Advanced Mathematical Met

During the past five years there has been an immense growth in the use of symbolic computing and mathematical software packages, such as Maple, which provide an invaluable tool for mathematicians and physical scientists alike. Enhancing and extending traditional mathematical methods, Maple's capabilities range from symbolic algebraic manipulation, finding symbolic values for integrals and solutions of some simple differential equations to modeling extremely complex engineering problems. Maple also has powerful numeric and graphic capabilities.

The first three chapters of this book provide a user-friendly introduction to computer-assisted algebra with Maple. The rest of the book then develops these techniques and demonstrates the use of this technology for deriving approximate solutions to differential equations (linear and nonlinear) and integrals, using methods such as perturbation and asymptotic expansions. In each case, the mathematical concepts are comprehensively introduced, with an emphasis on understanding how solutions behave and why various approximations can be used. The text is richly illustrated with examples from applied mathematics and physics and, where appropriate, integrates the use of Maple to extend the utility of traditional approximation techniques. Users of symbolic languages other than Maple will also benefit from the majority of the material which is not system-specific.

Topics covered encompass Padé approximants and methods of accelerating the convergence of sequences, Fourier series, perturbation methods, Green's functions and Sturm–Liouville theory, asymptotic analysis (including asymptotic expansions for some integrals and solutions of certain types of differential equations), Floquet theory and elements of nonlinear dynamics including and introduction to chaotic phenomena, and various approximation schemes such as perturbation theory and averaging methods.

Advanced Mathematical Methods with Maple is the ideal companion text for advanced undergraduate and graduate students of mathematics and the physical sciences. It incorporates over 1000 exercises with different levels of difficulty, for which solutions are provided on the internet at http://mcs.open.ac.uk/dr9

DEREK RICHARDS is Reader in Applied Mathematics at the Open University where he teaches courses in mathematical methods, nonlinear systems, and approximation methods. He is also the co-author of *Introduction to Dynamics* (1982).

Advanced Mathematical Methods
with Maple

Derek Richards

CAMBRIDGE
UNIVERSITY PRESS

PUBLISHED BY THE PRESS SYNDICATE OF THE UNIVERSITY OF CAMBRIDGE
The Pitt Building, Trumpington Street, Cambridge, United Kingdom

CAMBRIDGE UNIVERSITY PRESS
The Edinburgh Building, Cambridge CB2 2RU, UK
40 West 20th Street, New York, NY 10011–4211, USA
10 Stamford Road, Oakleigh, VIC 3166, Australia
Ruiz de Alarcón 13, 28014 Madrid, Spain
Dock House, The Waterfront, Cape Town 8001, South Africa

http://www.cambridge.org

First published 2002

Printed in the United Kingdom at the University Press, Cambridge

Typeface Times 10/13pt.   *System* LaTeX 2_ε [UPH]

A catalogue record for this book is available from the British Library

Library of Congress Cataloguing in Publication data
Richards, Derek.
Advanced mathematical methods with Maple / Derek Richards.
p.   cm.
Includes bibliographical references and index.
ISBN 0 521 77010 6 – ISBN 0 521 77981 2 (pb.)
1. Numerical analysis–Data processing. 2. Maple (Computer file) I. Title.
QA297.R495 2001
519.4′0285–dc21   2001025976 CIP

ISBN 0 521 77010 6 hardback
ISBN 0 521 77981 2 paperback

Contents

Preface   *page* xiii

1   Introduction to Maple   1
  1.1 Introduction   1
  1.2 Terminators   3
  1.3 Syntax errors   5
  1.4 Variables   6
      1.4.1   The value and the assignment of variables   7
      1.4.2   Evaluation and unassigning or clearing variables   10
      1.4.3   Quotes   18
      1.4.4   Complex variables   19
  1.5 Simple graphs   20
      1.5.1   Line graphs   21
      1.5.2   Parametric curves   22
      1.5.3   Implicit and contour plots   23
      1.5.4   Plotting functions of two variables: plot3d   25
  1.6 Some useful programming tools   26
      1.6.1   The seq command   27
      1.6.2   The **if** conditional   28
      1.6.3   Loop constructions   29
  1.7 Expression sequences   33
  1.8 Lists   34
      1.8.1   The use of lists to produce graphs   36
  1.9 Sets   38
  1.10 Arrays   40
      1.10.1 One-dimensional arrays   40
      1.10.2 Two-dimensional arrays and matrices   43
  1.11 Elementary differentiation   48
  1.12 Integration   51
      1.12.1 Numerical evaluation   53
      1.12.2 Definite integration and the assume facility   54
  1.13 Exercises   57

2   Simplification   63
  2.1 Introduction   63
  2.2 Expand   63

2.3	Factor	67
2.4	Normal	68
2.5	Simplify	70
2.6	Combine	72
2.7	Convert and series	74
2.8	Collect, coeff and coeffs	82
2.9	Map	85
2.10	Exercises	87

3 Functions and procedures 91

3.1	Introduction	91
3.2	Functions	91
	3.2.1 Composition of functions	100
	3.2.2 Piecewise functions	102
3.3	Differentiation of functions	103
	3.3.1 Total derivatives	105
	3.3.2 Differentiation of piecewise continuous functions	108
	3.3.3 Partial derivatives	108
3.4	Differentiation of arbitrary functions	109
3.5	Procedures	112
3.6	The SAVE and READ commands	122
	3.6.1 PostScript output of figures	124
3.7	Exercises	125

4 Sequences, series and limits 131

4.1	Introduction	131
4.2	Sequences	134
	4.2.1 The limit of a sequence	134
	4.2.2 The phrase 'of the order of'	136
4.3	Infinite series	138
	4.3.1 Definition of convergence	138
	4.3.2 Absolute and conditional convergence	138
	4.3.3 Tests for convergence	144
	4.3.4 The Euler–Maclaurin expansion	148
	4.3.5 Power series	150
	4.3.6 Manipulation of power series	156
4.4	Accelerated convergence	159
	4.4.1 Introduction	159
	4.4.2 Aitken's Δ^2 process and Shanks' transformation	160
	4.4.3 Richardson's extrapolation	163
4.5	Properties of power series	165
4.6	Uniform convergence	169
4.7	Exercises	173

5   Asymptotic expansions   179
  5.1 Introduction   179
  5.2 Asymptotic expansions   182
    5.2.1 A simple example   182
    5.2.2 Definition of an asymptotic expansion   184
    5.2.3 Addition and multiplication of asymptotic expansions 187
    5.2.4 Integration and differentiation of asymptotic expansions 188
  5.3 General asymptotic expansions   189
  5.4 Exercises   191

6   Continued fractions and Padé approximants   195
  6.1 Introduction to continued fractions   195
    6.1.1 Maple and continued fractions   198
  6.2 Elementary theory   199
    6.2.1 Canonical representation   203
    6.2.2 Convergence of continued fractions   204
  6.3 Introduction to Padé approximants   209
  6.4 Padé approximants   211
    6.4.1 Continued fractions and Padé approximants   217
    6.4.2 The hypergeometric function   219
  6.5 Exercises   223

7   Linear differential equations and Green's functions   227
  7.1 Introduction   227
  7.2 The delta function: an introduction   228
    7.2.1 Preamble   228
    7.2.2 Basic properties of the delta function   228
    7.2.3 The delta function as a limit of a sequence   232
  7.3 Green's functions: an introduction   238
    7.3.1 A digression: two linear differential equations   239
    7.3.2 Green's functions for a boundary value problem   244
    7.3.3 The perturbed linear oscillator   248
    7.3.4 Numerical investigations of Green's functions   251
    7.3.5 Other representations of Green's functions   252
  7.4 Bessel functions: Kapteyn and Neumann series   254
  7.5 Appendix 1: Bessel functions and the inverse square law   258
  7.6 Appendix 2: George Green   261
  7.7 Exercises   262

8   Fourier series and systems of orthogonal functions   267
  8.1 Introduction   267
  8.2 Orthogonal systems of functions   269
  8.3 Expansions in terms of orthonormal functions   270
  8.4 Complete systems   271
  8.5 Fourier series   273

8.6 Addition and multiplication of Fourier series ... 276
8.7 The behaviour of Fourier coefficients ... 277
8.8 Differentiation and integration of Fourier series ... 280
8.9 Fourier series on arbitrary ranges ... 284
8.10 Sine and cosine series ... 284
8.11 The Gibbs phenomenon ... 286
8.12 Poisson summation formula ... 291
8.13 Appendix: some common Fourier series ... 293
8.14 Exercises ... 293

9   Perturbation theory ... 301
9.1 Introduction ... 301
9.2 Perturbations of algebraic equations ... 302
    9.2.1 Regular perturbations ... 302
    9.2.2 Singular perturbations ... 305
    9.2.3 Use of Maple ... 308
    9.2.4 A cautionary tale: Wilkinson's polynomial ... 313
9.3 Iterative methods ... 315
    9.3.1 A nasty equation ... 319
9.4 Perturbation theory for some differential equations ... 320
9.5 Newton's method and its extensions ... 325
    9.5.1 Newton's method applied to more general equations ... 328
    9.5.2 Power series manipulation ... 331
9.6 Exercises ... 336

10   Sturm–Liouville systems ... 342
10.1 Introduction ... 342
10.2 A simple example ... 343
10.3 Examples of Sturm–Liouville equations ... 345
10.4 Sturm–Liouville systems ... 352
    10.4.1 On why eigenvalues are bounded below ... 360
10.5 Green's functions ... 361
    10.5.1 Eigenfunction expansions ... 365
    10.5.2 Perturbation theory ... 369
10.6 Asymptotic expansions ... 371
10.7 Exercises ... 375

11   Special functions ... 377
11.1 Introduction ... 377
11.2 Orthogonal polynomials ... 377
    11.2.1 Legendre polynomials ... 381
    11.2.2 Tchebychev polynomials ... 386
    11.2.3 Hermite polynomials ... 389
    11.2.4 Laguerre polynomials ... 395
11.3 Gaussian quadrature ... 400

11.4 Bessel functions 404
11.5 Mathieu functions 410
    11.5.1 An overview 411
    11.5.2 Computation of Mathieu functions 416
    11.5.3 Perturbation theory 421
    11.5.4 Asymptotic expansions 425
11.6 Appendix 429
11.7 Exercises 431

12 Linear systems and Floquet theory 437
12.1 Introduction 437
12.2 General properties of linear systems 438
12.3 Floquet theory 443
    12.3.1 Introduction 443
    12.3.2 Parametric resonance: the swing 444
    12.3.3 O Botafumeiro: parametric pumping in the middle ages 446
    12.3.4 One-dimensional linear systems 447
    12.3.5 Many-dimensional linear, periodic systems 448
12.4 Hill's equation 452
    12.4.1 Mathieu's equation 454
    12.4.2 The damped Mathieu equation 457
12.5 The Liénard equation 466
12.6 Hamiltonian systems 467
12.7 Appendix 471
12.8 Exercises 473

13 Integrals and their approximation 478
13.1 Introduction 478
13.2 Integration with Maple 482
    13.2.1 Formal integration 482
    13.2.2 Numerical integration 489
13.3 Approximate methods 492
    13.3.1 Integration by parts 493
    13.3.2 Watson's lemma and its extensions 497
13.4 The method of steepest descents 506
    13.4.1 Steepest descent with end contributions 514
13.5 Global contributions 516
13.6 Double integrals 520
13.7 Exercises 521

14 Stationary phase approximations 526
14.1 Introduction 526
14.2 Diffraction through a slit 528
14.3 Stationary phase approximation I 532
    14.3.1 Anger's function 532

    14.3.2 General first-order theory   539
14.4 Lowest-order approximations for many-dimensional integrals   541
14.5 The short wavelength limit   544
14.6 Neutralisers   545
14.7 Stationary phase approximation II   554
    14.7.1 Summary   555
14.8 Coalescing stationary points: uniform approximations   557
    14.8.1 Rainbows   557
    14.8.2 Uniform approximations   562
14.9 Exercises   569

15 Uniform approximations for differential equations   573
15.1 Introduction   573
15.2 The primitive WKB approximation   575
15.3 Uniform approximations   579
    15.3.1 General theory   579
    15.3.2 No turning points   581
    15.3.3 Isolated turning points   582
15.4 Parabolic cylinder functions   587
    15.4.1 The equation $y'' + (x^2/4 - a)y = 0$ 588
    15.4.2 The equation $y'' - (x^2/4 + a)y = 0$ 592
15.5 Some eigenvalue problems   594
    15.5.1 A double minimum problem   602
15.6 Coalescing turning points   606
    15.6.1 Expansion about a minimum   606
    15.6.2 Expansion about a maximum   609
    15.6.3 The double minimum problem   610
    15.6.4 Mathieu's equation   615
15.7 Conclusions   621
15.8 Exercises   623

16 Dynamical systems I   628
16.1 Introduction   628
16.2 First-order autonomous systems   634
    16.2.1 Fixed points and stability   634
    16.2.2 Natural boundaries   637
    16.2.3 Rotations   637
    16.2.4 The logistic equation   638
    16.2.5 Summary   641
16.3 Second-order autonomous systems   641
    16.3.1 Introduction   641
    16.3.2 The phase portrait   644
    16.3.3 Integrals of the motion   653
    16.3.4 Fixed points and stability   655
    16.3.5 Classification of simple fixed points   657

	16.3.6 The linearisation theorem	664
	16.3.7 The Poincaré index	665
16.4	Appendix: existence and uniqueness theorems	668
16.5	Exercises	669

17 Dynamical systems II: periodic orbits — 673

17.1 Introduction	673
17.2 Existence theorems for periodic solutions	674
17.3 Limit cycles	677
17.3.1 Simple examples	677
17.3.2 Van der Pol's equation and relaxation oscillations	678
17.4 Centres and integrals of the motion	684
17.5 Conservative systems	691
17.6 Lindstedt's method for approximating periodic solutions	697
17.6.1 Conservative systems	698
17.6.2 Limit cycles	705
17.7 The harmonic balance method	707
17.8 Appendix: On the Stability of the Solar System, by H. Poincaré	711
17.9 Appendix: Maple programs for some perturbation schemes	717
17.9.1 Poincaré's method for finding an integral of the motion	717
17.9.2 Lindstedt's method applied to the van der Pol oscillator	719
17.9.3 Harmonic balance method	721
17.10 Exercises	723

18 Discrete Dynamical Systems — 727

18.1 Introduction	727
18.2 First-order maps	728
18.2.1 Fixed points and periodic orbits	728
18.2.2 The logistic map	731
18.2.3 Lyapunov exponents for one-dimensional maps	739
18.2.4 Probability densities	742
18.3 Area-preserving second-order systems	746
18.3.1 Fixed points and stability	746
18.3.2 Periodic orbits	751
18.3.3 Twist maps	753
18.3.4 Symmetry methods	757
18.4 The Poincaré index for maps	762
18.5 Period doubling	764
18.6 Stable and unstable manifolds	767
18.6.1 Homoclinic points and the homoclinic tangle	772
18.7 Lyapunov exponents	774
18.8 An application to a periodically driven, nonlinear system	776
18.9 Exercises	779

19 **Periodically driven systems** 785

19.1 Introduction: Duffing's equation 785

19.2 The forced linear oscillator 786

19.3 Unperturbed nonlinear oscillations 789

 19.3.1 Conservative motion, $F = \mu = 0$: $\epsilon < 0$ 789

 19.3.2 Conservative motion, $F = \mu = 0$: $\epsilon > 0$ 793

 19.3.3 Damped motion: $F = 0$, $\mu > 0$ 796

19.4 Maps and general considerations 797

19.5 Duffing's equation: forced, no damping 807

 19.5.1 Approximations to T-periodic orbits 807

 19.5.2 Subharmonic response 813

19.6 Stability of periodic orbits: Floquet theory 815

19.7 Averaging methods 817

 19.7.1 Averaging over periodic solutions 817

 19.7.2 General theory 819

 19.7.3 Averaging over a subharmonic 823

19.8 Appendix: Fourier transforms 825

19.9 Exercises 828

20 **Appendix I: The gamma and related functions** 834

21 **Appendix II: Elliptic functions** 839

22 **References** 845

22.1 References 845

22.2 Books about computer-assisted algebra 853

Index 855

Preface

The purpose of this book is to teach a variety of mathematical methods useful for deriving approximate solutions of the differential equations and integrals found in the mathematical sciences. Except in the simplest of cases, solving equations that describe physical processes usually requires some prior understanding of how the solutions can behave, at least in some circumstances, and a full understanding of the system is normally developed by iterating between mathematical and physical ideas. Books on mathematical methods can describe only part of this process, so I have attempted to partially bridge this gap by explaining some of the history behind the methods and, in some cases, a few applications, so the reader will understand a little of why methods were developed. The second departure from convention is the inclusion of four chapters on nonlinear dynamics, included because nonlinear equations are at least as important as the linear equations emphasised in older books. The third, and most significant, departure from convention is the first three chapters teaching the use of the computer algebra package Maple; this technology is used to enhance the more traditional methods taught in the remainder of the book, and all aspects of the software, numerical, graphical and symbolic algebra, are used in the study of nonlinear systems, particularly in chapters 17 to 19.

Computers, for better or worse, have changed the way we behave and the way science is performed, and have increased the range of problems that can be tackled. The study of nonlinear systems blossomed with the invention of good graphical systems coupled to relatively cheap computers in the last quarter of the twentieth century. Electronic calculators have had a significant effect on mathematics education, though not one that is easily quantified. However, the fact that young children have been taught to do the sum $\frac{2}{5} + \frac{1}{4}$ by adding 0.4 to 0.25 on a calculator, seems to have affected their later ability to perform simple algebraic manipulations, such as simplifying $\frac{a}{b} + \frac{c}{d}$.

Computer algebra packages were developed on mainframe computers and used originally for large calculations that could be performed no other way, for example high-order perturbation expansions. Now such systems are available on PCs, offer reasonable graphical and numerical facilities and, because of the power of a PC, can perform very sophisticated and complicated algebra. This technology inevitably affects the way mathematics is taught even though we have no idea of its long-term effects. In my experience it is the synthesis of the new technology and the traditional methods that is most useful and this book is organised accordingly. I have found that computer assisted algebra can substantially increase the utility of traditional methods and help with complex algebraic manipulations. But these systems currently have severe limitations. I find these systems less help with the evaluation of integrals than some marketing material suggests — because most nontrivial integrals require a change of variables to be cast in a recognisable form and it is normally quicker to refer to a book of standard integrals than to switch on a computer. In addition there seems to be a fundamental difference between the pen–hand–paper–brain interaction and the keyboard (mouse)–screen–brain interaction that makes it easier to perform initial calculations of any complexity using traditional methods: in short a computer can hinder thinking. Computers are better, however, when patterns have been discerned and algorithms developed. The structure of this book reflects these opinions. Indeed, apart from the first three chapters most of this text is like a conventional mathematics text, and many of the exercises are designed to be done without a computer because I have tried to provide the reader with sufficient understanding and experience to make a synthesis between old and new methods.

Notwithstanding these reservations about computer assisted algebra, the ease with which simple numerical calculations can be performed and the simple but powerful graphical facilities significantly increase the efficiency of performing many tasks.

This book includes a range of material taught to students of mathematics and the mathematical sciences between the second year of a good honours degree and post-graduate level. Chapters 1–10 are based on an Open University MSc course and chapters 16–19 include material taught at level three and in an MSc course. I have assumed the reader to have a good working knowledge of calculus and of linear differential equations, and some material in chapter 13 assumes an understanding of complex variable theory. I have tried to introduce most topics via their historical context because this helps understand why ideas were developed and how subjects are connected, and I have also tried to explain why methods work, rather than present them as recipes.

The first three chapters are devoted to teaching Maple and start by assuming the reader has no knowledge of this language, but has access to it. Perforce, much Maple has been neglected, but the content of these three chapters is sufficient for most users and provides a good background for further development.

The remaining sixteen chapters are devoted to traditional mathematical methods, with a few unconventional additions. These chapters can be read as a normal text, though some exercises require a computer. All the Maple required in these chapters is included in chapters 1–3: however, I have written the bulk of the text so a reader with the equivalent knowledge of similar languages should be able to cope with the computing exercises. I have also included a few sample Maple programs; in my experience these are trivially translated into other languages, so readers of chapters 4–19 familiar with Reduce or Mathematica, for example, will not be significantly disadvantaged.

Topics are always introduced in the conventional manner and the use of a computer discouraged until simple problems can be done using the conventional paper and pencil technology. Failure to do this can, and frequently does, result in nonsense posing as authoritative computer output.

Arguments over the relative merits of rival systems of computer assisted algebra are frequent, but mostly sterile. In practice a choice between systems has to be made because time and effort is needed to learn such systems. I believe that there is little to chose between the major systems and that the differences are version dependent.

The choice of topics in such a book reflects the author's interests, but I have tried to include material that is generally applicable. In many cases I have extended the normal syllabus either because of enthusiasm for some unusual material, as in the treatment of Newton's method, section 9.5, or because important problems are frequently ignored, as in treatment of Fourier series, chapter 8. The use of computer assisted algebra changes the way we use some skills, so the introduction to the two chapters on integration, 13 and 14, provides a brief account of symbolic integration and of its limitations, but also shows how to combine computer algebra methods with traditional theory to derive useful asymptotic expansions for certain classes of integral. Similarly, the discussion of perturbation methods for approximating periodic solutions in chapter 17 goes beyond that normally found. It is hoped that undergraduates, postgraduates and mature scientists will find the subjects included of interest and value.

There are over 1000 exercises included in the text; solutions to these will be posted on my website at http://mcs.open.ac.uk/dr9.

Without the help and encouragement of numerous friends and colleagues the preparation of this book would have been far more difficult, in particular it is a pleasure to thank June Barrow-Green, Peter Hall and Francis Wright for their valuable help and advice. It would not have written at all without the encouragement of Helen, to whom this book is dedicated.

D. Richards
Milton Keynes
January 2001

1

Introduction to Maple

1.1 Introduction

Chapters 1–3 are designed to introduce Maple to new users. Readers with some experience of Maple could either start with chapter 4 or read selected parts of the first three chapters and/or do selected exercises. While preparing these three chapters I consulted a number of books dealing with computer algebra systems in general and Maple in particular and have provided a bibliography of these texts on page 853. The most comprehensive treatment of Maple is found in Heck (1996), but I also found the general discussion and problems in MacCallum and Wright (1991) and the treatment of physical problems in MacDonald (1994), both of which use Reduce, useful and illuminating. The reader requiring a detailed discussion of the mathematics of computer algebra systems should consult Gathen and Gerhard (1999) or Geddes *et al.* (1992), and a detailed comparision of systems, together with problems that cause difficulties and many interesting articles is provided by Wester (1999).

Maple is a large, complex system, only part of which is introduced in the first three chapters. Initially you may find the language and interface daunting; but do not be intimidated as it is necessary to learn only a few commands before quite complicated problems can be tackled. The initial stages of learning the system are worst as you will not know what to expect and when, inevitably and frequently, things go wrong you will not know whether there is a program bug or whether you have done something wrong and if so what it is; as with most computer systems Maple is not very forgiving of errors and the subsequent messages are rarely helpful to the new user. Thus initially learning Maple may be infuriating and time consuming, but you should find that progress is sufficiently rapid for many significant problems to be solved quicker and more reliably than by using traditional methods.

At this early stage, however, we must emphasise that Maple is a tool appropriate for some tasks but not others. You need to know enough about it to decide whether or not to use it, or when to stop using it and revert to more conventional methods. This is an important skill.

Maple contains a huge number of commands and functions of which we use only a small subset. This and the next two chapters are designed to introduce this subset and to enable you to use it in a variety of circumstances. There is a lot to remember, especially for users new to Maple; if this is the case, at first you should simply try to remember *what* can be done rather than *how* it is done by skimming this chapter quickly without using the computer, then read it slowly, typing in the commands as written and doing the exercises in order to build your confidence. The exercises at the end of the chapter should be attempted soon after reading it because only by using the language will you be able to remember it.

The syntax and structure of Maple can be described fairly precisely and logically. But for most people this description is of use only after considerable familiarity with the language has been developed; on the other hand it is necessary to understand some of the logical constructions and distinctions made by Maple before proceeding very far. So, in practice, one learns use and syntax simultaneously and the best way of doing this is via simple examples, which is what this chapter is devoted to.

We assume that you have managed to install Maple and can start the program running; moreover we assume that you are running it on a PC. After starting Maple the screen comprises a set of four menu bars above an empty *worksheet* which can be made to fill most of the screen. At the top left hand corner of the worksheet will be the input prompt — normally a > sign† — so the system is waiting for you to input a command. At this point you may find it useful to do the *New User's Tour* provided by Maple; this is accessed by clicking on *Help* in the top menu bar (or just type Alt-H), then click on *New User's Tour* (or just type N).

When Maple starts there are four bars at the top of the screen. On the first bar there is the title *Maple V Release 5* or *Maple 6* in versions 5 and 6 respectively. In the second bar are the usual pull down menus and below them, in the third bar, are some often used menu items which have been incorporated into buttons. But for frequently used commands you will find it far more efficient to use key-strokes rather than mouse-clicks. There is also an extra input line that seems to be most useful when typing mathematics into a text line.

The cursor is positioned by the prompt inside a worksheet ready to enter an expression for processing by Maple.

Maple has a particularly good *help* facility and you should familiarise yourself with this as it is an essential aid; indeed when a new Maple command is introduced we usually expect you to look it up rather than explain all its details in the text. In order to access the help files you need to click on *Help* in the top menu bar, or just type Alt-H, and then on either *Topic Search* (type T) or *Full Text Search* (type F), after which you will usually need to type in a name in a box labelled *Topic* or *Word(s)* respectively.

We suggest that you try this facility as soon as possible as you will find it an essential help both for learning new commands and as a reminder of syntax. For example in exercise 1.2, on page 5, we use the Maple command `ifactor`; go to the *Help* menu, then *Topic Search* and then type `ifactor` and *return* in the *Topic* box to obtain some information about this function. Alternatively, from *Help* go to *Full Text Search* before typing `ifactor` in the *Word(s)* box to obtain a list of help files which may be relevant. In most help files there is hypertext giving direct access to other relevant files. An alternative, and easier, method of accessing help on a particular topic, `ifactor` for

† The prompt sign can be changed using the **interface** command. For example to change it to a %, type **interface(prompt="%")**. The student version of Maple 4 has a different prompt that cannot be changed.

instance, is to type ?ifactor directly after the prompt: however, this assumes that the name and spelling are known exactly.

Beware that it is possible to accumulate many help-windows unless you explicitly close each one before opening another; note that under the *Windows* option in the top menu bar there is a facility for closing all help files.

If you find constant use of the mouse irritating then many mouse clicks can be avoided by using key-strokes. For instance, all the menus on the upper bar may be accessed using the Alt-key with the underlined letter, thus Alt-I brings down the *Insert* menu, then a section can be added simply by typing S. Such use of the key board is, after a little practice, far faster and more efficient than using the mouse. Some useful key board strokes are listed below:

- ctr-F S will save a worksheet and ctr-F A will allow you to save it to a particular file.
- ctr-K and ctr-J add another input line respectively before and after the cursor.
- F5 changes the input mode. The default is [> and a Maple command is expected. Pressing F5 changes this to the text input mode for which the prompt is [. This mode is used for all textual comments in your worksheet. When in text-input mode the F5 key will switch to mathematics input, which allows you to input mathematical expressions into the text.
- F4 joins input lines together, which means that they are executed together. F3 will separate them, which is useful for inserting new lines between joined lines.

Many exercises are provided in this chapter; you should attempt as many of these as you have time, because Maple can be learnt only by using it. The solutions to all exercises and any Maple worksheets mentioned in the text may be found on the Web site described in the preface.

All the Maple used in this text was written using version 5: Maple 6 was launched as the text was being finalised. This later version of Maple includes significant additions, the most significant being, perhaps, an enhanced capability for numerical computations involving large matrices, but these do not affect the contents of this text. There are also some changes of syntax that affect a few of the commands described below: where necessary I have described these differences. All the code included in the text and on the Web site works on both Maple 5 and 6 and in most cases the user need make no changes.

1.2 Terminators

After typing in a command you need to tell the computer that the command has ended *and* that it should be accepted. A command is ended by typing one of the two possible *terminators*; the usual terminator is a semi-colon and this should be followed by a *return* (or *enter*) to tell the computer to accept the command, as in the following simple arithmetic examples — note, the *return* is not shown.

```
>   4+3;
```
$$7$$

```
>   40*99;
```
$$3960$$

```
>   4!;
```
$$24$$

```
>   5!; exp(5); sin(Pi); sin(pi); cos(3*Pi/2);
```
$$120$$
$$e^5$$
$$0$$
$$\sin(\pi)$$
$$0$$

These commands should be self explanatory, but observe from the last example there can be as many commands as will fit on a single line provided they are each separated by a semi-colon (or, see below, a colon). Observe also that `Pi` and `pi` are different; `Pi` is the number $3.141\,592\,6\ldots$ whereas `pi` is just a variable; because Maple knows the Greek alphabet this variable is given as π in the Maple output. This example shows that Maple is case sensitive so one needs to be careful. Finally, note that multiplication must be made explicit with the $*$ symbol; thus 3π is input as `3*Pi` not as `3Pi`.

If you forget the semi-colon and just type *return* nothing happens except that the cursor is placed at the next prompt and, as you have not finished the command, a warning is issued. Either

- replace the cursor at the end of the line just typed and add the semi-colon followed by the *return*, or
- just type a semi-colon followed by the *return* at the new prompt, or
- if you had not finished your input, simply continue typing and end with a terminator. It is often useful to break up long commands in this manner.

The second method of ending an input command is with a colon followed by a *return*. Maple now accepts the command, provided it is syntactically correct, but does not print the output. For instance if we wish to define y to be $\sin(x^2)$ the colon and semi-colon terminators produce the following results.

```
>   y:=sin(x^2):
>   y:=sin(x^2);
```
$$y := \sin(x^2)$$

You might expect an $=$ sign where we used :=, that is a : followed by =. In Maple := and = have fundamentally different meanings that will be discussed soon.

Exercise 1.1
The probability of winning a large prize in the original UK national lottery with one ticket is $p = 1/x$ where
$$x = \frac{49!}{6!\,(49-6)!}.$$
What is p?

Exercise 1.2

Use the Maple function `ifactor` to find the factors of $n! + 1$ where $n = 3, 4, 5, 7, 11, 20$ and 27.

Note that if $n = 3, 11$ and 27, $n! + 1$ is a prime. Can you find any other values of n with this property? Do not spend too long on this last part.

Hint: the command `ifactor(7!+1)` will find the factors of $7! + 1$.

Exercise 1.3

Use the function `isprime` to determine which of the following numbers $p = 2^n - 1$, where $n = 30, 31, 32, 61$ and 67, are prime numbers.

For those that are, use the function `nextprime` to determine the next two largest primes and find out whether or not these can be expressed in the form $2^n - 1$ for some n.

Hint: for any integer N the command `isprime(N)` returns either `true` or `false`.

1.3 Syntax errors

Inevitably one makes syntax errors when typing commands, for instance forgetting a bracket or an operator. On making such a mistake an error message is given after the *return*; from this it may be clear what is wrong or, more frequently, the message may seem quite meaningless. In simple cases Maple will attempt to show you where the error is by placing the cursor at the point where it thinks the error occurs, but for complicated expressions this normally shows only where Maple has detected the error; for example, in the case of a missing bracket, the error may have occurred before this point. After accumulating some experience, as with any computer system, you should be able to guess what has caused the less comprehensible messages. Normally, upon correction of the error Maple will accept your input on re-typing *return* with the cursor on the command line.

Maple has a syntax-checker on the third tool bar, but I have not found this very useful.

A common error is to forget one or more brackets when typing in complex mathematical expressions. The occurrence of this type of error can be minimised by typing in the brackets first, to ensure that opening and closing brackets match. For instance when typing in the expression

$$(x-1)\left(\sqrt{3x-4} + (7x)^{3/4}\right) \sin\left(e^{2x} + \frac{1}{x}\right)$$

it is easier to type in the brackets and some of the operators first, as follows

```
( )*(sqrt() + ()^() ) * sin( );
```

and then go back and fill in the details before entering *return*.

There are also occasions when the cursor disappears: such behaviour may be induced by selecting a graph and deleting it, though, in this case, the cursor normally reappears after a short delay. On other occasions, not reproducible, I find that it returns after moving to another worksheet or a help file.

It is normally best to start a new worksheet with the `restart` command as this will reset the system so that a variable used in a previous worksheet does not carry its value over to the new one.

The Stop button, on the middle menu bar, is a useful feature on many occasions, for instance when one starts a computation that is taking far longer than it ought or when an error has been noted during a computation. Clicking on this will stop the process, though sometimes only after a considerable delay, particularly if the hard disk is being used extensively. Sometimes, however, Maple will not respond to this button and then more drastic action is required.

If none of these tricks work, and you are using Windows 95 or 98, then it is normally necessary to abort the session by using *control-Alt-delete* or, more drastically, by pressing the reset button on the computer. Windows NT, however, provides a more elegant method of ending a session using the *Task Manager*. Though such extreme measures are not often called for, when taken all work is lost — so you should save worksheets frequently, by clicking on either the *save* button (the key board strokes are Alt-F followed by S) or the *save as* button (Alt-F followed by A) in the *File* item on the top menu bar. Remember also that if Maple fails you may also lose other work that is running in other sessions outside Maple, so it is prudent to save this before opening Maple.

In our experience different users and/or different machines produce different patterns of errors, so it is difficult to give general advice. However, a typical situation when such action is necessary is when performing a large calculation and the computer has filled up its available memory and is proceeding by swapping in and out to the disk; then the system does not respond to the *stop* button. There are two classes of error which crash the system; these are *general exception fault* and *illegal execution*. In earlier versions of Maple it was easy to produce these, but versions 5 and 6 are more robust, though not infallible.

1.4 Variables

Variables in Maple are denoted by sequences of letters, digits and underscores, not beginning with a digit. For example

```
x, X, initial\_conditions, CO2, \_A
```

are all legitimate variables, but 2a, for instance, is not.

Not all combinations are allowed as some names are already used by Maple. For instance the symbols Pi and I are reserved for the constant $\pi = 3.141\,59\ldots$ and the square-root of -1; the lower-case symbols pi, e and i are not used by Maple. This illustrates the fact that Maple is case-sensitive so x and X, for instance, are treated as different variables. In older versions of Maple, and some other languages, the symbol E was reserved for the base of the natural logarithm $e = 2.718\,28\cdots$: in later versions the value of e is given by the exponential function evaluated at 1, that is e= exp(1): note that exp(1) returns a symbol e (which evaluates to the number e) but exp(1.0) returns the number $2.718\,281\,828$ directly.

All the standard mathematical functions, for example *sin, cos, sinh, cosh, sqrt, exp,* etc., have pre-defined names, as do many Maple procedures and commands, for example *series, int, diff, add, sum, coeff, rhs, lhs* and names of data types *set, list, matrix, array.* In addition Maple uses names starting with an underscore as output for some procedures, for instance the symbolic solution of some differential equations using `dsolve`, so these are best avoided.

Finally Maple has a number of reserved words, not accepted as valid names. Some of these are listed in table 1.1.

Table 1.1. *Maple Reserved Names*

and	by	do	done	elif	RETURN
else	end	fi	for	from	break
if	in	intersect	local	minus	next
mod	not	od	option	options	
or	proc	quit	read	save	
stop	then	to	union	while	

This list of proscribed names sounds formidable, but most are fairly obvious and you do not need to remember them as, if used inadvertently, an error message appears and it is easy to go back and change the name.

1.4.1 The value and the assignment of variables

Variables in Maple can be *free* or *unbounded* as in ordinary algebraic manipulations. A simple example is provided by finding the solution of the quadratic equation $ax^2 + bx + c = 0$; in Maple this is obtained as follows,

```
>  solve(a*x^2 + b*x + c=0,x);
```

$$\frac{1}{2}\frac{-b + \sqrt{b^2 - 4ac}}{a}, \frac{1}{2}\frac{-b - \sqrt{b^2 - 4ac}}{a}$$

Here *a, b* and *c* are parameters and *x* the unknown variable; the letters *a, b, c* and *x* are *free* or *unbounded* variables. This is in contrast to many programming languages, Basic, C and Fortran for instance, in which variables are usually used only after being assigned a specific *numerical* value.

A variable can be made *bound* or *assigned* a *value* using the symbol :=, that is a colon followed immediately by an equals sign. Thus we assign the variable *y* the value $ax^2 + bx + c$ by writing

```
>  y:=a*x^2+b*x +c:
```

so that until *y* is *cleared* or assigned another value, *y* is $ax^2 + bx + c$. Thus we can solve the equation $ax^2 + bx + c = c$ with the commands

```
>  y:=a*x^2 + b*x + c:
>  solve(y=c,x);
```

$$0, -\frac{b}{a}$$

One initial source of confusion is the difference between assign, :=, and equals, =. By assigning a variable y:=a*x+b, for instance, we are instructing the computer to replace y by a*x+b wherever it occurs. An equation in mathematics $y = ax + b$ can be used in this manner, but it also means that $y - b = ax$ or $x = (y - b)/a$; an assignment cannot be rearranged. On the other hand it can be more general than an equation as the next example will show.

In Maple the word *value*, as used above, is initially misleading; the variable y is not being given a numerical value, rather it is being made equivalent to an expression. We could, and often do, assign a variable to be an equation,

> y:=a*x^2+b*x+c=c:

so that the solution of the equation $ax^2 + bx + c = c$ is obtained with the command solve(y,x).

Suppose that we wish to solve the simultaneous equations

$$ax + by = A, \quad cx + dy = B$$

for x and y. First, we need to restart the session because y has already been assigned a value so needs to be cleared: more elegant methods of clearing variables will be discussed later. Next we assign z, say, to the set of equations and u to the set of unknown variables and use solve to create a set of solutions as follows

> restart;
> z:={a*x+b*y=A,c*x+d*y=B}:
> u:={x,y}:
> sol:=solve(z,u);

$$sol := \{x = -\frac{-dA + Bb}{ad - bc}, \, y = \frac{aB - Ac}{ad - bc}\}$$

Note the use of the curly braces to define a set; these will be discussed further in section 1.9. Having obtained the solution we may want to assign x and y to be these solutions; this is achieved using the assign command, as follows

> assign(sol);
> x,y;

$$\frac{-dA + Bb}{-da + bc}, \, -\frac{aB - Ac}{-da + bc}$$

This assignment works only if there is one solution; if the equations we are solving have more solutions then it is necessary to modify this method using other Maple constructions which are illustrated in subsequent examples. It is important to remember that after using the command assign neither x nor y are free variables.

It is not always necessary to assign x and y in this way. Suppose we require the distance of the intersection of these two lines from the origin, $\sqrt{x^2 + y^2}$, then we may use the substitute command, subs:

> dist:=subs(sol,sqrt(x^2+y^2));

$$dist := \sqrt{\frac{(-dA + Bb)^2}{(ad - bc)^2} + \frac{(aB - Ac)^2}{(ad - bc)^2}}$$

With this use x and y remain free variables.

Exercise 1.4

Use Maple to solve the equations

(i) $5x + 2y = 1$, $6x + 7y = 8$,
(ii) $x^3 - 6x^2 + 11x - 6 = 0$,
(iii) $5x^2 + 2y^2 = 1$, $2x^2 - y^2 = 2$,
(iv) $3x^2 + 2y^2 = 2$, $2x^2 - 2xy + y^2 = 2$.

In the last two questions you will need to use the `allvalues` command and proper use of this requires the use of lists, which will be dealt with later in this chapter.

A common problem can be created in the following manner. If we have assigned y to be $3x^2 + 4x + 5$ and wish to know the numerical value of y when $x = 1$ a method of doing this is to assign the value 1 to x so y automatically has the value $3 + 4 + 5 = 12$; this is done as follows

```
>   y:=3*x^2 + 4*x + 5;
```
$$y := 3x^2 + 4x + 5$$

```
>   x:=1;  y;
```
$$x := 1$$
$$12$$

Suppose that subsequently we need the integral of y with respect to x, that is $\int dx\,(3x^2 + 4x + 5)$. Using the Maple procedure for integration, `int`, it is natural to write

```
>   z:=int(y,x);
```

`Error, (in int) wrong number (or type) of arguments`

This is a common, but not a very helpful, error message until you have experienced it a few times. The problem is that both x and y are now numbers whereas `int` is expecting x to be a free variable. The better method is to use the `subs` command which allows us to set $x = 1$ locally without assigning the value 1 to x.

```
>   restart:                    # restart, because x is defined previously
>   y:=3*x^2 + 4*x + 5;
```
$$y := 3x^2 + 4x + 5$$

```
>   subs(x=1,y);
```
$$12$$

```
>   z:=int(y,x);
```
$$z := x^3 + 2x^2 + 5x$$

In this set of Maple commands we have also introduced use of the # after which all text is treated as a comment and not as a Maple command. There is an alternative method of substituting values into expressions involving the `eval` command, described in the next section.

Exercise 1.5

Use the `subs` command to determine the value of the cubic $y = 9x^3 + 5x^2 + 2x + 10$ at $x = 1, 4/3$ and 2.45. Notice that the value returned by `subs` is the same type of number as substituted for x.

The subs command can be used to make several simultaneous substitutions when used in either of the forms

```
subs(v1=r1,v2=r2,expression)
subs({v1=r1,v2=r2},expression)
```

in which the substitution is carried out from left to right, that is variable v1 is replaced by r1, variable v2 is replaced by r2. The order of substitution is sometimes important; any number of simultaneous substitutions may be made in this way. Note that the second version, in which the substituted variables are in a set, is that used above.

Exercise 1.6

Use Maple to evaluate the following integrals,

(i) $\int dx \; \sin x \sin 2x,$ (ii) $\int dx \; \frac{x+1}{x^2+4x+6},$

(iii) $\int dx \; \frac{1}{\sqrt{x+a}-\sqrt{x+b}},$ (iv) $\int dx \, x^4 \sin x.$

Use the result of part (iii) and the subs command to show that

$$\int_0^1 dx \; \frac{1}{\sqrt{x+2}-\sqrt{x+1}} = 2\sqrt{3} - \frac{2}{3}.$$

1.4.2 *Evaluation and unassigning or clearing variables*

Once a variable is assigned it normally remains assigned for the remainder of the session *and* if assigned in one worksheet it is assigned to the same value in all other worksheets of the same session. This means that when starting a new worksheet in the same session it is normally safest to start with the restart command which clears or unassigns all variables. This command can be used at any point of a worksheet.

Other methods of clearing variables involve *evaluation* so it is helpful to understand a little about this. When you use a name on the right hand side of an assignment Maple checks to see if it has been assigned a value; if it has, this value is substituted for the name: if this value itself has an assigned value Maple performs another substitution, and so on recursively until no more substitutions are possible. Consider the following simple example.

```
>   x:=y;
```
$$x := y$$

```
>   y:=z;
```
$$y := z$$

```
>   z:=5;
```
$$z := 5$$

Now Maple evaluates *x* fully, that is it substitutes *y* for *x*, *z* for *y* and finally 5 for *z*, so the value of *x* is,

```
>   x;
```

One common error message produced by recursive substitution is *recursive definition of a name*, a mistake frequently made by those used to programming in languages like Basic, C or Fortran. An easy way of producing this error is as follows.

```
>   w:=w+1;
```

```
Warning, recursive definition of name
```

$$w := w + 1$$

and if we try to evaluate an expression containing *w*,

```
>   sin(w);
```

```
Error, too many levels of recursion
```

Here Maple tried to find the value of *w*, but because

$$w = 1 + w = 1 + (1 + w) = 1 + (1 + (1 + w)) = 1 + (1 + (1 + (1 + w))) = \cdots,$$

is a sequence that never terminates, or rather finishes when Maple's stack is full, an error results. Having made this error it is necessary to *clear* the variable before proceeding; the methods of doing this will be described shortly. Sometimes, after such an error, it is also necessary to restart the session.

This error occurs only if w is free; if we set w:=5 and subsequently set w:=w + 1 the value of w becomes 6, as would be expected; the same result would be obtained in Basic, C or Fortran.

Evaluation can be delayed in one of two ways. Either we can use the evaln command, which simply returns the name (**eval**uate to a **n**ame)

```
>   evaln(x);
```

$$x$$

Or we can surround the name in single forward-quotes ' (on many keyboards ' is found on or near the top left hand key); the forward-quote ' *must not* be confused with the back-quote ` . It is easy to confuse these quotes, so it is best if you spend a little time now typing the next few Maple commands to determine which of your key-board buttons represents these quotes, because they are useful. The next subsection discusses the role of the various quotes in a little more detail; here we show how forward-quotes may be used to clear variables.

If x, for example, is assigned a value,

```
>   x:=cos(ln(u));
```

$$x := \cos(\ln(u))$$

then typing x simply returns this value:

```
>   x;
```

$$\cos(\ln(u))$$

But x may be returned unevaluated by enclosing it in forward-quotes

```
>   'x';                          # This statement returns x unevaluated
```

$$x$$

Hence x may be cleared by assigning x to its name,

```
>   x:='x';                                    # This statement clears x
```

$$x := x$$

Alternatively, rather than use quotes the evaln function may be used: thus if

```
>   z:=sinh(cos(u));
```

$$z := \sinh(\cos(u))$$

z may be cleared with the assignment

```
>   z:=evaln(z);                               # This statement clears z
```

$$z := z$$

```
>   z;
```

$$z$$

A third method of unassigning variables is with the unassign command. The arguments of unassign must be surrounded by forward-quotes otherwise either there is no result or an error message is produced, depending upon the type of name.

```
>   z:=exp(x);        a:=1;
```

$$z := e^x$$
$$a := 1$$

```
>   unassign('z,a'); z;a;
```

$$z$$
$$a$$

Another common use of quoted variables is in the sum and similar commands. Suppose we require the sum of the first N integers. The sum procedure requires the use of an index; suppose we have already assigned the variable k to be 5, but have forgotten this. Then an error is produced:

```
>   k:=5:      s1:=sum(k,k=1..N);
```

```
Error, (in sum) summation variable previously assigned,
    second argument evaluates to, 5 = 1 .. N
```

This problem can be cured in a number of ways. First we could just use another summation variable or simply clear k before the sum command. Or we can delay the evaluation of k inside the sum by surrounding every occurrence of k with forward-quotes,

```
>   s1:=sum('k','k'=1..N);    k;
```

$$s1 := \frac{1}{2}(N+1)^2 - \frac{1}{2}N - \frac{1}{2}$$

$$5$$

Note that this method does not change the original value of k. Alternatively we could first clear k, if we are sure that we do not need its original value:

```
>   k:='k':    s1:=sum(k,k=1..N);    k;
```

$$s1 := \frac{1}{2}(N+1)^2 - \frac{1}{2}N - \frac{1}{2}$$

$$k$$

and k remains a free variable after being used in sum. Note, also that sum does not give the simplest form for the sum of the first N integers; to obtain this we need to factor the given expression. We illustrate one use of the factor command to obtain the usual result.

```
>   s1:=factor(s1);
```
$$s1 := \frac{1}{2} N (N+1)$$

Maple is decidedly schizophrenic when requiring variables to be free or not. The sum procedure requires the index to be free but the similar add procedure does not, neither does the seq command nor do loops, as the following examples show.

```
>   k:=10;
```
$$k := 10$$

```
>   sum(k^3,k=1..4);
```

```
Error, (in sum) summation variable previously assigned,
  second argument evaluates to, 10 = 1 .. 4
```

```
>   add(k^3,k=1..5); print('Value of k is ',k);
```
$$225$$
$$\textit{Value of } k \textit{ is }, 10$$

```
>   seq(k^5,k=0..4); print('Value of k is ',k);
```
$$0, 1, 32, 243, 1024$$
$$\textit{Value of } k \textit{ is }, 10$$

Note that the text in the print command is enclosed in *back-quotes*, for reason that are discussed later. Now use k as a loop counter to form the sum of the first 5 cubes — we shall discuss the syntax of loops later in this chapter.

```
>   y:=0:
>       for k from 1 to 5 do;
>       y:=y + k^3;
>       od:
>   print('Sum of first 5 cubes is',y,'  and the value of k is',k);
```
$$\textit{Sum of first 5 cubes is}, 225, \quad \textit{and the value of } k \textit{ is}, 6$$

So the loop construction does not require the counter, here k, to be clear but it does reset it. Thus we can use the same counter in a subsequent loop, for instance to find the sum of the cubes of the first five odd numbers.

```
>   y:=0:
>       for k from 1 to 9 by 2 do;
>       y:=y + k^3;
>       od:
>   print('Sum of cubes of the first 5 odd numbers is',y,
>           'and the value of k is',k);
```
$$\textit{Sum of cubes of the first 5 odd numbers is}, 1225, \quad \textit{and the value of } k \textit{ is}, 11$$

In general, if in doubt, it is best to be safe and clear the index before use, in whatever context it is used.

From this example the add and sum procedures would seem to be the same but they are subtly different, as shown in the next exercise.

Exercise 1.7

Use the sum procedure to show that

$$1^3 + 2^3 + 3^3 + \cdots + n^3 = \frac{1}{4}n^2(n+1)^2.$$

Can you obtain the same result with the add procedure?

This exercise illustrates the difference between sum and add. The command sum will first attempt a symbolic summation and evaluate this expression if successful: if symbolic summation is not possible it will attempt to sum the series numerically: if these methods fail it simply returns the original expression. On the other hand, add attempts no symbolic summation but simply adds up a given number of terms. There are occasions when either sum or add can be used to obtain an answer, for instance $\sum_{k=1}^{10^6} k$ can be evaluated by either, but sum is quicker as it uses the formula $n(n+1)/2$ with $n = 10^6$, whereas add simply adds up $1\,000\,000$ numbers. But add determines a numerical value of $\sum_{k=1}^{10^2} \sqrt{k}$ quicker as sum first attempts to find a formula, which does not exist.

Exercise 1.8

Use the sum procedure to show that

(i) $$\sum_{k=0}^{n-1} a^k = \frac{a^n - 1}{a - 1},$$

(ii) $$\sum_{k=0}^{n} \sin kx = \frac{\sin(nx/2)}{\sin(x/2)} \sin(x(n+1)/2),$$

(iii) $$\sum_{k=0}^{\infty} \frac{k^2}{(k+1)!} = e - 1,$$

(iv) $$\sum_{k=0}^{\infty} \frac{2k+1}{(k+1)(k+2)(k+3)} = \frac{3}{4}.$$

Note that in part (ii) you may have difficulty getting Maple to give this form of the sum.

Exercise 1.9

This exercise is set in order to show what can go wrong. Evaluate the four sums using Maple and determine which is evaluated correctly.

$$\sum_{k=0}^{\infty} \frac{k^2 + k - 2}{(k+1)!}, \quad \sum_{k=0}^{\infty} \frac{k^2}{(k+1)!}, \quad \sum_{k=0}^{\infty} \frac{k}{(k+1)!}, \quad \sum_{k=0}^{\infty} \frac{1}{(k+1)!}.$$

Exercise 1.10

Use the command sum(a^k,k=0..infinity) to find the value of the sum

$$S = \sum_{k=0}^{\infty} a^k$$

given by Maple. Under what circumstances is the answer correct?

If you cannot remember whether or not a variable has been assigned a value the command assigned can be used. For instance k is clear and s1 is not, so

> assigned(k); assigned(s1);

false

true

It is also possible to determine all the assigned names with the `anames()` command.

```
>  restart;
>  z:=x:   w:=u*v + a:   n:=2:   m:=1.2:
>  anames();
```
$$m, n, w, z$$

Thus `anames()` returns a sequence of names that are currently assigned values other than their own name; this will include names used internally by Maple so the list may be unwieldy.

The `evaln` function, introduced on page 11, is one of many *evaluation* commands required. We shall introduce these as necessary and here consider `evalf` which evaluates an expression to a floating point number if possible. This command can have two arguments: the second optional argument, which must be a positive integer, determines the number of digits to which the result is evaluated. The following sequence of commands illustrates some uses of `evalf` .

```
>  u:=sin(1);
```
$$u := \sin(1)$$

This can be evaluated to a 10 digit floating point number using `evalf`,

```
>  evalf(u);
```
$$.8414709848$$

but if we require more or less accuracy the second argument can be used; for 4 and 20 digit accuracy,

```
>  evalf(u,4);   evalf(u,20);
```
$$.8415$$
$$.84147098480789650665$$

Alternatively we could use the fact that floating point arithmetic is normally contagious, as the following examples show†:

```
>  sin(2.0);
```
$$.9092974268$$

```
>  GAMMA(3.0); GAMMA(3);
```
$$2.000000000$$
$$2$$

```
>  2*ln(3.0);
```
$$2.197224578$$

where `GAMMA(z)` is the gamma function, usually denoted by $\Gamma(z)$; note this is one of the few Maple functions having the whole name capitalised, probably to distinguish it from gamma= γ, the name reserved for Euler's constant:

$$\gamma = \lim_{n\to\infty}\left(1 + \frac{1}{2} + \frac{1}{3} + \cdots + \frac{1}{n} - \ln n\right) = 0.577\,216\cdots.$$

† The accuracy given with this method is determined by the value of the variable **Digits**, the default value of which is 10.

This constant, like π, normally remains in symbolic form until `evalf` is used:

```
> gamma; evalf(gamma);
```

$$\gamma$$
$$.5772156649$$

```
> Pi; evalf(Pi); 2*Pi; 2.0*Pi;
```

$$\pi$$
$$3.141592654$$
$$2\,\pi$$
$$2.0\,\pi$$

Note that in last case `2.0*Pi` is *not* evaluated to a floating point number: for expressions involving `Pi` numerical evaluation is achieved only by using `evalf`. Maple usually, but not always, automatically evaluates numbers like $\sin(\pi)$ and $\cos(\pi)$ to their correct numerical value rather than leaving them in symbolic form,

```
> sin(Pi); cos(Pi);
```
$$0$$
$$-1$$

The default number of digits `evalf` computes is 10; this can be changed by re-defining `Digits` and this affects most subsequent calculations.

```
> Digits:=40;
```
$$Digits := 40$$

```
> s:=evalf(sqrt(2));
```
$$s := 1.414213562373095048801688724209698078570$$

On squaring this, using 45 digits, we obtain

```
> evalf(s^2, 45);
```
$$2.00000000000000000000000000000000000000092808$$

Here we have used the optional second argument of `evalf` so that 45 digits are used only in this calculation.

Exercise 1.11
Find the values of π, e, $\sqrt{2}$, $2^{1/5}$, $\pi/\sqrt{10}$, and any other numbers you fancy, to 20 significant figures. Where possible check the results by applying suitable 'inverse' functions, for example apply a few trigonometric functions to the value of π, take the log of the value of e, etc.

When numerically evaluating more complex expressions some care is needed, as with all numerical computations, because severe cancellation can result in the numerical accuracy of a computation sometimes being unrelated to the value of `Digits`, except that increasing `Digits` normally improves numerical accuracy.

For instance the difference $\sin(2 \times 10^i \pi + y) - \sin y$ is identically zero, but by evaluating it to a given accuracy we see that for large values of i a nonzero value is returned. In this example we evaluate the difference to four figures in order to exaggerate the problem.

```
> y:=0.2:
```
Print headings for the lists computed below

```
> printf("%2s %9s %22s \n%35s","i","x","difference","sin(x)-sin(y)");
```

Now compute the differences:

```
>   for i from 1 to 5 do;
>   x:=evalf( 2*10^i * Pi +y,4);
>   printf("%2.0f   %9.2f      %12.10f\n",i,x, evalf( sin(x)-sin(y) ));
>   od:
```

i	x	difference sin(x)-sin(y)
1	63.04	.0079778507
2	628.60	.0790980805
3	6284.00	.5288455487
4	62840.55	.7587278078
5	628400.78	-.4092089823

We take this opportunity of introducing `printf` and the quotes ", about which more will be said in the next section. In this calculation the result is nonsense when $i = 3$ because 0.2 added to 6284 gives 6284 if only four digits are used; increasing the number of digits in the calculation of x moves the point at which this calculation is inaccurate to larger values of i.

It was mentioned that the two commands `evaln` and `evalf` were members of a family of commands which evaluate expressions in a variety of ways. Another useful command of this type is `eval`, which evaluates expressions; to understand why this is necessary consider the following commands, first remembering that Maple previously set $\cos 0$ to 1 and $\sin \pi$ to zero.

```
>   y:=sin(Pi*cos(x));
```
$$y := \sin(\pi \cos(x))$$

```
>   z:=subs(x=0, y);
```
$$z := \sin(\pi \cos(0))$$

In this case Maple seems not to know that $\cos 0 = 1$. In fact it does, but has not evaluated it, as substitution is one of the circumstances in which variables are not fully evaluated. In these cases evaluation is forced by using the `eval` command, thus

```
>   eval(z);
```
$$0$$

A form of the `eval` command can also be used in place of `subs`; for instance if we wish to find the value of the expression

```
>   y:=4*x^4 + 12*x^3 - 9*x^2 + 5*x -6:
```

at $x = 2$ then either of the following commands may be used:

```
>   subs(x=2,y);
```
$$128$$

or

```
>   eval(y,x=2);
```
$$128$$

With this use it would seem that `eval` and `subs` are equivalent; in fact `eval` is far more general as will be demonstrated in later examples.

One of the useful commands that is affected by changing the value of `Digits` is `fsolve`, which is a function for finding numerical solutions of equations; it is one of the `solve` family which exists in many guises. The following exercises provide some illustrations of the use of `fsolve` and some of its options.

Exercise 1.12

The command

```
fsolve(sin(x)=x/2, x=0..Pi, avoid={x=0});
```

will find a numerical approximation to the solution of the equation $\sin x = x/2$ for x in the interval $0 < x \leq \pi$. In this example we have used the range option, `x=0..Pi`, and the avoid option, `avoid={x=0}`, which tells the procedure to avoid particular values of the variable.

Investigate the effect of changing `Digits` on the result produced by this command.

Exercise 1.13

All the roots of a polynomial may be found with `fsolve` using the `complex` option. If y is the name of a polynomial the command `fsolve(y,complex)` returns all roots. Use this to find the roots of the polynomials

$$y = 1 + x + x^2 + x^4 + x^8 + \cdots + x^{2^n}$$

for $n = 1, 2, \ldots, 4$. Can you see where these roots lie in the complex plane?

Exercise 1.14

Find the approximate numerical value for the root of

$$x^4 + y^4 = \frac{3}{4}, \quad \sin(xy) = \frac{1}{2},$$

for x in the interval $0.8 < x < 1$.

1.4.3 Quotes

There are three different types of quotes used by Maple. The forward-quote ' has already been met in the previous section and is used to delay evaluation and consequently also to clear variables.

The back-quote ` is used to form a symbol or a name. The simplest symbol is a single letter possibly followed by numbers, letters, underscores and with lower and upper case letters distinct. A name may also be formed by enclosing any sequence of characters in a pair of back-quotes ( ` ` ); for instance we can write `This is a name`:=2. Any valid name formed without using back-quotes is precisely the same as the name formed with back-quotes, so `y` and y refer to the same name. Back-quotes are often useful in conjunction with the `print` command, as illustrated on page 13.

The double-quote " — a single key stroke — is used to delimit strings, which is a sequence of characters having no value and which cannot be assigned to. Thus a:="This is a string" creates a string a and sin(a), for instance, is an invalid command. The `convert` command can be used to convert between names and strings and vice versa; we shall see one reason why this is useful in chapter 2, where graphs are annotated. Strings are also needed in the `printf` command, as illustrated on page 16.

Finally we mention the % sign, which is an operator used to denote the previously defined expression. The related operators %% and %%% denote the second last and third last expression defined.†

1.4.4 Complex variables

The square-root of -1 is denoted by I in Maple and with this complex arithmetic is automatic.

```
>  c:=(2+3*I) + (5+2*I);
```

$$c := 7 + 5I$$

```
>  c1:=1/c;
```

$$c1 := \frac{7}{74} - \frac{5}{74}I$$

```
>  conjugate(c);
```

$$7 - 5I$$

```
>  abs(c);
```

$$\sqrt{74}$$

```
>  Re(c);
```

$$7$$

```
>  Im(c);
```

$$5$$

```
>  argument(c);
```

$$\arctan\left(\frac{5}{7}\right)$$

The argument function gives an angle, θ, in the interval $-\pi < \theta \leq \pi$. Note that the abs function can be used for both real and complex values; similarly most other functions that have natural extensions into the complex plane can be used with complex arguments.

```
>  cs:=sin(c);
```

$$cs := \sin(7 + 5I)$$

This expression may be put in the canonical form $a + ib$, where a and b are real, using the procedure evalc, one of the evaluation family, that symbolically evaluates expressions assuming all variables in the expression are real and puts the result in the canonical form.

```
>  cs:=evalc(cs);
```

$$cs := \sin(7)\cosh(5) + I\cos(7)\sinh(5)$$

and to evaluate this to numeric form we need to use evalf

```
>  cn:=evalf(cs);
```

$$cn := 48.75494167 + 55.94196773I$$

or, if fewer digits are required,

† This operator changed between Maple versions 4 and 5.

```
>  cn:=evalf(cs,4);
```
$$cn := 48.76 + 55.94\,I$$

More directly we could use `cn:=evalf(sin(c),4)`.

Non-integer powers can cause problems because it is necessary to know which branch is required; for instance the Maple square-root function has its branch cut along the negative real axis, so that

$$\sqrt{-1 + i\epsilon} = \begin{cases} i + \epsilon/2 + i\epsilon^2/8 + \cdots, & \epsilon \geq 0, \\ -i - \epsilon/2 - i\epsilon^2/8 + \cdots, & \epsilon < 0. \end{cases}$$

Thus,

```
>  sqrt(-8);
```
$$2I\,\sqrt{2}$$

Exercise 1.15
Given that $z_1 = 1 + i\sqrt{3}$ and $z_2 = 4 + 3i$ find the moduli and arguments of $z_1 z_2$, z_2/z_1 and $\sin z_1 \sinh z_2$.

Exercise 1.16
If $z_1 = 2 + i$, $z_2 = -2 + 4i$ and

$$\frac{1}{z_3} = \frac{1}{z_1} + \frac{1}{z_2}$$

use `solve` to find z_3.

Exercise 1.17
Find the complex numbers z such that

$$|z + 2i - 3| = |z + 3i| \quad \text{and} \quad \arg z = \frac{\pi}{4}.$$

In order to solve this problem set $z = re^{i\pi/4}$ and assume $r > 0$.
Hint: it is easier to do this exercise using the `assume` command, introduced later, so we advise that this exercise is left for revision.

Exercise 1.18
It is often necessary to perform numerical calculations to higher precision than is required in the final result. The command `fnormal(e,d)` sets to zero all numerical values in the expression e that are smaller than 10^{-d}.

Find the values returned by the commands

```
fnormal(sqrt(-4+I*eps),5) and fnormal(sqrt(-4-I*eps),5)
```

where eps$= 1.0 \times 10^{-20}$, that is eps=1.0e-20, and explain the results.

1.5 Simple graphs

One of the very useful features of Maple is the ease with which many different types of graphs may be drawn; we shall make fairly extensive use of this facility but expect you to learn most about the graphics package from the help files. Here we provide an introduction to the simplest graphs; later in this chapter, when more Maple has been introduced, more complex graphs will be produced. It is advisable to try all the examples in this section†.

All graphs are drawn with variants of the plot command, but this has several forms and each has many optional arguments which determine the appearance of the resulting graph. A few of these forms and a few of the options will be introduced in this section; if, at any time, you wish to see a list of options type ?plot,options at the prompt and the relevant help page will appear.

1.5.1 Line graphs

The simplest graphs are those of functions of a single real variable; for example, the graph of the function $\sin(10/(1 + x^2))$ for $0 < x < 10$ is given with the command

```
>   plot(sin(10/(1+x^2)),x=0..10);
```

This graph is shown in figure 1.1. In this command the first argument is the formula for the graph and the second is the range of x; if we needed the graph for $10 < x < 100$ we should replace 0..10 by 10..100. On the other hand the command plot(sin(10/(1+x^2)),x) plots the graph over the default range $-10 \le x \le 10$.

The extent of the y variation is limited by adding a second range: thus if only the range $0 < y < 1$ is required we use the command

```
>   plot(sin(10/(1+x^2)), x=0..10, 0..1);
```

to give figure 1.2.

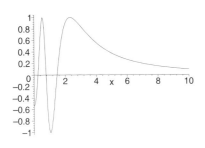

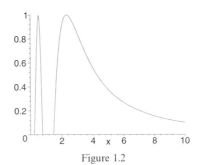

Figure 1.1   Figure 1.2

A graph may be assigned to a variable and used subsequently — we shall see why this is useful in section 1.10 — with the command

```
>   g1:=plot(sin(10/(1+x^2)), x=0..10):
```

Notice that with this type of command it is advisable to use a colon terminator.

In order to look at this graph, simply type g1;. You should try both these commands and examine the effect of replacing the terminating colon in the assignment of g1 by a semi-colon.

---

† Working versions of all the examples given in this section can be found on the web site described in the preface.

Exercise 1.19

Assign the expression $\sin x + 2\sin 2x + 3\sin 3x + 4\sin 4x$ to the variable y and use the command plot(y,x=0..2*Pi) to draw the graph of this function.

Exercise 1.20

Use the sum command to create the series

$$y = \sum_{k=1}^{n} \frac{\sin kx}{k}$$

for $n = 5$, 10, 15 and 20 and plot the graphs of these functions for $-2\pi \le x \le 2\pi$.

Simple changes to graphs can be made after they are drawn by placing the cursor in the vicinity of the figure and clicking the left mouse button. Then a new menu appears on the fourth bar which allows some changes: the extreme left hand box gives the cursor coordinates which is useful for finding approximate coordinates of, say, stationary points. The same changes can be made, possibly more easily, by right-clicking with the cursor in the vicinity of the figure to produce a more conveniently placed menu.

The default position of the figure is in the centre of the page; it may be either right- or left-justified by left-clicking on the figure and using the *Format* item in the upper menu bar.

This context-sensitive behaviour is a general feature of Maple, but it is best learnt by doing it because the commands are rather difficult to describe in words. However, a start can be made by typing in a function, say $\sin(5\sin(x))$ at the prompt, pressing *return*, to create some Maple output, then right-clicking on this Maple output to produce a menu from which a graph can be drawn. After playing with this try the function $\sin(3\sin x\cos y)$, to obtain a three-dimensional plot.

1.5.2 Parametric curves

Many interesting curves are described parametrically, that is by specifying the x- and y-coordinates as functions of a third variable, t. For instance the cycloid, the curve traced out by a point on a circle of radius a rolling without slipping along the x-axis, is

$$x = a(t - \sin t), \quad y = a(1 - \cos t).$$

The graph of this curve with $a = 1$, shown in figure 1.3, may be drawn with the commands

```
> x:=t-sin(t):  y:=1-cos(t):
> plot([x,y,t=0..2*Pi], scaling=constrained, colour=black);
```

Note that x, y and the independent variable, t, are separated by commas and all three are surrounded by square brackets. Here we have introduced the scaling options because we wanted the length of the y-axis to be in proportion to that of the x-axis. The alternative, scaling=unconstrained, is the default and you should try this to see how different the graph looks. Here we have also used the colour option; a list of colours available can be found by typing ?plot,colour at the prompt, where you can also find out how to define your own colours. Maple accepts both the English and American spellings of colour.

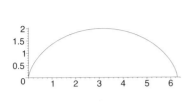

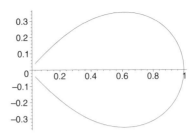

Figure 1.3   A cycloid.   Figure 1.4   The lemniscate of Bernoulli.

Curves expressed in polar coordinates may also be drawn using the option `coords=polar`; the default is `coords=cartesian` so in many cases this optional argument is unnecessary. An example is the lemniscate of Bernoulli, which can be expressed in terms of the polar coordinates (r, θ) by the equation $r^2 = a^2 \cos 2\theta$ and its graph, for $x > 0$, is given by

```
>   plot([sqrt(cos(2*t)),t,t=-Pi/4..Pi/4], coords=polar,
>   numpoints=1000,colour=black);
```

This curve is shown in figure 1.4. Here we need to use the `numpoints` option to insist that 1000 points are used, necessary in this case because equally spaced points in t, which Maple uses, are compressed towards the right hand side of the curve. The order in which options occur is irrelevant.

Exercise 1.21
One set of parametric curves are Lissajou's figures given by the equations $x = \sin nt$, $y = \cos mt$, $0 < t \le 2\pi$, where (n, m) are different, coprime positive integers. Investigate the shapes of some of these curves for various pairs (n, m).

1.5.3 Implicit and contour plots
Maple has facilities for drawing many types of curves and surfaces, some of which require the use of library packages which need to be loaded. There are many different sets of library packages, all of which are accessed in the same manner, that is by typing `with(name)`, name being the name of the required package: a list of these packages can be found by typing `?with` and clicking on `index[packages]` at the end of this help item. Here we shall use some of the procedures accessed by `with(plots)` — if you type `with(plots);` a list of available procedures will appear, as shown below; normally it is better to use the colon terminator. Here we consider `implicitplot` and `contourplot`. First type

```
>   restart; with(plots);
```

[*animate, animate3d, animatecurve, changecoords, complexplot, complexplot3d,*

  conformal, contourplot, contourplot3d, coordplot, coordplot3d, cylinderplot,

  densityplot, display, display3d, fieldplot, fieldplot3d, gradplot, gradplot3d,

  implicitplot, implicitplot3d, inequal, listcontplot, listcontplot3d, listdensityplot,

  listplot, listplot3d, loglogplot, logplot, matrixplot, odeplot, pareto, pointplot,

  pointplot3d, polarplot, polygonplot, polygonplot3d, polyhedra_supported,

  polyhedraplot, replot, rootlocus, semilogplot, setoptions, setoptions3d,

  spacecurve, sparsematrixplot, sphereplot, surfdata, textplot, textplot3d, tubeplot]

Suppose that a curve in the (x, y)-plane is defined implicitly by an equation $z(x, y) =$ constant. For instance if

$$z = 6x^2 + 8y^3 + 3(2x + y - 1)^2, \tag{1.1}$$

we first define the expression

```
>   z:=6*x^2 + 8*y^3 + 3*(2*x+y-1)^2:
```
and then the graph defined by $z(x, y) = 1$ in the region $-1/2 < x < 1$, $-1 < y < 1$, is produced by the command

```
>   implicitplot(z=1,x=-0.5..1,y=-1..1, grid=[50,50]):
```
Here the `grid` option is used because the default values often result in a poor resolution of the curve, there being too few points. Generally it is better to start with the default grid size because increasing the number of points slows the plotting procedure.

   If two curves, $z = 1$ and $z = 2$, say, are required then we must enclose the two equation in *curly* brackets (other types of brackets result in an error).

```
>   implicitplot({z=1,z=2}, x=-0.5..1,y=-1..1, grid=[50,50]);
```
This command produces the curves shown in figure 1.5.

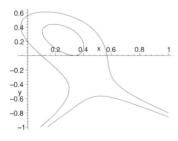

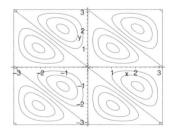

Figure 1.5   The curves $z = 1, 2$
with z given in equation 1.1.

Figure 1.6   Contours of
equation 1.2.

Contour plots are obtained equally easily: for example the contours of the function

$$z = \sin x \sin y \sin(x + y), \tag{1.2}$$

are plotted with the following sequence of commands:

```
>  z:=sin(x)*sin(y)*sin(x+y):
>  cl:=[-1,-0.8,-0.6,-0.4,-0.2,0,0.2,0.4,0.6,0.8,1]:
>  contourplot(z,x=-Pi..Pi,y=-Pi..Pi,grid=[50,50],contours=cl);
```

In this example we have used the option contours which defines the values of the function at which the contours are drawn; these values are best put in the list cl and then fed into the contour plotting procedure as shown.

You will notice that there is no indication in this contour plot, shown in figure 1.6, as to which are the high and low points; when drawn in colour on the computer screen the relative heights become clearer, but it is a defect of this routine that the values of the contours are not readily displayed. Thus it is sometimes helpful to draw some supplementary graphs, such as the graph of z on the lines $y = x$, or $y = -x$. Again contour plots devour computer time, so it is best to draw preliminary graphs with the default grid size.

1.5.4 Plotting functions of two variables: plot3d

Plotting graphs of functions of two variables is as easy as for functions of one variable, but now the plot3d procedure is used. For instance, suppose that the shape of the function $z = \sin(5\cos x \sin y)$ is required, then the relevant commands used to plot the surface shown in figure 1.7 are

```
>  z:=sin(5*cos(x)*sin(y)):
>  plot3d(z,x=-Pi..Pi,y=-Pi..Pi,axes=boxed,
>  style=patchcontour,orientation=[64,65]);
```

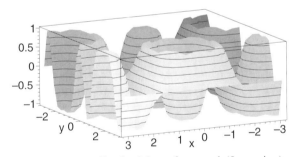

Figure 1.7   Graph of the surface $z = \sin(5\cos x \sin y)$.

The general form of this command is plot3d(z,x=a..b,y=c..d, options) but initially the options can be ignored as they may be chosen interactively. Try typing the command plot3d(z,x=-Pi..Pi,y=-Pi..Pi);, where z is defined above, or in general it may be any expression involving the variables whose ranges are listed in the argument list. After typing this command (and return) a form of the graph shown in figure 1.7 should appear on the screen; this can be modified in a variety of ways simply by using the menu that appears when you right-click in the vicinity of the figure — you should experiment with the items on this menu. The orientation of the figure is changed simply by using the left mouse button: press on the left button with the cursor in the vicinity of the figure and move the mouse while the button is depressed and you should find that the figure rotates in the direction of the cursor (unless the figure has been turned upside down) and the angles defining the direction from which the graph is being viewed are displayed in the top left hand corner of the screen.

If you wish to make many such plots in the same Maple session then the continual defining of the various options becomes tedious, consequently it is possible to change the default settings during a session using the `setoptions3d` command. Thus the options for figure 1.7 are fixed as follows:

```
>   with(plots):
>   setoptions3d(axes=boxed,style=patchcontour,
>                               orientation=[64,65]):
```

Note the need to invoke `with(plots)`. Now the same graph will be drawn with the simpler command `plot3d(z,x=-Pi..Pi,y=-Pi..Pi)`. There is a similar command `setoptions` which allows one to define options for the plot command. There is no need to re-define these options after a `restart` command.

Exercise 1.22
Plot the graph of the function

$$y = 15x^5 - 15x^4 + 20x^3 - 330x^2 + 600x + 2, \quad 0 < x < 3.$$

Find graphically, by carefully positioning the cursor, the approximate position and value of the stationary points of y.

Exercise 1.23
Draw graphs of the curves given by the parametric equations

$$x = 2\cos\theta + \cos n\theta, \quad y = 2\sin\theta - \sin n\theta, \quad 0 \le \theta < 2\pi,$$

for various integer values of $n \ge 2$.

Exercise 1.24
Draw graphs of the curves defined by the equation $z(x, y) = \pm 1$ for $-2 < x < 5, -2 < y < 5$, where $z(x, y) = y^3 + x^3 - 9xy + 1$.

Exercise 1.25
Use the `plot` command, with no options, to draw the graph of $y = e^x + \ln|4 - x|$ for $0 \le x \le 5$. Comment on the result you obtain.
Hint: what is the value of y at $x = 4$?

Exercise 1.26
Plot the graph of the function $z(x, y) = x(x^2 - 3y^2)$ for $-1 \le x, y \le 1$.

1.6 Some useful programming tools

One of the main advantages of computers is that repetitive operations are readily performed. Whilst Maple can be used as a very sophisticated calculator, it has far more potential because it is quite a powerful programming language. In this section we introduce three very useful tools, the `seq` command, the `if` conditional and `loops`.

1.6.1 The seq command

A particularly useful command is seq which creates sequences. For instance a sequence of the first 10 even numbers is generated by

```
seq(2*k,k=0..9);
```

and the 10 odd numbers following 101 by

```
seq(101+2*k,k=1..10);
```

and the 101th to 110th primes are

```
>  seq(ithprime(k),k=101..110);
```
$$547, 557, 563, 569, 571, 577, 587, 593, 599, 601$$

In general a command of the form `seq(f(k),k=m..n)` will produce the sequence `f(m),f(m+1),...,f(n-1),f(n)`; with this use it is necessary that $m \leq n$. The seq command can, however, be used for quite complicated problems, and here we give two illustrative examples.

The equation

$$\tan \pi x = x, \quad x > 0,$$

has an infinite set of roots, x_k, $k = 1, 2, \ldots$ with $k < x_k < k + 1/2$, as may be seen by drawing the graphs of $y = \tan \pi x$ and $y = x$. A subset of these may be found and put in a `list` using fsolve and seq. The command

```
fsolve(tan(Pi*x)=x,x=k..k+1/2);
```

will find a numerical approximation to the kth root, so combining this with seq in the form

```
>  z:=[seq(fsolve(tan(Pi*x)=x,x=k..k+1/2),k=1..5)];
```
$$z := [1.290109651, 2.373052972, 3.409179047, 4.429320704, 5.442155595]$$

gives numerical approximations for the first five roots. Note, the outer square brackets which are necessary to ensure that z is a list, see section 1.8. Another way of creating this list uses the `loop` construction described later in this section.

For the second example we create a list of plot structures; the reason why this is useful will be seen in section 1.8, page 34. Suppose we want to examine the shapes of the graphs $y = x^k(1 - x)^k$ as k varies between 1 and 10. We first create a `list` of graphs of the function for $k = 1, 2, \ldots, 10$ and $0 \leq x \leq 1$; this is achieved with the following command

```
g:=[seq(plot(x^k*(1-x)^k, x=0..1),k=1..10)]:
```

Note that we have used the colon terminator; with a semi-colon much unwanted output is produced. The graph of $y = x^5(1 - x)^5$ is obtained with the command g[5];. Other, more interesting ways of using such lists are considered in section 1.8.

Exercise 1.27
Form a list of the plot structures of the graphs $y = x^k(1 - x^k)$ for $0 \leq x \leq 1$ and $k = 1, 2, \ldots, 20$.

Exercise 1.28

Use the `fsolve` and `seq` commands to create a list of the first ten zeros of the equation

$$\tan \pi x = -\frac{1}{x}, \quad x > 0.$$

Hint: first show that the kth root, x_k, is in the interval $k - 1/2 < x_k < k$.

*1.6.2 The **if** conditional*

A useful construction is the `if` conditional, which has the elementary form

if *condition* **then** *command* **else** *command* **fi**;

Note that this conditional must be ended with `fi;`. You can look up this conditional statement under *Topic Search* and by typing *if* in the topic box, where other examples of its use may be found.

This conditional is normally used to control steps through a program and will be needed frequently in the problems encountered later: here we provide a trivial example, in order to illustrate its syntax.

In this example we test whether or not a given number is prime and print an appropriate message.

```
>  k:=98478389453934:
>  if isprime(k) then
>        printf("The number k=%15.0f is a prime.",k)
>        else
>        printf("The number  k=%15.0f is not a prime.",k)
>  fi;
```

```
The number  k= 98478389453934 is not a prime.
```

On the other hand, for a different number

```
>  k:=98478389453947:
>  if isprime(k) then
>        printf("The number k=%15.0f is a prime.",k)
>        else
>        printf("The number  k=%15.0f is not a prime.",k)
>  fi;
```

```
The number k= 98478389453947 is a prime.
```

`if` conditionals can be nested any number of times. For instance the following modification of the previous test also determines whether the given number is divisible by 3, if is not a prime, by the addition of another conditional.

```
>  k:=9874473:
>  if isprime(k) then
>  printf("The number k=%10.0f is a prime.",k)
>  else                                                      #(1)
>        if k mod 3 = 0 then                                 #(2)
>        printf("The number k=%10.0f is divisible by 3.",k)
>        else
>        printf("The number k=%10.0f is not a prime and is
                     not divisible by 3.",k)
>        fi;                                                 #(3)
>  fi;
```

```
The number k=   9874473 is divisible by 3.
```

In this type of nesting it is possible to contract the else in line (1) and the if in line (2) into an elif , line (4) below, provided the fi of line (3) is also removed. Thus the shortened code is:

```
>  k:=9874473:
>  if isprime(k) then
>  printf("The number k=%10.0f is a prime.",k)
>  elif                                                        #(4)
>      k mod 3 = 0 then
>      printf("The number k=%10.0f is divisible by 3.",k)
>      else
>      printf("The number k=%10.0f is not a prime and is
                    not divisible by 3.",k)
>  fi;
```

```
The number k=  9874473 is divisible by 3.
```

Other examples of this construction are given in the next section on loops.

Maple 6

In Maple 6 the if conditional can also be finished by typing end if — note the space — so the construction becomes:

$$\textbf{if } \textit{condition } \textbf{then } \textit{command } \textbf{else } \textit{command } \textbf{end if;}$$

Both constructions are valid in version 6.

1.6.3 Loop constructions

The for and while loops are programming devices for performing repetitive operations. Consider the simple task of multiplying the number 1234 by 1, 2, 3, 4 and displaying the results; this is accomplished with the following command:

```
>   for k from 1 to 4 do;
>   1234*k;
>   od;
```

$$1234$$
$$2468$$
$$3702$$
$$4936$$

Or, if we need the sum of all the odd numbers from 1 to 21,

```
>   sm:=0:
>       for k from 1 to 21 by 2 do;
>       sm:=sm + k;
>       od:
>   printf("Sum of odd numbers from 1 to 21 is %4.0f",sm);
```

```
Sum of odd numbers from 1 to 21 is  121
```

In this case we put a colon after the end command, od:, in order to suppress output from the loop interior. This use of the colon terminator is useful if there are many steps and the intermediate results of little interest.

These two examples show the general structure of the for loop:

for *counter* **from** *first counter value* **by** *increment* **to** *last counter value* **do**
             commands to be repeated
od;

It is important to remember that the last command must be od; or od:. If the increment is 1 it may be assumed, as in the first example of this section. Also if the value of the first counter is 1 it may be omitted:

```
>  sm:=0: for k to 121 by 2 do;   sm:=sm + k;   od:
>  printf("Sum of odd numbers from 1 to 121 is %4.0f",sm);

   Sum of odd numbers from 1 to 121 is 3721
```

And the default value of the last counter is infinity, so the construction

for *counter* **do;**   *commands* **od;**

is equivalent to

for k from 1 by 1 to infinity do; *commands* od;

Such a loop will run forever unless it is stopped in some manner, for example by using the break construction, described below.

Loops can also be controlled by conditional statements; for instance the product of all the odd numbers between -5 and 5 can be computed as follows.

```
>  prod:=1:
>      for k from -5 by 2 while k<6 do;
>      prod:=prod * k;
>      od:
>  printf("The product of the odd numbers from -5 to 5 is%5.0f",prod);

   The product of the odd numbers from -5 to 5 is -225
```

The body of a loop can also be controlled by the if conditional introduced above. Suppose we need the sum of all the integers between 1 and 100 which are not prime numbers, then we write,

```
>  sm:=0:
>      for k from 1 to 100 do;
>      if not isprime(k)  then sm:=sm + k fi;
>      od:
>  sm;
```
3990

Note that 1 is not a prime.

Another variant is the while loop having the form

while *condition* **do;** *commands* **od;**

Suppose we require the first integer larger than 10^{10} divisible by 109; this can be found using the modp function — which you should look up — and the commands,

```
>  n:=10^10:
>      while modp(n,109) >0 do;
>      n:=n+1
>      od:
>  printf("First number is %12.0f",n);

   First number is   10000000080
```

Care is necessary to ensure that such loops can end.

Sometimes it is necessary to either break out of a loop or to proceed directly to the next iteration; in Maple the commands break and next allow one to do this.† For instance the product of the odd numbers between −5 and 5 could be obtained as follows

```
>   prod:=1:
>      for k from -5 to 5 do;
>      if modp(k,2)=0 then next else prod:=prod*k fi;
>      od:
>   printf("The product of the odd numbers from -5 to 5 is %5.0f",prod);
```

```
The product of the odd numbers from -5 to 5 is -225
```

And if we require all the prime numbers up to 10,

```
>   for k do;
>   j:=ithprime(k);
>   if  j >= 10 then break else print(j) fi;
>   od:
```

$$2$$

$$3$$

$$5$$

$$7$$

Note that in many cases the seq command and a loop construction can be used for the same purpose, but the seq command normally leads to more elegant code. Thus the sequence of roots of $\tan \pi x = x$, $x > 0$, found in the previous section, is also given by the following commands:

```
>   restart;
>   z:=NULL:
>      for k from 1 to 6 do;
>      z:=z, fsolve(tan(Pi*x)=x,x=k..(k+1/2));
>      od:
>   z:=[z];
```

$z := [1.290109651, 2.373052972, 3.409179047, 4.429320704, 5.442155595,$

$6.451047259]$

The only new idea is the NULL sequence defined in the first line. The loop is used to build up the sequence of roots but, in this case, use of seq is neater.

The loop counter may also be taken from the elements of a list, as illustrated on page 36.

Loops may be embedded, as in the following simple example which evaluates the double sum

$$S = \sum_{r=1}^{10} \sum_{s=1}^{10} e^{-rs}.$$

```
>   sm:=0:
>      for r to 10 do;
>         for s to 10 do;
>         sm:=sm + evalf( exp(-r*s) );
>         od:
>      od:
>   printf("Value of double sum is %10.8f",sm);
```

```
Value of double sum is   .82020667
```

† The **continue** command seems to be synonymous with **next** but it is not mentioned in the help files.

A more elegant, and probably faster, method of evaluating this sum is to use two applications of the add command

```
>   sm:=add( add(evalf(exp(-r*s)), r=1..10),s=1..10);
```
$$sm := .8202066675$$

The sum is computed much faster if evalhf is used in place of evalf because it uses the underlying floating point arithmetic to double precision accuracy; on my computer this gives 16 significant figures. In this example the same computation is performed at least ten times faster with evalhf than with evalf, but there are restrictions on the use of evalhf which are described in the help files: for instance evalhf does not work in the sum command, though it does work with add.

Maple 6

In Maple 6 the for loop may also be finished by typing end do — note the space — rather than od, so the construction becomes:

for *counter* **from** *first counter value* **by** *increment* **to** *last counter value* **do**
                        commands to be repeated
end do;

Both constructions are valid in version 6.

Exercise 1.29
Use the loop construction to find the value of the sums

$$S_1 = \sum_{k=1}^{100} \frac{1}{\sqrt{k}}, \quad S_2 = \sum_{k=1}^{10} \frac{1}{1+k^2}, \quad S_3 = \sum_{n=1}^{10} \sum_{m=1}^{20} \frac{1}{\sqrt{n^2+m^2}}.$$

Exercise 1.30
The factorial function of the integer n is defined by the product

$$n! = n \times (n-1) \times (n-2) \times \cdots \times 3 \times 2 \times 1.$$

Write a loop to compute $n!$ and compute 10! and 21! and compare your results with the value given by the factorial function provided by Maple.

Exercise 1.31
The double factorial function for the integer n is defined to be

$$n!! = n \times (n-2) \times (n-4) \times \cdots \times \begin{cases} 4 \times 2 & \text{if } n \text{ is even} \\ 3 \times 1 & \text{if } n \text{ is odd} \end{cases}$$

Write a loop to compute $n!!$ and use it to show that $20!! = 3\,715\,891\,200$ and $31!! = 191\,898\,783\,962\,510\,625$.

Exercise 1.32
Find all the prime numbers between 100 and 150.

Exercise 1.33
In one loop using conditional statements determine all the numbers between 100 and 200, inclusive, that are divisible by 7, 11 or 13.

Exercise 1.34
Find the values of x_k, $k = 0, 1, 2, \ldots$, where

$$x_{k+1} = \frac{1}{2} \left(x_k + \frac{2}{x_k} \right), \quad x_0 = 1,$$

until $|x_{k+1} - x_k| < 10^{-4}$.

Exercise 1.35
Find the first two consecutive prime numbers such that the difference between them is divisible by $m = 10$. Next consider all values of m in the range $m = 2, 3, \ldots, 20$.

Exercise 1.36
For the real number x in the range $1 \leq x \leq 3$ find the integer $N(x)$ such that

$$\sum_{k=1}^{N(x)} \frac{1}{k^x} \geq 1.3.$$

1.7 Expression sequences

We have come to the point where it is necessary to examine some of the basic types of Maple objects, because without some notion of these types it can be difficult to use and understand Maple.

One of several basic Maple structures is the expression sequence, which is just a group of Maple expressions separated by commas.

```
>  1,2,3,4;
```
$$1, 2, 3, 4$$

```
>  a,b,c,x;
```
$$a, b, c, x$$

```
>  sin(x),cos(x),arccos(x);
```
$$\sin(x), \cos(x), \arccos(x)$$

Expression sequences are neither lists nor sets; they are a distinct data structure having their own properties. For instance, they preserve the order and repetition of their elements; items stay in the order they are entered, and should you enter an item twice both copies remain.

The concatenation operator is an elementary name creating operation that can be used with sequences. There are two forms of this operator, the first of which is written in Maple as '.' and is used as follows.

```
>  a.b;
```
$$ab$$

This is a binary operator requiring a name or a string as its left operand: the right

operand is evaluated and then concatenated to this name or string. The other form
uses the `cat` function,

```
>  cat(a,b);
```

$$ab$$

which can be used to concatenate any number of names or strings. In Maple 6 the
first form, `a.b`, has become `a||b`, and in this version the user is encouraged to use `cat`
function if possible.

When applied to a sequence this operator affects each element. For example, if `S` is
the sequence of numbers

```
>  S:=1,2,3,4:
```

we have

```
>  a.S;
```

$$a1, a2, a3, a4$$

The order is important: the commands `a.S` and `S.a` produce different effects. Also the
command `cat(a,S)` is different and produces `a1a2a3a4`. One use of this operator is in
the creation of power series,

```
>  n:=5:   y:=add(a.k * x^k, k=0..n);
```

$$y := a0 + a1\,x + a2\,x^2 + a3\,x^3 + a4\,x^4 + a5\,x^5$$

or, equivalently

```
>  y:=add(cat(a,k) * x^k,k=0..n):
```

It is worth noting here that this type of construction involves x^0 in the first term: if
x is subsequently set to zero this can, in some circumstances, cause an error so it is
always safer to use the construction

```
y:=a0 + add(cat(a,k) * x^k,k=1..n):
```

A variable assigned to an expression sequence can be cleared by any of the methods
described above.

1.8 Lists

A list is created by enclosing a Maple expression sequence in *square* brackets. At first
this may not seem a very useful construction but, as will be seen in this and the
next two chapters, it facilitates many useful operations. In this section we shall first
concentrate on the manipulation of lists before describing one of their common uses
for producing graphs. Other uses will become apparent later. Some lists are

```
>  data:=[1,2,3,10,20];
```

$$data := [1, 2, 3, 10, 20]$$

```
>  poly:=[1,x,(3*x^2-1)/2,(5*x^3-3*x)*x/2];
```

$$poly := [1,\, x,\, \frac{3}{2}x^2 - \frac{1}{2},\, \frac{1}{2}(5\,x^3 - 3\,x)\,x]$$

```
>  Fourier:=[1,sin(x),cos(x),sin(2*x),cos(2*x),sin(3*x),cos(3*x)];
```

$$Fourier := [1,\, \sin(x),\, \cos(x),\, \sin(2\,x),\, \cos(2\,x),\, \sin(3\,x),\, \cos(3\,x)]$$

Maple preserves the order and repetition in a list, so $[a, b, c]$, $[b, a, c]$ and $[a, a, b, c]$ are different; one can extract a particular element from a list by counting from the left hand end, starting from 1; thus

```
> data[4];
```

$$10$$

```
> Fourier[3];
```

$$\cos(x)$$

The number of elements in a list can be determined using the nops command,

```
> nops(Fourier);
```

$$7$$

An alternative method of extracting particular elements from a list, or making sub-lists, is with the op command, (which has a far wider range of uses)

```
> op(5,data);
```

$$20$$

```
> L1:=[op(4..7,Fourier)];
```

$$L1 := [\sin(2x), \cos(2x), \sin(3x), \cos(3x)]$$

The command op(Fourier) produces the whole expression sequence inside the list.
   Another, convenient, construction is

```
> L1:=Fourier[4..7];
```

$$L1 := [\sin(2x), \cos(2x), \sin(3x), \cos(3x)]$$

As with expression sequences, variables assigned to lists can be cleared using any of the methods described above, for example

```
> L1:=[2,3,4,t,r,s,te]; L1:='L1': L1;
```

$$L1 := [2, 3, 4, t, r, s, te]$$

$$L1$$

Individual elements of a list are most easily changed simply by re-assigning the particular element. For instance, if the fourth element of data should be 15, we set

```
> data[4]:=15: data;
```

$$[1, 2, 3, 15, 20]$$

The more general subsop command, which you should look up, can also be used.
   Lists can also be combined or concatenated using the op command; for example if

```
> n1:=[George, Henry, Nicholas]:
> n2:=[Natashia, Anastasia, Emma]:
```

the concatenated list is

```
> nc:=[op(n1),op(n2)];
```

$$nc := [George, Henry, Nicholas, Natashia, Anastasia, Emma]$$

and we can append or prepend elements to a list in a similar manner,

```
> n3:=[op(n2),Anna];
```

$$n3 := [Natashia, Anastasia, Emma, Anna]$$

and

> `n4:=[Peter,op(n1)];`

$$n4 := [Peter, George, Henry, Nicholas]$$

and elements can be inserted as follows:

> `n5:=[op(1..2,n2),Catherine, Olivia, op(n1)];`

$$n5 := [Natashia, Anastasia, Catherine, Olivia, George, Henry, Nicholas]$$

Elements in a list can also be used to control loops, as in the following simple example.

```
>   L:=[4,6,2,5,9,3,1,0,7]:
>      for i in L do;
>      if i > 5 then print(i) fi;
>      od;
```

$$6$$

$$9$$

$$7$$

Finally we note that the `seq` command provides a particularly convenient way of creating lists: thus the first 11 members of the list of numbers 2, $2^{1/2}$, $2^{1/4}$,... is given, to six decimal places, by the command

> `z:=[seq( evalf( 2^(2^(-k)),6),k=0..10)];`

$$z := [2., 1.41421, 1.18921, 1.09051, 1.04427, 1.02190, 1.01089, 1.00543, 1.00271,$$
$$1.00135, 1.00068]$$

Exercise 1.37

Elements of a list may be added or multiplied using the `convert` command, which we shall deal with more fully in chapter 2. Form the list `L:=[1,2,...,n]` for various specific values of n using the `seq` command and then use the commands `convert(L,'+')` and `convert(L,'*')`, not forgetting the back-quotes, to form the sum and product of the elements in the list. Recall that back-quotes are used to define names when special characters are used.

1.8.1 The use of lists to produce graphs

One important use of lists is producing certain types of graphs: here we give two illustrations. Suppose that we wish to compare graphically two functions, for example $\cos^{10} x$ and $\exp(-5x^2)$ for $0 < x < 1$. The command

`plot([cos(x)^10, exp(-5*x^2)],x=0..1, colour=[red, blue])`

will plot $\cos^{10} x$ as the red line and $\exp(-5x^2)$ as the blue line, because here we have put the two functions and the two required colours in a list. We have not shown this graph in the text as the difference between the two functions is too small to be noticeable. You should plot this graph yourself. The reason why these two functions are almost identical for $|x| < 1$ is given in exercise 3.54 (page 127).

In the next example we show how lists can be used to graph data from sequences of points. We plot a graph showing the difference between successive prime numbers starting at, say, the 100th prime and continuing for the next 100 primes.

This little program also contains some useful programming hints, which you should study. The object of the program is to create a list, s, of the difference between adjacent primes, and this is done by incrementally adding numbers to an expression sequence s, which is initially defined to be the NULL sequence, and keeping a list of the primes in an expression sequence x.

First we demonstrate this program for the difference of $N = 10$ primes.

```
>   s:=NULL:                        # Define the NULL sequence.
>   N:=10:                   # Define the number of primes to be used.
>   n:=ithprime(100): x:=n:           # Define x=n,  the 100th prime.
>     for k from 1 to N do;          # Start the loop round the primes
>       m:=nextprime(n);         # Define m to be the next prime after n
>       x:=x,m:                  # Expression sequence of all the primes
>       s:=s,(m-n);           # Append the difference to the sequence s
>       n:=m;                                        # Reset n to m
>     od:                                               # End of loop
>   s:=[s]; x:=[x];                        # Convert s and x to lists
```

$$s := [6, 10, 6, 6, 2, 6, 10, 6, 6, 2]$$

$$x := [541, 547, 557, 563, 569, 571, 577, 587, 593, 599, 601]$$

In order to draw a graph of these numbers we need a list of coordinates (x_k, s_k), in Maple these will be represented by [x[k],s[k]], and we form this list, denoted by s1, using the seq command

```
>   s1:=[ seq([x[k],s[k]],k=1..N) ];
```

$s1 := [[541, 6], [547, 10], [557, 6], [563, 6], [569, 2], [571, 6], [577, 10], [587, 6],$

$\qquad [593, 6], [599, 2]]$

Repeating this with $N = 100$, but suppressing the output, forms a longer sequence of points; the graph of these points, see figure 1.8, is drawn using the command

```
>   plot(s1);
```

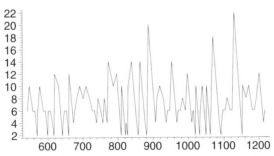

Figure 1.8   A graph of the differences between adjacent prime numbers.

You should type these commands into your computer to re-create this graph.† In section 1.10 we shall give an example of how arrays and lists can be used to produce

† If you experience difficulty this program can be found on the website described in the preface.

animated graphs. Here we simply note that if you wish to plot two, or more, arrays such as the above, s1 and s2 for instance, the appropriate command is plot([s1,s2]).

Exercise 1.38
Using the sequence command seq and the function ithprime create a list of the first 15 prime numbers and another list of the first 10 prime numbers greater than 99.

Exercise 1.39
Use the random number generator rand in the form rand(0..100) to create a list of 10 random integers in the interval $(0, 100)$ and then use sort to put this list in ascending order.

Exercise 1.40
Create a list of 100 random numbers between 1 and 100 and determine how many of these lie in the range $50 \leq n \leq 59$.

Repeat this count N times and compute the mean of the N numbers you obtain as a function of N for $1 \leq N \leq 50$. Note that this does not require a separate computation of the mean for each value of N.

Exercise 1.41
A sequence of numbers $x_1, x_2, \ldots$ is created using the equation

$$x_{k+1} = bx_k(1 - x_k), \quad 0 < b \leq 4,$$

where b is a given constant. Use Maple to form a list of the coordinates $[k, x[k]]$, $k = 1, 2, \ldots, 40$, where $x[k] = x_k$, starting with $x_0 = 0.2$ and with $b = 2$. Plot the graph of the sequence x_k against k.

Investigate the effects of changing the value of b, in particular try $b = 3.1$, 3.5, 3.6 and 3.84.

The equation relating x_{k+1} to x_k is known as the logistic map and is one of the simplest dynamical systems to exhibit chaotic motion: it is studied in detail in chapter 18.

1.9 Sets

Maple supports sets of the mathematical kind; a set is a sequence, separated by commas and enclosed by curly brackets.

```
>   data_set:={1,2,3,10,20};
```
$$data\_set := \{2, 3, 10, 20, 1\}$$

```
>   variables:={u,v,w,x,y,z};
```
$$variables := \{x, z, v, w, y, u\}$$

Maple does not preserve order and repetition in a set; moreover, Maple determines the order in which it prints the elements of a set and this order can be different in each session and can change during a session. The following sets are therefore identical $\{a, b, c\}$, $\{b, a, c\}$ and $\{a, a, b, c\}$, but the sets $\{2\}$ and $\{2, 2.0\}$ are not identical as the integer 2 is distinct from the floating point number 2.0. Maple knows about

$$ar_1 := 2$$
$$ar_2 := 3$$

It is not necessary to create the array first as we have done here; the whole array may be created in one go:

```
> cube:=array(1..3,[1,8,27]);
```

$$cube := [1, 8, 27]$$

The command `array([1,8,27])` is also valid.

In order to see the contents of an array use `print`, which displays results differently depending upon whether the first value of the index is 1 or not 1:

```
> print(ar);
```

$$\text{table}([$$
$$0 = 1$$
$$1 = 2$$
$$2 = 3$$
$$])$$

```
> print(cube);
```

$$[1, 8, 27]$$

In many respects lists and arrays with the index starting at 1 are the same, but you should be aware that internally Maple treats them as different types of structures. The structure that Maple names a `vector` is the same as an array with index starting at 1.

We shall find it useful to use array elements as coefficients of power series and there are several methods of constructing such series. First we can use the loop construction, discussed previously.

```
> n:=5:   a:=array(0..n):
> y:=a[0]:
>      for k from 1 to n do;
>      y:=y + a[k]*x^k;
>      od:
> y;
```

$$a_0 + a_1 x + a_2 x^2 + a_3 x^3 + a_4 x^4 + a_5 x^5$$

Alternatively, and more elegantly, the add command may be used:

```
> y:=a[0] + add(a[k]*x^k,k=1..n);
```

$$y := a_0 + a_1 x + a_2 x^2 + a_3 x^3 + a_4 x^4 + a_5 x^5$$

Notice that we have not included the term `a[0]` in the sum in order to avoid the construction x^0 which Maple interprets as 1 if $x \neq 0$ and 0 if $x = 0$; to see this consider the values of `subs(x=0,p=0,x^p)` and `subs(p=0,x=0,x^p)`. The command `sum` can also be used to create this expression.

Arrays may be cleared in the same manner as other variables; an array a is cleared by assigning the name to itself with the command `a:='a'`. The other methods for clearing variables, discussed in section 1.4.2 (page 10), also work with arrays.

An important use of one-dimensional arrays is in creating animated graphs; we illustrate this use with the simple example in which we show how the curve defined by the equation

$$y(\theta) = \cos(6\theta - 6x\sin\theta), \quad -\pi < \theta < \pi$$

changes as the parameter x changes from $\frac{1}{2}$ to $\frac{3}{2}$ in small steps. For this use of the graphics package it is necessary to invoke `with(plots)`.

```
>  restart; with(plots):
```

Now plot several graphs of $y(\theta)$ for the required values of x and simultaneously create an expression sequence of these plots; we name this sequence pg and change it into a list at the end, because a list is needed for the animation. First define the expression:

```
>  y:=cos(6*(t-x*sin(t))):
```

and then form a loop which creates 21 graphs with $0.5 \le x \le 1.5$ with increments of x being $1/20$.

```
>  pg:=NULL:                       # set pg to be the NULL sequence
>  N:=20:                          # N+1 is the number of graphs
>     for k from 0 to N do;        # beginning of the loop
>     x:=0.5 + k/N;                # Define x
>     pg:=pg,plot(y,t=-Pi..Pi);    # Append graph to the sequence
>     od:                          # End of loop
>  pg:=[pg]:                       # Convert the sequence to a list
```

Each graph has been assigned to an element of the array pg. This stage may take some time as 21 graphs are being constructed. The loop was ended with od: because the semi-colon terminator produces a vast amount of senseless output.

Finally use the `display` procedure, accessed using `with(plots)`, in the form

```
>  display(pg, insequence=true);
```

to display these graphs in the sequence of increasing x.

We can show no more on paper; you should repeat this sequence of commands on your computer and then click on any point near the graph to show the animation menu on the bar immediately above the worksheet; the *on* button will then start the animation so you will be able to see how the curve changes with x.† The context-sensitive right button can also be used to control the animation sequence.

Finally, note that this animation sequence may also be created more elegantly using the `seq` command:

```
>  restart; with(plots):
>  y:=cos(6*(t-x*sin(t))):
>  N:=20:
>  pg:=[seq(plot(eval(y,x=0.5+k/N),t=-Pi..Pi),k=0..N)]:
>  display(pg,insequence=true);
```

Exercise 1.46

Use the animation facility of Maple to examine the behaviour of the functions

$$\text{(i)} \quad f(x) = x - a\sin x, \quad -\pi \le x \le \pi, \quad -1 \le a \le 2,$$

$$\text{(ii)} \quad f(x) = x^4 - a^2 x^2 + \frac{1}{5}ax, \quad -1 \le x \le 1, \quad 0 \le a \le 1,$$

as the parameter a varies over the specified range.

---

† This sequence of commands can be found on the website given in the *Preface*.

1.10.2 Two-dimensional arrays and matrices

Arrays are not limited to the one-dimensional form introduced in the previous section but can be of any dimension, the upper limit depending only on the machine memory. Here, however, we consider only two-dimensional arrays and concentrate on the special form of these which allows matrix algebra to be performed. First, however, we consider a few more general aspects of two-dimensional arrays.

An example of a two-dimensional array is

```
> m1:=array(-1..1,0..1,[(-1,0)=a,(-1,1)=b,(0,0)=c,(0,1)=d]);
```

$$
\begin{aligned}
m1 := \text{array}(&-1..1, 0..1, [\\
&(-1, 0) = a\\
&(-1, 1) = b\\
&(0, 0) = c\\
&(0, 1) = d\\
&(1, 0) = m1_{1,0}\\
&(1, 1) = m1_{1,1}\\
&])
\end{aligned}
$$

There are two points to notice here. First, the indices are specified but can start with any integer value; here they start at -1 and 0. Second, not all the array elements need be specified.

A *matrix* is a two-dimensional array with both indices starting at 1, and may be created without mention of the array limits:

```
> m2:=array([[u,v],[x,y]]);
```

$$
m2 := \begin{pmatrix} u & v \\ x & y \end{pmatrix}
$$

or by using the alternative matrix command;

```
> m3:=matrix([[u,x],[v,y]]);
```

$$
m3 := \begin{pmatrix} u & x \\ v & y \end{pmatrix}
$$

The third way of inserting matrices into a worksheet is to use the palettes provided under the *View* option on the upper menu bar; these can be used to create templates of commonly used matrices and one may navigate between the blanks using the tab key.

Individual elements of an array can be accessed as one would expect; for example the $(1, 1)$ element of m2 is

```
> m2[1,1];
```

$$
u
$$

Note that square brackets are necessary.

In order to view the whole matrix it is necessary to use the evalm command, one of the eval family: the print command will display an array and a matrix, but is less useful for matrices. Simply typing m2 just returns the symbol m2: to see the value of the matrix it is necessary to type

> evalm(m2);

$$\left(\begin{array}{cc} u & v \\ x & y \end{array} \right)$$

This is an example of the fact that, by default, arrays are not fully evaluated; for arrays, Maple performs *last name evaluation*. One consequence of this is that if we wish to substitute matrix elements care is necessary. Suppose we need to form a new matrix m4 from m2 by replacing u by $a+b$; the substitution command has to be used in the following manner

> m4:=subs(u=a+b,evalm(m2));

$$m4 := \left(\begin{array}{cc} a+b & v \\ x & y \end{array} \right)$$

If the evalm function is forgotten then we simply obtain m4:=m2, with no substitution.

Matrices can be multiplied, but not with the ∗ operator which is the multiplication symbol reserved for commuting variables: for matrix multiplication the operator &∗ is used, so the product of m2 with m3 is given by

> evalm(m2 &* m3);

$$\left(\begin{array}{cc} u^2 + v^2 & u\,x + v\,y \\ u\,x + v\,y & x^2 + y^2 \end{array} \right)$$

Note that the evalm function is now necessary† — the eval function does not work in this situation. Similarly, matrices can be divided, added and subtracted; for example

> evalm(m3/m2);

$$\left(\begin{array}{cc} -\dfrac{u\,y}{-u\,y+v\,x} + \dfrac{x^2}{-u\,y+v\,x} & \dfrac{u\,v}{-u\,y+v\,x} - \dfrac{x\,u}{-u\,y+v\,x} \\[3ex] -\dfrac{v\,y}{-u\,y+v\,x} + \dfrac{y\,x}{-u\,y+v\,x} & \dfrac{v^2}{-u\,y+v\,x} - \dfrac{u\,y}{-u\,y+v\,x} \end{array} \right)$$

and

> evalm(m3 + m2);

$$\left(\begin{array}{cc} 2\,u & x+v \\ x+v & 2\,y \end{array} \right)$$

Multiplication by scalars is performed using the usual multiplication symbol

> m5:=a*evalm(m2);

$$m5 := a \left(\begin{array}{cc} u & v \\ x & y \end{array} \right)$$

> evalm(m5);

$$\left(\begin{array}{cc} a\,u & a\,v \\ a\,x & a\,y \end{array} \right)$$

---

† This is because &∗ is an inert operator, an idea that will be introduced in section 1.11.

or

> `m5:=evalm(a*m2);`

$$m5 := \begin{pmatrix} a\,u & a\,v \\ a\,x & a\,y \end{pmatrix}$$

If the elements of a matrix can be defined as a function of the array indices then it is a relatively elementary matter to input the matrix using the function notation of Maple — we shall introduce this more fully in chapter 3, but here its use is self-explanatory. Suppose that the (i, j), element of a 4×4 matrix is $i + j$ then we write

> `m6:=matrix(4,4,(i,j)->i+j);`

$$m6 := \begin{pmatrix} 2 & 3 & 4 & 5 \\ 3 & 4 & 5 & 6 \\ 4 & 5 & 6 & 7 \\ 5 & 6 & 7 & 8 \end{pmatrix}$$

Or if the (i, j) element depends upon another variable x, say, as $1/(i + j - x)$

> `m7:=matrix(4,4,(i,j)-> 1/(i+j-x));`

$$m7 := \begin{pmatrix} \dfrac{1}{2-x} & \dfrac{1}{3-x} & \dfrac{1}{4-x} & \dfrac{1}{5-x} \\[2ex] \dfrac{1}{3-x} & \dfrac{1}{4-x} & \dfrac{1}{5-x} & \dfrac{1}{6-x} \\[2ex] \dfrac{1}{4-x} & \dfrac{1}{5-x} & \dfrac{1}{6-x} & \dfrac{1}{7-x} \\[2ex] \dfrac{1}{5-x} & \dfrac{1}{6-x} & \dfrac{1}{7-x} & \dfrac{1}{8-x} \end{pmatrix}$$

The seemingly equivalent command `array(1..4,1..4,(i,j)->1/(i+j-x))` does not work.

Some special types of matrices can be created using the array command with the options *symmetric, antisymmetric, identity, diagonal* and *sparse*; for instance a 2×2 diagonal matrix is formed as follows:

> `m8:=array(diagonal,1..2,1..2):   evalm(m8);`

$$\begin{pmatrix} m8_{1,1} & 0 \\ 0 & m8_{2,2} \end{pmatrix}$$

The command `array(1..2,1..2,diagonal)` produces the same effect. For instance, a 2×2 symmetric matrix with unit off diagonal elements is given with the commands:

> `m9:=array(1..2,1..2,symmetric):  m9[1,2]:=1:   evalm(m9);`

$$\begin{pmatrix} m9_{1,1} & 1 \\ 1 & m9_{2,2} \end{pmatrix}$$

and a sparse one-dimensional array is given by

> `v1:=array(sparse,1..10):  v1[1]:=1:  v1[2]:=5:  v1[10]:=2:`

```
>  evalm(v1);
```
$$[1, 5, 0, 0, 0, 0, 0, 0, 0, 2]$$

You should experiment with the other options.

For any serious use of matrices with Maple you should load the linear algebra package, as this is specially designed for linear algebra computations and therefore has many features not available without it, for instance the determinant and trace functions.† A number of the exercises set below and at the end of this chapter will need some features from this package. As a simple example the determinants of m2 and m7 are obtained as follows:

```
>  with(linalg):

Warning, new definition for norm
Warning, new definition for trace

>  d2:=det(m2);
```
$$d2 := u\,y - v\,x$$

and

```
>  d7:=sort(det(m7));
```
$$d7 := \frac{144}{(x-2)\,(x-4)^3\,(x-6)^3\,(x-8)\,(x-7)^2\,(x-5)^4\,(x-3)^2}$$

Note the use of sort; without it the factors are written in an 'unnatural' form, $(-2+x)$ for instance.

Exercise 1.47

For the following matrices

$$A = \begin{pmatrix} 1 & 0 & 2 \\ 2 & -1 & 3 \\ 4 & 1 & 8 \end{pmatrix}, \quad B = \begin{pmatrix} -3 & 2 \\ 0 & 1 \\ 7 & 4 \end{pmatrix}$$

compute

(i)	A^{-1},	(ii)	$AA^{\top}$,
(ii)	$B^{\top}AB$,	(iv)	$(2A + BB^{\top})A^{\top}$,

where $A^{\top}$ is the transpose of A.

In order to compute the transpose of a matrix you will need to load the linear algebra package using with(linalg), then the trace and determinant of a matrix A are given by trace(A) and det(A).

Exercise 1.48

For the matrices

$$A = \begin{pmatrix} 1 & 2 & 3 \\ 2 & 3 & 0 \\ 4 & 8 & 5 \end{pmatrix}, \quad B = \begin{pmatrix} 3 & 8 & -2 \\ 4 & 7 & -1 \\ 0 & 3 & 5 \end{pmatrix}$$

compute $3BA - 4A^3$.

† In Maple 6 there is also the *Linear Algebra* package containing a group of linear algebra commands which are more powerful and efficient when calculating with large numeric matrices. The data structures used in this package are Matrices and Vectors, corresponding to the matrices and vectors of the linear algebra package: these types do not exist in Maple 5.

Exercise 1.49

Find all the 2×2 and 3×3 matrices that commute with

$$A = \begin{pmatrix} 2 & -1 \\ 3 & 5 \end{pmatrix}, \quad B = \begin{pmatrix} 1 & 2 & -1 \\ 3 & 0 & 1 \\ 5 & -1 & 1 \end{pmatrix}.$$

Exercise 1.50

Find by explicit computation the number of distinct integer powers of the following matrices:

$$\begin{pmatrix} 0 & 1 \\ 1 & 0 \end{pmatrix}, \quad \begin{pmatrix} 0 & 1 \\ -1 & 0 \end{pmatrix}, \quad \begin{pmatrix} 0 & i \\ -1 & 0 \end{pmatrix}.$$

Exercise 1.51

The Pauli spin matrices are defined as

$$\sigma_x = \begin{pmatrix} 0 & 1 \\ 1 & 0 \end{pmatrix}, \quad \sigma_y = \begin{pmatrix} 0 & -i \\ i & 0 \end{pmatrix}, \quad \sigma_z = \begin{pmatrix} 1 & 0 \\ 0 & -1 \end{pmatrix}.$$

Prove that

$$\sigma_x^2 = \sigma_y^2 = \sigma_z^2 = I$$

and that

$$\sigma_r \sigma_s - \sigma_s \sigma_r = 2i\sigma_t$$

where r, s and t are cyclic permutations of x, y and z.

Exercise 1.52

Solve the following linear simultaneous matrix equations for the matrices A and B:

(i) $\quad A - 2B = \begin{pmatrix} 1 & 2 \\ -1 & 1 \end{pmatrix}, \quad A - B = \begin{pmatrix} 2 & 1 \\ 1 & -1 \end{pmatrix},$

(ii) $\quad A - B = \begin{pmatrix} 1 & -2 \\ -1 & 3 \end{pmatrix}, \quad A + B = \begin{pmatrix} 3 & 0 \\ 3 & 1 \end{pmatrix}.$

Exercise 1.53

Use the `eigenvalues` procedure from the `with(linalg)` package to find the eigenvalues of the following matrices:

$$\begin{pmatrix} 1 & 1 \\ -1 & 2 \end{pmatrix}, \quad \begin{pmatrix} 0 & 2 & 4 \\ 1 & 1 & -2 \\ -2 & 0 & 5 \end{pmatrix}, \quad \begin{pmatrix} 1 & 1 & 1 & 1 \\ 0 & 1 & 2 & 1 \\ 1 & 0 & 4 & 1 \\ 1 & 1 & 0 & 3 \end{pmatrix}.$$

In the last, 4×4 matrix, what is the difference in the output of `eigenvalues` when (a) all the entries are integers, and (b) just one entry is a floating point number?

Exercise 1.54

Use the `eigenvectors` procedure from the `with(linalg)` package to show that the eigenvectors of

$$A = \begin{pmatrix} x & x^2 - \delta \\ x^2 + \delta & x^3 \end{pmatrix},$$

are

$$\mathbf{v}_1 = \left(x(1 - x^2) + \sqrt{x^2(x^2 + 1)^2 - 4\delta^2}, 2(x^2 + \delta) \right),$$

$$\mathbf{v}_2 = \left(x(1 - x^2) - \sqrt{x^2(x^2 + 1)^2 - 4\delta^2}, 2(x^2 + \delta) \right)$$

and that the scalar product of these vectors is

$$\mathbf{v}_1 . \mathbf{v}_2 = 8\delta(x^2 + \delta).$$

Exercise 1.55

Use the `eigenvectors` procedure from the `with(linalg)` package to show that the two real eigenvectors of

$$A = \begin{pmatrix} 1 & 2 & 1 & 1 \\ 0 & 1 & 2 & 1 \\ 1 & 0 & 4 & 1 \\ 1 & 1 & -1 & 3 \end{pmatrix}$$

are, to four significant figures,

$$\mathbf{v}_1 = (0.2871, 0.3586, 0.8776, -0.9505), \quad \mathbf{v}_2 = (0.4501, 0.4231, 0.7461, 0.07462).$$

1.11 Elementary differentiation

The basic differential operator for expressions is `diff`, which we illustrate with the simple examples

```
>  diff(sin(y),y);
```
$$\cos(y)$$

The first differential of $\sin\sqrt{x^2 + a^2}$ with respect to x is

```
>  z1:=diff(sin(sqrt(x^2+a^2)),x);
```
$$z1 := \frac{\cos(\sqrt{x^2 + a^2})\, x}{\sqrt{x^2 + a^2}}$$

and the differential of this is

```
>  diff(z1,x);
```
$$-\frac{\sin(\sqrt{x^2 + a^2})\, x^2}{x^2 + a^2} - \frac{\cos(\sqrt{x^2 + a^2})\, x^2}{(x^2 + a^2)^{3/2}} + \frac{\cos(\sqrt{x^2 + a^2})}{\sqrt{x^2 + a^2}}$$

or, in one go, using the sequence operator $:

```
>  diff(sin(sqrt(x^2+a^2)),x$2);
```
$$-\frac{\sin(\sqrt{x^2 + a^2})\, x^2}{x^2 + a^2} - \frac{\cos(\sqrt{x^2 + a^2})\, x^2}{(x^2 + a^2)^{3/2}} + \frac{\cos(\sqrt{x^2 + a^2})}{\sqrt{x^2 + a^2}}$$

Similarly, the fifth derivative of $\sin(x^2)$ is

> `diff(sin(x^2),x$5);`

$$32\cos(x^2)\,x^5 + 160\sin(x^2)\,x^3 - 120\cos(x^2)\,x$$

which is equivalent to

> `diff(sin(x^2),x,x,x,x,x):`

The operator `diff` is one of a few operators having an *inert* form, named `Diff`. This is essentially the same as `diff`, but does not evaluate the differential,

> `d:=Diff(sin(y),y);`

$$d := \frac{\partial}{\partial y}\sin(y)$$

and we can evaluate this using the `value` function,

> `value(d);`

$$\cos(y)$$

A tidy output may be produced in the following manner:

> `d:=Diff(sin(x)^3,x$6):    d=value(d);`

$$\frac{\partial^6}{\partial x^6}\sin(x)^3 = 546\sin(x)\cos(x)^2 - 183\sin(x)^3$$

Partial differentiation is performed by a straight forward extension. For example the three second derivatives of $\sin x \cos y$ are

> `f:=sin(x)*cos(y):`

> `d:=Diff(f,x$2):  d=value(d);`

$$\frac{\partial^2}{\partial x^2}\sin(x)\cos(y) = -\sin(x)\cos(y)$$

> `d:=Diff(f,x,y):  d=value(d);`

$$\frac{\partial^2}{\partial y\,\partial x}\sin(x)\cos(y) = -\cos(x)\sin(y)$$

> `d:=Diff(f,y$2):  d=value(d);`

$$\frac{\partial^2}{\partial y^2}\sin(x)\cos(y) = -\sin(x)\cos(y)$$

Exercise 1.56
For the functions defined below find the stationary point(s) and value(s) of the functions at these point(s), and use Maple to draw graphs of these functions over a range of x large enough to include all stationary points, or the range given,

(i) $y = x^2 + 4x + 7$,
(ii) $y = (4x^2 - 2)e^{-x^2/2}$,
(iii) $y = 12x^5 - 15x^4 + 20x^3 - 330x^2 + 600x + 2$,
(iv) $y = \tanh 3x - \dfrac{1}{3}\tanh^2 x, \quad x > 0$,
(v) $y = \sin(x^{\cos x}), \quad 0 \le x \le 2\pi$.

Exercise 1.57

Show that the function

$$u = \frac{x + 2y + 2}{x^2 + y^2 + 1}$$

has stationary values at the points $(1/5, 2/5)$, $(-1, -2)$ and determine the nature of these stationary points.

Use the procedure `contourplot`, remembering the `with(plots)` command, to draw the contours of the function $u(x, y)$.

Exercise 1.58

Show that the function

$$u = \frac{x + 2y + 1}{x^2 + y^2 + 1}$$

has two stationary points and determine the nature of these stationary points.

Note this formula is very similar to that considered in exercise 1.57, but Maple treats the root-finding in a slightly different manner and you may need the command `allvalues` in order to extract all the roots from the `RootOf` construction.

Use the procedure `contourplot` to draw the contours of the function $u(x, y)$.

Exercise 1.59

Use the `diff` operator to find the first five coefficients of the Taylor's series of $\sin(x + x^2)$ about $x = 0$ and $x = \pi/2$.

Hint: the command `eval( diff(f(x),x),x=u )` is a handy means of evaluating $f'(u)$ and also works if the Maple function `f(x)` is replaced by an expression. An alternative is `subs( x=u,diff(f(x),x) )`.

Exercise 1.60

By evaluating the first five derivatives at $x = 0$ and using Taylor's expansion, show that

$$\ln\left(\frac{1 + \sin x}{\cos x}\right) - 2\sin x + x\cos x = \frac{x^5}{15} + \cdots.$$

Exercise 1.61

Using combinations of the `sort`, `simplify` and the `normal` Maple commands show that:

if $\quad y = \dfrac{1 + \sin x}{1 + \cos x},\qquad \dfrac{dy}{dx} = \dfrac{\cos x + \sin x + 1}{1 + 2\cos x + \cos^2 x},$

if $\quad y = \ln\sqrt{\dfrac{1 + x}{1 - x}},\qquad \dfrac{dy}{dx} = -\dfrac{1}{(x + 1)(x - 1)},$

if $\quad y = \ln\left(\dfrac{(1 + x)^{1/2}}{(1 - x)^{1/3}}\right),\qquad \dfrac{dy}{dx} = \dfrac{1}{6}\dfrac{x - 5}{(x - 1)(x + 1)}.$

Exercise 1.62

Find the stationary points of the curve given parametrically by the equations

$$x = \sin t, \quad y = \cos^3 t, \quad -\pi \le t \le \pi.$$

Use Maple to plot the graph of this curve.

Exercise 1.63

If $z = \ln(x^2 + y^2)$ find the value of the constant a such that

$$\left(\frac{\partial z}{\partial x}\right)^2 + \left(\frac{\partial z}{\partial y}\right)^2 = a e^{-z}.$$

Exercise 1.64

Show that the function $f = 1/r$, where $r^2 = (x - a)^2 + (y - b)^2 + (z - c)^2$ and a, b and c are constants, is a solution of Laplace's equation

$$\frac{\partial^2 f}{\partial x^2} + \frac{\partial^2 f}{\partial y^2} + \frac{\partial^2 f}{\partial z^2} = 0.$$

1.12 Integration

Here we introduce the integration operator, leaving a more advanced discussion to chapter 13. Maple can perform many types of integration symbolically, and can provide numerical estimates of most integrals that cannot be evaluated symbolically as well as those that can, albeit rather slowly by comparison with other purpose built software. We have already introduced the simple command for indefinite integration, exercise 1.6, page 10: for example

```
>   i:=Int(x*exp(a*x^2),x):    i=value(i);
```

$$\int x e^{(a x^2)}\, dx = \frac{1}{2}\frac{e^{(a x^2)}}{a}$$

where we have used both the *inert* and *active* forms of int. Note that Maple does not include the arbitrary constant of integration, which means that the same function may be integrated to a different value depending upon how it is given to Maple; for instance

```
>   z1:=int((1+x)^2,x);
```

$$z1 := \frac{1}{3}(1 + x)^3$$

```
>   x1:=expand(z1);
```

$$x1 := \frac{1}{3} + x + x^2 + \frac{1}{3}x^3$$

```
>   z2:=int(1+2*x+x^2,x);
```

$$z2 := x + x^2 + \frac{1}{3}x^3$$

so the difference between z1 and z2 is a constant, z1 - z2 = 1/3.

Integration between limits is performed as follows;

```
>   i:=Int(x*exp(a*x^2),x=0..1):    i=value(i);
```

$$\int_0^1 x e^{(a x^2)}\, dx = \frac{1}{2}\frac{e^a - 1}{a}$$

and

```
>   i:=Int(1/sqrt(u-x),x=0..u):    i=value(i);
```

$$\int_0^u \frac{1}{\sqrt{u-x}}\, dx = 2\,\sqrt{u}$$

So the range of integration is given via the x=a..b construction which we have used before; here we see that a or b can be variables or numbers.

If Maple cannot evaluate an integral it simply returns it (after a short delay):

```
>   int(exp(x^3),x);
```

$$\int e^{(x^3)}\, dx$$

Maple sometimes produces quite horrendous expressions for integrals; for instance

```
>   i:=Int(x/(1+x^5),x):    i=value(i);
```

$$\int \frac{x}{1+x^5}\, dx = -\frac{1}{5}\ln(1+x) - \frac{1}{20}\ln(2\,x^2-x-\sqrt{5}\,x+2)\,\sqrt{5} + \frac{1}{20}\ln(2\,x^2-x-\sqrt{5}\,x+2)$$

$$+\frac{2}{5}\frac{\arctan\left(\dfrac{4\,x-1-\sqrt{5}}{\sqrt{10-2\,\sqrt{5}}}\right)\sqrt{5}}{\sqrt{10-2\,\sqrt{5}}} + \frac{1}{20}\ln(2\,x^2-x+\sqrt{5}\,x+2)\,\sqrt{5}$$

$$+\frac{1}{20}\ln(2\,x^2-x+\sqrt{5}\,x+2) - \frac{2}{5}\frac{\arctan\left(\dfrac{4\,x-1+\sqrt{5}}{\sqrt{10+2\,\sqrt{5}}}\right)\sqrt{5}}{\sqrt{10+2\,\sqrt{5}}}$$

and in this case it is more convenient to assign the integral to a variable

```
>   z:=int(x/(1+x^5),x=0..u):
```

and evaluate this as necessary,

```
>   z1:=evalf(subs(u=1,z));
```

$$z1 := .4069016344$$

In this and similar cases where only one or a few numbers are needed it may be quicker to evaluate the integral numerically using the methods discussed in the next section.

At this point you need to experiment with some simple integrals just to see what can be done.

Exercise 1.65
Use Maple to integrate the following functions, where a and b are constants, with respect to x:

(i) $\sqrt{e^x - 1}$,   (ii) $x^2(ax+b)^{5/2}$,   (iii) $\sinh(6x)\sinh^4(x)$,

(iv) $x^{10}e^x$,   (v) $1/\cosh(x)^6$,   (vi) $\sin(\ln x)$,

(vii) $\dfrac{\cos x}{2+\sin x+2\cos x}$,   (viii) $\dfrac{1}{8-x^{3/2}}$,   (ix) $\dfrac{1+x}{x\sqrt{x-1}}$,

(x) $\dfrac{1}{\sqrt{1+x}}$,   (xi) $\dfrac{\sqrt{x-1}}{x}$,   (xii) $\dfrac{1}{(1-x)\sqrt{1+x}}$.

Exercise 1.66
Use Maple to evaluate the following definite integrals:

(i) $\displaystyle\int_0^\infty dx\,\frac{1}{(1+x)(1+x^2)}$,

(ii) $\displaystyle\int_0^\infty dx\,\frac{e^{-x}}{\cosh x}$,

(iii) $\displaystyle\int_{\pi/4}^{\pi/3} dx\,\frac{1}{\sin x \cos^4 x}$,

(iv) $\displaystyle\int_{1/2}^1 dx\,\frac{1}{1+x^3}$,

(v) $\displaystyle\int_0^{1/2} dx\,\frac{1+x+x^2}{1-x+x^2}$,

(vi) $\displaystyle\int_0^{1/2} dx\,x\ln(1-x^3)$,

(vii) $\displaystyle\int_0^1 dx\,x^2\tan^{-1}x$,

(viii) $\displaystyle\int_0^{\pi/6} dx\,\frac{1}{\cos x + \cos^3 x}$,

(ix) $\displaystyle\int_0^{\pi/2} dx\,\frac{1}{1-2a\cos x + a^2}$, $a \neq 1$,

(x) $\displaystyle\int_0^1 dx\,\frac{1}{(1+x)\sqrt{1+x^2}}$.

From these exercises you will observe that Maple's ability to do quite nasty integrals is impressive, and that it easily evaluates almost all of the integrals found in elementary texts and most of the indefinite integrals given in standard tables. In chapter 13 we discuss the types of integral that Maple can and cannot do. There we shall see that Maple knows most of the tricks taught in elementary mathematics courses and can apply these to many types of integral. Unfortunately the majority of integrals do not yield to these methods.

1.12.1 Numerical evaluation
If Maple fails to find a symbolic answer, or if you know that the integral in question cannot be evaluated in closed form, then it may be necessary to evaluate it numerically. Maple does this by using the `evalf` function, as illustrated in the next example. First we try to find a symbolic result,

```
>   z:=int(exp(-t)*arcsin(t),t=0..1);
```

$$z := \int_0^1 e^{(-t)}\arcsin(t)\,dt$$

Maple just returns the integral, which means that it cannot evaluate the integral symbolically. The numerical value of the integral is found simply by evaluating z to a floating point number:

```
>   z:=evalf(z);
```

$$z := .2952205678$$

As with all numerical schemes this technique will work only if the integrand evaluates to a number: even replacing the integrand by $ae^{-t}\sin^{-1}t$, where a is a free variable, stops evaluation. These two operations can be combined: thus if only four digit accuracy is required,

```
> z:=evalf(Int(exp(-t)*arcsin(t),t=0..1),4);
```
$$z := .2952$$

In this case we used the inert form of `Int` because this saves Maple the futile attempt at evaluating the integral symbolically before the numerical evaluation.

Exercise 1.67

Use Maple to evaluate the following integrals numerically to four digits accuracy,

(i) $\displaystyle\int_0^1 dx \, \exp(x^3),$ (ii) $\displaystyle\int_0^{10} dx \, \frac{1}{\sqrt{1+x^4}},$

(iii) $\displaystyle\int_0^5 dx \, \sin(\exp(x/2)),$ (iv) $\displaystyle\int_0^{\pi/2} dx \, \frac{1}{\sin^{3/4} x \sqrt{1-0.9\sin^5 x}}.$

The integration routine provided with Maple seems able to accurately estimate the values of most well-behaved definite integrals. It does, however, fail in extreme cases, for instance when the integrand is very oscillatory; these types of problems are discussed in chapter 14.

1.12.2 Definite integration and the assume facility

So far we have discussed symbolic evaluation of indefinite integrals, but Maple knows a large variety of definite integrals and many of the special functions that can be defined in terms of definite integrals, some of which will be introduced later. Here we limit ourselves to a few examples. For instance the integral

```
> int(x^2 * ln(x)^2 /(1-x),x);
```
$$\int \frac{x^2 \ln(x)^2}{1-x} \, dx$$

cannot be evaluated in terms of elementary functions. But the related definite integral can be evaluated in terms of the zeta function,

```
> i:=Int(x^2 * ln(x)^2/(1-x),x=0..1):   i=value(i);
```
$$\int_0^1 \frac{x^2 \ln(x)^2}{1-x} \, dx = -\frac{9}{4} + 2\,\zeta(3)$$

where $\zeta(z)$ is the zeta function defined by the infinite series,

$$\zeta(z) = \sum_{n=1}^{\infty} \frac{1}{n^z}, \quad \Re(z) > 1. \tag{1.3}$$

Maple knows the zeta function by the name `zeta(z)`.

L. Euler
(1707–1783)

G. F. B. Riemann
(1826–1866)
The zeta function was known to Euler but most of its properties were discovered by Riemann in his investigations into prime numbers, hence it is often called the Riemann zeta function; it is discussed in chapter 4.

Many definite integrals contain parameters which must lie in specific ranges for the integral to exist. For example, if we try to evaluate the integral

$$\int_0^\infty dx\, e^{-xy}$$

the following error message is produced

```
>   int(exp(-x*y),x=0..infinity);

Definite integration: Can't determine if the integral is convergent.
Need to know the sign of --> y
Will now try indefinite integration and then take limits.
```

$$\lim_{x\to\infty} -\frac{e^{(-xy)}-1}{y}$$

Maple, quite rightly, gives an error message because this integral does not exist unless $y > 0$, or more generally $\Re(y) > 0$. We can define properties of variables using the assume facility, which you should look up; in this case we need to tell Maple that $y > 0$,

```
>   assume(y>0);
>   z:=int(exp(-y*x),x=0..infinity);
```

$$z := \frac{1}{y}$$

and now obtain the expected result.

A useful command, related to assume, is about which will tell you whether a variable has any properties associated with it and if so what they are. Thus if we require a to be real and positive, $0 < b < 1$ and c to be real then we type

```
>   assume(a>0); assume(b, RealRange(0,1)); assume(c,real);
```

or, alternatively

```
>   assume(a>0, b,RealRange(0,1), c,real);
```

and if we subsequently forget what properties we have associated with the variables a, b, c and p then we type

```
>   about(a,b,c,p);

Originally a, renamed a~:
  is assumed to be: RealRange(Open(0),infinity)

Originally b, renamed b~:
  is assumed to be: RealRange(0,1)

Originally c, renamed c~:
  is assumed to be: real

p:
  nothing known about this object
```

On trying these commands you will notice that any variable with associated properties has a tilde appended to it, as in the above Maple messages. If you wish to remove this then either click on the *Options* item in the top menu bar, and then on the *Assumed variables* entry, or type `interface(showassumed=0)` at the prompt; this setting can be saved from the *File* item in the top menu bar, then all subsequent Maple sessions will not show the tilde, in most circumstances. Incidentally, these, and other, settings are stored in the file *Maplev5.ini*, created when Maple is installed.

Another frequently recurring example is in the computation of Fourier components; suppose we need the *n*th Fourier components of the function x^4 on the interval $(-\pi, \pi)$ given by the integral

```
>  i:=Int(x^4 * cos(n*x),x=-Pi..Pi):    i=value(i);
```

$$\int_{-\pi}^{\pi} x^4 \cos n x \, dx =$$

$$2 \frac{n^4 \pi^4 \sin(\pi n) + 4 n^3 \pi^3 \cos(\pi n) - 12 n^2 \pi^2 \sin(\pi n) + 24 \sin(\pi n) - 24 \pi n \cos(\pi n)}{n^5}$$

This is clearly not very useful and simplifies considerably because $\sin(n\pi) = 0$ for all integer *n*; Maple has not set these factors to zero because it does not know that *n* is an integer:

```
>  assume(n,integer);
```

```
>  i:=Int(x^4 * cos(n*x),x=-Pi..Pi):    i=value(i);
```

$$\int_{-\pi}^{\pi} x^4 \cos(n x) \, dx = 8 \frac{\pi (-1)^n (n^2 \pi^2 - 6)}{n^4}$$

Not only has Maple set $\sin(n\pi) = 0$ but it has also set $\cos(n\pi) = (-1)^n$.

Maple can be made to forget assumptions about specific variables by clearing them.

You should be aware, however, that there are many definite integrals that can be evaluated but are not known by Maple; some examples are

$$\int_0^\infty dx \, \frac{x^2}{\cosh x} = \frac{\pi^3}{8}, \qquad \int_0^\infty dx \, \frac{x}{(1 + x^2) \sinh \pi x} = \ln 2 - \frac{1}{2},$$

$$\int_0^1 dx \, \frac{1 - x}{(1 + x)(1 + x^2) \ln x} = -\frac{1}{2} \ln 2, \qquad \int_{-\pi+u}^{\pi+u} dx \, \cos nx \cos mx = \pi \delta_{nm}.$$

In the last example *n* and *m* are integers and Maple returns zero for all *n* and *m*, unless *n* and *m* are explicitly made equal. Other examples can be found in the standard reference texts such as Gradshteyn and Ryzhik (1965). These examples are quoted as a warning: computer systems do not always produce correct results.

There is another class of integral which, at present, defeats a naive application of Maple. If *n* is an arbitrary positive integer, integrals like

$$\int dx \, x^n (ax + b)^{1/2} = 2 \frac{(ax + b)^{3/2}}{a^n} \sum_{k=0}^{n} {}^nC_r \frac{(ax + b)^{n-k} b^k}{2n - 2k + 3}$$

$$\int dx \, x^n e^{ax} = e^{ax} \left\{ \frac{x^n}{a} + \sum_{k=1}^{n} (-1)^k \frac{n(n-1) \cdots (n - k + 1)}{a^{k+1}} x^{n-k} \right\},$$

which can be evaluated by repeated integration by parts, are not evaluated by Maple unless specific values of n are chosen, in which case Maple can do the integrals.

Another problem with automatic integration is that the answer is sometimes not in the form required but given as a very complicated expression that is not easily simplified. Examples of this type of problem can be found when spherical geometry is used; a particular example is the integral

$$I = \int \frac{d\theta}{\sqrt{1 - \cos^2 \beta / \sin^2 \theta}} = \text{constant} - \sin^{-1}\left(\frac{\cos\theta}{\cos\beta}\right).$$

For this integral Maple gives

```
>  int(1/sqrt(1-cos(beta)^2/sin(x)^2),x);
```

$$\frac{1}{2} \frac{\sqrt{4}\sin(x)\sqrt{-\dfrac{\cos(x)^2 - 1 + \cos(\beta)^2}{\cos(x)^2 + 2\cos(x) + 1}}\,\arctan\left(\dfrac{\cos(x)}{(1+\cos(x))\sqrt{-\dfrac{\cos(x)^2 - 1 + \cos(\beta)^2}{\cos(x)^2 + 2\cos(x) + 1}}}\right)}{\sqrt{\dfrac{\cos(x)^2 - 1 + \cos(\beta)^2}{-1 + \cos(x)^2}}\,(-1 + \cos(x))}$$

But a simpler result is obtained by rewriting the integral in the form

$$I = \int d\theta \, \frac{\sin\theta}{\sqrt{\sin^2\theta - \cos^2\beta}} = \int d\theta \, \frac{\sin\theta}{\sqrt{\sin^2\beta - \cos^2\theta}} = \text{constant} - \sin^{-1}\left(\frac{\cos\theta}{\cos\beta}\right),$$

obtained by putting $\cos\theta = \sin x \sin\beta$. In general we observe that automatic simplification of trigonometric expressions is difficult and often it is easier to do this type of algebra either by hand or with partial help from Maple.

1.13 Exercises

Exercise 1.68

Marin Mersenne (1588–1648) conjectured that the only primes p for which $M_p = 2^p - 1$ is prime are 2, 3, 5, 7, 13, 17, 19, 31, 67, 127 and 257. It is now known that M_{67} and M_{257} are not prime, but that M_{61}, M_{89} and M_{107} are prime. Verify some of these assertions.

You may be interested in the complete list of the first 37 Mersenne primes: 2, 3, 5, 7, 13, 17, 19, 31, 61, 89, 107, 127, 521, 607, 1 279, 2 203, 2 281, 3 217, 4 253, 4 423, 9 689, 9 941, 11 213, 19 937, 21 701, 23 209, 44 497, 86 243, 110 503, 132 049, 216 091, 756 839, 859 433, 1 257 787, 1 398 269, 2 976 221, 3 021 377. The next known is 6972593, but it is not known whether this is the 38th: this Mersenne prime comprises 2 098 960 digits, about 3 km of type. These primes, and much more information, may be found on the web site: http://www.utm.edu/research/primes/mersenne.shtml

Exercise 1.69

If $y = x^a \ln x$ show that

$$x\frac{dy}{dx} = x^a + ay.$$

Exercise 1.70

Show that $y = \sin x/(1 - x^2)$ is a solution of the differential equation

$$(1 - x^2)\frac{d^2y}{dx^2} - 4x\frac{dy}{dx} - (1 + x^2)y = 0.$$

Exercise 1.71

If $y = e^{-kt}(a\cos\omega t + b\sin\omega t)$, where a, b, ω and k are constants, show that

$$\frac{d^2y}{dt^2} + 2k\frac{dy}{dt} + (\omega^2 + k^2)y = 0.$$

Exercise 1.72

If $y = \left(x + \sqrt{1 + x^2}\right)^a$, show that

$$\frac{dy}{dx} - \frac{ay}{\sqrt{1 + x^2}} = 0 \quad \text{and} \quad \sqrt{1 + x^2}\frac{d^2y}{dx^2} - a\frac{dy}{dx} + \frac{axy}{\sqrt{1 + x^2}} = 0.$$

Exercise 1.73

Show that the function $g = (z - c)/r$, where $r^2 = (x - a)^2 + (y - b)^2 + (z - c)^2$, and a, b and c are constants, is a solution of the equation

$$\frac{\partial^2 g}{\partial x \partial y} + \frac{A(x - a)}{r^2}\frac{\partial g}{\partial y} = 0, \quad \text{provided} \quad A = 3.$$

Exercise 1.74

If $y = (x^3 - 3x^2)e^{2x}$, show that

$$\frac{d^8y}{dx^8} = 256x(x + 3)(x + 6)e^{2x}.$$

Exercise 1.75

Show that if n is a positive integer,

$$\int dx \, \sinh(n + 1)x \, \sinh^{(n-1)} x = \frac{1}{n}\sinh^n x \sinh nx.$$

Exercise 1.76

Show that

$$\int_0^a \frac{dx}{\sqrt{a^4 - x^4}} = \frac{\pi^{3/2}\sqrt{2}}{4a\Gamma(3/4)^2}.$$

Note you may need to make the change of variable $x = ay$ before applying Maple, and then use `convert(*,GAMMA)`.

Exercise 1.77

Use Maple to show that, for all integers $n \geq 2$,

$$x^n - nx + n - 1 = (x - 1)(x^{n-1} + x^{n-2} + \cdots + x - n + 1)$$

and hence prove that $x^n + n - 1 \geq nx$ for $n \geq 1$ and $x \geq 0$.

Exercise 1.78

Two sequences of integers a_n and b_n, $n = 1, 2, \cdots$ are defined by the equations

$$a_{n+1} = a_n + 2b_n, \quad b_{n+1} = a_n + b_n, \quad a_1 = 3, \ b_1 = 2.$$

Use Maple to show that the ratio a_n/b_n appears to converge to $\sqrt{2}$ and is alternately larger and smaller than $\sqrt{2}$. Experiment with different initial conditions.

If $x_n = a_n/b_n$ prove that $|x_{n+1} - \sqrt{2}| < |x_n - \sqrt{2}|/2$ and hence that the above sequence converges to $\sqrt{2}$.

Exercise 1.79

Two ladders, of lengths 20 and 30 feet, cross at a point 8 feet above a lane, as shown in the diagram.

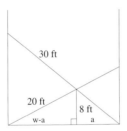

30 ft

20 ft

8 ft

w-a   a

Figure 1.9   Diagram showing the distances
defined.

Each ladder reaches from the base of one wall to some point on the opposite wall. Show that the width of the lane, w, satisfies the equation

$$\frac{1}{8} = \frac{1}{\sqrt{20^2 - w^2}} + \frac{1}{\sqrt{30^2 - w^2}}.$$

Use the `fsolve` procedure of Maple to find the width of the lane.

Exercise 1.80

Use Maple to show that

$$1^3 + 2^3 + \cdots + n^3 = (1 + 2 + \cdots + n)^2.$$

Hint: you will need to use the `sum` and `simplify` commands.

Exercise 1.81

If

$$\sigma_k(n) = \sum_{p=1}^{n} p^k$$

use Maple to show that

$$\sigma_1(n)^3 = \frac{1}{4}\sigma_3(n) + \frac{3}{4}\sigma_5(n),$$

$$\sigma_1(n)^5 = \begin{cases} \frac{1}{16}\sigma_5(n) + \frac{5}{8}\sigma_7(n) + \frac{5}{16}\sigma_9(n), \\ \frac{1}{16}\left(5\sigma_5(n) + 5\sigma_7(n) + 54\sigma_5(n)^2 - 48\sigma_3(n)\sigma_7(n)\right), \end{cases}$$

and also that

$$\sigma_5(n) = \frac{1}{3}\sigma_1(n)^2\{4\sigma_1(n) - 1\}, \qquad \sigma_7(n) = \frac{1}{3}\sigma_1(n)^2\{6\sigma_1(n)^2 - 4\sigma_1(n) + 1\},$$

$$\sigma_9(n) = \frac{1}{5}\sigma_1(n)^2(2\sigma_1(n) - 1)(8\sigma_1(n)^2 - 6\sigma_1(n) + 3).$$

Exercise 1.82
Factorise $x^n - 1$ for various integer values of n.

Exercise 1.83
Use matrix inversion to solve the following equations:

$$\text{(i)} \quad \begin{array}{rcl} 7x + 4y - 2z &=& 8 \\ 4x + 7y + 5z &=& 5 \\ 2x - 3y + 8z &=& 2 \end{array} \qquad \text{(ii)} \quad \begin{array}{rcl} 2x - 5y + z &=& a \\ 7x - y + 4z &=& b \\ 3x - 6y + 5z &=& 9. \end{array}$$

Exercise 1.84
For which values of x and y is the following matrix singular

$$\begin{pmatrix} 2 & 7 & x \\ 3 & y & 1 \\ 2 & 4 & 6 \end{pmatrix} ?$$

Exercise 1.85
If

$$C = \begin{pmatrix} 3 & 4 & 1 \\ 9 & 2 & 7 \\ 4 & 1 & 5 \end{pmatrix} \quad \text{and} \quad E = \begin{pmatrix} 5 & 6 & 2 \\ 9 & 4 & 7 \\ 2 & 1 & 1 \end{pmatrix}$$

solve the following matrix equations for A:

(i) $CAE = C - E,$
(ii) $C^2(A + 2E) = EA,$
(iii) $2AE^{-1} = A^2EC$ (assume A to be non-singular).

Exercise 1.86
If A is a square matrix, $q(x)$ and $p(x)$ are polynomials and $f(x) = p(x)/q(x)$, we define $f(A)$ to be $p(A)q(A)^{-1}$ provided $q(A)^{-1}$ exists. If

$$A = \begin{pmatrix} 5 & 6 \\ 2 & 8 \end{pmatrix}$$

evaluate the rational function $f(x) = (x^3 + 2)/(x^2 + 1)$ at A.
Show that if $\mathbf{x}$ is an eigenvector of A with eigenvalue λ then $f(A)\mathbf{x} = f(\lambda)\mathbf{x}$.

Exercise 1.87

Determine the values of η for which the equations

$$2x + 4y + \eta z = 3, \quad \eta x - y + z = 8, \quad 3x - \eta y - z = 12$$

have a unique solution.

Exercise 1.88

If a and b are real, $0 \le a \le 1$ and $b^2 = 2a(1 - a)$, show that the matrix

$$\begin{pmatrix} a & a-1 & b \\ a-1 & a & b \\ -b & -b & 2a-1 \end{pmatrix}$$

is orthogonal with determinant equal to unity. A square matrix A is orthogonal if $AA^\top = A^\top A = I$, where $A^\top$ denotes the transpose of A.

Exercise 1.89

Consider the $N \times N$ matrix $M^{(N)}$ for which all diagonal elements are $1 + x^2$ and the off-diagonal elements are

$$M^{(N)}_{i,i+1} = x, \ i = 1, 2, \ldots, N-1 \quad \text{and} \quad M^{(N)}_{i-1,i} = x, \ i = 2, 3, \ldots, N$$

and all other elements are zero.

Compute $\det(M^{(4)})$ and $\det(M^{(5)})$ and try to guess the form of $\det(M^{(N)})$; check your guess by evaluating other cases.

Exercise 1.90

Find the values of the following sums and products:

(i) $\displaystyle\sum_{n=1}^{\infty} \frac{1}{(n+1)(n+5)}$,

(ii) $\displaystyle\sum_{n=1}^{\infty} \frac{(-1)^n}{n(n+1)}$,

(iii) $\displaystyle\prod_{n=2}^{\infty} \left(1 - \frac{2}{n(n+1)}\right)$,

(iv) $\displaystyle\prod_{n=1}^{\infty} \left(1 + \frac{1}{n^3}\right)$.

Exercise 1.91

Plot the graph of $y = (\tanh x)^x$ for $x \ge 0$ and use `fsolve` to show that there is a minimum at $x \simeq 0.44068$.

Exercise 1.92

Plot the graph of

$$y = \left(\frac{x}{1 - x^2}\right)^{(1-x^2)/x}, \quad 0 \le x \le 1$$

and use `fsolve` to show that there is a maximum at $x \simeq 0.8328$ at which $y \simeq 1.4447$. Show also that

$$\int_0^1 dx \left(\frac{x}{1 - x^2}\right)^{(1-x^2)/x} \simeq 0.63674.$$

This integral causes Maple problems and it is advisable to use the `timelimit` command and then proceed with caution.

Exercise 1.93
Plot the graph of $y = x^{\sin x}$ for $x \geq 0$ and use `fsolve` to show that there are maxima at $x \approx 2.128,\ 7.9150,\ 14.1638,\ldots$.

Show that these are given approximately by the equation
$$x_k = \alpha_k + \frac{1}{\alpha_k \ln \alpha_k} + \cdots, \quad \alpha_k = \left(2k + \frac{1}{2}\right)\pi, \quad k = 0, 1, 2,\ldots.$$

Exercise 1.94
Evaluate the following integrals; in some cases you will need to assist Maple by making a transformation and in others numerical evaluation is necessary.

(i) $\displaystyle\int dx\, \frac{1}{\sqrt{a\sqrt{x}+b}}$ $\quad a, b > 0,$

(ii) $\displaystyle\int dx\, x^4 \tan^{-1} x,$

(iii) $\displaystyle\int_0^1 dx\, \frac{1}{(1+x)\sqrt{1+x^2}},$

(iv) $\displaystyle\int_0^{3/2} dx\, \sqrt{\cos x},$

(v) $\displaystyle\int_0^{1/2} dx\, \frac{\sin^{-1} x}{(1-x^2)^{3/2}},$

(vi) $\displaystyle\int_0^1 dx\, x^5 \exp\left(x^4\right).$

Exercise 1.95
Use Maple to find $f(x)$ if
$$\cosh\left(\int dx f(x)\right) = f(x).$$

Exercise 1.96
Where possible use Maple to show that

(i) $\displaystyle\int_0^{\pi/2} dx\, \ln(\sin x) = -\frac{\pi}{2}\ln 2,$

(ii) $\displaystyle\int_0^\infty dx\, \frac{\sin x}{x} = \frac{\pi}{2},$

(iii) $\displaystyle\int_0^\infty dx\, \left(\frac{\sin x}{x}\right)^2 = \frac{\pi}{2},$

(iv) $\displaystyle\int_0^\infty dx\, \left(\frac{\sin x}{x}\right)^5 = \frac{115}{384}\pi,$

(v) $\displaystyle\int_0^\infty dx\, \frac{(\ln x)^2}{1+x^2} = \frac{\pi^3}{8},$

(vi) $\displaystyle\int_0^\infty dx\, \frac{x^2 e^{-x}}{1+e^{-2x}} = \frac{\pi^3}{16},$

(vii) $\displaystyle\int_0^\infty dx\, \frac{x^{-a}}{1+x} = \frac{\pi}{\sin \pi a}, \qquad 0 < a < 1,$

(viii) $\displaystyle\int_0^\infty dx\, \frac{1}{1+x^n} = \frac{\pi/n}{\sin(\pi/n)}, \qquad n \text{ a positive integer,}$

(ix) $\displaystyle\int_{-\infty}^\infty dx\, \frac{1}{\cosh x + \cos \alpha} = \frac{2\alpha}{\sin \alpha}, \qquad 0 < \alpha < \pi.$

2

Simplification

2.1 Introduction

Maple rarely creates expressions in the form needed by the user, so some manipulation of the output is usually necessary to produce acceptable results. There are many facilities for 'simplifying' or manipulating expressions that can be learnt only with practice: often the process of transforming an expression to the 'correct' form is a matter of trial and error, though with experience the number of errors decreases. There is a natural tendency to attempt all the analysis using Maple; it is, however, sometimes easier and quicker to do part of it by hand. The purpose of this chapter is to describe some of the Maple commands that can be used for manipulating expressions. With increasing experience you will find many other facilities which can be useful; these may be found using the help files, which you should use to look up all the commands introduced in this chapter.

Some reflection should convince you that there can be no absolute rules for simplifying expressions, and consequently there is no definition of what 'simple' means. For example, it is sometimes 'simpler' to re-write $\cos 2x$ as either $2\cos^2 x - 1$ or $1 - 2\sin^2 x$ or as $\cos^2 x - \sin^2 x$; sometimes it is better to write $x^3 - y^3$ as $(x - y)(x^2 + xy + y^2)$ and sometimes it is better in the original form. The problem is that all mathematical expressions can be written in many equivalent forms; it is one problem to change between the equivalent forms and another to recognise which is the most convenient for subsequent use.

A related but more clearly defined problem is called the *zero equivalence* problem which is how to recognise whether a given expression — which may be only part of a larger expression — is zero. For example it is not immediately obvious that $\ln(\tan(x/2 + \pi/4)) - \sinh^{-1}(\tan x) = 0$.

In this chapter we introduce some Maple commands that change the form of expressions. There are very many such commands if one also includes the allowed options; since we cannot deal with all of these it is important that you spend a little time skimming through the help files. Despite the vast number of options available, you will find that most problems can be dealt with by remembering only a few commands.

2.2 Expand

The expand command does exactly what its name implies. Its operation on polynomial expressions is fairly obvious, as the following examples show:

```
>  expand( (x+1)*(x+z)^2 );
```

$$x^3 + 2x^2 z + x z^2 + x^2 + 2 x z + z^2$$

The two-argument form does not expand a sub-expression, in this case $x + 1$,

```
>  expand( (x+1)*(x+z)^2, x+1 );
```

$$(x + 1) x^2 + 2 (x + 1) x z + (x + 1) z^2$$

When applied to trigonometric functions expand uses the sum rules to remove multiple angles, replacing them with powers,

```
>  expand( cos(2*x) );
```

$$2 \cos(x)^2 - 1$$

and

```
>  expand(  sin(5*x)*cos(3*x)  );
```

$$64 \sin(x) \cos(x)^7 - 96 \sin(x) \cos(x)^5 + 40 \sin(x) \cos(x)^3 - 3 \sin(x) \cos(x)$$

Note that the alternative form, $\cos 2x = 1 - 2 \sin^2 x$, is not given, neither is it particularly easy to tell Maple to use this form as an alternative, at every occurrence, although the simplify command, discussed in section 2.5, can be used to change particular expressions. Some examples of the expand command are

```
>  expand( sin(x+y) );
```

$$\sin(x) \cos(y) + \cos(x) \sin(y)$$

```
>  expand( cos(2*x+y) );
```

$$2 \cos(y) \cos(x)^2 - \cos(y) - 2 \sin(y) \sin(x) \cos(x)$$

```
>  expand( cos(x+y+z) );
```

$$\cos(x) \cos(y) \cos(z) - \cos(x) \sin(y) \sin(z) - \sin(x) \sin(y) \cos(z) - \sin(x) \cos(y) \sin(z)$$

The opposite of this command, convert(*,trig), is dealt with in section 2.7 (page 74). Partial expansions, say $\cos 5x = \cos 4x \cos x - \sin 4x \sin x$, are obtained using tricks, such as

```
>   c:=cos(5*x):
>   c:=subs(5*x=x+y,c);
```

$$c := \cos(x + y)$$

```
>   c:=expand(c);
```

$$c := \cos(x) \cos(y) - \sin(x) \sin(y)$$

```
>   c:=subs(y=4*x,c);
```

$$c := \cos(x) \cos(4 x) - \sin(x) \sin(4 x)$$

The operation of expand on the exponential function is straightforward, but some care is needed with the logarithmic function and with powers because these are multivalued functions. For the exponential we have

```
>   expand( exp(x+y) );
```

$$e^x e^y$$

```
>   expand( exp(z*(x+y)) );
```

$$e^{(z\,x)} e^{(z\,y)}$$

But for the logarithm we have

> `expand( ln(x*y) );`

$$\ln(x\,y)$$

and similarly expand does not change $\ln(x/y)$ unless $x > 0$ or $y > 0$. Thus forcing x to be positive gives:

> `assume(x>0);   expand( ln(x*y) );`

$$\ln(x) + \ln(y)$$

Further, the function $\ln(x^y)$ expands only if both x and y are positive; thus

> `expand( ln(x^y) );`

$$\ln(x^y)$$

> `assume(y>0);   expand( ln(x^y) );`

$$y\,\ln(x)$$

Powers are also changed with the expand command: first we clear the variables x and y to remove the assumptions just imposed upon them,

> `x:='x': y:='y': expand( z^(x+y));`

$$z^x\,z^y$$

But $(xy)^z$ expands only if one of x or y is positive; thus

> `expand( (x*y)^z );`

$$(x\,y)^z$$

however, if $x > 0$ we have

> `assume(x>0);   expand( (x*y)^z );`

$$x^z\,y^z$$

Similarly, the command

> `expand( (x/y)^z);`

$$x^z\left(\frac{1}{y}\right)^z$$

works only if either x or y is positive.

These rules may seem complicated, but they are the natural rules obeyed by all complex variables; we shall see later that the `simplify` command can be made to have the same effect without the conditions on variables.

On the other hand expand treats square roots and fractional powers inconsistently, as shown in the following exercise.

Exercise 2.1
Show that the command `expand( sqrt(x)*sqrt(1/x) );` produces the result:

$$\sqrt{x}\,\sqrt{\frac{1}{x}}$$

whereas the mathematically equivalent Maple command `expand( x^(1/2)*x^(-1/2) );` gives 1. Re-apply these two commands after using `assume(x>0)` to make $x > 0$.

Exercise 2.2

Investigate how Maple deals with explicit fractional powers. Show that, for instance

```
>  expand( x^(1/2)*x^(3/4) );
```

combines the fractional powers as would be expected, to give $x^{5/4}$.

Now consider the effect on more general fractions, using the `assume` command to force p, q, r and t to be integers.

```
>  assume(p,integer, q,integer, r,integer, t,integer);
>  expand( x^(p/q)*x^(r/t) );
```

Now Maple does not combine the fractions as before, but gives:

$$x^{\left(\frac{p}{q}\right)} x^{\left(\frac{r}{t}\right)}$$

To be fair to Maple, in this last exercise we are using expand on a problem that it is not really designed for; powers should be combined using the `combine` command, introduced on page 72. But these exercise illustrate an important point: Maple is not very good at dealing with non-integer powers, though this appears to be a problem with most algebraic languages.

This example also highlights another feature of Maple: many operations can be performed by more than one command, though, in principle, there is a 'correct' command for each operation.

Expand also works on simple functions known to Maple, for example

```
>  expand( (n+1)! );
```

$$n!\,(n+1)$$

```
>  expand( binomial(n+1,k+1) );
```

$$\frac{(n+1)\,\mathrm{binomial}(n,\,k)}{k+1}$$

and expand knows the recurrence relations for Bessel functions,

```
>  expand( BesselJ(2,x) );
```

$$2\,\frac{\mathrm{BesselJ}(1,\,x)}{x} - \mathrm{BesselJ}(0,\,x)$$

The output, and use, of Bessel function can be made to look more natural using the `alias` command, that is described in more detail in chapter 3.

```
>  alias(J=BesselJ):
>  expand( J(2,x) );
```

$$2\,\frac{J(1,\,x)}{x} - J(0,\,x)$$

Thus Maple expands $J_n(x)$ in terms of $J_0(x)$ and $J_1(x)$ for any integer n. For example:

```
>  expand( J(4,x) );
```

$$48\,\frac{J(1,\,x)}{x^3} - 24\,\frac{J(0,\,x)}{x^2} - 8\,\frac{J(1,\,x)}{x} + J(0,\,x)$$

Exercise 2.3

Try the effect of expand on

$$x^{y^{a+b}}, \qquad (n+4)!, \quad \text{and} \quad \binom{n+2}{k+3}.$$

Exercise 2.4

Try the effect of expand on $\Gamma(m+x)$ for various explicit values of the integer m, both positive and negative.

Remember that the Maple name for $\Gamma(x)$ is GAMMA(x) and that the $\Gamma(x)$ is the natural extension of the factorial function to non-integer arguments, $z! = \Gamma(z+1) = z\Gamma(z)$; some of its properties are discussed in Appendix I.

Exercise 2.5

Try the effect of expand on BesselJ(n/2,x) for various odd and even values of the integer n.

Exercise 2.6

Use Maple to show that

$$
\begin{aligned}
\cos 5x &= 16\cos^5 x - 20\cos^3 x + 5\cos x, \\
\cos 8x &= 128\cos^8 x - 256\cos^6 x + 160\cos^4 x - 32\cos^2 x + 1, \\
\sinh 5x &= 16\sinh x \cosh^4 x - 12\sinh x \cosh^2 x + \sinh x \\
&= (4\cosh^2 x + 2\cosh x - 1)(4\cosh^2 x - 2\cosh x - 1)\sinh x.
\end{aligned}
$$

2.3 Factor

For polynomials the factor command is, in some respects, the inverse of expand; more generally factor is useful for factoring multivariate polynomials over the rationals. Consider the simple polynomials

```
>  f:=x^5 - x^4 - 7*x^3 + x^2 + 6*x:    factor(f);
```
$$x(x-1)(x-3)(x+2)(x+1)$$
```
>  g:=x^4 + 4*x^3*y - 7*x^2*y^2 - 22*x*y^3 + 24*y^4: factor(g);
```
$$(-y+x)(-2y+x)(3y+x)(4y+x)$$

factor deals with rational expressions, by factoring both numerator and denominator,

```
>  r:=f/g:    factor(r);
```
$$\frac{x(x-1)(x-3)(x+2)(x+1)}{(-y+x)(-2y+x)(3y+x)(4y+x)}$$
```
>  s:=(x^3-y^3)/(x^4-y^4):    factor(s);
```
$$\frac{x^2+xy+y^2}{(y+x)(x^2+y^2)}$$

In this last case both numerator and denominator contained the common factor $x-y$, which has been canceled.

Exercise 2.7

(i) Expand the function $f = (x + y + 1)^3$ and factorise $f - 1$.

(ii) Use the `expand` command on the function $g = \sin 5x$ and `factor` on the result. Explain what Maple is doing.

Exercise 2.8

Find the factors of $x^{2n} + x^n + 1$ for $n = 3, 4, \ldots, 23$. What do you notice about the factors when n is a prime?

2.4 Normal

The `normal` command is provided to simplify rational functions of the form f/g, where f and g are polynomials, although it also provides some simplification in other circumstances, as shown in the examples contained in the help files. For ratios of polynomials `normal` explicitly finds the greatest common divisor (gcd) of the numerator and denominator and factors it out, leaving the remainders unfactored; in some cases this leads to the same result given by `factor`, which finds factors and hence implicitly the greatest common divisor; exercise 2.9 gives an example of the similarities and differences. In general `normal` is faster than `factor` so is to be preferred. Consider a simple example: if

```
>   f:=x^2 + 3*x + 2: g:=x^2 + 5*x + 6: r:=f/g;
```

$$r := \frac{x^2 + 3x + 2}{x^2 + 5x + 6}$$

even though the gcd is $x+2$, Maple does not automatically simplify rational expressions to a relatively prime form, as it does with rational numbers, except when it immediately recognises common factors. We can factorise the numerator and denominator separately using the `numer` and `denom` commands

```
>   numer(r), denom(r);
```

$$x^2 + 3x + 2, \quad x^2 + 5x + 6$$

```
>   factor( numer(r) ), factor( denom(r) );
```

$$(x + 2)(x + 1), \quad (x + 3)(x + 2)$$

If only the greatest common divisor is needed it is better to use the `gcd` command

```
>   gcd(f,g);
```

$$x + 2$$

In order to express a rational polynomial as a ratio of relatively prime terms we use the `normal` command:

```
>   normal(r);
```

$$\frac{x + 1}{x + 3}$$

This command automatically divides out the gcd. You should confirm that, in this case, `factor` produces the same result.

In general, Maple will only automatically simplify rational functions when it immediately recognises common factors; for example, if

```
>   ff:=(x-1)*f;   gg:=(x-1)^2 * g;   rr:=ff/gg;
```

$$ff := (x-1)(x^2 + 3x + 2)$$

$$gg := (x-1)^2 (x^2 + 5x + 6)$$

$$rr := \frac{x^2 + 3x + 2}{(x-1)(x^2 + 5x + 6)}$$

and

```
>   normal(rr);
```

$$\frac{x+1}{(x+3)(x-1)}$$

In this case normal appears to have factored the result, but this is because the denominator was initially partially factored; to see the effect of this try the commands

```
r:=expand(ff)/expand(gg);   normal(r);
```

One difference between factor and normal is seen in the following example; if

```
>   f:=x^4 + 10*x^3 + 35*x^2 + 50*x + 24;
```

$$f := x^4 + 10x^3 + 35x^2 + 50x + 24$$

and

```
>   g:=x^5 + 12*x^4 + 54*x^3 + 112*x^2 + 105*x + 36;
```

$$g := x^5 + 12x^4 + 54x^3 + 112x^2 + 105x + 36$$

then the results of factor and normal on the ratio f/g are

```
>   factor(f/g);   normal(f/g);
```

$$\frac{x+2}{(x+3)(x+1)}$$

$$\frac{x+2}{x^2 + 4x + 3}$$

So, we see that normal reduces the expression to the simplest form, but does not factor the result.

Exercise 2.9
Show that in this case simplify(f/g) gives the same result as normal(f/g) and that simplify(f/g,factor) gives the same result as factor(f/g).

It is not always useful to factor a rational polynomial; for example the normal form of $(x^{1000} - 1)/(x-1)$ is much less easy to deal with than the original form.

The normal command also removes the greatest common divisor of multivariate polynomials,

```
>   f:=126*x*y+161*y^2+77+144*x^2*y+346*x*y^2+88*x+207*y^3+99*y:
```

```
>   g:=28*x^2+166*y+91+32*x^3+56*x*y+104*x+36*x^2*y+63*y^2:
```

```
>   r:=normal(f/g);
```

$$r := \frac{x+1}{x+3}$$

Further, `normal` also works on generalised polynomials such as

$$f = \frac{1 - \sin^2 x}{1 + \sin x} :$$

```
> f:=(1-sin(x)^2)/(1+sin(x)): normal(f);
```
$$-\sin(x) + 1$$

Exercise 2.10
Convert the expression

$$\frac{x^4 + x^3 - 4x^2 - 4x}{x^4 + x^3 - x^2 - x} \quad \text{into} \quad \frac{(x-2)(x+2)}{(x-1)(x+1)} \quad \text{and} \quad \frac{x^2 - 4}{x^2 - 1}.$$

2.5 Simplify

This is the general purpose simplification command which one often tries first, although it is not always the most appropriate command; the result of using `simplify` can be unpredictable so you are advised to keep a copy of the original. Normally, however, it produces a simpler result. `Simplify` applies the following rules:

$$\begin{aligned}
\sin^2 x &\rightarrow 1 - \cos^2 x \\
\sinh^2 x &\rightarrow -1 + \cosh^2 x \\
\exp(x)\exp(y) &\rightarrow \exp(x+y) \\
u^x u^y &\rightarrow u^{x+y} \\
u^{a/b} u^{c/d} &\rightarrow u^{(ad+bc)/bd}
\end{aligned}$$

and positive integer powers of trigonometric functions are simplified by these rules as far as is possible, for example,

```
> simplify(cosh(x)^4 - sinh(x)^4);
```
$$-1 + 2\cosh(x)^2$$

Note, `simplify` does not convert $\tan x$ to $\sin x / \cos x$; this is done using the `convert` command described below.

One useful feature of `simplify` is that it allows the user to define constraints or *side-relations*, which are defined using a second argument. For instance, the Maple expression

```
> f:=sin(x)^3 - sin(x)*cos(x)^2 + 3*cos(x)^3;
```
$$f := \sin(x)^3 - \sin(x)\cos(x)^2 + 3\cos(x)^3$$

can be simplified to:

```
> simplify(f);
```
$$-2\sin(x)\cos(x)^2 + 3\cos(x)^3 + \sin(x)$$

or, by putting $\cos^2 x = 1 - \sin^2 x$, we can obtain the alternative form, as follows

```
> side:={cos(x)^2=1-sin(x)^2}: simplify(f,side);
```
$$2\sin(x)^3 - 3\cos(x)\sin(x)^2 + 3\cos(x) - \sin(x)$$

Note that the side-relation, `side`, was enclosed in curly brackets.

Another very useful side-relation allows one to multiply polynomials and set to zero all terms of degree equal to or larger than some specified number. For example suppose

```
> f1:=add(x^k/(k+1),k=0..5);
```
$$f1 := 1 + \frac{1}{2}x + \frac{1}{3}x^2 + \frac{1}{4}x^3 + \frac{1}{5}x^4 + \frac{1}{6}x^5$$

we can truncate this to a third-order polynomial using the side-relation `{x^4=0}` as follows

```
> simplify(f1,{x^4=0});
```
$$1 + \frac{1}{2}x + \frac{1}{3}x^2 + \frac{1}{4}x^3$$

This facility is particularly useful if two truncated series are multiplied since there is no point in retaining terms which would have contributions from the truncated parts. Thus if

```
> f2:=add(x^k/(k^2+1),k=0..5);
```
$$f2 := 1 + \frac{1}{2}x + \frac{1}{5}x^2 + \frac{1}{10}x^3 + \frac{1}{17}x^4 + \frac{1}{26}x^5$$

the product of `f1` and `f2` up to and including the x^5 term is

```
> sort(simplify(f1*f2,{x^6=0}));
```
$$\frac{1847}{4420}x^5 + \frac{1021}{2040}x^4 + \frac{37}{60}x^3 + \frac{47}{60}x^2 + x + 1$$

Note the use of `sort` to put the terms in order.

Exercise 2.11

Use the power series expansion of e^x,

$$e^x = \sum_{n=0}^{\infty} \frac{x^n}{n!},$$

and the `add` command to construct the series expansions of $\exp x$ and $\exp x^2$ up to the x^8 term. By multiplying these together and using `simplify` find the series expansion of $\exp(x + x^2)$ up to the x^8 term.

Exercise 2.12

Consider the general cubic polynomial

$$f(x) = \frac{1}{3}ax^3 + bx^2 + cx + d$$

where a, b, c and d are real constants. If the stationary points of f are at x_1 and x_2, use

Maple to show that

$$f(x_1) - f(x_2) = -\frac{a}{6}(x_1 - x_2)^3$$

and that

$$f\left(\frac{x_1 + x_2}{2}\right) = \frac{1}{2}(f(x_1) + f(x_2)).$$

Can you find a method of deriving these results without using Maple?

Exercise 2.13
If

$$a + b + c = 3, \quad a^2 + b^2 + c^2 = 10 \quad \text{and} \quad a^3 + b^3 + c^3 = 25,$$

use `simplify` with side-relations to find the values of

$$a^n + b^n + c^n$$

for $n = 4$, 6 and 10.

One of the useful, but little publicised, options of the `simplify` command is `symbolic`, which sometimes forces the 'natural' simplification when multivalued functions are involved. Consider the problem of simplifing `sqrt(x)*sqrt(1/x)`, dealt with in section 2.2; using `simplify` on this produces 1 only if we use `assume` to make $x > 0$; otherwise we obtain

```
>   simplify(sqrt(x)*sqrt(1/x));
```

$$\sqrt{x}\sqrt{\frac{1}{x}}$$

But, the 'natural' result is obtained automatically if the `symbolic` option is included,

```
>   simplify(sqrt(x)*sqrt(1/x),symbolic);
```

$$1$$

Another example is

```
>   simplify(arctan(tan(x)));
```

$$\arctan(\tan(x))$$

which has no unique simplified form because $\tan(2\pi n + x) = \tan(x)$ for any integer n; but frequently it is not necessary to worry about this complication so we expect the result x when we attempt to simplify this expression. This natural simplification can be forced:

```
>   simplify(arctan(tan(x)),symbolic);
```

$$x.$$

2.6 Combine

The `combine` command is in some respects the opposite of `expand`; the following examples show how it can be used. Other examples are in the help files; note that the `combine` command is usually used with an optional argument.

```
>  combine( exp(x)*exp(y) );
```
$$e^{(x+y)}$$

```
>  combine( x^u * x^v, power);
```
$$x^{(u+v)}$$

```
>  combine( ln(x) + ln(y),ln);
```
$$\ln(x) + \ln(y)$$

In this case one can either invoke the symbolic option

```
>  combine( ln(x) + ln(y),ln,symbolic);
```
$$\ln(x\,y)$$

or make at least one of the variables positive,

```
>  assume(x>0):  combine( ln(x) + ln(y),ln);
```
$$\ln(x\,y)$$

The restriction is necessary as $\ln(z)$ is the principal branch logarithm so $\ln(z) = \ln|z| + i\arg(z)$ where $-\pi < \arg(z) \le \pi$ and the rule $\ln(z_1 z_2) = \ln(z_1) + \ln(z_2)$ is true only if $-\pi < \arg(z_1) + \arg(z_2) \le \pi$. The symbolic option overrides these nuances.

It is also worth noting that Maple treats e^x and $\exp(x)$ differently because e is not a reserved variable so is treated as any other number (other languages, such as Reduce, and older versions of Maple use E as a reserved variable for e). Further, the Maple screen output is different for e^x and $\exp(x)$. The following examples illustrate the differences.

```
>  combine( e^x * e^y ,exp);
```
$$e^x\, e^y$$

but

```
>  combine( e^x * e^y ,power);
```
$$e^{(x+y)}$$

In the text of this book we use standard notation and make *no* distinction between e^z and $\exp z$.

A useful feature of combine is trigonometric simplification because it transforms powers of the sine and cosine functions into their multiple angle expansions, for instance,

```
>  combine(sin(x)^4, trig);
```
$$\frac{3}{8} + \frac{1}{8}\cos(4\,x) - \frac{1}{2}\cos(2\,x)$$

```
>  combine(sin(x)^2 * (1-cos(x)^5), trig);
```
$$\frac{1}{2} + \frac{3}{64}\cos(5\,x) + \frac{1}{64}\cos(3\,x) - \frac{5}{64}\cos(x) - \frac{1}{2}\cos(2\,x) + \frac{1}{64}\cos(7\,x)$$

The combine command does not always immediately produce the required result. For example, if we define

$$f = a_0 + \sum_{k=1}^{3} a_k \sin kx$$

for some constants $a_0, \ldots, a_k$ the command `combine(f^5,trig);` will produce an expression containing products such as $\sin 3x \cos 4x$. For such expressions it is necessary to first use expand and then `combine`; thus the command

`combine( expand(f^5),trig );`

produces an expression with no products of trigonometric functions. You should try these commands, for various powers of f, to see for yourself what happens. The Maple output is not included here because it is too lengthy.

Exercise 2.14
Find the multiple angle forms for $\sin^n x$, $\cos^n x$, $\sinh^n x$ and $\cosh^n x$ for $n = 3, 4, \ldots, 10$.

2.7 Convert and series

The command `convert` changes the form of expressions and has many variants accessed via the second argument. Convert can be used to change from one data type to another, for instance from a `list` to a `set`,

> `l1:=[1,2,3,4,5]:  convert(l1,set);`

$$\{1, 2, 3, 4, 5\}$$

or a number to a continued fraction,

> `convert(3.142, confrac);`

$$[3, 7, 23, 1, 2]$$

where this list is to be interpreted as

$$3 + \cfrac{1}{7 + \cfrac{1}{23 + \cfrac{1}{1 + \cfrac{1}{2}}}}.$$

Some properties of continued fractions are explored in chapter 6.

The complete list of possible conversions is long and can be found in the help files, by typing `?convert` at the prompt.

Another use of `convert` is to change the form of expressions; for example $\tan x$ is replaced with $\sin x / \cos x$ using the `sincos` option

> `f:=convert((1+tan(x))/(1-tan(x)),sincos): normal(f);`

$$-\frac{\cos(x) + \sin(x)}{-\cos(x) + \sin(x)}$$

and the trigonometric functions are replaced by exponentials using the `exp` option,

> `f:=convert(sin(x),exp);`

$$f := -\frac{1}{2} I \left(e^{(I\,x)} - \frac{1}{e^{(I\,x)}} \right)$$

For example, a Fourier series can be readily expressed as a Laurent expansion in $z = \exp(ix)$, as follows

```
>  f:=a[0] + add(a[k]*sin(k*x),k=1..3);
```
$$f := a_0 + a_1 \sin(x) + a_2 \sin(2x) + a_3 \sin(3x)$$

```
>  f:=convert(f,exp);
```
$$f := a_0 - \frac{1}{2} I a_1 (e^{(I x)} - \frac{1}{e^{(I x)}}) - \frac{1}{2} I a_2 (e^{(2I x)} - \frac{1}{e^{(2I x)}}) - \frac{1}{2} I a_3 (e^{(3I x)} - \frac{1}{e^{(3I x)}})$$

The substitution $e^{ikx} = z^k$, for $k = 1$, 2 and 3 gives the Laurent expansion,

```
>  for k from 1 to 3 do;
>  f:=subs( exp(k*I*x)=z^k, f);
>  od: f;
```
$$a_0 - \frac{1}{2} I a_1 (z - \frac{1}{z}) - \frac{1}{2} I a_2 (z^2 - \frac{1}{z^2}) - \frac{1}{2} I a_3 (z^3 - \frac{1}{z^3})$$

Alternatively, the previous command can be replaced by:

```
>  s1:=seq(exp(I*k*x)=z^k,k=1..3);
```
$$s1 := e^{(I x)} = z, \, e^{(2I x)} = z^2, \, e^{(3I x)} = z^3$$

```
>  f:=subs(s1,f);
```
$$f := a_0 - \frac{1}{2} I a_1 (z - \frac{1}{z}) - \frac{1}{2} I a_2 (z^2 - \frac{1}{z^2}) - \frac{1}{2} I a_3 (z^3 - \frac{1}{z^3})$$

The conversion of exponentials to the trigonometric functions is achieved using the trig option:

```
>  convert(exp(x),trig);
```
$$\cosh(x) + \sinh(x)$$

One important use of convert is with the series command, a general purpose procedure usually returning a power series expansion of a function or an expression; we introduce this command here because it is useful and used frequently.

The series command gives all the usual power series expansions. It is normally used with three arguments, series(expr,x=a,n) where expr is the expression, involving a variable x, to be expanded about $x = a$ up to and including the $(n-1)$th power; for instance the expansion of $\tan^{-1}(x)$ about $x = 0$ up to and including the x^9 term is given by†

```
>  X:=series( arctan(x), x=0, 10);
```
$$X := x - \frac{1}{3} x^3 + \frac{1}{5} x^5 - \frac{1}{7} x^7 + \frac{1}{9} x^9 + O(x^{10}).$$

The series command returns a series data type as may be seen using the whattype command

```
>  whattype(X);
```
$$series$$

This data type cannot always be used until converted into a polynomial; for example it is not possible to plot the graph of X or multiply it with another expression of the

---

† The **series** command returns an expression ending with a term $O(x^n)$, for some n; this notation will be explained in chapter 4.

same type, though differentiation and integration of these types is possible. We change X to an ordinary polynomial using convert, as follows:

```
> Xs:=convert(X,polynom);  whattype(Xs);
```

$$Xs := x - \frac{1}{3} x^3 + \frac{1}{5} x^5 - \frac{1}{7} x^7 + \frac{1}{9} x^9$$
$$+$$

Sometimes it is convenient to use the default form of the series command, given in this case by series(arctan(x),x); which assumes the expansion to be about zero and the order to be 6, the default size of the Order environment variable.

At this point it is worth noting that Maple allows one to find an asymptotic expansion of some functions — asymptotic expansions are considered in chapter 5, but here we treat them as power series expansions in $1/x$,

```
> Y:=asympt(1/(1+x^2),x,10);  whattype(Y);
```

$$Y := \frac{1}{x^2} - \frac{1}{x^4} + \frac{1}{x^6} - \frac{1}{x^8} + O(\frac{1}{x^{10}})$$
$$+$$

Notice that this procedure produces a different data type, and now the $O(x^{-10})$ may be removed by substituting 0 (zero) for the O using the eval (or subs) command:

```
> Y:=eval(Y,O=0);
```

$$Y := \frac{1}{x^2} - \frac{1}{x^4} + \frac{1}{x^6} - \frac{1}{x^8}$$

This evaluation trick does not work with series data types. The convert(Y,polynom) command also works.

Exercise 2.15

Use the series command to find the power series for $\sin x$ and $\cos x$, about $x = 0$ up to the term x^9.

If s and c are the series data types representing these two series show that the command series(2*s*c,x=0,10) gives the series for $\sin 2x$.

The series command can also be used to evaluate some integrals in terms of series: for example the integral

$$\int_0^1 dy \, \exp(a \sin(xy))$$

cannot be expressed in terms of elementary functions, but if x is small its series representation can be useful, and this is found using the following command:

```
> series(  Int(exp(a*sin(x*y)),y=0..1),  x=0,  5);
```

$$1 + \frac{1}{2} a x + \frac{1}{6} a^2 x^2 + (-\frac{1}{24} a + \frac{1}{24} a^3) x^3 + (-\frac{1}{30} a^2 + \frac{1}{120} a^4) x^4 + O(x^5)$$

Note the use of Int rather than int as this saves the futile attempt at evaluating the integral in terms of known functions before finding the series representation.

Exercise 2.16

Use Maple to show that

$$\int_0^1 dy \, \sin\left(x\left(y^{1/2} + y^{1/3}\right)\right) = \frac{17}{12}x - \frac{1083}{1820}x^3 + \frac{565\,249}{6\,511\,680}x^5 + \cdots$$

$$\int_0^x dy \, \sin\left(x\left(y^{1/2} + y^{1/3}\right)\right) = \frac{3}{4}x^{7/3} + \frac{2}{3}x^{5/2} - \frac{1}{12}x^5 + \cdots.$$

For the second example, in order to avoid an error message from Maple, you will need to change the order in which you apply `series` and `int`.

A particularly useful feature of the `series` command is that, in conjunction with `solve`, it may be used to invert series. We illustrate this with a simple example: the series for $\tan^{-1} x$ about $x = 0$ can clearly be inverted to give the series for the tan function; thus if

$$y = \tan^{-1} x = x - \frac{x^3}{3} + \frac{x^5}{5} - \frac{x^7}{7} + \cdots, \tag{2.1}$$

then

$$x = \tan y = y + \frac{y^3}{3} + \frac{2y^5}{15} + \frac{17y^7}{315} + \cdots. \tag{2.2}$$

To find this with Maple we first form equation 2.1

```
>   eqn1:=y=series(arctan(x),x=0,15);
```

$$eqn1 := y = x - \frac{1}{3}x^3 + \frac{1}{5}x^5 - \frac{1}{7}x^7 + \frac{1}{9}x^9 - \frac{1}{11}x^{11} + \frac{1}{13}x^{13} + O(x^{15})$$

Now solve `eqn1` for x, using the `solve` command

```
>   x:=solve(eqn1,x);
```

$$x := y + \frac{1}{3}y^3 + \frac{2}{15}y^5 + O(y^6)$$

to give x as the required series in y; note, in this context it is assumed that `eqn1` is a series data type. The result is also a series data type, so may need to be converted to a polynomial,

```
>   x:=convert(x,polynom);
```

$$x := y + \frac{1}{3}y^3 + \frac{2}{15}y^5$$

The original series for $\tan^{-1} x$ contained terms to x^{13}, but the inverted series is truncated at y^5. In order to obtain further terms in the inversion it is necessary to reset the global variable `Order`; we now redo the inversion with `Order=15`, first using `restart` to clear all variables

```
>   restart; Order:=15:
>   eqn1:=y=series(arctan(x),x=0,Order);
```

$$eqn1 := y = x - \frac{1}{3}x^3 + \frac{1}{5}x^5 - \frac{1}{7}x^7 + \frac{1}{9}x^9 - \frac{1}{11}x^{11} + \frac{1}{13}x^{13} + O(x^{15})$$

Now invert this to give the series for $x = \tan y$,

```
>  x:=solve(eqn1,x);
```

$$x := y + \frac{1}{3} y^3 + \frac{2}{15} y^5 + \frac{17}{315} y^7 + \frac{62}{2835} y^9 + \frac{1382}{155925} y^{11} + \frac{21844}{6081075} y^{13} + O(y^{15})$$

In this example the use of `restart` was unnecessary as `forget(solve)` could have been used to reset the parameters used by `solve`; with this more gentle method, it would not have been necessary to re-define `eqn1`.

Exercise 2.17

Show that the inverse of the function

$$y = \tan^{-1}\left(\frac{x}{1-x}\right),$$

which is zero when $y = 0$, has the series representation

$$x = y - \frac{1}{2} y^2 + \frac{4}{3} y^3 - \frac{5}{3} y^4 + \frac{32}{15} y^5 - \frac{122}{45} y^6 + \frac{1088}{315} y^7 + \cdots .$$

Plot the graph of this series together with the graph of the exact inverse,

$$x = \frac{\tan y}{1 + \tan y} \quad \text{for} \quad 0 \le y \le 1.$$

Use the `convert` command with the `ratpoly` option to convert the above series to a ratio of polynomials, $X(y)$, and show, graphically, that this is a far better approximation than the original series. Note that `convert(s,ratpoly)` works only if s is a series data type.

This latter type of approximation, whereby a truncated power series is re-expressed as a ratio of polynomials, is dealt with in chapter 6. Frequently such ratios are better approximations than the original series.

Exercise 2.18

Use Maple to show that if

$$z = \int_0^1 dy \, \sinh(a \sin(xy)),$$

then for small values of z

$$x = 2u + \frac{2}{3}(1 - a^2)u^3 + \frac{2}{45}(13a^4 - 10a^2 + 13)u^5 + \cdots,$$

where $u = z/a$.

Text in graphs

It is often necessary to plot graphs of a function $f(x, a)$ for various values of the parameter a; it is then usually helpful to include the value of a in a title or caption. This is achieved by converting a, which we assume has a numerical value, to a string and concatenating this with other text. The convert command with the form `convert(expr,string)` changes expr into a `string`; the inverse function is `parse` and is needed in exercise 2.41 (page 89). If the value of expr is less than unity, say -0.01, the string returned by `convert` is in scientific notation, `-.1e-1`, and some effort is required to put this in a more readable form.

For instance, suppose that $f(x, a) = \exp(a \sin x)$ and that this is to be plotted with

$a = \pi/2$. The following set of instructions will draw this graph and add a title in a Times Roman font with size 12 pt (1pt=1/72 inch). Notice that when converting the numerical value of *a* to a string it was evaluated to a reasonable number of digits using the option in `evalf`.

```
>   b:=Pi/2:                         # Value of the constant to be used
>   f:=exp(a*sin(x)):                           # The function
>   t1:=cat("Value of a is ",convert(evalf(b,4),string)):
>                                               # The title string
>   tfont:=[TIMES,ROMAN,12]:          # Definition of the title font
>   plot(eval(f,a=b),x=0..2*Pi,title=t1,titlefont=tfont);
```

These commands produce the following graph.

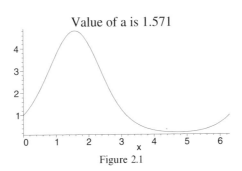

Figure 2.1

This technique is particularly useful in conjunction with the animation package, the feature of the graphics facility allowing one to view a sequence of graphs as a moving picture. Obviously we cannot show this feature in the text, but one frame of the animation is shown below.

Suppose we need to create a series of graphs of $y = \sin(n \sin x)$ for various values of *n* and each with a legend depending upon the value of *n*. The following sequence of commands will produce the graphs and put them in the array gr. In order to do this we need to add text to the figures, which is done using the `textplot` and `display` commands, as shown below.†

The following sequence of commands creates a list of plots; each command is explained in the text below. It is necessary to invoke `display` using `with(plots)` and we also find it better to overwrite the Maple default tick-marks using the two optional arguments of `plot`, `xtickmarks` and `ytickmarks`, combined in one expression, `ticks`.

```
>   restart; with(plots):
>   xtick:=[seq(k/2,k=0..6)]: ytick:=[seq(-1+k/2,k=0..4)]:
>   ticks:=xtickmarks=xtick,ytickmarks=ytick;
```

$$ticks := xtickmarks = [0, \frac{1}{2}, 1, \frac{3}{2}, 2, \frac{5}{2}, 3], ytickmarks = [-1, \frac{-1}{2}, 0, \frac{1}{2}, 1]$$

---

† It is necessary to use **textplot** here rather than use the title option because, in Maple 5 and subsequent versions, titles are not shown in animations.

```
>   fnt:=[TIMES,ROMAN,12]:                                      # 1
>   N:=20: gr:=NULL:                                            # 2
>   string1:="y=sin(n sin(x)) for n=":                          # 3
>       for k from 1 to N do;                                   # 4
>       string2:=convert(k,string);                             # 5
>       str:=cat(string1,string2);                              # 6
>       t:=textplot([1.5,1.1,str],font=fnt);                    # 7
>       p:=plot(sin(k*sin(x)),x=0..Pi, ticks);                  # 8
>       gr:=gr,display(p,t);                                    # 9
>       od:
>   gr:=[gr]:                                                   # 10
```

(1) Here the fonts used in `textplot` are defined; this is not necessary as `textplot` has default fonts, but a nicer result is obtained this way. In the list `fnt` the first two items define the font to be used and the third the size of the characters. More can be found out about font options by typing `?plot,options` at the command prompt.

In the Maple lines before the font definition we set up the tickmarks on the x- and y-axes, using the two commands `xtickmarks` and `ytickmarks` which, when used in this manner, provide more flexibility than the simpler `tickmarks=[n,m]` command and are better than the default settings, though it seems impossible to add small, intermediate tickmarks.

(2) In this line we define the number of graphs to be drawn, `N=20`, and define `gr`, which will eventually contain the graphs, to be the NULL sequence.

(3) This is the string of the first part of the title.

(4) Start of the loop that creates a figure on each pass through it.

(5) Convert the number k to the string `string2`.

(6) Concatenate the two strings to form the complete title in the form of a string.

(7) The title, in the string `str`, now has to be included in a graph; the first stage is to create a suitable object using the `textplot` function. You can find out more about this by typing `?textplot` at the command prompt. The first argument is a list containing the x and y coordinates of the string position and the third is the text to be included, which must be a string. Other options include the font to be used, here defined in the variable `fnt`, and the alignment of the text; here we have used the default.

This function is very useful when annotating graphs.

(8) The graph of the function $y = \sin(k \sin x)$, for $0 \le x \le \pi$, is drawn.

(9) In this line two operations are performed. The `display` command combines the plot created in line 8 with the text defined in line 7; then this plot is appended to the sequence of previous plots.

(10) Finally, the sequence `gr` is converted to a list.

This set of commands produces a list of graphs that can be viewed individually by typing gr[k] for specific values of k in the range 1 to $N(= 20)$ or by using the `display` command.† At this point you should type the above commands into your computer

---

† A version of this worksheet may be found on the web site given in the Preface.

and reproduce the fifth plot, shown below

> `gr[5];`

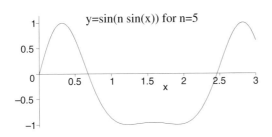

Figure 2.2   Graph of the function $y = \sin(5 \sin x)$.

The command `display(gr, insequence=true);` produces an animated sequence of plots. When the first graph is displayed on the screen left-click on it to activate the animation menu bar at the top of the screen and then use this to control the animation. The right mouse button gives a context sensitive menu.

Exercise 2.19
Use `convert` with the appropriate option to show that:

$$\sin x \quad = \quad \frac{2t}{1+t^2}, \quad t = \tan(x/2) \qquad \cos x \quad = \quad \frac{1-t^2}{1+t^2}, \quad t = \tan(x/2),$$

$$\sin^{-1} x \quad = \quad -i \ln(ix + \sqrt{1-x^2}), \qquad \sinh^{-1} x \quad = \quad \ln(x + \sqrt{1+x^2}),$$

$$\cos^{-1} x \quad = \quad -i \ln(x + i\sqrt{1-x^2}), \qquad \cosh^{-1} x \quad = \quad \ln(x + \sqrt{x^2-1}),$$

$$\tan^{-1} x \quad = \quad i \ln \sqrt{\frac{1-ix}{1+ix}}, \qquad \tanh^{-1} x \quad = \quad \ln \sqrt{\frac{1+x}{1-x}}.$$

Exercise 2.20
The `series` command sometimes produces some odd results, as the following examples show.

Determine the result of using the `series` command, expanding about $x = 0$, on the functions

$$\text{(i)} \quad f_1(x) = x^x, \qquad \text{(ii)} \quad f_2(x) = x^{x^x}, \qquad \text{(iii)} \quad f_3(x) = \sin(x^x).$$

Exercise 2.21
Compare, graphically, the series expansion of $\sin x$ about $x = 0$ up to x^n, for various values of n, and for $0 \leq x \leq \pi$. Animate your graphs for $n = 3, 5, \ldots, 9$.

How large does n need to be for the absolute value of the series approximation to be less than 0.01 at $x = \pi$ and $x = 2\pi$?

Exercise 2.22

Use the `series` command to show that

$$s_1 \quad = \quad \sin(\tan x) - \tan(\sin x) = -\frac{1}{30}x^7 + \cdots$$

$$s_2 \quad = \quad \sin^{-1}(\tan^{-1} x) - \tan^{-1}(\sin^{-1} x) = -\frac{1}{30}x^7 + \cdots$$

and find all the terms of the expansion up to, and including, the term x^{13}.
    Hence show that

$$\frac{s_1}{s_2} = 1 + \frac{5}{3}x^2 + \frac{1313}{1890}x^4 + \cdots.$$

Another example of this behaviour is considered in exercise 2.29, and the general case is dealt with in exercise 4.62 (page 174).

2.8 Collect, coeff and coeffs

We now consider some of the ways power series and other types of series can be manipulated and dissected. It is often useful to extract the coefficients of the various powers, and this is done singly using the `coeff` command or all together using `coeffs`. The `coeff` command can be used directly on `series` data types, the X defined below for instance, and on other data types. On other series, however, it is usually necessary to use `collect` prior to using `coeff`. For an elementary example we use `coeff` to extract the coefficient of x^5 from the power series expansion of $\tan^{-1} x$ by performing the following Maple commands:

```
>  X:=series(arctan(x),x=0,10);
```

$$X := x - \frac{1}{3}x^3 + \frac{1}{5}x^5 - \frac{1}{7}x^7 + \frac{1}{9}x^9 + \mathrm{O}(x^{10})$$

```
>  coeff(X,x,5);
```

$$\frac{1}{5}$$

If all of the coefficients are required, the command `coeffs` will return them as a sequence: the third optional argument of `coeffs`, `pow` in the example below, is an expression sequence of these powers. Since the ordering is not predictable[†] this second list is necessary:

```
>  Xs:=convert(X,polynom):
>  coeffs(Xs,x,'pow');   pow;
```

$$1, \frac{-1}{3}, \frac{1}{5}, \frac{-1}{7}, \frac{1}{9}$$

$$x, x^3, x^5, x^7, x^9$$

Note, `coeffs` does not work on `series` type data structures, and sometimes requires that the `collect` procedure be used first.

---

† The order of the coefficients produced by **coeffs** in the earlier releases of Maple was usually not predictable, but in version 5 it seems to be the natural order.

We illustrate the use of `collect` by showing how the Maple expression for the function

$$f = (x + y + 1)^3 (x + 3y + 5) \qquad (2.3)$$

can be converted to a polynomial in either x or y:

```
> f:=(x+y+1)^3 * (x+3*y+5);
```
$$f := (x + y + 1)^3 (x + 3y + 5)$$

```
> fx:=collect(f,x);
```

$$fx := x^4 + (6y + 8) x^3 + ((y + 1)(2y + 2) + (y + 1)^2 + (3y + 3)(3y + 5)) x^2$$
$$+ ((y + 1)^3 + ((y + 1)(2y + 2) + (y + 1)^2)(3y + 5)) x + (y + 1)^3 (3y + 5)$$

Alternatively, the coefficients of y^k may be collected *and* factored using the `factor` option:

```
> fy:=collect(f,y,factor);
```

$$fy := 3y^4 + (10x + 14) y^3 + 12 (x + 2)(x + 1) y^2 + 6 (x + 3)(x + 1)^2 y$$
$$+ (x + 1)^3 (x + 5)$$

Another way of factoring composite expressions is to use the `map` command, described in the next section.

The `collect` command can also be used on rational powers: for instance, if

$$g = \left(x + a\sqrt{x} + bx^{1/3} \right)^3 + (x + b\sqrt{x})^2,$$

the powers of x are collected as follows:

```
> g:=(x+a*sqrt(x)+b^(1/3))^3 + (x+b*sqrt(x))^2:
> collect(g,x);
```

$$x^3 + 3\,a\,x^{(5/2)} + (1 + 3\,b^{(1/3)} + 3\,a^2) x^2 + (2b + 4\,b^{(1/3)}\,a + a\,(2\,b^{(1/3)} + a^2)) x^{(3/2)}$$
$$+ (b^{(1/3)}\,(2\,b^{(1/3)} + a^2) + 2\,a^2\,b^{(1/3)} + b^{(2/3)} + b^2) x + 3\,b^{(2/3)}\,a\,\sqrt{x} + b$$

However, neither `coeff` nor `coeffs` work on functions containing non-integer powers.

The `collect` and `coeff` combination can also be used to extract other useful information, as we illustrate by finding the Fourier coefficients of the function

$$f = (a_0 + a_1 \sin x + a_2 \sin 2x + a_3 \sin 3x + a_4 \sin 4x)^2.$$

In these Maple commands we suppress the output to save space, but recommend that you repeat them without suppressing the output to see what happens. First define the Maple expression

```
> f:=(a[0] + add(a[k] * sin(k*x), k=1..4))^2;
```
$$f := (a_0 + a_1 \sin(x) + a_2 \sin(2x) + a_3 \sin(3x) + a_4 \sin(4x))^2$$

and then convert this to multiple angle form:

```
> f:=combine(f,trig):
```

The coefficients of $\sin(kx)$ for $k = 1, 2, \ldots, 5$ are found as follows:

```
>   for k from 1 to 5 do;
>   g:=collect(f,sin(k*x)):              # Collect the sin(k*x) terms,
>   s:=coeff(g,sin(k*x)):                # Put the coefficient in s,
>   print('For k=',k,' coefficient of sin(k*x) is    ',s);
>   od:
```

$$
\begin{aligned}
&\text{For } k =, 1, \quad \text{coefficient of } \sin(k * x) \text{ is} \quad , 2\,a_0\,a_1 \\
&\text{For } k =, 2, \quad \text{coefficient of } \sin(k * x) \text{ is} \quad , 2\,a_0\,a_2 \\
&\text{For } k =, 3, \quad \text{coefficient of } \sin(k * x) \text{ is} \quad , 2\,a_0\,a_3 \\
&\text{For } k =, 4, \quad \text{coefficient of } \sin(k * x) \text{ is} \quad , 2\,a_0\,a_4 \\
&\text{For } k =, 5, \quad \text{coefficient of } \sin(k * x) \text{ is} \quad , 0
\end{aligned}
$$

Note that in this context `collect` deals with each $\sin kx$ individually, whereas when used to collect powers of x it finds all powers simultaneously. More than one item can be collected by including all items in a set: this technique is illustrated by finding the coefficients of $\cos kx$, $k = 1, 2, \ldots, 3$:

```
>   sc:={seq(cos(k*x),k=1..3)}:          # Put cos(k*x),k=1..3 in a set.
>   g:=collect(f,sc):                    # Collect the items in this set.
```

Now isolate each coefficient and print it:

```
>   for k from 1 to 3 do;
>   s:=coeff(g,cos(k*x)):
>   print('For k=',k,' coefficient of cos(k*x) is    ',s);
>   od:
```

$$
\begin{aligned}
&\text{For } k =, 1, \quad \text{coefficient of } \cos(k * x) \text{ is} \quad , a_1\,a_2 + a_2\,a_3 + a_3\,a_4 \\
&\text{For } k =, 2, \quad \text{coefficient of } \cos(k * x) \text{ is} \quad , a_1\,a_3 - \frac{1}{2}\,a_1{}^2 + a_2\,a_4 \\
&\text{For } k =, 3, \quad \text{coefficient of } \cos(k * x) \text{ is} \quad , a_1\,a_4 - a_1\,a_2
\end{aligned}
$$

This technique finds all but the constant coefficient; it seems to be impossible to find this in the same manner so we revert to using the `remove` introduced in chapter 1, by removing all occurrences of sin and cos

```
>   c0:=remove(has,f,{sin,cos});
```

$$
c0 := a_0{}^2 + \frac{1}{2}\,a_1{}^2 + \frac{1}{2}\,a_4{}^2 + \frac{1}{2}\,a_3{}^2 + \frac{1}{2}\,a_2{}^2
$$

Note that c0 is just the mean value of f averaged over an interval of length 2π.

An alternative, and sometimes superior, method sets `sin=0` and `cos=0` using the `eval` or `subs` command, as follows

```
eval( subs(sin=0,cos=0,f) );
eval( eval(f,sin=0), cos=0);
```

This is the same trick as used to remove the 0-term from the result produced by asympt on page 76.

Exercise 2.23

If

$$
f = (a_0 + a_1 \sin x + a_2 \sin 2x + a_3 \sin 3x)^4,
$$

use Maple to find the mean value of f and the coefficients of $\sin kx$ and $\cos kx$ for $k = 1, 2$ and 3.

Hint: in this case it is necessary to expand before using combine: these instructions can be combined, thus combine(expand(f), trig).

2.9 Map

The map command is particularly useful when manipulating large expressions and in combination with the functions and procedures discussed in the next chapter. As with many Maple commands, a precise definition of map is not helpful, but the general idea is that map applies a given operation or function to elements of a compound expression, for instance the elements of a matrix or an array. We illustrate this with some simple examples.

One use is the factorisation of coefficients of a series. In the previous section it was shown how the collect command can be used to factor expressions; another method of doing this is to use the map command. Thus for the expression defined in equation 2.3 we may factor the coefficients of x with the command f:=collect(f,x) followed by f:=map(factor,f); combined these give

```
> fx:=map(factor, collect(f,x));
```

$$fx := x^4 + 2(3y+4)x^3 + 6(2y+3)(y+1)x^2 + 2(5y+8)(y+1)^2 x$$
$$+ (y+1)^3 (3y+5)$$

Here the use of map is not essential, but often it affords a far more elegant method than any alternative.

Consider the problem of evaluating $S = \dot{R}R^{-1}$, where R is the matrix

$$R = \begin{pmatrix} \cos t & \sin t \\ -\sin t & \cos t \end{pmatrix}.$$

First we form the differential of the R by applying the differential operator to each element,

```
> Rd:=map(diff,R,t);
```

$$Rd := \begin{pmatrix} -\sin t & \cos t \\ -\cos t & -\sin t \end{pmatrix}.$$

Observe the use of the third argument that defines the variable diff operates upon. The product is formed with the command:

```
> S:=evalm( Rd &* (1/R)):
```

the result of which has to be simplified, again using the map command:

```
> S:=map(simplify,S);
```

$$S := \begin{pmatrix} 0 & 1 \\ -1 & 0 \end{pmatrix}.$$

This shows that the matrix R is a solution of the equation $\dot{R} = SR$.

Matrices with numerical entries can be expressed with fewer significant figures using map and fnormal; thus if we evaluate the matrix R at $t = 0.874$,

```
>  B:=evalf( subs(t=0.874,evalm(R)));
```

$$B := \begin{pmatrix} .6417640810 & .7669021217 \\ -.7669021217 & .6417640810 \end{pmatrix}$$

This may be evaluated to three significant figures with the command

```
>  B:=map(fnormal,B,3);
```

$$B := \begin{pmatrix} .642 & .767 \\ -.767 & .642 \end{pmatrix}$$

Exercise 2.24

Show that the function

$$f(x) = \int_0^1 dt \, \exp\left(a(\sin(xt) + \sin(2xt))\right)$$

has the series expansion

$$f = 1 + \frac{3}{2}ax + \frac{3}{2}a^2x^2 + \frac{3}{8}(3a^2 - 1)ax^3$$
$$+ \frac{9}{40}(3a^2 - 4)a^2x^4 + \frac{1}{240}(81a^4 - 270a^2 + 11)ax^5 + \cdots .$$

Show, further, that the fourth-order expansion may be expressed as the ratio of polynomials,

$$f(x) = \frac{1 + a^2 + \frac{1}{10}a(1 + 3a^2)x + \frac{1}{20}(5 + 6a^2 + 3a^4)x^2}{1 + a^2 - \frac{1}{5}(7 + 6a^2)x + \frac{1}{20}(1 + 3a^2)(5 + 3a^2)x^2}.$$

Exercise 2.25

Show, by substitution, that the vectors

$$\mathbf{u} = \begin{pmatrix} \sin t \\ \cos t \end{pmatrix} e^{-t} \quad \text{and} \quad \mathbf{v} = \begin{pmatrix} -\cos t \\ \sin t \end{pmatrix} e^{t/2}$$

are solutions of the differential equation

$$\frac{d\mathbf{x}}{dt} = A(t)\mathbf{x}, \quad A(t) = \begin{pmatrix} -1 + \frac{3}{2}\cos^2 t & 1 - \frac{3}{4}\sin 2t \\ -1 - \frac{3}{4}\sin 2t & -1 + \frac{3}{2}\sin^2 t \end{pmatrix}.$$

Exercise 2.26

Show that if A is the skew-symmetric matrix

$$A = \begin{pmatrix} 0 & -c & b \\ c & 0 & -a \\ -b & a & 0 \end{pmatrix}$$

where $a^2 + b^2 + c^2 = 1$, then

$$e^{At} = \begin{pmatrix} a^2 + (1 - a^2)\cos t & ab(1 - \cos t) - c\sin t & b\sin t + ac(1 - \cos t) \\ c\sin t + ab(1 - \cos t) & b^2 + (1 - b^2)\cos t) & -a\sin(t) + cb(1 - \cos t) \\ -b\sin t + ca(1 - \cos t) & a\sin t + cb(1 - \cos t) & c^2 + (1 - c^2)\cos t \end{pmatrix}.$$

Hence show that if $\dot{x} = Ax$, $x(0) = (0, 0, 1)^\top$, then

$$x(t) = \left(b \sin t + ac(1 - \cos t), -a \sin(t) + cb(1 - \cos t), c^2 + (1 - c^2) \cos t\right)^\top.$$

Hint: the exponential of a matrix is computed using the `exponential` function in the `linalg` package.

2.10 Exercises

Exercise 2.27
Write a program to animate a sequence of graphs of the function

$$y = x^3 + cx, \quad -1 \le x \le 1,$$

as c is varied from -1 to 1 in steps of 0.2.

Use the appropriate 'button' on the menu bar to (a) make the graphs cycle through from beginning to end, (b) cycle slowly through the graphs backwards and (c) cycle through the graphs in a repeated loop.

How does the pattern of stationary points change with the parameter c?

Exercise 2.28
If $f(x, y) = (x + y + 1)^5 (x^2 + y + 2)^2 (x + y^2 + 3)$ show that the coefficients of x^3 and x^5 are

$$
\begin{aligned}
c_3 &= 2(y + 1)^2 (y + 2)(5y^4 + 16y^3 + 38y^2 + 63y + 56), \\
c_5 &= 5y^6 + 41y^5 + 151y^4 + 359y^3 + 597y^2 + 557y + 208.
\end{aligned}
$$

Exercise 2.29
Find the series expansions of the functions

$$
\begin{aligned}
f_1(x) &= \sin(\tan^{-1} x) - \tan^{-1}(\sin x), \\
f_2(x) &= \sin\left(\ln \sqrt{\frac{1 + x}{1 - x}}\right) - \ln \sqrt{\frac{1 + \sin x}{1 - \sin x}}.
\end{aligned}
$$

up to and including the x^9 term.

Can you see why in both cases the leading term is ax^7 for some constant a?

The generalisation of this result to arbitrary odd functions is considered in exercise 4.62 (page 174).

Exercise 2.30
Show that

$$\frac{1}{\cos^2 x} = \frac{27}{(3 - x^2)^3} + \frac{1}{135} x^6 + \cdots.$$

Exercise 2.31
Use the `limit` command to show that

$$\lim_{x \to 0} \left(\frac{a \sin bx - b \sin ax}{x^3}\right) = \frac{1}{6} ab(a^2 - b^2).$$

Exercise 2.32
Show that

$$\ln\left(\frac{1+\sin x}{\cos x}\right) = 2\sin x - x\cos x + \frac{x^5}{15} + \cdots,$$

and that

$$e^{\tan^{-1} x} = \left(1 - \frac{x^3}{3}\right)e^x + \frac{x^5}{5} + \cdots.$$

Exercise 2.33
For the 2π-periodic function

$$f(x) = (a_0 + a_1\cos x + a_2\cos 2x)^n$$

find the mean value of f,

$$\langle f \rangle = \frac{1}{2\pi}\int_{-\pi}^{\pi} dx\, f(x),$$

for $n = 2, 3, \ldots, 5$.
   For $n = 4$ show that the coefficient of $\cos 2x$ is

$$\frac{1}{2}a_1^4 + 3(a_1 a_0)^2 + 6a_0 a_1^2 a_2 + \frac{9}{4}(a_1 a_2)^2 + 4a_0^3 a_2 + 3a_0 a_2^3.$$

Remember to be careful to ensure that no products of trigonometric functions remain after applying `combine`.

Exercise 2.34
If $1 + x - \cos x = y$ show that

$$x = y - \frac{1}{2}y^2 + \frac{1}{2}y^3 - \frac{7}{12}y^4 + \frac{3}{4}y^5 + \cdots.$$

Exercise 2.35
Find x as a series in y about $y = 0$, if $y = x - \sin x$. Be careful because for small x, $y = x^3/6 + O(x^4)$.

Exercise 2.36
Use the `loop` construction to form the product

$$f = \prod_{k=1}^{N}(x + y + k)^k$$

and find the factorised form of the coefficient of x^N for $N = 1, 2, 3$ and 4.

Exercise 2.37
The `series` command will also express integrals as series, even if Maple cannot evaluate the integral. Define $f(x)$ by the integral

$$f(x) = \int_0^x dy\, \exp(y^3 + \sin 2y),$$

and use the `series` command to find a series approximation to $f(x)$ up to x^9.

Exercise 2.38

Define f to be the integral

$$f = \int_x^{ax} dy\; \ln(\sin y + \sin 2y), \quad a \geq 1,$$

and use the series command to find a series approximation to f up to x^9.

What is the difference between using int and Int in your Maple definition of f? Which is preferable?

Exercise 2.39

The sine function can be represented as the infinite product

$$\sin x = x \prod_{k=1}^{\infty} \left(1 - \frac{x^2}{k^2\pi^2}\right) = x\left(1 - \frac{x^2}{\pi^2}\right)\left(1 - \frac{x^2}{2^2\pi^2}\right)\cdots.$$

Use the loop construction to compute the first N terms of this product and express the result as the polynomial

$$s_N(x) = \sum_{k=1}^{N} c_{2k+1}(N) x^{2k+1}.$$

We know that

$$\lim_{N\to\infty} c_{2k+1}(N) = \frac{(-1)^k}{(2k+1)!}$$

so form the products $(2k+1)!\,c_{2k+1}(N)$ to examine how rapidly the first few coefficients of the series $s_N(x)$ approach their limit.

Compare graphically the functions $\sin x$ and $s_N(x)$ over a suitable range of x.

Exercise 2.40

Define the 3×3 matrices

$$E = \begin{pmatrix} a & 0 & 0 \\ 0 & 0 & 0 \\ 0 & 0 & b \end{pmatrix}, \quad S = \begin{pmatrix} c & 0 & s \\ 0 & 1 & 0 \\ -s & 0 & c \end{pmatrix}, \quad R = \begin{pmatrix} \cos t & -\sin t & 0 \\ \sin t & \cos t & 0 \\ 0 & 0 & 1 \end{pmatrix}$$

and use Maple to form the product $Z = R^{-1}S^{-1}ESR$. We now wish to find the mean value of Z over the interval $0 < t \leq 2\pi$, assuming all other variables are constant, using the two methods discussed in the text, page 84.

Using the combine command, in the form map(combine,Z,trig) transform the trigonometric functions in Z into their multiple angle form; you should obtain, $Z_{13} = sc(a - b)\cos t/(s^2 + c^2)$, for example. Next use both methods mentioned above to try and find the mean value of Z.

This may seem an artificial problem, but is useful in finding an approximate solution to the dynamical problem $\dot{x} = A(t)x$, where $A(t)$ is a particular 3×3 anti-symmetric matrix representing a rapid rotation about a slowly moving axis; the average is over the rapid rotations. This type of averaging approximation is considered in chapter 19.

Exercise 2.41

The digits $1, 2, 3, \ldots, 9$ are arranged in numerical order, each appearing once and only once: a $+$ sign or a $-$ sign is inserted in front of one or more of the digits and the numerical value, N, of the resulting sum is found. The numbers $-12345 + 6789$ and

$$-1 + 234 - 5 + 6 - 7 + 89 \;\; = \;\; N = 316$$
$$1 + 2 + 3 + 4 + 5678 - 9 \;\; = \;\; 5679$$

are all legitimate arrangements.

Find all the combinations such that $N = 100$. Repeat the computation for the digits arranged in reverse order. You should write the Maple program to be able to deal with any reasonable value of N.

Note: one method of doing this question uses the `parse` command, the inverse of `convert(expr,string)`, and the `loop` construction discussed in chapter 1. A version of this question was set as homework to the final year of a primary school class. It is easier using Maple.

Exercise 2.42

The numbers N defined in the previous exercise take integer values in the interval $(-M, M)$, $M = 123\,456\,789$. For $0 \le n \le 100$ determine the number of times the number N has the value n and display your results graphically.

3

Functions and procedures

3.1 Introduction

Maple can be used as a collection of 'black boxes' which perform calculations only once. But it is often more useful used as a high-level programming language with repeated applications, either numeric or symbolic, of the same sequence of commands. We have already seen such usage in the loop construction. The writing of user defined functions and procedures takes us further down this road, and allows the encoding of complex sequences of instructions into a single command that may be saved and recalled in subsequent sessions.

This chapter will introduce the two main programming devices of Maple: functions and procedures. Maple functions will be treated as if they are the same as the ordinary functions of one or more variables, so here we shall also introduce the D operator designed for differentiating functions and returning another function. Procedures are more general sets of commands, and are introduced in section 3.5: formally a Maple function is a procedure, but it is clearer to make a distinction.

In writing a function or a procedure some effort is involved in testing it to ensure that it is correct, so there are facilities in Maple allowing one to save and retrieve them for use in future sessions, hence saving duplication of effort; we introduce these in section 3.6.

3.2 Functions

Most of the well known functions, all the trigonometric functions, exponential functions and logarithms are pre-defined by Maple; in addition numerous special functions are defined. A list of such functions is obtained by typing ?inifcn at the prompt. Nevertheless, it is frequently convenient to define your own functions and this is achieved using the *arrow* notation; for instance the function

$$f(x) = ax^3 + 2x + 1,$$

where a is some constant, can be defined in Maple as follows:

```
>  f:= x-> a*x^3 + 2*x + 1;
```
$$f := x \rightarrow a x^3 + 2x + 1$$

The arrow is typed in as a 'minus sign' followed by a 'greater than sign'. In this case our function involves a constant a which may need to be assigned a value before, for instance, drawing the graph of $f(x)$, though in many cases it may be left as a free variable. Values of f are obtained using the Leibniz notation; so the value of f at $x = 1$, $f(1)$, is

```
>  f(1);
```
$$a + 3$$

Similary

```
>  f(u);
```
$$a u^3 + 2 u + 1$$

and

```
>  f(w+1);
```
$$a (w + 1)^3 + 2 w + 3$$

though it would normally be better to expand this function:

```
>  expand(f(w+1));
```
$$a w^3 + 3 a w^2 + 3 a w + a + 2 w + 3$$

but if one needs to express this as a polynomial in w it is best to just collect all the coefficients of w together,

```
>  collect(f(w+1),w);
```
$$a w^3 + 3 a w^2 + (2 + 3 a) w + a + 3$$

This may be expressed as another function, $g(w)$, as follows

```
>  g:=w->collect(f(w+1),w);
```
$$g := w \rightarrow collect(f(w + 1), w)$$

which gives,

```
>  g(z);
```
$$a z^3 + 3 a z^2 + (2 + 3 a) z + a + 3$$

Such functions may be integrated and differentiated: for the function $f(w)$ defined above,

```
>  diff(f(w),w);
```
$$3 a w^2 + 2$$

```
>  int(f(w),w);
```
$$\frac{1}{4} a w^4 + w^2 + w$$

Note that in both cases it is necessary to include the argument of the function, and also that the expression returned is *not* a Maple function.

A common error made by new users is to try and define a function, $h(x) = x^3 + 2x$, for instance, using the Maple command

```
>  h(x):=x^3 + 2*x;
```
$$h(x) := x^3 + 2 x$$

There are algebraic computing languages in which this is a legitimate method of defining functions, but it is important to remember that Maple is *not* one of these. All this command does is to assign the symbol 'h(x)' the value $x^3 + 2x$: the function $h(w)$ is not defined, as is shown by the following command

```
> h(w);
```
$$h(w)$$

which simply returns the different symbol 'h(w)', that has not yet been assigned a value.

Care is needed in using the notation $g(x)$ when the function has not been previously defined, as Maple recognises this string as a function in only a few circumstances. Two of these are the `diff` and `int` commands which recognise the string 'g(w)' as the image of the value w under the undefined function g. For instance, if we try to differentiate or integrate the string $g(w)$ this is what we get

```
> diff(g(w),w);
```
$$\frac{\partial}{\partial w} g(w)$$

```
> int(g(w),w);
```
$$\int g(w)\, dw$$

In these circumstances, Maple even knows the chain rule

```
> diff(g(h(w)),w);
```
$$D(g)(h(w))\left(\frac{\partial}{\partial w} h(w)\right)$$

for example

```
> diff(g(w^2),w);
```
$$2\, D(g)(w^2)\, w$$

and the product rule

```
> diff(g(w)*h(w),w);
```
$$\left(\frac{\partial}{\partial w} g(w)\right) h(w) + g(w)\left(\frac{\partial}{\partial w} h(w)\right)$$

The results of these formal manipulations are expressed in terms of the differential operator D, which is an important extension of the `diff` command and will be dealt with in section 3.3. We shall show how the D operator can be used to perform purely formal manipulations in section 3.4.

Whilst both `diff(g(w),w)` and `int(g(w),w)` recognise that 'g(w)' means that $g(w)$ is some function of w even if it has not yet been defined, the similar commands `diff(f,w)` and `int(f,w)`, where f has no prior definition simply return 0 and wf respectively because here f is treated as a constant.

The function construction is convenient to use but is not always necessary because often an expression sequence is adequate, and then it is better programming practice to use this simpler Maple construction. When defining a Maple function, or more generally a procedure, you should ensure that it is necessary; needless to say, there are no strict rules and often it is more convenient to use a function even if not strictly necessary. Several of the examples used in this chapter can be done in other ways, but I have chosen to use procedures in order to illustrate their use; the next example is an instance of such misuse as the same graph could equally well be drawn using an expression.

Graphs of user defined functions are readily drawn; for example if $h(x) = \sin(5\sin(\pi x))$,

```
>  h:=x->sin(5*sin(Pi*x));
```

$$h := x \to \sin(5\sin(\pi x))$$

its graph is produced with the command

```
>  tick:=xtickmarks=[0.5,1.0,1.5,2.0],ytickmarks=[-1,-0.5,0,0.5,1]:
>  plot(h(z),z=0..2,title="Graph of h(z)",tick,colour=black);
```

which gives the following graph:

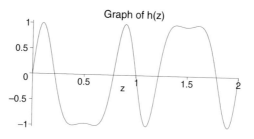

Figure 3.1   Graph of $\sin(5\sin\pi x)$.

The alternative command

```
plot(h,0..2,title="Graph of h(z)");
```

in which the arguments are omitted, can be used, though now there is no label on the axis.

Functions of two or more variables are defined by a straightforward extension; for example the function

$$H(x, y) = \frac{1}{2}y^2 + \frac{1}{2}x^2 - \frac{1}{3}ax^3$$

where a is some constant, is defined as follows:

```
>  H:=(x,y)-> y^2/2 + x^2/2 - a*x^3/3;
```

$$H := (x, y) \to \frac{1}{2}y^2 + \frac{1}{2}x^2 - \frac{1}{3}ax^3$$

and its value at the point $(1, 1)$ is,

```
>  H(1,1);
```

$$1 - \frac{1}{3}a$$

The differentials of $H(x, y)$ may also be obtained using the `diff` command, for instance the second differential with respect to x is

```
>  Diff(H(x,y),x$2)=diff(H(x,y),x$2);
```

$$\frac{\partial^2}{\partial x^2}\left(\frac{1}{2}y^2 + \frac{1}{2}x^2 - \frac{1}{3}ax^3\right) = 1 - 2ax$$

Exercise 3.1

Define the function

$$f(x) = \frac{1}{2}\left(x + \frac{2}{x}\right)$$

and use the loop construction to form the sequence of numbers

$$x_k = f(x_{k-1}), \quad k = 1, 2, \ldots, 5, \quad \text{with} \quad x_0 = 1.$$

Plot the graph of the sequence $[k, x_k]$ for $k = 0, 1, \ldots, 10$ in the case $x_0 = 10$.
What is the value of x_k as $k \to \infty$?

Exercise 3.2

Define the function

$$f(x) = \left(\frac{x}{1 + x^2}\right)^{x/(1-x)}$$

and plot the graph of the real parts of $f(x)$ and $f'(x)$ for $-1 \le x \le 1$.
Use the Maple `limit` function to find

$$\lim_{x \to 0} f(x) \quad \text{and} \quad \lim_{x \to 0} f'(x).$$

Exercise 3.3

Define the function

$$f(x) = \sin x + \frac{3}{2}\sin 2x$$

and by drawing the graphs of $y = x$ and $y = f(x)$ for $0 \le x \le \pi$ show that the equation $x = f(x)$ has a root near $x = 1.5$.

Use the `fsolve` facility to find a numerical approximation, a, to this root to 5 digit accuracy and check that $f(a) - a \simeq 0$.

Hint: since the root is near 1.5 and from the graph is clearly in the range $(1, 2)$, using `fsolve` in the form

```
Digits:=5:a:=fsolve(f(x)=x,x=1..2); Digits:=10:
```

gives the required solution to an accuracy of 5 digits. Note that `Digits` is reset to the default value after the calculation.

The arrow notation does not always produce the desired result; for instance if we wish to define a function which is the second differential of

$$g(x) = \exp(x^2), \tag{3.1}$$

it would be natural to write

```
>  G:=x-> diff(exp(x^2),x$2);
```
$$G := x \to \operatorname{diff}(e^{(x^2)}, x \$ 2)$$

but when we try to find the value of $G(1)$ an error is returned

```
>  G(1);
Error, (in G) wrong number (or type) of parameters in
function diff
```

because Maple has substituted 1 for x in the `diff` operator, and hence found a number where it was expecting a free variable. The same error would occur if we tried to plot the graph of $G(x)$ when defined this manner. There are three ways of overcoming this problem; one is to use the `unapply` command, another is dealt with in exercise 3.6, (page 99), and the third using the `D` operator is discussed in the next section.

Here we concentrate on the `unapply`† command, which is particularly useful when you wish to convert an expression to a function. Suppose we have created the expression

```
>   expr:=a*x*sin(1/(b^2 + x^2)) + exp(-x):
```

and subsequently decide that this needs to be a function of x, $F(x)$ say. This is achieved with the command

```
>   F:=unapply(expr,x);
```

$$F := x \rightarrow a\,x\,\sin(\frac{1}{b^2 + x^2}) + e^{(-x)}$$

or if we wish to also treat b as a variable,

```
>   F:=unapply(expr,x,b);
```

$$F := (x,\,b) \rightarrow a\,x\,\sin(\frac{1}{b^2 + x^2}) + e^{(-x)}$$

Returning to the function G, for the second differential of $\exp(x^2)$ we simply write

```
>   G:=unapply(diff(exp(x^2),x$2),x);
```

$$G := x \rightarrow 2\,e^{(x^2)} + 4\,x^2\,e^{(x^2)}$$

so that $G(1)$ is

```
>   G(1);
```

$$6\,e$$

The Maple function `G(x)` can now be used to draw graphs.

A more complicated use of the `unapply` command is illustrated in the following example. The roots of the cubic equation,

$$x^3 - (a-1)x^2 + a^2x - a^3 = 0, \quad a \ge 0, \tag{3.2}$$

will be functions of the real parameter a; if $a > 1/(1 + \sqrt{3})$ only one of these is real, as is shown in exercise 3.5, and the objective is to define a function of a which is this root.

First we solve the equation

```
>   sol:=[solve(x^3-(a-1)*x^2+a^2*x-a^3=0,x)];
```

---

† It is reasonable to ask about the origin of the term **unapply**; I have yet to see a convincing explanation.

$$sol := [\frac{1}{6}\%1^{(1/3)} - 6\%2 + \frac{1}{3}a - \frac{1}{3},$$

$$-\frac{1}{12}\%1^{(1/3)} + 3\%2 + \frac{1}{3}a - \frac{1}{3} + \frac{1}{2}I\sqrt{3}(\frac{1}{6}\%1^{(1/3)} + 6\%2),$$

$$-\frac{1}{12}\%1^{(1/3)} + 3\%2 + \frac{1}{3}a - \frac{1}{3} - \frac{1}{2}I\sqrt{3}(\frac{1}{6}\%1^{(1/3)} + 6\%2)]$$

$$\%1 := 80a^3 + 12a^2 + 24a - 8 + 12\sqrt{48a^6 + 24a^5 - 12a^3 + 33a^4}$$

$$\%2 := \frac{\frac{2}{9}a^2 + \frac{2}{9}a - \frac{1}{9}}{\%1^{(1/3)}}$$

If you repeat these commands on your computer the Maple output may not look like the above; in particular the expression may not contain the Maple-defined variables %1 and %2, with the consequence that it may look far more complicated. Symbols like %1 are known as labels and information about them can be found by typing ?labeling, or ?labelling; they are available only if the Maple output option is set to typeset notation. You can change to this option using the menu items *Options* (Alt-O), *Output Display* (O) then choose *Typeset Notation* (T); clicking on *Save Settings* in the *File* menu saves this choice for future use.

Visual inspection of the above list shows that the real root is the first of the three components of the list sol. In order to convert this element of the list into the required function we set

> ```
r:=unapply(sol[1],a);
```

$$r := a \rightarrow \frac{1}{6}(80a^3 + 12a^2 + 24a - 8 + 12\sqrt{48a^6 + 24a^5 - 12a^3 + 33a^4})^{(1/3)}$$

$$- 6\frac{\frac{2}{9}a^2 + \frac{2}{9}a - \frac{1}{9}}{(80a^3 + 12a^2 + 24a - 8 + 12\sqrt{48a^6 + 24a^5 - 12a^3 + 33a^4})^{(1/3)}} + \frac{1}{3}a - \frac{1}{3}$$

This form of the solution is not very illuminating, so it is sensible to look at its graph

> ```
plot(Re(r(a)),a=0..1);
```

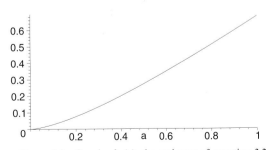

Figure 3.2   Graph of $r(a)$, the real root of equation 3.2.

This graph suggests that the root increases linearly with a; in fact, for large a the methods developed in chapter 9 can be used to show, directly from the original cubic equation, that $r(a) \simeq a - 1/2 + 1/(4a)$.

Exercise 3.4

Use Maple to construct the function $r(a)$ and find its numerical values for various values of x in the range $(0.2, 0.3)$.

The results of this exercise show that for some values of a numerical errors produce small complex parts to $r(a)$. This is the reason why we plotted the real part of the function in figure 3.2. If we simply plot $r(a)$ the graphical routine leaves a gap in the curve at those values of a at which it thinks $\mathrm{Im}(r) \neq 0$; we remove these gaps by explicitly plotting the real part of r. When doing this you must be sure that the nonzero complex parts are due to numerical errors.

Exercise 3.5

Show analytically that if $a > 1/(1 + \sqrt{3}) \simeq 0.366$ the cubic equation

$$x^3 - (a-1)x^2 + a^2 x - a^3 = 0$$

has one and only one real root.

Use Maple to refine this estimate by showing that if $a > 0.277\ldots$ the equation has only one real root.

Functions can be added, subtracted and composed. In Maple the notation

$$(f+g)(x) \quad \text{means} \quad f(x) + g(x), \quad \text{and}$$
$$(-f)(x) \quad \text{means} \quad -f(x).$$

In the special case that f is a number (an integer, a real or a complex number), in Maple $f(x)$ means f, thus

```
>  2(x);  2.2(x);  (2+3*I)(x);
```

$$2$$
$$2.2$$
$$2 + 3I$$

which allows us to write $(f+1)(x)$ for $f(x) + 1$, for instance.

In all the examples considered so far the arguments of the functions have been one or more real, or complex, variables and the value of the output of the function has been a single number. Some generalisations of this format are possible, for instance we can create a list of functions of the same variable(s) as follows,

```
>  h:=x->[2*x, 3*x^3, 4*x^4];
```

$$h := x \rightarrow [2x, 3x^3, 4x^4]$$

so that

```
>  h(1);  h(2);
```

$$[2, 3, 4]$$
$$[4, 24, 64]$$

and

```
>  diff(h(x),x);
```

$$[2, 9\,x^2,\ 16\,x^3]$$

We may also use lists as arguments for functions; suppose we need a function which sums all the elements of a list:

```
>  f:=ls -> sum(ls[k],k=1..nops(ls));
```

$$f := ls \to \sum_{k=1}^{\text{nops}(ls)} ls_k$$

so if

```
>  ls:=[x,x^2,x^3,x^4,x^5];
```

$$ls := [x,\ x^2,\ x^3,\ x^4,\ x^5]$$

we have

```
>  f(ls);
```

$$x + x^2 + x^3 + x^4 + x^5$$

The map command can also be combined with function notation to give a very useful facility. For instance, if A is a matrix, array or list the command `map(x->x^2, A)` will square all elements of A: and `map(x->evalf(x,4),A)` will convert all numbers in A to floating point form with four significant figures, which is sometimes useful because `evalf(x,N)` is one of the operations that does not always work with map.

Exercise 3.6
Show that the function $G(x) = g''(x)$ can be defined by either of the commands

```
G:=x-> subs(u=x,diff(g(u),u$2)):
G:=x-> eval(diff(g(u),u$2),u=x):
```

By using the function g defined in equation 3.1, page 95, show that these methods of defining the function G allow one to draw the graph of G and to evaluate $G(x)$ at particular values of x.

Exercise 3.7
Define a function which forms the sum of the squares of the elements of a list using (*a*) the method described in the text above, (*b*) the convert command in the form `convert( list,'+')` and (*c*) the add command. Test your function on suitable examples.

Exercise 3.8
Define a function to compute the scalar product of two lists of the same length, that is if

$$\mathbf{x} = [x_1, x_2, \ldots, x_N] \quad \text{and} \quad \mathbf{y} = [y_1, y_2, \ldots, y_N]$$

then

$$f(\mathbf{x}, \mathbf{y}) = x_1 y_1 + x_2 y_2 + \cdots + x_N y_N.$$

Exercise 3.9

Define the function

$$f(t, n) = -2 \sum_{r=1}^{n} \frac{(-1)^r}{r} \sin rt$$

and plot the graph of this function for $-\pi \le t \le \pi$ and for $n = 4$, 8 and 16.

Exercise 3.10

The function $x(y, a)$ is defined to be the inverse of

$$y = e^{a \sin x} - 1, \quad |x| < \frac{\pi}{2},$$

for some nonzero constant a, and can be represented as a power series in y; express the first five terms of this series as a Maple function.

Exercise 3.11

Define the difference function

$$d(x, h) = \frac{1}{h} \left(f(x + h/2) - f(x - h/2) \right)$$

for an arbitrary function f. Plot the graphs of $d(x, h)$ and $f'(x)$ when $f(x) = \sin x$ and $\exp(x)$ and for $h = 2$, 1 and 0.5.

Exercise 3.12

If a function $f(x)$ has a series representation

$$f(x) = \sum_{k=0}^{\infty} a_k x^k$$

and A is a square matrix, the matrix $f(A)$ is defined by the series

$$f(A) = \sum_{k=0}^{\infty} a_k A^k$$

if it exists.

If A is the matrix

$$A = \begin{pmatrix} 2 & -1 & 1/2 \\ 6 & -4 & 3 \\ 9 & -7 & 11/2 \end{pmatrix}$$

obtain $\sin A$ and $\cos A$ to $O(A^{12})$ in floating point form, accurate to 12 digits. Using `map` and `fnormal` show that to six figures $\cos^2 A + \sin^2 A = I$.

3.2.1 Composition of functions

Composition of functions of a single variable is also allowed and is achieved using the @ symbol so $f@g$ means $f \circ g$, or in Leibniz notation $(f@g)(x)$ means $f(g(x))$: in the first use $f@g$ no brackets are necessary, but in the second $(f@g)(x)$ both sets of brackets are needed. Thus if, for some constant b,

```
>  f:=x->b*x*(1-x);    g:=x->x^2;
```

$$f := x \rightarrow b\, x\, (1 - x)$$

$$g := x \rightarrow x^2$$

we have the composition $g \circ f$,

> `gf:=g@f;`

$$gf := g@f$$

or explicitly $g(f(x))$

> `gf(x);`

$$b^2 x^2 (1-x)^2$$

and $f \circ g$,

> `fg:=f@g;`

$$fg := f@g$$

and explicitly

> `fg(x);`

$$b x^2 (1-x^2)$$

Note that we define the *function* $f \circ g$ with the command $f@g$, whereas the command $(f@g)(x)$ simply returns an *expression*,

> `(g@f)(x);`

$$b^2 x^2 (1-x)^2$$

Multiple compositions of the same function are obtained using $@@n$, n being an integer; thus the three-fold composition of f, $(f \circ f \circ f)(x)$, is given by

> `(f@@3)(x);`

$$b^3 x (1-x)(1-b x(1-x))(1-b^2 x(1-x)(1-b x(1-x)))$$

Exercise 3.13
If $f(x) = x^{1/2}$ for $x \geq 0$, draw the graphs of

$$\underbrace{f \circ f \circ f \circ \cdots \circ f}_{n \text{ times}}(x)$$

over the interval $0 \leq x \leq 10$ for $n = 2, 4, \ldots, 10$.

Exercise 3.14
If $f(x) = x^{1/2} + x^{-1/2}$, $x > 0$, draw the graphs of $f^{(2)}(x) = (f \circ f)(x)$, $f^{(4)}(x)$ and $f^{(6)}(x)$ for $1 < x < 100$.
Use Maple to show that

$$
\begin{aligned}
f^{(2)}(x) &= x^{1/4} + x^{-1/4} + \frac{1}{2} x^{-3/4} + \cdots \\
f^{(4)}(x) &= x^{1/16} + x^{-1/16} + \frac{1}{2} x^{-3/16} + \cdots \\
f^{(6)}(x) &= x^{1/64} + x^{-1/64} + \frac{1}{2} x^{-3/64} + \cdots.
\end{aligned}
$$

In order to do this you will need the command `asympt`, but for each case you will need to first make a substitution of the form $x = u^\alpha$, otherwise the expressions obtained from `asympt` will not be very illuminating. The value of α is different in each case and should be chosen after examining the leading term using ordinary methods.

An extension of this result is derived in exercise 3.60 (page 129).

3.2.2 *Piecewise functions*

Piecewise continuous functions can also be defined using the arrow notation, for example the function

$$h(x) = \begin{cases} \exp(-x^2), & x \le 0, \\ \cos(x), & x > 0, \end{cases} \qquad (3.3)$$

is best represented in Maple using the `piecewise` function which, in this case, becomes

```
>  h:=x->piecewise(x<=0, exp(-x^2), cos(x));
```
$$h := x \rightarrow \text{piecewise}(x \le 0, e^{(-x^2)}, \cos(x)).$$

For the general syntax of `piecewise` you should look in the help files where more examples are given.

Graphs of such functions can be drawn with the usual command,

```
>  plot(h(z),z=-2..2*Pi);
```

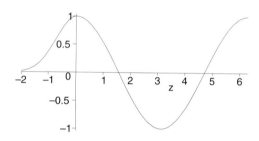

Figure 3.3   Graph of the piecewise function defined in
equation 3.3.

Such functions can be differentiated with `diff` to give an expression; a method that returns a function directly is described in the next section.

Exercise 3.15
Use the command

```
hp:=diff(h(u),u);
```

to form an expression for the differential of the function $h(u)$ defined above using the piecewise command, and plot the graph of $h'(u)$ using

```
plot(hp,u=-2..2*Pi);
```

Using similar commands plot the graphs of d^2h/du^2 and d^3h/du^3.

Compare the results of using the two commands `subs` and `eval(hp,x=1)` to find the value of the differential at $x = 1$ and $x = 0$.

Piecewise functions can be composed with themselves and other functions; for example the composition of $h(x)$ with itself is

```
>  h2:=h@@2;
```
$$h2 := h^{(2)}$$

and the graph of $h(h(y))$ is

```
> plot(h2(y),y=-2..2*Pi);
```

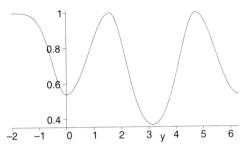

Figure 3.4   Graph of the composition of the piecewise
function defined in equation 3.3 with itself.

In principle we can construct $h \circ h \circ h$ etc., but in practice computations with this type
of function can take a long time.

Maple functions defined with `piecewise` may also be integrated symbolically, provided the constituent parts can be evaluated. Thus for the function $h(x)$ defined in 3.3,
and the subsequent Maple commands, the integral

$$g(x) = \int_0^x dt\, h(t)$$

is produced with the command

```
> g:=unapply( int(h(t),t=0..x),x );
```

$$g := x \rightarrow \frac{1}{2}\sqrt{\pi}\,\mathrm{erf}(x) - \frac{1}{2}\sqrt{\pi}\,\mathrm{erf}(x)\,\mathrm{Heaviside}(x) + \sin(x)\,\mathrm{Heaviside}(x)$$

If, however, $\exp(-x^2)$ is replaced by $\exp(-x^3)$, so the integral for $x < 0$ cannot be
evaluated symbolically, then Maple simply returns the integral unevaluated.

Exercise 3.16
If $h(x)$ is the function defined in equation 3.3 and $f(x) = h(x-1)$, plot the graphs of $h(\sin x)$,
$h(f(x))$ and $f(h(x))$ for $-4 < x < 4\pi$.

Exercise 3.17
Define a function $s(x)$ in Maple which is 1 for $-1 \le x \le 1$ and zero elsewhere. Plot the
graph of $s(x)$ for $-2 \le x \le 2$ and of $s(4\sin(x^2))$ for $0 \le x \le 2\pi$.

3.3 Differentiation of functions

In chapter 1 we introduced the `diff` command which can be used to differentiate
formulae, expressions or functions, but always returns an expression. The operator
`D` computes the derivative of a function (or procedure) and returns a function (or
procedure), as is shown in the following simple example; let $f(x)$ be the function

```
> f:=x-> x^n * exp(cos(x));
```

$$f := x \rightarrow x^n\, e^{\cos(x)}$$

of the single variable x, then its derivative, $g = f'$, is given with the command

>   `g:=D(f);`

$$g := x \to \frac{x^n \, n \, e^{\cos(x)}}{x} - x^n \sin(x) \, e^{\cos(x)}$$

The command `D(f)(w)` simply returns the *expression* for $g(w)$, so should not be used if a function is required; thus we have

>   `g(w);`

$$\frac{w^n \, n \, e^{\cos(w)}}{w} - w^n \sin(w) \, e^{\cos(w)}$$

and

>   `D(f)(w);`

$$\frac{w^n \, n \, e^{\cos(w)}}{w} - w^n \sin(w) \, e^{\cos(w)}$$

The argument of `D(f)(w)` does not need to be a free variable; the value of a derivative at a particular point, say $x = \pi/2$, is given by

>   `D(f)(Pi/2);`

$$2 \frac{(\frac{1}{2}\pi)^n \, n}{\pi} - (\frac{1}{2}\pi)^n$$

In practice `D(f)` is equivalent to `unapply(diff(f(x),x),x)`.

It is important to remember that `D` operates on functions and returns functions, whereas `diff` operates on expressions or functions and returns expressions. In the following example we illustrate this by finding the derivative of the sine function

>   `D(sin);`

$$\cos$$

and the second and third derivatives are obtained using the @ symbol introduced above,

>   `(D@@2)(sin);   (D@@2)(cos);`

$$-\sin$$

$$-\cos$$

Thus in Maple the function $G(x)$, defined in equation 3.1 (page 95), is

>   `G:=(D@@2)(x->exp(x^2));`

$$G := x \to 2 \, e^{(x^2)} + 4 \, x^2 \, e^{(x^2)}$$

so the value of the function $G(x)$ at $x = 1$ is

>   `G(1);`

$$6 \, e$$

The operator `D` should not be used on expressions because commands like

>   `z:=D(cos(u));`

$$z := D(\cos(u))$$

do not produce useful or meaningful results because, in this context, Maple does not consider $\cos(u)$ to be the cosine of u.

Exercise 3.18
Define a Maple function for $f(x) = \sin(\exp(ax))$ and use the D operator to find Maple functions and expressions for its first and second derivatives.

3.3.1 Total derivatives

The D operator can also be applied to expressions containing several free variables and then it returns the total derivative of the expression with respect to a dummy variable; for instance if

$$f = \frac{xy}{x^2 + y^2}$$

and if both x and y are functions of the independent variable s, then

$$\frac{df}{ds} = \frac{y}{x^2 + y^2}\frac{dx}{ds} + \frac{x}{x^2 + y^2}\frac{dy}{ds} - \frac{2xy}{(x^2 + y^2)^2}\left(x\frac{dx}{ds} + y\frac{dy}{ds}\right).$$

In Maple we have

```
>   f:=x*y/(x^2 + y^2);
```
$$f := \frac{x\,y}{x^2 + y^2}$$

and

```
>   D(f);
```
$$\frac{D(x)\,y}{x^2 + y^2} + \frac{x\,D(y)}{x^2 + y^2} - \frac{x\,y\,(2\,D(x)\,x + 2\,D(y)\,y)}{(x^2 + y^2)^2}$$

Here x and y are being treated as functions, so $D(x)$ is the differential of the function x.

This use of D works only on expressions; it fails if applied to Maple functions of two or more variables because then D computes partial derivatives and needs to know which variable is to be differentiated; this aspect of differentiation is discussed in section 3.3.3.

If the differential of a function of f, say $\sin\left(f^2\right)$, is required then we use the composition rule,

```
>   g:=D( sin@(f^2) );
```

$$g := \cos@\!\left(\frac{x^2\,y^2}{(x^2 + y^2)^2}\right)(2\,\frac{D(x)\,x\,y^2}{(x^2 + y^2)^2} + 2\,\frac{x^2\,D(y)\,y}{(x^2 + y^2)^2} - 2\,\frac{x^2\,y^2\,(2\,D(x)\,x + 2\,D(y)\,y)}{(x^2 + y^2)^3})$$

Generally it is most convenient to treat this result as an expression sequence involving x and y.

Implicit differentiation can also be carried out with the D operator, using a method best understood through examples. Suppose x and y are related through the implicit equation

$$x^3 + y^4 = \text{constant}$$

and we require dy/dx; first define a variable eq to be the equation

```
>   eq:=x^3 + y^4 =c:
```

Now operate on this equation with D to obtain another equation, containing the derivatives of *x*, *y* and *c*, for we have not yet told Maple that *c* is a constant,

> eq1:=D(eq);

$$eq1 := 3\,D(x)\,x^2 + 4\,D(y)\,y^3 = D(c)$$

This can be solved for D(y)

> dy:=solve(eq1,D(y));

$$dy := -\frac{1}{4}\,\frac{3\,D(x)\,x^2 - D(c)}{y^3}$$

but *c* is a constant so D(c)=0, and by definition D(x)=1, hence the required derivative, yp, is obtained by substitution

> yp:=subs(D(c)=0,D(x)=1,dy);

$$yp := -\frac{3}{4}\,\frac{x^2}{y^3}$$

An alternative, neater and more convenient way of doing this is to use the assume command to tell Maple that *c* is a constant and to define the variable *x* to be the function *x* → *x*: we write

> x:=x->x;

$$x := x \rightarrow x$$

> assume(c,constant):   eq:=x^3+y^4=c:
> eq1:=D(eq);

$$eq1 := 3\,x^2 + 4\,D(y)\,y^3 = 0$$

which, as before, we solve for D(y).

> yp:=solve(eq1,D(y));

$$yp := -\frac{3}{4}\,\frac{x^2}{y^3}$$

Similarly, the second derivative can be found:

> eq2:=(D@@2)(eq);

$$eq2 := 6\,x + 4\,(D^{(2)})(y)\,y^3 + 12\,D(y)^2\,y^2 = 0$$

which can again be solved for (D@@2)(y), after first substituting yp for D(y),

> y2p:=solve(eval(eq2,D(y)=yp),(D@@2)(y));

$$y2p := -\frac{3}{16}\,\frac{x\,(8\,y^4 + 9\,x^3)}{y^7}$$

Exercise 3.19
The velocity of a wave of wavelength *l* in a certain medium of density *d* and temperature *T* is

$$v = \sqrt{a + \frac{2\pi T}{ld}},$$

where *a* is a constant. If the relative decrease in the temperature is 2%, that is $T \rightarrow T + \Delta T$

where $\Delta T = -0.02T$, and the relative increase in l is 1%, show that

$$\frac{\Delta v}{v} = -\frac{0.03\pi T}{dlv^2}.$$

What small relative changes in T and l keep v constant?

Exercise 3.20

In this exercise we demonstrate an alternative method of dealing with implicit functions using the `alias` command. The fact that y is to be considered a function of x is revealed to Maple by making an equivalence between y and $y(x)$ as follows:

```
>  alias(y=y(x));
```
$$I, y$$

Now whenever we type y Maple interprets it as y(x); thus `diff(y,x)` gives

```
>  diff(y,x);
```
$$\frac{\partial}{\partial x}\, y$$

whereas without the alias zero would be returned. Now proceed as before with the following commands to find $\frac{dy}{dx}$:

```
assume(c,constant):
eq:=x^3 + y^4=c:
eq1:=diff(eq,x):
dy:=solve(eq1,diff(y,x)):
```

Exercise 3.21

If $\ln(x^2 + y^2) = 2\tan^{-1}(y/x)$ show that

$$\frac{dy}{dx} = \frac{x+y}{x-y}, \qquad \frac{d^2y}{dx^2} = \frac{2(x^2+y^2)}{(x-y)^3}, \qquad \frac{d^3y}{dx^3} = \frac{4(2y+x)(x^2+y^2)}{(x-y)^5}.$$

Exercise 3.22

Another use of the `alias` command is to create a convenient short-hand for any of the special functions defined in Maple as these tend to have long names. We illustrate this use with the ordinary Bessel function, conventionally denoted by $J_n(x)$ but in Maple by `BesselJ(n,x)`: you can find out what Maple knows about Bessel functions and how to call them by typing `?Bessel`, or `?bessel`, at the prompt. The command

```
>  alias(J=BesselJ):
```

gives $J_n(x)$ the Maple name J(n,x). Thus the graph of $J_2(x)$, for $0 < x < 10$ is obtained with the command `plot(J(2,x),x=0..10)`:

Use the `alias` command to plot the graphs of $J_n(nx)$ for $n = 2, 4, \ldots, 10$ for $0 < x < 3$, and comment upon the position of the first maximum of these functions. The reason for the behaviour you should observe is given in chapter 14.

You might also consider creating the alias `Ai` for the Airy function, $Ai(x)$, denoted in Maple by `AiryAi(x)`, and examining the nature of this function graphically.

3.3.2 Differentiation of piecewise continuous functions

The D operator can also be used to differentiate piecewise continuous functions; for instance the first and second differentials of the function defined in equation 3.3 (page 102), are simply

```
> hp:=D(h);
```

$$hp := x \rightarrow \text{piecewise}(x \le 0, -2\,x\,e^{(-x^2)}, -\sin(x))$$

```
> h2p:=(D@@2)(h);
```

$$h2p := x \rightarrow \text{piecewise}(x \le 0, -2\,e^{(-x^2)} + 4\,x^2\,e^{(-x^2)}, -\cos(x))$$

and the graphs of these functions are given as follows:

```
> plot([hp(x),h2p(x)], x=-2..2*Pi);
```

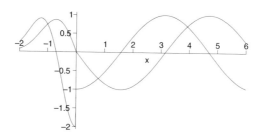

Figure 3.5   Graphs of $h'(x)$ and $h''(x)$.

Exercise 3.23

If

$$f(x) = \begin{cases} 0, & |x| > 1/2, \\ 1 - 2x, & 0 \le x \le 1/2, \\ 1 + 2x, & -1/2 \le x \le 0, \end{cases}$$

construct a Maple function for f and draw the graphs of $f(x)$, $(f \circ f)(x)$ and $f'(x)$.

3.3.3 Partial derivatives

The D operator can also be used to compute partial derivatives. Suppose $f(x, y, z)$ is a function of the three variables x, y and z, then D[1](f) is the derivative with respect to the first argument, x, D[2](f) the derivative with respect to the second argument, y, etc. Thus, if

```
> f:=(x,y,z)-> 1/sqrt(x^2 + y^2 + z^2);
```

$$f := (x, y, z) \rightarrow \frac{1}{\sqrt{x^2 + y^2 + z^2}}$$

then $\partial f / \partial x$ is given by

```
> D[1](f);
```

$$(x, y, z) \rightarrow -\frac{x}{\sqrt{x^2 + y^2 + z^2}^3}$$

and the second differential with respect to x is denoted by D[1,1](f)=D[1](D[1](f)) with the obvious generalisation; thus $\partial^4 f / \partial x^2 \partial y \partial z$ is

```
>  D[1,1,2,3](f);
```
$$(x, y, z) \rightarrow 105 \frac{z\, y\, x^2}{\sqrt{x^2 + y^2 + z^2}^9} - 15 \frac{z\, y}{\sqrt{x^2 + y^2 + z^2}^7}$$

Although the D operator can be used to obtain any partial derivative it does not know about the chain rule for partial derivatives.

Finally, the value of partial derivatives at particular points is given by the direct extension of the rule for ordinary derivatives given on page 104. Thus the value of $\partial^4 f / \partial x^2 \partial y \partial z$ at (1,2,3) is

```
>  D[1,1,2,3](f)(1,2,3);
```
$$-\frac{45}{38416} \sqrt{14}$$

Exercise 3.24
Define the Maple function $f(x, y, z) = (x^2 + y^2 + z^2)^{-1/2}$, using the exponential form of the square root rather than sqrt as in the above example, and find the partial derivatives $\partial f / \partial x$ and $\partial^4 f / \partial x^2 \partial y \partial z$. Note how these differ from those found previously.

Exercise 3.25
If a, b and c are positive constants use Maple to determine the coordinates of the minimum of the function
$$f(x, y, z) = \frac{\exp(ax + by + cz)}{xyz}$$

and its value at this point.

Exercise 3.26
Find and classify the stationary points of the functions

(i) $z = (x^3 + y^3) - 2(x^2 + y^2) + 3xy,$

(ii) $z = x^4 + 64y^4 - 2(x + 8y)^2.$

Use plot3d and the contour plotting command contourplot to plot graphs of each function in a region of the (x, y)-plane sufficiently large to show all stationary points; remember to use with(plots) before contourplot.
Hint: in these questions, having found the stationary points, it is useful to use the Maple allvalues function.

3.4 Differentiation of arbitrary functions

In most of the examples dealt with in the preceding two sections we differentiated functions or expressions which had explicit definitions, for example $f(x) = \sin(5 \cos x)$. It sometimes happens that we wish to consider arbitrary functions which satisfy more general relations. The following problems provide some illustrative ideas.

(1) If $h(x, y, z)$ is a function of three variables, then a function $z(x, y)$ may be defined by the implicit equation $h(x, y, z) = \text{constant}$, and the partial derivatives of $z(x, y)$ may be expressed in terms of the partial derivatives of h.

(2) Another, related, class of problems is illustrated by the following specific example. If $f(w)$ is an arbitrary function and $z(x, y)$ is defined implicitly by the equation $x + y + z = f(x - z)$, then it can be shown (exercise 3.28) that z satisfies the partial differential equation

$$\frac{\partial z}{\partial x} - 2\frac{\partial z}{\partial y} = 1.$$

(3) If ϵ is a small parameter and $\omega(x)$ is a positive function such that it and its derivatives are $O(1)$, then the function

$$y(t) = \frac{1}{\sqrt{\omega(\epsilon t)}} \exp\left(\pm i \int_0^t dz\, \omega(\epsilon z)\right)$$

can be shown (exercise 3.31) to be an approximate solution of the differential equation

$$\frac{d^2 y}{dt^2} + \omega(\epsilon t)^2 y = 0.$$

These types of problem are relatively common and fairly easily treated using the usual methods of calculus, but it is not entirely clear how they can be tackled using Maple.

The crucial step in using Maple is to first define an arbitrary function with the desired properties; this we do in the intuitive manner using the arrow notation. As is often the case, these ideas are best understood by studying specific examples.

Suppose that we wish to show that $z = x^a f(y/x)$ is a solution of the partial differential equation

$$x\frac{\partial z}{\partial x} + y\frac{\partial z}{\partial y} = az,$$

then the first step is to tell Maple that z is a function having the given form,

```
>  z:=(x,y)-> x^a * f(y/x);
```

$$z := (x,\, y) \rightarrow x^a\, \mathrm{f}\left(\frac{y}{x}\right)$$

First find the partial derivative of z with respect to x,

```
>  zx:=D[1](z)(x,y);
```

$$zx := \frac{x^a\, a\, \mathrm{f}\left(\frac{y}{x}\right)}{x} - \frac{x^a\, \mathrm{D}(f)\left(\frac{y}{x}\right) y}{x^2}$$

and y,

```
>  zy:=D[2](z)(x,y);
```

$$zy := \frac{x^a\, \mathrm{D}(f)\left(\frac{y}{x}\right)}{x}$$

Now form the differential equation from these two expressions,

```
>   d:=x*zx + y*zy - a*z(x,y);
```

$$d := x\left(\frac{x^a\, a\, f\left(\frac{y}{x}\right)}{x} - \frac{x^a\, D(f)\left(\frac{y}{x}\right)y}{x^2}\right) + \frac{y\, x^a\, D(f)\left(\frac{y}{x}\right)}{x} - a\, x^a\, f\left(\frac{y}{x}\right)$$

Naturally Maple does not produce the result of this sum in the simplest form, so we need to use the simplify command,

```
>   simplify(d);
```

$$0$$

Thus the required equation is satisfied.

All the examples listed at the beginning of this section can be treated in the same manner. The same result is obtained using the diff command instead of the D operator.

Exercise 3.27
Repeat the above steps in Maple using the diff command in place of the D operator.

Exercise 3.28
If the function $z(x, y)$ is defined implicitly by the equation $x + y + z = f(x - z)$, where f is an arbitrary differentiable function, show that

$$\frac{\partial z}{\partial x} - 2\frac{\partial z}{\partial y} = 1.$$

Hint: it helps to use the command alias(z=z(x,y)).

Exercise 3.29
If $z = f(x - 2y) + g(3x + y)$, where f and g are arbitrary functions, and if

$$\frac{\partial^2 z}{\partial x^2} + a\frac{\partial^2 z}{\partial x \partial y} + b\frac{\partial^2 z}{\partial y^2} = 0,$$

find the constants a and b.

Exercise 3.30
If $f = \exp k(r - x)$, where k is a constant and $r^2 = x^2 + y^2$, prove that

$$\text{(i)} \qquad \left(\frac{\partial f}{\partial x}\right)^2 + \left(\frac{\partial f}{\partial y}\right)^2 + 2fk\frac{\partial f}{\partial x} = 0,$$

$$\text{(ii)} \qquad \frac{\partial^2 f}{\partial x^2} + \frac{\partial^2 f}{\partial y^2} + 2k\frac{\partial f}{\partial x} = \frac{fk}{r}.$$

Exercise 3.31
Use Maple to find the first and second derivatives of the function

$$y(t) = \frac{1}{\sqrt{f(\epsilon t)}}\sin\left(\int_0^t ds\, f(\epsilon s)\right),$$

where f is a positive function and ϵ a constant.

Hence, or otherwise, show that

$$\frac{d^2 y}{dt^2} + yf(\epsilon t)^2 = \epsilon^2 A(t),$$

where A is some function of t, and deduce that for small ϵ, $y(t)$ is an approximate solution of the equation

$$\frac{d^2 z}{dt^2} + z f(\epsilon t)^2 = 0.$$

An extension of this idea is given in exercise 3.61 (page 130).

3.5 Procedures

Maple procedures comprise sets of Maple commands which one needs to use repeatedly. A procedure could be used to define a function — and often is if the function is too complicated to write in the usual form of a Maple function — or it may be used to return a matrix, a graph or a logical value, for instance. For those familiar with programming, a Maple procedure is similar to a user written function or subroutine in Fortran, Basic or C.

As usual it is best to start with some simple examples which illustrate syntax and use. First consider a relatively simple function which, nevertheless, is a little too complex to write conveniently as a Maple function;

$$f(x) = y^2 + \frac{1}{1 + y^2} + \sin y, \quad y = 2 + \sin x. \tag{3.4}$$

For this the Maple procedure is:

```
>  f:=proc(x)
>  local y;
>  y:=2 + sin(x);
>  y^2 + 1/(1+y^2) + sin(y);
>  end;
```

After typing this set of commands in Maple returns the following output,

$$f := \mathbf{proc}(x) \ \mathbf{local} \ y; \ y := 2 + \sin(x) \ ; \ y^2 + 1/(1 + y^2) + \sin(y) \ \mathbf{end}$$

Normally we should not print this output, which is suppressed by replacing the last line, end; with end:. Note that the semicolon prior to the end statement is not necessary. In this example we can treat f as a function. For example the numerical value of $f(x)$ at $x = 1$ is

```
>  evalf(f(1));
```
$$8.479799297$$

The same result is obtained, in this case, by typing f(1.0), rather than f(1). The main points to notice are:

(1) The procedure starts with proc() and ends with end; or end:, the latter suppressing the output printed above. In between proc() and end; is the body of the procedure.

(2) The argument of this procedure is x; as with functions there may be many arguments or none. This is the main method of passing information into a procedure. The values of arguments *must not* be changed in the procedure body.

(3) The value returned by the procedure is the value of the last statement, that is the statement immediately preceding end;. This 'value' need not be a number: it may be a graph, a matrix or any Maple expression for example.

(4) The use of a local variable. In this case the local variable is y and its value outside the procedure is unchanged — you should check this by assigning y a value before using the procedure and then checking that its value is unchanged afterwards.

Maple forces all variables used only in the body of the procedure to be local variables. The command local y; is therefore optional, but if left out a message is produced stating that the variable y is implicitly declared local, though the procedure will work. You should try the procedure with and without the local command to see what happens.

Variables, a and b for example, may be made global with the global a,b; command: this is an alternative method of passing information to and from procedures.

Exercise 3.32
Use Maple to draw the graph of the function $f(x)$ defined in equation 3.4 for $0 \le x \le 2\pi$.

The function which is the derivative of $f(x)$ can be found using the D operator:

```
>   fp:=D(f);
```
and the Maple response is

$fp := \mathbf{proc}(x)$
    local y, yx;
        $yx := \cos(x)\,;\, y := 2 + \sin(x)\,;\, 2 * y * yx - 2 * y * yx/(1 + y^2)^2 + \cos(y) * yx$
    end

Note how Maple has been clever enough to add another local variable yx, the derivative of $y(x)$; the equivalent expression is obtained using diff; you should try both methods. For more complicated procedures it may not be possible to find the derivative in this manner.

Exercise 3.33
Use Maple to plot graphs of $f'(x)$ and $f''(x)$, where $f(x)$ is defined in equation 3.4.

In the next example we write a procedure which returns the graph of the function

$$g(x, a) = y^4 + \exp(y) + \sin(2y), \quad y = a + \sin x, \tag{3.5}$$

for $0 \le x \le 2\pi$ and for given values of the parameter a. This procedure is essentially the same as the previous one but there are more local variables, the last statement is different and the argument of the procedure is a.

```
>  gr:=proc(a) local y,x,f;
>  y:=a + sin(x);
>  f:=y^4 + exp(y) + sin(2*y);
>  plot(f,x=0..2*Pi);
>  end;
```

The Maple command gr(5); will then plot the graph of $g(x, 5)$.

Exercise 3.34

Type in the above procedure and plot the graphs of $g(x, a)$ for $a = 1$ and 5.

Procedures are frequently useful for evaluating complicated expressions involving conditional statements, but for these types of procedures care is needed when plotting graphs: for example, the piecewise smooth function defined in equation 3.3 can be defined with the procedure

```
>  h:=proc(x);
>  if x <= 0 then exp(-x^2) else cos(x) fi;
>  end:
```

Using this function in plot now results in the error message

```
>  plot(h(x),x=-2..6);
```

    Error, (in h) cannot evaluate boolean

that can be avoided by evaluating h(x) to its name in plot using forward-quotes or the evaln procedure: either of the following commands will successfully plot the graph of h(x):

```
>  plot('h(x)',x=-2..6);
>  plot(evaln(h(x)),x=-2..6);
```

Whilst the D operator can be used on procedures, it appears that neither the int nor Int operators can be used, though these can be used on the piecewise command described above.

In these examples the object returned by the procedure was the last evaluated statement. This is sometimes a limitation and can be overcome in several ways. If several numbers or objects are required they can be put in a list and extracted from the list, as and when needed. In the following example we create a list comprising the next two prime numbers greater than a given integer n.

```
>  np:=proc(n::integer) local a, b;
>  a:=nextprime(n);
>  b:=nextprime(a);
>  [a,b];
>  end;
```

So the next two primes after 1000 are given by

```
>  np(1000);
```

$$[1009, 1013]$$

In this procedure the local variable b is not necessary, for we could omit the line b:=nextprime(a) and replace [a,b] with [a,nextprime(a)]. Note that the input declares n to be an integer; if you type np(x) where x is a real number, π for example, an error message is returned. You should try this.

Exercise 3.35

Re-write the preceding procedure to ensure that if the input integer is a prime then it is the first member of the list.

Exercise 3.36

Write a procedure to create a list of the first N prime numbers greater than a given number n. This procedure will have two integer arguments, n and N.

Another method of extracting results from a procedure is via the RETURN statement. The command

```
RETURN(sequence)
```

causes an immediate return from the procedure and the value of the sequence becomes the value obtained by invoking the procedure; this is useful mainly in conditional statements. The word RETURN must be in capitals.

An example of this technique is shown in the following procedure that returns the first even number from a given list

```
> first_even:=proc(L::list) local el;
>     for el in L do;
>         if type(el,even) then RETURN(el) fi;
>     od;
> end:
```

which gives, for example,

```
> L:=[1,3,5,11,101,421,2,10,13,14]:
> first_even(L);
```

$$2$$

We illustrate these features with an example that will be particularly useful in later chapters. It is frequently necessary to find numerical solutions to equations like $f(x) = 0$: if values of $f(x)$ are given by simple expressions then fsolve suffices, but for more complicated functions this either will not work or does not provide sufficient control: for instance if $f(x)$ is defined by the solution of a differential equation, or the iterates of a map, as in chapter 18, fsolve cannot be used. The following procedure can, with minor modifications, be used in these circumstances. It is based on the False Position Method, described in detail by Press *et al.* (1987, section 9.2), and the procedure given here is a direct translation of the Fortran subroutine they provide. This is not the most rapidly converging method, but it is very robust. The general idea is to start with two values of x such that $f(x_1)f(x_2) < 0$, so we know that there is a root between x_1 and x_2: it is assumed that x_1 and x_2 have been chosen with sufficient care for there to be only one root. A first estimate is given by linear interpolation between these points, and this estimate replaces one of x_1 or x_2; then the process is repeated to obtain a new estimate. This process continues until a value of $x = x_r$ that satisfies $|f(x_r)| < \epsilon$, for some small pre-defined number ϵ, is obtained. The only minor complication in this method is in identifying which of the limits gives the smallest value of $f(x)$. In this procedure $f(x)$ is represented by the Maple function or procedure fun(x) and it is assumed that this evaluates to a real number; notice that inside procedures other functions and procedures are assumed global.

```
>   Root:=proc(X1,X2)
>   local FL,FH,XL,XH,swap,dx,k,rtf,F,del,eps;
>   eps:=1.0e-08:                                                    # 1
>   FL:=fun(X1); FH:=fun(X2);
>   if FL*FH > 0 then RETURN('No root in interval') fi;             # 2
>     if FL < 0 then                                                # 3
>     XL:=X1; XH:=X2
>       else
>     XL:=X2; XH:=X1; swap:=FL; FL:=FH; FH:=swap;
>   fi;                                                             # 4
```
Loop round iterations
```
>   dx:=XH-XL;
>   for k from 1 to 50 do;                                          # 5
>   rtf:=XL + dx*FL/(FL-FH); F:=fun(rtf);                           # 6
>     if F < 0 then                                                 # 3
>     del:=XL-rtf: XL:=rtf; FL:=F;
>       else
>     del:=XH-rtf; XH:=rtf; FH:=F;
>     fi;                                                           # 4
>   dx:=XH-XL;
>   if abs(del) < eps or abs(F) < eps  then break fi;              # 7
>   od;
>   [rtf, k, fun(rtf)]; end:                                        # 8
```

(1) This line defines the parameter ϵ that determines when the iteration should stop.

(2) This is a check to ensure that there is a root in the interval (x_1, x_2): no account is taken of the possibility of there being two or more roots. The RETURN command is used here to produce an error message if the inequality is not satisfied.

(3) and (4) This check, repeated later on, ensures that XL corresponds to the end of the range at which $f < 0$.

(5) Start of the iterative loop: a loop structure is used here to ensure that at most 50 iterations are used. A check of this nature is essential.

(6) A linear interpolation between the upper and lower limits, to give the next estimate of the root, rtf.

(7) This checks that either the successive estimates of the roots are closer than ϵ or that $|f(x_t)| < \epsilon$.

(8) A successful computation produces a list; the first element is the estimate of the root; the second the number of iterates required; the third the value of the function at the estimated root.

As an example consider the simple function $f(x) = \sin x - e^{-x}$, which has an infinity of positive roots: for $0 < x < 2\pi$ a graph shows that there are roots near $x = 0.6$ and 3.1. Thus, for the smaller root we have

```
>   Root(0.5,0.7);
```

$$[.5885327493, 5, .74\ 10^{-8}]$$

found in five iterations. For the second root,

```
>   Root(2.4,3.4);
```

$$[3.096363932, 4, .39\ 10^{-9}]$$

Exercise 3.37
Use the procedure Root to show that three solutions of the equation $e^{-x^2} = \sin \pi x$ are $x = -1.09704$, 0.34694 and 0.833677.

Exercise 3.38
Write a procedure to solve Kepler's equation,

$$x = f(u) = u - \epsilon \sin u, \quad 0 \le \epsilon < 1,$$

for $u(x)$ and any given x in the range $(0, 2\pi)$, using Newton's method, which gives the root as the limit of the sequence of iterates

$$u_{n+1} = u_n - \frac{u_n - \epsilon \sin u_n - x}{1 - \epsilon \cos u_n}, \quad u_0 = x + \epsilon \sin x.$$

Your procedure $RN(x, \epsilon, \delta)$ should stop either after 30 iterations or when $|f(u_k) - x| < \delta$, where δ is a small number, whichever is the sooner, and should return a list [u,k,f(u)-x], where u is the estimate of the root and k is the number of iterations used.

Use your procedure to determine the number of iterations $k(x)$ needed to find $u(x)$ when $\epsilon = 0.999$ and $\delta = 10^{-15}$ for x in the range $0.001 \le x \le 0.02$. Illustrate your results by plotting the graph of $k(x)$.

The results obtained should show that Newton's method is not reliable when ϵ is close to unity and x close to 0, or 2π; this is because $f'(u)$ is zero when $u = 0$ and $\epsilon = 1$. The little known Halley's method is designed to overcome this problem and is considered in exercise 3.62 (page 130) at the end of this chapter.

Both Newton's and Halley's methods are derived in chapter 9.

These examples provide some idea of the syntax and use of procedures; it should be clear that there are many varied uses of procedures and that here we have provided only an introduction.

The general form of a procedure is

proc(*parameter sequence*)
    [**local** *name sequence*;]
    [**option** *name sequence*;]
    [**global** *name sequence*;]
  procedure body
end;

with the items enclosed in square brackets being optional.

We have not, so far, mentioned either the option or the global options; here we illustrate the use of the option facility with a recursive procedure. Other uses can be found in the help file, you should look under procedure and then the click on the hypertext label for option at the end of that file. Information about the global option can be found in the same manner.

The Fibonacci numbers are defined by the following recurrence relation

$$f_n = f_{n-1} + f_{n-2}, \quad f_0 = 0, \quad \text{and} \quad f_1 = 1. \tag{3.6}$$

The following procedure computes these numbers for any integer n.

```
>  restart;
>  fib:=proc(n::integer)
>  if n <= 1 then n else fib(n-1) + fib(n-2) fi;
>  end;
```

This is an example of recursive procedure which calls itself until a condition is satisfied: here the condition is $n \le 1$.

Exercise 3.39

Type in this procedure and try it out for the first few Fibonacci numbers, which are: 0, 1, 1, 2, 3, 5, 8, 13, 21, 34, 55.

It is worth spending a little time trying to understand how this recursive procedure works; some help is provided by the trace facility which, when switched on, traces the path through the procedure. If fib(1) is called then 1 is returned and nothing else happens

```
>  trace(fib):  fib(1);
{--> enter fib, args = 1
                                    1
<-- exit fib (now at top level) = 1}
                                    1
```

But calling fib(2) calls fib(1) and fib(0), in that order,

```
>  fib(2);
{--> enter fib, args = 2
{--> enter fib, args = 1
                                    1
<-- exit fib (now in fib) = 1}
{--> enter fib, args = 0
                                    0
<-- exit fib (now in fib) = 0}
                                    1
<-- exit fib (now at top level) = 1}
                                    1
```

This shows the path followed by the repeated calls to the procedure. The various values of fib(n) are stored in a stack until the last call and then combined, so the number of levels of recursions allowed is limited by the size of the stack. It is instructive to repeat the above for different arguments on your machine, but beware that if you call fib with a large argument trace will produce a lot of output. The trace facility is switched off using untrace(fib).

The problem with this procedure for the Fibonacci numbers is that it is very inefficient, and hence slow: this is how long it takes to compute f_{20} and f_{25} on my computer:

```
>  t1:=time():
>  print('F(20)= ',fib(20),' Time taken ',time()-t1,' seconds');
```

$$F(20) = , 6765, \quad \textit{Time taken} , .165, \textit{seconds}$$

```
> t1:=time():
> print('F(25)= ',fib(25),' Time taken ',time()-t1,' seconds');
```

$$F(25) = ,\ 75025,\quad \textit{Time taken},\ 1.717,\ \textit{seconds}$$

That is, increasing n from 20 to 25 increases the computational time by a factor of about 10: for f_{50} I gave up waiting. The reason it takes so long to compute the large n Fibonacci numbers is that this procedure re-calculates the same result over and over again; to find f_6 we require f_5 and f_4, and to find f_5 we need f_4 and f_3, etc. A way of dealing with this is to use the remember option which tells Maple to store the intermediate results and hence to save re-computing them. Thus we re-write the procedure as,

```
> fib:=proc(n::integer) option remember;
> if n <= 1 then n else fib(n-1) + fib(n-2) fi;
> end;
```

to find a dramatic increase in speed:

```
> t1:=time():

> print('F(25)= ',fib(25),' Time taken ',time()-t1,' seconds');
```

$$F(25) = ,\ 75025,\quad \textit{Time taken},\ 0,\ \textit{seconds}$$

```
> t1:=time():

> print('F(200)= ',fib(200),' Time taken ',time()-t1,' seconds');
```

$$F(200) = ,\ 280571172992510140037611932413038677189525,\quad \textit{Time taken},\ .005,$$
$$\textit{seconds}$$

Exercise 3.40

Write a procedure to compute the Legendre polynomials $P_n(x)$, which can be generated from the following recurrence relation:

$$P_n(x) = x\frac{2n-1}{n}P_{n-1}(x) - \frac{n-1}{n}P_{n-2}(x),\quad n \geq 2,$$

where $P_0(x) = 1$ and $P_1(x) = x$.

Plot the graphs of the first few Legendre polynomials for $-1 \leq x \leq 1$.

We end this section with a more complicated problem which provides a good exercise in writing procedures but also introduces some of the useful Maple facilities for numerically solving differential equations.

Consider the innocuous looking equation

$$\frac{dy}{dx} = \cos(\pi xy),\quad y(0) = a, \tag{3.7}$$

where a is a positive number. We wish to understand how this solution depends upon a. This differential equation is nonlinear and has no solution that can be expressed in terms of elementary functions, so it is necessary to solve it either numerically or using some other approximate method. Here we consider a numerical solution.

There are two useful Maple commands which help; first dsolve can be used to create a procedure which can then be used inside the command odeplot to produce a graph of the solution. In the following set of Maple commands we show only one example of the use of these commands; other uses may be found in the help files.

First it is necessary to load the graphics library in order to use `odeplot`:

```
>  with(plots):
```

Now we assign the differential equation to a variable `eq` — this is not necessary as the differential equation can be inserted directly into the argument of `dsolve`, but it is clearer to do it this way. Note the use of the D operator and the function `y(x)`: the `diff` command could also be used.

```
>  eq:=D(y)(x)=cos(Pi*x*y(x));
```

$$eq := D(y)(x) = \cos(\pi \, x \, y(x))$$

The initial conditions are also more easily dealt with by assigning them to a variable, here `ic`, and we set $a = 0.9$.

```
>  ic:=y(0)=0.9;
```

$$ic := y(0) = .9$$

It is important to note that in this numerical application of `dsolve` it is necessary to set $y(0)$ to be a specific number, not a free variable. Finally we use `dsolve` to set up a procedure which will numerically solve the differential equation:

```
>  p:=dsolve({eq,ic},y(x), type=numeric);
```

$$p := \mathbf{proc}(rkf45\_x) \ldots \mathbf{end}$$

This may be used to compute specific values of $y(x)$ at prescribed values of x, for example at $x = 0.1$,

```
>  z1:=p(0.1);
```

$$z1 := [x = .1, \, y(x) = .9984456777707413]$$

Sometimes this use of `dsolve` is essential, for instance if the difference between $y(x)$ and some other function is required, then it is necessary to extract the value of $y(x)$ from this list. This is best done using the `select` command, because in some applications of `dsolve` the order in the output list is not predictable. The relevant command comprises three parts which we first show separately:

```
>  z2:=select(has,z1,y(x));
```

$$z2 := [y(x) = .9984456777707413]$$

This is a list, and we need the first (and only) component:

```
>  z3:=z2[1];
```

$$z3 := y(x) = .9984456777707413$$

z3 is an equation and we need its right hand side:

```
>  z:=rhs(z3);
```

$$z := .9984456777707413$$

These three commands are better combined:

```
>  z:=rhs(select( has,y1,y(x))[1]);
```

$$z := .9984456777707413$$

If, however, we require just the graph of $y(x)$, then the command

```
>  odeplot(p,[x,y(x)],0..5,numpoints=50):
```

will plot† the graph for $0 < x < 5$, using 50 points.

Since we require graphs of $y(x)$ for a variety of values of a it is far better to put all these commands in procedure, which also allows the upper limit, X, of the range of x to be varied.

```
>  gr:=proc(a,X) local eq,ic,p;
>  eq:=D(y)(x)=cos(Pi*x*y(x));
>  ic:=y(0)=a;
>  p:=dsolve({eq,ic}, y(x), type=numeric);
>  odeplot(p,[x,y(x)],0..X,numpoints=100);
>  end:
```

Thus for $a = 1$ and $X = 5$ we type gr(1,5) to obtain

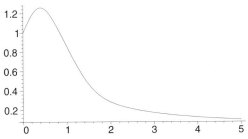

Figure 3.6   Solution of equation 3.7 with $y(0) = a = 1$.

The display command can be used to combine several of these graphs:

```
>  display(gr(1,6),gr(2,6),gr(3,6));
```

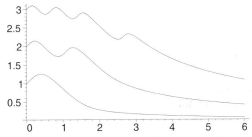

Figure 3.7   Solutions of equation 3.7 with $a = 1$ (lower curve),
2 and 3 (upper curve).

In addition to the Maple procedures discussed here there is a whole range of other procedures in the DEtools package. We introduce some of these facilities in chapter 16 when they are needed.

---

† More information about the arguments of **odeplot** and other examples of its use is obtained by typing **?odeplot** at the prompt.

Exercise 3.41

Write a procedure which draws graphs of the solutions of equation 3.7 for a set of initial conditions, contained in a list, and for $0 \leq x \leq X$, X being a given number.

Use this command to re-create figure 3.7 and draw a similar graph for $a = 0.2k$, $k = 1, 2, \ldots, 15$.

Exercise 3.42

Write a procedure to solve the differential equation

$$\frac{dy}{dx} = \sin(\pi y \sin x), \quad y(0) = a,$$

and to plot the graphs of $y(x)$ for several values of a over the interval $(0, X)$ for some given value of X.

Use your procedure to draw the graphs for $a = 0.2k$, $k = 1, 2, \ldots, 15$ and $0 < x < 2\pi$.

Exercise 3.43

A particle of unit mass moves in the time-dependent potential $V(x,t) = \frac{1}{4}tx^4$ and starts at the origin at time $t = 0$ with a positive velocity. Its equation of motion is

$$\frac{d^2x}{dt^2} + tx^3 = 0, \quad x(0) = 0, \quad \dot{x}(0) = v > 0,$$

which has solutions $x(t,v)$. The values of v for which the particle returns to the origin at time $t = 6$ are required, that is the roots of $g(v) = x(6,v) = 0$.

Write a Maple procedure to compute $g(v)$ for a given v by solving the equation of motion numerically and plot the graph of $g(v)$ for $0.1 < v < 5$. Use the `Root` procedure given on page 116 to show that for

$$v = 0.129\,38,\ 0.498\,79,\ 1.0853,\ 1.8774,\ 2.8678,\ 4.0507$$

the particle returns to the origin at time $t = 6$.

3.6 The SAVE and READ commands

Having written and tested Maple programs for functions and procedures, it is natural to want them saved for possible future use: the `save` and `read` commands are for this purpose.

For example, suppose that the function with the name *my_function* has been written and that it is to be saved under the file name *fred.fun*. Then the command

```
save my_function, "fred.fun" ;
```

will save this function in the file *fred.fun* in a particular default directory (called a folder in Windows 95 or 98); we discuss the location of this directory and how to change it below. If, in a subsequent session, the function *my_function* is needed, it may be retrieved by issuing the command

```
read "fred.fun" ;
```

In both cases the quotes round the file name are necessary, otherwise some characters, in this case the dot, will not be read.

As an example consider a procedure, `gra(x,n)`, which plots the graph of the function $\cos(nu - x\sin u)$ on the interval $0 \le u \le 2\pi$.

```
>  gra:=proc(x,n); plot( cos(n*u-x*sin(u)),u=-Pi..Pi); end:
```

We save this procedure in the file *graph.map* as follows:

```
>  save gra, "graph.map";
```

It would be good to know where this has been put; Maple is unforthcoming on this matter, so if this is the first time you have saved a file it is necessary to search for the file *graph.map*. You will probably find that it has been put in a sub-directory `Bin.wnt`; on my computer and with the version of Maple I am using to write this (which is probably an earlier version than the one you are using) the full address of the destination is

`c:\maplev5\Bin.wnt\graph.map`

though the actual address will depend upon how Maple was installed. This directory is where many Maple executable files are kept, so it is *not* a very good idea to keep other files here. Thus you will need to save procedures in other directories. To place the file in the directory

`c:\maplev5\examples`

either of the following commands may be used:

```
>  save gra,"c:/maplev5/examples/graph1.map";
>  save gra, "c:\\maplev5\\examples\\graph2.map";
```

In the first case the usual DOS backslash \ has been replaced by a /, because \ is used as a Maple command; you can find out how by typing `?\` at the prompt. The second alternative allows one to use the backslash by typing it twice. The reason for this confusion is probably historical, as Maple initially worked with the Unix operating system which uses / to define a directory tree.

The following commands will retrieve the procedure `gra(z,n)`, though any comments added to help you remember what the procedure did are not retrieved.

```
>  read "c:/maplev5/examples/graph1.map";
```
$$gra := \mathbf{proc}(x, n)\, \text{plot}(\cos(n \times u - x \times \sin(u)),\, u = -\pi..\pi)\, \mathbf{end}$$

or

```
>  read "c:\\maplev5\\examples\\graph2.map":
```

Another useful command is `readdata`, which allows one to read data files from other sources into Maple; these sources may be other computers, experimental results or just different software running on the same computer. The `writedata` command writes data out of a Maple session into a file. We illustrate the use of these commands by first writing data to a file and then reading it back into Maple.

The following commands evaluate the integral

$$f(x) = \frac{1}{\pi} \int_0^\pi dt\, \cos(2t - 2\sin t - x\sin 2t).$$

for $0 \le x \le 10$, in steps of 0.2, and writes the pairs $(x_k, f(x_k))$ to the file `c:\maplev\examples\data01.dat`.

```
>  restart:
>  file:="c:/maplev/examples/data01.dat":                          #1
>  fd:=fopen(file,WRITE):                                           #2
>      for x from 0 to 10.0 by 0.2 do;
>      f:=evalf( Int(cos(2*(t-sin(t)) - x*sin(2*t)),t=0..Pi)/Pi);   #3
>      ar:=array([[x,f]]):                                          #4
>      writedata(fd,ar,float):                                      #5
>      od:
>  fclose(file);                                                    #6
```

(1) This line names the output file details to be `file`: this is not necessary but saves subsequent typing.
(2) This command opens the named file and defines its mode to be WRITE: other modes are READ and APPEND.
(3) This command evaluates the required integral.
(4) For each k the pair $(x_k, f(x_k))$ is put in the 1×2 array `ar`. Note the double square brackets.
(5) The array `ar` is written to `data01.dat`.
(6) Finally the file is closed. Without this command the file `data01.dat` cannot be accessed unless the Maple session is ended.

The first and last two lines of the file `data01.dat` are:

$$0 \qquad .3528340285$$
$$.2 \qquad .369979158$$
$$\vdots \qquad \vdots$$
$$9.8 \qquad .0246469776$$
$$10.0 \qquad .0133432156$$

Now suppose that in another Maple session we need the data in this file. It is read into the variable X with the command

```
>  X:=readdata("c:/maplev/examples/data01.dat",2):
```

because there are two columns in the file `data01.dat`. With this format X is a list of the lists $[x_k, f(x_k)]$, $k = 1, 2, \ldots, 102$, the first two entries being

```
>  X[1];
```

$$[0, .3528340285]$$

```
>  X[2];
```

$$[.2, .369979158]$$

3.6.1 PostScript output of figures
It is sometimes useful to save the figures produced in a Maple session as a PostScript file for use elsewhere. Indeed, this is how all the figures in this book were produced.

The easiest way to do this is to export the Maple session; either use the menu buttons, *File, ExportAs, LaTeX*, or more conveniently the keystrokes alt-F, E, L, to produce a LaTeX version of the whole worksheet together with separate EPS files, one for each figure in the worksheet; the *.tex file thus produced is useful only with appropriate software, but the PostScript files can be used separately. These PostScript figures will not necessarily be an exact replica of the figures on the screen, but will be similar to the default figure that first appears: in particular they will be square and sometimes the tickmarks are different.

More control is obtained using the `plotsetup` options, listed below. Here it is assumed that the figure has been assigned to the variable p1 and that the EPS file of this plot is to be named `c:\temp\plot.eps`.

```
> with(plots):
> pltop:='width=240pt,height=120pt,portrait,
>                                      leftmargin=0pt,bottommargin=0pt';
> file:="c:/temp/plot.eps";                      # File name
> plotsetup(ps,plotoptions=pltop,plotoutput=file);
> display(p1);                                 # output to PS file
> plotsetup(default);        # Return to default values, i.e. the screen.
```

The variable `pltop` defines the size and the position of the figure: here the units are points (1 pt= 1/72 inch), but other units may be used. The variable `file` defines the file name and this definition is followed by a command setting up the plotting parameters: the plot device used here is ps for postscript, but other devices are available. With these parameters the command `display(p1)` produces the required file. It is wise to return to the default settings at the end. The resulting PS-figure is usually not an exact replica of the screen version, so it is often necessary to edit the PS file to align text better or to change the relative thickness of lines, for instance.

3.7 Exercises

Exercise 3.44
Use Maple to derive the given expressions, or similar expressions, for the differentials of the following functions:

$$f(x) = (\sin rx)^m e^{cx}, \qquad f'(x) = [c \sin rx + mr \cos rx]\, e^{cx}(\sin rx)^{m-1},$$

$$f(x) = \tan^{-1}\left(\frac{x}{1+x^2}\right), \qquad f'(x) = \frac{1-x^2}{1+3x^2+x^4},$$

$$f(x) = \ln\left(\frac{\cos x - \sin x}{\cos x + \sin x}\right), \qquad f'(x) = \frac{2}{1-2\cos^2 x},$$

$$f(x) = \frac{(x^2-1)^{3/2}}{(x^4+1)^{1/3}}, \qquad f'(x) = \frac{x\sqrt{x^2-1}(5x^4+4x^2+9)}{3(1+x^4)^{4/3}}.$$

Exercise 3.45
Differentiate the function

$$f(x) = \left(\frac{x}{x^2+1}\right)^{x/(1-x^2)}$$

and plot graphs of $f(x)$ and $f'(x)$ for $0 \le x \le 1$.

Exercise 3.46

Differentiate the function

$$f(x) = \sin(x^{\cos x})$$

and plot graphs of $f(x)$ and $f'(x)$ for $0 \le x \le 2\pi$.

Exercise 3.47

Use Maple to draw the graph of the curve $y = 4\cos x - 2\sin 3x + 3\cos 5x$ for $0 \le x \le \pi$. Find the values of x at which $y(x)$ has a maximum or a minimum and the values of y at these points.

Exercise 3.48

The equation of a curve $y(x)$ is defined implicitly by the equation

$$y^3 + x^3 - 9xy + 1 = 0$$

and (x_1, y_1) is a point on this curve at which the tangent is parallel to the x-axis. Prove that at this point

$$\frac{d^2 y}{dx^2} = \frac{18}{27 - x_1^3}.$$

Prove also that the stationary values of $y(x)$ occur at the points for which

$$x = (27 \pm 3\sqrt{78})^{1/3}$$

and determine which of these gives a maximum value of $y(x)$ and which a minimum.

Exercise 3.49

If the function $y(x)$ is defined implicitly by the equation $x^a + y^a = 1$, where a is a positive constant, show that

$$\frac{dy}{dx} = -\left(\frac{x}{y(x)}\right)^{a-1}.$$

Find an expression for $d^2 y/dx^2$.

Exercise 3.50

If the function $y(x)$ is defined implicitly by the equation $xy^2 - \ln y = x + \sin x$, show that

$$y(0) = 1, \quad \left.\frac{dy}{dx}\right|_{x=0} = -1, \quad \left.\frac{d^2 y}{dx^2}\right|_{x=0} = -3, \quad \text{and} \quad \left.\frac{d^3 y}{dx^3}\right|_{x=0} = 0.$$

Exercise 3.51

If $y = \tan^{-1}(\sinh x)$, show that

$$\frac{d^2 y}{dx^2} + \left(\frac{dy}{dx}\right)^2 \tan y = 0.$$

Exercise 3.52

If $y = \cosh a\theta$ and $x = \cosh \theta$, show that

$$(x^2 - 1)\frac{d^2 y}{dx^2} + x\frac{dy}{dx} - a^2 y = 0.$$

Exercise 3.53
Write a Maple function $s(n)$, with integer argument, to evaluate the ratio of Stirling's approximation to $n!$, that is

$$n! \simeq s(n) = \sqrt{2\pi n}\,(n/e)^n,$$

to the exact value of $n!$, $r(n) = s(n)/n!$. The value of this ratio should be expressed as a floating point number.

Plot the graph of $r(n)$ for $1 \le n \le 30$ and determine the values of n for which $s(n)$ is within 2% and 1% of $n!$.

Exercise 3.54
One of the definitions of the exponential function is

$$e^x = \lim_{n \to \infty} f(x,n), \quad \text{where} \quad f(x,n) = \left(1 + \frac{x}{n}\right)^n.$$

Write a Maple function for $f(x,n)$ and compare, graphically, this approximation with e^x for $0 < x < 2$ and various values of n.

Define the function

$$g(x,y) = \left(1 + \frac{x}{y}\right)^y \exp(-x)$$

and use the `implicitplot` procedure in the form

```
implicitplot(g(x,y)=0.9, x=1..7, y=1..200, grid=[50,50]);
```

to provide an estimate of the value of $n(x)$ at which the ratio of approximation $f(x,n)$ to e^x is about 0.9.

Show that for small values of x and large n

$$\cos^n x \simeq \exp\left(-\frac{nx^2}{2}\right)$$

and examine, graphically, the accuracy of this approximation for $|x| < 1$ and various values of n.

Exercise 3.55
If

$$z(x) = \int_0^{f(x)} dt\, g(x,t),$$

use Maple to show that

$$\frac{dz}{dx} = f'(x)g(x,f(x)) + \int_0^{f(x)} dt\, \frac{\partial g}{\partial x}.$$

Exercise 3.56
Plot the graphs of

$$c_n(x) = \underbrace{c \circ c \circ \cdots \circ c(x)}_{n \text{ times}}, \quad \text{where} \quad c(x) = \cos x$$

and

$$s_n(x) = \underbrace{s \circ s \circ \cdots \circ s(x)}_{n \text{ times}}, \quad \text{where} \quad s(x) = \sin x$$

for various values of n and $0 \le x \le 2\pi$. Discuss the behaviour of these functions as n increases.

In particular, show that

$$\lim_{n \to \infty} c_n(x) = c, \quad \text{where } c \text{ is the first positive root of } \cos c = c,$$

and that for large n $c_n(x) = c + \delta_n$, where $\delta_{n+1} = -\delta_n \sin c$.

Also show that the sequence $s_n(x)$ converges to zero, but that the convergence is much slower and is given approximately by

$$s_n(x) \simeq \sqrt{\frac{3x^2}{3 + nx^2}},$$

for small x. You will need the small x-expansion of $\sin x$ to obtain this result.

Exercise 3.57

Write a procedure which returns the value of the skew symmetric tensor $\mathscr{E}_{ijk}$, where i, j and k are integers taking the values 1, 2 and 3, and where

$$\mathscr{E}_{ijk} = \begin{cases} 0 & \text{if any index is repeated} \\ 1 & \text{if } i, j \text{ and } k \text{ are even permutations of } (1, 2, 3) \\ -1 & \text{if } i, j \text{ and } k \text{ are odd permutations of } (1, 2, 3), \end{cases}$$

so, for example, $\mathscr{E}_{123} = 1$ and $\mathscr{E}_{213} = -1$.

Use this procedure in another procedure which computes the vector product of two three-dimensional vectors $\mathbf{x} = (x_1, x_2, x_3)$ and $\mathbf{y} = (y_1, y_2, y_3)$,

$$\mathbf{x} \times \mathbf{y} = (x_2 y_3 - x_3 y_2, x_3 y_1 - x_1 y_3, x_1 y_2 - x_2 y_1),$$

which can be written in the form

$$(\mathbf{x} \times \mathbf{y})_i = \sum_{ijk} \mathscr{E}_{ijk} x_j y_k.$$

Exercise 3.58

Simpson's N-point rule for the numerical evaluation of the integral

$$I = \int_a^b dx \, f(x)$$

is, for even values of N,

$$I \simeq I_s = \frac{h}{3} \left[f_0 + f_N + 2 \left(f_2 + f_4 + \cdots + f_{N-2} \right) + 4 \left(f_1 + f_3 + \cdots + f_{N-1} \right) \right],$$

where $f_k = f(a + kh)$ and $h = (b - a)/N$. The error is approximately

$$|I - I_s| \sim \frac{(b - a)^5}{180 N^4} f^{(4)}(\zeta)$$

for some ζ in (a, b).

Write a procedure $\texttt{sim(f,a,b,N)}$ which evaluates the integral of a function f over the interval (a, b) using N points; your procedure should check that N is even and increment it by unity if it is not.

For the function $f(x) = \sqrt{x}$, $a = 0$ and $b = 1$, plot the graph of $\log(I/I_s)$ for $N = 4, 6, \ldots, 20$.

Use the save command to save your procedure.

Exercise 3.59

If $f(x)$ is a periodic function of x with period X, so $f(x + X) = f(x)$ for all x, normally the most efficient numerical method of evaluating the integral of f over one period,

$$I = \int_y^{y+X} dx\, f(x) = \int_0^X dx\, f(x), \quad \text{for all } y$$

is by an equally spaced trapezoidal rule,

$$I \simeq I_N = \frac{X}{N}(f_0 + f_1 + \cdots + f_{N-1}), \quad f_k = f\left(\frac{kX}{N}\right).$$

If $f(x)$ has a continuous pth derivative, the errors are less than CN^{-p} for some positive constant C. If all derivatives of $f(x)$ exist then the errors decrease faster than any inverse power of N, for instance as e^{-N}. The reasons for this are given in chapter 8.

Write a procedure to evaluate I_N for any f and N. Use your procedure to compute the mean value of

$$f(x) = \left(\sin^{20} x + \cos^{15} x\right)^2$$

for $N = 6, 7, \ldots, 25$, and plot a graph which compares these values with the exact value computed using the method discussed in chapter 2.

Use your program for evaluating Simpson's rule to compute the same mean value and compare, graphically, the two methods for different values of N.

Exercise 3.60

This exercise is an extension of exercise 3.14. Consider the functions

$$f(x) = x^{1/2} + x^{-1/2} \quad \text{and} \quad f^{(n+1)} = f^{(n)} \circ f \quad \text{with} \quad f^{(0)} = f,$$

and define

$$g_n(u) = f^{(n)}\left(u^{2^n}\right).$$

Use Maple to find $g_k(u)$ for $k = 2, 3, \ldots, 6$ and use the asympt command to show that

$$g_2(u) - g_3(u) = -\frac{1}{4u^7} + \cdots$$

$$g_3(u) - g_4(u) = -\frac{1}{8u^{15}} + \cdots$$

$$g_4(u) - g_5(u) = -\frac{1}{16u^{31}} + \cdots$$

$$g_5(u) - g_6(u) = -\frac{1}{32u^{63}} + \cdots.$$

These results are an example of a functional iteration which converges to a limiting function.

Exercise 3.61

In this question we find an approximate solution to the equation

$$\frac{d^2 y}{dt^2} + \omega(\epsilon t)^2 y = 0, \quad |\epsilon| \ll 1.$$

From exercise 3.31 we see that one approximate solution has the form

$$y(t) = \frac{1}{\sqrt{\omega(\epsilon t)}} \exp\left(i \int_0^t ds\, \omega(\epsilon s)\right),$$

and we now improve upon this by considering a modification of this solution of the form

$$y(t) = \frac{1}{\sqrt{\omega(\epsilon t)}} \exp\left(i \int_0^t ds\, \omega(\epsilon s) + i\epsilon h(\epsilon t)\right),$$

where $h(\epsilon t)$ is an unknown function.

Use Maple to substitute this approximate solution into the original equation, and expanding in powers of ϵ, choose h to eliminate the largest terms. Hence show that

$$h(z) = \int_0^z ds \left(\frac{3}{8\omega(s)^3} \left(\frac{d\omega}{ds} \right)^2 - \frac{1}{4\omega(s)^2} \frac{d^2\omega}{ds^2} \right)$$

$$= -\frac{1}{4\omega(z)^2} \frac{d\omega}{dz} - \frac{1}{8} \int_0^z ds \frac{1}{\omega^3} \left(\frac{d\omega}{ds} \right)^2 + \text{constant}.$$

Exercise 3.62

Repeat exercise 3.38 (page 117) using Halley's method for the solution of the equation

$$x = f(u) = u - \epsilon \sin u,$$

which gives the solution as the limit of the sequence

$$u_{n+1} = u_n + \frac{2f'(u_n)(x - f(u_n))}{2f'(u_n)^2 + (x - f(u_n))f''(u_n)}.$$

The derivation of this sequence will be given in chapter 9, on perturbation theory, where the derivations of both Newton's and Halley's method are given.

4

Sequences, series and limits

4.1 Introduction

A sequence is a collection of elements, in a given order, such as $1, \frac{1}{2}, \frac{1}{3}, \frac{1}{4}, \ldots$ or $1, x, x^2, x^3, \ldots$; an infinite sequence contains infinitely many elements. It is more difficult to explain what is meant by an infinite series, but for the moment you may think of it as an extension of the idea of a sum, for instance $1 + \frac{1}{2} + \frac{1}{3} + \frac{1}{4} + \cdots$. Sequences and series arise in diverse circumstances and their behaviour can be very subtle, even bizarre, and difficult to understand when first seen. Most of the analysis discussed in this chapter helps explain the behaviour of sequences and series and was developed in the nineteenth century by a variety of Continental mathematicians. A consequence of the lapse of time is that the problems which gave rise to this rather technical subject have been largely forgotten outside the specialist history books. So, before starting we briefly review the state of calculus in the eighteenth century and the problems that caused the fundamental changes which created some of the ideas presented here.

Post Newton and Leibniz the notion of a function was not clearly defined, but was considered to be a formula, x, x^2 and polynomials, for instance. There were problems in understanding the nature of a differential, which was defined in terms of the limit

$$\frac{df}{dx} = \lim_{\delta \to 0} \frac{f(x + \delta) - f(x)}{\delta},$$

because both numerator and denominator were zero at $\delta = 0$; these ideas worked in simple applications but problems arose when this new subject was applied to some physical problems.

In 1742 d'Alembert† derived a partial differential equation describing the motion of a vibrating string; in modern notation this is

$$\frac{\partial^2 y}{\partial x^2} - \frac{1}{c^2} \frac{\partial^2 y}{\partial t^2} = 0. \tag{4.1}$$

In addition d'Alembert obtained general solutions of this equation which highlighted many problems with the understanding of the character of functions at that time. The nature of the solutions of this equation led to a detailed inspection and clarification of the whole of eighteenth century analysis; the theory of functions, convergence of series and the physical interpretation of solutions. Two examples suffice to illustrate the problems.

---

† Jean le Rond d'Alembert (1717–1783) was an orphan abandoned on the steps of the church of St Jean le Rond, near Notre Dame, in Paris.

L. Euler
(1707–1783)

First, calculus was considered to comprise operations on algebraic expressions, the operations being differentiation and integration. The implicit assumption was that all operations should be applied to what we now call differentiable functions. But problems arose with Euler's solution, in 1749, of d'Alembert's wave equation 4.1; this solution has, again using modern notation, the general form $y(x,t) = F(ct-x) + G(ct+x)$, where the two functions F and G are expressed in terms of the initial shape and velocity of the string

$$y(x,0) = h(x), \qquad \frac{\partial}{\partial t} y(x,t)\Big|_{t=0} = v(x),$$

as follows:

$$F(-x) = \frac{1}{2}h(x) - \frac{1}{2c}\int dx\, v(x), \quad G(x) = \frac{1}{2}h(x) + \frac{1}{2c}\int dx\, v(x),$$

giving the solution

$$y(x,t) = \frac{1}{2}\{h(x-ct) + h(x+ct)\} + \frac{1}{2c}\int_{x-ct}^{x+ct} dz\, v(z).$$

However, a string may be plucked; then its initial speed is zero, $v(x) = 0$, and the gradient of the displacement, $h(x)$, is discontinuous at the point(s) where it is plucked — in modern terms it is piecewise differentiable — so $h(x)$ was not part of the class of functions dealt with by the existing calculus.

D. Bernoulli
(1700–1782)

The second problem was created by D. Bernoulli in 1753 by extending Taylor's work; he wrote the solution of the wave equation 4.1, for a string of length l fixed at both ends and initially stationary, in the form of an infinite series of functions:

$$y(x,t) = \sum_{k=1}^{\infty} a_k \sin\frac{k\pi x}{l} \cos\frac{k\pi ct}{l}, \quad y(x,0) = h(x), \qquad (4.2)$$

where the a_k, $k = 1, 2, \ldots$ are constants. This form of solution, now quite familiar, was unheard of prior to 1753, for it implies that *any* initial function, $h(x)$, defining the shape of the string, can be expressed as an infinite sum of sine functions,

$$h(x) = \sum_{k=1}^{\infty} a_k \sin\frac{k\pi x}{l}.$$

J. B. J. Fourier
(1768–1830)

This revolutionary idea was later to be developed by Fourier into the general theory of Fourier series, which will be discussed in chapter 8. The change from finite to infinite sums of functions can be very subtle: for example, it is not at all obvious how the above sum of continuous and infinitely differentiable functions can represent a function which is only piecewise smooth, as in the example of the plucked string.

Another class of problems associated with infinite series is illustrated by the series

$$S = 1 - \frac{1}{2} + \frac{1}{3} - \frac{1}{4} + \cdots,$$

which can be shown to sum to the value $S = \ln 2$; but when rearranged in the form

$$W = 1 + \left(\frac{1}{3} - \frac{1}{2} + \frac{1}{5}\right) + \left(\frac{1}{7} - \frac{1}{4} + \frac{1}{9}\right) + \left(\frac{1}{11} - \frac{1}{6} + \frac{1}{13}\right) + \cdots,$$

which contains exactly the same terms, though in a different order, sums to $W = \frac{3}{2}\ln 2$. Worse, the same terms can be rearranged to sum to *any* given value (exercise 4.8, page 140). Such examples show that finite and infinite series need not behave in the same manner. The behaviour of infinite series of functions is likely to be as bizarre as that of infinite series of numbers.

These problems struck at the very heart of the current theories of analysis, and it took most of the nineteenth century to resolve the fundamental problems and create the modern analysis that we now use. Only at the end of the nineteenth century did most of these developments impinge upon English mathematics; indeed, as late as the 1890s the work of Cantor, Weierstrass and Dedekind was largely unknown at Cambridge, and only by travelling to the USA and Europe did Russell learn sufficient to embark on the work necessary for *Principia Mathematica* (Monk, 1992, pages 45 and 113). It is interesting to note that much of the blame for this inadequacy has been attributed to the examination system in place at Cambridge during the nineteenth century; the articles by Forsyth (1935), Hardy (1948), and also Gandy (1973) provide an interesting view of that examination system and a useful comparison with the current debate in the UK; an alternative view of the old Cambridge examination systems is given by Pearson (1936).

The modern applied mathematician can, as a consequence of the efforts of earlier mathematicians and the power of modern computers, sometimes ignore many of the difficulties that created modern analysis; for example, we now know when it is safe to use infinite series of functions, whilst Fourier, and related, series have become standard tools. Nevertheless, it is still necessary to understand why the methods we use work, and this chapter, particularly sections 4.2 and 4.3, is largely about this basic theory. It is also helpful to understand the mathematical history of this area, and I recommend Boyer (1968) for an overall picture, and Gratton-Guinness (1970) for a more detailed discussion.

Some of the simplest sequences occur when finding the zeros of a simple function, $f(x) = 0$; for instance Newton's method gives the sequence

$$x_{k+1} = x_k - \frac{f(x_k)}{f'(x_k)}, \quad k = 0, 1, 2, \ldots,$$

which often converges to the required zero if the initial guess, x_0, is good enough. Understanding whether and how fast this sequence converges is clearly important.

Series provide a natural method of representing or approximating functions: for example the exponential function can be defined by the series

$$e^x = 1 + x + \frac{x^2}{2!} + \frac{x^3}{3!} + \cdots = \sum_{n=0}^{\infty} \frac{x^n}{n!},$$

and from this all its well known properties can, with some ingenuity, be derived, see for instance Whittaker and Watson (1965, appendix). Most other common functions and solutions of many linear differential equations can be represented as series. Such series normally converge, in a sense to be defined in section 4.3, for at least some values of x.

However, extracting numerical values from infinite series is not always simply a matter of sequentially adding terms. The value of an infinite series is obtained by a limiting process which can be quite subtle and needs to be understood; for example the two series for S and W, discussed above, behave non-intuitively for reasons which will be discussed in section 4.2.

In addition the direct summation of an infinite series may not be practical: for instance, the series

$$\sum_{k=2}^{\infty} \frac{\ln k}{k^2}$$

converges to the value 0.937548.., but the first five decimal places of this value cannot be obtained by direct summation except by adding about 10^7 terms. Methods of accelerating the convergence of sequences are therefore necessary and are discussed in section 4.4.

Other types of series exist whose behaviour is more difficult to understand; these are the divergent series discussed in chapter 5. As an example, the series

$$f(x) = \frac{1}{x} - \frac{1!}{x^2} + \frac{2!}{x^3} - \frac{3!}{x^4} + \frac{4!}{x^5} - \cdots, \quad x > 0, \tag{4.3}$$

converges for no finite value of x but, nevertheless, if truncated after a finite number of terms it will give a good approximation to the value of the integral

$$g(x) = \int_0^{\infty} dt \, \frac{e^{-t}}{x+t}, \quad x > 0, \tag{4.4}$$

provided x is sufficiently large.

It transpires that very many problems have solutions which are more usefully expressed in terms of divergent rather than convergent series. We shall discover some of the reasons for this in later chapters.

In the following we provide some necessary theory of sequences and series. I have assumed that much of the material covered in sections 4.2 and 4.3 is known, at least partially, to the reader, so have covered it fairly briskly in order to spend more time on the newer material in later sections. There are numerous books covering this early material, such as Rudin (1976), Apostol (1963) and Whittaker and Watson (1965), which provide more examples and detail.

4.2 Sequences

4.2.1 The limit of a sequence

A sequence is a denumerable set, z_k, $k = 1, 2, \ldots$, of real or complex numbers in a specified order. The value of z_k can behave in a variety of ways as $k \to \infty$; for instance if $z_k = 1/k$ it is intuitively clear that z_k converges to zero; if $z_k = \exp(ak)$ the limit depends upon the value of a and will tend to zero if $\Re(a) < 0$, infinity if $\Re(a) > 0$ and be indeterminate if $\Re(a) = 0$. In other cases it is not so clear how the sequence behaves; for instance if

$$z_k = 1 + \frac{1}{2} + \frac{1}{3} + \cdots + \frac{1}{k} \quad \text{or} \quad z_k = 1 - \frac{1}{2} + \frac{1}{3} - \cdots + \frac{(-1)^{k-1}}{k},$$

the behaviour as $k \to \infty$ is not obvious. In these circumstances it is essential to have some clear definitions of what we mean by convergence and to have tests that determine whether or not a given sequence converges.

The most easily understood sequences are those that converge to some finite limit as $k \to \infty$; we define this notion rigorously in the following manner.

If there exists a number l such that for every positive number ϵ, no matter how small, a finite number n_0 can be found such that

$$|z_k - l| < \epsilon \quad \text{for all } k > n_0,$$

the sequence z_k is said to tend to the limit l as k tends to infinity; this statement is written as

$$\lim_{k \to \infty} z_k = l.$$

If the sequence is such that, given an arbitrary number X, no matter how large, we can find an n_0 such that $|z_k| > X$ for all values of k greater than n_0, we say that '$|z_k|$ tends to infinity' and write

$$|z_k| \to \infty.$$

A sequence of real numbers x_k tends to infinity if for $k > n_0$, $x_k > X$ for any $X > 0$ and to $-\infty$ if $-x_k > X$ for $k > n_0$. If a sequence of real numbers does not tend to a limit or to $\pm\infty$ it is said to oscillate.

Frequently sequences comprise sums and products of other sequences; if u_k and v_k are two sequences and $u_k \to u$ and $v_k \to v$ as $k \to \infty$, then the following results hold:

(i) if a and b are numbers independent of k, then $au_k + bv_k \to au + bv$ as $k \to \infty$;
(ii) $u_k v_k \to uv$ as $k \to \infty$;
(iii) if $v \neq 0$ then $u_k/v_k \to u/v$ as $k \to \infty$;
(iv) if $u_k < v_k$ for all $k > N$, then $u \leq v$ as $k \to \infty$.

Exercise 4.1
Determine the limits of the following sequences as $k \to \infty$:

(i) $\quad z_k = \dfrac{k}{1+k}$, (ii) $\quad z_k = \dfrac{1}{1+k^2}$, (iii) $\quad z_k = \dfrac{k^2}{1+k}$,

(iv) $\dfrac{(ak+b)^2}{ck^2+d}$, (v) $\quad z_k = i^k$, (vi) $\dfrac{k}{k^2} + \dfrac{k+1}{k^2} + \cdots + \dfrac{2k}{k^2}$,

where a, b, c and d are positive constants.

Exercise 4.2
Describe the behaviour of a^n as $n \to \infty$ when $a < -1$, $-1 < a < 0$, $0 < a < 1$ and $a > 1$.

Sketch the graph of the function $f(x) = x^n$ for $0 \leq x \leq 1$ and for some large positive values of n. State the values of $\lim_{n \to \infty} f(x)$ for $x > 0$.

Exercise 4.3
If $a > 1$ and $a^{1/n} = 1 + x$, show that $0 < x < a/n$. Deduce that

$$\lim_{n \to \infty} a^{1/n} = 1.$$

What are the corresponding results if (i) $a = 1$, (ii) $0 < a < 1$?
Hint: for part (ii) consider $b = 1/a$.

Exercise 4.4
If $y > 1$ and $y^{1/y} = 1 + x$, show that $0 < x < \sqrt{2/(y-1)}$. Deduce that

$$\lim_{y \to \infty} y^{1/y} = 1.$$

4.2.2 The phrase 'of the order of'

It is frequently necessary to compare the magnitude of two sequences, particularly when making approximations, and we need a convenient notation allowing such comparisons: for example it is clear that, for large enough n, $w_n = 1 + n^2$ is larger than $u_n = 100^{100}n$ and smaller than $v_n = 10^{-6}n^{5/2}$, but how can we conveniently express this idea?

If z_k and ζ_k are two sequences and a number n_0 exists such that

$$\left| \frac{\zeta_n}{z_n} \right| < K \quad \text{whenever} \quad n > n_0,$$

where K is *independent* of n, we say that ζ_n is *of the order of* z_n and write

$$\zeta_n = O(z_n) \quad \text{as } n \to \infty.$$

For instance, if $\zeta_n = (10n + 2)/(n^2 + 1)$ and $z_n = 1/n$ then

$$\frac{\zeta_n}{z_n} = \frac{10n^2 + 2n}{n^2 + 1} = 10 \frac{1 + 1/5n}{1 + 1/n^2} < 11$$

for all $n > n_0 = 1$. Thus we write

$$\zeta_n = \frac{10n + 2}{n^2 + 1} = O(z_n) = O\left(\frac{1}{n}\right) \quad \text{as } n \to \infty.$$

Frequently the phrase 'as $n \to \infty$', or the equivalent, is omitted as it is clear from the context. Similarly, if $w_n = 1 + n^2$ and $u_n = 100^{100}n$ then $w_n/u_n = O(n)$ as $n \to \infty$, despite the very large numerical factor, which means that $u_n > w_n$ for $n < 100^{100}$. This example emphasises that the comparisons may only be useful for very large n; normally, however, the numerical coefficients are not so large, and the comparisons are meaningful at smaller values of n.

Note that with this definition, since $n < n^2$ for all $n \geq 2$, we have $n = O(n^2)$ and $n = O(n^{10})$ for example, so it would be perfectly correct to write

$$3n^2 + 5n + 1 = O(n^4) \quad \text{as} \quad n \to \infty.$$

Generally, in order to convey as much information as possible, we put the smallest possible term in the order symbol, so in the last equation the right hand side would be written as $O(n^2)$.

This notation can also be used to compare the size of two functions $f(x)$ and $g(x)$ of the real variable x: we say that '$f(x)$ is of the order of $g(x)$ as $x \to y$', or

$$f(x) = O(g(x)) \quad \text{as} \quad x \to y, \quad \text{if} \quad \lim_{x \to y} \left| \frac{f(x)}{g(x)} \right| < K$$

for some K independent of y. Observe that if $f(x) \simeq g(x)$ or if $f(x) \ll g(x)$ as $x \to y$ then $f(x) = O(g(x))$. Thus we write, for instance,

$$\sin x = x - \frac{x^3}{3!} + O(x^5) \quad \text{as } x \to 0,$$

to signify that the difference $\sin x - (x - x^3/3!)$ behaves like Kx^5 for small $|x|$.

In many applications we need to know the order of a function $f(x)$ as x tends either to zero or infinity. Thus if $f(x) = (a+x)^b - a^b$, a being some positive number, then for small x the binomial expansion can be used to give

$$(a + x)^b = a^b \left(1 + \frac{bx}{a} + O(x^2) \right), \quad \text{hence} \quad f(x) = O(x) \quad \text{as } x \to 0.$$

In this type of analysis Taylor's series, which will be discussed in section 4.3, is often useful, and in the limit $x \to \infty$ it is sometimes helpful to define a new variable $w = 1/x$ and use the Taylor expansion about $w = 0$.

Typically, if we require the behaviour of a function as $x \to 0$ we need to know whether it behaves as some power, that is $f(x) = O(x^\alpha)$ where α is a real number, or as a logarithm, $f(x) = O(\ln(x))$, or as some other function, for instance $x^\alpha \ln x$ or $e^{-1/x}$. Some simple examples will show how these ideas can be applied.

Since $\sin x = x - x^3/3! + \cdots$, we have

$$\sin x = O(x) \quad \text{as } x \to 0, \quad \text{because} \quad \frac{\sin x}{x} = 1 - \frac{x^2}{6} \to 1.$$

Similarly, since $\cos x = 1 - x^2/2! + \cdots$, we have $\cos x = O(1)$ as $x \to 0$. Functions of functions can be dealt with by working outwards from the inner function; thus if $f(x) = \ln(\sin x)$ then since $\sin x = O(x)$ we may replace $\sin x$ by x to obtain $\ln(\sin x) = O(\ln(x))$ as $x \to 0$.

Exercise 4.5

Determine the order of the following expressions as $x \to 0$:

(i) $\sqrt{x(1-x)}$,   (ii) $10000x^{1/2}$,   (iii) $\ln(1 + \sin x)$,

(iv) $\dfrac{1 - \cos x}{1 + \cos x}$,   (v) $\dfrac{x^{3/2}}{1 - \sin x}$,   (vi) $\dfrac{x^{5/4}}{1 - \cos x}$,

(vii) $\exp(\tanh x)$,   (viii) $1 + \dfrac{1 + 2x}{x(1 - 2x)}$,   (ix) $\ln \left(1 + \dfrac{\ln \left(\frac{1+2x}{x} \right)}{1 - 2x} \right)$.

You should be able to do these calculations 'by hand', but an easy way is to use the Maple command `series(f,x,3)`, where `f` is the expression representing the given function, then the first term of the given series is normally the order of the expression. Note that the `series` command uses the order notion to provide an idea of the magnitude of the ignored terms.

Exercise 4.6
Determine the order of the following functions as $x \to \infty$:

(i) $\dfrac{x}{x^2 - 1}$, (ii) $x^{1/x} - 1$, (iii) $\sqrt{x^2 + x} - x$,

(iv) $(x + 1)^a - x^a$, $a > 0$, (v) $(a + x)^b - a^b$, $a > 0$.

4.3 Infinite series

4.3.1 Definition of convergence

If z_k, $k = 1, 2, \ldots$, is an infinite sequence of real or complex numbers, we can form another sequence S_n from the partial sums of z_k,

$$S_n = z_1 + z_2 + \cdots + z_n = \sum_{k=1}^{n} z_k.$$

If the sequence S_n tends to a limit S, the infinite series $z_1 + z_2 + z_3 + z_4 + \cdots$ is said to be *convergent* and to converge to the sum S, the value of which is unique; then we write

$$S = z_1 + z_2 + \cdots = \sum_{k=1}^{\infty} z_k. \tag{4.5}$$

Otherwise the infinite series is said to be *divergent*. The number S is called the sum of the infinite series, but it should be understood as the limit of a sequence and not obtained simply by addition, although in practice we often find approximations to the value of series by summing a finite number of terms.

When the series converges the expression

$$R_n = S - S_n = z_{n+1} + z_{n+2} + \cdots$$

is named the *remainder* after n terms.

A necessary condition for the sequence S_n to converge is that z_n tends to zero as n tends to infinity; that this condition is not sufficient is seen by studying the harmonic series

$$S = 1 + \frac{1}{2} + \frac{1}{3} + \frac{1}{4} + \frac{1}{5} + \cdots, \tag{4.6}$$

which is shown to diverge in exercise 4.22 (page 147).

The reader could be excused for thinking that only convergent series are useful; we shall see in the next chapter that divergent series are both useful and important.

4.3.2 Absolute and conditional convergence

When doing arithmetic we make use of rules which always hold when finite sets of numbers are added or multiplied, and it is reasonable to ask whether these rules remain true when applied to infinite series. As an example, the order in which any finite set of numbers is added does not affect the value of the sum, so the finite series

$$S_n = 1 - \frac{1}{2} + \frac{1}{3} - \frac{1}{4} + \cdots + \frac{(-1)^{n-1}}{n} \tag{4.7}$$

has a unique value regardless of the order of summation. But, the value of the infinite series

$$S = 1 - \frac{1}{2} + \frac{1}{3} - \frac{1}{4} + \frac{1}{5} - \cdots \tag{4.8}$$

may be changed by the order in which the summation is performed.

Exercise 4.7
The infinite series 4.8 can be re-ordered in the form

$$W = 1 + \left(\frac{1}{3} - \frac{1}{2} + \frac{1}{5}\right) + \left(\frac{1}{7} - \frac{1}{4} + \frac{1}{9}\right) + \cdots + \left(\frac{1}{4k-1} - \frac{1}{2k} + \frac{1}{4k+1}\right) + \cdots,$$

so W contains the same terms as S but in a different order. Define the two partial sums

$$S_n = \sum_{k=1}^{4n+1} \frac{(-1)^{k-1}}{k},$$

$$W_n = 1 + \sum_{k=1}^{n} \left(\frac{1}{4k-1} - \frac{1}{2k} + \frac{1}{4k+1}\right).$$

Use Maple to evaluate these sums and plot the graph of the ratio W_n/S_n over a suitable range of n to show that $W_n/S_n \simeq 3/2$.

This elementary example, examined in more detail later on, shows that care is needed in passing from finite to infinite sums. The notion of absolute convergence, defined below, is necessary to distinguish those infinite series which behave like finite series.

In order that the series of real or complex terms

$$S = \sum_{k=1}^{\infty} z_k \tag{4.9}$$

should converge it is *sufficient* that the series

$$\sum_{k=1}^{\infty} |z_k| \tag{4.10}$$

should converge. If the series 4.10 converges then the series 4.9 is said to be *absolutely convergent*. Series that are convergent but not absolutely convergent are said to be *conditionally* convergent.

Absolute convergence is important for two reasons. First, if the two series

$$\begin{aligned} S &= u_1 + u_2 + u_3 + \cdots \\ T &= v_1 + v_2 + v_3 + \cdots \end{aligned}$$

are absolutely convergent, then the product of these series

$$P = (u_1 + u_2 + u_3 + \cdots)(v_1 + v_2 + v_3 + \cdots) = u_1 v_1 + u_2 v_1 + u_1 v_2 + \cdots,$$

formed by the products of their terms written in *any* order, is also absolutely convergent and has the value ST. On the contrary, the product of two conditionally convergent series is not unique; its value depends upon the order of summation, and may diverge. In these circumstances we need a definition of the product; one definition is

$$P = \sum_{n=2}^{\infty} w_n, \quad w_n = \sum_{k=1}^{n-1} u_k v_{n-k}, \quad n = 1, 2, \ldots. \tag{4.11}$$

This is the definition used by computer algebra systems; for real series the nth term involves $n-1$ products, so the sum of the series P to N terms requires roughly N^2 multiplications.

Second, the sum of an absolutely convergent series is unaffected by the order of the terms. But for a conditionally convergent real series it may be shown that the terms can be rearranged to converge to *any* real number.

This result may seem strange and to conflict with the notion of convergence to a unique limit defined by equation 4.5. But there is no contradiction because different rearrangements produce different sequences; for instance the partial sums S_n and W_n defined in exercise 4.7 are not the same.

A conditionally convergent series

$$S = u_1 + u_2 + u_3 + \cdots + u_n + \cdots$$

can be re-arranged to

$$R = u_1' + u_2' + u_3' + \cdots + u_n' + \cdots,$$

where R is any prescribed real number, in the following manner. Suppose, with no loss of generality, that $R > 0$: for u_1' take the first non-negative term in S; if this is less than R for u_2' take the second non-negative term. Proceed in this way until the sum exceeds R. For the next term take the first negative term of S; if the sum still exceeds R continue taking the successive negative terms until the sum is less than R.

Continue in this way, taking alternate groups of non-negative and negative terms of S until the sum becomes respectively greater and less than R. It can be proved that this process converges to give a re-arrangement of S summing to R.

Exercise 4.8
Write a Maple procedure which implements this method to sum the sequence

$$1 - \frac{1}{2} + \frac{1}{3} - \frac{1}{4} + \frac{1}{5} + \cdots,$$

not necessarily in that order, to produce an approximation to any given number R from a finite number, N, of terms of the series. You will need to restrict the magnitude of R otherwise very many terms are needed, that is, N will need to be commensurately large.

Finally, we return to the series discussed in exercise 4.7; recall the two series:

$$S = 1 - \frac{1}{2} + \frac{1}{3} - \frac{1}{4} + \frac{1}{5} + \cdots, \tag{4.12}$$

$$W = 1 + \left(\frac{1}{3} - \frac{1}{2} + \frac{1}{5}\right) + \left(\frac{1}{7} - \frac{1}{4} + \frac{1}{9}\right) + \left(\frac{1}{11} - \frac{1}{6} + \frac{1}{13}\right) + \cdots, \tag{4.13}$$

formed from the same terms, but in a different order. We shall see below that the sum of moduli of these terms diverges, so, at best, these two series are conditionally convergent.

Let S_n and W_n denote the sums of the first n terms — note that this definition is different from that used in exercise 4.7 — and set

$$\sigma_n = 1 + \frac{1}{2} + \cdots + \frac{1}{n},$$

so that

$$
\begin{aligned}
\sigma_{2n} - \sigma_n &= \left(1 + \frac{1}{2} + \frac{1}{3} + \cdots + \frac{1}{2n}\right) - \left(1 + \frac{1}{2} + \frac{1}{3} + \cdots + \frac{1}{n}\right), \\
&= 1 + \left(\frac{1}{2} - 1\right) + \frac{1}{3} + \left(\frac{1}{4} - \frac{1}{2}\right) + \cdots \\
&\qquad + \frac{1}{2k-1} + \left(\frac{1}{2k} - \frac{1}{k}\right) + \cdots + \left(\frac{1}{2n} - \frac{1}{n}\right), \\
&= 1 - \frac{1}{2} + \frac{1}{3} - \frac{1}{4} + \cdots + \frac{1}{2k-1} - \frac{1}{2k} + \cdots - \frac{1}{2n}.
\end{aligned}
$$

Hence $S_{2n} = \sigma_{2n} - \sigma_n$. Similarly,

$$
\begin{aligned}
W_{3n} &= 1 + \frac{1}{3} + \cdots + \frac{1}{4n-1} - \left(\frac{1}{2} + \frac{1}{4} + \cdots + \frac{1}{2n}\right) = \sigma_{4n} - \frac{1}{2}\sigma_{2n} - \frac{1}{2}\sigma_n \\
&= \sigma_{4n} - \sigma_{2n} + \frac{1}{2}(\sigma_{2n} - \sigma_n) = S_{4n} + \frac{1}{2}S_{2n}.
\end{aligned}
$$

On letting $n \to \infty$ we obtain $W = \frac{3}{2}S$, the result obtained numerically in exercise 4.7.

We now discuss a few elementary examples of series.

Geometric series

A *geometric series* is defined to be a series of the form

$$S = a + az + az^2 + az^3 + \cdots + az^n + \cdots,$$

z and a being complex numbers. The sum of the first $n+1$ terms of this series is given by the formula

$$S_{n+1} = a\left(1 + z + z^2 + \cdots + z^n\right) = a\frac{1 - z^{n+1}}{1 - z}.$$

The series S is absolutely convergent if $|z| < 1$, as may be shown by considering the partial sum of moduli

$$1 + |z| + |z|^2 + |z|^3 + \cdots + |z|^n = \frac{1}{1 - |z|} - \frac{|z|^{n+1}}{1 - |z|}.$$

If $|z| < 1$ the second term on the right tends to zero as $n \to \infty$ and the geometric series is absolutely convergent to

$$S = \frac{a}{1 - z}, \quad |z| < 1.$$

If $|z| > 1$ the second term increases without bound and the series diverges. The case $|z| = 1$ is dealt with in the following exercise.

Exercise 4.9
Consider the geometric series

$$S = 1 + z + z^2 + \cdots + z^n + \cdots.$$

Show that the sum of the first $n + 1$ terms can be written in the form

$$S_{n+1} = \frac{z^{(n+1)/2} - z^{-(n+1)/2}}{z^{1/2} - z^{-1/2}} z^{n/2},$$

and by putting $z = e^{i\theta}$, for some real θ, deduce that

$$S_{n+1}(\theta) = \sum_{k=0}^{n} e^{ik\theta} = \frac{\sin((n+1)\theta/2)}{\sin(\theta/2)} e^{in\theta/2}.$$

Show that

$$\lim_{\theta \to 0} S_{n+1}(\theta) = n + 1, \quad \lim_{\theta \to \pi} S_{n+1}(\theta) = \begin{cases} 1, & \text{if } n \text{ is even,} \\ 0, & \text{if } n \text{ is odd.} \end{cases}$$

Plot the graph of $e^{-in\theta/2} S_{n+1}(\theta)$ for $-\pi \le \theta \le \pi$ for various values of n.

Exercise 4.10
Show that

$$\sum_{k=-N}^{N} e^{ik\theta} = \frac{\sin\left(N + \frac{1}{2}\right)\theta}{\sin(\theta/2)}.$$

Exercise 4.11
By considering the partial sums, show that if

$$S_n = \sum_{r=1}^{n} \frac{1}{r(1+r)} \quad \text{then} \quad \lim_{n \to \infty} S_n = 1.$$

The zeta function and harmonic series
The Riemann zeta function is defined by the infinite series

$$\zeta(z) = \sum_{k=1}^{\infty} \frac{1}{k^z}, \quad \Re(z) > 1. \tag{4.14}$$

This function was known to Euler, but many of its remarkable properties were discovered by Riemann. One of the more tantalising aspects of this function is its occurrence in diverse fields of mathematics and physics. Euler connected it to the prime numbers via the relation

G. F. B. Riemann
(1826–1866)

$$\frac{1}{\zeta(s)} = \prod_{\text{primes}} \left(1 - \frac{1}{p^s}\right).$$

More recently the zeros of the zeta function have been connected with the energy levels of a quantised dynamical system whose classical analogue is chaotic. This connection comes via the Riemann hypothesis that all the zeros of $\zeta(s)$ other than $s = -2, -4, \ldots$, have real part $\frac{1}{2}$, that is if $\zeta(s_k) = 0$ then $s_k = \frac{1}{2} + iE_k$, where E_k is real. It was pointed out by Hilbert and Pólya that this conjecture would be true if the E_k could be shown to be the eigenvalues of a Hermitian operator. Since the Hamiltonian of a dynamical system is represented, in quantum mechanics, by such an operator, this, together with other analysis, see for instance Keating (1993) and Berry (1987), suggests a connection between energy levels and the zeros of the zeta function. Such reasoning suggests that the spacing between adjacent zeros, $S = s_{k+1} - s_k$ suitably normalised, has the probability density† given by $\rho(S) = 32S^2 \exp(-4S^2/\pi)/\pi^2$. This empirical rule agrees well with numerical calculations using the first 10^5 zeros; other statistical results are given by Berry (1987, page 192).

Returning to more elementary problems, it is known that if z is an even integer, $z = 2n$, then $\zeta(z) = c\pi^{2n}$, where c is a rational number; a reason for this is given in chapter 8. Specific values of the zeta function are

$$\zeta(2n) = \frac{(2\pi)^{2n}}{2(2n)!}|B_{2n}|, \quad \zeta(-2n) = 0 \quad \text{and} \quad \zeta(1 - 2n) = -\frac{B_{2n}}{2n},$$

where B_n are the Bernoulli numbers, defined in the introduction to chapter 5, see also section 4.3.4.

The series defining the zeta function, equation 4.14, can be shown to converge if $\Re(z) > 1$, exercise 4.22. If $\Re(z) \leq 1$ other representations of $\zeta(z)$ are more useful, some of which are given in Whittaker and Watson (1965, chapter 13) and Abramowitz and Stegun (1965, chapter 23). One remarkable formula, due to Riemann, is

$$2\Gamma(z)\zeta(z)\cos\left(\frac{1}{2}z\pi\right) = (2\pi)^z\zeta(1 - z).$$

If $z = x$ is real, the series 4.14 converges if $x > 1$ and diverges if $x \leq 1$. If $x = 1$ it reduces to the harmonic series, equation 4.6 (page 138), and it can be shown that the harmonic series diverges as $\ln n$, or, more accurately,

$$\lim_{n \to \infty}\left[1 + \frac{1}{2} + \frac{1}{3} + \cdots + \frac{1}{n} - \ln n\right] = \gamma, \tag{4.15}$$

where $\gamma = 0.577\,215\,664\,9\ldots$; this limit defines Euler's constant. In Maple the numerical value of γ is given by `evalf(gamma)`.

If $s < 1$ each term of the series for the zeta function diminishes more slowly than when $s = 1$, so the series diverges.

Exercise 4.12
Prove, by changing the order of summation, that

$$\text{(i)} \quad \sum_{n=2}^{\infty}\left(\zeta(n) - 1\right) = 1, \qquad \text{(ii)} \quad \sum_{n=2}^{\infty}(-1)^n\left(\zeta(n) - 1\right) = \frac{1}{2}.$$

---

† This means that the relative number of differences in the interval (x_1, x_2) is $\int_{x_1}^{x_2} dS\,\rho(S)$.

Exercise 4.13

Plot graphs of $\zeta(x)$ for $-20 < x < -14$, $-10 < x < -1$ and $2 < x < 6$.

4.3.3 Tests for convergence

It is clearly important to know whether or not a sequence or series converges, consequently a variety of tests have been developed for this purpose. Here we list a few of the most useful and popular methods.

Comparison test

The series $S = z_1 + z_2 + z_3 + \cdots$ is absolutely convergent if there exists an absolutely convergent series $v_1 + v_2 + v_3 + \cdots$ and a constant C, independent of n, such that $|z_n| \leq C|v_n|$ for all n. For example, the series

$$S = \cos x + \frac{1}{2^2}\cos 2x + \frac{1}{3^2}\cos 3x + \cdots, \quad x \text{ real},$$

is absolutely convergent because $V = 1 + 2^{-2} + 3^{-2} + \cdots = \zeta(2) = \pi^2/6$ is absolutely convergent and

$$|z_k| = \left|\frac{\cos kx}{k^2}\right| \leq \frac{1}{k^2} = v_k.$$

Exercise 4.14

Show that if α and β are positive numbers, the series

$$\sum_{n=1}^{\infty} \frac{(n+1)^\alpha - n^\alpha}{n^\beta}$$

is convergent if $\beta > \alpha$. Determine the behaviour of the series if $\beta = \alpha$.

Exercise 4.15

Show that for all real $\alpha < 1$ the series

$$S_n = 1 + \frac{1}{2^\alpha} + \frac{1}{3^\alpha} + \cdots + \frac{1}{n^\alpha} \quad \text{is divergent.}$$

Cauchy's test

If

$$\lim_{n\to\infty} |z_n|^{1/n} < 1, \quad \text{the series} \quad \sum_{n=1}^{\infty} z_n$$

converges absolutely. But if

$$\lim_{n\to\infty} |z_n|^{1/n} > 1,$$

z_n does not tend to zero so the series does not converge. If the limit is unity this test provides no information.

Exercise 4.16

Prove that if α, a and x are positive real numbers, the following series converges if $x < 1$:

$$\sum_{n=0}^{\infty} \frac{x^n}{a + n^\alpha}.$$

D'Alembert's ratio test

The series

$$\sum_{n=1}^{\infty} z_n$$

is absolutely convergent if

$$\lim_{n\to\infty} \left| \frac{z_{n+1}}{z_n} \right| < 1 \quad \text{and divergent if} \quad \lim_{n\to\infty} \left| \frac{z_{n+1}}{z_n} \right| > 1.$$

As with Cauchy's test, this test provides no information if the limit is unity. But the series is absolutely convergent if a positive number c exists such that

$$\lim_{n\to\infty} n \left\{ \left| \frac{z_{n+1}}{z_n} \right| - 1 \right\} = -1 - c.$$

This means that the ratio $|z_{n+1}/z_n|$ must not approach 1 from below too fast. Thus if

$$\left| \frac{z_{n+1}}{z_n} \right| = 1 + \frac{a}{n} + O\left(n^{-2} \right),$$

where a is independent of n, then the series is absolutely convergent if $a < -1$. This is sometimes named Raabe's test. The series diverges if $a \geq 1$.

Exercise 4.17

Determine the conditions on the numbers p and q, satisfying $0 < q < p$, such that the following series converge:

$$S_1 = \sum_{n=1}^{\infty} p^n n^p, \quad S_2 = \sum_{n=2}^{\infty} \frac{1}{n^p - n^q}, \quad S_3 = \sum_{n=1}^{\infty} \frac{1}{p^n - q^n}.$$

Exercise 4.18

If a_n are non-negative numbers and $\sum_{n=1}^{\infty} a_n$ converges, prove that $\sum_{n=1}^{\infty} \sqrt{a_n}/n$ also converges.

Exercise 4.19

Determine the conditions on the real numbers r and k for the following series to converge:

$$\sum_{n=1}^{\infty} n^r \exp\left(-k \sum_{m=1}^{n} \frac{1}{m} \right).$$

Exercise 4.20

Prove that the following series converges:

$$\sum_{n=2}^{\infty} \left(n^{1/n} - 1\right)^n.$$

Dirichlet's test

If a_n, $n = 1, 2, \ldots$, is a sequence of numbers such that

$$\left| \sum_{n=1}^{p} a_n \right| < N,$$

where N is a number independent of p; and if another sequence of numbers f_n satisfy $f_n \geq f_{n+1} > 0$, for all n, and $\lim_{n \to \infty} f_n = 0$, then the sum

$$\sum_{n=1}^{\infty} a_n f_n$$

converges. For instance, if we set $f_n = \frac{1}{n}$ and $a_n = (-1)^{n+1}$ we see that the series 4.8 (page 139), converges.

Exercise 4.21

Show that the following series converges:

$$\sum_{n=2}^{\infty} \frac{(-1)^n}{\ln n}.$$

Integral test

If $f(x)$ is a bounded, real, decreasing function for $x \geq 1$, the series

$$s_n = \sum_{k=1}^{n} f(k)$$

is bounded by the inequalities

$$\int_{1}^{n+1} dx\, f(x) \leq s_n \leq f(1) + \int_{1}^{n} dx\, f(x). \tag{4.16}$$

We can interpret this geometrically by observing that the sum s_n represents two different areas.

Consider the bound on s_5. The lower bound is obtained by interpreting s_5 as the area under the rectangles shown in figure 4.1; the area under the smooth curve from $x = 1$ to $x = 6$ must be less than (or equal to) the area of these rectangles. The upper bound is found from the area of the rectangles shown in figure 4.2: the area under these is clearly less than (or equal to) the area under the smooth curve from $x = 1$ to $x = 5$ plus the area of the first rectangle. We could consider the area under the curve from

$x = 0$ to $x = 5$, but as $f(x)$ is frequently infinite at $x = 0$, for example if $f(n) = n^{-2}$, this may result in an undefined integral.

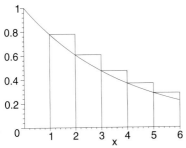

Figure 4.1   Lower bound of sum.

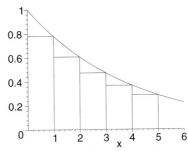

Figure 4.2   Upper bound of sum.

It may be shown that the sequence s_n and the integral $\int_1^X dx\, f(x)$ both converge or both diverge as $n \to \infty$ and $X \to \infty$ respectively. In either case,

$$\lim_{n \to \infty} \left(s_n - \int_1^n dx\, f(x) \right) \quad \text{exists.}$$

Exercise 4.22
By setting $f(x) = x^{-\alpha}$ show that the series

$$s_n = \sum_{k=1}^n \frac{1}{k^\alpha}$$

converges as $n \to \infty$ if $\alpha > 1$ and diverges otherwise.

By putting $\alpha = 1$, show that the nth partial sum of the harmonic series given in equation 4.6 (page 138) satisfies the inequalities

$$\ln(n + 1) \le s_n \le 1 + \ln n.$$

Deduce that the series representation of the zeta function,

$$\zeta(z) = \sum_{k=1}^\infty k^{-z},$$

converges if $\Re(z) > 1$ and that the harmonic series diverges as $\ln n$.

Exercise 4.23
Show that the series

$$S = \sum_{n=2}^\infty \frac{1}{n(\ln n)^\alpha}$$

converges if $\alpha > 1$ and in this case

$$\frac{1}{(\alpha - 1)(\ln 2)^{\alpha-1}} \le S \le \frac{1}{2(\ln 2)^\alpha} + \frac{1}{(\alpha - 1)(\ln 2)^{\alpha-1}}.$$

Leibniz's criterion

The infinite alternating series

$$S = \sum_{k=1}^{\infty}(-1)^k a_k$$

formed from the sequence of real numbers a_k, which, for sufficiently large k, are positive and decreasing, converges provided $\lim_{k\to\infty} a_k = 0$. The proof of this result can be found in Rudin (1976, page 71).

It can also be shown that

$$S < S_n + a_{n+1},$$

where S_n is the nth partial sum, showing that the error in truncating the series after n terms is less than the first ignored term, a_{n+1}.

Use of this criterion shows that the alternating harmonic series

$$S = 1 - \frac{1}{2} + \frac{1}{3} - \frac{1}{4} + \cdots + \frac{(-1)^{n-1}}{n} + \cdots$$

converges as does $\sum_{n=1}^{\infty}(-1)^{n-1}n^{-\alpha}$ for $\alpha > 0$.

4.3.4 The Euler–Maclaurin expansion

The integral test discussed above uses integrals to bound sums; a more powerful and useful result, with which a sum can be approximated by an integral, is obtained using the Euler–Maclaurin expansion:

$$\int_a^{a+x} dz\, f(z) \;=\; \frac{1}{2}x\,[f(a) + f(a+x)] \tag{4.17}$$

$$-\sum_{k=1}^{N} \frac{B_{2k}x^{2k}}{(2k)!}\left[f^{(2k-1)}(a+x) - f^{(2k-1)}(a)\right] + R_{N+1},$$

where $R_{N+1} = O(x^{2N+2})$ is a remainder term and B_k are the Bernoulli numbers, some of which are

$$B_2 = \frac{1}{6}, \quad B_4 = -\frac{1}{30}, \quad B_6 = \frac{1}{42}, \quad B_8 = -\frac{1}{30}.$$

These numbers are small, but $|B_{2n}| > 1$ for $n = 7$ and increases very rapidly for larger n. The values of B_k may be obtained using the Maple command `bernoulli(k)`, and typing `?bernoulli` takes you to the help file containing information about these numbers; their definition and some of their properties are given in the introduction to chapter 5. A proof of equation 4.17, together with an expression for the remainder, R_N, may be found in Whittaker and Watson (1965, page 127), but note this book uses a different notation for B_k from that used here.

On putting $x = 1$ and $a = r$, for some integer r, the expression 4.17 becomes

$$\int_r^{r+1} dz\, f(z) \;=\; \frac{1}{2}\,[f(r) + f(r+1)]$$

$$-\sum_{k=1}^{N} \frac{B_{2k}}{(2k)!}\left[f^{(2k-1)}(r+1) - f^{(2k-1)}(r)\right] + R_{N+1}.$$

Now form $m - n$ similar expressions by putting $r = n, n + 1, \ldots, m - 1$, and add these together to give

$$\int_n^m dz\, f(z) = \sum_{r=n}^m f(r) - \frac{1}{2}\left[f(n) + f(m)\right] - \sum_{k=1}^N \frac{B_{2k}}{(2k)!}\left[f^{(2k-1)}(m) - f^{(2k-1)}(n)\right],$$

where we have ignored the remainder terms. Finally rearrange this to give the following approximation to the sum:

$$\sum_{k=n}^m f(k) = \int_n^m dz\, f(z) + \frac{1}{2}\left[f(n) + f(m)\right]$$

$$- \sum_{k=1}^N \frac{B_{2k}}{(2k)!}\left[f^{(2k-1)}(n) - f^{(2k-1)}(m)\right] + R'_{N+1}, \qquad (4.18)$$

R'_{N+1} being the remainder. Since the B_k increase rapidly and without bound† as k increases, this approximation is generally useful only for small values of N, unless the sum terminates as when $f(z)$ is a polynomial.

The Euler–Maclaurin expansion may also be used to obtain numerical estimates of integrals. If $x_k = x_0 + kh$, $k = 0, 1, \ldots, N$, repeated application of equation 4.17 with $x = h$ gives

$$\int_{x_0}^{x_N} dx\, f(x) = h\left\{\frac{1}{2}f(x_0) + f(x_1) + f(x_2) + \cdots + f(x_{N-1}) + \frac{1}{2}f(x_N)\right\}$$

$$+ \sum_{k=1}^M \frac{B_{2k}h^{2k}}{(2k)!}\left(f^{(2k-1)}(x_N) - f^{(2k-1)}(x_0)\right) + R_{2M}, \qquad (4.19)$$

where $R_{2M} = O(Nh^{2M+3})$. Note that if $f(x)$ is periodic, with period $x_N - x_0$ and $f^{(p)}(x_0) = f^{(p)}(x_N)$, for all p, then all terms in the second sum are zero: in this case the remainder is smaller than R_{2M}, for the reasons discussed in section 8.7.

Exercise 4.24
Use equation 4.18 to find expressions for

$$\sum_{k=1}^m k^2 \quad \text{and} \quad \sum_{k=1}^m k^5.$$

Exercise 4.25
Write a Maple procedure to use the Euler–Maclaurin expansion, equation 4.18, to find an approximation to the sum

$$S = \sum_{k=n}^m f(k),$$

assuming that $f(x)$ is a decreasing function and the integral can be expressed as a closed formula. Test your procedure in the cases $f(k) = k$ and k^2.

† An approximation for B_{2n} valid in the large n limit is given on page 179.

Use this procedure to evaluate an approximation to the sum

$$S = \sum_{k=n}^{m} \frac{\ln k}{k^2}$$

using various values of N in equation 4.18.

Exercise 4.26

In the discussion of continued fractions, chapter 6, the infinite product

$$P = \prod_{k=1}^{\infty} \left(1 + \frac{1}{k(k+2)} \right)^{\frac{\ln k}{\ln 2}}$$

occurs. Show that if $S = \ln P$ then

$$S = \frac{1}{\ln 2} \sum_{k=1}^{\infty} \ln \left(1 + \frac{1}{k(k+2)} \right) \ln k$$

and show that this sum converges. Use the Euler–Maclaurin expansion to show that $P \simeq 2.685452$.

Hint: a direct application of the Euler–Maclaurin expansion provides poor accuracy because B_k increases too rapidly. This problem may be circumvented by writing

$$\sum_{k=1}^{\infty} f(k) = \sum_{k=1}^{n-1} f(k) + \sum_{k=n}^{\infty} f(k)$$

for some n, evaluating the first sum directly and using the Euler–Maclaurin expansion to estimate the second sum. Because $f^{(2k-1)}(n) \ll f^{(2k-1)}(1)$ this ensures that the sum over k in equation 4.18 is relatively small, provided N is not too large.

4.3.5 Power series

A series of the form

$$f(z) = a_0 + a_1 z + a_2 z^2 + \cdots + a_n z^n + \cdots, \tag{4.20}$$

in which the a_k, $k = 0, 1, \ldots$, are numbers, possibly complex, and are independent of the complex variable z, is called a power series: the number a_k is named the coefficient of z^k. Often we use a truncated form of power series and write this as

$$f(z) = a_0 + a_1 z + a_2 z^2 + \cdots + a_n z^n + O(z^{n+1}),$$

to show that the true value of f differs from the first $n+1$ terms by, in this case, a quantity of order z^{n+1}.

Such series are of fundamental importance in applied mathematics as they may be used to represent many useful functions. From a numerical viewpoint, however, power series are not always the most useful representation of a function, though they frequently provide a useful starting point for other approximations.

The convergence of such series may depend upon the value of z, and we now show that if it converges for a particular value z_0 then it will converge absolutely for all values of z inside the circle $|z| = |z_0|$ in the complex plane.

Since $\sum_{k=0}^{\infty} a_k z_0^k$ converges, $|a_k z_0^k| \to 0$ as $k \to \infty$, so there is a number M, independent of k, such that $|a_k z_0^k| < M$ and

$$|a_k z^k| = \left| a_k z_0^k \left(\frac{z}{z_0} \right)^k \right| < M \left| \frac{z}{z_0} \right|^k.$$

Thus every term of the series $\sum_{k=0}^{\infty} |a_k z^k|$ is less than the corresponding term in the geometric series

$$M \sum_{k=0}^{\infty} \left| \frac{z}{z_0} \right|^k,$$

which is absolutely convergent provided $|z| < |z_0|$.

A circle of radius r which includes all the values of z for which the series converges, and excludes all values for which it diverges, is named the circle of convergence, and r is named the *radius of convergence*. The value of the radius of convergence may be shown to be given by either of the limits

$$r = \lim_{k \to \infty} \left| \frac{a_k}{a_{k+1}} \right|, \qquad r = \lim_{k \to \infty} |a_k|^{-1/k}, \qquad (4.21)$$

if they exist.

A power series will converge for all values of the variable z provided the coefficients a_k decay to zero fast enough; for example, the series representations of the basic trigonometric functions

$$\sin z = z - \frac{z^3}{3!} + \frac{z^5}{5!} + \cdots + \frac{(-1)^k (z)^{2k+1}}{(2k+1)!} + \cdots, \qquad (4.22)$$

$$\cos z = 1 - \frac{z^2}{2!} + \frac{z^4}{4!} + \cdots + \frac{z^{2k}(-1)^k}{(2k)!} + \cdots, \qquad (4.23)$$

converge over the whole complex plane, because the factorial function increases so rapidly.

On the other hand, power series associated with rational functions, for instance

$$\frac{1}{1-z} = 1 + z + z^2 + z^3 + \cdots + z^n + \cdots,$$

have a finite radius of convergence, unity in this case, given by the minimum distance between the origin and the zeros of the denominator.

Exercise 4.27
Can you think of any other common functions represented by power series which (*a*) converge for all x, (*b*) converge for x on a finite interval?

The radius of convergence may be zero; for instance, in the series

$$1 + 1!z + 2!z^2 + 3!z^3 + \cdots + n!z^n + \cdots$$

the ratio of successive terms is

$$\left| \frac{u_{k+1}}{u_k} \right| = (n+1)|z|,$$

and tends to infinity for any nonzero value of z, so, using the ratio test, the series is divergent except at the point $z = 0$.

It is important to note that a power series does not necessarily converge *on* the circumference of the circle of convergence. Consider the series

$$f(z) = \sum_{k=1}^{\infty} \frac{z^k}{k^s}, \quad s \text{ real},$$

whose radius of convergence is unity since

$$\left| \frac{a_k}{a_{k+1}} \right| = \left(\frac{1+k}{k} \right)^s \to 1 \text{ as } k \to \infty.$$

At the point $z = 1$ the series converges if $s > 1$ but diverges if $s \leq 1$: but at $z = -1$ the series converges when $s = 1$.

Exercise 4.28

Determine the radius of convergence of the series

$$y = \sum_{k=0}^{\infty} \frac{(k+1)^{k-1}}{k!} x^k.$$

Hint: the approximation $(1 - x/n)^n = e^{-x} + O(1/n)$ is useful.

Generally, a power series is unbounded at only a finite number of points on its circle of convergence, but the series

$$f(z) = 1 + z^2 + z^4 + z^8 + \cdots + z^{2^n} + \cdots \tag{4.24}$$

provides a counter example. Clearly this series converges for $|z| < 1$ and diverges as $z \to 1$. But since $f(z) = z^2 + f(z^2)$ the function is unbounded at the two points at which $z^2 = 1$: similarly $f(z) = z^2 + z^4 + f(z^4)$, so $f(z)$ is unbounded at the four points at which $z^4 = 1$. In general f is unbounded at any root of any of the equations

$$z = 1, \quad z^2 = 1, \quad z^4 = 1, \quad \ldots, \quad z^{2^n} = 1, \ldots$$

that is, at the points $z = \exp(2\pi i 2^n / p)$, $p = 1, 2, \ldots, 2^n$, $n = 0, 1, 2 \ldots$.

Exercise 4.29

Use Maple to compute the complex zeros of the functions

$$f(z, n) = 1 + z^2 + z^4 + z^8 + \cdots + z^{2^n}$$

for various values of n — use the Maple command `fsolve(f(z,n)=0,z,complex)` — and use `pointplot` to plot these points in the complex plane; you should also draw the graph of the unit circle on the same figure.

In addition examine the nature of the function on the boundary, $|z| = 1$, by plotting the graph of $\Re(f(\exp i\theta, n))$ for some values of n and $0 \leq \theta < 2\pi$.

Hint: to draw a circle of radius r centred at (a, b) use the commands

```
with(plottools):     c:=circle([a,b,],r):
```

and use the `display` command to plot c together with your plot containing the zeros.

The power series defined in equation 4.20 is an expansion about the origin because it comprises the powers z^k, $k = 1, 2, \ldots$, which all vanish at $z = 0$. The series, formed by expanding about the point z_0,

$$g(z) = a_0 + a_1(z - z_0) + a_2(z - z_0)^2 + \cdots + a_n(z - z_0)^n + \cdots$$

behaves similarly, but the circle of convergence is centred at $z = z_0$ and the series converges inside the circle $|z - z_0| < r$, where r is given by either limit of 4.21.

Taylor's series

Taylor's series are the source of many power series. If a real function $f(x)$ has a finite $(N+1)$th derivative in the interval $a < x < b$, provided the Nth derivative is continuous in $a \le x \le b$ and if $a \le x_0 \le b$, then

$$f(x) = f(x_0) + \sum_{k=1}^{N} \frac{f^{(k)}(x_0)}{k!}(x - x_0)^k + R_{N+1}, \quad a < x < b, \tag{4.25}$$

where the remainder term is

$$R_{N+1} = \frac{f^{(N+1)}(\xi)}{(N + 1)!}(x - x_0)^{N+1}$$

for some (unknown) ξ in the interval $a \le \xi \le b$. If we set $R_{N+1} = 0$ an Nth degree polynomial approximation to $f(x)$ is obtained; this is sometimes called the Taylor polynomial of degree N. The magnitude of $|R_{N+1}|$ provides an error estimate for this polynomial which can be found if a suitable bound on $|f^{N+1}(\xi)|$ over $[a, b]$ is known. Taylor polynomials are usually inefficient approximations, for reasons that will be discussed in chapter 8.

The Taylor's series for a function having infinitely many derivatives is obtained by taking the limit $N \to \infty$,

$$f(x) = \sum_{k=0}^{\infty} \frac{f^{(k)}(x_0)}{k!}(x - x_0)^k, \quad a < x_0, x < b. \tag{4.26}$$

The necessary and sufficient conditions for this series to exist and to sum to the function $f(x)$ is that $R_N \to 0$ as $N \to \infty$. In the case that $x_0 = 0$ the series 4.26 is sometimes called a Maclaurin's series.† Maclaurin's series for the sine and cosine functions are given in equations 4.22 and 4.23 (page 151).

C. Maclaurin
(1698–1746)

---

† The formal series in equation 4.26 was first published by Brook Taylor in 1715; the result obtained by putting $x_0 = 0$ was discovered by Stirling in 1717 and published by Maclaurin in 1742.

The Taylor series about $x = c$ to order N may be computed using the Maple command `taylor(f,x=c,N+1)`. For functions that possess a Taylor's series the command `series(f,x=c,N+1)` produces the same result.

A.-L. Cauchy
(1789–1857)

Cauchy noted that the series 4.26 may converge, but not necessarily to the original function $f(x)$; for example if

$$f(x) = \begin{cases} \exp(-1/x^2), & x \neq 0, \\ 0, & x = 0, \end{cases}$$

then it may be shown that $f^{(k)}(x)$ exists and $f^{(k)}(0) = 0$ for all $k > 0$. Thus the Taylor's series 4.26 for f exists and sums to zero. Similarly, the series for $g(x) = e^{-x} + f(x)$ converges to e^{-x} rather than $g(x)$. In this case the remainder term, R_N, defined to be the difference between $f(x)$ and the first N terms of the series, does not vanish with increasing N. Whilst such behaviour is rare, it is necessary to be aware of it.

Taylor's series also exist for the more restrictive class of complex functions which are analytic in the neighbourhood of a point a in the complex z-plane: if $\mathscr{C}$ is a circle, centred at a, which does not contain any singular points† of $f(z)$, then

$$f(z) = \sum_{k=0}^{\infty} \frac{f^{(k)}(a)}{k!}(z - a)^k. \tag{4.27}$$

The radius of convergence, r, of this series is the distance from a to the nearest singularity of $f(z)$ and the circle of convergence is the circle of radius r centred at $z = a$.

From this it follows that the radius of convergence of the Taylor's series of an analytic function about $x = a$ is the distance between a and the nearest singularity of $f(x)$. For example, the function $g(z) = 1/(1 + z^2)$ has poles at $z = \pm i$, so the Taylor's series about $x = 0$ for the real function

$$\frac{1}{1 + x^2} = 1 - x^2 + x^4 - x^6 + \cdots$$

can converge only for $|x| < 1$.

The similarity of the two series 4.26 and 4.27 can be a little misleading because a real function with infinitely many derivatives in an interval is not necessarily analytic. For example, the function

$$f(x) = \begin{cases} -1, & x \leq -\pi/2, \\ \sin x, & |x| < \pi/2, \\ 1, & x \geq \pi/2, \end{cases}$$

is infinitely differentiable for $|x| < \pi/2$ and its Taylor's series about the origin converges to f on this interval, but not elsewhere, though it converges to the analytic function $\sin x$ for all x.

---

† We shall not attempt to define the elusive term singularity, but hope that when used the context makes its meaning clear. For example, the function $f(z) = (a - z)^\alpha$ has a singularity at $z = a$, provided α is not a positive integer; generally most of the differentials of a function do not exist at a singularity.

Differentiation and integration of power series

A convergent power series may be differentiated and integrated term by term to give other convergent series having the same radius of convergence as the original series.

Thus if

$$f(z) = \sum_{k=0}^{\infty} a_k z^k$$

is a power series with radius of convergence r, then

$$f_1(z) = \sum_{k=1}^{\infty} k a_k z^{k-1} \quad \text{and} \quad f_2(z) = \sum_{k=0}^{\infty} \frac{a_k}{k+1} z^{k+1}$$

are also convergent series having the same radius of convergence, and for $|z| < r$, $f_1(z) = f'(z)$ and $f_2(z) = \int_0^z dw\, f(w)$.

Exercise 4.30

Use the definition

$$\tan^{-1} x = \int_0^x dt\, \frac{1}{1+t^2},$$

by expanding the integrand and integrating term by term, to derive Gregory's expansion for $\tan^{-1} x$,

$$\tan^{-1} x = x - \frac{x^3}{3} + \frac{x^5}{5} - \cdots = \sum_{k=0}^{\infty} \frac{(-1)^k}{2k+1} x^{2k+1}.$$

From this derive Gregory's series for π,

$$\frac{\pi}{4} = 1 - \frac{1}{3} + \frac{1}{5} - \frac{1}{7} + \frac{1}{9} + \cdots.$$

Exercise 4.31

Using the definition

$$\sin^{-1} x = \int_0^x dt\,(1-t^2)^{-1/2},$$

derive the series expansions

$$\sin^{-1} x = \sum_{k=0}^{\infty} \frac{(2k)!}{4^k k!^2} \frac{x^{2k+1}}{2k+1},$$

$$\sin^{-1}(1-x) = \frac{\pi}{2} - \sqrt{2x}\left(1 + \sum_{k=1}^{\infty} \frac{1.3.5.\cdots.(2k-1)}{4^k (2k+1)k!} x^k\right).$$

Exercise 4.32

Use Maple to find the first five nonzero terms of Taylor's series of the function

$$f(x) = \frac{1+x}{x^2}\left\{\frac{2+2x}{1+2x} - \frac{\ln(1+2x)}{x}\right\}.$$

4.3.6 Manipulation of power series

Power series may be manipulated in many ways, but without some computer assistance the amount of algebra required normally increases so rapidly that only the most intrepid are willing to go beyond the second-order term. For example,

$$\tanh x = x - \frac{x^3}{3} + \frac{2x^5}{15} - \frac{17x^7}{315} + \cdots \quad \text{and} \quad e^x = 1 + x + \frac{x^2}{2!} + \frac{x^3}{3!} + \cdots,$$

so we could compute the power series of $e^{\tanh x}$ simply by substituting the first series into each term of the second and collecting the coefficients of x^k for all the necessary values of k; it is relatively trivial to find the first few terms but the algebra rapidly becomes lengthy, tedious and error prone.

Even elementary operations such as multiplication, division and inversion of series, for which there are simple rules relating the series coefficients, are tedious to apply by hand. For example, the rule for multiplying the two series

$$S = \sum_{k=0}^{\infty} a_k \quad \text{and} \quad T = \sum_{k=0}^{\infty} b_k$$

is

$$ST = \sum_{k=0}^{\infty} c_k \quad \text{where} \quad c_n = \sum_{k=0}^{n} a_k b_{n-k}, \tag{4.28}$$

and for the division of these two series,

$$S/T = \sum_{k=0}^{\infty} c_k, \quad \text{where} \quad c_n = \frac{1}{b_0}\left(a_n - \sum_{j=0}^{n-1} c_j b_{n-j}\right), \quad b_0 \neq 0.$$

But such algebra is ideally suited to the computer and various packages have been developed to help. When using these methods it is important to remember that the computer manipulates series as algebraic objects with no regard to the convergence, or even the meaning, of the resulting series and it is incumbent on the user to determine the analytic properties of the results, see for instance exercise 4.69 (page 176).

We have already used the `series` command, which is a general purpose procedure for finding series; for Taylor's series the command `taylor`, having the same syntax, can also be used. There are more specialised procedures for creating and manipulating power series when formulae for the coefficients are known; here we provide a brief introduction to some of these facilities.

The `powseries` library package provides facilities to create and to manipulate power series, expanded about the origin, when a formula defining the coefficients is known; it needs to be invoked using the `with` command

> `with(powseries);`

[*compose, evalpow, inverse, multconst, multiply, negative, powadd, powcos, powcreate,*
    powdiff, powexp, powint, powlog, powpoly, powsin, powsolve, powsqrt, quotient,
    reversion, subtract, tpsform]

We used the semi-colon terminator in order to list the procedures available; typing
`?compose`, for instance, at the prompt will access the help file on `compose`.

In order to create a power series, $\sum_{k=0}^{N} f_k x^k$, it is necessary to define the coefficients
f_k; thus for the exponential function we put the coefficients in `f(k)` with the command

> `powcreate(f(k)=1/k!);`

and produce the required series with the `tpsform` command — short for truncated
power series **form** — which has a similar syntax to the `series` command,

> `e:=tpsform(f,x,6);`

$$e := 1 + x + \frac{1}{2}x^2 + \frac{1}{6}x^3 + \frac{1}{24}x^4 + \frac{1}{120}x^5 + O(x^6)$$

the variable e is of series type, so may be reduced to a polynomial using the command
`convert(e,polynom)` if necessary. However, Maple has a range of procedures which
manipulate these series without the need to convert. We consider some of these
procedures, but first need to define another series; the coefficients of the power series
of $\ln(1 + x)$ are

> `powcreate(g(n)=(-1)^(n-1)/n, g(0)=0);`

Note that in this case we added the value of `g(0)` as a special case because the formula
does not exist at $n = 0$; in general we may add as many special cases as required and
these take precedence over the formula. The series for $\ln(1 + x)$ is

> `L:=tpsform(g,x,6);`

$$L := x - \frac{1}{2}x^2 + \frac{1}{3}x^3 - \frac{1}{4}x^4 + \frac{1}{5}x^5 + O(x^6)$$

Various operations on these series, or rather the coefficients, may be performed to
create other series. Addition and multiplication, for example, are performed as follows:
addition is given by

> `a:=powadd(f,g): tpsform(a,x,8);`

$$1 + 2x + \frac{1}{2}x^3 - \frac{5}{24}x^4 + \frac{5}{24}x^5 - \frac{119}{720}x^6 + \frac{103}{720}x^7 + O(x^8)$$

and multiplication by

> `m:=multiply(f,g): tpsform(m,x,8);`

$$x + \frac{1}{2}x^2 + \frac{1}{3}x^3 + \frac{3}{40}x^5 - \frac{7}{144}x^6 + \frac{23}{504}x^7 + O(x^8)$$

More complicated operations may be performed on some series; for example $e^{-x} = 1/e^x$ and this is given by `inverse`.†

---

† There is potential confusion over the meaning of the word inverse. Here the inverse of $y = f(x)$ is the
function $1/f(x)$. In other circumstances it can mean the function $x = g(y)$, where $y = f(g(y))$; in the
context of **powseries** this latter operation is named **reversion**.

```
>  i:=inverse(f): tpsform(i,x,6);
```

$$1 - x + \frac{1}{2}x^2 - \frac{1}{6}x^3 + \frac{1}{24}x^4 - \frac{1}{120}x^5 + O(x^6)$$

but this operation will produce an error message when applied to the power series, L, for $\ln(1 + x)$, because $\ln(1 + x) = x - x^2/2 + \cdots$ and $1/\ln(1 + x)$ behaves as $1/x$ for small $|x|$, therefore cannot be represented as a power series.

Composition is performed using compose ; thus if $A(x)$ and $B(x)$ are two power series with coefficients $a(n)$ and $b(n)$, $n = 0, 1, \ldots$, the command compose(a,b) gives the coefficients of the function $C = A(B(x))$ when expressed as a series in x:

$$C(x) = a_0 + a_1 B(x) + a_2 B(x)^2 + \cdots.$$

Note that the constant term of this series is the infinite sum $\sum_{k=0}^{\infty} a_k b_0^k$, which may not exist, so a necessary condition for the use of compose(a,b) is that $b_0 = 0$, as then the constant term is guaranteed to exist. The series for $e^{\ln(1+x)}$ is clearly $1 + x$, and compose gives

```
>  c1:=compose(f,g): tpsform(c1,x,10);
```

$$1 + x + O(x^{10})$$

On the other hand, for the reason given above, the composition $\ln(1 + e^x)$ cannot be determined by this command, even though a convergent power series exists.

Other operators are described in the help file — under *Help, Full Text Search* type *power series* — but we mention only the useful reversion command which inverts power series provided the constant term is zero *and* the coefficient of x is unity; if

$$y = x + a_2 x^2 + a_3 x^3 + \cdots,$$

then x may be expressed as a similar power series in y. For instance, if $y = \ln(1+x)$ then $x = e^y - 1$ and the power series for this function is determined using the coefficients $g(n)$ for $\ln(1 + x)$ defined above:

```
>  r:=reversion(g): z:=tpsform(r,y,6);
```

$$z := y + \frac{1}{2}y^2 + \frac{1}{6}y^3 + \frac{1}{24}y^4 + \frac{1}{120}y^5 + O(y^6)$$

which is $e^y - 1$ as expected. More generally, if

$$y = a_0 + a_1 x + a_2 x^2 + a_3 x^3 + \cdots, \quad a_0 \neq 1, \ a_1 \neq 1,$$

we may define a related power series,

$$z = \frac{y - a_0}{a_1} = x + b_2 x^2 + b_3 x^3 + \cdots, \quad b_k = \frac{a_k}{a_1},$$

and proceed as before to determine x as a power series in z and hence $(y - a_0)$.

Exercise 4.33

If

$$y = x + x^2 + x^3 + \cdots + x^n + \cdots,$$

use Maple to show that

$$x = y - y^2 + y^3 + \cdots - (-y)^{n-1} + \cdots.$$

Verify this result by finding a closed expression for $y(x)$ and using conventional algebra.

Exercise 4.34

If

$$y = x + 2x^2 + 3x^3 + \cdots + nx^n + \cdots,$$

use Maple to show that

$$x = y - 2y^2 + 5y^3 - 14y^4 + 42y^5 + \cdots.$$

Find the radius of convergence of the series for $y(x)$ and use the coefficients obtained using `reversion` to estimate the radius of convergence of the series for $x(y)$ using the ratio test. The convergence of this ratio is very slow so do not attempt to find an accurate limit; we shall see how to improve the accuracy of this calculation in the next section.

Find a formula for $x(y)$ and the exact value of the radius of convergence of the series for $x(y)$.

Exercise 4.35

Use the Maple command `powsolve` to find a series solution to $O(x^{10})$ for the differential equation

$$\frac{d^2y}{dx^2} - xy = 0, \quad y(0) = 1, \; y(1) = b$$

and plot the graphs of these solutions, using the animate facility, for $b = 0.6$ to 1.6 in steps of 0.1.

Hint: remember that to animate graphs you need to put the graphs in a list, p say, and use the command `display(p,insequence=true)`, and that this needs the command `with(plots)`.

4.4 Accelerated convergence

4.4.1 Introduction

This chapter has been concerned mainly with the problem of determining whether or not a given sequence converges. This is important, but even if a series converges it may be a non-trivial task to find a numerical approximation to its limit if the convergence is very slow. An example of this very slow convergence was seen in exercise 4.34; another is the evaluation of the series

$$S = 1 - \frac{1}{2} + \frac{1}{3} - \frac{1}{4} + \cdots$$

which requires about 6000 and 200 000 terms to obtain an approximation accurate to four and six decimal places respectively, when the sum is evaluated directly.

In this section we describe two common methods for speeding the convergence of sequences. These are not the only methods, and often the application of some ingenuity and the use of known results will help. A simple example is the series for $\ln(1+x)$,

$$\ln(1+x) = x - \frac{x^2}{2} + \frac{x^3}{3} + \cdots = -\sum_{n=1}^{\infty} \frac{(-x)^n}{n}.$$

For x close to unity this converges as n^{-1} which is too slow for practical use. Multiply both sides of the equation by $(1+ax)$ and rearrange the series to the form

$$(1+ax)\ln(1+x) = x + \sum_{n=2}^{\infty} \frac{(-x)^n}{n(n-1)} \{1 - n(1-a)\}.$$

Now choose $a = 1$ so that the term proportional to n inside the curly brackets of the numerator is zero; this gives

$$(1+x)\ln(1+x) = x + \sum_{n=2}^{\infty} \frac{(-x)^n}{n(n-1)}.$$

Now the series converges as n^{-2}. In this example the special form of the series and a little ingenuity were used to obtain more rapid convergence; clearly this method can be extended as shown in exercise 4.70 (page 176). In general, however, we require methods that do not depend sensitively on special features of the series.

In this section two common extrapolation techniques are introduced. The aim is to construct a new sequence converging faster than the original sequence, but to the same limit. These methods have a long history and have important applications in a variety of areas, for instance the solution of ordinary differential equations, see Press *et al.* (1987, section 15.4). A brief, but interesting, history of extrapolation methods and their connection with Padé approximants, the subject of chapter 6, may be found in Brezinski (1981, 1985 and 1996).

4.4.2 *Aitken's Δ^2 process and Shanks' transformation*

Suppose that the dominant behaviour of the nth term of a sequence A_n is described by the rule

$$A_n \simeq A + \alpha q^n, \quad |q| < 1, \quad \left(\text{thus} \quad \lim_{n\to\infty} A_n = A \right), \tag{4.29}$$

where q and α are (unknown) numbers, independent of n. Three adjacent values of A_n can be used to determine the three constants A, α and q. If we use A_n, A_{n+1} and A_{n+2} the expression for A is

$$A = \frac{A_{n+2}A_n - A_{n+1}^2}{A_{n+2} - 2A_{n+1} + A_n} = A_{n+1} - \frac{(A_{n+2} - A_{n+1})(A_{n+1} - A_n)}{(A_{n+2} - A_{n+1}) - (A_{n+1} - A_n)}, \tag{4.30}$$

the latter form being preferable, being less prone to rounding errors.

Consider, for example, the geometric series

$$A_n = \sum_{k=0}^{n} (-z)^k = \frac{1 - z^{n+1}}{1 + z},$$

which has the same form as equation 4.29 with $A = 1/(1+z)$, $\alpha = -z/(1+z)$ and $q = z$: we see that

$$A = \frac{A_{n+2}A_n - A_{n+1}^2}{A_{n+2} - 2A_{n+1} + A_n} = \frac{1}{1+z}, \quad \text{for all } n,$$

which is the exact value of $\lim_{n\to\infty} A_n$ because, in this example, the form assumed in equation 4.29 for A_n is exact.

In general, given a sequence A_k, the transformation

$$B_n = S(A_n) = A_{n+1} - \frac{(A_{n+2} - A_{n+1})(A_{n+1} - A_n)}{(A_{n+2} - A_{n+1}) - (A_{n+1} - A_n)}, \tag{4.31}$$

defines a new sequence, B_k, which in many (but not all) cases approaches the same limit as the original sequence but faster. The transformation $S(A_n)$ was given by Aitken in 1926 and is normally referred to as *Aitken's Δ^2 process*.† Aitken used the transformation to determine approximations to the dominant root of a polynomial, but noted that von Nagelsbach used the same method when approximating the roots of nonlinear equations in 1876. Maxwell in 1892 also used the technique for extrapolating experimental data, and the Japanese mathematician Kowa used it to compute the value of π to 11 decimal places. The transformation $S(A_n)$ is sometimes called *Shanks' transformation* after Shanks, who generalised equation 4.31, see equation 4.32 below.

A. C. Aitken (1895–1967)

J. C. Maxwell (1831–1879)

S. Kowa (c. 1642–1708)

Exercise 4.36
Write a Maple procedure to find the nth partial sum

$$A_n = 1 - \frac{1}{2} + \frac{1}{3} + \cdots + \frac{(-1)^{n-1}}{n}, \quad \left(\lim_{n\to\infty} A_n = \ln 2 = 0.693147\right).$$

Write another procedure which gives a new sequence $A_n^{(1)} = S(A_n)$, equation 4.31. Using these procedures show that $A_{10} = 0.6456$ and $A_{10}^{(1)} = 0.693066$, having errors of 7% and 0.012% respectively.

Now apply the transform to the sequence $A_n^{(1)}$ to obtain another sequence $A_n^{(2)} = S(A_n^{(1)}) = S^2(A_n)$, and then $A_n^{(3)} = S^3(A_n)$, to show that

$$1 - \frac{A_{10}^{(2)}}{\ln 2} = 4.3 \times 10^{-7}, \quad 1 - \frac{A_{10}^{(3)}}{\ln 2} = 5.7 \times 10^{-9}.$$

Recall that A_{6000} is accurate to only four figures.

Note that in this case $S(A_n) = A_{n+1} - (-1)^n/(2n+3)$.

Exercise 4.37
Show that the transformation defined in equation 4.31 applied to a sequence satisfying the two-term linear recurrence relation

$$a_0(A_n - A) + a_1(A_{n+1} - A) = 0$$

gives $S(A_n) = A$, provided $a_0 a_1 \neq 0$ and $a_0 + a_1 \neq 0$.

By putting $A_n = A + c\beta^n$, show also that the general solution of the recurrence relation is

$$A_n = A + \left(-\frac{a_0}{a_1}\right)^n (A_0 - A).$$

---

† In numerical analysis Δ denotes the forward difference operator, $\Delta(f_n) = f_{n+1} - f_n$, so the denominator of $S(A_n)$ is $\Delta^2(A_n)$.

Shanks (1955) discussed a generalisation of Aitken's Δ^2 process for which a sequence satisfying the m-term linear recurrence relation

$$a_0(A_n - A) + a_1(A_{n+1} - A) + \cdots + a_m(A_{n+m} - A) = 0, \quad \text{for all } n,$$

where $a_0 a_m \neq 0$ and $a_0 + a_1 + \cdots + a_m \neq 0$, is reduced to a number independent of n. The transformation is

$$S^{(m)}(A_n) = \frac{\det\left(N^{(m)}(A_n)\right)}{\det\left(D^{(m)}(A_n)\right)}, \tag{4.32}$$

where the numerator and denominator are respectively the $m \times m$ and $(m-1) \times (m-1)$ square matrices

$$N^{(m)} = \begin{pmatrix} A_n & A_{n+1} & \cdots & A_{n+m} \\ A_{n+1} & A_{n+2} & \cdots & A_{n+m+1} \\ \vdots & \vdots & & \vdots \\ A_{n+m} & A_{n+m+1} & \cdots & A_{n+2m} \end{pmatrix}, \quad D^{(m)} = \begin{pmatrix} \Delta^2 A_n & \cdots & \Delta^2 A_{n+m-1} \\ \vdots & & \vdots \\ \Delta^2 A_{n+m} & \cdots & \Delta^2 A_{n+2m-2} \end{pmatrix},$$

and Δ is the forward difference operator:

$$\Delta A_n = A_{n+1} - A_n, \quad \Delta^2 A_n = \Delta(\Delta A_n) = A_{n+2} - 2A_{n+1} + A_n.$$

Notice that the evaluation of $S^{(m)}(A_n)$ requires the values of $(A_n, A_{n+1}, \cdots, A_{n+m})$.

In 1956 Wynn showed that the successive values of $S^{(m)}(A_n)$, $m = 1, 2, 3, \cdots$ could be obtained by iterating the sequence

$$T_n^{(k+1)} = T_{n+1}^{(k-1)} + \frac{1}{T_{n+1}^{(k)} - T_n^{(k)}}, \quad \text{where} \tag{4.33}$$

$$T_n^{(-1)} = 0, \quad T_n^{(0)} = S(A_n) = S^{(1)}(A_n),$$

since $S^{(m)}(A_n) = T_n^{(2m)}$. This is important because it allows evaluation of $S^{(m)}$ without computing determinants, which can be sensitive to rounding errors.

Exercise 4.38
Show that $S^{(1)}(A_n) = S(A_n)$, where S is defined in equation 4.31.

Exercise 4.39
Apply the algorithm defined in equation 4.33 to the sequence A_n defined in exercise 4.36 and show that

$$1 - \frac{S^{(1)}(A_{10})}{\ln 2} = 4.3 \times 10^{-7}, \qquad 1 - \frac{S^{(2)}(A_{10})}{\ln 2} = 3.1 \times 10^{-9},$$

$$1 - \frac{S^{(3)}(A_{10})}{\ln 2} = 3.2 \times 10^{-11}, \qquad 1 - \frac{S^{(4)}(A_{10})}{\ln 2} = 4.2 \times 10^{-13}.$$

Exercise 4.40
By putting $A_n = A + c\beta^n$ show that the general solution of the three-term recurrence relation

$$a_0(A_n - A) + a_1(A_{n+1} - A) + a_2(A_{n+2} - A) = 0$$

is

$$A_n = A + c_1\beta_1^n + c_2\beta_2^n, \quad \beta_{1,2} = \frac{-a_1 \pm \sqrt{a_1^2 - 4a_0a_2}}{2a_2},$$

where c_k, $k = 1$ and 2, depend upon the initial conditions.

4.4.3 Richardson's extrapolation

The Aitken's process and Shanks' method assume that the sequence A_n tends to its limit A in a particular way, equation 4.29, so it will fail or be less efficient if the elements of the sequence depend upon n in other ways. Richardson's extrapolation determines an approximation to the limit when

$$A_n \simeq A + \frac{B_1}{n} + \frac{B_2}{n^2} + \frac{B_3}{n^3} + \cdots \quad \text{for large } n. \tag{4.34}$$

This type of convergence is common when computing the radius of convergence using the ratio test, page 145, as frequently $|a_{n+1}/a_n| = A + O(n^{-1})$. For sequences having this dependence upon n Aitken's Δ^2 process is not helpful, as shown in the next exercise.

Exercise 4.41
Show that if

$$A_n = A + \frac{B_1}{n} + \frac{B_2}{n^2} + O(n^{-3})$$

Aitken's Δ^2 process applied to A_n gives

$$S(A_n) = A + \frac{B_1}{2n} + \frac{B_2}{2n^2}.$$

In this case the original and transformed sequence converge at the same rate.

If A_n behaves as in equation 4.34, we assume that it may be expressed as a polynomial in n^{-1}, and write it in the form

$$A_n = A + \frac{B_1}{n} + \frac{B_2}{n^2} + \frac{B_3}{n^3} + \cdots + \frac{B_N}{n^N} \tag{4.35}$$

for some N, and then use $N+1$ values $(A_n, A_{n+1}, \ldots, A_{n+N})$ to determine the coefficient A.

Consider first the simple case $N = 1$, which leads to the two equations

$$\begin{aligned} A_n &= A + B_1/n \\ A_{n+1} &= A + B_1/(n+1), \end{aligned}$$

which may be solved for A and B_1,

$$A = (n+1)A_{n+1} - nA_n, \quad B_1 = n(n+1)(A_n - A_{n+1}). \tag{4.36}$$

In general, writing A_n as an Nth degree polynomial in n^{-1} necessitates the solution of the $N+1$ equations

$$A_n = A + \frac{B_1}{n} + \frac{B_2}{n^2} + \cdots + \frac{B_N}{n^N},$$

$$A_{n+1} = A + \frac{B_1}{n+1} + \frac{B_2}{(n+1)^2} + \cdots + \frac{B_N}{(n+1)^N},$$

$$\vdots$$

$$A_{n+N} = A + \frac{B_1}{n+N} + \frac{B_2}{(n+N)^2} + \cdots + \frac{B_N}{(n+N)^N}. \tag{4.37}$$

These equations look unpleasant, but there is an elegant solution for A, and B_1, derived in exercises 4.74 to 4.76 (page 177). The solution for A is

$$A = \sum_{k=0}^{N} (-1)^{k+N} \frac{A_{n+k}(n+k)^N}{k!(N-k)!}. \tag{4.38}$$

The values of B_k, $k = 1, 2, \ldots, N$ are of no importance here. Equation 4.38 is the basic formula of Richardson's extrapolation method.

Because A_k are positive for sufficiently large k, the sum 4.38 is oscillatory, and if n and N are large there may be severe cancellation when evaluating the sum numerically. In practice it is therefore often best to use relatively small values of N.

Exercise 4.42

Consider the series

$$A = 1 + \frac{1}{2^2} + \frac{1}{3^2} + \frac{1}{4^2} + \cdots = \frac{\pi^2}{6}.$$

Use the integral test of equation 4.16 (page 146) to infer that the nth partial sum behaves as $A_n = A + O(n^{-1})$ and use equation 4.36 to derive the results listed in the following table.

Table 4.1. *Table of the partial sums A_n and the estimate A, given by equation 4.36, as a ratio of the exact limit, $\pi^2/6$.*

n	$6A_n/\pi^2$	$6A/\pi^2$
10	0.9421	0.9974
15	0.9608	0.9988
20	0.9704	0.9993

Exercise 4.43

Use equation 4.38 to estimate the ratio $6A_n/\pi^2$, where A_n is defined in exercise 4.42, with $n = 10$ and $N = 1, 2, \ldots, 6$.

Exercise 4.44

Consider the series

$$S = \sum_{k=1}^{\infty} \frac{1}{k^2\sqrt{k+1}}$$

and show that $S_n = S + O(n^{-3/2})$.

Despite this behaviour, compute the estimate of S given by equation 4.38 for $n = 10$, $N = 1, 2, \ldots, 10$ and $n = 20$, $N = 1, 2, \ldots, 10$. Describe the behaviour of your results.

Can you think of a way to approximate this series using ζ-functions?

Exercise 4.45

In exercise 4.34 you found the inverse of the series

$$y = x + 2x^2 + 3x^3 + \cdots + nx^n + \cdots$$

using the Maple command `reversion` to determine the coefficients of the power series representing $x(y)$. Use these together with the ratio test and Richardson's extrapolation to compute the radius of convergence of this series.

Repeat this exercise for the series

$$y = x + 2^\alpha x^2 + 3^\alpha x^3 + \cdots + n^\alpha x^n + \cdots$$

for $\alpha = 2, 3, 4$ and some non-integer values of α in between.

The standard application of Richardson's method assumes that the nth member of the sequence has an expansion like that in equation 4.34, and then the limit is approximated by equation 4.38. If different functional forms are assumed the equation for the limit cannot necessarily be expressed as a neat formula. But there are circumstances in which the functional form given in equation 4.34 is not appropriate, as we saw in exercise 4.44. The following exercise explores a modification of Richardson's method.

Exercise 4.46

Consider the series

$$S = \sum_{k=2}^{\infty} \frac{\ln k}{k^2}.$$

If S_n is the nth partial sum, use the integral test to show that

$$\frac{1}{2}(1 + \ln 2) - \frac{1 + \ln(n+1)}{n+1} \le S_n \le \frac{1}{2}\left(1 + \frac{3}{2}\ln 2\right) - \frac{1 + \ln n}{n},$$

and to infer that

$$S_n = S + a\frac{\ln n}{n} + \frac{b}{n} + \text{higher order terms},$$

where a and b are unknown constants. Find an equation for S in terms of S_n, S_{n+1} and S_{n+2} and use this with $n = 500$ and 1000 to show that $S \simeq 0.9375$. How many terms of the original series would be needed to obtain this accuracy?

It is difficult to extend this method further if one relies upon algebraic methods of solving the equations for S, but since the equations are linear in the unknown constants we can easily solve for S numerically. This option is explored in exercise 4.73, (page 177).

4.5 Properties of power series

Equation 4.21 (page 151) shows that the radius of convergence of a power series can be deduced from the limiting behaviour of the coefficients. It is natural to ask whether or not other properties of the function represented by the power series can be inferred from the behaviour of these coefficients. Here we show how the nature of the singularity on the circle of convergence can sometimes be deduced from the behaviour of these coefficients. The method is not infallible, but works well if the function is not too complicated — in a sense which will become clear after we have considered a few examples. Further discussion of the methods discussed here may be found in Hinch (1991, chapter 8).

Consider the power series given by the binomial expansion

$$\left(1 + \frac{x}{a}\right)^{\alpha} = \sum_{n=0}^{\infty} c_n x^n, \quad a^n c_n = \binom{\alpha}{n} = \frac{\Gamma(1 + \alpha)}{\Gamma(1 + n)\Gamma(1 + \alpha - n)}. \tag{4.39}$$

If α is a positive integer, $\alpha = m > 0$, then

$$a^n c_n = \binom{m}{n} = \frac{m!}{n!(n - m)!}, \quad 0 \le n \le m,$$

and the series terminates at $n = m$. We assume that α is not a positive integer, so the function has a singularity[†] at $x = -a$; it is convenient to call α the power of this singularity. Since the radius of convergence is clearly a, we must have $\lim_{n\to\infty} |c_n/c_{n-1}| = 1/a$, but since the coefficients depend upon α we expect the rate at which this limit is reached to depend upon α (and n). The ratio of coefficients is

$$\frac{c_n}{c_{n-1}} = \frac{1}{a} \frac{\Gamma(n)}{\Gamma(n+1)} \frac{\Gamma(2 + \alpha - n)}{\Gamma(1 + \alpha - n)};$$

using a basic property of the Gamma function, $\Gamma(z + 1) = z\Gamma(z)$, this ratio becomes

$$\frac{c_n}{c_{n-1}} = -\frac{1}{a}\left(1 - \frac{1 + \alpha}{n}\right). \tag{4.40}$$

The power of the singularity on the circle of convergence, α, arises in the second term.

Now suppose that we have a function of the form $f(x) = g(x)(1 + x/a)^{\alpha}$, where g is some function having a power series expansion with radius of convergence exceeding a. Since f has the same type of singularity as the series 4.39 at $x = a$, we expect the coefficients, d_n, of the power series expansion of f to behave in a similar manner to c_n, provided n is large enough and that the singularities of g do not interfere with that at $x = -a$. We can test this idea by using Maple to compute the ratio d_n/d_{n-1} using simple examples.

Exercise 4.47
Consider the power series expansion of the functions

(i) $\quad f(x) = (a + x)^{0.33}(1 + x)^{0.79}, \quad a = 3, 2, 1.5, 1, 0.9,$ and 0.5,
(ii) $\quad f(x) = (1 + x)^{\mu} \sin x, \quad \mu = 0.76, -0.33, -2.89,$ and -4.

Write a Maple procedure that computes the coefficients, d_n, of the Taylor's series of a given function and creates a sequence of the ratios d_n/d_{n-1}, $n = 1, 2, \ldots, N$, then use Richardson's extrapolation to deduce the radius of convergence, r, of the series.

Form a second sequence,

$$\alpha_n = n\left(1 - r\left|\frac{d_n}{d_{n-1}}\right|\right) - 1,$$

and find an approximation to $\alpha = \lim_{n\to\infty} \alpha_n$.

---

† We use the term 'singularity' rather loosely here; if α is negative the function is infinite at $x = -a$; if α is positive and not an integer, derivatives of the function are infinite here.

In the following table is shown the results of the calculation for the first function: these values were computed using 30 terms of the series expansion and Richardson's extrapolation with five terms.

a	0.5	0.6	0.7	0.8	0.9
r	0.5000	0.5999	0.7003	0.8004	0.9079
α	0.3315	0.3455	0.2890	0.2311	−0.6636
a	1.0	1.5	2.0	3.0	
r	1.0000	0.9989	1.0000	1.0000	
α	1.12	0.9697	0.7808	0.7891	

If $|a| < 1$ the function $f(x) = (a + x)^{0.33}(1 + x)^{0.79}$ is, for small $|x|$, dominated by the $(a + x)^{0.33}$ factor, so has radius of convergence a and the order of the singularity is $\alpha = 0.33$. If $a \gg 1$ the factor $(1 + x)^{0.79}$ dominates and the radius of convergence is 1 and the order of the singularity is 0.79. The numerical results, shown in the table, agree with this simple analysis; also if $a = 1$, $f(x) = (1 + x)^{1.12}$ and the radius of convergence is 1 and the order of the singularity is 1.12, as computed. But if $a \simeq 1$ the two singularities in $f(x)$ are close and the analysis leading to the estimate of the singularity order is invalid.

To summarise, we have found that in some circumstances the limiting behaviour of the ratio c_n/c_{n-1} of the coefficients of a power series provides valuable information about the nature of the singularity of the series. The first example treated in exercise 4.47 suggests that the method is accurate provided the singularities of the function are not too close.

The next exercise shows how useful this method can be, but it also shows that when $\alpha = -1$, equation 4.40, it fails; after this exercise we shall remedy this defect.

Exercise 4.48
Consider the function $f(x)$ defined by the integral

$$f(x) = \int_0^1 dt \, \frac{1}{(1 + xt^2)^\beta}.$$

By expanding the integrand as a power series and integrating term-by-term, show that the power series of $f(x)$ is

$$f(x) = \frac{1}{\Gamma(\beta)} \sum_{n=0}^\infty \frac{(-1)^n \Gamma(n + \beta) x^n}{(2n + 1)n!}.$$

Show that its radius of convergence is unity and that $f(x)$ has a singularity of order

$\alpha = 1 - \beta$. Hence deduce that for positive non-integer values of β,

$$f(x) \simeq \frac{A}{(1+x)^{\beta-1}} \quad \text{for } x \text{ near } -1.$$

Can you find a method of checking this result?

In the cases $\beta = 1$ and 2 evaluate the integral algebraically to show that

$$f(x) \quad \simeq \quad \frac{1}{2} \ln\left(\frac{4}{1+x}\right) \quad \text{for } x \simeq -1 \quad \text{and} \quad \beta = 1,$$

$$f(x) \quad \simeq \quad \frac{1}{2(1+x)} + O(\ln(1+x)) \quad \text{for } x \simeq -1 \quad \text{and} \quad \beta = 2,$$

and compare this behaviour with that suggested by equation 4.40.

The last exercise illustrates the utility of the method, but it also shows that it fails if $\beta = 1$, when the integral diverges as $\ln(1+x)$ for $x \simeq -1$.

Consider the expansion of the functions

$$g(x) = (1-x)^m \ln(1-x), \quad m = 0, 1, 2, \dots . \tag{4.41}$$

The power series expansion of this function clearly has unit radius of convergence and is given by multiplying the series for $(1-x)^m$ and $\ln(1-x)$,

$$g(x) = -\left[\sum_{r=0}^{m} \binom{m}{r} (-x)^r \right] \sum_{k=1}^{\infty} \frac{x^k}{k}$$

to give the coefficient, c_n, of x^n as the finite series

$$c_n \quad = \quad \sum_{r=0}^{m} \frac{(-1)^r}{n-r} \binom{m}{r} \quad \text{if} \quad n > m$$

$$= \quad \frac{1}{n} \sum_{r=0}^{m} (-1)^r \binom{m}{r} \left(1 + \frac{r}{n} + \frac{r^2}{n^2} + \cdots \right).$$

Now define S_p to be the series

$$S_p = \sum_{r=0}^{m} (-1)^r r^p \binom{m}{r},$$

then $S_p = 0$ for $0 \le p < m$ and $S_m = (-1)^m m!$, exercise 4.74 (page 177), so

$$c_n = \frac{(-1)^m m!}{n^{m+1}} + O(n^{-m-2}) \quad \text{and} \quad \frac{c_n}{c_{n-1}} \simeq \left(\frac{n-1}{n}\right)^{m+1} = \left(1 - \frac{m+1}{n} + \cdots \right). \tag{4.42}$$

This analysis suggests that if the limiting form of the ratio c_n/c_{n-1} behaves as in equation 4.40 with $\alpha = 0$, then the singularity may have the form $\ln(1-x/a)$. Applying this result to the problem in exercise 4.48 with $\beta = 1$, we see that $m = 0$ and the suggested behaviour of f is $\ln(1+x)$ for x near -1, as given by the exact evaluation.

Exercise 4.49
The complete elliptic integrals of the first kind is defined by the integral

$$K(x) = \int_0^{\pi/2} d\theta \, \frac{1}{\sqrt{1 - x^2 \sin^2 \theta}},$$

and it is clear that $K(x) \to \infty$ as $x^2 \to 1$. Show that for $x^2 \simeq 1$

$$K(x) \simeq -A \ln(1 - x^2)$$

for some constant A.

Hint: you will need the integral

$$\int_0^{\pi/2} d\theta \, \sin^{2a} \theta = \frac{\sqrt{\pi}}{2} \frac{\Gamma(a + 1/2)}{\Gamma(a + 1)}, \quad a > -\frac{1}{2}.$$

4.6 Uniform convergence

In this section we extend our discussion of infinite series to include those with elements which are functions of a real variable x; the power series discussed in section 4.3.5 are examples of such series. Specifically, we consider the infinite series

$$u_1(x) + u_2(x) + \cdots + u_n(x) + \cdots = \sum_{k=1}^{\infty} u_k(x), \qquad (4.43)$$

where each $u_k(x)$ is a differentiable function of x in some interval $[a, b]$. From this series we form a sequence of functions, $f_n(x)$, comprising the partial sums

$$f_n(x) = u_1(x) + u_2(x) + \cdots + u_n(x),$$

and for every $a \le x \le b$ *define* the function $f(x)$ in terms of the limit of this sequence,

$$f(x) = \lim_{n \to \infty} f_n(x), \quad a \le x \le b. \qquad (4.44)$$

With this definition we say that the sequence $f_n(x)$ converges *pointwise* to $f(x)$. The main problem is to determine whether the properties of the members of this sequence, for example continuity and differentiability, are preserved when this limit is taken.

In order to understand some of the problems that can arise, consider the series

$$f(x) = x^2 + \frac{x^2}{1 + x^2} + \frac{x^2}{(1 + x^2)^2} + \cdots + \frac{x^2}{(1 + x^2)^n} + \cdots,$$

so the terms $u_k(x)$ of equation 4.43 are given by $u_k(x) = x^2/(1+x^2)^k$. Since $u_{n+1}(x)/u_n(x) = 1/(1 + x^2) < 1$ for $x \ne 0$, the series is absolutely convergent for all real values of x, except possibly $x = 0$; but here all partial sums are zero, so $f(0) = 0$. This series is a geometric series and so the nth partial sum is

$$f_n(x) = x^2 \sum_{k=0}^{n-1} \frac{1}{(1 + x^2)^k} = 1 + x^2 - \frac{1}{(1 + x^2)^{n-1}},$$

giving the limit

$$f(x) = \lim_{n \to \infty} f_n(x) = 1 + x^2, \quad x \ne 0.$$

But, we have established that $f(0) = 0$, so the limit function is not continuous at $x = 0$, even though each term in the sum and each partial sum is continuous. Graphs of $f_n(x)$ for various values of n are shown in figure 4.3; these show how the discontinuity develops as $n \to \infty$.

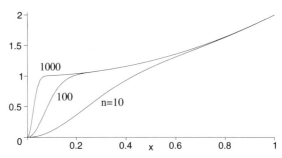

Figure 4.3   Graphs of $f_n(x)$, for $n = 10$, 100 and 1000.

To put this another way, recall that continuity is defined as a limit: a function is continuous at a point x if

$$\lim_{t \to x} f(t) = f(x),\tag{4.45}$$

or, more fully, given any $\epsilon > 0$ we can find a $\delta > 0$ such that if $|x - t| < \delta$ then $|f(x) - f(t)| < \epsilon$.

But the definition of the function $f(x)$, equation 4.44, involves the limit $n \to \infty$, so in asking whether a limit of the sequence of partial sums is continuous we are implicitly asking whether

$$\lim_{t \to x} \lim_{n \to \infty} f_n(t) = \lim_{n \to \infty} \lim_{t \to x} f_n(t),$$

that is, whether the two limiting processes are interchangeable. Two simple examples show that it is not always possible to interchange limits.

First consider the double sequence

$$S_{mn} = \frac{m}{m + n}.$$

For every fixed n

$$\lim_{m \to \infty} S_{mn} = 1, \quad \text{giving} \quad \lim_{n \to \infty} \lim_{m \to \infty} S_{mn} = 1.$$

But for every fixed m,

$$\lim_{n \to \infty} S_{mn} = 0, \quad \text{giving} \quad \lim_{m \to \infty} \lim_{0 \to \infty} S_{mn} = 0.$$

For the second example consider the sequence of functions

$$f_n(x) = n^2 x (1 - x^2)^n, \quad 0 \le x \le 1, \quad n = 1, 2, \dots.$$

If $0 \le x \le 1$ we have $\lim_{n \to \infty} f_n(x) = 0$, and so

$$\int_0^1 dx \left(\lim_{n \to \infty} f_n(x) \right) = 0.$$

But also

$$\int_0^1 dx\, n^2 x (1 - x^2)^n = \frac{n^2}{2(n + 1)} \quad \text{so that} \quad \lim_{n \to \infty} \int_0^1 dx\, f_n(x) = \infty.$$

In other words the two limits are not interchangeable,

$$\lim_{n\to\infty}\left(\int_0^1 dx\, f_n(x)\right) \neq \int_0^1 dx\, \left(\lim_{n\to\infty} f_n(x)\right).$$

Exercise 4.50

Sketch a few representative samples of the functions $f_n(x) = n^2 x(1 - x^2)^n$ and explain geometrically why the integral diverges.

Exercise 4.51

Consider the sequence of functions

$$f_n(x) = \frac{\sin nx}{\sqrt{n}}, \quad x \text{ real}, \quad n = 1, 2, \ldots$$

and show that

$$\frac{d}{dx}\left(\lim_{n\to\infty} f_n(x)\right) \neq \lim_{n\to\infty} \frac{d}{dx} f_n(x).$$

These examples show that care needs to be exercised when interchanging the order of limits.

Until the middle of the nineteenth century it had been generally assumed that if an infinite series converged to a continuous and differentiable function, $f(x)$, then the series obtained by differentiating the series term by term converged to $f'(x)$. Several mathematicians, Stokes, Seidel and Weierstrass, independently developed the notion of *uniform convergence* which is needed to be sure that various limits may be interchanged.

G. G. Stokes (1819–1903)
P. L. V. Seidel (1821–1896)
K. Weierstrass (1815–1897)

A sequence of functions $f_n(x)$, $n = 1, 2, \ldots$, is said to converge uniformly to the function $f(x)$ on an interval (a, b) if for every $\epsilon > 0$ there is an integer $N(\epsilon)$, independent of x, such that if $n \geq N$ then $|f(x) - f_n(x)| < \epsilon$ for every x in the interval (a, b).

It is clear that uniformly convergent sequences are pointwise convergent. The explicit difference between these ideas is that with pointwise convergence, for every x in the interval and every $\epsilon > 0$ there is an $N(\epsilon, x)$ such that $|f(x) - f_n(x)| < \epsilon$ if $n > N(\epsilon, x)$. If the sequence converges uniformly, then for each ϵ it is possible to find an N *independent* of x such that $|f(x) - f_n(x)| < \epsilon$.

In summary we have seen, using the example described at the beginning of this section, that the infinite sum of continuous functions need not be continuous everywhere; at first sight this is a strange and disturbing result and shows that the passage from finite to infinite sums is subtle. The idea of uniform convergence is introduced to guarantee that infinite sums of a set of continuous functions is continuous; this class of series is easier to handle than non-uniformly convergent series. In practice, however, non-uniformly convergent series are common and useful; for example in chapter 5 we shall use series which do not even converge to obtain useful approximations.

A sufficient, but not necessary, condition for the series 4.43 to be uniformly convergent is that there exist numbers M_k such that $|u_k(x)| \leq M_k$, for all x in the interval, and that the series $\sum_{k=1}^{\infty} M_k$ is convergent. This condition is due to Weierstrass and is usually referred to as the Weierstrass M-test. Note that this condition, if satisfied, also shows that the original series is absolutely convergent.

Thus the series

$$f(x) = \cos x + \frac{1}{2^2}\cos^2 x + \frac{1}{3^2}\cos^3 x + \cdots$$

is uniformly convergent because we may take $M_k = 1/k^2$ and $\sum_{k=1}^{\infty} k^{-2}$ converges (to $\pi^2/6$). Similarly, all power series are uniformly convergent inside their circle of convergence.

N. H. Abel
(1802–1829)
A more sensitive test for uniform convergence has been given by Abel. If

$$f(x) = \sum_{k=0}^{\infty} u_k(x) \quad \text{and} \quad u_k(x) = a_k g_k(x),$$

where $\sum_{k=0}^{\infty} a_k$ converges and the $g_k(x)$ satisfy $g_{k+1}(x) < g_k(x)$ and $0 \le g_k(x) \le M$ for all $a \le x \le b$, then the series for $f(x)$ converges uniformly in $[a, b]$.

Finally we state, without proof, two important results which are due to the fact that for uniformly convergent series limits may be interchanged. If the series

$$f(x) = u_1(x) + u_2(x) + \cdots + u_n(x) + \cdots = \sum_{k=1}^{\infty} u_k(x)$$

is uniformly convergent for $a \le x \le b$, then we may differentiate and integrate term by term:

$$\frac{df}{dx} = \sum_{k=1}^{\infty} \frac{du_k}{dx},$$

$$\int_{\alpha}^{\beta} dx\, f(x) = \sum_{k=1}^{\infty} \int_{\alpha}^{\beta} dx\, u_k(x), \quad \text{where} \quad a \le \alpha < \beta \le b.$$

The differential of a uniformly convergent series need not be uniformly convergent, though the integral is uniformly convergent, reflecting the fact that integration is a 'smoothing' operation.

These two properties of uniformly convergent series are important because they show when it is completely safe to treat infinite series as ordinary functions. Many series, however, are not uniformly convergent but, nevertheless, can be differentiated provided care is exercised; some important examples of such series are discussed in the next chapter.

The distinction between uniform and absolute convergence
Uniform convergence does not imply absolute convergence and absolute convergence does not imply uniform convergence.

For instance, as we saw above, the series

$$\sum_{k=0}^{\infty} \frac{x^2}{(1+x^2)^k}$$

is absolutely, but not uniformly, convergent near $x = 0$.

Conversely, the series

$$\frac{1}{1+x^2} - \frac{1}{2+x^2} + \frac{1}{3+x^2} + \cdots = \sum_{k=1}^{\infty} \frac{(-1)^{k-1}}{k+x^2}$$

is only conditionally convergent but, nevertheless, is uniformly convergent, see exercise 4.64.

Absolutely convergent series behave like series comprising a finite number of terms in that we can multiply them and transpose the order of summation.

Uniformly convergent series behave like series comprising a finite number of terms in that they are continuous if each term in the series is continuous.

4.7 Exercises

Exercise 4.52
Determine the behaviour of the following sequences as $k \to \infty$:

(i) $z_k = (a^k + b^k)^{1/k}$,   (ii) $z_k = \dfrac{1}{1 + k^2}$,   (iii) $z_k = k \sin(1/k)$,

(iv) $z_k = (-1)^k (1 + \dfrac{1}{k})$,   (v) $z_k = k(-1)^k$,   (vi) $z_k = k^2 \sin(k\pi/2)$.

Exercise 4.53
Evaluate the following limits as $n \to \infty$:

$$(i) \quad \frac{3^n + (-2)^n}{3^n - 2^n}, \qquad (ii) \quad \frac{a^n - 1}{a^n + 1}, \quad a > 0.$$

Exercise 4.54
Show by geometric arguments that the sequence

$$s_{n+1} = \sqrt{2 + \sqrt{s_n}}, \quad s_0 > 0,$$

converges to a positive root of $s^4 - 4s^2 - s + 4 = 0$.

Exercise 4.55
Show that each of the following series converges, and by expressing each term as a partial fraction show that they converge to the limit shown:

$$S_1 = \sum_{k=1}^{\infty} \frac{1}{(k+1)(k+3)(k+5)} = \frac{23}{480},$$

$$S_2 = \sum_{k=1}^{\infty} \frac{3k-2}{k(k+1)(k+2)} = 1.$$

Exercise 4.56
Discuss the convergence of the following series:

$$S_1 = \frac{1}{1} - \frac{2}{3} + \frac{3}{5} - \frac{4}{7} + \cdots,$$

$$S_2 = \left(\frac{a}{1} - \frac{b}{2}\right) + \left(\frac{a}{3} - \frac{b}{4}\right) + \cdots + \left(\frac{a}{2r-1} - \frac{b}{2r}\right) + \cdots,$$

where a and b are positive constants.

Exercise 4.57

For which values of the positive constants α and β does the following series converge

$$\sum_{n=3}^{\infty} \frac{1}{n(\ln n)^{\alpha}(\ln \ln n)^{\beta}} \ ?$$

Exercise 4.58

Find the radius of convergence of each of the following series:

$$\text{(i)} \ \sum_{k=0}^{\infty} k^3 x^k, \quad \text{(ii)} \ \sum_{k=0}^{\infty} \frac{2^k}{k!} x^k, \quad \text{(iii)} \ \sum_{k=0}^{\infty} \frac{5^k}{k^{3/2}} x^k, \quad \text{(iv)} \ \sum_{k=0}^{\infty} \frac{k^4}{5^k} x^k.$$

Exercise 4.59

Find the radii of convergence of the series:

$$\text{(i)} \ \sum_{k=0}^{\infty} \frac{x^k}{(3+(-1)^k)^k}, \quad \text{(ii)} \ \sum_{k=0}^{\infty} \frac{x^k}{2^{(k+(-1)^k)}}.$$

Exercise 4.60

Determine the radius of converence of the series:

$$S = \frac{x}{a+1} + \frac{x^2}{a+\sqrt{2}} + \frac{x^3}{a+\sqrt{3}} + \cdots + \frac{x^n}{a+\sqrt{n}} + \cdots,$$

where a is a real positive constant, and determine whether or not the series is uniformly convergent.

Exercise 4.61

If $E(x)$ denotes the sum of the power series

$$E(x) = \sum_{k=0}^{\infty} \frac{x^k}{k!},$$

show that $E(x)E(y) = E(x+y)$.

Exercise 4.62

In this exercise we generalise the results of exercises 2.22 and 2.29 (pages 82 and 87).

If $f(x)$ and $g(x)$ are odd functions having powers series expansions about $x = 0$, under what circumstances does

$$d(x) = f(g(x)) - g(f(x)) = ax^7 + O(x^9)$$

for some constant a? If $x = F(y)$ and $x = G(y)$ are respectively the inverses of $y = f(x)$ and $y = g(x)$, so $x = F(f(x))$, show that when these conditions are satisfied

$$D(y) = F(G(y)) - G(F(y)) = ay^7 + O(y^9).$$

Exercise 4.63

Consider the series

$$f(x) = \sum_{k=1}^{\infty} \frac{1}{1+k^2 x}.$$

For what values of x does the series converge absolutely? On which intervals does it converge uniformly? Is $f(x)$ continuous wherever the series converges?

Exercise 4.64

Show that the series

$$\sum_{k=0}^{\infty} \frac{(-1)^k}{k+x^2}$$

is uniformly and conditionally convergent.

Exercise 4.65

Prove that the series

$$\sum_{k=1}^{\infty} (-1)^k \frac{k+x^2}{k^2}$$

converges uniformly in every bounded interval, but does not converge absolutely for any real value of x.

Exercise 4.66

Show that the binomial expansion gives

$$\frac{x}{1-x} = x + x^2 + x^3 + \cdots + x^n + \cdots,$$

$$\frac{1}{1-1/x} = 1 + \frac{1}{x} + \frac{1}{x^2} + \cdots + \frac{1}{x^n} + \cdots.$$

By adding these results show that

$$0 = \sum_{k=-\infty}^{\infty} x^k.$$

This result was first derived by Euler. Why has this absurdity occurred?

Exercise 4.67

Show that

$$\int_0^1 dx \, \frac{\ln(1+x)}{x} = \frac{1}{2}\zeta(2),$$

$$\int_0^1 dx \, \frac{\ln(1-x)}{x} = -\zeta(2),$$

$$\int_0^1 dx \, \frac{[\ln(1-x)]^p}{x} = (-1)^p \Gamma(1+p)\zeta(1+p), \quad p > 0.$$

In the last example it helps to use the substitution $x = 1 - e^{-t}$ and to use the result

$$\int_0^{\infty} dt \, t^k e^{-\lambda t^m} = \frac{1}{m\lambda^{(k+1)/m}} \Gamma\left(\frac{k+1}{m}\right),$$

for suitable values of the constants k, λ and m.

Exercise 4.68
Show that

$$\sum_{k=1}^{\infty} \frac{x^{k-1}}{(1-x^k)(1-x^{k+1})} = \begin{cases} \dfrac{1}{(1-x)^2}, & |x| < 1, \\[2mm] \dfrac{1}{x(x-1)^2}, & |x| > 1. \end{cases}$$

Exercise 4.69
Show that the series

$$1 - \frac{1}{\sqrt{2}} + \frac{1}{\sqrt{3}} - \frac{1}{\sqrt{4}} + \cdots$$

is conditionally convergent. Show also that its square, constructed using the rule defined in equation 4.28 (page 156), is

$$S = 1 - \sqrt{2} + \left(\frac{2}{\sqrt{3}} + \frac{1}{2}\right) - \left(1 + \frac{2}{\sqrt{6}}\right) + \cdots + (-1)^n s_n + \cdots,$$

where

$$S_n = \sum_{k=0}^{n} \frac{1}{\sqrt{(k+1)(n+1-k)}}, \qquad n = 0, 1, 2, \ldots.$$

Further, show that S converges to a negative number.

Exercise 4.70
Show that the convergence of the series for $\ln(1 + x)$ can be improved by rewriting the series in the form

$$(1 + 3x + 3x^2 + x^3)\ln(1 + x) = x + \frac{5}{2}x^2 + \frac{11}{6}x^3 + 6\sum_{k=4}^{\infty} \frac{(-1)^k x^k}{k(k-1)(k-2)(k-3)}.$$

How many terms of this series are needed to estimate the value of $\ln 2$ to six decimal places?

Exercise 4.71
Write Gregory's series for $\pi/4$, exercise 4.30 (page 155), as the limit of the partial sums

$$S_n = 2\sum_{k=1}^{n} \frac{1}{(4k-1)(4k-3)}$$

and, by showing that $S_n = \frac{\pi}{4} + \frac{A_1}{n} + \frac{A_2}{n^2} + \cdots$, use Richardson's extrapolation to determine π to 10 decimal places. Remember to increase `Digits` to a sufficiently large value.

Exercise 4.72
A variant of the extrapolation methods considered in this chapter can be used to improve the accuracy of numerical integration.

Suppose that an N-point rule is used to approximate the integral $I = \int_a^b dx\, f(x)$ by the number I_N and that I and I_N are related by

$$I = I_N + aN^{-p} + \text{higher order terms}$$

for some positive integer p. Show that a better approximation is given by

$$I \simeq \frac{2^p I_{2N} - I_N}{2^p - 1}.$$

Use this method together with Simpson's rule, exercise 3.58 (page 128), for which $p = 4$, to evaluate $\int_0^3 dx \, \exp(\sqrt{x})$ to 2 decimal places.

Exercise 4.73

In exercise 4.46 we estimated the value of the sum

$$S = \sum_{k=2}^{\infty} \frac{\ln k}{k^2}$$

by assuming the partial sums have the form $S_n = S + (a/n)\ln n + (b/n)$. Now assume that

$$S_n = S + \left(\frac{a}{n} + \frac{b}{n^2}\right)\ln n + \left(\frac{c}{n} + \frac{d}{n^2}\right),$$

and, by solving the five linear equations for the unknown constants (S, a, b, c, d) numerically, using `fsolve`, show that $S \simeq 0.937548\ldots$.

Exercise 4.74

If S_{mp} is defined by the series

$$S_{mp} = \sum_{r=0}^{m} (-1)^r r^p \binom{m}{r},$$

show that $S_{mp} = 0$ for $0 \le p \le m-1$ and that $S_{mm} = (-1)^m m!$.
Hint: consider the binomial expansion of $(1-x)^m$ and repeatedly operate on both sides by $x\frac{d}{dx}$.

Exercise 4.75

Using the solution of the previous exercise, show that the function

$$f_p(x) = \sum_{k=0}^{N} (-1)^{k+N} \frac{(x+k)^p}{k!(N-k)!}$$

has the values

$$f_p(x) = \begin{cases} 1 & \text{for all } x \text{ if } p = N, \\ 0 & \text{for all } x \text{ if } 0 \le p \le N-1. \end{cases}$$

Further, show that $f_{N+1}(-N) = -f_{N+1}(0)$ and by integration that

$$f_{N+1}(x) = (N-1)x + f_{N+1}(0) \quad \text{and hence that} \quad f_{N+1}(x) = (N+1)(N+2x)/2.$$

Exercise 4.76

Use the results obtained in the previous exercise to show that the equations

$$R_n = \beta_0 + \frac{\beta_1}{n} + \frac{\beta_2}{n^2} + \cdots + \frac{\beta_N}{n^N}$$

$$R_{n+1} = \beta_0 + \frac{\beta_1}{n+1} + \frac{\beta_2}{(n+1)^2} + \cdots + \frac{\beta_N}{(n+1)^N}$$

$$\vdots$$

$$R_{n+N} = \beta_0 + \frac{\beta_1}{n+N} + \frac{\beta_2}{(n+N)^2} + \cdots + \frac{\beta_N}{(n+N)^N}$$

for β_k, $k = 0,\ldots,N$ may be solved to give

$$\beta_0 = \sum_{k=0}^{N} \frac{(-1)^{N+k} R_{n+k}(n+k)^N}{k!(N-k)!}$$

and

$$\beta_1 = \sum_{k=0}^{N} \frac{(-1)^{N+k+1} R_{n+k}(n+k)^N}{k!(N-k)!} - \frac{1}{2}\beta_0(N+1)(N+2n).$$

Exercise 4.77
Show that if

$$I_{nm} = \int_0^1 dx\, x^n (\ln x)^m \quad \text{then} \quad I_{nm} = -\frac{m}{n+1} I_{n\,m-1}$$

and hence that $I_{nm} = (-1)^m m!/(n+1)^{m+1}$. Use this result to show that

$$\int_0^1 dx\, x^{ax} = \sum_{k=0}^{\infty} \frac{(-1)^k a^k}{(k+1)^{k+1}}.$$

Exercise 4.78
Show that

$$\sum_{k=0}^{\infty} \frac{(-1)^k}{\alpha+k} = \int_0^1 dx\, \frac{x^{\alpha-1}}{1+x}, \quad \alpha > 0.$$

Exercise 4.79
By showing that

$$\int_0^1 dx\, \frac{1-x^m}{1+x^n} = \sum_{k=0}^{\infty} (-1)^k \left(\frac{1}{nk+1} - \frac{1}{nk+m+1} \right)$$

and by choosing $n = 6$ and $m = 4$, show that

$$1 - \frac{1}{5} - \frac{1}{7} + \frac{1}{11} + \frac{1}{13} - \frac{1}{17} + \cdots = \frac{1}{\sqrt{3}} \ln(2 + \sqrt{3}).$$

5

Asymptotic expansions

5.1 Introduction

In the previous chapter we dealt with convergent series; there are, however, series that do not converge anywhere, and these are often more useful than convergent series. This may seem paradoxical, or even bizarre, so before proceeding with the general theory we discuss a few examples.

The use of infinite series goes back to the time of Newton, though the mathematicians I. Newton
(1642–1727) of the eighteenth century were often more interested in the formal manipulation of these series than in rigorous proofs. Nevertheless some of the results obtained were remarkable. In 1730 Stirling gave an infinite series for $\ln(m!)$, which we write as J. Stirling
(1692–1770)

$$\ln(m!) = x \ln x - x + \frac{1}{2} \ln(2\pi) + \sum_{k=1}^{\infty} \frac{B_{2k}(\frac{1}{2})}{(2k-1)\, 2k\, x^{2k-1}}, \quad x = m + \frac{1}{2}, \quad (5.1)$$

where $B_n(x)$ is Bernoulli's polynomial defined implicitly by the relation†

$$\frac{z e^{xz}}{e^z - 1} = \sum_{k=0}^{\infty} B_k(x) \frac{z^k}{k!}. \quad (5.2)$$

De Moivre subsequently gave a similar formula in terms of the Bernoulli number B_{2n}, A. De Moivre
(1667–1754) the value of the Bernoulli polynomial at $x = 0$,

$$\ln(m!) = \left(m + \frac{1}{2} \right) \ln m - m + \frac{1}{2} \ln(2\pi) + \sum_{k=1}^{\infty} \frac{B_{2k}}{(2k-1)\, 2k\, m^{2k-1}}. \quad (5.3)$$

But for large n we have‡

$$B_{2n} = B_{2n}(0) \simeq (-1)^{n-1} 4n \sqrt{\frac{2n-1}{2\pi}} \left(\frac{2n-1}{2\pi e} \right)^{2n-1},$$

so it is clear that the second of these series does not converge for any value of m; it can also be shown that Stirling's series does not converge. Nevertheless, Stirling was able to calculate $\log(1000!)$ to ten decimal places using only the first few terms of his series; De Moivre's approximation provides the same accuracy.

The theory which put Stirling's work on a rigorous basis was eventually developed by

---

† This definition of Bernoulli polynomials is not universal. For instance, a different definition is used by Whittaker and Watson (1965, chapter 7) and this leads to a different definition of the Bernoulli numbers: the nth number of Whittaker, $\bar{B}_n$, is related to that used here by $\bar{B}_n = (-1)^{n-1} B_{2n}$.

‡ The derivation of this approximation is given in exercise 13.32.

H. Poincaré
(1854–1912)

Poincaré, and this general theory, described here, is the foundation of a very important tool widely used in applied mathematics.

Exercise 5.1

Write a Maple procedure f(m,N) to compute the first N terms of the series 5.1 and compare graphically the values of $f(m, N)/\ln(m!)$ for $m = 3$, 4 and 5 for various values of N, up to about 50.

Hint: you may find it convenient to use the command alias(B=bernoulli), so that subsequently B(n,x) is the value of the nth Bernoulli polynomial at x, $B_n(x)$ and B(n) is the value of the nth Bernoulli number, $B_n = B_n(0)$.

In this exercise you should observe that for small N the ratio $f(m, N)/\ln(m!)$ is close to unity but at some critical value of N, dependent upon m, the ratio begins to increase.

P. S. Laplace
(1749–1827)

Another example comes from Laplace's *Théorie Analytique des Probabilités*, published in 1812. Here he showed that the error function

$$\mathrm{erf}(x) = \frac{2}{\sqrt{\pi}} \int_0^x dt\, e^{-t^2} \tag{5.4}$$

can be represented by the uniformly convergent Taylor's series

$$\begin{aligned}\mathrm{erf}(x) &= \frac{2}{\sqrt{\pi}} \sum_{k=0}^{\infty} (-1)^k \frac{x^{2k+1}}{(2k+1)k!} \\ &= \frac{2}{\sqrt{\pi}} \left(x - \frac{1}{3}x^3 + \frac{1}{10}x^5 - \frac{1}{42}x^7 + \cdots \right),\end{aligned} \tag{5.5}$$

obtained by expanding the integrand and integrating term by term. This series converges uniformly for all x. At $x = 1$ eight terms of this series gives a relative accuracy of 10^{-6}; for $x = 2$, 17 terms are required, and at $x = 6$, 104 terms are needed to obtain the same degree of accuracy. Moreover, the large magnitude of the individual terms means that in practice, rounding errors limit the magnitude of x for which this type of series can be used; this problem is explored in exercise 5.16 (page 191). Laplace also derived the approximation, exercise 5.22 (page 192),

$$\mathrm{erf}(x) = 1 - \frac{e^{-x^2}}{x\sqrt{\pi}} \left\{ 1 - \frac{1}{2x^2} + \frac{1.3}{(2x^2)^2} - \frac{1.3.5}{(2x^2)^3} + \cdots \right\}, \quad x > 0, \tag{5.6}$$

which diverges for all finite x. However, at $x = 2$ the first three terms in the curly bracket provides a relative accuracy of less than 10^{-4}, and for $x > 3$ the first two terms give a relative accuracy of less than 10^{-6}.

Exercise 5.2

Write Maple functions to evaluate the series, truncated at the nth term, defined in equations 5.5 and 5.6. Compare the results obtained using these series, truncated at some suitable point, with the exact values given by the Maple function erf(x), for $0 < x < 4$. For the Taylor's series use ten terms and for the second series use four terms. Investigate the effect of varying the number of terms.

Hint: both these series can produce large values, the Taylor's series at large values of x and the second series at small values (because of the x^{-1} term), so in the plot command it is helpful to use the view option in the form view=[0..4,-2..2] for instance.

At this stage it is natural to ask why such series exist and why they are so important. Clearly there is no simple, complete answer to this type of question, but one of the main sources of asymptotic expansions is differential equations with the form

$$\epsilon^2 \frac{d^2 y}{dx^2} + g(x)y = 0,$$

where ϵ is a small parameter and $g(x)$ some known function. Such equations arise naturally in the description of wave motion — such as sound, light or water, and in the quantum description of matter, in which the wavelength is relatively small — and also in fluid flow where viscous effects are significant only close to a boundary. Since ϵ is small we should expect some kind of small ϵ expansion to be useful; but when $\epsilon = 0$ the equation reduces to $g(x)y = 0$, suggesting that the behaviour of the solutions as $\epsilon \to 0$ is far from simple. It transpires that this type of small ϵ expansion is usually an asymptotic expansion, though to explain why would take us too far from our main theme; we shall see examples of this behaviour in chapters 14 and 15, and simple cases are treated in exercise 5.3 below.

Integrals of the form

$$\int_0^\infty dt \, \exp(ih(t)/\epsilon), \quad |\epsilon| \ll 1,$$

where $h(t)$ is real, also give rise to asymptotic expansions, because in many cases the dominant contribution comes from a particular value of t and the expansion about this point usually estimates the exponentially small contributions from other values of t incorrectly. Such approximations are considered in chapter 14.

In this chapter we introduce the basic theory, leaving the derivation of asymptotic expansions for particular problems to later chapters, particularly 13 to 15. If you wish to study this subject and its applications further, there are a number of good books; Whittaker and Watson (1965, chapter 8) gives a good introductory account; asymptotic methods are used by de Bruijn (1961) in a variety of interesting problems and by Erdélyi (1956) and Murray (1984) to evaluate integrals and find solutions of differential equations; Copson (1967) and Bleistein and Handelsman (1975) provide comprehensive studies of the evaluation of integrals using asymptotic approximations; Olver (1974) provides a detailed account of the application of asymptotic analysis to special functions.

Exercise 5.3

Solve the differential equation

$$\epsilon \frac{dx}{dt} = 1 - x, \quad x(0) = a > 0, \quad 0 < \epsilon \ll 1,$$

and plot the graphs of the solution for $a = 0.1$ and $a = 2$ with $\epsilon = 0.1$ and 0.01. Observe that, in this case, for $t \gg \epsilon$, $x(t) \simeq 1$.

Next solve the equation

$$\epsilon^2 \frac{d^2 x}{dt^2} + x = 0, \quad x(0) = 1, \quad \dot{x}(0) = 0,$$

and plot the graphs of the solution for $\epsilon = 0.1$ and 0.01 over the range $0 < t < 1$.

Exercise 5.4

Use equation 5.2 to show that

$$B_1(x) = -\frac{1}{2} + x, \quad B_2(x) = \frac{1}{6} - x + x^2.$$

Use Maple and equation 5.2 to derive the next eight Bernoulli polynomials.

5.2 Asymptotic expansions

In the introduction we gave two examples of series, equations 5.1 and 5.6, which did not converge on any interval, but which nevertheless provided useful approximations if truncated. We now examine the nature of this type of series and introduce the theory which justifies their use. Methods of deriving asymptotic expansions will be discussed later, particularly in chapters 13 to 15.

5.2.1 A simple example

We start with the specific example of the function $g(x)$ defined by the integral

$$g(x) = \int_0^\infty dt \, \frac{e^{-t}}{x+t}, \quad x > 0. \tag{5.7}$$

This function is related to the exponential integral, defined as

$$E_1(x) = \int_x^\infty dt \, \frac{e^{-t}}{t}, \tag{5.8}$$

by the relation

$$g(x) = e^x E_1(x). \tag{5.9}$$

Maple knows about $E_1(x)$, which it denotes by $\texttt{Ei(1,x)}$, and you may find it helpful to plot the graph of $g(x)$.

From the definition of $g(x)$ we expect $g(x) \to 0$ as $x \to \infty$ but wish to determine exactly how it approaches zero; integrating by parts gives the following expression:

$$g(x) = \frac{1}{x} - \int_0^\infty dt \, \frac{e^{-t}}{(x+t)^2}. \tag{5.10}$$

The integral here is clearly smaller than the original if $x > 1$, which suggests repeating this operation. Using the identity

$$\int_0^\infty dt \, \frac{e^{-t}}{(x+t)^n} = \frac{1}{x^n} - n \int_0^\infty dt \, \frac{e^{-t}}{(x+t)^{n+1}}, \tag{5.11}$$

obtained by integrating by parts, we find that†

$$g(x) = \frac{1}{x} - \frac{1}{x^2} + \frac{2!}{x^3} + \cdots + \frac{(-1)^{n-1}(n-1)!}{x^n} + (-1)^n n! \int_0^\infty dt \, \frac{e^{-t}}{(x+t)^{n+1}}. \tag{5.12}$$

---

† This operation can be automated and performed using Maple.

Exercise 5.5
Write two Maple functions, one to evaluate the function $g(x)$ using equation 5.9, and the other to evaluate the partial sums

$$g_n(x) = \sum_{k=1}^{n} \frac{(-1)^{k-1}(k-1)!}{x^k}.$$

Use these to plot graphs of the ratio $g_n(m)/g(m)$ as a function of n for fixed $x = m$ with n in the range $1 \le n \le 2m+1$ and for $m = 5$, 10 and 20.

Use Stirling's approximation, in the form $k! \simeq \sqrt{2\pi k}\,(k/e)^k$, to show that

$$\frac{(k-1)!}{x^k} \simeq \sqrt{\frac{2\pi}{k}} \left(\frac{k}{ex} \right)^k,$$

and use this to explain your graphical results.

The results obtained in exercise 5.5 show that for any sufficiently large x the partial sum of the series provides a good approximation if n is not too large. In this case we can see why this is so by considering the remainder,

$$g(x) - g_n(x) = (-1)^n n! \int_0^\infty dt\, \frac{e^{-t}}{(x+t)^{n+1}}.$$

Replacing $(x+t)^{-n-1}$ by the upper bound x^{-n-1} gives

$$|g(x) - g_n(x)| \le \frac{n!}{x^{n+1}} \int_0^\infty dt\, e^{-t} = \frac{n!}{x^{n+1}} \simeq \frac{\sqrt{2\pi n}}{x} \left(\frac{n}{xe} \right)^n,$$

where we have used Stirling's approximation, given in the exercise above, to derive the last expression. Consider the function $h(x,n) = \sqrt{2\pi n}(n/ex)^n/x$, for some large fixed value of x, as n varies. If $n > ex$, $h \gg 1$ and if $n < ex$, h is small. Further,

$$\frac{dh}{dn} = \left(\frac{1}{2n} + \ln n - \ln x \right) h(x,n),$$

so if x is large $h(x,n)$ has a minimum near $n = x$, where $dh/dn \simeq 0$. This behaviour is shown in figure 5.1, where the variation of $\ln(h(x,n))$ with n for $x = 5$, 10 and 15 is shown.

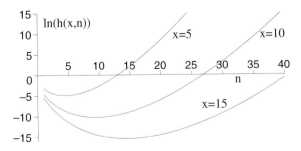

Figure 5.1   Representative graphs of $\ln(h(x,n))$ for fixed x.

From these graphs we see that for any given x the bound $h(x, n)$ of the difference $|g(x) - g_n(x)|$ at first decreases with increasing n, reaching a minimum at $n \simeq x$ after which the difference increases. Also, the figure shows that the larger x the smaller the value of $h(x, n)$ at the minimum; in fact the minimum value is approximately $\sqrt{2\pi n}e^{-n}$ at $n \simeq x$. Thus, in contrast to the behaviour of a Taylor's series, we cannot keep adding terms to improve the accuracy. This is one of the main characteristics of this type of series; the infinite series

$$\frac{1}{x} - \frac{1}{x^2} + \frac{2!}{x^3} + \cdots + \frac{(-1)^{n-1}(n-1)!}{x^n} + \cdots$$

is said to be an *asymptotic expansion* for the function $g(x)$; in the next section we make these ideas more precise.

Exercise 5.6

Figure 5.1 depicts graphs of the bound $h(x, n)$. In this exercise you will show that the exact difference behaves in the same manner by using the fact that the integral for the difference $g(x) - g_n(x)$ can be expressed in terms of known functions.

Show that if the function $\mathrm{Ei}(n, x)$ is defined by the integral

$$\mathrm{Ei}(n, x) = \int_1^\infty dt\, \frac{e^{-xt}}{t^n},$$

then

$$|g(x) - g_n(x)| = \frac{n!\,e^x}{x^n}\,\mathrm{Ei}(n + 1, x).$$

Use the Maple function $\mathrm{Ei}(n,x)$ to reproduce figure 5.1 using this exact difference, and compare, graphically, this exact function with $h(n, x)$, used above.

5.2.2 Definition of an asymptotic expansion

A series

$$a_0 + \frac{a_1}{x} + \frac{a_2}{x^2} + \cdots + \frac{a_n}{x^n} + \cdots \tag{5.13}$$

with the sum of the first $n + 1$ terms being denoted by $f_n(x)$,

$$f_n(x) = a_0 + \frac{a_1}{x} + \frac{a_2}{x^2} + \cdots + \frac{a_n}{x^n}, \tag{5.14}$$

is said to be an *asymptotic expansion* of a function $f(x)$ if the expression

$$R_n(x) = x^n \{f(x) - f_n(x)\} \tag{5.15}$$

satisfies the limiting condition

$$\lim_{x \to \infty} R_n(x) = 0 \quad (n \text{ fixed}). \tag{5.16}$$

Often the series 5.13 will diverge and, despite the limit 5.16,

$$\lim_{n \to \infty} R_n(x) = \infty \quad (x \text{ fixed}). \tag{5.17}$$

Nevertheless, if the first limit is satisfied we can make $|R_n(x)|$ arbitrarily small by fixing n and taking x sufficiently large. We denote the fact that the series is an asymptotic expansion of $f(x)$ by writing

$$f(x) \sim a_0 + \frac{a_1}{x} + \frac{a_2}{x^2} + \cdots + \frac{a_n}{x^n} + \cdots,$$

though in some texts the $\sim$ is replaced by $=$.

In the example defined in equation 5.7 (page 182),

$$|R_n| = x^n |g(x) - g_n(x)| < n! x^{-1},$$

which satisfies condition 5.16, for if we fix n the value of R_n can be made arbitrarily small by increasing x. But if x is fixed R_n increases without bound as n increases.

The definition just given is due to Poincaré (1886a) and is the simplest and most common type of asymptotic expansion.

If an expression for the remainder term, $|f(x) - f_n(x)|$, is known, then, for each x, it is possible to determine a priori the number of terms of the sum which minimises the magnitude of the remainder. In practice, however, a general expression for the remainder is rarely known and we normally resort to the simple device of continuing to add terms while their magnitude is decreasing. Thus, for the series defined in equation 5.13, we would add the term $a_{n+1} x^{-n-1}$ to the nth partial sum f_n only if $|a_{n+1} x^{-n-1}| < |a_n x^{-n}|$, that is $|a_{n+1}| \leq |xa_n|$; the sum is truncated at the first value of n at which $|a_{n+1}| > |xa_n|$. For the series 5.12 this criterion suggests truncating the series at $n \simeq x$.

Exercise 5.7
Write a Maple procedure which evaluates the asymptotic expansion of the error function, given in equation 5.6 (page 180), and uses the criterion discussed above to truncate the series.

Use your procedure to compute an approximation to $\mathrm{erf}(x)$ for $x = 1, 2$ and 3, comparing your values with the exact value of the function.

Exercise 5.8
Show that

$$\frac{1}{x} + \frac{1}{x^2} + \cdots + \frac{1}{x^n} + \cdots \tag{5.18}$$

is an asymptotic expansion of the function $f(x) = 1/(x-1)$, for $x > 1$, and that in this case

$$R_n(x) = \frac{1}{x} + O(x^{-2}) \quad \text{as } x \to \infty.$$

Show also that the series 5.18 converges for all $x > 1$.

In this example the asymptotic expansion is a convergent series. In fact it is simply the Taylor's series of the function expanded about $x = \infty$, as may be seen by setting $x = 1/w$ and finding Taylor's series of $f(x(w))$ about $w = 0$.

Exercise 5.9

Show that

$$\int_0^\infty dt\, \frac{e^{-xt}}{1+t^2} \sim \frac{1}{x} - \frac{2!}{x^3} + \cdots = \sum_{k=1}^\infty (-1)^k \frac{(2k)!}{x^{2k+1}},$$

and that the remainder after $n+1$ terms of the sum, $R_{n+1}(x)$, is bounded by

$$|R_{n+1}(x)| < \frac{(2n+2)!}{x^2}.$$

Hint: use the results

$$\frac{1}{1+t^2} = 1 - t^2 + t^4 + \cdots + (-1)^n t^{2n} - \frac{(-1)^n t^{2n+2}}{1+t^2}, \qquad \int_0^\infty dz\, e^{-z} z^n = n!, \quad n = 0, 1, 2, \cdots.$$

Different functions can have the same asymptotic expansions. For example, there are functions $L(x)$ which are represented asymptotically by a series all of whose terms are zero, that is $\lim_{x\to\infty} x^n L(x) = 0$ for every positive integer n; an example of such a function is $L(x) = e^{-x}$ for $x > 0$, so an asymptotic expansion of $f(x)$ is also an asymptotic expansion of $f(x) + L(x)$.

On the other hand the asymptotic expansion of a *given* function is unique, for if

$$f(x) \sim \sum_{k=0}^\infty a_k x^{-k} \quad \text{and} \quad f(x) \sim \sum_{k=0}^\infty b_k x^{-k}$$

then, by equation 5.16, for *all* positive integers n,

$$\lim_{x\to\infty} x^n \sum_{k=0}^n (a_k - b_k) x^{-k} = 0, \quad n = 1, 2, \cdots,$$

which is possible only if $a_k = b_k$ for $k = 0, 1, \cdots$.

Maple has a command asympt, with syntax similar to that of series and taylor, which computes the asymptotic expansion of expressions, where possible, of many standard functions. Thus the asymptotic expansion of the function

$$f = \frac{x^2 + 3}{x^3 + 5x^2}$$

up to and including the term $O(x^5)$ is given by the command

>   f1:=asympt(f,x,6);

$$f1 := \frac{1}{x} - \frac{5}{x^2} + \frac{28}{x^3} - \frac{140}{x^4} + \frac{700}{x^5} + O(\frac{1}{x^6}),$$

and the first few terms of the asymptotic expansion of $(2x)!/(x!)^2$ are

>   f2:=asympt((2*x)! / (x!)^2, x,3);

$$f2 := \left(\frac{\sqrt{\frac{1}{x}}}{\sqrt{\pi}} - \frac{1}{8}\frac{(\frac{1}{x})^{3/2}}{\sqrt{\pi}} + O((\frac{1}{x})^{5/2}) \right) 4^x$$

Expressions returned by `asympt` are not of series type, though the order term may be removed using the `convert` command,

> `convert(f2,polynom);`

But, because the expression is not of series type, it is sometimes convenient to simply substitute zero for O,

> `eval(f2,O=0);`

$$\left(\frac{\sqrt{\frac{1}{x}}}{\sqrt{\pi}} - \frac{1}{8}\frac{(\frac{1}{x})^{(3/2)}}{\sqrt{\pi}} \right) 4^x$$

alternatively the command `eval( subs(O=0,f2) )` may be used.

Exercise 5.10
Use Maple to determine the first term in the asymptotic expansion of

$$g_n(x) = \frac{(nx)!}{(x!)^n}$$

for various integer values of *n*.

5.2.3 Addition and multiplication of asymptotic expansions
If the functions $f(x)$ and $g(x)$ have the asymptotic expansions

$$f(x) \sim \sum_{k=0}^{\infty} a_k x^{-k} \quad \text{and} \quad g(x) \sim \sum_{k=0}^{\infty} b_k x^{-k},$$

then the asymptotic expansion of the sum $h(x) = \alpha f(x) + \beta g(x)$, where α and β are constants, is

$$h(x) \sim \sum_{k=0}^{\infty} (\alpha a_k + \beta b_k) x^{-k}.$$

The asymptotic expansion of the product $\phi(x) = f(x)g(x)$ is

$$\phi(x) \sim \sum_{k=0}^{\infty} c_k x^{-k}, \quad \text{where} \quad c_n = \sum_{s=0}^{n} a_s b_{n-s}. \tag{5.19}$$

The proof of both these statements can be found in Whittaker and Watson (1965, page 153).

It follows from this last result that any positive integral power of $f(x)$ possesses an asymptotic expansion, as does any polynomial in $f(x)$. Further, if $a_0 \neq 0$ then

$$\frac{1}{f(x)} \sim \frac{1}{a_0} + \sum_{k=1}^{\infty} d_k x^{-k}, \tag{5.20}$$

for some coefficients d_k which may be expressed in terms of the coefficients a_k, as in exercise 5.14 (page 190).

More generally, any rational function of $f(x)$ has an asymptotic power series expansion provided the denominator does not tend to zero as $x \to \infty$, that is as $f(x) \to a_0$.

5.2.4 Integration and differentiation of asymptotic expansions

If

$$f(x) \sim \sum_{k=2}^{\infty} \frac{a_k}{x^k},$$

then it may be shown, for instance in Whittaker and Watson (1965, page 153), that

$$\int_x^{\infty} dx\, f(x) \sim \sum_{k=2}^{\infty} \frac{a_k}{(k-1)x^{k-1}}. \tag{5.21}$$

Note that for the integral to exist it is necessary to start the sum at $k = 2$.

On the other hand it is not always permissible to differentiate an asymptotic expansion term by term to obtain the asymptotic expansion of the differential of the original function, as may be seen by considering an example. Suppose that

$$f(x) = \sum_{k=0}^{\infty} \frac{a_k}{x^k} + e^{-x} \sin(e^x),$$

so that the asymptotic expansion of f is

$$f(x) \sim \sum_{k=0}^{\infty} \frac{a_k}{x^k}.$$

Differentiation gives

$$f'(x) = -\sum_{k=1}^{\infty} \frac{k a_k}{x^{k+1}} + \cos(e^x) - e^{-x} \sin(e^x).$$

Since $\cos(e^x) = O(1)$ the asymptotic expansion of $f'(x)$ does not exist.

In general, however, for a function $f(x)$ that is differentiable *and* if both $f(x)$ and $f'(x)$ possess asymptotic expansions, then if

$$f(x) \sim \sum_{k=0}^{\infty} \frac{a_k}{x^k} \quad \text{it follows that} \quad f'(x) \sim -\sum_{k=1}^{\infty} \frac{k a_k}{x^{k+1}}. \tag{5.22}$$

We end this section by giving two of the better known asymptotic expansions. The first example is the gamma function and the related factorial function, $\Gamma(x+1) = x!$. One form of this result, due to Stirling, was quoted in the introduction, equation 5.1; the more useful form, which will be derived in chapter 14, is

$$n! \sim \sqrt{2\pi n} \left(\frac{n}{e}\right)^n \left(1 + \frac{1}{12n} + \frac{1}{288n^2} - \frac{139}{51\,840 n^3} - \frac{571}{2\,488\,320 n^4} \cdots\right).$$

This expansion is remarkable for the fact that the first term is a reasonable approximation even for quite small values of n; thus the first term gives $1! \sim 0.922$, and $3! \sim 5.84$: the first two terms give $1! \sim 0.998\,98$ and $3! \sim 5.998$.

The second example is the prime number theorem which states that if $\pi(x)$ is the number of primes not exceeding x, then

$$\pi(x) = \int_2^x \frac{dt}{\ln t} + O(x \exp(-c\sqrt{\ln x})),$$

where c is a positive constant; normally the integral is replaced by its large x approximation so we have $\pi(x) \sim x/\ln x$. Strictly, an asymptotic expansion of the integer valued function $\pi(x)$ does not exist, because there is no known relation for the difference between adjacent primes, though an approximation to the statistical distribution of these differences is known; it is the same as the probability density for the difference between the zeros of the zeta function, quoted on page 143 of chapter 4. A proof of the above asymptotic formula can be found in Apostol (1976, chapter 13), and for further discussion of the prime number theorem see the interesting article by Shiu (1996).

Exercise 5.11
Using integration by parts show that the asymptotic expansion of

$$g_1(x) = \int_0^\infty dt \, \frac{e^{-t}}{(x+t)^2}$$

is

$$g_1(x) \sim \frac{1}{x^2} - \frac{2}{x^3} + \frac{2.3}{x^4} + \cdots + (-1)^n \frac{(n-1)!}{x^n} + \cdots.$$

Show also that $g_1(x) = -g'(x)$, where $g(x)$ is defined in equation 5.7 (page 182), and hence confirm equation 5.22 for this example.
Hint: you will find equation 5.11 useful.

5.3 General asymptotic expansions

The asymptotic expansion defined by equation 5.14 is only one of a class of asymptotic expansions. It is necessary to generalise our notion of asymptotic expansions because there are many functions whose asymptotic behaviour cannot be described in the manner of equation 5.14. For instance, the asymptotic expansion of $1/(\sqrt{x} - 1)$ is, on using the result of exercise 5.8,

$$\frac{1}{\sqrt{x} - 1} \sim \frac{1}{x^{1/2}} + \frac{1}{x} + \frac{1}{x^{3/2}} + \cdots.$$

Clearly this behaviour cannot be described by an expression of the form given in equation 5.14.

In addition we have assumed that an approximation as $x \to \infty$ is required; whilst this is often the case we sometimes require an asymptotic expansion about $x = 0$, or other points.

In order to resolve these problems we generalise the ideas introduced in the previous section by using *asymptotic sequences* of functions. A sequence of functions $\phi_k(x)$, $k = 0, 1, 2, \ldots$, is an asymptotic sequence as $x \to a$ if

$$\lim_{x \to a} \frac{\phi_1(x)}{\phi_0(x)} = 0, \quad \lim_{x \to a} \frac{\phi_2(x)}{\phi_1(x)} = 0, \quad \cdots \quad \lim_{x \to a} \frac{\phi_{n+1}(x)}{\phi_n(x)} = 0 \cdots, \tag{5.23}$$

and we assume that $\phi_k(x)$ has no zeros in the neighbourhood of $x = a$. If, for each value of N,

$$f(x) = \sum_{k=0}^{N} a_k \phi_k(x) + O(\phi_{N+1}(x)) \quad \text{as} \quad x \to a, \tag{5.24}$$

then $f(x)$ has an asymptotic expansion in the sequence $\phi_k(x)$, and as before we write this as

$$f(x) \sim \sum_{k=0}^{\infty} a_k \phi_k(x). \tag{5.25}$$

Exercise 5.12
Show that the functions $\phi_k(x) = x^{-k}$ are an asymptotic sequence as $x \to \infty$.

Some other examples of asymptotic sequences are

$$
\begin{aligned}
(x-a)^n, & \quad x \to a, \\
x^{-\alpha_n}, & \quad x \to \infty, \quad \Re(\alpha_{n+1}) > \Re(\alpha_n), \\
(\ln x)^{-n}, & \quad x \to \infty, \\
e^x x^{-\beta_n}, & \quad x \to \infty, \quad \beta_n \text{ real and } \beta_{n+1} > \beta_n,
\end{aligned} \tag{5.26}
$$

where $n = 0, 1, 2, \dots$.

If an asymptotic expansion of $f(x)$ exists for a given asymptotic sequence $\phi_k(x)$, then it is unique and the coefficients a_k are uniquely determined by the following rules:

$$
\begin{aligned}
a_0 &= \lim_{x \to a} \frac{f(x)}{\phi_0(x)}, \\
a_1 &= \lim_{x \to a} \frac{f(x) - a_0 \phi_0(x)}{\phi_1(x)} \\
\vdots \quad &\vdots \qquad \qquad \vdots \\
a_n &= \lim_{x \to a} \frac{f(x) - \sum_{k=0}^{n-1} a_k \phi_k(x)}{\phi_n(x)},
\end{aligned} \tag{5.27}
$$

which are a consequence of equation 5.24 and the conditions 5.23.

Exercise 5.13
Use the fact that equation 5.24 holds for all N to derive equation 5.27.

Exercise 5.14
Use the rules defined by equations 5.27 to show that the first three coefficients of the asymptotic expansion of $1/f(x)$, with respect to the asymptotic sequence x^{-k}, $k = 0, 1, \dots$, with

$$f(x) \sim \sum_{k=0}^{\infty} a_k x^{-k}, \quad a_0 \neq 0,$$

are given by

$$d_0 = \frac{1}{a_0}, \quad d_1 = -\frac{a_1}{a_0^2}, \quad d_2 = \frac{a_1^2}{a_0^3} - \frac{a_2}{a_0^2}.$$

Use Maple to determine the next three coefficients.

The asymptotic expansion of a function depends upon the sequence $\phi_k(x)$ chosen, consequently a function may have several asymptotic expansions. For example, consider the power series expansion of the function $f(x) = 1/(x-1)$; for $|x| > 1$ we have

$$\frac{1}{x-1} \sim \sum_{k=1}^{\infty} \frac{1}{x^k} \quad \text{as} \quad x \to \infty,$$

but on replacing x by x^2 this becomes

$$\frac{1}{x^2-1} \sim \sum_{k=1}^{\infty} \frac{1}{x^{2k}},$$

and so we also have

$$\frac{1}{x-1} \sim (x+1) \sum_{k=1}^{\infty} \frac{1}{x^{2k}} \quad \text{as} \quad x \to \infty.$$

These two (convergent) asymptotic expansions are different as they are based on the two different asymptotic sequences x^{-k} and x^{-2k}, $k = 0, 1, \ldots$, respectively.

As before, the asymptotic expansion, with respect to a given sequence, of a sum of functions is the sum of their asymptotic expansions. But in general the product of two functions need not have an asymptotic expansion, for it may not be possible to arrange the products $\phi_n(x)\phi_m(x)$ as an asymptotic sequence; the asymptotic sequence x^{-k} is an exception and gives the rule of equation 5.19, see also exercise 5.28.

Exercise 5.15
Show that the sequence $\phi_k(x) = e^{ix}x^{-k}$, $k = 0, 1, 2, \ldots$, is an asymptotic sequence as $x \to \infty$ and that if

$$F(x) = \int_x^{\infty} dt\, \frac{\cos t}{t},$$

then

$$F(x) \sim \left(\frac{1}{x^2} - \frac{3!}{x^4} + \frac{5!}{x^6} + \cdots \right) \cos x - \left(\frac{1}{x} - \frac{2!}{x^3} + \frac{4!}{x^5} - \cdots \right) \sin x.$$

5.4 Exercises

Exercise 5.16
Use Stirling's approximation to show that the modulus of the nth term of the Taylor series of $\sin x$, $x^{2n+1}/(2n+1)!$ is small for $n > ex/2$, and that for large $|x|$ the largest term in the Taylor's series is about $e^x/\sqrt{2\pi x}$.

Show further that with arithmetic accurate to N significant figures the series approximation for $\sin x$ can be used directly to find values of $\sin x$ accurate to $M < N$ significant figures only for $x < (N - M)\ln 10$. Check this behaviour using Maple.

Exercise 5.17

Show that the Bernoulli polynomials, defined implicitly by equation 5.2 (page 179), satisfy the relations

(i) $B'_n(x) = nB_{n-1}(x)$,   (ii) $B_n(x+1) - B_n(x) = nx^{n-1}$,   (iii) $B_n(0) = (-1)^n B_n(1)$.

Exercise 5.18

If $f'(x) \to 0$ as $x \to \infty$, does it follow that $f \sim c$ as $x \to \infty$, where c is a constant? Hint: consider the function $f(x) = \sqrt{x}$.

Exercise 5.19

The function

$$f(x) = \frac{x}{1+x} \cos x \quad \text{behaves as} \quad \cos x \quad \text{when } x \to \infty.$$

Show that if

$$F(x) = \int_0^x dt \, \frac{t \cos t}{1+t}$$

then it is not true that

$$F(x) \sim \int_0^x dt \, \cos t = \sin x \quad \text{as} \quad x \to \infty.$$

Exercise 5.20

Show that if $a > 1$ then

$$\int_0^\infty dt \, \frac{e^{-xt}}{1+t^a} \sim \sum_{k=0}^\infty (-1)^k \frac{\Gamma(1+ak)}{x^{1+ak}}.$$

Exercise 5.21

Show that

$$y(x) = \int_0^\infty dt \, \frac{e^{-t}}{1+xt} \sim 1 - x + 2!x^2 - 3!x^3 + \cdots (-1)^N N! x^N + \cdots \quad \text{as} \quad x \to 0$$

and that the remainder of the series summed to N terms is

$$R_N = (-1)^N (N)! x^{N+1} \int_0^\infty dt \, \frac{e^{-t}}{(1+xt)^{N+1}},$$

and hence that limit 5.16 (page 184) is satisfied.

Exercise 5.22

The error function is defined by the integral

$$\text{erf}(x) = \frac{2}{\sqrt{\pi}} \int_0^x dt \, e^{-t^2}.$$

Show that this can be re-written in the form

$$\text{erf}(x) = 1 - \frac{2}{\sqrt{\pi}} \int_x^\infty dt \, e^{-t^2}.$$

By expressing the integral in terms of the new variable, $y = t^2$, and repeated integration by parts, show that the asymptotic expansion of $\text{erf}(x)$ is

$$\text{erf}(x) = 1 - \frac{e^{-x^2}}{x\sqrt{\pi}} \left(1 - \frac{1}{2x^2} + \frac{1.3}{(2x^2)^2} - \frac{1.3.5}{(2x^2)^3} + \cdots \right).$$

Exercise 5.23
The Sine integral is defined by

$$\text{Si}(x) = \int_0^x dt \, \frac{\sin t}{t}.$$

Show that the asymptotic expansion as $x \to \infty$ is

$$\text{Si}(x) \sim \frac{\pi}{2} - \frac{1}{x} \left(1 - \frac{2!}{x^2} + \frac{4!}{x^4} - \frac{6!}{x^6} + \cdots \right) \cos x - \frac{1}{x^2} \left(1 - \frac{3!}{x^2} + \frac{5!}{x^4} - \frac{7!}{x^6} + \cdots \right) \sin x,$$

and that the Taylor's series about the origin is

$$\text{Si}(x) = \sum_{k=0}^{\infty} \frac{(-1)^k x^{2k+1}}{(2k+1)(2k+1)!}.$$

Exercise 5.24
Show that the sequences defined in equation 5.26 (page 190), satisfy the conditions of equation 5.23.

Exercise 5.25
Show that the sequence $\ln(1 + x^k)$ is an asymptotic sequence as $x \to 0$.

Exercise 5.26
Show that $\phi_k(x) = x^k(a + \sin(x^{-k}))$, $k = 0, 1, \ldots$, is an asymptotic sequence as $x \to 0$, but that $d\phi_k/dx$ is not.

Exercise 5.27
If a_1, a_2, b_1 and b_2 are nonzero constants and

$$g(x) \sim a_1 x^{b_1} + a_2 x^{b_2}, \quad \text{with} \quad b_1 < b_2,$$

find the values of these constants which ensure that $g(x)$ gives the first two terms in the asymptotic expansion of the following functions as $x \to 0$:

(i) $f = \left(1 + \frac{1}{\cos x} \right)^{3/4}$,   (ii) $f = \sinh \sqrt{1 + xy}$,

(iii) $f = (1 + xy)^{1/x}$,   (iv) $f = \int_0^x dy \, \sin(\sqrt{x} + xy)$,

(v) $f = \sum_{k=1}^{\infty} \left(\frac{1}{3} \right)^k \sin \left(\frac{x}{k} \right)$,   (vi) $f = \int_0^{\pi} dt \, \frac{\cos t}{(t+x)^{3/2}}$,

where y is any positive real number.

Exercise 5.28

If $\phi_k(x)$, $k = 0, 1, \ldots$, is an asymptotic sequence with the property $\phi_k(x)\phi_l(x) = \phi_{k+l}(x)$, and

$$f(x) \sim \sum_{k=0}^{\infty} a_k \phi_k(x), \quad g(x) \sim \sum_{k=0}^{\infty} b_k \phi_k(x),$$

show that the asymptotic expansion of the product $f(x)g(x)$ is

$$f(x)g(x) \sim \sum_{k=0}^{\infty} c_k \phi_k(x), \quad \text{where} \quad c_n = \sum_{k=0}^{n} a_k b_{n-k}.$$

Exercise 5.29

Show that the functions e^{-kx}, $k = 1, 2, \ldots$, form an asymptotic sequence as $x \to \infty$.

Show further, using Maple if necessary, that

$$y(x) = ae^{-x} - \left(ae^{-x}\right)^2 + 4\left(ae^{-x}\right)^3 - \frac{88}{3}\left(ae^{-x}\right)^4 + \cdots$$

is an approximate solution of the equation

$$\left(\frac{d^2y}{dx^2}\right)^2 = \frac{dy}{dx} + y.$$

Use Maple to continue this expansion in order to estimate whether or not it converges for any finite value of x.

6

Continued fractions and Padé approximants

6.1 Introduction to continued fractions

Continued fractions are a useful method of representing numbers as fractions and, more generally, functions as ratios of polynomials. They are useful in many areas of mathematics because they provide the 'best' method of approximating numbers; this means that if $\frac{a}{b}$ is a continued fraction that approximates a real number α, given any other fraction $\frac{c}{d}$ with a smaller denominator, $0 < d \le b$, then $|\frac{a}{b} - \alpha| < |\frac{c}{d} - \alpha|$. This is the reason why the approximation $\frac{22}{7}$ for π is better than the Egyptian approximation $\frac{19}{6}$. In this section we shall give an introductory account of continued fractions; more details can be found in the excellent book by Khinchin (1964).

A continued fraction is a number (or function) that can be represented in the form

$$c = a_0 + \cfrac{b_1}{a_1 + \cfrac{b_2}{a_2 + \cfrac{b_3}{a_3 \cdots}}} \tag{6.1}$$

where a_0 and (a_k, b_k), $k = 1, 2, \ldots$, may be numbers, complex, real or integers, or functions of one or more variables. The continued fraction is simpler if $b_k = 1$ for all k and the a_k are positive integers:

$$c = a_0 + \cfrac{1}{a_1 + \cfrac{1}{a_2 + \cfrac{1}{a_3 \cdots}}} \tag{6.2}$$

We are primarily interested in this case, although many of the results quoted are not restricted in this manner.

The numbers a_k are named the *elements* of the given continued fraction and there may be an infinite or a finite number of elements, in which case the continued fraction terminates at the nth element,

$$c = a_0 + \cfrac{1}{a_1 + \cfrac{1}{a_2 + \cfrac{1}{\cdots \cfrac{1}{a_n}}}} \tag{6.3}$$

and is called a continued fraction of order n; so an nth-order continued fraction has $n+1$ elements. In this case, if the elements are positive integers the continued fraction is

195

a rational number. For obvious reasons it is convenient to write the continued fraction of equation 6.2 in the form

$$c = [a_0; a_1, a_2, \ldots],$$ (6.4)

or for an nth-order continued fraction

$$c = [a_0; a_1, a_2, \ldots, a_n].$$ (6.5)

Another convenient, commonly used, notation is

$$c = a_0 + \frac{1}{a_1+} \frac{1}{a_2+} \frac{1}{a_3+} + \cdots.$$

John Wallis (1616–1703)

Euclid (c. 330 BC)

Continued fractions have a very long history, although John Wallis first used the name in a book published in 1655. It is generally agreed that Euclid's algorithm for finding the gcd of two integers is the first technique that can be expressed in terms of continued fractions, though Euclid did not use the algorithm in this way. If a and b are two positive integers, with $a > b$, and the sequence of integers r_k is defined by the relation

$$r_k = r_{k+1} q_k + r_{k+2}, \quad r_0 = a, \quad r_1 = b,$$ (6.6)

with $0 \leq r_{k+1} \leq r_k$ and q_k an integer, then for some N, $r_{N+2} = 0$, so $r_N = r_{N+1} q_N$, and r_{N+1} is the gcd of a and b. A proof of this statement may be found in Apostol (1976, chapter 1).

The formula 6.6 may be re-written as

$$\frac{r_k}{r_{k+1}} = q_k + \frac{1}{r_{k+1}/r_{k+2}},$$ (6.7)

so

$$\frac{a}{b} = q_0 + \frac{1}{r_1/r_2} = q_0 + \frac{1}{q_1 + \frac{1}{r_2/r_3}} = \cdots = [q_0; q_1, q_2, \ldots, q_N].$$

Many subsequent applications that are now expressed in terms of continued fractions revolve around the approximation of numbers by simpler fractions. A well-known example is the approximation of π by 22/7 and 355/113, exercise 6.4. Another ancient problem, originally connected with the appearance of constellations, is the solution of the diophantine equation,

$$ax + by = c,$$ (6.8)

Diophantus (c. 3rd century BC)

Aryabhata (born AD 476)

where a, b and c are positive integers and where integers values of x and y are required. This type of equation is named after Diophantus, who found a rational solution. The general solution, reproduced below, was discovered by the Indian mathematician Aryabhata.

If a and b are relatively prime, then from Euclid's algorithm

$$\frac{a}{b} = [q_0; q_1, q_2, \ldots, q_N],$$

and we define another rational number by

$$\frac{p}{q} = [q_0; q_1, q_2, \ldots, q_{N-1}],$$

then it follows from equation 6.22 (page 206), that $bp - aq = (-1)^N$. Hence

$$ax + by = c(bp - aq)(-1)^N \quad \text{or} \quad \frac{1}{b}(x + (-1)^N cq) = \frac{1}{a}(cp(-1)^N - y) = t,$$

giving the general solution

$$x = bt - (-1)^N cq, \quad y = (-1)^N cp - at. \tag{6.9}$$

Giving t integer values produces solutions to equation 6.8. This solution was first discovered by Aryabhata and re-discovered in Europe by Méziriac in 1612.

More generally any equation to be solved in integers is named a diophantine equation. For example, if $f(x)$ is a polynomial of degree greater than three, it may be shown that the diophantine equation $y^2 = f(x)$ has at most a finite number of solutions. Pell's equation, $x^2 = 1 + ny^2$, named after, but not discovered by, John Pell is another example.

The original diophantine equation arose from a type of problem that also produces difficulties in the construction of orreries, mechanical models of the Solar System, and planetariums: for these mechanical devices the accurate approximation of numbers by fractions with small denominators is required. This led Huygens to his studies of continued fractions. We illustrate this problem using modern data. In its orbit round the Sun the Earth covers $360°$ in 365.2564 days.† The period of Saturn is 10 759.22 days, so while the Earth rotates through $360°$ Saturn rotates through

$$\alpha = \frac{365.2564}{10\,759.22} \times 360° = 12.221\,36°.$$

In order for a mechanical planetarium to accurately reproduce this ratio it is necessary to approximate it by a fraction with a small denominator, in order that the gears have as few teeth as possible. The continued fraction development of this ratio is $\alpha = [12; 4, 1, 1, 13, 1, 3]$ which gives a series of fractions of increasing accuracy:

$$[12; 4] = \frac{49}{4} = 12.25, \qquad [12; 4, 1] = \frac{61}{5} = 12.20,$$

$$[12; 4, 1, 1] = \frac{110}{9} = 12.2\dot{2}, \qquad [12; 4, 1, 1, 13] = \frac{1491}{122} = 12.221\,31.$$

After Huygens the subject flourished, and most of the major mathematicians of the eighteenth century were involved in the development of the theory, with the most

C. G. B.
de Méziriac
(1581–1638)

J. Pell
(1610–1685)

Christiaan
Huygens
(1629–1695)

---

† This is the orbit relative to the fixed stars, and defines the siderial year. The passage from perihelion to perihelion takes slightly longer, 365.2596 days, because the semi-major axis of the Kepler ellipse precesses through $0.003\,15°$ per siderial year, in the direction of the Earth's orbit, that is anti-clockwise when looking from the pole star onto the Earth's orbit. A day is defined to be 86 400 seconds, the second being defined by reference to an atomic clock: this is approximately the same as the mean solar day which is 86 400.002 seconds. The orbital direction of the Earth's rotation round the Sun may be deduced from two facts: (1) that the Sun rises in the east, and (2) that the solar day (rotation of the Earth relative to the Sun) is longer than the mean period (relative to the fixed stars), 86 164.101 seconds. This data is taken from the *Observer's Handbook* (1995).

L. Euler
(1707–1783)

significant developments being made by Euler, who also gave the first clear exposition of the subject.

Some modern applications of continued fractions are given in the booklet by Moore (1964) and a series of *American Journal of Physics* articles by Srinivasan (1992), Krantz (1998), Pickett (1997), Sullivan (1978) (see also Bryant (1995)), Phelps (1995), English and Winters (1997) (see also the comments by Holstein (1997) and Dean (1999)).

6.1.1 *Maple and continued fractions*

Maple can deal with continued fractions. There are two methods of converting a real number or fraction to a continued fraction. The simplest is to use convert with the confrac option: thus the continued fraction of *e* is

```
> convert( evalf(exp(1),12), confrac);
```
$$[2, 1, 2, 1, 1, 4, 1, 1, 6, 1, 1, 8, 1, 1, 10]$$

Here the value of *e* has been computed to twelve digits — with fewer digits the last element is different: in this example the number of elements computed depends upon the value of Digits and the second argument of evalf. Care is normally needed to ensure that the number of elements requested is consistent with the accuracy of the number provided: there is no simple relation giving the size of Digits needed to produce an accurate continued fraction of a given length. Note also that the list of elements has the form $[a_0, a_1, \ldots, a_n]$, which is another common notation for continued fractions.

For more serious use it is better to use the procedure cfrac accessed with the command with(numtheory). This may be used in the form

```
cfrac(evalf(x),n,'quotients','con')
```

which finds the first $n + 1$ convergents of *x* in the form of a list put in con. In the argument list con is enclosed in forward-quotes in order to ensure that it is unevaluated; alternatively evaln(con) may be used or con may be cleared before using it; not clearing this argument produces an error if con is bound. The optional argument 'quotients' controls the form of the output: without it the continued fraction is in the extended form of equation 6.3 (page 195), which is very unwieldy if *n* is large.

For instance, if $x = \pi$, we obtain

```
> cfrac(evalf(Pi,11),7,'quotients',evaln(con));
```
$$[3, 7, 15, 1, 292, 1, 1, 1, \ldots]$$

Here we evaluated π to eleven digits: ten digits gives $[3; 7, 15, 1, 293, 11, 1, 1, \ldots]$. Some of the convergents are:

$$\mathrm{con}[2] = \frac{22}{7}, \quad \mathrm{con}[3] = \frac{333}{106}, \quad \mathrm{con}[4] = \frac{355}{133},$$

which give π to 3, 5 and 7 digits, respectively.

The reverse procedure, obtaining numbers from lists, is achieved using the nthconver command, also accessed using with(numtheory). For the continued fraction $c =$

$[1; 2, 3, 4, 5, \ldots]$, the second and fifth convergents, $[1; 2, 3]$ and $[1; 2, 3, 4, 5, 6]$ respectively, are

```
>   c:=[seq(k,k=1..10)]:
>   c2:=nthconver(c,2):     c2=evalf(c2);
```

$$\frac{10}{7} = 1.428571429$$

```
>   c5:=nthconver(c,5):     c5=evalf(c5);
```

$$\frac{1393}{972} = 1.433127572$$

Exercise 6.1

Use Euler's algorithm to find the greatest common divisors of

$$(105, 96), \quad (1005, 98) \quad \text{and} \quad (81313, 106361).$$

Exercise 6.2

Show that the equations

$$\text{(i)} \quad 31x + 11y = 2, \qquad \text{(ii)} \quad 33x + 19y = 5,$$

have the solutions $(x, y) = (11n - 12, 34 - 31n)$ and $(19n - 20, 35 - 33n)$, where n is an integer.

Exercise 6.3

Show that the continued fraction of the angle 12.22136 covered by Saturn in a siderial year is $[12; 4, 1, 1, 13, 1, 3]$ and that inclusion of more elements is not warranted.

Exercise 6.4

Find the first four convergents of π and determine their relative accuracy, that is $E = (c/\pi - 1)$, c being the continued fraction approximation.

6.2 Elementary theory

Before proceeding further, we consider how continued fractions are evaluated and computed, as without some experience in manipulating these objects the subsequent algebra will be difficult to understand. Please note that whilst Maple will perform all these calculations it is important that you do not use it, other than as a hand calculator if necessary, because it is unlikely that you will understand how continued fractions work unless you perform some computations by hand.

Consider the specific finite continued fraction

$$c = [1; 2, 2, 2] = 1 + \cfrac{1}{2 + \cfrac{1}{2 + \frac{1}{2}}}$$

We evaluate this 'backwards':

$$c = 1 + \cfrac{1}{2 + \frac{2}{5}} = 1 + \frac{5}{12} = \frac{17}{12}.$$

This process will be formalised when we consider the canonical representation of continued fractions.

Exercise 6.5

Evaluate the continued fractions

$$[1; 1, 2], \quad [1; 1, 2, 1, 2], \quad [1; 1, 2, 1, 2, 1, 2],$$
$$[1; 2, 2], \quad [1; 2, 2, 2, 2], \quad [1; 2, 2, 2, 2, 2, 2].$$

If these patterns were continued, can you guess what the limits would be?

Now consider the inverse problem of finding the continued fraction, with positive integer elements, from a given real number, and as an example consider the number $e = 2.718\,281\,83$ which we wish to approximate as a sequence of finite continued fractions. Clearly the first element is 2, so write

$$e = [2; a_1, a_2, \ldots] = 2 + \cfrac{1}{a_1 + \cfrac{1}{a_2 + \cdots}} = 2 + \frac{1}{r_1}, \quad \text{with} \quad r_1 > 1,$$

where a_k, $k = 1, 2, \ldots$, are positive integers and we have put

$$r_1 = a_1 + \cfrac{1}{a_2 + \cfrac{1}{a_3 \cdots}} = [a_1; a_2, a_3, \ldots].$$

Thus $r_1 = \frac{1}{e-2} = 1.39\cdots$ and a_1 will be the integer part of this, just as a_0 was the integer part of e. Thus†

$$a_1 = \text{int}\left(\frac{1}{e-2}\right) = 1.$$

Now repeat this process, but starting with r_1 instead of e, to give

$$r_1 = a_1 + \frac{1}{r_2} \quad \text{with} \quad r_2 = a_2 + \cfrac{1}{a_3 + \cfrac{1}{a_4 \cdots}}$$

so that

$$r_2 = \frac{1}{r_1 - a_1} \quad \text{and hence} \quad a_2 = \text{int}(r_2) = 2.$$

The general formulae are

$$r_{k+1} = \frac{1}{r_k - a_k} \quad \text{and} \quad a_{k+1} = \text{int}(r_{k+1}). \tag{6.10}$$

In practice this involves performing the following sequence on a calculator. For the number $e = 2.718\,281\,83$:

(1) subtract the integer part ($a_0 = 2$) to form the next number $x_1 = e - 2$, $0 \le x_1 < 1$;
(2) define $r_1 = 1/x_1 > 1$ and subtract the integer part ($a_1 = \text{int}(r_1) = 1$) to form the next number $x_2 = r_1 - a_1$, $0 \le x_2 < 1$;

---

† We use the notation int(x) to mean the integer part of a positive number x, the largest integer less than or equal to x; int is the same as the Maple functions **floor** and **trunc** for positive arguments.

(3) define $r_2 = 1/x_2 > 1$ and subtract the integer part ($a_2 = \text{int}(r_2) = 2$) to form the next number $0 \le x_3 = r_2 - a_2 < 1$; etc.

Note that for positive arguments the Maple functions for int(x) are floor(x) and trunc(x). The function floor(x) gives the integer less than or equal to x, thus floor(4.4)=4 and floor(-4.4)=-5: if $x > 0$, floor(x)=trunc(x), but if $x < 0$, floor(x)=trunc(x)-1.

Using this algorithm we obtain

$$e \simeq 2.718\,281\,83 = [2; 1, 2, 1, 1, 4, 1, 1, 6].$$

In practice we know the original number to only a certain finite number of decimal places, which limits the number of elements that can be found. Any elements that are obtained beyond this point are essentially meaningless. As an example my calculator, which is accurate to 12 decimal places, gives

$$e = [2; 1, 2, 1, 1, 4, 1, 1, 6, 1, 1, 8, 1, 1, 10, \overbrace{2, 10, 1, 1, 5, 1}^{\text{wrong}}, \ldots],$$

$$\pi = [3; 7, 15, 1, 292, 1, 1, 1, 2, 1, \overbrace{4, 1, 2, 14, 16, 13, 1}^{\text{wrong}}, \ldots].$$

Using Maple with a sufficiently high accuracy shows that the designated last few elements are incorrect.

Exercise 6.6
With a hand calculator compute the fifth-order continued fractions for

$$e^2, \quad \pi, \quad \sqrt{2}, \quad \sqrt{3}, \quad 2^{1/3}, \quad \text{and} \quad e^{\pi}.$$

We now return to the general discussion of continued fractions.

The kth *convergent* of a continued fraction $c = [a_0; a_1, a_2, \ldots]$ is defined to be the kth-order continued fraction

$$c_k = [a_0; a_1, a_2, \ldots, a_k]. \tag{6.11}$$

Hence an nth-order continued fraction has $n + 1$ convergents,

$$a_0, \quad [a_0; a_1], \quad [a_0; a_1, a_2], \quad \ldots, \quad [a_0; a_1, a_2, \ldots, a_n].$$

The *remainder* is defined to be

$$r_k = [a_k; a_{k+1}, a_{k+2}, \ldots]. \tag{6.12}$$

For finite continued fractions we have

$$[a_0; a_1, a_2, \ldots, a_n] = [a_0; a_1, a_2, \ldots, a_{k-1}, r_k], \quad \text{for all} \quad 1 \le k \le n, \tag{6.13}$$

and for an infinite continued fraction

$$[a_0; a_1, a_2, \ldots] = [a_0; a_1, a_2, \ldots, a_{k-1}, r_k], \quad \text{for all} \quad k \ge 1. \tag{6.14}$$

Interesting types of continued fractions are those in which the elements are all the same, $a_k = a$ for $k = 1, 2,\ldots$ for some positive number a, not necessarily an integer. In this case the continued fraction $c = [1; a, a,\ldots]$ can be written in one of the forms

$$c = [1; r], \quad c = [1; a, r], \quad c = [1; a, a, r], \quad \text{etc.,}$$

where $r = [a; a, a,\ldots]$. The first two of these expressions can be written as

$$c = 1 + \frac{1}{r} \quad \text{and} \quad c = 1 + \cfrac{1}{a + \cfrac{1}{r}}$$

and eliminating c gives a quadratic equation in r: the positive root yields an explicit expression for c,

$$c = 1 + \frac{\sqrt{a^2 + 4} - a}{2}.$$

Thus if $a = 2$ we have $c = [1; 2, 2,\ldots] = \sqrt{2}$, and if $a = 1$, $c = [1; 1, 1,\ldots] = (1 + \sqrt{5})/2$.

Exercise 6.7
Consider the continued fraction

$$c = [1; a, b, a, b,\ldots],$$

where a and b are positive numbers, and show that

$$c = 1 - \frac{b}{2} + \frac{\sqrt{a^2 b^2 + 4ab}}{2a}.$$

Hence show that $[1; 1, 2, 1, 2,\ldots] = \sqrt{3}$.

From your solution of this exercise it should be clear that any periodic continued fraction c, that is one in which a particular finite sequence of integer elements repeats indefinitely, can be expressed in the form $c = (m_1 + \sqrt{m_2})/m_3$ for some integers m_1, m_2 and m_3.

Exercise 6.8
If

$$x_n = \underbrace{[2; 2, 2,\ldots, 2]}_{n+1 \text{ terms}}, \quad n = 1, 2,\ldots,$$

show that

$$x_k = 2 + \frac{1}{x_{k-1}}, \quad x_1 = \frac{5}{2},$$

and hence that $x_k \to 1 + \sqrt{2}$ as $k \to \infty$. By defining $y_k = x_k - 1$ deduce that the sequence

$$y_{n+1} = \frac{y_n + 2}{y_n + 1}, \quad n = 1, 2,\ldots, \quad y_1 = \frac{3}{2},$$

converges to $\sqrt{2}$ through the continued fraction convergents.

Find a similar sequence which converges to $\sqrt{p}$ and show that

$$\left(y_{n+2} - \sqrt{p} \right) \leq \left(\frac{\sqrt{p} - 1}{\sqrt{p} + 1} \right)^2 \left(y_n - \sqrt{p} \right).$$

Exercise 6.9

Use Maple to show that the iterative sequence

$$x_{n+1} = \frac{1}{2}\left(x_n + \frac{2}{x_n}\right), \quad x_0 = 1,$$

converges to $\sqrt{2}$ and that x_1, x_2, x_3 and x_4 correspond to the continued fraction

$$z_m = \underbrace{[1;2,2,\ldots,2]}_{m}$$

for $m = 2$, 4, 8 and 16 respectively.

Exercise 6.10

Compute the first six convergents of e, e_k, $k = 1, 2, \ldots, 6$, and find the relative error

$$E_k = 1 - \frac{e_k}{e}.$$

Repeat for the first five convergents of π and the first six convergents of $2^{1/3}$.

From the results of this exercise you should notice that (a) the convergents approach the limit very rapidly, and faster if the denominators of the convergents are large, and (b) the convergents are alternately larger and smaller than the limit. We show later that these results are quite general and give reasons for the rapid convergence.

6.2.1 Canonical representation

In most of the preceding examples the elements have been positive integers and the convergents reduce naturally to unique fractions, but if the elements are not integers there is no unique form for the continued fraction. For instance, consider the number

$$c = [1;\pi,\pi,\pi] = 1 + \cfrac{1}{\pi + \cfrac{1}{\pi + \cfrac{1}{\pi}}} = \frac{1.291\,493\cdots}{1}$$

$$= \frac{1 + 2\pi + \pi^2 + \pi^3}{2\pi + \pi^3} = \frac{48.159\,066}{37.289\,462}.$$

For reasons that will soon become apparent it is convenient to express the kth-convergent as a ratio p_k/q_k, so we need a prescription which defines unique values of p_k and q_k: this is named the canonical form.

The zeroth-order continued fraction is

$$[a_0] = a_0,$$

so we take the canonical representation of this to be $p_0/q_0 = a_0/1$, that is $p_0 = a_0$ and $q_0 = 1$. We now use induction and assume that the canonical representation of the continued fractions of all orders less than n are defined. Using equation 6.13, with $k = 1$, we can write

$$[a_0;a_1,a_2,\ldots,a_n] = [a_0;r_1] = a_0 + \frac{1}{r_1}, \quad (6.15)$$

where $r_1 = [a_1; a_2, a_3, \cdots, a_n]$ is an $(n-1)$th-order continued fraction. By definition we know the canonical representation of this, and write it as

$$r_1 = \frac{p'}{q'}, \quad \text{hence from equation 6.15} \quad [a_0; a_1, a_2, \ldots, a_n] = \frac{a_0 p' + q'}{p'},$$

so the canonical representation of the nth-order continued fraction is known. Thus for a finite continued fraction the canonical representation is obtained by working 'backwards' through the sequence, as in the example on page 199. The example given at the beginning of this section should make this process clear.

$$c = [1; \pi, \pi, \pi] = [1; r_1], \quad r_1 = [\pi; \pi, \pi] = [\pi; r_2],$$
$$r_2 = [\pi; \pi] = \pi + \frac{1}{\pi} = \frac{\pi^2 + 1}{\pi}.$$

Therefore,

$$r_1 = \pi + \frac{1}{r_2} = \pi + \frac{\pi}{\pi^2 + 1} = \frac{\pi^3 + 2\pi}{\pi^2 + 1}, \quad \text{and}$$
$$c = 1 + \frac{1}{r_1} = 1 + \frac{\pi^2 + 1}{\pi^3 + 2\pi} = \frac{\pi^3 + \pi^2 + 2\pi + 1}{\pi^3 + 2\pi},$$

the latter expression being the canonical form for c.

Exercise 6.11

Find the canonical representations of all the convergents of the number

$$c = [1; \pi, \pi^{\frac{1}{2}}, \pi^{\frac{1}{3}}],$$

and show that its canonical representation is

$$c = \frac{\pi^{\frac{11}{6}} + 2\pi^{\frac{5}{6}} + \pi^{\frac{1}{3}} + 1}{\pi^{\frac{11}{6}} + \pi^{\frac{5}{6}} + \pi^{\frac{1}{3}}} = 1.294365.$$

6.2.2 Convergence of continued fractions

We now come to one of the most important results in the theory of continued fractions, which relates the numerators and denominators of successive convergents. From this result, equation 6.17 below, we shall be able to deduce many properties of the convergents.

Consider the finite continued fraction

$$c_n = [a_0; a_1, a_2, \ldots, a_n], \tag{6.16}$$

and suppose that p_k/q_k is the canonical representation of the kth-convergent, $c_k = p_k/q_k$. Then for any $k \geq 2$,

$$\begin{aligned} p_k &= a_k p_{k-1} + p_{k-2} \\ q_k &= a_k q_{k-1} + q_{k-2}. \end{aligned} \tag{6.17}$$

The proof of this relation is by induction. First we show, by direct computation, that the result is true for $k = 2$:

$$\frac{p_0}{q_0} = [a_0], \quad \text{so} \quad p_0 = a_0, \quad q_0 = 1,$$

$$\frac{p_1}{q_1} = [a_0; a_1] = p_0 + \frac{1}{a_1}, \quad \text{so} \quad p_1 = p_0 a_1 + 1, \quad q_1 = a_1,$$

$$\frac{p_2}{q_2} = [a_0; a_1, a_2] = \frac{a_2(q_1 p_0 + 1) + p_0}{a_2 q_1 + 1},$$

which gives the required result since $p_1 = p_0 a_1 + 1 = p_0 q_1 + 1$.

We now proceed by induction, by assuming the results to be true for all $k < m$ and then proving it true for $k = m$. Consider the $(m-1)$th-order continued fraction $\alpha = [a_1; a_2, a_3, \ldots, a_m]$ and denote its rth-order convergent by α_r / β_r, then

$$\frac{p_k}{q_k} = [a_0; a_1, a_2, \ldots, a_k] = \left[a_0; \frac{\alpha_{k-1}}{\beta_{k-1}} \right] = a_0 + \frac{\beta_{k-1}}{\alpha_{k-1}}$$

$$= \frac{a_0 \alpha_{k-1} + \beta_{k-1}}{\alpha_{k-1}} \quad \text{for all } k = 1, 2, \ldots, m-1, \quad (6.18)$$

so we have

$$\begin{aligned} p_k &= a_0 \alpha_{k-1} + \beta_{k-1} \\ q_k &= \alpha_{k-1}. \end{aligned} \quad (6.19)$$

But we have assumed that equations 6.17 hold for α_{m-1} and β_{m-1}, but with a_m rather than a_{m-1}, because α begins with a_1 not a_0, thus

$$\begin{aligned} \alpha_{m-1} &= a_m \alpha_{m-2} + \alpha_{m-3} \\ \beta_{m-1} &= a_m \beta_{m-2} + \beta_{m-3}. \end{aligned}$$

On substituting these equations into equation 6.17 for p_m we obtain

$$\begin{aligned} p_m &= a_0 (a_m \alpha_{m-2} + \alpha_{m-3}+) + a_m \beta_{m-2} + \beta_{m-3} \\ &= a_m (a_0 \alpha_{m-2} + \beta_{m-2}) + (a_0 \alpha_{m-3} + \beta_{m-3}) \\ &= a_m p_{m-1} + p_{m-2}, \end{aligned}$$

where equation 6.18 with $k = m - 1$ and $m - 2$ has been used to obtain the last line. Similarly,

$$\begin{aligned} q_m &= a_m \alpha_{m-2} + \alpha_{m-3} \\ &= a_m q_{m-1} + q_{m-2}. \end{aligned}$$

Thus if equations 6.17 are true for $k = m - 1$ they are true for $k = m$; since they are true for $k = 2$ they are true for all $k \geq 2$.

This proof is also valid for infinite continued fractions and does not assume that the elements are integers. Finally we note that by defining $p_{-1} = 1$ and $q_{-1} = 0$ the equations for the convergents $c_n = [a_0; a_1, a_2, \ldots, a_n] = p_n / q_n$ become

$$p_k = a_k p_{k-1} + p_{k-2}, \quad p_{-1} = 1, \quad p_0 = a_0, \quad k = 1, 2, \ldots, \quad (6.20)$$

$$q_k = a_k q_{k-1} + q_{k-2}, \quad q_{-1} = 0, \quad q_0 = 1. \quad (6.21)$$

From these relations we can determine some properties of the convergents. Multiply equation 6.20 by q_{k-1}, 6.21 by p_{k-1} and subtract to obtain

$$q_k p_{k-1} - q_{k-1} p_k = -(q_{k-1} p_{k-2} - q_{k-2} p_{k-1}), \quad k \geq 1,$$

and since $q_0 p_{-1} - q_{-1} p_0 = 1$ this gives

$$q_k p_{k-1} - q_{k-1} p_k = (-1)^k, \quad k \geq 1.$$

Dividing by q_k and q_{k-1} gives

$$\frac{p_{k-1}}{q_{k-1}} - \frac{p_k}{q_k} = \frac{(-1)^k}{q_k q_{k-1}}, \quad k \geq 1. \tag{6.22}$$

Similarly, by multiplying equations 6.20 by q_{k-2}, 6.21 by p_{k-2} we obtain

$$\frac{p_{k-2}}{q_{k-2}} - \frac{p_k}{q_k} = \frac{(-1)^{k-1} a_k}{q_k q_{k-2}}, \quad k \geq 2. \tag{6.23}$$

Exercise 6.12
Prove equation 6.23.

This last result, 6.23, is illuminating because it shows that even-order convergents form an increasing sequence and those of odd-order a decreasing sequence, if $a_k > 0$ for $k \geq 1$.

Also equation 6.22 shows that every odd-order convergent is greater than the following even-order convergent.

Thus for a finite-order continued fraction α every even-order convergent is less than α and every odd-order convergent is greater than α — except for the last convergent which equals α. For infinite continued fractions there is an infinite sequence of convergents p_k/q_k, $k = 0, 1, 2, \ldots$; if this sequence converges to a unique number α this is defined to be the value of the continued fraction. We can write this result in the form

$$\frac{p_0}{q_0} < \frac{p_2}{q_2} < \frac{p_4}{q_4} < \cdots < \alpha < \cdots < \frac{p_5}{q_5} < \frac{p_3}{q_3} < \frac{p_1}{q_1}.$$

A necessary and sufficient condition for this sequence to converge is that the sum

$$\sum_{k=1}^{\infty} a_k$$

diverges, which is certainly the case when the a_k are positive integers. A proof of this result is given in Khinchin (1964, page 10).

For an infinite continued fraction with value α it can be shown, Khinchin (1964, pages 9–15), that

$$\frac{1}{q_k(q_k + q_{k+1})} < \left| \alpha - \frac{p_k}{q_k} \right| < \frac{1}{q_k q_{k+1}},$$

which bounds the difference $|\alpha - p_k/q_k|$, showing that it depends upon the magnitude of the denominators, q_k, so some idea of the rate at which these grow with k is useful. We have, from the second line of equation 6.17,

$$q_k = a_k q_{k-1} + q_{k-2} > q_{k-1} + q_{k-2} \geq 2q_{k-2}$$

since a_k is a positive integer, and the second of equations 6.18 shows that $q_{k-1} \geq q_{k-2}$. Successive applications of this result gives

$$q_{2k} > 2^k q_0, \qquad q_{2k+1} > 2^k q_1 > 2^k.$$

Thus the denominators of the convergents grow at least as rapidly as the terms of a geometric progression, and the convergents approach the limit exponentially fast,

$$\left| \alpha - \frac{p_{2k}}{q_{2k}} \right| < 4^{-k}.$$

Exercise 6.13

Find the continued fraction approximations of the numbers

$$\sqrt{2}, \quad \sqrt{3}, \quad \frac{\sqrt{5}-1}{2}, \quad \frac{\sqrt{17}-4}{2}, \quad \frac{\sqrt{35}-5}{2}$$

using the command

```
cfrac( evalf(a,20), 20,'quotients', 'con')
```

where a is one of the above numbers. Remember to use with(numtheory).

For each number plot a graph of the sequence [k,c[k]], $k = 1, 2, \ldots, N-1$, where $N =$nops(con) and

```
c[k]:=log10( abs(a-con[k]) )
```

What do you notice about the elements for these numbers, and can you see any correlation between the behaviour of the graphs and the magnitude of the elements?

Repeat this exercise for the numbers

$$\pi, \quad e, \quad 2^{1/3}, \quad 2^{1/7}, \quad \frac{5^{1/3}-1}{2}.$$

Do you notice any differences between these results and those of the previous set?

The patterns you have just observed are contained in a theorem due to Lagrange: a real number c has an infinite continued fraction expansion which eventually repeats if and only if c is of the form $(p + q\sqrt{m})/r$ for some integers p, q, r and m with $m \geq 2$ and not a square. Further results of this nature are examined in exercise 6.15.

It can be shown that every real number has a unique continued fraction representation. From a practical viewpoint continued fractions are not always useful as, for instance, there are no convenient rules for doing arithmetic; even adding two continued fractions using their elements is difficult. The main practical use of continued fractions stems from the fact that they provide the 'best' method, as defined in section 6.1, of approximating real numbers by fractions.

Exercise 6.14

Use Maple to compute the first 50 elements of the number

$$x = 13^{\frac{3}{11}} = [a_0; a_1, a_2, \ldots, a_{50}, \ldots]$$

and find the value of the geometric mean

$$\left[\prod_{k=1}^{N} a_k\right]^{1/N}, \quad N = 50.$$

Repeat this computation for different values of x, say $x = 7^{k/11}$ and $x = 13^{k/11}$, $k = 1, 2, \ldots, 8$.

You should find that all your results are close to the value of the limit

$$\prod_{r=1}^{\infty}\left(1 + \frac{1}{r(r+2)}\right)^{\frac{\ln r}{\ln 2}} \simeq 2.68\ldots.$$

The results obtained in the previous exercise are a consequence of the theorem of Khinchin (1964, page 86) which states that if $f(r)$ is a non-negative function of the integer argument r such that $f(r) < Cr^{\frac{1}{2}-\delta}$, $r = 1, 2, \ldots$ for some positive constants C and δ then for almost all α in the interval $(0, 1)$ if

$$\alpha = [0; a_1, a_2, \ldots, a_n, \ldots]$$

then

$$\lim_{N \to \infty} \frac{1}{N} \sum_{k=1}^{N} f(a_k) = \frac{1}{\ln 2} \sum_{r=1}^{\infty} f(r) \ln\left(1 + \frac{1}{r(r+2)}\right). \tag{6.24}$$

On setting $f(r) = \ln r$ the following result is obtained:

$$(a_1 a_2 \cdots a_N)^{1/N} \to \prod_{r=1}^{\infty}\left(1 + \frac{1}{r(r+2)}\right)^{\frac{\ln r}{\ln 2}} \simeq 2.6854\cdots.$$

This product converges very slowly so it is difficult to obtain the limit of the left hand side numerically — over 500 000 terms are needed to achieve the result quoted above by direct multiplication. Methods of estimating such quantities more efficiently were discussed in chapter 4, particularly exercise 4.26 (page 150).

Finally, we use this theorem to derive another interesting result. Choose an integer M and set

$$f(r) = \begin{cases} 1 & r = M \\ 0 & r \neq M \end{cases}$$

so that the sum on the left hand side of equation 6.24,

$$\sum_{j=1}^{N} f(a_j),$$

is just the number of times the integer M occurs amongst the first N elements. Thus the ratio

$$d_M = \lim_{N \to \infty} \frac{1}{N} \sum_{j=1}^{N} f(a_j) = \frac{\ln\left((M+1)^2\right) - \ln\left(M(M+2)\right)}{\ln 2} \sim \frac{1}{M^2 \ln 2} + O(M^{-3}) \tag{6.25}$$

is just the frequency of occurrence of the integer M in the continued fraction expansion of *most* numbers in the interval $(0, 1)$, and it is independent of the actual number. That is, a given positive integer M occurs in the continued fraction expansion of most numbers with the same frequency, given by equation 6.25.

Exercise 6.15
Find the continued fraction representation of the numbers $\sqrt{k}$ for $2 \leq k \leq 99$, excluding the perfect squares, and show that they all have the form

$$\sqrt{k} = [a_0; a_1, a_2, \ldots, a_{n-1}, 2a_0, a_1, \ldots]$$

where $a_{pn} = 2a_0$, $a_{pn+r} = a_r$, $r = 1, 2, \ldots, n-1$, $p = 2, 3, \ldots$ and n is some integer satisfying $n \geq 1$.

The number n is named the period of this type of continued fraction. Use Maple to show that for the integers $2 \leq k \leq 500$, excluding the perfect squares, the number of numbers with period N are as shown in the table.

period N	1	2	3	4	5	6	7	8	9	10
number of non-square integers between 2 and 500	22	95	5	69	17	52	8	38	7	0

If a number x has an n-periodic continued fraction show that an equation for x is

$$x = [a_0; a_1, a_2, \ldots, a_{n-1}, a_0 + x]$$

and that this is a quadratic equation for x.

If $n = 2$ show that x is a perfect square provided a_0 is divisible by a_1. If $n = 3, 5$ or 7 show that if $a_i = a_{n-i}$, $i = 1, \ldots, (n-1)/2$ then x may be a perfect square. For the cases $n = 4, 6$ or 8 show that the same result holds if $a_i = a_{n-i}$, $i = 1, \ldots, n/2$.

6.3 Introduction to Padé approximants

Power series and, in particular, Taylor's series do not usually provide practical approximations to functions. For example, the Taylor's series approximation to $\sin x$ requires about 12 terms to find the value of $\sin 2\pi \, (= 0)$ to better than 10^{-6} even though the series converges for all values of x. There are a variety of superior methods for approximating functions of a single variable, and in this section we investigate a method using sequences of functions comprising ratios of polynomials; for simplicity we shall call such ratios *rational functions*. That such approximations will sometimes be vastly superior to a power series is seen from the simple example

$$\frac{1}{1 - x} = 1 + x + x^2 + \cdots + x^n + \cdots$$

in which the right hand side converges only for $|x| < 1$ and the left hand side is defined for all x other than $x = 1$. We should like to know when a given power series can be represented more efficiently by a simpler rational approximation.

There is no unique method of approximating a given function by a rational function, so before we embark on our discussion of the method of Padé we examine a particular

H. E. Padé
(1863–1953)

example and approximate it by various rational functions. Consider the function

$$f(x) = \left(\frac{1+4x}{1+x}\right)^{1/2}, \quad x \geq 0, \tag{6.26}$$

and its Taylor's series

$$f(x) = 1 + \frac{3}{2}x - \frac{21}{8}x^2 + \frac{87}{16}x^3 - \frac{1677}{128}x^4 + \cdots. \tag{6.27}$$

The function $f(x)$ varies smoothly from 1, at $x = 0$, to 2 as $x \to \infty$, and has singularities, at $x = -1$ and $x = -\frac{1}{4}$. The latter singularity determines that the Taylor's series converges only for $|x| < \frac{1}{4}$. The 'problem' singularity at $x = -\frac{1}{4}$ can be negated by applying the Euler transformation which, in this example, means changing to the new variable defined by

$$y = \frac{x}{1+4x}, \quad 0 \leq y \leq \frac{1}{4}, \quad \text{or} \quad x = \frac{y}{1-4y},$$

which moves this singularity at $x = -\frac{1}{4}$ to infinity and gives

$$f(x(y)) = \frac{1}{\sqrt{1-3y}} \tag{6.28}$$

$$= 1 + \frac{3}{2}y + \frac{27}{8}y^2 + \frac{135}{16}y^3 + \frac{2835}{128}y^4 + \cdots. \tag{6.29}$$

Since this last series has radius of convergence $\frac{1}{3} (> \frac{1}{4})$ we can produce a sequence of approximations, valid for all $x > 0$, simply by truncating the series for $f(x(y))$ at any given order and substituting for $y(x)$. The first few rational approximations are

$$f_1 = \frac{2+11x}{2(1+4x)}, \quad f_2 = \frac{8+76x+203x^2}{8(1+4x)^2}, \quad f_3 = \frac{16+216x+1014x^2+1759x^3}{16(1+4x)^3},$$

obtained by truncating the series 6.29 at y, y^2 and y^3 respectively. This sequence of approximations gives the following estimates of $f(\infty) = 2$:

$$1, \quad 1.375, \quad 1.586, \quad 1.718, \quad 1.804, \quad 1.863, \quad 1.903, \quad 1.931,$$

from which we observe that the convergence to 2 is rather slow. A better idea of the accuracy of this type of approximation is given by plotting the graphs of $f(x)$ against $f_k(x)$, for $k = 1, 2$ and 3, as shown in figure 6.1.

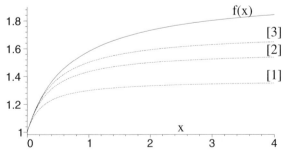

Figure 6.1   Graphs of the functions $f(x)$, equation 6.27, and the approximations $f_k(x)$, $k = 1, 2$ and 3.

In this example we have replaced a power series approximation, equation 6.27, which converges over only part of the range, by another series converging over the whole range by making an Euler transformation that takes into account the singularity of $f(x)$ nearest the point of expansion. From this new series we have created a sequence of rational approximations. One lesson to be learned from this simple example is that understanding the nature of the singularities of a function is an important aspect of finding good approximations.

6.4 Padé approximants

Padé approximants are a particular type of rational approximation to functions and are constructed to match the Taylor's series approximation of the original function as far as possible. Before dealing with the general theory we shall consider an example. One of the simplest rational functions is

$$\frac{a + bx}{c + dx},$$

where a, b, c and d are constants, and on setting this equal to the truncated Taylor's series of $f(x)$, equation 6.27, we see that to ensure equality at $x = 0$ we need $a = c$ and that the only remaining free parameters are b/a and d/a, which are obtained by forcing the equation

$$1 + \frac{3}{2}x - \frac{21}{8}x^2 = \frac{1 + bx/a}{1 + dx/a}$$

to be satisfied up to and including terms $O(x^2)$; this gives $\frac{b}{a} = \frac{13}{4}$ and $\frac{d}{a} = \frac{7}{4}$, and hence

$$\sqrt{\frac{1 + 4x}{1 + x}} \simeq \frac{1 + \frac{13}{4}x}{1 + \frac{7}{4}x}. \tag{6.30}$$

This is one of the Padé approximants of $f(x)$. The value of the approximation as $x \to \infty$ is $\frac{13}{7} = 1.857\ldots$, which is closer to the exact value, 2, than any of the approximations shown in figure 6.1; in order to exceed this accuracy using equation 6.29 we need the expansion to $O(y^5)$. Equation 6.31, derived in the next exercise, is another Padé approximant of the function $f(x)$.

Exercise 6.16
Use a rational function of the form

$$\frac{1 + ax + bx^2}{1 + cx + dx^2}$$

and the method just described in the text to show that the function defined in equation 6.26, expanded to $O(x^4)$, can be approximated by

$$f(x) \simeq \frac{1 + \frac{23}{4}x + \frac{121}{16}x^2}{1 + \frac{17}{4}x + \frac{61}{16}x^2}. \tag{6.31}$$

Compare the graphs of the exact and approximate functions, given in equations 6.30 and 6.31, for $0 \le x \le 10$.

This problem is more easily done using Maple, but you should do most of the analysis 'by hand' in order to see how the algebra works because this will help you understand the general case considered later.

Exercise 6.17
Show that the function defined in equation 6.28 is approximated by the Padé approximants

$$\frac{1 - \frac{3}{4}y}{1 - \frac{9}{4}y}, \quad \frac{1 - \frac{9}{4}y + \frac{9}{16}y^2}{1 - \frac{15}{4}y + \frac{45}{16}y^2}.$$

Show further that by substituting $y = x/(1 + 4x)$ in each of these expressions the results quoted in equation 6.30 and 6.31 are regained.
Hint: in this example you are advised to use Maple which will save some effort. The relevant Maple commands, which will be discussed later in this section, are

```
g:=1/sqrt(1-3y):
n:=3: gs:=taylor(g,y=0,n);
gp:=convert(gs,ratpoly);
```

and setting n:=3 and 5.

The general idea of a Padé approximant is to replace a power series $\sum_{n=0}^{\infty} a_n x^n$ by a sequence of ratios of polynomials of the form

$$P_M^N(x) = \frac{\sum_{n=0}^{N} A_n x^n}{\sum_{n=0}^{M} B_n x^n}, \quad B_0 = 1, \tag{6.32}$$

where the coefficient B_0 is chosen, without loss of generality, to be unity, and where the numerator and denominator have no common factors. In this approximation there are $N + 1$ coefficients in the numerator and M in the denominator, which are chosen so that the first $N + M + 1$ terms of the Taylor's series of $P_M^N(x)$ match the first $N + M + 1$ terms of the original power series. The resulting function is named a Padé
F. G. Frobenius   approximant; when it exists Frobenius (1881) and Padé (1892) proved it to be unique.
(1849–1917)   It is convenient to name this the [N/M] Padé approximant.

Some Padé approximants of the function $\sqrt{(1 + 4x)/(1 + x)}$ are given in equations 6.30 and 6.31. This example illustrates the property that a Padé approximant can be valid over a wider range than the original power series from which it is obtained. At first sight this may seem strange, but if we consider the problem the other way round and start from a rational function, say $1/(1 - x)$ whose power series about $x = 0$ converges only for $|x| < 1$, we see that in these circumstances we may consider the Padé approximant as an attempt to re-construct the original function from its power series representation.

If a function, $f(x)$, can be represented by the power series $\sum a_n x^n$ and $P_M^N(x)$ are Padé approximants derived from this series, then in many circumstances $P_M^N(x) \to f(x)$ as $N \to \infty$ and $M \to \infty$. In general N and M need not be the same, and one can consider the convergence of the sequence $P_r^{J+r}(x)$ as r increases, for some fixed value of J; here we normally consider only the diagonal approximants, $N = M$, in which the numerator and denominator polynomials have the same order.

Exercise 6.18
Show that the [0/1] and [1/1] Padé approximants of the series

$$f(x) = \sum_{n=0}^{\infty} a_n x^n \quad \text{are} \quad P_1^0(x) = \frac{a_0}{1 - x a_1/a_0} \quad \text{and} \quad P_1^1(x) = \frac{a_0 + (a_1^2 - a_0 a_2)x/a_1}{1 - a_2 x/a_1}.$$

Construction of the Padé approximant

The general evaluation of the Padé approximant is achieved via a sequence of simple operations. The fundamental equation which defines the constants A_k and B_k of equation 6.32 is

$$\frac{\sum_{j=0}^{N} A_j x^j}{\sum_{j=0}^{M} B_j x^j} = \sum_{k=0}^{N+M} a_k x^k + O(x^{N+M+1}).$$

On multiplying by the denominator of the left hand side this becomes

$$\sum_{j=0}^{N} A_j x^j = \left[\sum_{k=0}^{N+M} a_k x^k \right] \left(1 + B_1 x + B_2 x^2 + \cdots + B_M x^M \right).$$

Expanding the product on the right hand side, equating the coefficients of the powers of x and assuming that $M = N$, we obtain the following $2N + 1$ linear equations for the A and B coefficients:

$$
\begin{aligned}
a_0 &= A_0 \\
a_1 + B_1 a_0 &= A_1 \\
a_2 + B_1 a_1 + B_2 a_0 &= A_2 \\
&\ \ \vdots \\
a_N + B_1 a_{N-1} + B_2 a_{N-2} + \cdots + B_N a_0 &= A_N \\[1em]
a_{N+1} + B_1 a_N + B_2 a_{N-1} + \cdots + B_N a_1 &= 0 \\
a_{N+2} + B_1 a_{N+1} + B_2 a_N + \cdots + B_N a_2 &= 0 \\
&\ \ \vdots \\
a_{2N} + B_1 a_{2N-1} + B_2 a_{2N-2} + \cdots + B_N a_N &= 0.
\end{aligned}
\tag{6.33}
$$

In the more general case, $M \neq N$, the equations are similar but care is needed with the boundary terms; these equations are given in Baker (1975, page 6). The first $N + 1$ of these equations define the A coefficients in terms of the B coefficients; the remaining N equations are linear and involve only the N, B coefficients. We assume a solution of these equations exists to give the [N/N] Padé approximant to the power series. More generally if $N \neq M$ the same method gives the [N/M] approximant.

There are two simple but important consequences following from the construction of $P_M^N(x)$. First, if the [N/M] Padé approximant of $f(x)$ is $A_N(x)/B_M(x)$, then the [M/N] Padé approximant of $1/f(x)$ is $B_M(x)/A_N(x)$, provided $f(0) \neq 0$.

Second, if the $[M/M]$ diagonal Padé approximant of $f(x)$ is $P_M^M(x) = A_M(x)/B_M(x)$ and if x and y are related by the Euler transformation

$$x = \frac{\alpha y}{1 + \beta y} \tag{6.34}$$

for some constants α and β, then the $[M/M]$ approximant of $g(y) = f(x(y))$ is $A_M(x(y))/B_M(x(y))$. This generalises the result found in exercise 6.17.

The transformation 6.34 can be used to move any point, other than $y = 0$, to say $x = 1$. Consequently, all points other than $y = 0$ are equivalent in the Padé approximant. Hence we expect the domain of convergence of a Padé approximant to depend on the intrinsic analytic structure of the function rather than the artificially restricted radius of convergence of its Taylor's series representation.

Exercise 6.19

If $A_M(x)/B_M(x)$ is the $[M/M]$ Padé approximant of the function $f(x)$ show that the $[M + J/M]$ approximant of $g(x) = x^J f(x)$ is $x^J A_M(x)/B_M(x)$.

Maple has built-in procedures for finding Padé approximants. One is accessed using the convert command, with the option ratpoly, operating on series data types produced using the series or taylor commands, for instance; this automatically produces the diagonal Padé approximant if the order of the series is even. Thus

```
>  f:=((1+16*x)/(1+x))^(1/4);
```

$$f := \left(\frac{1 + 16\,x}{1 + x}\right)^{1/4} \tag{6.35}$$

```
>  ftay:=taylor(f,x=0,7);
```

$$ftay := 1 + \frac{15}{4}\,x - \frac{795}{36}\,x^2 + \frac{29505}{128}\,x^3 - \frac{5169405}{2048}\,x^4 + \frac{247579545}{8192}\,x^5$$
$$- \frac{25056627135}{65536}\,x^6 + \mathrm{O}(x^7)$$

It is worth noting here that the taylor command does not always, on the first application, produce a series to the required order; sometimes it seems to be necessary to apply it twice, for unknown reasons.

Now convert this to a rational polynomial:

```
>  fpad:=convert(ftay, ratpoly);
```

$$fpad := \frac{1 + \dfrac{219}{8}\,x + \dfrac{6921}{32}\,x^2 + \dfrac{228857}{512}\,x^3}{1 + \dfrac{189}{8}\,x + \dfrac{4881}{32}\,x^2 + \dfrac{118487}{512}\,x^3}$$

Note that fpad is not a series type as was ftay. Now compare these results graphically:

```
>  plot([f,fpad],x=0..4,color=black);
```

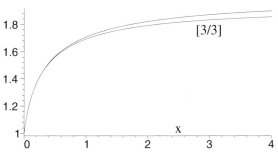

Figure 6.2   The upper line depicts the function defined in
equation 6.35, the lower line its [3/3] Padé approximant.

If the order n of the series is not even then `convert(ftay,ratpoly)`, by default, produces $P^{(n+1)/2}_{(n-1)/2}(x)$. If other approximants are required, then the two optional arguments
can be used as described in the help files — just type `?ratpoly` at the prompt.

An alternative, and more direct, method of obtaining Padé approximants is to use
the `pade` command, which needs to be accessed using `with(numapprox)`. Thus the
[2/3] Padé approximant of the expression f, defined in equation 6.35, is obtained using
the commands:

```
>  with(numapprox):      pade(f,x,[2,3]);
```

$$\frac{1 + \dfrac{190201}{9038} x + \dfrac{27994591}{289216} x^2}{1 + \dfrac{312617}{18076} x + \dfrac{16422781}{289216} x^2 - \dfrac{15946875}{1156864} x^3}$$

More can be found out about the `pade` command by typing `?pade` at the prompt.

The accuracy of the example shown in figure 6.2 is not always achieved. For instance
make a simple change to the function defined in 6.35:

```
>  f:=((1+16*x)/(1+x^2))^(1/4);
```

$$f := \left(\frac{1 + 16\,x}{1 + x^2}\right)^{(1/4)} \tag{6.36}$$

The [3/3] Padé approximant of this function is:

```
>  fpad:=pade(f,x,[3,3]);
```

$$fpad := \frac{1 + \dfrac{35058952603}{1225735952} x + \dfrac{10503242331}{43776284} x^2 + \dfrac{5269065384497}{9805887616} x^3}{1 + \dfrac{30156008795}{1225735952} x + \dfrac{50797711731}{306433988} x^2 + \dfrac{2430511050791}{9805887616} x^3}$$

```
>  plot([f,fpad],x=0..2,color=black);
```

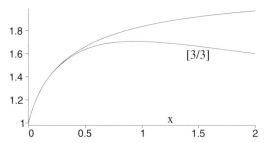

Figure 6.3   The upper line depicts the function 6.36, the lower
curve its [3/3] Padé approximant.

Now the approximation is not so good, although it is far better than the Taylor's series approximation and can be improved by increasing the order of the Padé approximant. You should try increasing the order N of the $[N/N]$ approximant; when $N = 8$ the resulting approximation is reasonable for $0 < x < 6$, but in other cases, $N = 5$ for instance, the denominator of the Padé approximant has a zero at $x = 0.5335$. These two examples show that Padé approximants can be very accurate but that care is needed as seemingly slight changes in the original function can change the accuracy of the approximation. In addition this last example shows that changing the order of the approximation may introduce zeros in the denominator, and hence poles in the approximation.

It is clear from these few examples that the accuracy with which a Padé approximant represents a given function depends on, amongst other things, the position and nature of the singularities of the function. One interesting aspect of the Padé approximant is the manner in which it represents singularities like $\sqrt{1+x}$ or $(1+x)^{-1/4}$, since the approximation itself contains only poles of the form $(x - a)^{-n}$, where a is a real or complex number and n a positive integer. In figure 6.4 we show how some of poles of the $[N/N]$, $N = 10j + 1$, $j = 1, 2$ and 3, Padé approximants of $\sqrt{1+x}$ are arranged: in this example all the poles are on the real axis to the left of -1, so to distinguish the different cases we have added an imaginary amount $0.01j$ to each pole and have shown only those poles in the interval $(-2, -1)$.

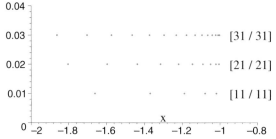

Figure 6.4   Diagram of some poles of the Padé approximants
of $\sqrt{1+x}$.

This figure shows that as j, the order of the approximation, increases, more and more poles accumulate to the left of $x = -1$. An understanding of this example is obtained by examining the analytic structure of the complex function $(1 + z)^{1/2}$, see for example Baker (1975, page 128). In general it is found that Padé approximants represent these types of singularity by a sequence of poles, that is zeros in the denominator, bunching near the singularity.

This concludes our introduction to the basic ideas of Padé approximants; more detail may be found in Baker (1975), which is devoted to the study of these approximations, and in Bender and Orszag (1978, chapter 8).

Exercise 6.20
Compute the $[10j + 1/10j + 1]$ Padé approximants for the function $(1 + x)^{-1/2}$, for $j = 1, 2$ and 3, and produce the equivalent to figure 6.4.

Exercise 6.21
Consider the variations of the Padé approximants of the real function

$$f(x) = \frac{1}{\sqrt{(x - 1)^2 + a^2}},$$

for various values of the real parameter a.

Set $a = 1, 1/2, 1/4$ and $1/8$, graphically compare the $[N/N]$ approximants with the original function for $0 < x < 2$ and describe the behaviour you observe: in particular show that for small values of a the Padé approximant behaves more like $(x - 1)^{-1}$ than $f(x)$.

Compute and graph the positions of the zeros of the denominator of various Padé approximants for these values of a and discuss your results.

6.4.1 Continued fractions and Padé approximants

Padé approximants and continued fractions are closely connected, and the connection is significant because many well known functions can be represented as continued fractions. In order to make this connection we need to use the more general form of the continued fraction defined in equation 6.1, for convenience reproduced here:

$$c = a_0 + \cfrac{b_1}{a_1 + \cfrac{b_2}{a_2 + \cfrac{b_3}{a_3 + \cdots}}} \tag{6.37}$$

This reduces to the form considered previously if we set $b_k = 1$, $k = 1, 2, \ldots$. If the nth truncation is the fraction A_n/B_n it may be shown that

$$\frac{A_0}{B_0} = \frac{a_0}{1}, \quad \frac{A_1}{B_1} = \frac{b_1 + a_0 a_1}{a_1}, \quad \frac{A_2}{B_2} = \frac{b_1 a_2 + b_2 a_0 + a_0 a_1 a_2}{b_2 + a_1 a_2},$$

and the general recurrence relations for A_n and B_n, $n \geq 0$, are

$$A_{n+1} = a_{n+1} A_n + b_{n+1} A_{n-1}, \quad A_{-1} = 1, \quad A_0 = a_0,$$

$$B_{n+1} = a_{n+1} B_n + b_{n+1} B_{n-1}, \quad B_{-1} = 0, \quad B_0 = 1,$$

which are the equivalent of equation 6.17.

Now set $a_k = 1$, $k = 1, 2, \ldots$, $a_0 = b_0$ and replace b_k by xb_k, where the b_k are numbers, and the continued fraction becomes the function

$$f(x) = b_0 + \cfrac{xb_1}{1 + \cfrac{xb_2}{1 + \cfrac{xb_3}{1 + \cdots}}} \qquad (6.38)$$

It can be shown, see Baker (1975, page 47), that if the coefficients converge to the limit b, that is, $\lim_{n \to \infty} b_n = b$, then this continued fraction converges provided $bx > -1/4$.

The first few convergents are

$$\frac{A_0}{B_0} = \frac{b_0}{1}, \quad \frac{A_1}{B_1} = \frac{b_0 + b_1 x}{1}, \quad \frac{A_2}{B_2} = \frac{b_0 + (b_1 + b_0 b_2)x}{1 + b_2 x},$$

and the recurrence relations for the numerators and the denominators are

$$\begin{aligned} A_{n+1}(x) &= A_n(x) + xb_{n+1} A_{n-1}(x), \quad A_{-1} = 1, \quad A_0 = b_0, \\ B_{n+1}(x) &= B_n(x) + xb_{n+1} B_{n-1}(x), \quad B_{-1} = 0, \quad B_0 = 1. \end{aligned} \qquad (6.39)$$

This scheme represents the function $f(x)$ as a ratio of two polynomials which we relate to the Padé approximant of the function after connecting the continued fraction representation to the power series of $f(x)$. The first few terms of the power series of the continued fraction 6.38 are

$$\begin{aligned} f(x) &= b_0 + b_1 x - b_1 b_2 x^2 + b_1 b_2 (b_2 + b_3) x^3 - \\ & \quad b_1 b_2 (b_2 (b_2 + 2b_3) + b_3 (b_3 + b_4)) x^4 + \cdots \\ &= \sum_{n=0}^{\infty} f_n x^n. \end{aligned} \qquad (6.40)$$

From this expansion we see that the coefficient b_m appears in f_n only if $n \geq m$ and has no effect on the coefficients $(f_0, f_1, \ldots, f_{m-1})$. Conversely, the coefficient f_n depends only upon the numbers $(b_0, b_1, b_2, \ldots, b_n)$.

The functions $A_n(x)$ and $B_n(x)$ will be polynomials in x, and we see from equation 6.39 that the degrees of successive polynomials are related by

$$\deg(A_{n+1}) = \max \left(\deg(A_n), 1 + \deg(A_{n-1}) \right),$$

with a similar equation for B_n. Since $\deg(A_0) = \deg(B_0) = 0$, $\deg(A_1) = 1$ and $\deg(B_1) = 0$, this gives

$$\begin{aligned} \deg(A_{2n}) &= n, & \deg(B_{2n}) &= n, \\ \deg(A_{2n+1}) &= n + 1, & \deg(B_{2n+1}) &= n. \end{aligned}$$

Thus the convergent $A_{2N}(x)/B_{2N}(x)$ is a ratio of N-degree polynomials which depend only upon the coefficients $(f_0, f_1, \ldots, f_{2N})$ of the power series representation of $f(x)$. It follows from the uniqueness of the Padé approximant that this convergent is the $[N/N]$ Padé approximant of $f(x)$.

Hence the continued fraction representation of a function is equivalent to a Padé approximant, an important result because many well known functions can be expressed naturally in terms of continued fractions. We explain why in the next section.

6.4.2 The hypergeometric function

Many common functions can be represented as continued fractions because they may be represented in terms of the hypergeometric function. The connection with continued fractions is algebraicly complicated, so we advise you to skim through this section first before reading it in detail.

One of the most important special functions of mathematical physics is the hypergeometric function, defined by the series

$$F(a,b;c;z) = 1 + \frac{ab}{1.c}z + \frac{a(a+1)\,b(b+1)}{1.2\,c(c+1)}z^2 + \frac{a(a+1)(a+2)\,b(b+1)(b+2)}{1.2.3\,c(c+1)(c+2)}z^3$$

$$+ \cdots \frac{a(a+1)\cdots(a+n-1)\,b(b+1)\cdots(b+n-1)}{n!\,c(c+1)\cdots(c+n-1)}z^n + \cdots \qquad (6.41)$$

$$= \frac{\Gamma(c)}{\Gamma(a)\Gamma(b)} \sum_{k=0}^{\infty} \frac{\Gamma(a+k)\Gamma(b+k)}{k!\,\Gamma(c+k)}z^k,$$

where a and b can be any complex numbers and c should not equal $0, -1, -2, \ldots$. If either a or b is a negative integer, $-N$, the hypergeometric function is a polynomial of degree N. This series converges for all $|z| < 1$, as can be proved using the ratio test; within this disk in the complex plane the series is an analytic function.

Exercise 6.22
Show that $F(a,b;c;x) = F(b,a;c;x)$, $F(a,0;c;x) = 1$ and that

$$F\left(a,b;a;\frac{x}{b}\right) = 1 + x + \sum_{k=2}^{\infty} \frac{x^k}{k!} \prod_{r=1}^{k-1} \left(1 + \frac{r}{b}\right).$$

Maple knows about the hypergeometric function, which is accessed with the command `hypergeom([a,b],[c],z)`, and you can check many of the relations given in equation 6.43 below using this built-in function and the `simplify` command; for example the command

```
simplify(hypergeom([1,1],[2],-x),hypergeom);
```

will produce `ln(1+x)/x`. The series 6.41 can be reproduced, to $O(x^5)$, using the command

```
series(hypergeom([a,b],[c],x),x=0,6);
```

for example. More information can be found in the help files under *hypergeom*.

The hypergeometric function is important because it satisfies the linear second-order differential equation

$$z(1-z)\frac{d^2u}{dz^2} + \{c - (a+b+1)z\}\frac{du}{dz} - abu = 0, \qquad (6.42)$$

which, for various values of the constants a, b and c, arises in many problems. One common source of such equations is the solution of second-order partial differential equations obtained using the method of separating variables. For example, if the system being studied is spherically symmetric, using spherical polar coordinates the equation describing variations with the latitude θ is given by setting $a = -n$, $b = n + 1$ and $z = (1 - x)/2$, where n is an integer, to obtain

$$(1 - x^2)\frac{d^2u}{dx^2} - 2x\frac{du}{dx} + n(n + 1)u = 0,$$

which is the equation satisfied by the Legendre polynomial $P_n(x)$, discussed in chapter 11: in physical approximations the variable x is normally related to the latitude θ, $0 \le \theta \le \pi$, in spherical polar coordinates by $x = \cos\theta$, and this accounts for the zero in the coefficient of $\frac{d^2u}{dx^2}$ at $x = \pm 1$.

There are many other special values of these constants for which the hypergeometric function reduces to well known functions; some of these are:

$$
\begin{aligned}
xF(1, 1; 2; -x) &= \ln(1 + x), \\
F\left(\frac{1}{2}, \frac{1}{2}; \frac{3}{2}; x^2\right) = \sqrt{1 - x^2}\, F(1, 1; \frac{3}{2}; x^2) &= x^{-1}\sin^{-1} x, \\
xF\left(\frac{1}{2}, 1; \frac{3}{2}; x^2\right) &= \tanh^{-1} x = \frac{1}{2}\ln\left(\frac{1 + x}{1 - x}\right), \\
F(a, b; b; x) &= (1 - x)^{-a}, \\
\lim_{b \to \infty} F(1, b; 1; x/b) &= e^x, \\
F\left(a, 1 - a; \frac{1}{2}; \sin^2 x\right) &= \frac{\cos(2a - 1)x}{\cos x}, \\
F\left(a, \frac{1}{2} + a; \frac{1}{2}; -\tan^2 x\right) &= \cos^{2a} x \cos(2ax).
\end{aligned}
\tag{6.43}
$$

Other special cases are listed in Abramowitz and Stegun (1965, page 556), which also describes many other properties of the hypergeometric function. A more detailed discussion of the analytic properties of these functions can be found in Whittaker and Watson (1965, chapters 10, 14 and 15).

Exercise 6.23
Show that the series defined in equation 6.41 satisfies the differential equation 6.42.

Exercise 6.24
Show that

$$\frac{d}{dz}F(a, b; c; z) = \frac{ab}{c}F(a + 1, b + 1; c + 1; z).$$

The reason that the hypergeometric function is related to continued fractions is that it is possible to express the ratio

$$\frac{F(a, b + 1; c + 1; x)}{F(a, b; c; x)}$$

as a continued fraction with known coefficients. Because many elementary functions can be expressed in terms of this ratio, we have a direct method of constructing their continued fraction representation. The required result is

$$\frac{F(a,b+1;c+1;x)}{F(a,b;c;x)} = \cfrac{1}{1 + \cfrac{b_1 x}{1 + \cfrac{b_2 x}{1 + \cfrac{b_3 x}{1 + \cdots}}}} \qquad (6.44)$$

where

$$b_{2n+1} = -\frac{(n+a)(n+c-b)}{(2n+c)(2n+c+1)}, \qquad b_{2n} = -\frac{(n+b)(n+c-a)}{(2n+c-1)(2n+c)}. \qquad (6.45)$$

Since $\lim_{n\to\infty} b_n = -\frac{1}{4}$ this continued fraction converges for $x < 1$.

We now derive this result, though you are advised to leave this proof until your second or third reading of this section; the proof ends at equation 6.51 below.

Consider the difference

$$F(a,b+1;c+1;x) - F(a,b;c;x).$$

The coefficient of x^n is

$$\frac{a(a+1)\cdots(a+n-1)}{n!} \left\{ \frac{(b+1)\cdots(b+n)}{(c+1)\cdots(c+n)} - \frac{b(b+1)\cdots(b+n-1)}{c(c+1)\cdots(c+n-1)} \right\}$$

$$= \frac{a(a+1)\cdots(a+n-1)}{n!} \frac{(b+1)\cdots(b+n)}{(c+1)\cdots(c+n)} \left\{ 1 - \frac{b(c+n)}{c(b+n)} \right\}$$

$$= \frac{a(c-b)}{c(c+1)} \times \frac{(a+1)\cdots(a+n-1)}{(n-1)!} \frac{(b+1)+\cdots(b+n-1)}{(c+2)\cdots(c+n)},$$

which is $a(c-b)/c(c+1)$ times the coefficient of x^n in $xF(a+1,b+1;c+2;x)$. Thus we have

$$F(a,b+1;c+1;x) - F(a,b;c;x) = \frac{a(c-b)}{c(c+1)} xF(a+1,b+1;c+2;x). \qquad (6.46)$$

But, since $F(a,b;c;x) = F(b,a;c;x)$, this gives

$$F(a+1,b;c+1;x) - F(a,b;c;x) = \frac{b(c-a)}{c(c+1)} xF(a+1,b+1;c+2;x), \qquad (6.47)$$

derived by interchanging the first two arguments in 6.46 and then re-labelling, $a \leftrightarrow b$. Now rewrite 6.46 in the form

$$\frac{F(a,b+1;c+1;x)}{F(a,b;c;x)} = \frac{1}{1 - x \dfrac{a(c-b)}{c(c+1)} \dfrac{F(a+1,b+1;c+2;x)}{F(a,b+1;c+1;x)}} \qquad (6.48)$$

and equation 6.47 in the form

$$\frac{F(a+1,b;c+1;x)}{F(a,b;c;x)} = \frac{1}{1 - x \dfrac{b(c-a)}{c(c+1)} \dfrac{F(a+1,b+1;c+2;x)}{F(a+1,b;c+1;x)}},$$

or, on making the replacements $b \to b+1$ and $c \to c+1$,

$$\frac{F(a+1,b+1;c+2;x)}{F(a,b+1;c+1;x)} = \frac{1}{1 - x\dfrac{(b+1)(c+1-a)}{(c+1)(c+2)}\dfrac{F(a+1,b+2;c+3;x)}{F(a+1,b+1;c+2;x)}}. \tag{6.49}$$

But this is exactly the correct expression to re-substitute into the denominator of equation 6.48. Further, we can also substitute equation 6.48 into the denominator of 6.49. Thus by alternately substituting 6.49 into 6.48 and vice versa we obtain the continued fraction representation,

$$\frac{F(a,b+1;c+1;x)}{F(a,b;c;x)} = \cfrac{1}{1 + \cfrac{b_1 x}{1 + \cfrac{b_2 x}{1 + \cfrac{b_3 x}{1 + \cdots}}}} \tag{6.50}$$

where

$$b_1 = -\frac{a(c-b)}{c(c+1)}, \qquad b_2 = -\frac{(b+1)(c+1-a)}{(c+1)(c+2)},$$

$$b_3 = -\frac{(a+1)(c+1-b)}{(c+2)(c+3)}, \qquad b_4 = -\frac{(b+2)(c+2-a)}{(c+3)(c+4)},$$

or in general

$$b_{2n+1} = -\frac{(n+a)(n+c-b)}{(2n+c)(2n+c+1)}, \qquad b_{2n} = -\frac{(n+b)(n+c-a)}{(2n+c-1)(2n+c)}. \tag{6.51}$$

Now consider some special cases: since $F(a,0;c;x) = 1$ we have from equation 6.44 and by replacing $c+1$ by c,

$$F(a,1;c;x) = \cfrac{1}{1 + \cfrac{r_1 x}{1 + \cfrac{r_2 x}{1 + \cfrac{r_3 x}{1 + \cdots}}}} \tag{6.52}$$

with

$$r_{2n+1} = -\frac{(n+a)(n+c-1)}{(2n+c-1)(2n+c)}, \qquad r_{2n} = -\frac{n(n+c-a-1)}{(2n+c-2)(2n+c-1)}. \tag{6.53}$$

Specialising further, set $a = 1$, $c = 2$ and $x \to -x$ and use equation 6.43 to obtain

$$\ln(1+x) = \cfrac{x}{1 + \cfrac{x/2}{1 + \cfrac{x/6}{1 + \cfrac{x/3}{1 + \cfrac{x/5}{1 + \cfrac{3x/10}{1 + \cdots}}}}}} \tag{6.54}$$

with

$$r_{2n+1} = \frac{n+1}{2(2n+1)}, \qquad r_{2n} = \frac{n}{2(2n+1)}.$$

Exercise 6.25

Use the relation

$$\tanh^{-1} x = x F\left(\frac{1}{2}, 1; \frac{3}{2}; x^2\right)$$

to show that

$$\tanh^{-1} x = \cfrac{x}{1 - \cfrac{x^2/3}{1 - \cfrac{4x^2/15}{1 - \cfrac{9x^2/35}{1 - \cfrac{16x^2/63}{1 - \cdots}}}}}$$

and that the general coefficients are

$$r_{2n+1} = -\frac{(2n+1)^2}{(4n+1)(4n+3)}, \quad r_{2n} = -\frac{4n^2}{16n^2 - 1}.$$

Exercise 6.26

Show that

$$(1-x)^{-a} = \cfrac{1}{1 - \cfrac{ax}{1 + \cfrac{(a-1)x/2}{1 - \cfrac{(a+1)x/6}{1 + \cfrac{(a-2)x/6}{1 - \cfrac{(a+2)x/10}{1 + \cfrac{(a-3)x/10}{1 + \cdots}}}}}}}$$

and that the general coefficients are

$$r_{2n+1} = -\frac{n+a}{2(2n+1)}, \quad r_{2n} = -\frac{n-a}{2(2n-1)}, \quad n > 0.$$

Note: special care is needed to find the value of r_1.

6.5 Exercises

Exercise 6.27

Use Maple to show that the continued fraction representation of $e = \exp(1)$ is

$$e = [2; 1, 2, 1, 1, 4, 1, 1, 6, 1, 1, 8, 1, \ldots];$$

a result obtained by Cotes in 1714. To how many decimal places does this represent e? Can you find more elements of this continued fraction?

R. Cotes (1682–1716)

Exercise 6.28

Use Maple to find the first 30 elements of the continued fraction representation of $\sqrt{e}$.

Exercise 6.29

For any positive integer n find the continued fraction representation of $\sqrt{n^2 + 1}$.
Hint: use the identity $(\sqrt{n^2 + 1} - n)^{-1} = \sqrt{n^2 + 1} + n$.

Exercise 6.30

Expand the generalised continued fractions

$$t_1(x) = \cfrac{1}{x^{-1} - \cfrac{1}{3x^{-1}}} \quad \text{and} \quad t_2(x) = \cfrac{1}{x^{-1} - \cfrac{1}{3x^{-1} - \cfrac{1}{5x^{-1}}}}$$

as a power series in x. Do you recognise these series?

Exercise 6.31

Show that the value of the periodic continued fraction $x = [1; a, b, c, a, b, c, \ldots]$ is

$$x = 1 - \frac{c}{2} - \frac{a - b}{2(1 + ab)} + \frac{\sqrt{[c(1 + ab) + a + b]^2 + 4}}{2(1 + ab)}.$$

Exercise 6.32

Given the finite continued fraction $x = [a_0; a_1, \ldots, a_N]$ for some given N, write a procedure to compute all the convergents p_k/q_k, $k = 0, 1, \ldots, N$ and to output these as a list of length $N + 1$, assuming the input to be the list $x = [a_0, a_1, \ldots, a_N]$; your procedure should also output lists containing p_k and q_k for $k = 1, 2, \ldots, N$.

Exercise 6.33

An alternative method of evaluating continued fractions is to mimic the way they are evaluated 'by hand', that is backwards starting with the element with the largest index. Write a Maple procedure to perform this operation.

Exercise 6.34

Use equation 6.22 (page 206), and the method described in section 6.1 to show that the general solution of the diophantine equation 6.8 (page 196), is given by 6.9.

Exercise 6.35

Show that the $[L/0]$ Padé approximant of the infinite series $f(x) = \sum_{k=0}^{\infty} a_k x^k$ is just the Lth partial sum.

Exercise 6.36

Recall from equation 5.12 (page 182), that the asymptotic expansion of

$$g(x) = e^x E_1(x) \quad \text{is} \quad g(x) \sim -\sum_{k=1}^{\infty} (-1)^k \frac{(k-1)!}{x^k}.$$

Use Maple to find the Padé approximant of the Nth partial sum of this series by first substituting $y = 1/x$ to convert it to a power series, and then compare, graphically, the exact function, $g(x)$, with the Nth partial sum and the derived Padé approximant for $0 < x < 5$.

Exercise 6.37

Use Maple to construct Taylor polynomial approximations, of various orders, and the consequent Padé approximants of the function

$$f(x) = \ln(1 + \sin x),$$

and graphically compare these approximations with the exact function for $0 < x < \pi$.

Exercise 6.38
Use Maple to form Padé approximations of Stirling's approximation to the factorial function, $x!$, and graphically compare both approximations with the exact values of $x!$.

Exercise 6.39
Use Maple to construct Taylor polynomial approximations, of various orders, and the consequent Padé approximants of the function

$$f(x) = \ln \cosh x,$$

and graphically compare these approximations with the exact function for $0 < x < 10$.

Exercise 6.40
Use the result of exercise 6.26 to show that

$$2^{1/3} \simeq \cfrac{1}{1 - \cfrac{1/3}{1 + \cfrac{2/3}{1 + \cfrac{1/9}{1 + \cfrac{7}{18}}}}}$$

and evaluate this fraction and the relative error.

Exercise 6.41
Use the relation

$$F\left(\frac{1}{2}, \frac{1}{2}; \frac{3}{2}; z^2\right) = \left(1 - z^2\right)^{1/2} F\left(1, 1; \frac{3}{2}; z^2\right)$$

to derive the result

$$\frac{\sin^{-1} x}{\sqrt{1 - x^2}} = \cfrac{x}{1 - \cfrac{x^2/6}{1 - \cfrac{2x^2/15}{1 - \cfrac{3x^2/35}{1 - \cfrac{4x^2/21}{1 - \cdots}}}}}$$

Exercise 6.42
Use the relation

$$g(x) = \frac{(1+x)^w - (1-x)^w}{(1+x)^w + (1-x)^w} = wx \frac{F\left(\frac{1}{2}(1-w), 1 - \frac{w}{2}; \frac{3}{2}; x^2\right)}{F\left(\frac{1}{2}(1-w), -\frac{w}{2}; \frac{1}{2}; x^2\right)}$$

together with equation 6.51 to show that

$$g(x) = \cfrac{wx}{1 - \cfrac{(1 - w^2)x^2/3}{1 - \cfrac{(4 - w^2)x^2/15}{1 - \cfrac{(9 - w^2)x^2/35}{1 - \cfrac{(16 - w^2)x^2/63}{1 - \cdots}}}}}$$

and in general

$$a_{2n+1} = -\frac{(2n+1)^2 - k^2}{(4n+1)(4n+3)}, \qquad a_{2n} = -\frac{4n^2 - k^2}{16n^2 - 1}.$$

By putting $x = i \tan \phi$ show that

$$g(x) = i \tan w\phi$$

and hence that

$$\tan w\phi = \cfrac{w \tan \phi}{1 + \cfrac{(1 - w^2) \tan^2 \phi/3}{1 + \cfrac{(4 - w^2) \tan^2 \phi/15}{1 + \cfrac{(9 - w^2) \tan^2 \phi/35}{1 + \cfrac{(16 - w^2) \tan^2 \phi/63}{1 + \cdots}}}}}$$

Show that if $w = 2$ this reduces to the usual identity $\tan 2\phi = 2 \tan \phi/(1 - \tan^2 \phi)$. Find the equivalent formula for $\tan 3\phi$ in terms of $\tan \phi$.

Exercise 6.43

In this question you will investigate the statistics of the number of elements needed to exactly represent a finite decimal number of the form $x = 0.d_1 d_2 \cdots d_N$ for various values of N, where x is uniformly distributed in $(0, 1)$. One reason for wanting to know this is the possibility of using continued fractions to compress data. If fewer digits are needed to store a number as a continued fraction this might be an efficient method of storing numerical data; as will be seen, this is not the case.

Write a Maple program to generate an x, for any given N, using the `rand` function. For a large sample of such numbers, x_k, $k = 1, 2, \ldots, M$ with $M \gg 1$, for each number find its continued fraction representation and the number n_k of elements required. Use this data to compute the means

$$\langle n \rangle = \frac{1}{M} \sum_{k=1}^{M} n_k, \qquad \langle n^2 \rangle = \frac{1}{M} \sum_{k=1}^{M} n_k^2$$

and the probability $P(m)$ that n has the value m. For $N = 3, 4, \ldots, 10$ find the values of $\langle n \rangle$ and $\langle n^2 \rangle$; for the first few values of N they are shown in the table.

Table 6.1. *Values of $\langle n \rangle$ and $\langle n^2 \rangle$ for some representative values of N, obtained using a sample of M = 1500 numbers.*

N	3	5	7	9
$\langle n \rangle$	5.41	9.19	13.1	16.9
$\langle n^2 \rangle$	32.7	90.28	179.3	295.4

Show, graphically, that for sufficiently large M

$$P(m) \simeq \frac{1}{\sigma\sqrt{2\pi}} \exp\left(-\frac{(m - \langle n \rangle)^2}{2\sigma^2}\right), \qquad \sigma^2 = \left(\langle n^2 \rangle - \langle n \rangle^2\right).$$

7

Linear differential equations and Green's functions

7.1 Introduction

In this chapter we introduce the Green's function, a device allowing one to write the solution of particular linear differential equations in terms of integrals; for instance the solution of the initial value problem

$$\frac{d^2x}{dt^2} + \omega^2 x = f(t), \quad x(0) = \dot{x}(0) = 0, \tag{7.1}$$

can be written as the integral

$$x(t) = \int_0^t ds \, G(t,s) f(s), \quad G(t,s) = \frac{1}{\omega} \sin \omega(t-s). \tag{7.2}$$

Here $G(t,s)$ is the Green's function which depends only upon the properties of the differential operator $\left(\frac{d^2}{dt^2} + \omega^2\right)$ and the initial values, but not upon the function $f(t)$. Green's functions, when they exist, may be regarded as the inverse of the differential operator.

Green's functions exist for partial as well as the ordinary differential equations treated here and are used extensively in quantum mechanics and the theory of electromagnetism; indeed Green's original essay, Green (1828), had the title *An Essay on the Application of Mathematical Analysis to the Theories of Electricity and Magnetism*. Green's functions are important because they enable the unknown function, $x(t)$ in the above equation, to be expressed explicitly in terms of known quantities, and this sometimes allows one to determine properties of the solution without actually finding it. They also provide a useful tool for obtaining perturbation type expansions for solutions.

Of equal interest is the story of Green himself, for this is one of the more unusual stories in the history of mathematics and physics; a very brief synopsis of this story is provided in appendix 2 of this chapter, but much more detail is given in the delightful book by Cannell (1993).

George Green (1793–1841)

The theory of Green's functions can be developed in a variety of ways: here we choose to base the development on delta functions because this provides a transparent physical picture of why solutions of linear differential equations can be written in the form of integrals like that in equation 7.2. In addition delta functions are useful in a variety of other applications; one arcane use is described in section 7.4.

Thus we start this chapter with an introduction to delta functions, followed by their

227

application to the theory of Green's functions, the theory of which is developed further in chapter 10.

7.2   The delta function: an introduction

7.2.1 Preamble

The delta function, $\delta(x)$, sometimes called the Dirac delta function, is a strange object, useful in many branches of applied mathematics. We introduce this 'function' of a real variable in a heuristic and non-rigorous manner, with the object of explaining its properties and showing how to use it in some circumstances. It will quickly become apparent that the delta function is not a function in the usual sense, but it is convenient to treat it as if it were because it is a useful device, sometimes allowing complicated algebra to be circumvented; this is similar to the way we sometimes treat the differential $\frac{dy}{dx}$ as a fraction, for instance when changing variables in an integral, knowing full well that it is not a fraction.

We first 'define' the delta function using its fundamental property, equation 7.3, and from this derive many of its other properties simply by treating it as an ordinary function. This approach is intended to provide you with an idea of how the delta function behaves and how it may be used; it is not rigorous although all the results obtained here can be proved rigorously. Following this heuristic introduction we show how the delta function can be considered as the limit of a sequence of ordinary, well behaved functions; such sequences sometimes provide useful alternative representations of the delta function.

7.2.2 Basic properties of the delta function

In this section we shall assume that all the ordinary functions used — generally denoted here by $f(x)$ or $g(x)$ — are sufficiently well behaved for the final expressions, which do not involve delta functions, to exist.

The basic property of the delta function, $\delta(x)$, is that for *all* sufficiently well behaved functions $f(x)$

$$\int_{-\infty}^{\infty} du\, \delta(u-x)f(u) = f(x) \quad \text{for all } x. \tag{7.3}$$

Since this relation is true for a sufficiently wide class of functions and for all x it can be shown that $\delta(u-x) = 0$ for all $u \neq x$. Hence $\delta(u-x)$ is zero everywhere except the single point where its argument is zero, $u = x$. In addition the property 7.3 holds if $f(u) = 1$ and hence $\int_{-\infty}^{\infty} du\, \delta(u-x) = 1$, for all x, so it should be clear that the delta function is not a function in the ordinary sense. The result 7.3 also holds on finite intervals,

$$\int_{a}^{b} du\, \delta(u-x)f(u) = f(x) \quad \text{provided} \quad a < x < b.$$

In Maple the delta function is named `Dirac(x)`.

The dimensions of $\delta(x)$ are the inverse of the dimensions of x because, from equation 7.3, the quantity $du\,\delta(u-x)$ must be dimensionless.

Before justifying the use of delta functions we shall assume that the usual operations of calculus apply and derive some of its properties. All the following results are important and may be justified rigorously, though to do so would take us far beyond the scope of this book; more details can be found in the books by Lighthill (1962), Jones (1966) and Hoskins (1979), for example.

One simple consequence of equation 7.3 is the relation

$$\delta(x-u)g(x) = \delta(x-u)g(u), \tag{7.4}$$

where $g(x)$ is any function continuous at $x = u$, because both sides of this equation when inserted into equation 7.3 lead to the same result.

This illustrates an important result which we use frequently. If

$$\int_{-\infty}^{\infty} du\, f(u)\, g_1(u)\delta(u-x) = \int_{-\infty}^{\infty} du\, f(u)\, g_2(u)\delta(u-x)$$

for all sufficiently well behaved functions f, then

$$g_1(u)\delta(u-x) = g_2(u)\delta(u-x).$$

This result is needed in exercise 7.3.

Exercise 7.1
Prove equation 7.4.

We shall now generalise equation 7.3 to show how the derivatives of the delta function may be defined; the relation, known to Maple, is

$$\int_{-\infty}^{\infty} du\, f(u)\frac{d^n}{du^n}\delta(u-x) = (-1)^n\frac{d^n}{dx^n}f(x) \quad \text{for all } x \text{ and } n = 0, 1,\dots. \tag{7.5}$$

This follows by repeated integration by parts. Consider the case $n = 1$ and assume that integration by parts is allowed,

$$\int_{-\infty}^{\infty} du\, f(u)\frac{d}{du}\delta(u-x) = [\delta(u-x)f(u)]_{-\infty}^{\infty} - \int_{-\infty}^{\infty} du\, \delta(u-x)\frac{d}{du}f(u)$$

$$= -\frac{df}{dx}, \tag{7.6}$$

the last relation following from equation 7.3. In this, and other similar relations, we normally assume that the functions involved, here $f(x)$, are sufficiently well behaved that the boundary terms can be ignored.

Exercise 7.2
Use repeated integration by parts to prove equation 7.5.

Exercise 7.3
Use equation 7.3 to show that $\delta(x)$ and $\delta(-x)$ have the same effect. This shows that $\delta(x)$ may be treated as an even function.

Show also that the nth derivative of $\delta(x)$ is odd or even according as n is odd or even. Further, by changing variables in equation 7.3 show that for any nonzero constant a

$$\delta(ax) = \frac{1}{|a|}\delta(x). \tag{7.7}$$

Another useful function is the Heaviside step function defined by the relation

$$H(x) = \begin{cases} 1, & x > 0, \\ 0, & x < 0, \end{cases} \tag{7.8}$$

so its differential is zero for all $x \neq 0$; note that $H(x)$ is not defined at $x = 0$. In Maple this function is denoted by `Heaviside(x)`.

One of the properties of the Heaviside function is that $H'(x) = \delta(x)$, which is proved by considering the function $F(x)$ defined by the equation

$$F(x) = \int_{-\infty}^{\infty} du\, H(u - x) f(u) = \int_{x}^{\infty} du\, f(u);$$

the differential of F is given by

$$\begin{aligned}
\frac{dF}{dx} &= \frac{d}{dx} \int_{-\infty}^{\infty} du\, H(u - x) f(u) = \int_{-\infty}^{\infty} du\, f(u) \frac{d}{dx} H(u - x) \\
&= -\int_{-\infty}^{\infty} du\, f(u) \frac{d}{du} H(u - x),
\end{aligned} \tag{7.9}$$

where we have used the rule for differentiating under the integral† on the first line and the chain rule on the second line. We also have

$$\frac{dF}{dx} = \frac{d}{dx} \int_{x}^{\infty} du\, f(u) = -f(x). \tag{7.10}$$

Thus, on equating 7.9 and 7.10, we find that for all sufficiently well behaved functions

$$\int_{-\infty}^{\infty} du\, f(u) \frac{d}{du} H(u - x) = f(x). \tag{7.11}$$

Comparing this relation with equation 7.3, we deduce that

$$\delta(x) = \frac{d}{dx} H(x). \tag{7.12}$$

Another important result concerns the interpretation of a delta function of an ordinary function: for any sufficiently well behaved functions f and g,

$$\int_{-\infty}^{\infty} dx\, g(x) \delta(f(x)) = \sum_{k=1}^{N} \frac{g(x_k)}{|f'(x_k)|}, \tag{7.13}$$

where x_k, $k = 1, 2, \ldots, N$, are the simple zeros of $f(x) = 0$, that is $f(x_k) = 0$ and $f'(x_k) \neq 0$; if $f'(x_k) = 0$ for some x_k the integral is not defined. We demonstrate the validity of this useful result by finding the contributions to the integral from intervals which include only one zero of $f(x)$ and on which $f(x)$ is monotonic. Consider the interval (a, b) containing only one zero, x_k, such that in it $f(x)$ is monotonic, so the

---

† The general rule for differentiating under the integral sign is given in chapter 13 (page 479).

relation $u = f(x)$ may be inverted to give a single valued function $x(u)$ such that $x(0) = x_k$. The contribution to the whole integral from this interval is

$$\int_a^b dx\, g(x)\delta(f(x)) = \int_{a_1}^{b_1} du\, \frac{g(x(u))}{f'(x(u))}\delta(u) = \frac{g(x_k)}{|f'(x_k)|},$$

where $a_1 = f(a)$, $b_1 = f(b)$ and, since $f(x_k) = 0$, a_1 and b_1 have opposite signs. Adding these contributions leads to the required result. Observe that equation 7.7 is a special case of equation 7.13 obtained by putting $f(x) = ax$. The contribution to the integral from x_k is large if the gradient of $f(x)$ is small at $x = x_k$ and small otherwise.

In the theory of Green's functions discontinuities are important. A function $f(x)$ is discontinuous at $a = x$ if the right and left hand limits differ, that is,

$$\lim_{\epsilon \to 0_+} f(a - \epsilon) \neq \lim_{\epsilon \to 0_+} f(a + \epsilon).$$

At $x = a$ the first derivative behaves like a delta function. The exact behaviour is examined in the following two exercises.

Exercise 7.4
Show that the function

$$f(x) = \begin{cases} f_2(x), & x > a, \\ f_1(x), & x < a, \end{cases}$$

where $f_k(x)$ and $f_k'(x)$, $k = 1, 2$, are defined at $x = a$, can be written in terms of the Heaviside step function as follows:

$$f(x) = f_1(x)H(a - x) + f_2(x)H(x - a),$$

and hence that its first derivative is

$$f'(x) = f_1'(x)H(a - x) + f_2'(x)H(x - a) + [f_2(a) - f_1(a)]\,\delta(x - a),$$

assuming that f_1 and f_2 are defined at $x = a$.
Deduce that if ϵ is a small positive quantity,

$$\int_{a-\epsilon}^{a+\epsilon} dx\, f'(x) = f_2(a) - f_1(a) + O(\epsilon).$$

Exercise 7.5
If $f(x)$ is the discontinuous function

$$f(x) = \begin{cases} -x, & x < 0, \\ 1 + x, & x \geq 0, \end{cases} \quad \text{and} \quad F(x) = \int_{-1}^x du\, f(u),$$

show that

$$f'(x) = \begin{cases} 1, & x > 0, \\ -1, & x < 0, \end{cases} \quad \text{and} \quad F(x) = \begin{cases} \dfrac{1}{2}(1 - x^2), & x < 0, \\ \dfrac{1}{2}(1 + x)^2, & x \geq 0, \end{cases}$$

and plot the graphs of $f(x)$ and $F(x)$. Note that the integral $F(x)$ is continuous at $x = 0$.

This analysis seems plausible – we have taken an object, $\delta(x)$, with the property defined in equation 7.3 and have treated as if it were an ordinary function. So far the results obtained seem reasonable, but care is needed. Objects like $\delta(x)$ and $H(x)$ do not always behave like the ordinary functions that arise in applied mathematics, so they are usually referred to as *generalised* functions and there is a well developed theory for them, see for example Lighthill (1962), Jones (1966) and Hoskins (1979). Generalised functions do not always behave like ordinary functions, but behave according to rules which are a restricted subset of those satisfied by ordinary functions. For instance there is generally no meaning attached to powers or products of generalised functions; therefore, care and a bit of common sense are needed when using them.

7.2.3 *The delta function as a limit of a sequence*

We now consider how the above ideas may be made plausible by defining the delta function as the limit of a sequence of ordinary, well behaved functions. The main difficulty with this approach is in proving that the various limits exist; we ignore this problem but note that it is not insurmountable. The advantage is that one can develop an intuitive notion of how the delta function works without taking this limit. There are many sequences which produce the delta function, some of which will be given later, equations 7.18 to 7.20: first, for illustrative purposes, we consider the sequence of functions

$$\delta_n(x) = \sqrt{\frac{n}{\pi}} e^{-nx^2}, \quad n = 1, 2, \ldots, \tag{7.14}$$

as n becomes very large. In figures 7.1 and 7.2 we show the graphs of $\delta_n(x)$ for various values of n, and observe that as n increases the function is more and more concentrated around $x = 0$.

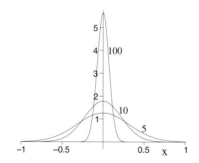

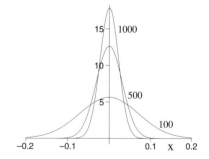

Figure 7.1   Graphs of the function 7.14 for $n = 5$, 10 and 100.

Figure 7.2 $n = 100$, 500 and 1000. Note that the scales of this and figure 7.1 differ.

These graphs strongly suggest that for large n, provided $f(x)$ exists at $x = 0$, is sufficiently slowly varying and if $e^{-nx^2} f(x) \to 0$ as $|x| \to \infty$ sufficiently rapidly,

$$\int_{-\infty}^{\infty} dx \, \delta_n(x) f(x) \simeq f(0) \int_{-\infty}^{\infty} dx \, \delta_n(x), \quad n \gg 1.$$

Since the area under $\delta_n(x)$ is unity this argument suggests that the sequence $\delta_n(x)$ tends to a delta function as $n \to \infty$. We prove this result after finding the area under $\delta_n(x)$.

Exercise 7.6
Use the properties of the gamma function, given in Appendix I, in particular equations 20.14, to show that

$$\int_{-\infty}^{\infty} dx\, \delta_n(x) = 1 \quad \text{for all } n.$$

These approximations suggest that if n is large enough $\delta_n(x)$ behaves like $\delta(x)$. In fact it is not too difficult to show that

$$\lim_{n \to \infty} \int_{-\infty}^{\infty} dx\, \delta_n(x) f(x) = f(0), \tag{7.15}$$

for we have, using the result of exercise 7.6,

$$
\left| f(0) - \int_{-\infty}^{\infty} dx\, \sqrt{\frac{n}{\pi}}\, e^{-nx^2} f(x) \right| = \sqrt{\frac{n}{\pi}} \left| \int_{-\infty}^{\infty} dx\, e^{-nx^2} \{f(x) - f(0)\} \right|
$$

$$
\leq \max |f'(x)| \sqrt{\frac{n}{\pi}} \int_{-\infty}^{\infty} dx\, |x| e^{-nx^2}
$$

$$
= \frac{1}{\sqrt{n\pi}} \max |f'(x)| \to 0 \text{ as } n \to \infty.
$$

The second line follows from the mean value theorem, which states that there is a w in the range $0 \leq w \leq x$ such that

$$f(x) - f(0) = x f'(w) \leq x \max(f').$$

The graphs of $\delta_n(x)$ show why this result holds for any reasonable function $f(x)$; essentially, for large values of n, $\delta_n(x)$ is very small except for $|x| < A/\sqrt{n}$, for some positive number A, and in this small interval we may approximate $f(x)$ by its value at $x = 0$.

For most functions $f(x)$ the integral $\int_{-\infty}^{\infty} dx\, \delta_n(x) f(x)$ cannot be evaluated in terms of known functions, but there are a few cases for which the integral may be performed and the limit taken. The integrals given in Appendix I allow the evaluation for polynomials; other examples are treated in the next exercise.

Exercise 7.7
By evaluating the integrals examine the behaviour of

$$F_1 = \int_{-\infty}^{\infty} dx\, \delta_n(x) e^{-a(x-b)^2} \quad \text{and} \quad F_2 = \int_{-\infty}^{\infty} dx\, \delta_n(x) e^{ikx},$$

where a, b and k are real with $a > 0$. In particular show that

$$F_1 = e^{-ab^2} \left\{ 1 + \frac{a}{n}\left(ab^2 - \frac{1}{2}\right) + \cdots \right\}, \quad F_2 = 1 - \frac{k^2}{4n} + \cdots,$$

and hence confirm that the limit 7.15 holds.

The sequence $\delta_n(x)$ defining the delta function can also be used to examine the behaviour of $\delta'(x)$ and higher derivatives. For instance, we have

$$\delta'_n(x) = -2nx\sqrt{\frac{n}{\pi}}e^{-nx^2},$$

which is an odd function of x and, by integrating by parts,

$$\int_{-\infty}^{\infty} dx\,\delta'_n(x)f(x) = [\delta_n(x)f(x)]_{-\infty}^{\infty} - \int_{-\infty}^{\infty} dx\,\delta_n(x)f'(x)$$

$$\longrightarrow \quad -f'(0) \quad \text{as} \quad n \to \infty. \tag{7.16}$$

The geometric reason for this result is seen in the figures below showing the graphs of $\delta'_n(x)$ and $\delta''_n(x)$ for various values of n.

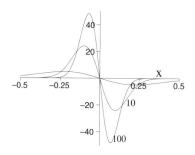

Figure 7.3   Graphs of $\delta'_n(x)$ for $n = 10, 50$ and 100.

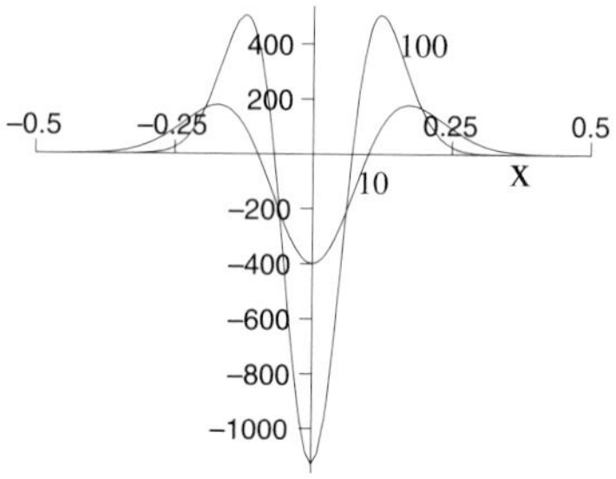

Figure 7.4   Graphs of $\delta''_n(x)$ for $n = 50$ and 100.

Since

$$\delta''_n(x) = 2n\sqrt{\frac{n}{\pi}}(2nx^2 - 1)e^{-nx^2},$$

the stationary points of $\delta'_n(x)$ are at $x = \pm 1/\sqrt{2n}$, and at these points $\delta'_n(x) = \mp n\sqrt{2/\pi e}$. Thus the integral picks out adjacent values of f, either side of zero, and subtracts them to form the differential.

It is clear that the technique used to derive equation 7.16 can be used to deal with any differential.

Exercise 7.8

Use integration by parts to prove that

$$\int_{-\infty}^{\infty} dx\,f(x)\frac{d^k}{dx^k}\delta_n(x) = (-1)^k \int_{-\infty}^{\infty} dx\,\delta_n(x)\frac{d^k}{dx^k}f(x)$$

$$\longrightarrow \quad (-1)^k f^{(k)}(0) \quad \text{as } n \to \infty.$$

The integral of the sequence $\delta_n(x)$ also defines a sequence of functions which, from equation 7.12, we should expect to behave like the Heaviside function.

Exercise 7.9

Use Maple to plot the graphs of the function

$$H_n(x) = \int_{-\infty}^{x} dw \, \delta_n(w) \quad \text{for} \quad n = 10 \text{ and } 100.$$

What is the value of $H_n(0)$? State, without proof, the value of the limit

$$\lim_{n \to \infty} H_n(x - a).$$

Another useful generalised function is the sgn function defined by

$$\text{sgn}(x) = \begin{cases} -1, & x < 0, \\ 1, & x > 0. \end{cases} \tag{7.17}$$

This is the `signum` function in Maple. Clearly we can write $\text{sgn}(x) = -1 + 2H(x)$ to give $\text{sgn}'(x) = 2\delta(x)$; alternatively $\text{sgn}(x) = H(x) - H(-x)$. Note that $\text{sgn}(x)$ is not defined at the origin.

Exercise 7.10

Show that

$$\text{sgn}(x) = \frac{d}{dx}|x|.$$

Try evaluating the right hand side with Maple.

Exercise 7.11

Show by direct integration that

$$\int_{-\infty}^{\infty} dx \, \text{sgn}(x - a) f'(x) = -2f(a)$$

and, by integrating by parts, that

$$\int_{-\infty}^{\infty} dx \, f(x) \frac{d}{dx} \text{sgn}(x - a) = 2f(a),$$

provided all integrals exist. Deduce that $\text{sgn}'(x - a) = 2\delta(x - a)$.

There are numerous other sequences which converge to the delta function; three are

$$\delta_L(x) = \begin{cases} 0, & |x| > 1/2L, \\ L, & |x| < 1/2L, \end{cases} \tag{7.18}$$

$$\delta_L(x) = \frac{L}{\pi(1 + (Lx)^2)}, \tag{7.19}$$

$$\delta_L(x) = \frac{1}{2\pi} \int_{-L}^{L} du \, e^{ixu} = \frac{\sin Lx}{\pi x}. \tag{7.20}$$

All these sequences converge to the delta function as $L \to \infty$, in the sense that

$$\lim_{L \to \infty} \int_{-\infty}^{\infty} dx \, \delta_L(x) f(x) = f(0).$$

Exercise 7.12

Plot the graphs of each of the functions $\delta_L(x)$ defined in equations 7.18 to 7.20 for suitable values of L.

For all three functions show that

$$\int_{-\infty}^{\infty} dx\, \delta_L(x) = 1 \quad \text{and} \quad \lim_{L\to\infty} \int_{-\infty}^{\infty} dx\, \delta_L(x) f(x) = f(0)$$

for all sufficiently well behaved functions $f(x)$. For the function defined in equation 7.20, the last result follows from the analysis given in Whittaker and Watson (1965, section 9.7).

For these sequences it is fairly clear that $\delta_L(x)$ has the desired property because $\delta_L(0) \propto L$, but $\delta_L(x)$ is large only over a narrow range of x which decreases as L increases in such a manner that the area under $\delta_L(x)$ is unity.

There are other types of sequence which have the required limit, as shown by the following example:

$$\delta_N(x) \;=\; \frac{1}{2\pi} + \frac{1}{\pi}\sum_{k=1}^{N} \cos(kx), \quad -\pi \le x \le \pi, \tag{7.21}$$

$$=\; \frac{1}{2\pi}\sum_{k=-N}^{N} e^{-ikx}. \tag{7.22}$$

In this case $\delta_N(x)$ is a finite Fourier series, with each of the N coefficients equal, so each member of the sequence is 2π-periodic, $\delta_N(x+2\pi) = \delta_N(x)$ for all x. Clearly $\delta_N(2n\pi) = (N+\frac{1}{2})/\pi$ for all $n = 0, \pm 1, \ldots$, but it is not so clear why the function should be small for other values of x. In fact the behaviour of this function is not so simple, as is seen by noting that equation 7.22 is a geometric series with the sum

$$\delta_N(x) = \frac{\sin(N+\frac{1}{2})x}{2\pi\sin(x/2)}, \tag{7.23}$$

the proof of which is outlined in exercise 4.10 (page 142). This function is large only if $\sin(x/2)$ is small; at $x = 2n\pi$ both numerator and denominator are zero but we have, by l'Hospital's rule,

$$\delta_N(2n\pi) = \lim_{x\to 2n\pi} \frac{\sin(N+\frac{1}{2})x}{2\pi\sin(x/2)} = \lim_{x\to 2n\pi} \frac{(N+\frac{1}{2})\cos(N+\frac{1}{2})x}{\pi\cos(x/2)} = \frac{1}{\pi}\left(N+\frac{1}{2}\right).$$

It can be shown that the sequence $\delta_N(x)$ converges to the sum of delta functions,

$$\frac{1}{2\pi}\sum_{k=-\infty}^{\infty} e^{-ikx} = \sum_{m=-\infty}^{\infty} \delta(x - 2\pi m). \tag{7.24}$$

Exercise 7.13

Plot graphs of

$$\delta_N(x) = \frac{1}{2\pi}\sum_{k=-N}^{N} e^{-ikx} = \frac{\sin(N+\frac{1}{2})x}{2\pi\sin(x/2)}$$

for various values of N over a sufficiently large range of x to see the peaks of equation 7.24 develop and to see how $\delta_N(x)$ behaves between these peaks.

The graphs obtained in this exercise and equation 7.23 show that $\delta_N(x)$ is strongly peaked at $x = 2n\pi$, $n = 0, \pm 1, \ldots$, but that it does not tend to zero at all other points, for instance $\delta_N(\pi) = (-1)^N/2\pi$; this is not implied by a literal interpretation of equation 7.24. Moreover, all the local maxima in $(-\pi, \pi)$ exceed $1/2\pi$ and all the local minima are less than $-1/2\pi$ and $\delta_N(x)$ does not tend to zero with increasing N for most values of x. Instead the function oscillates more as N increases, and it is these oscillations that provide the cancellation to give the limits

$$\lim_{N \to \infty} \int_a^b dx\, f(x)\delta_N(x) = 0, \quad 0 < a < b < 2\pi,$$

and

$$\lim_{N \to \infty} \int_{-\pi}^{\pi} dx\, f(x)\delta_N(x) = f(0).$$

It is in this sense that each of the sequences 7.21 and 7.22 tends to a delta function.

These limits are a consequence of the more general result, proved in Whittaker and Watson (1965, section 9.41), that if $\int_a^b dx\, g(x)$ exists and if $g(x)$ has bounded variation in $a < x < b$ then

$$\lim_{\lambda \to \infty} \int_a^b dx\, g(x) \sin \lambda x = 0.$$

Thus taking $g(x) = f(x)/\sin(x/2)$ gives the first result, and for the second we write

$$\int_{-\pi}^{\pi} dx\, f(x)\delta_N(x) = \int_{-\pi}^{\pi} dx\, \frac{f(x) - f(0)}{\sin(x/2)} \sin\big((N + 1/2)x\big) + f(0) \int_{-\pi}^{\pi} dx\, \delta_N(x)$$

to obtain the required limit if $f(x)$ is sufficiently well behaved.

This analysis is important because it shows that an integral containing rapidly oscillating functions that do not necessarily tend to zero can behave like delta functions. In these circumstances the statement

$$\delta_N(x) \to \delta(x) \quad \text{means that} \quad \int_a^b dx\, f(x)\delta_N(x) \to \int_a^b dx\, f(x)\delta(x).$$

If you have difficulty understanding the behaviour of delta functions it usually helps to think of them inside integrals.

We have considered the behaviour of the sequence in the interval $(-\pi, \pi)$, but since each member of the sequence is 2π-periodic it follows that

$$\lim_{N \to \infty} \int_{-\infty}^{\infty} dx\, f(x) \frac{1}{2\pi} \sum_{k=-N}^{N} e^{-ikx} = \int_{-\infty}^{\infty} dx\, f(x) \sum_{m=-\infty}^{\infty} \delta(x - 2\pi m)$$

$$= \sum_{m=-\infty}^{\infty} f(2\pi m).$$

This limit is the meaning of equation 7.24.

Exercise 7.14
Form the function

$$H_N(x) = \int_{-\pi}^{x} dw\, \delta_N(w), \quad -\pi \le x \le \pi,$$

where $\delta_N(x)$ is defined in equation 7.21, to show that for $-\pi \leq x \leq \pi$ the Heaviside function can be represented in the form

$$H(x) = \frac{x+\pi}{2\pi} + \frac{1}{\pi}\sum_{k=1}^{\infty}\frac{\sin kx}{k}.$$

Use Maple to plot the graph of $H_N(x)$ for various values of N and comment on the convergence of $H_N(x)$ to $H(x)$.

The sequence 7.22 converges to a sum of delta functions because of cancellation between the oscillatory components of the sum. An illuminating alternative expression is

$$\frac{1}{2\pi}\sum_{k=-\infty}^{\infty}e^{ik(x-y)} = \delta(x-y), \quad |x-y| < 2\pi. \tag{7.25}$$

Here each term is the product of the oscillatory function, e^{ikx}, and its complex conjugate evaluated at a different point. If $x = y$ this product is positive so the sum necessarily diverges, otherwise no matter how small $|x-y|$ there is always a k such that the signs of the real parts of e^{ikx} and e^{-iky} differ, and then there is destructive interference between the components of the sum which tends to zero.

There are many other sets of oscillatory functions, some of which are discussed in chapter 11, so we should expect other similar relations. For instance, consider the Legendre polynomials, $P_n(x)$; these can be computed using the recurrence relation given in chapter 3, or using the library functions provided in Maple, invoked with the `with(orthopoly)` command.

Exercise 7.15
Use Maple to plot the graph of the function

$$f_N(\theta, \phi) = \sum_{l=0}^{N}(2l+1)P_l(\cos\theta)P_l(\cos\phi), \quad 0 < \theta, \phi < \pi,$$

for a fixed ϕ and for $0 < \theta < \pi$, and for various values of N, say 10 to 20. Beware, this computation can take a long time if N is too large.

In this case it can be shown that

$$\sum_{l=0}^{\infty}(2l+1)P_l(\cos\theta)P_l(\cos\phi) = 2\delta(\cos\theta - \cos\phi).$$

Another example of this behaviour is considered in exercise 7.42 (page 263).

7.3 Green's functions: an introduction

In this section the theory of Green's functions is introduced: it will be developed further in chapter 10. The Green's function can be regarded as the inverse of a linear differential equation, and some analogous ideas may be helpful in putting it in context. The vector equation $A\mathbf{y} = \mathbf{f}$, where A is an $N \times N$ non-singular matrix and $\mathbf{y}$ and $\mathbf{f}$ are N-dimensional vectors, has the solution $\mathbf{y} = A^{-1}\mathbf{f}$; the inverse, A^{-1}, is independent of **f** and can be used to find the solution **y** for any vector **f**. Similarly the differential equation

$$\frac{dy}{dx} = f(x) \quad \text{has an inverse} \quad y(x) = \int dx\, f(x),$$

so the integral $\int dx$ may be regarded as an inverse of the operator $\frac{d}{dx}$, which can be made unique by specifying an initial condition. In both these examples we find the unknown by first determining the inverse of a linear operator, and then applying it to a specific case.

Green's functions may be regarded as the inverse of certain linear differential operators. For example, the solution of the linear differential equation

$$\frac{d}{dx}\left(p(x)\frac{dy}{dx}\right) + q(x)y = f(x), \quad y(a) = y(b) = 0, \tag{7.26}$$

can, for sufficiently well behaved functions $p(x)$ and $q(x)$, be written as the integral

$$y(x) = \int_a^b du\, G(x, u) f(u), \tag{7.27}$$

where the Green's function, G, is *independent* of the function f. The form taken by the Green's function depends upon *both* the differential operator *and* the boundary or initial conditions. The relation 7.27 is very similar to the solution of the vector equation discussed above which, in component form, is $y_k = \sum_j (A^{-1})_{kj} f_j$; in both cases the solution is a linear function of f and is a sum or integral over the components of f. In exercises 7.26 and 7.27, at the end of this section, we develop this analogy a little further.

Green's functions are important because they enable the unknown function $y(x)$ to be expressed explicitly in terms of known quantities. The analogy with integration suggests that Green's functions can be found explicitly for only a relatively small class of problems; fortunately this includes many important equations, including some partial differential equations.

Green's functions exist only for linear equations; the crucial property being that the sum of two or more solutions is also a solution. Before considering the more general applications we digress to a discussion of two simple physical models which, in part, explain why linearity enables the solutions to be expressed in the relatively simple manner of equation 7.27.

7.3.1 A digression: two linear differential equations
Motion of a particle
A crucial property of Newton's equations of motion is that forces add vectorially, so the equations of motion are linear in the force. In addition for most mechanical systems near stable equilibrium, where the system is stationary and where small departures from equilibrium remain small, the equations of motion are approximately linear; this means that if individual forces $\mathbf{F}_1$ and $\mathbf{F}_2$ give rise to the motions $\mathbf{r}_1(t)$ and $\mathbf{r}_2(t)$ then

the combined force $\mathbf{F}_1 + \mathbf{F}_2$ produces the motion $\mathbf{r}_1(t) + \mathbf{r}_2(t)$. The theory of Green's functions rests upon this property.

Linear equations of motion arise naturally when considering the equilibrium positions of mechanical systems, which include for example the motion of a violin string, the airflow in a trumpet bell and the motion of a wave on the surface of deep water. Also, in very many circumstances, the electric field produced by several electrons is the linear sum of the fields produced by each individual electron, so the equations describing these electromagnetic fields depend linearly upon the sources. Green's function methods can be applied in all these circumstances, and in all cases the basic ideas are the same and can be illustrated with very simple problems.

In our first illustration we consider a particle of unit mass moving along the x-axis under the influence of a force $F(t)$ depending only upon the time t and which, for the sake of simplicity, we assume to be zero when $t \leq 0$. For the same reason we assume the particle to be stationary at the origin at $t = 0$, so Newton's equation of motion and the initial conditions are

$$\ddot{x} = F(t), \quad x(0) = \dot{x}(0) = 0. \tag{7.28}$$

Obviously we can find $x(t)$ by direct integration, but here we tackle the problem using a less direct method.

The equation of motion is linear, so if $x_k(t)$ is the motion due to a force $f_k(t)$ the motion due to the sum of forces $\sum f_k(t)$ is $\sum x_k(t)$. Moreover, we know the motion of a free particle, $f = 0$, and it is relatively easy to find the motion due to a force which is nonzero for only a very short time. This suggests that we split the function $F(t)$ into elementary rectangles of width $\delta\tau$, one of which is shown in the figure.

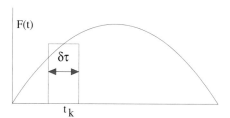

Figure 7.5   An elementary rectangle in the
decomposition of $F(t)$.

The kth rectangle is centred at $t = t_k = (k + \frac{1}{2})\delta\tau$ and has height $F(t_k)$; the sum of the areas of the first n rectangles is approximately the area under the curve $F(t)$ for $0 \leq t < t_n$:

$$\int_0^{t_n} dt\, F(t) \simeq \delta\tau \sum_{j=0}^{n} F(t_j).$$

We wish to write the force as a sum of impulses such that the kth impulse is zero everywhere except on an interval of width $\delta\tau$ around $t = t_k$. To this end we define the function

$$\Delta(t - t_k) = \begin{cases} 0, & |t - t_k| > \delta\tau/2, \\ \delta\tau^{-1}, & |t - t_k| < \delta\tau/2, \end{cases} \quad k = 0, 1, \ldots, \tag{7.29}$$

so that the graph of $\Delta(t - t_k)$ is simply a rectangle centred at $t = t_k$, of width $\delta\tau$, height $1/\delta\tau$ and with unit area. Comparison with equation 7.18 (page 235), shows that this sequence tends to $\delta(t - t_k)$ as $\delta\tau \to 0$.

Now form the sum

$$\sum_{\text{all } k} \Delta(t_k - t)F(t_k)\delta\tau \quad \simeq \quad F(t) \tag{7.30}$$

$$\longrightarrow \int_0^\infty d\tau \, F(\tau)\delta(\tau - t) \quad \text{as} \quad \delta\tau \to 0.$$

The motion $x(t)$ due to $F(t)$ is just the sum of the motions due to the impulses $\Delta(t_k - t)F(t_k)\delta\tau$, occuring at t_k, for $t_k < t$, that is $k = 0, \ldots, t/\delta\tau$. The motion $x_k(t)$, due to the impulse at t_k, is given by the solution of the equation

$$\ddot{x}_k(t) = \Delta(t_k - t)F(t_k)\delta\tau, \quad x_k(0) = \dot{x}_k(0) = 0.$$

But the right hand side is nonzero only for a short interval $\delta\tau$ around t_k, and it may be assumed that during this time the particle velocity changes but its position does not, and at all other times it moves as a free particle. Thus for $t < t_k - \delta\tau/2$ it is stationary at the origin, and for times $t > t_k + \delta\tau/2$, it is moving at a constant speed V away from the origin, that is $x(t) = (t - t_k)V + \beta\delta\tau$, where β is some constant, present because of the finite duration of the impulse: we ignore this term because it tends to zero with $\delta\tau$. The change in velocity V is given by integrating the equation of motion across the impulse:

$$V = \int_{t_k - \delta\tau/2}^{t_k + \delta\tau/2} dt \, \ddot{x}_k = \int_{t_k - \delta\tau/2}^{t_k + \delta\tau/2} dt \, \Delta(t_k - t)F(t_k)\delta\tau = F(t_k)\delta\tau.$$

Effectively, at the time $t = t_k$ the particle starts moving from rest with the speed $F(t_k)\delta\tau$ and subsequently moves as a free particle. Thus we have, for sufficiently small $\delta\tau$,

$$x_k(t) = \begin{cases} 0, & t < t_k, \\ (t - t_k)F(t_k)\delta\tau, & t > t_k. \end{cases}$$

The impulse at $t = t_k$ affects the motion only for $t > t_k$, as would be expected.

Since $F(t)$ can be decomposed into the sum of elementary impulses, equation 7.30, the motion due to F is just the sum of the effects of each impulse,

$$x(t) \quad \simeq \quad \sum_{\text{all } k} x_k(t) = \sum_{t_k < t} (t - t_k)F(t_k)\delta\tau$$

$$\longrightarrow \int_0^t d\tau \, (t - \tau)F(\tau) \quad \text{as} \quad \delta\tau \to 0.$$

Note that the sum is limited to values of $t_k < t$ because $x_k(t)$ is zero for $t_k > t$.

This analysis has shown that the solution of the equation

$$\ddot{x} = F(t), \quad x(0) = \dot{x}(0) = 0,$$

can be written in the general form of equation 7.27, that is

$$x(t) = \int_0^\infty du\, G(t,u)F(u), \quad \text{where} \quad G(t,u) = H(t-u)(t-u) \tag{7.31}$$

is the Green's function.

In this simple example the same result may be obtained by directly integrating the equation $\ddot{x} = F(t)$ twice and taking into account the initial conditions to give

$$
\begin{aligned}
x(t) &= \int_0^t du_1 \int_0^{u_1} du\, F(u) \\
&= \int_0^t du\, F(u) \int_u^t du_1 = \int_0^t du\, (t-u)F(u),
\end{aligned}
$$

where we have changed the order of integration to derive the second line.

Exercise 7.16
Verify that equation 7.31 satisfies the differential equation 7.28 and has the correct initial conditions.

Exercise 7.17
Solve the equation

$$\frac{d^2x}{dt^2} = e^{-at}, \quad x(0) = \dot{x}(0) = 0, \quad a > 0,$$

by direct integration and confirm that the same result is obtained by applying equation 7.31.

A stationary, stretched string
For the second example we consider a static string of length L lying along the x-axis. If it is disturbed by a steady force acting perpendicular to its axis it will bend into a curve $y(x)$, such as shown in figure 7.6.

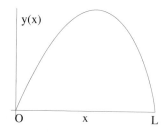

Figure 7.6   Diagram showing a string displaced by a force $f(x)$.

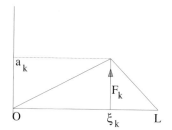

Figure 7.7   Diagram showing a string displaced by a localised force $F_k\Delta(x-\xi_k)$.

Suppose that the string is stationary and is fixed at each end, so $y(0) = y(L) = 0$, then if the force per unit length is $f(x)$ and if $y(x)$ is not too large the equation describing the displacement is, in appropriate units,

$$\frac{d^2y}{dx^2} = -f(x), \quad y(0) = y(L) = 0,$$ (7.32)

where the force is positive in the direction of increasing y. The only difference between this and equation 7.28 is the boundary conditions. Again, this is a linear equation, so we approximate the force $f(x)$ as a sum,

$$f(x) = \int_0^L d\xi \, \delta(\xi - x)f(\xi) \simeq \sum_k \Delta(\xi_k - x)f(\xi_k)\delta\xi,$$

with each component of the sum representing a force localised at one point, as if a stick of width $\delta\xi$ were used to prod the string at $x = \xi_k$, as shown in figure 7.7. On ignoring the small width of each component, it is clear that the shape of the string due to the localised force at ξ_k is the two straight lines shown in the figure, that is

$$y_k(x) = \begin{cases} \dfrac{xa_k}{\xi_k}, & 0 \le x \le \xi_k, \\[2mm] \dfrac{(L-x)a_k}{L-\xi_k}, & \xi_k \le x \le L, \end{cases}$$ (7.33)

for some constant a_k, the displacement of the string at $x = \xi_k$. The value of the constant a_k is unknown but, as in the previous case, it can be determined by integrating the equation for $y_k(x)$ across the localised force. The equation for the displacement due to the force at ξ_k is

$$\frac{d^2y_k}{dx^2} = -\Delta(\xi_k - x)f(\xi_k)\delta\xi, \quad y_k(0) = y_k(L) = 0.$$ (7.34)

Integrating this equation over a small interval that includes the whole of $\delta\xi$ gives the change in gradient of the string due to the force,

$$\Delta\left(\frac{dy_k}{dx}\right) = \int_{\xi_k - \delta\xi/2}^{\xi_k + \delta\xi/2} dx \, \frac{d^2y_k}{dx^2} = -f(\xi_k)\delta\xi.$$

But from equation 7.33 we see that the change in the gradient at $x = \xi_k$ is

$$\Delta\left(\frac{dy_k}{dx}\right) = -\frac{a_k}{L-\xi_k} - \frac{a_k}{\xi_k}.$$

These two expressions for the change in slope give $a_k = \xi_k(L - \xi_k)f(\xi_k)\delta\xi/L$.

Since the equation for y is linear the response to the force $f(x)$ is just the sum of the response to each component,

$$\begin{aligned} y(x) &= \sum_k y_k(x) = \frac{x}{L}\sum_{\xi_k > x}(L-\xi_k)f(\xi_k)\delta\xi + \frac{L-x}{L}\sum_{\xi_k < x}\xi_k f(\xi_k)\delta\xi \\ &\to \frac{x}{L}\int_x^L d\xi \, f(\xi)(L-\xi) + \frac{L-x}{L}\int_0^x d\xi \, f(\xi)\xi \quad \text{as } \delta\xi \to 0. \end{aligned}$$ (7.35)

This solution can be written in the canonical form of equation 7.27 by defining the Green's function,

$$G(x, \xi) = \frac{x(L - \xi)}{L} H(\xi - x) + \frac{\xi(L - x)}{L} H(x - \xi).$$

This Green's function and that given in equation 7.31 belong to the same operator, $\frac{d^2}{dx^2}$, but are different because the first arose from an initial value problem and the second from a boundary value problem.

Exercise 7.18
Verify that equation 7.35 satisfies the differential equation 7.32 and the correct boundary conditions.

7.3.2 Green's functions for a boundary value problem

The method of constructing Green's functions for more general equations is essentially the same as the method used in the above two examples; we merely replace the sequence defined in equation 7.29 by a delta function, which takes us directly to the end result, equations 7.31 and 7.35. Here we deal only with specific examples, leaving the general theory to chapter 10.

Consider the simple linear boundary value problem

$$\frac{d^2 y}{dx^2} = h(x), \quad y(0) = A, \quad y(L) = B, \tag{7.36}$$

where $h(x)$ is some function and A, B and $L > 0$ are constants. Observe first that because the system is linear the solution may be written as the sum of the solution to the homogeneous equation with inhomogeneous boundary conditions,

$$\frac{d^2 y}{dx^2} = 0, \quad y(0) = A, \quad y(L) = B, \tag{7.37}$$

and the inhomogeneous equation with homogeneous boundary conditions,

$$\frac{d^2 y}{dx^2} = h(x), \quad y(0) = 0, \quad y(L) = 0. \tag{7.38}$$

The solutions to *both* these equations can be obtained from the general solution of the homogeneous equation $d^2 y/dx^2 = 0$, which, in this case, is $\alpha x + \beta$, α and β being constants. The solution of equation 7.37 is obtained by choosing the constants α and β to satisfy the boundary conditions:

$$y = A + (B - A)\frac{x}{L}.$$

Now concentrate on the inhomogeneous equation and attempt to express its solution in the form of equation 7.27. In order to do this we solve the related equation

$$\frac{d^2}{dx^2} G(x, u) = \delta(x - u), \quad G(0, u) = G(L, u) = 0, \quad 0 < u < L, \tag{7.39}$$

which is similar to equation 7.34 in the sense that the right hand side is zero for most values of x. The function $G(x,u)$ is a function of u besides x but, for the present, we treat u as a constant. Despite its appearance, it is easier to solve this equation than to solve the original inhomogeneous equation because the right hand side is zero everywhere except at $x = u$; we construct this solution shortly. First, suppose that $G(x,u)$ is the solution and define the function

$$y(x) = \int_0^L du\, G(x,u)h(u). \tag{7.40}$$

That this is the required solution of equation 7.38 is seen by differentiating twice with respect to x,

$$\frac{d^2y}{dx^2} = \frac{d^2}{dx^2} \int_0^L du\, G(x,u)h(u) = \int_0^L du\, \frac{d^2}{dx^2} G(x,u)h(u)$$

$$= \int_0^L du\, \delta(x-u)h(u)$$

the last relation following from the definition of $G(x,u)$, equation 7.39. Thus we obtain

$$\frac{d^2y}{dx^2} = h(x), \quad \text{from equation 7.3 (page 228).}$$

This analysis shows that the solution of the inhomogeneous equation with homogeneous boundary conditions can be written in terms of a Green's function.

Exercise 7.19
Show that the function defined in equation 7.40 satisfies the boundary conditions $y(0) = y(L) = 0$.

The solution of the original equation, 7.36, can be found simply by adding these two solutions to give

$$y(x) = A + (B - A)\frac{x}{L} + \int_0^L du\, G(x,u)h(u). \tag{7.41}$$

It follows that if $G(x,u)$ can be found, the required solution is given by the integral 7.41. The existence of the function $G(x,u)$, which is *independent* of $h(x)$, means that the solution of the original differential equation has been cast in the form of an integral, which is an important simplification.

Construction of the Green's function
We now show that the Green's function may be constructed using the general solution to the homogeneous equation. Divide the interval $0 < x < L$ into two parts. In the first, $0 \le x < u$, we set $G = G_1(x,u)$ to give

$$\frac{d^2G_1}{dx^2} = 0, \quad G_1(0,u) = 0 \quad \Longrightarrow \quad G_1(x,u) = \alpha(u)x,$$

where $\alpha(u)$ is an unknown function of u.
In the second interval, $u < x \le L$, we set $G = G_2(x,u)$ to give

$$\frac{d^2G_2}{dx^2} = 0, \quad G_2(L,u) = 0 \quad \Longrightarrow \quad G_2(x,u) = \beta(u)(L - x),$$

where $\beta(u)$ is another unknown function of u. These two expressions for G_1 and G_2 are essentially the same as the expressions for $y_k(x)$ given in equation 7.33 with $\alpha(u)$ playing the role of a_k/ξ_k and $\beta(u)$ the role of $a_k/(L - \xi_k)$. The arbitrary functions $\alpha(u)$ and $\beta(u)$ are determined by matching G_1 and G_2 and their gradients at $x = u$.

The change in the gradient at $x = u$ is given by integrating the differential equation 7.39 over the interval $u - \epsilon \le x \le u + \epsilon$ for $0 < \epsilon \ll 1$,

$$\int_{u-\epsilon}^{u+\epsilon} dx \, \frac{d^2 G}{dx^2} = \int_{u-\epsilon}^{u+\epsilon} dx \, \delta(x - u) = 1,$$

giving the condition

$$\lim_{\epsilon \to 0} \left[\frac{d}{dx} G(x, u) \right]_{x=u-\epsilon}^{u+\epsilon} = 1$$

for the change in the gradient of $G(x, u)$ across $x = u$. Since ϵ can be made arbitrarily small, the gradient of $G(x, u)$ is discontinuous at $x = u$ and with unit step.

Now we require the relation between $G_1(x, u)$ and $G_2(x, u)$ at $x = u$. In the physical examples considered above it was obvious that the solutions were continuous at $x = u$, that is $G_1(u, u) = G_2(u, u)$. This is always true, as shown by using the result derived in exercise 4 (page 231): if $G(x, u)$ were not continuous at $x = u$ then the differential $\partial G/\partial x$ would contain a component $\delta(x - u)$ and then $\partial^2 G/\partial x^2$ would have the term $\delta'(x - u)$, which is a contradiction.

These two conditions are sufficient to determine the two unknown functions $\alpha(u)$ and $\beta(u)$ because they are true for *all* values of u in $(0, L)$. At $x = u$ we have $G(u, u) = u\alpha(u)$ and $G(u, u) = \beta(u)(L - u)$, and continuity gives

$$\alpha(u)u = \beta(u)(L - u).$$

The discontinuity in the derivative gives

$$\lim_{\epsilon \to 0} \left[\frac{d}{dx} G(x, u) \right]_{u-\epsilon}^{u+\epsilon} = -\beta(u) - \alpha(u) = 1.$$

Solving these two equations gives

$$\alpha(u) = -\frac{L - u}{L}, \quad \beta(u) = -\frac{u}{L},$$

and hence the final expression for $G(x, u)$ is

$$G(x, u) = \begin{cases} G_1(x, u) = -\dfrac{x(L - u)}{L}, & 0 \le x \le u, \\[2mm] G_2(x, u) = -\dfrac{(L - x)u}{L}, & u \le x \le L. \end{cases} \tag{7.42}$$

Observe that $G_2(u, x) = G_1(x, u)$, which is a general property of Green's functions for boundary value problems. It is sometimes more convenient to write the Green's function using the Heaviside function, defined in equation 7.8,

$$G(x,u) = G_1(x,u)H(u-x) + G_2(x,u)H(x-u). \tag{7.43}$$

The shape of this function is shown in the figure below: note that $-G(x,u)$ is plotted because this shows its shape more clearly.

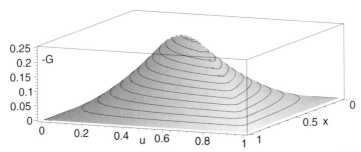

Figure 7.8   Three-dimensional plot of the Green's function defined in equation 7.42.

The general solution of the equation

$$\frac{d^2y}{dx^2} = h(x), \quad y(0) = A, \quad y(L) = B,$$

can now be written in the form

$$\begin{aligned}
y(x) &= A + (B-A)\frac{x}{L} + \int_0^L du\, G(x,u)h(u), \\
&= A + (B-A)\frac{x}{L} - \frac{L-x}{L}\int_0^x du\, u h(u) - \frac{x}{L}\int_x^L du\,(L-u)h(u),
\end{aligned}$$

which should be compared with equation 7.35. There are a few dubious steps in this analysis, but all can be made rigorous. Normally it is easier to solve the equations for $G(x,u)$ than the original inhomogeneous equation 7.36.

The function $G(x,u)$ is called the Green's function for the operator $\frac{d^2}{dx^2}$ with the boundary conditions of equation 7.38.

Exercise 7.20
Use the Green's function found above to find the solution of the equation

$$\frac{d^2y}{dx^2} = x^2, \quad y(0) = y(1) = 0.$$

Exercise 7.21
Show that the Green's function for the equation

$$\frac{d^2y}{dx^2} = h(x), \quad y(a) = 0, \quad y'(b) = 0, \quad \text{is} \quad G(x,u) = \begin{cases} -(x-a), & x \le u, \\ -(u-a), & x \ge u. \end{cases}$$

Hence find the solution of the differential equation

$$\frac{d^2y}{dx^2} = \sin x, \quad y(a) = A, \quad y'(b) = B.$$

Exercise 7.22

Show that the Green's function, $G(x, u)$, for the equation

$$\frac{d^3 y}{dx^3} = h(x), \quad y(0) = y'(0) = y(1) = 0$$

satisfies the condition

$$\lim_{\epsilon \to 0} \left[\frac{\partial^2 G}{\partial x^2} \right]_{x=u-\epsilon}^{u+\epsilon} = 1$$

and that $G(x, u)$ and $\frac{\partial G}{\partial x}$ are continuous at $x = u$.

Hence show that

$$G(x, u) = \begin{cases} -\dfrac{1}{2}(u-1)^2 x^2, & 0 \le x \le u, \\[2mm] -\dfrac{1}{2}(1-x)(2x - u - ux), & u \le x \le 1. \end{cases}$$

7.3.3 *The perturbed linear oscillator*

The linear oscillator is one of the fundamental systems in dynamics because it describes the motion of a dynamical system near a point of stable equilibrium. The one-dimensional forced linear oscillator is an important variant because it describes the motion of a system, S_1, near stable equilibrium when it interacts with another dynamical system, S_2, in such a manner that S_1 has a negligible effect on S_2. For instance, a pendulum, S_1, hanging in a train will not significantly affect the motion of the train, S_2, but the motion of the train does affect the pendulum.

If the configuration of the system S_1 can be described by a single variable, x, — the pendulum constrained to move in a vertical plane, for instance — near equilibrium, the equation of motion will often have the form

$$\frac{d^2 x}{dt^2} + \omega^2 x = f(t),$$

where $f(t)$ represents the force, and depends only upon t and not x, and ω is some positive constant. This type of equation is normally valid only if $|x(t)|$ remains small by comparison to some scale. If $f(t) = 0$ the equation has the general solution $x(t) = A \sin \omega(t - t_0)$, showing that ω is the angular frequency of the unperturbed motion: the constants A and t_0 depend upon the initial conditions; for this linear system the amplitude A is independent of the frequency ω.

Typically this is an initial value problem because the system normally starts from a known state at a given time, that is, $x(t)$ and $\dot{x}(t)$ are known at some instant $t = t_0$ and we require the solution for $t > t_0$. In order to be specific we put $t_0 = 0$ and assume that $x(0) = \dot{x}(0) = 0$, thus we need to solve the equation

$$\frac{d^2 x}{dt^2} + \omega^2 x = f(t), \quad x(0) = \dot{x}(0) = 0. \tag{7.44}$$

We can express the solution of this equation as an integral over $f(t)$ by finding the

Green's function, that is, the solution of the equation

$$\frac{d^2}{dt^2}G(t,\tau) + \omega^2 G(t,\tau) = \delta(t-\tau), \quad 0 = G(0,\tau) = \frac{d}{dt}G(t,\tau)\,|_{t=0}. \qquad (7.45)$$

This is found in exactly the same manner as the solution of the simpler equation 7.36, the main differences being that here we have an initial value rather than a boundary value problem and that the solutions of the homogeneous equation are different. As before, we consider the two time intervals $0 \le t < \tau$ and $t > \tau > 0$, and match the two solutions at the boundary $t = \tau$.

The result obtain in exercise 7.4 (page 231), shows that $G(t,\tau)$ is continuous at $t = \tau$. Since the Green's function is bounded, direct integration of equation 7.45 across the interval $\tau - \epsilon < t < \tau + \epsilon$, with $\epsilon > 0$, shows that

$$\lim_{\epsilon \to 0} \left[\frac{dG}{dt}\right]_{t=\tau-\epsilon}^{\tau+\epsilon} = 1.$$

These two conditions allow the unique determination of the Green's function.

For $0 \le t < \tau$ the Green's function satisfies the equation

$$\frac{d^2}{dt^2}G(t,\tau) + \omega^2 G(t,\tau) = 0, \qquad 0 = G(0,\tau) = \frac{d}{dt}G(t,\tau)\,|_{t=0},$$

and the only solution is $G(t,\tau) = 0$, $t < \tau$. This result should not be surprising, for the system is at rest at the equilibrium position at time $t = 0$ and remains undisturbed until the impulse at time $t = \tau > 0$.

For $t > \tau$ the equation is

$$\frac{d^2}{dt^2}G(t,\tau) + \omega^2 G(t,\tau) = 0,$$

and this has the general solution

$$G(t,\tau) = A(\tau)\sin\omega(t-\tau) + B(\tau)\cos\omega(t-\tau),$$

where A and B are functions of τ, determined by matching the two solutions at $t = \tau$. Continuity of the Green's function at $t = \tau$ shows that $B(\tau) = 0$. Integrating across $t = \tau$ gives $A(\tau) = 1/\omega$. Thus the Green's function is

$$G(t,\tau) = \frac{1}{\omega}H(t-\tau)\sin\omega(t-\tau), \qquad (7.46)$$

and the solution of the original equation is

$$x(t) = \frac{1}{\omega}\int_0^t d\tau\, f(\tau)\sin\omega(t-\tau). \qquad (7.47)$$

You should check, by differentiation, that this function satisfies the original equation. The shape of this Green's function is shown in the following figure.

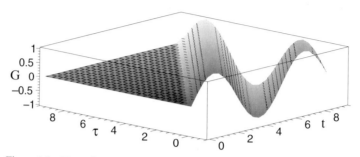

Figure 7.9   Three-dimensional plot of the Green's function defined in equation 7.46.

Exercise 7.23

The energy $E(t)$ of the system defined in equation 7.44 is defined to be

$$E(t) = \frac{1}{2}\dot{x}^2 + \frac{1}{2}\omega^2 x^2.$$

Show that if $f(t)$ is real the energy of the oscillator as $t \to \infty$ is

$$E_f = \frac{1}{2}\left|\int_0^\infty d\tau\, f(\tau)e^{-i\omega\tau}\right|^2.$$

Hint: in the solution, equation 7.47, write $\sin\omega(t-\tau) = \Im e^{i\omega(t-\tau)}$.

Exercise 7.24

Show that the Green's function satisfying the equation

$$\frac{d^2}{dx^2}G(x,u) = \delta(x-u), \qquad G(0,u) = \left.\frac{d}{dx}G(x,u)\right|_{x=0} = 0$$

is

$$G(x,u) = (x-u)H(x-u).$$

Hence show that the solution of the equation

$$\frac{d^2 y}{dx^2} = h(x), \quad y(0) = y'(0) = 0,$$

can be written in the form

$$y(x) = \int_0^x du\,(x-u)h(u). \tag{7.48}$$

By directly integrating the equation for y twice show that an alternative representation of the solution is

$$y(x) = \int_0^x dt_2 \int_0^{t_2} dt_1\, h(t_1),$$

and deduce the identity

$$\int_0^x du\,(x-u)h(u) = \int_0^x dt_2 \int_0^{t_2} dt_1\, h(t_1).$$

The generalisation of this result to the n-dimensional integrals is considered in exercise 7.43 (page 264).

7.3.4 Numerical investigations of Green's functions

We have not justified the use of the delta function because a rigorous development of the necessary theory would be out of place here, but we can numerically examine the behaviour of the 'Green's' function $G_n(x, u)$ obtained by replacing the delta function by a member of the sequence $\delta_n(x - u)$ which defines a delta function; we chose the exponential sequence defined in equation 7.14, page 232. To be specific, we consider the initial value problem defined in exercise 7.24,

$$\frac{d^2}{dx^2} G_n(x, u) = \delta_n(x - u), \quad G_n(0, u) = \frac{d}{dx} G_n(x, u)\bigg|_{x=0} = 0. \tag{7.49}$$

The result obtained in exercise 7.24, equation 7.48, allows us to write the solution of this equation as the integral

$$\begin{aligned} G_n(x, u) &= \int_0^x dw\, (x - w)\delta_n(w - u) \\ &\longrightarrow (x - u)H(x - u) \quad \text{as } n \to \infty. \end{aligned} \tag{7.50}$$

If n is finite we can evaluate the integral 7.50 using Maple and compare its value with the limiting form. Such comparisons are shown in figures 7.10 to 7.12 for various values of n and for $u = \frac{1}{2}$.

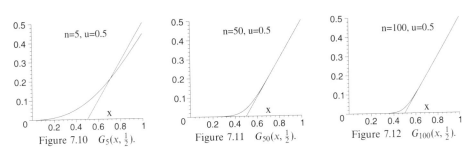

Figure 7.10 $G_5(x, \frac{1}{2})$.   Figure 7.11 $G_{50}(x, \frac{1}{2})$.   Figure 7.12 $G_{100}(x, \frac{1}{2})$.

This numerical comparison shows clearly how the limit is approached; notice that the Green's function is continuous at $x = u$ for all values of n used.

Exercise 7.25

Solve the boundary value problem

$$\frac{d^2}{dx^2} G_n(x, u) = \delta_n(x - u), \quad G_n(0, u) = G_n(1, u) = 0,$$

where $\delta_n(u)$ is defined in equation 7.14, and examine graphically the behaviour of $G_n(x, u)$ as a function of x, for various values of u and as n increases.

7.3.5 Other representations of Green's functions

We end this section on Green's functions by first showing how the Green's function is related to the linear algebraic equation $A\mathbf{y} = \mathbf{f}$, discussed in the introduction to this section, and second by showing how alternative representations of the Green's function may be obtained, as a pre-cursor to the more general theory to be developed in chapter 10.

First consider the algebraic equation $A\mathbf{y} = \mathbf{f}$; the connection between this and the Green's function can be made explicit by approximating all functions by their values at a discrete set of equally spaced points. If for example $0 < x < 1$, the function $f(x)$, which is zero at $x = 0$, is approximated by the N-dimensional vector $\mathbf{f} = (f_1, f_2, \ldots, f_N)$ with $f_k = f(k\delta)$, $N\delta = 1$.

A simple example is the equation $\frac{dy}{dx} = f(x)$, $y(0) = 0$. We first discretise the value of the differential at $x_k = k\delta$ by $\frac{dy}{dx} \rightarrow (y_k - y_{k-1})/\delta$ so the differential operator $\frac{dy}{dx}$ is approximated by the matrix operator D/δ, where

$$
D = \begin{pmatrix}
1 & 0 & 0 & 0 & \cdots & 0 & 0 & 0 \\
-1 & 1 & 0 & 0 & \cdots & 0 & 0 & 0 \\
0 & -1 & 1 & 0 & \cdots & 0 & 0 & 0 \\
\vdots & \vdots & \vdots & \vdots & & \vdots & \vdots & \vdots \\
0 & 0 & 0 & 0 & \cdots & -1 & 1 & 0 \\
0 & 0 & 0 & 0 & \cdots & 0 & -1 & 1
\end{pmatrix}, \tag{7.51}
$$

consequently the differential equation becomes $D\mathbf{y} = \delta\mathbf{f}$ and has the approximate solution $\mathbf{y} = \delta D^{-1}\mathbf{f}$. It is important to note that the form of the matrix D depends on the operator $\frac{dy}{dx}$ *and the initial conditions*.

Exercise 7.26

Use Maple to find the inverse of the matrix D, for suitable values of N, and use this to find an approximate solution of the equation

$$
\frac{dy}{dx} = x^p, \quad y(0) = 0,
$$

for various values of p, comparing your answers with the exact result $x^{p+1}/(p+1)$.

Find the discrete approximation to the Green's function for this operator.

It takes a long time for Maple to invert a large matrix, but it is faster if the elements of the matrix are floating point numbers, so we suggest that initially you choose N to be between 5 and 10, making it larger only when you are confident that the program works. Note, also that inverting D is probably the most inefficient method of solving these linear equations, which are trivially solved 'by hand', but here we are simply illustrating an idea.

Exercise 7.27

Find the discrete approximation to the Green's function satisfying the equation

$$
\frac{d^2G}{dx^2} + G = \delta(x-y), \quad G(0, y) = \frac{d}{dx}\,G(x, y)|_{x=0} = 0,
$$

on the interval $0 \le x \le \pi$.

Use the matrix D of equation 7.51 to approximate $\frac{d^2}{dx^2}$ by $E = (D/\delta)^2$, $N\delta = \pi$, so the differential equation becomes the matrix equation $G_k = (E + I)^{-1}\mathbf{d}_k$, where $\mathbf{d}_k$ is an N-dimensional vector all components of which are zero except the kth which is $1/\delta$.

Use Maple to find G for suitable values of N and k and compare your result graphically with the exact solution given in equation 7.46.

A given function can be approximated by a finite set of numbers in many different ways, the above discretisation is only one method and is not particularly efficient. A finite power series expansion, $f(x) = \sum a_n x^n$, such as a Taylor polynomial, can be regarded as a way of approximating $f(x)$ by the vector $(a_0, a_1, \ldots)$. Another, more commonly used, approach is to expand the relevant functions in terms of a complete set of functions† satisfying the boundary conditions. This method also converts a differential equation into an algebraic problem, and is illustrated in the next exercise.

Exercise 7.28
Consider the boundary value problem

$$\frac{d^2 y}{dx^2} + \omega^2 y = f(x), \quad y(0) = y(\pi) = 0,$$

with ω not an integer. Since the set of functions $\sin nx$, $n = 1, 2, \ldots$, is complete on the space with these boundary conditions both $f(x)$ and $y(x)$ can be expressed in the form

$$f(x) = \sum_{k=1}^{\infty} f_k \sin kx, \quad y(x) = \sum_{k=1}^{\infty} y_k \sin kx,$$

where f_k are known real constants. Show that $y_k = f_k/(\omega^2 - k^2)$. Note that finding this solution is particularly simple because $\frac{d^2}{dx^2} \sin kx = -k^2 \sin kx$.

Now extend this technique using the following series representation of the delta function:

$$\delta(x - y) = \frac{2}{\pi} \sum_{k=1}^{\infty} \sin kx \sin ky$$

to show that the Green's function for this problem may be expressed as the infinite series

$$G(x, y) = \frac{2}{\pi} \sum_{k=1}^{\infty} \frac{\sin kx \sin ky}{\omega^2 - k^2}.$$

Show also the standard approach described in the text gives

$$G(x, y) = \begin{cases} -\dfrac{\sin \omega(\pi - y) \sin \omega x}{\omega \sin \pi \omega}, & x \leq y, \\[4mm] -\dfrac{\sin \omega(\pi - x) \sin \omega y}{\omega \sin \pi \omega}, & x \geq y. \end{cases}$$

By truncating the sum of the infinite series representation after N terms, use Maple to compare these two formulae graphically as functions of x for a few values of y and N.

---

† The idea of a complete set of functions will be developed in chapter 8. Here you may assume that it is a set of functions $\phi_k(x)$, $k = 0, 1, \ldots$, such that any suitable function can be represented as a linear combination of the $\phi_k(x)$.

7.4 Bessel functions: Kapteyn and Neumann series

The theory presented next provides an illustration of the power of the delta function by deriving results involving series of ordinary Bessel functions; here delta functions replace some fairly complicated analysis with a few lines of algebra. The method works because the integral definition of the ordinary Bessel has a suitable form. The theory presented here uses Fourier series, which are introduced in the next chapter.

The ordinary Bessel function of integer order, $J_n(x)$, occurs in many applications of mathematics; it may be defined by the integral†

$$J_n(x) = \frac{1}{2\pi} \int_{-\pi}^{\pi} d\theta \, \exp i(n\theta - x\sin\theta) \tag{7.52}$$

$$= \frac{1}{\pi} \int_0^{\pi} d\theta \, \cos(n\theta - x\sin\theta).$$

In words, $J_n(x)$ is the nth harmonic in the Fourier series development of $e^{-ix\sin\theta}$. The ordinary Bessel function is also given by the solution of the differential equation

$$x^2 \frac{d^2 y}{dx^2} + x\frac{dy}{dx} + \left(x^2 - n^2\right) y = 0, \tag{7.53}$$

which is regular at the origin. It is easily seen that the function defined by the integral 7.52 satisfies this equation. The series representation of the ordinary Bessel function is

$$J_n(x) = \left(\frac{x}{2}\right)^n \sum_{k=0}^{\infty} (-1)^k \frac{(x^2/4)^k}{k!(n+k)!}.$$

This series converges for all (complex) values of x.

A great deal is known about these functions, much of which is described in the comprehensive treatise of Watson (1966). They have occurred in the work of Euler (1764, in the vibrations of a stretched membrane), Lagrange (1770, in the theory of planetary motion), Fourier (1822, in his theory of heat flow), Poisson (1823, in the theory of heat flow in spherical bodies) and by Bessel (1824, who studied these functions in detail); Watson (1966) abandons his attempt to delineate the chronological order of the study after Bessel as 'After the time of Bessel investigations on the functions become so numerous...'.

Exercise 7.29

Use Maple to plot the graphs of $J_n(x)$ for $-20 \le x \le 20$ for $n = 0, \pm 1, \ldots, \pm 4$. From these graphs what can you deduce about the relation between $J_n(x)$ and $J_{-n}(x)$ and between $J_n(x)$ and $J_n(-x)$? Use the integral definition 7.52 to prove your observations.

By plotting the graphs of $J_n(x)$, $n = 1, 2, \ldots, 10$ for $0 \le x \le 20$ can you see any pattern in the behaviour of the position of the first maximum of $J_n(x)$?

Plot the graphs of $J_n(nx)$ for $n = 5, 10, 20$ and 50 for $0 \le x \le 2$. Comment upon the position of the first maximum of $J_n(nx)$.

You may find it convenient to use the Maple `alias`   command in the form

---

† This integral definition of the ordinary Bessel function is valid only if n is an integer; for the generalisation see for instance Whittaker and Watson (1965, sections 17.1 and 17.231).

`alias(J=BesselJ)` to save typing. For instance, with this definition the graph of $J_3(x)$ for $0 < x < 5$ is given by `plot(J(3,x),x=0..5)`.

The graphs drawn in the last part of exercise 7.29 show that $J_n(nx)$ is oscillatory for $x > 1$ but decreases rapidly to zero as $x \to 0$. In fact we have

$$J_n(nx) \simeq \sqrt{\frac{2}{\pi n x}} \cos\left(nx - \left(\frac{1}{2}n + \frac{1}{4}\right)\pi\right), \quad x \gg 1, \tag{7.54}$$

$$\simeq \frac{1}{\sqrt{2\pi n}}\left(\frac{ex}{2}\right)^n \quad 0 < x \ll 1. \tag{7.55}$$

These and other results can be deduced from the integral 7.52 using the methods described in chapters 13 to 15.

Kapteyn series

Series of the form

$$\sum_{n=0}^{\infty} a_n J_n(nz)$$

are known as Kapteyn series after the Dutch mathematician W. Kapteyn, who made the first systematic study of their properties as a function of the complex variable z; here we assume that z is real. Such series arise because of the connection between Bessel functions and planetary motion and, as a consequence, in the theory of radiation from accelerating electrons; the connecting link being the inverse square law of gravitational and electric forces. The details of these connections are not relevant to the mathematical behaviour of these series, but are provided in the first appendix to this chapter for the interested reader; that analysis provides an alternative method of deriving some of the results obtained here.

W. Kapteyn
Born 1849

Consider the specific series

$$S(x) = \sum_{n=1}^{\infty} J_n(nx) = -\frac{1}{2} + \frac{1}{2}\sum_{n=-\infty}^{\infty} J_n(nx). \tag{7.56}$$

If $|x| \ll 1$ there are effectively only a finite number of terms in this infinite series because for fixed $|x| \ll 1$, $J_n(nx)$ decreases exponentially with increasing n, equation 7.55; in this case $S(x)$ can be shown to be analytic in a region of the complex x-plane. But if $x > 1$, $J_n(nx)$ is oscillatory and its magnitude decreases only very slowly with increasing n, equation 7.54, and the series $S(x)$ is more like a trigonometric series and does not converge off the real axis. Thus the function $S(x)$ changes its character at $x = 1$; this change is reflected in expressions we obtain for the sum.

Using the integral definition of the Bessel function, equation 7.52, and interchanging the order of summation and integration, we obtain

$$S(x) = -\frac{1}{2} + \frac{1}{4\pi}\int_{-\pi}^{\pi} d\theta \sum_{n=-\infty}^{\infty} \exp\left(in(\theta - x\sin\theta)\right)$$

$$= -\frac{1}{2} + \frac{1}{2}\int_{-\pi}^{\pi} d\theta\, \delta(\theta - x\sin\theta), \tag{7.57}$$

where we have used equation 7.22 to evaluate the sum in terms of the delta function. If $x < 1$ the equation $\theta - x \sin \theta = 0$ has only one real root which is at $\theta = 0$, so the integral over the delta function is, on using equation 7.13 (page 230),

$$S(x) = -\frac{1}{2} + \frac{1}{2(1-x)} = \frac{x}{2(1-x)}, \quad x < 1. \tag{7.58}$$

This result can also be derived using standard methods, as shown in the appendix.

But if $x > 1$ the equation $\theta - x \sin \theta = 0$ has three real roots, one at $\theta = 0$ and others at $\theta = \pm \phi$, where $0 < \phi < \pi$ is the positive root of $\theta = x \sin \theta$; thus again using equation 7.13 we have

$$S(x) = -\frac{1}{2} + \frac{1}{2(x-1)} + \frac{1}{1 - x \cos \phi}, \quad \phi = x \sin \phi. \tag{7.59}$$

This last result is not so easily derived using standard methods and is not quoted in the usual tables of series, see for instance Gradshteyn and Ryzhik (1965). Putting these results together, we find

$$\sum_{n=1}^{\infty} J_n(nx) = \begin{cases} \dfrac{x}{2(1-x)}, & x < 1, \\[2mm] \dfrac{2-x}{2(x-1)} + \dfrac{1}{1 - x \cos \phi(x)}, & x > 1. \end{cases}$$

Exercise 7.30

Use the Bessel functions provided with Maple to evaluate the finite series

$$S_N(x) = \sum_{n=1}^{N} J_n(nx)$$

for various values of N and $0 \le x < 1$, and graphically compare $S_N(x)$ with $S(x) = x/2(1-x)$.

Can you make the same direct comparison for $x > 1$? If not, can you think of a method of numerically checking equation 7.59?

Exercise 7.31

Show that for $0 \le x < 2\pi$

$$\sum_{k=1}^{\infty} \frac{J_k'(kx)}{k} = \begin{cases} \dfrac{1}{2} + \dfrac{x}{4}, & 0 \le x \le 1, \\[2mm] \cos \alpha - \dfrac{1}{2} + \dfrac{x}{4}, & 1 \le x < 7.79\ldots, \end{cases}$$

where α is the real root of $\theta = x \sin \theta$ in the range $0 \le \alpha < \pi$. Use Maple to examine this result graphically.

Note the differential $J_k'(z)$ can be computed from the recurrence relation

$$J_k'(z) = \frac{1}{2} (J_{k-1}(z) - J_{k+1}(z)),$$

and the result found in exercise 7.14 (page 237) will be useful.

Neumann series

Neumann series have the form

$$\sum_{n=0}^{\infty} a_n J_n(x).$$

Here we consider the one simple case, leaving others to be dealt with as exercises,

$$S(x) = \sum_{n=-\infty}^{\infty} J_n(x) = J_0(x) + 2\sum_{n=1}^{\infty} J_{2n}(x).$$

On using the integral 7.52 this can be re-written as

$$S(x) = \frac{1}{2\pi} \int_{-\pi}^{\pi} d\theta\, e^{-ix\sin\theta} \sum_{n=-\infty}^{\infty} e^{in\theta} = \int_{-\pi}^{\pi} d\theta\, \delta(\theta) e^{-ix\sin\theta} = 1.$$

Thus

$$J_0(x) + 2\sum_{n=1}^{\infty} J_{2n}(x) = 1 \quad \text{for all } x. \tag{7.60}$$

Exercise 7.32

Extend the result of the previous example to show that

$$\sum_{n=-\infty}^{\infty} J_n(x) e^{-iny} = e^{-ix\sin y}.$$

Deduce equation 7.60 as a special case and derive the two real forms

$$\cos(x\sin y) = J_0(x) + 2\sum_{k=1}^{\infty} J_{2k}(x)\cos 2ky,$$

$$\sin(x\sin y) = 2\sum_{k=0}^{\infty} J_{2k+1}(x)\sin(2k+1)y.$$

Exercise 7.33

Show that for $|x| < 2\pi$

(i)
$$1 + 2\sum_{k=1}^{\infty} J_0(kx) = \frac{2}{|x|},$$

(ii)
$$\sum_{k=1}^{\infty} J_{2n}(kx) = \begin{cases} \dfrac{1}{|x|}, & x \neq 0,\ n \neq 0, \\ 0, & x = 0. \end{cases}$$

Exercise 7.34

Show that for $|x| < 1$

$$2\sum_{k=1}^{\infty} k^{2n} J_k(kx) = (-1)^n \frac{d^{2n}}{dw^{2n}} \left(\frac{1}{1 - x\cos\theta(w)} \right), \quad \text{where} \quad w = \theta - x\sin\theta,$$

where the expression on the right hand side is evaluated at $w = 0$. Hence show that

$$\sum_{k=1}^{\infty} k^2 J_k(kx) = \frac{x}{2(1-x)^4}, \quad |x| < 1.$$

Write a Maple procedure to evaluate this differential for any given n and use this to show that, for $|x| < 1$,

$$\sum_{k=1}^{\infty} k^{10} J_k(kx) = \frac{x(1 + 1008x + 50166x^2 + 457200x^3 + 893025x^4)}{2(1-x)^{16}}.$$

7.5 Appendix 1: Bessel functions and the inverse square law

J.-L. Lagrange
(1736–1813)

The properties of the function $J_n(nx)$ for integer values of n are intimately connected with the motion of a particle in an inverse square force, for example the motion of a planet round a sun or an electron round its nucleus. Indeed, one of the first occurrences of the Bessel function was in the work of Lagrange describing the motion of a planet. Here we briefly describe the background of this theory for historical interest and to provide some idea of the connections between the various subjects. First we provide a simple description of the motion of a particle moving in an inverse square force, such as the gravitational attraction between the Sun and the Earth. A full analysis of this motion can be found in almost any text on dynamics, for example Whittaker (1964, section 47), Goldstein (1980, chapter 3) or Landau and Lifshitz (1965a, chapter 2).

A particle with cartesian coordinates (x, y, z) which is attracted to a fixed point, chosen to be the origin, by the force $\mathbf{F} = -G\hat{\mathbf{r}}/r^2$, $r^2 = x^2 + y^2 + z^2$, moves in a plane (because the force is spherically symmetric so angular momentum is conserved) and on an ellipse, in this plane, with a focus at the origin. We can choose coordinates such that Ox lies along the major axis and Oy parallel to the minor axis of the ellipse, as shown in the figure.

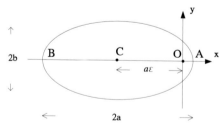

Figure 7.13

The point of closest approach A is named the perihelion and the most distant point, B, the aphelion. The perihelion is fixed in space because the inverse square law has special properties.

The parameter which defines the size of the ellipse is a, the length of the semi-major axis, which can take any positive value. The parameter which defines the shape of the ellipse is the eccentricity ϵ, which lies in the interval $0 \le \epsilon \le 1$; the distance between the focus and the centre, OC, is $a\epsilon$. The length of the semi-minor axis is $b = a\sqrt{1 - \epsilon^2}$. The ellipse degenerates to a circle of radius a when $\epsilon = 0$ and a straight line of length $2a$ when $\epsilon = 1$.

You can draw some ellipses using Maple; the following sequence of commands will draw three ellipses with $a = 1$ and eccentricity $\epsilon = \frac{3}{4}, \frac{1}{2}$ and $\frac{1}{4}$, all with the origin at the ellipse centre, not as in figure 7.13.

```
>   restart;  with(plots): with(plottools):
>   ecc:=[0.75, 0.5, 0.25]:
>   ell:=seq(ellipse([0,0],1,sqrt(1-e^2)),e=ecc):
>   display(ell);
```

Returning to the motion of a particle in the inverse square force, it can be shown that the particle moves on the ellipse shown in figure 7.13. From the equations of motion it can be shown that the relations between the coordinates and the time can be written in the form

$$x = a(\cos u - \epsilon), \quad y = a\sqrt{1 - \epsilon^2} \sin u, \tag{7.61}$$

where the eccentric anomaly u is related to the time by:

$$\omega t = u - \epsilon \sin u \quad \text{(Kepler's equation)}, \tag{7.62}$$

where ω is the frequency of the motion, and depends only upon a (and the force constant G and the reduced mass). In fact $\omega \propto a^{-3/2}$, which is Kepler's third law. When u increases by 2π the time t increases by $2\pi/\omega$ and for $0 \le \epsilon \le 1$, t is a monotonic increasing function of u. Note that the function $u - \epsilon \sin u$ is similar to the function in the exponent of the integral definition of the Bessel function, equation 7.52.

The geometric significance of u is shown in the figure, which is a view of planetary motion from the pole star. The planets of the Solar System move round the Sun in the direction of increasing u.

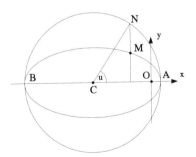

Figure 7.14   Diagram showing the geometric significance of the eccentric anomaly u; the planet M is moving on the ellipse AMB.

Around the ellipse draw a circle of radius a centred at the ellipse centre. If the particle is at a point M on the ellipse, project perpendicularly from the x-axis on to this circle, to the point N. Then u is the angle OCN; twice in every period u coincides with the polar angle of the particle, at $u = 0$ and π.

Now equations 7.61 and 7.62 giving (x, y) in terms of u are satisfyingly simple, but the wrong way round for most purposes. In practice we would need to know the position at a given time, which means solving Kepler's equation $\omega t = u - \epsilon \sin u$ for $u(t)$; this is a transcendental equation so its solution cannot be expressed simply in terms of known functions; various approximations have been used and this is where Bessel functions arise. Modern work tends to emphasise numerical ways of solving Kepler's equation; for small values of ϵ Newton's method works well but this fails for $\epsilon \simeq 1$ (the orbits of comets often have eccentricities near unity) so in general Halley's method is preferable, see for example exercises 3.38 (page 117) and 3.62 (page 130). These methods are discussed in chapter 9.

An important observation is that we can write $u = \theta + P(\theta, \epsilon)$, $\theta = \omega t$, where P is a 2π-periodic function of θ, which can therefore be expressed as a Fourier sine series (since it is an odd function of θ). It transpires that

$$P(\theta, \epsilon) = \epsilon \sin u(\theta) = 2 \sum_{n=1}^{\infty} \frac{J_n(n\epsilon)}{n} \sin n\theta. \tag{7.63}$$

This relation can be proved directly using the integral form of the Bessel function, but when first derived by Lagrange he essentially defined the Bessel function of order n by this Fourier series. Thus there is a close relation between Bessel functions and the motion of a particle in an inverse square law, which accounts for the similarity between Kepler's equation, 7.62, and the exponent in the integral definition of Bessel functions, equation 7.52.

It is also possible to express functions of x, y and $r = \sqrt{x^2 + y^2}$ as Fourier series in $\theta = \omega t$, and in many cases the coefficients can be expressed in terms of Bessel functions, as would be expected. Some examples are

$$\frac{r}{a} = 1 + \frac{1}{2}\epsilon^2 - 2\epsilon \sum_{n=1}^{\infty} \frac{J_n'(n\epsilon)}{n} \cos n\theta \tag{7.64}$$

$$\frac{a}{r} = 1 + 2 \sum_{n=1}^{\infty} J_n(n\epsilon) \cos n\theta \tag{7.65}$$

$$\frac{x}{a} = -\frac{3}{2}\epsilon + 2 \sum_{n=1}^{\infty} \frac{J_n'(n\epsilon)}{n} \cos n\theta \tag{7.66}$$

$$\frac{y}{a} = 2\sqrt{1 - \epsilon^2} \sum_{n=1}^{\infty} \frac{J_n(n\epsilon)}{n\epsilon} \sin n\theta. \tag{7.67}$$

All these relations can be derived using equations 7.61 and 7.62 and the integral form of the Bessel function, equation 7.52.

The original practical value of these relations resides in the small value of the eccentricity of all planets. For instance, the eccentricity of the Earth's orbit is $\epsilon \simeq 0.0167$ whilst the largest eccentricity is that of Pluto's orbit, 0.250, with Mercury's orbit having the next largest at 0.2056. Thus for the Earth the ratio of successive terms in the expansion of $1/r$, $J_{n+1}((n+1)\epsilon)/J_n(n\epsilon)$ is roughly constant at 0.02, so successive terms decrease very rapidly.

From these relations we can obtain various simple identities. For instance, when $u = 0$ we have $\theta = 0$ and $x/a = r/a = 1 - \epsilon$ and $y = 0$, so from equation 7.65

$$\sum_{n=1}^{\infty} J_n(n\epsilon) = \frac{\epsilon}{2(1-\epsilon)}, \quad 0 \le \epsilon < 1,$$

and from equation 7.66 we obtain part of the result quoted in exercise 7.31 (page 256). Most of the series involving $J_n(nx)$, with $|x| < 1$, arise from such expansions or their derivatives. For instance, on differentiating equation 7.65 twice with respect to θ we obtain

$$2\sum_{n=1}^{\infty} n^2 J_n(n\epsilon) = -\frac{d^2}{d\theta^2} \left. \frac{1}{1 - \epsilon \cos u} \right|_{\theta=0},$$

which is the result derived in exercise 7.34.

7.6 Appendix 2: George Green

The story of George Green is unusual and I provide a very brief summary in the hope of inspiring sufficient interest in the reader that s/he will want to discover more; should this be the case, the book by Cannell (1993) is highly recommended.

George Green was born in 1793, the only son of a Nottingham miller, also named George Green. When seven years old he started school, but stayed only four terms and at nine started work in his father's bakery. Little is known of Green's life between then and 1823 when he joined the Bromely House Subscription Library: this was just five years before the publication in 1828 of his first major paper *An Essay on the Application of Mathematical Analysis to the Theories of Electricity and Magnetism*. Bromely was a meeting place for the intellectual activities of Nottingham: when Green joined it had about 100 subscribers and the president was Rev. R. W. Almond, a mathematics graduate from Queens' College, Cambridge. This library gave Green access to a number of books and scientific journals, besides lively debates on scientific matters, which were then a matter of general interest.

In the 1828 paper Green generalised the work of Poisson on magnetism and used the mathematics of Laplace, from his *Mécanique Céleste*, in order to derive what are now known as Green's theorem and Green's functions. This essay was revolutionary and largely ignored until after Green's death, when its full value was realised. It was published by private subscription and probably only about 100 copies were produced. The work was remarkable not only because Green worked alone, but also because the mathematics used was not widely known in England at that time; Green's ideas were all built upon developments in mathematics originating in Europe and ignored in England. How Green was able to develop such original ideas whilst working full-time as a miller is unknown, though Cannell provides an illuminating story of the background that made it possible. One significant source of material, however, was the English translation of Book I of Laplace's *Mécanique Céleste* by the Rev J Toplis in 1814, also a Cambridge mathematics graduate and a member of the Bromely Library.

Of all the subscribers to the first essay E F Bromhead was probably the most influential, and when Green's second work, *Mathematical Investigations Concerning the Laws of the Equilibrium of Fluids analogous to the Electric Fluid, with other similar Researches*, was finished Bromhead forwarded it to the Cambridge Philosophical Society, and it was duly published in 1833. A number of other significant works were published by the Cambridge Philosophical Society and the Royal Society of Edinburgh, in rapid succession.

On 6th October 1833, at the age of forty, Green was admitted as an undergraduate to Caius College, Cambridge. It is difficult to see how he would have fitted into life at Cambridge; Green was considerably older than most college tutors and had already produced a number of original works, so the emphasis on examination problems may have been difficult to cope with. Nevertheless, he passed his BA examination, in Latin, Greek and ecclesiastical history, and proceeded to become Fourth Wrangler. In October 1839 he was awarded a Perse Fellowship, £10 per year, so could reside in College indefinitely. Despite this he returned home to Nottingham six months later, never to return, and died from influenza on 31st May 1841, having published a total of eight original papers and leaving Jane Smith, his common law wife, with seven children.

Green's work was subsequently 'discovered' only a few years after his death by William Thomson, later to become Lord Kelvin, one of the leading scientists of the nineteenth century. Thompson was about to start research on electromagnetism, and in preliminary reading came across a reference to 'the ingenious Essay by Mr Green of Nottingham' by R. Murphy, a Fellow at Caius contemporary with Green. Thomson had difficulty finding a copy of this Essay, of which only a few copies were printed, but tracked one down the night before setting off to Paris to work on these problems. Green, it transpired, had many years previously solved several of the problems on which Thomson and his colleagues had planned to work. Green's work created some excitement on the Continent and was published in three parts in 1850, 1852 and 1854, in English, and later translated into German. Green's first Essay was not published in England until 1871.

Green's mathematics now permeate most of applied mathematics and physics and the most significant applications are in subjects unknown in Green's time.

7.7 Exercises

Exercise 7.35
Show that the two generalised functions $x\delta'(x)$ and $-\delta(x)$ are identical.

Exercise 7.36
Verify that the sequence

$$\delta_n(x) = \begin{cases} 0, & x < 0, \\ ne^{-nx}, & x > 0, \end{cases}$$

tends to a delta function.

Exercise 7.37

For the sequence

$$\delta_n(x) = \frac{n}{2 \cosh^2 nx}$$

show that $\int_{-\infty}^{\infty} dx\, \delta_n(x) = 1$. Define $h_n(x) = \int_{-\infty}^{x} dx\, \delta_n(x)$ and show that $h_n(x) \to H(x)$ as $n \to \infty$.

Exercise 7.38

For the sequence

$$\delta_L(x) = \frac{L}{\pi(1 + (Lx)^2)}$$

show that

$$\int_{-\infty}^{\infty} dx\, e^{ikx} \delta_L(x) = 1 - \frac{k}{L} + O(L^{-2}).$$

Exercise 7.39

Use the sequence

$$g_L(x) = \int_0^L du\, e^{iux}$$

to show that

$$\int_0^{\infty} du\, e^{iux} = \pi\delta(x) + \frac{i}{x}.$$

Hint: in an integral if L is large $\sin^2(xL/2)$ may be replaced by its mean, $\frac{1}{2}$.

Exercise 7.40

By differentiating the result found in the previous exercise, show that

$$\int_0^{\infty} du\, u^n e^{iux} = \pi \frac{d^n}{dx^n} \delta(x) + \frac{n!}{x^{n+1}} e^{i(n+1)\pi/2}.$$

Exercise 7.41

Express the function $f(x, y) = \min(x, y)$ in terms of Heaviside functions and hence show that

$$\frac{\partial^2 f}{\partial x \partial y} = \delta(x - y).$$

For $0 \le x, y \le T$ the function $f(x, y)$ can be represented by the Fourier series

$$f = \frac{8T}{\pi^2} \sum_{n=1}^{\infty} \frac{1}{(2n+1)^2} \sin\left(\frac{2n+1}{2T}\pi x\right) \sin\left(\frac{2n+1}{2T}\pi y\right).$$

Use this to obtain another series representation for the delta function and use Maple to confirm, graphically, that truncated forms of these series are consistent with the exact functions.

Exercise 7.42

In equation 7.25 (page 238), we used the sine function to produce a series representation of the delta function. Such series can be constructed from many other functions, and here we shall use functions closely related to the Hermite polynomials, $H_n(x)$, $n = 0, 1, 2, \ldots$; these polynomials can be found in Maple by invoking with(orthopoly), then H(n,x) is the nth order Hermite polynomial.

Define the new functions

$$u_n(x) = \frac{1}{\sqrt{2^n n! \sqrt{\pi}}} e^{-x^2/2} H_n(x), \quad n = 0, 1, \ldots,$$

and examine their behaviour graphically; you should observe that u_n is oscillatory for $|x| < \sqrt{2n}$ and exponentially decaying for larger $|x|$. Construct the series

$$\delta_n(x, y) = \sum_{n=0}^{N} u_n(x) u_n(y)$$

and examine, graphically, its behaviour as a function of x for fixed y, for N sufficiently large, say 20 to 50.

It can be shown that $\delta_n(x, y) \to \delta(x - y)$ as $n \to \infty$. Note that, as $u_n(x)$ is odd or even according as n is odd or even, you need only consider positive values of x and y.

Exercise 7.43

Consider the initial value problem

$$\frac{d^n y}{dx^n} = h(x), \quad y(0) = \left. \frac{d^k y}{dx^k} \right|_{x=0} = 0, \quad k = 1, 2, \ldots, n-1.$$

By integrating n times show that the solution can be written as the n-fold integral

$$y(x) = \int_0^x dt_n \int_0^{t_n} dt_{n-1} \cdots \int_0^{t_3} dt_2 \int_0^{t_2} dt_1 h(t_1).$$

Alternatively, show that the Green's function that satisfies the equation

$$\frac{d^n}{dx^n} G(x, u) = \delta(x - u), \quad u > 0, \quad G(0, u) = \left. \frac{d^k}{dx^k} G(x, u) \right|_{x=0} = 0, \quad k = 1, 2, \ldots, n-1,$$

is

$$G(x, u) = \frac{1}{(n-1)!} (x - u)^{n-1} H(x - u)$$

and hence that an alternative expression for $y(x)$ is

$$y(x) = \frac{1}{(n-1)!} \int_0^x dt \, (x - t)^{n-1} h(t).$$

Deduce the identity

$$\frac{1}{(n-1)!} \int_0^x dt \, (x - t)^{n-1} h(t) = \int_0^x dt_n \int_0^{t_n} dt_{n-1} \cdots \int_0^{t_3} dt_2 \int_0^{t_2} dt_1 h(t_1). \tag{7.68}$$

Exercise 7.44

Find the Green's function for the equation

$$\frac{d^2 y}{dx^2} - k^2 y = h(x), \quad y(0) = y(1) = 0,$$

where k is a positive constant. Hence show that the solution is

$$y(x,k) = -\frac{\sinh k(1-x)}{k \sinh k} \int_0^x du\, h(u) \sinh ku - \frac{\sinh kx}{k \sinh k} \int_x^1 du\, h(u) \sinh k(1-u).$$

For the function

$$h(x) = \begin{cases} -1, & x \le \frac{1}{4} \quad \text{and} \quad x \ge \frac{3}{4}, \\ 0, & \text{otherwise,} \end{cases}$$

use Maple to evaluate these integrals and to plot the graph of $y(x,k)$ for $0 < x < 1$ and $0 \le k \le 10$.

Exercise 7.45

Consider the forced linear oscillator

$$\ddot{x} + \omega^2 x = \frac{FT^2}{T^2 + t^2},$$

which is initially undisturbed so that in the distant past, $t \ll -T$, the right hand side of the equation may be ignored and $x(t) = a \sin(\omega t + \phi)$ for some phase ϕ.

Use the Green's function obtained in exercise 7.21 (page 247) to show that for large t

$$x(t) = a \sin(\omega t + \phi) + \frac{FT^2}{\omega} \int_{-\infty}^{\infty} d\tau\, \frac{\sin \omega(t-\tau)}{T^2 + \tau^2} = a' \sin(\omega t + \phi'),$$

where

$$a' = a\sqrt{1 + 2\beta \cos \phi + \beta^2}, \quad \beta = \frac{\pi F T}{a\omega} e^{-\omega T}.$$

The energy E is defined by the relation given in exercise 7.21. Show that the energy change $\Delta E = E(\infty) - E(-\infty)$ is given by

$$\Delta E = \frac{1}{2} \omega^2 \beta (\cos \phi + \beta).$$

Exercise 7.46

For the differential equation

$$p(x)\frac{d^2 y}{dx^2} + q(x)y = 0, \quad y(0) = y(L) = 0,$$

where $p(x) > 0$ for $0 < x < L$, and both p and q are continuous in this range, show that the Green's function satisfies

(i) $\qquad \lim_{\epsilon \to 0} \frac{\partial}{\partial x} G(x,u) \Big|_{x=u-\epsilon}^{u+\epsilon} = \frac{1}{p(u)},$

(ii) $\qquad \lim_{\epsilon \to 0} G(x,u) \Big|_{x=u-\epsilon}^{u+\epsilon} = 0.$

Exercise 7.47

Find the Green's function for the equations

(i) $\qquad \frac{d}{dx}\left((1-x^2)\frac{dy}{dx} \right) = 0, \quad y(0) = 0, \quad y'(0) = \alpha,$

(ii) $\qquad \frac{d^2 y}{dx^2} = 0, \quad y(0) = 0, \quad y(1) + \alpha y'(1) = 0,$

where α is a constant.

Exercise 7.48

Use the integral form of the ordinary Bessel function,

$$J_n(x) = \frac{1}{2\pi} \int_{-\pi}^{\pi} d\theta \, \exp(i(n\theta - x\sin\theta))$$

together with equation 7.52 to derive the addition formula

$$\sum_{n=-\infty}^{\infty} J_n(x)J_{n+k}(y) = J_k(y - x).$$

Exercise 7.49

Show that

(i) $$\sum_{k=1}^{\infty} (2k)^2 J_{2k}(x) = \frac{1}{2}x^2,$$

(ii) $$\sum_{k=1}^{\infty} kJ_k'(kx) = \begin{cases} \dfrac{1}{2(1-x)^2}, & 0 \le x < 1, \\[2mm] \dfrac{1}{2(1-x)^2} + \dfrac{\cos\alpha - x}{(1-x\cos\alpha)^3}, & x > 1, \end{cases}$$

(iii) $$\sum_{k=1}^{\infty} k^2 J_{2k}(2kx) = \frac{x^2}{4(1-x^2)^4}, \quad 0 \le |x| < 1,$$

where, for $x > 1$, α is the real root of $\theta = x\sin\theta$ in the range $(0, \pi)$.

8

Fourier series and systems of orthogonal functions

8.1 Introduction

In chapter 4 we introduced Taylor's series, a representation in which the function is expressed in terms of the value of the function and all its derivatives at the point of expansion. In general such series are useful only close to the point of expansion, and far from here it is usually the case that very many terms of the series are needed to give a good approximation. An example of this behaviour is shown in figure 8.1, where we compare various Taylor polynomials of $\sin x$ about $x = 0$ over the range $0 < x < 2\pi$.

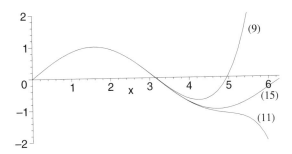

Figure 8.1   Taylor polynomials of $\sin x$ with degree 9, 11 and 15.

It is seen that many terms are needed to obtain a good approximation at $x = 2\pi$. Such behaviour should not be surprising, for the nth term of the Taylor's series is $(-1)^n x^{2n+1}/(2n + 1)!$ and at $x = 2\pi$ this is larger than 1 if $n \leq 6$. Some values of the magnitude of this coefficient for various n at $x = 2\pi$ and 3π are shown in figure 8.2.

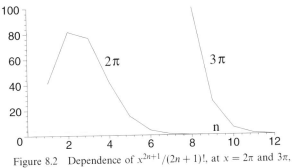

Figure 8.2   Dependence of $x^{2n+1}/(2n + 1)!$, at $x = 2\pi$ and 3π, upon n.

267

A Taylor's series approximation is most accurate near the point of expansion, $x = a$, and its accuracy generally decreases as $|x - a|$ increases, so this type of approximation suffers from the defect that it is not usually uniformly accurate over the required range of x, although the Padé approximant of the Taylor's series, introduced in chapter 6, can often provide improved accuracy.

Fourier series eliminate these problems by approximating functions in a quite different manner. The essential idea is very simple. Suppose we have a set of functions $\phi_k(x)$, $k = 1, 2, \ldots,$ (which may be complex) defined on some interval $a \le x \le b$; these could be, for instance, the polynomials $\phi_k(x) = x^k$, or the trigonometric functions $\phi_k(x) = \sin kx$. Then we attempt to approximate a function $f(x)$ as the sum

$$f(x) \simeq f_N(x) = \sum_{k=1}^{N} c_k \phi_k(x)$$

by choosing the coefficients c_k to minimise the mean square difference

$$E_N = \int_a^b dx \left| f(x) - \sum_{k=1}^{N} c_k \phi_k(x) \right|^2 .$$

This is a more democratic method of approximation because no point in the interval is picked out for favoured treatment, as in a Taylor's series. In order to put this idea into practice we need to know how to choose the functions $\phi_k(x)$ and to understand how the approximation converges with increasing N. For this it is necessary to introduce the notion of a complete set of functions together with some connected technical details.

Exercise 8.1

Use Stirling's approximation to show that the modulus of the nth term of the Taylor's series of $\sin x$, $x^{2n+1}/(2n+1)!$, is small for $n > ex/2$, and that for large $|x|$ the largest term in the Taylor's series has a magnitude of about $e^x/\sqrt{2\pi x}$.

In addition show that with arithmetic accurate to N significant figures the series approximation for $\sin x$ can be used directly to find values of $\sin x$ accurate to $M(< N)$ significant figures only for $x < (N - M) \ln 10$. Check this behaviour using Maple.

Exercise 8.2

This exercise is about the approximation of $\sin x$ on the interval $(0, \pi)$ using the functions $\phi_1 = 1$, $\phi_2 = x$ and $\phi_3 = x^2$.

Use Maple to evaluate the integral

$$E(a, b, c) = \int_0^\pi dx \left(a + bx + cx^2 - \sin x \right)^2$$

to form the function $E(a, b, c)$ which is quadratic in a, b and c. Find the position of the minimum of this function by solving the three equations $\frac{\partial E}{\partial a} = 0$, $\frac{\partial E}{\partial b} = 0$ and $\frac{\partial E}{\partial c} = 0$ for a, b and c. Hence show that

$$\sin x \simeq 12 \frac{\pi^2 - 10}{\pi^3} - 60 \frac{\pi^2 - 12}{\pi^5} x(\pi - x), \quad 0 \le x \le \pi.$$

Compare your approximation graphically with the exact function and various Taylor polynomials.

Note that with Maple it is relatively easy to include higher order polynomials in this expansion: it is worth exploring the effect of doing this.

8.2 Orthogonal systems of functions

Here we consider complex functions of the real variable x on an interval $a \le x \le b$. The *inner product* of two such functions $f(x)$ and $g(x)$ is denoted by (f, g) and is defined by the integral

$$(f, g) = \int_a^b dx \, f^*(x) g(x), \qquad (8.1)$$

where f^* denotes the complex conjugate of f; note that $(g, f) = (f, g)^*$. The inner product of a function with itself is real, positive and $\sqrt{(f, f)}$ is named the *norm*. A function whose norm is unity is said to be *normalised*.

Two functions f and g are *orthogonal* if their inner product is zero, $(f, g) = 0$. A system of normalised functions $\phi_k(x)$, $k = 1, 2, \ldots$, every pair of which is orthogonal is named an *orthogonal system*. If, in addition, each function is normalised, so

$$(\phi_r, \phi_s) = \delta_{rs} = \begin{cases} 1, & r = s, \\ 0, & r \ne s, \end{cases}$$

the system is called an *orthonormal system*. The symbol δ_{rs} introduced here is named the *Kronecker delta*. An example of a real orthonormal system on the interval $(0, 2\pi)$, or more generally any interval of length 2π, is the set of functions

$$\frac{1}{\sqrt{2\pi}}, \quad \frac{\cos x}{\sqrt{\pi}}, \quad \frac{\cos 2x}{\sqrt{\pi}}, \quad \ldots, \quad \frac{\cos kx}{\sqrt{\pi}}, \quad \ldots.$$

On the same interval the set of complex functions

$$\phi_k(x) = \frac{e^{ikx}}{\sqrt{2\pi}}, \quad k = 0, \pm 1, \pm 2, \quad \ldots$$

is also an orthonormal system.

Exercise 8.3
Find the appropriate value of the constant A that makes the norm of each of the functions

$$\phi_1(x) = Ax, \quad \phi_2(x) = A(3x^2 - 1), \quad \phi_3(x) = Ax(5x^2 - 3)$$

on the intervals $-1 \le x \le 1$ and $0 \le x \le 1$, unity. For each interval determine the matrix of inner products (ϕ_i, ϕ_j) for $i, j = 1, 2, 3$.

Exercise 8.4
By evaluating (h, h), where $h(x) = (f, g)f(x) - (f, f)g(x)$, and using the fact that $(h, h) \ge 0$, prove the Schwarz inequality,

$$|(f, g)|^2 \le (f, f)(g, g), \qquad (8.2)$$

Exercise 8.5

Show that the functions

$$\phi_k(x) = \sqrt{\frac{2}{\alpha}} \sin\left(\frac{2\pi}{\alpha}kx\right), \quad k = 1, 2, \ldots,$$

are orthogonal on any interval of length α.

8.3 Expansions in terms of orthonormal functions

Suppose that $\phi_k(x)$, $k = 1, 2, \ldots$, is a system of orthonormal functions on the interval $a \le x \le b$, then we expand a given real function $f(x)$ in terms of these functions by writing

$$f(x) \simeq f_N(x) = \sum_{k=1}^{N} c_k \phi_k(x), \quad a \le x \le b, \tag{8.3}$$

and choose the N coefficients c_k — which may be complex — to minimise the square of the norm of $(f_N - f)$. That is, we minimise the function

$$
\begin{aligned}
F(\mathbf{c}) &= \int_a^b dx \left| \sum_{k=1}^{N} c_k \phi_k(x) - f(x) \right|^2 \\
&= \sum_{k=1}^{N} |c_k - (\phi_k, f)|^2 - \sum_{k=1}^{N} |(\phi_k, f)|^2 + (f, f).
\end{aligned}
\tag{8.4}
$$

It follows from this last expression that $F(\mathbf{c})$ has its only minimum when the first term is made zero by choosing

$$c_j = (\phi_j, f), \quad j = 1, 2, \ldots, N. \tag{8.5}$$

The numbers $c_j = (\phi_j, f)$ are the expansion coefficients of f with respect to the orthogonal system $\{\phi_1, \phi_2, \ldots\}$. This type of approximation is called an *approximation in the mean*.

An important consequence follows from the trivial observation that

$$\int_a^b dx \left| \sum_{k=1}^{N} c_k \phi_k(x) - f(x) \right|^2 \ge 0,$$

and by expanding the integrand and integrating term by term to give

$$\sum_{k=1}^{N} |c_k|^2 - \sum_{k=1}^{N} \left[c_k(\phi_k, f)^* + c_k^*(\phi_k, f) \right] + (f, f) \ge 0, \quad \text{for any } c_k.$$

Hence

$$\sum_{k=1}^{N} |c_k|^2 \le (f, f) \quad \text{if} \quad c_k = (\phi_k, f) \quad \left[\text{and} \quad (\phi_k, \phi_k) = 1 \right].$$

Since (f,f) is independent of N it follows that

$$\sum_{k=1}^{\infty}|c_k|^2 \le (f,f), \qquad \text{Bessel's inequality.} \tag{8.6}$$

Bessel's inequality is true for every orthogonal system: it shows that the sum of squares of the coefficients always converges, provided the norm of f exists. From this inequality it follows from section 4.3.1 that $c_k \to 0$ as $k \to \infty$.

Exercise 8.6
Derive equation 8.4.

Exercise 8.7
Equations 8.5, for the expansion coefficients, and Bessel's inequality, equation 8.6, were both derived for an orthonormal system for which $(\phi_j,\phi_k) = \delta_{jk}$. Show that if the ϕ_k are orthogonal, but not necessarily orthonormal, these relations become

$$c_j = \frac{(\phi_j,f)}{(\phi_j,\phi_j)} \quad \text{and} \quad \sum_{k=1}^{\infty}\frac{|(\phi_k,f)|^2}{(\phi_k,\phi_k)} \le (f,f).$$

8.4 Complete systems

A *complete orthogonal system*, $\phi_k(x)$, $k = 1, 2, \ldots$, has the property that any function, taken from a given particular set of functions, can be approximated in the mean to any desired accuracy by choosing N large enough. In other words, for any $\epsilon > 0$, no matter how small, we can find an $N(\epsilon)$ such that for $M > N(\epsilon)$

$$\int_a^b dx \left| f(x) - \sum_{k=1}^{M}c_k\phi_k(x) \right|^2 < \epsilon, \quad \text{where} \quad c_k = \frac{(\phi_k,f)}{(\phi_k,\phi_k)}. \tag{8.7}$$

That is, the mean square error can be made arbitrarily small. Notice that this definition needs the function f to belong to a given set, normally the set of integrable functions. For a complete orthogonal system it can be proved that Bessel's inequality becomes the equality

$$(f,f) = \sum_{k=1}^{\infty}(\phi_k,\phi_k)|c_k|^2 = \sum_{k=1}^{\infty}\frac{(\phi_k,f)^2}{(\phi_k,\phi_k)}, \tag{8.8}$$

and this is known as the *completeness relation*. It may also be shown that a sufficient condition for an orthogonal system to be complete is that this completeness relation holds; a proof of this statement is given in Courant and Hilbert (1965, page 52).

There are three points worthy of note.

- First, the functions $f(x)$ being approximated need not be continuous or differentiable at every point in the interval of approximation, as required by Taylor's series.

- Second, the fact that

$$\lim_{N\to\infty} \int_a^b dx \left| f(x) - \sum_{k=1}^{N} c_k \phi_k(x) \right|^2 = 0, \quad c_k = \frac{(\phi_k, f)}{(\phi_k, \phi_k)}, \tag{8.9}$$

does *not* imply pointwise convergence, that is,

$$\lim_{N\to\infty} \sum_{k=1}^{N} c_k \phi_k(x) = f(x), \quad a \le x \le b. \tag{8.10}$$

If the limit 8.9 holds we say that the sequence of functions

$$f_N(x) = \sum_{k=1}^{N} c_k \phi_k(x)$$

converge to $f(x)$ in the mean. If the limit 8.10 holds $f_N(x)$ converges pointwise to $f(x)$, section 4.6. If, however, the series $f_N(x)$ converges uniformly then convergence in the mean implies pointwise convergence.
- Third, if two piecewise continuous functions have the same expansion coefficients with respect to a complete system of functions then it may be shown that they are identical, see for example Courant and Hilbert (1965, page 54).

Finally, we note that systems of functions can be complete even if they are not orthogonal. Examples of such complete systems are the polynomials $\phi_k(x) = x^k$,

$$1, \; x, \; x^2, \ldots, x^n, \ldots,$$

which form a complete system in any closed interval $a \le x \le b$, for the approximation theorem of Weierstrass states that any function continuous in the interval $a \le x \le b$ may be approximated uniformly by polynomials in this interval. This theorem asserts uniform convergence, not just convergence in the mean, but restricts the class of functions to be continuous. A proof may be found in Powell (1981, chapter 6).

Another set of functions is

$$\frac{1}{x + \lambda_1}, \quad \frac{1}{x + \lambda_2}, \quad \cdots \quad \frac{1}{x + \lambda_n}, \cdots,$$

where $\lambda_1, \lambda_2, \ldots, \lambda_n, \ldots$ are positive numbers which tend to infinity with increasing n; this set is complete in every finite positive interval. An example of the use of these functions is given in exercise 8.28 (page 294), and another set of complete functions is given in exercise 8.41 (page 298).

In this and the preceding sections we have introduced the ideas of:

- inner product and norm;
- orthogonal functions;
- complete orthonormal systems.

You should ensure that you understand these ideas before passing to the next section.

8.5 Fourier series

In modern mathematics a Fourier series is an expansion of a function in terms of a set of complete functions. Originally and in many modern texts the same name is used in the more restrictive sense to mean an expansion in terms of the trigonometric functions

$$1, \cos x, \sin x, \cos 2x, \sin 2x, \ldots, \cos nx, \sin nx, \ldots \tag{8.11}$$

or their complex equivalents

$$\phi_k = e^{-ikx}, \quad k = 0, \pm 1, \pm 2, \ldots, \tag{8.12}$$

which are complete and orthogonal on the interval $(-\pi, \pi)$, or any interval of length 2π; the interval $(0, 2\pi)$ is often used. Series of this type are named *trigonometric series* if it is necessary to distinguish them from more general Fourier series: in the remainder of this chapter we treat the two names as synonyms. Trigonometric series are one of the simplest of this class of Fourier expansions — because they involve the well understood trigonometric functions — and have very many applications.

Any sufficiently well behaved function $f(x)$ may be approximated by the *trigonometric series*

$$F(x) = \sum_{k=-\infty}^{\infty} c_k e^{-ikx}, \quad c_k = \frac{(\phi_k, f)}{(\phi_k, \phi_k)} = \frac{1}{2\pi} \int_{-\pi}^{\pi} dx\, e^{ikx} f(x), \tag{8.13}$$

where we have used the first result of exercise 8.7. We restrict our attention to real functions, in which case c_0 is real, and is just the mean value of the function $f(x)$, and $c_{-k} = c_k^*$. The constants c_k are named the *Fourier coefficients*.

The Fourier series $F(x)$ is often written in the real form

$$F(x) = \frac{1}{2} a_0 + \sum_{k=1}^{\infty} a_k \cos kx + \sum_{k=1}^{\infty} b_k \sin kx, \tag{8.14}$$

where $a_k = 2\Re(c_k)$, $b_k = 2\Im(c_k)$, or

$$a_0 = \frac{1}{\pi} \int_{-\pi}^{\pi} dx\, f(x), \quad \binom{a_k}{b_k} = \frac{1}{\pi} \int_{-\pi}^{\pi} dx \binom{\cos kx}{\sin kx} f(x), \quad k = 1, 2, \ldots. \tag{8.15}$$

The constants a_k and b_k are also called Fourier coefficients. It is often more efficient and elegant to use the complex form of the Fourier series, though in special cases, see for instance exercise 8.10, the real form is more convenient.

One of the main questions to be settled is how the Fourier series $F(x)$ relates to the original function $f(x)$. This relation is given by Fourier's theorem, discussed next, but before you read this it will be helpful to do the following two exercises.

Exercise 8.8
Show that the Fourier series of the function $f(x) = |x|$ on the interval $-\pi \le x \le \pi$ is

$$F(x) = \frac{\pi}{2} - \frac{4}{\pi} \sum_{k=1}^{\infty} \frac{\cos(2k-1)x}{(2k-1)^2}.$$

Use Maple to compare graphically the Nth partial sum, $F_N(x)$, of the above series with $f(x)$ for $N = 1, 2, 6$ and 10 over the range $-2\pi \le x \le 2\pi$.

Further, show that the mean square error, defined in equation 8.7, of the Nth partial sum decreases as N^{-3}. Show also that

$$F_N(0) = \frac{1}{\pi N} + O(N^{-2}).$$

In this example we notice that even the first two terms of the Fourier series provide a reasonable approximation to $|x|$ for $-\pi \le x \le \pi$, and for larger values of N the graphs of $F_N(x)$ and $|x|$ are indistinguishable over most of the range. A close inspection of the graphs near $x = 0$, where f has no derivative, shows that here more terms are needed to obtain the same degree of accuracy as elsewhere.

For $|x| > \pi$, $f(x)$ and $F(x)$ are different. This is not surprising as $F(x)$ is a periodic function with period π — generally this type of Fourier series is 2π-periodic but here $F(x)$ is even about $x = \pi$.

Finally, note that for large k the Fourier coefficients c_k are $O(k^{-2})$; we shall see that this behaviour is partly due to $f(x)$ being even about $x = 0$.

Now consider a function which is piecewise smooth on $(-\pi, \pi)$ and discontinuous at $x = 0$.

Exercise 8.9
Show that the Fourier series of the function

$$f(x) = \begin{cases} x/\pi, & -\pi < x < 0, \\ 1 - x/\pi, & 0 \le x < \pi, \end{cases}$$

is

$$F(x) = \frac{4}{\pi^2} \sum_{k=1}^{\infty} \frac{\cos(2k-1)x}{(2k-1)^2} + \frac{2}{\pi} \sum_{k=1}^{\infty} \frac{\sin(2k-1)x}{2k-1}.$$

Use Maple to compare graphically the Nth partial sum of this series with $f(x)$ for $N = 1, 4,$ 10 and 20. Make your comparisons over the interval $(-2\pi, 2\pi)$ and investigate the behaviour of the partial sums of the Fourier series in the range $-0.1 < x < 0.1$ in more detail.

Find the values of $f(0)$ and $f(\pm\pi)$ and show that

$$F(0) = \frac{1}{2} \quad \text{and} \quad F(\pm\pi) = -\frac{1}{2}.$$

Hint: remember that the piecewise function can be defined in Maple using the command
`x->piecewise(x<0, x/Pi, 1-x/Pi);`.

In this comparison there are four points to notice:

- First, observe that for $-\pi < x < \pi$, but x not too near 0 or $\pm\pi$, $F_N(x)$ is close to $f(x)$ but that $F(x) \neq f(x)$ at $x = 0$ and $\pm\pi$.
- Second, as in the previous example, F and f are different for $|x| > \pi$ because $F(x)$ is a periodic extension of $f(x)$.
- Third, in this case the convergence of the partial sums to $f(x)$ is slower because now $c_k = O(k^{-1})$; again this behaviour is due to the nature of $f(x)$, as will be seen later.
- Finally, we see that for $x \simeq 0$, $F_N(x)$ oscillates about $f(x)$ with a period that decreases with increasing N but with an amplitude that does not decrease: this important phenomenon is due to the discontinuity in $f(x)$ at $x = 0$ and will be discussed in section 8.11.

Some of the observations made above are summarised in the following theorem, which gives *sufficient* conditions for the Fourier series of a function to coincide with the function.

Fourier's theorem

Let $f(x)$ be a function given on the interval $-\pi \le x < \pi$ and defined for all other values of x by the equation

$$f(x + 2\pi) = f(x)$$

so that $f(x)$ is 2π-periodic. Assume that $\int_{-\pi}^{\pi} dx\, f(x)$ exists and that the complex Fourier coefficients c_k are defined by the equations

$$c_k = \frac{1}{2\pi} \int_{-\pi}^{\pi} dx\, f(x) e^{ikx}, \quad k = 0, \pm 1, \pm 2, \ldots,$$

then if $-\pi < a \le x \le b < \pi$ and if in this interval $|f(x)|$ is bounded, the series

$$F(x) = \sum_{k=-\infty}^{\infty} c_k e^{-ikx} \tag{8.16}$$

is convergent and has the value

$$F(x) = \frac{1}{2}\left(\lim_{\epsilon \to 0_+} f(x + \epsilon) + \lim_{\epsilon \to 0_+} f(x - \epsilon) \right). \tag{8.17}$$

If $f(x)$ is continuous at a point $x = w$ the limit reduces to $F(w) = f(w)$.

The conditions assumed here are normally met by functions found in practical applications. In the example treated in exercise 8.9 we have, at $x = 0$,

$$\lim_{\epsilon \to 0_+} f(\epsilon) = 1 \quad \text{and} \quad \lim_{\epsilon \to 0_+} f(-\epsilon) = 0,$$

so equation 8.17 shows that the Fourier series converges to $\frac{1}{2}$ at $x = 0$, as found in the exercise.

Fourier's theorem gives the general relation between the Fourier series and $f(x)$. In addition it can be shown that if the Fourier coefficients have bounded variation and $|c_k| \to 0$ as $k \to \infty$ the Fourier series converges uniformly in the interval $0 < |x| < \pi$. At $x = 0$ care is sometimes needed as the real and imaginary parts of the series behave differently; convergence of the real part depends upon the convergence of the sum $c_0 + c_1 + c_2 + \cdots$, whereas the imaginary part is zero, since c_0 is real for real functions (Zygmund, 1990, chapter 1).

The completeness relation, equation 8.8 (page 271), modified slightly because the functions used here are not normalised, gives the identity

$$\sum_{k=-\infty}^{\infty} |c_k|^2 = \frac{1}{2\pi} \int_{-\pi}^{\pi} dx\, f(x)^2, \tag{8.18}$$

or, for the real form of the Fourier series,

$$\frac{1}{2} a_0^2 + \sum_{k=1}^{\infty} (a_k^2 + b_k^2) = \frac{1}{\pi} \int_{-\pi}^{\pi} dx\, f(x)^2. \tag{8.19}$$

These relations are known as *Parseval's theorem*. It follows from this relation that if the integral exists $|c_k|$ tends to zero faster than $|k|^{-1/2}$ as $|k| \to \infty$.

There is also a converse of Parseval's theroem: the Riesz–Fischer theorem, which states that if numbers c_k exist such that the sum in equation 8.18 exists then the series defined in equation 8.13 (page 273), is the Fourier series of a square integrable function. A proof of this theorem can be found in Zygmund (1990, chapter 4).

In the appendix to this chapter some common Fourier series are listed.

Exercise 8.10

Show that if $f(x)$ is a real function even about $x = 0$ then its Fourier series, on the interval $(-\pi, \pi)$, contains only cosine terms, but that if $f(x)$ is an odd function its Fourier series contains only sine terms.

Exercise 8.11

Let $F(x)$ be the Fourier series of the function $f(x) = x^2$ on the interval $0 \le x \le 2\pi$. Sketch the graph of $F(x)$ in the interval $(-2\pi, 4\pi)$. What are the values of $F(2n\pi)$ for integer n? Note, you are not expected to find an explicit form for $F(x)$.

Exercise 8.12

Use the relation

$$\ln\left(\frac{1}{1-z}\right) = z + \frac{1}{2}z^2 + \frac{1}{3}z^3 + \cdots + \frac{1}{n}z^n + \cdots, \quad |z| \le 1, z \ne 1,$$

$$= \ln\left|\frac{1}{1-z}\right| - i\arg(1-z)$$

to show that for $-\pi \le \theta < \pi$,

$$\frac{1}{2}\ln\left(\frac{1}{1+r^2-2r\cos\theta}\right) = \sum_{k=1}^{\infty} \frac{r^k}{k}\cos k\theta, \quad r \le 1,$$

$$\tan^{-1}\left(\frac{r\sin\theta}{1-r\cos\theta}\right) = \sum_{k=1}^{\infty} \frac{r^k}{k}\sin k\theta, \quad r \le 1,$$

and that

$$\sum_{k=1}^{\infty} \frac{\cos k\theta}{k} = -\ln\left|2\sin\frac{\theta}{2}\right|, \quad \sum_{k=1}^{\infty} \frac{\sin k\theta}{k} = \begin{cases} \frac{1}{2}(\pi - \theta), & 0 < \theta < \pi, \\ -\frac{1}{2}(\pi + \theta), & -\pi < \theta < 0. \end{cases}$$

Observe that for $r < 1$ the Fourier coefficients tend to zero faster than exponentially with increasing k, in contrast to the Fourier coefficients obtained in exercises 8.8 and 8.9, for example.

8.6 Addition and multiplication of Fourier series

The Fourier coefficients of the sum and difference of two functions are given by the sum and difference of the constituent coefficients, as would be expected. Thus if

$$f_1(x) = \sum_{k=-\infty}^{\infty} c_k e^{-ikx}, \quad \text{and} \quad f_2(x) = \sum_{k=-\infty}^{\infty} d_k e^{-ikx}, \tag{8.20}$$

then, for any constants a and b,

$$af_1(x) + bf_2(x) = \sum_{k=-\infty}^{\infty} (ac_k + bd_k) e^{-ikx}.$$

The Fourier series of the product of the two functions is, however, more complicated; suppose that

$$f_1(x)f_2(x) = \sum_{k=-\infty}^{\infty} D_k e^{-ikx}, \qquad (8.21)$$

then if the Fourier series for f_1 and f_2 are absolutely convergent we also have

$$f_1(x)f_2(x) = \sum_{k=-\infty}^{\infty} \sum_{l=-\infty}^{\infty} c_k d_l e^{-i(k+l)x},$$

$$= \sum_{p=-\infty}^{\infty} e^{-ipx} \sum_{k=-\infty}^{\infty} c_k d_{p-k}. \qquad (8.22)$$

On comparing equations 8.21 and 8.22 we see that the nth Fourier coefficient of the product is

$$D_n = \sum_{k=-\infty}^{\infty} c_k d_{n-k} = \sum_{k=-\infty}^{\infty} c_{n-k} d_k. \qquad (8.23)$$

Exercise 8.13
Show that the nth Fourier coefficient of $f_1(x)^2$, where the Fourier series of f_1 is given in equation 8.20, is

$$\sum_{k=-\infty}^{\infty} c_k c_{n-k}.$$

Use the Fourier series for x to deduce that

$$\sum_{\substack{k=-\infty \\ k \neq 0, n}}^{\infty} \frac{1}{k(k-n)} = \frac{2}{n^2}.$$

Another consequence of equation 8.22 is the addition formula for Bessel functions, exercise 8.31 (page 295).

8.7 The behaviour of Fourier coefficients

In general it is difficult to make a priori estimates of the asymptotic behaviour of Fourier components, although we know from Bessel's inequality that $\lim_{k \to \infty} c_k = 0$, and that $|c_k|$ must decay to zero faster than $k^{-1/2}$ if $f(x)$ is square integrable. For the function treated in exercise 8.8 we have $c_k = O(k^{-2})$, and in exercise 8.9 $c_k = O(k^{-1})$; exercise 8.12 provides an example for which $c_k \to 0$ exponentially. It is clearly important to know how rapidly $|c_k|$ decreases to zero because this determines the number of terms needed to achieve a given accuracy. Here we present a few elementary observations.

Consider a function having N continuous derivatives on $(-\pi, \pi)$. The integral for the Fourier components, equation 8.13 (page 273), can be integrated by parts N times; the first two integrations give

$$
c_k = \frac{(-1)^k}{2\pi i k}[f(\pi) - f(-\pi)] - \frac{1}{2\pi i k}\int_{-\pi}^{\pi} dx\, f'(x)e^{ikx} \tag{8.24}
$$

$$
= \frac{(-1)^k}{2\pi i k}[f(\pi) - f(-\pi)] + \frac{(-1)^k}{2\pi k^2}[f'(\pi) - f'(-\pi)] - \frac{1}{2\pi k^2}\int_{-\pi}^{\pi} dx\, f''(x)e^{ikx}.
$$

Clearly this process can be continued until the Nth differential appears in the integral, but useful information can be gleaned from these expressions.

If $f(x)$ is even, $f(\pi) = f(-\pi)$ and it follows that $c_k = O(k^{-2})$. The same result holds if f is not even but $f(\pi) = f(-\pi)$. If the function is odd then $c_k = O(k^{-1})$, unless $f(\pi) = 0$.

If $f(x)$ is 2π-periodic then $f^{(r)}(\pi) = f^{(r)}(-\pi)$, $r = 0, 1, \dots, N$, and after further integration by parts we obtain

$$
c_k = \frac{1}{2\pi}\left(\frac{i}{k}\right)^N \int_{-\pi}^{\pi} dx\, f^{(N)}(x)e^{ikx},
$$

since all the boundary terms are now zero. But

$$
\left|\int_{-\pi}^{\pi} dx\, f^{(N)}(x)e^{ikx}\right| \le \left|\int_{-\pi}^{\pi} dx\, f^{(N)}(x)\right|,
$$

so $c_k = O(k^{-N})$. If $f(x)$ is periodic and *all* derivatives exist then c_k will tend to zero faster than any power of $1/k$, for instance as e^{-k}, as in the example of exercise 8.12.

One important consequence of this last result is that the numerical estimate of the mean of a sufficiently well behaved periodic function over a period obtained using N equally spaced points converges faster than any power of N^{-1}; this is faster than most other numerical procedures. We prove this using the Fourier series of the function: suppose, for simplicity, that the function is 2π-periodic so possesses the Fourier series

$$
f(x) = \sum_{k=-\infty}^{\infty} C_k e^{-ikx},
$$

where the coefficients C_k are unknown and C_0 is the required mean value of $f(x)$. The mean of $f(x)$ over $(0, 2\pi)$ can be approximated by the sum over N equally spaced points,

$$
\frac{1}{2\pi}\int_0^{2\pi} dx\, f(x) \simeq \frac{1}{N}\sum_{j=1}^{N} f\left(\frac{2\pi j}{N}\right). \tag{8.25}
$$

Using the Fourier series of $f(x)$ the sum can be written in the alternative form

$$
\sum_{j=1}^{N} f\left(\frac{2\pi j}{N}\right) = \sum_{j=1}^{N}\sum_{k=-\infty}^{\infty} C_k \exp\left(-i\frac{2\pi k j}{N}\right)
$$

$$
= \sum_{k=-\infty}^{\infty} C_k \sum_{j=1}^{N} \exp\left(-i\frac{2\pi k j}{N}\right).
$$

But we have the relation

$$\sum_{j=1}^{N} e^{-izj} = R(z)\exp\left(-i(N+1)z/2\right), \quad R(z) = \frac{\sin(Nz/2)}{\sin(z/2)},$$

which follows because the left hand side is a geometric series. Now $z = 2\pi k/N$, and

$$R\left(\frac{2\pi k}{N}\right) = \frac{\sin \pi k}{\sin(\pi k/N)}.$$

This is zero unless $k = Np$ for some integer p, that is $z = 2\pi p$, in which case we can find the value of R by taking the limit, or more easily by noting that the original sum becomes

$$\sum_{j=1}^{N} \exp\left(-i\frac{2\pi k j}{N}\right) = \sum_{j=1}^{N} \exp\left(-i2\pi pj\right) = N, \quad k = Np.$$

Thus

$$\frac{1}{N}\sum_{j=1}^{N} f\left(\frac{2\pi j}{N}\right) = C_0 + \sum_{\substack{p=-\infty \\ p\neq 0}}^{\infty} C_{Np}. \tag{8.26}$$

If all derivatives of f exist then, since $f(x)$ is periodic, $|C_{Np}| \to 0$ faster than any power of $(Np)^{-1}$, so that the numerical estimate of the mean on the left hand side converges to C_0 faster than any power of N^{-1}. This result is of practical value.

The ideas presented in this section can sometimes be put to good use in speeding the convergence of Fourier series. Consider a function $f(x)$ continuous on $-\pi < x \leq \pi$ but with $f(\pi) \neq f(-\pi)$, so its Fourier coefficients $c_k = O(k^{-1})$, as $k \to \infty$. Define a new function $g(x) = f(x) - \alpha x$ with the constant α chosen to make $g(\pi) = g(-\pi)$, that is

$$\alpha = \frac{1}{2\pi}(f(\pi) - f(-\pi)),$$

so Fourier components of g behave as $O(k^{-2})$.

As an example consider the odd function $f(x) = \sin\sqrt{2}x$ with the Fourier series

$$\sin\sqrt{2}x = \sum_{k=1}^{\infty} b_k \sin kx, \tag{8.27}$$

where

$$b_k = \frac{2k(-1)^{k-1}\sin\pi\sqrt{2}}{\pi(k^2-2)} = \frac{2(-1)^{k-1}}{k\pi}\sin\pi\sqrt{2} + O(k^{-2}).$$

In this case $\alpha = (\sin\pi\sqrt{2})/\pi$ and using the Fourier series of x, given in the appendix of this chapter, we see that the Fourier coefficients, G_k, of $g(x) = f(x) - \alpha x$, are

$$G_k = b_k - \frac{2(-1)^{k-1}}{k\pi}\sin\pi\sqrt{2} = \frac{4}{\pi}\frac{(-1)^{k-1}}{k(k^2-2)}\sin\pi\sqrt{2} = O(k^{-3}).$$

Hence for $-\pi \le x \le \pi$ we may write

$$f(x) = \sin \sqrt{2}x = \frac{x}{\pi}\sin \pi\sqrt{2} + \frac{4}{\pi}\sin \pi\sqrt{2}\sum_{k=1}^{\infty}\frac{(-1)^{k-1}}{k(k^2-2)}\sin kx. \qquad (8.28)$$

In the following two figures we compare the function, $f(x) = \sin \sqrt{2}x$, with ten terms of the original Fourier series 8.27, on the left, and just two terms of the modified series 8.28, on the right; in the second case with more terms the two functions are practically indistinguishable. The main point to notice is that by removing the discontinuity in the Fourier series at $x = \pm\pi$ a far more rapidly converging approximation has been obtained that also converges *pointwise* over the whole range.

It is clear that further corrections that produce a more rapidly converging Fourier series may be added. This idea is developed further by Lanczos (1966, section 16).

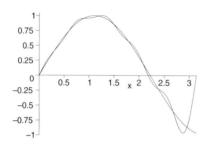

Figure 8.3   Comparison of $f(x)$ with ten terms of the original Fourier series 8.27.

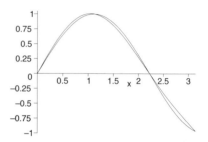

Figure 8.4   Comparison of $f(x)$ with two terms of the modified Fourier series 8.28.

Exercise 8.14
Show that the Fourier series of $f(x) = \sinh x$ on $(-\pi, \pi)$ is

$$\sinh x = \frac{2\sinh \pi}{\pi}\sum_{k=1}^{\infty}\frac{(-1)^{k-1}k}{1+k^2}\sin kx.$$

Show further that

$$\sinh x = \frac{x}{\pi}\sinh \pi - 2\frac{\sinh \pi}{\pi}\sum_{k=1}^{\infty}\frac{(-1)^{k-1}}{k(1+k^2)}\sin kx.$$

Use Maple to compare various partial sums of these two series with $\sinh x$ and hence demonstrate that the latter is a more useful approximation.

8.8 Differentiation and integration of Fourier series

The Fourier series of a function $f(x)$ is uniformly convergent in an interval (a, b), where $-\pi < a \le x \le b < \pi$, if $f(x)$ is continuous in this interval. Then the term by term differentiation and integration of the series are respectively the differential and integral of $f(x)$. Thus if

$$f(x) = \sum_{k=-\infty}^{\infty} c_k e^{-ikx} \tag{8.29}$$

is uniformly convergent for $a \le x \le b$, then

$$\frac{df}{dx} = -i \sum_{k=-\infty}^{\infty} k c_k e^{-ikx}, \tag{8.30}$$

and the integral is

$$\int dx\, f(x) = A + xc_0 + i \sum_{\substack{k=-\infty \\ k \ne 0}}^{\infty} \frac{c_k}{k} e^{-ikx},$$

where A is the constant of integration and the sum does not include the $k = 0$ term. The expressions for the differential and integral of a function given above are not entirely satisfactory: the series for $\int dx\, f(x)$ is not a Fourier series, and if $c_k = O(k^{-1})$ as $k \to \infty$ the convergence of the series given for $f'(x)$ is problematical and the series is certainly not useful.

Integration
The first difficulty may be cured by forming a definite integral and using the known Fourier series for x. The definite integral is

$$\int_0^x dx\, f(x) = xc_0 + i \sum_{\substack{k=-\infty \\ k \ne 0}}^{\infty} \frac{c_k}{k} \left(e^{-ikx} - 1 \right),$$

but

$$x = 2 \sum_{k=1}^{\infty} \frac{(-1)^{k-1}}{k} \sin kx = i \sum_{\substack{k=-\infty \\ k \ne 0}}^{\infty} \frac{(-1)^{k-1}}{k} e^{-ikx}.$$

Substituting this expression for x gives

$$\int_0^x dx\, f(x) = 2 \sum_{k=1}^{\infty} \frac{\Im(c_k)}{k} + i \sum_{\substack{k=-\infty \\ k \ne 0}}^{\infty} \left(\frac{c_k - (-1)^k c_0}{k} \right) e^{-ikx}. \tag{8.31}$$

If $f(x)$ is real the alternative, real, form for the integral is

$$\int_0^x dx\, f(x) = \sum_{k=1}^{\infty} \frac{b_k}{k} + \sum_{k=1}^{\infty} \left\{ \frac{a_k - (-1)^k a_0}{k} \sin kx - \frac{b_k}{k} \cos kx \right\}, \tag{8.32}$$

where $2c_k = a_k + ib_k$, $k \ge 1$ and $a_0 = 2c_0$.

Exercise 8.15
Use equation 8.32 to integrate the Fourier series for x to show that

$$x^2 = \frac{\pi^2}{3} - 4 \sum_{k=1}^{\infty} \frac{(-1)^{k-1} \cos kx}{k^2}.$$

Explain why the Fourier series for x on $(-\pi, \pi)$ depends only on sine functions and that for x^2 only upon cosine functions.

Differentiation

For many differentiable functions the leading term in the asymptotic expansion of c_k is $O(k^{-1})$; the reason for this is given in the previous section, particularly equation 8.24. Then the convergence of the series 8.30 for $f'(x)$ is questionable even though $f'(x)$ may have a Fourier series expansion. Normally $c_k = O(k^{-1})$ because the periodic extension of $f(x)$ is discontinuous at $x = \pm \pi$, then we expect the Fourier coefficients of $f'(x)$ to also be $O(k^{-1})$, not $O(1)$ as suggested by equation 8.30. We now show how this is achieved.

The analysis of section 8.7 leads us to assume that for large k

$$c_k = (-1)^k \frac{c_\infty}{k} + O(k^{-2}), \quad c_\infty = \frac{1}{2\pi i} \left(f(\pi) - f(-\pi) \right),$$

so we define a constant c by the limit

$$c = \lim_{k \to \infty} (-1)^k k c_k$$

and consider the Fourier series

$$g(x) = -i \sum_{k=-\infty}^{\infty} \left(k c_k - (-1)^k c \right) e^{-ikx}$$

which is clearly related to the series 8.30. With the assumed behaviour of c_k, the Fourier components of this series are $O(k^{-1})$. Now integrate this function using equation 8.31:

$$\int_0^x du\, g(u) = f(x) - 2 \sum_{k=1}^{\infty} \frac{1}{k} \Re(k c_k - (-1)^k c).$$

Thus $g(x) = f'(x)$ and we have found the Fourier series of $f'(x)$ that converges at the same rate as the Fourier series of $f(x)$, that is

$$\frac{df}{dx} = g(x) = -i \sum_{k=-\infty}^{\infty} \left(k c_k - (-1)^k c \right) e^{-ikx}. \tag{8.33}$$

If $f(x)$ is real, $\Re(c) = 0$ and the real form of the Fourier series is

$$\frac{df}{dx} = -\frac{b}{2} + \sum_{k=1}^{\infty} \left\{ (k b_k - (-1)^k b) \cos kx - k a_k \sin kx \right\},$$

where $b = -2ic = \lim_{k \to \infty} (-1)^k k b_k$ and $a_k + i b_k = 2 c_k$.

Exercise 8.16

Using the Fourier series

$$x^3 = 2\sum_{k=1}^{\infty}(-1)^{k-1}\frac{k^2\pi^2 - 6}{k^3}\sin kx, \quad |x| < \pi,$$

and the above expression for the differential of a Fourier series, show that

$$x^2 = \frac{\pi^2}{3} - 4\sum_{k=1}^{\infty}\frac{(-1)^{k-1}}{k^2}\cos kx, \quad |x| \le \pi.$$

Explain why the Fourier coefficients for x^3 and x^2 are respectively $O(k^{-1})$ and $O(k^{-2})$ as $k \to \infty$.

Exercise 8.17

Consider the function

$$f(x) = \sum_{k=2}^{\infty}(-1)^k\frac{k}{k^2-1}\sin kx = \frac{i}{2}\sum_{|k|\ge 2}(-1)^k\frac{k}{k^2-1}e^{-ikx}.$$

Show that the Fourier series of $f'(x)$ is

$$f'(x) = -\frac{1}{2} + \cos x + \sum_{k=2}^{\infty}(-1)^k\frac{\cos kx}{k^2-1}.$$

Find also the series for $f''(x)$ and hence show that $f(x) = \frac{1}{4}\sin x + \frac{1}{2}x\cos x$.

Exercise 8.18

(i) If $f(x) = e^{ax}$, use the fact that $f'(x) = af(x)$ together with equation 8.33 to show that the Fourier coefficients of $f(x)$ on the interval $(-\pi, \pi)$ are

$$c_k = \frac{i(-1)^k c}{a + ik}$$

for some number c.

Show directly that the mean of $f'(x)$ is $(f(\pi) - f(-\pi))/2\pi$ and use this to determine the value of c, hence showing that

$$e^{ax} = \frac{\sinh \pi a}{\pi}\sum_{k=-\infty}^{\infty}\frac{(-1)^k}{a + ik}e^{-ikx}.$$

(ii) By observing that $e^{2ax} = e^{ax}e^{ax}$, or otherwise, use the Fourier series for e^{ax} to show that

$$1 + 2a^2\sum_{k=1}^{\infty}\frac{1}{a^2 + k^2} = \frac{\pi a}{\tanh \pi a},$$

and by considering the small a expansion deduce that, for some numbers f_n,

$$\sum_{k=1}^{\infty}\frac{1}{k^{2n}} = f_n\pi^{2n}, \quad n = 1, 2,\dots.$$

8.9 Fourier series on arbitrary ranges

The Fourier series of a function $f(x)$ over a range $a \leq x \leq b$, different from $(-\pi, \pi)$, is obtained using the preceding results and by defining a new variable

$$w = \frac{2\pi}{b-a}\left(x - \frac{b+a}{2}\right), \quad x = \frac{b+a}{2} + \frac{b-a}{2\pi}w, \tag{8.34}$$

which maps $-\pi \leq w \leq \pi$ onto $a \leq x \leq b$. Then it follows that the function $f(x(w))$ of w can be expressed as a Fourier series on $-\pi \leq w \leq \pi$,

$$f(x(w)) = \sum_{k=-\infty}^{\infty} c_k e^{-ikw} \quad \text{and so} \quad f(x) = \sum_{k=-\infty}^{\infty} d_k \exp\left(-i\frac{2\pi k}{b-a}x\right),$$

where

$$d_k = \frac{1}{b-a}\int_a^b dx\, f(x)\exp\left(\frac{2\pi ik}{b-a}x\right), \tag{8.35}$$

this last relation following from the definition of c_k and the transformation 8.34.

Exercise 8.19
Show that the Fourier series of $f(x) = \cosh x$ on the interval $(-L, L)$ is

$$\cosh x = \frac{\sinh L}{L}\left(1 + 2\sum_{k=1}^{\infty}\frac{(-1)^k L^2}{L^2 + \pi^2 k^2}\cos\left(\frac{\pi k x}{L}\right)\right), \quad -L \leq x \leq L.$$

From the form of this Fourier series we observe that the Fourier components are small only if $k \gg L$; thus if L is large many terms of the series are needed to provide an accurate approximation, as may be seen by comparing graphs of partial sums of the series with $\cosh x$ on $-L < x < L$ for various values of L.

8.10 Sine and cosine series

Fourier series of even functions, $f(x) = f(-x)$, comprise only cosine terms, and Fourier series of odd functions, $f(x) = -f(-x)$, comprise only sine terms. This fact can be used to produce cosine or sine series of *any* function over a given range which we shall take to be $(0, \pi)$.

For any function $f(x)$ an odd extension, $f_o(x)$, may be produced as follows:

$$f_o(x) = \begin{cases} -f(-x), & x < 0, \\ f(x), & x \geq 0. \end{cases}$$

Figure 8.6 shows the odd extension of the function $f(x)$, depicted in figure 8.5.

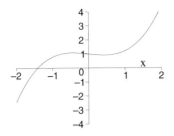

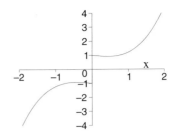

Figure 8.5 Original function.　　　　　　　　Figure 8.6 Odd extension.

We would normally use this extension when $f(0) = 0$ so that $f_o(x)$ is continuous at $x = 0$. Then the Fourier series of $f_o(x)$ on $(-\pi, \pi)$ contains only sine functions and we have, from equation 8.15,

$$f(x) = \sum_{k=1}^{\infty} b_k \sin kx, \quad 0 < x \le \pi, \quad \text{with} \quad b_k = \frac{2}{\pi} \int_0^{\pi} dx\, f(x) \sin kx, \quad k = 1, 2, \ldots,$$

the latter relation following from equation 8.13. This series is often called the *half-range Fourier sine series*.

Similarly we can extend $f(x)$ to produce an even function,

$$f_e(x) = \begin{cases} f(-x), & x < 0, \\ f(x), & x \ge 0, \end{cases}$$

an example of which is shown in figure 8.8.

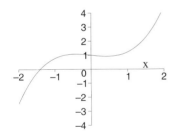

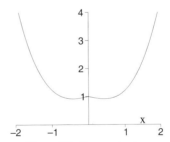

Figure 8.7 Original function.　　　　　　　　Figure 8.8 Even extension.

The even extension produces a Fourier series containing only cosine functions, to give the *half-range Fourier cosine series*

$$f(x) = \frac{1}{2}a_0 + \sum_{k=1}^{\infty} a_k \cos kx, \quad 0 < x \le \pi,$$

where

$$a_k = \frac{2}{\pi} \int_0^{\pi} dx\, f(x) \cos kx, \quad k = 0, 1, \ldots.$$

Exercise 8.20

Show that

$$x = 2 \sum_{k=1}^{\infty} \frac{(-1)^{k-1}}{k} \sin kx, \quad 0 \le x < \pi,$$

and

$$x = \frac{\pi}{2} - \frac{4}{\pi} \sum_{k=1}^{\infty} \frac{\cos(2k-1)x}{(2k-1)^2}, \quad 0 \le x \le \pi.$$

Explain why the even extension converges more rapidly.

Use Maple to compare, graphically, the partial sums of these two approximations with the exact function.

Exercise 8.21

Show that

$$\sin x = \frac{4}{\pi} \left\{ \frac{1}{2} - \sum_{k=1}^{\infty} \frac{\cos 2kx}{4k^2 - 1} \right\}, \quad 0 \le x \le \pi,$$

and deduce that

$$\sum_{k=1}^{\infty} \frac{(-1)^{k-1}}{4k^2 - 1} = \frac{\pi}{4} - \frac{1}{2}.$$

8.11 The Gibbs phenomenon

The Gibbs phenomenon occurs in the Fourier series of any discontinuous function, such as that considered in exercise 8.9, but before discussing the mathematics behind this strange behaviour we give the following description of its discovery, taken from Lanczos (1966, section 10).

The American physicist Michelson† invented many physical instruments of very high precision. In 1898 he constructed a harmonic analyser that could compute the first 80 Fourier components of a function described numerically; this machine could also construct the graph of a function from the Fourier components, thus providing a check on the operation of the machine because, having obtained the Fourier components from a given function, it could be reconstructed and compared with the original. Michelson found that in most cases the input and synthesised functions agreed well.

---

† A. A. Michelson (1852–1931) was born in Prussia. He moved to the USA when two years old and is best known for his accurate measurements of the speed of light, which was his life long passion; in 1881 he determined this to be 186 329 miles/sec and in his last experiment, finished after his death, obtained 186 280 miles/sec, the current value for the speed of light in a vacuum is 186 282.397 miles/sec ($2.997\,924\,58 \times 10^8$ m/sec). In 1887, together with Morley, he published results showing that there was no measurable difference between the speed of light when the Earth was moving towards or away from a source: this result was crucial in falsifying the theories of the aether and in Einstein's formulation of the special theory of relativity.

But when he tried the machine on a square wave the synthesised function agreed
well with the square wave input, except near the point of discontinuity at which the
synthesis had a peculiar protuberance not present in the original function, as shown
in figure 8.9 below. Michelson was puzzled and thought that there was a defect in the
machine: he wrote to Gibbs, the eminent mathematical physicist, asking his opinion, J. W. Gibbs
                                                                                     (1839–1903)
who explained it in terms of the non-uniform convergence of the Fourier series in the
vicinity of a point of discontinuity, and published his findings in *Nature* (Gibbs, 1898,
1899).

In figure 8.9 we show graphs of the type of result confronting Michelson; this shows
the Heaviside unit function, $H(x)$, defined in equation 7.8 (page 230), and the 10th, 20th
and 40th partial sums of its Fourier decomposition, on a small interval surrounding
the discontinuity at $x = 0$. The Nth partial sum is

$$H_N(x) = \frac{1}{2} + \frac{2}{\pi} \sum_{k=1}^{N} \frac{\sin(2k-1)x}{2k-1}. \tag{8.36}$$

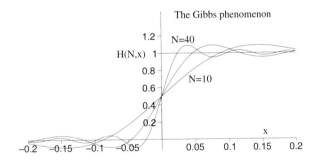

Figure 8.9   Graphs of $H_N(x)$, equation 8.36, for $N = 10$, 20 and 40.

Observe the 'wiggles' at small $|x|$ in this graph and note how they compress towards
$x = 0$ as N increases but that their amplitude remains constant. It is illuminating
to watch an animation showing how these oscillations change with N. The following
Maple commands will create such an animation, which is worth watching because it
helps one to understand the phenomenon.

```
>  restart; with(plots):
```
First define the series defined in equation 8.36.

```
>  f:=proc(x,n) local k;
>  1/2 + 2*add(evalf( sin((2*k-1)*x)/(2*k-1) ),k=1..n)/Pi;
>  end:
```
Now create a sequence of plots for the animation; here N increases from 5 to 300.

```
>   fnt1:=[TIMES,ROMAN,14]:                                    # Define a font
>   fnt2:=[TIMES,ROMAN,12]:                                    # Define a font
>   t3:=textplot([0,1.3,"The Gibbs phenomenon"],font=fnt1):
>
>   pl:=NULL:                                                  # Title text
>     for k from 10 to 300 by 5  do;                           # Start loop
>     t1:=convert(N=k, string);               # Label giving value of N used
>     t2:=textplot([0.1,0.8,t1],align=RIGHT,font=fnt2);
>     p:=plot(f(x,k),x=-0.2..0.2,-0.1..1.1,numpoints=100):
>                                               # This line plots the graph
>     p:=display([p,t2,t3],view=[-0.2..0.2,-0.1..1.3]):
>                                               # Add captions to graph
>     pl:=pl,p;                        # Form the expression sequence of plots
>     od:                             # End loop: NOTE the colon terminator
```

Finally display these graphs as an animation

```
>   display(pl,insequence=true);
```

In this animation you should see the first, large oscillation move towards the y-axis as N increases but with unchanging amplitude. Our problem is to understand this strange phenomenon.

There are two stages to this analysis. The first is quite general and not limited to any specific function; in the second stage we apply this analysis to the Heaviside function and use known properties of the Sine integral.

Our first objective is to find a convenient expression for the Nth partial Fourier series of the function $f(x)$,

$$f_N(x) = \sum_{k=-N}^{N} c_k e^{-ikx}. \tag{8.37}$$

Replace the Fourier coefficients c_k by their integral definition, equation 8.13 (page 273), and change the order of summation and integration to obtain

$$f_N(x) = \frac{1}{2\pi} \int_{-\pi}^{\pi} dy\, f(y) \sum_{k=-N}^{N} e^{ik(y-x)}. \tag{8.38}$$

The sum in this integral is just a geometric series and we have

$$s_N(z) = \sum_{k=-N}^{N} e^{ikz} = \frac{\sin(N+1/2)z}{\sin(z/2)}. \tag{8.39}$$

This shows that $s_N(0) = 2N + 1$, that the width of the maximum at $z = 0$ is about π/N and that for most other values of z, $s_N(z)$ is relatively small. In the following two figures are shown graphs of $s_{10}(z)$ and $s_{100}(z)$; note that the range of z is quite different in these two cases. The function $s_N(z)$ is, of course, 2π-periodic.

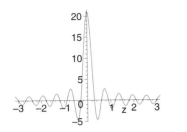

Figure 8.10 $s_{10}(z)$.

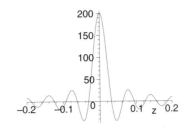

Figure 8.11 $s_{100}(z)$.

You will recall from equation 7.24 (page 236) that

$$\frac{1}{2\pi} \sum_{k=-\infty}^{\infty} e^{ikx} = \sum_{n=-\infty}^{\infty} \delta(x - 2n\pi).$$

One can understand this limit intuitively from the above graphs. Similarly, one can see that if $f(x)$ is continuous the result of taking the limit $N \to \infty$ in equation 8.38 is to give $f_N(x) \to f(x)$: a proof of this result may be found in Whittaker and Watson (1965, section 9.42). Fejér extended this result to functions that are not continuous; further details of this method can be found in Lanczos (1966, section 11).

We are interested in the case where $f(x)$ is discontinuous and want to know how $f_N(x)$ behaves near a discontinuity of $f(x)$. As an example put $f(x) = H(x)$, the Heaviside unit function, then equations 8.38 and 8.39 give the Nth partial sum of $H(x)$ in the form

$$
\begin{aligned}
H_N(x) &= \frac{1}{2\pi} \int_0^\pi dy \, \frac{\sin(N + 1/2)(y - x)}{\sin\left(\frac{y-x}{2}\right)}, \\
&= \frac{1}{2\pi} \int_{-x}^{\pi - x} d\xi \, \frac{\sin(N + 1/2)\xi}{\sin(\xi/2)}.
\end{aligned}
\tag{8.40}
$$

This equation is exact, but to proceed further approximations are necessary. Assume that $|x|$ is small — because we are interested only in the behaviour near the discontinuity at the origin — and that N is large so the integrand is peaked at $\xi = 0$. Then we can replace $\sin(\xi/2)$ by $\xi/2$ to give

$$H_N(x) \simeq \frac{1}{\pi} \int_{-x}^{\pi - x} d\xi \, \frac{\sin(N + 1/2)\xi}{\xi}.$$

Now use the Sine integral, denoted in Maple by Si(z),

$$\mathrm{Si}(z) = \int_0^z du \, \frac{\sin u}{u}, \qquad \mathrm{Si}(-z) = -\mathrm{Si}(z),$$

to write $H_N(x)$ in the form

$$H_N(x) = \frac{1}{\pi} \left[\mathrm{Si}((N + 1/2)(\pi - x)) + \mathrm{Si}((N + 1/2)x) \right].
\tag{8.41}$$

Integrating the expression for Si(z) repeatedly by parts† shows that the asymptotic

† This method is dealt with in chapters 5 and 13, and in particular exercise 5.23.

expansion of Si(z) as $z \to \infty$ is

$$\mathrm{Si}(z) \sim \frac{\pi}{2} - \frac{1}{z}\left(1 - \frac{2!}{z^2} + \frac{4!}{z^4} + \cdots\right)\cos z - \frac{1}{z^2}\left(1 - \frac{3!}{z^2} + \frac{5!}{z^4} + \cdots\right)\sin z.$$

Thus, if N is large and $|x| \ll 1$ the first term in the square brackets of 8.41 is approximately $\pi/2$, and the equation reduces to

$$H_N(x) \simeq \frac{1}{2} + \frac{1}{\pi}\mathrm{Si}\left(\left(N + \tfrac{1}{2}\right)x\right), \quad N \gg 1, \quad |x| \ll 1.$$

$$\simeq 1 - \frac{\cos\left(N + \tfrac{1}{2}\right)x}{\left(N + \tfrac{1}{2}\right)x}, \quad N \gg 1, \quad |x| \ll 1 \quad \text{and} \quad Nx \gg 1.$$

Exercise 8.22
Derive equation 8.41 and use it to find the value of $H_N(0)$ for large N.

The graph of $\mathrm{Si}(z)/\pi$, figure 8.12, shows that for $z > 2$ the Sine integral oscillates about its asymptotic value of $\pi/2$. The first maximum is at $z = \pi$ and here $\mathrm{Si}(z)/\pi = 0.5895$.

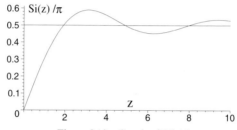

Figure 8.12   Graph of $\mathrm{Si}(z)/\pi$.

Since $H_N(x)$ depends upon the product $(N + \tfrac{1}{2})x$, rather than N and x separately, we see that for *every* finite N, $H_N(x)$ always overshoots the value of $H(x)$ by the same amount, namely 0.0895, at the point $x \simeq \pi/N$, so the oscillations in $H_N(x)$ have an amplitude independent of N although for any fixed $x > 0$, $|H(x) - H_N(x)| < (\pi(N + \tfrac{1}{2})x)^{-1}$ as $N \to \infty$. A similar result holds for $-\pi < x < 0$.

This analysis explains the results seen in figure 8.9, in Michelson's synthesis and in the Maple animation. It is also applicable to other discontinuous functions whose left and right derivatives exist at the discontinuity; we simply write such a function as the sum of a continuous function and a Heaviside function, as follows. Suppose that the function $f(x)$ is discontinuous at $x = x_d$, that

$$\lim_{\epsilon \to 0_+} (f(x_d + \epsilon) - f(x_d - \epsilon)) = D$$

and that the left and right derivatives at $x = x_d$ exist, that is, the limits

$$\lim_{\epsilon \to 0_+} \frac{f(x_d + \epsilon) - f(x_d)}{\epsilon}, \quad \lim_{\epsilon \to 0_+} \frac{f(x_d) - f(x_d - \epsilon)}{\epsilon}$$

exist. Then the new function

$$g(x) = f(x) - DH(x - x_d)$$

is continuous at $x = x_d$, so the Fourier series of $g(x)$ converges pointwise. The Fourier series of $f(x)$ is the same as the Fourier series of $g(x)$ plus the Fourier series of $DH(x - x_d)$, and the latter we have just found.

8.12 Poisson summation formula

The Poisson summation formula is an important transformation allowing one to find different representations of functions that are expressed as infinite series. Suppose that $f(x)$ is a continuous function defined on the whole real axis; from it we may form the 2π-periodic function $F(x)$ by the simple device of adding copies of $f(x)$,

$$F(x) = \sum_{n=-\infty}^{\infty} f(x + 2\pi n). \tag{8.42}$$

It is easy to see that $F(x)$ is 2π-periodic because, for all x,

$$F(x + 2\pi) = \sum_{n=-\infty}^{\infty} f(x + 2\pi(n + 1)) = \sum_{m=-\infty}^{\infty} f(x + 2\pi m) = F(x).$$

Since $F(x)$ is periodic it can be represented by a Fourier series,

$$F(x) = \sum_{k=-\infty}^{\infty} c_k e^{-ikx}, \tag{8.43}$$

and we can write the Fourier component c_k in terms of the integral

$$
\begin{aligned}
c_k &= \frac{1}{2\pi} \int_{-\pi}^{\pi} dx\, F(x) e^{ikx} \\
&= \frac{1}{2\pi} \int_{-\pi}^{\pi} dx \sum_{n=-\infty}^{\infty} f(x + 2\pi n) e^{ikx} \\
&= \frac{1}{2\pi} \sum_{n=-\infty}^{\infty} \int_{(2n-1)\pi}^{(2n+1)\pi} dy\, f(y) e^{iky}, \quad \text{where} \quad y = x + 2\pi n \\
c_k &= \frac{1}{2\pi} \int_{-\infty}^{\infty} dy\, f(y) e^{iky}. \tag{8.44}
\end{aligned}
$$

Now substitute this expression for c_k back into the original series to obtain

$$F(x) = \frac{1}{2\pi} \sum_{k=-\infty}^{\infty} e^{-ikx} \int_{-\infty}^{\infty} dy\, f(y) e^{iky},$$

and since $F(x)$ is itself a sum, equation 8.42, we obtain the following identity between two infinite series:

$$\sum_{n=-\infty}^{\infty} f(x + 2\pi n) = \frac{1}{2\pi} \sum_{k=-\infty}^{\infty} e^{-ikx} \int_{-\infty}^{\infty} dy\, f(y) e^{iky}. \tag{8.45}$$

This is the Poisson summation formula. On setting $x = 0$ we obtain a slightly less general result,

$$\sum_{n=-\infty}^{\infty} f(2\pi n) = \frac{1}{2\pi} \sum_{k=-\infty}^{\infty} \int_{-\infty}^{\infty} dy\, f(y)e^{iky}. \tag{8.46}$$

At first sight it would appear that little has been achieved and that we have simply regained the original expression via some rather tortuous algebra. This is not the case; consider the simple example of the infinite series

$$S(\lambda) = \sum_{n=-\infty}^{\infty} e^{-\lambda n^2}, \quad \lambda > 0. \tag{8.47}$$

If $0 < \lambda \ll 1$ very many terms are needed to evaluate this sum accurately, as shown in table 8.1. In order to apply equation 8.46 we define

$$f(x) = e^{-\mu x^2}, \quad \text{giving} \quad \sum_{n=-\infty}^{\infty} f(2\pi n) = \sum_{n=-\infty}^{\infty} e^{-\lambda n^2} \quad \text{with} \quad 4\pi^2 \mu = \lambda.$$

With this definition of $f(x)$ the integral on the right hand side of equation 8.46 is

$$\int_{-\infty}^{\infty} dy\, e^{-\mu y^2 + iky} = \sqrt{\frac{\pi}{\mu}} \exp\left(-\frac{k^2}{4\mu}\right),$$

and hence

$$\sum_{n=-\infty}^{\infty} e^{-\lambda n^2} = \sqrt{\frac{\pi}{\lambda}} \sum_{k=-\infty}^{\infty} \exp\left(-\frac{\pi^2 k^2}{\lambda}\right). \tag{8.48}$$

If λ is small, $\exp(-\pi^2 k^2/\lambda) \ll 1$ provided $k \neq 0$ and so the left hand side of this equation converges very slowly and the right hand side very rapidly. This behaviour is illustrated in the table, where we show some partial sums, $S_N(\lambda)$, of the left hand side of equation 8.47 with $\lambda = 0.01$. In this case the $k = 0$ term of the right hand side of 8.48 gives $17.7245\cdots$ and the remaining terms are negligible. Further, the right hand side of equation 8.48 shows that the leading term of the asymptotic expansion of $S(\lambda)$ as $\lambda \to 0$ is $\sqrt{\pi/\lambda}$.

Table 8.1. *Some values of the partial sums of the left hand side of 8.47 when $\lambda = 0.01$. To three decimal places the exact value is 17.725.*

N	1	2	5	10	20	30
$S_N(\lambda)$	2.980	4.901	9.991	15.292	17.658	17.724

Exercise 8.23

Use the function

$$f(x) = \frac{1}{1 + (ax)^2}, \quad a > 0,$$

and the Poisson summation formula to show that

$$\sum_{n=-\infty}^{\infty} \frac{1}{1+(2\pi an)^2} = \frac{1}{2a}\left\{1+2\sum_{n=1}^{\infty}\exp(-n/a)\right\} = \frac{1}{2a\tanh(1/2a)}.$$

Hint: use the integral

$$\int_{-\infty}^{\infty} dw \, \frac{e^{-iaw}}{1+w^2} = \pi e^{-|a|}, \quad a \text{ real}.$$

Another example of the Poisson summation formula is the sum of delta functions introduced in chapter 7:

$$\sum_{n=-\infty}^{\infty} \delta(x+2\pi n) = \frac{1}{2\pi}\sum_{n=-\infty}^{\infty} e^{-ikx},$$

obtained from equation 8.45 by replacing $f(x)$ by $\delta(x)$.

8.13 Appendix: some common Fourier series

Here we list a few of the simpler Fourier series; in all cases the interval is $-\pi < x \leq \pi$.

$$2\sum_{k=1}^{\infty} \frac{(-1)^{k-1}}{k}\sin kx = x,$$

$$2\sum_{k=1}^{\infty} \frac{\sin kx}{k} = \begin{cases} -\pi - x, & -\pi \leq x < 0, \\ \pi - x, & 0 \leq x < \pi, \end{cases}$$

$$\frac{4}{\pi}\sum_{k=1}^{\infty} \frac{\sin(2k-1)x}{2k-1} = \begin{cases} -1, & -\pi \leq x < 0, \\ 1, & 0 \leq x < \pi, \end{cases}$$

$$\sum_{k=1}^{\infty} \frac{\cos kx}{k} = -\ln\left(2\sin\left(\frac{|x|}{2}\right)\right),$$

$$2\sum_{k=1}^{\infty} \frac{\cos(2k-1)x}{2k-1} = -\ln\left(\tan\left(\frac{|x|}{2}\right)\right),$$

$$\frac{\pi^2}{3} - 4\sum_{k=1}^{\infty} \frac{(-1)^{k-1}}{k^2}\cos kx = x^2,$$

$$\sum_{k=1}^{\infty} \frac{(-1)^{k-1}}{k^3}\sin kx = \frac{x}{12}\left(\pi^2 - x^2\right),$$

$$1 + 2\sum_{k=1}^{\infty}\left(\frac{-a}{1+\sqrt{1-a^2}}\right)^k \cos kx = \frac{\sqrt{1-a^2}}{1+a\cos x}, \quad 0 \leq a < 1.$$

8.14 Exercises

Exercise 8.24

Let

$$f(x) = \sum_{k=0}^{N} c_k x^k,$$

where $\mathbf{c} = (c_0, c_1, \ldots, c_N)$ are real constants. Write a Maple procedure to evaluate the integral

$$F(\mathbf{c}) = \int_0^1 dx \, (f(x) - e^x)^2$$

for a given value of N, to form the function $F(\mathbf{c})$, quadratic in $\mathbf{c}$, and which finds the position of the minimum of this function by solving the $N + 1$ equations $\frac{\partial F}{\partial c_k} = 0$, for $\mathbf{c}$.

In the cases $N = 2$ and 3 show that

$$e^x \simeq (39e - 105) + (588 - 216e)x + (210e - 570)x^2$$

$$e^x \simeq (536e - 1456) + (16800 - 6180e)x + (15120e - 41100)x^2 + (27020 - 9940e)x^3$$

and compare, graphically, these approximations with the exact function and various Taylor polynomials.

Exercise 8.25
Show that the Fourier series of the functions e^x and $e^{-|x|}$ are

$$e^x = \frac{\sinh \pi}{\pi} + \frac{2 \sinh \pi}{\pi} \sum_{k=1}^{\infty} \frac{(-1)^k}{1 + k^2} (\cos kx - k \sin kx), \quad -\pi < x \le \pi,$$

$$e^{-|x|} = \frac{1 - e^{-\pi}}{\pi} + \frac{2}{\pi} \sum_{k=1}^{\infty} \frac{1 - e^{-\pi}(-1)^k}{1 + k^2} \cos kx, \quad -\pi < x \le \pi,$$

and explain why the second series converges more rapidly than the first.

Exercise 8.26
A 2π-periodic function is defined by the conditions $f(x + 2\pi) = f(x)$ and

$$f(x) = \exp\left(-x^2\right), \quad -\pi \le x \le \pi.$$

Sketch the graph of $f(x)$ on the interval $-3\pi \le x \le 3\pi$ and state which terms are present in the Fourier series development of $f(x)$ on the interval $-\pi \le x \le \pi$.

Exercise 8.27
Show that Fourier series for the function

$$f(x) = \begin{cases} 0, & -h < x \le 0, \\ e^x - 1, & 0 \le x \le h, \end{cases}$$

on the interval $-h < x \le h$ is

$$f(x) = \frac{e^h - 1}{2h} - \frac{1}{2} + \sum_{k=1}^{\infty} \frac{h(e^h(-1)^k - 1)}{h^2 + \pi^2 k^2} \cos\left(\frac{\pi k x}{h}\right)$$

$$+ \sum_{k=1}^{\infty} \left[\frac{(-1)^k - 1}{\pi k} - \frac{\pi k}{h^2 + \pi^2 k^2} \left(e^h(-1)^k - 1\right) \right] \sin\left(\frac{\pi k x}{h}\right).$$

Exercise 8.28
Investigate the use of the complete system

$$\phi_k(x) = \frac{1}{x + k}, \quad k = 1, 2, 3, \ldots$$

to approximate the function $f(x) = e^{-x}$ on the interval $0 \le x \le 1$.

Show that the minimum of the function

$$F(\mathbf{c}) = \int_0^1 dx \left(\sum_{k=1}^N c_k \phi_k(x) - e^{-x} \right)^2,$$

where $\mathbf{c} = (c_1, c_2, \ldots, c_N)$ are N real variables, is given by the solution of the linear equations

$$\sum_{k=1}^N c_k M_{kj} = r_j,$$

where

$$r_j = \int_0^1 dx \frac{e^{-x}}{x+j} = e^j (E_1(j) - E_1(j+1)),$$

$$M_{kj} = \int_0^1 \frac{dx}{(x+k)(x+j)} = \begin{cases} \dfrac{1}{k-j} \ln\left(\dfrac{k(j+1)}{j(k+1)} \right), & k \neq j, \\[2ex] \dfrac{1}{k(k+1)}, & k = j, \end{cases}$$

where $E_1(x)$ is the exponential integral defined in equation 5.8 (page 182). Use Maple to solve these equations for $\mathbf{c}$ for various values of N and compare, graphically, the resulting approximation with the exact function.

Exercise 8.29
Use the Fourier series of $f(x) = x$ and Parseval's theorem to show that

$$\zeta(2) = \sum_{k=1}^\infty \frac{1}{k^2} = \frac{\pi^2}{6} \quad \text{and hence that} \quad \sum_{k=1}^\infty \frac{(-1)^{k-1}}{k^2} = \frac{\pi^2}{12}.$$

Exercise 8.30
Use the Fourier series for x^2 to show that

$$\zeta(4) = \sum_{k=1}^\infty \frac{1}{k^4} = \frac{\pi^4}{90}, \quad \zeta(6) = \sum_{k=1}^\infty \frac{1}{k^6} = \frac{\pi^6}{945}, \quad \zeta(8) = \sum_{k=1}^\infty \frac{1}{k^8} = \frac{\pi^8}{9450}.$$

From the solution to this exercise you will observe that the process can be continued to find $\zeta(N)$ for any even value of N and that $\zeta(2n) = c_{2n}\pi^{2n}$ for some rational number c_{2n}.

Exercise 8.31
The ordinary Bessel function $J_n(x)$ can be defined by the integral

$$J_n(x) = \frac{1}{2\pi} \int_{-\pi}^\pi d\theta \exp i(n\theta - x\sin\theta).$$

That is, $J_n(x)$ is the nth Fourier coefficient of the periodic function $e^{-ix\sin\theta}$.
  Use this definition to show that

$$J_n(x+y) = \sum_{k=-\infty}^\infty J_k(x) J_{n-k}(y).$$

Exercise 8.32

Use the integral given in the previous question to show that

$$e^{-ix\sin\theta} = \sum_{n=-\infty}^{\infty} J_n(x)e^{-in\theta}.$$

Hence, or otherwise, show that

$$J_0(x) + 2J_2(x) + 2J_4(x) + \cdots = 1 \quad \text{and} \quad J_0(x)^2 + 2\sum_{k=1}^{\infty} J_k(x)^2 = 1.$$

Exercise 8.33

The Legendre polynomials are, to within a multiplicative constant, the bounded solutions of the differential equation

$$\frac{d}{dx}\left((1-x^2)\frac{dP_n}{dx}\right) + n(n+1)P_n = 0, \quad |x| \le 1, \quad n = 0, 1, 2, \dots.$$

The first few of which are

$$P_0(x) = 1, \quad P_1(x) = x, \quad P_2(x) = \frac{1}{2}(3x^2 - 1), \quad P_3(x) = \frac{1}{2}x(5x^2 - 3).$$

Using the differential equation, show that

$$\int_{-1}^{1} dx\, P_n(x)P_m(x) = c_n\delta_{nm},$$

for some constant c_n, and hence that the Legendre polynomials form an orthogonal system on $(-1, 1)$. In fact, Legendre polynomials form a complete system.

Exercise 8.34

Let a and b be any positive numbers, with a/π not an integer, and let λ_j, $j = 0, \pm1, \dots$ be the set of all roots of the equation

$$\tan a\lambda = b\lambda.$$

Show that the functions

$$\phi_k(x) = \sin \lambda_k x, \quad k = 0, \pm1, \dots$$

form an orthogonal system on the interval $(-a, a)$.

Exercise 8.35

The Weierstrass approximation theorem states that on any interval (a, b) the polynomials $f_k(x) = x^k$, $k = 0, 1, \dots$, are complete, but it is clear that this system of functions is not orthogonal. In this exercise we show how a new set of orthogonal polynomials $\phi_k(x)$ can be constructed. In order to be specific we consider only the interval $(-1, 1)$.

On the interval $(-1, 1)$, $f_0 = 1$ and $f_1 = x$ are orthogonal, so set $\phi_0 = f_0$ and $\phi_1 = f_1$ and define

$$\phi_2(x) = x^2 + a_1\phi_1(x) + a_0\phi_0(x).$$

Find values of a_0 and a_1 that make ϕ_2 orthogonal to ϕ_1 and ϕ_0, that is, solve the equations

$$0 = \int_{-1}^{1} dx \, \phi_0(x)\phi_2(x) = \int_{-1}^{1} dx \left(a_0 + a_1 x + x^2\right) = 2a_0 + \frac{2}{3},$$

$$0 = \int_{-1}^{1} dx \, \phi_1(x)\phi_2(x) = \int_{-1}^{1} dx \, x \left(a_0 + a_1 x + x^2\right) = \frac{2}{3}a_1,$$

for a_0 and a_1.

Similarly, find constants b_k, $k = 1$, 2 and 3 such that

$$\phi_3(x) = x^3 + b_2\phi_2(x) + b_1\phi_1(x) + b_0\phi_0(x)$$

is orthogonal to ϕ_k, $k = 1$, 2 and 3.

In general, define

$$\phi_n(x) = x^n + \sum_{k=0}^{n-1} b_{nk}\phi_k(x) \quad \text{and show that} \quad b_{nk} = -\frac{(x^n, \phi_k)}{(\phi_k, \phi_k)}.$$

Deduce that $\phi_{2k}(x)$ is an even function of x and $\phi_{2k+1}(x)$ is odd. Write a Maple program to find $\phi_n(x)$ for any n.

The functions $\phi_k(x)$, $k = 0, 1, 2, \ldots$, are orthogonal polynomials, and $\phi_k(x)$ has degree k and is proportional to the kth Legendre polynomial defined in exercise 8.33. This set of polynomials is also complete.

This procedure is a particular application of the Gram–Schmidt orthogonalisation process whereby from a system of linearly independent vectors v_k, $k = 1, 2, \ldots, N$, one may construct a system of N linearly independent orthogonal vectors, see for instance Courant and Hilbert (1965).

Exercise 8.36
Use the Gram–Schmidt process defined in the previous exercise to find the first ten orthonormal polynomials on the interval $(0, 1)$. The first four are:

$$1, \quad (2x - 1)\sqrt{3}, \quad (6x^2 - 6x + 1)\sqrt{5}, \quad (20x^3 - 30x^2 + 12x - 1)\sqrt{7}.$$

Exercise 8.37
If $f(t)$ is a periodic function with period $2\pi/\Omega$ and with Fourier series

$$f(t) = \sum_{k=-\infty}^{\infty} c_k e^{-ik\Omega t},$$

show that the solution to the differential equation

$$\frac{d^2 y}{dt^2} + \omega^2 y = f(t)$$

can be written in the form

$$y(t) = \frac{c_0}{\omega^2} + ae^{i\omega t} + be^{-i\omega t} - \sum_{\substack{k=-\infty \\ k \neq 0}}^{\infty} \frac{c_k}{(k\Omega)^2 - \omega^2} e^{-ik\Omega t},$$

provided there are no terms in the series with $k\Omega = \pm\omega$, for some integer k.

Exercise 8.38

In chapter 7 it was shown that the sequence

$$\delta_n(x) = \sqrt{\frac{n}{\pi}} e^{-nx^2}$$

tends to the delta function as $n \to \infty$. If $f_n(x)$ is the 2π-periodic function which coincides with $\delta_n(x)$ for $|x| < \pi$, show that for large n

$$f_n(x) = \frac{1}{2\pi} \exp\left(-\frac{1}{4n}\right) + \frac{1}{\pi} \sum_{k=1}^{\infty} \exp\left(-\frac{k^2}{4n}\right) \cos kx$$

and write down the Fourier series of $f_\infty(x)$.

Use Maple to plot the graphs of the partial sums of $f_n(x)$ for various values of n and for $-\pi < x < \pi$.

Exercise 8.39

The function

$$f(x, \epsilon) = \begin{cases} 0, & -\pi < x < -\epsilon, \\ (2\epsilon)^{-1}, & -\epsilon \le x \le \epsilon, \\ 0, & \epsilon < x < \pi \end{cases}$$

is one of the sequences whose limit as $\epsilon \to 0$ is the delta function. Show that the Fourier series of the periodic extension of this function is

$$f(x, \epsilon) = \frac{1}{2\pi} \left(1 + 2 \sum_{k=1}^{\infty} \frac{\sin k\epsilon}{k\epsilon} \cos kx\right),$$

and show that this reduces to the Fourier series of the delta function as $\epsilon \to 0$.

Exercise 8.40

The Fourier series of the function $f(x) = 1/(1 + a\cos x)$, $0 \le a < 1$, is

$$\frac{1}{1 + a\cos x} = \frac{1}{\sqrt{1 - a^2}} \left(1 + 2 \sum_{k=1}^{\infty} (-1)^k \left(\frac{a}{1 + \sqrt{1 - a^2}}\right)^k \cos kx\right).$$

Show that the nth Fourier component decreases to zero exponentially as $n \to \infty$, and explain why the Fourier components of this function decay faster than those of polynomials such as x and x^2.

The function $f(x)$ is singular at $x = \pm\pi$ when $a = 1$, yet, because of the factor $(1 - a^2)^{-1/2}$, the Fourier series appears to tend to infinity for all x when $a = 1$. Examine the behaviour of the Fourier series as $a \to 1$ and resolve this paradox.

Exercise 8.41

It can be proved, Courant and Hilbert (1965, page 102), that the infinite sequence of powers $1, x^{\lambda_1}, x^{\lambda_2}, \ldots$ with positive exponents, λ_k, which approach infinity as $k \to \infty$ is complete in the interval $0 \le x \le 1$ if and only if $\sum_{k=1}^{\infty} 1/\lambda_k$ diverges.

Show that the least squares approximation to the function $f(x) = x^a$, $a > 0$,

$$x^a \simeq f_N(x) = \sum_{k=0}^{N} c_k x^{\lambda_k}, \quad 0 \le x \le 1,$$

is given by those c_k which minimise the function

$$F(\mathbf{c}) = -2 \sum_{k=0}^{N} \frac{c_k}{1 + a + \lambda_k} + \sum_{k=0}^{N} \frac{c_k^2}{1 + 2\lambda_k} + 2 \sum_{k=0}^{N-1} \sum_{j=k+1}^{N} \frac{c_k c_j}{1 + \lambda_k + \lambda_j}.$$

In the case $\lambda_k = \sqrt{1+k}$ and $a = \frac{3}{2}$ use Maple to find these values of $\mathbf{c}$ and plot the graphs of $x^a - f_N(x)$ for $N = 2$ and 3. Also consider other positive values of a.

Exercise 8.42
Use the Fourier series for the function x^2 to show that

$$\sum_{\substack{k = -\infty \\ k \ne 0, n}}^{\infty} \frac{1}{k^2 (n-k)^2} = \frac{2\pi^2}{3n^2} - \frac{6}{n^4}, \quad n \ne 0.$$

Exercise 8.43
A mechanical device that produces an approximate saw-tooth action is shown in the following diagram.

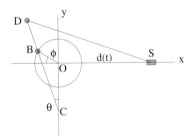

Here the point B moves on the circle of radius a with uniform angular speed ω, so the angle ϕ between OB and the negative x-axis is $\phi = \omega t$. The rod CD is pivoted at C and B and has length $L_2 > L_1 + 2a$, where $L_1 = OC$. The slide S is free to move along the axis Ox and is attached to D by the rod DS of length L, pivoted at D and S.

Show that the coordinates of D in the Oxy-plane are

$$(-L_2 \sin\theta, L_2 \cos\theta - L_1), \quad \text{where} \quad \sin\theta = \frac{a}{\sqrt{a^2 + L_1^2 + 2aL_1 \sin\phi}} \cos\phi.$$

Hence show that the distance OS is

$$d(\phi, a) = \sqrt{L^2 - (L_2 \cos\theta(\phi) - L_1)^2} - L_2 \sin\theta(\phi).$$

Write a procedure to compute $d(\phi, a)$ for $L_1 = 60$, $L_2 = 120$ and $L = 150$. Use this to plot the graph of $d(\phi, a)$ over one period, and show that as $a \to L_1$ the path of S approximates that of a saw-tooth.

Use the numerical method outlined in section 8.7, equation 8.25, to compute the Fourier components of $d(\phi)$ for particular values of a. Show that for $a = 10$ the motion is dominated by one harmonic,

$$d(\phi) \simeq 137.8 - 19.9 \sin \phi - 0.350 \cos 2\phi + 1.66 \sin 2\phi - 0.207 \sin 3\phi,$$

and also that as a increases, all other parameters being fixed, the number of harmonics needed to faithfully represent the motion increases.

9

Perturbation theory

9.1 Introduction

Perturbation theory is a collection of methods for obtaining approximate solutions to equations, involving a small parameter, which we normally denote by ϵ or δ; the equations may be of any kind, algebraic, differential or integral for instance, but often the success, or otherwise, of the application depends upon the ingenuity applied to the problem — a judicious change of variable can turn a seemingly intractable problem into one that is more manageable.

In this chapter we provide an introduction to the general principles of perturbation theory by applying the method to simple cases. There are applications to many other types of problem than cannot be mentioned here, but in most cases the applications are non-trivial and require knowledge of the system being studied so cannot be included in an introduction. Two such applications are described in chapter 17, where we show how to approximate periodic solutions of some types of differential equations; other applications will be described in later chapters.

The main assumption underlying perturbation theory is that the solution is a well-behaved function of ϵ, the small parameter. A *regular* perturbation problem is defined as one whose perturbation series is a power series in ϵ with a non-vanishing radius of convergence; for these problems the perturbed solution smoothly approaches the solution of the unperturbed equation — that is, the equation with $\epsilon = 0$ — as $\epsilon \to 0$.

A *singular* perturbation problem is one whose perturbation series either does not take the form of a power series, or if it does the power series has a vanishing radius of convergence; for example it could be an asymptotic expansion. In these problems the equation normally has a different character when $\epsilon = 0$ than when $\epsilon \neq 0$. Such problems are very common and we shall meet them in several different guises; an example is given in the introduction to chapter 5. It is surprising that even in these unpromising circumstances perturbation expansions are still useful.

Perturbation theory was developed long before computers existed; now that computers can be routinely used to easily solve a wide variety of numerical problems it is necessary to explain why the older methods are still useful. First, numerical solutions are not always required, nor are they always the most useful form of approximation; this is particularly true if the equations to be solved contain many parameters and we require the solution as a function of these parameters. Second, it is sometimes more convenient to have an approximate solution in terms of a formula than to have an accurate numerical result. Finally, with symbolic processing it is often possible to find

301

perturbation expansions to much higher order, and hence accuracy, than is possible by hand, thus increasing their utility.

In practice perturbation and numerical methods are complementary; for instance we frequently use perturbation theory to check numerical programs or to provide a first guess for an iterative method, and numerical solutions may be used to check for algebraic or programming errors.

In general, perturbation methods are algebraically messy and it is unusual for a perturbation series of a real problem to be computed to more than $O(\epsilon^2)$, simply because the algebra becomes too unwieldy. The use of computer assisted algebra to some extent removes this block and sometimes allows many higher terms to be computed; we shall illustrate how this can be done in the next few sections and chapters 13 and 14. However, be warned; computer assisted algebra is not a substitute for thought and in using these tools it is still necessary to understand the mathematical processes involved: doing algebra without understanding can produce all sorts of silly solutions.

There are many good books dealing exclusively with perturbation methods in general and applied to particular types of problems. One of the most readable and interesting elementary texts is Hinch (1991); more advanced books are Holmes (1995) and Bender and Orszag (1978, Part III), the latter being well worth reading: more specialised texts are Nayfeh (1973), Nayfeh and Mook (1979) and Giacaglia (1972).

9.2 Perturbations of algebraic equations

9.2.1 Regular perturbations

We illustrate perturbation theory by applying it to particular, simple problems that can be solved exactly. We then consider more complex problems, not having exact solutions, and show how Maple can be used to considerably ease the algebraic effort and find more accurate solutions.

In general we wish to solve equations of the form $f(x, \epsilon) = 0$, where f is some function of x and the small parameter ϵ: the function $f(x, \epsilon)$ may also depend upon other parameters. It is assumed that the *unperturbed equation*, $f(x, 0) = 0$, obtained by setting $\epsilon = 0$, has a known *unperturbed solution*, x_0, and that the solution of the perturbed equation, $f(x, \epsilon) = 0$, is close to this solution and can be expanded as a power series in ϵ, or possibly some function of ϵ, such as $\epsilon^{\frac{1}{2}}$.

We begin by finding an approximation to the roots of the simple quadratic equation

$$x^2 - ax + \epsilon = 0, \quad |\epsilon| \ll 1, \tag{9.1}$$

as a function of the small parameter ϵ and the constant a. This simple example provides an illustration of all the salient ideas but also allows us to compare our approximate solutions with the exact solutions. An alternative, but equivalent, approach is discussed in exercise 9.47 (page 336).

The *unperturbed equation* is obtained by putting $\epsilon = 0$,

$$x^2 - ax = 0, \tag{9.2}$$

and this has roots $x = x_0$ with $x_0 = 0$, a. When $|\epsilon| \ll 1$ we should expect the roots of equation 9.1 to be close to these unperturbed roots and express this assumption by writing them as a power series in ϵ,

$$x = x_0 + \epsilon x_1 + \epsilon^2 x_2 + \cdots, \tag{9.3}$$

where $x_0 = 0$ or a but x_k, $k = 1, 2, \ldots$, are unknown coefficients, to be determined: these depend upon the value taken by x_0, but are independent of ϵ. In many cases this perturbation series will have a finite radius of convergence, but it may also be an asymptotic expansion; it is normally difficult to determine a priori the nature of this series, especially when the coefficients x_k depend upon parameters. Nevertheless, we assume that the series is sufficiently well behaved that all the operations subsequently carried out on it are allowed.

In order to find the coefficients, substitute the series 9.3 into the original equation 9.1 and collect all terms of each power of ϵ. In this example terms up to $O(\epsilon^2)$ are included, so the original equation becomes

$$\left(x_0 + \epsilon x_1 + \epsilon^2 x_2\right)^2 - a\left(x_0 + \epsilon x_1 + \epsilon^2 x_2\right) + \epsilon + O(\epsilon^3) = 0, \tag{9.4}$$

which can be rearranged to give

$$x_0^2 - ax_0 + \epsilon\left(2x_0 x_1 - ax_1 + 1\right) + \epsilon^2\left(2x_0 x_2 + x_1^2 - ax_2\right) + O(\epsilon^3) = 0. \tag{9.5}$$

Now comes a crucial step. The parameter ϵ is considered to be a *variable*, not a fixed constant. Therefore equation 9.5 is valid for a range of values† of ϵ: this means that *all* the coefficients of ϵ^k, $k = 0, 1, \ldots$, must be identically zero. Thus equation 9.5 leads to the three equations

$$x_0^2 - ax_0 = 0, \quad \text{(zero order, } \epsilon^0\text{)}, \tag{9.6}$$
$$2x_0 x_1 - ax_1 + 1 = 0, \quad \text{(first order, } \epsilon\text{)},$$
$$2x_0 x_2 + x_1^2 - ax_2 = 0, \quad \text{(second order, } \epsilon^2\text{)},$$

which can be solved successively for x_0, x_1 and x_2. The first of these, equation 9.6, is just the unperturbed equation, which is a useful check upon this part of the analysis. From the other two equations we obtain

$$x_1 = \frac{1}{a - 2x_0}, \quad \text{and} \quad x_2 = \frac{x_1^2}{a - 2x_0} = \frac{1}{(a - 2x_0)^3}. \tag{9.7}$$

The zero-order equation has the two solutions $x_0 = 0$ and a, and on substituting these into equation 9.7 we obtain the approximate solutions

$$x = \frac{\epsilon}{a} + \frac{\epsilon^2}{a^3} + O(\epsilon^3), \quad x = a - \frac{\epsilon}{a} - \frac{\epsilon^2}{a^3} + O(\epsilon^3). \tag{9.8}$$

The application of perturbation theory to this problem is harder than finding the exact solutions, which are just

$$x = \frac{a}{2}\left(1 \pm \sqrt{1 - 4\epsilon/a^2}\right). \tag{9.9}$$

---

† In some cases the series will be an asymptotic expansion, with zero radius of convergence. The theory discussed in chapter 5 shows that the following steps are still admissible.

Expanding the square root as a Taylor's series gives

$$x = \frac{a}{2} \pm \left(\frac{a}{2} - \frac{\epsilon}{a} - \frac{\epsilon^2}{a^3} + \cdots \right),$$

agreeing with perturbation theory.

There are five stages involved in this type of analysis.

S 1: Identification of a small parameter ϵ — easy in this case but not always obvious. Further, there is often no unique choice for this parameter, with different choices giving series formally the same but numerically different when truncated, see for instance equations 9.11 and 9.12.

S 2: Solving the unperturbed equation, that is, the $\epsilon = 0$ limit. If these solutions cannot be found then it is necessary to reformulate the problem.

S 3: Expansion of the solution as a power series in ϵ.

S 4: Substitution of this expansion into the equation and construction of a set of equations, one for each coefficient of ϵ^k, $k = 1, 2, \ldots, N$ for some suitable N, which depends upon the accuracy required.

S 5: Solution of these N equations.

For more difficult problems, each of these stages can be difficult; the art of using perturbation theory is to find an unperturbed equation with a solution close to the exact solution and for which the successive equations in stage 4 can be solved.

There are no general rules which make the application of these methods automatic, but problems can often be simplified by changing variables. For instance, in the simple example treated above, if we define a new variable w by $x = \beta w$, where β is some constant, the original equation 9.1 becomes, after some rearrangement,

$$w^2 - \frac{a}{\beta} w + \frac{\epsilon}{\beta^2} = 0,$$

so on putting $\beta = a$, $\delta = \epsilon/a^2$, the new equation is

$$w^2 - w + \delta = 0, \quad x = aw,$$

which has only one small parameter, δ. In this case such a rescaling merely simplifies the analysis by decreasing the number of variables, which is always useful; in other cases more profound simplifications can be made.

Changing variables may also give different expansions: formally these are just different representations of the same function but will be different when the series are truncated, and this may lead to improved accuracy. For instance, the equation

$$f(x) = x^3 - (1 + \epsilon)x - \epsilon = 0, \quad 0 \le \epsilon \ll 1, \tag{9.10}$$

has roots near $x = 0$ and ± 1. The root near $x = 1$ has the expansion

$$x = r_1 = 1 + \epsilon - \epsilon^2 + O(\epsilon^3). \tag{9.11}$$

The substitution $y = x/\sqrt{1 + \epsilon}$ transforms the equation to

$$y^3 - y = \delta = \frac{\epsilon}{(1 + \epsilon)^{3/2}},$$

and the expansion of the root of this equation near $y \simeq x = 1$ gives

$$x = \sqrt{1+\epsilon}\left(1 + \frac{\epsilon}{2(1+\epsilon)^{3/2}} - \frac{3\epsilon^2}{8(1+\epsilon)^3} + \cdots\right). \tag{9.12}$$

The infinite series 9.11 and 9.12 are different representations of the same function, but when truncated are different functions. If $\epsilon = 0.1$ the exact root is, to six decimal places, $x = 1.091\,608$. Using the series to the order quoted, the ϵ-expansion, 9.11, gives $x = 1.09$, a relative error of 0.15%, and $f(1.09) = -0.004$; whereas the δ-expansion, 9.12, gives $x = 1.091\,31$, with 0.03% error, and $f(1.091\,31) = -0.000\,74$.

Exercise 9.1

Find a perturbation expansion for the real root of

$$x^3 - 2\epsilon x - 1 = 0, \quad |\epsilon| \ll 1,$$

up to and including terms $O(\epsilon^2)$.

Exercise 9.2

Introduce a small parameter into the equation

$$x^{1/3} - 4.001x + 0.003 = 0,$$

and hence find the root near $1/8$ with an error less than 10^{-6}.

9.2.2 Singular perturbations

In all the above examples the addition of the small perturbation did not significantly alter the equation or the form of the solution. We name this a *regular* perturbation, and usually the solution can be expressed as a power series in ϵ having a non-vanishing radius of convergence: the exact solution smoothly approaches the unperturbed solution as $\epsilon \to 0$.

In many interesting cases, however, the 'small' perturbation may change the nature of the solution. For instance, the quadratic equation

$$\epsilon x^2 + x - 1 = 0 \tag{9.13}$$

has one solution, $x = 1$, when $\epsilon = 0$ but two solutions when $\epsilon \neq 0$. This is an example of a *singular* perturbation in which the solutions of the equation have a different character when $\epsilon = 0$ than when $\epsilon \neq 0$.

For this example, if $\epsilon \neq 0$ the product of the roots is $-1/\epsilon$, and if $0 < |\epsilon| \ll 1$ one root is near 1 so the other must be approximately $-1/\epsilon$.

The two roots are given by the usual expression,

$$x_1(\epsilon) = \frac{2}{1 + \sqrt{1+4\epsilon}} = 1 - \epsilon + 2\epsilon^2 - 5\epsilon^3 + 14\epsilon^4 + O(\epsilon^5),$$

$$x_2(\epsilon) = -\left(\frac{1 + \sqrt{1+4\epsilon}}{2\epsilon}\right) = -\frac{1}{\epsilon} - 1 + \epsilon - 2\epsilon^2 + 5\epsilon^3 + O(\epsilon^4),$$

where the expansions converge for $|\epsilon| < 1/4$. The first root can be obtained from a direct application of perturbation theory and $x_1(\epsilon) \to 1$ as $\epsilon \to 0$. The second root is $O(1/\epsilon)$, and tends to infinity as $\epsilon \to 0$ and is not a root of the unperturbed equation and cannot therefore be given by a direct application of perturbation theory.

This example illustrates one aspect of singular perturbations, but it also shows that the distinction between regular and singular perturbations is not as clear cut as our definitions suggest. For, in this case, we can find a new variable in which the perturbation problem becomes regular, though we should emphasise that such tricks are not normally possible. If $x \sim \epsilon^{-1}$ both the ϵx^2 and the x terms of equation 9.13 are $O(\epsilon^{-1})$, suggesting that we define a new variable $y = \epsilon x$ to obtain the equation

$$y^2 + y - \epsilon = 0, \tag{9.14}$$

the solutions of which can be obtained using regular perturbation theory. Thus by rescaling the variable a singular perturbation has been converted into a regular perturbation. It must be emphasised that this is rarely possible.

Exercise 9.3
Find the first three terms of the series solution for all three roots of the equation

$$\epsilon^2 x^3 + 2x^2 + x + \epsilon = 0.$$

Hint: when $\epsilon = 0$ the roots are $x = 0$ and $-1/2$ and series expansions of the roots that reduce to these may be found. There are many methods of finding the third root, but one is to note that the sum of all roots is $-2\epsilon^{-2}$.

Exercise 9.4
Find the perturbation series for both solutions of equation 9.14, and hence 9.13, correct to $O(\epsilon^2)$.

Another simple example is provided by the apparently innocent equation

$$f(x) = (1 - \epsilon)x^2 - 2x + 1 = 0. \tag{9.15}$$

By re-writing this in the form $(1 - x)^2 = \epsilon x^2$ and examining the graphs of the functions $y = (1 - x)^2$ and $y = \epsilon x^2$, shown in figure 9.1, we see immediately that there are two real roots for $\epsilon > 0$, one for $\epsilon = 0$ and none for $\epsilon < 0$, so a power series expansion in ϵ cannot exist. This type of behaviour is quite common and is caused by the two roots of the quadratic equation being complex for $\epsilon < 0$, real and different for $\epsilon > 0$ and real and identical for $\epsilon = 0$.

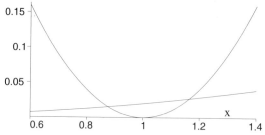

Figure 9.1   Graphs of the functions $y = (1 - x)^2$ and ϵx^2 near $x = 1$ in the case $\epsilon = 0.02$.

In this case the exact solutions, for $\epsilon \geq 0$, are easily obtained by taking the square root of the rearranged equation,

$$x = \frac{1}{1 \mp \sqrt{\epsilon}} = \frac{1 \pm \sqrt{\epsilon}}{1 - \epsilon} = \begin{cases} 1 + \epsilon^{1/2} + \epsilon + \epsilon^{3/2} + O(\epsilon^2), \\ 1 - \epsilon^{1/2} + \epsilon - \epsilon^{3/2} + O(\epsilon^2). \end{cases}$$

Thus for this problem the solution can be expressed as a power series in $\epsilon^{1/2}$ rather than ϵ.

This is all very well, but because the result was derived from the exact solution it is of little value. In general it is necessary to guess the correct expansion parameter, and there are no simple rules applicable to all cases, though the iterative method described in the next section can often be used to suggest the correct expansion parameter.

In this instance the graph in figure 9.1 shows that, for $\epsilon > 0$, the root is near $x = 1$ where $|1 - x|$ is relatively small; thus we can approximate the original equation by $1 - x = \pm x\sqrt{\epsilon} \simeq \sqrt{\epsilon}$ (because $x \simeq 1$) which gives $x \simeq 1 \pm \sqrt{\epsilon}$ as a first approximation, suggesting that $\epsilon^{1/2}$ is the correct expansion parameter.

An alternative, more algebraic, approach is to assume that the root has the expansion

$$x = 1 + x_1 \epsilon^{\alpha}, \tag{9.16}$$

for some numbers x_1 and $0 < \alpha < 1$, and attempt to find x_1 and α by the usual method. On substituting this expansion into equation 9.15 we obtain

$$f(x) = -\epsilon + x_1^2 \epsilon^{2\alpha} - 2x_1 \epsilon^{1+\alpha} - x_1^2 \epsilon^{1+2\alpha} + \text{higher order terms} = 0.$$

We now have to chose α and x_1 to make the leading term of this equation vanish. In other words we need to make one of the terms $x_1^2 \epsilon^{2\alpha}$, $2x_1 \epsilon^{1+\alpha}$ and $x_1^2 \epsilon^{1+2\alpha}$ cancel the $-\epsilon$ term in such a manner that the other two terms are $O(\epsilon^{1+\eta})$ for some $\eta > 0$. In this case it is clear that the only choice is $\alpha = \frac{1}{2}$ and $x_1 = \pm 1$, to give

$$x = 1 \pm \epsilon^{1/2} + O(\epsilon).$$

Exercise 9.5
Assume a series solution of the form

$$x = 1 + x_1 \epsilon^{1/2} + x_2 \epsilon + O(\epsilon^{3/2})$$

to find a perturbation solution for equation 9.15.

Exercise 9.6
By sketching graphs of the left and right hand sides of the equation

$$1 - \epsilon x = \sin x, \quad 0 < \epsilon \ll 1,$$

show that there are two roots near $x = \frac{\pi}{2}$ and that they are given by the expansions

$$x = \frac{\pi}{2} \pm \sqrt{\pi \epsilon} + \epsilon \pm \frac{\pi^2 + 12}{24\sqrt{\pi}} \epsilon^{3/2} + O(\epsilon^2).$$

Exercise 9.7

Show that the equation

$$1 + (x^2 + \epsilon^2)^{1/2} = e^x, \quad 0 < \epsilon \ll 1,$$

has a real root near $x = 0$. In addition show that the leading term in the perturbation expansion is $x = \epsilon^{2/3}$ and that a better approximation is given by

$$x \simeq \frac{\epsilon^{2/3}}{(1 + \frac{7}{12}\epsilon^{2/3})^{1/3}}.$$

Hint: this equation is tricky: first note that if $\epsilon = 0$ it becomes $1 + |x| = e^x$, having $x = 0$ as a root. When $\epsilon > 0$ expanding the original equation requires care because it is not known a priori whether $x > \epsilon$ or $x < \epsilon$, so first write the equation in the form $x^2 + \epsilon^2 = (\exp(x) - 1)^2$, expand this in powers of x and then use an iterative method as discussed in section 9.3.

9.2.3 Use of Maple

We now consider more complicated algebraic equations which do not, in general, have solutions that can be expressed in terms of a simple finite formula. We shall use regular perturbation theory to show that a unique series expansion can, in principle, be constructed and shall use Maple to construct it.

Consider the general equation

$$f(x) = \epsilon g(x), \tag{9.17}$$

for the arbitrary, sufficiently well behaved, functions f and g; we assume that

(a) the equation $f(x) = 0$ has a simple root at $x = 0$, so $f(0) = 0$ and $f'(0) \neq 0$,
(b) $g(0) \neq 0$, and
(c) both $f(x)$ and $g(x)$ have Taylor expansions about the origin with a finite radius of convergence.

If the root of $f(x) = 0$ is at $x = a \neq 0$, then we simply define a new variable $y = x - a$ and replace the original equation by $F(y) = \epsilon G(y)$, where $F(y) = f(y + a)$ and $G(y) = g(y + a)$.

We seek an approximation to the root $r(\epsilon)$ of equation 9.17, which reduces to zero at $\epsilon = 0$, in the form of the series expansion

$$r(\epsilon) = \epsilon r_1 + \epsilon^2 r_2 + \cdots + \epsilon^n r_n + \cdots, \tag{9.18}$$

and where the parameter r_n is determined from the previous $n - 1$ parameters $(r_1, r_2, \ldots, r_{n-1})$ and where $r_1 = g(0)/f'(0)$.

Substituting this expression into equation 9.17 and making a Taylor's expansion about the origin gives

$$\sum_{p=1}^{\infty} \frac{\epsilon^p}{p!}(r_1 + \epsilon r_2 + \cdots)^p f^{(p)}(0) = \sum_{p=0}^{\infty} \frac{\epsilon^{p+1}}{p!}(r_1 + \epsilon r_2 + \cdots)^p g^{(p)}(0). \tag{9.19}$$

The first two terms of this equation are

$$\epsilon \left[r_1 f'(0) - g(0) \right] + \epsilon^2 \left[r_2 f'(0) + \frac{1}{2} r_1^2 f''(0) - r_1 g'(0) \right] + O(\epsilon^3) = 0,$$

and on equating the coefficients of ϵ and ϵ^2 to zero we obtain

$$r_1 = \frac{g(0)}{f'(0)} \quad \text{and} \quad r_2 = \frac{g(0)}{2f'(0)^3} \left[2g'(0)f'(0) - g(0)f''(0) \right].$$

These computations are readily extended to higher orders using Maple.

Inspection of equation 9.19 shows that to order ϵ^N we may write it in the form

$$\epsilon c_1(r_1) + \epsilon^2 c_2(r_1, r_2) + \cdots + \epsilon^N c_N(r_1, \ldots, r_N) = 0, \tag{9.20}$$

where the coefficient c_n is a function of $(r_1, r_2, \ldots, r_n)$, but does not depend upon the coefficients r_k, $k \geq n+1$. This is the case because r_k always appears on the left of equation 9.19 in the combination $(\epsilon^k r_k)^m$ and on the right in the combination $\epsilon(\epsilon^k r_k)^m$ for $m = 1, 2, \ldots$. This also means that the coefficient c_n can be written in the simpler form

$$c_n = A_n r_n - \bar{c}_n(r_1, \ldots, r_{n-1}) \tag{9.21}$$

for some constant A_n and function $\bar{c}_n$ which does not depend upon r_n. As before we use the fact that equation 9.20 holds for a range of values of ϵ, and therefore the coefficients of ϵ^k, $k = 1, 2, \ldots, N$ must all individually be zero and hence, since r_1 is known, we can find r_n, $n = 2, 3, \ldots$ using equation 9.21.

This analysis shows that it is possible to find a unique expansion of the root of $f(x) = \epsilon g(x)$ near a simple root of $f(x) = 0$. But in practice only the first few terms of the series can be found 'by hand' as the amount of algebra quickly becomes unwieldy. With Maple, however, it is a relatively trivial matter to determine the perturbation expansion to a very high order, the upper limit of which is determined mainly by the computer memory available.

There are two ways of tackling this problem using Maple and we illustrate both. The first is essentially an implementation of the above analysis; the second uses the ability of Maple to invert series, which is easier to use, but less general; for instance the first method can be generalised to solve coupled equations.

We illustrate these methods with a simple example: consider the equation

$$\sin x = \epsilon e^{-x}, \quad x > 0, \quad 0 \leq \epsilon \ll 1. \tag{9.22}$$

From the sketch of the graphs of $y = \sin x$ and $y = \epsilon e^{-x}$ shown in figure 9.2 it is seen that this equation has an infinite number of positive roots near $x = n\pi$, $n = 0, 1, \ldots$.

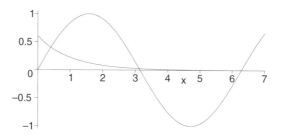

Figure 9.2   Graphs of the functions $y = \sin x$ and $y = \epsilon e^{-x}$ for
$\epsilon = 0.6$.

In order to find the root near $x = \pi$, change variables by defining $y = \pi - x$, so that $|y|$ is small; since $\sin x = \sin y$ the equation satisfied by y is

$$\sin y = \delta e^{y}, \quad \delta = \epsilon e^{-\pi}, \tag{9.23}$$

and the new expansion parameter is δ. When $\delta = 0$ the equation becomes $\sin y = 0$ and the relevant solution is $y = 0$, therefore we assume that the solution can be represented by the series

$$y = \delta r_1 + \delta^2 r_2 + \delta^3 r_3 + \cdots. \tag{9.24}$$

The value of r_1 can be found by inspecting equation 9.23, or equivalently by using the relation $r_1 = g(0)/f'(0)$ derived previously. Since $|y| \ll 1$ we may replace $\sin y$ by y and e^y by 1 to give $y \simeq \delta$ and hence $r_1 = 1$. The remaining coefficients we determine using the following Maple procedure.

```
>   ser:=proc(n::integer) local k,y,ys,c,r;
>   if n <= 1 then RETURN else
>                y:=d+add(r[k]*d^k,k=2..n) fi;           #1
>   ys:=series(sin(y)-d*exp(y),d=0,n+1);                 #2
>   c:=[seq(coeff(ys,d,k),k=1..n)];                      #3
>       for k from 2 to n do;
>       r[k]:=solve(c[k]=0,r[k]);                        #4
>       od;
>   eval(y); end;
```

(1) defines the series of equation 9.18, with $r_1 = 1$, after first ensuring that n is large enough.

(2) expresses $\sin y - \delta e^y$ as a power series in δ. The Maple command `taylor` could also be used here.

(3) puts the coefficient of δ^k into the list element `c[k]`; we know that this depends only upon `r[q]`, q = 1, 2,...,k, in the form of equation 9.21 and hence can be solved to give a unique value of `r[k]`.

(4) solves equation 9.21, and the following line outputs the series solution of the equation. Note that the use of `eval` is essential in the next line; without it the procedure simply outputs the expression for `y` defined in line 1, without substituting for the values of `r[k]`.

If you cannot see how this procedure works, type in the individual commands starting from line 2, after first assigning n a suitable value, say n:=3, and inspect the output of each line.

For the case $n = 10$ we obtain, by typing y:=ser(10);

$$y = \delta + \delta^2 + \frac{5}{3}\delta^3 + \frac{10}{3}\delta^4 + \frac{221}{30}\delta^5 + \frac{52}{3}\delta^6 + \frac{5\,365}{126}\delta^7$$
$$+ \frac{6\,800}{63}\delta^8 + \frac{141\,245}{504}\delta^9 + \frac{2\,102\,152}{2\,835}\delta^{10}. \tag{9.25}$$

This procedure will find series with more terms; on a Pentium 133MHz with 32 Mb of RAM the procedure works with $n = 15$, but with $n = 18$ it starts to swap in and out of the disk and becomes interminably slow. Nevertheless, this method is useful because it sometimes works in circumstances where the simpler method described below cannot be used, as in exercise 9.18.

The alternative method requires fewer commands and will, in practice, provide more terms in the series. It uses the Taylor expansion of the expression $\sin y - \delta e^y$ about $y = 0$ and then assumes that the equation

$$\sin y - \delta e^y = 0 = -\delta + (1 - \delta)y - \frac{1}{2}\delta y^2 + O(y^3) \tag{9.26}$$

can be solved by expressing y as a power series in δ. This inversion is accomplished using the command solve, as described in chapter 2.

A Maple procedure which performs these operations up to and including terms of $O(n)$ is given below:

```
>   ser2:=proc(n::integer) local eq,y,Old_order,u;
>   Old_order:=Order; Order:=n+1;                          #1
>   eq:=0=series(sin(y)-d*exp(y),y=0,Order);               #2
>   u:=sort(  convert(solve(eq,y),polynom)  );             #3
>   Order:=Old_order; u;                                   #4
>   end:                                                   #4
```

(1) redefines the global Order variable, keeping the old value; this is necessary to invert the series – line 3 – to the same order as the original series. The Order variable is reset in line 4.

(2) defines the series 9.26 up to and including terms $O(\delta^n)$.

(3) compress three commands into one. The inner command, solve, inverts the series obtained in line 2; the middle command converts this to a polynomial and the outer command, sort, arranges the powers in order.

Thus the fifth-order perturbation expansion is given by the command

```
>   y:=ser2(5);
```

$$y := \frac{221}{30}d^5 + \frac{10}{3}d^4 + \frac{5}{3}d^3 + d^2 + d.$$

This method can be used for larger values of n; the same computer as described above has no problem in finding the 30th-order expansion.

Exercise 9.8

Show that the perturbation expansion of the root of the equation 9.22 near $x = n\pi$, for some positive integer n, is given by equation 9.25 but with $\delta = (-1)^{n-1}\epsilon e^{-n\pi}$.

Exercise 9.9

Show that the equation

$$\tan x = \frac{\epsilon}{1 + ax}, \quad 0 \le \epsilon \ll 1, \quad a > 0,$$

has positive roots near $x = n\pi$, $n = 0, 1, \ldots$, and that the series expansion of the root of the equation near $x = 0$ is

$$x = \epsilon - a\epsilon^2 + \left(2a^2 - \frac{1}{3}\right)\epsilon^3 + a\left(\frac{4}{3} - 5a^2\right)\epsilon^4 + O(\epsilon^5).$$

In addition show that the root near 2π is

$$x = 2\pi + \frac{\epsilon}{2\pi a + 1} - \frac{\epsilon^2}{(2\pi a + 1)^3} + O(\epsilon^3),$$

and continue this expansion to $O(\epsilon^6)$.

Finding a series approximation is only part of the solution: we also need to understand how this series behaves, so in the following exercise you will explore the behaviour of this series and examine how its nature reflects the behaviour of the original equation, $\sin y = \delta e^y$.

Exercise 9.10

(i) Use one of the procedures given above to determine the ratio r_{k-1}/r_k, $k = 2, 3, \ldots$, where r_k is the coefficient of δ^k in the series solution of $\sin y = \delta e^y$.

(ii) Use this data to estimate δ_c, the radius of convergence of the series, and use Richardson's extrapolation, described in chapter 4, to show that $\delta_c \simeq 0.3224$. Remember that if you use the first procedure you will not be able to use values of n much in excess of 15. Using the theory of section 4.5, show that for $\delta \simeq \delta_c$, $y = A + a\sqrt{\delta_c - \delta}$ for some constants a and A.

(iii) By examining the graphs of the functions $w = \sin y$ and $w = \delta e^y$ for $\delta \simeq \delta_c$ and $0 < y < \pi$, explain why the series fails to converge at $\delta = \delta_c$ and show that the values of δ_c is given by the solution of the simultaneous equations

$$\delta = e^{-y}\sin y, \quad \delta = e^{-y}\cos y,$$

that is, $\delta = e^{-\pi/4}/\sqrt{2}$.

Exercise 9.11

This exercise is a continuation of the last. By expanding about the critical point, $\delta = \delta_c$, show that near here an approximation to the root is

$$y = \frac{\pi}{4} - \sqrt{\epsilon}\left(1 - \frac{\sqrt{\epsilon}}{3} + \frac{7\epsilon}{36}\right) + O(\epsilon^2), \quad \epsilon = 1 - \frac{\delta}{\delta_c}, \quad \delta_c = \frac{e^{-\pi/4}}{\sqrt{2}}.$$

Exercise 9.12

Use perturbation theory to show that the real root of

$$e^{ax} \ln x = \epsilon, \quad a > 0, \quad 0 \le \epsilon \ll 1,$$

is

$$x = 1 + \delta + \left(\frac{1}{2} - a\right)\delta^2 + O(\delta^3), \quad \delta = \epsilon e^{-a}.$$

For the case $a = 1$ use Maple to find $x_s(\delta)$, the series approximate root to $O(\delta^{12})$, and convert this to the $[6/6]$ Padé approximant, $x_p(\delta)$. Plot the graphs of

$$y(\delta) = \log\left(\left|e^{x(\delta)} \ln x(\delta) - \delta\right| + 10^{-D}\right),$$

where D is the value of `Digits` and $x(\delta) = x_p$ and x_s, for $0 < \delta \le 5$. Hence show that over most of this range of δ the Padé approximant is a far better approximation than the original series.

Exercise 9.13

Compute the perturbation series approximation to the root of $\sin y = \delta e^y$ up to and including terms of $O(\delta^{10})$, and from this series derive the $[5/5]$ Padé approximant. If these two expressions are denoted by $y_s(\delta)$ and $y_p(\delta)$, compute the two functions

$$f_s(\delta) = \log\left(|\sin y_s - \delta e^{y_s}| + 10^{-D}\right), \quad f_p(\delta) = \log\left(|\sin y_p - \delta e^{y_p}| + 10^{-D}\right),$$

where D is the values of `Digits`, and compare them graphically for $0 \le \delta < \delta_c$.

These results, and those of the previous exercise, show that, in some circumstances, the accuracy of a perturbation expansion can be improved by using Padé approximants.

Exercise 9.14

Consider the simultaneous equations

$$x^2 + 2y^2 = 3 + \epsilon \cos(\pi x y),$$
$$xy = 1 + \epsilon \sin\left(\frac{\pi(2y + ax)}{2 + a}\right).$$

Show that when $\epsilon = 0$ there are four real roots and that one of these is at $x = y = 1$, which has the expansion

$$x = 1 + \frac{\epsilon}{2} + \epsilon^2\left(\frac{7}{8} - \pi + \frac{4\pi}{2 + a}\right) + O(\epsilon^3)$$

$$y = 1 - \frac{\epsilon}{2} - \epsilon^2\left(\frac{5}{8} - \frac{\pi}{2} + \frac{2\pi}{a + 2}\right) + O(\epsilon^3),$$

and extend this series to $O(\epsilon^4)$.

9.2.4 *A cautionary tale: Wilkinson's polynomial*

No discussion of perturbation theory is complete without a discussion of the example described by Wilkinson, and now named Wilkinson's polynomial. This concerns the behaviour of the roots of high degree polynomials, and in particular the 20 degree polynomial

$$p(x, \epsilon) \;=\; \prod_{k=1}^{20}(x - k) + \epsilon x^{19} \tag{9.27}$$

$$=\; x^{20} - (210 - \epsilon)x^{19} + 26\,615x^{18} - 1\,256\,850x^{17} + \cdots + 20!\,.$$

This is a regular perturbation problem because the perturbation, ϵx^{19}, does not alter the degree of the polynomial which has 20 roots for all ϵ. When $\epsilon = 0$ these roots are $(1, 2, \ldots, 20)$; we wish to understand how these change as ϵ increases.

The perturbation affects only the coefficient of x^{19}, and if $\epsilon = 10^{-9}$ the relative change in the magnitude of this coefficient is $10^{-9}/210 \simeq 4.8 \times 10^{-12}$; this is exceedingly small, so we might be justified in thinking that the roots are affected little. The following results belie this. In figure 9.3 we show a diagram of the position of the roots of $p(x, \epsilon) = 0$ in the complex plane for $\epsilon = 10^{-9}$ (the ellipses), for $\epsilon = 10^{-9}/3$ (boxes) and for $\epsilon = 10^{-9}/5$ (diamonds).

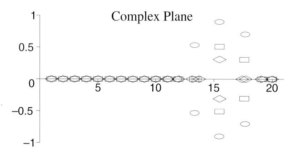

Figure 9.3   Position of the roots in the complex plane of equation 9.27 when $\epsilon = 10^{-9}$, $10^{-9}/3$ and $10^{-9}/5$. See text for the legend.

This diagram shows that for these very small values of ϵ the smallest and the largest roots are hardly affected, but that for $\epsilon = 10^{-9}$ six roots have moved a significant way into the complex plane. Given the small relative change in the magnitude of the coefficient of x^{19} this is surprising. But, first-order perturbation theory provides a fairly simple explanation of this behaviour by giving an estimate of the critical value of ϵ at which pairs of roots coalesce.

Exercise 9.15
Show that the first-order approximation to the root x_k of $p(x, \epsilon) = 0$ is

$$x_k \;=\; k - \epsilon k^{19}\left[\prod_{\substack{r=1 \\ r \neq k}}^{20}(k - r) \right]^{-1}, \quad k = 1, 2, \ldots, 20,$$

$$=\; k - \frac{(-1)^k k^{19}\epsilon}{(k-1)!(20-k)!}.$$

Hence show that the pairs of roots $(1, 2), (3, 4), \ldots, (19, 20)$ approach each other as ϵ increases from zero.

Plot the graph of the magnitude of the perturbation $|x_k - k|$ and show that it is largest for $k = 16$, and using this first-order approximation show that the value of ϵ at which the unperturbed roots $x = 15$ and 16 coincide is $\epsilon \simeq 0.22 \times 10^{-9}$. Use `fsolve` to check this estimate.

This exercise shows that as ϵ increases from zero to 10^{-9} pairs of roots approach each other, eventually coalesce and move into the complex plane as complex conjugate pairs. The example illustrates two important points. First, that the roots of high degree polynomials may be extraordinarily sensitive to small changes in their coefficients, even though the perturbation is regular. Second, the results obtained from perturbation theory can be misleading unless one understands the nature of the problem; in this case a naive application of perturbation theory could be misleading, but an examination of the behaviour of the perturbation solutions shows that roots coalesce and this suggests that the expansion is likely to fail because the roots may move into the complex plane.

9.3 Iterative methods

Iterative methods provide an alternative to perturbation expansions. Normally these methods are best suited to numerical approximation, but in some cases they can provide simple analytic approximations to the solutions of equations. The general idea is to rewrite the equation in the form $x = F(x)$ for some function $F(x)$ and to replace this equation by an iterative scheme

$$x = F(x) \quad \longrightarrow \quad x_{k+1} = F(x_k), \quad k = 0, 1, 2, \ldots, \tag{9.28}$$

for the sequence $x_0, x_1, x_2, \cdots$. This method is often more convenient than perturbation theory if the sequence converges to the required root of $x = F(x)$.

Some idea of whether the sequence converges and, if it does, how rapidly, can be obtained by assuming x_k is close to the required root for some k and by linearising about η, the root. This is achieved by setting $x_k = \eta + \delta_k$ and by assuming that all terms $O(\delta_j^2)$, can be ignored. The kth iterate is then

$$\eta + \delta_k = F(\eta + \delta_{k-1}) = F(\eta) + \delta_{k-1} F'(\eta) + O(\delta_{k-1}^2).$$

Since, by definition, $\eta = F(\eta)$, this gives

$$\delta_k = F'(\eta)\delta_{k-1} \quad \text{or} \quad \delta_N = \delta_0 F'(\eta)^N,$$

which shows that $\delta_N \to 0$ geometrically, provided $|F'(\eta)| < 1$. If $-1 < F'(\eta) < 0$ the sign of the error δ_k alternates, so successive iterates bracket the limit. This analysis assumes that the starting point x_0 is sufficiently close to the root, in which case we can obtain an estimate of the rate of convergence by replacing the unknown η by x_0 in the multiplier $F'(\eta)$.

It is sometimes useful to describe the iterative process defined by equation 9.28 geometrically; we show how this is done using the logistic map, $F(x) = bx(1 - x)$, so that the equation $x = F(x)$ has roots at $x = 0$ and $1 - 1/b$ and at this latter root $F'(x) = 2 - b$, so the iterations should converge if $1 < b < 3$. The graphs of $y = x$ and $y = F(x)$ are shown in figure 9.4 by the dashed lines; these lines intersect at the root we are trying to find. This figure also shows the iteration diagramatically, with the solid lines. Starting on the x-axis at some point x_0 we draw a vertical line to the point $(x_0, F(x_0))$ and then a horizontal line to the curve $y = x$, that is the point $(x_1, F(x_0))$ where $x_1 = F(x_0)$. Next we move vertically to the point $(x_1, F(x_1))$ and then again horizontally to the line $y = x$, that is the point $(x_2, F(x_1))$, where $x_2 = F(x_1)$. In this manner we can follow the iterate geometrically. The final result, in the case $b = 2.5$ and $x_0 = 0.95$, is shown in figure 9.4, and we can see the iterations converging on the root $\eta = 3/5$.

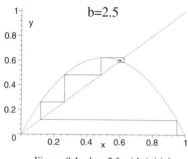

Figure 9.4 $b = 2.5$ with initial
condition $x_0 = 0.95$.

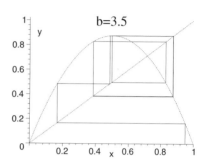

Figure 9.5 $b = 3.5$ with initial
condition $x_0 = 0.95$.

In figure 9.5 the iterations for the case $b = 3.5$ and $x_0 = 0.95$ are shown; in this case there is a real root, at $x = \eta = 5/7$, but $|F'(\eta)| > 1$ and the iterations do not converge to it, but eventually alternate between the two roots of $F(F(x)) = x$; the reasons and consequences of this are discussed in chapter 18.

It is interesting to watch these iterations develop: the following Maple procedure produces an animation of N iterations, starting at $x = x_0$, $0 < x_0 < 1$, with a given value of b, which must be in the interval $0 < b \le 4$.

```
>   with(plots): with(plottools):
>   anim := proc(b,x0,N) local k,pg,z1,u,z,lg,F;
>   F := (x,b) -> b*x*(1-x);
>   # Set up the graph of F(x,b) as a background plot:
>   pg := plot([x,F(x,b)], x=0..1, color=black, linestyle=4,
>                 title=cat("b = ", convert(b,string)));
>   # Set up the iteration trajectory as an animation:
>   u := x0;   z1 := [u,0];
>   lg[0] := CURVES([z1]);
>       for k from 1 to N do
>       z := F(u,b);   z1 := z1, [u,z], [z,z];   u := z;
>       lg[k] := CURVES([z1])
>       od;
>   display([seq(lg[k], k=0..N)], insequence=true);
>   # Finally, display the animation and background together:
>   display({pg, %})
>   end:
```

The command `anim(2.5,0.9,10)` will produce an animation of figure 9.4. For this problem variations in x_0 produce no interesting changes, but varying the value of b changes the nature of the iterations dramatically.

There are many uses for iterative methods because, like perturbation theory, they can be applied to a wide variety of equation types, not just the algebraic equations considered here. Iterative methods can be applied to matrix equations, differential and integral equations as well as more general operator equations: some unusual examples are discussed in section 9.5.

In most cases the equation(s) to be solved do not come in a form suited for iterations and need to be rearranged. Normally there are many ways of rewriting equations only some of which are suitable for iteration; some rearrangements will diverge, some will lead to illegal operations such as taking the square root of a negative number. Some examples of these problems are discussed at the end of this section. Usually a certain amount of trial and error is required before finding a suitable form.

For a first simple example consider the iterative scheme obtained from equation 9.15 (page 306), by rewriting it in the form

$$(x-1)^2 = \epsilon x^2 \quad \text{or} \quad x = 1 \pm x\sqrt{\epsilon}, \quad \epsilon \geq 0,$$

which suggests the iterative scheme

$$x_{k+1} = 1 \pm x_k\sqrt{\epsilon} \quad \text{with} \quad x_0 = 1. \tag{9.29}$$

By comparing this with equation 9.28, remembering that the root η will be near unity, we see that $F'(\eta) \simeq \sqrt{\epsilon}/2$, and according to our previous discussion the iterative scheme should converge for small ϵ. The first three iterations give

$$\begin{aligned}
x_1 &= 1 \pm \epsilon^{1/2} \\
x_2 &= 1 \pm \left(1 \pm \sqrt{\epsilon}\right)\sqrt{\epsilon} = 1 \pm \sqrt{\epsilon} + \epsilon \\
x_3 &= 1 \pm \left(1 \pm \sqrt{\epsilon} + \epsilon\right)\sqrt{\epsilon} = 1 \pm \epsilon^{1/2} + \epsilon \pm \epsilon^{3/2},
\end{aligned}$$

which is just the series obtained using perturbation theory. But there are two points worth noting. First, in this case the iterative scheme was much easier to apply. Second, this method automatically determines the expansion parameter – in this case $\epsilon^{1/2}$.

Exercise 9.16
Can you estimate the values of the positive constant a for which the sequence $x_{k+1} = 1 + \sqrt{ax_k}$, $k = 0, 1, \ldots$, converges if $x_0 > 0$?
Hint: use the diagrammatic method illustrated in figure 9.4.

Exercise 9.17
Rewrite equation 9.1 in the form $x = a - \epsilon/x$ to form an iterative scheme with $x_0 = a$ and find the following iterates:

$$x_2 = \frac{a(a^2 - 2\epsilon)}{a^2 - \epsilon}, \quad x_3 = \frac{a^4 - 3a^2\epsilon + \epsilon^2}{a(a^2 - 2\epsilon)}, \quad x_4 = \frac{a(a^2 - \epsilon)(a^2 - 3\epsilon)}{a^4 - 3a^2\epsilon + \epsilon^2}.$$

Exercise 9.18

Write a Maple recursive procedure which performs the iteration defined in equation 9.29 and expresses the final result as a power series in $\sqrt{\epsilon}$ up to and including the term of order $\epsilon^{N/2}$ for some specified N.

Hint: it is easiest to define a new parameter $\delta = \sqrt{\epsilon}$.

Successful iteration often depends upon re-casting the original equation in a suitable form. In order to see what can go wrong, consider the equation

$$x = -\ln x \quad \text{with the iterative scheme} \quad x_{k+1} = -\ln x_k.$$

Graphs of the functions $y = x$ and $y = -\ln x$ show that there is a real root between 0 and 1; on setting $x_0 = 0.5$ we obtain $x_1 = 0.693$, $x_2 = 0.365$ and $x_3 = -1.0$, at which point the iteration fails. In this example $F(x) = -\ln x$ and $|F'(x)| = 1/x > 1$ if $0 < x < 1$, so this scheme is unstable and clearly a bad choice.

Now re-write the equation in the form

$$x = e^{-x} \quad \text{with the iterative scheme} \quad x_{k+1} = e^{-x_k},$$

for which $F(x) = e^{-x}$ and $|F'(x)| = e^{-x}$, and this scheme converges if $x > 0$; with $x_0 = 0.5$ we obtain, after 16 iterations, $x_{16} \simeq 0.5671$ as an approximation to the root. This example shows that the stability of the iterations near a root is not an intrinsic property of the root, but depends upon the iteration scheme.

Another problem is that an iterative scheme may converge to a root other than that being sought. For instance, the quadratic equation $2x^2 - 4x + 1 = 0$ has roots $1 \pm 1/\sqrt{2}$ and the equations can be re-written as the iterative scheme

$$x_{k+1} = \sqrt{2x_k - \frac{1}{2}}.$$

If $x_0 = 1 - 1/\sqrt{2} + \delta$, for any $\delta > 0$, the iterations converge to the larger root $1 + 1/\sqrt{2}$.

Exercise 9.19

Use the geometric description of the iterative process shown in figure 9.4 to explain the behaviour just described. What happens if the initial point is in the interval $(\frac{1}{4}, 1 - \frac{1}{\sqrt{2}})$?

Exercise 9.20

Show that the iterative scheme $x_{k+1} = e^{-x_k}$, with $x_0 = 0.5$, requires about 21 iterations to obtain an approximation to the real root of the equation $x = e^{-x}$ accurate to six decimal places, but that six iterations and two applications of the Aitken's process provide similar accuracy.

Exercise 9.21

Consider the following iterative schemes, with the given values of x_1,

(i) $x_{k+1} = \cos x_k$, $x_1 = 0.01, 1.4,$
(ii) $x_{k+1} = 2 \sin x_k$, $x_1 = 1.5, 2.9,$
(iii) $x_{k+1} = (x_k + 1)^{1/4}$, $x_1 = 0.1, 3.0,$
(iv) $x_{k+1} = x_k^4 - 1$, $x_1 = 0.5, 1.3.$

Comment upon the convergence and illustrate the iterative process graphically; you may find it convenient to use a version of the program provided above.

Exercise 9.22
Write a Maple program to iterate the equations

$$x_{k+1} = 1 - 2y_k^2/3, \quad y_{k+1} = 3x_k^2 - \frac{3}{4}.$$

Starting with initial values $(x, y) = (\frac{1}{2}, \frac{1}{2})$ show that $(0.037\,47, -0.745\,75)$ is a solution of these equations. Determine, graphically, the approximate region contained in the rectangle $|x| < 1.5$, $|y| < 2$ that iterates to this solution.

9.3.1 A nasty equation
It is often necessary to solve, or understand the solutions of, equations involving powers and exponentials, and we illustrate the problems that this sort of equation can produce by tackling the relatively simple example

$$xe^{-x} = \epsilon, \quad 0 < \epsilon \ll 1. \tag{9.30}$$

The function $y = x\exp(-x)$ has a single local maximum at $x = 1$ with value $\max(y) = e^{-1}$; also $y(x)$ is zero at $x = 0$ and as $x \to \infty$. Thus the equation $xe^{-x} = \epsilon$ has two positive real roots provided $0 < \epsilon < e^{-1}$, as shown in the figure.

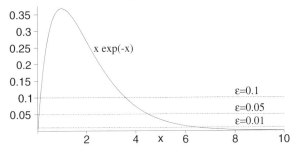

Figure 9.6   Graph of $y = x\exp(-x)$ with the dashed lines showing $y = \epsilon$ for $\epsilon = 0.1$, 0.05 and 0.01.

The smaller root can be found either by a perturbation expansion or an iterative scheme; this is left to exercise 9.51 (page 337). The larger root presents a more difficult problem.

In order to find an approximation to the larger root we find, after some trial and error, that it is convenient to re-cast the equation in the form

$$e^x = \frac{x}{\epsilon} \quad \Longrightarrow \quad x = \ln(1/\epsilon) + \ln x. \tag{9.31}$$

If x is large then $x \gg \ln x$ — which is the reason why the equation was re-written in this form — so a first approximation to the solution is $x = \ln(1/\epsilon)$ because if ϵ is sufficiently small $\ln(1/\epsilon) \gg \ln\ln(1/\epsilon)$. However, the slow rate of increase of the logarithm with decreasing ϵ means that it has to be very small indeed for the ratio of the first to the second term to be very large; for instance if $\epsilon = 10^{-3}$ then $\ln(1/\epsilon)/\ln\ln(1/\epsilon) \simeq 3.6$ and if $\epsilon = 10^{-6}$ the ratio is still only 5.3. This is the reason why good perturbation-type approximations to this root are very difficult to obtain and why the series we derive below converges so slowly.

Equation 9.31 suggests the iterative scheme

$$x_{k+1} = \ln X + \ln x_k \quad \text{with} \quad x_0 = \ln X \quad \text{and} \quad X = \frac{1}{\epsilon}. \tag{9.32}$$

If η is the root and $x_k = \eta + \delta_k$, an expansion about η gives

$$\delta_{k+1} \simeq \frac{\delta_k}{\eta}, \quad \text{and since} \quad \eta \simeq \ln(1/\epsilon), \quad \delta_k \simeq \frac{\delta_0}{[\ln(1/\epsilon)]^k},$$

showing that the convergence of this iterative scheme increases very slowly as ϵ decreases. For example for $\epsilon = 0.01$, δ decreases by a factor of about 5 at each iteration; if ϵ decreases by a factor of 10 to 0.001 this factor increases only a small amount to 7. Therefore good numerical accuracy is obtained only for very small values of ϵ or very many iterations.

The first three iterations with $x_0 = \ln(1/\epsilon)$ are

$$\begin{aligned}
x_1 &= \ln X + \ln\ln X, & X = \ln(1/\epsilon), \\
x_2 &= \ln X + \ln(\ln X + \ln\ln X) \\
x_3 &= \ln X + \ln\left[\ln X + \ln(\ln X + \ln\ln X)\right].
\end{aligned}$$

Exercise 9.23
Define Maple functions which represent the first four iterates, $x_k(\epsilon)$, $k = 1, 2, \ldots, 4$, defined by equation 9.32. Plot the graphs of the functions $g_k(\epsilon) = x_k(\epsilon) - \epsilon e^{x_k(\epsilon)}$ for $\epsilon = 10^{-8}, 10^{-7}, \ldots, 10^{-1}$.

Exercise 9.24

(i) Find the first three terms of an iteration sequence for the positive root of the equation

$$\exp(-ax^2) = \epsilon x, \quad a > 0, \quad |\epsilon| \ll 1.$$

(ii) Use Maple to find the root numerically and hence find the approximate values of a and ϵ for which your solution has a relative accuracy of 1%.

9.4 Perturbation theory for some differential equations

Perturbation theory can be applied to differential equations, but is more difficult and usually requires some qualitative understanding of the behaviour of the solutions. It is technically more difficult because integrals are involved; these usually become more complex with each order and eventually may become intractable. Worse, minor changes to an equation may change a regular into a singular expansion. Thus more care is needed when dealing with differential equations. For these reasons different methods have been developed for particular types of differential equations, and these require a specialised study: two useful methods are described in chapter 17. Here we discuss methods which are the direct generalisations of those used above for algebraic equations: such methods are strictly algebraic and 'dumb' in that they take no regard of possible differences in the behaviour of the perturbed and the unperturbed solutions. Nevertheless, they can sometimes yield useful approximations and provide a useful introduction to the subject.

We illustrate the method with the simple example of a first-order differential equation,

$$\frac{dx}{dt} = x + \epsilon x^2, \quad 0 \le \epsilon \ll 1, \quad x(0) = A > 0, \tag{9.33}$$

and attempt to solve this using a direct extension of the above perturbation theory. This example is chosen because it can be solved in terms of elementary functions and because it illustrates some of the problems that can arise. First, expand the solution as a power series in ϵ:

$$x(t) = x_0(t) + \epsilon x_1(t) + \epsilon^2 x_2(t) + \cdots,$$

where each $x_k(t)$ is an unknown function of the independent variable, t, rather than a number. The initial condition can be satisfied most simply by imposing the constraints $x_0(0) = A$ and $x_k(0) = 0$ for $k \ge 1$. Now proceed exactly as before by substituting this expansion into the original equation, collecting together the coefficients of ϵ^k, $k = 0, 1, \ldots$, and setting these to zero, to obtain the following differential equations for the unknown functions:

$$\frac{dx_0}{dt} = x_0, \quad x_0(0) = A, \tag{9.34}$$

$$\frac{dx_1}{dt} = x_1 + x_0(t)^2, \quad x_1(0) = 0, \tag{9.35}$$

$$\frac{dx_2}{dt} = x_2 + 2x_0(t)x_1(t), \quad x_2(0) = 0, \tag{9.36}$$

$$\frac{dx_3}{dt} = x_3 + x_1^2 + 2x_0 x_2, \quad x_3(0) = 0. \tag{9.37}$$

Exercise 9.25
Derive equations 9.34 to 9.37.

The solution of the unperturbed equation 9.34 is

$$x_0(t) = Ae^t,$$

so the equation for $x_1(t)$ becomes

$$\frac{d}{dt}\left(e^{-t}x_1\right) = A^2 e^t \quad \Longrightarrow \quad x_1 = A^2 e^t \left(e^t - 1\right).$$

Similarly,

$$x_2 = A^3 e^t \left(e^t - 1\right)^2 \quad \text{and} \quad x_3 = A^4 e^t \left(e^t - 1\right)^3,$$

giving the second-order approximation

$$x(t) = Ae^t \left[1 + \epsilon A \left(e^t - 1 \right) + \epsilon^2 \left[A \left(e^t - 1 \right) \right]^2 + \cdots \right]. \tag{9.38}$$

Observe that each term in the outer square brackets is $\epsilon A(e^t - 1)$ times the preceding term. Thus, for a given accuracy, the first-order expansion $x_0(t) + \epsilon x_1(t)$ is useful for smaller t than the second-order expansion which, itself, is useful for smaller t than the third-order expansion. Conversely, for a given accuracy higher orders are needed the larger t is. These are common properties of perturbation expansions for differential equations, and are illustrated in figure 9.7.

The differential equation 9.33 can be integrated directly to give

$$x(t) = \frac{Ae^t}{1 - \epsilon A(e^t - 1)}. \tag{9.39}$$

Expanding this in powers of ϵ gives the series 9.38. The solution $x(t)$ should be regarded as a function of the two variables t and ϵ; in this example the dependence upon these two variables is clear, but in more complex problems the dependence upon the independent variable and the perturbation parameter is not so transparent. Here the exact solution becomes infinite at $t = T_c$ given by

$$\epsilon A(e^{T_c} - 1) = 1 \quad \text{or} \quad T_c = \ln \left(\frac{1 + \epsilon A}{A\epsilon} \right),$$

which is the radius of convergence of the series 9.38 when considered as a function of t; the perturbation expansion is valid only for $t < T_c$. The behaviour of the unperturbed solution, $x(t) = x_0(t) = Ae^t$, is different in that it is finite for all finite t.

The reason for this is to be found in the different qualitative behaviour of the perturbed and unperturbed equations. For $x > 0$ we have $\dot{x} > 0$, and so $x(t)$ is always increasing. Initially $x \gg \epsilon x^2$ so the solution will behave like that of the unperturbed equation, $\dot{x} = x$, that is $x \simeq Ae^t$. But there is a value of t where $\epsilon x(t) \simeq 1$, and for larger t the equation will be like $\dot{x} = \epsilon x^2$ which has the general solution $x(t) = 1/(b - \epsilon t)$, for some constant b, and this is infinite at $t = b/\epsilon$.

In other words at some critical value of t the behaviour of the unperturbed and the perturbed solutions differ qualitatively, and here the perturbation expansion breaks down. Thus the 'small' perturbation, ϵx^2, has qualitatively changed the large t behaviour of the solution. This type of behaviour is common; one of its manifestations, in more complicated equations, is chaos. Suffice to say that in these circumstances perturbation approximations are useful only for a limited range of the dependent variable, t.

If $t \ll T_c$ the perturbation expansion is quite accurate, as is shown in figure 9.7, where we show, with the solid line, the graph of the exact solution divided by the unperturbed solution,

$$g(t) = \frac{x(t)}{Ae^t} = \frac{1}{1 - \epsilon A(e^t - 1)},$$

and, with the dashed lines, the approximate quotient equivalent to the Nth-order perturbation expansion,

$$g_N(t) = \sum_{k=0}^{N} \left(\epsilon A(e^t - 1) \right)^k,$$

in the case $\epsilon = 0.1$, $A = 1$ and for $N = 1, 2, 4$ and 10. For these parameters $T_c \simeq 2.4$.

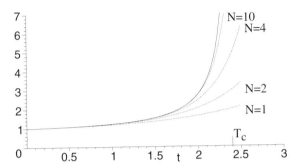

Figure 9.7   Comparison of the exact solution of equation 9.33, the solid
line, with the various order perturbation approximations.

It is worth noting that, in this case, an estimate of T_c may be obtained directly from the unperturbed solution: we expect the perturbation expansion to be accurate when $x < \epsilon x^2$ and substituting the unperturbed solution into this inequality gives $A\epsilon e^{T_c} \sim 1$, which is close to the previous estimate if $A\epsilon \ll 1$.

This example is very simple, but it illustrates the method and some of the difficulties of applying perturbation to differential equations. The following exercises will provide some practice in these methods, some idea of the difficulties involved in implementing the method and also show how useful the method can be for some equations.

Exercise 9.26
Use perturbation theory to obtain the exact solution of the equation

$$\frac{dx}{dt} = \epsilon x, \quad x(0) = A > 0,$$

in the form of the series

$$x(t) = A \sum_{k=0}^{\infty} \frac{(\epsilon t)^k}{k!} = A e^{\epsilon t}.$$

Exercise 9.27
Show that the first-order perturbation expansion of the solution to

$$\frac{dx}{dt} = x^{1/2} + \epsilon x^{1/4}, \quad x(0) = A > 0, \quad |\epsilon| \ll 1,$$

is

$$x(t) = u + 4\epsilon \left(u^{1/4} - A^{1/4} \right) \sqrt{u} + O(\epsilon^2),$$

where $u(t) = \left(\sqrt{A} + t/2 \right)^2$. Use Maple to find the second- and third-order terms.

Exercise 9.28

The function $y(x)$ satisfies the second-order linear differential equation

$$\frac{d^2y}{dx^2} = \epsilon y f(x), \quad y(0) = a, \quad y'(0) = b.$$

Show that the terms of the perturbation expansion $y(x) = \sum_{k=0}^{\infty} \epsilon^k y_k(x)$ can be chosen to satisfy the equations

$$y_0(x) = a + bx, \quad \frac{d^2y_k}{dx^2} = y_{k-1}f(x), \quad y_k(0) = y_k'(0) = 0, \quad k \geq 1,$$

and hence that

$$y_k = \int_0^x dt\, (x-t)f(t)y_{k-1}(t), \quad k \geq 1.$$

Use this result and Maple to show that the solution of

$$\frac{d^2y}{dx^2} = \epsilon y e^{-x}, \quad y(0) = a, \quad y'(0) = b,$$

correct to $O(\epsilon^2)$ is

$$y(x) = a + bx + \left[(a+b)x - a - 2b + (bx + a + 2b)e^{-x}\right]\epsilon +$$
$$\left[\frac{1}{4}(b+2a)x - \frac{1}{4}(5a + 3b) + ((a+b)x + a)e^{-x} + \frac{1}{4}(bx + a + 3b)e^{-2x}\right]\epsilon^2,$$

and use your Maple program to extend this result to include the term $O(\epsilon^3)$ and show that for large x the solution behaves as

$$y \sim a - (a + 2b)\epsilon - \frac{1}{4}(5a + 3b)\epsilon^2 - \left(\frac{5a}{18} + \frac{11b}{108}\right)\epsilon^3$$
$$+ \left(b + (a+b)\epsilon + \frac{1}{4}(2a + b)\epsilon^2 + \frac{1}{36}(3a + b)\epsilon^3\right)x, \quad x \gg 1.$$

Exercise 9.29

Use perturbation theory to show that the equation

$$\frac{d^2x}{dt^2} + \Omega^2 x + \epsilon x^3 = F \cos t, \quad |\epsilon| \ll 1, \quad \Omega < 1,$$

has an approximate 2π-periodic solution

$$x(t) = -\frac{F}{1 - \Omega^2}\cos t - \frac{\epsilon F^3}{4(1 - \Omega^2)^3}\left(\frac{3\cos t}{1 - \Omega^2} + \frac{\cos 3t}{9 - \Omega^2}\right) + O(\epsilon^2).$$

Use Maple to check this result, to determine the next term in the series and to compare, graphically, this perturbation solution with the relevant numerical solution of the differential equation for various values of ϵ and F.

Hint: for the numerical comparison you will need to use dsolve and odeplot or DEtools.

Exercise 9.30

Consider the linear delay-equation

$$\frac{dx}{dt} = -\lambda x + \epsilon x(t - a), \quad x(0) = 1, \quad |\epsilon| \ll 1,$$

where a and λ are positive constants. Here $x(t-a)$ means the function $x(t)$ evaluated at $t-a$.

Show that, if $x(t)$ is a solution of this equation, then successive terms of the perturbation expansion

$$x(t) = x_0(t) + \epsilon x_1(t) + \epsilon^2 x_2(t) + \cdots$$

are related by the equation

$$x_n(t) = e^{-\lambda t} \int_0^t ds\, e^{\lambda s} x_{n-1}(s-a),$$

with $x_0(t) = e^{-\lambda t}$. Hence show that for small values of ϵ the approximate solution is

$$x(t) = e^{-\lambda t} + \epsilon t e^{-\lambda(t-a)} + \frac{1}{2}\epsilon^2 t(t-2a)e^{-\lambda(t-2a)} + O(\epsilon^3).$$

Show that

$$x_6(t) = \frac{t(t-6a)^5}{6!} e^{-\lambda(t-6a)} \quad \text{and} \quad x_8(t) = \frac{t(t-8a)^7}{8!} e^{-\lambda(t-8a)}$$

and find expressions for $x_n(t)$ for other values of n. These results suggest that

$$x_n(t) = \frac{t(t-na)^{n-1}}{n!} e^{-\lambda(t-na)}.$$

Prove that this is a solution of the equation for $x_n(t)$.

9.5 Newton's method and its extensions

A well known iterative method of solving algebraic equations is Newton's method — sometimes called the Newton–Raphson method — which is a scheme produced using a first-order Taylor expansion about a point close to the zero. The method was introduced in an elementary form by Newton in *Methods of Fluxions*, a treatise written in 1671 but not published until 1736, where it is explained by its application to the cubic $x^3 - 2x - 5 = 0$. In 1690 Raphson published a method closely resembling Newton's but superior in that each successive step was obtained from the same equation. More details of these differences can be found in Cajori (1991, pages 191–203); Raphson's is the familiar modern form. This iterative scheme is cleverer than the simple-minded approach of equation 9.28 (page 315), because it uses knowledge of the function at each iterate to help guess the next: the price paid for this is the need to compute values of the differential besides the function at each iteration; the gain is much faster convergence.

I. Newton
(1642–1727)

J. Raphson
(1648–1715)

Suppose that $(x_0, x_1, \ldots, x_k, \ldots)$ is a sequence of iterates obtained from the initial guess x_0. We put $x_{k+1} = x_k + \delta$, assume that $f(x_{k+1}) = 0$ and use Taylor's expansion

$$0 = f(x_k + \delta) = f(x_k) + \delta f'(x_k) + O(\delta^2) \quad \text{or} \quad \delta \simeq -\frac{f(x_k)}{f'(x_k)}. \tag{9.40}$$

Thus the next approximation to the root is

$$x_{k+1} = x_k - \frac{f(x_k)}{f'(x_k)}. \tag{9.41}$$

If the initial guess, x_0, is close to the solution, $x_0 \simeq \eta$, where $f(\eta) = 0$, then if this sequence converges it does so exponentially, as can be seen by putting $x_k = \eta + \delta_k$ into equation 9.41 and expanding up to $O(\delta_k^2)$ to give

$$\delta_{k+1} = \frac{f''(\eta)}{2f'(\eta)}\delta_k^2. \tag{9.42}$$

Thus if δ_k is $O(10^{-p})$ then $\delta_{k+1} = O(10^{-2p})$ and $\delta_{k+m} = O(10^{-mp})$, showing that the number of accurate significant figures doubles with each iteration. Newton's scheme should be compared with that defined in equation 9.28 (page 315), for which $\delta_{k+1} = F'(\eta)\delta_k$; this gives geometric convergence, $\delta_N \sim F'(\eta)^N \delta_0$, as opposed to the exponential convergence of Newton's scheme, $\delta_N \sim u^N \delta_0^{2N}$, where $u = f''(\eta)/2f'(\eta)$, and in both cases η is the root.

In order to see this behaviour in a particular example, consider the equation

$$f(x) = 2\sin x - x,$$

which has a root η in the interval $\pi/2 < \eta < \pi$. The scheme defined by equation 9.41 is

$$x_{k+1} = x_k - \frac{2\sin x_k - x_k}{2\cos x_k - 1}, \tag{9.43}$$

and the iterates starting with the initial value $x_0 = \pi/2$ are shown in the table below. Here we also show, for comparison, the results obtained using a conventional iterative scheme introduced in section 9.3,

$$x_{k+1} = 2\sin x_k, \quad x_0 = \frac{\pi}{2}, \tag{9.44}$$

which requires 38 iterations before $\delta_k < 10^{-8}$.

	Newton's method equation 9.43		Conventional iteration equation 9.44	
k	x_k	$\delta_k = \lvert x_k - \eta \rvert$	x_k	δ_k
1	2.000 000	1.0×10^{-1}	2.000 000	1.0×10^{-1}
2	1.900 995	5.5×10^{-3}	1.818 595	7.7×10^{-2}
3	1.895 511	1.7×10^{-5}	1.938 909	4.3×10^{-2}
4	1.895 494	1.7×10^{-10}	1.866 016	3.0×10^{-2}
5	$x_4 + O(10^{-10})$	6.1×10^{-20}	1.913 476	1.8×10^{-2}

The results shown in this table confirm that for Newton's method the value of δ_k decreases as $\delta_{k+1} \simeq 0.55\,\delta_k^2$ and for the conventional method $\delta_{k+1} \simeq 0.64\,\lvert\delta_k\rvert$, which produces the far slower convergence seen in the table.

Exercise 9.31
Show that when applied to the equation $\tan x = 1 - x^2$, Newton's scheme becomes

$$x_{k+1} = x_k - \frac{\tan x_k + x_k^2 - 1}{1 + \tan^2 x_k + 2x_k}.$$

Use this scheme with the initial values $x_0 = 0$ and $x_0 = \pi$ to find the positive roots $x = 0.583\,248$ and $1.924\,880$.

Exercise 9.32

Use Newton's method to find the positive root of the equation

$$J_0(x) = x$$

which lies between zero and the first zero of the zero-order Bessel function at $x \simeq 2$. Remember that it is convenient to use `alias(J=BesselJ)` so `J(n,x)` becomes the Maple symbol for the nth-order Bessel function.

Hint: you may find the following result useful:

$$2J'_n(x) = J_{n-1}(x) - J_{n+1}(x).$$

Exercise 9.33

Prove that the values of the function at each iteration, $f_k = f(x_k)$, are related by

$$f_{k+1} = f_k^2 \frac{f''_k}{2\left(f'_k\right)^2}.$$

It is natural to ask whether or not a scheme giving an even faster rate of convergence can be found. It is clear from the derivation of equations 9.41 and 9.42 that the quadratic convergence is a consequence of using a first-order Taylor expansion to approximate δ in equation 9.40, so we should expect the inclusion of higher-order terms in this expansion to lead to faster convergence. On retaining the second-order term equation 9.40 becomes

$$0 = f(x_k) + \delta f'(x_k) + \frac{1}{2}\delta^2 f''(x_k) + O(\delta^3), \tag{9.45}$$

which is now a quadratic equation for δ; but we expect the solution of this to be close to that given in equation 9.40, so we re-write this quadratic equation in the form

$$\delta = -\frac{2f(x_k)}{2f'(x_k) + \delta f''(x_k)},$$

and since $\delta \simeq -f(x_k)/f'(x_k)$, we may substitute this value of δ into the right hand side to obtain Halley's iterative scheme:

$$x_{k+1} = x_k - \frac{2f(x_k)f'(x_k)}{2f'(x_k)^2 - f''(x_k)f(x_k)}. \tag{9.46}$$

Exercise 9.34

Derive equation 9.46 and show that the equivalent of equation 9.42 is

$$\delta_{n+1} = -\frac{2f'(\eta)f'''(\eta) - 3f''(\eta)^2}{12f'(\eta)^2}\delta_n^3.$$

You may find it easier to use Maple to obtain this result.

The result derived in this exercise shows that Halley's scheme converges even faster than Newton's method, the number of accurate decimal places *tripling* with each iteration. The penalty paid is the need to evaluate $f''(x_n)$ as well as $f(x_n)$ and $f'(x_n)$ and in some cases this extra effort outweighs the advantage of the faster convergence. A comparison between these methods was made in exercise 3.62 (page 130); that example was chosen because it also showed that there are circumstances in which Halley's method converges when Newton's method does not.

Exercise 9.35

Use Halley's method to find the root of the equation $2\sin x - x = 0$ near $\pi/2$, starting with the initial guess $x_0 = \pi/2$, and compare the rate of convergence with that of Newton's method.

9.5.1 Newton's method applied to more general equations
Systems of algebraic equations

Formally Newton's method is trivially extended to deal with systems of algebraic equations; thus for the set of N equations

$$f_j(\mathbf{x}) = 0, \quad j = 1, 2, \dots, N, \quad \mathbf{x} = (x_1, x_2, \dots, x_N), \tag{9.47}$$

the iterative scheme equivalent to that of equation 9.41 is

$$\mathbf{x}_{n+1} = \mathbf{x}_n - M(\mathbf{x}_n)^{-1}\mathbf{f}(\mathbf{x}_n), \quad \text{where} \quad M_{ij}(\mathbf{z}) = \frac{\partial f_i}{\partial z_j}, \tag{9.48}$$

and where $\mathbf{f}(\mathbf{x})$ is the N-dimensional vector of functions $\mathbf{f} = (f_1, f_2, \dots, f_N)$. Here $M(\mathbf{x})$ is an $N \times N$ matrix, and so each iteration involves finding the inverse of this matrix, though in practice it is usually more efficient to find $M^{-1}\mathbf{f}$ by solving the linear equations $\mathbf{f} = M\mathbf{y}$ for $\mathbf{y}$. Because this is a computationally intensive operation, it is often more efficient to replace $M(\mathbf{x}_n)^{-1}$ by its value at the initial guess, $M(\mathbf{x}_0)^{-1}$, which need be found only once at the beginning of the calculation.

Although Newton's method generalises relatively easily from one to many variables you should not think that it is almost as easy to find roots of systems of nonlinear equations as for single equations. The statement 'There are *no* good, general methods for solving systems of more than one nonlinear equation', made and justified by Press *et al.* (1987, chapter 9, page 269), should be adequate warning.

Exercise 9.36
Derive equation 9.48.

Exercise 9.37
By plotting suitable graphs show that the simultaneous equations $f_1(x, y) = 0$, $f_2(x, y) = 0$, where

$$f_1(x, y) = x^4 + y^3 - 16 \quad \text{and} \quad f_2(x, y) = e^{3x/2} - e^y - 1,$$

have a root near $(1.5, 2.2)$. Write two Maple procedures which implement the Newton scheme defined in equation 9.48, one using $M(\mathbf{x}_n)$ and the other $M(\mathbf{x}_0)$; use both to find a better approximation to the root.

A differential equation

The above generalisation from one to many variables immediately suggests that Newton's method can be used for a much wider class of problems, because a function, $y(x)$, of a continuous variable x can be approximated by a finite dimensional vector $\mathbf{y} = (y_1, y_2, \ldots, y_N)$, where $y_k = y(x_k)$ and $x_k = (k-1)\delta$ for some suitably small δ. This suggests that Newton's method could be adapted to solve equations whose solutions are functions rather than numbers. The same idea is used in exercise 7.26 (page 252), where the differential operator $\frac{dy}{dx}$ becomes a matrix multiplication $D\mathbf{y}$ and a differential equation becomes N equations in y_k, $k = 1, 2, \ldots, N$.

As a specific example we follow Sawyer (1967, page 110) and consider the nonlinear differential equation

$$\frac{dy}{dx} + y^2 = \frac{1}{(1+x)^2}, \quad y(0) = 1. \tag{9.49}$$

In principle we could replace the function $y(x)$ by the vector $\mathbf{y}$ and the operator $\frac{d}{dx}$ by the matrix D, as described in chapter 7, and then apply Newton's method to the N coupled nonlinear equations; but this method seems likely to result in rather messy equations, so we proceed by analogy working directly from the original differential equation.

In order to produce an iterative scheme assume that we have an approximate solution, $w_k(x)$, which satisfies the initial conditions, $w_k(0) = 1$, and, as in equation 9.40, assume that the exact solution is $w_{k+1}(x) = w_k(x) + g(x)$ for some small function $g(x)$; if both $w_{k+1}(x)$ and $w_k(x)$ satisfy the initial conditions then $g(0) = 0$. An approximate equation for $g(x)$, in terms of $w_k(x)$, is found by substituting w_{k+1} into the differential equation 9.49 and ignoring terms $O(g^2)$:

$$\frac{dg}{dx} + 2w_k(x)g = F_k(x), \quad g(0) = 0, \quad \text{where} \quad F_k(x) = \frac{1}{(1+x)^2} - \frac{dw_k}{dx} - w_k^2, \tag{9.50}$$

which is the equivalent of equation 9.48 and, as in that case, we need to invert the linear equation $g' + 2w_k g = F_k$, the equivalent of the matrix equation $M(\mathbf{x}_k)^{-1} = \mathbf{f}$. If w_k is a solution of equation 9.49, then $F_k = 0$ and $g(x) = 0$. The linear operator $\frac{d}{dx} + 2w_k(x)$ defining $g(x)$ involves $w_k(x)$, which changes with each iteration — as does the matrix $M(\mathbf{x}_k)$ — and as before it is expeditious to replace $w_k(x)$ by the zero-order estimate to the solution $w_0(x)$, which we take to be the solution of

$$\frac{dy}{dx} + y^2 = 0, \quad y(0) = 1, \tag{9.51}$$

that is,

$$y(x) = w_0(x) = \frac{1}{1+x}.$$

Thus the required approximation to equation 9.50 is

$$\frac{dg}{dx} + \frac{2g}{1+x} = F_k(x), \quad g(0) = 0, \tag{9.52}$$

which can be re-written as

$$\frac{1}{(1+x)^2} \frac{d}{dx} \left[(1+x)^2 g\right] = F_k(x), \quad g(0) = 0,$$

and then integrated directly to give

$$g_k(x) = \frac{1}{(1+x)^2} \int_0^x ds \, F_k(s)(1+s)^2, \tag{9.53}$$

where the function $F_k(s)$ depends upon $w_k(s)$, the previous estimate of the solution, through the relation given in equation 9.50. Thus the first iterate is

$$
\begin{aligned}
w_1(x) &= w_0(x) + \frac{1}{(1+x)^2} \int_0^x ds \, (1+s)^2 \left[\frac{1}{(1+s)^2} - \frac{dw_0}{ds} - w_0(s)^2\right] \\
&= \frac{1}{1+x} + \frac{1}{(1+x)^2} \int_0^x ds = \frac{1}{1+x} + \frac{x}{(1+x)^2}.
\end{aligned}
$$

The general iteration is

$$w_{k+1}(x) = w_k(x) + \frac{1}{(1+x)^2} \int_0^x ds \, (1+s)^2 \left[\frac{1}{(1+s)^2} - \frac{dw_k}{ds} - w_k^2(s)\right]. \tag{9.54}$$

This may be simplified by integrating by parts the second term in the square brackets:

$$w_{k+1}(x) = \frac{1}{1+x} + \frac{1}{(1+x)^2} \int_0^x ds \, \left[2(1+s)w_k(s) - (1+s)^2 w_k(s)^2\right]. \tag{9.55}$$

The first few iterates yield the following approximations to the solution:

$$
\begin{aligned}
w_0(x) &= \frac{1}{1+x}, \\
w_1(x) &= \frac{1}{1+x} + \frac{x}{(1+x)^2}, \\
w_2(x) &= \frac{1}{1+x} + 2\frac{\ln(1+x)}{(1+x)^2} - \frac{x}{(1+x)^3}.
\end{aligned}
$$

Other terms can be found, but it is easiest to write a Maple procedure to perform the necessary integrations. In figure 9.8 we show a comparison of $w_2(x)$ with the numerical solution of the equation, the solid line, using the results derived in the next exercise; in figure 9.9 we show the differences between the numerical solution of the differential equation and the second and fourth iterates.

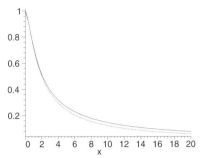

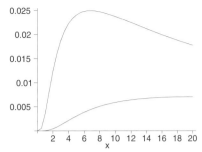

Figure 9.8   Graph showing the numerical solution of equation 9.49, the solid line, and the second iterate.

Figure 9.9   Graph showing the differences between the exact numerical solution of equation 9.49 and the second and fourth iterates.

Exercise 9.38

Write a Maple procedure that determines $w_{n+1}(x)$ given $w_n(x)$ and graphically compare the first five iterates of the solution with the exact solution generated using `dsolve` for $0 < x < 20$ and with $w_0 = 1/(1+x)$.

This example has been given in order to show that Newton's method can be extended to other types of equation, and is not limited to the numerical solutions of algebraic equations. In practice it is not easy to solve differential equations in this manner because the solution of the equivalent of equation 9.52 and the consequent generalisation of the iterative scheme 9.54 yields, in most cases, intractable integrals. In addition it is much more difficult to understand the conditions required for convergence of this type of expansion.

In the next exercise the method is applied to a generalisation of equation 9.49 and the resulting integrals provide some idea of the problems involved.

Exercise 9.39

Apply Newton's method to the generalisation of equation 9.49, namely

$$\frac{dy}{dx} + F(y) = f(x), \quad y(0) = A,$$

where $f(x)$ and $F(y)$ are suitably well behaved functions of x and y respectively, to show that an iterative scheme is given by

$$w_{k+1} = \frac{AF(u(x))}{F(A)} + F(u(x)) \int_0^x ds \, \frac{1}{F(u(s))} \left[w_k(s)\frac{dF}{du} + f(s) - F(w_k(s)) \right],$$

where $u(x)$ is a function defined to be the solution of the unperturbed equation

$$\frac{du}{dx} + F(u) = 0, \quad u(0) = A.$$

9.5.2 *Power series manipulation*

The iterative schemes discussed in the previous sections can also be used for the formal manipulation of series to find, for example, the power series representing the inverse or square root of given power series. We introduce this novel use of Newton's method by showing how it may be used to compute the inverse of a series; more details of these ideas can be found in Geddes *et al.* (1992, chapter 4). Recall from chapter 4 that if

$$y(x) = a_0 + a_1 x + a_2 x^2 + \cdots + a_n x^n + \cdots, \quad a_0 \neq 0, \tag{9.56}$$

then

$$z(x) = \frac{1}{y(x)} = b_0 + b_1 x + b_2 x^2 + \cdots + b_n x^n + \cdots, \tag{9.57}$$

where

$$b_0 = \frac{1}{a_0}, \quad b_k = -\frac{1}{a_0} \sum_{j=0}^{k-1} b_j a_{k-j}.$$

With this direct method of finding the inverse, the computation of b_n requires n multiplications, given $(b_0, b_1, \ldots, b_{n-1})$, so the calculation of $z(x)$ to $O(x^n)$ requires about n^2 multiplications and the computational time increases as $O(n^2)$. The method described here reduces this to a time $O(n \ln n)$.

First, consider the simpler problem of the evaluation of the inverse of a real number using Newton's method, because this shows how an approximation to a division may be obtained with only multiplications. Suppose we require the inverse of 5, that is $1/5$ or equivalently the zero of the function

$$f(x) = 5 - \frac{1}{x}. \tag{9.58}$$

For this equation Newton's iterative scheme, equation 9.41 (page 325), is

$$x_{n+1} = x_n - \frac{5 - 1/x_n}{1/x_n^2} = x_n(2 - 5x_n), \tag{9.59}$$

which requires only multiplications and no division and was the method sometimes used in early computers that had no hardware division implemented. Starting with $x_0 = 0.1$, we obtain

$$x_1 = 0.15, \quad x_2 = 0.1875, \quad x_3 = 0.199\,219, \quad x_4 = 0.199\,997.$$

The second iterate, x_2, is accurate to one decimal place, the third to two decimal places, the fourth to five decimal places and the fifth iterate, not shown here, is accurate to ten decimal places. Very roughly there is a doubling of accuracy with each iteration. In this example Halley's scheme involves division and gives $x_1 = 1/5$ independent of the starting value.

Now consider a generalisation of this method which, in the first instance, we use to compute the inverse of a power series. Let $y(x)$ and $z(x)$ be the power series defined in equations 9.56 and 9.57 and define

$$f = y(x) - \frac{1}{z(x)},$$

by analogy with equation 9.58. Thus the value of $z(x)$ is the value at which $f = 0$ and, by analogy with equation 9.59, the tentative iterative scheme is

$$z_{n+1} = z_n - \frac{f(z_n)}{f'(z_n)}, \tag{9.60}$$

which reduces to

$$z_{n+1} = z_n[2 - z_n y(x)]. \tag{9.61}$$

Now assume that the polynomial $z_n(x)$ has degree $p_n - 1$ and gives an approximation to $1/y(x)$ with an error $O(x^{p_n})$ for some integer p_n, that is, $z_n(x)y(x) - 1 = O(x^{p_n})$. Then on using equation 9.61 we see that

$$y(x)z_{n+1}(x) - 1 = (z_n(x)y(x) - 1)^2 = O(x^{2p_n}).$$

Hence $p_{n+1} = 2p_n$, and the iterative scheme 9.61 doubles the number of terms in the series with each iteration.

Taking $z_0(x) = 1/a_0$, then $z_1 = 1/a_0 - a_1 x/a_0^2$ so $p_1 = 2$; substituting z_1 into equation 9.61 then gives

$$z_2(x) = \frac{1}{a_0} - \frac{a_1}{a_0^2}x + \frac{a_1^2 - a_2 a_0}{a_0^3}x^2 + \frac{2a_0 a_1 a_2 - a_0^2 a_3 - a_1^3}{a_0^4}x^3,$$

and we see directly that $p_2 = 4$; in general $p_n = 2^n$. Since z_n is a polynomial of degree $2^n - 1$ we write equation 9.61 in the form

$$z_{n+1} = z_n[2 - z_n y(x)] \quad \mathrm{mod}\ x^{2^{n+1}}, \tag{9.62}$$

which means that all terms on the right hand side $O(x^{2^{n+1}})$ are ignored in computing z_{n+1}.

In order to see this method working consider the series

$$y(x) = 1 - 2x + 3x^2 + 4x^3 - 5x^5 + x^6 - 2x^7 + 10x^9.$$

Using the scheme of equation 9.62 and with $z_0 = 1$ we obtain, using Maple for the second and third iterate,

$$\begin{aligned} z_1 &= 2 - (1 - 2x) \quad \mathrm{mod}(x^2) \\ &= 1 + 2x, \end{aligned}$$

and

$$\begin{aligned} z_2 &= (1 + 2x)\left[2 - (1 + 2x)(1 - 2x + 3x^2 + 4x^3)\right] + O(x^4) \\ &= 1 + 2x + x^2 - 8x^3. \end{aligned}$$

Note that $y(x)$ is needed only to $O(x^3)$ in this part of the calculation. For the third iterate

$$\begin{aligned} z_3 &= (1 + 2x + x^2 - 8x^3)\left[2 - (1 + 2x + x^2 - 8x^3)(1 - 2x + 3x^2 \right. \\ &\quad \left. + 4x^3 - 5x^5 + x^6 - 2x^7)\right] + O(x^8) \\ &= 1 + 2x + x^2 - 8x^3 - 27x^4 - 29x^5 + 64x^6 + 328x^7. \end{aligned}$$

This example shows the number of terms doubling with each iteration; because of this it can be shown, Geddes *et al.* (1992, chapter 4), that the computational cost of computing z_n increases as $n \ln n$ rather than n^2, as for the direct method.

Exercise 9.40
Write a Maple procedure to implement the scheme defined in equation 9.62; reproduce the results quoted above and find $z_4(x)$ and $z_5(x)$.

Can you estimate how the time that the Maple command `series(1/y, x=0, N)` takes to compute depends upon N?

Note: it is necessary to use a different series y for each value of N because Maple remembers previous calculations and simply increasing N with the same series results in a false, low time.

The method can clearly be extended to find other functions of power series, and as an example we determine the square root of the series $y(x) = \sum_{k=0}^{\infty} a_k x^k$ with $a_0 > 0$. In this case we iteratively solve the equation

$$f = y(x) - z^2, \tag{9.63}$$

and the iterative sequence is

$$
\begin{aligned}
z_{n+1} &= z_n - \frac{f(z_n)}{f'(z_n)} \\
&= \frac{1}{2} z_n + \frac{y}{2z_n} \quad \mathrm{mod} \left(x^{2^{n+1}} \right)
\end{aligned}
\tag{9.64}
$$

with $z_0 = \sqrt{a_0}$. This scheme is more complicated because it involves finding $1/z_n$, though as $y - z_n^2 = O(x^{2^n})$ we need only know this to $O(x^{2^n})$, which is the order to which $z_n(x)$ is known.

Exercise 9.41
Write a Maple procedure using Newton's method to find a series representing the square root of

$$y(x) = 4 + 2x + 3x^3 + 6x^4 + x^5 + 10x^6 + x^7 + 9x^8 + 5x^9$$

to the order of the original series.

Exercise 9.42
Consider the generalisation of equation 9.63,

$$f = y(x) - g(z),$$

and show that the power series representing the function $g^{-1}(y(x))$ is given formally by the iterative sequence

$$z_{n+1} = z_n + \frac{y - g(z_n)}{g'(z_n)}.$$

Further, show that

$$y(x) - g(z_{n+1}) = -\frac{g''(z_n)}{2g'(z_n)^2} (y - g(z_n))^2 + O\left((y - g(z_n))^3 \right),$$

and deduce that if $y(x) - g(z_n(x)) = O(x^{p_n})$ then, with a suitable choice of z_0, $p_n = 2^n$.

The use of Newton's method to manipulate polynomials suggests that Halley's method, equation 9.46 (page 327), might produce even faster convergence. As we saw above, Halley's method does not help in finding the inverse, but for other functions it does provide a more rapidly converging scheme at the cost of some extra complexity. We illustrate the idea by finding the square root of a series; using the function f defined in equation 9.63, together with the scheme of equation 9.46, we obtain

$$z_{n+1} = z_n + \frac{2z_n(y - z_n^2)}{y + 3z_n^2},$$ (9.65)

which gives

$$y - z_{n+1}^2 = \frac{(y - z_n^2)^3}{(y + 3z_n^2)^2}.$$

Thus if $y(x) - z_n^2(x) = O(x^{q_n})$ it follows, as before, that $q_{n+1} = 3q_n$ and that if we set $z_0 = \sqrt{a_0}$, $a_0 > 0$, $q_n = 3^n$, which means that the number of terms triples at each iteration, in contrast to Newton's method in which it doubles. The cost of this more rapid convergence is that we now need to determine $y + 3z_n^2$ and then its inverse, instead of just the inverse of z_n.

Exercise 9.43
Write a Maple procedure for performing the iteration defined by equation 9.65. Use your procedure to find a power series $z(x)$, to $O(x^9)$, such that

$$z(x)^2 = 4 + 2x + 3x^3 + 6x^4 + x^5 + 10x^6 + x^7 + 9x^8 + 5x^9, \quad z(0) = 2,$$

and show directly that $q_{n+1} = 3q_n$.

Exercise 9.44
Generalise the result of exercise 9.42 to show that using Halley's scheme, equation 9.46 (page 327), and the function $f = y - g(z)$,

$$z_{n+1} = z_n + \frac{2(y - g(z_n))g'(z_n)}{2g'(z_n)^2 + g''(z_n)(y - g(z_n))},$$

and that

$$y(x) - g(z_{n+1}) = O\left((y - g(z_n))^3\right).$$

Exercise 9.45
If A is an $N \times N$ real matrix and x a real number, find iterative schemes equivalent to Newton's and Halley's methods to find (i) the inverse of $I + xA$, and (ii) the square root of $I + xA$, as a power series in x with the coefficients of x^k being $N \times N$ matrices.

Use Maple to apply both methods to find approximations to the square root, B, of the matrix $A = \begin{pmatrix} 1 & x \\ x & 1 \end{pmatrix}$. In both cases use three iterations to approximate B, compute the matrix $C = B^2 - A$ and compare the magnitude of

$$f(x) = \sqrt{C_{11}^2 + C_{12}^2 + C_{21}^2 + C_{22}^2}$$

for various values of x.

9.6 Exercises

Exercise 9.46
Use sixth-order perturbation theory to find an approximation to all the real roots of the equations

$$\text{(i)} \quad x^2 + 2x + 6\epsilon = 0, \qquad \text{(ii)} \quad x^3 - \epsilon x - 1 = 0.$$

Exercise 9.47
Another method of solving some algebraic equations, $f(x, \epsilon) = 0$, is to assume that the solution $x(\epsilon)$ has a Taylor's series expansion about $\epsilon = 0$ and to use the equation $f(x, \epsilon) = 0$ to directly compute the required differentials; for instance the first differential is obtained from

$$\frac{\partial f}{\partial x}\frac{dx}{d\epsilon} + \frac{\partial f}{\partial \epsilon} = 0$$

by setting $\epsilon = 0$ and x to the unperturbed root.

Use this method to show that for the equation $f(x, \epsilon) = x^2 - ax + \epsilon = 0$,

$$\frac{d^n \epsilon}{d\epsilon^n} = \frac{2^{n-1} \, 1.2.3. \cdots .(2n - 1)}{(a - 2x)^{2n-1}}$$

and use this develop the perturbation solution for equation 9.1.

Exercise 9.48
Show that a perturbation series expansion for the root of the equation

$$x^{1/3} - (a + \epsilon)x^{1/5} + b\epsilon = 0, \qquad a, \, b > 0,$$

near the real root $x = a^{15/2}$ is

$$x = a^{15/2} - \frac{15}{2}a^5 \left(b - a^{3/2} \right) \epsilon + \frac{15}{8}a^{5/2} \left(7b^2 + 13a^2 - 20ba^{3/2} \right) \epsilon^2 + O(\epsilon^3),$$

and find the next two terms of the expansion.

Exercise 9.49
Consider the equation

$$(x - 1)^n = \epsilon x, \qquad |\epsilon| \ll 1,$$

where n is a positive integer.

Show that if n is even there are two positive real roots near unity if $\epsilon > 0$ but none if $\epsilon < 0$, and that if n is odd there is always one real root near unity regardless of the sign of ϵ.

Show that an iterative scheme for these roots is

$$x_{k+1} = 1 \pm \delta x_k^{1/n}, \quad n \text{ even}, \quad \delta = \epsilon^{1/n}, \quad \epsilon > 0,$$
$$x_{k+1} = 1 + \delta x_k^{1/n}, \quad n \text{ odd}, \quad \delta = \text{sgn}(\epsilon)|\epsilon|^{1/n},$$

and hence find the first five terms in the perturbation expansion of these roots.

Exercise 9.50
Determine how the roots of the following polynomials behave as $\epsilon \to 0$:

$$\text{(i)} \quad \epsilon x^7 - \epsilon^2 x^5 + x - 1 = 0, \qquad \text{(ii)} \quad \epsilon^2 x^8 - \epsilon x^6 + x - 4 = 0.$$

Exercise 9.51

Show that the perturbation expansion of the real root of the equation $xe^{-x} = \epsilon$ that tends to zero as $\epsilon \to 0$ is

$$x = \epsilon + \epsilon^2 + \frac{3}{2}\epsilon^3 + \frac{8}{3}\epsilon^4 + \frac{125}{24}\epsilon^5 + \frac{54}{5}\epsilon^6 + O(\epsilon^7).$$

Exercise 9.52

Consider the real equation

$$1 + (x^\beta + \epsilon)^{1/\beta} = e^x, \quad x, \beta > 0, \quad 0 \le \epsilon \ll 1.$$

Show that there is a root near $x = 0$ which is given approximately by $x = \delta$, where

$$\delta = \left(\frac{2\epsilon}{\beta}\right)^{1/(1+\beta)}.$$

Show that the first three terms of the perturbation expansion are

$$x = \delta - \frac{1 + 3\beta}{12(1 + \beta)}\delta^2 - \frac{3\beta^3 - 18\beta^2 - 13\beta - 4}{288(1 + \beta)^2}\delta^3 + \cdots$$

and find the next two terms of this series.

Exercise 9.53

Consider the equation

$$1 - \epsilon x^2 = \sin x, \quad 0 < \epsilon \ll 1.$$

Show graphically that there are two real roots near $x = \pi/2$ and show that the first three terms of the series expansion of these roots is

$$x = \frac{\pi}{2} \pm \pi\sqrt{\frac{\epsilon}{2}} + \pi\epsilon \pm \frac{\pi}{96}(\pi^2 + 96)\epsilon\sqrt{2\epsilon},$$

and find the next two terms of this expansion.

Exercise 9.54

Consider the equation

$$A + \epsilon \cos x = A \sin(x - \epsilon), \quad A > 0, \quad 0 \le \epsilon \ll 1.$$

Show graphically that there are two real roots near $x = \pi/2$ and that the first two terms of their series expansion are

$$x = \frac{\pi}{2} + \left(\frac{B \mp 1}{B \pm 1}\right)\epsilon \pm \frac{B^3 \mp 3B^2 + B \pm 1}{6(B \pm 1)^2}\epsilon^3 + \cdots,$$

where $B = \sqrt{2A + 1}$, and find the next two terms in this series.

Exercise 9.55

Examine the nature of the roots of the equation

$$f(x) = x^3 - (1 - \epsilon)x^2 - x + 1 = 0, \quad |\epsilon| \ll 1.$$

Show that there are roots near ± 1 and that the series expansions of these roots are

$$x = -1 - \frac{1}{4}\epsilon - \frac{1}{16}\epsilon^2 + \cdots,$$

$$x = 1 \pm \sqrt{-\frac{\epsilon}{2} - \frac{3}{8}\epsilon \pm \frac{13}{128}(-\epsilon)^{3/2}},$$

and find the next four terms of each of these series.

Exercise 9.56
Show that the second-order approximation to the positive roots of

$$\sin x \cos y = e^{-x}, \quad \cos x \sin y = \frac{1}{x^2 + y^2}$$

are

$$x = n\pi + (-1)^{n+m}e^{-n\pi}\epsilon - e^{-2n\pi}\epsilon^2,$$

$$y = m\pi + \frac{(-1)^{n+m}}{\pi^2(n^2 + m^2)}\epsilon - \frac{2\left(m + n\pi^2 e^{-n\pi}(1 + n^2)\right)}{\pi^5(n^2 + m^2)^3}\epsilon^2,$$

where n and m are positive integers.

Exercise 9.57
One of the famous equations of celestial mechanics is Kepler's equation for the motion of a planet round the Sun,

$$t = u - \epsilon \sin u,$$

where t is the time, in suitable units, u one of the orbital angles defining the position on the elliptical path of the planet, and ϵ is the eccentricity of the ellipse, see figure 7.14 (page 259). Normally ϵ is small — for the Earth $\epsilon = 0.0167$ — and one requires u in terms of t. Show that the fourth-order perturbation expansion for $u(t)$ is

$$u(t) = t + \epsilon \sin t + \frac{\epsilon^2}{2}\sin 2t + \frac{\epsilon^3}{8}(3\sin 3t - \sin t)$$

$$+ \frac{\epsilon^4}{6}(2\sin 4t - \sin 2t) + \epsilon^5\left(\frac{1}{192}\sin t - \frac{27}{128}\sin 3t + \frac{125}{384}\sin 5t\right) + O(\epsilon^6).$$

Exercise 9.58
The function

$$f(x) = a^2 - x^2 + \epsilon x^4, \quad |\epsilon| \ll 1,$$

has two real roots $x = \pm b(a)$ with $b(a) \simeq a$ for small ϵ. Show that the integral

$$F(a, \epsilon) = \int_0^{b(a)} dx \, \frac{1}{\sqrt{a^2 - x^2 + \epsilon x^4}}$$

can be re-cast in the form

$$F(a, \epsilon) = \int_0^{\pi/2} d\phi \, \frac{1}{\sqrt{1 - \epsilon b^2(1 + \sin^2 \phi)}}.$$

Hence, or otherwise, find a perturbation expansion for $F(a, \epsilon)$ up to and including terms $O(\epsilon^6)$. A more general variant of this integral is dealt with in exercise 15.31 (page 623).

Hint: in the original integral change variables with the transformation $x = b \sin \phi$ and then use the identity $a^2 = b^2 - \epsilon b^4$.

Exercise 9.59

Show that an approximate solution, accurate to $O(\epsilon)$, of the equation

$$\frac{dx}{dt} = \sin x + \epsilon x, \quad x(0) = A, \quad |\epsilon| \ll 1,$$

where $0 \leq A \leq \pi$, is

$$x = u + \left\{ [A \cot A - \ln \sin A] \sin u - u \cos u + (\sin u) \ln \sin u \right\} \epsilon,$$

where $u(t) = 2 \tan^{-1}(e^t \tan(A/2))$.

Exercise 9.60

Show that an approximate solution, accurate to $O(\epsilon^2)$, of the equation

$$\frac{dx}{dt} = x - \epsilon x^\alpha, \quad 0 \leq \alpha < 1, \quad 0 \leq \epsilon \ll 1,$$

with the initial condition $x(0) = A > \epsilon^{1/(1-\alpha)}$, is

$$x(t) = Ae^t + \frac{A^\alpha \epsilon}{1 - \alpha} \left(e^{\alpha t} - e^t \right) + \frac{\alpha A^{2\alpha-1} \epsilon^2}{2(1 - \alpha)^2} \left(e^t + e^{(2\alpha-1)t} - 2e^{\alpha t} \right).$$

Why is the condition on A necessary?

Exercise 9.61

Show that the perturbation expansion $y = \sum_{k=0}^{\infty} \epsilon^k y_k(x)$ of the solution for the second-order equation

$$\frac{d^2 y}{dx^2} = \epsilon y^2 f(x), \quad y(0) = a, \quad y'(0) = b,$$

is given by

$$y_0(x) = a + bx,$$

$$y_n(x) = \int_0^x dt \, (x - t) f(t) \sum_{k=0}^{n-1} y_k(t) y_{n-1-k}(t), \quad n \geq 1.$$

In the case $f(x) = e^{-x}$ show that

$$y = a + bx + \left\{ x^2 b^2 + (2ab + 4b^2)x + a^2 + 4ab + 6b^2)e^{-x} \right.$$
$$\left. + (2b^2 + a^2 + 2ab)x - 4ab - 6b^2 - a^2 \right\} \epsilon + O(\epsilon^3).$$

Find the asymptotic form of the solution for large x to $O(\epsilon^4)$.

Exercise 9.62

Show that the first-order perturbation approximation to the solution of the nonlinear boundary value problem

$$\frac{d^2 y}{dx^2} + y = \frac{\cos x}{3 + \epsilon y^2}, \quad y(0) = y(\pi/2) = 0,$$

where ϵ is a small positive parameter, is

$$y(x) = \frac{2x - \pi}{12} \sin x + \epsilon y_1(x), \quad \text{where}$$

$$y_1(x) = -\frac{1}{62208} (2x - \pi) \left(4x^2 - 4\pi x + \pi^2 - 6\right) \sin x$$

$$-\frac{1}{82944} \left(16x^2 - 16\pi x + 7 - 2\pi^2\right) \cos x$$

$$-\frac{1}{82944} \left(8x^2 - 8\pi x - 7 + 2\pi^2\right) \cos 3x + \frac{1}{13824} (2x - \pi) \sin 3x.$$

Exercise 9.63

Consider the nonlinear eigenvalue problem

$$\frac{d^2 y}{dx^2} + \lambda y = \epsilon x y^2, \quad y(0) = y(\pi) = 0, \quad \int_0^\pi dx \, y(x)^2 = 1,$$

where $|\epsilon| \ll 1$.

Show that the unperturbed eigenvalues are $\lambda = n^2$, $n = 1, 2, \ldots$, and use perturbation theory to show that the first-order approximation to the nth eigenvalue is

$$\lambda_n = n^2 - \epsilon \frac{4(-1)^n}{3n} \sqrt{\frac{2}{\pi}},$$

and that to the nth eigenfunction is:

$$y_n(x) = \sqrt{\frac{2}{\pi}} \sin nx + \frac{\epsilon}{n^2 \pi} \left\{ x \left(1 - \frac{4(-1)^n}{3} \cos nx + \frac{1}{3} \cos 2nx\right) \right.$$

$$\left. + \frac{2}{9n} (5(-1)^n \sin nx - 2 \sin 2nx) \right\}.$$

Notice that in contrast to the linear system, $\epsilon = 0$, the value of the eigenvalue λ depends upon both the boundary conditions *and* the normalisation condition.

Exercise 9.64

Consider the equation

$$\frac{dx}{dt} = -\lambda x + x(t + \epsilon \sin t), \quad x(0) = A,$$

where A is a positive constant, $\lambda > 1$ and ϵ is a small parameter. Note that $x(t + \epsilon \sin t)$ means the function $x(t)$ evaluated at $t + \epsilon \sin t$.

By expressing this in the form suitable for the use of perturbation theory, show that an approximate solution is

$$x(t) = Ae^{-\mu t} - A\epsilon \mu (1 - \cos t) e^{-\mu t}$$

$$+ \frac{1}{4} A \epsilon^2 \mu^2 (3 - 4\cos t + \cos 2t) e^{-\mu t} + \frac{1}{8} A \epsilon^2 \mu (\mu - 2)(2t - \sin 2t) e^{-\mu t},$$

where $\mu = \lambda - 1$.

Exercise 9.65

Consider the linear integro-differential equation

$$\frac{dx}{dt} = -\lambda x + \epsilon \int_0^t ds\, e^{-a(t-s)} x(t-s), \quad x(0) = A > 0, \quad |\epsilon| \ll 1,$$

where λ and a are positive constants.

If $z(t)$ is defined by the relation $x = e^{-\lambda t} z$, show that

$$\frac{dz}{dt} = \epsilon e^{\lambda t} \int_0^t ds\, e^{-\mu s} z(s), \quad \mu = \lambda + a,$$

and deduce that if $z(t) = z_0 + \epsilon z_1(t) + \epsilon^2 z_2(t) + \cdots$ then the successive coefficients of ϵ^k are related by

$$\frac{dz_{n+1}}{dt} = e^{\lambda t} \int_0^t ds\, e^{-\mu s} z_n(s), \quad \text{that is,} \quad z_{n+1}(t) = \frac{1}{\lambda} \int_0^t ds\, e^{-\mu s} \left(e^{\lambda t} - e^{\lambda s} \right) z_n(s),$$

with $z_0 = A$.

Hence show that an approximate solution is

$$x(t) = Ae^{-\lambda t} \left(1 + \epsilon \left(\frac{\lambda e^{-at} + ae^{-\lambda t} - \lambda - a}{\lambda a(\lambda + a)} \right) + \cdots \right),$$

and find the next two terms.

10

Sturm–Liouville systems

10.1 Introduction

J. C. F. Sturm
(1803–1855)

J. Liouville
(1809–1882)

The theory presented in this chapter was created in a series of articles between 1829 and 1840 by Sturm and Liouville: their work, later known as Sturm–Liouville theory, created a whole new subject in mathematical analysis. In particular, it allows the generalisation of the idea of trigonometric series and extends the use of Green's functions to a wider range of differential equations. The theory deals with the general linear second-order differential equation

$$\frac{d}{dx}\left(p(x)\frac{dy}{dx}\right) + \left(q(x) + \lambda w(x)\right) y = 0, \tag{10.1}$$

where the real variable is confined to an interval, $a \leq x \leq b$, which may be the whole real line or just $x \geq 0$. The real functions $p(x)$, $q(x)$ and $w(x)$ are known and satisfy certain, not very restrictive, conditions that will be delineated in section 10.4. In addition to the differential equation, boundary conditions are specified, with the consequence that solutions exist for only particular values of the constant λ, which are named *eigenvalues*, and the solution $y_\lambda(x)$ is called the *eigenfunction* for the eigenvalue λ. At this stage we shall not specify any boundary conditions because different types of problems produce different types of conditions. Equation 10.1, together with any necessary boundary conditions, is known as a Sturm–Liouville system, or problem.

Sturm–Liouville problems are important partly because they arise in diverse circumstances and partly because the properties of the eigenvalues and eigenfunctions are well understood. For example, we find that there is an infinite set of *real* eigenvalues λ_k, $k = 0, 1, 2,\ldots$, and that the set of eigenfunctions $y_k(x)$, $k = 0, 1, 2,\ldots$, is complete, and that these functions may be used to form generalised Fourier series, section 10.4. Further, we know the leading term in the asymptotic expansion of both the eigenvalues and eigenfunctions as $k \to \infty$, section 10.6.

The achievements of Sturm and Liouville are more impressive when seen in the context of early nineteenth century mathematics. Prior to 1820, work on differential equations was concerned with finding solutions in terms of finite formulae; but for the general equation 10.1 Sturm could not find an expression for the solution and instead obtained information about the properties of the solution from the equation itself. This was the first qualitative theory of differential equations, and anticipated Poincaré's work on nonlinear differential equations developed at the end of that century; a discussion of this work and the relation to Sturm–Liouville theory can be found at the beginning of chapter 16. Today the work of Sturm and Liouville is intimately interconnected:

however, though lifelong friends who discussed their work prior to publication, this theory emerged from a series of articles published separately by each author during the period 1829 to 1840. More details of this history may be found in Lützen (1990, chapter 10).

Before introducing the general theory, section 10.4, in section 10.2 we consider the solutions of a very simple Sturm–Liouville problem, and in section 10.3 we show, in some detail, how this type of differential equation arises from some common partial differential equations. The simple example is discussed because almost all solutions of Sturm–Liouville equations behave in a similar manner, only the details vary. The derivation of Sturm–Liouville systems is provided because there are subtle variations in the type of differential equations that occur and it is helpful to know how and why these differences arise.

The general Sturm–Liouville theory is discussed in section 10.4, after which we continue the development of Green's function theory, started in chapter 7. Finally, in section 10.6 we determine the asymptotic expansions of the eigenvalues and eigenvectors in the limit of large order and show how the solutions of most Sturm–Liouville systems behave in a similar manner in this limit.

10.2 A simple example

It is helpful at this stage to consider a particular example, because the solutions of most Sturm–Liouville equations behave in a similar manner, independent of the forms taken by the functions p, q and w. Consider the elementary Sturm–Liouville system

$$\frac{d^2 y}{dx^2} + \lambda y = 0, \quad y(0) = y(\pi) = 0. \tag{10.2}$$

Here $p = w = 1$, $q = 0$ and the interval is taken, for convenience, to be $(0, \pi)$. If $\lambda < 0$ the general solution is $y = A \sinh(\sqrt{-\lambda}x) + B \cosh(\sqrt{-\lambda}x)$ and we see immediately that only one boundary condition can be satisfied. If $\lambda = 0$ only the trivial solution $y = 0$ satisfies both the equation and the boundary conditions. If $\lambda = \omega^2$, with $\omega > 0$, the general solution is $y = A \sin \omega x + B \cos \omega x$, which can be made to fit the boundary condition at $x = 0$ by choosing $B = 0$ and the condition at $x = \pi$ by making ω an integer, $\omega = 1, 2, \dots$. Because the equation is linear the value of the constant A is indeterminate, and we take as the solutions

$$y_n(x) = \sin nx, \quad \lambda_n = n^2, \quad n = 1, 2, \dots.$$

The eigenvalues are real, $\lambda_n = n^2$, and the eigenfunctions are $y_n(x) = \sin nx$. In this example each eigenvalue is associated with a unique eigenfunction. The boundary conditions, $y(0) = y(\pi) = 0$, determine that there are a countable number of eigenvalues and their dependence upon the index; thus, for instance, the boundary conditions $y'(0) = y'(\pi)$ will give different eigenvalues and eigenfunctions, see exercise 10.1.

Graphs of $y_n(x)$, and $y_{n+1}(x)$ for $n = 1$, 4 and 8 are shown in figures 10.1 to 10.3 respectively, with $y_{n+1}(x)$ depicted by the heavier line

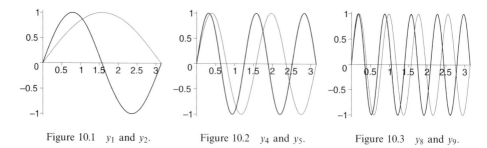

Figure 10.1 y_1 and y_2.   Figure 10.2 y_4 and y_5.   Figure 10.3 y_8 and y_9.

These figures suggest that $y_n(x)$ has $n-1$ zeros in the interval $0 < x < \pi$ and the zeros interlace, with a single zero of $y_{n+1}(x)$ between adjacent zeros of $y_n(x)$.

There are several important properties of the eigenvalues and eigenfunctions of this simple system that are common to the solutions of many Sturm–Liouville systems, so it is useful to list them here.

- The eigenvalues are real; this is a fundamental property and is a consequence of the form of the differential equation.
- The smallest eigenvalue is unity, but there is no largest eigenvalue. Further, $\lambda_n/n^2 = O(1)$ as $n \to \infty$. In general there is a smallest but no largest eigenvalue and λ_n increases as n^2 for large n: reasons for this are given in section 10.4.1.
- For each eigenvalue there is one eigenfunction. This is true for a wide variety of Sturm–Liouville problems, but is not universally true, as will be seen in exercise 10.1, part (iv), and the examples given in the next section.
- The nth eigenfunction has $n-1$ zeros in $0 < x < \pi$. For the general problem with the interval $[a, b]$ the nth eigenfunction has $n-1$ zeros in $a < x < b$.
- The zeros interlace, so there is one and only one zero of $y_{n+1}(x)$ between adjacent zeros of $y_n(x)$.
- The inner product, equation 8.1, of two eigenfunctions with *distinct* eigenvalues is zero,

$$\int_0^\pi dx\, y_n(x)y_m(x) = \int_0^\pi dx\, \sin nx \sin mx = 0, \quad n \neq m.$$

Again this is a universal property of the solutions of Sturm–Liouville systems and stems from the form of the differential equation, as shown in section 10.4, though in general it is necessary to use a modified inner product, equation 10.19 (page 356).

- The eigenfunctions $y_n(x) = \sin nx$ form a complete orthogonal system, so any sufficiently well behaved function $F(x)$ may be approximated in the mean by the series

$$F(x) = \sum_{k=1}^{\infty} a_n y_n(x)$$

for some coefficients a_k, equation 10.23 (page 357). This property generalises the theory of Fourier series developed in chapter 8, and many of the properties of Fourier series carry over to these more general series.

Exercise 10.1

Find *all* the eigenvalues and eigenfunctions of the Sturm–Liouville systems defined by the differential equation

$$\frac{d^2 y}{dx^2} + \lambda y = 0,$$

and the boundary conditions

(i) $y'(0) = y'(\pi) = 0,$
(ii) $y(0) = y'(\pi) = 0,$
(iii) $y(0) = 0,\ y(\pi) - y'(\pi) = 0,$
(iv) $y(x)$ is 2π-periodic on $-\pi \le x \le \pi.$

What do you notice about the eigenfunctions of the fourth example?

In each case show that the eigenfunctions belonging to distinct eigenvalues are orthogonal.

Exercise 10.2

Show that the general, second-order, linear differential equation

$$a_2(x)\frac{d^2 y}{dx^2} + a_1(x)\frac{dy}{dx} + a_0(x)y = 0$$

may be put in the form

$$\frac{d}{dx}\left(p(x)\frac{dy}{dx} \right) + q(x)y = 0,$$

where

$$p(x) = \exp\left(\int dx\, \frac{a_1(x)}{a_2(x)} \right) \qquad \text{and} \qquad q(x) = \frac{a_0(x)}{a_2(x)}p(x).$$

Exercise 10.3

There are many ways of transforming equation 10.1 into other forms. Show that the change of independent variable

$$\xi = \int_a^x \frac{dx}{p(x)}$$

removes the first differential to give the equation

$$\frac{d^2 y}{d\xi^2} + p(\xi)\left(q(\xi) + \lambda w(\xi) \right) y = 0.$$

10.3 Examples of Sturm–Liouville equations

In this section we show how various types of Sturm–Liouville problems arise. This material is not essential, but it is recommended that you read it, and do the exercises at the end of the section.

The original work of Sturm appears to have been motivated by the problem of heat conduction. One example he discussed is the temperature distribution in a one-dimensional bar, described by the linear partial differential equation

$$h(x)\frac{\partial u}{\partial t} = \frac{\partial}{\partial x}\left(p(x)\frac{\partial u}{\partial x}\right) - l(x)u, \tag{10.3}$$

where $u(x,t)$ denotes the temperature at a point x of the bar at time t, and h, p and l are positive functions. If the surroundings of the bar are held at constant temperature and the ends of the bar, at $x = 0$ and $x = L$, are in contact with large bodies at a different temperatures, then the boundary conditions are

$$\begin{aligned}
p(x)\frac{\partial u}{\partial x} + \alpha u(x,t) &= 0, \qquad \text{at } x = 0,\\[2mm]
p(x)\frac{\partial u}{\partial x} + \beta u(x,t) &= 0, \qquad \text{at } x = L,
\end{aligned} \tag{10.4}$$

for some constants α and β. Finally, the initial temperature of the bar needs to be specified, so $u(x,0) = f(x)$ where $f(x)$ is the known initial temperature.

Sturm attempted to solve this equation by first substituting a function of the form $u(x,t) = X(x)e^{-\lambda t}$ which yields the ordinary differential equation

$$\frac{d}{dx}\left(p(x)\frac{dX}{dx}\right) + \left(\lambda h(x) - l(x)\right)X = 0$$

for $X(x)$ in terms of the unknown constant λ, together with the boundary conditions

$$p(0)X'(0) + \alpha X(0) = 0, \quad p(L)X'(L) + \beta X(L) = 0.$$

This is an eigenvalue problem. Assuming that there are solutions $X_k(x)$ with eigenvalues λ_k, for $k = 1, 2, \dots$, Sturm used linearity of the original equation to write a general solution as the sum

$$u(x,t) = \sum_{k=1}^{\infty} A_k X_k(x)e^{-\lambda_k t},$$

where the coefficients A_k are arbitrary. This solution formally satisfies the differential equation and the boundary conditions, but not the initial condition $u(x,0) = f(x)$, which will be satisfied only if

$$f(x) = \sum_{k=1}^{\infty} A_k X_k(x).$$

Thus the problems reduces to that of finding the values of the A_k satisfying this equation.

J. Fourier
(1768–1830)

S. D. Poisson
(1781–1840)

Fourier and Poisson found expressions for the coefficients A_k for particular functions $h(x)$, $p(x)$ and $l(x)$, but Sturm and Liouville determined the general solution.

Typically, Sturm–Liouville equations arise when the method of separating variables is used to solve linear partial differential equations, some examples of which are

$$\nabla^2 \psi + k^2 \psi = 0, \tag{10.5}$$

$$\nabla^2\psi - \frac{1}{c^2}\frac{\partial^2\psi}{\partial t^2} = 0, \quad \text{wave equation,} \tag{10.6}$$

$$\frac{1}{\beta(x)}\frac{\partial}{\partial x}\left(\beta(x)\frac{\partial\psi}{\partial x}\right) - \frac{1}{c^2}\frac{\partial^2\psi}{\partial t^2} = 0, \quad \text{canal or horn equation,} \tag{10.7}$$

where c is a constant representing the speed of propagation of small disturbances in the media, k is a constant, $\beta(x)$ some positive function of x and

$$\nabla^2\psi = \frac{\partial^2\psi}{\partial x^2} + \frac{\partial^2\psi}{\partial y^2} + \frac{\partial^2\psi}{\partial z^2}.$$

The first of these equations arises in the solution of Poisson's equation — viz., $\nabla^2\psi = -F(\mathbf{r})$, for a scalar field $\psi(\mathbf{r})$ in terms of the source $F(\mathbf{r})$ — in the solution of the time-independent Schrödinger equation and the wave equation. The second equation, 10.6, is the wave equation for propagation of small disturbances in an isotropic medium and describes a variety of wave phenomena such as electromagnetic radiation, water and air waves, waves in strings and membranes. The last equation is a variant of the previous wave equation and in this form was derived by Green (1838) to describe waves on a canal of rectangular cross section but with a width varying along its length; a similar equation describes, approximately, the air pressure in a horn, though in many instruments the flare is sufficiently rapid for the longitudinal and radial modes to couple, so it is necessary to use the two-dimensional version of 10.7 in which the variation of the air pressure along the length of the pipe and in the radial direction are included.

The many applications of these equations have different ranges of validity and degrees of accuracy. Thus there are circumstances in which the results of calculations using Schrödinger's equation and Poisson's equation cannot be distinguished from experimental results. But in some mechanical applications effects not included in the derivation of these equations are significant and the solutions then provide only a first estimate of the system behaviour. An example is the piano wire: the frequency of the harmonics of the lowest note do not increase linearly with the order of the harmonic, as predicted by the wave equation, but somewhat faster, see Blackham (1978); in addition it would seem that there is coupling between the various wires so that a single note will excite other notes, resulting in a more complex response than suggested by a simple wave equation for an isolated wire.

Sturm–Liouville systems are derived from partial differential equations by separating variables, and it is often the boundaries that determine the coordinates to use, and these consequently determine important features of the resultant Sturm–Liouville equations. Thus we now derive two representative equations that illustrate different features. We consider, for illustrative purposes, the wave equation describing the vibrations of a perfectly flexible membrane,

$$\frac{\partial^2\psi}{\partial x^2} + \frac{\partial^2\psi}{\partial y^2} - \frac{1}{c^2}\frac{\partial^2\psi}{\partial t^2} = 0, \quad \psi(\mathbf{r}, t) = 0 \text{ on the boundary.} \tag{10.8}$$

Here the xy-plane is horizontal and the undisturbed membrane is assumed to lie in this plane, that is, gravity has a negligible effect by comparison to the tension in the membrane. The function $\psi(\mathbf{r},t)$ is the vertical displacement from this equilibrium position at time t and is assumed to be small. We consider rectangular and circular membranes.

For a rectangular membrane it is natural to use the cartesian coordinates defined by two edges of the membrane which has sides of lengths a and b, that is, $0 \le x \le a$ and $0 \le y \le b$. We start by looking for oscillations with an unspecified frequency ω, so assume a solution of the form $\psi(\mathbf{r},t) = F(x,y)e^{-i\omega t}$. On substituting this into equation 10.8 we obtain an equation for $F(x,y)$, independent of the time,

$$\frac{\partial^2 F}{\partial x^2} + \frac{\partial^2 F}{\partial y^2} + \frac{\omega^2}{c^2}F = 0, \quad F(\mathbf{r}) = 0 \text{ on the boundary.}$$

Because the membrane is rectangular it is sensible to assume a solution of the form $F(\mathbf{r}) = X(x)Y(y)$, where X is a function of x only and Y a function of y only. On substituting this into the above equation and dividing by F we arrive at the equation

$$\frac{1}{X}\frac{d^2 X}{dx^2} + \frac{1}{Y}\frac{d^2 Y}{dy^2} + \frac{\omega^2}{c^2} = 0.$$

The only way that this equation can be satisfied is if each term is a constant. Hence we can set $X''/X = -\lambda_x$ and $Y''/Y = -\lambda_y$ to obtain the following two equations for the spatial components:

$$\begin{aligned}
\frac{d^2 X}{dx^2} + \lambda_x X &= 0, & X(0) = X(a) = 0, \\
\frac{d^2 Y}{dy^2} + \lambda_y Y &= 0, & Y(0) = Y(b) = 0,
\end{aligned} \tag{10.9}$$

where the two eigenvalues λ_x and λ_y are related to the frequency by the equation

$$\lambda_x^2 + \lambda_y^2 = \frac{\omega^2}{c^2}.$$

These are the equations dealt with in section 10.2, only the range has changed. Thus the eigenvalues are $\lambda_x = (n\pi/a)^2$ and $\lambda_y = (m\pi/b)^2$, where n and m are positive integers, and this solution is

$$F(\mathbf{r}) = \sin\left(\frac{n\pi}{a}x\right)\sin\left(\frac{m\pi}{b}y\right).$$

The time-dependent solution for motion at this frequency is thus

$$\psi(\mathbf{r},t) = (A\sin\omega_{nm}t + B\cos\omega_{nm}t)\sin\left(\frac{n\pi}{a}x\right)\sin\left(\frac{m\pi}{b}y\right), \tag{10.10}$$

$$\frac{\omega_{nm}^2}{c^2} = \left(\frac{n\pi}{a}\right)^2 + \left(\frac{m\pi}{b}\right)^2,$$

so the frequency depends upon the values of the integers n and m.

You should note how the Sturm–Liouville equations arose from the original partial differential equation, and the role of the boundary conditions in this process. Note also that for each eigenvalue λ_x there is a unique eigenfunction $X(x)$, and similarly for λ_y.

Do not be deceived by the apparent simplicity of these solutions; if the length of the sides a and b are commensurate the nodal patterns of the oscillations of particular frequency can be quite complicated, as shown in exercise 10.7.

Equation 10.10 is not, of course, the complete solution. As for the heat equation treated above, we may write the general solution of this linear equation as a sum over all of these eigenfunctions,

$$\Psi(\mathbf{r}, t) = \sum_{n=1}^{\infty} \sum_{m=1}^{\infty} \sin\left(\frac{n\pi}{a}x\right) \sin\left(\frac{m\pi}{b}y\right) (A_{nm} \sin \omega_{nm} t + B_{nm} \cos \omega_{nm} t),$$

where the constants A_{nm} and B_{nm} are determined by the initial shape and velocity of the membrane. Thus if the membrane is initially stationary with a prescribed shape, $\Psi(\mathbf{r}, 0) = f(\mathbf{r})$ and $\partial\Psi/\partial t = 0$ at $t = 0$, we have $A_{nm} = 0$ and

$$f(\mathbf{r}) = \sum_{n=1}^{\infty} \sum_{m=1}^{\infty} B_{nm} \sin\left(\frac{n\pi}{a}x\right) \sin\left(\frac{m\pi}{b}y\right),$$

which is just a double Fourier series, so

$$B_{nm} = \frac{4}{ab} \int_0^a dx \int_0^b dy\, f(\mathbf{r}) \sin\left(\frac{n\pi}{a}x\right) \sin\left(\frac{m\pi}{b}y\right).$$

Exercise 10.4
Consider a drum initially flat, $\Psi(\mathbf{r}, 0) = 0$, and with velocity $\partial\Psi/\partial t = g(\mathbf{r})$ at $t = 0$, the initial conditions produced by an impulse. Show that with these initial conditions the subsequent motion is

$$\Psi(\mathbf{r}, t) = \sum_{n=1}^{\infty} \sum_{m=1}^{\infty} A_{nm} \sin\left(\frac{n\pi}{a}x\right) \sin\left(\frac{m\pi}{b}y\right) \sin \omega_{nm} t,$$

where

$$\omega_{nm} A_{nm} = \frac{4}{ab} \int_0^a dx \int_0^b dy\, g(\mathbf{r}) \sin\left(\frac{n\pi}{a}x\right) \sin\left(\frac{m\pi}{b}y\right).$$

The same problem, but with a different shaped boundary, produces different types of equations. For a circular boundary it is more natural to use the polar coordinates $x = r \cos\theta$, $y = r \sin\theta$, so the boundary conditions become $\psi(r, \theta, t) = 0$ for $r = r_a$ and $\psi(r, \theta, t)$ must be a 2π-periodic function of θ, that is, $\psi(r, \theta + 2\pi, t) = \psi(r, \theta, t)$ for all θ, which simply means that $\psi(\mathbf{r}, t)$ is a single valued function of the position; a condition tacitly assumed for the rectangular case.

In polar coordinates

$$\frac{\partial^2 \psi}{\partial x^2} + \frac{\partial^2 \psi}{\partial y^2} = \frac{1}{r}\frac{\partial}{\partial r}\left(r\frac{\partial \psi}{\partial r}\right) + \frac{1}{r^2}\frac{\partial^2 \psi}{\partial \theta^2}$$

and, as above, we search for the solution representing motion oscillating with the angular frequency ω, and consider a solution with the form $\psi = U(r, \theta)e^{-i\omega t}$. Substituting this into the equation gives

$$r\frac{\partial}{\partial r}\left(r\frac{\partial U}{\partial r}\right) + \frac{\partial^2 U}{\partial \theta^2} + r^2\frac{\omega^2}{c^2} U = 0,$$

which is solved by assuming a separable form $U = R(r)\Theta(\theta)$, to give the two equations

$$\frac{d^2\Theta}{d\theta^2} + \lambda\Theta = 0, \qquad \Theta(\theta + 2\pi) = \Theta(\theta),$$

$$\frac{d}{dr}\left(r\frac{dR}{dr}\right) + \left(r\omega^2 - \frac{\lambda}{r}\right) = 0, \qquad R(r_a) = 0,$$

(10.11)

where λ is the unknown separation constant, which is also the eigenvalue of the angular equation.

At this point it is worth noting the difference between these equations and 10.9 for the rectangular membrane. Now we have only one separation constant, λ, rather than two, and the boundary conditions are different. In the angular equation $p = w = 1$ and $q = 0$, as before, but the boundary conditions are different and give the eigenvalues $\lambda = n^2$, $n = 0, \pm 1, \pm 2, \ldots$ with the eigenfunctions

$$\Theta_n(\theta) = e^{in\theta}, \quad n = \pm 1, \pm 2, \ldots, \quad \Theta_0 = 1,$$

or, in real form, $\{\cos\theta, \sin\theta\}$. In this case all but one of the eigenvalues has *two* eigenfunctions, $\{\cos nx, \sin nx\}$: such behaviour is typical of periodic boundary conditions and should be contrasted with the behaviour of the general rectangular membrane in which each eigenvalue has a unique eigenfunction. When an eigenvalue has more than one eigenfunction the system is said to be *degenerate*, otherwise it is *non-degenerate*.

The radial equation is also different. The coordinate r is restricted to the range $0 \le r \le r_a$ and now $p(r) = w(r) = r$, $q(r) = -n^2/r$; the eigenvalue of this equation has become the square of the frequency, ω^2, in contrast to the previous case in which the frequency was determined by two eigenvalues, λ_x and λ_y. Finally, there is a *single* boundary condition at $r = r_a$: at the other boundary, $r = 0$, the function $p(r) = r = 0$ and in these circumstances the boundary condition at $r = 0$ is replaced by the condition that the solution be bounded for $0 \le r \le r_a$. Systems for which $p(r)$ is zero at one or both ends of the range are named *singular Sturm–Liouville* systems and have to be treated with a certain amount of caution, though they are common.

The zero in $p(r)$ at $r = 0$ is a property of the coordinate system used, not the physical system being studied. At $r = 0$ the coordinate system is singular because all values of θ label the same point. In three-dimensional spherical polar coordinates (r, θ, ϕ) there are similar problems at the north and south poles, where the value of ϕ is irrelevant and in this case for the θ-equation $p(\theta) = \sin\theta$. We should not, on physical grounds, expect that the Sturm–Liouville solution, which is singular by virtue of the coordinates used, to be significantly different from that of an ordinary Sturm–Liouville system.

To continue with the analysis of the circular membrane we see that the radial equation can be written as

$$\frac{d^2R}{dr^2} + \frac{1}{r}\frac{dR}{dr} + \left(\omega^2 - \frac{n^2}{r^2}\right)R = 0, \quad R(r_a) = 0 \quad \text{and } R(r) \text{ bounded.}$$

This is Bessel's equation with the two solutions $J_n(\omega r)$ and $Y_n(\omega r)$: since $|Y_n(\omega)|$ is unbounded at $r = 0$ the only relevant solution is $J_n(\omega r)$. The condition at $r = r_a$ determines the allowed values of the eigenvalue ω. For each value of n the eigenvalues

ω_{nj}, $j = 1, 2, \ldots$, are given by the positive solutions of the equation $J_n(\omega r_a) = 0$, as shown in exercise 10.5.

Thus the solutions that oscillate with frequency ω_{nj} are linear combinations of

$$J_n(\omega_{nj} r) e^{i(n\theta + \omega_{nj} t)} \quad \text{and} \quad J_n(\omega_{nj} r) e^{-i(n\theta - \omega_{nj} t)},$$

or in real form

$$\psi_{nj} = J_n(\omega_{nj} r) \Big\{ \cos n\theta \left[A_{nj} \cos \omega_{nj} t + B_{nj} \sin \omega_{nj} t \right]$$
$$+ \sin n\theta \left[C_{nj} \cos \omega_{nj} t + D_{nj} \sin \omega_{nj} t \right] \Big\}.$$

Exercise 10.5
The positive zeros of the ordinary Bessel function $J_v(x)$, $v \geq 0$, are traditionally denoted by $j_{v,s}$, $s = 1, 2, \ldots$, with $j_{v,1}$ the first positive zero. Use the Maple command `BesselJZeros` to find $j_{1,k}$, $k = 1, 2, \ldots, 10$. Plot graphs of $J_1(j_{1,k} x)$ for $0 < x < 1$ and various k to show numerically that:

(i) $J_1(j_{1,k} x)$ has $k - 1$ zeros for $0 < x < 1$, and
(ii) that the zeros of $J_1(j_{1,k+1} x)$ interlace those of $J_1(j_{1,k} x)$. Compare this behaviour with that of the example treated in section 10.2.

Repeat these comparisons for $J_2(j_{2,k} x)$.

Exercise 10.6
Use Maple to plot the graphs of $J_0(x)$, $J_{\pm3}(x)$ and $J_{\pm6}(x)$ for $0 < x < 10$, and infer that $\omega_{nj} = \omega_{-nj}$, for these values of n and all j. Use the integral representation,

$$J_n(x) = \frac{1}{2\pi} \int_{-\pi}^{\pi} du \, e^{i(nu - x \sin u)},$$

to show that $J_n(x) = (-1)^n J_{-n}(x)$ and $J_n(x) = (-1)^n J_n(-x)$. Show that $\omega_{nj} = \omega_{-nj}$ for all n and j.

Exercise 10.7

Nodal lines
It has been shown, equation 10.10, that a rectangular membrane will oscillate with the frequency given by

$$\frac{\omega_{nm}^2}{c^2} = \left(\frac{n\pi}{a}\right)^2 + \left(\frac{m\pi}{b}\right)^2 \quad \text{and shape} \quad f(\mathbf{r}) = \sin \frac{n\pi x}{a} \sin \frac{m\pi y}{b},$$

where n and m are integers.

The *nodal lines* of a solution are the lines along which the solution is zero, for all time, $f(\mathbf{r}) = 0$; for the membrane they are the curves along which the membrane is at rest. The nodal lines of the above solution are simply lines parallel to the axes.

The real numbers a and b are said to be *commensurable* if there are two relatively prime integers p and q such that $\frac{a}{b} = \frac{p}{q}$; otherwise they are *incommensurable*.

If the sides of the rectangle are commensurable, $b = aq/p$, then

$$\left(\frac{\omega a}{\pi c}\right)^2 q^2 = n^2 q^2 + m^2 p^2.$$

and there several solutions of the diophantine equation $n^2 q^2 + m^2 p^2 = N^2$. For instance, if the membrane is square, $p = q = 1$, and if $n = n_1$ and $m = n_2$ are solutions of $n^2 + m^2 = N^2$, so are $n = n_2$ and $m = n_1$. In this case the multiplicity of the eigenvalue is given by the solution to the number-theoretic question: in how many ways can a number N^2 be represented as the sum of two squares?

In such a case the nodal lines are more complicated and depend upon the exact linear combination of the component eigenfunctions. Thus in the case $a = b = 1$ and $n_1 = 3$, $n_2 = 1$ two distinctly different patterns are shown in the figures: other patterns are given by different values of A and B.

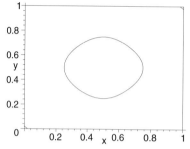

Figure 10.4   Nodal lines for $A = B = 1$.

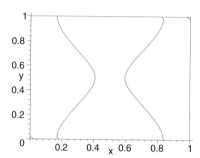

Figure 10.5   Nodal lines for $A = 1$, $B = -\frac{2}{3}$.

For the case $n = 4$, $m = 1$ produce an animated sequence of nodal curves for $A = 1$, $B = k/20$, $k = 0, 1, \ldots, 20$. Also consider the case $n = 5$, $m = 2$ with the same values for A and B.

In summary, we have shown that the equation of an oscillating membrane gives rise to a variety of types of Sturm–Liouville equation depending upon the shape of the boundary. For the rectangular boundary we obtained two regular non-degenerate equations for the spatial modes; for the circular boundary we obtained an angular equation that was degenerate and a radial equation that was singular. But in both cases the equations were of Sturm–Liouville type.

10.4 Sturm–Liouville systems

In this section the results illustrated in the previous sections are put on a more formal basis. A *regular Sturm–Liouville* system comprises the linear second-order differential equation†

$$\frac{d}{dx}\left(p(x)\frac{dy}{dx}\right) + \left(q(x) + \lambda w(x)\right) y = 0, \qquad (10.12)$$

---

† There is no agreed convention for the signs in this equation. For instance, in Courant and Hilbert (1965) the sign in front of $q(x)$ is negative and in Körner (1988) the signs in front of $q(x)$ and λ are negative. Since the behaviour of the eigenvalues can depend upon the sign of $q(x)$, see for instance exercise 10.16, care is needed when using different sources.

defined on a finite interval of the real axis $a \leq x \leq b$, together with the homogeneous boundary conditions

$$A_1 y(a) + A_2 y'(a) = 0, \quad B_1 y(b) + B_2 y'(b) = 0 \tag{10.13}$$

where

- the functions $p(x)$, $w(x)$ and $q(x)$ are real and continuous for $a \leq x \leq b$;
- $p(x)$ and $w(x)$ are positive for $a \leq x \leq b$;
- $p'(x)$ exists and is continuous for $a \leq x \leq b$;
- A_1, A_2, B_1 and B_2 are real constants and the trivial cases $A_1 = A_2 = 0$ and $B_1 = B_2 = 0$ are both excluded.

The examples treated in exercise 10.1 (page 345), parts (i), (ii) and (iii), are regular Sturm–Liouville problems with $p = w = 1$ and $q = 0$; but separation of variables in polar coordinates, equation 10.11, produces equations with periodic boundary conditions or $p(x) = 0$ at one or both end points. Thus we see that the above definition is too restrictive and excludes many commonly occurring problems; we partly rectify this by introducing singular Sturm–Liouville systems.

If $p(x)$ is positive for $a < x < b$ but vanishes at one or both ends, then a *singular Sturm–Liouville* system comprises the differential equation 10.12, with $w(x)$ and $q(x)$ satisfying the same conditions as for a regular system, and

- the solution $y(x)$ is bounded for $a < x < b$;
- at an end point at which $p(x)$ does not vanish $y(x)$ satisfies a boundary condition of the type 10.13.

For instance, the system

$$\frac{d}{dr}\left(r\frac{dy}{dr}\right) + \left(r\lambda - \frac{1}{r}\right)y = 0, \quad 0 \leq r \leq a, \tag{10.14}$$

with boundary condition $B_1 y(a) + B_2 y'(a) = 0$, $|B_1| + |B_2| \neq 0$, and $y(x)$ bounded is a singular Sturm–Liouville system. On defining $x = r\sqrt{\lambda}$ it becomes

$$x^2 \frac{d^2 y}{dx^2} + x\frac{dy}{dx} + (x^2 - 1)y = 0,$$

which is just Bessel's equation with the two linearly independent solutions $J_1(x)$ and $Y_1(x)$. But, $Y_1(x)$ is unbounded — near the origin $Y_1(x) \sim \ln x$ — thus $J_1(x)$ is the only bounded solution. This behaviour is typical; a singular Sturm–Liouville equation will often have two linearly independent solutions only one of which is bounded.

Exercise 10.8
Show that the eigenvalues of the equation 10.14 are given in terms of the roots of

$$B_1 J_1(au) + u B_2 J_1'(au) = 0, \quad \text{by} \quad \lambda = u^2.$$

In the case $B_1 = B_2 = 1$, $a = 2$, use Maple to show that the first four eigenvalues are approximately 1.37, 2.85, 4.38 and 5.94. Plot the graphs of these eigenfunctions $\phi_k(r) =$

$J_1(r\sqrt{\lambda_k})$ for $0 < r < 2$ and show numerically that

$$\int_0^2 dr\, r\phi_k(r)\phi_j(r) = h_k\delta_{kj},$$

where h_k are positive numbers.

The differential equation 10.12 looks special, but the result of exercise 10.2 (page 345), shows that any second-order differential equation with the form

$$a_2(x)\frac{d^2 y}{dx^2} + a_1(x)\frac{dy}{dx} + a_0(x)y + \lambda w(x)y(x) = 0$$

can be cast in this form provided $a_1(x)$ and $a_2(x)$ are sufficiently well behaved.

Many of the nice properties of the eigenvalues and eigenfunctions of a Sturm–Liouville system, regular and singular, stem from the special form of the differential equation 10.12; the first important result is *Lagrange's identity*,

$$vLu - uLv = \frac{d}{dx}\left[p(x)\left(v\frac{du}{dx} - u\frac{dv}{dx}\right)\right] \qquad (10.15)$$

where L is the real, linear differential operator

$$Lf = \frac{d}{dx}\left(p(x)\frac{df}{dx}\right) + q(x)f$$

and u and v are *any* functions for which both sides of the identity exist.

Exercise 10.9
Prove Lagrange's identity, equation 10.15.

The inner product† of two functions f and g was defined in chapter 8 to be the integral

$$(f,g) = \int_a^b dx\, f(x)^* g(x). \qquad (10.16)$$

It follows from Lagrange's identity that

$$(Lu, v) = (u, Lv), \qquad (10.17)$$

where $u(x)$ and $v(x)$ are two functions, which may be complex, for which Lu and Lv exist *and* which satisfy any boundary conditions. Note that u and v need not be solutions of the Sturm–Liouville equation.

An operator satisfying relation 10.17 is said to be *self-adjoint*. In order to prove this important relation we note that

$$\begin{aligned}(Lu, v) - (u, Lv) &= \int_a^b dx\left[vLu^* - u^*Lv\right]\\ &= \left[p(x)\left(v(x)\frac{du^*}{dx} - u(x)^*\frac{dv}{dx}\right)\right]_a^b,\end{aligned}$$

---

† There is no agreed version of the inner product notation. That adopted here is normally used in physics, particularly in quantum mechanics, but in mathematics texts the integrand is often taken to be $f(x)g(x)^*$. Provided one definition is used consistently the difference is immaterial.

where we have used the relation $(Lu)^* = Lu^*$, true because L is real. The boundary term vanishes at both $x = a$ and $x = b$, whether the system be regular or singular. For a regular system, at $x = a$ and assuming $A_2 \neq 0$, we have

$$
\begin{aligned}
A_1 u(a)^* &+ A_2 u'(a)^* &= 0 \\
A_1 v(a) &+ A_2 v'(a) &= 0
\end{aligned}
\quad\Longrightarrow\quad
\frac{u'(a)^*}{u(a)^*} = \frac{v'(a)}{v(a)},
$$

showing that the boundary term is zero at $x = a$; a similar analysis shows it to be zero at $x = b$. If $A_2 = 0$ then $u(a) = v(a) = 0$ and the same result follows.

For a singular system, if $p(a) = p(b) = 0$ the result is obvious. If, for instance, $p(a) = 0$ and $p(b) \neq 0$ then we use the above algebra at $x = b$. Thus in all cases $(Lu, v) = (u, Lv)$ and the operator L is self-adjoint. It is important to note that both the form of the differential operator *and* the boundary conditions determine whether or not an operator is self-adjoint, see exercises 10.12 and 10.17.

Exercise 10.10
Prove that if $p(x)$ is periodic and the boundary conditions are periodic then L is self-adjoint. Note: periodic boundary conditions are examples of mixed boundary conditions in which the values of the function, and possibly its derivative, at the two ends of the range are non-trivially related. Normally mixed boundary conditions produce operators that are not self-adjoint, exercise 10.17.

Exercise 10.11
In this chapter the operators considered are real, but complex operators are often useful.

Show that on the space of differentiable functions for which $\int_{-\infty}^{\infty} dx \, |u(x)|^2$ exists the real operator $L = \frac{d}{dx}$ is not self-adjoint, but that the complex operator $\mathscr{L} = iL$ is self-adjoint.

Exercise 10.12
Show that the operator L defined by

$$
Ly = \frac{d^2 y}{dx^2} + \alpha y = 0, \quad y(0) = A, \quad y'(\pi) = B,
$$

where α, A and B are nonzero constants, is not self-adjoint.
This exercise shows why the boundary conditions need to be homogeneous. In exercise 10.17 it will be seen that mixed boundary conditions may also produce non-self-adjoint operators.

The eigenvalues of a self-adjoint operator are real. If $\phi(x)$ is an eigenfunction corresponding to an eigenvalue λ, then $L\phi = -\lambda w \phi$ and

$$
(L\phi, \phi) = (-\lambda w \phi, \phi) = -\lambda^* (w\phi, \phi).
$$

Also

$$
(\phi, L\phi) = (\phi, -\lambda w \phi) = -\lambda (\phi, w\phi)
$$

and hence, since $w(x)$ is real,

$$
0 = (L\phi, \phi) - (\phi, L\phi) = (\lambda - \lambda^*) \int_a^b dx \, w(x) |\phi(x)|^2.
$$

Since $w(x) > 0$ and $|\phi(x)|^2 > 0$, for almost all x, the right hand side can be zero only if $\lambda = \lambda^*$, that is the eigenvalues of a Sturm–Liouville system are real: this proof is valid for regular and singular systems and if the boundary conditions are periodic.

Now consider two eigenfunctions $\phi(x)$ and $\psi(x)$ corresponding to distinct eigenvalues λ and μ, respectively, that is $L\phi = -\lambda w\phi$ and $L\psi = -\mu w\psi$. By the self-adjoint property,

$$
\begin{aligned}
0 &= (L\phi, \psi) - (\phi, L\psi) \\
&= -\lambda(w\phi, \psi) + \mu(\phi, w\psi) \\
&= (\mu - \lambda) \int_a^b dx\, w(x)\phi(x)^* \psi(x).
\end{aligned}
$$

Since we have assumed that $\mu \neq \lambda$ it follows that

$$
\int_a^b dx\, w(x)\phi(x)^* \psi(x) = \left(\sqrt{w}\phi, \sqrt{w}\psi\right) = 0. \tag{10.18}
$$

In other words, the weighted eigenfunctions of distinct eigenvalues are orthogonal, with respect to the inner product 10.16.

Alternatively, it is more convenient to re-define the inner product to make the eigenfunctions orthogonal. The modified inner product is, for suitable functions f and g,

$$
(f, g) = \int_a^b dx\, w(x)f(x)^* g(x), \tag{10.19}
$$

where $w(x) > 0$ is named the *weight* function and is the same function that multiplies the eigenvalue in the original Sturm–Liouville equation. We use the same notation for this inner product as that used above because there is rarely confusion between the two.

Thus far we have shown, very easily, that, should they exist, the eigenvalues are real and the eigenfunctions are orthogonal. Next we quote a theorem giving several important results: a proof of this may be found in Young (1988, chapter 11) or Hartman (1964, chapter 11).

The Sturm–Liouville theorem
Solutions of the regular Sturm–Liouville system

$$
\frac{d}{dx}\left(p(x)\frac{dy}{dx}\right) + (q(x) + \lambda w(x))\, y = 0 \tag{10.20}
$$

with the homogeneous boundary conditions

$$
A_1 y(a) + A_2 y'(a) = 0, \quad B_1 y(b) + B_2 y'(b) = 0, \tag{10.21}
$$

have the following properties.

- There is an infinite sequence of real eigenvalues λ_k, $k = 1, 2, \ldots$, and $\lambda_k \to \infty$ as $k \to \infty$: this means that there can be at most a finite number of negative eigenvalues.
- For each eigenvalue there is a unique — up to a multiplicative constant — eigenfunction $\phi_k(x)$, that is, the system is non-degenerate.

- The eigenfunctions are orthogonal with respect to the weight function $w(x)$,

$$\int_a^b dx\, w(x)\phi_k(x)\phi_j(x) = 0, \quad k \neq j, \tag{10.22}$$

and form a complete set of functions, which means that any function $f(x)$, for which $\int_a^b dx\, |f(x)|^2$ exists, can be represented by the infinite series

$$f(x) = \sum_{k=1}^{\infty} a_k \phi_k(x), \quad a_k = \frac{\int_a^b dx\, w(x)\phi_k(x)^* f(x)}{\int_a^b dx\, w(x)|\phi_k(x)|^2} = \frac{(\phi_k, f)}{(\phi_k, \phi_k)}, \tag{10.23}$$

which converges in the mean to $f(x)$,

$$\lim_{N \to \infty} \int_a^b dx\, \left| f(x) - \sum_{k=1}^{N} a_k \phi_k(x) \right|^2 = 0.$$

For most well behaved functions the series is also pointwise convergent. In particular the series converges pointwise to $f(x)$ at the end point $x = a$ if *either* $A_2 \neq 0$ or $f(a) = 0$; similarly at $x = b$.

- The nth eigenfunction, $\phi_n(x)$, has precisely $n - 1$ zeros in the interval $a < x < b$, and each zero of $\phi_{n+1}(x)$ lies between adjacent zeros of $\phi_n(x)$; this latter result is often known as the *oscillation theorem*; for other versions of this result see exercises 10.40 and 10.41 (pages 375 and 376).

A casual reading of this last paragraph might suggest that (*a*) the zeros of $\phi_n(x)$ never coincide with those of $\phi_m(x)$, for $m \neq n$, and that (*b*) the $n - 1$ zeros, $x_{n,k}$, $k = 1, 2, \ldots, n - 1$, of $\phi_n(x)$ are, if $n \gg 1$, approximately uniformly spread over the interval $a < x < b$. Neither of these statements is true. For the simple example treated in section 10.2 the nth eigenfunction is $\phi_n(x) = \sin nx$ with zeros $x_{n,k} = k\pi/n$, $k = 1, 2, \ldots, n - 1$, and there are many coincident zeros of ϕ_m and ϕ_n provided $m \neq n \pm 1$. For this simple example, $\phi_n = \sin nx$, the zeros $x_{n,k} = k\pi/n$ are uniformly spread between 0 and π; such behaviour is exceptional and is a consequence of the simple form of the particular differential equation; in exercise 10.36 we shall find an approximation to the spacing between adjacent zeros for large n and shall see that they are not normally evenly distributed in $a < x < b$.

Exercise 10.13
If $\phi_k(x)$, $k = 1, 2, \ldots$, are orthogonal with respect to the weight function $w(x)$ on the interval $a \leq x \leq b$, show that the mean square error,

$$E_N = \int_a^b dx\, w(x) \left| f(x) - \sum_{k=1}^{N} a_k \phi_k(x) \right|^2$$

is minimised by setting

$$a_k = \frac{\int_a^b dx\, w(x)\phi_k^*(x)f(x)}{\int_a^b dx\, w(x)|\phi_k(x)|^2} = \frac{(\phi_k, f)}{(\phi_k, \phi_k)}.$$

Exercise 10.14

In exercise 10.1 (page 345), you found the following sets of eigenfunctions, $y_n(x)$, and eigenvalues $\lambda_n = \omega_n^2$,

(i) $y_n(x) = \cos \omega_n x$, $\omega_n = n$, $n = 0, 1, \ldots$;

(ii) $y_n(x) = \sin \omega_n x$, $\omega_n = n + \frac{1}{2}$, $n = 0, 1, \ldots$;

(iii) $y_n(x) = \sin \omega_n x$, where $\tan \pi \omega_n = \omega_n$, $n = 1, 2, \ldots$, and $y_0 = \sinh v_0 x$ with $\tanh v_0 \pi = v_0$, and $\lambda_0 = -v_0^2$,

for the equation $d^2 y/dx^2 + \lambda y = 0$ with three different boundary conditions.

The Sturm–Liouville theorem shows that each of these sets of functions is complete on $(0, \pi)$. Use this to show that

$$x = \frac{\pi}{2} - \frac{4}{\pi} \sum_{k=0}^{\infty} \frac{\cos(2k+1)x}{(2k+1)^2}$$

$$x = \frac{2}{\pi} \sum_{k=0}^{\infty} \frac{(-1)^k}{(k+1/2)^2} \sin\left(k + \frac{1}{2}\right) x$$

$$x = -\frac{2(\pi-1)\cosh \pi v_0}{v_0(\pi - \cosh^2 \pi v_0)} \sinh v_0 x - 2(\pi-1) \sum_{k=1}^{\infty} \frac{\cos \omega_k \pi \sin \omega_k x}{\omega_k(\pi - \cos^2 \omega_k \pi)},$$

and use Maple to construct the partial sums of this series and to make a graphical comparisons in all three cases.

The expansion in equation 10.23 is a direct generalisation of Fourier series expansions dealt with in chapter 8. Indeed in modern texts the name *Fourier series* applies to expansions using complete orthogonal sets of functions; if a distinction is necessary the original Fourier series are named *trigonometric* series. The coefficients a_k in equation 10.23 are named *Fourier coefficients*.

One of the useful results of chapter 8 that generalises is *Parseval's theorem*, equation 8.18, which becomes

$$(f, f) = \int_a^b dx\, w(x) f(x)^2 = \sum_{k=1}^{\infty} (\phi_k, \phi_k) |a_k|^2. \tag{10.24}$$

The Sturm–Liouville theorem is powerful, but has one major flaw: it is valid only for regular problems, yet we have seen that many important cases arising in practice are singular or have periodic boundary conditions. The extension to general singular systems is problematical because not all such systems have eigenvalues, exercise 10.15, however, almost all the differential equations occurring in physical applications arise from particular types of variational principles, which can also be used to show that the resultant Sturm–Liouville system is complete and that the properties of the eigenfunctions are similar to those of a regular Sturm–Liouville system. Such methods are beyond the scope of this text but may be found in Courant and Hilbert (1965, chapter 6).

Decomposition of the delta function

One of the expansions required in the next section is that of the delta function, $\delta(x-\xi)$. Because this function is invariant under the interchange of x and ξ, the expansion must be of the form

$$\delta(x-\xi) = \sum_{k=1}^{\infty} a_k \phi_k(x) \phi_k(\xi),$$

where $\phi_k(x)$ is an eigenfunction of a Sturm–Liouville system; here we assume, for convenience, that all eigenfunctions are real. The coefficient a_n is found in the usual manner, by multiplying by $w(x)\phi_n(x)$ and integrating,

$$
\begin{aligned}
\int_a^b dx\, w(x)\phi_n(x)\delta(x-\xi) &= w(\xi)\phi_n(\xi) \\
&= \sum_{k=1}^{\infty} a_k \phi_k(\xi) \int_a^b dx\, w(x)\phi_n(x)\phi_k(x) \\
&= a_n(\phi_n, \phi_n)\phi_n(\xi),
\end{aligned}
$$

whence $a_n = w(\xi)/(\phi_n, \phi_n)$. Because $w(\xi)$ is independent of n we may write the final result in the symmetric form

$$\delta(x-\xi) = \sqrt{w(x)w(\xi)} \sum_{k=1}^{\infty} \frac{\phi_k(x)\phi_k(\xi)}{(\phi_k, \phi_k)}. \tag{10.25}$$

This decomposition of the delta function will be used later to express the Green's function in terms of a series of eigenfunctions.

This section has concentrated on some formal aspects of Sturm–Liouville systems and, apart from elementary cases involving trigonometric functions, no other examples of eigenfunctions have been given. In fact most of the special functions of mathematical physics are solutions of Sturm–Liouville systems, so many other examples exist. We shall describe some of these in the next chapter.

Exercise 10.15
Show that the singular Sturm–Liouville system

$$\frac{d}{dx}\left(x^2 \frac{dy}{dx}\right) + \lambda y = 0, \quad y(1) = 0,$$

with $y(x)$ bounded for $0 \le x \le 1$, has no eigenvalues.
Hint: find a particular solution using a function of the form $y(x) = x^a$, and then find the general solution by setting $y = x^a v(x)$.

Exercise 10.16
Consider the Sturm–Liouville problem with the boundary conditions

$$y(a) = y(b) = 0 \quad \text{or} \quad y(a) = y'(b) = 0 \quad \text{or} \quad y'(a) = y(b) = 0.$$

If $\phi_k(x)$ is the normalised eigenfunction, $(\phi_k, \phi_k) = 1$, of the eigenvalue λ_k, use the identity

$$\lambda_k = \int_a^b dx\, \lambda_k w(x)\phi_k(x)^2$$

to show that

$$\lambda_k = \int_a^b dx \left[p(x) \left(\frac{d\phi_k}{dx} \right)^2 - q(x)\phi_k(x)^2 \right] - \left[p(x)\phi_k \frac{d\phi_k}{dx} \right]_a^b.$$

Deduce that if $q(x) \leq 0$ for $a < x < b$ all the eigenvalues are positive.

Exercise 10.17

Consider the following system with mixed boundary conditions:

$$\frac{d^2 y}{dx^2} + \lambda y = 0, \quad y(0) = 0, \quad y(\pi) = ay'(0), \quad a > 0.$$

Show that if $0 < a < \pi$ there are a finite number of real eigenvalues given by the real roots of the equation $\sin \omega \pi = a\omega$, $(\omega_1, \omega_2, \ldots, \omega_N)$, with $\lambda = \omega^2$ and with eigenfunctions $y_k(x) = \sin \omega_k x$ and $N \sim 1/2a$.

Are these eigenfunctions orthogonal?

Show also that there are an infinite number of complex eigenvalues, also given by the zeros of the equation $\sin \omega \pi = a\omega$, and find the asymptotic expansion of these eigenvalues, in the limit where the imaginary part is large.

Exercise 10.18

Find the eigenvalues and eigenfunctions of the following Sturm–Liouville systems:

(i) $y'' + \lambda y = 0, \quad y'(0) = 0, \quad y(\pi) = 0,$
(ii) $y'' + \lambda y = 0, \quad y(0) = 0, \quad y(\pi) - 3y'(\pi) = 0,$
(iii) $y'' + y' + \lambda y = 0, \quad y(0) = 0, \quad y'(2) = 0.$

10.4.1 On why eigenvalues are bounded below

The Sturm–Liouville theorem states that the eigenvalues λ_k are bounded below but not above: at first sight this asymmetry seems anomalous so here we provide a heuristic explanation.

Suppose that $y(x)$ is an eigenfunction of the Sturm–Liouville equation with eigenvalue λ. We may assume y to be normalised to unity, so

$$\lambda = \lambda \int_a^b dx \, w(x)y(x)^2 = -\int_a^b dx \left(\frac{d}{dx} \left(p \frac{dy}{dx} \right) + qy \right) y,$$

the second equation following from the substitution of wy using the original equation 10.20. Now integrate the first term by parts to rewrite this in the form

$$\lambda = \int_a^b dx \left(p \left(\frac{dy}{dx} \right)^2 - qy^2 \right) - \left[py \frac{dy}{dx} \right]_a^b.$$

Supposing now that the boundary terms vanish, as they do in many cases, exercise 10.16, this gives

$$\lambda = \int_a^b dx \left(p \left(\frac{dy}{dx} \right)^2 - qy^2 \right). \tag{10.26}$$

If $q(x) \le 0$ then $\lambda > 0$. If $0 < q(x) < Q$ for $a \le x \le b$ then

$$\int_a^b dx\, qy^2 \le Q \int_a^b dx\, y^2 \le \frac{Q}{\min(w(x))},$$

so λ is bounded below. A more rigorous and general treatment of this problem may be found in Courant and Hilbert (1965, chapter 6, section 5).

Expression 10.26 for λ also shows why the eigenvalues are not bounded above. The nth eigenvalue has $n-1$ zeros in $a < x < b$, and since it is normalised to unity $|y_n'(x)|$ is proportional to n, consequently the equation 10.26 shows that $\lambda_n \sim n^2$. More refined estimates for both the eigenvalues and eigenfunctions in the large n limit are derived in section 10.6.

10.5 Green's functions

In chapter 7 Green's functions were introduced using simple examples. Here the same theory is applied to Sturm–Liouville systems, but first we recall the essential ideas of the Green's function by finding the solution of the elementary linear, boundary value problem

$$\frac{d^2y}{dx^2} = f(x), \quad y(a) = A, \quad y(b) = B. \tag{10.27}$$

The first stage is to use linearity and write the solution as the sum of two functions $y(x) = y_1(x) + y_2(x)$, where $y_1(x)$ satisfies the homogeneous equation with inhomogeneous boundary conditions

$$\frac{d^2y_1}{dx^2} = 0, \quad y_1(a) = A, \quad y_1(b) = B, \tag{10.28}$$

and where $y_2(x)$ satisfies the inhomogeneous equation with homogeneous boundary conditions

$$\frac{d^2y_2}{dx^2} = f(x), \quad y_2(a) = 0, \quad y_2(b) = 0. \tag{10.29}$$

The distinction between these two types of equation is important in the following theory, because only the second is a self-adjoint system.

The solution of the homogeneous equation is

$$y_1(x) = A + \frac{B-A}{b-a}(x-a).$$

The solution of the inhomogeneous equation is obtained with the help of the Green's function, $G(x, \xi)$, the solution of

$$\frac{d^2G}{dx^2} = \delta(x - \xi), \quad G(a, \xi) = 0, \quad G(b, \xi) = 0. \tag{10.30}$$

For $x < \xi$ and $x > \xi$ the right hand side of this equation is zero and the solution has the form $\alpha(\xi) + \beta(\xi)x$. The functions $\alpha(\xi)$ and $\beta(\xi)$ — which are different for $x < \xi$

and $x > \xi$ — are obtained by using the fact that $G(x, \xi)$ is continuous at $x = \xi$ and that

$$\lim_{\epsilon \to 0_+} \left[\frac{\partial G}{\partial x}\right]_{x=\xi-\epsilon}^{\xi+\epsilon} = 1. \tag{10.31}$$

These two conditions define the Green's function uniquely,

$$G(x, \xi) = \begin{cases} -\dfrac{(x-a)(b-\xi)}{b-a}, & x \le \xi, \\[2mm] -\dfrac{(b-x)(\xi-a)}{b-a}, & x \ge \xi, \end{cases} \tag{10.32}$$

and allow us to write the solution in the form

$$\begin{aligned} y(x) &= y_1(x) + \int_a^b d\xi \, G(x, \xi) f(\xi) \\ &= A + \frac{B-A}{b-a}(x-a) - \frac{b-x}{b-a}\int_a^x d\xi \, (\xi - a)f(\xi) - \frac{x-a}{b-a}\int_x^b d\xi \, (b-\xi)f(\xi). \end{aligned}$$

At this point you should do the following exercise to ensure that you can reproduce all the steps leading to this last equation.

Exercise 10.19
Show that the Green's function for the system

$$\frac{d^2y}{dt^2} + 2\frac{dy}{dt} + 2y = f(t), \quad y(0) = y'(0) = 0,$$

is $G(t, \tau) = H(t - \tau)e^{-(t-\tau)}\sin(t - \tau)$, so that the solution can be written as

$$y(t) = \int_0^t d\tau \, f(\tau)e^{-(t-\tau)}\sin(t - \tau).$$

Now consider the more general linear differential equation

$$Ly = \frac{d}{dx}\left(p(x)\frac{dy}{dx}\right) + q(x)y = f(x), \tag{10.33}$$

with the unmixed homogeneous boundary conditions

$$A_1 y(a) + A_2 y'(a) = 0, \quad B_1 y(b) + B_2 y'(b) = 0. \tag{10.34}$$

The Green's function for this system is the function $G(x, \xi)$, satisfying the equation

$$LG = \frac{d}{dx}\left(p(x)\frac{dG}{dx}\right) + q(x)G = \delta(x - \xi), \quad a < \xi < b, \tag{10.35}$$

and the boundary conditions 10.34. The Green's function satisfies the homogeneous equation $LG = 0$ for $x < \xi$ and $x > \xi$ and the solutions either side of ξ are related by the conditions:

- $G(x, \xi)$ is continuous at $x = \xi$;
- The first derivative of $G(x, \xi)$ is discontinuous at $x = \xi$:

$$\lim_{\epsilon \to 0_+} \left[\frac{\partial G}{\partial x} \right]_{x = \xi - \epsilon}^{\xi + \epsilon} = \frac{1}{p(\xi)}. \tag{10.36}$$

This is the generalisation of the condition 10.31.

The first condition is the same as was derived in chapter 7; the second condition is derived simply by integrating equation 10.35 across $x = \xi$: thus if $0 < \epsilon \ll 1$,

$$\int_{\xi - \epsilon}^{\xi + \epsilon} dx \left[\frac{d}{dx} \left(p(x) \frac{dG}{dx} \right) + q(x)G \right] = \int_{\xi - \epsilon}^{\xi + \epsilon} dx \, \delta(x - \xi) = 1$$

which gives

$$\left[p(x) \frac{\partial G}{\partial x} \right]_{\xi - \epsilon}^{\xi + \epsilon} + O(\epsilon) = 1.$$

Taking the limit $\epsilon \to 0$ yields the condition 10.36 because $p(x)$ is continuous.

The Green's function is a solution of 10.35, and since the operator is self-adjoint $(LG, y) = (G, Ly) = (G, f)$, that is,

$$\int_a^b d\xi \, LG \, y(\xi) = \int_a^b d\xi \, G(x, \xi) f(\xi),$$

and since $LG = \delta(x - \xi)$ this gives the solution of the original equation 10.33,

$$y(x) = \int_a^b d\xi \, G(x, \xi) f(\xi).$$

The problem now is to find the Green's function. It is normally easier to do this than to directly integrate the original equation because the Green's function comprises terms that satisfy only one of the two boundary conditions. Let $u_1(x)$ be any function that satisfies the equation $Lu_1 = 0$ and the boundary condition at $x = a$, so that the most general solution of this type is $c_1 u_1(x)$, c_1 being a nonzero constant — note that we have again used the linearity of the equation. Similarly, let $u_2(x)$ satisfy $Lu_2 = 0$ and the boundary condition at $x = b$, so $c_2 u_2(x)$ is the most general such solution. The Green's function may therefore be written in the form

$$G(x, \xi) = \begin{cases} c_1(\xi) u_1(x), & x \le \xi, \\ c_2(\xi) u_2(x), & x \ge \xi, \end{cases} \tag{10.37}$$

where now c_1 and c_2 are functions, unknown at present, of ξ.

From Lagrange's identity, equation 10.15 (page 354), we have $u_1 u_2' - u_2 u_1' = c/p(x)$ for some constant c. There are two possibilities, either $c = 0$ or $c \ne 0$. If $c = 0$ then $u_1'/u_1 = u_2'/u_2$ for all x and hence $u_2(x)$ is proportional to $u_1(x)$, that is, the solutions are linearly dependent; we shall treat this as an exceptional case and deal with it later.

If $c \ne 0$ then $u_1 u_2' - u_2 u_1' \ne 0$ for all x and the curves $c_1 u_1(x)$ and $c_2 u_2(x)$ can never touch, for if there was such a point $x = \eta$ then both the equations $c_1 u_1(\eta) = c_2 u_2(\eta)$

and $c_1 u_1'(\eta) = c_2 u_2'(\eta)$ would be satisfied and this contradicts the condition. Thus if $c \neq 0$ the two solutions are linearly independent.

Assuming that the two functions $u_1(x)$ and $u_2(x)$ are linearly independent, we may continue with the construction of the Green's function. Continuity at $x = \xi$ is produced by setting $c_1(\xi) = d u_2(\xi)$ and $c_2(\xi) = d u_1(\xi)$ for some nonzero d. The left and right derivatives of G are then

$$\lim_{\epsilon \to 0_+} p(x) \left. \frac{\partial G}{\partial x} \right|_{x = \xi - \epsilon} = d\, p(\xi) u_2(\xi) u_1'(\xi)$$

$$\lim_{\epsilon \to 0_+} p(x) \left. \frac{\partial G}{\partial x} \right|_{x = \xi + \epsilon} = d\, p(\xi) u_1(\xi) u_2'(\xi),$$

so that condition 10.36 gives

$$\frac{1}{d} = p(\xi) W(\xi), \quad \text{where} \quad W(\xi) = u_1(\xi) u_2'(\xi) - u_2(\xi) u_1'(\xi), \tag{10.38}$$

which exists because $u_1(\xi)$ and $u_2(\xi)$ are linearly independent. The function $W(\xi)$ is known as the *Wronskian* or Wronskian determinant† of the functions $u_1(\xi)$ and $u_2(\xi)$, and it is important precisely because the two functions are linearly independent whenever their Wronskian is not identically zero, exercise 10.21; in general the Wronskian can be defined for any number of functions, see for instance Arfken and Weber (1995, chapter 8) or Kreider *et al.* (1966, chapter 3).

It follows that the Green's function may be written in the form

$$G(x, \xi) = \frac{H(\xi - x) u_2(\xi) u_1(x) + H(x - \xi) u_1(\xi) u_2(x)}{p(\xi) W(\xi)}, \tag{10.39}$$

and the solution to equation 10.33 satisfying the boundary conditions 10.34 is

$$y(x) = u_2(x) \int_a^x d\xi\, \frac{f(\xi) u_1(\xi)}{p(\xi) W(\xi)} + u_1(x) \int_x^b d\xi\, \frac{f(\xi) u_2(\xi)}{p(\xi) W(\xi)}. \tag{10.40}$$

This is the generalisation of the result obtained in section 7.3.5.

Exercise 10.20
Use equation 10.40 to write the solution of the equation

$$\frac{d^2 y}{dx^2} + \omega^2 y = f(x), \quad y(0) = y(\pi),$$

where ω is positive but not an integer, in terms of a suitable Green's function.

Exercise 10.21
The Wronskian
Two functions $y_1(x)$ and $y_2(x)$ are linearly independent whenever their Wronskian determinant

$$W(x) = y_1 \frac{dy_2}{dx} - \frac{dy_1}{dx} y_2$$

is not *identically* zero. The converse is not true: that is if $W(x) \equiv 0$ then it is not necessarily true that the functions are linearly dependent, as the following example shows.

---

† The Wronskian is often denoted by $W[u_1, u_2]$ in order to emphasis its dependence on the functions.

The two functions $y_1 = x^3$ and $y_2 = |x|^3$ are linearly independent for $-1 < x < 1$. Show that their Wronskian is identically zero.

This is clearly a pathological example and a few restrictions on the nature of the two functions suffices to allow the inverse. If $y_1(x)$ and $y_2(x)$ are solutions of the homogeneous linear differential equation

$$a_2(x)\frac{d^2y}{dx^2} + a_1(x)\frac{dy}{dx} + a_0(x)y = 0, \quad a \le x \le b,$$

and if $y_1 y_2' - y_1' y_2 \equiv 0$ for $a \le x \le b$ then y_1 and y_2 are linearly dependent, see exercise 10.38 (page 375).

10.5.1 Eigenfunction expansions

In this section we shall, for simplicity, assume that all eigenfunctions are real; this is not restrictive for the reason given in exercise 10.37 (page 375). We have shown how to construct the Green's function, $G(x, \xi)$, for the self-adjoint differential operator

$$Ly = \frac{d}{dx}\left(p(x)\frac{dy}{dx}\right) + q(x)y,$$

with the homogeneous boundary conditions defined in equation 10.34, using two linearly independent solutions of $Lu = 0$ each satisfying only one boundary condition. In that analysis it was assumed that a non-trivial solution of $Lu = 0$ satisfying *both* boundary conditions did not exist. Now we find an alternative representation in terms of the eigenfunctions of L which, besides being useful, provides a method of dealing with the case where $u_1(x)$ and $u_2(x)$ are linearly dependent and the previous method fails.

Suppose that $\phi_k(x)$ are the eigenfunctions of the Sturm–Liouville system, $L\phi_k + \lambda_k w \phi_k = 0$, with eigenvalues λ_k, $k = 1, 2, \ldots$, and that none of the λ_k are zero. Then it follows from the eigenfunction representation of the delta function, equation 10.25 (page 359), that the Green's function may be expressed in terms of the series

$$G(x, \xi) = -\sum_{k=1}^{\infty} \frac{\phi_k(x)\phi_k(\xi)}{(\phi_k, \phi_k)\,\lambda_k}. \tag{10.41}$$

Exercise 10.22
By applying the operator L to each term of the sum 10.41 and using the expansion 10.25, show that the function $G(x, \xi)$ defined by the infinite series is the required Green's function.

Exercise 10.23
Use the eigenfunctions of the Sturm–Liouville system

$$\frac{d^2y}{dx^2} + \lambda y = 0, \quad y(0) = y(\pi) = 0,$$

to construct the Green's function for the equation $\frac{d^2G}{dx^2} = \delta(x - \xi)$. Hence show that

$$-\frac{2}{\pi} \sum_{k=1}^{\infty} \frac{\sin kx \sin k\xi}{k^2} = \begin{cases} -\dfrac{x(\pi - \xi)}{\pi}, & 0 \le x \le \xi, \\[2mm] -\dfrac{(\pi - x)\xi}{\pi}, & \xi \le x \le \pi. \end{cases}$$

A minor generalisation of equation 10.41, which is useful when we consider small perturbations in the next section, is obtained by expressing the Green's function of the operator

$$Hy = Ly + \mu w(x)y = \frac{d}{dx}\left(p(x)\frac{dy}{dx}\right) + \left(q(x) + \mu w(x)\right)y \qquad (10.42)$$

in terms of the eigenvalues and eigenfunctions of L. We have

$$G(x, \xi) = \sum_{k=1}^{\infty} \frac{\phi_k(x)\phi_k(\xi)}{(\mu - \lambda_k)} \frac{1}{(\phi_k, \phi_k)}, \qquad (10.43)$$

which is valid provided μ is not equal to one of the eigenvalues. This generalises the result found in exercise 7.28 (page 253).

Exercise 10.24
Prove equation 10.43.

The expansion 10.41 for the Green's function is valid provided zero is not an eigenvalue, that is provided that there is no non-trivial solution, $u_0(x)$, of the equation $Lu_0 = 0$. But this was just the condition required for the construction of the Green's function of equation 10.39. Recall that we required the functions $u_1(x)$ and $u_2(x)$ to be linearly independent, and that if they were linearly dependent then there exists a solution of $Lu = 0$ satisfying both boundary conditions.

The zero-eigenvalue problem is connected with the fact that the equation $Lu = f$ has a unique solution if and only if $u_0(x)$, the solution of $Lu_0 = 0$, is orthogonal (with weight function unity) to $f(x)$, that is

$$(u_0, f) = \int_a^b dx\, u_0(x)f(x) = 0.$$

We may see that this is necessary by taking the inner product of the equation $Lu = f$ with u_0, $(u_0, Lu) = (u_0, f)$ but since L is self-adjoint $(u_0, Lu) = (Lu_0, u) = 0$ and hence $(u_0, f) = 0$. We show that it is sufficient by constructing the solution.

Consider the equation

$$Ly = \frac{d}{dx}\left(p(x)\frac{dy}{dx}\right) + q(x)y = 0 \qquad (10.44)$$

with the boundary conditions defined in equation 10.34, and assume that zero is an eigenfunction of L, with nonzero function $\phi_0(x)$. Let $(\lambda_0, \lambda_1, \ldots)$ and $(\phi_0(x), \phi_1(x), \ldots)$, with $\lambda_0 = 0$, be the eigenvalues and functions of L, $L\phi_k = -\lambda_k w \phi_k$. The expansion 10.41, valid in the absence of a zero eigenvalue, suggests defining the generalised Green's function,

$$\overline{G}(x,\xi) = -\sum_{k=1}^{\infty} \frac{\phi_k(x)\phi_k(\xi)}{\lambda_k(\phi_k,\phi_k)}, \qquad (10.45)$$

as a sum over all but one of the eigenfunctions. Then

$$
\begin{aligned}
L\overline{G}(x,\xi) &= w(x)\sum_{k=1}^{\infty} \frac{\phi_k(x)\phi_k(\xi)}{(\phi_k,\phi_k)}, \\
&= \delta(x-\xi) - \frac{\phi_0(x)\phi_0(\xi)}{(\phi_0,\phi_0)} w(x). \qquad (10.46)
\end{aligned}
$$

From this last equation it follows that the solution of

$$Ly = f \quad \text{is} \quad y(x) = \int_a^b d\xi \, f(\xi)\overline{G}(x,\xi),$$

as may be seen directly by applying L to the second equation:

$$
\begin{aligned}
Ly &= \int_a^b d\xi \, f(\xi)L\overline{G} \\
&= f(x) - \frac{w(x)\phi_0(x)}{(\phi_0,\phi_0)} \int_a^b d\xi \, \phi_0(\xi)f(\xi).
\end{aligned}
$$

But since $(\phi_0, f) = 0$ the result follows.

Equation 10.46 may be regarded as the equation defining $\overline{G}(x,\xi)$. However, it does not determine $\overline{G}$ uniquely for if $\overline{G}$ is a solution so is $\overline{G}(x,\xi) + a(\xi)\phi_0(x)$, for some function $a(\xi)$. The solution is made unique and consistent with the series 10.45 by choosing it to be orthogonal to $\phi_0(x)$. Then $\overline{G}(x,\xi)$ is named the generalised Green's function and is the necessary modification if the system has a zero eigenvalue.

In order to show how this theory works, consider the system with periodic boundary conditions

$$Ly = \frac{d^2y}{dx^2}, \qquad y(-\pi) = y(\pi), \ y'(-\pi) = y'(\pi), \qquad (10.47)$$

dealt with in exercise 10.1. There it was shown that the eigenvalues are $\lambda_n = n^2$, $n = 0, \pm 1, \pm 2, \ldots$, with the eigenfunctions

$$\phi_0 = a, \quad \phi_n(x) = \{\sin nx, \cos nx\}, \quad n = 1, 2, \ldots,$$

where a is a nonzero constant. This system is degenerate and has zero for an eigenvalue. It is also the system that generates the orthogonal functions that provide the basis for the conventional Fourier series, so is not pathological.

Since the eigenfunction associated with the zero eigenvalue is a constant, it follows from the general theory that the equation

$$\frac{d^2u}{dx^2} = A = \text{constant}, \qquad u(-\pi) = u(\pi), \ u'(-\pi) = u'(\pi), \qquad (10.48)$$

has no solution unless $A = 0$, for only then is A orthogonal to $\phi_0(x) = a \neq 0$.

For this system $\phi_0(x) = a$ and $w(x) = 1$ so

$$w(x)\frac{\phi_0(x)\phi_0(\xi)}{(\phi_0, \phi_0)} = \frac{1}{2\pi},$$

and equation 10.46 for the generalised Green's function becomes

$$\frac{d^2\overline{G}}{dx^2} + \frac{1}{2\pi} = \delta(x - \xi). \tag{10.49}$$

For $x \neq \xi$ the general solution of this equation is

$$\overline{G}(x, \xi) = \begin{cases} -\dfrac{1}{4\pi}(x - \xi)^2 + A(\xi)(x - \xi) + B(\xi), & x < \xi, \\[2mm] -\dfrac{1}{4\pi}(x - \xi)^2 + C(\xi)(x - \xi) + D(\xi), & x > \xi, \end{cases}$$

for some functions A, B, C and D. The conditions at $x = \xi$ give

$$\begin{aligned} B(\xi) - D(\xi) &= 0, \quad \text{continuity,} & (10.50) \\ C(\xi) - A(\xi) &= 1, \quad \text{discontinuity in the derivative.} & (10.51) \end{aligned}$$

Also the boundary conditions give

$$-\frac{1}{4\pi}(\pi + \xi)^2 - (\pi + \xi)A(\xi) = -\frac{1}{4\pi}(\pi - \xi)^2 - (\pi - \xi)C(\xi) \tag{10.52}$$

$$\frac{1}{2\pi}(\pi + \xi) + A(\xi) = -\frac{1}{2\pi}(\pi - \xi) + C(\xi). \tag{10.53}$$

Equation 10.53 is the same as 10.51. Equation 10.52 can be solved to give $C = -A = \frac{1}{2}$, so the solution is

$$\overline{G} = -\frac{1}{4\pi}(x - \xi)^2 + \frac{1}{2}|x - \xi| + B(\xi)$$

for some $B(\xi)$. This solution is orthogonal to $\phi_0 = $ constant if $\int_{-\pi}^{\pi} dx\, \overline{G}(x, \xi) = 0$, which gives $B = -\pi/6$. Hence the solution which is orthogonal to $\phi_0(x)$ is

$$\overline{G}(x, \xi) = \frac{1}{2}|x - \xi| - \frac{(x - \xi)^2}{4\pi} - \frac{\pi}{6}. \tag{10.54}$$

Exercise 10.25
Show explicitly that equation 10.48 has a non-trivial solution only if $A = 0$.

Exercise 10.26
Show, by direct integration, that the solution of the system

$$\frac{d^2y}{dx^2} = \sin x, \qquad y(-\pi) = y(\pi),\ y'(-\pi) = y'(\pi)$$

is $y = -\sin x + A$, where A is a constant.

Use the generalised Green's function defined in equation 10.54 to obtain the solution

$$y(x) = \int_{-\pi}^{\pi} d\xi\, \overline{G}(x, \xi)\sin \xi = -\sin x.$$

Explain why these two solutions differ.

Exercise 10.27

Use equation 10.54 and the expansion 10.45 to show that

$$\frac{1}{2}|x-\xi| - \frac{(x-\xi)^2}{4\pi} = \frac{\pi}{6} - \frac{1}{\pi}\sum_{k=1}^{\infty}\frac{\cos k(x-\xi)}{k^2},$$

and use Maple to plot some representative graphs of the left hand side and partial sums of the right hand side.

10.5.2 Perturbation theory

The ideas of the previous section allow the development of a simple first-order perturbation theory for eigenvalues and eigenfunctions. Consider the operator $Ly = (py')' + qy$ with the eigenvalues λ_k and eigenfunctions satisfying $L\phi_k = -\lambda_k w\phi_k$ with the boundary conditions defined in equation 10.34 (page 362); for simplicity assume that the eigenfunctions are real, non-degenerate and that they have been normalised, $\int_a^b dx\, w\phi_k^2 = 1$: neither assumption is necessary.

First we concentrate upon the effect on the eigenvalues of changing either $q(x)$ or $p(x)$ by a small amount. Suppose that $q(x)$ is replaced by $q(x) + \epsilon h(x)$ with $|\epsilon| \ll 1$ and $h(x)$ an integrable function. Then we require the new eigenvalues μ_k and eigenfunctions $\psi_k(x)$ satisfying the equation $(L + \epsilon h)\psi_k = -\mu_k w\psi_k$ with the same boundary conditions as the original equation. The equation for the nth eigenfunction may be re-written in the form

$$(L + \lambda_n w)\psi_n = (\lambda_n - \mu_n)w\psi_n - \epsilon h\psi_n. \tag{10.55}$$

But we noted above that the self-adjoint system $\mathscr{L}u = f$ has a unique solution only if the solution u_0 of $\mathscr{L}u_0 = 0$ is orthogonal to f. In the present context this means that equation 10.55 has a unique solution only if the solution of $(L + \lambda_n w)u = 0$, that is, ϕ_n, is orthogonal to the right hand side. Thus

$$(\lambda_n - \mu_n)\int_a^b dx\, w(x)\phi_n(x)\psi_n(x) = \epsilon \int_a^b dx\, \phi_n(x)h(x)\psi_n(x).$$

But if $|\epsilon h(x)|$ is small we should expect $|\psi_n - \phi_n|$ and $|\lambda_n - \mu_n|$ both to be $O(\epsilon)$ — an expectation that may be proved, see for instance Courant and Hilbert (1965, chapter 6) — and thus to first order we may replace ψ_n by ϕ_n to obtain the first-order estimate of the change in the eigenvalue,

$$\mu_n = \lambda_n - \epsilon \int_a^b dx\, h(x)\phi_n(x)^2, \quad (\phi_n, \phi_n) = 1. \tag{10.56}$$

In a similar manner it may be shown that changing $p(x)$ to $p(x) + \epsilon h(x)$ produces the change

$$\mu_n = \lambda_n + \epsilon \int_a^b dx\, h(x)\left(\frac{d\phi_n}{dx}\right)^2, \tag{10.57}$$

and changing $w(x)$ to $w(x)(1 + \epsilon h(x))$ produces the change

$$\mu_n = \lambda_n \left(1 - \epsilon \int_a^b dx\, w(x) h(x) \phi_n(x)^2\right). \tag{10.58}$$

These results show, in particular, that

- if the change in either $q(x)$ or $w(x)$ is everywhere non-negative all eigenvalues *decrease*;
- if the change in $p(x)$ is everywhere non-negative all eigenvalues *increase*.

Exercise 10.28
Prove equations 10.57 and 10.58.

One physical interpretation of these results is found in the wave equation for the stretched string, of length L and fixed at both ends. The equation of motion is

$$\frac{\partial^2 \psi}{\partial x^2} - \frac{1}{c^2}\frac{\partial^2 \psi}{\partial t^2} = 0, \quad c^2 = \frac{T}{\rho}, \quad \psi(0, t) = \psi(L, t) = 0,$$

where T is the tension and ρ the density. It is a well known empirical fact that pulling a string tighter (increasing T) increases the frequency of vibration; we now show that both the exact solution of this equation and the above perturbation theory agree with this observation.

First consider the exact solution: setting $\psi = X(x)e^{-i\omega t}$ gives the eigenvalue equation for X, as in section 10.3,

$$\frac{d^2 X}{dx^2} + \frac{\omega^2}{c^2} X = 0, \quad \text{or} \quad LX = -\frac{\lambda}{c^2} X,$$

where L is the operator $\frac{d^2}{dx^2}$ and the weight function is $w(x) = c^{-2}$. The eigenvalues are $\lambda_n = \omega_n^2$ with $\omega_n = n\pi c/L$.

Increasing the string tension T increases the speed of propagation c and hence the frequency. Alternatively, increasing c decreases the weight function, w, and in equation 10.58 $h < 0$, showing that $\mu_n > \lambda_n$.

Exercise 10.29
Using the exact solution and perturbation theory, show that if the tension in the vibrating string changes from T to $T + \delta T$, to first order the eigenvalue λ_n becomes

$$\mu_n = \lambda_n \left(1 + \frac{\delta T}{T}\right).$$

Now consider the change in the eigenfunctions when q changes to $q + \epsilon h$. The exact eigenfunctions are given by the solution of 10.55. A convenient approximation is obtained using the generalised Green's function for the operator on the left hand side of this equation, which is given by combining equations 10.43 and 10.45,

$$\bar{G}(x, \xi) = \sum_{k \neq n} \frac{\phi_k(x)\phi_k(\xi)}{\mu_n - \lambda_k},$$

so the component of the solution $\psi_n(x)$ which is orthogonal to $\phi_n(x)$ is

$$\sum_{k \neq n} \frac{\phi_k(x)}{\mu_n - \lambda_k} \int_a^b d\xi \, \phi_k(\xi) \left[(\lambda_n - \mu_n) w(\xi) \psi_n(\xi) - \epsilon h(\xi) \psi_n(\xi) \right],$$

and the component parallel to $\phi_n(x)$ we take to be $\alpha \phi_n(x)$ for some constant α. As before, use the fact that a perturbation $O(\epsilon)$ produces changes of a similar order and replace ψ_k by ϕ_k to obtain the first-order correction to the eigenfunction,

$$\psi_n(x) = \phi_n(x) - \epsilon \sum_{k \neq n} \frac{\phi_k(x)}{\lambda_n - \lambda_k} \int_a^b d\xi \, \phi_k(\xi) h(\xi) \phi_n(\xi), \tag{10.59}$$

where the constant α has been set to unity to ensure that $(\psi_n, \psi_n) = 1 + O(\epsilon^2)$.

Exercise 10.30
If $q(x)$ changes to $q(x) + \alpha w(x)$ for some constant α and where $w(x)$ is the weight function, use perturbation theory to show that the eigenfunctions are unchanged and that the new eigenvalues are $\mu_n = \lambda_n - \alpha$. Show also that in this case these results are exact.

Exercise 10.31
Show that if the function $p(x)$ is changed by a small amount, to $p(x) + \epsilon h(x)$, then the nth eigenfunction becomes

$$\psi_n(x) = \phi_n(x) + \epsilon \sum_{k \neq n} \frac{\phi_k(x)}{\lambda_n - \lambda_k} \int_a^b d\xi \, \frac{d\phi_k}{d\xi} h(\xi) \frac{d\phi_n}{d\xi}.$$

10.6 Asymptotic expansions

The Sturm–Liouville theorem, page 356, shows that the eigenfunctions and eigenvalues of Sturm–Liouville systems all behave in a similar manner when the eigenvalue is large: in particular the eigenvalues λ_n tend to infinity with n and the eigenfunction $\phi_n(x)$ has $n - 1$ zeros. We now consider the large n limit in a little more detail and obtain explicit formulae for the leading term in the asymptotic expansion of both the eigenvalues and eigenfunctions. The results presented here were first obtained by Liouville by changing both the dependent and independent variables to transform the Sturm–Liouville equation into a more suitable form: this transformation is essential if further terms in the asymptotic expansion are required, but here we present a simpler treatment that illuminates the basic principles more clearly.

The solutions of the Sturm–Liouville equation

$$\frac{d}{dx} \left(p(x) \frac{dy}{dx} \right) + (q(x) + \lambda w(x)) \, y = 0, \quad a \leq x \leq b, \tag{10.60}$$

are oscillatory when λ is large, so it is natural to attempt to approximate the solutions with a function of the form

$$\psi(x) = \sin \left(\sqrt{\lambda} g(x) + \alpha \right), \tag{10.61}$$

where $g(x)$ is some unknown functions of x and α a constant. The eigenvalue λ appears in the form $\sqrt{\lambda}$ because of the need to differentiate $\psi(x)$ twice: we have

$$\frac{d}{dx}\left(p(x)\frac{d\psi}{dx}\right) = \sqrt{\lambda}\frac{d}{dx}\left(p(x)\frac{d\psi}{dx}\right)\cos\left(\sqrt{\lambda}g(x)+\alpha\right) - \lambda p\left(\frac{dg}{dx}\right)^2\psi. \qquad (10.62)$$

Since $\lambda \gg 1$ equation 10.60 becomes

$$\lambda\left(w - p\left(\frac{dg}{dx}\right)^2\right)\psi + O(\sqrt{\lambda}) = 0,$$

and the dominant term is zero if

$$g(x) = \int_a^x du\,\sqrt{\frac{w(u)}{p(u)}}. \qquad (10.63)$$

This provides a first estimate to the eigenfunctions and the eigenvalues if the approximate solution is made to fit the boundary conditions. By way of an example suppose that the boundary conditions are $y(a) = y(b) = 0$, then since $g(a) = 0$ we must set $\alpha = 0$, and then the condition at $x = b$ gives

$$\sqrt{\lambda}g(b) = n\pi \quad\Longrightarrow\quad \lambda_n = \left(\frac{n\pi}{g(b)}\right)^2,$$

showing that λ_n increases as n^2 independent of the form of w and p and also that asymptotically the eigenvalues and eigenfunctions are independent of $q(x)$. Notice also that with this value of λ_n for $a < x < b$, $\psi_n(x)$ has $n-1$ zeros.

Exercise 10.32
Show that with the boundary condition

$$A_1 y(a) + A_2 y'(a) = 0, \quad B_1 y(b) + B_2 y'(b) = 0,$$

the asymptotic limit of the eigenvalue is given by the solution of the equation

$$\tan\left(\sqrt{\lambda}g(b)\right) = \frac{\sqrt{\lambda}(B_1 A_1 g'(a) - A_1 B_2 g'(b))}{A_1 B_1 + \lambda A_2 B_2 g'(a)g'(b)}.$$

The approximation 10.61 has constant amplitude, which is clearly not correct. An improvement may be obtained by adding another unknown function to modulate the amplitude, thus we take as the next approximation

$$\psi(x) = A(x)\sin\left(\sqrt{\lambda}g(x)+\alpha\right), \qquad (10.64)$$

where $A(x)$ is the unknown amplitude. Substituting this into the equation gives

$$\frac{d}{dx}\left(p(x)\frac{d\psi}{dx}\right) + (q(x) + \lambda w(x))\,\psi = \frac{d}{dx}\left(p(x)\frac{d\psi}{dx}\right)\sin\left(\sqrt{\lambda}g(x)+\alpha\right) + q\psi$$

$$+\sqrt{\lambda}\left[p\frac{dg}{dx}\frac{dA}{dx} + \frac{d}{dx}\left(pA\frac{dg}{dx}\right)\right]\cos\left(\sqrt{\lambda}g(x)+\alpha\right)$$

$$+\lambda\left[w - p\left(\frac{dg}{dx}\right)^2\right]\psi.$$

As before choose $g(x)$ to remove the dominant λ term of the right hand side; but now the term $O(\sqrt{\lambda})$ may also be eliminated by defining $A(x)$ to satisfy the equation

$$p\frac{dg}{dx}\frac{dA}{dx} + \frac{d}{dx}\left(pA\frac{dg}{dx}\right) = 0 \quad\Longrightarrow\quad A(x) = \frac{\beta}{(pw)^{1/4}}, \tag{10.65}$$

for some constant β. The approximation to the eigenvalue remains the same and $q(x)$ still plays no role to this order in the expansion. Thus our approximation becomes

$$\phi_n(x) \sim \psi(x) = \frac{\beta}{(pw)^{1/4}} \sin\left(\sqrt{\lambda_n}\int_a^x du\,\sqrt{\frac{w(u)}{p(u)}} + \alpha_n\right). \tag{10.66}$$

The multiplicative constant β may be defined using the normalisation condition. Suppose $(\phi_n, \phi_n) = 1$, then

$$1 \simeq \beta^2 \int_a^b dx\,\frac{w(x)}{\sqrt{w(x)p(x)}}\sin^2\left(\sqrt{\lambda_n}g(x) + \alpha_n\right).$$

Since $n \gg 1$ the eigenfunction has many zeros and the square of the sine function may be replaced by it mean value, $\frac{1}{2}$, to give $\beta = \sqrt{2/g(b)}$.

For regular Sturm–Liouville systems this approximation shows that if $n \gg 1$, $\lambda_n \sim n^2$, exercise 10.32, and the value of α_n may be obtained from the boundary conditions. Examples of this approximation applied to a regular system are given in the next chapter, for example figures 11.7 and 11.8 (page 393).

For singular Sturm–Liouville systems, for which $p(x)$ is zero at one or both of the end points, the approximation is singular, but more seriously the eigenvalues can not be obtained by the method used in exercise 10.32. Nevertheless, it is still possible to show that, provided all integrals exist,

$$\lambda_n \sim \left(\frac{n\pi + v}{g(b)}\right)^2 \tag{10.67}$$

where v is a constant independent of n, but which depends upon the form of the boundary conditions; a method of deriving this result is outlined in exercise 10.34. Examples are given in the next chapter, in particular exercise 11.9 (page 386) and figures 11.11 and 11.12 (page 399).

Another feature of this simple approximation is that it allows integrals of the type

$$F_{nm} = \int_a^b dx\,w(x)f(x)\phi_n(x)\phi_m(x), \tag{10.68}$$

where n and m are large but $|n-m|/n \ll 1$, which occur in quantum mechanics, to be approximated by a Fourier-like integral,

$$F_{nm} \simeq \frac{1}{\pi}\int_0^\pi dt\,f(x(t))\cos\left((n-m)\left(t + \frac{d\alpha_n}{dn}\right)\right), \quad t(x) = \frac{\pi g(x)}{g(b)}. \tag{10.69}$$

The derivation of this approximation is outlined in exercise 10.35. In practical applications the function $x(t)$ is often fairly simple, so the approximation affords simple approximations to rather unpleasant integrals. Some examples of this approximation are considered in the next chapter, in particular exercises 11.9 and 11.27.

Exercise 10.33

Derive equation 10.65 for $A(x)$.

Exercise 10.34

The functions

$$\psi_n(x) = \frac{\beta_n}{(pw)^{1/4}} \sin\left(\sqrt{\lambda_n}g(x) + \alpha\right), \quad g(x) = \int_a^x du \sqrt{\frac{w(u)}{p(u)}},$$

where β_n, α and λ_n are constants, are approximate solutions of the Sturm–Liouville equation 10.60 even if the system is singular, provided $g(x)$ exists. If I_{nm} is the inner product of ψ_n and ψ_m (with weight function $w(x)$), show that

$$I_{nm} = \frac{1}{2}\beta_n\beta_m g(b) \int_0^1 du \cos\left[\left(\sqrt{\lambda_n} - \sqrt{\lambda_m}\right)g(b)u\right]$$

$$+ \frac{1}{2}\beta_n\beta_m g(b) \int_0^1 du \cos\left[\left(\sqrt{\lambda_n} + \sqrt{\lambda_m}\right)g(b)u + 2\alpha\right],$$

where $u = g(x)/g(b)$.

If $\lambda_n \gg 1$ the integrand of the second integral oscillates many times as u increases from 1 to -1, so this integral is relatively small and may be ignored. Use this approximation to show that if $\psi_n(x)$ is normalised to unity,

$$\beta_n = \beta = \sqrt{\frac{2}{g(b)}},$$

and hence that

$$\int_a^b dx\, w(x)\psi_n(x)\psi_m(x) = \int_0^1 du \cos\left[\left(\sqrt{\lambda_n} - \sqrt{\lambda_m}\right)g(b)u\right].$$

Use this last result to deduce that

$$\lambda_n = \left(\frac{\pi n + v}{g(b)}\right)^2$$

for some constant v.

Exercise 10.35

If

$$F_{nm} = \int_a^b dx\, w(x)f(x)\phi_n(x)\phi_m(x),$$

where $\phi_n(x)$ are eigenfunctions of the Sturm–Liouville system 10.60, using the approximation 10.66 show that

$$F_{nm} \simeq \frac{2}{g(b)} \int_a^b dx\, g(x)f(x) \sin\left(\sqrt{\lambda_n}g(x) + \alpha_n\right) \sin\left(\sqrt{\lambda_m}g(x) + \alpha_m\right).$$

Use the method outlined in exercise 10.34 to write this as

$$F_{nm} \simeq \frac{1}{g(b)} \int_a^b dx\, g'(x)f(x) \cos\left((n - m)\left(\frac{d\sqrt{\lambda_n}}{dn}g(x) + \frac{d\alpha_n}{dn}\right)\right),$$

and by changing variables derive equation 10.69.

Exercise 10.36

Use the approximation 10.66 to show that the difference between adjacent zeros of $\phi_n(x)$ is, approximately,

$$x_{i+1} - x_i \sim \frac{\pi}{\sqrt{\lambda_n}} \sqrt{\frac{p(\bar{x}_i)}{w(\bar{x}_i)}}, \quad \bar{x}_i = \frac{1}{2}(x_i + x_{i+1}).$$

10.7 Exercises

Exercise 10.37

If L is the real differential operator defined in equation 10.20 (page 356), and if λ is an eigenvalue with the complex eigenfunction $\phi(x)$, which is not simply proportional to a real eigenfunction, show that $\phi(x)^*$ is also an eigenfunction with the eigenvalue λ.

Deduce that (a) the system must be degenerate, and (b) it is possible to find a pair of real eigenfunctions with eigenvalue λ.

Exercise 10.38

Let $y_1(x)$ and $y_2(x)$ be two linearly independent solutions of the equation

$$a_2(x)\frac{d^2y}{dx^2} + a_1(x)\frac{dy}{dx} + a_0(x)y = 0.$$

The Wronskian of these two functions is defined to be $W(x) = y_1 y_2' - y_1' y_2$. By differentiating this with respect to x and forming a first-order differential equation for W, show that

$$W(x) = W(x_0)\exp\left(-\int_{x_0}^{x} ds\, \frac{a_1(s)}{a_2(s)}\right).$$

Note that this result shows that if $W(x) \neq 0$ for an x it is not zero for any x, provided a_1 and a_2 are sufficiently well behaved.

Exercise 10.39

Use the fact that two functions $f(x)$ and $g(x)$ are linearly independent in an interval $a \leq x \leq b$ if the Wronskian, $W = fg' - f'g$, is not identically zero on this interval, to determine whether the following functions are linearly independent for all real x.

    (i) $f = e^x$, $g = e^{-x}$,   (ii) $f = e^x$, $g = \sin x$,

    (iii) $f = x$, $g = x^2$,   (iv) $f = \sin x$, $g = \cos(x + \alpha)$.

Exercise 10.40

Let $y_1(x)$ and $y_2(x)$ be two linearly independent solutions of the equation

$$a_2(x)\frac{d^2y}{dx^2} + a_1(x)\frac{dy}{dx} + a_0(x)y = 0,$$

and let $a < b$ be two adjacent zeros of $y_1(x)$. Using the fact that the Wronskian never vanishes show that $y_2(x)$ must have one and only one zero for $a < x < b$.

Deduce that the functions $a_1 \sin x + a_2 \cos x$ and $b_1 \sin x + b_2 \cos x$ have alternating zeros if $a_1 b_2 \neq a_2 b_1$

Exercise 10.41

In this exercise you will show that if y_1 and y_2 are non-trivial solutions of the equations

$$\frac{d^2y}{dx^2} + p_1(x)y = 0, \quad \frac{d^2y}{dx^2} + p_2(x)y = 0, \quad a \le x \le b,$$

where $p_1(x) \ge p_2(x)$ for $x \in [a, b]$, then between any two adjacent zeros of y_2 there is at least one zero of y_1.

Let α and β be two adjacent zeros of y_2 and assume that $y_1 \ne 0$ for $\alpha \le x \le \beta$; it may therefore be assumed that both y_1 and y_2 have the same sign. If $W(x) = y_1 y_2' - y_1' y_2$ is the Wronskian, show that $W(\alpha) \ge 0$ and $W(\beta) \le 0$ and that

$$\frac{dW}{dx} = (p_1 - p_2)y_1 y_2 \ge 0.$$

Deduce that $y_1(x) = 0$ for some $x \in [\alpha, \beta]$.

Exercise 10.42

Show that if the weight function $w(x)$ is changed by a small amount, to $w(x)(1 + \epsilon h(x))$, then the nth eigenfunction becomes

$$\psi_n(x) = \phi_n(x) - \epsilon \lambda_n \sum_{k \ne k} \frac{\phi_k(x)}{\lambda_n - \lambda_k} \int_a^b d\xi \, w(\xi)\phi_k(\xi)h(\xi)\phi_n(\xi).$$

11

Special functions

11.1 Introduction

The special functions of mathematical physics are important because they arise from the solutions of linear differential equations that describe many different phenomena. Most are solutions of Sturm–Liouville equations, the notable exceptions being the gamma and zeta functions which are not defined as solutions of differential equations with rational coefficients. Here we introduce a few of the more important examples. Because these functions have been known about for well over a hundred years and have been used in a wide variety of problems a great deal is known about them, and any initial immersion can be rather daunting because there are many formulae and results. In the following I have attempted to provide some of the salient results and, in some cases, to explain where they come from or why they are important, though this is not always possible in such a brief summary. Maple knows about most special functions — to find out which type `?inifcn`: the orthogonal polynomials are accessed using the `with(orthopoly)` command. You may find out more from many sources, for instance, Whittaker and Watson (1965), Watson (1966), Arfken and Weber (1995), whilst Abramowitz and Stegun (1965) and Gradshteyn and Ryzhik (1965) provide compendious lists of properties.

11.2 Orthogonal polynomials

Orthogonal polynomials are systems of polynomials $p_n(x)$ of degree n, $n = 0, 1, 2, \ldots,$ that are orthogonal with respect to a weight function $w(x)$ on the interval $a \leq x \leq b$,

$$\int_a^b dx \, w(x) p_n(x) p_m(x) = h_n \delta_{nm}, \tag{11.1}$$

with $w(x) > 0$ for $a < x < b$; $w(x)$ may be zero at either $x = a$, $x = b$ or both and the interval may be infinite. Weierstrass's approximation theorem, page 272, shows that the set of polynomials $\{1, x, x^2, \ldots, x^n, \ldots\}$ is complete on any interval $a \leq x \leq b$, and the Gram–Schmidt orthogonalisation process, exercise 8.35 (page 296), shows how to construct $p_n(x)$ for all n. The theory of orthogonal polynomials can thus be developed without reference to Sturm–Liouville theory since it is part of the theory of linear spaces, in particular the space of orthogonal polynomials form a Hilbert space, see for example Powell (1981) and Young (1988). In many respects this modern approach presents a more unified approach to the subject. However, the traditional approach, though treating each set of orthogonal polynomials as a special case, by using complex variable theory allows these polynomials to be related to other types of functions, besides enabling many other inter-relations to be found. The reader interested in the latter approach should consult Whittaker and Watson (1965). Historically, most of the commonly used orthogonal polynomials arose from the solutions of differential equations.

Before considering some orthogonal polynomials in detail we provide a short catalogue of results which are common to all these functions. In the following table are listed the orthogonal polynomials that occur most frequently, together with their weight functions and other relevant details; notice that in three of these examples the interval $a < x < b$ is infinite.

Table 11.1. *Table of some common orthogonal polynomials and various associated constants; the constant h_n is defined in equation 11.1; the value of h_0 is given only if it is inconsistent with the formula for h_n.*

function label	name of polynomial	a	b	$w(x)$	h_n	h_0
$P_n(x)$	Legendre	-1	1	1	$\dfrac{2}{2n+1}$	
$T_n(x)$	Tchebychev of first kind	-1	1	$1/\sqrt{1-x^2}$	$\pi/2$	π
$U_n(x)$	Tchebychev of second kind	-1	1	$\sqrt{1-x^2}$	$\pi/2$	
$H_n(x)$	Hermite	$-\infty$	∞	e^{-x^2}	$2^n n!\sqrt{\pi}$	
$L_n(x)$	Laguerre	0	∞	e^{-x}	1	
$L_n^{(\alpha)}(x)$	generalised Laguerre	0	∞	$x^\alpha e^{-x}, \ \alpha > -1$	$\dfrac{\Gamma(n+\alpha+1)}{n!}$	
$P_n^{(\alpha,\beta)}(x)$	Jacobi	-1	1	$(1-x)^\alpha(1+x)^\beta$	see text	

For the Jacobi polynomials the normalisation constant is

$$h_n = \frac{2^{\alpha+\beta+1}}{2n+\alpha+\beta+1} \frac{\Gamma(n+\alpha+1)\Gamma(n+\beta+1)}{n!\Gamma(n+\alpha+\beta+1)}, \quad \alpha > -1, \ \beta > -1.$$

All sets of orthogonal polynomials have the following properties.

- Each set of orthogonal polynomials satisfies the orthogonality relation 11.1. If the value of h_n is not specified the set p_n is defined to within a multiplicative constant. Normally this is standardised by fixing $p_n(x)$ to have a given value at one end of the range, which then fixes h_n. Different authors often standardise the functions in different ways.
- The polynomial $p_n(x)$ is an nth-degree polynomial with leading terms

$$p_n(x) = k_n x^n + l_n x^{n-1} + \cdots,$$

where k_n and l_n are numbers depending upon n and the choice of h_n.
- The polynomial $p_n(x)$ satisfies a second-order linear differential equation

$$g_2(x)\frac{d^2y}{dx^2} + g_1(x)\frac{dy}{dx} + \alpha_n y = 0, \tag{11.2}$$

where $g_2(x)$ and $g_1(x)$ are independent of n and α_n a constant, some values are given in table 11.2.

Table 11.2. *Table of the functions and constants occurring in the differential equation 11.2.*

function	$g_2(x)$	$g_1(x)$	α_n
$P_n(x)$	$1-x^2$	$-2x$	$n(n+1)$
$T_n(x)$	$1-x^2$	$-x$	n^2
$U_n(x)$	$1-x^2$	$-3x$	$n(n+2)$
$H_n(x)$	1	$-2x$	$2n$
$L_n(x)$	x	$1-x$	n
$L_n^{(\alpha)}(x)$	x	$1+\alpha-x$	n
$P_n^{(\alpha,\beta)}(x)$	$1-x^2$	$\beta-\alpha-x(2+\alpha+\beta)$	$n(n+\alpha+\beta+1)$

- The polynomials $p_n(x)$ satisfies the three-term recurrence relation

$$p_{n+1}(x) = (a_n + xb_n)\,p_n(x) - c_n p_{n-1}(x),\qquad(11.3)$$

where a_n, b_n and c_n are independent of x, and are given by

$$b_n = \frac{k_{n+1}}{k_n},\quad a_n = b_n\left(\frac{l_{n+1}}{k_{n+1}} - \frac{l_n}{k_n}\right),\quad c_n = \frac{k_{n+1}k_{n-1}h_n}{k_n^2 h_{n-1}}.$$

Table 11.3. *Table of the functions and constants occurring in the recurrence relation 11.3.*

function	a_n	b_n	c_n
$P_n(x)$	0	$\dfrac{2n+1}{n+1}$	$\dfrac{n}{n+1}$
$T_n(x)$	0	2	1
$U_n(x)$	0	2	1
$H_n(x)$	0	2	$2n$
$L_n(x)$	$\dfrac{2n+1}{n+1}$	$-\dfrac{1}{n+1}$	$\dfrac{n}{n+1}$
$L_n^\alpha(x)$	$\dfrac{2n+\alpha+1}{n+1}$	$-\dfrac{1}{n+1}$	$\dfrac{n+\alpha}{n+1}$

- The function $p_n(x)$ can be generated by the following formula, named Rodrigues' formula,

$$p_n(x) = \frac{1}{e_n w(x)} \frac{d^n}{dx^n}\left[w(x)g(x)^n\right],\qquad(11.4)$$

where $g(x)$ is a polynomial independent of n, e_n is a constant and $w(x)$ is the weight function. This formula was first derived for Legendre polynomials by the French economist Rodrigues. It is this special case, equation 11.8, which is most often referred to by the name.

O. Rodrigues
(1794–1851)

Table 11.4. *Table of the functions and constants occurring in Rodrigues' formula 11.4.*

function	e_n	$w(x)$	$g(x)$
$P_n(x)$	$(-1)^n 2^n n!$	1	$1 - x^2$
$T_n(x)$	$(-1)^n 2^{n+1} \Gamma(n + 1/2)/\sqrt{\pi}$	$(1 - x^2)^{-1/2}$	$1 - x^2$
$U_n(x)$	$(-1)^n 2^{n+1} \Gamma(n + 3/2)/((n+1)\sqrt{\pi})$	$(1 - x^2)^{1/2}$	$1 - x^2$
$H_n(x)$	$(-1)^n$	$\exp(-x^2)$	1
$L_n(x)$	$n!$	$\exp(-x)$	x
$L_n^{(\alpha)}(x)$	$n!$	$x^\alpha \exp(-x)$	x

- **Generating functions:** A function $g(x, t)$ exist such that $p_n(x)$ is proportional to the coefficient of t^n in the Taylor expansion of $g(x, t)$ about $t = 0$, that is,

$$g(x, t) = \sum_{n=0}^{\infty} a_n p_n(x) t^n \tag{11.5}$$

for some constants a_n. The function $g(x, t)$ is named the *generating function* and provides an alternative definition of $p_n(x)$. The generating functions for some orthogonal polynomials are listed in table 11.5. Generating functions exist for many special functions, not just the orthogonal polynomials: the generating function for Bernoulli polynomials is given in equation 5.2 (page 179), and that for Bessel functions in equation 11.63.

Table 11.5. *Table of generating functions for various orthogonal polynomials $p_n(x)$: here $R^2 = 1 - 2xt + t^2$.*

function	a_n	$g(x, t)$
$P_n(x)$	1	R^{-1}
$T_n(x)$	2	$(1 - t^2)R^{-2} + 1$
$U_n(x)$	1	R^{-2}
$H_n(x)$	$1/n!$	$\exp(2xt - x^2)$
$H_{2n}(x)$	$(-1)^n/(2n)!$	$\cos(2xt)e^{t^2}$
$H_{2n+1}(x)$	$(-1)^n/(2n + 1)!$	$\sin(2xt)e^{t^2}$
$L_n(x)$	1	$(1 - t)^{-1} \exp\left(xt/(t - 1)\right)$
$L_n^{(\alpha)}(x)$	1	$(1 - t)^{-1-\alpha} \exp\left(xt/(t - 1)\right)$

- **Christoffel–Darboux formula**: The orthogonal polynomials satisfy the sum formula

$$\sum_{m=0}^{n} \frac{1}{h_m} p_m(x)p_m(y) = \frac{k_n}{k_{n+1}h_n} \frac{p_{n+1}(x)p_n(y) - p_n(x)p_{n+1}(y)}{x - y}. \tag{11.6}$$

- The set of differentials $\{p_1'(x), p_2'(x), \dots, p_n'(x), \dots\}$ is also a set of orthogonal polynomials, though with a different weight function, see for instance exercise 11.8 and also 11.55 (page 431).

11.2.1 Legendre polynomials

The Legendre polynomials occur frequently because they arise as the angular components when variables are separated in many common partial differential equations; they are particularly important in the quantum mechanical description of angular momentum, see for example Edmonds (1974) or Brink and Satchler (1971).

Legendre polynomials may be defined as suitably normalised, polynomial solutions of the singular Sturm–Liouville problem

$$\frac{d}{dx}\left((1-x^2)\frac{dy}{dx}\right) + \lambda y = 0, \quad -1 \le x \le 1. \tag{11.7}$$

In many applications $x = \cos\theta$, $\pi/2 - \theta$ being the latitude of a point on a sphere. This is a singular Sturm–Liouville problem because $p(x) = 1 - x^2$ is zero at both ends of the range. The bounded solutions of this equation are polynomials of degree n if $\lambda = n(n+1)$, and if, in addition, we demand that $y(1) = 1$ these solutions are the Legendre polynomials $y = P_n(x)$.

Graphs of the first few Legendre polynomials are shown in the following figures.

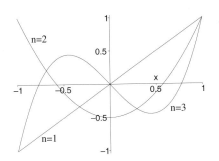

Figure 11.1   Graphs of $P_n(x)$ for $n = 1, 2$ and 3.

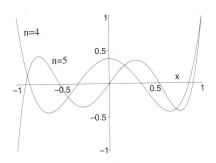

Figure 11.2   Graphs of $P_n(x)$ for $n = 4$ and 5.

Observe that although $|P_n(\pm 1)| = 1$ the value of $|P_n(x)|$ away from $x = \pm 1$ is smaller: in fact we shall see, exercise 11.9, that for x not near ± 1 the height of the local maxima of $|P_n(x)|$ are $O(n^{-1/2})$. Compare this behaviour with that of the Tchebychev polynomials shown in figure 11.3 (page 387); we see later that this difference is important in some applications.

The Legendre polynomials may be computed in a variety of ways; one is to use Rodrigues' formula,

$$P_0(x) = 1, \quad P_n(x) = \frac{1}{n!\,2^n}\frac{d^n}{dx^n}(x^2 - 1)^n, \quad n \ge 0. \tag{11.8}$$

With this definition, $P_n(1) = 1$, $P_n(-1) = (-1)^n$ and $P_n(x)$ satisfies the differential equation: proofs of these results are are set as exercises for the reader.

Equation 11.7 shows that the weight function is unity, and use of Rodrigues' formula shows that

$$\int_{-1}^{1} dx \, P_n(x) P_m(x) = \frac{2\delta_{nm}}{2n+1}. \tag{11.9}$$

The proof is outlined in exercise 11.63 (page 434).

There are other solutions of equation 11.7, though these are not polynomials, and we digress for a moment to show briefly how these arise. Cauchy's integral for derivatives expresses the nth differential of an analytic function $f(z)$ by the contour integral

$$\frac{d^n f}{dz^n} = \frac{n!}{2\pi i} \oint_{\mathscr{C}} dt \, \frac{f(t)}{(t-z)^{n+1}}, \tag{11.10}$$

where $\mathscr{C}$ is a contour enclosing the point z and no singularities of $f(t)$, and is traversed counter-clockwise. Proof of this result may be found in Whittaker and Watson (1965, chapter 6) or Arfken and Weber (1995, chapter 6). This formula allows Rodrigues' formula to be written as the contour integral

$$P_n(z) = \frac{1}{2^{n+1}\pi i} \oint_{\mathscr{C}} dt \, \frac{(t^2-1)^n}{(t-z)^{n+1}}, \tag{11.11}$$

which is Schläfli's integral formula for the Legendre polynomial. Two useful results that follow from this integral are

$$\frac{1}{\pi} \int_0^\pi d\phi (1 + a \cos \phi)^n = (1-a^2)^{n/2} P_n \left(\frac{1}{\sqrt{1-a^2}} \right), \quad |a| \le 1,$$

$$\frac{1}{\pi} \int_0^\pi d\phi \frac{1}{(1 + a \cos \phi)^{n+1}} = \frac{1}{(1-a^2)^{(n+1)/2}} P_n \left(\frac{1}{\sqrt{1-a^2}} \right), \quad |a| < 1,$$

where n is a positive integer. These are known respectively as Laplace's first and second integrals and proofs may be found in Whittaker and Watson (1965, chapter 15). It is interesting to note that the second integral is a direct corollary of the first because $P_{-n-1}(z) = P_n(z)$, for positive integers n, a result that follows by expressing $P_n(z)$ in terms of the hypergeometric function, $P_n(z) = F(n+1, -n; 1; \frac{1}{2} - \frac{z}{2})$.

Schläfli's integral allows the restriction that n be an integer to be lifted and for it to be replaced by a real number ν. Provided sufficient care is taken in defining the contour $\mathscr{C}$, it may be shown, Whittaker and Watson (1965), that $P_\nu(z)$ is a solution of equation 11.7 with $\lambda = \nu(\nu + 1)$, but the orthogonality condition 11.9 is satisfied only for integer values of ν. This generalisation from integer to non-integer values of n is similar to the extension of the factorial function $n!$ to non-integer values of n using the gamma function, $\nu! = \Gamma(\nu + 1)$.

Moreover, a different choice of contour gives the Legendre functions of the second kind; these are denoted by $Q_\nu(z)$ and are not polynomials even if ν is an integer, neither are they bounded at $x = \pm 1$. For instance, if $|x| < 1$ we have

$$Q_0(x) = \frac{1}{2} \ln \left(\frac{1+x}{1-x} \right), \quad Q_1(x) = \frac{x}{2} \ln \left(\frac{1+x}{1-x} \right) - 1,$$

and for $n \geq 1$

$$Q_n(x) = \frac{1}{2} P_n(x) \ln \left(\frac{1+x}{1-x} \right) - \sum_{k=1}^{n} \frac{1}{k} P_{k-1}(x) P_{n-k}(x).$$

Return now to our main topic, the behaviour of the polynomial $P_n(x)$. From Rodrigues' formula 11.8 we see that $P_n(x)$ is a polynomial of degree n; the first few are $P_0(x) = 1$ and:

$$P_1(x) = x, \qquad\qquad P_2(x) = \frac{1}{2}(3x^2 - 1), \qquad P_3(x) = \frac{x}{2}(5x^2 - 3),$$

$$P_4(x) = \frac{1}{8}(35x^4 - 30x^2 + 3), \qquad P_5(x) = \frac{x}{8}(63x^4 - 70x^2 + 15).$$

The completeness of Legendre polynomials together with the above orthogonality relation allows any suitably behaved function, defined on the interval $-1 \leq x \leq 1$, to be expressed in terms of the Fourier series

$$f(x) = \sum_{k=0}^{\infty} a_k P_k(x), \qquad a_n = \left(n + \frac{1}{2} \right) \int_{-1}^{1} dx\, f(x) P_n(x). \tag{11.12}$$

This expansion has properties similar to that of trigonometric series, discussed in chapter 8; for instance it exhibits the Gibbs phenomenon and behaves similarly at discontinuities, exercise 11.61 (page 433).

Legendre polynomials may also be defined by the generating function

$$\frac{1}{(1 - 2xt + t^2)^{1/2}} = \sum_{n=0}^{\infty} P_n(x)\, t^n, \quad |t| < 1, \quad |x| \leq 1. \tag{11.13}$$

The origin of this expression is the ubiquitous inverse square law: for example a point charge e at $(0, a)$ in the Oxz-plane, as shown in the diagram below, produces the potential

$$V_p = \frac{e}{d} = \frac{e}{\sqrt{r^2 + a^2 - 2ar \cos \theta}}$$

at the point P a distance r from the origin, where θ is defined in the diagram and where the cosine rule is used to express the distance d in terms of r, a and θ.

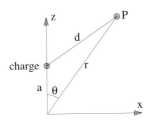

Expanding the potential in powers of a/r, assuming $r > a$, gives

$$V_P = \frac{e}{r} \sum_{n=0}^{\infty} \left(\frac{a}{r} \right)^n P_n(\cos \theta),$$

which is similar to equation 11.13. Since $\nabla^2(V_P) = 0$ it follows, by separating variables in spherical polar coordinates, that the functions $P_n(\cos\theta)$, *defined* by equation 11.13, satisfy Legendre's equation, exercise 11.64, (page 434).

In the following nine exercises some of the well known results involving Legendre polynomials are derived; many of the following relations are special cases of more general results satisfied by all orthogonal polynomials, and almost all can be derived in many ways. Other similar results may be found in Abramowitz and Stegun (1965, chapters 8 and 22).

Exercise 11.1
Show that $P_{2n}(x)$ is an even function and that $P_{2n+1}(x)$ is an odd function. Use Maple to plot the graph of $P_n(x)$ for some large values of n, say $n \sim 20$. What do you observe about the distribution of the roots of $P_n(x) = 0$?

Exercise 11.2
Use the generating function for Legendre polynomials to show that

$$P_n(1) = 1, \qquad P_n'(1) = \frac{1}{2}n(n+1),$$

$$P_{2n}(0) = (-1)^n \frac{\Gamma(n+1/2)}{n!\sqrt{\pi}}, \qquad P_{2n+1}'(0) = (2n+1)P_{2n}(0),$$

and use Stirling's formula to show that $P_{2n}(0) \sim (-1)^n/\sqrt{n\pi}$, $P_{2n+1}'(0) \sim 2(-1)^n\sqrt{n/\pi}$.

Exercise 11.3
Show that

$$\frac{d}{dx}P_{n+1}(x) - \frac{d}{dx}P_{n-1}(x) = (2n+1)P_n(x), \quad n \geq 1,$$

and $P_1'(x) = P_0(x)$.

Hint: note that the left hand side is a polynomial of degree n, so may be expressed as a sum of the first $n+1$ Legendre polynomials, with coefficients given by equation 11.12.

Exercise 11.4
Prove that the nth degree polynomial

$$y = \frac{d^n}{dx^n}\left[(x^2 - 1)^n\right]$$

satisfies the differential equation 11.7 if $\lambda = n(n+1)$.

Hint: expand the term $(1 - x^2)^n$ and then differentiate.

Exercise 11.5
Recurrence relation
(i) If $F_m(x)$ is a polynomial of degree m show that it may be represented by a sum of the first $m+1$ Legendre polynomials,

$$F_m(x) = \sum_{k=0}^{m} a_k P_k(x),$$

and that $F_m(x)$ is orthogonal to $P_n(x)$ if $n \geq m+1$.

(ii) By noting that $xP_n(x)$ is a polynomial of degree $n+1$, using the result derived in part (i) and equation 11.12, show that

$$xP_n(x) = a_{n+1}P_{n+1}(x) + a_n P_n(x) + a_{n-1}P_{n-1}(x)$$

and use the result of exercise 11.1 to show that $a_n = 0$.

(iii) Using the values of $P_n(1)$ and $P_n'(1)$ deduce the recurrence relation

$$(n+1)P_{n+1}(x) = (2n+1)xP_n(x) - nP_{n-1}(x). \tag{11.14}$$

Another method of deriving this recurrence relation, using the generating function 11.13, is given in exercise 11.57 (page 432).

Exercise 11.6
Recurrence relation

By differentiating the generating function with respect to x, show that

$$P_{n+1}'(x) + P_{n-1}'(x) = xP_n'(x) + P_n(x)$$

and use the recurrence relation 11.14 to prove that

$$nP_n(x) = xP_n'(x) - P_{n-1}'(x).$$

By differentiating this with respect to x, show that

$$(1 - x^2)\frac{dP_n}{dx} = nP_{n-1}(x) - nxP_n(x).$$

Exercise 11.7
Delta function decomposition

(i) Show that

$$\delta(x - \xi) = \frac{1}{2}\sum_{k=0}^{\infty}(2k+1)P_k(x)P_k(\xi).$$

(ii) Use the recurrence relation 11.14 to show that

$$(2k+1)xP_k(\xi)P_k(x) = (k+1)P_k(\xi)P_{k+1}(x) + kP_k(\xi)P_{k-1}(x),$$
$$(2k+1)\xi P_k(x)P_k(\xi) = (k+1)P_k(x)P_{k+1}(\xi) + kP_k(x)P_{k-1}(\xi),$$

and deduce the Christoffel–Darboux formula,

$$\sum_{k=0}^{n}(2k+1)P_k(x)P_k(\xi) = \frac{n+1}{x-\xi}\left[P_{n+1}(x)P_n(\xi) - P_n(x)P_{n+1}(\xi)\right]. \tag{11.15}$$

(iii) Deduce that

$$\lim_{n\to\infty}(n+1)\left[\frac{P_{n+1}(x)P_n(\xi) - P_n(x)P_{n+1}(\xi)}{x-\xi}\right] = 2\delta(x-\xi).$$

Exercise 11.8
Show that the polynomials $o_n(x) = dP_n/dx$, $n = 1, 2, \ldots$, are orthogonal with respect to the weight function $w(x) = (1 - x^2)$.

Hint: find the differential equation satisfied by $o_n(x)$. An extension of this result is discussed in exercise 11.55 (page 431).

Exercise 11.9

Use the method outlined in exercise 10.34, and assume $v = \pi/2$, to show that the asymptotic expansion of $P_n(x)$ can be written in the form

$$P_n(x) \sim \frac{\beta_n}{(1-x^2)^{1/4}} \sin\left(\left(n+\frac{1}{2}\right)\phi + \alpha_n\right), \quad x = \sin\phi,$$

where α_n and β_n are constants. By normalising this approximation, equation 11.9, show that $\beta_n^2 = 2/(\pi(n+1/2))$, and by using the values of $P_{2n}(0)$ and $P'_{2n+1}(0)$, derived in exercise 11.2, show that

$$P_n(x) \sim \sqrt{\frac{2}{\pi(n+\frac{1}{2})\sqrt{1-x^2}}} \cos\left(\left(n+\frac{1}{2}\right)\phi - \frac{n}{2}\pi\right), \quad x = \sin\phi.$$

Make graphical comparisons of this approximation with some exact values of $P_n(x)$.

Use this approximation to show that

(i) the zeros of $P_n(x)$ are approximately

$$x_k = \sin\left(\frac{2k+1-n}{2n+1}\pi\right) \quad k = 0, 1, \ldots, n-1,$$

(ii) and that if $|n - m| \ll n$ then

$$\int_{-1}^{1} dx\, f(x) P_n(x) P_m(x) \simeq \frac{1}{\pi\sqrt{(n+\frac{1}{2})(m+\frac{1}{2})}} \int_{0}^{\pi} d\theta\, f(\cos\theta) \cos(n-m)\theta.$$

11.2.2 Tchebychev polynomials

Tchebychev polynomials are closely related to trigonometric functions. Starting with the integral

$$\int_{0}^{\pi} d\theta\, \cos n\theta \cos m\theta = h_n \delta_{nm}, \quad h_n = \begin{cases} \pi, & n = 0, \\ \pi/2, & n \geq 1, \end{cases} \tag{11.16}$$

and changing the integration variable to $x = \cos\theta$ gives

$$\int_{-1}^{1} dx\, \frac{T_n(x) T_m(x)}{\sqrt{1-x^2}} = h_n \delta_{nm},$$

where

$$T_n(x) = \cos\left(n\cos^{-1} x\right). \tag{11.17}$$

The functions $T_n(x)$ are therefore orthogonal on the interval $-1 \leq x \leq 1$ with weight function $w(x) = (1 - x^2)^{-1/2}$. Further, they are nth degree polynomials in x, as may be seen by noting that $\cos n\theta$ may be expressed as an nth degree polynomial in $\cos\theta$, exercise 11.11.

The equation satisfied by the Tchebychev polynomials, in self-adjoint form is, exercise 11.12,

$$\frac{d}{dx}\left(\sqrt{1-x^2}\frac{dy}{dx}\right) + \frac{n^2}{\sqrt{1-x^2}}y = 0,$$

and the first few of these polynomials are:

$$T_0(x) = 1, \qquad T_1(x) = x, \qquad T_2(x) = 2x^2 - 1,$$
$$T_3(x) = x(4x^2 - 3), \qquad T_4(x) = 8x^4 - 8x^2 + 1, \qquad T_5(x) = x(16x^4 - 20x^2 + 5).$$

Equation 11.17 defines the Tchebychev polynomial† of order n, named after their founder, the Russian mathematician Tchebychev‡.

From the definition we see that

P. L. Tchebychev
(1821-1894)

$$|T_n(x)| \le 1 \quad \text{for} \quad -1 \le x \le 1,$$

that $T_n(1) = 1$ and $T_n(-1) = (-1)^n$ and that the n roots of

$$T_n(x) = 0 \quad \text{are at} \quad x_j = \cos\left(\frac{j-\frac{1}{2}}{n}\right)\pi, \quad j = 1, 2, \ldots, n.$$

Moreover, between each pair of roots, $T_n(x)$ achieves either its maximum, $+1$, or its minimum, -1, value. The graphs of the first few of these polynomials are shown in the figures.

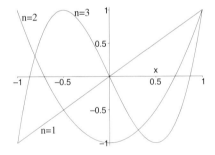

Figure 11.3 $T_n(x)$ for $n = 1, 2$ and 3.

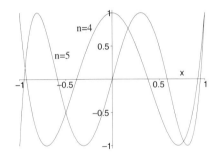

Figure 11.4 $T_n(x)$ for $n = 4$ and 5.

Comparing these graphs and those in figures 11.1 and 11.2, showing the Legendre polynomials of the same order, reveals an important difference. The Tchebychev polynomials reach their global maximum and minimum values between adjacent zeros, and because of this the least-squares fit of a function $f(x)$ by a finite sum of Tchebychev polynomials is very nearly the same as the *minimax polynomial* which (among all polynomials of the same degree) has the smallest maximum deviation from the original function, $f(x)$. Further discussion of this important property can be found in Press *et al.* (1987), Acton (1970, chapter 12) and Powell (1981).

† There seems to be no agreed convention for the normalisation of Tchebychev polynomials.
‡ There are several different anglicised forms of this name: others are Chebychev, Chebichev, Tschebytschew and Tchebycheff.

The discussion of this section started with the integral 11.16; a similar integral is

$$\int_0^\pi d\theta \, \sin n\theta \sin m\theta = h_n \delta_{nm},$$

and by making the same substitution it is seen that the functions

$$V_n(x) = \sin(n \cos^{-1} x) \tag{11.18}$$

satisfy the same orthogonality relation,

$$\int_{-1}^1 dx \, \frac{V_n(x)V_m(x)}{\sqrt{1-x^2}} = h_n \delta_{nm}.$$

These functions are not, however, polynomials although they do satisfy the same Sturm–Liouville equation, exercises 11.11 and 11.12.

Exercise 11.10

Use the trigonometric addition rules to prove that the following recurrence relations hold:

$$\begin{aligned}
T_{n-1}(x) + T_{n+1}(x) &= 2x T_n(x), \\
T_{n-m}(x) + T_{n+m}(x) &= 2 T_n(x) T_m(x), \\
T_{n-m}(x) - T_{n+m}(x) &= 2 V_n(x) V_m(x).
\end{aligned}$$

Show also that $V_n(x)$ satisfy the relations

$$\begin{aligned}
V_{n+1}(x) + V_{n-1}(x) &= 2x V_n(x), \\
V_{n+m}(x) + V_{n-m}(x) &= 2 V_n(x) T_m(x), \\
V_{n+m}(x) - V_{n-m}(x) &= 2 T_n(x) V_m(x).
\end{aligned}$$

Exercise 11.11

The simplest method of expressing $T_n(x)$ as polynomials in x and of seeing why the other choice $V_n(x) = \sin(n \cos^{-1} x)$ does not produce a polynomial in x is to consider the complex function

$$\begin{aligned}
z_n(x) = \cos n\theta + i \sin n\theta &= e^{in\theta}, \quad \text{with} \quad x = \cos\theta \\
&= \left(x + i\sqrt{1-x^2}\right)^n.
\end{aligned}$$

Then if $T_n(x) = \cos(n \cos^{-1} x)$ and $V_n(x) = \sin(n \cos^{-1} x)$, we see that

$$T_n(x) = \Re\left(x + i\sqrt{1-x^2}\right)^n \quad \text{and} \quad V_n(x) = \Im\left(x + i\sqrt{1-x^2}\right)^n.$$

Use these expressions to find the first few Tchebychev polynomials and show that

$$\begin{aligned}
V_0(x) &= 0, & V_1(x) &= \sqrt{1-x^2}, \\
V_2(x) &= 2x\sqrt{1-x^2}, & V_3(x) &= (4x^2 - 1)\sqrt{1-x^2}.
\end{aligned}$$

Exercise 11.12

By differentiating equation 11.17 show that the Tchebychev polynomial $T_n(x)$ satisfies the equation

$$(1 - x^2)\frac{d^2 y}{dx^2} - x\frac{dy}{dx} + n^2 y = 0,$$

and that $V_n(x)$ satisfies the same equation.

Exercise 11.13

Tchebychev polynomials of the second kind can be defined by

$$U_n(x) = \frac{\sin(n+1)\theta}{\sin\theta}, \quad x = \cos\theta, \quad n = 0, 1, \ldots. \tag{11.19}$$

(i) Show that $U_n(x)$ is a polynomial of degree n in x and that

$$U_0(x) = 1, \quad U_1(x) = 2x, \quad U_2(x) = 4x^2 - 1,$$
$$U_3(x) = 4x(2x^2 - 1), \quad U_4(x) = 16x^4 - 12x^2 + 1.$$

(ii) Show that $U_n(x)$ satisfy the orthogonality relation

$$\int_{-1}^{1} dx \sqrt{1 - x^2}\, U_n(x)U_m(x) = h_n \delta_{nm}, \quad h_n = \frac{\pi}{2},$$

that they satisfy the differential equation

$$(1 - x^2)\frac{d^2y}{dx^2} - 3x\frac{dy}{dx} + n(n+2)y = 0$$

and that they are related to the Tchebychev polynomials by

$$\frac{dT_{n+1}}{dx} = (n+1)U_n(x).$$

Exercise 11.14

Show that the zeros of the nth-order Legendre and Tchebychev polynomials are given by

$$x_k \simeq -\cos\left(\frac{2k+3/2}{2n+1}\right)\pi, \quad z_k = -\cos\left(\frac{2k+1}{2n}\right)\pi, \quad k = 0, 1, \ldots, n-1,$$

respectively. Compare the first of these approximations with the exact roots of $P_n(x) = 0$, computed numerically, and compare the differences $x_k - z_k$, $0 \le k \le n-1$ when n is large.

11.2.3 Hermite polynomials

Hermite polynomials are different from Legendre or Tchebychev polynomials because they are defined on the whole real axis, so the weight function, $w(x)$, must decrease faster than $|x|^n$ for any integer n, as $|x| \to \infty$. The simplest function with this property is $w(x) = \exp(-x^2)$, though sometimes $\exp(-x^2/2)$ is used. With this weight function it is possible to use the Gram–Schmidt process, exercise 8.35, to construct the Hermite polynomials, $H_n(x)$ $n = 0, 1, \ldots$, which are complete on $(-\infty, \infty)$. But it is more convenient to define the nth polynomial to be a suitably normalised nth degree polynomial solution of the differential equation

$$\frac{d^2y}{dx^2} - 2x\frac{dy}{dx} + 2ny = 0, \quad n = 0, 1, 2, \ldots. \tag{11.20}$$

If n is not an integer the two linearly independent solutions of this equation can be represented by infinite series, exercise 11.23, one of which collapses to an nth degree polynomial if n is an integer. Henceforth we assume that n is an integer. The self-adjoint form of this equation is

$$\frac{d}{dx}\left(e^{-x^2}\frac{dy}{dx}\right) + 2ne^{-x^2}y = 0, \tag{11.21}$$

showing that $p(x) = w(x) = \exp(-x^2)$ and that the eigenvalues are $\lambda = 2n$. There are many ways of making the solutions of this equation unique; here we adopt the convention that the coefficient of x^n is 2^n. Then the orthogonality relation 11.1 becomes

$$\int_{-\infty}^{\infty} dx\, e^{-x^2} H_n(x) H_m(x) = h_n \delta_{nm}, \qquad h_n = 2^n n! \sqrt{\pi}, \tag{11.22}$$

as will be shown below, equation 11.27.

Equation 11.20 has a particularly simple form that allows the solution to be expressed in terms of a simple contour integral, from which many of the properties of the Hermite polynomials follow almost trivially. The general analysis of this type of equation is described briefly in the appendix to this chapter; for the particular case of equation 11.20 we obtain

$$y_n(x) = e^{x^2} \oint_{\mathscr{C}} dz\, \frac{e^{-z^2}}{(z-x)^{n+1}}, \qquad n \text{ a non-negative integer}, \tag{11.23}$$

where $\mathscr{C}$ is any contour enclosing the pole at $z = x$. Comparing this with Cauchy's integral, equation 11.10 (page 382), suggests putting $f(z) = \exp(-z^2)$ to give

$$y_n(x) = a e^{x^2} \frac{d^n}{dx^n} \left(e^{-x^2} \right),$$

where a is an arbitrary constant. Clearly $y_n(x)$ is an nth degree polynomial and is even or odd according as n is even or odd. The term of highest degree is $a(-2x)^n$, and to obtain the conventional Hermite polynomial we set $a = (-1)^n$ to give

$$H_n(x) = (-1)^n e^{x^2} \frac{d^n}{dx^n} \left(e^{-x^2} \right). \tag{11.24}$$

This is Rodrigues' formula for the Hermite polynomial. The first few polynomials are:

$$
\begin{aligned}
H_0(x) &= 1, & H_1(x) &= 2x, \\
H_2(x) &= 4x^2 - 2, & H_3(x) &= 8x^3 - 12x, \\
H_4(x) &= 16x^4 - 48x^2 + 12, & H_5(x) &= 32x^5 - 160x^3 + 120x.
\end{aligned}
$$

Exercise 11.15
Use Rodrigues' formula 11.24 to show that:

(i) $H_{n+1}(x) = 2x H_n(x) - 2n H_{n-1}(x),$ (11.25)

(ii) $H_n'(x) = 2n H_{n-1}(x),$ (11.26)

(iii) $H_{2n}(0) = (-1)^n (2n)!/n!, \quad H_{2n+1}(0) = 0,$

(iv) $H_n(x) = (-1)^n H_n(-x).$

The Hermite polynomial $H_n(x)$ tends to $\pm\infty$ as $|x| \to \infty$ and has n zeros which, as will be seen below, are in the interval $|x| < \sqrt{2n+1}$: outside this interval $|H_n(x)|$ increases rapidly, as $|x|^n$, but inside this interval the function oscillates between large values; for graphical comparisons it is therefore more convenient to draw graphs of

$$y_n(x) = e^{-x^2/2} H_n(x)/\sqrt{h_n}, \qquad h_n = 2^n n! \sqrt{\pi},$$

some of which are shown in figures 11.5 and 11.6.

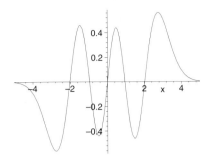

Figure 11.5 $y_5(x)$.

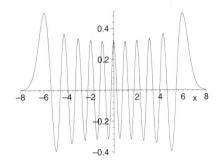

Figure 11.6 $y_{20}(x)$.

The value of h_n in the orthogonalisation integral 11.22 is given by replacing one of the $H_n(x)$ in this integral by Rodrigues' formula 11.24,

$$h_n = (-1)^n \int_{-\infty}^{\infty} dx \, H_n(x) \frac{d^n}{dx^n} \left(e^{-x^2} \right),$$

and integrating by parts n times,

$$h_n = \int_{-\infty}^{\infty} dx \, e^{-x^2} \frac{d^n H_n}{dx^n}.$$

Because $H_n(x) = (2x)^n + O(x^{n-1})$, $d^n H_n/dx^n = 2^n n!$ and hence

$$h_n = \int_{-\infty}^{\infty} dx \, e^{-x^2} H_n(x)^2 = 2^n n! \sqrt{\pi}. \tag{11.27}$$

Hermite polynomials may also be defined using the generating function

$$g(x, t) = \exp(-t^2 + 2xt) = \sum_{n=0}^{\infty} \frac{t^n}{n!} H_n(x), \tag{11.28}$$

the derivation of which is outlined in exercise 11.19. This relation can be useful for evaluating some integrals, exercises 11.20 and 11.21.

One of the reasons that Hermite polynomials are important is that the related functions

$$\phi_n(x) = \exp\left(-x^2/2\right) H_n(x) \tag{11.29}$$

satisfy the differential equation, sometimes named the Weber equation,

$$\frac{d^2\phi}{dx^2} + (v^2 - x^2)\phi = 0, \quad v = \sqrt{2n+1}. \tag{11.30}$$

This equation arises in the quantum mechanical description of the simple harmonic oscillator — the motion of a particle under the action of an attractive force increasing in magnitude in proportion to the distance from a point — which is of fundamental significance because it describes the forces between many molecular systems near equilibrium. This is the quantum mechanical equivalent of the classical linear oscillator.

In order to obtain a simple approximation, accurate when $v \gg 1$, we first rescale the independent variable, $x = vz$, to cast the equation in the form

$$\frac{d^2\phi}{dz^2} + v^4(1 - z^2)\phi = 0.$$

The solution is oscillatory for $|z| < 1$ and the required solution tends to zero exponentially fast as $|z| \to \infty$, so the method used in section 10.6 suggests an approximation of the form

$$\phi(z) \sim \begin{cases} A(z)e^{iv^2 S(z)}, & |z| < 1, \\ 0, & |z| > 1. \end{cases}$$

Substituting this into the equation for ϕ gives

$$\frac{d^2 A}{dz^2} + \frac{iv^2}{A}\frac{d}{dz}\left(A^2\frac{dS}{dz}\right) + Av^4\left\{-\left(\frac{dS}{dz}\right)^2 - (1 - z^2)\right\} = 0. \qquad (11.31)$$

For $v \gg 1$ an approximation is obtained by setting the coefficients of v^4 and v^2 to zero. Then integration of the two ensuing equations gives

$$\phi_n(x) \sim \frac{1}{(v^2 - x^2)^{1/4}} \exp\left(\pm i \int_0^x dx\, \sqrt{v^2 - x^2}\right), \quad |x| \ll v = \sqrt{2n + 1}.$$

This approximation is singular at $x = \pm v$ because here the assumptions made in deriving the approximation are invalid, namely that the magnitude of the coefficient of ϕ in equation 11.30 is large.

Using the facts that $\phi_{2n}(x)$ is even and $\phi_{2n+1}(x)$ odd, together with the known normalisation, equation 11.27, an approximation to the magnitude of $\phi_n(x)$ is easily obtained. The sign of this function is, however, a convention determined by the choice of the coefficient of x^n, see page 390; we account for this by ensuring that the value of $\phi_n(0)$, for even n, of $\phi_n'(0)$ for odd n, is correct. This gives the approximation:

$$\phi_n(x) \sim \sqrt{2}\left(\frac{2n}{e}\right)^{n/2}\left(\frac{2n}{v^2 - x^2}\right)^{1/4}\begin{cases} (-1)^{n/2}\cos S(x), & n \text{ even}, \\ (-1)^{(n-1)/2}\sin S(x), & n \text{ odd}, \end{cases} \quad |x| \ll v.$$
$$(11.32)$$

where

$$S(x) = \frac{v^2}{2}\sin^{-1}\left(\frac{x}{v}\right) + \frac{x}{2}\sqrt{v^2 - x^2}, \quad v = \sqrt{2n + 1},$$

and Stirling's approximation has been used to simplify h_n.

In the next two figures these approximations are compared with the exact values of $\phi_n(x)$, the thicker lines, in the cases $n = 5$ and 10.

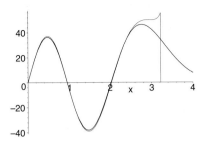

Figure 11.7 $n = 5$.   Figure 11.8 $n = 10$.

A more accurate asymptotic expansion, valid over the entire range of x, is obtained in chapter 15 (page 593).

Exercise 11.16
Derive the approximation 11.32.

Exercise 11.17
Use the asymptotic approximation 11.32 to show that for large n the difference between adjacent zeros of $H_n(x)$, $x_{i+1} > x_i$, is given approximately by

$$x_{i+1} - x_i = d(\bar{x}_i) \simeq \frac{\pi}{\sqrt{2n + 1 - \bar{x}_i^2}}, \quad \bar{x}_i = \frac{1}{2}(x_i + x_{i+1}).$$

Exercise 11.18
Show that

$$\int_{-\infty}^{\infty} dx\, e^{-x^2} x^m H_n(x) = 0$$

if $m \leq n - 1$ is an integer.

Exercise 11.19
Define the function $g(x,t)$ by the series

$$g(x,t) = \sum_{n=0}^{\infty} \frac{t^n}{n!} H_n(x)$$

and use equation 11.26 to show that $\frac{\partial g}{\partial x} = 2tg$ and hence that $g(x,t) = e^{2tx - t^2}$.

Exercise 11.20
Use the generating function found in the previous exercise to show that:

(i) $$\int_{-\infty}^{\infty} dx\, e^{-x^2/2} H_n(x) = \begin{cases} \sqrt{2\pi}\, n!/(n/2)!, & n \text{ even}, \\ 0, & n \text{ odd}, \end{cases}$$

(ii) $$\int_{-\infty}^{\infty} dx\, x e^{-x^2/2} H_n(x) = \begin{cases} 0, & n \text{ even}, \\ 2\sqrt{2\pi}\, \dfrac{n!}{((n-1)/2)!}, & n \text{ odd}. \end{cases}$$

Exercise 11.21

By considering the product of two generating functions, show that

$$e^{2ts} \int_{-\infty}^{\infty} du\, (u+t+s)^p e^{-u^2} = \sum_{n=0}^{\infty} \sum_{m=0}^{\infty} \frac{t^n s^m}{n!m!} \int_{-\infty}^{\infty} dx\, x^p e^{-x^2} H_n(x) H_m(x).$$

If $p = 1$ show that this reduces to

$$(t+s)e^{2ts} = \frac{1}{\sqrt{\pi}} \sum_{n=0}^{\infty} \sum_{m=0}^{\infty} \frac{t^n s^m}{n!m!} \int_{-\infty}^{\infty} dx\, xe^{-x^2} H_n(x) H_m(x),$$

and hence that

$$\frac{1}{\sqrt{h_n h_m}} \int_{-\infty}^{\infty} dx\, xe^{-x^2} H_n(x) H_m(x) = \sqrt{\frac{\max(n,m)}{2}} \left\{ \delta_{mn-1} + \delta_{mn+1} \right\},$$

where $h_n = 2^n n! \sqrt{\pi}$ is the normalisation constant.

Show also that

$$\frac{1}{\sqrt{h_n h_m}} \int_{-\infty}^{\infty} dx\, x^2 e^{-x^2} H_n(x) H_m(x) = \delta_{mn} \frac{2n+1}{2} + \delta_{mn-2} \frac{\sqrt{n(n-1)}}{2}$$
$$+ \delta_{mn+2} \frac{\sqrt{(n+1)(n+2)}}{2}.$$

Hint: the integrals required for the first part of this exercise may be found in Appendix I. For the second result it is easier to use the recurrence relation 11.25 to replace $xH_n(x)$ by a linear combination of H_n and H_{n-1}.

Exercise 11.22

Using the differential equation 11.20, show that

$$H_{2n}\left(\frac{x}{2\sqrt{n}}\right) = \sqrt{2}\left(-\frac{4n}{e}\right)^n \cos x, \quad \text{as } n \to \infty,$$

$$H_{2n+1}\left(\frac{x}{2\sqrt{n+1/2}}\right) = 2\sqrt{2n+1}\left(-\frac{4n}{e}\right)^n \sin x, \quad \text{as } n \to \infty.$$

Plot graphs to compare these limits with the asymptotic expansion given in equation 11.32.

Exercise 11.23

Substitute the series

$$y = a_0 + a_1 x + a_2 x^2 + \cdots + a_r x^r + \cdots$$

into equation 11.20 to show that the coefficients a_k satisfy the recurrence relation

$$(k+1)(k+2)a_{k+2} = (k-n)a_k, \quad k = 0, 1, \ldots .$$

By assigning the initial conditions

$$\text{(i) } a_0 = 1, \quad a_1 = 0, \qquad \text{(ii) } a_0 = 0, \quad a_1 = 1,$$

show that two linearly independent solutions are

$$y_1(x) = 1 + \sum_{m=1}^{\infty} \frac{n(n-2)\cdots(n+2-2m)}{(2m)!}(-2x^2)^m,$$

$$y_2(x) = x \sum_{m=0}^{\infty} \frac{(n-1)(n-3)\cdots(n+1-2m)}{(2m+1)!}(-2x^2)^m,$$

and that these solutions converge for all x.

Note that if n is an integer one of these solutions is an nth degree polynomial and coincides with the solution defined by Rodrigues' formula, equation 11.24.

11.2.4 Laguerre polynomials

Laguerre polynomials, denoted by $L_n(x)$, $n = 0, 1, \ldots$, are complete on the positive real axis, $x \geq 0$. The weight function $w(x)$ must therefore decrease faster than x^n, for any $n > 0$, as $x \to \infty$, and the simplest function with this property is $w(x) = e^{-x}$. The nth Laguerre polynomial is the suitably normalised polynomial solution of the differential equation

$$x\frac{d^2y}{dx^2} + (1-x)\frac{dy}{dx} + ny = 0. \tag{11.33}$$

In contrast to the equation for Hermite polynomials, this equation is singular at the origin and only one of the two linearly independent solutions is finite at the origin. The general theory of such equations, see for instance Ince (1956, chapter 7), shows that the regular solution can be expressed as a power series and that the other solution has a logarithmic singularity at the origin. We are interested only in the first of these which becomes a finite series when n is an integer, exercise 11.28. Equation 11.33 has the self-adjoint form

$$\frac{d}{dx}\left(xe^{-x}\frac{dy}{dx}\right) + ne^{-x}y = 0, \tag{11.34}$$

showing that $w(x) = e^{-x}$, $p(x) = xe^{-x}$ and that the eigenvalues are $\lambda = n$. The multiplicative constant of the solution is chosen† so that $L_n(0) = 1$, and then the orthogonality relation 11.1 becomes

$$\int_0^{\infty} dx\, e^{-x}L_n(x)L_m(x) = \delta_{nm}, \quad (L_n(0) = 1). \tag{11.35}$$

Equation 11.33 has a similar structure to that defining the Hermite polynomials, equation 11.20, consequently the same method, outlined in the appendix to this chapter, shows that the solution can be written in terms of the contour integral

$$L_n(x) = \frac{1}{2\pi i} \oint_{\mathscr{C}} d\zeta \left(\frac{\zeta-1}{\zeta}\right)^n \frac{e^{x\zeta}}{\zeta}, \tag{11.36}$$

where the contour $\mathscr{C}$ encloses the origin in the ζ-plane. The external factor is chosen to ensure that $L_n(0) = 1$.

† Other conventions are also used; for instance sometimes the coefficient of x^n is defined to be $(-1)^n$.

Two transformations of the integration variable re-cast this integral into more convenient forms from which other important results may be derived. First, set $\zeta = z/(1-z)$ to give

$$L_n(x) = \frac{1}{2\pi i} \oint_{\mathscr{C}} dz \, \frac{e^{-xz/(1-z)}}{(1-z)z^{n+1}}, \tag{11.37}$$

where the contour $\mathscr{C}$ now encloses the origin in the z-plane.

The second transformation is to the variable u, defined by

$$\frac{xz}{1-z} = u - x \quad \text{or} \quad z = \frac{u-x}{u},$$

which gives

$$L_n(x) = \frac{e^x}{2\pi i} \oint_{\mathscr{D}} du \, \frac{u^n e^{-u}}{(u-x)^{n+1}}, \tag{11.38}$$

where $\mathscr{D}$ is a curve enclosing the pole at $u = x$.

This relation, together with Cauchy's theorem, equation 11.10, gives Rodrigues' formula for Laguerre polynomials,

$$L_n(x) = \frac{e^x}{n!} \frac{d^n}{dx^n} \left(x^n e^{-x} \right). \tag{11.39}$$

Using this formula it is easy to see that the first few polynomials are $L_0(x) = 1$ and

$L_1(x) = 1 - x,$ $\qquad\qquad\qquad$ $L_2(x) = 1 - 2x + \frac{1}{2}x^2,$

$L_3(x) = 1 - 3x + \frac{3}{2}x^2 - \frac{1}{6}x^3,$ $\qquad$ $L_4(x) = 1 - 4x + 3x^2 - \frac{2}{3}x^3 + \frac{1}{24}x^4,$

$L_5(x) = 1 - 5x + 5x^2 - \frac{5}{3}x^3 + \frac{5}{24}x^4 - \frac{1}{120}x^5.$

Exercise 11.24

Use equation 11.39 and Leibniz's rule for the differentiation of products to show that

$$L_n(x) = n! \sum_{k=0}^{n} \frac{(-x)^k}{k!^2(n-k)!}.$$

Deduce that the coefficients of x, x^2 and x^n are respectively $-n$, $n(n-1)/4$ and $(-1)^n/n!$.

The Laguerre polynomial $L_n(x)$ behaves as $(-x)^n$ as $x \to \infty$ and has n zeros which, as will be seen below, are all in the interval $0 < x < 4n + 2$: outside this interval $|L_n(x)|$ increases as $|x|^n$, but inside this interval the function oscillates between large values, so for graphical comparisons it is more convenient to re-normalise the functions and plot the graphs of

$$\phi_n(x) = e^{-x/2} L_n(x).$$

Graphs of these functions, for $n = 5$ and 10, are shown in the following figure:

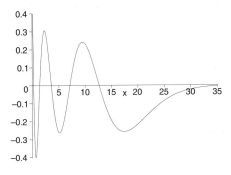

Figure 11.9 $\phi_5(x) = \exp(-x/2)L_5(x)$.

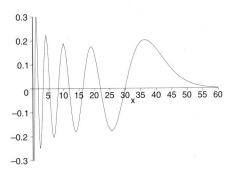

Figure 11.10 $\phi_{10}(x) = \exp(-x/2)L_{10}(x)$.

Cauchy's theorem (page 382) applied to equation 11.37 shows that $L_n(x)$ is $n!$ times the nth differential of the function $(1-t)^{-1}e^{-xt/(1-t)}$, so the generating function representation is

$$\frac{e^{-xt/(1-t)}}{1-t} = \sum_{n=0}^{\infty} t^n L_n(x), \quad |t| < 1. \tag{11.40}$$

This relation can be used to evaluate certain integrals, as in exercise 11.26, and to derive some recurrence relations, as in the next exercise.

Exercise 11.25
By differentiating equation 11.40 with respect to x, show that

$$L_n(x) = L'_n(x) - L'_{n+1}(x).$$

By differentiating the same equation with respect to t, show that

$$(n+1)L_{n+1}(x) = (2n+1-x)L_n(x) - nL_{n-1}(x). \tag{11.41}$$

Use these two results and the original differential equation to show that

$$xL'_n(x) = nL_n(x) - nL_{n-1}(x). \tag{11.42}$$

The elementary asymptotic expansion of the Laguerre polynomials can be found in a similar manner to that of the Hermite polynomials, equation 11.32, though because the defining equation is singular there is a subtlety that can only be resolved using the methods developed in chapter 15. The function $\phi_n(x) = e^{-x/2}L_n(x)$ satisfies the equation

$$x\frac{d^2\phi}{dx^2} + \frac{d\phi}{dx} + \left(v - \frac{x}{4}\right)\phi = 0, \quad v = n + \frac{1}{2}. \tag{11.43}$$

When the coefficient of ϕ is large and positive, $x < 4v$, the solution of this equation is oscillatory, and when this coefficient is large and negative, $x > 4v$, it is either exponentially increasing or decreasing. The solution we require decreases exponentially and in order to find an approximation in the oscillatory region we first rescale the independent variable, $z = x/v$, and then assume a solution of the form

$$\phi(z) = \begin{cases} A(z)\exp(ivS(z)), & v = n + \tfrac{1}{2}, & 0 < z < 4, \\ 0, & & z > 4. \end{cases}$$

Substituting this into the transformed equation gives

$$z\frac{d^2A}{dz^2} + \frac{dA}{dz} + i\frac{v}{A}\left\{z\frac{dW}{dz} + W\right\} - v^2A\left\{z\left(\frac{dS}{dz}\right)^2 - \left(1 - \frac{z}{4}\right)\right\} = 0, \qquad (11.44)$$

where $W(z) = A^2 dS/dz$. If $v \gg 1$ an approximate solution is obtained by equating the coefficients of v and v^2 to zero and solving the two equations obtained to give

$$S(z) = \int_0^z dz\sqrt{\frac{1}{z} - \frac{1}{4}} + \alpha, \quad A(z) = \frac{B}{(z(4-z))^{1/4}},$$

where α and B are constants. These give

$$\phi_n(x) \sim \frac{B}{(x(4v - x))^{1/4}}\sin\left(\int_0^x \frac{dx}{\sqrt{x}}\sqrt{v - \frac{x}{4}} + \alpha\right), \quad 0 < x < 4v = 4n + 2. \quad (11.45)$$

The constant B may be obtained from the normalisation condition if we note that when $n \gg 1$ the sine function oscillates rapidly and that its square may be approximated by its mean value, $\tfrac{1}{2}$. Thus

$$1 = \int_0^\infty dx\, e^{-x}L_n(x)^2 \simeq \frac{B^2}{2}\int_0^{4v} dx\,\frac{1}{\sqrt{x(4v-x)}} = \frac{1}{2}\pi B^2.$$

Hence $B = \sqrt{2/\pi}$.

The approximation 11.45 is singular at the origin where it behaves as $x^{-1/4}$, unless $\alpha = m\pi$, m being an integer. This singularity is a direct consequence of the original equation being singular at the origin, so that near $x = 0$ the first two terms of 11.44 dominate and the approximation is invalid.

In order to find the value of α it is necessary to consider the solution in the neighbourhood of $x = 4v$ where the coefficient of ϕ in equation 11.43 changes sign, so the solution changes from being oscillatory when $x < 4v$ to decaying exponentially for $x > 4v$. The method of dealing with this type of problem is discussed in chapter 15 and there we shall show that $\alpha = \pi/4$. Thus we obtain, for $0 < x < 4n + 2$,

$$\phi_n(x) \sim \frac{\sqrt{2}}{\sqrt{\pi}\,(x(4v-x))^{1/4}}\sin\left(2v\sin^{-1}\sqrt{\frac{x}{4v}} + \frac{1}{2}\sqrt{x(4v-x)} + \frac{\pi}{4}\right), \qquad (11.46)$$

where $v = n + 1/2$.

In the next two figures this approximation is compared with the exact values of $e^{-x/2}L_n(x)$, the thicker lines, for $n = 5$ and 10. Note that the weak singularity in the approximation is significant only very close to the origin and is not noticeable in these graphs.

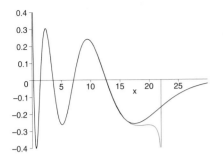

Figure 11.11 $n = 5$.

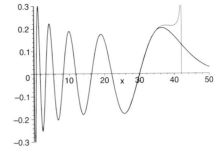

Figure 11.12 $n = 10$.

Exercise 11.26

Use the generating function for Laguerre polynomials, equation 11.40, to show that

$$\frac{1}{(1-t)(1-s)} \exp\left(-\frac{x(1-st)}{(1-t)(1-s)}\right) = \sum_{s=0}^{\infty}\sum_{t=0}^{\infty} s^m t^n e^{-x} L_n(x) L_m(x),$$

and use this to confirm that

$$\int_0^{\infty} dx\, e^{-x} L_n(x) L_m(x) = \delta_{nm}.$$

Show also that

$$\int_0^{\infty} dx\, e^{-x} x L_n(x) L_m(x) = \begin{cases} 2n+1, & n=m, \\ -n-1, & m=n+1, \\ -n, & m=n-1, \\ 0, & \text{otherwise}. \end{cases}$$

and that

$$\int_0^{\infty} dx\, e^{-x} x^2 L_n(x) L_m(x) = \begin{cases} 6n^2+6n+2, & n=m, \\ -4n^2, & m=n-1, \\ -4(n+1)^2, & m=n+1, \\ n(n-1), & m=n-2, \\ (n+1)(n+2), & m=n+2, \\ 0, & \text{otherwise}. \end{cases}$$

Hint: for this result you may find it easier to use the recurrence relation 11.41 to replace $xL_n(x)$ by a linear combination of L_n and L_{n-1}.

Exercise 11.27

Use the asymptotic approximation given in equation 11.46 to show that if $n \simeq m \gg 1$,

$$F_{nm} = \int_0^{\infty} dx\, e^{-x} f(x) L_n(x) L_m(x) \simeq \frac{1}{\pi} \int_0^{\pi} dt\, f(x(t)) \cos(n-m)t,$$

where $x(t) = 4\bar{v}\sin^2(t/2)$ and $\bar{v} = (n+m+1)/2$ is the mean value of $v_n = n+1/2$ and $v_m = m+1/2$. For the case $f(x) = x$ show that this approximation is exact. For $f(x) = x^2$

show that the approximation gives

$$
F_{nm} \simeq \begin{cases} 6(n+1/2)^2, & m = n, \\ -4n^2, & m = n-1, \\ -4(n+1)^2, & m = n+1, \end{cases} \qquad F_{nm} \simeq \begin{cases} (n-1/2)^2, & m = n-2, \\ (n+3/2)^2, & m = n+2, \\ 0, & \text{otherwise.} \end{cases}
$$

Hint: it is helpful to write the approximation for $\phi_n(x)$ in the form

$$
\phi_n(x) = \sqrt{\frac{2}{\pi}} \frac{1}{(x(4v_n - x))^{1/4}} \sin\left(\int_0^x \frac{du}{\sqrt{u}} \sqrt{v_n - \frac{u}{4}} + \frac{\pi}{4}\right),
$$

to use the identity $2 \sin A \sin B = \cos(A - B) + \cos(A + B)$ and to note that the second term is highly oscillatory so may be ignored. Finally, use the approximation $g(v_n, x) - g(v_m, x) \simeq (v_n - v_m)\partial g/\partial v$.

Exercise 11.28
Substitute the series

$$
y(x) = x^c \sum_{r=0}^{\infty} a_r x^r
$$

into the differential equation

$$
x\frac{d^2 y}{dx^2} + (1 - x)\frac{dy}{dx} + \alpha y = 0, \quad \alpha \geq 0,
$$

and show that $c^2 a_0 = 0$.
  Consider the solution with $c = 0$, $a_0 = 1$ and show that

$$
a_p = (-1)^p \frac{\alpha(\alpha - 1)\cdots(\alpha - p + 1)}{(p!)^2}
$$

and that the radius of convergence of this solution is infinite. Show also that if α is an integer, $\alpha = n$, this series solution is an nth degree polynomial.

11.3 Gaussian quadrature

A practical use of orthogonal polynomials is the approximation of integrals using the *Gaussian quadrature* method. We introduce this method using Legendre polynomials and then show how other sets of orthogonal polynomials may be similarly used to approximate integrals with various types of singularities at the end points.
  The general idea is to approximate an integral as a sum over N terms, in the form

$$
\int_{-1}^{1} dx\, f(x) = \sum_{k=1}^{N} c_k f(x_k) + R_N, \tag{11.47}
$$

where $f(x)$ is a sufficiently well behaved function and where the weights c_k and abscissae x_k are $2N$ numbers and R_N is the remainder term. We now show that these $2N$ constants may be chosen to make the approximations *exact* when $f(x)$ is a polynomial of degree $2N - 1$ or less. This contrasts with the behaviour of equally spaced N-point rules that are exact only for either N or $N - 1$ degree polynomials.

On the interval $-1 \leq x \leq 1$ the function $f(x)$ may be expanded as a series of Legendre polynomials,

$$f(x) = \sum_{k=0}^{\infty} a_k P_k(x),$$

with the Fourier coefficients given by equations 11.12 (page 383); the first coefficient is related to the integral we require:

$$a_0 = \frac{1}{2} \int_{-1}^{1} dx\, f(x).$$

Substituting the expansion into the integral 11.47 gives

$$\int_{-1}^{1} dx\, f(x) \simeq \sum_{j=1}^{N} c_j f(x_j) = \sum_{k=0}^{\infty} a_k \sum_{j=1}^{N} c_j P_k(x_j), \tag{11.48}$$

where the weights c_k and abscissae x_k remain to be chosen. Now $P_N(x)$ has exactly N zeros in $-1 \leq x \leq 1$, so if the abscissae are taken to be these zeros, $P_N(x_j) = 0$, $j = 1, 2, \ldots, N$, the coefficient of a_N vanishes. Now choose the N weights c_j to satisfy the N relations

$$\sum_{j=1}^{N} c_j P_0(x_j) = \sum_{j=1}^{N} c_j = 2, \tag{11.49}$$

$$\sum_{j=1}^{N} c_j P_k(x_j) = 0, \quad k = 1, 2, \ldots, N-1, \tag{11.50}$$

so that

$$\sum_{i=1}^{N} c_i f(x_i) = 2a_0 + R_N, \quad R_N = \sum_{k=N+1}^{\infty} a_k \sum_{j=1}^{N} c_j P_k(x_j). \tag{11.51}$$

Finally, use the equation for a_0 to give

$$\int_{-1}^{1} dx\, f(x) = \sum_{i=1}^{N} c_i f(x_i) - R_N. \tag{11.52}$$

In order to use this formula the weights c_i and abscissae x_i are needed. The abscissae need to be computed numerically: there are tables of these numbers, for instance Abramowitz and Stegun (1965, chapter 25), and Maple will find all roots using the command `fsolve(P(n,x)=0,x,complex)`, provided n is not too large and the variable `Digits` is increased if $n > 20$. It may be shown that the weights are given by

$$c_i = \frac{2}{(1 - x_i^2)P_N'(x_i)^2}. \tag{11.53}$$

Observe that if $f(x)$ is a polynomial of degree $N - 1$, or less, it follows from the above construction that $R_N = 0$; that is, for this case the sum is exact. However, the remarkable feature of these methods that makes them so useful is that the sum is exact if $f(x)$ is a polynomial of degree $2N - 1$ or less. This is seen by noting that such a polynomial may be decomposed into the sum of two terms,

$$f(x) = P_N(x)Q_1(x) + Q_2(x)$$

where Q_1 and Q_2 are polynomials of degree $N - 1$ or less, both of which must be orthogonal to $P_N(x)$, exercise 11.5 (page 384). Then the integral is

$$\int_{-1}^{1} dx\, f(x) = \int_{-1}^{1} dx\, P_N(x)Q_1(x) + \int_{-1}^{1} dx\, Q_2(x).$$

But $P_N(x)$ and $Q_1(x)$ are orthogonal so the first term is identically zero, and the second integral is given exactly by the N-point quadrature sum. Hence the remainder is zero.

In general the remainder term may be written in the form

$$R_N = \left(\frac{N!^2}{(2N)!}\right)^2 \frac{2^{2N+1}}{(2N+1)} \frac{f^{(2N)}(\xi)}{(2N)!} \simeq \frac{\pi}{4^N} \frac{f^{(2N)}(\xi)}{(2N)!}, \quad -1 < \xi < 1. \tag{11.54}$$

The last ratio is the coefficient of u^{2N} in the Taylor's series of $f(\xi + u)$. In practice this formula for the remainder is of little value because of the difficulty in computing the high order differential: a more practical alternative estimate is given by Lanczos (1966, chapter 6). However, if the Taylor's series of $f(x)$ about $x = \xi$ has radius of convergence r, then $r^{2N} f^{(2N)}(\xi)/(2N)! \simeq 1$, which provides the estimate $R_N \simeq \pi(2r)^{-2N}$, for $N \gg 1$.

Exercise 11.29
Show that for the arbitrary interval $a \le x \le b$, equation 11.47 becomes

$$\int_{a}^{b} dx\, f(x) = \frac{1}{2}(b - a) \sum_{k=1}^{N} c_k f(y_k) \quad \text{with} \quad y_i = \frac{1}{2}(b + a) + \frac{1}{2}(b - a)x_i,$$

where x_i are the roots of $P_N(x) = 0$.

Exercise 11.30
Use Maple to find the weights and abscissae for $N = 10$. Use these to evaluate the integrals

$$\int_{-1}^{1} dx\, x^{2n}$$

for $n = 1, 2, \ldots, 20$, and for each case determine the relative error of the approximation. Consider also the effects of changing `Digits` and N and observe how the accuracy changes as n increases through N.

Perform the same calculations on the integral $\int_{-1}^{1} dx\, \exp(-ax)$ for $1 \le a \le 50$.

General theory
For an integral over the interval (a, b) with weight function $w(x)$ we use the same general theory but with the polynomials $p_n(x)$ orthogonal on (a, b) with weight function $w(x)$. A similar analysis gives

$$\int_{a}^{b} dx\, w(x)f(x) \simeq \sum_{k=1}^{N} c_k f(x_k), \tag{11.55}$$

where the abscissae x_j, $j = 1, 2, \ldots, N$, are the zeros of $p_N(x)$ and the N weights c_j are chosen to satisfy

$$\sum_{j=1}^{N} c_j = \frac{h_0}{p_0^2}, \tag{11.56}$$

$$\sum_{j=1}^{N} c_j p_k(x_j) = 0, \quad k = 1, 2, \ldots, N-1, \tag{11.57}$$

where h_0 is defined in equation 11.1 and p_0 is a constant.

Different types of integral may be evaluated using various polynomials. Thus for integrals of the type

$$\int_0^\infty dx\, e^{-x} f(x), \qquad \int_{-\infty}^\infty dx\, e^{-x^2} f(x),$$

we use, respectively, Laguerre polynomials and Hermite polynomials; weights and abscissae of these and other examples are given by Abramowitz and Stegun (1965, chapter 25). For integrals with square-root singularities at the end points we use Tchebychev polynomials. Thus for integrals with weight function $w(x) = (1 - x^2)^{-1/2}$ the weights are $w_j = \pi/N$ and

$$\int_{-1}^{1} dx\, \frac{f(x)}{\sqrt{1 - x^2}} \simeq \frac{\pi}{N} \sum_{j=1}^{N} f(x_j), \quad x_j = \cos\left(\frac{j - \frac{1}{2}}{N} \right) \pi. \tag{11.58}$$

On putting $x = \cos\theta$ the integral can be cast in the form

$$\frac{1}{\pi} \int_0^\pi g(\theta) = \frac{1}{N} \sum_{j=1}^{N} g\left(\frac{j - \frac{1}{2}}{N} \pi \right), \quad g(\theta) = f(\cos\theta).$$

The left hand side is the mean over one period of the 2π-periodic function $g(\theta)$ approximated by a sum over equally spaced points. We saw in section 8.7 that this sum converged very rapidly to the integral with increasing N provided all differentials of g exist.

Exercise 11.31
Show that for the arbitrary interval $a \leq x \leq b$ equation 11.58 becomes

$$\int_a^b dx\, \frac{f(x)}{\sqrt{(x - a)(b - x)}} \simeq \frac{\pi}{N} \sum_{j=1}^{N} f(y_j), \quad y_j = \frac{1}{2}(b + a) + \frac{1}{2}(b - a)\cos\left(\frac{j - \frac{1}{2}}{N} \right)\pi.$$

Exercise 11.32
For integrals along the whole positive real axis Laguerre polynomials may be used, so we have

$$\int_0^\infty dx\, e^{-x} f(x) \simeq \sum_{k=1}^{N} c_k f(x_k),$$

where x_k are the N roots of $L_N(x) = 0$ and the weights are

$$c_k = \frac{x_k}{(N+1)^2 L_{N+1}(x_k)^2}.$$

Write a Maple procedure to use this method for the integral

$$\int_0^\infty dx\, F(x),$$

where $F(x)$ is an arbitrary function. Use your procedure, with various values of N, to evaluate the infinite integral when

$$F(x) = \frac{1}{1+x^2}, \quad \frac{\sin x}{x}, \quad \frac{1}{1+x^2+x^5}, \quad \text{and} \quad \frac{\sin x}{1+x^3}.$$

Compare your results with either the exact value of the integral or the value given by Maple.

Exercise 11.33
Show that the set of functions $T_{2n+1}(x)$, $n = 0, 1, \ldots$, are complete on $(0, 1)$ and use these functions to develop the quadrature

$$\int_0^1 dx\, \sqrt{\frac{x}{1-x}}\, f(x) \simeq \sum_{k=1}^N w_k f(x_k), \quad \text{where} \quad x_i = \cos^2 \frac{2i-1}{2N+1} \frac{\pi}{2}, \quad w_i = \frac{2\pi x_i}{2N+1}.$$

11.4 Bessel functions

Bessel functions are probably the most commonly occurring of the special functions. They occur when separating variables in a variety of situations, and also in the analysis of the motion of a particle moving in an inverse square force, see appendix 1 of chapter 7 which gives a brief description of their origin.

One form of Bessel's equation is

$$x^2 \frac{d^2 y}{dx^2} + x \frac{dy}{dx} + (x^2 - v^2)y = 0, \tag{11.59}$$

where both x and v may be complex variables; here, however, we shall assume x to be real and shall concentrate on the most important values of v, that is when it is an integer, $v = n$, or a half integer, $v = n + \frac{1}{2}$, $n = 0, \pm 1, \pm 2, \ldots$. Some of the results quoted are valid for all v; so we adopt the convention that n denotes an integer, positive or negative, and v a real number, possibly in a restricted range, and results quoted for $J_n(x)$ are not necessarily valid for $J_v(x)$.

Equation 11.59 is the standard form of Bessel's equation, but there are many other equivalent equations, two of which are derived in exercise 11.39.

When v is an integer there are two linearly independent solutions denoted by $Y_n(x)$ and $J_n(x)$ — in Maple by BesselJ(n,x) and BesselY(n,x) — with $J_n(x)$ bounded for all x and $Y_n(x)$ unbounded at the origin:

$$J_v(x) \sim \frac{(x/2)^v}{\Gamma(1+v)}, \quad v \neq -1, -2, \ldots, \quad J_{-n}(x) = (-1)^n J_n(x), \tag{11.60}$$

$$Y_0(x) \sim \frac{2}{\pi} \left(\gamma + \ln \left(x/2 \right) \right),$$

$$Y_v(x) \sim -\frac{\Gamma(v)}{\pi (x/2)^v}, \quad \Re(v) > 0, \quad Y_{-n}(x) = (-1)^n Y_n(x). \tag{11.61}$$

The series representation of $J_v(x)$ is

$$J_v(x) = \left(\frac{x}{2} \right)^v \sum_{k=0}^{\infty} \frac{\left(-x^2/4 \right)^k}{k! \, \Gamma(v+k+1)}, \tag{11.62}$$

and is uniformly convergent for all x; the series for $Y_n(x)$ is more complicated and is given in Abramowitz and Stegun (1965, 9.1.11). Although this series converges for all x it is of practical value only for small x, typically $|x| < |v|$, because the terms alternate in sign, just like those of the sine function: the problems associated with the practical summation of the sine series are considered in exercise 5.16 (page 191), and the same problems occur in the summation of 11.62. One way of avoiding this problem is to use the recurrence relation 11.75 as shown in exercise 11.69 (page 435).

The generating function for $J_n(x)$ is

$$\exp \left(\frac{x}{2} \left(t - \frac{1}{t} \right) \right) = \sum_{k=-\infty}^{\infty} t^k J_k(x), \quad t \neq 0, \tag{11.63}$$

and useful versions of this are obtained by setting $t = e^{-i\theta}$:

$$e^{-ix \sin \theta} = \sum_{k=-\infty}^{\infty} e^{-ik\theta} J_k(x), \tag{11.64}$$

$$\cos(x \sin \theta) = J_0(x) + 2 \sum_{k=1}^{\infty} J_{2k}(x) \cos 2k\theta \tag{11.65}$$

$$\sin(x \sin \theta) = 2 \sum_{k=0}^{\infty} J_{2k+1}(x) \sin(2k+1)\theta \tag{11.66}$$

$$\cos(x \cos \theta) = J_0(x) + 2 \sum_{k=1}^{\infty} (-1)^k J_{2k}(x) \cos 2k\theta \tag{11.67}$$

$$\sin(x \cos \theta) = 2 \sum_{k=0}^{\infty} (-1)^k J_{2k+1}(x) \cos(2k+1)\theta. \tag{11.68}$$

The first of these expressions shows that $J_n(x)$ is just the nth Fourier component in the Fourier series of $e^{-ix \sin \theta}$, which is a direct reflection of their origin, discussed in the first appendix of chapter 7.

Exercise 11.34
Use the generating function to show that

$$1 = J_0(x) + 2 \sum_{k=1}^{\infty} J_{2k}(x),$$

$$\cos x = J_0(x) - 2J_2(x) + 2J_4(x) - 2J_6(x) + \cdots, \quad \text{and}$$

$$\sin x = 2J_1(x) - 2J_3(x) + 2J_5(x) - \cdots.$$

The series for $J_\nu(x)$ are infinite, but if $\nu = n + \frac{1}{2}$ the spherical Bessel function,

$$j_n(x) = \sqrt{\frac{\pi}{2x}} J_{n+\frac{1}{2}}(x), \quad y_n(x) = \sqrt{\frac{\pi}{2x}} Y_{n+\frac{1}{2}}(x), \tag{11.69}$$

that satisfy the equation

$$x^2 \frac{d^2 y}{dx^2} + 2x \frac{dy}{dx} + (x^2 - n(n+1))y = 0, \quad n = 0, \pm 1, \pm 2, \ldots, \tag{11.70}$$

can be represented by the finite series

$$
\begin{aligned}
j_0(x) &= \frac{\sin x}{x}, \quad j_1(x) = \frac{\sin x}{x^2} - \frac{\cos x}{x}, \\
j_2(x) &= \left(\frac{3}{x^3} - \frac{1}{x} \right) \sin x - \frac{3}{x^2} \cos x,
\end{aligned}
$$

and in general

$$j_n(x) = f_n(x) \sin x + (-1)^{n+1} f_{-n-1}(x) \cos x,$$

where $f_0(x) = 1/x$, $f_1(x) = 1/x^2$ and $f_n(x)$ are given by the recurrence relation

$$f_{n-1} + f_{n+1} = \frac{2n+1}{x} f_n, \quad n = 0, \pm 1, \pm 2, \ldots.$$

For small $|x|$ the series expansions are useful,

$$j_n(x) = \left\{ 1 - \frac{\frac{1}{2}x^2}{1!\,(2n+3)} + \frac{\left(\frac{1}{2}x^2\right)^2}{2!\,(2n+3)(2n+5)} + \cdots \right\},$$

whilst $x^{n+1} y_n(x) \to -1.3.5. \cdots .(2n-1)$ as $x \to 0$.

The other important variants of the Bessel functions are the Airy functions $\mathrm{Ai}(x)$ and $\mathrm{Bi}(x)$ — in Maple `AiryAi(x)` and `AiryBi(x)` — which satisfy the equation

$$\frac{d^2 y}{dx^2} - xy = 0. \tag{11.71}$$

The Airy function that occurs most often can be defined by the integral, derived in the appendix to this chapter,

$$
\begin{aligned}
\mathrm{Ai}(x) &= \frac{1}{\pi} \int_0^\infty dt \, \cos\left(\frac{1}{3} t^3 + xt \right), \tag{11.72} \\
&= \frac{1}{2\pi} \int_{-\infty}^\infty dt \, \exp i \left(\frac{1}{3} t^3 + xt \right).
\end{aligned}
$$

The integrand does not tend to zero as $|t| \to \infty$, but the integrals converge because the rapidly increasing oscillations induce convergence. The function $\mathrm{Ai}(x)$ is the solution of 11.71 that tends to zero as $x \to \infty$. The linearly independent solution $\mathrm{Bi}(x)$ can be represented by the integral

$$\mathrm{Bi}(x) = \int_0^\infty dt \left[\exp\left(-\frac{1}{3} t^3 + xt \right) + \sin\left(\frac{1}{3} t^3 + xt \right) \right]. \tag{11.73}$$

This function tends to infinity exponentially as $x \to \infty$. For $x < 0$ both solutions oscillate in the same manner, as shown in the figures.

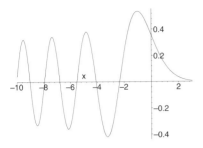

Figure 11.13   Ai(x).

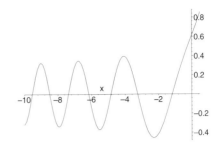

Figure 11.14   Bi(x).

The asymptotic expansions of the Airy functions, which will be derived in chapter 15, clearly reflect this behaviour:

$$\text{Ai}(x) \sim \begin{cases} \dfrac{1}{2\sqrt{\pi}x^{1/4}}e^{-\zeta} \\[3mm] \dfrac{1}{\sqrt{\pi}|x|^{1/4}}\sin\left(\zeta+\dfrac{\pi}{4}\right) \end{cases}, \quad \text{Bi}(x) \sim \begin{cases} \dfrac{1}{\sqrt{\pi}x^{1/4}}e^{\zeta}, & x\to\infty \\[3mm] \dfrac{1}{\sqrt{\pi}|x|^{1/4}}\cos\left(\zeta+\dfrac{\pi}{4}\right), & x\to-\infty \end{cases}$$

where $\zeta = 2|x|^{3/2}/3$. Both Airy functions are related to ordinary Bessel functions of order $\frac{1}{3}$, see for instance Abramowitz and Stegun (1965, 10.4.15).

The first Airy function, Ai(x), is probably the most important and it is named after Airy, the Astronomer Royal between 1835 and 1881. It is important because it may be used to find good approximations to some integrals and differential equations that occur frequently in mathematical physics, particularly in the descriptions of wave motion with short wavelength; indeed it was in his description of rainbows that Airy introduced these functions. This theory will be discussed in chapters 14 and 15, but see also exercise 11.40 for an example.

J. B. Airy (1801–1892)

The behaviour of the ordinary Bessel functions, for real values of the argument, is quite simple and is shown in the figures for $n = 0, 1, 3, 5$ and 10.

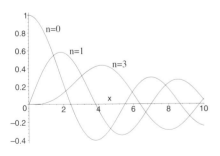

Figure 11.15   The Bessel functions $J_n(x)$ for $n = 0, 1$ and 3.

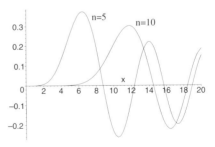

Figure 11.16 $J_n(x)$ for $n = 5$ and 10.

Apart from $J_0(x)$ these Bessel functions increase from zero at $x = 0$ to the first maximum at $x \sim n$, after which they oscillate with a period of approximately 2π and with an amplitude that decreases as $1/\sqrt{|x|}$. The behaviour of $J_0(x)$ differs only because there is no initial rise and $J_0(0) = 1$; the asymptotic behaviour is similar. For $n \geq 1$, this behaviour is seen most clearly by plotting the graphs of $J_n(nx)$, as these functions are weakly dependent upon n for $x \gg 1$. In the figure these functions are shown for $n = 1$, 3, 5 and 10; the different curves may be identified by noting that the larger n the nearer the abscissa of the first maximum to unity.

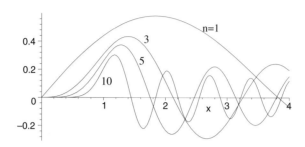

Figure 11.17 $J_n(nx)$ for $n = 1$, 3, 5 and 10.

Many important properties of $J_n(x)$ can be derived from its integral representation

$$J_n(x) = \frac{1}{2\pi} \int_{-\pi}^{\pi} d\theta \, e^{i(n\theta - x\sin\theta)}, \tag{11.74}$$

which is another way of writing equation 11.64. From this representation the following recurrence relations can be derived:

$$J_{n-1}(x) + J_{n+1}(x) = \frac{2n}{x} J_n(x), \tag{11.75}$$

$$J_{n-1}(x) - J_{n+1}(x) = 2J_n'(x). \tag{11.76}$$

The addition formula

$$J_n(x + y) = \sum_{k=-\infty}^{\infty} J_k(x) J_{n-k}(y) \tag{11.77}$$

also follows directly from this integral using the rules for the product of Fourier series, exercise 8.31 (page 295). Also, using methods developed in chapter 14, the following crude asymptotic expansion is easily derived from the same integral:

$$J_n(nx) \sim \sqrt{\frac{2}{\pi n \sqrt{x^2 - 1}}} \cos\left(n\cos^{-1}\left(\frac{1}{x}\right) - \sqrt{x^2 - 1} + \frac{\pi}{4}\right), \quad x > 1, \tag{11.78}$$

$$\sim \sqrt{\frac{2}{\pi n x}} \cos\left(x - \left(\frac{n}{2} + \frac{1}{4}\right)\pi\right), \quad x \gg 1.$$

More accurate approximations that are not singular at $x = 1$ can be found: an example is given in exercise 11.40 and the theory behind this type of approximation is described in chapter 15. Notice that the singularity at $x = 1$ is similar to that in the approximations to the Legendre, Hermite and Laguerre polynomials. Apart from the phase $\pi/4$ this approximation is readily derived from Bessel's differential equation 11.59.

From this asymptotic expansion it is clear that $J_n(x)$ has infinitely many, approximately equally spaced zeros; the sth zero of $J_n(x)$ is traditionally denoted by $j_{n,s}$, and it is clear from equation 11.78 that for large n, $j_{n,s} \sim (2n+1)\pi/4 + (2p-1)\pi/2$ for some p. Exact values of $j_{n,s}$ and $y_{n,s}$ are given by the Maple functions `BesselJZeros` and `BesselYZeros`. The asymptotic expansion of $j_{v,s}$ is, for $s \gg v$,

$$j_{v,s} \sim \beta - \frac{\mu - 1}{\beta} - \frac{4(\mu - 1)(7\mu - 31)}{3(8\beta)^2} + \cdots, \quad \beta = \left(s + \frac{v}{2} - \frac{1}{2}\right)\pi, \ \mu = 4v^2.$$

It is this oscillatory behaviour of the Bessel functions that makes the set $J_n(j_{n,k}\,x)$, $k = 1, 2, \ldots$, orthogonal, and allows the Fourier expansions of equation 10.23 (page 357), and exercises 11.36 and 11.37 below. In practice these expansions are of less value than trigonometric series, partly because it is more difficult to compute the values of Bessel functions and partly because it is usually difficult to compute the Fourier coefficients.

Exercise 11.35
Use the integral representation of the ordinary Bessel function, equation 11.74, to derive the recurrence relations 11.75, 11.76 and the addition formula 11.77.

Exercise 11.36
Show that the equation

$$\frac{d}{dx}\left(x\frac{dy}{dx}\right) + \left(\lambda^2 x - \frac{n^2}{x}\right)y = 0, \quad y(b) = 0,$$

where $b > 0$, has the bounded solutions $J_n(j_{n,k}\,x/b)$, $k = 1, 2, \ldots$.
   Using the result

$$\int_0^1 dt\, t J_n(j_{n,k}\,t)\, J_n(j_{n,l}\,t) = \frac{1}{2}J_n'(j_{n,k})^2\delta_{kl},$$

find an integral expression for the Fourier coefficients a_k, where

$$f(x) = \sum_{k=1}^{\infty} a_k J_n(j_{n,k}\,x/b).$$

Exercise 11.37
Show that

$$\delta(x - \xi) = 2\sqrt{x\xi}\sum_{k=1}^{\infty}\frac{J_n(j_{n,k}\,x)J_n(j_{n,k}\,\xi)}{J_n'(j_{n,k})^2}, \quad 0 \le x, \xi \le 1.$$

Use Maple to evaluate this sum for $n = 1$ and 2 truncated to ten terms.

Exercise 11.38
Show that the Heaviside function can be represented by the Bessel function series

$$H(x - \xi) = \sum_{k=1}^{\infty} a_k(\xi)J_n(j_{n,k}\,x), \quad 0 < x, \xi < 1,$$

where

$$a_k(\xi) = \frac{2}{J_n'(j_{n,k})^2}\int_\xi^1 du\, u J_n(j_{n,k}\,x),$$

and where $j_{n,k}$ is the kth positive zero of $J_n(x)$.

In the case $n = 1$ use Maple to find $j_{n,k}$, write a procedure to numerically evaluate the coefficient $a_k(\xi)$, for any k and ξ, and plot the graph of the partial sum

$$H_N(x - \xi) = \sum_{k=1}^{N} a_k(\xi) J_n(j_{n,k} x),$$

for sufficiently large N, typically between 10 and 20, to see Gibbs' oscillations develop.

Exercise 11.39

Show that the following differential equations have the solutions given:

$$\frac{d^2w}{dx^2} + \left(\lambda^2 e^{2x} - v^2\right) w = 0, \qquad w = J_v(\lambda e^x),$$

$$\frac{d^2w}{dx^2} + \left(\lambda^2 - \frac{v^2 - \frac{1}{4}}{x^2}\right) w = 0, \qquad w = x^{1/2} J_v(\lambda x).$$

Exercise 11.40

Use Maple to make a graphical comparison of the following Airy function approximation for $J_n(nx)$:

$$J_n(nx) \simeq \left(\frac{4\zeta(x)}{1 - x^2}\right)^{1/4} n^{-1/3} \mathrm{Ai}\left(n^{2/3}\zeta(x)\right), \quad x \geq 0,$$

where the function $\zeta(x)$ is positive for $0 \leq x < 1$, negative for $x > 1$, and defined by

$$\frac{2}{3}\zeta^{3/2} = \ln\left(\frac{1 + \sqrt{1 - x^2}}{x}\right) - \sqrt{1 - x^2}, \quad 0 \leq x < 1,$$

$$\frac{2}{3}(-\zeta)^{3/2} = \sqrt{x^2 - 1} - \cos^{-1}(1/x), \quad x \geq 1.$$

Consider the cases $n = 1, 4$ and 10 for $0 < x < 5$.

Note: formally this approximation is the first term in a type of asymptotic expansion, valid for large n, and will be derived from the integral representation in chapter 14 and from the differential equation in chapter 15. This asymptotic approximation is quite accurate even for $n = 1$.

This exercise provides an illustration of the use of Airy functions; this type of approximation can be found for the solutions of a large class of differential equations and will be discussed in chapter 15, see also Abramowitz and Stegun (1965, 10.4.111).

11.5 Mathieu functions

Mathieu functions are the π- and 2π-periodic solutions of Mathieu's equation

$$\frac{d^2y}{dz^2} + (a - 2q \cos 2z)y = 0, \tag{11.79}$$

where a and q are constants. This is a Sturm–Liouville system with periodic boundary conditions and solutions exist only for particular values of the eigenvalue a which depend upon q; other solutions exist for every pair (a, q), but here we consider only the periodic solutions, leaving the others to the next chapter. The most significant difference between Mathieu's equation and other equations defining special functions is that the coefficient of y is a periodic function of the independent variable, so it is the simplest case of the more general equation

$$\frac{d^2 y}{dz^2} + (a + p(z))y = 0,\qquad(11.80)$$

where $p(z)$ is a periodic function of z. This is called Hill's equation, after the American astronomer who derived this type of equation in his investigations of the stability of lunar motion, see for instance Barrow-Green (1996). G. W. Hill (1838–1914)

The periodic solutions of equation 11.79 seem to have been first discussed by Mathieu in the context of the vibrations of an elliptic membrane. Then the equation arises from the separation of the equation E. Mathieu (1835–1890)

$$\frac{\partial^2 u}{\partial x^2} + \frac{\partial^2 u}{\partial y^2} + k^2 u = 0$$

in the elliptic coordinates (ξ, η) defined by $x = h \cosh \xi \cos \eta$, $y = h \sinh \xi \sin \eta$, where $(\pm h, 0)$ are the coordinates of the ellipse foci, see for instance Whittaker and Watson (1965, chapter 12). One of the subsequent equations can be written in the form of equation 11.79, and the periodicity requirement follows because the solution must be single valued.

However, Mathieu's equation and its generalisations are more important than this single application would suggest. The motion of an electron in a periodic array of atoms is important in understanding electrical conduction and gives rise to equations like 11.80. Some larger molecules comprise parts that can rotate relative to the other parts, for instance the CH_3 and CF_3 parts of the molecule H_3CCF_3 may rotate about their common axis: this type of motion is describe by equations like Mathieu's equation, see for instance Townes and Schawlow (1975, chapter 12) in which many similar examples are given. In this case the eigenvalues are closely related to the molecular energies. The study of the stability of the periodic orbits of nonlinear systems also produces equations like 11.80; this application will be discussed in chapter 19.

By comparison to most other special functions, the behaviour of Mathieu functions and their eigenvalues is relatively rich and consequently more difficult to understand. The next section provides a succinct summary of the main properties of these quantities, and this is followed by a section describing two simple methods of computing Mathieu functions, which can also be used to find the periodic solution of Hill's equation. Some asymptotic expansions are given in the section 11.5.4 in order to help understand the behaviour of some Mathieu functions; this analysis is extended in chapter 15.

11.5.1 An overview
The behaviour of Mathieu functions is fairly complicated, particularly as we need to understand both the z and q dependence of all eigenfunctions. Thus, before delving into details we provide a brief overview, which also serves to introduce some notation. More details may be found in Abramowitz and Stegun (1965, chapter 20) and Gradshteyn and Ryzhik (1965).

The variable q is a parameter and we need to obtain the behaviour of both the eigenvalues and eigenfunction as a function of q. In the special case $q = 0$ there are periodic solutions only if $a = n^2$, $n = 0, 1, 2, \ldots$, and these solutions are

$$\begin{array}{cccc} 1, & \cos z, & \cos 2z, & \cdots & \text{(even solutions)}, \\ & \sin z, & \sin 2z, & \cdots & \text{(odd solutions)}. \end{array} \tag{11.81}$$

Observe that these are alternately 2π- and π-periodic functions and that each pair $\{\cos nz, \sin nz\}$ has the same eigenvalue, and each has n zeros in the interval $\pi < z \leq \pi$. The Mathieu functions that reduce to these as $q \to 0$ are labelled by

$$\begin{array}{cccc} ce_0(z, q), & ce_1(z, q), & ce_2(z, q), & \cdots & \text{(even solutions)}, \\ & se_1(z, q), & se_2(z, q), & \cdots & \text{(odd solutions)}. \end{array} \tag{11.82}$$

If $q \neq 0$ each of these eigenfunctions has a distinct eigenvalue; for each eigenvalue there is at most one solution of period π or 2π (apart from multiplicative constants), and each pair $\{ce_n(z, q), se_n(z, q)\}$ has n zeros in the interval $\pi < z \leq \pi$. These solutions may be made unique in a variety of ways: one is to choose the coefficient of $\cos nz$ in $ce_n(z, q)$ and of $\sin nz$ in $se_n(z, q)$ to be unity, as in Whittaker and Watson (1965, chapter 19); the other is to fix the normalisation constant h_n, equation 11.83, as in Abramowitz and Stegun (1965, equation 20.5.3), with the sign of the functions determined by making the sign of $\cos nz$ or $\sin nz$ positive. In all results presented here we use the latter convention with $h_n = \pi$.

The eigenvalue associated with the even solutions, $ce_k(z, q)$ is labelled by $a_k(q)$, $k = 0, 1, 2, \ldots$, and that associated with the odd Mathieu function, $se_k(z, q)$, by $b_k(q)$, $k = 1, 2, \cdots$. The parity and period of these functions is summarised in the following table.

Table 11.6. *Parity and periods of Mathieu functions.*

Function	period	parity about $z = 0$	parity about $z = \pi/2$	eigenvalues	
$ce_{2r}(z, q)$	π		even	$a_{2r}(q)$,	$r = 0, 1, 2, \ldots$
		even			
$ce_{2r+1}(z, q)$	2π		odd	$a_{2r+1}(q)$,	$r = 0, 1, 2, \ldots$
$se_{2r}(z, q)$	π		odd	$b_{2r}(q)$,	$r = 1, 2, 3, \ldots$
		odd			
$se_{2r-1}(z, q)$	2π		even	$b_{2r-1}(q)$,	$r = 1, 2, 3, \ldots$

Mathieu's equation is a Sturm–Liouville system with $p(z) = 1$, $q(z) = -2q \cos 2z$, $w(z) = 1$ and with periodic boundary conditions. The set of eigenfunctions $\{ce_r(z, q),$ $se_r(z, q)\}$ is complete on $-\pi \leq z \leq \pi$. In addition each of the sets $\{ce_r(z, q)\}$ and $\{se_r(z, q)\}$ is complete on $0 \leq z \leq \pi$ and each of the sets $\{ce_{2r}(z, q)\}$, $\{ce_{2r+1}(z, q)\}$, $\{se_{2r}(z, q)\}$ and $\{se_{2r+1}(z, q)\}$ is complete on $0 \leq z \leq \pi/2$. The Mathieu functions are orthogonal:

$$\int_{-\pi}^{\pi} dz \, ce_m(z, q) ce_n(z, q) = h_n \delta_{nm},$$

$$\int_{-\pi}^{\pi} dz\, se_m(z,q)se_n(z,q) = h_n\delta_{nm},$$ (11.83)

$$\int_{-\pi}^{\pi} dz\, ce_m(z,q)se_n(z,q) = 0.$$

Here we use the convention that $h_n = \pi$. The eigenvalues are ordered as follows:

$$a_0(q) < b_1(q) < a_1(q) < b_2(q) < a_2(q) < \cdots \quad (q \neq 0),$$

but the relation between adjacent eigenvalues is more complicated than this simple ordering suggests. It is helpful to divide the eigenvalues into three types; there are those with values much less than $2q$, those with values much larger that $2q$ and those in between.

For the large eigenvalues — for which $a - 2q\cos 2z$ is always positive — we have $a_r(q) \to b_r(q) \to r^2$ as $q \to 0$, and for fixed q and large r the difference between a_r and b_r is exponentially small,

$$a_r(q) - b_r(q) = O\left(\frac{q^r}{r^{r-1}}\right) \quad \text{as} \quad r \to \infty.$$ (11.84)

These large r eigenvalues behave as

$$a_r(q) \simeq b_r(q) \sim r^2 + \frac{q^2}{2r^2} + \cdots.$$

This approximation is derived later, exercise 11.46 (page 425).

Conversely, if $q \gg 1$ and $a \ll q$ — so $a - 2q\cos 2z$ changes sign — $b_{r+1}(q)$ and $a_r(q)$ are close; more precisely, the difference is given by

$$b_{r+1}(q) - a_r(q) \sim \sqrt{\frac{2}{\pi}} \frac{2^{4r+5}}{r!} q^{r/2+3/4} e^{-4\sqrt{q}} \quad \text{as} \quad q \to \infty,$$ (11.85)

and the magnitude by

$$a_r(q) \simeq b_{r+1}(q) \sim -2q + 2R\sqrt{q} - \frac{R^2+1}{8} + \cdots \quad \text{as} \quad q \to \infty,$$

where $R = 2r + 1$. More terms of this series may be found in Abramowitz and Stegun (1965, equation 20.2.30). The second of these equations will be derived in section 11.5.4.

For small $|q|$ the eigenvalues can be expanded as power series in q, which may be found using the perturbation method discussed in section 11.5.3. But a clearer picture of the relation between the a_r and b_r is seen in the following figure, in which the values of the first 11 eigenvalues are shown for $0 < q < 20$: this figure was produced using the method described in section 11.5.2. The ordering of lines in this figure may be understood by noting that $a_0 = -q^2/2 + O(q^4)$, that for $r \geq 1$, $a_r(0) = b_r(0) = r^2$ and that for $q > 0$, $a_r(q) > b_r(q)$ and for $q \gg 1$, $a_r(q) \simeq b_{r+1}(q)$.

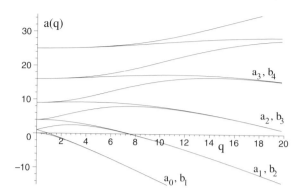

Figure 11.18   Eigenvalues as a function of q.

This figure shows clearly the change in the relation between eigenvalues of the even Mathieu functions, a_r, and the odd functions, b_r, as q increases: this behaviour is also apparent in the relation between the corresponding eigenfunctions, shown in figures 11.19 to 11.22.

If $q < 0$ the variable change $z \to \frac{\pi}{2} - z$ changes the sign of $\cos 2z$ in Mathieu's equation, and it may be shown that the following relations hold:

$$a_{2r}(-q) = a_{2r}(q), \qquad b_{2r}(-q) = b_{2r}(q), \qquad a_{2r+1}(-q) = b_{2r+1}(q)$$

and

$$
\begin{aligned}
ce_{2r}(z, -q) &= (-1)^r ce_{2r}(\pi/2 - z, q), \\
ce_{2r+1}(z, -q) &= (-1)^r se_{2r+1}(\pi/2 - z, q), \\
se_{2r+1}(z, -q) &= (-1)^r ce_{2r+1}(\pi/2 - z, q), \\
se_{2r}(z, -q) &= (-1)^{r-1} se_{2r}(\pi/2 - z, q).
\end{aligned}
$$

Henceforth we assume that $q \geq 0$.

We now consider the behaviour of the eigenfunctions. In the next few graphs are shown some representative Mathieu functions when q has the relatively large value of 20: all these figures were computed using the numerical method and Maple program outlined in section 11.5.2.

For $q = 20$ the lowest eigenvalue is $a_0 = -31.313\,390$ and the next largest is $b_1 = -31.313\,386$, which are consistent with equation 11.85. In figures 11.19 and 11.20 are shown respectively the even, π-periodic eigenfunction $ce_0(z, 20)$ and the odd, 2π-periodic function $se_1(z, 20)$. Observe that both functions have strong local extrema at $z = \pm\frac{\pi}{2}$ and are relatively small elsewhere and also that, with the sign convention chosen, $se_1(z, q) \simeq \pm ce_0(\pm z, q)$, for large $|q|$; however, $ce_r(0, q) \neq 0$ and $ce_r(\pm\pi, q) \neq 0$ for $q \neq 0$.

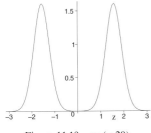

Figure 11.19 $ce_0(z, 20)$,
$a_0 = -31.31 \ll 2q$.

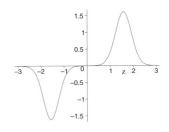

Figure 11.20 $se_1(z, 20)$,
$a_1 = -31.31 \ll 2q$.

The next two eigenvalues are $a_1 = -14.4913$ and $b_2 = -14.4911$. The π-periodic, even function, $ce_1(z, 20)$ is shown in figure 11.21, the related $se_2(z, 20) \simeq \mp ce_1(\pm z, 20)$ is not shown. In figure 11.22 both $ce_2(z, 20)$ and $se_3(z, 20)$, with eigenvalues $a_2 = 1.1543$ and $b_2 = 1.1607$, are shown; note that now the approximate relation $se_2(z, 20) \simeq \pm ce_3(\pm z, 20)$ is not so good near the origin.

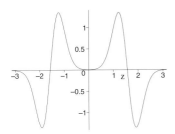

Figure 11.21 $ce_1(z, 20)$, $a_1 = -14.49 \ll 2q$.

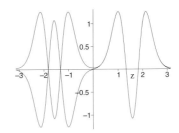

Figure 11.22 $ce_2(z, 20)$, $(a_2 = 1.154)$ and $se_3(z, 20)$, $(b_2 = 1.161)$.

For the next three examples we show the eigenfunctions for which $a \simeq 2q = 40$, that is, $ce_5(z, 20)$, $se_6(z, 20)$ and $ce_6(z, 20)$. In these cases, particularly $ce_5(z, 20)$ and $ce_6(z, 20)$, the magnitude of the eigenfunctions is more strongly peaked at $z = 0$ and $\pm\pi$, in contrast to the behaviour seen in figure 11.19. Such behaviour is seen more clearly if the square of the Mathieu function is plotted.

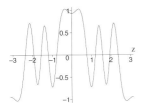

Figure 11.23 $ce_5(z, 20)$,
$a_5 = 36.64$.

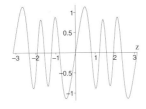

Figure 11.24 $se_6(z, 20)$,
$b_6 = 40.59$.

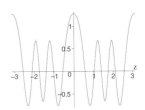

Figure 11.25 $ce_6(z, 20)$,
$a_6 = 44.06$.

The eigenfunctions with eigenvalues $a > 2q$ are not so interesting and simply oscillate fairly uniformly. There is, however, a subtle change in the relation between the odd and even functions, for if $a \gg 2q$, $b_r(q) \simeq a_r(q)$, whereas if $a \ll 2q$, $a_{r+1}(q) \simeq b_r(q)$. Figure 11.26 depicts $se_8(z, 20)$ and figure 11.27 shows $ce_8(z, 20)$.

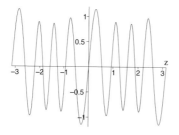

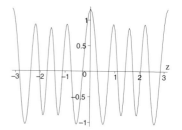

Figure 11.26 $se_8(z, 20)$, $b_8 = 67.25$.   Figure 11.27 $ce_8(z, 20)$, $a_8 = 67.35$.

11.5.2 Computation of Mathieu functions

In this section we outline briefly two methods of computing Mathieu functions and their eigenvalues; both methods may also be used to find solutions of the more general Hill's equation. The first method is simply numerical and can be implemented on any computer provided suitable linear algebra routines are available; in Maple the procedure `eigenvectors(M)`, where M is a square matrix, suffices.

Periodic solutions with periods π and 2π can be approximated using the truncated Fourier series

$$y(z) = \sum_{k=-N}^{N} c_k e^{ikz}, \tag{11.86}$$

in which the $(2N + 1)$ complex coefficients c_k, $k = -N, -N + 1, \ldots, N$ are unknown. Substituting this solution into Mathieu's equation gives

$$\sum_{k=-N}^{N} k^2 c_k e^{ikz} + q \left(\sum_{k=-N+2}^{N} c_{k-2} e^{ikz} + \sum_{k=-N}^{N-2} c_{k+2} e^{ikz} \right) = a \sum_{k=-N}^{N} c_k e^{ikz}, \tag{11.87}$$

and equating the coefficients of e^{ikz} to zero we obtain the eigenvalue equation $M\mathbf{c} = a\mathbf{c}$, where $\mathbf{c} = (c_{-N}, \ldots, c_{N-1}, c_N)$ is the $(2N + 1)$-vector of Fourier coefficients and M is the $(2N + 1) \times (2N + 1)$ banded matrix:

$$M = \begin{pmatrix} N^2 & 0 & q & 0 & 0 & \cdots & & \cdots & & 0 \\ 0 & (N-1)^2 & 0 & q & 0 & \cdots & & \cdots & & 0 \\ q & 0 & (N-2)^2 & 0 & q & \cdots & & \cdots & & 0 \\ \vdots & \vdots & \vdots & \vdots & \vdots & & \vdots & & \vdots & \vdots \\ 0 & \cdots & & & q & 0 & (N-2)^2 & 0 & & q \\ 0 & \cdots & & & 0 & q & 0 & (N-1)^2 & & 0 \\ 0 & \cdots & & & 0 & 0 & q & 0 & & N^2 \end{pmatrix}. \tag{11.88}$$

The eigenvalues of M approximate the first $2N + 1$ eigenvalues $\{a_0, b_1, a_1, \ldots, b_N, a_N\}$ of Mathieu's equation.

If all the first $2N + 1$ eigenvalues are required this method is probably the easiest and was used to produce figure 11.18. But since the time needed to compute the eigenvalues and eigenvectors of M is proportional to the cube of the matrix size, if only particular Mathieu functions are required it is far quicker to take advantage of the known symmetries and halve the size of the matrix, hence decreasing the computational time by a factor of about eight.

The relevant Fourier series and associated eigenvalues are, for $n \geq 0$,

$$c_{2n}(z, q) \;=\; c_0 + \sum_{k=1}^{N-1} c_k \cos 2kz, \qquad \text{even, } \pi \text{ periodic}, \qquad a_{2n}(q),$$

$$s_{2n+1}(z, q) \;=\; \sum_{k=1}^{N} c_k \sin(2k-1)z, \qquad \text{odd, } 2\pi \text{ periodic}, \qquad b_{2n+1}(q),$$

$$c_{2n+1}(z, q) \;=\; \sum_{k=1}^{N} c_k \cos(2k-1)z, \qquad \text{even, } 2\pi \text{ periodic}, \qquad a_{2n+1}(q),$$

$$s_{2n+2}(z, q) \;=\; \sum_{k=1}^{N} c_k \sin 2kz, \qquad \text{odd, } \pi \text{ periodic}, \qquad b_{2n+2}(q).$$

We consider the first of these in detail, leaving the other cases as exercises.

Substituting the approximation for $c_{2n}(z, q)$ into Mathieu's equation and equating the coefficients of $\cos 2kz$ to zero gives the eigenvalue equation $M\mathbf{c} = a\mathbf{c}$, where M is the $N \times N$ real, banded matrix,

$$M = \begin{pmatrix} 0 & q & 0 & 0 & 0 & \cdots & 0 & 0 & 0 \\ 2q & 2^2 & q & 0 & 0 & \cdots & 0 & 0 & 0 \\ 0 & q & 4.2^2 & q & 0 & \cdots & 0 & 0 & 0 \\ \vdots & \vdots & \vdots & \vdots & \vdots & \vdots & \vdots & \vdots & \vdots \\ 0 & 0 & 0 & 0 & 0 & \cdots & 4(N-3)^2 & q & 0 \\ 0 & 0 & 0 & 0 & 0 & \cdots & q & 4(N-2)^2 & q \\ 0 & 0 & 0 & 0 & 0 & \cdots & 0 & q & 4(N-1)^2 \end{pmatrix}. \tag{11.89}$$

The eigenvalues of M are approximations to $\{a_0(q), a_2(q), \ldots, a_{2(N-1)}(q)\}$ and the eigenvectors of M can be used to approximate $c_{2n}(z, q)$. In practice, the size of the matrix M must be kept within reasonable bounds in order to keep the computation time sensible. Further, most procedures that determine eigenvalues and eigenvectors numerically create results in an unpredictable order; the Maple procedure eigenvectors will also produce results in a different order in sequential applications even when the matrix is unchanged. Thus, specific eigenvectors can be found only after some additional work; a method of finding specific Mathieu functions is described below.

It is also necessary to correctly normalise the Mathieu functions. Suppose that the numerical routine returns the coefficients c_k with the normalisation† $\sum_{k=0}^{N-1} |c_k|^2 = C$. But

$$\int_{-\pi}^{\pi} dz \, ce_{2n}(z, q) = 2\pi c_0^2 + \pi \sum_{k=1}^{N-1} c_k^2 = \pi(C + c_0^2),$$

so it is necessary to divide all c_k of the eigenvector for $ce_{2n}(z, q)$ by $\sqrt{C + c_0^2} \, \text{sgn}(c_n)$ to ensure that $h_n = \pi$, and that the coefficient of $\cos 2nz$ is positive. Finally, note that Mathieu functions corresponding to the eigenvectors of M with the largest eigenvalues will not be given accurately by this method; thus the value of N needs to be chosen appropriately.

The procedure `eigenvectors` is accessed using `with(linalg)`. For this problem it is advisable to increase the size of `Digits`. In the following command we have set $q = 0.5$ and $N = 10$ — to obtain a visually manageable output — so M is a 10×10 matrix, initially set to zero. The size of N is determined by the magnitude of q and the accuracy required; also if the required eigenfunction or eigenvector has large order, $n \gg 1$, it is not always necessary to start the truncated Fourier series at $k = 1$. All the figures in the previous section were obtained using a similar method to that outlined here with $N \simeq 20$.

```
>   restart; with(plots):with(linalg):Newdigits:=15:
>   Digits:=Newdigits:
>   q:=1.0:                                    # Define the value of q
>   N:=10:                          # The number of Fourier components is N-1
>   M:=matrix(N,N,(i,j)->0):                        # Initialise the matrix
```

Define diagonal elements of M,

```
>       for j from 1 to N do;
>       if j=1 then M[j,j]:=0 else M[j,j]:=4*(j-1)^2; fi;
>       od:
```

and the upper and lower diagonals, note that M_{21} has a special value.

```
>       for j from 1 to N do;
>       if j <= N-1 then M[j,j+1]:=q; fi;
>       if j >= 2 then M[j,j-1]:=q; fi;
>       od:
>   M[2,1]:=2*q:
```

Compute the eigenvalues and eigenvectors using the procedure `eigenvectors`,

```
>   a:=[eigenvectors(M)]:
>   Digits:=10:
```

The variable a is a list of N lists each with the form $[\lambda_k, m, \{\mathbf{v}_k\}]$, $k = 1, \ldots, N$, with λ_k an eigenvalue, m the multiplicity of the eigenvector (which should be unity for this problem) and $\mathbf{v}_k$ a list of the components of the associated eigenvector. The order of this list of lists is not predictable, and a typical output is shown below where the eigenvalues, the second column, and the largest components of the eigenvectors are printed out. Notice that for this relatively small value of q each eigenvector is

---

† The procedure **eigenvectors** seems to return most, but not all, eigenvectors with $C = 1$.

dominated by one component.

k	λ_k	$\mathbf{v}_k$
1,	64.01,	$[0, 0, 0, 0.03569, 0.9990, -0.02777, 0, 0, 0, 0]$
2,	16.03,	$[0, 0.08357, 0.9953, -0.04990, 0, 0, 0, 0, 0]$
3,	144.0,	$[0, 0, 0, 0, 0, -0.02272, -0.9996, 0.01923, 0, 0]$
4,	256.0,	$[0, 0, 0, 0, 0, 0, 0, -0.01666, -0.9998, 0.01470]$
5,	36.01,	$[0, 0, 0.04995, 0.9981, -0.03569, 0, 0, 0, 0, 0]$
6,	324.0,	$[0, 0, 0, 0, 0, 0, 0, 0, -0.01470, -0.9999]$
7,	-0.4551,	$[0.9099, -0.4141, 0.02521, 0, 0, 0, 0, 0, 0, 0]$
8,	100.0,	$[0, 0, 0, 0, -0.02777, -0.9994, 0.02272, 0, 0, 0]$
9,	4.371,	$[0.2269, 0.9919, -0.08553, 0, 0, 0, 0, 0, 0, 0]$
10,	196.0,	$[0, 0, 0, 0, 0, 0, 0.01923, 0.9997, -0.01666, 0]$

Because the output of `eigenvectors` has no predictable order it is necessary to write a procedure that enables one to pick out particular eigenvalues: the following procedure produces the list $[[k_1, \lambda_{k_1}], [k_2, \lambda_{k_2}], \ldots, [k_N, \lambda_{k_N}]]$, where $\lambda_{k_1} < \lambda_{k_2} < \cdots < \lambda_{k_N}$ are the ordered eigenvalues and where k_m is the position of λ_{k_m} in the original list of the lists, $[k, a[k][1]]$, contained in the variable ev used in the argument of the following procedure.

```
>   st:=proc(ev) local ev1,lis1,lis2,k,j;
>   lis1:=sort([seq(ev[k][2],k=1..nops(ev))]);              # 1
>   lis2:=NULL:
>       for k from 1 to nops(ev) do;                        # 2
>           for j from 1 to nops(ev) do;
>           if lis1[k] = ev[j][2]   then lis2:=lis2,j  else next fi;
>           od;
>       od;
>   [seq([lis2[k],lis1[k]],k=1..nops(ev))];
>   end:
```

(1) An ordered list of eigenvalues.

(2) A loop that re-orders the original list, keeping track of the change in order.

Using this procedure we can re-order eigenvalues, putting them into the variable **v**.

```
>   ev:=[seq([k,a[k][1]],k=1..nops(a))]:          # Original list
>   v:=st(ev):                            # ordered list of eigenvalues
```

Now write a procedure to plot a graph of $ce_{2n}(z, q)$ and to give the truncated Fourier series for a given n. This uses the list **v**, a global variable. Because vectors and matrices have indices starting at unity, the eigenvalue $a_{2k}(q)$ is the $k + 1$th element of the list, thus $ce_0(z, q)$ is obtained by setting the argument of the following procedure to one, rather than zero. The output of this procedure is a list: `gr(k)[1]` is the graph of the Mathieu function $ce_{2(k-1)}(z, q)$ and `gr(k)[2]` is the truncated Fourier series of this function.

```
>   gr:=proc(n) local qs,t1,t2,ne,z,zn,p,pg,k,cn,conds,ticks;
>   global N,q,w,v,a;
>   qs:=convert(q,string);
>   t1:=substring(convert(v[n][2], string),1..6);
>   t2:=cat("q= ",qs,",  Eigenvalue= ",t1);                        # 1
>   ne:=v[n][1]: z:=convert(a[ne][3][1],list);                     # 2
>   zn:=add(z[k]^2,k=1..nops(z));                                  # 3
>   cn:= evalf( sqrt(z[1]^2 + zn)*signum(z[n])):
>   z:=[seq(evalf(z[k]/cn),k=1..nops(z))]:                         # 4
>   p:=z[1] + add(cos(2*(k-1)*x)*z[k],k=2..nops(z));               # 5
>   p:=fnormal(p,5);
>   ticks:=xtickmarks=[seq(k,k=-3..3)],ytickmarks=[seq(-2+0.5*k,k=0..8)];
>   conds:=colour=[black],ticks,title=t2,titlefont=[TIMES,ROMAN,13]:
>   pg:=plot([p],x=-Pi..Pi,conds);
>   [pg,eval(p,x=w)];
>   end:
```

(1) t1 and t2 are titles for the graph.

(2) Here ne is the number of the required eigenvalue in the original list a.

(3) zn is the length of the eigenvector and cn the normalisation constant.

(4) Re-normalise the Fourier coefficients.

(5) Form the truncated Fourier series and, in the next line, remove all small components.

Thus approximations to $ce_0(z,q)$ and $ce_2(z,q)$, with $q = 1$, are obtained by typing gr(1)[2] and gr(2)[2] respectively:

$$ce_0(z,1) \simeq 0.672\,99 - 0.306\,30\cos 2z + 0.018\,646\cos 4z,$$
$$ce_2(z,1) \simeq 0.2169 + 0.9483\cos 2z - 0.081\,77\cos 4z + 0.002\,59\cos 6z.$$

Exercise 11.41

Use the matrix 11.88 and the Maple procedure eigenvalues to compute the difference $d_r(q) = b_{r+1}(q) - a_r(q)$. For $1 < q < 20$ plot the graphs of this function together with the graphs of the approximate difference given in equation 11.84 (page 413), for $r = 1, 2, \ldots, 4$. Note: for $q = 20$ the smallest eigenvalues are computed sufficiently accurately if $N = 15$.

Exercise 11.42

Write a Maple procedure, as outlined above, to compute the Mathieu functions $ce_{2n+1}(z,q)$ and $se_{2n}(z,q)$ and plot some representative graphs.

Exercise 11.43

Consider the equation

$$\frac{d^2y}{dz^2} + (a + 2q\cos z)y = 0, \quad -\pi < z \le \pi.$$

Using the truncated Fourier series

$$y(z) = \sum_{k=0}^{N-1} c_k \cos kz,$$

show that approximate even, 2π-periodic eigenfunctions and their eigenvalues may be obtained from the eigenvectors and eigenvalues of the matrix equation $M\mathbf{c} = a\mathbf{c}$, where

$$
M = \begin{pmatrix}
0 & -q & 0 & 0 & \cdots & 0 & 0 & 0 \\
-2q & 1 & -q & 0 & \cdots & 0 & 0 & 0 \\
0 & -q & 2^2 & -q & \cdots & 0 & 0 & 0 \\
\vdots & \vdots & \vdots & \vdots & \vdots & \vdots & \vdots & \vdots \\
0 & 0 & 0 & 0 & \cdots & (N-3)^2 & -q & 0 \\
0 & 0 & 0 & 0 & \cdots & -q & (N-2)^2 & -q \\
0 & 0 & 0 & 0 & \cdots & 0 & -q & (N-1)^2
\end{pmatrix}.
$$

Use the eigenvectors of M to construct approximations to the eigenfunctions of the original equation that are normalised according to $\int_{-\pi}^{\pi} dz\, y_k(z) y_j(z) = \pi \delta_{ij}$.

In the case $q = 60$ show that there are about ten eigenvalues less than $2q$, that the lowest eigenvalue is about -112.7, and plot graphs of all the eigenvectors with eigenvalues less than 140.

11.5.3 Perturbation theory

If $|q|$ is small Mathieu's equation may be solved using a form of perturbation theory. This is readily implemented in Maple, which allows the computation of many terms in the perturbation expansion of both the eigenvalues and Mathieu functions. The method can also be applied to Hill's equation.

Write Mathieu's equation in the form

$$
\frac{d^2 y}{dz^2} + a(q)y = 2qy \cos 2z \tag{11.90}
$$

and assume that both $a(q)$ and $y(z, q)$ may be expressed as power series in q,

$$
\begin{aligned}
a(q) &= a_0 + a_1 q + a_2 q^2 + \cdots \\
y(z, q) &= y_0(z) + y_1(z)q + y_2(z)q^2 + \cdots,
\end{aligned}
$$

where all the $y_k(z)$ are 2π-periodic and are either all even, for the $ce_r(z, q)$ functions, or all odd, for $se_r(z, q)$. The approximation to ce_n, $n \geq 0$, is obtained by setting $a_0 = n^2$ and $y_0(z) = \cos nz$; and for se_n, $n \geq 1$, we set $a_0 = n^2$ and $y_0(z) = \sin nz$.

Substituting the series into the differential equation and equating the coefficients of q^k, $k = 0, 1, 2, \ldots$, to zero gives the following set of linear differential equations:

$$
\begin{aligned}
\frac{d^2 y_0}{dz^2} + n^2 y_0 &= 0, \\
\frac{d^2 y_1}{dz^2} + n^2 y_1 &= 2y_0(z) \cos 2z - a_1 y_0(z), \tag{11.91} \\
&\quad\vdots \quad \vdots \quad \vdots \\
\frac{d^2 y_k}{dz^2} + n^2 y_k &= 2y_{k-1}(z) \cos 2z - \sum_{j=1}^{k} a_j y_{k-j}(z).
\end{aligned}
$$

The equation for $y_k(z)$ depends only upon $\{y_0, y_1, \ldots, y_{k-1}\}$ and $\{a_0, a_1, \ldots, a_k\}$ and may be solved for *both* a_k and $y_k(z)$ by imposing the periodic boundary conditions and the parity requirement. The method is similar to Lindstedt's method, described in chapter 17, which is used to find periodic solutions to particular types of nonlinear differential equations. In order to see how this scheme works we shall first use it to find an approximation to the smallest eigenvalue $a_0(q)$ and its eigenfunction $ce_0(z, q)$, which is a special case, and then for a more typical example, $se_3(z, q)$.

If $n = 0$ the only non-trivial, periodic solution of the equation for y_0, apart from an arbitrary multiplicative constant, is $y_0(z) = 1$, and the equation for $y_1(z)$ becomes

$$\frac{d^2 y_1}{dz^2} = 2\cos 2z - a_1.$$

This has a periodic solution only if $a_1 = 0$, and then $y_1(z) = -\frac{1}{2}\cos 2z$. The equation for $y_2(z)$ is then

$$\begin{aligned}
\frac{d^2 y_2}{dz^2} &= 2y_1 \cos 2z - a_1 y_1 - a_2 y_0 \\
&= -\frac{1}{2} - a_2 - \frac{1}{2}\cos 4z.
\end{aligned}$$

For $y_2(z)$ to be periodic we need $a_2 = -\frac{1}{2}$, and then $y_2(z) = \frac{1}{32}\cos 4z$. The third equation is

$$\begin{aligned}
\frac{d^2 y_3}{dz^2} &= 2y_2(z)\cos 2z - a_1 y_2(z) - a_2 y_1(z) - a_3 y_0(z) \\
&= \frac{1}{32}\cos 6z - \frac{15}{32}\cos 2z - a_3,
\end{aligned}$$

and hence

$$a_3 = 0, \quad \text{and} \quad y_3(z) = \frac{7}{128}\cos 2z - \frac{1}{1152}\cos 6z.$$

Clearly this method may be carried to any order and the algorithm readily implemented using Maple, which gives

$$a_0(q) = -\frac{q^2}{2} + \frac{7q^4}{128} - \frac{29q^6}{2304} + \frac{68\,687q^8}{18\,874\,368} - \frac{123\,707q^{10}}{104\,857\,600} + \cdots \tag{11.92}$$

and

$$ce_0(z, q) = 1 - \frac{q}{2}\cos 2z + \frac{q^2}{32}\cos 4z + \frac{q^3}{128}\left(7\cos 2z - \frac{1}{9}\cos 6z\right) + \cdots. \tag{11.93}$$

The norm of this approximation is

$$\frac{1}{\pi} \int_{-\pi}^{\pi} dq\, ce_0(z, q)^2 = 2\left[1 + \frac{q^2}{8} + O(q^4)\right]^2.$$

Thus if we require a Mathieu function normalised with $h_n = \pi$, equation 11.83, it is necessary to divide the above approximation by the square root of the right hand side of this equation, to give

$$ce_0(z,q) = \frac{1}{\sqrt{2}}\left[1 - \frac{q}{2}\cos 2z - \frac{q^2}{32}(2 - \cos 4z)\right.$$

$$\left. + q^3\left(\frac{11}{128}\cos 2z - \frac{1}{1152}\cos 6z\right) + O(q^4)\right]. \quad (11.94)$$

Exercise 11.44
Write a Maple procedure to compute the series expansion of both $a_0(q)$ and $ce_0(z,q)$ to any given order. Use your procedure to confirm equations 11.92 and 11.93 and find $ce_0(z,q)$ to $O(q^5)$, normalised as in 11.83. In addition compute the series for $a_0(q)$ to a sufficiently large order to show that the radius of convergence of the series is $q_R \simeq 1.5$ and confirm this estimate by finding the poles in the Padé approximant of the series.

Use the numerical procedure developed in the previous section to find the lowest eigenvalues and eigenvectors and compare, graphically, these results with those of the perturbation expansion for $0 < q < 2$.

The same type of analysis can be used to find approximations to all other Mathieu functions and eigenvalues, but the algebra is slightly different when $n \neq 0$ because the equations 11.91 for $y_k(z)$ are different. For instance, consider the expansions of $se_3(z,q)$ and $b_3(q)$, for which $a_0 = 3$ and $y_0(z) = \sin 3z$. The equation for $y_1(z)$ is

$$\frac{d^2 y_1}{dz^2} + 9y_1 = 2\sin 3z \cos 2z - a_1 \sin 3z$$

$$= \sin z + \sin 5z - a_1 \sin 3z.$$

A periodic solution exists only if $a_1 = 0$, for if $a_1 \neq 0$ the solution would contain the non-periodic term proportional to $a_1 z \cos 3z$. Thus we must set $a_1 = 0$ to obtain

$$y_1(z) = \frac{1}{8}\sin z - \frac{1}{16}\sin 5z + c_1 \cos 3z + c_2 \sin 3z,$$

where c_1 and c_2 are constants. But $se_3(z,q)$ is an odd function and $\cos 3z$ is even, so $c_1 = 0$. Further, it is unnecessary to include the $\sin 3z$ term because this harmonic is already included in $y_0(z)$; to include it is equivalent to multiplying the solution by a constant. Thus we can also choose $c_2 = 0$.

The subsequent equation for $y_2(z)$ is

$$\frac{d^2 y_2}{dz^2} + 9y_2 = 2y_1(z)\cos 2z - a_2 \sin 3z$$

$$= -\frac{1}{8}\sin z + \left(\frac{1}{16} - a_2\right)\sin 3z - \frac{1}{16}\sin 7z,$$

which has an odd periodic solution only if $a_2 = 1/16$, and then as before the appropriate solution is

$$y_2(z) = -\frac{1}{64}\sin z + \frac{1}{640}\sin 7z.$$

Continuing in this manner, we obtain

$$b_3(q) = 9 + \frac{q^2}{16} - \frac{q^3}{64} + \frac{13q^4}{20\,480} + \frac{5q^5}{16\,384} - \frac{1\,961q^6}{23\,592\,960} + \frac{609q^7}{104\,857\,600} + \cdots \quad (11.95)$$

and

$$se_3(z,q) \simeq \sin 3z + q\left(\frac{1}{8}\sin z - \frac{1}{16}\sin 5z\right) - q^2\left(\frac{1}{64}\sin z - \frac{1}{640}\sin 7z\right) + \cdots. \quad (11.96)$$

The conventionally normalised function is,

$$se_3(z,q) \simeq \sin 3z + q\left(\frac{1}{8}\sin z - \frac{1}{16}\sin 5z\right)$$
$$-q^2\left(\frac{1}{64}\sin z + \frac{5}{512}\sin 3z - \frac{1}{640}\sin 7z\right) + \cdots.$$

This method may be used for any specific value of n and to any order; but for general n it produces a series, exercise 11.46, in q^2 in which the coefficient of q^{2r} contains the factor $(n-r)$ in the denominator; therefore when using this series for a specific value of n it must be truncated at the $q^{2(n-1)}$ term, and the last few terms of this series are incorrect, as seen by comparing the results derived in exercises 11.45 and 11.46.

Exercise 11.45

Write Maple procedures to compute $a_r(q)$ and $b_r(q)$. Use your procedure to check equations 11.95 and 11.96 and to show that

$$a_0(q) = -\frac{q^2}{2} + \frac{7q^4}{128} - \frac{29q^6}{2304} + \frac{68\,687q^8}{18\,874\,368} - \frac{123\,707q^{10}}{104\,857\,600} + \cdots,$$

$$\left.\begin{array}{c} b_1(q) \\ a_1(-q) \end{array}\right\} = 1 - q - \frac{q^2}{8} + \frac{q^3}{64} - \frac{q^4}{1\,536} - \frac{11q^5}{36\,864} + \frac{49q^6}{589\,824}$$
$$- \frac{55q^7}{9\,437\,184} - \frac{83q^8}{35\,389\,440} + \frac{12\,121q^9}{15\,099\,494\,400} + \cdots,$$

$$b_2(q) = 4 - \frac{q^2}{12} + \frac{5q^4}{13\,824} - \frac{289q^6}{79\,626\,240} + \frac{21\,391q^8}{458\,647\,142\,400} - \cdots,$$

$$a_2(q) = 4 + \frac{5q^2}{12} - \frac{763q^4}{13\,824} + \frac{1\,002\,401q^6}{79\,626\,240} - \frac{1\,669\,068\,401q^8}{458\,647\,142\,400} + \cdots,$$

$$\left.\begin{array}{c} b_3(q) \\ a_3(-q) \end{array}\right\} = 9 + \frac{q^2}{16} - \frac{q^3}{64} + \frac{13q^4}{20\,480} + \frac{5q^5}{16\,384} - \frac{1\,961q^6}{23\,592\,960}$$
$$+ \frac{609q^7}{104\,857\,600} + \frac{4\,957\,199q^8}{2\,113\,929\,216\,000} + \cdots,$$

$$b_4(q) = 16 + \frac{q^2}{30} - \frac{317q^4}{864\,000} + \frac{10\,049q^6}{2\,721\,600\,000} - \frac{93\,824\,197q^8}{2\,006\,581\,248\,000\,000} + \cdots,$$

$$a_4(q) = 16 + \frac{q^2}{30} + \frac{433q^4}{864\,000} - \frac{5\,701q^6}{2\,721\,600\,000} - \frac{112\,236\,997q^8}{2\,006\,581\,248\,000\,000} + \cdots.$$

Also show that the leading terms of the difference $a_r(q) - b_r(q)$ for $r = 2, 3, \ldots, 8$ are, respectively,

$$\frac{q^2}{2}, \quad \frac{q^3}{2^5}, \quad \frac{q^4}{2^7 3^2}, \quad \frac{q^5}{2^{13} 3^2}, \quad \frac{q^6}{2^{15} 3^2 5^2}, \quad \frac{q^7}{2^{19} 3^4 5^2}, \quad \frac{q^8}{2^{21} 3^4 5^2 7^2}.$$

Exercise 11.46
Use perturbation theory to show that for $n > 5$,

$$\left.\begin{array}{c} a_n(q) \\ b_n(q) \end{array}\right\} = n^2 + \frac{q^2}{2(n^2-1)} + \frac{(5n^2+7)q^4}{32(n^2-4)(n^2-1)^3} + \frac{(9n^4+58n^2+29)q^6}{64(n^2-9)(n^2-4)(n^2-1)^5}$$

$$+ \frac{(1469n^{10}+9144n^8-140\,354n^6+64\,228n^4+827\,565n^2+274\,748)q^8}{8192(n^2-16)(n^2-9)(n^2-4)^3(n^2-1)^7}.$$

For $n = 2$, 3 and 4 compare the first few terms of this series, up to the divergent term, with
the relevant series found in the previous exercise.

Exercise 11.47
Show that the three eigenvalues of the equation

$$\frac{d^2y}{dx^2} + (a - \alpha q\cos x - \beta q\cos 2x)y = 0, \quad -\pi \le x \le \pi,$$

where $y(x)$ is 2π-periodic, belonging to the solutions that reduce to $\{1, \cos x, \sin x\}$ when
$q = 0$, have the series expansions

$$a_0 = -\frac{1}{2}\left(\alpha^2 + \frac{1}{4}\beta^2\right)q^2 + \frac{3}{8}\alpha^2\beta q^3 + \left(\frac{7}{32}\alpha^4 - \frac{5}{36}\alpha^2\beta^2 + \frac{7}{2048}\beta^4\right)q^4 + \cdots,$$

$$a_1 = 1 - \frac{\beta}{2}q - \left(\frac{\alpha^2}{12} + \frac{\beta^2}{32}\right)q^2 + \left(\frac{7\alpha^2\beta}{288} + \frac{\beta^3}{512}\right)q^3 + \cdots,$$

$$a_2 = 1 + \frac{\beta}{2}q - \left(\frac{5\alpha^2}{12} - \frac{\beta^2}{32}\right)q^2 - \left(\frac{121\alpha^2\beta}{288} - \frac{\beta^3}{512}\right)q^3 + \cdots,$$

and find the associated eigenfunction to this order.
    What difference is made if the $\cos x$ term is replaced by $\sin x$?

Exercise 11.48
Show that the four smallest eigenvalues of the equation

$$\frac{d^2y}{dx^2} + (a - 2q\cos 4x)y = 0,$$

where $y(x)$ is 2π-periodic, have the series expansions

$$\lambda_0(q) = -\frac{1}{8}q^2 + \frac{7}{8192}q^4 - \frac{29}{2\,359\,296}q^6,$$

$$\lambda_1(q) = \lambda_2(q) = 1 - \frac{1}{6}q^2 + \frac{11}{4320}q^4 - \frac{47}{544\,320}q^6,$$

$$\lambda_3(q) = \lambda_4(-q) = 4 - q - \frac{1}{32}q^2 + \frac{1}{1024}q^3 - \frac{1}{98\,304}q^4 - \frac{11}{9\,437\,184}q^6.$$

Also find the associated eigenfunctions to the same order.

Exercise 11.49
Show that with $y_0 = e^{iz}$ the perturbation expansion 11.91 cannot yield a periodic solution.

11.5.4 Asymptotic expansions
The figures of section 11.5.1 show that Mathieu functions behave in a variety of different
ways depending on the relative values of a and q. It is not immediately obvious why

this should be the case, and most schemes used to compute these functions and their eigenvalues do not help explain their behaviour. Asymptotic methods, however, are helpful, so here we provide some simple approximations valid when $q \gg 1$ and when either $a \gg 2q$ or $a \ll 2q$; more accurate approximations, particularly for $a \sim 2q$, will be considered in chapter 15.

The simplest case is $q \gg 1$ and $a > 2q$ because then $a - 2q \cos 2z > 0$ for all z. Define $v = a/2q > 1$ and assume a solution of the form $y \sim A(z) \exp(i\sqrt{2q}S(z))$, so equation 11.79 for the Mathieu function becomes

$$\frac{d^2A}{dz^2} + i\frac{\sqrt{2q}}{A}\frac{d}{dz}\left(A^2\frac{dS}{dz}\right) + 2qA\left\{-\left(\frac{dS}{dz}\right)^2 + (v - \cos 2z)\right\} = 0. \qquad (11.97)$$

Equating the coefficients of the two highest powers of $\sqrt{2q}$ to zero gives the approximate solutions

$$y_{\pm}(z) = \frac{1}{(a - 2q\cos 2z)^{1/4}} \exp\left(\pm i\int_0^z du\,\sqrt{a - 2q\cos 2u}\right). \qquad (11.98)$$

Approximate eigenvalues are obtained by imposing the periodic boundary conditions on these solutions to give the implicit equation

$$\int_0^\pi du\,\sqrt{a - 2q\cos 2u} = n\pi, \quad n \text{ a positive integer.} \qquad (11.99)$$

We shall see that even values of n give the π-periodic solutions and odd values of n the 2π-periodic solutions. A lower bound on n below which the approximation is invalid is given by putting $a = 2q$,

$$n > \frac{\sqrt{2q}}{\pi}\int_0^\pi du\,\sqrt{1 - \cos 2u} = \frac{4\sqrt{q}}{\pi}, \quad \text{or approximately} \quad n^2 > q.$$

For $a \gg 2q$, that is, large n, we obtain the simple approximation $a = n^2$, and hence $a_n = b_n = n^2$, for the eigenvalues. For smaller values of n a form of perturbation theory may be used: putting $\cos 2u = 2\cos^2 u - 1$ in 11.99 gives

$$\sqrt{a + 2q}\,E(k) = \frac{n\pi}{2}, \quad k^2 = \frac{4q}{a + 2q} < 1 \quad \text{for} \quad a > 2|q|, \qquad (11.100)$$

where $E(k)$ is the complete elliptic integral† of the second kind.

Equation 11.100 for the eigenvalues may be solved numerically for a, given q and n, or as the perturbation series

$$\left.\begin{array}{c} a_n(q) \\ b_n(q) \end{array}\right\} \simeq n^2\left(1 + \frac{\delta}{2} + \frac{5\delta^2}{32} + \frac{9\delta^3}{64} + \frac{1469\delta^4}{8192} + \cdots\right), \quad \delta = \frac{q^2}{n^4}. \qquad (11.101)$$

In this approximation the two eigenvalues $a_r(q)$ and $b_r(q)$ are identical: the coefficients of δ^k may be obtained from the series obtained in exercise 11.46 by taking the first term in the $1/n$ asymptotic expansion of each coefficient. In fact $a_r(q)$ and $b_r(q)$ are not identical, though the difference $a_r - b_r$ is exponentially small as $r \to \infty$, equation 11.84; approximations to this difference will be derived in equation 15.94.

---

† The elliptic integrals are discussed in Appendix II. In Maple the two functions are denoted by **EllipticE(k)** and **EllipticK(k)** respectively.

The odd and even eigenfunctions, $se_n(z,q)$ and $ce_n(z,q)$, are normalised linear combinations of the basic solutions, given by equation 11.98,

$$ce_n(z,q) \quad \sim \quad \frac{\alpha_n}{(a_n - 2q\cos 2z)^{1/4}} \cos\left(\int_0^z du \sqrt{a_n - 2q\cos 2u}\right),$$

$$se_n(z,q) \quad \sim \quad \frac{\alpha_n}{(a_n - 2q\cos 2z)^{1/4}} \sin\left(\int_0^z du \sqrt{a_n - 2q\cos 2u}\right),$$

(11.102)

where α_n is the normalisation constant, chosen to satisfy equation 11.83,

$$\alpha_n^2 = \frac{\pi\sqrt{a_n + 2q}}{2K(k_n)}, \quad k_n^2 = \frac{4q}{a_n + 2q},$$

where $K(k)$ is the complete elliptic integral of the first kind.

In the next two figures we compare the eigenfunctions of the two lowest eigenvalues for which the approximations 11.102 (together with 11.101) are valid when $q = 5$, that is $ce_3(z,5)$ ($a_3 = 11.6$) in figure 11.28 and $se_4(z,5)$ ($b_4 = 16.5$) in figure 11.29; the approximation 11.100 gives $a = 10.59$ and $b = 16.81$ respectively. In these figures the thick line denotes the approximation 11.102. It is seen that the approximation is qualitatively reasonable for the first of these and fairly accurate for the second, in which case the differences are barely noticeable on the scale of the graph. For the next highest pair of eigenvalues the differences are smaller than those seen in figure 11.29.

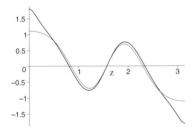

Figure 11.28 $ce_3(z,5)$.

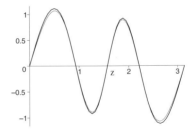

Figure 11.29 $se_4(z,5)$.

In the next two figures are compared the eigenfunctions of the two lowest eigenvalues for $q = 10$, that is $se_5(z,10)$ ($b_5 = 26.8$) in figure 11.30 and $ce_5(z,10)$ ($a_5 = 27.7$) in figure 11.31.

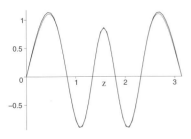

Figure 11.30 $se_5(z,10)$.

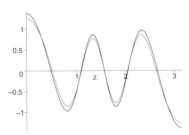

Figure 11.31 $ce_5(z,10)$.

Exercise 11.50
Derive equations 11.98 and 11.102.

Exercise 11.51
By writing equation 11.100 in the form

$$\frac{\pi k}{2E(k)} = \epsilon = \frac{2\sqrt{q}}{n},$$

and expanding the left hand side as a power series in k, derive the series 11.101.

Now consider the opposite extreme $a \ll 2q$, and find approximations for the lowest few eigenvalues and eigenfunctions when $q \gg 1$. The graphs in figure 11.19 show that these eigenfunctions are concentrated around $z = \pm\pi/2$ and are practically zero at $z = 0$ and $\pm\pi$, and since the eigenfunctions are either odd or even we concentrate on the interval $0 < z < \pi$. The form of the graph in figure 11.19 (page 415) suggests expanding about $z = \pi/2$: thus we define $x = z - \pi/2$, so $\cos 2z = -1 + 2\sin^2 x$ and Mathieu's equation becomes

$$\frac{d^2 y}{dx^2} + (a + 2q - 4q \sin^2 x)y = 0.$$

For $z \simeq \pi/2$, $|x| \ll 1$ so we may expand $\sin^2 x$ and re-scale the variables to obtain the equation

$$\frac{d^2 y}{d\xi^2} + (v^2 - \xi^2)y = 0, \quad \xi = (4q)^{1/4}x, \quad v^2 = \frac{a + 2q}{2\sqrt{q}}, \tag{11.103}$$

which is just Weber's equation 11.30 (page 391).

If $q \gg 1$, $z = 0$ corresponds to $\xi \ll 1$ and $z = \pi$ to $\xi \gg 1$, so we require a solution that tends to zero as $\xi \to \pm\infty$. These are just $y = e^{-\xi^2}H_n(\xi)$, where $v^2 = 2n + 1$. Thus for $z > 0$ approximations to the lowest order Mathieu functions are

$$\left.\begin{array}{c} ce_n(z,q) \\ se_{n+1}(z,q) \end{array}\right\} \simeq Ae^{-\xi^2/2}H_n(\xi), \quad \xi = (4q)^{1/4}\left(z - \frac{\pi}{2}\right), \quad 0 < z < \pi, \tag{11.104}$$

where the external constant

$$A^2 = \frac{\sqrt{2\pi}q^{1/4}}{2^{n+1}n!},$$

is chosen to ensure the correct normalisation, equation 11.83 with $h_n = \pi$, though the overall sign of this approximation may differ from the exact function. The corresponding eigenvalues are

$$a_n \simeq b_{n+1} \simeq -2q + 2(2n + 1)\sqrt{q}, \quad n = 0, 1, \dots.$$

In the following figures are compared the approximation 11.104 (the thicker lines) with $ce_n(z,q)$ for $q = 20$ and $n = 0, 1$ and 2.

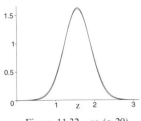

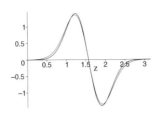

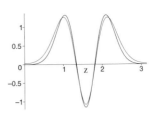

Figure 11.32 $ce_0(z, 20)$.   Figure 11.33 $ce_1(z, 20)$.   Figure 11.34 $ce_2(z, 20)$.

11.6 Appendix

Solutions of the linear nth-order differential equations having the special form

$$(a_n z + b_n)\frac{d^n y}{dz^n} + (a_{n-1}z + b_{n-1})\frac{d^{n-1}y}{dz^{n-1}} + \cdots + (a_1 z + b_1)\frac{dy}{dz} + (a_0 z + b_0)y = 0,$$

where (a_k, b_k), $k = 0, 1, \ldots, n$, are constants, may be expressed in terms of an integral by using an extension of the Laplace transform method. Assume that the solution may be written as the complex integral

$$w(z) = \int_{\mathscr{C}} d\zeta\, v(\zeta)e^{z\zeta},$$

where both the contour $\mathscr{C}$ and the analytic function $v(\zeta)$ are unknown. Substituting this function into the equation gives

$$\int_{\mathscr{C}} d\zeta\, v(\zeta)e^{z\zeta} \sum_{k=0}^{n}(a_k z + b_k)\zeta^k = 0. \tag{11.105}$$

But

$$\int_{\mathscr{C}} d\zeta\, v(\zeta)\zeta^k z e^{z\zeta} = \left[v(\zeta)\zeta^k e^{z\zeta}\right]_{\mathscr{C}} - \int_{\mathscr{C}} d\zeta\, e^{z\zeta}\frac{d}{d\zeta}\left(v(\zeta)\zeta^k\right),$$

and consequently equation 11.105 may be re-written in the form

$$\left[v(\zeta)e^{z\zeta}P(\zeta)\right]_{\mathscr{C}} - \int_{\mathscr{C}} d\zeta\, e^{z\zeta}\left\{P(\zeta)\frac{dv}{d\zeta} - v(\zeta)Q(\zeta)\right\} = 0, \tag{11.106}$$

where

$$P(\zeta) = \sum_{k=0}^{n} a_k \zeta^k \quad \text{and} \quad Q(\zeta) = \sum_{k=0}^{n}\left(b_k \zeta^k - ka_k\zeta^{k-1}\right).$$

If the contour $\mathscr{C}$ can be chosen so that the boundary term vanishes and the function $v(\zeta)$ such that the term in curly braces vanishes, then $w(z)$ will be a solution of the original equation. The equation for $v(\zeta)$ is the first-order differential

$$\frac{1}{v}\frac{dv}{d\zeta} = \frac{Q(\zeta)}{P(\zeta)},$$

and since both $P(\zeta)$ and $Q(\zeta)$ are polynomials in ζ its solution may, in principle, be expressed in terms of known functions; further analysis of this type of equation may be found in Ince (1956, chapter 18).

Hermite polynomials

For Hermite polynomials equation 11.20 gives $a_2 = a_0 = 0$, $a_1 = -2$, $b_0 = 2n$, $b_1 = 0$ and $b_2 = 1$, giving $P = -2\zeta$ and $Q = \zeta^2 + 2(n+1)$. Thus $v(\zeta) = e^{-\zeta^2/4}\zeta^{-n-1}$. The boundary term is now $\left[e^{-\zeta^2/4}\zeta^{-n}\right]_{\mathscr{C}}$, and because n is an integer this may be made zero by choosing $\mathscr{C}$ to be a closed curve surrounding the origin in the ζ-plane. Now change variables, $\zeta = 2(x - z)$, to write the solution, apart from an arbitrary multiplicative constant, in the form

$$w_n(x) = e^{x^2} \int_{\mathscr{C}} dz \, \frac{e^{-z^2}}{(z - x)^{n+1}}, \tag{11.107}$$

where $\mathscr{C}$ is any contour enclosing the pole at $z = x$. The remaining part of the analysis can be found in the main text and leads to equation 11.24 (page 390).

Laguerre polynomials

For Laguerre polynomials equation 11.33 gives $a_0 = 0$, $a_1 = -1$, $a_2 = 1$, $b_0 = n$, $b_1 = 1$ and $b_2 = 0$ so that $P(\zeta) = \zeta(\zeta - 1)$ and $Q(\zeta) = n + 1 - \zeta$. Thus $v(\zeta) = (1 - 1/\zeta)^n/\zeta$ and

$$w(z) = \oint_{\mathscr{C}} d\zeta \, \left(\frac{\zeta - 1}{\zeta}\right)^n \frac{e^{z\zeta}}{\zeta},$$

and the boundary term is made zero by choosing $\mathscr{C}$ to be a closed curve surrounding the origin in the ζ-plane.

Airy functions

For Airy functions equation 11.71 shows that $a_0 = -1$, $a_1 = a_2 = b_0 = b_1 = 0$ and $b_2 = 1$, so that $P(\zeta) = -1$ and $Q(\zeta) = \zeta^2$, giving $v(\zeta) = \exp(-\zeta^3/3)$ and

$$w(z) = A \int_{\mathscr{C}} d\zeta \, \exp\left(z\zeta - \frac{1}{3}\zeta^3\right),$$

where A is a constant. The boundary term is $\left[\exp\left(z\zeta - \frac{1}{3}\zeta^3\right)\right]_{\mathscr{C}}$, and this is zero if $\mathscr{C}$ goes to infinity at both ends in regions where $\Re(\zeta^3) > 0$, which are the shaded regions shown in the diagram.

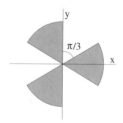

A contour on the edge of this region is $\zeta = iy$, where y is real, and this gives

$$w(z) = iA \int_{-\infty}^{\infty} dy \, \exp i \left(zy + \frac{1}{3}y^3 \right) = 2iA \int_{0}^{\infty} dy \, \cos \left(zy + \frac{1}{3}y^3 \right),$$

for the solution with real z and which tends to zero as $z \to \infty$.

11.7 Exercises

Exercise 11.52
Show that if $\{p_0, p_1, \ldots, p_n, \ldots\}$ are a set of orthogonal polynomials on $a \le x \le b$ with $p_n(x)$ of degree n and with the orthogonality relation 11.1, then they satisfy the three-term recurrence relation

$$p_{n+1}(x) = (a_n + xb_n)p_n(x) - c_n p_{n-1}(x),$$

where the parameters a_n, b_n and c_n do not depend upon x.

Exercise 11.53
The system of polynomials

$$p_n(x) = k_n x^n + l_n x^{n-1} + \cdots,$$

orthogonal with respect to the weight function $w(x)$, satisfy the recurrence relation derived in the previous exercise. Use this relation to show that

$$b_n = \frac{k_{n+1}}{k_n}, \quad a_n = b_n \left(\frac{l_{n+1}}{k_{n+1}} - \frac{l_n}{k_n} \right),$$

and use the orthogonality relation 11.1 (page 377) to show that $c_n = k_{n+1}k_{n-1}h_n/(k_n^2 h_{n-1})$.

Exercise 11.54
Use the results of exercises 11.52 and 11.53 to verify the Christoffel–Darboux formula, equation 11.6 (page 380).

Exercise 11.55
Suppose that the orthogonal polynomials $f_n(x)$, $n = 0, 1, 2, \ldots$, satisfy the differential equation

$$g_2(x)\frac{d^2 y}{dx^2} + g_1(x)\frac{dy}{dx} + \alpha_n y = 0, \quad a \le x \le b,$$

where α_n is independent of x. Show that the weight function is

$$w(x) = \frac{p(x)}{q_2(x)}, \quad p(x) = \exp \left(\int_{a}^{x} du \, \frac{g_1(u)}{g_2(u)} \right) \quad \text{and that} \quad \int_{a}^{b} dx \, w f_n f_m = h_n \delta_{nm}$$

for some constants h_n.

Show also that the polynomials df_n/dx, $n = 1, 2, \ldots$, are orthogonal with respect to the weight function $p(x)$, that is,

$$\int_{a}^{b} dx \, p(x) \frac{df_n}{dx} \frac{df_m}{dx} = h'_n \delta_{nm}.$$

Exercise 11.56

Using Rodrigues' formula show that the Fourier coefficient, a_n, defined in equation 11.12 (page 383), may be written in the form

$$a_n = \frac{n + \frac{1}{2}}{2^n n!} \int_{-1}^{1} dx \, (1 - x^2)^n \frac{d^n f}{dx^n},$$

provided the nth differential of $f(x)$ exists.

Exercise 11.57

Recurrence relation for Legendre polynomials

Using the generating function $g(x, t) = (1 - 2xt + t^2)^{-1/2}$ for the Legendre polynomials, show that

$$\frac{\partial g}{\partial t} = \frac{x - t}{(1 - 2xt + t^2)^{3/2}} = \sum_{n=0}^{\infty} n P_n(x) t^{n-1},$$

Hence show that

$$(1 - 2xt + t^2) \sum_{n=0}^{\infty} n P_n(x) t^{n-1} = (x - t) \sum_{n=0}^{\infty} P_n(x) t^n,$$

and by finding the coefficient of t^n derive the recurrence relation

$$(2n + 1) x P_n(x) = (n + 1) P_{n+1}(x) + n P_{n-1}(x), \quad n = 1, 2, \ldots.$$

Exercise 11.58

Use the factorisation $1 - 2t \cos \theta + t^2 = (1 - te^{i\theta})(1 - te^{-i\theta})$ and the generating function for Legendre polynomials, equation 11.13 (page 383), to show that

$$P_{2m}(\cos \theta) = \frac{1}{\pi} \left(\frac{\Gamma(m + \frac{1}{2})}{m!} \right)^2 + \frac{2}{\pi} \sum_{p=1}^{m} \frac{\Gamma(m + p + \frac{1}{2}) \Gamma(m - p + \frac{1}{2})}{(m + p)! (m - p)!} \cos 2p\theta,$$

$$P_{2m+1}(\cos \theta) = \frac{2}{\pi} \sum_{p=0}^{m} \frac{\Gamma(m + p + \frac{3}{2}) \Gamma(m - p + \frac{1}{2})}{(m + p + 1)! (m - p)!} \cos(2p + 1)\theta.$$

Exercise 11.59

Use the recurrence relation found in exercise 11.3 (page 384) to show that if

$$f(x) = \sum_{k=0}^{\infty} a_n P_n(x),$$

then

$$\int_x^1 dx \, f(x) = a_0 + \frac{1}{3} a_1 - \sum_{k=0}^{\infty} \left(\frac{a_{n-1}}{2n - 1} - \frac{a_{n+1}}{2n + 1} \right) P_n(x).$$

Exercise 11.60

Generating functions can also be used to solve certain types of coupled linear differential equations. The following is an example of a particularly simple system having an elegant solution. Equations such as this occur in quantum mechanics.

If the functions $x_k(t)$, $k = 0, \pm 1, \pm 2, \ldots$, are the solutions of the infinite set of equations

$$\frac{dx_k}{dt} = A(x_{k-1} - x_{k+1}), \quad x_0(0) = 1, \quad x_k(0) = 0, \ k \neq 0,$$

where the A is a real constant. Show that the generating function

$$G(\theta, t) = \sum_{k=-\infty}^{\infty} x_k(t)e^{-ik\theta}$$

satisfies the differential equation

$$\frac{\partial G}{\partial t} = -2iAG(\theta, t)\sin\theta,$$

and hence that

$$x_k(t) = \frac{1}{2\pi}\int_{-\pi}^{\pi} d\theta \, e^{i(k\theta - 2At\sin\theta)} = J_k(2At).$$

Using a similar method, show that the solution of the equations

$$\frac{dx_k}{dt} = A\left(x_{k-1}e^{-i\omega t} - x_{k+1}e^{i\omega t}\right), \quad x_0(0) = 1, \quad x_k(0) = 0, \ k \neq 0,$$

where ω is also a positive constant, is

$$x(t) = e^{-i\omega t/2}J_k\left(\frac{4A}{\omega}\sin(\omega t/2)\right).$$

Explain the effect of the oscillating terms $e^{\pm i\omega t}$ on the solution.

Exercise 11.61
If $H(z)$ is the Heaviside unit function, show that

$$H(x - a) = \frac{1}{2}(1 - a) + \frac{1}{2}\sum_{k=1}^{\infty} P_k(x)\left(P_{k-1}(a) - P_{k+1}(a)\right).$$

Show also that the value of the nth partial sum at $x = a$ is

$$S_n(a) = \frac{1}{2} - \frac{1}{2}P_n(a)P_{n+1}(a) \quad \text{and that} \quad \lim_{n\to\infty} S_n(a) = \frac{1}{2}.$$

Draw graphs of some representative partial sums of the series for $H(x - a)$ and demonstrate explicitly that this type of expansion exhibits the Gibbs' phenomenon.

Exercise 11.62
The functions

$$P_1(x) = x \quad \text{and} \quad Q_0(x) = \frac{1}{2}\ln\left(\frac{1+x}{1-x}\right)$$

are both solutions of the equation

$$\frac{d}{dx}\left((1 - x^2)\frac{dy}{dx}\right) + n(n + 1)y = 0,$$

with $n = 0$ and 1 respectively.

Show that $\int_{-1}^{1} dx\, P_1(x)Q_0(x) \neq 0$ and explain why the proof of orthogonality, given in section 10.4, breaks down.

Exercise 11.63

Use Rodrigues' formula and integration by parts to show that

$$\int_{-1}^{1} dx\, P_n(x)^2 = \frac{2}{2n+1}.$$

The following identities may be useful:

$$2\int_{0}^{\pi/2} d\theta\, \cos^{2n+1}\theta = \frac{\Gamma(1/2)\Gamma(n+1)}{\Gamma(n+3/2)}, \quad \Gamma(1/2) = \sqrt{\pi},$$

$$\Gamma(n+1/2) = \frac{1.3.5.\cdots.(2n-1)}{2^n}\Gamma(1/2).$$

Exercise 11.64

If

$$\nabla^2 f = \frac{\partial^2 f}{\partial x^2} + \frac{\partial^2 f}{\partial y^2} + \frac{\partial^2 f}{\partial z^2},$$

show that if $V(r)$ is a function of r only, then $\nabla^2 V = V''(r) + 2V'(r)/r$ and hence that if $\nabla^2 V = 0$ then $V = \alpha/r$ for some constant α.

In spherical polar coordinates $x = r\sin\theta\cos\phi$, $y = r\sin\theta\cos\phi$ and $z = r\cos\theta$,

$$\nabla^2 f = \frac{1}{r^2}\frac{\partial}{\partial r}\left(r^2\frac{\partial f}{\partial r}\right) + \frac{1}{r^2\sin\theta}\frac{\partial}{\partial\theta}\left(\sin\theta\frac{\partial f}{\partial\theta}\right) + \frac{1}{r^2\sin^2\theta}\frac{\partial^2 f}{\partial\phi^2}.$$

Use this result and the expansion

$$V(r) = \frac{1}{\sqrt{r^2 + a^2 - 2ar\cos\theta}} = \frac{1}{r}\sum_{n=0}^{\infty}\left(\frac{a}{r}\right)^n P_n(\cos\theta)$$

to show that the functions $P_n(\cos\theta)$, defined by this relation, satisfy the equations

$$\frac{d}{dw}\left((1-w^2)\frac{dP_n}{dw}\right) + n(n+1)P_n = 0, \quad n = 0, 1, \ldots, \qquad w = \cos\theta.$$

Exercise 11.65

Show that the following differential equations have the solutions given:

$$\frac{d^2 w}{dx^2} + \left(\frac{\lambda^2}{4x} - \frac{v^2 - 1}{4x^2}\right)w = 0, \quad w = x^{1/2}J_v(\lambda x^{1/2}),$$

$$\frac{d^2 w}{dx^2} + \lambda^2 x^{p-2}w = 0, \quad w = x^{1/2}J_{1/p}(2\lambda x^{p/2}/p),$$

$$\frac{d^2 w}{dx^2} - \frac{2v-1}{x}\frac{dw}{dx} + \lambda^2 w = 0, \quad w = x^v J_v(\lambda x).$$

Exercise 11.66

Use Maple, or otherwise, to show that

$$x = \begin{cases} P_1(x) \\ T_1(x) \\ \frac{1}{2}U_1(x) \end{cases} \qquad x^2 = \begin{cases} \frac{1}{3} + \frac{2}{3}P_2(x) \\ \frac{1}{2} + \frac{1}{2}T_2(x) \\ \frac{1}{4} + \frac{1}{4}U_2(x) \end{cases}$$

$$x^3 = \begin{cases} \frac{3}{5}P_1(x) + \frac{2}{5}P_3(x) \\ \frac{3}{4}T_1(x) + \frac{1}{4}T_3(x) \\ \frac{1}{4}U_1(x) + \frac{1}{8}U_3(x) \end{cases} \qquad x^4 = \begin{cases} \frac{1}{5} + \frac{1}{7}P_2(x) + \frac{8}{35}P_4 \\ \frac{3}{8} + \frac{1}{2}T_2(x) + \frac{1}{8}T_4(x) \\ \frac{1}{8} + \frac{3}{8}U_2(x) + \frac{1}{16}U_4(x) \end{cases}$$

and find similar expressions involving Laguerre and Hermite polynomials.

Exercise 11.67

Use the change of variable $x = \alpha w^2$ in the differential equation for the Laguerre polynomial $L_n(x)$ with a suitable choice of α to show that

$$L_n(x) \to J_0(2\sqrt{nx}) \quad \text{as} \quad n \to \infty,$$

where $J_0(x)$ is the zero order Bessel function. Compare graphically this asymptotic form with the exact values of $L_n(x)$ and with the asymptotic expansion given in equation 11.46.

Exercise 11.68

Use the integral

$$\int_0^1 dx\, x^\rho P_n(x) = \frac{\sqrt{\pi} 2^{-\rho-1}\Gamma(1+\rho)}{\Gamma\left(1 + \frac{\rho}{2} - \frac{n}{2}\right)\Gamma\left(\frac{3}{2} + \frac{\rho}{2} + \frac{n}{2}\right)}, \quad \Re(\rho) > -1,$$

to find the Legendre series approximation to the function

$$f(x) = \begin{cases} 0, & -1 \le x \le 0, \\ x^\rho, & 0 < x \le 1. \end{cases}$$

Use Maple to make a graphical comparison between an N-term series and the exact function for various values of ρ, negative and positive, and N.

Exercise 11.69

The accurate computation of Bessel functions for arbitrary order and argument is a non-trivial task. The Taylor's series 11.62 may be used only if $x < n$ and the asymptotic expansion is accurate only if $x \gg n$. However, the recurrence relation

$$J_{n-1}(x) + J_{n+1}(x) = \frac{2n}{x}J_n(x)$$

may be used to compute $J_k(x)$, $k = 0, 1, \ldots$, if care is taken.

Because $J_n(x)$ is, for $x < n$, a decreasing function of n, the recurrence relation is stable in the direction of *decreasing* n, but unstable for increasing n. Thus we proceed as follows.

Suppose that we need to compute $J_k(1.75)$ for $k = 0, 1, 2$ and 3, then we start with two initial guesses for $J_{11}(1.75)$ and $J_{10}(1.75)$, $J_{11}(1.75) = 0$ and $J_{10}(1.75) = 1$, and use these to compute trial values for $J_k(1.75)$, $k = 8, 7, \ldots, 0$, then use the identity

$$J_0(x) + 2J_2(x) + 2J_4(x) + \cdots = 1$$

to re-normalise these values. Thus the trial values for $J_0(1.75)$ and $J_1(1.75)$ are, respectively, 485 176.88 and 762 746.78, and the normalisation factor is 1 314 726.4, which gives $J_0(1.75) = 0.3690$, $J_1(1.75) = 0.5802$ and $J_2(1.75) = 0.2940$. Write a Maple procedure that implements this method.

Exercise 11.70

(i) Show that the generalised Green's function for the differential equation

$$\frac{d}{dx}\left((1-x^2)\frac{dy}{dx}\right) = f(x), \quad -1 \le x \le 1,$$

where $y(x)$ is bounded for $-1 \le x \le 1$, can be written in terms of the infinite sum

$$\overline{G}(x,u) = -\sum_{n=1}^{\infty} \frac{n+\frac{1}{2}}{n(n+1)} P_n(x)P_n(u),$$

where $P_n(x)$ is a Legendre polynomial.

(ii) By truncating this series suitably show that an approximate bounded solution of the equation

$$\frac{d}{dx}\left((1-x^2)\frac{dy}{dx}\right) = \cos \pi x, \quad -1 \le x \le 1,$$

is

$$y(x) \simeq \frac{7(2\pi^2 - 81)}{32\pi^4} - \frac{105(2\pi^2 - 27)}{16\pi^4}x^2 + \frac{315(2\pi^2 - 21)}{32\pi^4}x^4.$$

(iii) Consider the two cases $f(x) = x^2$ and $f(x) = x^2 - \frac{1}{3}$ and explain why the generalised Green's function gives a solution in the second case but not in the first.

12

Linear systems and Floquet theory

12.1 Introduction

In this chapter we continue the study of linear differential equations, but now consider equations of the form

$$\frac{d\mathbf{x}}{dt} = A(t)\mathbf{x}, \tag{12.1}$$

where $A(t)$ is a real, non-singular, $n \times n$ matrix and $\mathbf{x} = (x_1, x_2, \ldots, x_n)$ an n-dimensional vector. The elements of $A(t)$ are functions of the independent variable t, and our main interest is the case where these functions are periodic, though we start with a brief description of the properties of more general systems. All the equations dealt with in chapter 11 may be cast in this form, with $n = 2$, though normally this re-formulation is not helpful because the methods used to understand the special functions of that chapter are usually different from those used to understand the solutions of 12.1.

Linear equations like 12.1 arise naturally in the study of nonlinear equations, $\dot{\mathbf{x}} = \mathbf{X}(\mathbf{x}, t)$, where $\mathbf{X}(\mathbf{x}, t) = (X_1(\mathbf{x}, t), X_2(\mathbf{x}, t), \ldots, X_n(\mathbf{x}, t))$ is an n-dimensional vector function of $\mathbf{x}$ and possibly the time t. If $\mathbf{X}$ is not a function of the time then an expansion about the fixed points, defined by the equations $\mathbf{X}(\mathbf{x}) = 0$, normally gives rise to equations like 12.1 with constant matrix A, and the study of these linear equations provides important information about the stability of these fixed points. This type of problem is dealt with in some detail in chapter 16. Alternatively, if the nonlinear system admits a periodic solution $\boldsymbol{\xi}(t)$, so $\boldsymbol{\xi}(t + T) = \boldsymbol{\xi}(t)$ for all t and some $T > 0$, then an expansion about $\boldsymbol{\xi}(t)$ leads to equations like 12.1 but with $A(t)$ a T-periodic function of time; the study of these linear equations helps us understand the nature of the original periodic orbit. Because linear equations are far easier to deal with than nonlinear equations, these techniques are important. A simple example of this latter application is described in section 12.5.

Other important problems that give rise to equations like 12.1 arise from Schrödinger's equation when the potential is spatially periodic, as in a crystal,

$$\frac{\hbar^2}{2\mu}\nabla^2 \psi + (E - V(\mathbf{x}))\,\psi = 0, \quad V(\mathbf{x} + \mathbf{L}_k) = V(\mathbf{x}) \text{ for all } \mathbf{x},$$

where $\mathbf{x}$ is an N-dimensional vector and the $\mathbf{L}_k$ are N linearly independent vectors, or when the potential is a periodic function of time,

$$i\hbar\frac{\partial \psi}{\partial t} = \frac{\hbar^2}{2\mu}\nabla^2 \psi - V(\mathbf{x}, t)\psi, \quad V(\mathbf{x}, t + T) = V(\mathbf{x}, t) \text{ for all } t,$$

437

as when an atom or molecule is influenced by a periodic electric field, such as a laser.

12.2 General properties of linear systems

The equation

$$\frac{d\mathbf{x}}{dt} = A(t)\mathbf{x}, \tag{12.2}$$

where $A(t)$ is a real, non-singular, $n \times n$ matrix and $\mathbf{x} = (x_1, x_2, \ldots, x_n)$ an n-dimensional vector, is linear, and this imposes stringent constraints on the behaviour of the solutions. Here we list some important definitions and properties: proofs may be found in Arnold (1973), Cesari (1963), Hartman (1964), Hirsch and Smale (1974) or Jordan and Smith (1999), for example.

Existence: if the elements of $A(t)$ are piecewise continuous, with a finite number of discontinuous points and integrable at each discontinuity, the solution of the equation $\dot{\mathbf{x}} = A(t)\mathbf{x}$ with initial condition $\mathbf{x}(0) = \mathbf{x}_0$ exists and is unique.

Linearity: if $\{\mathbf{x}_1(t), \mathbf{x}_2(t), \ldots, \mathbf{x}_m(t)\}$ are m solutions of 12.2, real or complex, then the sum

$$\mathbf{y}(t) = \sum_{k=1}^{m} \alpha_k \mathbf{x}_k(t),$$

where α_k are constants, real or complex, is also a solution.

Linear dependence: if $\{\mathbf{z}_1(t), \mathbf{z}_2(t), \ldots, \mathbf{z}_m(t)\}$ are any m vector functions (real or complex), continuous on some interval of t, none being identically zero, and if there exists constants α_k, $k = 1, 2, \ldots, m$, not all of which are zero, such that

$$\sum_{k=1}^{m} \alpha_k \mathbf{z}_k(t) = 0,$$

then the functions $\mathbf{z}_k(t)$, $k = 1, 2, \ldots, m$, are said to be *linearly dependent*. Otherwise they are *linearly independent*.

Linear dependence of solutions: any $n + 1$ non-trivial solutions of the n equations 12.2 are linearly dependent. Conversely there exists a set of n linearly independent solutions of 12.2.

  Further, if $\{\boldsymbol{\phi}_1(t), \boldsymbol{\phi}_2(t), \ldots, \boldsymbol{\phi}_n(t)\}$ is a set of n linearly independent solutions, *every* solution of 12.2 is a linear combination of these solutions.

Fundamental matrix: if $\{\boldsymbol{\phi}_1(t), \boldsymbol{\phi}_2(t), \ldots, \boldsymbol{\phi}_n(t)\}$ is a set of n linearly independent solutions, the matrix

$$\Phi(t) = \left(\boldsymbol{\phi}_1(t), \boldsymbol{\phi}_2(t), \ldots, \boldsymbol{\phi}_n(t)\right) = \begin{pmatrix} \phi_{11}(t) & \phi_{12}(t) & \cdots & \phi_{1n}(t) \\ \phi_{21}(t) & \phi_{22}(t) & \cdots & \phi_{2n}(t) \\ \vdots & \vdots & & \vdots \\ \phi_{n1}(t) & \phi_{n2}(t) & \cdots & \phi_{nn}(t) \end{pmatrix},$$

the kth column of which is the vector $\boldsymbol{\phi}_k(t)$, is named a *fundamental matrix*.

For any system there are an infinity of fundamental matrices, each satisfying the matrix differential equation

$$\frac{d\Phi}{dt} = A(t)\Phi.$$

Different sets of linearly independent solutions give rise to different fundamental matrices, but since the components of one set may be expressed as linear combinations of the components of any other set, any two fundamental matrices $\Phi_1(t)$ and $\Phi_2(t)$ are related by $\Phi_2(t) = \Phi_1(t)C$, where C is a constant, non-singular matrix.

The special fundamental matrix satisfying $\Phi(0) = I$ is sometimes named the *matrizant* of equation 12.1. If the matrix $A(t)$ is T-periodic the *monodromy matrix* is the value of the matrizant at $t = T$, that is $\Phi(T)$ where $\Phi(0) = I$. This matrix plays an important role in the theory of periodic systems that follows.

Determinant of a fundamental matrix: if $X(t)$ is an $n \times n$ matrix of the solutions $\{\mathbf{x}_1(t), \mathbf{x}_2(t), \ldots, \mathbf{x}_n(t)\}$, then either

(i) $\det(X(t)) \neq 0$, for all t, in which case the $\mathbf{x}_k(t)$ are linearly independent and $X(t)$ is a fundamental matrix, or

(ii) $\det(X(t)) = 0$, for all t, in which case the $\mathbf{x}_k(t)$ are linearly dependent. Conversely, if $\mathbf{x}_k(t)$ are linearly dependent, $\det(X(t)) = 0$.

The Wronskian of a set of n linearly independent solutions is defined to be the determinant of the fundamental matrix,

$$W(t) = \det(\Phi(t)),$$

and can be shown, see for instance Hartman (1964, chapter 4), to satisfy the linear differential equation

$$\frac{dW}{dt} = W(t)\text{Tr}(A(t)) \quad \text{with solution} \quad W(t) = W(t_0)\exp\left(\int_{t_0}^{t} ds\, \text{Tr}(A(s))\right).$$
$$(12.3)$$

Thus if $\text{Tr}(A) = 0$, $\det(\Phi(t)) = $constant. This result is originally due to Liouville. This definition of the Wronskian is directly equivalent to that given in chapter 10 if the kth row of Φ is the $(k-1)$th differential of the first row.

Solutions in terms of a fundamental matrix: the solution of 12.2 with the initial condition $\mathbf{x}(t_0) = \mathbf{x}_0$ can be expressed as

$$\mathbf{x}(t) = \Phi(t)\Phi(t_0)^{-1}\mathbf{x}(t_0),$$

where $\Phi(t)$ is *any* fundamental matrix, the proof of this is outlined in exercise 12.33 (page 473).

The matrix $U(t_2, t_1) = \Phi(t_2)\Phi(t_1)^{-1}$, depending upon the two times t_1 and t_2, is named the *propagator* because it propagates the solution from a time t_1 to a time t_2, which can precede t_1: for autonomous systems $U(t_2, t_1)$ depends only upon the difference $t_2 - t_1$. Because the system is linear the propagator has the multiplicative property

$$U(t_3, t_1) = U(t_3, t_2)U(t_2, t_1).$$

This is sometimes useful in allowing the long-time evolution to be constructed from the product of many short-time propagators that are easier to obtain, because less happens during small time intervals. An important example of this use is described in Feynman and Hibbs (1965, chapter 2).

In general it is not possible to find solutions of the equation $\dot{\mathbf{x}} = A(t)\mathbf{x}$ in terms of known functions. When the matrix A is constant, however, the solution in any particular case can be expressed in terms of trigonometric or hyperbolic functions and products of these and polynomials in t.

Consider the system $\dot{\mathbf{x}} = A\mathbf{x}$, where A is now a constant, real, non-singular $n \times n$ matrix. On putting $\mathbf{x} = \mathbf{r}e^{\lambda t}$, where λ is a constant and $\mathbf{r}$ a constant vector, we obtain the eigenvalue equation

$$A\mathbf{r} = \lambda\mathbf{r}.$$

The vector $\mathbf{r}$ will be non-trivial ($\mathbf{r} \neq 0$) only if λ is an eigenvalue of A, that is, λ must satisfy the nth degree polynomial

$$\det(A - \lambda I) = 0.$$

This equation has n solutions $\lambda_1, \lambda_2, \ldots, \lambda_n$, which may not all be distinct, and will be real or, if complex, will occur in complex conjugate pairs (because A is real). For any given eigenvalue λ_p there will be a corresponding eigenvector $\mathbf{r}_p$ satisfying the equation $A\mathbf{r}_p = \lambda_p\mathbf{r}_p$. Because this is a homogeneous equation $\mathbf{r}_p$ is undefined to within a multiplicative constant – that is the length of $\mathbf{r}_p$ is arbitrary.

If the matrix A has n linearly independent eigenvectors $\{\mathbf{r}_1, \mathbf{r}_2, \ldots, \mathbf{r}_n\}$ corresponding to the eigenvalues $\lambda_1, \lambda_2, \ldots, \lambda_n$, which need not all be distinct, the general solution of the equation

$$\frac{d\mathbf{x}}{dt} = A\mathbf{x} \quad \text{is} \quad \mathbf{x}(t) = \sum_{k=1}^{n} c_k\mathbf{r}_k e^{\lambda_k t},$$

where the c_k are constants.

If an eigenvalue λ_p has multiplicity $m \leq n$, so $\det(A - \lambda I)$ has a factor $(\lambda - \lambda_p)^m$, there are m linearly independent eigenvectors associated with the eigenvector λ_p. In this case there are m solutions of the form

$$\mathbf{p}_1(t)e^{\lambda_p t}, \quad \mathbf{p}_2(t)e^{\lambda_p t}, \quad \ldots, \quad \mathbf{p}_m(t)e^{\lambda_p t},$$

where the $\mathbf{p}_k(t)$ are vector polynomials in t of degree $m - 1$, or less.

For two-dimensional systems, $n = 2$, the various combinations of eigenvalues and eigenvectors that occur give rise to ten distinctly different types of solution: these are discussed in detail in chapter 16.

If $n \geq 3$ there are too many possibilities for a useful classification scheme, and each problem needs to be treated individually. If, however, the eigenvalues of A satisfy the condition $\Re(\lambda_k) \leq 0$, then the solutions of $\dot{\mathbf{x}} = A\mathbf{x}$ and 12.4 are bounded. In addition, if all the solutions of $\dot{\mathbf{x}} = A\mathbf{x}$ are bounded then all the solutions of

$$\frac{d\mathbf{x}}{dt} = (A + C(t))\mathbf{x} \quad \text{are bounded for } t > t_0 \text{ if } \quad \sum_{i=1}^{n}\sum_{j=1}^{n}\int_{t_0}^{\infty} dt\,|C_{ij}(t)| \tag{12.4}$$

is bounded, Cesari (1963, section 3.3).

Exercise 12.1
Find the eigenvalues and eigenvectors of the matrices

$$\begin{pmatrix} 0 & 2 & 4 \\ 1 & 1 & -2 \\ -2 & 0 & 5 \end{pmatrix}, \quad \begin{pmatrix} 1 & 0 & 0 \\ 1 & 3 & 2 \\ 1 & 2 & 3 \end{pmatrix}, \quad \begin{pmatrix} 6 & 24 & 12 \\ 20 & -34 & -32 \\ -43 & 56 & 58 \end{pmatrix}.$$

Exercise 12.2
Determine whether or not the following vectors are linearly dependent:

(i) $(1, -1, 1)$, $(2, 1, 1)$, $(0, 1, -3)$,
(ii) $(t, 2t)$, $(3t, 5t)$, $(4t, 6t)$,
(iii) $(\sin(t - \alpha), \cos(t - \alpha))$, $(\sin(t - \beta), \cos(t - \beta))$.

Exercise 12.3
Find the fundamental matrix with the initial value $\Phi(0) = I$ and construct the propagator for the systems

$$\text{(i)} \quad \frac{d\mathbf{x}}{dt} = \begin{pmatrix} 0 & 1 \\ -1 & -2 \end{pmatrix}\mathbf{x}, \qquad \text{(ii)} \quad \frac{d\mathbf{x}}{dt} = \begin{pmatrix} -1 & 0 & 0 \\ 1 & 1 & 1 \\ 0 & -1 & 0 \end{pmatrix}\mathbf{x}.$$

Exercise 12.4
If $\Phi(t)$ is a fundamental matrix of the system $\dot{\mathbf{x}} = A(t)\mathbf{x}$, by making the change of variables $\mathbf{x} = \Phi(t)\mathbf{y}$ show that a solution of the equation

$$\frac{d\mathbf{x}}{dt} = A(t)\mathbf{x} + \mathbf{F}(t),$$

where $\mathbf{F}(t)$ is a vector function of t only, is

$$\mathbf{x}(t) = \Phi(t)\int ds\, \Phi(s)^{-1}\mathbf{F}(s).$$

Deduce that the solution of the equation with the initial conditions $\mathbf{x}(t_0) = \mathbf{x}_0$ is

$$\mathbf{x}(t) = \Phi(t)\Phi(t_0)^{-1}\mathbf{x}_0 + \Phi(t)\int_{t_0}^{t} ds\, \Phi(s)^{-1}\mathbf{F}(s).$$

Exercise 12.5
By defining $y = \dot{x}$ convert the system $\ddot{x} + \omega^2 x = 0$ to the form of equation 12.2 and show that the matrizant is

$$\Phi(t) = \begin{pmatrix} \cos\omega t & \omega^{-1}\sin\omega t \\ -\omega\sin\omega t & \cos\omega t \end{pmatrix}.$$

Using this and the result of the previous exercise show that the solution of the equation

$$\frac{d^2x}{dt^2} + \omega^2 x = F(t), \quad x(0) = a, \quad \dot{x}(0) = b,$$

is

$$x(t) = a\cos\omega t + \frac{b}{\omega}\sin\omega t + \frac{1}{\omega}\int_0^t ds\, F(s)\sin\omega(t-s).$$

Note: the last integral contains the Green's function for this system.

Exercise 12.6

By defining suitable auxiliary variables, as in exercise 12.5, write the equation

$$\frac{d^3x}{dt^3} - \frac{d^2x}{dt^2} + 4\frac{dx}{dt} - 4x = 0$$

in the form of equation 12.2, find a fundamental matrix and show that the propagator is

$$U(t_2, t_1) = \frac{1}{5}\begin{pmatrix} e + 4c + 2s & -10s & 4e - 4c + 8s \\ e - c + 2s & 5c & 4e - 4c - 2s \\ e - c - \frac{1}{2}s & \frac{5}{2}s & 4e + c - 2s \end{pmatrix},$$

where $e = \exp\tau$, $c = \cos 2\tau$, $s = \sin 2\tau$, and $\tau = t_2 - t_1$.

Exercise 12.7

Find the fundamental matrix, with the initial condition $\Phi(0) = I$, of the system of equations

$$\frac{dx}{dt} = -y, \quad \frac{d^2y}{dt^2} = -x - y + \frac{dy}{dt}.$$

Deduce that the only solutions that are bound for all t have the initial conditions $(x(0), y(0), \dot{y}(0)) = (0, a, 0)$ for any constant a.

Exercise 12.8

Find the eigenvalues and eigenvectors of the matrix

$$A = \begin{pmatrix} 1 & 0 & 0 \\ 1 & 3 & 2 \\ 1 & 2 & 3 \end{pmatrix},$$

and hence show that the solution of the equation

$$\dot{\mathbf{x}} = A\mathbf{x}, \quad \text{with} \quad \mathbf{x}(0) = (2, 2, 1) \quad \text{is} \quad \mathbf{x}(t) = \left(2e^t, 2e^{5t}, 2e^{5t} - e^t\right).$$

Exercise 12.9

Find the eigenvalues and eigenvectors of the matrix

$$A = \begin{pmatrix} 1 & 1 & 0 \\ 0 & 1 & 1 \\ 0 & -2 & 2 \end{pmatrix},$$

and hence find the solution of the equation $\dot{\mathbf{x}} = A\mathbf{x}$, $\mathbf{x}(0) = (1, 1, -3)$ and show that

$$x_3(t) = -e^{-3t/2}\left(\sqrt{7}\sin\left(\frac{t\sqrt{7}}{2}\right) + 3\cos\left(\frac{t\sqrt{7}}{2}\right)\right).$$

Exercise 12.10

Show that the matrix

$$A = \begin{pmatrix} 0 & 1 \\ -b^2 & 2b \end{pmatrix},$$

where b is a constant, has an eigenvalue $\lambda = b$, with multiplicity 2, and a single eigenvector $\mathbf{r}_1 = (1, b)^T$. If $\mathbf{r}_2$ is another vector and c_1 and c_2 are arbitrary constants, show that

$$\mathbf{x} = c_1 \mathbf{r}_1 e^{bt} + c_2 (\mathbf{r}_2 + t\mathbf{r}_1) e^{bt}$$

is a solution of the differential equation $\dot{\mathbf{x}} = A\mathbf{x}$ if $\mathbf{r}_2$ satisfies the equation $(A - bI)\mathbf{r}_2 = \mathbf{r}_1$. Hence show that the solution with the initial conditions $\mathbf{x}(0) = (\alpha, \beta)$ is

$$\mathbf{x}(t) = (\alpha + (\beta - b\alpha)t, \beta + bt (\beta - b\alpha)) e^{bt}.$$

Exercise 12.11

Express the differential equations

$$\frac{dx_1}{dt} - 3\frac{dx_2}{dt} = 2x_1 + x_2, \qquad 2\frac{dx_1}{dt} + \frac{dx_2}{dt} = 4x_1 - 3x_2,$$

in the form $\dot{\mathbf{x}} = A\mathbf{x}$ and hence find the matrizant $\Phi(t)$.

Exercise 12.12

Show that the equation

$$\frac{d^2x}{dt^2} - \frac{2}{t}\frac{dx}{dt} + x = 0$$

has solutions $x = (1 \mp it)e^{\pm it}$.

This result shows that the solutions of the equation formed by taking the limit of the equation as $t \to \infty$ are not always given by the same limit of the solutions of the original equation. Does this contradict the statement made about equation 12.4?

12.3 Floquet theory

12.3.1 Introduction

We now turn to the linear system with periodic coefficients,

$$\frac{d\mathbf{x}}{dt} = A(t)\mathbf{x}, \qquad A(t + T) = A(t) \quad \text{for all } t, \tag{12.5}$$

where $A(t)$ is a real, non-singular $n \times n$ matrix with elements that are T-periodic functions of t. The French mathematician Floquet first developed the general theory of linear, periodic systems and provided a systematic study of such systems. In general G. Floquet (1847–1920) the solutions of such equations cannot be expressed in terms of known functions, but linearity and the periodicity of $A(t)$ means that the behaviour of a solution for *all* times can be deduced from the general solution on a finite interval of length T. This unusual property means that the behaviour of the solutions as $t \to \infty$ can often be deduced from approximate or numerical solutions.

In the introduction to this chapter we described a number of important circumstances described by equations like 12.5. Such equations, and nonlinear equivalents, also occur in the description of dynamical systems with parameters that vary periodically with time; for instance the vertical pendulum with a point of support that moves periodically, as in exercise 12.38 (page 475).

In order to provide some idea of how such systems can behave, we start this section with a description of two systems. Both examples are essentially vertical pendulums and in each case the length is made to vary periodically. With the correct choice of this period it is shown that the energy (that is the amplitude of the swing) can be made to increase rapidly. The method of supplying energy to a system whereby system parameters are varied periodically is known as *parametric pumping*.

12.3.2 Parametric resonance: the swing

The first example of parametric resonance with which you are bound to be familiar is a child on a playground swing. Very careful observation of the child shows that the amplitude of the motion is increased by the rhythmical bending and straightening of the child's body with the effect that the centre of mass is raised as the swing passes through its lowest point and lowered when the swing reaches its highest point. An idealisation of this motion is obtained by treating the swing and child as a vertical pendulum with shifts in the centre of mass taking place instantaneously at the lowest and highest points, as shown in the diagram. The advantage of this approximation is that we can understand the motion without solving any differential equations, although conservation of angular momentum and energy are needed.

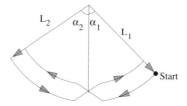

Figure 12.1

During each quarter oscillation, when the length L of the pendulum is constant, the swing behaves like the vertical pendulum for which the energy is

$$E = \frac{1}{2}mL^2\dot{\theta}^2 - mgL\cos\theta,$$

where θ is the angle between the swing and the downward vertical. Suppose that the system is released from rest at an angle $\theta = \alpha_1$ with length L_1, then the angular velocity $\dot{\theta} = \omega_1$ at the bottom, where $\theta = 0$, is obtained using the energy equation

$$\frac{1}{2}L_1^2\omega_1^2 = gL_1(1 - \cos\alpha_1) \quad \text{or} \quad \omega_1^2 = \frac{4g}{L_1}\sin^2(\alpha_1/2).$$

At the bottom the length changes instantaneously from L_1 to $L_2 < L_1$, and since the velocity change is towards the point of support, angular momentum, $mL^2\dot{\theta}$, is conserved and immediately after this length change the angular velocity, ω_2, is larger and given by

$$L_1^2 \omega_1 = L_2^2 \omega_2 \quad \text{or} \quad \omega_2 = \frac{L_1^2}{L_2^2} \omega_1 > \omega_1. \tag{12.6}$$

The amplitude, α_2, of the next quarter swing can be related to ω_2 using the energy equation again,

$$\frac{1}{2} L_2^2 \omega_2^2 - g L_2 = -g L_2 \cos \alpha_2 \quad \text{or} \quad \omega_2^2 = \frac{4g}{L_2} \sin^2 \left(\frac{\alpha_2}{2} \right).$$

But since ω_1 and ω_2 are related by equation 12.6 we can obtain a relation between successive amplitudes and the lengths L_1 and L_2,

$$L_1^3 \sin^2 \left(\frac{\alpha_1}{2} \right) = L_2^3 \sin^2 \left(\frac{\alpha_2}{2} \right). \tag{12.7}$$

When the swing reaches its maximum amplitude, $\theta = \alpha_2$, its length is returned instantaneously to L_1. Again angular momentum is conserved, but at this point on the swing the angular velocity is zero so remains unchanged. Thus the swing starts its next half cycle with length L_1 but from the larger amplitude α_2.

This procedure can be performed on each swing, so after passing through the lowest point N times we have, on setting $L_1 = L_2 + h$,

$$\sin \left(\frac{\alpha_{N+1}}{2} \right) = \left(1 + \frac{h}{L_2} \right)^{3N/2} \sin \left(\frac{\alpha_1}{2} \right) \simeq \exp \left(\frac{3Nh}{2L_2} \right) \sin \left(\frac{\alpha_1}{2} \right).$$

Thus the amplitude increases *exponentially* with N. After a finite number of swings the right hand side of this equation becomes larger than unity and the theory is invalid: this happens when the system has sufficient energy for the swing to rotate round the support. In practice a real swing does not increase its amplitude so rapidly because the changes in length take place more gradually and not precisely at the optimum point.

Another simple example of parametric resonance is the child's toy, shown in the diagram, comprising a disc suspended on a loop of thread which can be made to spin alternately in opposite directions by pulling the loops twice in each complete cycle.

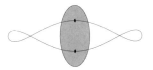

In both these examples large amplitude motion is produced by changing the system parameters with a frequency of *twice* the natural frequency of the system. This is typical, but the frequency does not have to be exactly twice the natural frequency, ω_0: the system can be parametrically pumped with a nearby frequency, and one important problem is to determine the width of the frequency band round $2\omega_0$ in which resonances occur. Parametric pumping can also occur near the frequencies $2\omega_0/n$, for integer n; in the example of the swing this is fairly obvious as we need only pump on every alternate pass through the bottom, for instance.

12.3.3 O Botafumeiro: parametric pumping in the middle ages

One of the oldest recorded examples of parametric pumping is swinging the giant censer, O Botafumeiro, in the cathedral of Santiago in Santiago de Compostela, a town in Galicia in northwest Spain. This cathedral was a pilgrims' shrine, famous throughout Christendom during the middle ages. The censer, with coals, weighs about 57 kg and hangs in the transept; it swings on a rope about 21m long with a maximum amplitude of about 80° and has a period of about 10 seconds; near the bottom of the swing it is travelling at about 68 km/hr (about 40 mph) half a metre above the floor.

Producing and maintaining such motion is far from trivial, but the rite of pumping O Botafumeiro appears to be about 700 years old, for the cathedral was built between the years 1078 to 1211, on the site of an older one destroyed by Almanzor, the military commander of the Moorish caliphate of Córdoba, in AD 997. The first recorded use of the censer is a 14th century margin note in the Codex, Liber Sancti Jacobi donated to the cathedral in about 1150: so we know that the rite of pumping O Botafumeiro started between 1150 and 1325, that is at least four centuries before the pendulum was studied scientifically.

The motion is started by moving the censer off the vertical to let it swing like a pendulum; then a team of men cyclically pull at cords attached to the upper end of the rope in order to decrease and increase its length as it passes through the lowest and highest points of the motion, as shown schematically in figure 12.1. Here the similarity with the child's swing ends. In that example of parametric pumping no formal rules are needed; a child never formulates the mechanism by which it pumps the swing. But O Botafumeiro needs a team effort, the chief verger calling orders where required; to obtain motion with an amplitude of about 80° roughly 17 pumping cycles are required and the total time taken is about 80 seconds. There is some evidence that at some point the rules for pumping became understood and transmitted to other local cathedrals; there are records of the cathedrals at Orense and Tuy (respectively 100 km SE and S of Santiago), but there are no records of the large gold censer at old St Peter's in Rome ever having been set in motion.

The motion of O Botafumeiro is said to be an impressive sight. It clearly imposes significant strains on the supports, with the result that several accidents have occurred. When pumping, the highest tensions in the ropes occur at the top and bottom of the swing, so it is here that the rope is most likely to break; the recorded accidents support this view. In 1622 the rope broke and the censer fell vertically, just missing the men pulling at the rope, suggesting that the break occurred near the highest point. In 1499 the chains attached to the censer broke and it landed at the side of the transept, crushing the door about 30 m from the centre of the swing; this could happen only if the amplitude of the motion was large and the break occurred near the bottom of the swing.

The dynamics of O Botafumeiro is complicated; in particular the amplitude of the motion is too large for the linear approximation to be made, the rope is heavy so its mass needs to be taken into account, partly because at the highest point of the swing it is far from straight, and air resistance needs to be included. Nevertheless this system has been studied, Sanmartín (1984), and the theoretical motion agrees well with its actual motion.

12.3.4 One-dimensional linear systems

First-order linear systems are relatively easy to understand and serve as a useful stepping stone to more complicated systems. The most general homogeneous, first-order system has the form

$$\frac{dx}{dt} = a(t)x, \tag{12.8}$$

where $a(t)$ is a T-periodic function of time, $a(t + T) = a(t)$ for all t. This equation can be solved by direct integration,

$$x(t) = x(t_0)\exp\left(\int_{t_0}^{t} ds\, a(s)\right), \tag{12.9}$$

but it is more helpful to use methods that can easily be generalised to higher-dimensional systems that cannot be integrated.

The linearity of the equations and the periodicity of the coefficient impose constraints upon the possible types of solution: if $x(t)$ is a solution then the function $y(t) = x(t+T)$ is also a solution because

$$\frac{dy}{dt} = \frac{d}{dt}x(t + T) = \frac{dx(t + T)}{d(t + T)} = a(t + T)x(t + T)$$
$$= a(t)y(t).$$

Hence $x(t + T)$ and $x(t)$ satisfy the same equation. They are, however, *not* necessarily the same function, but because the differential equation is first-order and linear there is only one linearly independent solution, so we must have

$$x(t + T) = cx(t), \tag{12.10}$$

where c is a constant, independent of t.

Exercise 12.13
Show that for any integer n,

$$x(t_0 + nT) = c^n x(t_0) \tag{12.11}$$

and use equation 12.9 to deduce that

$$c = \exp\left(\int_{0}^{T} dt\, a(t)\right).$$

What property of $a(t)$ is required for $x(t)$ to be periodic?

Equation 12.11 has a particularly simple interpretation. If $c > 1$ then c^n grows exponentially with increasing n, so the solution $x(t)$ also increases exponentially. But if $c < 1$ then $c^n \to 0$ as $n \to \infty$, so $x(t) \to 0$ as $t \to \infty$. In the special case $c = 1$, $x(t + T) = x(t)$, for all t, and the solution is periodic: this occurs only when the mean value of $a(t)$ is zero.

As an example consider the 2π-periodic function

$$a(t) = A + B \cos t, \tag{12.12}$$

where A and B are constants. In this case equation 12.9 becomes

$$x(t) = x(0) \exp\left(At + B \sin t\right)$$

and

$$x(t + 2\pi) = e^{2\pi A} x(t), \quad \text{giving} \quad c = e^{2\pi A}.$$

If $A < 0$ then $c < 1$ and $x(t) \to 0$ as $t \to \infty$: if $A > 0$ then $c > 1$ and the solution grows without bound. When $A = 0$, $x(t) = \exp(B \sin t)$ and the solution is 2π-periodic. Examples of a growing, a periodic and a decaying solution are shown in the following graphs.

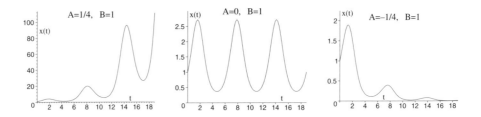

Fig. 12.2. Graphs of $x(t) = \exp(At + B \sin t)$ for $B = 1$ and various values of A.

Exercise 12.14
Consider the equation

$$\frac{dx}{dt} = x \sin t \sin vt, \quad x(0) = 1, \quad 0 < v \le 1,$$

with $v = p/q$, where p and q are coprime positive integers. Show that the solution is periodic with period $T = \pi q$. Show also that the solution is

$$x(t) = \exp\left(\frac{\sin(v-1)t}{2(v-1)} - \frac{\sin(v+1)t}{2(v+1)}\right),$$

and that as v varies over the interval $0 < v < 1$ the maximum value of this solution is unbounded.

12.3.5 *Many-dimensional linear, periodic systems*
The general theory of n-dimensional linear, periodic systems is very similar to the theory of one-dimensional systems presented above, but is generally more difficult to apply since the solutions of the equations of motion cannot normally be expressed in terms of integrals.

Consider the nth-order, linear homogeneous equation

$$\frac{d\mathbf{x}}{dt} = A(t)\mathbf{x}, \quad A(t + T) = A(t) \text{ for all } t, \tag{12.13}$$

where $A(t)$ is a real, $n \times n$ matrix whose elements are T-periodic functions. Assuming that this equation has n linearly independent solutions $\{x_1(t), x_2(t), \ldots, x_n(t)\}$, we may form a fundamental matrix

$$\Phi(t) = \begin{pmatrix} x_{11}(t) & x_{12}(t) & \cdots & x_{1n}(t) \\ x_{21}(t) & x_{22}(t) & \cdots & x_{2n}(t) \\ \vdots & \vdots & & \vdots \\ x_{n1}(t) & x_{n2}(t) & \cdots & x_{nn}(t) \end{pmatrix}$$

satisfying the matrix equation

$$\frac{d\Phi}{dt} = A(t)\Phi.$$

The vectors $y_k(t) = x_k(t + T)$ are also solutions of the original differential equation, for we have

$$\frac{dy_k}{dt} = \frac{d}{dt}x_k(t + T) = \frac{dx_k(t + T)}{d(t + T)} = A(t + T)x_k(t + T) = A(t)y_k.$$

Thus $y_k(t)$ must be a linear combination of the $x_j(t)$, $j = 1, 2, \ldots, n$, that is,

$$y_k(t) = \sum_{j=1}^{n} x_j(t)e_{jk}, \tag{12.14}$$

for some constants e_{jk}. These new solutions may be used to form the fundamental matrix $\Phi(t + T)$, and from 12.14 we have

$$\Phi(t + T) = \Phi(t)E,$$

where E is the matrix with elements e_{ij}. Since $\det(\Phi(t + T)) = \det(\Phi(t))\det(E)$ and $\det(\Phi) \neq 0$, E is non-singular. In the special case $\Phi(0) = I$ the matrix E is named the *monodromy matrix* of equation 12.13 and then $E = \Phi(T)$.

The eigenvalues and eigenvectors of E are important. If λ is an eigenvalue and a the associated eigenvector, $Ea = \lambda a$, then the solution

$$z(t) = \Phi(t)a \quad \text{has the property} \quad z(t + T) = \lambda z(t) \quad \text{for all } t.$$

This follows because

$$z(t + T) = \Phi(t + T)a = \Phi(t)Ea = \lambda\Phi(t)a = \lambda z(t), \quad \text{for all } t.$$

The eigenvalues of the matrix E are therefore named the *characteristic multipliers* or *numbers* of the system 12.13. The eigenvalues of E are independent of the choice of fundamental matrix so are a property of the system, not of any particular solution. This follows because if the different fundamental matrices $\Phi_1(t)$ and $\Phi_2(t)$ give rise to the matrices E_1 and E_2 then since, from section 12.2, $\Phi_2(t) = \Phi_1(t)C$ for some constant matrix C, we have

$$\Phi_2(t + T) = \Phi_1(t + T)C = \Phi_1(t)E_1C = \Phi_2(t)C^{-1}E_1C \implies E_2 = C^{-1}E_1C.$$

Thus E_1 and E_2 are connected by a similarity transformation and have the same eigenvalues.

It is convenient to write the eigenvalues in the form

$$\lambda_k = e^{T\rho_k},$$

where ρ_k is made unique by choosing its imaginary part to satisfy $-\pi < \Im(T\rho_k) \le \pi$, so we may write

$$\rho_k = \frac{1}{T}\ln(\lambda_k) \quad \text{or} \quad T\rho_k = \ln|\lambda_k| + i\arg(\lambda_k), \quad -\pi < \arg(\lambda_k) \le \pi,$$

$\ln z$ being the principal branch† of the natural logarithm. The ρ_k are named the *characteristic exponents*.

If E has n distinct eigenvalues, λ_k, then equation 12.13 has n linearly independent solutions, which may be written in the form

$$\mathbf{z}_k(t) = \mathbf{p}_k(t)e^{\rho_k t}, \tag{12.15}$$

where $\mathbf{p}_k(t)$ is a T-periodic function of time. This follows because

$$\begin{aligned}
\mathbf{p}_k(t+T) &= \mathbf{z}_k(t+T)e^{-\rho_k(t+T)} \\
&= \mathbf{z}_k(t)e^{-\rho_k t} = \mathbf{p}_k(t), \quad \text{for all } t.
\end{aligned}$$

Equation 12.15 shows that the long-time behaviour of the solution $\mathbf{z}_k(t)$ is determined solely by the magnitude of ρ_k:

- if $|\rho_k| > 1$, $\mathbf{z}_k(t)$ is unbounded as $t \to \infty$;
- if $|\rho_k| < 1$, then $\mathbf{z}_k(t) \to 0$ as $t \to \infty$;
- if $|\rho_k| = 1$, then $\mathbf{z}_k(t)$ oscillates between finite bounds for all t, though it is not necessarily periodic.

An arbitrary solution will be a linear combination of the $\mathbf{z}_k(t)$, so its long-time behaviour will also depend upon the initial conditions. But if *all* $|\rho_k| > 1$ then all solutions are unbounded: if *all* $|\rho_k| < 1$ then all solutions tend to zero as $t \to \infty$, and if *all* $|\rho_k| = 1$ all solutions are bound for all t.

Since the values of ρ_k are determined by the matrix $E = \Phi(t)^{-1}\Phi(t+T)$, which depends only upon n linearly independent solutions on the interval $(t, t+T)$, we see that linear, periodic systems are special because the long-time behaviour of all solutions depends only upon a finite number of solutions on a finite interval.

The matrix E is important and may be computed using the result $\Phi(t+T) = \Phi(t)E$. Numerical integration from t to $t+T$ with the initial conditions $\Phi(t) = I$ gives $E = \Phi(t+T)$. A translation in time is equivalent to forming a new fundamental matrix, giving a different matrix E', related to the original matrix by a similarity transformation, so we normally set the initial value of t to zero and compute the monodromy matrix. One useful check on these numerical computations of E is equation 12.3 (page 439).

---

† The principal branch of the logarithm of a complex number $z = r\exp(i\theta)$, $-\pi < \theta \le \pi$, $r > 0$, is defined to be $\ln z = \ln r + i\theta$. Equivalently, $\ln z = \int_0^z dt\, 1/t$, where the integration path does not pass through the origin and does not cross the negative real axis. The general logarithm function is the many-valued function $\zeta = \ln z + 2\pi i n$, n being an integer, and is the solution of the equation $z = \exp(\zeta)$, and is sometimes denoted by $\mathrm{Ln}(z)$.

This theory is encapsulated and made more precise by the Floquet–Lyapunov theorem.

Floquet–Lyapunov theorem: for the equation $\dot{\mathbf{x}} = A(t)\mathbf{x}$, where the elements of the $n \times n$ matrix $A(t)$ are T-periodic, piecewise continuous functions of t, with a finite number of discontinuities on $(-\infty, \infty)$ and integrable at each discontinuity, a fundamental matrix may be expressed in the form

$$\Phi(t) = P(t)e^{Kt},$$

where $P(t)$ is a T-periodic, $n \times n$ matrix, non-singular for all t and with elements continuous with integrable, piecewise continuous derivatives. Also K is a constant $n \times n$ matrix.

If $A(t)$ is real then $P(t)$ and K are real and

$$P(t + T) = P(t)R,$$

for some real matrix R satisfying $R^2 = I$ and which commutes with K, $KR = RK$. If $R = I$ then $P(t + T) = P(t)$, but otherwise $P(t + 2T) = P(t)$; the proof of this result is outlined in exercise 12.40. Other, important special cases are dealt with in section 12.6.

In a problem requiring numerical integration with particular initial conditions for $0 \le t \le mT$, if $m < n$ it is generally faster to integrate the equations for the whole time, but if $m > n$ it may be more efficient to compute a fundamental matrix by integrating n initial conditions over one period.

Exercise 12.15
Show that the matrizant $\Phi(t)$ satisfies $\Phi(t + T) = \Phi(t)\Phi(T)$.

Exercise 12.16
If E has an eigenvalue λ that is an mth root of unity, $\lambda^m = 1$, m being a positive integer, with $\mathbf{a}$ the associated eigenvector, show that the solution $\mathbf{z}(t) = \Phi(t)\mathbf{a}$ has period mT.

Exercise 12.17
Use equation 12.3 to show that the characteristic multipliers of the system λ_k satisfy the relation

$$\lambda_1 \lambda_2 \cdots \lambda_n = \exp\left(\int_0^T dt\, \mathrm{Tr}(A(t)) \right). \tag{12.16}$$

Exercise 12.18
Write a Maple procedure to compute the monodromy matrix $E(\omega, a)$ for the equation

$$\frac{d^2x}{dt^2} + \frac{dx}{dt}\sin t + (\omega^2 + a\cos t)x = 0,$$

and use this to plot the graph of $\mathrm{Tr}(E)$ for $0 < \omega < 4$ when $a = 1/2$, 1, 2 and 4.

In the case $a = 1$ show numerically that the system is unstable for $0.2845 \le \omega \le 0.7770$, but is otherwise stable, except possibly when ω is close to $n/2$, n being an integer larger than 2.

When $\omega \gg a$ show that an approximate solution to the equation is

$$x(t) = (A \cos \omega t + B \sin \omega t) \exp \left(-\sin^2(t/2)\right),$$

where A and B are constants, and that with this approximation $\text{Tr}(E) = 2 \cos 2\pi\omega$, independent of a.

Show further that when $\omega = 0$ and $a = 1$ the solution is

$$x(t) = x_0 e^{\cos t} + \dot{x}_0 e^{\cos t} \int_0^t ds\, e^{-\cos s} \simeq (x_0 + \dot{x}_0 C t)\, e^{\cos t},$$

where $C = \langle e^{-\cos t} \rangle \simeq 1.27$ is the mean of $e^{-\cos t}$ over one period. Deduce that this solution is unstable if $\dot{x}_0 \neq 0$.

Make numerical comparisons of these approximations with exact solutions.

12.4 Hill's equation

Hill's equation, section 11.5, may be written in the form

$$\frac{d^2 x}{dt^2} + (a + p(t))x = 0, \tag{12.17}$$

where a is a constant and $p(t)$ a T-periodic function. Mathieu's equation, also introduced in chapter 11, is a special case with $p(t) = -2q \cos 2t$ and $T = \pi$. By defining $y = \dot{x}$ we may write Hill's equation in the standard matrix form,

$$\frac{d\mathbf{r}}{dt} = A(t)\mathbf{r}, \quad \mathbf{r} = \begin{pmatrix} x(t) \\ y(t) \end{pmatrix}, \quad A = \begin{pmatrix} 0 & 1 \\ -a - p(t) & 0 \end{pmatrix}. \tag{12.18}$$

Since $\text{Tr}(A) = 0$, equation 12.3 shows that $\det(\Phi) = $ constant and hence $\det(E) = 1$ and the product of the eigenvalues is unity. The eigenvalues of E are therefore given by

$$\lambda^2 - \text{Tr}(E)\lambda + 1 = 0, \quad \Longrightarrow \quad 2\lambda = \text{Tr}(E) \pm \sqrt{\text{Tr}(E)^2 - 4}. \tag{12.19}$$

Hence the long-time behaviour of the solutions is determined mainly by the single real number $\text{Tr}(E)$; this type of system is easier to understand than those for which both $\text{Tr}(E)$ and $\det(E)$ depend upon the system parameters.

Consider the two independent solutions of 12.17 that satisfy the initial conditions

$$\eta_1(0) = 1, \quad \dot{\eta}_1(0) = 0, \qquad \eta_2(0) = 0, \quad \dot{\eta}_2(0) = 1, \tag{12.20}$$

so $\Phi(0) = I$ and

$$\text{Tr}(E) = \eta_1(T) + \dot{\eta}_2(T). \tag{12.21}$$

In the following these initial conditions are assumed.

There are five separate cases according to the value of $\text{Tr}(E)$.

(1) $\text{Tr}(E) > 2$. The eigenvalues are positive, different, not equal to $+1$ and satisfy $0 < \lambda_1 < 1 < \lambda_2$. The characteristic exponents are $\pm\rho$, where

$$T\rho = \ln \lambda_2 > 0,$$

and two linearly independent solutions are

$$\xi_1(t) = e^{-\rho t} p_1(t), \quad \xi_2(t) = e^{\rho t} p_2(t), \tag{12.22}$$

where $p_k(t)$ are T-periodic functions.

(2) $\text{Tr}(E) = 2$. The eigenvalues are identical and equal to $+1$, hence $\rho = 0$. Now the behaviour of the solutions depends upon the number of independent eigenvectors of E:

(2a) E has two linearly independent eigenvectors: there are two T-periodic solutions and as in 12.22,

$$\xi_1(t) = p_1(t), \quad \xi_2(t) = p_2(t),$$

where $p_k(t)$ are T-periodic functions. This case is comparatively rare.

(2b) E has one linearly independent eigenvector (the normal case): the two independent solutions of 12.17 are

$$\xi_1(t) = p_1(t), \quad \xi_2(t) = tp_1(t) + p_2(t), \tag{12.23}$$

where $p_k(t)$ are T-periodic functions. The first solution is bounded; the amplitude of the second solution increases linearly with t, so there are one stable and one unstable solution. An example of this behaviour is shown in figures 12.7 and 12.8 (page 456).

Proofs of these statements are given in the appendix to this chapter.

(3) $|\text{Tr}(E)| < 2$. The eigenvalues of E are complex and may be written in the form $\lambda = e^{\pm i\theta}$, $0 < \theta < \pi$, with the characteristic exponents $T\rho = \pm i \cos^{-1}(\text{Tr}(E)/2)$. Now the two independent solutions are

$$\xi_1(t) = e^{i\rho t} p_1(t), \quad \xi_2(t) = e^{-i\rho t} p_2(t), \tag{12.24}$$

where $p_k(t)$ are T-periodic functions. In this case, all solutions are bounded for all times.

(4) $\text{Tr}(E) = -2$. The eigenvalues are identical and equal to -1. Again the behaviour of the solutions depends upon the number of independent eigenvectors of E:

(4a) E has two linearly independent eigenvectors: there are two $2T$-periodic solutions because $T\rho = i\pi$, and the two independent solutions are

$$\xi_1(t + T) = -\xi_1(t), \quad \xi_2(t + T) = -\xi_2(t).$$

Again this case is comparatively rare.

(4b) E has one linearly independent eigenvector (the normal case): the two independent solutions of 12.17 are

$$\xi_1(t) = p_1(t), \quad \xi_2(t) = tp_1(t) + p_2(t), \tag{12.25}$$

where $p_k(t)$ are two $2T$-periodic functions. The first solution is bounded; the amplitude of the second solution increases linearly with t, so there is one stable and one unstable solution. An example of this behaviour is shown in figures 12.5 and 12.6 (page 455).

(5) $\mathrm{Tr}(E) < -2$. The eigenvalues are negative, different, not equal to -1 and satisfy $\lambda_2 < -1 < \lambda_1 < 0$. The characteristic exponents are

$$T\rho = \pm \ln(-\lambda_2) + i\pi$$

and the two linearly independent solutions are as in 12.22, with $\phi_1(t)$ decreasing and $\phi_2(t)$ increasing as $t \to \infty$.

This characterisation of stability gives the impression that there is a sharp distinction between stable and unstable solutions. This is true, but only for long times, or formally as $t \to \infty$. If one observes a solution for a finite time the distinctions are not so clear, because the solutions are generally well-behaved functions of the system parameters. For instance, if $\mathrm{Tr}(E) = -2$ the amplitude of one solution increases linearly with t: if $\mathrm{Tr}(E) = -2 + \epsilon^2$ the amplitude of this solution will also initially increase linearly with t and may, if ϵ is sufficiently small, reach a relatively large value before decreasing. In general the time at which a clear distinction between unstable and stable solutions becomes apparent increases as ϵ decreases. An example of this behaviour is examined in exercise 12.19.

12.4.1 Mathieu's equation
In order to illustrate the behaviour outlined above we re-visit Mathieu's equation,

$$\frac{d^2x}{dt^2} + (a - 2q \cos 2t)x = 0, \tag{12.26}$$

fix $q = 2$ and investigate how the solutions depend upon a by computing $\mathrm{Tr}(E)$. Sturm–Liouville theory shows that for $0 \le t < 2\pi$ there are π- and 2π-periodic solutions at discrete values of a, which depend upon q: these are ordered according to the convention $a_0(q) < b_1(q) < a_1(q) < \cdots$, $q \ne 0$, with a_k being associated with the even solutions and b_k the odd solutions. Here we consider the solutions for other values of a.

Figure 12.3 shows the graph of $\mathrm{Tr}(E)$ for $-2 < a < 12$, which was computed numerically using dsolve to integrate the equations with the initial conditions 12.20 from $t = 0$ to $t = \pi$, the period of the matrix $A(t)$.

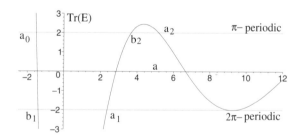

Figure 12.3   A graph of $\mathrm{Tr}(E(a))$ for equation 12.26 with $q = 2$.

The general shape of $\mathrm{Tr}(E)$ shown in this figure is typical of this type of problem: observe the steep gradient of $\mathrm{Tr}(E)$ for $a < 0$. For large a (not shown on figure) the curve is almost tangent to the lines $\mathrm{Tr}(E) = \pm 2$.

The points where $\mathrm{Tr}(E) = 2$ give the eigenvalues $a_0(q)$, $\{b_{2k}(q), a_{2k}(q)\}$, $k = 1, 2, \ldots$, corresponding to the π-periodic Mathieu functions $x = ce_0(t, q)$, $\{se_{2k}(t, q), ce_{2k}(t, q)\}$. The points where $\mathrm{Tr}(E) = -2$ give the eigenvalues $\{b_{2k+1}(q), a_{2k+1}(q)\}$, $k = 0, 1, \ldots$, corresponding to the 2π-periodic Mathieu functions $\{se_{2k+1}(t, q), ce_{2k+1}(t, q)\}$. Notice that for $a \simeq 9.2$ the graph of $\mathrm{Tr}(E)$ appears to be tangential to the line -2: in fact the results quoted in equation 11.84 (page 413) show that the large roots of $\mathrm{Tr}(E) = \pm 2$ are exponentially close. In addition the negative roots of $\mathrm{Tr}(E) = \pm 2$, corresponding to $a_0(q)$ and $b_1(q)$, are also very close, equation 11.85, because here $\mathrm{Tr}(E)$ is changing very rapidly.

The figure also shows that if a is in the interval $b_r(q) \le a \le a_r(q)$ the solutions are unstable; otherwise they are stable. These features are succinctly summarised in the graphs of $a_r(q)$ and $b_r(q)$ shown in the following figure.

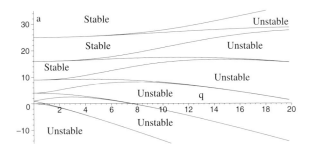

Figure 12.4   Regions of stability in the (q, a)-plane.

If q is large and $a \ll 2q$ the eigenvalues appear in pairs with $a_{r+1} \simeq b_r$ and the difference $a_{r+1} - b_r$ being exponentially small, equation 11.85. This is due to the symmetries in Mathieu's equation and has the consequence that $\mathrm{Tr}(E)$ changes rapidly with a when $a \ll 2q$.

Conversely, if $a \gg 2q$ the difference $a_r(q) - b_r(q)$ is exponentially small, equation 11.84, so the graph of $\mathrm{Tr}(E)$ is almost tangent to the lines ± 2.

In the following two figures are shown the two independent solutions at $a = a_1(2) = 2.379 \ldots$ where $\mathrm{Tr}(E) = -2$. Figure 12.5 is the 2π-periodic solution with the initial conditions $\xi(0) = 1$, $\dot{\xi}(0) = 0$. Figure 12.6 is the solution with the initial conditions $\xi(0) = 0$, $\dot{\xi}(0) = 1$, whose amplitude increases linearly with t.

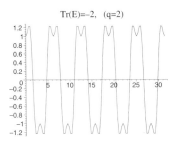

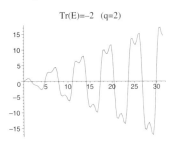

Figure 12.5 2π-periodic solutions for $a = a_1(2) = 2.379 \ldots$, corresponding to $se_1(t, q)$.

Figure 12.6   The non-periodic solution for $a = 2.379 \ldots$.

The next two figures show the two independent solutions at $a = b_2(q) = 3.672\ldots$, where $\mathrm{Tr}(E) = 2$. Figure 12.7 is the π-periodic solution with the initial conditions $\xi(0) = 0$, $\dot{\xi}(0) = 1$. Figure 12.8 is the solution with the initial conditions $\xi(0) = 1$, $\dot{\xi}(0) = 0$, whose amplitude increases linearly with t.

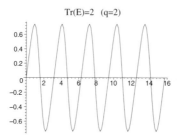

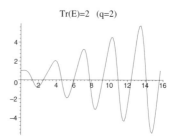

Figure 12.7 π-periodic solutions for $a = 3.672\ldots$, corresponding to $se_2(t, q)$.

Figure 12.8   The non-periodic solution for $a = b_2 = 3.672\ldots$.

Exercise 12.19

When $q = 2$ some eigenvalues of Mathieu's equation are

$$
\begin{aligned}
a_0 &= -1.513\,956\,9, & \mathrm{Tr}(E) &= 2, \\
b_1 &= -1.390\,676\,5, & \mathrm{Tr}(E) &= -2, \\
a_1 &= 2.379\,199\,9, & \mathrm{Tr}(E) &= -2, \\
b_2 &= 3.672\,232\,7, & \mathrm{Tr}(E) &= 2.
\end{aligned}
$$

Write a Maple procedure to plot the graph of the solution of Mathieu's equation with given initial conditions over a given interval $0 < t < T$. Use this procedure to examine the solution with initial conditions $(1, 0)$ in the stable regions $a_0 + \epsilon$, $b_1 - \epsilon$, $a_1 + \epsilon$ and $b_2 - \epsilon$, for $0 < \epsilon \ll 1$, to observe that for stable solutions the amplitude can be large and that for small times it increases linearly with t.

Note: for large T accurate graphs need `numpoints` and `maxfun`, an optional argument of `dsolve` to be large; see for instance the help files found by typing `?classical`.

Exercise 12.20

Write Maple procedures to find the values of a in the interval $0 < a < 16$ for which there are 2π- and 4π-periodic solutions of the equation

$$
\frac{d^2 x}{dt^2} + (a - 4\cos 2t - 2\cos t)\, x = 0.
$$

Note that accurate roots of the equation $\mathrm{Tr}(E) = \pm 2$ may be found using the false position method described in chapter 3.

Show, numerically, that if $a = 2.049\,46\ldots$ there is a 2π-periodic solution, and plot the graph of this and the associated unbounded solution. Show also that if $a = 2.182\,37\ldots$ there is a 4π-periodic solution, and plot the graph of this and the associated unbounded solution.

12.4.2 *The damped Mathieu equation*

Here we briefly consider the effects of adding a small, linear damping term to Mathieu's equation, which now becomes

$$\frac{d^2x}{dt^2} + v\frac{dx}{dt} + (a - 2q\cos 2t)x = 0, \quad v \geq 0. \tag{12.27}$$

In this discussion we shall suppose that v is small. When $v = 0$ instabilities occur when the parametric term, $2qx\cos 2t$, drives the linear oscillator, $\ddot{x} + ax$, appropriately, so the energy of the oscillator increases without bound: for small q we see from figure 12.4 that the regions of instability emanate from the points $a = n^2$ on the a-axis. The damping term, $v\dot{x}$, removes energy from the system so is in competition with the resonant driving term. Thus we expect the presence of damping to decrease the area in the (q, a)-plane associated with unstable motion. Further, if $x(t)$ is even $\dot{x}$ is odd, and vice versa, so the solutions of the damped equation cannot have a definite parity as do those of the undamped equation.

The damped Mathieu equation has the canonical form

$$\frac{d}{dt}\begin{pmatrix} x \\ y \end{pmatrix} = A(t)\begin{pmatrix} x \\ y \end{pmatrix}, \quad y = \frac{dx}{dt}, \quad A(t) = \begin{pmatrix} 0 & 1 \\ 2q\cos 2t - a & -v \end{pmatrix}. \tag{12.28}$$

Thus, using equation 12.3 (page 439), we see that the determinant of the monodromy matrix, E, is (since $T = \pi$)

$$\det(E) = e^{-v\pi}.$$

Setting $\mathrm{Tr}(E) = 2\theta$, the eigenvalues of E, that is the characteristic multipliers, are the solutions of

$$\lambda^2 - 2\theta\lambda + e^{-v\pi} = 0 \quad \Longrightarrow \quad \lambda_{\pm} = \theta \pm \sqrt{\theta^2 - e^{-v\pi}}.$$

The condition for stability is $|\lambda_{\pm}| \leq 1$, and since $\lambda_{+}\lambda_{-} = e^{-v\pi} < 1$ we see that it is now possible for both multipliers to be real *and* for the solutions to be stable.

Stability boundaries

If $\theta > 0$ then $0 < \lambda_{-} < \lambda_{+}$, so the stability boundary is given by the condition $\lambda_{+} = 1$, that is,

$$\theta = \frac{1}{2}\left(1 + e^{-v\pi}\right).$$

If $\theta < 0$ then $\lambda_{-} < \lambda_{+} < 0$, so the stability boundary is given by the condition $\lambda_{-} = -1$, that is,

$$\theta = -\frac{1}{2}\left(1 + e^{-v\pi}\right).$$

It follows that the condition for stability is

$$|\mathrm{Tr}(E)| < 1 + e^{-v\pi}. \tag{12.29}$$

If $\mathrm{Tr}(E) = 1 + e^{-v\pi}$ then $\lambda_+ = 1$ and $\lambda_- = e^{-v\pi}$ and the general theory shows that two independent solutions of 12.27 are

$$p_+(t) \quad \text{and} \quad p_-(t)e^{-vt},$$

where $p_\pm(t)$ are π-periodic functions of t; examples of such solutions are shown in figures 12.10 and 12.11.

If $\mathrm{Tr}(E) = -(1 + e^{-v\pi})$ then $\lambda_- = -1$ and $\lambda_+ = -e^{-v\pi}$ and two independent solutions of 12.27 are

$$q_-(t) \quad \text{and} \quad q_+(t)e^{-vt},$$

where $q_\pm(t)$ are 2π-periodic functions of t, figures 12.12 and 12.13.

A numerical example

The effect of damping is seen by plotting the graph of $\mathrm{Tr}(E)$ as a function of a, for fixed q and v: for $q = 1$ and $v = 0.2$ such a graph is shown below.

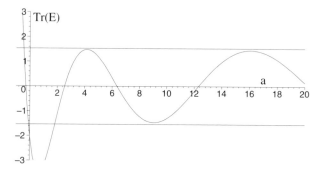

Figure 12.9   Graph of $\mathrm{Tr}(E)$ for $q = 1$ and $v = 0.2$.

In this example the boundary between the stable and unstable region is $|\mathrm{Tr}(E)| = 1.533$, and closer examination of the region near $\max(|\mathrm{Tr}(E)|)$ shows that there are only three intersections: for later use we give these roots:

$$a = -0.451\,679, \quad -0.088\,303\,7, \quad 1.837\,843.$$

Thus, for $q = 1$ and $v = 0.2$ the only unstable regions are:

$$a < -0.452 \quad \text{and} \quad -0.0883 < a < 1.838.$$

Knowledge of the values of a at which $\mathrm{Tr}(E) = 1 + e^{-v\pi}$ enables the two solutions $p_+(t)$ and $e^{-vt}p_-(t)$ to be computed using the eigenvectors of E to define the initial conditions. For the case shown in figure 12.9, at $a = -0.4517$ where $\mathrm{Tr}(E) = 1 + e^{-v\pi}$, we have

$$E = \begin{pmatrix} 0.6601 & 0.0404 \\ 1.067 & 0.8734 \end{pmatrix}, \quad \mathbf{e}_+ = \begin{pmatrix} 0.1178 \\ 0.9930 \end{pmatrix} \quad \mathbf{e}_- = \begin{pmatrix} -0.3037 \\ 0.9528 \end{pmatrix},$$

with $\lambda_+ = 1$ and $\lambda_- = 0.5335$. Hence the initial conditions $(x(0), \dot{x}(0))^\top = \mathbf{e}_\pm$ give, respectively, a π-periodic function $p_+(t)$ and an exponentially decaying function $e^{-vt}p_-(t)$.
Graphs of these two functions are shown below.

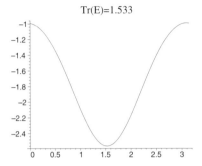

Tr(E)=1.533

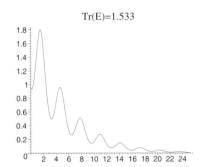

Tr(E)=1.533

Figure 12.10   Graph of a π-periodic solution
$p_+(t)$ for $a = -0.4517$.

Figure 12.11   Graph of a decaying solution
$e^{-vt}p_-(t)$ for $a = -0.4517$.

The perturbation approximation to $p_+(t)$ with the appropriate initial condition is easily
derived from the solution found in exercise 12.24, and gives

$$p_+(t) \simeq -1.695 + 0.7462\cos(2t) - 0.05296\cos(4t) - 0.08474\sin(2t).$$

At the boundary $\mathrm{Tr}(E) = -(1 + e^{v\pi})$, one solution is $a = 1.8378$ and here

$$E = \begin{pmatrix} -0.6568 & -0.0385 \\ -1.0992 & -0.8767 \end{pmatrix}, \quad \mathbf{e}_- = \begin{pmatrix} 0.1115 \\ 0.9938 \end{pmatrix}, \quad \mathbf{e}_+ = \begin{pmatrix} 0.2980 \\ -0.9546 \end{pmatrix},$$

with $\lambda_- = -1$ and $\lambda_+ = -0.5335$. Graphs of the two solutions are shown in the
following figures.

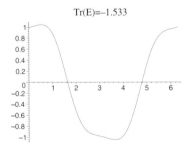

Tr(E)=−1.533

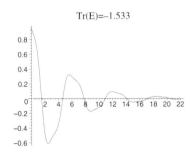

Tr(E)=−1.533

Figure 12.12   Graph of a 2π-periodic
solution $q_-(t)$ for $a = 1.8378$.

Figure 12.13   Graph of a decaying
solution $e^{-vt}q_+(t)$ for $a = 1.8378$.

Exercise 12.21
Write a Maple procedure to compute the monodromy matrix for equation 12.28 given
values of (v, a, q). Use this to draw the graphs of $\mathrm{Tr}(E)$ for $-2 < a < 10$, $q = 2$ and various
values of v in the range $0 < v \le 0.4$.

Exercise 12.22
For $q = 1$, $v = 0.2$ and various a near -0.4517 find the monodromy matrix and use this to determine the initial conditions of the solution associated with the eigenvalue $\lambda_+ \simeq 1$. Use these initial conditions to draw graphs of this solution for a variety of a values.
        Repeat for $a \simeq 1.8378$ and $\lambda_- \simeq -1$.

Perturbation theory
The perturbation expansion of periodic solutions in the presence of damping is more complicated than when damping is absent, except if $a \simeq 0$ or $a \simeq 1$.
        When $a \simeq n \geq 2$ and $v = 0$, $a_n(q) - b_n(q) \sim q^n$ and the unstable region near $a = n$ has a width $O(q^n)$. Since damping can only decrease this width the term $v\dot{x}$ affects the perturbation expansion only at the nth order, or beyond. This is consistent with the perturbation expansion for $v = 0$, where we saw that $a_n(q)$ and $b_n(q)$ differ only at the nth order, exercise 11.45.
        Consequently the analysis needed for the perturbation solution is algebraically messy, so in this section we describe only the bare outline of the method, for particular cases, leaving the reader to complete the calculations, which are most easily done using Maple. We assume that both v and q are small and initially set $v = \mu q$ in order to derive a perturbation expansion in q. The equation is then most conveniently written in the form

$$\frac{d^2x}{dt^2} + a(q)x = q\left(2x\cos 2t - \mu\frac{dx}{dt}\right), \quad v = \mu q.$$

Now write both $x(t)$ and $a(q)$ as power series in q,

$$
\begin{aligned}
x(t) &= x_0(t) + x_1(t)q + x_2(t)q^2 + \cdots, \\
a(q) &= a_0 + a_1 q + a_2 q^2 + \cdots,
\end{aligned}
$$

to give the infinite set of equations:

$$\frac{d^2x_0}{dt^2} + a_0 x_0 = 0,$$

$$\frac{d^2x_1}{dt^2} + a_0 x_1 = -a_1 x_0 + 2x_0 \cos 2t - \mu\frac{dx_0}{dt},$$

$$\frac{d^2x_2}{dt^2} + a_0 x_2 = -a_2 x_0 - a_1 x_1 + 2x_1 \cos 2t - \mu\frac{dx_1}{dt},$$

$$\vdots \qquad\qquad\qquad \vdots$$

$$\frac{d^2x_n}{dt^2} + a_0 x_n = -\sum_{k=1}^{n} a_k x_{n-k} + 2x_{n-1}\cos 2t - \mu\frac{dx_{n-1}}{dt}.$$

This scheme can be used if $a_0 = 0$ or 1 but, for the reasons discussed below, fails if $a_0 \geq 2$. Here consider the case $n = 1$, and hence find an approximation for $q_-(t)$, the 2π-periodic solution. Then the solution to the equation for $x_0(t)$ is

$$x_0(t) = A_0\cos t + B_0\sin t,$$

where (A_0, B_0) are constants to be determined. Substituting this into the right hand side of the equation for x_1 gives, after some simplification,

$$\frac{d^2 x_1}{dt^2} + x_1 = (A_0(1 - a_1) - \mu B_0) \cos t + (B_0(1 + a_1) - \mu A_0) \sin t$$
$$+ A_0 \cos 3t + B_0 \sin 3t. \tag{12.30}$$

This equation has periodic solutions only if the coefficients of both $\cos t$ and $\sin t$ are zero,

$$A_0(1 - a_1) - \mu B_0 = 0, \quad B_0(1 + a_1) - \mu A_0 = 0. \tag{12.31}$$

In the undamped limit, $\mu = 0$, these equations have two solutions:

(i) $\quad a_1 = -1, \quad A_0 = 0 \quad \Longrightarrow \quad x = B_0 \sin t \quad$ leading to $\quad se_1(q, t)$,

(ii) $\quad a_1 = 1, \quad B_0 = 0 \quad \Longrightarrow \quad x = A_0 \cos t \quad$ leading to $\quad ce_1(q, t)$,

where A_0 and B_0 are chosen to satisfy appropriate normalisation conditions.

If $\mu > 0$ these homogeneous equations have a unique solution only if the determinant of the coefficients is zero, that is if

$$a_1^2 = 1 - \mu^2, \tag{12.32}$$

which leads to the two solutions,

(i) $\quad a_1 = -\sqrt{1 - \mu^2}, \quad A_0 = \mu B_0 / (1 + \sqrt{1 - \mu^2})$,

(ii) $\quad a_1 = \sqrt{1 - \mu^2}, \quad B_0 = \mu A_0 / (1 + \sqrt{1 - \mu^2})$.

Since $a_0 = 1$ equation 12.32 gives, in terms of the original variable $v = \mu q$,

$$a = 1 \pm \sqrt{q^2 - v^2} \quad \text{or} \quad q^2 - (a - 1)^2 = v^2, \tag{12.33}$$

which is the equation of a hyperbola in the (q, a)-plane; this is shown in the following figure, which also shows the region in which solutions are unstable. The asymptotes of this hyperbola, $a = 1 \pm q$, are the boundaries between the stable and unstable regions in the undamped limit, $v = 0$, shown in figure 12.4,

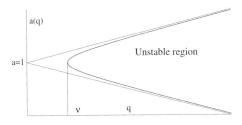

Figure 12.14   One branch of the hyperbola
$(a - 1)^2 = q^2 - v^2$.

This analysis shows that if $0 \le q < v$ and $a \simeq 1$ all solutions are stable; thus for small q, solutions that are unstable when $v = 0$ become stable when $v > 0$.

Now return to the derivation of the perturbation expansion. Unique solutions are obtained by fixing the coefficient of either $\cos t$ or $\sin t$. The solution that reduces to $se_1(t, q)$ when $v = 0$ is obtained by setting $a_1 = -\sqrt{1 - \mu^2}$, $B_0 = 1$ *and* by ensuring that there is no $\sin t$ term in $x_k(t)$, $k \geq 1$. Similarly, the solution that reduces to $ce_1(t, q)$ when $v = 0$ is obtained by setting $a_1 = \sqrt{1 - \mu^2}$, $A_0 = 1$ *and* by ensuring that there is no $\cos t$ term in $x_k(t)$, $k \geq 1$.

In the following example we find the second solution by setting $A_0 = 1$ and $a_1 = \sqrt{1 - \mu^2}$, though the latter substitution is best done at the end of the calculation. Equation 12.30 for $x_1(t)$ now becomes

$$\frac{d^2 x_1}{dt^2} + x_1 = \cos 3t + B_0 \sin 3t, \quad B_0 = \frac{\mu}{1 + \sqrt{1 - \mu^2}},$$

having the solution

$$x_1(t) = -\frac{1}{8} \cos 3t - \frac{B_0}{8} \sin 3t + B_1 \sin t$$

for some constant B_1. Substituting this into the equation for $x_2(t)$ and setting the coefficients of $\cos t$ and $\sin t$ to zero gives equations for a_2 and B_1, the solutions of which are

$$B_1 = 0, \quad a_2 = -\frac{1}{8}.$$

Now the equation for $x_2(t)$ becomes

$$\frac{d^2 x_2}{dt^2} + x_2 = \frac{1}{8}(a_1 + 3\mu B_0)\cos 3t + \frac{1}{8}(a_1 B_0 - 3\mu)\sin 3t - \frac{1}{8}(\cos 5t - B_0 \sin 5t).$$

The relevant solution of this equation is

$$x_2(t) = B_2 \sin t - \frac{1}{64}(a_1 + 3\mu B_0)\cos 3t - \frac{1}{64}(a_1 B_0 - 3\mu)\sin 3t$$
$$+ \frac{1}{192}(\cos 5t + B_0 \sin 5t).$$

Substituting this into the equation for $x_3(t)$ and collecting the coefficients of $\cos t$ and $\sin t$ gives equations for B_2 and a_3 having the solutions:

$$B_2 = \frac{3\mu}{64\sqrt{1 - \mu^2}(1 + \sqrt{1 - \mu^2})}, \quad a_3 = -\frac{1 + 2\mu^2}{64\sqrt{1 - \mu^2}}.$$

Hence, to $O(q^3)$ we have

$$a(q) = 1 + \sqrt{1 - \mu^2}\, q - \frac{q^2}{8} - \frac{1 + 2\mu^2}{64\sqrt{1 - \mu^2}} q^3 + O(q^4), \tag{12.34}$$

and to $O(q)$,

$$x(t) = \cos t + \frac{\mu \sin t}{1 + \sqrt{1 - \mu^2}} - \frac{q}{8}\left(\cos 3t + \frac{\mu \sin 3t}{1 + \sqrt{1 - \mu^2}}\right) + O(q^2). \tag{12.35}$$

These series reduce to the series for $a_1(q)$ and $ce_1(t, q)$ when $\mu = 0$. For the values illustrated in figure 12.12, $q = 1$, $\mu = v = 0.2$, the first series gives $a = 1.8376$, compared with the exact value of 1.8378, and the perturbation solution is, to $O(q^2)$, and with $x(0) = 1$,

$$x(t) = 1.1667 \cos t + 0.1238 \sin t - 0.0059 \sin 3t - 0.1731 \cos 3t + 0.0064 \cos 5t.$$

Graphs of the exact and this approximate solutions are indistinguishable.

Exercise 12.23
Show that the other solution for $a_0 = 1$, that is,

$$B_0 = 1, \quad a_1 = -\sqrt{1 - \mu^2}, \quad A_0 = \frac{\mu}{1 + \sqrt{1 - \mu^2}},$$

is

$$a = 1 - \sqrt{1 - \mu^2}\, q - \frac{q^2}{8} + \frac{q^3}{64} \frac{1 + 2\mu^2}{\sqrt{1 - \mu^2}} + O(q^4)$$

$$x(t) = \sin t - \frac{\mu \cos t}{1 + \sqrt{1 - \mu^2}} - \frac{q}{8}\left(\sin 3t + \frac{\mu \cos 3t}{1 + \sqrt{1 - \mu^2}}\right) + O(q^2).$$

For the case $q = 1$, $v = 0.2$, show that the first series gives $a \simeq -0.087\,57$ and inclusion of the second-order term in the series for $x(t)$ gives

$$x(t) = \sin t + 0.1059 \cos t - 0.1087 \sin 3t - 0.0205 \cos 3t + 0.0052 \sin 5t.$$

Using this value of a and this approximate solution to define the initial conditions, make a graphical comparison with the exact solution.

Exercise 12.24
If $a_0 = 0$ show that

$$a_0(q, \mu) = -\frac{q^2}{2} + \frac{q^4}{8}\left(\frac{7}{16} + \mu^2\right) - \frac{q^6}{32}\left(\frac{29}{72} + \frac{11}{64}\mu^2 + \mu^4\right)$$

$$+ \frac{q^8}{128}\left(\frac{68\,687}{147\,456} + \frac{7685}{2592}v^2 + \frac{1023}{256}v^4 + \mu^6\right) + O(q^{10}),$$

where $v = \mu q$, and that the associated solution is

$$x(t) = 1 - \frac{q}{2}\cos 2t + \frac{q^2}{32}(\cos 4t + 8\mu \sin 2t)$$

$$+ \frac{q^3}{8}\left(\left(\frac{7}{16} + \mu^2\right)\cos 2t - \frac{1}{144}\cos 6t - \frac{3\mu}{16}\sin 4t\right) + O(q^4).$$

Show also that at $q = 1$, $v = 0.2$ the 14th-order expansion gives $a_0 = -0.451\,75$: note that the exact value of v is $-0.451\,68\cdots$.

Exercise 12.25
Apply the above perturbation scheme with the starting value $a_0 = 4$ to show that forcing $x_1(t)$ to be periodic gives the equation $a_1^2 + 4\mu^2 = 0$. Deduce that this perturbation scheme is invalid. Show, more generally, that a similar result is obtained if $a_0 = n^2$, $n \geq 2$.

This last exercise shows that a naive application of perturbation theory fails when $n \geq 2$. The reason for this is discussed in the introduction to this section and the remedy is to set $\mu = vq^n$ when $a_0 = n^2$, for then the damping term does not appear until the equation for $x_n(t)$ and the first $n-1$ equations of the perturbation scheme are the same as for the undamped case, although the required solution is *not* the same to this order.

Consider, for example, the case $n = 2$, and look for a solution that reduces to $ce_2(t, q)$ as $\mu \to 0$; we start with the general solution for $x_0(t)$:

$$a_0 = 4, \quad x_0(t) = A_0 \cos 2t + B_0 \sin 2t,$$

where A_0 and B_0 are constants to be determined. We subsequently seek solutions for $x_k(t)$, $k \geq 1$, that do not contain a $\cos 2t$ term.

On substituting $x_0(t)$ into the equation for $x_1(t)$ we find that it has a periodic solution only if $a_1 = 0$ and then

$$x_1(t) = \frac{A_0}{4} - \frac{A_0}{12} \cos 4t + B_1 \sin 2t - \frac{B_0}{12} \sin 4t.$$

Substituting this expression into the equation for $x_2(t)$ and setting the coefficients of $\cos 2t$ and $\sin 2t$ to zero gives the following equations:

$$A_0 \left(a_2 - \frac{5}{12} \right) + 2\mu B_0 = 0, \qquad \left(a_2 + \frac{1}{12} \right) B_0 - 2\mu A_0 = 0, \qquad (12.36)$$

for a_2 and B_0. If $\mu = 0$ the only non-trivial solutions are $a_2 = 5/12$, $A_0 = 1$, $B_0 = 0$ and $a_2 = -1/12$, $A_0 = 0$, $B_0 = 1$, leading to $ce_2(t, q)$ and $se_2(t, q)$ respectively. If $\mu > 0$ we may eliminate A_0 and B_0 to give an equation relating a_2 to μ: remembering that $a = 4 + a_2 q^2$ and $v = \mu q^2$, this gives the following equations for $a(q)$:

$$4v^2 + \left(a - 4 - \frac{5q^2}{12} \right) \left(a - 4 + \frac{q^2}{12} \right) = 0, \qquad (12.37)$$

which is the equivalent of equation 12.33. When $v = 0$ this reduces to the two lines

$$a = 4 - \frac{q^2}{12}, \quad a = 4 + \frac{5q^2}{12}, \qquad (12.38)$$

corresponding to $b_2(q)$ and $a_2(q)$ respectively. When $v > 0$ the equation gives the curve drawn in the following figure, which also shows the asymptotes defined in 12.38. In this figure $v = 0.03$.

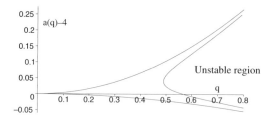

Figure 12.15   Boundary between stable and unstable solutions for $v = 0.03$ and $a(q) \simeq 4$.

The curve defined by equation 12.37 is closest to the a-axis when $dq/da = 0$, that is when $q = \sqrt{8v}$: for small v this is further from the a-axis than the equivalent point when $a_0 = 1$, figure 12.14.

Equations 12.36 have the two solutions

$$B_0 = \frac{2\mu A_0}{a_2 + \frac{1}{12}}, \qquad a_2 = \frac{1}{6} + \frac{1}{4}\sqrt{1 - 64\mu^2}, \quad (ce_2(t, q) \quad \text{when} \quad \mu = 0),$$

$$A_0 = -\frac{2\mu B_0}{a_2 - \frac{5}{12}}, \qquad a_2 = \frac{1}{6} - \frac{1}{4}\sqrt{1 - 64\mu^2}, \quad (se_2(t, q) \quad \text{when} \quad \mu = 0),$$

which are real if $0 \leq \mu \leq \frac{1}{8}$. In the following we chose the first of these and set $A_0 = 1$.

Proceeding with the perturbation expansion, the solution for $x_2(t)$ is

$$x_2(t) = \frac{1}{384}(\cos 6t + B_0 \sin 6t) + B_2 \sin 2t - \frac{B_1}{12}\sin 4t,$$

where B_1 and B_2 are unknown constants.

Substituting $x_2(t)$ into the equation for $x_3(t)$ and again setting the coefficients of $\cos 2t$ and $\sin 2t$ to zero gives

$$B_1 = a_3 = 0.$$

A periodic solution for $x_3(t)$ may then be found which, when substituted into the equation for $x_4(t)$, gives as before the two linear equations

$$a_4 + 2\mu B_2 = -\frac{\mu B_0}{36} - \frac{19a_2}{144}$$

$$a_4 B_0 + \left(a_2 + \frac{1}{12}\right)B_2 = \frac{\mu}{36} - \frac{B_0}{4608} - \frac{B_0 a_2}{144}.$$

These may be solved and the proceedure continued. The series obtained for $a(q)$ is

$$a = 4 + \frac{q^2}{12}\left(2 + 3\sqrt{1 - 64\mu^2}\right) - q^4\left(\frac{763}{13\,824} + \frac{128}{9}\mu^4 + \cdots\right) + O(q^6).$$

Here the coefficient of q^4 has been expanded in powers of μ, in order to simplify the presentation and to show that when $\mu = 0$, $a_2(q)$ is regained. The first three terms of this solution are

$$x(t) = \cos 2t + \frac{8\mu}{C}\sin 2t + q\left(\frac{1}{4} - \frac{\cos 4t}{12} - \frac{2\mu}{3C}\sin 2t\right)$$

$$+ q^2\left(\frac{\cos 6t}{384} + \frac{\mu}{48C}\sin 6t + \frac{(17C - 576\mu^2)\sin 2t}{36\left[(1 - 32\mu^2)C - 32\mu^2\right]}\right) + \cdots,$$

where $C = 1 + \sqrt{1 - 64\mu^2}$.

Exercise 12.26
Use perturbation theory to find the solution of the damped Mathieu equation that reduces to $se_2(t, q)$ at $\mu = 0$. For this solution use the choice

$$B_0 = 1, \quad A_0 = -\frac{2\mu}{a^2 - \frac{5}{12}}, \quad a_2 = \frac{1}{6} - \frac{1}{4}\sqrt{1 - 64\mu^2}$$

to obtain

$$a(q) = 4 + \frac{q^2}{12}\left(2 - 3\sqrt{1 - 64\mu^2}\right) + q^4\left(\frac{5}{13\,824} + \frac{128}{9}\mu^4 + \frac{8192}{9}\mu^6 + \cdots\right) + O(q^6),$$

with the solution

$$
\begin{aligned}
x(t) &= \sin 2t + \frac{8\mu}{C}\cos 2t + q\left(\frac{2\mu}{C} - \frac{\sin 4t}{12} - \frac{2\mu}{3C}\cos 4t\right) \\
&\quad + q^2\left(\frac{(C - 576\mu^2)\cos 2t}{36(32\mu^2 + (32\mu^2 - 1)C)} + \frac{\sin 6t}{384} + \frac{\mu\cos 6t}{48C}\right) + \cdots.
\end{aligned}
$$

Exercise 12.27

(i) Show that when $a \simeq 9$ for the perturbation expansion to be of use it is necessary to write $v = \mu q^3$, where v is the coefficient of the damping term, and with this choice the expansion coefficients for a are $a_1 = 0$ and $a_2 = \frac{1}{16}$.

(ii) Show also that the third-order equations of the expansion lead to the equation

$$9v^2 + \left(a - 9 - \frac{q^2}{16} - \frac{q^3}{64}\right)\left(a - 9 - \frac{q^2}{16} + \frac{q^3}{64}\right) = 0$$

for the boundary between stable and unstable regions. Use this expression to show that in the (q, a)-plane the curve defined by this equation is parallel to the a-axis at $q = (192v)^{1/3}$.

(iii) Further, show that at the third-order term in the expansion the solutions for (A_0, B_0) are

$$A_0 = 1, \qquad B_0 = \frac{3\mu}{a_3 + \frac{1}{64}}, \qquad a_3 = \frac{1}{64}\sqrt{1 - (192\mu)^2},$$

$$B_0 = 1, \qquad A_0 = -\frac{3\mu}{a_3 - \frac{1}{64}}, \qquad a_3 = -\frac{1}{64}\sqrt{1 - (192\mu)^2},$$

and that the first choice leads to

$$a = 9 + \frac{q^2}{16} + \frac{Cq^3}{64} + \frac{13q^3}{20\,480} - \frac{q^5(3 + 2C^2)}{16\,398\,C} + O(q^6),$$

where $C = \sqrt{1 - (192\mu)^2}$ and

$$x(t) = \cos 3t + \frac{192\mu\sin 3t}{1 + C} + q\left(\frac{1}{8}\cos t + \frac{24\mu\sin t}{1 + C} - \frac{\cos 5t}{16} - \frac{12\mu\sin 5t}{1 + C}\right) + O(q^2).$$

12.5 The Liénard equation

One important use of Floquet theory is to determine whether or not periodic solutions of nonlinear equations are stable. We illustrate this application with the relatively simple example of Liénard's equation,

$$\frac{d^2x}{dt^2} + f(x)\frac{dx}{dt} + g(x) = 0. \tag{12.39}$$

This equation possesses isolated periodic solutions in a variety of circumstances, some of which are discussed in chapter 17. Here we assume that a periodic orbit exists.

Let $p(t)$ be a T-periodic solution and define a new variable y by $x = p(t) + y$. Substituting this into the Liénard equation and expanding to first order in y gives

$$\frac{d^2 y}{dt^2} + f(p)\frac{dy}{dt} + \left[f'(p)\frac{dp}{dt} + g'(p) \right] y = 0 \tag{12.40}$$

or, in matrix form,

$$\frac{d\mathbf{z}}{dt} = \left(\begin{array}{cc} 0 & 1 \\ -\dot{p}f'(p) - g'(p) & -f(p) \end{array} \right) \mathbf{z}, \quad \mathbf{z} = \left(\begin{array}{c} y \\ \dot{y} \end{array} \right). \tag{12.41}$$

The original equation is autonomous, so if τ is any constant $p(t+\tau)$ is also a solution. Setting $x = p(t+\tau) = p(t) + \tau\dot{p}(t) + O(\tau^2)$ it follows that $\dot{p}(t)$ is one solution of 12.40 and hence that one of the multipliers is unity. The fact that $\dot{p}(t)$ is a solution may be verified by differentiating the original equation. On setting $\lambda_1 = 1$ and using equation 12.16 (page 451), we obtain

$$\lambda_2 = \exp\left(-\int_0^T dt f(p) \right) = \exp(-T \langle f(p) \rangle) > 0.$$

Thus the nature of the orbit depends on the mean of $f(p)$: it is stable if $\lambda_2 < 1$, that is $\langle f(p) \rangle > 0$, and unstable if $\langle f(p) \rangle < 0$; no conclusion can be drawn if $\langle f(p) \rangle = 0$. It is interesting to note that whilst the existence of the periodic orbit depends upon both $f(x)$ and $g(x)$ its stability depends directly only upon $f(x)$.

Exercise 12.28
Consider the real, autonomous n-dimensional system

$$\frac{d\mathbf{x}}{dt} = \mathbf{X}(\mathbf{x}), \quad \mathbf{x} = (x_1, x_2, \ldots, x_n).$$

If $\mathbf{p}(t)$ is a periodic solution of this system, by defining $\mathbf{x} = \mathbf{p}(t) + \mathbf{y}$ and assuming $|\mathbf{y}|$ to be small, show that $\mathbf{y}$ satisfies the linear equations

$$\frac{d\mathbf{y}}{dt} = A(t)\mathbf{y},$$

where $A(t)$ is a real, $n \times n$ matrix with the periodic coefficients $A_{ij}(t) = \partial X_i/\partial x_j$ evaluated at $\mathbf{x} = \mathbf{p}(t)$. Show that one solution of this linear equation is $\dot{\mathbf{p}}$.

Exercise 12.29
Use the fact that $\dot{p}$ is a solution of 12.40 and the substitution $y = z\dot{p}$ to show that the other solution is

$$y(t) = A\dot{p} \int dt \, \frac{1}{\dot{p}(t)^2} \exp\left(-\int dt \, f(p(t)) \right),$$

where A is a constant.

12.6 Hamiltonian systems

Hamiltonian systems are sufficiently important to warrant special attention. They describe a wide variety of physical phenomena, but here we are concerned only with the special structure of the equations of motion and the consequences that follow.

No prior knowledge of Hamiltonian systems is assumed; our analysis starts with equation 12.45 below, the preamble being a brief introduction to this equation.

Hamiltonian systems are special because

(i) the coordinates describing the state of the system occur in pairs, frequently denoted by (q_k, p_k), $k = 1, \ldots, N$, named conjugate variables, and

(ii) the equations of motion are determined by a *Hamiltonian function*, $H(\mathbf{q}, \mathbf{p}, t)$, via *Hamilton's equations*:

$$\frac{dq_k}{dt} = \frac{\partial H}{\partial p_k}, \quad \frac{dp_k}{dt} = -\frac{\partial H}{\partial q_k}, \quad k = 1, 2, \ldots, N.$$

The special form of these equations of motion constrains the behaviour of the system in many ways, one of which is explored in this section.

If there is a T-periodic solution of Hamilton's equations, $(\mathbf{Q}(t), \mathbf{P}(t))$, where $Q_k(t+T) = Q_k(t)$, $P_k(t+T) = P_k(t)$, for all t and $k = 1, \ldots, N$, its stability is examined by changing the origin to a point moving on the periodic orbit

$$q_k = Q_k(t) + x_k, \quad p_k = P_k(t) + y_k, \quad k = 1, 2, \ldots, N.$$

It may be shown that the Hamiltonian describing the motion in the moving coordinate system is

$$K(\mathbf{x}, \mathbf{y}, t) = H(\mathbf{Q} + \mathbf{x}, \mathbf{P} + \mathbf{y}, t) + \mathbf{x} \cdot \dot{\mathbf{P}} - \mathbf{y} \cdot \dot{\mathbf{Q}},$$

and that the equations of motion are

$$\frac{dx_k}{dt} = \frac{\partial K}{\partial y_k}, \quad \frac{dx_k}{dt} = -\frac{\partial K}{\partial x_k}, \quad k = 1, 2, \ldots, N. \tag{12.42}$$

If $H(\mathbf{q}, \mathbf{p})$ does not depend explicitly upon t then $K(\mathbf{x}, \mathbf{y}, t)$ is T-periodic in t. Further, the expansion of $K(\mathbf{x}, \mathbf{y}, t)$ about $\mathbf{x} = 0$ and $\mathbf{y} = 0$ involves no terms linear in $\mathbf{x}$ or $\mathbf{y}$ and the first non-trivial term is a quadratic form in $\mathbf{x}$ and $\mathbf{y}$; thus, to this order, the equations of motion are linear. At this point it is convenient to change notation and define $Q_{i+n} = P_i$ and $x_{i+n} = y_i$, $i = 1, 2, \ldots, n$. Expanding the Hamiltonian $K(\mathbf{x}, \mathbf{y}, t)$ about the periodic orbit then gives

$$K = \frac{1}{2} \sum_{i=1}^{2N} \sum_{j=1}^{2N} B_{ij}(t) x_i x_j, \quad B_{ij}(t) = \frac{\partial^2 H}{\partial Q_i Q_j} = B_{ji}(t), \tag{12.43}$$

where the coefficients B_{ij} are independent of $\mathbf{x}$, and T-periodic functions of t: $B_{ij}(t+T) = B_{ij}(t)$ for all t. In the remainder of this section $B(t)$ is the matrix with elements $B_{ij}(t)$.

The subsequent equations of motion may be cast in a more convenient form by introducing the skew-symmetric, real $2N \times 2N$ matrix

$$J_{2N} = \begin{pmatrix} 0 & -I_N \\ I_N & 0 \end{pmatrix}, \tag{12.44}$$

where I_N is the $N \times N$ identity matrix. The linear equations of motion for the variation can now be written in the standard matrix form

$$\frac{d\mathbf{x}}{dt} = -J_{2N} B(t) \mathbf{x}. \tag{12.45}$$

Exercise 12.30

Show that $J_{2N}^2 = -I_{2N}$ and hence that $J_{2N}^{\mathsf{T}} = -J_{2N} = J_{2N}^{-1}$. Show also that $\text{Tr}(J_{2N} B) = 0$.
Hint: it is easiest to do this by noting that J_{2N} is a partitioned matrix. An $n \times m$ matrix A is partitioned when put in the form

$$A = \left(\begin{array}{cc|cc}
\begin{array}{cccc}
a_{11} & a_{12} & \cdots & a_{1q} \\
a_{21} & a_{22} & \cdots & a_{2q} \\
\vdots & \vdots & \vdots & \vdots \\
a_{p1} & a_{p2} & \cdots & a_{pq}
\end{array} &
\begin{array}{ccc}
a_{1\,q+1} & \cdots & a_{1n} \\
a_{2\,q+1} & \cdots & a_{2n} \\
\vdots & \vdots & \vdots \\
a_{p\,q+1} & \cdots & a_{pn}
\end{array} \\
\hline
\begin{array}{cccc}
a_{p+1\,1} & a_{p+1\,2} & \cdots & a_{p+2\,q} \\
a_{p+2\,1} & a_{p+2\,2} & \cdots & a_{p+2\,q} \\
\vdots & \vdots & \vdots & \vdots \\
a_{m1} & a_{m2} & \cdots & a_{mq}
\end{array} &
\begin{array}{ccc}
a_{p+1\,q+1} & \cdots & a_{p+1\,n} \\
a_{p+2\,q+1} & \cdots & a_{p+2\,n} \\
\vdots & \vdots & \vdots \\
a_{m\,q+1} & \cdots & a_{mn}
\end{array}
\end{array} \right).$$

It may then be regarded as a 2×2 matrix with matrices for elements,

$$A = \left(\begin{array}{cc} A_{11} & A_{12} \\ A_{21} & A_{22} \end{array} \right).$$

The product of two similarly partitioned matrices A and B is obtained as if the matrix elements of A and B were numbers:

$$AB = \left(\begin{array}{cc} A_{11}B_{11} + A_{12}B_{21} & A_{11}B_{12} + A_{12}B_{22} \\ A_{21}B_{11} + A_{22}B_{21} & A_{21}B_{12} + A_{22}B_{22} \end{array} \right).$$

Exercise 12.31

Show that the Hamiltonian for Hill's equation is

$$H = \frac{1}{2}(a + p(t)) x_1^2 + \frac{1}{2} x_2^2, \quad \text{where} \quad x_1 = x, \quad x_2 = \dot{x},$$

and that the elements of the matrix $B(t)$, defined in equation 12.43, are

$$B_{11} = a + p(t), \quad B_{22} = 1, \quad B_{12} = B_{21} = 0.$$

If $X(t)$ is the fundamental matrix for equation 12.45 that satisfies the initial condition $X(0) = I_{2N}$, then

$$X(t)^{\mathsf{T}} J_{2N} X(t) = J_{2N}. \tag{12.46}$$

This follows because the matrix $X(t)$ satisfies the equation

$$J_{2N} \frac{dX}{dt} = BX, \quad X(0) = I_{2N},$$

so that

$$\frac{d}{dt}\left(X^{\top}J_{2N}X\right) = \frac{dX^{\top}}{dt}J_{2N}X + X^{\top}J_{2N}\frac{dX}{dt},$$
$$= X^{\top}BJ_{2N}^{2}X + X^{\top}BX = 0.$$

The result follows because $X(0) = I_{2N}$.

The monodromy matrix is just $E = X(T)$, and by rearranging equation 12.46 we see that E and $(E^{\top})^{-1}$ are related by the similarity transformation

$$E = J_{2N}^{-1}\left(E^{\top}\right)^{-1}J_{2N}, \tag{12.47}$$

that is, the eigenvalues of E and $(E^{\top})^{-1}$ are identical. But for any matrix M the eigenvalues of M and $M^{\top}$ are identical and if λ is an eigenvalue of M then $1/\lambda$ is an eigenvalue of M^{-1}. Hence the characteristic numbers of the system 12.45 comprise the sets $\{\lambda_k\}$ and $\{\lambda_k^{-1}\}$, $k = 1, 2, \ldots, N$. It follows that the characteristic polynomial

$$\det(E - \lambda I) = \lambda^{2N} + a_{2N-1}\lambda^{2N-1} + a_{2N-2}\lambda^{2N-2} + \cdots + a_1\lambda + 1$$

must be reflexive, that is, $a_k = a_{2N-k}$, $k = 1, 2, \ldots, N$, and if μ is a root so is μ^{-1}.

Exercise 12.32
Explain why the coefficient of λ^{2N} and the constant term of the characteristic polynomial are both unity.

For real systems the coefficients of the characteristic polynomial are real, so if λ is a complex eigenvalue, so is λ^*. It follows that the complex multipliers of the linear Hamiltonian system $d\mathbf{x}/dt = -J_{2N}B(t)\mathbf{x}$ may be grouped into the quadruples $\{\lambda, 1/\lambda, \lambda^*, 1/\lambda^*\}$, as illustrated in the left hand diagram. Eigenvalues on the unit circle, $|\lambda| = 1$, must occur in pairs, as in the centre diagram: real eigenvalues also occur in pairs, with one larger and one smaller than unity, as in the centre and right hand diagrams.

Fig. 12.16. Diagram showing possible positions of the eigenvalues in the complex λ-plane. The circle has unit radius.

For Hill's equation $n = 1$ so there can be only two eigenvalues which are constrained by the above rules to both lie either on the real axis or on the unit circle in the complex plane.

If all the eigenvalues of E are distinct and lie on the unit circle in the complex plane — note that this implies that $\lambda \neq \pm 1$ — the characteristic exponents are purely imaginary and the system is stable. Furthermore, because the eigenvalues are distinct, small perturbations of the system, that preserve the Hamiltonian structure, cannot make the system unstable. This is proved by enclosing each of the $2N$ eigenvalues by non-intersecting neighbourhoods: a small change in the parameters of the system will change E to E_1, and since the eigenvalues of a finite matrix depend continuously upon the elements of the matrix, the eigenvalues of E_1 are close to those of E and for sufficiently small changes will lie inside the chosen neighbourhoods. Therefore eigenvalues of E_1 also lie on the unit circle, for if one, μ say, did not then there would be another at $1/\mu$, and since both μ and $1/\mu$ must lie in the given neighbourhood there would be more than $2N$ eigenvalues, which is impossible.

An eigenvalue of E can leave the unit circle only by colliding with another eigenvalue, as shown in the following diagrams.

If the collision occurs off the real axis two pairs of eigenvalues must collide, as in the figure on the left; collisions on the real axis result in pairs of real roots, as shown on the right. For a system with $n = 1$, Hill's equation for instance, only the latter option is possible. Collisions of eigenvalues do not necessarily lead to instabilities, because it is possible for them to pass through each other and remain on the unit circle: this possibility is discussed in more detail by Arnold (1978, section 42) and Yakubovich and Starzhinskii (1975, chapter 3).

12.7 Appendix

Here we consider the linear, second-order system with T-periodic coefficients,

$$\frac{d^2x}{dt^2} + a_1(t)\frac{dx}{dt} + a_2(t)x = 0, \quad a_k(t + T) = a_k(t), \tag{12.48}$$

which is equivalent to the matrix equation

$$\frac{d}{dt}\begin{pmatrix} x \\ y \end{pmatrix} = \begin{pmatrix} 0 & 1 \\ -a_2(t) & -a_1(t) \end{pmatrix}\begin{pmatrix} x \\ y \end{pmatrix}, \quad y = \frac{dx}{dt}. \tag{12.49}$$

We are interested only in the special case where the matrix E, equation 12.14, has two identical, real eigenvalues, so $\text{Tr}(E)^2 = 4\det(E)$.

If λ is this eigenvalue then there is a solution of 12.48 satisfying

$$x_1(t + T) = \lambda x_1(t) \quad \text{for all } t. \tag{12.50}$$

Let $x_2(t)$ be any solution of 12.48 which is linearly independent of $x_1(t)$. Since $x_2(t+T)$ is also a solution of 12.48 there are constants d_1 and d_2 such that

$$x_2(t + T) = d_1x_1(t) + d_2x_2(t). \tag{12.51}$$

The value of d_2 is found by evaluating the Wronskian of these two solutions,

$$W(t) = x_1(t)\dot{x}_2(t) - \dot{x}_1(t)x_2(t).$$

By substituting equations 12.50 and 12.51 into the equivalent expression for $W(t+T)$ we obtain

$$W(t+T) = \lambda d_2 W(t).$$

But, equation 12.3 (page 439) gives

$$W(t+T) = W(t)\exp\left(\int_0^T dt\,\mathrm{Tr}(A)\right) = W(t)\exp\left(-\int_0^T dt\,a_1(t)\right)$$

and hence

$$\lambda d_2 = \exp\left(-\int_0^T dt\,a_1(t)\right).\tag{12.52}$$

Now construct E using the two linearly independent solutions with the initial conditions

$$\xi_1(0) = 1, \quad \dot{\xi}_1(0) = 0, \qquad \xi_2(0) = 0, \quad \dot{\xi}_2(0) = 1,$$

so that

$$E = \begin{pmatrix} \xi_1(T) & \xi_2(T) \\ \dot{\xi}_1(T) & \dot{\xi}_2(T) \end{pmatrix}$$

and the multiplier, λ, satisfies the equation

$$\lambda^2 - \left(\xi_1(T) + \dot{\xi}_2(T)\right)\lambda + \exp\left(-\int_0^T dt\,a_1(t)\right) = 0.\tag{12.53}$$

But we have assumed that the eigenvalues are identical so this equation has a double root, and hence

$$\lambda^2 = \exp\left(-\int_0^T dt\,a_1(t)\right)$$

and hence $d_2 = \lambda$. Thus equation 12.51 becomes

$$x_2(t+T) = d_1 x_1(t) + \lambda x_2(t),$$

and there are two cases to consider.

Case 1 If $d_1 = 0$ we have $x_2(t+T) = \lambda x_2(t)$. If $\lambda = 1$, ($\det(E) = 1$ and $\mathrm{Tr}(E) = 2$), both linearly independent solutions are T-periodic. If $\lambda = -1$, ($\det(E) = 1$ and $\mathrm{Tr}(E) = -2$), both linearly independent solutions are $2T$-periodic

This situation occurs only if E has two linearly independent eigenvectors, that is if

$$\dot{\xi}_1(T) = \xi_2(T) = 0 \quad \text{and} \quad \xi_1(T) = \dot{\xi}_2(T).$$

These three conditions are relatively rare.

Case 2 If $d_1 \neq 0$ we define two functions,

$$
\begin{aligned}
p_1(t) &= e^{-\rho t} x_1(t), \quad \lambda = e^{\rho T} \\
p_2(t) &= e^{-\rho t} x_2(t) - \frac{d_1 t}{T \lambda} p_1(t).
\end{aligned}
$$

It is clear that $p_1(t)$ is T-periodic; we now show that $p_2(t)$ is also T-periodic:

$$
\begin{aligned}
p_2(t + T) &= e^{-\rho(t+T)} (d_1 x_1(t) + \lambda x_2(t)) - \frac{d_1}{T\lambda}(t + T) p_1(t) \\
&= e^{-\rho t} x_2(t) - \frac{d_1 t}{T\lambda} p_1(t) = p_2(t).
\end{aligned}
$$

Hence we have

$$
x_2(t) = e^{\rho t} \left\{ p_2(t) + \frac{d_1 t}{T\lambda} p_1(t) \right\}, \tag{12.54}
$$

showing that the amplitude of $x_2(t)$ increases linearly with t. This situation occurs if E has just one eigenvector and is the normal case.

If $\det(E) = 1$ the double root of the eigenvalue equation is ± 1, then, in both cases, the period of the periodic solution(s) is T if $\lambda = 1$, $(\text{Tr}(E) = 2)$, and is $2T$ if $\lambda = -1$, $(\text{Tr}(E) = -2)$.

12.8 Exercises

Exercise 12.33
If $\Phi(t)$ is any fundamental matrix of the system $\dot{\mathbf{x}} = A(t)\mathbf{x}$, show, by differentiation, that the vector $\mathbf{x}(t) = \Phi(t)\Phi(t_0)^{-1}\mathbf{x}_0$ satisfies the equation

$$
\frac{d\mathbf{x}}{dt} = A(t)\mathbf{x}, \quad \mathbf{x}(t_0) = \mathbf{x}_0.
$$

Exercise 12.34
The exponential of a matrix B can be defined by the series

$$
e^B = \sum_{n=1}^{\infty} \frac{B^n}{n!} = I + B + \frac{B^2}{2!} + \cdots.
$$

Using this definition show that if $\dot{\mathbf{x}} = A\mathbf{x}$, where A is a constant matrix, the solution is

$$
\mathbf{x}(t) = e^{At}\mathbf{x}(0),
$$

provided all series exist. Hence show that a fundamental matrix is $\Phi(t) = e^{At}$.

Exercise 12.35
By solving the equation

$$
\frac{d\mathbf{x}}{dt} = C\mathbf{x}, \quad \text{where} \quad C = \begin{pmatrix} 0 & 1 & 0 \\ 0 & 0 & 1 \\ 0 & -1 & 0 \end{pmatrix},
$$

and finding the matrizant, show that

$$e^{Ct} = \begin{pmatrix} 1 & \sin t & 1 - \cos t \\ 0 & \cos t & \sin t \\ 0 & -\sin t & \cos t \end{pmatrix}.$$

Exercise 12.36

Consider the linear system

$$\frac{d^2 x}{dt^2} + \omega^2(t)x = 0,$$

where the parameter $\omega(t)$ is the T-periodic function

$$\omega(t) = \begin{cases} a^2, & 0 \leq t < \tau, \\ 0, & \tau \leq t < T. \end{cases}$$

Find the two appropriate solutions to show that the monodronomy matrix is

$$E = \begin{pmatrix} \cos \gamma - (\beta - \gamma)\sin \gamma & \dfrac{1}{a}(\sin \gamma + (\beta - \gamma)\cos \gamma) \\ -a \sin \gamma & \cos \gamma \end{pmatrix}, \quad \gamma = a\tau, \quad \beta = aT.$$

Further, show that $\mathrm{Tr}(E) = 2\cos \gamma - (\beta - \gamma)\sin \gamma$, and plot the zones in the (β, γ)-plane in which the motion is stable and the lines on which T- and $2T$-periodic solutions exist. Remember that only the region $\beta > \gamma$ is significant.

Exercise 12.37

Meissner's equation may be written in the form

$$\frac{d^2 x}{dt^2} + (a + 2q\, p(t))x = 0,$$

where $p(t)$ is the π-periodic function

$$p(t) = \begin{cases} 1, & 0 \leq t < \pi/2, \\ -1, & \pi/2 \leq t < \pi. \end{cases}$$

Show that the trace, T, of the monodromy matrix is

$$\begin{aligned} T &= 2\cos \beta_+ \cos \beta_- - \frac{2a}{\sqrt{a^2 - 4q^2}} \sin \beta_+ \sin \beta_-, \quad a > 2q, \\ &= 2\cos \beta_+ \cosh \gamma_- - \frac{2a}{\sqrt{4q^2 - a^2}} \sin \beta_+ \sinh \gamma_-, \quad a < 2q, \end{aligned}$$

where

$$\beta_\pm = \frac{\pi}{2}\sqrt{a \pm 2q}, \quad \gamma_- = \frac{\pi}{2}\sqrt{2q - a}.$$

For $q = 1$ plot the graph of $T(a)$ for $a > 2$ and show, graphically, that $T(a) = 2$ for $a \simeq (2n)^2$ and $T(a) = -2$ for $a \simeq (2n-1)^2$, where n is an integer.

If $a > 2q$ show that the stability boundaries defined by the condition $T = 2$ are given approximately by

$$a = \begin{cases} 4n^2 - \dfrac{q^2}{4n^2} + \dfrac{\pi^2 n^2 - 9}{192 n^6} q^4 + \cdots, \\ 4n^2 + \dfrac{3q^2}{4n^2} - \dfrac{\pi^2 n^2 + 21}{192 n^6} q^4 + \cdots, \end{cases}$$

For large n the width of the unstable region is $\Delta a \sim q^2/n^2$, whereas for Mathieu's equation it decreases as q^n. What is the reason for this difference?

Exercise 12.38
The equation of motion of a vertical pendulum comprising a light, stiff rod of length l and mass m, fixed at its end, and whose point of support oscillates vertically with frequency Ω and amplitude a, is

$$\frac{d^2\theta}{dt^2} + \omega^2 \left(1 + k^2 \sin \Omega t\right) \sin \theta = 0, \quad \omega^2 = \frac{g}{l}, \quad k^2 = \frac{a}{l}\frac{\Omega^2}{\omega^2},$$

where θ is the angle between the downward vertical and the pendulum. Here ω is the frequency of the unperturbed, small amplitude motion, that is, $k = 0$ and $|\theta| \ll 1$.

By putting $\theta = \pi - \phi$ and expanding about $\phi = 0$, show that the equation of motion can be written in the form

$$\frac{d^2\phi}{d\tau^2} + (A - 2Q \cos 2\tau)\phi = 0, \quad A = -\frac{4\omega^2}{\Omega^2}, \quad Q = \frac{2a}{l}.$$

Deduce that for $\Omega/\omega > l\sqrt{2}/a$ the upward vertical position is stable.

A more general analysis of this problem that takes account of the nonlinearity gives the same result, but also the maximum possible amplitude of stable oscillations.

Exercise 12.39
The adjoint system corresponding to the real, linear system $\dot{\mathbf{x}} = A(t)\mathbf{x}$ is defined to be $\dot{\mathbf{y}} = -A(t)^\top\mathbf{y}$, where $A^\top$ is the transpose of the matrix A. Show that if $X(t)$ and $Y(t)$ are respectively fundamental matrices of these systems then $X(t)^\top Y(t)$ is a constant matrix.

Show also that if $\mathbf{x}(t)$ is any solution of $\dot{\mathbf{x}} = A(t)\mathbf{x}$ and $\mathbf{y}(t)$ any solution of $\dot{\mathbf{y}} = -A(t)^\top\mathbf{y}$ then $\mathbf{x}(t)^\top\mathbf{y}(t) = \text{constant}$.

Exercise 12.40
It can be shown, Yakubovich and Starzhinskii (1975, chapter 2), that if X is a real non-singular matrix there exist matrices K and R such that

$$e^K = XR, \quad \text{where} \quad R^2 = I,$$

and R commutes with both X and K, $XR = RX$ and $KR = RK$.

If, in addition, X has no negative [positive] real eigenvalues then there exists a real matrix $K = \ln(X)$ $[K = \ln(-X)]$ such that $e^K = X$ $[e^K = -X]$.

If $\Phi(t)$ is a real, fundamental matrix and $\Phi(t + T) = \Phi(t)E$, it follows from the above that there exists a matrix K such that $e^{KT} = ER$, where $R^2 = I$ and R commutes with both K and E. Show that $R = R^{-1}$ and that $Ee^{-TK} = R$. By defining the matrix $F(t) = \Phi(t)e^{-Kt}$, show that $F(t + T) = F(t)R$ and $F(t + 2T) = F(t)$.

Exercise 12.41
In this exercise examples of logarithms of 2×2 matrices are considered.

(i) The matrix

$$X = \begin{pmatrix} 1 & -1 \\ 2 & 4 \end{pmatrix}$$

has real positive eigenvalues, $\lambda = 2$ and 3. Show that if

$$K = \begin{pmatrix} \ln(4/3) & \ln(2/3) \\ \ln(9/4) & \ln(9/2) \end{pmatrix} \quad \text{then} \quad X = \exp(K).$$

(ii) If X is the matrix

$$X = \begin{pmatrix} \cos\theta & \sin\theta \\ -\sin\theta & \cos\theta \end{pmatrix},$$

where θ is a real variable with complex eigenvalues $\exp(\pm i\theta)$, show that if

$$K = \begin{pmatrix} 0 & \theta \\ -\theta & 0 \end{pmatrix} \quad \text{then} \quad X = \exp(K).$$

(iii) The matrix

$$X = \begin{pmatrix} -7 & -5 \\ 10 & 8 \end{pmatrix}$$

has eigenvalues $\lambda = -2$ and 3, and the matrix

$$R = \begin{pmatrix} -\sqrt{1 - bc} & b \\ c & \sqrt{1 - bc} \end{pmatrix} \quad \text{satisfies} \quad R^2 = I.$$

Find values of the variables b and c such that $XR = RX$ and that XR has positive eigenvalues. Hence show that $XR = \exp(K)$, where K is defined in part (i) of this exercise.

Exercise 12.42
Consider the equation

$$\frac{d^2 x}{dt^2} + a_1(t)\frac{dx}{dt} + a_2(t)x = 0,$$

where $a_k(t)$ are T-periodic functions of t.

(i) Show that the characteristic multipliers satisfy the relation

$$\lambda_1 \lambda_2 = \exp\left(-\int_0^T dt\, a_1(t)\right).$$

(ii) By making the transformation $x(t) = u(t)f(t)$, to a new dependent variable $u(t)$, show that by choosing

$$f(t) = \exp\left(-\frac{1}{2}\int_0^t ds\, a_1(s)\right)$$

the equation for $u(t)$ becomes

$$\frac{d^2 u}{dt^2} + p(t)u = 0, \quad p(t) = a_2(t) - \frac{1}{4}a_1(t)^2 - \frac{1}{2}\frac{da_1}{dt}.$$

(iii) If $\Phi_1(t)$ is a fundamental matrix of the equation for $x(t)$, show that the related fundamental matrix for the u-equation is

$$\Phi_2(t) = \frac{1}{f(t)}\begin{pmatrix} 1 & 0 \\ a_1(t)/2 & 1 \end{pmatrix}\Phi_1(t).$$

Deduce that the characteristic multipliers, $\bar{\lambda}_k$, of this equation are given by

$$\lambda_k = f(T)\bar{\lambda}_k = \bar{\lambda}_k \exp\left(-\frac{1}{2}\int_0^T dt\, a_1(t)\right), \quad k = 1, 2.$$

Exercise 12.43

Consider the Hill's equation, $\ddot{x} + p(t)x = 0$, where $p(t)$ is a T-periodic function and $p(t) < 0$ for all t. Integrate the equation to show that

$$\frac{dx}{dt} = \dot{x}(0) + \int_0^t ds\,[-p(s)]x(s),$$

and use this to show that the solutions with initial conditions $(x(0), \dot{x}(0)) = (1, 0)$ and $(0, 1)$ increase monotonically. Does this mean that all solutions increase monotonically?

Exercise 12.44

Use perturbation theory to show that for small $|q|$ the 2π-periodic solution of

$$\frac{d^2x}{dt^2} + 2q\mu\sin(t + \delta)\frac{dx}{dt} + (a - 2q\cos 2t)x = 0$$

that reduces to $a = 1$, $x(t) = \cos t$ when $q = 0$ is

$$x(t) = \cos t + q\left(\frac{1}{3}\mu\cos(2t + \delta) - \frac{1}{8}\cos 3t + \mu\cos\delta\right) + O(q^2),$$

and

$$a(q) = 1 + q - q^2\left(\frac{1}{8} - \frac{2\mu^2}{8}\right) - q^3\left(\frac{1}{64} + \frac{4}{9}\mu^2 + \frac{1}{3}\mu^2\cos 2\delta\right) + O(q^4).$$

13

Integrals and their approximation

13.1 Introduction

Integrals arise frequently in most science and engineering problems and in many other applications of mathematics. In a first calculus course, for instance Apostol (1963, chapter 9), integration is introduced in two ways: as a limit of a sum and as the inverse of differentiation. The Fundamental Theorem of Calculus shows that these two definitions are equivalent. This connection between integration and differentiation is slightly puzzling because the rules of differentiation, viz.,

$$\frac{d}{dx}(\alpha f(x) + \beta g(x)) = \alpha f'(x) + \beta g'(x),$$

$$\frac{d}{dx}(f(x)g(x)) = f'(x)g(x) + f(x)g'(x),$$

$$\frac{d}{dx}\left(\frac{f(x)}{g(x)}\right) = \frac{f'(x)g(x) - f(x)g'(x)}{g(x)^2},$$

$$\frac{d}{dx}f(g(x)) = g'(x)f'(g),$$

where $f(x)$ and $g(x)$ are both differentiable functions of x, $f'(x)$ and $g'(x)$ being their differentials, and α and β are constants, allow us to treat differentiation as an algorithmic procedure. Thus, knowing the derivatives of a set of functions allows the computation of any finite combination of these functions.

Integration is different. It would seem that integration is performed using a collection of devices and special cases, and these can be used only on a small subset of integrals. For combinations of functions, other than addition, there are no simple rules; for instance we know the integral of the functions $\exp x$ and x^3, but not the composition $\exp(x^3)$. There are several methods available for evaluating integrals, the main ones being

$$\text{linearity}: \quad \int dx\,[\alpha f(x) + \beta g(x)] = \alpha \int dx\,f(x) + \beta \int dx\,g(x),$$

$$\text{change of variables}: \quad \int_a^b dx\,f(x) = \int_{g^{-1}(a)}^{g^{-1}(b)} dt\,f(g(t))g'(t),$$

$$\text{integration by parts}: \quad \int_a^b dx\,f(x)g'(x) = [f(x)g(x)]_a^b - \int_a^b dx\,f'(x)g(x),$$

provided all integrals exist. Even these rules need to be treated with some respect; for example the integrals $\int_0^{\pi/2} dx/x$ and $\int_0^{\pi/2} dx/\sin x$ do not exist but their difference $\int_0^{\pi/2} dx\,(1/x - 1/\sin x)$ does.

Another useful, but less well known, method is differentiation under the integral

478

sign. If $G(z, x)$ is a sufficiently well behaved function of z and x, $a(x)$ and $b(x)$ are differentiable functions of x and if $F(x)$ is defined by the integral

$$F(x) = \int_{a(x)}^{b(x)} dz\, G(z, x), \qquad (13.1)$$

the differential of $F(x)$ is

$$F'(x) = b'(x)G(b(x), x) - a'(x)G(a(x), x) + \int_{a(x)}^{b(x)} dz\, \frac{\partial G(z, x)}{\partial x}, \qquad (13.2)$$

provided both integrals exist. Some examples of the use of this method are given in exercises 13.1 and 13.2.

These are the main tools available for the evaluation of indefinite integrals in terms of known functions. But frequently the obvious direct application of these methods fails and it is necessary to use other methods for recalcitrant integrals. There are four basic ways of tackling an integral.

(1) Find a way of converting it to an integral having a 'standard' form, the integral of which can be expressed in terms of known functions.
(2) Use algorithmic methods, as implemented in Maple for instance; these methods generally define integration as the inverse of differentiation and work on only a subset of the limited class of integrals for which the first method can be used.
(3) Evaluate the integral numerically; these methods approximate an integral by a finite sum, to give a numerical answer.
(4) Find an approximation to the integral in terms of known functions.

Of all these techniques the first is the 'best', but this requires much practice and often considerable ingenuity; unfortunately most integrals cannot be evaluated by these means.

The use of algorithmic methods is also of limited value: essentially this method can deal only with integrals that the standard methods will evaluate, however, there are advantages and disadvantages of using these methods. Algorithmic methods are particularly good at evaluating complicated integrals without making silly algebraic errors. For instance the integral $I = \int dx\, \ln^4(a - x)$ can clearly be evaluated using substitution and the integration by parts rule,

$$\int dz\, \ln^n(z) = z\ln^n(z) - n \int dz\, \ln^{n-1}(z).$$

But four applications of this rule is tedious and error prone, so it is easier to use Maple to obtain

$$I = -u\left[\ln^4(u) - 4\ln^3(u) + 12\ln^2(u) - 24\ln(u) + 24\right], \quad u = a - x.$$

Also these methods will sometimes evaluate integrals that only the most intrepid will attempt, for instance†

$$\int dx \; \frac{x\left[\left(x^2 e^{2x^2} - \ln^2(x+1)\right)^2 + 2xe^{3x^2}\left[x - (2x^3 + 2x^2 + x + 1)\ln(x+1)\right]\right]}{(x+1)\left(\ln^2(x+1) - x^2 e^{2x^2}\right)^2}$$

$$= x - \ln(x+1) - \frac{xe^{x^2}\ln(x+1)}{\ln^2(x+1) - x^2 e^{2x^2}}$$

$$+ \tfrac{1}{2}\ln\left(\ln(x+1) + xe^{x^2}\right) - \tfrac{1}{2}\ln\left(\ln(x+1) - xe^{x^2}\right)$$

and

$$\int dx \left(\frac{2x^3 - 2x^2 - 2x + 3}{(x-1)^2}\right) x^2 e^{-x^2} = \frac{x^3 e^{-x^2}}{1-x}.$$

On the other hand, complicated algorithms attract programming errors, and whilst these will gradually be removed as systems develop it is necessary to be on guard against such errors. In order to illustrate these problems I give two examples which occurred in recent versions of Maple. The first, now corrected, using symbolic integration, gave

$$\int_1^2 dx \; \sin^{-1}\left(\sqrt{1 - 1/x}\right) = -1,$$

which is clearly wrong as the integrand is positive throughout the range; the correct result is $\pi/2 - 1$. The second example is

$$\int_a^{2\pi+a} dx \; \cos nx \cos mx = 0, \quad \text{both } n \text{ and } m \text{ being integers.}$$

This result is correct only if $n \neq m$: both Maple 5 and 6 give this integral incorrectly.

Wrong answers are rare; these examples are given only to alert you to the fact that all large computer programs have bugs and that care is always needed. A more common problem is that the computer will produce a correct result but not in a useful form. This often happens when trigonometric functions are integrated, as is shown by the following example which occurs in the description of three-dimensional motion in a central potential; the integral

$$F(x) = \int dx \; \sqrt{A^2 - B^2/\sin^2 x}, \quad \text{with} \quad 0 < |B| < A,$$

is evaluated by Maple but the result is not given in a useful form, even though the

---

† These examples are a little phoney, being originally constructed by differentiating the right hand sides. Nevertheless, they do demonstrate the power of these algebraic systems which happily perform the integrals.

substitution $B \cot x = \sqrt{A^2 - B^2} \sin \phi$ transforms the integral to the form

$$
\begin{aligned}
F(x) &= B\phi - A^2 B \int d\phi \, \frac{1}{A^2 \sin^2 \phi + B^2 \cos^2 \phi} \\
&= B\phi(x) - A \tan^{-1} \left(\frac{A \cot x}{\sqrt{A^2 - B^2 / \sin^2 x}} \right).
\end{aligned}
$$

Similar examples are treated in exercise 13.3.

These few examples show that computer systems are useful but do not explain why these systems work or provide an indication of their limitations. In section 13.2.1 we shall provide a very brief description of the methods used and some idea of their limitations.

An additional complication is that there exists a range of very powerful techniques, often involving the use of complex variable theory, for evaluating *definite* integrals which cannot be used for indefinite integrals. For instance the result

$$
F(u, v) = \int_{-\infty}^{\infty} dz \, \frac{e^{iuz}}{v^2 + z^2} = \frac{\pi e^{-v|u|}}{2v} \quad u, v \text{ real with } v > 0,
$$

is obtained using standard complex variable methods (and is given incorrectly by Maple 5 and 6). However, changing either integration limit produces an integral which cannot be expressed in terms of elementary functions. Maple, and similar languages, recognise many definite integrals of this type.

Furthermore, many 'special' functions† can be *defined* in terms of definite integrals, which means that there exists another class of integrals which can be evaluated in terms of these special functions. Again, computer algebra systems are able to recognise some of these integrals.

The numerical evaluation of integrals is normally relatively straightforward provided care is taken with any singularities, although modern 'black box' integrators, such as those provided with Maple, cope well with singulaities. Numerical methods are particularly suitable if a single value is needed; but if the integrand, or integration limits, are functions of one or more variables and it is necessary to understand how the integral depends upon these variables, then numerical methods are of limited value.

In these circumstances, and when methods 1 and 2 fail, it is often possible to approximate the integral using a variety of techniques, most of which were developed before computers made numerical methods so convenient and before the advent of computer algebra systems. These methods fall into two categories. In the first are those that approximate integrals dominated by contributions from the neighbourhood of isolated points; integrals like

$$
\int_0^{\infty} dt \, \frac{e^{-x\sqrt{t}}}{1 + t^{\sqrt{2}}} \quad \text{and} \quad \int_{-\pi}^{\pi} dt \, \exp i \, (t - x \sin t + y \sin 2t),
$$

where x and y are large, fall into this category. The methods for dealing with some of

---

† Maple contains approximations to many special functions and information about some of their properties: to determine which type **?inifcn** at the prompt.

these integrals are discussed in sections 13.3.2, 13.4 and the next chapter. In the second category are integrals whose dominant contribution comes not from an isolated neighbourhood but a range of the integration variable for which such expansion techniques fail; some methods of dealing with these problems are discussed in section 13.5.

Exercise 13.1
Use the integral

$$\int_0^\infty dx \, \frac{\sin x}{x} = \frac{\pi}{2} \quad \text{to show that} \quad f(a) = \int_0^\infty dx \, \frac{\sin ax}{x} = \frac{\pi}{2} \operatorname{sgn}(a),$$

and apply equation 13.2 to the function

$$g(a) = \int_0^\infty dx \, \frac{\sin^2 ax}{x^2}$$

to show that $g(a) = \pi|a|/2$.

Exercise 13.2
By noting that

$$\int_0^\infty dx \, e^{-ax} \sin x = \frac{1}{1+a^2}, \quad a > 0,$$

show, by differentiation with respect to a, that

$$f(a) = \int_0^\infty dx \, e^{-ax} \frac{\sin x}{x} = \frac{\pi}{2} - \tan^{-1} a.$$

Exercise 13.3
Show that

(i) $\displaystyle \int \frac{dx}{\sqrt{1 - (\cos^2 a / \sin^2 x)}} = -\sin^{-1}\left(\frac{\cos x}{\sin a}\right),$

(ii) $\displaystyle \int dx \, \frac{\cos a}{\sin^2 x \sqrt{1 - (\cos^2 a / \sin^2 x)}} = -\sin^{-1}(\cot a \cot x),$

and compare these results with those given by Maple.

13.2 Integration with Maple

13.2.1 Formal integration

The problem of indefinite integration is easy to state: given a function $f(x)$ of the real variable x find a function $g(x)$ satisfying the equation

$$\frac{dg}{dx} = f(x). \tag{13.3}$$

If the integral $g(x)$ can be found, we write

$$\int dx \, f(x) = g(x) \tag{13.4}$$

or $g(x) + c$, where c is a constant; here we ignore this constant.

In practice $g(x)$ is found by a collection of standard methods, integration by parts, substitution, looking in tables; alternatively some common functions $f(x)$ for which $g(x)$ 'cannot' be found by standard methods are often used to define new functions in terms of integrals. For instance,

$$\text{Si}(x) = \int_0^x dt\,\frac{\sin t}{t}, \quad \text{erf}(x) = \frac{2}{\sqrt{\pi}}\int_0^x dt\,e^{-t^2}, \quad \text{E}_n(x) = x^{n-1}\int_x^\infty dt\,\frac{e^{-t}}{t^n}, \quad (13.5)$$

where n is a positive integer in the last example; these functions have been tabulated, their analytic properties determined and they have now become 'standard' functions and can be used to evaluate other integrals. In short these, and other, functions have become part of the standard tool kit used to evaluate integrals in terms of formulae. This process is not confined to functions defined only by integrals: most of the special functions of mathematical physics, for example Bessel functions and Legendre polynomials, were originally defined by linear differential equations, but also have integral representations that can often be used.

It is a legitimate question to ask why none of the above three integrals can be expressed in terms of more elementary functions, such as polynomials, trigonometric functions, exponentials or logarithms, but it would be difficult to find an answer in any standard book on integration as this type of question requires mathematics quite different from the analysis normally used to evaluate or understand the properties of these functions. Some of these ideas will be described here: further detailed discussion can be found in Davenport (1981) and Davenport *et al.* (1993).

There is no reliable systematic approach to integration. The only case in which some form of algorithmic approach always provides an exact finite formula is when $f(x)$, the integrand, is a rational function, that is, $f(x) = P(x)/Q(x)$, where $P(x)$ and $Q(x)$ are polynomials; then $f(x)$ can be decomposed into sums of simpler functions using partial fractions, and the resulting integrals are known. For simple polynomials such analysis is easily performed by hand, but it is a more difficult task to write a satisfactory algorithm which will work efficiently for arbitrary polynomials, and only relatively recently have such algorithms been written; a detailed discussion of this problem is given by Davenport *et al.* (1993, chapter 5). The Horowitz method that you will observe Maple using in exercise 13.4(viii) is one of these methods.

Maple, and most computer algebra systems, start their search for an integral in a manner similar to that of a human. First, the program will test to see if the integrand is a polynomial or a sum of inverse powers; next it will check to determine whether the integrand is one of a number of simple functions it knows about, for instance trigonometric or exponential functions; next it will look for specific types, for example integrands of the form

$$p(x)e^{ax+b}\sin(cx+d),$$

for constants a, b, c and d and a polynomial $p(x)$, which it can evaluate by integration by parts. Finally it attempts to determine whether or not the integrand has the form $f(x) = h'(x)/h(x)$ for some function $h(x)$.

Some idea of how Maple tackles various integrals is obtained by setting the `infolevel` command before evaluation, as follows:†

```
>  restart;
>  infolevel[int]:=5;
```

$$infolevel_{int} := 5$$

```
>  Int((1+x^(1/4))^(1/3)*x^(-1/2),x)=int((1+x^(1/4))^(1/3)*x^(-1/2),x);
```

```
int/indef1:    first-stage indefinite integration
int/algebraic2/algebraic:   algebraic integration
int/algebraic2/algebraic:   applying algebraic substitution
int/indef1:    first-stage indefinite integration
int/indef1:    first-stage indefinite integration
int/algebraic2/algebraic:   algebraic integration
int/indef1:    first-stage indefinite integration
```

$$\int \frac{(1+x^{1/4})^{1/3}}{\sqrt{x}}\, dx = \frac{12}{7}\,(1+x^{(1/4)})^{(7/3)} - 3\,(1+x^{(1/4)})^{(4/3)}$$

Exercise 13.4
Set `infolevel[int]:=5` to see what Maple does when evaluating the indefinite integrals of the following functions.

(i)	$\sin x$	(ii)	$e^x \cos x$	(iii)	$\sin x \cos^5 x$
(iv)	$\dfrac{e^x}{\sqrt{1-e^{2x}}}$	(v)	$\dfrac{x^2}{\sqrt{1+x^{1/3}}}$	(vi)	$\dfrac{1}{(1+\sqrt{1+x})^4}$
(vii)	$\dfrac{1}{x+x^3}$	(viii)	$\dfrac{1}{a+x+x^3}$	(ix)	$\dfrac{\ln(x-4)}{x}$
(x)	$\dfrac{\cos x}{x^2}$	(xi)	$\dfrac{x^5}{\sqrt{x^{12}+1}}$	(xii)	$\dfrac{x^5}{\sqrt{x^{12}-1}}$
(xiii)	$\dfrac{e^x}{\sqrt{1+x^{1/5}}}$	(xiv)	$\dfrac{e^{x^2}}{\sqrt{1+x^{1/5}}}$	(xv)	$e^{ax+b}\sin(cx+d)\ln x$

To see how hard Maple needs to work for some integrals, consider the example

$$\int_0^\infty dx\,\frac{1}{x}\left(\frac{\ln(2+ax)}{\sqrt{1+ax}} - \frac{\ln(2+bx)}{\sqrt{1+bx}}\right) = \ln(b/a)\ln 2, \quad a,b>0.$$

An easier way of evaluating this is given in equation 13.64 (page 520).

A surprising number of integrals can be evaluated by these heuristic methods but this is an unsatisfactory technique, for if it fails to produce a result it is not clear whether this is because no integral of the assumed form exists or because the methods are not good enough to find the result. This problem was first stated clearly and studied systematically by Liouville:

J. Liouville
(1809–1882)

The integration problem: given two classes, A and B, of functions of x, the integration

---

† Maple 5 and 6 produce slightly different outputs in this example: the output quoted is from Maple 5.

problem from A to B is to find an algorithm which, for every member $a(x)$ of A, either gives an element $b(x)$ of B such that $a(x) = db/dx$, or proves that there is no element $b(x)$ of B such that $a(x) = db/dx$.

The choice of classes A and B is difficult because integration is one of the means of generating new types of functions, as shown by the definitions in equation 13.5; generally, but not always, if $f(x)$ is a function of a particular type — a polynomial, a ratio of polynomials, an exponential, for instance — then $f'(x)$ is of the same, or similar, type whereas $\int dx\, f(x)$ may be different. Thus the rational polynomial $1/x$ differentiates to another rational polynomial, $-x^{-2}$, but its integral cannot be expressed in terms of rational polynomials. The theory that deals with such classification problems was created in the middle of the nineteenth century by Liouville, building on the work of Laplace and Abel.

P.-S. de Laplace
(1749–1827)

N. H. Abel
(1802–1829)

Exercise 13.5

Prove that

$$\int dx\, \frac{1}{x} \neq \frac{P(x)}{Q(x)},$$

where $P(x)$ and $Q(x)$ are polynomials with no common factors.

Hint: differentiate and show that $Q(x)$ has a factor x^n for some integer $n \geq 1$ and hence derive a contradiction.

Consider how to define a suitable class of functions. We naturally start with the simplest functions x^k, $k = 0, 1, 2, \ldots$, and finite sums of such functions, that is, polynomials; this class is closed under addition, multiplication, differentiation and integration. To obtain closure under division it is necessary to extend this class to include all rational functions, that is ratios of polynomials; this class of functions is usually denoted by $\mathbf{Q}(x)$. It is obvious that $\mathbf{Q}(x)$ is closed under addition, multiplication, division and differentiation. It is also closed under less obvious operations: if the two rational functions $f(x)$ and $g(x)$ can be represented by the infinite series

$$f(x) = \sum_{k=0}^{\infty} f_k x^k, \qquad g(x) = \sum_{k=0}^{\infty} g_k x^k, \tag{13.6}$$

then the Hadamard product $f \circ g$ and the idempotency operator $f \to \hat{f}$,

$$(f \circ g)(x) = \sum_{k=0}^{\infty} g_k f_k x^k, \qquad \hat{f}(x) = \sum_{k=0}^{\infty} \mathrm{sgn}(f_k) x^k, \tag{13.7}$$

are also rational functions (Melzak, 1973, page 206), see also exercise 13.70 (page 525).

Clearly $\mathbf{Q}(x)$ is closed under differentiation, but not integration, for the logarithm occurs naturally as the integral of the rational function $1/x$. Thus the next natural extension is the logarithm, $\ln(x)$, and logarithms of polynomials, $\ln(p(x))$; we also add the inverse of the logarithm, the exponential function, e^x and $e^{f(x)}$, where $f(x) \in \mathbf{Q}(x)$, to our class of functions. These two extensions include the trigonometric functions and some of their inverses because they can be expressed in terms of the exponential and logarithmic functions with complex arguments, for instance

$$\sin x = \frac{1}{2i}\left(e^{ix} - e^{-ix}\right), \qquad \cos x = \frac{1}{2}\left(e^{ix} + e^{-ix}\right),$$

$$\tan^{-1} x = \frac{i}{2}\ln\left(\frac{1-ix}{1+ix}\right).$$

Finally, we observe that the trigonometric functions can also be defined in terms of integrals like

$$\sin^{-1} x = \int_0^x dt\,\frac{1}{\sqrt{1-t^2}}, \quad\text{and}\quad \cos^{-1} x = \int_x^1 dt\,\frac{1}{\sqrt{1-t^2}} = \frac{\pi}{2} - \sin^{-1} x,$$

so it is expedient to add to our list of functions *algebraic functions*, which are explicit, for example

$$y = x^{10/3}, \qquad\qquad y = \left(\frac{x^2 + x + 4}{2x + 1}\right)^{3/2}, \quad y = \frac{(1+x)^{1/2} - (1-x)^{1/3}}{(1+x)^{1/2} + (1-x)^{1/3}},$$

$$y = \sqrt{x + \sqrt{x + \sqrt{x}}}, \qquad y = \sqrt{x} + \sqrt{x + \sqrt{x}},$$

$$\text{(13.8)}$$

and can be represented in terms of any finite combination of the four elementary operations $+, -, \times, \div$ and any finite number of root extractions. More generally, a function $f(x)$ is an algebraic function of x, of degree n, if it is a root of an irreducible polynomial† of degree n, $P(f) = 0$, where P is

$$P(z) = z^n + a_{n-1}(x)z^{n-1} + \cdots + a_1(x)z + a_0(x),$$

where the coefficients $a_k(x)$ are ratios of polynomials of x with coefficients that may be integers, real or complex numbers; there is no loss of generality in assuming that $a_n = 1$. Since general polynomials of degree greater than four do not have roots that can be expressed in terms of surds, most algebraic functions are not explicit and these are named *implicit algebraic functions*. Functions like $x^{\sqrt{2}}$, $x^{1/2+i}$, e^x and $\ln x$ are not algebraic functions. An algebraic function of degree one is a rational function. Further discussion of algebraic functions can be found in Hardy (1966, 1967).

Exercise 13.6
Determine which of the following polynomials are irreducible:

$$y^4 - x^4 \quad\text{and}\quad y^2 + x.$$

Exercise 13.7
Find the polynomials satisfied by the explicit algebraic functions defined in equations 13.8.

Exercise 13.8
If y is an algebraic function of x show that x is an algebraic function of y.

In this manner we arrive at the class of *elementary functions* comprising functions of the following type:

† An irreducible polynomial cannot be expressed as the product of two other nonzero degree polynomials over Q, the set of rational numbers.

- rational functions,
- the logarithm function, $\ln(x)$,
- the exponential function, e^x,
- algebraic functions,
- all functions which can be defined by means of any finite combination of the above functions using the four elementary operations $+, -, \times, \div$ and root extraction.

Laplace conjectured, and Abel proved, that the integral of an algebraic function $y(x)$, if algebraic, will contain the algebraic function of the integrand and no other algebraic function. A number of such results are described by Hardy (1966), for instance if $y(x)$ is an algebraic function of degree n and if $\int dx\, y(x)$ is an algebraic function, it has the form

$$\int dx\, y(x) = R_0(x) + R_1(x)y + R_2(x)y^2 + \cdots + R_{n-1}(x)y^{n-1}, \tag{13.9}$$

where $R_r(x)$ are rational functions of x, Hardy (1966, pages 36–41); an important special case of this result is considered in exercises 13.51 and 13.52, (page 521). More generally, Liouville proved:

Liouville's Principle If $y_k(x)$, $k = 1, 2, \ldots$, are functions of x whose differentials are algebraic functions of x and $y_k(x)$, if $F(x, y_1, y_2, \ldots)$ is an algebraic function and if

$$\int dx\, F(x, y_1, y_2, \ldots, y_n)$$

is an elementary function, then it has the form

$$u_0(x) + \sum_i A_i \ln u_i(x),$$

where the $u_i(x)$ are algebraic functions of x and $y_k(x)$, $k = 1, 2, \ldots$, and the A_i are constants.

This result applies to functions like $F(x, y_1, y_2)$, where $y_1 = e^x$ and $y_2 = \sin x$ for which $y_1' = y_1$ and $y_2' = \sqrt{1 - y_2^2}$ respectively, are algebraic functions of y_1 and y_2, for instance. Other examples are discussed in Hardy (1966, page 60). Here the notion of algebraic has to be generalised to include more variables: z is an algebraic function of $(x, y_1, y_2, \ldots)$ if it satisfies an equation of the form

$$P_n(x, y_1, y_2, \ldots)z^n + P_{n-1}(x, y_1, y_2, \ldots)z^{n-1} + \cdots + P_0(x, y_1, y_2, \ldots) = 0,$$

where $P_k(x, y_1, y_2, \ldots)$ are polynomials in all variables.

These results only provide part of the story, for they do not give a method either of determining whether the integral exists or of finding it when it does. For rational functions it is always true that the integral can be written in this form, and there are algorithms for determining the constants A_i and the functions $u_i(x)$; these are described in Geddes *et al.* (1992, chapter 12). For elementary transcendental functions, that is, elementary functions without algebraic functions, Risch (1969) proved that there is an algorithm that, given such a function $f(x)$, either gives an elementary function which is the integral of $f(x)$, or proves that $f(x)$ has no elementary integral; subsequently Davenport (1981) extended this result to allow $f(x)$ to be elementary. Descriptions of some of these algorithms can be found in Davenport (1981).

Summary

In this section we have tried to give you some idea of what to expect from packages that perform indefinite integration symbolically. There have been huge improvements in these methods during the past thirty years and a good summary of this history and the developments made can be found in the article by Moses (1971) and also the introduction to the book by Geddes *et al.* (1992). In summary, these packages appear to be able to evaluate most of the standard indefinite integrals that one would expect a good honours mathematics graduate to manage; the packages will probably evaluate most of the more complicated integrals more accurately, but those involving trigonometric functions seem to pose problems for algorithms and the results are often not given in the most convenient form.

Is it necessary, therefore, to know the techniques of integration? If all integrals were of the form amenable to algorithmic methods, then there would seem to be little point in learning such skills because the computer can produce the result quicker and more reliably. But almost all integrals are not of this form and quite different techniques are required. In practice one requires either a few numerical values, in which case it is usually best to evaluate the integral numerically, although sometimes it is important to know that the result is, say, π rather than 3.14159265, or one needs to know how an integral depends upon parameters; for example, how do

$$\int_0^1 du \, \frac{1}{(u^5 + yu^3 + x)^{1/3}}, \quad \int_0^\pi du \, \cos\left(\frac{1}{5}u^5 + xu\right), \quad \int_0^\infty du \, u^y e^{-xu^2},$$

behave as functions of x and y? Such questions require different techniques requiring a good practical understanding of the methods needed to evaluate simpler integrals. Some of the more general analytic techniques required to evaluate these integrals are described in this and the next chapter. In short, without the skills required to evaluate the simple integrals that algorithmic methods can tackle, all integrals become impenetrable except to numerical methods.

Finally, we end this section by noting that many of the classic curves, derived geometrically, give rise to algebraic equations because they involve distances which, via Pythagoras' theorem, give rise to square roots. The geometric interpretation is sometimes useful as it shows clearly the behaviour of all real solutions of some algebraic equations, and if there are parameters in the equation then graphs of the solutions provide a convenient way of seeing how the solutions change with the parameter. A simple example is the Cissoid of Diocles, defined to be the curve traced out by the point Q, figure 13.1, where the lengths OQ and PT are equal; here O is the origin, OB the diameter of the circle of radius a and BT the tangent at B. It is easily seen that if OB is the x-axis the curve satisfies the equation $y^2(2a - x) = x^3$, hence $y(x)$ is an algebraic function of degree two, giving the curves shown.

Cissoid of Diocles

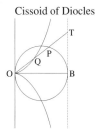

Conchoid of Nicomedes

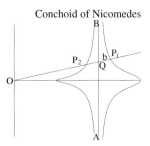

Figure 13.1 Diagram showing the
construction of the Cissoid of Diocles.

Figure 13.2 Diagram showing the
construction of the Conchoid of Nicomedes.

In this case the two solutions are relatively simple and the only parameter, the radius of the circle, does not change the shapes of the curves.

Exercise 13.9
The Conchoid of Nicomedes is depicted by the curves in figure 13.2. Here the line AB is perpendicular to the x-axis, crossing it at a distance $a > 0$ from the origin. The point Q moves along AB. If OQ is the straight line joining O to Q the points P_1 and P_2 are the two points on this line a distance b from Q. The Conchoid of Nicomedes is the loci of P_1 and P_2.

Show that the equation satisfied by these points is

$$(x^2 + y^2)(x - a)^2 = b^2 x^2$$

and that these curves have different behaviour according as $b > a$ or $b < a$.
Fix $a = 1$ and create an animation showing how these curves change with b.

13.2.2 Numerical integration
Maple has a black-box numerical integrator which provides numerical estimates of most definite integrals that exist, even those with integrable singularities, with highly oscillatory integrands and over infinite ranges, though the computational time can be excessive in extreme cases. The simplest method of numerically evaluating an integral is to use the evalf(Int(..)) combination. Thus the value of the integral

$$I = \int_0^1 dx \, \frac{1}{1 + x^2 + x^{1/2}}$$

is given by the commands:

```
>   f:=1+x^2 + sqrt(x):
>   Int(1/f,x=0..1)=evalf(Int(1/f,x=0..1));
```

$$\int_0^1 \frac{1}{1 + x^2 + \sqrt{x}} \, dx = .5371959283$$

Notice that evalf(Int(..)) and not evalf(int(...)) was used. If higher accuracy is required, say 20 digits, we invoke the second optional argument of evalf,

```
>  Int(1/f,x=0..1)=evalf(Int(1/f,x=0..1),20);
```

$$\int_0^1 \frac{1}{1+x^2+\sqrt{x}}\, dx = .53719592826258221156$$

Exercise 13.10

What is the difference between the commands `evalf(Int(..))` and `evalf(int(...))` and when is the former to be preferred?

Note: this difference was discussed in section 1.12.1.

The default method used by Maple is Clenshaw–Curtis quadrature, and an attempt is made to evaluate the integral to a relative error of 5×10^{-D}, where D is the second parameter in `evalf`; if this is not specified D is the global value of `Digits`; a description of the methods used can be found in Harding and Quinney (1989). If singularities are detected in or near the interval of integration then some techniques of symbolic analysis are used to deal with them.

It is possible, if you wish, to force Maple to use one of three methods for numerically evaluating the integral. In this case the general form of the command is `evalf(Int(f,x=a..b,N,flag))`, where N is the number of digits required and `flag` is one of the three values:

(1) `_CCquad` forces Maple to use only the Clenshaw–Curtis quadrature method and to avoid the singularity-handling routines.
(2) `_Dexp` forces Maple to use only the adaptive double-exponential method and to avoid the Clenshaw–Curtis quadrature and the singularity-handling routines.
(3) `_NCrule` forces Maple to use only the adaptive Newton–Cotes rule, which is a fixed-order method and hence is not effective for high precision, typically if `Digits`> 15.

Thus we evaluate the above integral, to 30 digits, using the double-exponential routine as follows†

```
>  Int(1/f,x=0..1)=evalf(Int(1/f,x=0..1,30,_Dexp));
```

$$\int_0^1 \frac{1}{1+x^2+\sqrt{x}}\, dx = .537195928262582211564543889199$$

When specifying the method to be used it is necessary to use `Int` rather than `int`. In this example the other two methods fail. In practice it seems to be better to let Maple decide which method to use, and it seems to be impossible to determine the actual method used.

If a definite integral can be evaluated symbolically, it is natural to assume that symbolic evaluation is preferable to numerical integration; this is not always the case. Consider the integral

$$I = \int_1^2 dx\, \frac{1}{1+x^7}.$$

† With Maple 6 this command fails if $N \ge 30$.

The integrand is a rational function so we know that the integral can be evaluated symbolically, and as a first attempt we try

> ```
>   f:=1/(1+x^7):    int(f,x=1..2);
> ```

$$\int_1^2 \frac{1}{1+x^7} \, dx$$

For some curious reason Maple 5 fails† to evaluate this, even though the equivalent indefinite integral is evaluated; so we integrate over the interval (a, b), a and b being variables,‡

> ```
>   z:=int(f,x=a..b);
> ```

$$z := \frac{1}{7}\ln(1+b) + \left(\sum_{\_R=\%1} \_R \ln(b+7\_R)\right) - \frac{1}{7}\ln(1+a) - \left(\sum_{\_R=\%1} \_R \ln(a+7\_R)\right)$$

$$\%1 := \text{RootOf}(117649\_Z^6 + 16807\_Z^5 + 2401\_Z^4 + 343\_Z^3 + 49\_Z^2 + 7\_Z + 1)$$

Now convert this to a function and evaluate the function at $a = 1$ and $b = 2$,

> ```
>   z:=unapply(z,a,b):    evalf(z(1,2));
> ```

$$.1163017045$$

Alternatively we could have used the `subs` command. In this case it is easier to simply evaluate the integral numerically,

> ```
>   evalf(Int(f,x=1..2));
> ```

$$.1163017044$$

which agrees with the previous number to nine decimal places. The command

```
int(f,x=1..2,'continuous')
```

which instructs `int` not to look for discontinuities, will also reproduce the above expression for z with $a = 1$ and $b = 2$.

I have found that the Maple numerical integration package efficiently provides numerical estimates of the required accuracy for a wide variety of integrals; but have noticed that this facility can fail when the integrand is highly oscillatory, especially when the range of integration extends to $\pm\infty$; also when the integrand is close to a non-integrable singularity the computational time becomes excessive. However, one should not be using such numerical methods for these types of integrals, but should try to approximate the integral using methods described in this and the next chapter. The following exercises provide some idea of the versatility of this package and also of integrals that it cannot manage.

† Maple 6 can cope with this integral.
‡ In this example the form of output produced by Maple depends upon the *Output Display* option used. Labels, here %1, are not used in the *Editable Math* option, or *Standard Math Notation* in Maple 6. Labels help avoid clutter on the screen.

Exercise 13.11

The function $f(x, a) = (x - 1/2)^2 + a^2$, $a \geq 0$, has zeros at $x = \frac{1}{2} \pm ia$, so that the integral

$$g(a) = \int_0^1 \frac{dx}{f(x, a)^2}$$

exists only if $a > 0$. Evaluate this integral numerically for a range of values of a in the interval $0 \leq a \leq 0.0001$, comparing the result obtained with the value found by symbolic integration.

Exercise 13.12

Use `evalf(Int(...))` to evaluate numerically each of the following integrals using `Digits`$= 8, 9, \ldots, 15$, and determine the time taken in each case.

(i) $\displaystyle\int_0^1 dx \, \frac{1}{1 + x^3 + x^{1/3} + \sin \pi x}$,

(ii) $\displaystyle\int_0^8 dx \, \sin(x^4)$,

(iii) $\displaystyle\int_0^1 dx \, \exp\left(\sin^{-1}(1 - x)^a\right)$, $\quad a = \frac{1}{2}, \frac{1}{4}$,

(iv) $\displaystyle\int_0^\infty dx \, \frac{\cos x}{x^{1/3}}$.

Note: in the second example you should evaluate the given integral and also that obtained by making the transformation $y = x^4$. In the third integral use of the Maple function `timelimit` is advised. Finally, I have observed that Maple 5 and Maple 6 perform differently on some of these integrals.

Exercise 13.13

Numerically evaluate the integral

$$F(x) = \int_0^x dt \, \sin(t^3)$$

with `Digits=5` and for $x = 5, 10, 20, 50$ and 100 if possible. Other ways of approximating this integral will be discussed later in this chapter.

Exercise 13.14

Numerically evaluate the integral

$$F(x) = \int_0^\pi dt \, \sin(t - x \sin t)$$

with `Digits=5` and 8 and for $x = 2, 10, 500, 1000$ and 5000, if possible, and compare your values with those given by the first term in the asymptotic expansion

$$F(x) \sim \frac{\sqrt{2\pi}}{(x^2 - 1)^{1/4}} \sin\left(\frac{\pi}{4} + \cos^{-1}(1/x) - \sqrt{x^2 - 1}\right), \quad x \gg 1,$$

derived in the next chapter.

13.3 Approximate methods

There are several methods of finding analytic approximations to integrals; none work for all integrals and there are many integrals for which no method works. The available techniques can only be used for certain classes of integral, so for a given integral the art is to determine the appropriate method or, if none, to transform the integral into a form in which one of the standard methods is applicable; this is largely a matter of experience and trial and error because there are few general rules. In this section we introduce techniques that are useful when the integral is dominated by contributions from the neighbourhood of one, or a few, isolated points.

13.3.1 Integration by parts

One of the standard methods of evaluating integrals is to use integration by parts, a method which relies upon splitting the integrand into the product of two functions, $f(x)$ and $g(x)$, and using the relation

$$\int dx\, f(x)g(x) = g(x)\left[\int dx\, f(x)\right] - \int dx\, \frac{dg}{dx}\left[\int dx\, f(x)\right]. \tag{13.10}$$

Repeating this operation N times gives the more general relation

$$\int dx\, f(x)g(x) = \sum_{r=0}^{N-1}(-1)^r f^{(-r-1)}(x)g^{(r)}(x) + (-1)^N \int dx\, f^{(-N)}(x)g^{(N)}(x), \tag{13.11}$$

where $g^{(k)}$ is the kth differential of g if $k \geq 1$ and the kth integral if $k \leq -1$, and $g^{(0)} = g$; the function $g^{(-k)}(x)$ contains an arbitrary polynomial of degree $k - 1$, the choice of which depends upon the context, but often all integration constants are set to zero.

Exercise 13.15
Prove equation 13.11.

In the simplest cases the integral on the right hand side of 13.11 can be evaluated directly for some N, usually when $g(x)$ is a polynomial, thus if $g(x) = x^2$ and $f(x) = \sin ax$, then since $g^{(2)}(x) = 2$ we have

$$\int dx\, x^2 \sin ax = g\, f^{(-1)} - g^{(1)} f^{(-2)} + g^{(2)} \int dx\, f^{(-2)}(x)$$

$$= -\frac{x^2}{a}\cos ax + \frac{2x}{a^2}\sin ax + \frac{2}{a^3}\cos ax.$$

Another simple application is when the integral on the right hand side of 13.11 is proportional to the original integral on the left hand side for some value of N, and then an exact expression for the integral can be obtained. Normally these types of integrals involve trigonometric and/or exponential functions, as in the next exercise.

Exercise 13.16
Using integration by parts twice, show that

$$\int_x^\infty dt\, e^{-at}\sin bt = \frac{b\cos bx + a\sin bx}{a^2 + b^2}e^{-ax}, \quad a > 0.$$

Another simple use of equation 13.10 is in obtaining 'reduction formulae', best explained by example. Suppose that

$$I_n = \int dx \, x^n e^{ax}, \quad \text{where } n \text{ is a positive integer,}$$

then using equation 13.10 with $g(x) = e^{ax}$ and $f(x) = x^n$ we obtain

$$aI_n = x^n e^{ax} - nI_{n-1}.$$

Repeated application of this result, which is equivalent to using equation 13.11 with $N = n$, and using the fact that $I_0 = e^{ax}/a$, gives

$$\int dx \, x^n e^{ax} = \left\{ \frac{x^n}{a} - \frac{nx^{n-1}}{a^2} + \frac{n(n-1)x^{n-2}}{a^3} - \cdots + \frac{(-1)^n n!}{a^{n+1}} \right\} e^{ax}.$$

Exercise 13.17
One of the better-known relations that can be derived using integration by parts is Wallis'
formula:

$$\frac{2}{\pi} \int_0^{\pi/2} dx \, \sin^{2n} x = \frac{2}{\pi} \int_0^{\pi/2} dx \, \cos^{2n} x \quad = \quad \frac{1 \cdot 3 \cdot 5 \cdots (2n-1)}{2 \cdot 4 \cdot 6 \cdots (2n)}, \quad n \text{ an integer,}$$

$$= \quad \frac{\Gamma(n + \frac{1}{2})}{\sqrt{\pi}\,\Gamma(n+1)}, \quad n \text{ a real number.}$$

Prove the first of these relations using integration by parts.

These examples are special in that the process terminates after a finite number of applications and provides an exact result. Our main interest in this section, however, is in cases where the process does not terminate but where the remaining integral on the right of equation 13.11 is smaller than the last term of the series, so that repeated integration by parts yields an infinite series. We have already seen an example of this type in chapter 5, where we used repeated integration by parts to derive asymptotic expansions. In that chapter we introduced the exponential integral, defined by†

$$E_1(x) = \int_x^\infty dt \, \frac{e^{-t}}{t}, \quad x > 0, \tag{13.12}$$

which cannot be evaluated in terms of elementary functions, and there showed that

$$E_1(x) \sim e^{-x} \sum_{k=1}^\infty \frac{(-1)^{k-1}(k-1)!}{x^k}.$$

In this example it is fairly easy to derive this result from equation 13.11 if we set $g(t) = 1/t$ and $f(t) = e^{-t}$ and let $N \to \infty$, though in more complicated cases the algebra can become overwhelming. But equation 13.11 is clearly an ideal candidate for the use of Maple.

For this technique to be useful it is necessary to make a judicious choice of $f(x)$ and $g(x)$ in order to ensure that (a) $f^{(-k)}(x)$ and $g^{(k)}(x)$ can be computed, and (b) that the remainder integral is small. If such functions cannot be found then the method is of

---

† This definition of the exponential integral is also valid for complex x provided $|\arg(x)| < \pi$; here we assume that x is real.

little value, but if such functions exist then the method can be used to derive useful series. Typically such methods can be used when $f(x)$ is e^{-ax} or e^{iax}, for then repeated integration of $f(x)$ is possible.

Exercise 13.18
Write a Maple procedure `int_part:=proc(f,g,N::integer)` which returns equation 13.11 for given expressions f and g and some integer N.

By choosing f and g appropriately, use this procedure to show that
$$e^x E_1(x) = \frac{1}{x} - \frac{1}{x^2} + \frac{2}{x^3} - \frac{6}{x^4} + \frac{24}{x^5} - \frac{120}{x^6} + \frac{720}{x^7} - \frac{5040}{x^8} + O(x^{-9}).$$

Exercise 13.19
Use Maple to show that
$$\int_0^\infty dt \, \frac{e^{-xt}}{\sqrt{1+t+t^2}} \sim \frac{1}{x} - \frac{1}{2x^2} - \frac{1}{4x^3} + \frac{21}{8x^4} - \frac{111}{16x^5} - \frac{345}{32x^6} + O(x^{-7}).$$

Using the expansion to $O(x^{-9})$ find the [3, 5] Padé approximant and compare, graphically, the values given by these two approximations with the numerical evaluation of the integral for $1 < x < 6$.

Exercise 13.20
By changing variables show that
$$\int_x^\infty dt \, e^{-t^4} \sim e^{-x^4} \left(\frac{1}{4x^3} - \frac{3}{16x^7} + \frac{21}{64x^{11}} + O(x^{-15}) \right).$$

Laplace integrals
Integrals of the form
$$F(x) = \int_0^\infty dt \, e^{-xt} f(t) \tag{13.13}$$
are called *Laplace integrals* and occur as solutions of some linear differential equations, normally initial value problems. The variable x may be complex but is normally restricted to the region $\Re(x) \geq a \geq 0$. If f is integrable over any finite interval $(0, T)$ and $f(t) = O(e^{at})$ for some constant a as $t \to \infty$, then the integral is absolutely convergent and represents an analytic function of x in the half-plane $\Re(x) > a$. Unless otherwise stated we shall assume that x is real.

If the integral exists and $f(t)$ is N times continuously differentiable in the neighbourhood of the origin, then the application of equation 13.11 gives
$$F(x) = \sum_{k=0}^{N} f^{(k)}(0) x^{-k-1} + \frac{(-1)^{N+1}}{x^{N+1}} \int_0^\infty dt \, f^{(N+1)}(t) e^{-xt}. \tag{13.14}$$

Provided all the differentials $f^{(k)}(0)$ exist the asymptotic expansion of $F(x)$ is the sum
$$F(x) = \int_0^\infty dt \, e^{-xt} f(t) \sim \sum_{k=0}^{\infty} f^{(k)}(0) x^{-k-1} \quad \text{as} \quad x \to \infty. \tag{13.15}$$

Fourier integrals

Integrals of the form

$$F(x) = \int_\alpha^\beta dt\, e^{ixt} f(t) \tag{13.16}$$

are called *Fourier integrals* and occur in very many circumstances, often involving wave motion or boundary value problems. In practice $|\alpha|$ and β can both be infinite, but in this section we shall assume that they are finite. An important application, discussed in chapter 8, is the representation of periodic functions in terms of Fourier series; then the Fourier coefficients are given by integrals of this form but with x taking discrete values.

The ordinary Bessel function of integer order, $J_n(x)$, can be defined by such an integral,

$$J_n(x) = \frac{1}{2\pi} \int_{-\pi}^\pi dt\, e^{i(nt - x\sin t)}, \tag{13.17}$$

but also two of the less well known real special functions, Anger's function $\mathbf{J}_\nu(x)$ and Weber's function $\mathbf{E}_\nu(x)$, are defined by similar integrals,

$$\mathbf{J}_\nu(x) + i\mathbf{E}_\nu(x) = \frac{1}{\pi} \int_0^\pi dt\, e^{i(\nu t - x\sin t)}. \tag{13.18}$$

The existence of the Fourier integral 13.16 for all real x is guaranteed by the Riemann–Lebesgue lemma, see for instance Whittaker and Watson (1965, page 172) or Apostol (1963, page 469). A slightly more general result states that if $f(t)$ is integrable over the range (a, b), if x is a real variable and if $a \le a' < b' \le b$, then

$$\int_{a'}^{b'} dt\, e^{ixt} f(t) \to 0 \quad \text{as} \quad x \to \pm\infty,$$

and the convergence is uniform in a' and b', Zygmund (1990, page 46).

The infinite integral $\int_\alpha^\infty dt\, e^{ixt} f(t)$ is a special case of the integral $\int_\alpha^\infty dt\, \phi(t) f(t)$ which can be shown to exist, Whittaker and Watson (1965, page 72), if the following two conditions are satisfied:

(1) the function $f(t)$ can be bounded by a function that decreases monotonically to zero as $t \to \infty$,

(2) and

$$\left| \int_a^w dt\, \phi(t) \right| \quad \text{is bounded as} \quad w \to \infty.$$

For Fourier integrals, $\phi(t) = e^{ixt}$, and $\int_a^w dt\, \exp(ixt)$ is bounded, so provided $f(t)$ satisfies condition (1) the integral 13.16 with $\beta = \infty$ exists. For instance, if $a > 0$,

$$\int_a^\infty dt\, \frac{\sin t}{t} = \frac{\pi}{2} \quad \text{exists though} \quad \lim_{T\to\infty} \int_a^T \frac{dt}{t} \quad \text{does not.}$$

If f and its derivatives are known the identity 13.11 can clearly be used to rewrite the integral for $F(x)$. Suppose that the first N derivatives of $f(t)$ exist but that $f^{(N+1)}(t)$ does not exist everywhere; then equation 13.11 gives

$$\int_\alpha^\beta dt\, e^{ixt} f(t) = -\left[e^{ixt} \sum_{r=0}^N \left(\frac{i}{x}\right)^{r+1} f^{(r)}(t) \right]_{t=\alpha}^\beta + \left(\frac{i}{x}\right)^{N+1} \int_\alpha^\beta dt\, f^{(N+1)}(t) e^{ixt}. \quad (13.19)$$

Exercise 13.21
Write a Maple procedure to evaluate the sum defined in equation 13.19 which will accept any suitable expression for $f(t)$ and arbitrary values of α, β and N. Use this procedure to show that the asymptotic expansion of the function defined by the integral

$$F(x) = \int_x^\infty dt\, \frac{\sin t}{\sqrt{t}}$$

is

$$F(x) \sim \frac{1}{x^{1/2}} \left(1 - \frac{3}{4x^2} + \frac{105}{16x^4} + \cdots\right) \cos x + \frac{1}{2x^{3/2}} \left(1 - \frac{15}{4x^2} + \frac{945}{16x^4} + \cdots\right) \sin x.$$

Extend this to $O(x^{11})$ and hence show that a Padé approximant is

$$F(x) = \frac{4(2280 + 489x^2 + 8x^4)}{x^{1/2}(10\,395 + 1980x^2 + 32x^4)} \cos x + \frac{2(10\,512 + 1641x^2 + 20x^4)}{x^{3/2}(63\,063 + 6864x^2 + 80x^4)} \sin x.$$

Compare the accuracy of this and the original asymptotic expansion for $1 < x < 6$.

Exercise 13.22
If the function $F(x)$ is defined by the integral

$$F(x) = \int_0^1 dt\, t^{-a} \sin(xt), \quad 0 < a < 1,$$

show that the asymptotic expansion of $F(x)$ for large x is

$$F(x) \sim \frac{A}{x^{1-a}} - \left(\frac{1}{x} - \frac{a(a+1)}{x^3}\right) \cos x - \frac{a}{x^2} \left(1 - \frac{(a+1)(a+2)}{x^2}\right) \sin x + O(x^{-5}),$$

where A is the number $\int_0^\infty dy\, y^{-a} \sin y$.
Hint: change the integration variable to $y = xt$ and rearrange the integral.

13.3.2 *Watson's lemma and its extensions*
The application of integration by parts to Laplace integrals often provides an asymptotic expansion, but is rather mechanical and can disguise the essential behaviour of the integrand that leads to the asymptotic expansion 13.14. Furthermore, direct application of this method fails to work in many important cases; the example in equation 13.21 overpage is one such case.

Consider the Laplace-type integral

$$F(x) = \int_0^A dt\, e^{-xt} f(t), \quad x > 0. \quad (13.20)$$

If the function $f(t)$ does not increase too rapidly and if A and x are sufficiently large, because e^{-xt} decays rapidly as t increases it is clear that the dominant contribution comes from the region around $t = 0$. In order to see how this can help we discuss the particular case

$$F(x) = \int_0^3 dt \, \frac{\sqrt{t}e^{-xt}}{1+\sqrt{t}},$$

(13.21)

for which the direct application of equation 13.14 fails because none of the derivatives of $f(t) = \sqrt{t}/(1+\sqrt{t})$ exist at the origin. The graphs of $f(t)$ and e^{-xt}, for various values of x, are shown in the figure.

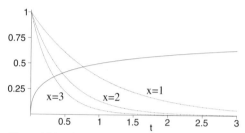

Figure 13.3 Graphs of $y = e^{-xt}$ for $x = 1, 2$ and 3, the dotted lines, and $y = \sqrt{t}/(1+\sqrt{t})$.

From this graph it is clear that for large values of x the behaviour of $f(t)$ at large t is irrelevant, so we should be able to replace $f(t)$ by its series expansion about the origin, $t^{1/2} - t + t^{3/2} + \cdots$, and then replace the upper limit of the integral by infinity (in order that the subsequent integrals can be evaluated) to give

$$F(x) \sim \int_0^\infty dt \, t^{1/2}e^{-xt} - \int_0^\infty dt \, te^{-xt} + \int_0^\infty dt \, t^{3/2}e^{-xt} + \cdots.$$

Using the definition of the Gamma function (equation 20.4, page 834), this becomes

$$F(x) \sim \frac{\sqrt{\pi}}{2x^{3/2}} - \frac{1}{x^2} + \frac{3\sqrt{\pi}}{4x^{5/2}} + \cdots.$$

(13.22)

Clearly it would be easy to find more terms of this series: the table below gives a numerical comparison between some exact values of $F(x)$, obtained by numerical integration, and those given by the above series.

Table 13.1. *Comparison of the approximation 13.22 with the exact integral.*

x	5	10	15	20	25
$F(x)$ (exact)	0.053 72	0.020 90	0.011 92	0.007 97	0.005 82
$F(x)$ (eqn 13.22)	0.063 05	0.022 23	0.012 34	0.008 15	0.005 92

In deriving this approximation we have performed many inadmissible operations justified on the grounds that for large values of x the dominant contribution to the integral comes from the region near $t = 0$. Consider what has been done. First we approximated the integral by replacing the upper limit by infinity,

$$F(x) \sim F_1(x) = \int_0^\infty dt \, \frac{\sqrt{t} e^{-xt}}{1 + \sqrt{t}}.$$

This causes no problems because we can write

$$F_1(x) = F(x) + R(x), \quad \text{where} \quad R = \int_3^\infty dt \, \frac{\sqrt{t} e^{-xt}}{1 + \sqrt{t}} < \frac{e^{-3x}}{x},$$

so $|F(x) - F_1(x)| \to 0$ exponentially as $x \to \infty$; it follows that the asymptotic expansion of $F_1(x)$ is also an asymptotic approximation of $F(x)$.

In the next stage of the approximation, however, we used the series approximation $1/(1 + \sqrt{t}) = 1 - t^{1/2} + t + \cdots$, which is valid only for $|t| < 1$, in the integral in which t ranged from 0 to ∞. Thus there is no justification for assuming that the series 13.22 provides an approximation to the original integral.

The proof that this approximation is a valid asymptotic expansion is provided by *Watson's lemma*, which emphasises the fact that if $x \gg 1$ the dominant contribution comes from the neighbourhood of $t = 0$.

Watson's lemma If $f(t)$ has the asymptotic expansion

$$f(t) \sim t^\alpha \sum_{k=0}^\infty a_k t^{k\beta}, \quad \text{valid as } t \to 0, \tag{13.23}$$

where $\beta > 0$ and $\alpha > -1$, conditions necessary for the existence of the integral 13.19, then the asymptotic expansion of $F(x)$ is

$$F(x) = \int_0^A dt \, e^{-xt} f(t) \sim \sum_{k=0}^\infty a_k \frac{\Gamma(\alpha + k\beta + 1)}{x^{\alpha + k\beta + 1}}, \quad \text{valid as } x \to \infty. \tag{13.24}$$

The proof of this result can be found in Copson (1967, chapter 6).

Notice that if $f(t)$ has a Taylor's series expansion at $t = 0$, $\alpha = 0$ and β is an integer, normally 1, or 2 if f is even, then only integer powers of $1/x$ appear in the asymptotic expansion of $F(x)$. In this case 13.24 is the same as 13.15.

The essence of this approximation is that for $x \gg 1$ the dominant contribution to the integral comes from near $t = 0$, consequently

$$F(x) = \int_0^A dt \, e^{-xt} f(t) = \int_0^\infty dt \, e^{-xt} f(t) + \text{exponentially small terms},$$

and replacing f in the last integral by its asymptotic expansion, interchanging the order of summation and integration gives the asymptotic expansion of the integral. This algebra does not prove equation 13.24, but it does show how the approximation is produced.

For an example of the application of Watson's lemma consider the integral representation of the modified Bessel function $K_0(x)$,

$$K_0(x) = \int_1^\infty dt \, \frac{e^{-xt}}{(t^2 - 1)^{1/2}}. \tag{13.25}$$

You may plot the graph of this function using its Maple representation `BesselK(n,x)` with n set to zero. Watson's lemma cannot be applied directly to this integral because the limits are wrong; but with the simple change of variables $s = t + 1$, it may be put in the form of equation 13.20,

$$K_0(x) = \frac{e^{-x}}{\sqrt{2}} \int_0^\infty ds \, \frac{e^{-sx}}{s^{1/2}(1 + s/2)^{1/2}}.$$

Now use the binomial theorem to expand the denominator in powers of s,

$$s^{-1/2}(1 + s/2)^{-1/2} = \frac{1}{\sqrt{\pi}} \sum_{k=0}^\infty s^{k-(1/2)} \frac{(-1)^k \Gamma(k + \frac{1}{2})}{2^k k!}, \quad |s| < 2.$$

Although this expansion is valid only for $|s| < 2$ we can still use it because it is valid as $s \to 0$; by comparison with the series 13.23 we see that $\alpha = -\frac{1}{2}$ and $\beta = 1$ and hence that the asymptotic expansion is

$$K_0(x) \sim \frac{e^{-x}}{\sqrt{2\pi x}} \sum_{k=0}^\infty (-1)^k \frac{\Gamma(k + \frac{1}{2})^2}{k!(2x)^k}, \quad x \to \infty.$$

In this example the similarity of the original integral to the basic form, defined in equation 13.20, was fairly obvious. In the next example, the complement to the error function, the relation is not so transparent. The integral representation of this function is

$$\text{erfc}(x) = 1 - \text{erf}(x) = \frac{2}{\sqrt{\pi}} \int_x^\infty dt \, e^{-t^2}. \tag{13.26}$$

We convert this into the canonical form by defining a new integration variable $t = x + w$ which transforms the integral into

$$\text{erfc}(x) = \frac{2e^{-x^2}}{\sqrt{\pi}} \int_0^\infty dw \, e^{-2xw} e^{-w^2}.$$

The Taylor expansion of e^{-w^2} is

$$e^{-w^2} = \sum_{n=0}^\infty \frac{(-1)^n}{n!} w^{2n},$$

and by comparing with equation 13.23 we see that $\alpha = 0$, $\beta = 2$, and in the integral replace x by $2x$, so that Watson's lemma gives

$$\text{erfc}(x) = \frac{e^{-x^2}}{x\sqrt{\pi}} \sum_{n=0}^\infty (-1)^n \frac{(2n-1).(2n-3). \cdots .3.1}{(2x^2)^n},$$

$$= \frac{e^{-x^2}}{x\sqrt{\pi}} \left\{ 1 - \frac{1}{2x^2} + \frac{1.3}{(2x^2)^2} - \frac{1.3.5}{(2x^2)^3} + \cdots \right\}. \tag{13.27}$$

Exercise 13.23
Show that the application of Watson's lemma to the integral 13.21 gives

$$F(x) \sim x^{-3/2} \sum_{k=0}^\infty (-1)^k \frac{\Gamma(k/2 + 3/2)}{x^{k/2}}.$$

Using the criterion discussed on page 185 and the approximation $\Gamma(az + b) \sim z^b \Gamma(az)$, $z \to \infty$, show that this series provides the best approximation when truncated at $k \simeq x$. Using this criterion evaluate the asymptotic expansion for $5 \le x \le 25$ and compare with the exact values quoted in table 13.1.

Exercise 13.24
Use Watson's lemma to show that if $A \gg 1$,

$$\int_0^A dt \, \frac{e^{-xt}}{1 + t^2} \sim \frac{1}{x} - \frac{2!}{x^3} + \frac{4!}{x^5} + \cdots + (-1)^n \frac{(2n)!}{x^{2n+1}} + \cdots .$$

Exercise 13.25
Show that if $a > 0$ and $b > 0$,

$$\int_a^\infty dt \, t^{-b} e^{-xt} \sim \frac{e^{-ax}}{xa^b} \sum_{k=0}^\infty \frac{(-1)^k}{(ax)^k} \frac{\Gamma(k + b)}{\Gamma(b)} .$$

Extensions to Watson's lemma
There are a number of extensions to Watson's lemma, most of which involve a change of variables to cast integrals into the required form of 13.20. It is worth studying these as they show very clearly how the behaviour of the integrand near $t = 0$ determines the asymptotic form of the integral.

The first extension removes the restriction on the form of the asymptotic expansion of $f(t)$, equation 13.23, which requires that the ratio of all successive terms behaves as y^β. This restriction is removed by the following result, the proof of which may be found in Erdélyi (1956, page 34); strictly, it is proved only for Laplace integrals with $A = \infty$. If the Laplace integral exists for some x and if λ_k are a set of real numbers with the order $0 < \lambda_1 < \lambda_2 < \cdots < \lambda_N$, and if $f(t)$ has the asymptotic expansion

$$f(t) \sim \sum_{k=1}^N a_k t^{\lambda_k - 1} \quad \text{as } t \to 0,$$

then

$$F(x) \sim \sum_{k=1}^N a_k \Gamma(\lambda_k) x^{-\lambda_k} \quad \text{as } x \to \infty. \tag{13.28}$$

For the second extension we observe that Watson's lemma can sometimes be used for integrals of the form

$$F(x) = \int_0^A dt \, f(t) e^{-h(t,x)}, \tag{13.29}$$

provided the function $h(t, x)$ is sufficiently simple; in some cases $h(t, x)$ is the product $h(t, x) = x h_1(t)$, where $h_1(t)$ is independent of x, and then the Laplace integral is regained if $h(t, x) = xt$.

The basic idea behind all the following analysis is to change the integration variable in order to ensure that the dominant contributions come from integrals of the form of equation 13.20 (page 497). There are a number of circumstances where this is possible and here we deal with two common cases.

The simplest case is when $h(t, x)$ is a monotonic increasing function of t on $0 \le t < A$ for all x in some range; this is the case when $h(x, t) = xt$. Introduce a new variable $s(t) = h(t, x) - h(0, x)$ so that for each fixed value of x, $s(t)$ is an increasing function of t and the inverse function $t(s)$ exists. Using s for the integration variable, the integral 13.29 becomes

$$F(x) = e^{-h(0,x)} \int_0^B ds \, g(s) e^{-s}, \quad \text{where} \quad g(s, x) = f(t(s)) \frac{dt}{ds}, \quad (13.30)$$

and $B = h(A, x) - h(0, x) \gg 1$. Watson's lemma may be applied to this integral; the only practical difficulty being the inversion of the function $s = h(t, x) - h(0, x)$ to find $t(s)$, but since we require only a series approximation to dt/ds and $f(t(s))$ about $s = 0$, this difficulty can often be overcome by using Maple.

Exercise 13.26
If the function $F(x)$ is defined by the integral

$$F(x) = \int_0^\infty dt \, f(t) e^{-x \tanh t}, \quad x > 0,$$

and

$$f(t) = t^{1/2} \sum_{k=0}^\infty f_k t^k \quad \text{as} \quad t \to 0,$$

use Maple to show that

$$F(x) \sim \frac{\sqrt{\pi}}{x^{3/2}} \left(\frac{f_0}{2} + \frac{3f_1}{4x} + \frac{35f_0 + 30f_2}{16x^2} + \frac{315f_1 + 210f_3}{32x^3} + \cdots \right).$$

Exercise 13.27
Show that if $x > 0$,

$$\int_0^{\pi/2} dt \, (1+t)^{1/4} \exp\left(-\frac{xt}{(1+t)^{1/4}} \right) \sim \frac{1}{x} \left\{ 1 + \frac{3}{4x} - \frac{15}{64x^3} + \frac{9}{16x^4} - \frac{945}{1024x^5} + \cdots \right\}.$$

In the above two exercises the exponent $h(t, x)$ was monotonic increasing for all t in the range of integration. In other cases this may not be true, but if there is a single maximum, at which $dh/dt = 0$, but no global minima, then a similar method may be used if the integration range is divided in two.

Exercise 13.28
If

$$F(x) = \int_0^1 dt \, f(t) e^{-xt(1-t)}, \quad x > 0,$$

and if $f(t)$ has a Taylor's series expansion about both $t = 0$ and $t = 1$, show that

$$F(x) \sim \sum_{k=1}^{\infty} \frac{a_k}{x^k},$$

where

$$
\begin{aligned}
a_0 &= f(0) + f(1), \quad a_1 = 2\left[f(0) + f(1)\right] + f'(0) + f'(1), \\
a_2 &= 12\left[f(0) + f(1)\right] + 6\left[f'(0) + f'(1)\right] + 2\left[f''(0) + f''(1)\right],
\end{aligned}
$$

and find expressions for a_3 and a_4.

In the case where $f(t) = t^\alpha$, $\alpha > 0$, show that

$$
\begin{aligned}
F(x) \quad \sim \quad & \frac{1}{x^{1+\alpha}} \left(\Gamma(1+\alpha) + \frac{(2+\alpha)\Gamma(2+\alpha)}{x} + \frac{\alpha(\alpha+3)\Gamma(\alpha+3)}{2x^2} + \cdots \right) \\
& + \frac{1}{x}\left(1 + \frac{2-\alpha}{x} + \frac{(\alpha-3)(\alpha-4)}{x^2} + \cdots \right).
\end{aligned}
$$

If the function $h(t, x)$, appearing in the integral 13.29, is not a monotonic increasing function of t and has minima, we expect the dominant contributions to come from the region of t around the global minimum of $h(t, x)$ for $0 \le t < A$. We illustrate the method used in this circumstance by finding the asymptotic expansion of the factorial function, $x!$, where x is real, that is, the generalisation of Stirling's formula for non-integer arguments. Because $x! = \Gamma(x+1)$ we may use the integral definition of the gamma function (equation 20.4, page 834), to obtain

$$x! = \int_0^\infty dt\, t^x e^{-t} = \int_0^\infty dt\, \exp(x \ln t - t), \quad x > -1. \tag{13.31}$$

The function $h(t, x) = t - x \ln t$ has a single stationary point at $dh/dt = 1 - x/t = 0$, that is, $t = x$, and since $d^2 h/dt^2 = x/t^2 > 0$ this is a minimum.

A seemingly sensible strategy is to expand about the minimum at $t = x$ to give

$$h(t, x) = x - x \ln x + \frac{(t-x)^2}{2x} + O((t-x)^3),$$

and to approximate the integral in the form

$$x! \simeq e^{-x+x\ln x} \int_{-\infty}^{\infty} dt\, \exp\left(-\frac{(t-x)^2}{2x} \right) = \sqrt{2\pi x} \left(\frac{x}{e} \right)^x.$$

This is the correct first term in the asymptotic expansion.

But how are the higher-order terms of the expansion obtained? An obvious method is to continue the expansion of $h(t, x)$ to higher order, $h(t, x) = x - x \ln x + \frac{(t-x)^2}{2x} - \frac{(t-1)^3}{3x} + \cdots$, but when substituted into equation 13.31 this produces an integral that cannot be evaluated in terms of elementary functions.

A more elegant method is to take advantage of the fact that h has a local minimum and near this minimum $h(t, x)$ has a shape similar to a parabola, as shown in figure 13.4.

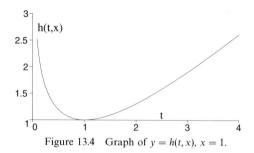

Figure 13.4   Graph of $y = h(t, x)$, $x = 1$.

The shape of this graph suggests defining a new variable z:

$$
\begin{aligned}
z^2 &= h(t, x) - h(x, x), \\
&= t - x - x(\ln t - \ln x),
\end{aligned}
\tag{13.32}
$$

so that $z(t)$ is zero at the position of the minimum, $t = x$. The signs of the square root are defined by the conditions

$$
z(t) > 0 \quad \text{for} \quad t > x \qquad \text{and} \qquad z(t) < 0 \quad \text{for} \quad t < x,
$$

so the relation between z and t is one-to-one for $0 < t < \infty$ and

$$
\lim_{t \to 0} z(t) = -\infty, \qquad \lim_{t \to \infty} z(t) = \infty.
$$

A graph of the function $t(z)$ is shown in figure 13.5.

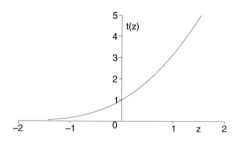

Figure 13.5   Graph of the function $t(z)$, with $x = 1$,
defined implicitly by equation 13.32.

The relation 13.32 can be inverted to give the function $t(z)$, and the integral 13.31 becomes

$$
x! = e^{-x + x \ln x} \int_{-\infty}^{\infty} dz \, \frac{dt}{dz} e^{-z^2}.
\tag{13.33}
$$

This equation is exact, but is useful only if $\frac{dt}{dz}$ can be expressed as a function of z. Equation 13.32 defines $z(t)$ uniquely, but to find $t(z)$ it is necessary to invert the Taylor's series of $z(t)$ about the minimum at $t = x$. The Taylor's series has a finite radius of convergence, so use of its inverse in 13.33 gives an asymptotic expansion for the integral, because exponentially small contributions from large $|z|$ are given incorrectly.

The expansion of 13.32 can be written as

$$z(t)^2 = \frac{(t-x)^2}{2x}\left[1 - \frac{2(t-x)}{3x} + \frac{(t-x)^2}{2x^2} + O((t-x)^3)\right].$$

On taking the square root this gives

$$z(t) = \frac{t-x}{\sqrt{2x}}\left[1 - \frac{t-x}{3x} + \frac{7(t-x)^2}{36x^2} + O((t-x)^3)\right], \tag{13.34}$$

where the sign is chosen to ensure that $z > 0$ when $t > 0$. Finally, we can invert this series to give $t - x$ as a series in z,

$$t = x + z\sqrt{2x} + \frac{2}{3}z^2 + \frac{\sqrt{2}}{18\sqrt{x}}z^3 + O(z^4).$$

Using this to determine dt/dz and substituting into equation 13.33 gives

$$x! \sim e^{-x+x\ln x}\int_{-\infty}^{\infty} dz\, e^{-z^2}\left(\sqrt{2x} + \frac{4}{3}z + \frac{\sqrt{2}}{6\sqrt{x}}z^2 + O(z^3)\right).$$

The integrals over the odd powers of z are zero, and using 20.13 (page 837), we obtain

$$x! \sim \sqrt{2\pi x}\left(\frac{x}{e}\right)^x\left(1 + \frac{1}{12x} + O(x^{-2})\right). \tag{13.35}$$

This procedure has given the first two terms in the asymptotic expansion. Other terms can be obtained simply by carrying this analysis to higher terms, and though this results in lengthy algebraic expressions, with Maple these computations are relatively easy.

Summary

In this section we have described methods for finding the asymptotic expansion of functions defined by integrals of the form

$$F(x) = \int_0^A dt\, f(t)e^{-h(t,x)}, \tag{13.36}$$

where x is a real variable and h and f suitably well behaved real functions. It has been shown that the behaviour of $F(x)$ depends crucially upon the shape of $h(t,x)$, considered as a function of t for fixed x, and we have also considered two possible cases.

First, if $h(t,x)$, regarded as a function of t, has no minima, then we may choose one (or two, see exercise 13.28) new variables $s = h(t,x) + \text{constant}$ so that over a range of t the function $h(t,x)$ is monotonic and the equation for $s(t)$ is invertible. In this case the integral reduces to either a Laplace-type integral or a sum of two Laplace-type integrals to which Watson's lemma may be applied. In the case $h(t,x) = xh_1(t)$ the dominant term in the asymptotic expansion of $F(x)$ normally behaves as x^{-1}.

In the second case $h(t,x)$ has one or more minima at $(t_1, t_2, \ldots, t_N)$, which may depend upon x; provided these are sufficiently far apart each of these minima may be regarded as separate, and by defining new variables $z^2 = h(t, x) - h(t_k, x)$, and expressing $t(z) - t_k$ as a power series in z, the asymptotic expansion for each contribution may be found. In the special case that $h(t, x) = x^2 h_1(t)$ the dominant term in the asymptotic expansion of $F(x)$ normally behaves as $x^{-1/2}$.

Exercise 13.29

Use the method just discussed to show that

$$\left(\frac{e}{x}\right)^x \frac{x!}{\sqrt{2\pi x}} \sim 1 + \frac{1}{12x} + \frac{1}{228x^2} - \frac{139}{51\,840x^3}$$
$$- \frac{571}{2\,488\,320x^4} + \frac{163\,879}{209\,018\,880x^5} + \frac{5\,246\,819}{75\,246\,796\,800x^6}$$
$$- \frac{534\,703\,531}{902\,961\,561\,600x^7} - \frac{4\,483\,131\,259}{86\,684\,309\,913\,600x^8} + O(x^{-9}).$$

Exercise 13.30

Show that if $a > 0$ and $x \gg 1$,

$$\int_{-\pi/2}^{\pi/2} dt \, \exp(-x \sin^2 t + a \cos t) \sim \frac{e^a \sqrt{2\pi}}{\sqrt{2a + x}} \left(1 + \frac{(8a + x)}{8(2a + x)^2} + \cdots\right).$$

Exercise 13.31

Show that the asymptotic expansion of the function

$$F(x, a) = \int_{-\infty}^{\infty} dt \, e^{-xh(t)}, \quad h(t) = t^2(t^2 - a^2) + \frac{a^4}{4} \quad \text{with } a > 0$$

is, for large x,

$$F(x, a) \sim \frac{1}{a} \sqrt{\frac{2}{\pi x}} \sum_{k=0}^{\infty} \frac{\Gamma(2k + 1/2)\Gamma(k + 1/2)}{(2k)!} \left(\frac{4}{xa^4}\right)^k.$$

Show also that $F(x, 0) = \Gamma(1/4)/(2x^{1/4})$.

Exercise 13.32

The Bernoulli number B_{2n} can be expressed in terms of the integral

$$B_{2n} = (-1)^{n-1} 4n \int_0^{\infty} dt \, \frac{t^{2n-1}}{e^{2\pi t} - 1}.$$

Use this to obtain the asymptotic expansion

$$B_{2n} = (-1)^{n-1} 4n \sqrt{\frac{2n-1}{2\pi}} \left(\frac{2n-1}{2\pi e}\right)^{2n-1} \left(1 + \frac{1}{12(2n-1)} + \frac{1}{288(2n-1)^2} + \cdots\right).$$

13.4 The method of steepest descents

The method of steepest descents is a general method for finding the asymptotic expansion of complex integrals having the form

$$F(\lambda) = \int_{\mathscr{C}} dz \, g(z) e^{\lambda h(z)}, \tag{13.37}$$

where λ is a large positive real variable, $g(z)$ and $h(z)$ are analytic functions and $\mathscr{C}$ a contour in the complex z-plane. The method was originally introduced by Riemann in a paper, published posthumously in 1892, in which he found an asymptotic expansion for the hypergeometric function, $F(n-c, n+1+a; 2n+2+a+b; s)$, defined in chapter 7, which has the integral representation

G. F. B. Riemann (1826–1866)

$$F = \frac{\Gamma(2n+2+a+b)}{\Gamma(n+1+a)\Gamma(n+1+b)} \int_0^1 dz \, z^a (1-z)^b (1-sz)^c \left(\frac{z(1-z)}{1-sz} \right)^n. \tag{13.38}$$

Riemann used this representation to find an approximation valid in the limit of large n. This integral can be put in the canonical form of equation 13.37 by setting $n = \lambda$, $g(z) = (1-sz)^c z^a (1-z)^b$ and $h(z) = \ln(z(1-z)) - \ln(1-sz)$. The method was subsequently developed by the chemist Debye in his analysis of the asymptotic expansion of the Bessel function, exercise 13.34.

P. J. W. Debye (1884–1966)

The method of steepest descent is slightly more complicated than the methods discussed previously because it involves understanding some complex variable theory, in particular we require Cauchy's theorem, stating that if $f(z)$ is analytic at all points on and inside a closed curve $\mathscr{D}$ then $\oint_{\mathscr{D}} dz \, f(z) = 0$; this allows integration paths to be moved to more convenient locations. Detailed discussions of this important theorem can be found in Whittaker and Watson (1965, chapter 6) and Apostol (1963, chapter 16).

The essence of the method is to deform the integration path $\mathscr{C}$ so that the value of the integral defining $F(\lambda)$ is dominated by a contribution from a neighbourhood of a single point, or a few isolated points. The most suitable path depends upon the nature of the function $h(z)$ — the behaviour of $g(z)$ normally being irrelevant — thus in practice it is necessary to understand the behaviour of $h(z)$ in the complex plane before choosing the most suitable path. This is usually the difficult part of the analysis. Once the correct path has been chosen the approximation to $F(\lambda)$ becomes fairly automatic; it is often a matter of using Watson's lemma, though the consequent algebra may be complicated.

The analysis is most easily accomplished by decomposing $h(z)$ into its real and imaginary parts,

$$h(z) = u(x, y) + iv(x, y), \quad z = x + iy, \tag{13.39}$$

where $u(x, y)$ and $v(x, y)$ are real functions of the real variables x and y. In deforming the path $\mathscr{C}$ the objective is to find points at which the largest value of $|e^{\lambda h(z)}| = e^{\lambda u}$ is concentrated on the shortest length of path; this suggests making the path pass through the stationary points of $h(z)$ and choosing it so that $h(z)$ decreases fastest when moving away from this point. In order to do this we need to understand a little how the contours of $u(x, y)$ and $v(x, y)$ can behave.

First consider a specific example: let

$$h(z) = \frac{1}{6} z^3 + \frac{1}{4} z^2 - z,$$

so $h'(z) = \frac{1}{2}(z-1)(z+2) = 0$ when $z = -2$ and 1. The contours of $u(x, y)$ and $v(x, y)$ are shown in figures 13.6 and 13.7.

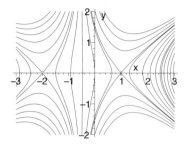

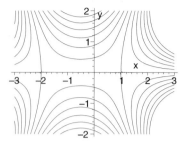

Figure 13.6   Contours of $u = \Re(h)$.   Figure 13.7   Contours of $v = \Im(h)$.

There are two important points to notice in the figures.

(1) At the stationary points, $z = -2$ and 1, neither $u(x, y)$ nor $v(x, y)$ have extrema; both are saddle points.
(2) At these saddle points two contour lines cross; we shall see below that they intersect at right angles.

Now return to the general case and consider an analytic function with a stationary point at $z = z_s$, so $h'(z_s) = 0$, and suppose also that the first $N(\geq 1)$ derivatives are zero, $h^{(r)}(z_s) = 0$ for $r = 1, \ldots, N$. The point z_s is named a *critical point* of the function $h(z)$. Near such a critical point the function may be approximated by the first two non-zero terms in the Taylor's series,

$$h(z_s + w) = h(z_s) + \frac{1}{(N+1)!}h^{(N+1)}(z_s)w^{N+1} + O(w^{N+2}).$$

Thus as we move round z_s on the small circle $w = re^{i\theta}$, the real and imaginary parts of $h(z)$ change according to

$$\begin{aligned}
\Re(h(z_s + w)) &= \Re(h(z_s)) + A\cos((N+1)\theta + \alpha) \\
\Im(h(z_s + w)) &= \Im(h(z_s)) + A\sin((N+1)\theta + \alpha),
\end{aligned} \tag{13.40}$$

where A and α are constants defined by

$$A = r^{N+1}|h^{(N+1)}(z_s)|/(N+1)!, \quad \alpha = \arg(h^{(N+1)}(z_s)).$$

The values of these constants are not relevant for the analysis that follows. As θ increases from 0 to 2π the values of both $u(x, y)$ and $v(x, y)$ oscillate about their values at the stationary point and equal this value at $N + 1$ equally spaced values of θ. It follows that stationary points of analytic functions cannot be extrema and that there are $N + 1$ contour lines passing through each stationary point, the angle between adjacent lines being $\pi/(N + 1)$. In addition these radial contour lines of $u(x, y)$ and $v(x, y)$ alternate and the angle between a radial contour of u and the adjacent contour of v is $\pi/(2N+2)$. In the example depicted in figures 13.6 and 13.7 the second derivative is not zero at either critical point, $N = 1$, and the contours through each point cross at right angles.

At an ordinary point, where $h'(z) \neq 0$, the contours of $u(x, y)$ and $v(x, y)$ are perpendicular. To see this, consider adjacent points z and $z + w$ and set $N = 0$ in equation 13.40, to give

$$\begin{aligned}
\Re(h(z + w)) &= \Re(h(z)) + A\cos(\theta + \alpha) \\
\Im(h(z + w)) &= \Im(h(z)) + A\sin(\theta + \alpha),
\end{aligned} \qquad (13.41)$$

where A and α are defined appropriately. On a contour of $u = \Re(h)$ passing through z we have $\theta + \alpha = \pm\pi/2$ and on a contour of $v = \Im(h)$, $\theta + \alpha = 0, \pi$; thus the two contours are perpendicular. This result is needed below.

Clearly it is the real part of $h(z)$ that governs the magnitude of the integrand, so if the contour is deformed to pass through a saddle we want it to coincide with the line along which $u(x, y)$ decreases most rapidly from its value at the saddle. It follows from equations 13.41 that $\mathscr{C}$ should be perpendicular to the contour lines of $u(x, y)$, and hence a contour of $v(x, y)$.

In summary, the path $\mathscr{C}$ should pass through the stationary points of $h(z)$ and coincide with one of the contours of $v(x, y) = \Im(h(z))$. Along this path $|e^{\lambda h(z)}| = e^{\lambda u(x, y)}$ decreases most rapidly and the phase of $e^{\lambda h(z)}$ is constant. If the integral is to be evaluated numerically this is the best path to choose for numerical efficiency because along it the integrand does not oscillate.

The saddle point method

The preceding analysis has shown that the dominant contribution to the integral will come from one or more of the saddle points, and has also shown which path to use through each saddle point. The general theory uses this, or an equivalent, path but is often difficult to apply, so we first describe a simpler method which provides the first term in the asymptotic expansion of the integral with very little effort.

This method is usually named the *saddle point* method. Suppose that the integration path $\mathscr{C}$ can be deformed to pass through the saddle point — we shall assume for the sake of simplicity that there is only one saddle point — at $z = z_s$ and that $h''(z_s) \neq 0$. The Taylor expansion about this point gives

$$h(z_s + w) = h(z_s) + \frac{1}{2}w^2 h''(z_s) + O(w^3), \quad h''(z_s) \neq 0, \qquad (13.42)$$

and we use this approximation for $h(z)$ in the integral. It is now necessary to choose an appropriate path through the critical point. Since $h''(z_s) \neq 0$ there are two lines of constant phase through z_s, crossing at right angles: a schematic diagram showing this is given in figure 13.8, where the critical point is at the origin. The expansion 13.40 shows that on one of these lines, represented by CD in the figure, $w^2 h''(z_s)$ increases with $|w|$ and on the other, AB, it decreases.

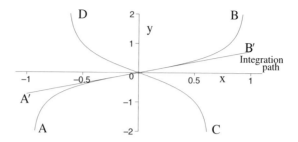

Figure 13.8   Diagram showing the two contours, AB and CD,
crossing at the critical point.

Thus we choose the integration path to be the straight line $A'B'$ tangent to AB at z_s, so that on this line $w^2 h''(z_s) \leq 0$. The direction of integration along this line is determined by the original contour so cannot be decided in this general analysis. Here we assume the integration is in the direction AB; the opposite direction merely changes the sign of the resultant formula. With this path and approximating $h(z)$ using the Taylor's series 13.42 we obtain

$$\int_{\mathscr{C}} dz\, g(z) e^{\lambda h(z)} \sim g(z_s) \exp\left(\lambda h(z_s) + i\frac{\pi}{2} - \frac{i}{2}\arg(h''(z_s))\right) \sqrt{\frac{2\pi}{\lambda |h''(z_s)|}}. \qquad (13.43)$$

It is important to remember that the overall sign of this expression is not determined by this general analysis but that it depends upon the original contour.

Exercise 13.33
Derive equation 13.43.

If there is more than one saddle point, and all are well separated, then the asymptotic expansion of $F(\lambda)$ comprises the sum of similar expressions, one for each saddle.

For an illustration of the method consider the calculation of the asymptotic expansion of the Airy function, which may be defined by the integral

$$\text{Ai}(u) = \frac{1}{\pi} \int_0^\infty ds\, \cos\left(\frac{1}{3}s^3 + us\right), \qquad (13.44)$$

where u is any real number. A graph of the Airy function is shown in figure 11.13 (page 407), and from this you will observe that for $u > 0$, $\text{Ai}(u)$ decreases monotonically to zero. We use the saddle point method to approximate this behaviour.

The integral may be re-cast into the canonical form, equation 13.37, by the change of variables, $s = u^{1/2}t$ and $\lambda = u^{3/2}$, and by expressing the cosine in terms of exponentials,

$$\text{Ai}(u) = \frac{\lambda^{\frac{1}{3}}}{2\pi} \int_{-\infty}^\infty dt\, \exp i\lambda\left(\frac{1}{3}t^3 + t\right), \qquad \lambda = u^{3/2}, \qquad (13.45)$$

so that $g(z) = 1$, $h(z) = i(z^3/3 + z)$ and the path $\mathscr{C}$ is the whole of the real line.

The function $h(z)$ is stationary at $z = \pm i$, and since $h(i) = -\frac{2}{3}$ and $h(-i) = \frac{2}{3}$ we deform the path to pass through $z = i$. Expanding about $z = i$ gives

$$h(z) = -\frac{2}{3} - w^2 + O(w^3), \qquad w = z - i,$$

so that

$$
\text{Ai}(u) \sim \frac{\lambda^{1/3} e^{-2\lambda/3}}{2\pi} \int_{-\infty}^{\infty} dw \, e^{-\lambda w^2}, \quad \lambda = u^{3/2},
$$

$$
= \frac{e^{-2\lambda/3}}{2\lambda^{1/6}\pi^{1/2}} = \frac{1}{2u^{1/4}\sqrt{\pi}} \exp\left(-\frac{2}{3}u^{3/2}\right).
$$

Exercise 13.34
The ordinary Bessel function of order n can be defined by the integral

$$
J_n(x) = \frac{1}{2\pi} \int_{-\pi}^{\pi} dt \, e^{i(nt - x \sin t)}.
$$

Show that if $0 < x \ll n$ and $n \gg 1$ then

$$
J_n(x) \sim \frac{1}{\sqrt{2\pi\sqrt{n^2 - x^2}}} \left(\frac{x e^{\sqrt{1 - x^2/n^2}}}{n + \sqrt{n^2 - x^2}} \right)^n, \quad n \gg x > 0.
$$

Method of steepest descent for the Airy function

The saddle point method is a local expansion about the stationary point and usually yields the first term in the asymptotic expansion with ease. Higher-order terms are more difficult to obtain because they require more terms in the Taylor's series 13.42 and involve understanding the behaviour of $h(z)$ sufficiently well that the contour can be chosen appropriately. This is more difficult, and we proceed by considering the asymptotic expansion of the Airy function in order to show the type of analysis that is necessary.

As in the saddle point method, the integration path is moved into the complex plane, but because the whole asymptotic expansion is required the whole path must be considered, so it is necessary to determine the region of z for which the integrand decreases to zero as $|z| \to \infty$. Put $z = Re^{i\theta}$ with R large and positive, so

$$
\left| e^{i\lambda z^3/3} \right| = \exp\left(-\frac{1}{3}\lambda R^3 \sin 3\theta\right) \to 0 \quad \text{as} \quad R \to \infty \quad \text{if} \quad \sin 3\theta > 0,
$$

that is, if

$$
0 < \theta < \frac{1}{3}\pi, \quad \frac{2}{3}\pi < \theta < \pi \quad \text{or} \quad \frac{4}{3}\pi < \theta < \frac{5}{3}\pi.
$$

These are the shaded regions in the diagram.

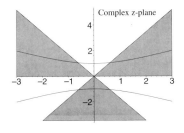

The stationary points of $h(z)$ are at the roots of

$$h'(z) = i(z^2 + 1) = 0, \quad \text{that is,} \quad z = \pm i.$$

At these points $h(i) = -\frac{2}{3}$ and $h(-i) = \frac{2}{3}$. We need to deform the path from the real axis to pass through one of these points. From the previous discussion the magnitude of the integrand will decrease most rapidly along a line given by $\Im(h) = $constant, and since $h(\pm i) = \mp\frac{2}{3}$ is real the equation for the line is $\Im(h(z)) = 0$, that is,

$$\Im(h) = \frac{1}{3}x(x^2 - 3y^2 + 3) = 0.$$

This equation defines three lines: one is the imaginary axis, $x = 0$; a second lies in the upper half plane and is the hyperbola $y = \sqrt{1 + x^2/3}$; the third lies in the lower half plane and is the hyperbola $y = -\sqrt{1 + x^2/3}$. The two branches of the hyperbola are shown by the solid curved lines in the diagram. The upper hyperbola has asymptotes $z = r\exp(i\pi/6)$ and $z = r\exp(5i\pi/6)$ as $r \to \infty$ which lie inside the shaded region, whilst on the asymptotes of the lower hyperbola $|e^{\lambda h(z)}|$ grows exponentially. Thus the path we require passes through the saddle at $z = i$.

On this path we have

$$\Re(h) = \frac{1}{3}y(y^2 - 3x^2 - 3),$$

and on the upper parabola, $x^2 = 3(y^2 - 1)$, $y > 1$, we have $\Re(h) = 2y(3 - 4y^2)/3$ which is negative and decreasing as y increases, as would be expected.

This behaviour is seen in the contour plots of $\Im(h)$ in figure 13.9 and of $\Re(h)$ in figure 13.10; note that the contours in the two graphs are perpendicular.

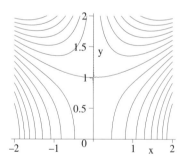

Figure 13.9   Contours of $v = \Im(h)$.

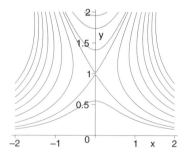

Figure 13.10   Contours of $u = \Re(h)$.

The values of the functions $\Im(h(z))$ and $\Re(h(z))$ are shown in figures 13.11 and 13.12, which give a more graphic picture of the saddle in both functions at $z = i$. Remember that $\Re(h)$ appears in an exponential as $\exp[\lambda\Re(h(z))]$ and for large λ the value of this exponential will decrease very rapidly as we move away from the saddle.

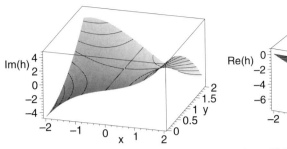

Figure 13.11   Surface defined by $v = \Im(h(z))$.

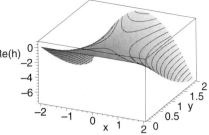

Figure 13.12   Surface defined by $u = \Re(h(z))$.

The original integral 13.45 (page 510) along this path can be written in the form

$$\frac{2\pi}{\lambda^{1/3}}\mathrm{Ai}(\lambda^{2/3}) = \int_i^{Z_1} dz\, e^{\lambda h(z)} - \int_i^{Z_2} dz\, e^{\lambda h(z)} = I_1 - I_2, \tag{13.46}$$

where $Z_1 = Re^{i\pi/6}$, $Z_2 = Re^{5i\pi/6}$ and where we shall let R tend to infinity. On Z_1, $\Re(z) > 0$, and on Z_2, $\Re(z) < 0$. Each of these integrals is evaluated using a variant of Watson's lemma, as described in section 13.3.2.

Consider the first integral I_1: define a new variable t in such a manner that when $z \simeq i$, $t \simeq z - i$, which, in this case, means setting

$$t^2 = h(i) - h(z) = (z - i)^2 - \frac{i}{3}(z - i)^3.$$

On the curve of steepest descent $\Im(h) =$ constant and since $h(i)$ is real the variable t is real on this curve and the square root is defined so that $t \geq 0$. Then

$$I_1 = e^{-2\lambda/3} \int_0^\infty dt\, e^{-\lambda t^2} \frac{dz}{dt}.$$

We now need to express $z - i$ as a power series in t; there is an elegant method of obtaining all terms of this series explicitly, but we leave this to the end of chapter, exercise 13.61 (page 523), as this analysis would distract from the story. The first few terms of this series are

$$z - i = t + \frac{it^2}{6} - \frac{5t^3}{72} - \frac{it^4}{27} + \cdots, \tag{13.47}$$

so that

$$I_1 = e^{-2\lambda/3} \int_0^\infty dt\, e^{-\lambda t^2} \left(1 + \frac{it}{3} - \frac{5t^2}{24} - \frac{4it^3}{27} + \cdots\right).$$

For the integral I_2 we make a similar substitution, but now choose t to be negative to give

$$I_2 = e^{-2\lambda/3} \int_0^{-\infty} dt\, e^{-\lambda t^2} \left(1 + \frac{it}{3} - \frac{5t^2}{24} - \frac{4it^3}{27} + \cdots\right),$$

and hence

$$\frac{2\pi}{\lambda^{1/3}}\,\mathrm{Ai}(\lambda^{2/3}) \;=\; e^{-2\lambda/3}\int_{-\infty}^{\infty} dt\, e^{-\lambda t^2}\left(1 - \frac{5t^2}{24} + \cdots\right),$$

$$=\; e^{-2\lambda/3}\sqrt{\frac{\pi}{\lambda}}\left(1 - \frac{5}{48\lambda} + \cdots\right).$$

Thus, remembering that $u = \lambda^{2/3}$, we finally obtain

$$\mathrm{Ai}(u) = \frac{1}{2u^{1/4}\sqrt{\pi}}\exp\left(-\frac{2u^{3/2}}{3}\right)\left(1 - \frac{5}{48u^{3/2}} + \cdots\right). \qquad (13.48)$$

All the other terms in this series are determined in exercise 13.61.

13.4.1 Steepest descent with end contributions

The example just considered was special in that the asymptotic expansion was given by an expansion about the saddle. When one or more of the integration limits is finite there are usually significant contributions from the end points of the integral which need to be taken into account. The effect of end points is illustrated in the following example.

Consider the function

$$F(\lambda) = \int_0^1 dt\, e^{i\lambda t^3}, \quad x \gg 1. \qquad (13.49)$$

By comparison with equation 13.37 we see that $h(t) = it^3$ and this is stationary at $t = 0$. The paths along which $e^{\lambda h(t)}$ decrease fastest away from $t = 0$ are the rays $t = re^{i\pi/6}$, $t = re^{5i\pi/6}$ and $t = re^{3i\pi/2}$, none of which pass through $t = 1$. The integrand is exponentially small for $|t| \gg 1$ in the segment $0 < \arg(t) < \pi/3$.

Thus the contour along which we integrate starts at the origin and goes to infinity along the ray $t = re^{i\pi/6}$, $\mathscr{C}_1$, and returns to $t = 1$ along a suitable path, $\mathscr{C}_2$, chosen below, equation 13.50, as shown in the figure.

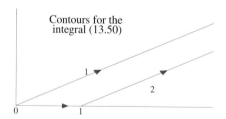

Figure 13.13   Sketch of the contours used to evaluate 13.49.

Thus the integral can be written as the sum of the two integrals

$$F(\lambda) = F_1 - F_2 = \int_{\mathscr{C}_1} dt\, e^{i\lambda t^3} - \int_{\mathscr{C}_2} dt\, e^{i\lambda t^3}. \qquad (13.50)$$

Consider each integral separately. The first integral, along the ray $t = re^{i\pi/6}$ away from the origin, may be evaluated exactly,

$$F_1(\lambda) = e^{i\pi/6} \int_0^\infty dr \, e^{-\lambda r^3} = \frac{\Gamma(4/3)}{\lambda^{1/3}} e^{i\pi/6}.$$

This contribution decreases relatively slowly with increasing λ, as a direct consequence of the t^3 term in the exponent of the integral, see exercise 13.37.

The second integral is evaluated using Watson's lemma, but we first need to choose a suitable path. This must start at $t = 1$ and go to infinity in the segment $0 < \arg(t) < \pi/3$ and it would be helpful if t^3 were a linear function of the new real integration variable. So we choose $it^3 = i - au$ for some a; then $t = 1$ when $u = 0$, and if we set $a = 1$, t approaches $\mathscr{C}_1$ as $u \to \infty$. On changing variables the second integral becomes

$$F_2(\lambda) = \frac{ie^{i\lambda}}{3} \int_0^\infty du \, \frac{e^{-\lambda u}}{(1 + iu)^{2/3}}.$$

This is exact and its asymptotic expansion can be found using Watson's lemma:

$$F_2(\lambda) \sim -\frac{e^{i\lambda}}{3\Gamma(2/3)} \sum_{k=0}^\infty \frac{\Gamma(k + 2/3)}{(i\lambda)^{k+1}}, \tag{13.51}$$

and hence

$$F(\lambda) = \int_0^1 dt \, e^{i\lambda t^3} \sim \frac{\Gamma(4/3)}{\lambda^{1/3}} e^{i\pi/6} + \frac{e^{i\lambda}}{3\Gamma(2/3)} \sum_{k=0}^\infty \frac{\Gamma(k + 2/3)}{(i\lambda)^{k+1}}. \tag{13.52}$$

For large λ the first term, the contribution from the saddle, is dominant, but there are also contributions $O(\lambda^{-1})$ from the end point at $t = 1$.

Exercise 13.35
Prove equation 13.52.

Exercise 13.36
Use the result given in equation 13.52 to show that

$$\int_0^x dt \, \sin(t^3) \sim \frac{1}{2}\Gamma(4/3) + \frac{1}{3\Gamma(2/3)} \sum_{k=0}^\infty \frac{\Gamma(k + 2/3)}{x^{3k+2}} \cos\left(x^3 - k\frac{\pi}{2}\right) \quad \text{as} \quad x \to \infty.$$

Exercise 13.37
Show that if $\lambda \gg 1$ and $\mu \geq 1$,

$$\int_0^1 dt \, e^{i\lambda t^\mu} \sim \frac{\Gamma(1 + 1/\mu)}{\lambda^{1/\mu}} e^{i\pi/2\mu} + \frac{e^{i\lambda}}{\mu\Gamma(1 - 1/\mu)} \sum_{k=0}^\infty \Gamma(k + 1 - 1/\mu) \left(-\frac{i}{\lambda}\right)^{k+1}.$$

Examine the limiting case $\mu = 1$.

Exercise 13.38
Show that

$$\int_0^\infty dt \, \exp\left[i\lambda\left(\frac{1}{3}t^3 + t\right)\right] \sim \frac{i}{\lambda} \sum_{p=0}^\infty \frac{(3p)!}{p! \, 3^p \lambda^{2p}} + \frac{e^{-2\lambda/3}}{4\lambda^{1/6}\sqrt{\pi}} \left(1 + \frac{i}{2\lambda} + \cdots\right), \quad \lambda \gg 1.$$

Why does this function decay to zero with increasing λ much more slowly than the related function defined by the integral 13.45?

13.5 Global contributions

In the previous two sections the dominant contributions to the integral came from a single point, or isolated points, and approximations are obtained by expanding about them. Not all integrals behave like this. Sometimes it is not possible to find such convenient points, and then there are no guaranteed methods of obtaining estimates of the integral and each example has to be treated individually. Nevertheless, there are several techniques which are often useful in extracting information from such integrals, some of which are changing variables, typically a re-scaling, splitting the integration range and subtracting from the integrand 'similar' functions that can be integrated. We illustrate the use of these methods in two 'typical' examples. In practice the successful conclusion of this type of analysis is frequently a matter of trial and error.

For our first example we examine the behaviour of the integral

$$I(\epsilon) = \int_0^1 \frac{du}{(\epsilon + u)^\alpha}, \quad 0 \le \alpha < 1, \quad 0 \le \epsilon \ll 1, \tag{13.53}$$

as $\epsilon \to 0$. This integral is finite for all $\epsilon \ge 0$, because $\alpha < 1$, but the derivative $I'(\epsilon)$ clearly does not exist at $\epsilon = 0$. One of the properties of $I(\epsilon)$ that we require is the form of the singularity in $I'(\epsilon)$ at $\epsilon = 0$.

This example has been chosen because the integral can be evaluated, and for this reason much of the following analysis may seem rather pointless: you should remember that it is being done simply to illustrate general methods and that in most examples the results are not so obvious. The integral is

$$
\begin{aligned}
I(\epsilon) &= \frac{1}{1 - \alpha} \left[(1 + \epsilon)^{1-\alpha} - \epsilon^{1-\alpha} \right], \quad 0 < 1 - \alpha \le 1, \\
&= \frac{1}{1 - \alpha} - \frac{\epsilon^{1-\alpha}}{1 - \alpha} + \epsilon - \frac{\alpha}{2} \epsilon^2 + O(\epsilon^3).
\end{aligned} \tag{13.54}
$$

For small ϵ it is clear that $I \simeq \frac{1}{1-\alpha}$ and $I'(\epsilon) = -\epsilon^{-\alpha} + 1 + O(\epsilon)$; we need to know how this behaviour arises and to be able to obtain it as a first term in an approximation.

Often it is helpful to expand the integrand about a suitable point, but in this case a simple expansion of the integrand $(\epsilon + u)^{-\alpha}$ over the entire integration range is not possible; the binomial expansion gives

$$
\begin{aligned}
(\epsilon + u)^{-\alpha} &= \epsilon^{-\alpha} \left(1 - \frac{\alpha u}{\epsilon} + O(u^2/\epsilon^2) \right), \quad u < \epsilon, \\
&= u^{-\alpha} \left(1 - \frac{\alpha \epsilon}{u} + O(\epsilon^2/u^2) \right), \quad u > \epsilon.
\end{aligned}
$$

These two expansions suggest splitting the range of integration into a short interval $(0, \epsilon)$ and a longer interval $(\epsilon, 1)$ and approximating the integrand by

$$(\epsilon + u)^{-\alpha} \simeq \begin{cases} \epsilon^{-\alpha}, & 0 \le u < \epsilon, \\ u^{-\alpha}, & \epsilon < u \le 1, \end{cases}$$

to give

$$I(\epsilon) \simeq \epsilon^{-\alpha} \int_0^\epsilon du + \int_\epsilon^1 du\, u^{-\alpha} = \epsilon^{1-\alpha} + \frac{1}{1-\alpha}\left(1 - \epsilon^{1-\alpha}\right)$$

$$\simeq \frac{1}{1-\alpha} - \frac{\alpha\epsilon^{1-\alpha}}{1-\alpha}, \tag{13.55}$$

showing that the dominant contribution comes from the integral $\int_0^1 du\, u^{-\alpha}$ obtained by setting $\epsilon = 0$, though we cannot be sure of this until we have estimated the size of the remainder. Observe that the second, smaller, term is given incorrectly by this simple approximation.

Since the dominant contribution to I comes from the integral $\int_0^1 du\, u^{-\alpha} = (1-\alpha)^{-1}$, we add and subtract this term to produce an equivalent integral with an integrand which is numerically smaller than the original over most of the integration range. That is, we write the integral 13.53 in the form

$$I(\epsilon) = \frac{1}{1-\alpha} - I_1(\epsilon), \quad I_1(\epsilon) = \int_0^1 du \left\{ \frac{1}{u^\alpha} - \frac{1}{(\epsilon+u)^\alpha} \right\}. \tag{13.56}$$

Now the problem is to estimate the dominant contribution to I_1. First, we re-scale the integration variable by defining a new variable v by $u = \epsilon v$ to give

$$I_1(\epsilon) = \epsilon^{1-\alpha} \int_0^{1/\epsilon} dv \left\{ \frac{1}{v^\alpha} - \frac{1}{(1+v)^\alpha} \right\}.$$

This idea can frequently be used to cast an integral into a more useful form. In this integral ϵ appears both in the external factor $\epsilon^{1-\alpha}$, which tends to zero with ϵ, and in the upper integration limit, which tends to infinity as $\epsilon \to 0$. In the $\epsilon = 0$ limit the integral exists because the integrand behaves as $v^{-1-\alpha}$, so now we introduce another useful but standard trick and write I_1 in the equivalent form

$$I_1(\epsilon) = \epsilon^{1-\alpha} \left[\int_0^\infty dv \left\{ \frac{1}{v^\alpha} - \frac{1}{(1+v)^\alpha} \right\} - \int_{1/\epsilon}^\infty dv \left\{ \frac{1}{v^\alpha} - \frac{1}{(1+v)^\alpha} \right\} \right], \tag{13.57}$$

which is useful because the first integral is independent of ϵ and the second is small. The second integral of equation 13.57 can be evaluated simply by expanding the integrand in inverse powers of v,

$$\frac{1}{v^\alpha} - \frac{1}{(1+v)^\alpha} = \frac{\alpha}{v^{1+\alpha}} - \frac{\alpha(1+\alpha)}{2v^{2+\alpha}} + \frac{\alpha(1+\alpha)(2+\alpha)}{6v^{3+\alpha}} + \cdots,$$

giving

$$I_1(\epsilon) = A(\alpha)\epsilon^{1-\alpha} - \epsilon + \frac{\alpha}{2}\epsilon^2 - \frac{1}{6}\alpha(1+\alpha)\epsilon^3 + \cdots, \tag{13.58}$$

where A is the number

$$A(\alpha) = \int_0^\infty dv \left\{ \frac{1}{v^\alpha} - \frac{1}{(1+v)^\alpha} \right\} = \frac{1}{1-\alpha}.$$

On combining this with equations 13.58 and 13.56 equation 13.54 is regained.

We have spent some time deriving this approximation to $I(\epsilon)$ because the analysis illustrates several techniques available for understanding such problems. In other similar situations the details will vary considerably but the basic methods are the same, though often some ingenuity is necessary for their application. These types of problem can be solved only by understanding the behaviour of the integrand and using this knowledge to produce a simpler problem that can be dealt with.

Exercise 13.39

Show that if $0 \leq \alpha < 1$,

$$\int_0^\infty dv \left\{ \frac{1}{v^\alpha} - \frac{1}{(1+v)^\alpha} \right\} = \frac{1}{1-\alpha}.$$

Can you obtain this result using Maple?

Exercise 13.40

Show that for $0 \leq x < 1$ and $0 \leq \alpha < \frac{1}{2}$,

$$\int_0^1 dt \frac{1}{(x+t^2)^\alpha} = \frac{1}{1-2\alpha} - A(\alpha)x^{\frac{1}{2}-\alpha} - \frac{x^{-\alpha}}{\Gamma(\alpha)} \sum_{k=1}^\infty \frac{(-1)^k}{2k-1} \frac{\Gamma(k+\alpha)}{k!} x^k,$$

where

$$A(\alpha) = \int_0^\infty dv \left\{ \frac{1}{v^{2\alpha}} - \frac{1}{(1+v^2)^\alpha} \right\}.$$

We now apply these methods to a different example for which the integral cannot be expressed as a closed expression. Consider the behaviour of the integral

$$I(\epsilon) = \int_0^1 dx \frac{\ln x}{\epsilon + x}, \quad 0 < \epsilon < 1. \tag{13.59}$$

At $\epsilon = 0$ the singularity in the integrand at $x = 0$ is not integrable, so $I(\epsilon) \to \infty$ as $\epsilon \to 0$: we need to know the form of the singularity in $I(\epsilon)$.

Before obtaining an expansion we estimate the dominant term for small ϵ using the same approximation as used to derive equation 13.55, that is, splitting the integration range into $(0, \epsilon)$ and $(\epsilon, 1)$ and making appropriate approximations to the integrand in each,

$$I(\epsilon) \simeq \frac{1}{\epsilon} \int_0^\epsilon dx \ln x + \int_\epsilon^1 dx \frac{\ln x}{x} = \ln \epsilon - 1 - \frac{1}{2}(\ln \epsilon)^2 \simeq -\frac{1}{2}(\ln \epsilon)^2,$$

where we have used the result $\int dx \, (\ln x)/x = (\ln x)^2/2$ and ignored the smaller first term. The dominant contribution comes from almost the entire integral. Since $|\ln \epsilon|$ grows very slowly as $\epsilon \to 0$, the above will be a reasonable approximation only for very small ϵ, so we need more terms in the expansion.

The form of the integrand suggests the substitution $x = \epsilon y$, which gives

$$\begin{aligned} I(\epsilon) &= \ln \epsilon \int_0^{1/\epsilon} dy \frac{1}{1+y} + \int_0^{1/\epsilon} dy \frac{\ln y}{1+y} \\ &= \ln(1 + 1/\epsilon) \ln \epsilon + I_1(\epsilon), \quad \text{where} \quad I_1(\epsilon) = \int_0^{1/\epsilon} dy \frac{\ln y}{1+y}. \end{aligned} \tag{13.60}$$

This looks more promising, because ϵ now occurs only in the upper limit, but $\ln(1 + 1/\epsilon) \ln \epsilon \sim -(\ln \epsilon)^2$ so the above approximate analysis suggests that $I_1(\epsilon) \sim (\ln \epsilon)^2/2$. In this integral the ϵ-dependence comes from the upper limit so, as in equation 13.57, we write the integral in the form

$$I_1(\epsilon) = \int_0^1 dy \, \frac{\ln y}{1 + y} + \int_1^{1/\epsilon} dy \, \frac{\ln y}{1 + y}.$$

Now the $(1 + y)^{-1}$ factor in the second integrand can be expanded in inverse powers of y. Let the value of the first integral be the number A, so

$$\begin{aligned}
I_1(\epsilon) &= A + \sum_{p=0}^{\infty} (-1)^p \int_1^{1/\epsilon} dy \, \frac{\ln y}{y^{p+1}} \\
&= A + \sum_{p=1}^{\infty} \frac{(-1)^p}{p^2} + \frac{1}{2}(\ln \epsilon)^2 - \ln \epsilon \ln(1 + \epsilon) - \sum_{p=1}^{\infty} \frac{(-1)^p}{p^2} \epsilon^p, \quad (13.61)
\end{aligned}$$

since

$$\int dy \, \frac{\ln y}{y^{p+1}} = -\frac{1}{p^2 y^p} - \frac{\ln y}{p y^p}, \quad p \geq 1.$$

But, in exercise 13.63 (page 524) it is shown that

$$A = \int_0^1 dy \, \frac{\ln y}{1 + y} = \sum_{p=1}^{\infty} \frac{(-1)^p}{p^2} = -\frac{\pi^2}{12},$$

so we finally arrive at the result

$$I(\epsilon) = \int_0^1 dx \, \frac{\ln x}{\epsilon + x} = -\frac{\pi^2}{6} - \frac{1}{2}(\ln \epsilon)^2 - \sum_{p=1}^{\infty} \frac{(-1)^p}{p^2} \epsilon^p, \quad 0 < \epsilon < 1. \quad (13.62)$$

Exercise 13.41
Use Maple to evaluate the integral 13.59 in terms of known functions and use this result to derive equation 13.62.

Exercise 13.42
Using the methods described above, show that

$$\int_0^1 du \, \frac{1}{\sqrt{u^2 + \delta^2}} \simeq \frac{1}{\delta} \int_0^{\delta} du + \int_{\delta}^1 \frac{du}{u} \simeq -\ln \delta$$

as $\delta \to 0$ and compare your result with the exact value of the integral.

Exercise 13.43
If $0 \leq \delta < 1$ and a, α and β are all positive real numbers with $0 < \alpha + \beta < 1$, show that

$$\int_0^a dx \, \frac{1}{x^{\alpha}(\delta + x)^{\beta}} = \frac{a^{1-\alpha-\beta}}{1 - \alpha - \beta} - A(\alpha, \beta)\delta^{1-\alpha-\beta} - \frac{a^{1-\alpha-\beta}}{\Gamma(\beta)} \sum_{k=1}^{\infty} \frac{(-1)^k \Gamma(k + \beta)}{k!(k + \alpha + \beta - 1)} \left(\frac{\delta}{a}\right)^k,$$

where

$$A(\alpha, \beta) = \int_0^{\infty} \frac{du}{u^{\alpha}} \left(\frac{1}{u^{\beta}} - \frac{1}{(1 + u)^{\beta}}\right).$$

13.6 Double integrals

We end this chapter with a description of an elegant method, given by Melzak (1973, chapter 5), for evaluating some integrals. The method uses the fact that for a function $f(x, y)$ of two variables (x, y) the order of the double integral over a rectangular region is often immaterial. Specifically, if $\mathcal{R}$ is the rectangle $x_0 \leq x \leq x_1$, $y_0 \leq y \leq y_1$, then

$$I = \int_{x_0}^{x_1} dx \left[\int_{y_0}^{y_1} dy \, f(x, y) \right] = \int_{y_0}^{y_1} dy \left[\int_{x_0}^{x_1} dx \, f(x, y) \right]. \tag{13.63}$$

Exercise 13.44

A sufficient condition for equation 13.63 to be valid is that the function $f(x, y)$ is defined and bounded in $\mathcal{R}$ and on the boundary, Apostol (1963, theorem 10-21).
    The function

$$g(x, y) = \frac{x - y}{(x + y)^3}$$

is not defined at the origin; show that

$$\int_0^1 dx \left[\int_0^1 dy \, g(x, y) \right] = \frac{1}{2}, \quad \int_0^1 dy \left[\int_0^1 dx \, g(x, y) \right] = -\frac{1}{2}.$$

In the remainder of this section we assume that equation 13.63 holds and that all integrals exist. In some circumstances this allows the computation of an expression in two different ways to produce interesting identities.
    First put $f(x, y) = F'(xy)$ and consider the rectangle $0 \leq x \leq X$, $a \leq y \leq b$ to obtain

$$\int_0^X dx \left[\int_a^b dy \, F'(xy) \right] = \int_0^X \frac{dx}{x} \left[\int_{xa}^{xb} dw \, F'(w) \right], \quad \text{where } w = xy$$

$$= \int_0^X dx \, \frac{F(xb) - F(xa)}{x}.$$

By changing the order of integration this can also be written in the form

$$\int_a^b dy \left[\int_0^X dx \, F'(xy) \right] = \int_a^b \frac{dy}{y} \left[F(Xy) - F(0) \right]. \tag{13.64}$$

If $F(u) \to 0$ as $u \to \infty$ these two results give the identity

$$\int_0^\infty dx \, \frac{F(xb) - F(xa)}{x} = -F(0) \ln(b/a). \tag{13.65}$$

The result quoted on page 484 of the introduction to this chapter is given by choosing

$$F(u) = \frac{\ln(2 + u)}{\sqrt{1 + u}}.$$

Exercise 13.45

Use equation 13.63 with $f(x, y) = x^y$ on the rectangle $0 \leq x \leq 1$, $a \leq y \leq b$ to show that

$$\int_0^1 dx \, \frac{x^b - x^a}{\ln x} = \ln \left(\frac{b + 1}{a + 1} \right).$$

Exercise 13.46

Use equation 13.65 with $F(u) = -e^{-u}$ to show that

$$\int_0^\infty dx \, \frac{e^{-ax} - e^{-bx}}{x} = \ln(b/a).$$

13.7 Exercises

Exercise 13.47

Using the results quoted in exercise 13.2 (page 482), show that

(i) $\quad \displaystyle\int_0^\infty dx \, \frac{\sin x}{x} \left(e^{-ax} - e^{-bx}\right) \;=\; \tan^{-1} b - \tan^{-1} a,$

(ii) $\quad \displaystyle\int_0^\infty dx \, \frac{\sin wx}{x^2} \left(e^{-ax} - e^{-bx}\right) \;=\; \frac{\pi}{2}(b-a) - \left[b \tan^{-1}\left(\frac{b}{w}\right) - a \tan^{-1}\left(\frac{a}{w}\right)\right]$

$$+ \frac{w}{2} \ln\left(\frac{w^2 + b^2}{w^2 + a^2}\right).$$

Exercise 13.48

Show that the following integral exists:

$$\int_0^\infty dt \, \frac{\sin(t^3 - at)}{t^\alpha}, \quad 0 < \alpha < 1.$$

Exercise 13.49

Show that

$$\int_0^\infty dt \, \frac{\cos at \sin bt}{t} = \frac{\pi}{4} \left[\operatorname{sgn}(a+b) + \operatorname{sgn}(b-a)\right].$$

Hint: use the addition formula $\cos At \sin Bt = \frac{1}{2} \sin(B+A)t + \frac{1}{2} \sin(B-A)t$.

Exercise 13.50

Prove that a periodic function $f(x)$ of fundamental period X, that is, $f(x+X) = f(x)$ for all x, cannot be represented as a rational function of x.
Hint: for some constant a, $f(x) - a$ has infinitely many zeros.

Exercise 13.51

If $y(x)$ is an algebraic function satisfying the equation $y'' - R(x) = 0$, where $R(x)$ is a rational function of x, use equation 13.9 and the fact that an algebraic function of order n satisfies no non-trivial equation of degree less than n to show that if $\int dx \, y(x)$ is an elementary function, then

$$\int dx \, y(x) = y(x)P(x),$$

where $P(x)$ is a rational function of x.

Exercise 13.52

If y is the algebraic function $y = (x^2 - 1)^{-1/3}$, show, by differentiating the result obtained in the previous exercise and by examining both sides of the equation at the singularities and at infinity, that $\int dx \, y(x)$ is not an algebraic function.

Exercise 13.53

If $J_n = \int_0^\infty x^n e^{-x}$, show that $J_n = n!$.

Exercise 13.54

If

$$I_n = \int dx\, \frac{1}{(x^3 + a^3)^n},$$

show that

$$3(n-1)a^3 I_n = \frac{x}{(x^3 + a^3)^{n-1}} + (3n-4)I_{n-1}.$$

Exercise 13.55

If

$$I_n = \int_0^\pi dx\, e^{-x} \sin^n x, \quad \text{prove that} \quad I_n = \frac{n(n-1)}{n^2 + 1} I_{n-2}, \quad n \ge 2.$$

Show that $I_0 = 1 - e^{-\pi}$ and deduce that

$$I_{2n} = \frac{(2n)!\,(1 - e^{-\pi})}{\prod_{k=1}^{n}(4k^2 + 1)}.$$

Exercise 13.56

If

$$I_n = \int_0^{\pi/4} dx\, \tan^n x, \quad \text{show that} \quad (n-1)(I_n + I_{n-2}) = 1,$$

and hence show that

$$
\begin{aligned}
I_{2n} &= \frac{1}{2n-1} - \frac{1}{2n-3} + \cdots + \frac{(1-)^n}{3} - (-1)^n(1 - I_0), \quad I_0 = \frac{\pi}{4}, \\
I_{2n+1} &= \frac{1}{2n} - \frac{1}{2n-2} + \cdots + \frac{(-1)^{n-1}}{2} + (-1)^n \ln \sqrt{2}.
\end{aligned}
$$

Exercise 13.57

Use integration by parts to show that if n and m are positive integers,

$$I_{nm} = \int_0^1 dx\, x^n (\ln x)^m$$

satisfies the relation

$$I_{nm} = -\frac{m}{n+1} I_{n\,m-1}, \quad \text{and hence that} \quad I_{nn} = (-1)^n \frac{n!}{(n+1)^{n+1}}.$$

Deduce that

$$\int_0^1 dx\, x^x = 1 - \frac{1}{2^2} + \frac{1}{3^3} - \frac{1}{4^4} + \cdots + (-1)^n \frac{1}{n^n} + \cdots.$$

Exercise 13.58
Show that if $a > 1$,

$$\int_x^\infty dt \, \exp(-t^a) \sim \frac{xe^{-x^a}}{\alpha\Gamma(1 - 1/\alpha)} \sum_{k=0}^\infty \frac{(-1)^k \Gamma(k + 1 - 1/\alpha)}{x^{(1+\alpha)k}}, \quad \text{as} \quad x \to \infty.$$

Exercise 13.59
Find the first five terms of the asymptotic expansions as $x \to \infty$ of the functions

(i) $\displaystyle\int_0^{\pi/2} dt \, \exp(-x \tan t),$ (ii) $\displaystyle\int_{-\pi/2}^{\pi/2} dt \, (t + 2) \exp(-x \cos t),$

(iii) $\displaystyle\int_{-\pi/2}^{\pi/2} dt \, \exp(-x \sin^2 t).$

Exercise 13.60
Show that as $x \to \infty$,

$$\int_0^\infty dt \, e^{-xt} e^{-1/t} \sim \frac{e^{-2y}\sqrt{\pi}}{y^2}\left(1 + \frac{3}{16y} - \frac{15}{512y^2} + \frac{105}{8192y^3} + \cdots\right).$$

Exercise 13.61
In this exercise you will obtain the general term of the series 13.47 (page 513), using the fact that $u = z - i$ is an analytic function of t, defined implicitly by the equation $t^2 = u^2 - iu^3/3$, and that

$$a_n = \frac{u^{(n)}(0)}{n!} = \frac{1}{2\pi i}\oint_\mathscr{C} dt \, \frac{u(t)}{t^{n+1}},$$

where $\mathscr{C}$ is a suitable contour surrounding the origin.
By changing variables show that

$$a_n = \frac{1}{2\pi i}\oint_{\mathscr{C}'} \frac{du}{u^n} \frac{(1 - iu/2)}{(1 - iu/3)^{n/2+1}} = \left(\frac{i}{3}\right)^{n-1} \frac{\Gamma\left(3n/2 - 1\right)}{n!\,\Gamma\left(n/2\right)}, \quad n \geq 1.$$

Deduce that

$$\text{Ai}(u) \sim \frac{\exp(-2u^{3/2}/3)}{2\pi u^{1/4}} \sum_{p=0}^\infty (-1)^p \frac{\Gamma(3p + 1/2)}{3^{2p}(2p)!} \frac{1}{(u)^{3p/2}}.$$

Exercise 13.62
Use the saddle point method to show that

$$\frac{1}{2\pi}\int_{-\infty}^\infty dt \, \exp\left[i\lambda\left(\frac{1}{5}u^5 + u\right)\right] \sim \frac{1}{\sqrt{2\pi\lambda}} \exp\left(-\frac{4\lambda}{5\sqrt{2}}\right)\cos\left(\frac{4\lambda}{5\sqrt{2}} - \frac{\pi}{8}\right).$$

Re-write this integral as a real integral which decreases to zero exponentially far from the origin, and use this to obtain numerical approximations to $F(\lambda)$ and compare these with the values given by your approximation.

Exercise 13.63

Prove that

$$\int_0^1 dt \, \frac{\ln t}{1+t} = -\frac{\pi^2}{12}.$$

Exercise 13.64

Show that for $0 \le \delta < 1$,

$$f(\delta) = \int_0^1 du \, (u^4 + \delta^4)^{-1/4} = A + \ln(1/\delta) + \frac{1}{4\Gamma(1/4)} \sum_{k=1}^{\infty} (-1)^{k-1} \frac{\Gamma(k+1/4)}{k \, k!} \delta^{4k},$$

where A is the number

$$A = \int_0^1 dt \, \frac{1}{(1+t^4)^{1/4}} + \int_1^{\infty} dt \left(\frac{1}{(1+t^4)^{1/4}} - \frac{1}{t} \right) = 0.9126\ldots .$$

Exercise 13.65

Show that as $x \to 1$ from below,

$$f(x) = \int_0^{\pi/2} du \, \frac{1}{\sqrt{1 - x^2 \sin^2 u}} \sim \ln \left(\frac{1}{\sqrt{1-x^2}} \right).$$

Exercise 13.66

Show that if $0 < \alpha < 1$ and $0 \le \delta < 1$

$$\int_0^{\infty} dt \, \frac{1}{(1+t)(\delta+t)^{\alpha}} = \frac{\pi}{\sin \pi \alpha} - \delta^{1-\alpha} \left(\frac{1}{1-\alpha} + \frac{\delta}{(1-\alpha)(2-\alpha)} \right.$$

$$\left. + \frac{2\delta^2}{(1-\alpha)(2-\alpha)(3-\alpha)} + \frac{6\delta^3}{(1-\alpha)(2-\alpha)(3-\alpha)(4-\alpha)} + \cdots \right).$$

Exercise 13.67

Show that if $0 < \alpha < 1$ and $0 < \epsilon \ll 1$,

$$\int_0^1 dt \, \frac{(-\ln t)^{\alpha}}{\epsilon + t} \sim \frac{1}{1+\alpha} \ln^{\alpha+1} \left(\frac{1}{\epsilon} \right).$$

Exercise 13.68

Show that if $\alpha > 1$,

$$\int_0^1 dx \, \frac{\ln x}{(\epsilon + x)^{\alpha}} \sim \frac{\ln \epsilon}{(\alpha - 1)\epsilon^{\alpha-1}} \quad \text{as} \quad \epsilon \to 0.$$

Exercise 13.69

Show that if I is defined by the integral

$$I = \int_0^{\infty} dx \, f \left((cx - a/x)^2 \right), \quad \text{where} \quad a, c > 0,$$

then

$$I = \frac{1}{c} \int_0^\infty dy \, f(y^2).$$

Use this result to show that

(i) with $f(u) = e^{-u}$,

$$I = \int_0^\infty dx \, \exp\left(-c^2x^2 - a^2/x^2\right) = \frac{\sqrt{\pi}}{2c} e^{-2ac},$$

(ii) with $f(u) = 1/(1+u)$,

$$I = \int_0^\infty dx \, \frac{x^2}{c^2x^4 + (1-2c)x^2 + a^2} = \frac{\pi}{2c}.$$

Exercise 13.70
Use the rational functions

$$f(x) = \frac{1}{1 + x + 3x^2}, \quad g(x) = \frac{1+x}{1 + x^2 + x^3}$$

and Maple to confirm that the Hadamard product and the idempotency product, equation 13.7 (page 485), are rational functions.

14

Stationary phase approximations

14.1 Introduction

In this chapter we consider approximations to integrals with highly oscillatory integrands; the typical integral is

$$F(x, v) = \int_a^b dt \, f(t, x) e^{ivh(t,x)}, \quad |v| \gg 1, \tag{14.1}$$

where $f(x, t)$ and $h(t, x)$ are real functions of t. The real variable x has been included as a reminder that in many applications the large number v is fixed and we need to understand how $F(x, v)$ changes with other variables; in order to avoid clutter, we often omit mention of x. The parameter v is real and large, in the sense that the change in the phase, $vh(t)$, over the integration range is much larger than π and consequently the integrand oscillates many times. It is assumed that $f(t, x)$ changes relatively little during one oscillation of the exponential. In the limit $v \to \infty$ the integrand may not exist and the approximation we obtain to $F(x, v)$ will normally be an asymptotic expansion. A typical example of such an integral is Anger's function,

$$J_v(x) = \frac{1}{\pi} \int_0^\pi dt \, \cos(vt - x \sin t), \tag{14.2}$$

the large $|v|$ behaviour of which is examined in section 14.3.

If $h(t)$ is stationary in the integration range then the asymptotic expansion of $F(x, v)$ is dominated by the contribution from an expansion about these stationary points; hence the class of approximations considered here are named *stationary phase approximations*, where $vh(t)$ is the *phase*, although we shall also consider more general integrals.

The need to understand the behaviour of $F(x, v)$ often means that the accurate numerical evaluation of such integrals is less useful than simple analytic approximations. Moreover, if v is very large the rapid oscillations in the integrand make accurate numerical evaluation of such integrals difficult.

The physical origin of oscillatory integrals is usually different from that of the integrals considered in chapter 13. They, and their many-dimensional equivalents, occur frequently in the theory of wave propagation, be it air, water or electromagnetic radiation, and then the variable v is normally inversely proportional to the wavelength. Thus in optical problems v will typically be the ratio of the wavelength of light, $\lambda \simeq 5 \times 10^{-5}$ cm, and the size, a, of an object causing a shadow, for instance if $a \sim 1$ cm then $v \sim 10^4$. In optics, as $v \to \infty$, the ray limit is obtained. In wave mechanics the de Broglie wavelength of a particle with momentum p is $\lambda = h/p$, where h is Planck's constant,† and this sets the relevant scale.

The numerical evaluation of oscillatory integrals is difficult and requires care because of cancellation between the positive and negative contributions. The following exercise illustrates some of the numerical problems that arise.

Exercise 14.1
Evaluate the integral

$$F(x) = \int_0^\pi dt \, \cos(xt^2)$$

numerically using the Maple command

```
y:=(x,N)->evalf( Int( cos(x*t^2), t=-Pi..Pi), N)
```

for various values of x and N. Compare the times of computation for $x = 50k, k = 1, 2, \ldots, 10$ for $N = 10$ and $N = 12$ and check the accuracy of the numerical integration against the value given by expressing the integral in terms of the Fresnel integral,

$$F(x) = \sqrt{\frac{2\pi}{x}} C(\sqrt{2\pi x}).$$

Hint: you may find it useful to use the Maple procedure `timelimit` to set an upper limit on the time that can be taken. In practice one should not attempt the numerical evaluation of such an integral in this manner. This example is designed only to show that the accurate numerical evaluation of integrals having highly oscillatory integrands presents problems.

The theory behind estimates of oscillatory integrals is quite simple, though the implementation is sometimes complicated. The essential idea is that dominant contributions to the asymptotic expansion as $|v| \to \infty$ come from the neighbourhood of points, which we name *critical points*. If these can be isolated by making appropriate changes of variable, the contribution from each critical point is obtained using integration by parts. In most cases the critical points are the stationary points of $h(t)$ and the end points, $t = a$ and b, of the integrand. It is relatively easy to obtain the lowest order approximation, but complete asymptotic expansions require a lot more effort: the elementary theory is given in section 14.3 and an introduction to the subject could end there.

Our discussion starts with a description of a simple physical example that produces an oscillatory integral. In this example the value of the integral is 'obvious' from simple physical considerations, which also show why the dominant contribution comes from the region where the phase, $vh(t, x)$, is stationary. This section is not essential reading. The following section, 14.3, provides a simple heuristic derivation of the lowest order approximation; the extension of this simple theory to multi-dimensional integrals is discussed in section 14.4.

The derivation of the complete asymptotic expansion, or even just a few more terms, is more difficult and it is easy to lose sight of the essential simplicity of the theory because of the morass of algebraic detail. The important additions for this extension are neutralisers, defined in section 14.6, which allow one to isolate each critical point

† $h = 6.63 \times 10^{-34}$ joule seconds $= 6.63 \times 10^{-27}$ g cm^2 sec^{-1}.

and to compute its contribution to the asymptotic expansion. Neutralisers also allow a rigorous development of the theory for one-dimensional integrals.

All this theory assumes that the critical points are, in some sense, well separated. But there are important physical phenomena for which critical points are too close and these approximations fail; the rainbow is a well known example. Thus we end this chapter with a discussion of the physical origins of the rainbow, which continues the discussion of section 14.2, and this leads to the development of uniform approximations, which are applied to differential equations in the following chapter.

14.2 Diffraction through a slit

The wave length, λ, of ordinary light varies between 3.9×10^{-5} cm (violet) and 7.8×10^{-5} cm (red), which means that light emitted from a point source casts a distinct shadow behind a large† object in its path. An example is shown in figure 14.1, where a light source illuminates part of a screen CD by passing through a slit in another screen AB. Also shown, on the extreme right, is the light intensity on CD: we assume that the distance between AB and CD is relatively small.

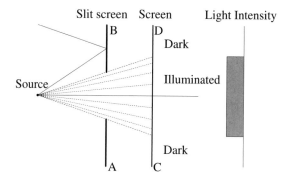

Figure 14.1 Schematic diagram showing the bright and dark areas behind a slit illuminated by a point source S according to the ray picture of light propagation.

This figure shows the common *ray* description of light propagation in which light is assumed to move along straight lines, the rays, from the source to observer. With this description of light shadows are sharp, and we see that the light intensity on the screen is approximately uniform provided a straight line can be drawn from the source to the screen without intersecting the slit. If the light source has finite size, then the edges of a shadow are not distinct and a penumbra of a shadow is produced, as seen for instance in an eclipse of the Sun. But of interest to us are the departures from figure 14.1 due to the wave length of light being small, but nonzero.

The mathematical treatment of scattering of waves is complicated; here we provide the simplest introduction with the sole intention of showing how oscillatory integrals

---

† Large means that that the size of the object is far larger than the wave length, λ, of light.

arise. Full descriptions of the wave theory of light may be found in the books by Longhurst (1967) and Mandel and Wolf (1995).

The first essential addition is that with each ray is associated a *phase*, the value of which is proportional to the path length, r, between the source and the observer. If there is only one ray between the source and the observer the amplitude of the light ϕ has the form $\phi = a(r)e^{i\omega(r-ct)/c}$, where $(r-ct)\omega/c$ is the phase of the ray and $a(r)$ a function that varies slowly by comparison with the oscillatory term. Here ω is the angular frequency of the light, which is related to the wave length, λ, by $\lambda\omega = 2\pi c$, c being the speed of light.† The intensity of the light is the square of the modulus of the amplitude, that is, $I = |\phi|^2 = a^2$, and in this case is independent of the phase.

But if there are two rays contributing to the signal having path lengths r_1 and r_2, then the amplitude is the sum

$$\phi = a_1 e^{i(\omega t - kr_1)} + a_2 e^{i(\omega t - kr_2)} = e^{i\omega t}\left\{a_1 e^{-ikr_1} + a_2 e^{-ikr_2}\right\}, \quad k = \frac{2\pi}{\lambda}.$$

Now the intensity is

$$I = |\phi|^2 = a_1^2 + a_2^2 + 2a_1 a_2 \cos\left(2\pi(r_1 - r_2)/\lambda\right).$$

The last term represents interference between the two rays due to the phase difference along them. The sum, $a_1^2 + a_2^2$, is the intensity without the interference; it is named the *incoherent sum*, and is the intensity given by the ray description.

Consider the simple case when $a_1 = a_2 = a$. If the difference in the path length is $r_1 - r_2 = (p + \frac{1}{2})\lambda$, where $p = 0, \pm 1, \pm 2, \ldots$, there is complete *destructive interference* between the two rays and $I = 0$. Whereas if $r_1 - r_2 = p\lambda$ there is *constructive interference* and $I = 4a^2$. If the light is not monochromatic but comprises contributions from a large enough range of frequencies, the intensity will be the sum over all these components, and then the interference part generally sums to a negligible quantity leaving only the incoherent sum, $a_1^2 + a_2^2$.

The difference $r_1 - r_2$ can often be made to vary by changing the point of observation. For instance the colours seen in a soap or oil film are due to this type of interference because one sees light reflected from both sides of the film, and whether the interference is constructive or destructive depends upon the thickness of the film and the wavelength of the light. The thickness of a soap bubble will vary across its surface, being thinnest at the top, so different colours are enhanced by constructive interference at different points on its surface. The colours of some butterfly wings are also a consequence of this type of interference. These and other interference effects are described in more detail in the excellent book by Walker (1977).

The second addition is due, in its elementary form, to Huygens, who explained wave propagation through the aether by assuming that each point on a wave front becomes a source of secondary waves.‡ This simple description is sufficient to explain many

C. Huygens (1629–1695)

---

† The frequency v is related to the angular frequency ω by $\omega = 2\pi v$ and the period of oscillation is $T = 1/v = 2\pi/\omega$, so $\lambda v = c$, where $c \simeq 3 \times 10^{10}$ cm sec^{-1} is the speed of light. Thus for red light $\lambda = 7.8 \times 10^{-5}$ cm and $v = 3.8 \times 10^{14}$ hertz, that is, cycles per second, or $\omega = 2.4 \times 10^{15}$ radians sec^{-1}.

‡ A particularly interesting account of Huygen's work, which encompassed many areas of science, and its relation to modern mathematics is given by Arnold (1990).

phenomena, including the classical laws of reflection and refraction. The modern version of this, based upon the solution of the wave equation satisfied by all electromagnetic waves, is due to Kirchhoff, see Longhurst (1967, section 10.6) for instance.

G. R. Kirchhoff
(1824–1887)

These ideas can be used to obtain an approximation to the amplitude of the light in the circumstances shown in figure 14.1. Consider the scheme shown in figure 14.2 where a line source S (perpendicular to the page) illuminates part of a screen CD by passing through a slit AB, both screens being perpendicular to the page and the axis SN. The slit extends a distance a_2 above and a_1 below the axis SN: the distance between the source and the slit is D_1 and that between the slit and the screen is D_2. For simplicity we assume u to be small.

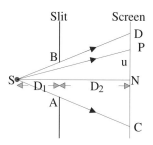

Figure 14.2   Diagram showing the configuration of the source S, slit and screen. The light amplitude at a point P a distance u from the axis is given approximately by equation 14.3.

The amplitude of light $\psi(u)$ at a point P on the screen at a distance u from the axis can be obtained from the rules briefly described above. For light with wave length λ it may be shown that

$$\psi(u) = A \int_{-a_1}^{a_2} dx \, \exp\left[-iv\left(x - \frac{uD_1}{D_1 + D_2}\right)^2\right], \quad v = \frac{\pi(D_1 + D_2)}{\lambda D_1 D_2}, \quad (14.3)$$

where the distances D_1 and D_2 are defined in figure 14.2 and A is a constant. Typically $\lambda \sim 5 \times 10^{-5}$ cm, and if $D_1 \sim D_2 \sim 10$ cm then $v \sim 10^4$, and the integrand is highly oscillatory provided a_1 and a_2 are not too small.

If one were confronted with this integral with no prior knowledge of its origin it would be difficult to know how $\psi(u)$ behaves. But because the ray description of light propagation is a good first approximation we should expect $\psi(u)$ to be almost zero above D and below C on the screen, that is if $u > a_2(D_1 + D_2)/D_1$ or $u < -a_1(D_1 + D_2)/D_1$, and for $|\psi|$ to be almost constant between these limits, in CD.

In the case of only one edge we may set $a_1 = 0$ and $a_2 = \infty$ and re-write the integral 14.3 in the form

$$\psi(u) = B\left\{\frac{1}{2}(1 - i) + \left(C(z) - iS(z)\right) \operatorname{sgn} u\right\}, \quad \text{where} \quad z = |u|\sqrt{\frac{2D_1}{\lambda D_2(D_1 + D_2)}}, \quad (14.4)$$

where B is some constant and

$$C(z) = \int_0^z dv \, \cos\left(\frac{\pi}{2}v^2\right), \quad S(z) = \int_0^z dv \, \sin\left(\frac{\pi}{2}v^2\right),$$

are the Fresnel integrals. For the case $z = 14|u|$, corresponding roughly to $D_1 = D_2 = 100$ cm, $\lambda = 5 \times 10^{-5}$ cm (for which $v \simeq 1260$), a graph of $|\psi|^2$ is shown in the following figure.

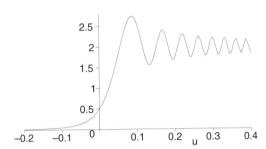

Figure 14.3   Graph of $|\psi|^2$, equation 14.4, for $z = 14|u|$.

According to the ray description of light the intensity would be zero for $u < 0$, and for $u > 0$ the intensity should decrease roughly in proportion to u. The figure shows that $|\psi|^2$ behaves approximately like this, being very small for $u < 0$, with a sharp increase near $u = 0$ and oscillations about a constant value for $u > 0$. The incorrect behaviour for $u \gg 1$ is due to 14.3 failing to take properly into account the distance between the screen and the source, and is not important. What is interesting is the way the integral mimics the step function at $u = 0$, which becomes sharper as v increases. A clue to the origin of this behaviour is given by looking at the shape of the integrand.

Exercise 14.2
If $D_1 = D_2 = d$ the integrand of the real part of the amplitude, equation 14.3, is proportional to

$$f(x,u) = \cos\left(v\left(x - \frac{1}{2}u\right)^2\right), \quad v = \frac{2\pi}{\lambda d}.$$

For $v = 50$ and 100 use the animate facility of Maple to plot graphs of $f(x,u)$, $0 \le x \le 1$, for a sequence of values of u in the range $-1/2 < u < 1$. Using these graphs, explain why $|\psi|^2 \simeq 0$ for $u < 0$ and why the largest maximum of $|\psi(u)|^2$ occurs for small positive values of u. The value of v is decreased from 1260 for reasons that will become obvious when you draw the graphs.

These graphs show that for $u < 0$ the oscillations in f are so rapid that the value of the integral is small because of cancellation. For $u > 0$ there is a broad maximum at $x = u/2$, so the value of the integral is much larger. This maximum in f is at the point where the exponent in equation 14.3 is stationary, that is, $x = uD_1/(D_1 + D_2)$, or $x = u/2$ if $D_1 = D_2$ as in the case treated above, which is precisely the point at which the ray from S to P in figure 14.2 intersects the screen. That is, the dominant contribution to the intensity at P comes from the neighbourhood of the ray from S to P, as would be expected.

In this section we have used the example of light passing through a slit to show that, in some cases, a complicated oscillatory integral can be approximated quite simply. This is because in the short wave length limit oscillations produced by the changing phase of the light interfere destructively except along the rays used in the simplest of descriptions; on these rays the phase is stationary. In other circumstances a simple physical picture is missing, but the general principle remains the same; however, the mathematical treatment often obscures this simplicity with complex algebra.

Exercise 14.3
Derive equation 14.4.

14.3 Stationary phase approximation I

In this section we derive the lowest-order approximation to oscillatory integrals using approximations based on intuition rather than rigor. In the first part we consider a particular example, Anger's function, after which the general case is dealt with. The ideas introduced here will be generalised and made rigorous in subsequent sections; in particular we show how higher-order approximations may be found. In practice, however, the simple approximations derived in this section are often all that is needed; moreover, for the many dimensional integrals of section 14.4 this simple theory is generally all that is available.

14.3.1 Anger's function

C. T. Anger
(1803–1858)

Anger studied the function defined by the integral†

$$\mathscr{J}_v(x) = \frac{1}{\pi} \int_0^\pi dt \, \cos(vt - x \sin t) \tag{14.5}$$

in his studies of the Bessel function and their generalisations. When $v = n$ is an integer this function reduces to an ordinary Bessel function, $\mathscr{J}_n(x) = J_n(x)$, equation 11.74 (page 408). Here v is real and we shall consider the case in which it is large. For our purposes it is more convenient to study the related function

$$\mathscr{J}_v(vx) = \frac{1}{\pi} \int_0^\pi dt \, \cos\left(v(t - x \sin t)\right). \tag{14.6}$$

The graphs of $\mathscr{J}_v(vx)$ for $v = 5$ and 5.5 are compared in the next figure, from which we see that for $|x| > 1$, $\mathscr{J}_v(vx)$ is oscillatory.

---

† The Maple name for the Anger function $\mathscr{J}_v(x)$ is **AngerJ(v,x)**.

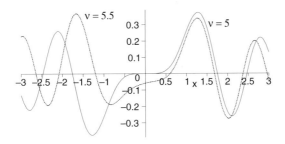

Figure 14.4   Graphs of $\mathscr{J}_v(vx)$ for $v = 5$ and 5.5.

The phase of the integrand is $vh(t)$, where $h(t) = t - x\sin t$, and in figure 14.5 we show graphs of $h(t)/\pi$, the thick line, together with the integrand $\cos(vh(t))$ for $v = x = 5$ and $-\pi \le t \le \pi$. The reason for showing this extended range of t will soon become clear.

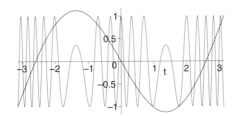

Figure 14.5   Graphs of the phase $h(t)/\pi$ (thick line) and
the integrand $\cos(vh(t))$ for $v = x = 5$.

When both x and v are fairly large the integrand, $\cos(vh(t))$, oscillates rapidly except near the stationary point of the phase, that is at the positive root of $\cos t = 1/x$, $t = t_s \simeq 1.37$; in this example there is also a stationary point at $t = -t_s$, not in the integration range. The dominant contribution to the integral will clearly come from the region around t_s, where $\cos(vh(t))$ varies more slowly.

If x changes both stationary points $\pm t_s(x)$ move, with t_s decreasing to zero as $x \to 1$. When $x = 1$ the two stationary points coalesce at $t = 0$ and when $|x| < 1$ they are complex. When x is slightly larger than unity the two stationary points are close together, the oscillations between them disappear and the graphs look like those in the next figure, in which $x = 1.2$ and $v = 5$.

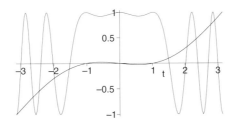

Figure 14.6   Graphs of the phase $h(t)/\pi$ (thick line) and
the integrand $\cos(vh(t))$ for $v = 5$ and $x = 1.2$.

From this figure it would seem that the value of $\mathscr{J}_v(vx)$ for $x = 1.2$ will be larger than for $x = 5$ because of the broad maximum in the integrand near $t = 0$. By doing the next exercise you will observe that as x increases through unity the width of this maximum increases, becoming broadest at $x \simeq 1.2$, after which the two stationary points $\pm t_s(x)$ are sufficiently far apart for oscillations to appear in the interval $(-t_s, t_s)$. The graphs plotted in this exercise suggest that $\mathscr{J}_v(vx)$ will have a global maximum at $x \simeq 1.2$; in general, when there are two stationary phase points the integral has a maximum when they are real and close, not when they coincide. The general theory of close and coalescing stationary points is described in section 14.8

Exercise 14.4
Create an animation of the curves $y = \cos(v(t - x \sin t))$, $-\pi \le t \le \pi$, with $v = 5$ and with x varying between 0.5 and 4 in suitably small steps.
   Using these graphs estimate the value of x, close to unity, that produces a maximum in $\mathscr{J}_5(5x)$. Plot the graph of $\mathscr{J}_5(5x)$ for $0 \le x \le 2$ and check your prediction, and consider other values of v.

For $|x| < 1$ there are no real stationary points of $h(t)$ and the graphs of the phase and integrand are as shown in the next figure.

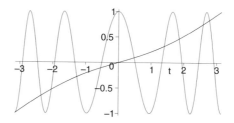

Figure 14.7   Graphs of the phase $h(t)/\pi$ (thick line) and the integrand $\cos(vh(t))$ for $v = 5$ and $x = 0.4$.

In this case the phase, $vh(t)$, is practically linear and oscillations of the integrand tend to cancel, so the integral has a small magnitude, equation 14.13 (page 539).
   This discussion suggests that for $|x| > 1$ the dominant contribution comes from the neighbourhood of the stationary points of $h(t) = t - x \sin t$, that is, the roots of

$$\frac{dh}{dt} = 1 - x \cos t = 0, \quad 0 < t < \pi.$$

In the range of integration there is only one stationary point, but when x is close to unity the other stationary point at $-t_s$ affects the behaviour of the function in the neighbourhood of $t = 0$ and cannot be ignored; we postpone this problem until section 14.8, because it cannot be dealt with using the methods described here.
   It is easier to deal with exponentials than trigonometric functions, so we define

$$F_v(x) = \frac{1}{\pi} \int_0^\pi dt \, e^{ivh(t)}, \quad h(t) = t - x \sin t, \quad \text{hence} \quad \mathscr{J}_v(vx) = \Re(F_v(x)). \tag{14.7}$$

Because the dominant contribution to the integral comes from the neighbourhood of t_s, a natural approximation is obtained by expanding $h(t)$ about this point; thus, assuming $x > 1$, we have

$$
\begin{aligned}
h(t) &= h(t_s) + \frac{1}{2}(t - t_s)^2 h''(t_s) + O((t - t_s)^3) \\
&\simeq \cos^{-1}\left(\frac{1}{x}\right) - \sqrt{x^2 - 1} + \frac{1}{2}\sqrt{x^2 - 1}\,(t - t_s)^2, \quad (x > 1).
\end{aligned}
\tag{14.8}
$$

Substituting this approximation into the integral 14.7 gives another integral which is an approximation to $F_v(x)$. But this cannot be evaluated in terms of simple known functions, so a further approximation is necessary.

Because $|v| \gg 1$ and x is sufficiently large, the contribution from each of the intervals $0 < t \ll t_s$ and $t_s \ll t < \pi$ is small because the many oscillations cause cancellation. Thus a first approximation is obtained by extending the range of integration to the whole real axis because this produces an integral that can be evaluated:

$$
F_v(x) \simeq \frac{1}{\pi} \exp\left(iv\left(\cos^{-1}\left(\frac{1}{x}\right) - \sqrt{x^2 - 1}\right)\right) \int_{-\infty}^{\infty} du \, \exp\left(i\frac{1}{2}v\sqrt{x^2 - 1}\,u^2\right).
\tag{14.9}
$$

Integrals of this type are dealt with in Appendix I, particularly page 837; using the result

$$
\int_{-\infty}^{\infty} dt \, e^{iat^2} = \sqrt{\frac{\pi}{|a|}} \exp\left(i\frac{\pi}{4}\operatorname{sgn} a\right)
$$

we obtain

$$
\mathcal{J}_v(vx) \simeq \sqrt{\frac{2}{\pi|v|\sqrt{x^2 - 1}}} \cos\left(v\left(\cos^{-1}\left(\frac{1}{x}\right) - \sqrt{x^2 - 1}\right) + \frac{\pi}{4}\operatorname{sgn} v\right), \quad x \gg 1.
\tag{14.10}
$$

In the next two figures we compare this approximation, the thicker line, with $\mathcal{J}_v(vx)$ for $x > 1$ and $v = 2$ and 5. It is seen that even for $v = 2$ — recall that this is a large $|v|$ expansion — the approximation is good, for $x > 2$, but the singularity at $x = 1$ causes large errors for $x \simeq 1$.

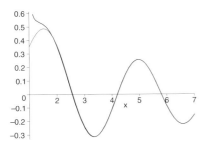

Figure 14.8 $\mathcal{J}_v(vx)$ for $v = 2$.

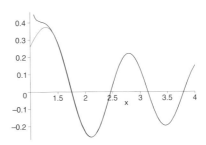

Figure 14.9 $\mathcal{J}_v(vx)$ for $v = 5$.

Exercise 14.5
Derive equation 14.10 and show that the general result, valid for positive and negative values of x, is

$$\mathcal{J}_v(vx) \simeq \sqrt{\frac{2}{\pi|v|\sqrt{x^2-1}}} \cos\left(v\left(\cos^{-1}\left(\frac{1}{x}\right) - \sqrt{x^2-1}\,\operatorname{sgn}x\right) + \frac{\pi}{4}\operatorname{sgn}(xv)\right), \quad |x| \gg 1.$$

There are three points to notice about the approximation 14.10.

(1) The magnitude of $\mathcal{J}_v(vx)$ decreases as $v^{-1/2}$, which is rather slow.
(2) As x varies $\mathcal{J}_v(vx)$ oscillates with a wavelength roughly proportional to v^{-1}.
(3) The approximation is singular at $x = 1$, which is clearly wrong because the integral 14.7 is bounded for all finite x. As $x \to 1$, from above, the two stationary points, $\pm t_s(x)$, approach zero, coalesce at $t = 0$ when $x = 1$, and here $h''(t_s) = 0$. Consequently the approximation 14.8 is inappropriate. At $x = 1$ we need to use the approximation

$$h(t) \simeq \frac{1}{6}t^3 + O(t^5), \quad (x = 1),$$

which provides the asymptotic expansion, in v, at $x = 1$, exercise 14.38. However, for x close to unity we have to use

$$h(t) = h(t_s(x), x) + \sqrt{\frac{x-1}{2}}(t - t_s)^2 + \frac{1}{6}(t - t_s)^3 + O((t - t_s)^4),$$

because neither the $(t-t_s)^2$ nor the $(t-t_s)^3$ term may be ignored: in section 14.8 we shall show how to find an approximation, accurate for $x \simeq 1$ and which reduces to the asymptotic form 14.10 for $|x| \gg 1$.

Next we consider the accuracy of the approximation 14.10 with varying v and fixed x. The following figure shows the difference between this approximation and the exact values of $\mathcal{J}_v(2v)$ for $2 < v < 10$.

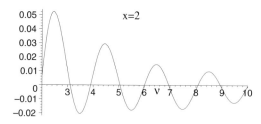

Figure 14.10   Difference between $\mathcal{J}_v(2v)$ and the approximation of equation 14.10.

This graph suggests that the approximation 14.10 is best when v is close to an integer. The reason for this is that the next term in the asymptotic expansion comes from the end points of the integral which cancel when the integrand is periodic, that is, when v is an integer. In general, as will be shown in the next section, the inclusion of more terms in the expansion of $h(t)$ about the stationary point leads to corrections $O(v^{-3/2})$ but end-point effects give corrections $O(v^{-1})$.

The contribution from the end points is obtained using a simple method, equivalent to that used to derive equation 14.10. Higher-order terms, from both sources, will be obtained using the more general methods described and justified in the next section.

No matter how rapidly the integrand oscillates, there cannot be complete cancellation near an end point, as implicitly assumed when deriving equation 14.10. The contribution, F_0, to the integral 14.7 from the neighbourhood of $t = 0$ is obtained by expanding $h(t)$ about $t = 0$ and retaining only the linear terms, $h(t) = -(x-1)t + O(t^2)$, and extending the range of integration to infinity:

$$F_0 = \frac{1}{\pi} \int_0^\infty dt\, e^{-iv(x-1)t} = -\frac{i}{\pi v(x-1)}, \quad |x| > 1,$$

where we have used the integral 20.15 (page 837). This contribution is purely imaginary so does not contribute to $\mathcal{J}_v(vx)$.

The contribution from the neighbourhood of $t = \pi$, F_π, is obtained similarly, by expanding about $t = \pi$. Since $h(\pi - s) = \pi - (x+1)s + O(s^2)$, we have

$$F_\pi = \frac{e^{i\pi v}}{\pi} \int_0^\infty ds\, e^{-iv(x+1)s} = -\frac{ie^{i\pi v}}{\pi v(x+1)}, \quad |x| > 1.$$

The sum of contributions from the stationary point, equation 14.9, and the end point $x = \pi$ give the first two terms in the asymptotic expansion,

$$\mathcal{J}_v(vx) \simeq \sqrt{\frac{2}{\pi |v| \sqrt{x^2-1}}} \cos\left(v\left(\cos^{-1}\left(\frac{1}{x}\right) - \sqrt{x^2-1}\,\mathrm{sgn}\,x\right) + \frac{\pi}{4}\mathrm{sgn}(xv)\right)$$
$$+ \frac{\sin \pi v}{\pi v(x+1)}. \tag{14.11}$$

Notice that if v is an integer the last term is identically zero. In the following figure we have redrawn figure 14.10 (the thin lines) together with the difference between $\mathcal{J}_v(2v)$ and the above approximation (thick lines) for $2 < v < 10$; this shows how contributions from the end points have improved the accuracy.

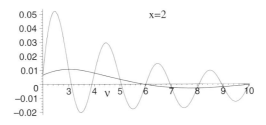

Figure 14.11   Differences between $\mathcal{J}_v(2v)$, and the approximation of equations 14.10, thin line, and 14.11.

Finally, for completeness, we briefly consider the behaviour of $\mathcal{J}_v(vx)$ for $0 < x < 1$ and $v > 0$, when the stationary points are complex; the equation for these points is still $\cos t_s = 1/x$, but now has the solutions

$$t_s = p\pi + i \ln\left(\frac{1 + \sqrt{1 - x^2}}{|x|}\right),$$

where p is an even integer if $0 < x < 1$ and an odd integer if $-1 < x < 0$. For this analysis the integral 14.6 (page 532) is easier to handle when put in the form

$$\mathcal{J}_v(vx) = \frac{1}{2\pi} \int_{-\pi}^{\pi} dt\, e^{iv(t - x \sin t)},$$

so the contour may be deformed off the real axis to pass through the stationary point on the imaginary axis, $t_s = i \ln\left(\frac{1 + \sqrt{1 - x^2}}{x}\right)$, at which

$$iv(t_s - x \sin t_s) = -v\left(\ln\left(\frac{1 + \sqrt{1 - x^2}}{x}\right) - \sqrt{1 - x^2}\right) < 0 \quad \text{for} \quad 0 < x < 1.$$

Deform the contour to the three sides of the rectangle $ABCD$, shown in the figure; the original integral is along AD, the deformed integral is along $ABCD$ and the dot on BC represents the stationary point.

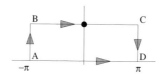

Figure 14.12   Diagram of the integration
contour, $x, v > 0$.

Thus the integral becomes

$$
\begin{aligned}
\mathcal{J}_v(vx) \;=\; & \frac{1}{2\pi}\left(\frac{x e^{\sqrt{1-x^2}}}{1 + \sqrt{1 - x^2}}\right)^v \int_{-\pi}^{\pi} du\, \exp\left(-2v\sqrt{1 - x^2}\,\sin^2\left(\frac{u}{2}\right) + iv(u - \sin u)\right) \\
& + \frac{\sin v\pi}{\pi} \int_0^{\tau_s} d\tau\, e^{-v(\tau + x \sinh \tau)},
\end{aligned}
\tag{14.12}
$$

where $\tau_s = \cosh^{-1}(1/x) = \ln\left(\frac{1 + \sqrt{1 - x^2}}{|x|}\right)$; note that this expression is exact. The first integral is the contribution from BC and the second the contributions from AB and CD. The first integral is exponentially small and may be evaluated by the method of steepest descent, chapter 13. The second term is zero if v is an integer, when the Anger function reduces to an ordinary Bessel function, but otherwise because $\tau_s \gg 1$ it may be estimated by expanding about $\tau = 0$, as in the derivation of 14.11:

$$\int_0^{\tau_s} d\tau\, e^{-v(\tau + x \sinh \tau)} \sim \int_0^{\infty} d\tau\, e^{-v(1 + x)\tau} = \frac{1}{v(1 + x)}.$$

The full asymptotic expansion of this integral may be found using the extensions to Watson's lemma, discussed in section 13.3.2.

For $0 < x < 1$ and $v > 0$ the first two terms in the asymptotic expansion are

$$\mathcal{J}_v(vx) \sim \sqrt{\frac{1}{2\pi v \sqrt{1-x^2}}} \left(\frac{xe^{\sqrt{1-x^2}}}{1+\sqrt{1-x^2}} \right)^v + \frac{\sin v\pi}{\pi v(1+x)}. \qquad (14.13)$$

For non-integer v the second term is dominant: if $v = n$ is an integer the second term is zero and the expression reduces to the first term in the asymptotic expansion of the Bessel function $J_n(nx)$.

Exercise 14.6
Weber's function is defined by the integral

$$E_v(x) = \frac{1}{\pi} \int_0^\pi dt \, \sin(vt - x \sin t).$$

Show that the first two terms of its asymptotic expansion are

$$\begin{aligned} E_v(x) &= \sqrt{\frac{2}{\pi \sqrt{x^2 - v^2}}} \sin\left(v \cos^{-1}\left(\frac{v}{x}\right) - \operatorname{sgn}(x)\sqrt{x^2 - v^2} + \frac{\pi}{4}\operatorname{sgn}(x) \right) \\ &\quad -\frac{1}{\pi(x-v)} - \frac{\cos \pi v}{\pi(x+v)}, \qquad |x| > |v|. \end{aligned}$$

Exercise 14.7
Show that if $x \gg 1$ and $v = n$ is an integer, equation 14.10 reduces to the simpler expression for the asymptotic expansion of the Bessel function,

$$J_n(w) \simeq \sqrt{\frac{2}{\pi w}} \cos\left(w - \frac{\pi}{2}\left(n + \frac{1}{2}\right) \right).$$

Compare, graphically, this approximation with that given in equation 14.10 and with the exact values of $J_n(w)$, for various values of n and w.

14.3.2 *General first-order theory*
The particular example treated above suggests that the largest contributions to the integral normally come from the end points of the integration and the points where the phase is stationary, the latter being dominant if they exist. In many applications only these stationary phase points need be considered. Thus, before we study the nuances of higher-order corrections we derive, heuristically, the first two terms in the asymptotic expansion of the general integral

$$F(v) = \int_a^b dt \, f(t) e^{ivh(t)}, \qquad |v| \gg 1, \qquad (14.14)$$

where both $f(t)$ and $h(t)$ are at least twice differentiable in t and $h(t)$ has N stationary points t_k, with $a < t_1 < t_2 < \cdots < t_N < b$ and $h''(t_k) \neq 0$, for all k. In the limit $|v| \to \infty$ each stationary point may be considered in isolation, that is, the integrand oscillates many times between each pair of stationary phase points and in the intervals (a, t_1),

(t_N, b). Then the contribution $F_k(v)$ from the neighbourhood of t_k is obtained in exactly the same manner that 14.10 was obtained. Write

$$h(t) = h(t_k) + (t - t_k)^2 h''(t_k)/2 + \cdots$$

to give

$$F_k(v) \sim e^{ivh(t_s)} f(t_k) \int_{-\infty}^{\infty} ds \, \exp\left(ivs^2 h''(t_k)/2\right),$$

where we have assumed that $f(t)$ varies slowly by comparison with the exponential term. Adding the contributions from all the stationary points and evaluating the integrals gives the approximation

$$F(v) \sim \sqrt{\frac{2\pi}{|v|}} \sum_{k=1}^{N} \frac{f(t_k)}{\sqrt{|h''(t_k)|}} \exp i \left(vh(t_k) + \frac{\pi}{4} \operatorname{sgn}\left(vh''(t_k)\right)\right), \quad |v| \gg 1. \qquad (14.15)$$

This formula is possibly the most useful part of this section and can be used in many circumstances to obtain the first term in the asymptotic expansion of many oscillatory integrals.

If either a or b is finite and $f(t)$ is not zero at the end points, the next term in the asymptotic expansion comes from the end points and is easily found using similar heuristics. The contribution from $x = a$, $F_a(v)$ for example, is obtained by setting $t = a + s$ and expanding to the lowest order, to give

$$F_a(v) \simeq e^{ivh(a)} f(a) \int_0^{\infty} ds \, e^{ivh'(a)s} = \frac{if(a)e^{ivh(a)}}{vh'(a)}.$$

It will be shown in the next section how this technique can be made rigorous. The contribution from the other end, $t = b$, is obtained similarly to give the first two terms of the asymptotic expansion,

$$F(v) \quad \sim \quad \sqrt{\frac{2\pi}{|v|}} \sum_{k=1}^{N} \frac{f(t_k)}{\sqrt{|h''(t_k)|}} \exp i \left(vh(t_k) + \frac{\pi}{4} \operatorname{sgn}\left(vh''(t_k)\right)\right)$$

$$+ \frac{i}{v} \left\{ \frac{e^{ivh(a)} f(a)}{h'(a)} - \frac{e^{ivh(b)} f(b)}{h'(b)} \right\} \quad \text{as} \quad |v| \to \infty. \qquad (14.16)$$

If both $h(t)$ and $f(t)$ are $(b - a)$-periodic the boundary term vanishes.

The reader will have observed that the method used here is dubious. Moreover, it is not clear how these ideas can be used to derive higher-order terms or how to deal with awkward cases, for instance when $f(t)$ has an integrable singularity at an end point, $f(t) = (t - a)^{-1/2}$ for example. These problems are addressed in section 14.6.

Exercise 14.8
Derive equation 14.16. Show also that if the only stationary point of $h(t)$ is at $t = a$ and $f(a) \neq 0$, then

$$F(v) \sim \sqrt{\frac{\pi}{2|v|}} \frac{f(a)}{\sqrt{|h''(a)|}} \exp i \left(vh(a) + \frac{\pi}{4} \operatorname{sgn}\left(vh''(a)\right)\right) - \frac{if(b)\exp(ivh(b))}{vh'(b)}.$$

Exercise 14.9
The Airy function, Ai(x), may be defined by the integral:

$$\text{Ai}(-x) = \frac{1}{2\pi} \int_{-\infty}^{\infty} dt \, \exp i \left(\frac{1}{3} t^3 - xt \right).$$

Show that if $x > 0$ there are two real stationary phase points and that if x is sufficiently large,

$$\text{Ai}(-x) \sim \frac{1}{\pi^{1/2} x^{1/4}} \sin \left(\frac{2}{3} x^{3/2} + \frac{\pi}{4} \right).$$

Exercise 14.10
Show that

$$\int_{-1}^{1} dt \, \frac{e^{ivt^2}}{1 + t^2} \sim \sqrt{\frac{\pi}{|v|}} \exp \left(i \frac{\pi}{4} \text{sgn}(v) \right) - \frac{ie^{iv}}{2v} + \cdots.$$

Exercise 14.11
Show that

$$\int_{0}^{\pi} dt \, \frac{\cos^n t \, e^{iv \sin^2 t}}{1 + t^2} \sim \sqrt{\frac{\pi}{4|v|}} \left(1 + \frac{(-1)^n}{1 + \pi^2} \right) \exp \left(i \frac{\pi}{4} \text{sgn}(v) \right) + \cdots.$$

Exercise 14.12
If $f(t)$ is an infinitely differentiable function for $-1 \le t \le 1$ and $F(v; x, y)$ is defined by the integral

$$F = \int_{-1}^{1} dt \, f(t) \exp iv \left(xt + y\sqrt{1 - t^2} \right),$$

show that

$$F \sim \sqrt{\frac{2\pi}{v}} \frac{y}{(x^2 + y^2)^{3/4}} f \left(\frac{x}{\sqrt{x^2 + y^2}} \right) \exp \left(iv\sqrt{x^2 + y^2} - i\frac{\pi}{4} \right), \quad x > 0, \, y > 0,$$

and find expressions for the asymptotic expansion of F when (x, y) is in other quadrants.

14.4 Lowest-order approximations for many-dimensional integrals

Most mathematical methods do not easily generalise from one to many dimensions, but the first-order stationary phase approximation is one example where the extension is relatively easy and useful, though the higher-order terms are far more difficult to obtain. Here we consider the N-dimensional integral

$$F(v) = \int \cdots \int d\mathbf{y} \, f(\mathbf{y}) \exp(ivh(\mathbf{y})), \quad |v| \gg 1, \tag{14.17}$$

where $\mathbf{y} = (y_1, y_2, \ldots, y_N)$, is an N-dimensional real variable and $f(\mathbf{y})$ and $h(\mathbf{y})$ are real functions of $\mathbf{y}$, g being continuous and f at least twice differentiable. The integral is over a sufficiently large region of $\mathbf{R}^N$ to contain one or more stationary points of h and it is assumed that all stationary points are far from the boundary.

With these assumptions the dominant contributions are from the neighbourhood of the stationary points of h, that is, the real roots of the N equations

$$\frac{\partial h}{\partial y_k} = 0, \quad k = 1, 2, \ldots, N. \tag{14.18}$$

It is necessary to assume that each stationary point is isolated, in the sense that between every pair of stationary points there are many oscillations of the integrand. Let $\boldsymbol{\alpha} = (\alpha_1, \alpha_2, \ldots, \alpha_N)$ be a stationary point and define new variables $\mathbf{z} = \mathbf{y} - \boldsymbol{\alpha}$, so the first two terms of the Taylor's series of h are

$$h(\mathbf{y}) = h(\boldsymbol{\alpha}) + \frac{1}{2}\mathbf{z}^T A \mathbf{z} + \cdots, \tag{14.19}$$

where A is a real, symmetric $N \times N$ matrix with elements $A_{rs} = \partial^2 h/\partial y_r \partial y_s$ evaluated at $\mathbf{y} = \boldsymbol{\alpha}$. We also need to assume that the stationary point is not degenerate, that is $\det(A) \neq 0$; this is the usual case and is equivalent to assuming that $h''(\alpha) \neq 0$ in the one-dimensional case. The contribution from the neighbourhood of this stationary point can therefore be written in the form

$$F_{\alpha}(v) = f(\boldsymbol{\alpha})e^{ivh(\boldsymbol{\alpha})} \int_{-\infty}^{\infty} \cdots \int_{-\infty}^{\infty} d\mathbf{z} \exp\left(\frac{1}{2}iv\mathbf{z}^T A \mathbf{z}\right). \tag{14.20}$$

Because A is real, symmetric and non-singular its eigenvalues, $\lambda_k, k = 1, 2, \ldots, N$, are real, and there is a matrix B defining a linear transformation $\mathbf{z} = B\mathbf{w}$ such that the matrix $D = B^T A B$ is diagonal with $D_{rr} = \lambda_r$. Thus, when expressed in terms of $\mathbf{w}$ the quadratic form $\mathbf{z}^T A \mathbf{z}$ becomes the sum of squares,

$$\mathbf{z}^T A \mathbf{z} = \mathbf{w}^T B^T A B \mathbf{w} = \sum_{k=1}^{N} \lambda_k w_k^2.$$

The Jacobian determinant of the transformation between $\mathbf{z}$ and $\mathbf{w}$ is the constant, $\partial(\mathbf{z})/\partial(\mathbf{w}) = |\det(B)|$, so the N-dimensional integral 14.20 becomes the product of N one-dimensional integrals,

$$\begin{aligned}
F_{\alpha}(v) &= |\det(B)|f(\boldsymbol{\alpha})e^{ivh(\boldsymbol{\alpha})} \prod_{k=1}^{N} \int_{-\infty}^{\infty} dw_k \exp\left(\frac{1}{2}iv\lambda_k w_k^2\right) \\
&= \left(\frac{2\pi}{|v|}\right)^{N/2} |\det(B)|f(\boldsymbol{\alpha}) \frac{\exp i\left(vh(\boldsymbol{\alpha}) + \frac{\pi}{4}\mathrm{sgn}(v)\sum_{k=1}^{N}\mathrm{sgn}\,\lambda_k\right)}{\prod_{k=1}^{N}\sqrt{|\lambda_k|}}.
\end{aligned} \tag{14.21}$$

But,

$$\prod_{k=1}^{N} \lambda_k = \det(D) = \det(B^T A B) = \det(A)\det(B)^2$$

and hence

$$F_{\alpha}(v) = \left(\frac{2\pi}{|v|}\right)^{N/2} \frac{f(\boldsymbol{\alpha})}{\sqrt{|\det(A)|}} \exp i\left(vh(\boldsymbol{\alpha}) + \frac{\pi}{4}\mathrm{sgn}(v)\mathrm{sig}(A)\right). \tag{14.22}$$

The quantity $\mathrm{sig}(A) = \sum_{k=1}^{N}\mathrm{sgn}\,\lambda_k$ is just the number of positive eigenvalues minus the number of negative eigenvalues and is named the *signature* of the matrix A. At a minimum $\text{sig}(A) = N$ and at a maximum $\text{sig}(A) = -N$.

If there are several, well separated, stationary points, $\boldsymbol{\alpha}_j$, inside the domain of integration, then the first term in the asymptotic expansion of $F(\lambda)$ is the sum of the individual terms,

$$F(v) \sim \left(\frac{2\pi}{|v|}\right)^{N/2} \sum_j \frac{f(\boldsymbol{\alpha}_j)}{\sqrt{|\det(A_j)|}} \exp i \left(vh(\boldsymbol{\alpha}_j) + \frac{\pi}{4} \text{sgn}\, v \, \text{sig}(A_j)\right). \qquad (14.23)$$

Higher-order terms of this expansion are more difficult to obtain. If the integration domain is finite the next largest term normally comes from the boundary; for two-dimensional integrals this contibution is derived by Mandel and Wolf (1995, section 3.3.3).

Exercise 14.13
Show that

$$\iint_{\mathscr{D}} dx\, dy \left(x^2 + \frac{y^2}{2}\right) \exp iv \left(y \sin \frac{\pi x}{2} - \frac{y^2}{2}\right) \sim \frac{12}{|v|} \exp i \left(\frac{v}{2} - \frac{\pi}{2} \text{sgn}(v)\right),$$

where $\mathscr{D}$ is a circle of radius $\frac{3}{2}$ centred at the origin.

Exercise 14.14
Snell's law: Consider two transparent media A and B, their interface being the xy-plane and B being in the region $z > 0$, as shown in the left hand figure. A light source S in medium A has coordinates $(\alpha, 0, -\beta)$, with $\alpha > 0$ and $\beta > 0$, and the observer O in medium B has coordinates $(0, 0, h)$.

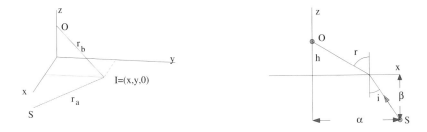

Supposing light travels in straight lines in each medium with speeds v_a and v_b and the amplitude, ψ, of the light at O is given, approximately, by the integral over the Oxy plane,

$$\psi = e^{i\omega t} \int_{-\infty}^{\infty} \int_{-\infty}^{\infty} dx\, dy\, a(\mathbf{r}) \exp\left(-ir_a(x, y)k_a - ir_b(x, y)k_b\right),$$

where $k_a = \omega/v_a$ and $k_b = \omega/v_b$, ω being the frequency of the light, and r_a the distance SI and r_b the distance IO, I being the point where the ray intersects the xy-plane.

Show that the stationary phase point of the integral is at $y = 0$ and

$$\frac{\sin i}{\sin r} = \frac{v_a}{v_b},$$

W. van Snel
(1580–1626)

where i is the angle of incidence and r the angle of refraction, as shown in the right hand diagram. This relation between the angles of incidence and refraction is known as Snell's law, after the Dutch astronomer and mathematician Willebrord van Snel van Royen (also called Snellius), first published by Huygens. The stationary phase path from S to O is also the path along which light travels in the shortest time: this is an illustration of Fermat's principle, see for instance Longhurst (1967).

14.5 The short wavelength limit

In the limit of very large $|v|$, if the stationary phase points are real the function

$$F(v, x) = \int_{-\infty}^{\infty} dt\, f(t, x) e^{ivh(t,x)} \tag{14.24}$$

contains terms like $e^{iv\alpha(x)}$, where $\alpha(x)$ and its derivative are $O(1)$. Thus $F(v, x)$ will, itself, be a rapidly oscillating function of both v and x. In many physical situations v is fixed and x varies — for instance in optical problems v is the inverse of a wavelength which may be relatively small. Observations are never precise and will generally involve an average over a small range of x which, for very large v, will include many oscillations. If such an observation was made on $F(v, x)$ the result would be zero, but normally we observe real quantities such as $|F(v, x)|^2$; for instance in the optics example F is the light amplitude and the intensity $|F|^2$ is the observed quantity; in quantum mechanics F is the probability amplitude and $|F|^2$ the probability. Here we obtain an approximation for $|F(v, x)|^2$, valid in the limit of large v, that removes the highly oscillatory components. If there is only one stationary point, at t_s, then

$$F \sim f(t_s) \sqrt{\frac{2\pi}{|vh''(t_s)|}} \exp i\left(vh(t_s) + \frac{\pi}{4}\mathrm{sgn}(h''(t_s v))\right)$$

and

$$|F(v, x)|^2 \simeq \frac{2\pi}{|v|}\frac{f(t_s)^2}{|h''(t_s)|} = 2\pi \int_{-\infty}^{\infty} dt\, f(t, x)^2 \delta\left(vh'(t, x)\right), \tag{14.25}$$

where we have used the properties of the delta function, in particular equation 7.13 (page 230), to express $|F|^2$ in terms of the delta function.

This result can also be derived directly from the original integral; the derivation is instructive and also useful because it is not restricted to integrals having only one stationary point. The magnitude of F may be expressed as the double integral,

$$|F(v, x)|^2 = \int_{-\infty}^{\infty} dw \int_{-\infty}^{\infty} dt\, f(w)f(t) \exp\left(iv(h(w) - h(t))\right).$$

The integrand is rapidly oscillating except in the neighbourhood of the line $w = t$, which suggests the change of variables $w = t + s$ and that we expand about $s = 0$:

$$|F(v, x)|^2 \simeq \int_{-\infty}^{\infty} dt \int_{-\infty}^{\infty} ds\, f(t + s)f(t) \exp\left(ivsh'(t) + O(s^2)\right),$$

$$\simeq \int_{-\infty}^{\infty} dt\, f(t)^2 \int_{-\infty}^{\infty} ds\, e^{ivsh'(t)} = \frac{2\pi}{|v|} \int_{-\infty}^{\infty} dt\, f(t)^2 \delta(h'(t)). \tag{14.26}$$

This result looks the same as that in equation 14.25, however, it is valid when there are many stationary points. Thus if $h'(t) = 0$ at $t = t_k$, $k = 1, 2, \ldots, N$,

$$|F(v, x)|^2 \simeq \frac{2\pi}{|v|} \sum_{k=1}^{N} \frac{f(t_k)^2}{|h''(t_k)|}. \tag{14.27}$$

The same result may be obtained from equation 14.15 (page 540), by ignoring the cross terms, which are oscillatory so will be small in any average over x.

Equation 14.26 provides an integral representation for $|F|^2$ in the limit of large v. In practical applications, if the phase $h(t, x)$ is a complicated function of x and t it may be easier to numerically compute the integral, using a suitable approximation for the delta function, than to evaluate the stationary phase approximation before taking the average because the latter involves finding all solutions of $h'(t) = 0$.

This is especially true for many-dimensional integrals. In this case, suppose that $F(v)$ is defined by the N-dimensional integral

$$F(v) = \int d\mathbf{t}\, f(\mathbf{t}) e^{iv h(\mathbf{t})} \tag{14.28}$$

and $h(\mathbf{t})$ is stationary at the points $\mathbf{t} = \mathbf{t}_k$, $k = 1, 2, \ldots, M$. A simple generalisation of the preceding analysis gives

$$|F(v)|^2 \simeq \left(\frac{2\pi}{|v|} \right)^N \sum_{j=1}^{M} \frac{f(\boldsymbol{\alpha}_j)^2}{|\det(A_j)|}, \tag{14.29}$$

which is just the square of the modulus of equation 14.23 with the cross terms neglected.

14.6 Neutralisers

In the examples treated above an estimate of the dominant part of the integral was obtained by expanding about the stationary and the end points of the integral. At each of these critical points the phase was expanded to obtain integrals like $\int_{-\infty}^{\infty} dt\, f(t) e^{ixt^2}$ at the stationary points and like $\int_0^{\infty} dt\, f(t) e^{ixt}$ at the end points; the idea being that the dominant contribution comes from the neighbourhood of these points. This idea can be developed to yield higher-order terms using two simple ideas. First, the contribution from each critical point must be isolated and dealt with by transforming to a new integration variable: this is achieved using *neutralisers*. Second, the asymptotic expansion of the resulting integrals is obtained by integrating by parts.

First consider the asymptotic expansion of

$$F(v) = \int_a^b dt\, f(t) h(vt), \tag{14.30}$$

where $f(t)$ is infinitely differentiable for $a \le t \le b$ and v a large, positive real number. In some cases this may be obtained using repeated integration by parts using the formula 13.11 (page 493), because each integration of $h(vt)$ introduces a factor v^{-1}.

This gives

$$F(v) = \sum_{r=0}^{N} \left[\frac{(-1)^r}{v^{r+1}} f^{(r)}(t) h^{(-r-1)}(vt) \right]_a^b - \frac{(-1)^N}{v^{N+1}} \int_a^b dt \, f^{(N+1)}(t) h^{(-N-1)}(vt), \qquad (14.31)$$

where the symbols $f^{(n)}(t)$ and $f^{(-n)}(t)$ denote respectively the nth differential and the nth repeated integral of f, so

$$f^{(0)} = f, \quad f^{(n)} = \frac{d}{dt} f^{(n-1)}, \quad \text{and} \quad f^{(-n+1)} = \frac{d}{dt} f^{(-n)}.$$

The function $f^{(-n)}(t)$ is defined only to within n integration constants; in many particular problems the choice of these is obvious, though in general the correct choice requires some care.

In simple cases this formula automatically provides an asymptotic expansion. For instance, if $h(s) = e^{is}$ the functions $h^{(-n)}(vt)$ may be chosen to form the asymptotic sequence

$$h^{(-n)}(vt) = \frac{1}{(iv)^n} e^{ivt},$$

and if all the derivatives of $f(t)$ exist we obtain

$$\int_a^b dt \, f(t) e^{ivt} \sim -\frac{i}{v} \sum_{r=0}^{\infty} \left(\frac{i}{v} \right)^r \left[e^{ivb} f^{(r)}(b) - e^{iva} f^{(r)}(a) \right]. \qquad (14.32)$$

This is a power series in $1/v$, and it is easy to see that its radius of convergence is zero if the radius of convergence of the expansion of $f(t)$ about either $t = a$ or b is finite. Suppose that $f(t)$ has a Taylor expansion about $t = a$ with radius of convergence T, then

$$|f^{(n)}(a)| \sim \sqrt{2\pi n} \left(\frac{n}{eT} \right)^n, \quad n \gg 1,$$

so the nth term of the asymptotic expansion from $t = a$ behaves as

$$\sqrt{2\pi n} \left(\frac{n}{eT|v|} \right)^n.$$

This function grows very rapidly as $n \to \infty$, but has a local minimum at $n \simeq Tv$ — a similar analysis was performed in chapter 5, see figure 5.1 (page 183). It follows that the maximum number of terms that should be used in the series 14.32 is about $T|v|$.

Exercise 14.15
By changing the integration variable to $s = t^\beta$ show that for $v \gg 1$ and $\beta > 0$,

$$\int_0^1 dt \, e^{ivt^\beta} \sim \frac{\Gamma(1/\beta)}{\beta} \left(\frac{i}{v} \right)^{1/\beta} - \frac{ie^{iv}}{\beta v} \sum_{k=0}^{\infty} \left(\frac{i}{v} \right)^k A_k,$$

where

$$A_k = (-1)^k \prod_{r=1}^{k} \left(r - \frac{1}{\beta} \right) = (-1)^k \frac{\Gamma(k+1-1/\beta)}{\Gamma(1-1/\beta)} = \frac{\Gamma(1/\beta)}{\Gamma(1/\beta - k)}.$$

Use the asymptotic expansion $\Gamma(n+a) \sim \sqrt{2\pi n} \left(n/e\right)^n n^a$ to show that the optimum number of terms in the sum is about v.

Exercise 14.16
Use equation 14.32 to show that if n is a large integer,

$$\int_0^1 dt \, \frac{\cos n\pi t}{1+t^2} \sim \frac{(-1)^{n-1}}{\pi^2 n^2} \left(\frac{1}{2} - \frac{15}{(n\pi)^4} + \frac{11\,340}{(n\pi)^8} - \frac{48\,648\,600}{(n\pi)^{12}} \cdots \right).$$

For each n estimate the number of terms required to give the most accurate values. Compare the values of this approximation with exact values obtained by numerical integration for some values of n.

If, for some m, $f^{(m)}(t)$ does not exist at either a or b then the series 14.31 cannot be carried beyond $r = m - 1$. For instance, the integral $\int_0^1 dt \, t^{1/2} e^{ivt}$ cannot be treated in this manner. Similar problems arise if some derivative of $f(t)$ does not exist at an interior point of the integral, $t = c$, where $a < c < b$, but then we may split the integral into two parts over the intervals (a, c) and (c, b).

The crucial step in dealing with such integrals is to isolate the contribution from each critical point so each may be dealt with using appropriate transformations of the independent variable. This is achieved using *neutralisers*, which are infinitely differentiable, real functions $q(t, \alpha, \beta)$ such that if $\alpha < \beta$,

$$q(t, \alpha, \beta) = \begin{cases} 0, & x \le \alpha, \\ 1, & x \ge \beta, \end{cases} \quad \text{or} \quad q(t, \alpha, \beta) = \begin{cases} 1, & x \le \alpha, \\ 0, & x \ge \beta. \end{cases} \tag{14.33}$$

These are the only properties of neutralisers required, but to show that such functions exist we construct an example. The function

$$p(x) = \begin{cases} 0, & x \le 0, \\ \exp(-1/x), & x > 0, \end{cases}$$

is infinitely differentiable and increases from 0 to 1 as x increases from 0 to ∞. The function

$$q_2(t, \alpha, \beta) = \begin{cases} 0, & x \le \alpha, \\ \dfrac{p(x - \alpha)}{p(x - \alpha) + p(\beta - x)}, & x > \alpha, \end{cases} \tag{14.34}$$

is a neutraliser satisfying the first of equations 14.33. A neutraliser satisfying the second of equations 14.33 is $q_1(t, \alpha, \beta) = 1 - q_2(t, \alpha, \beta)$. Graphs of these neutralisers are shown in figure 14.13.

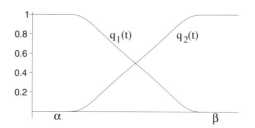

Figure 14.13   Sketch of the neutralisers q_1 and
$q_2 = 1 - q_1$.

If $a < \alpha < \beta < b$ we may write the integral 14.30 as the sum $F = F_a + F_b$ where

$$F_a(v) = \int_a^b dt\, f(t)q_1(t)h(t,v), \quad F_b(v) = \int_a^b dt\, f(t)q_2(t)h(t,v), \qquad (14.35)$$

thus isolating the contribution from each end point. Here, and henceforth, we assume that the integrand has no interior singularities and also omit mention of the parameters α and β, assuming them to be chosen appropriately.

If $f(t)$ is infinitely differentiable, each of $F_a(v)$ and $F_b(v)$ may be integrated by parts using the formula 14.32 with $f(t)$ replaced by $f(t)q_k(t)$, $k = 1$ and 2. Then the series for $F_a(v)$ will contain no contribution from $t = b$, the series for $F_b(v)$ no contribution from $t = a$, and equation 14.32 is regained.

Now consider an example where $f(t)$ is not differentiable at an end point:

$$F(v) = \int_a^b dt\, (t-a)^{\alpha-1} g(t) e^{ivt}, \quad 0 < \alpha < 1, \quad v \gg 1, \qquad (14.36)$$

$g(t)$ being an infinitely differentiable. Now the series 14.32 fails at the first term if we choose $f(t) = (t-a)^{\alpha-1} g(t)$. Proceed by introducing two neutralisers and writing $F(v) = F_a(v) + F_b(v)$, where

$$F_a(v) \quad = \quad \int_a^b dt\, q_1(t)(t-a)^{\alpha-1} g(t) e^{ivt}, \qquad (14.37)$$

$$F_b(v) \quad = \quad \int_a^b dt\, q_2(t)(t-a)^{\alpha-1} g(t) e^{ivt}, \qquad (14.38)$$

so $F_a(v)$ gives the contribution from the neighbourhood of $t = a$ and $F_b(v)$ the contribution from near $t = b$.

The integral for F_b presents no problems; setting $f(t) = q_2(t)g(t)(t-a)^{\alpha-1}$ gives

$$F_b(v) \sim -\frac{i e^{ivb}}{v} \sum_{r=0}^{\infty} \left(\frac{i}{v}\right)^r \frac{d^r}{db^r}\left[g(b)(b-a)^{\alpha-1}\right]. \qquad (14.39)$$

The evaluation of $F_a(v)$ is not so easy and we use a method, due to Erdélyi (1956, section 2.9), see also Bleistein and Handelsman (1975, page 90). However, the algebraic complexity of this rigorous method obscures its essential simplicity, so we first derive the expansion using a heuristic method that is easier to apply, particularly if Maple is used, and gives the same result.

In the integral for $F_a(v)$ set $t = s + a$ to give

$$F_a(v) = e^{iva} \int_0^\infty ds \, s^{\alpha-1} g(a+s) q_1(a+s) e^{ivs}.$$

Now expand $g(a+s)q_1(a+s)$ about $s = 0$, remembering that $q_1(a) = 1$ and that all derivatives of q_1 at $x = a$ are zero, and interchange the order of summation and integration to write the integral in the form

$$F_a(v) = e^{iva} \sum_{n=0}^\infty \frac{g^{(n)}(a)}{n!} \int_0^\infty ds \, s^{n+\alpha-1} e^{ivs}.$$

Using the integral

$$\int_0^\infty ds \, s^\beta e^{ivs} = \Gamma(1+\beta) \left(\frac{i}{v}\right)^{1+\beta}, \quad \beta > -1,$$

we obtain the asymptotic expansion

$$F_a(v) = e^{iva} \sum_{n=0}^\infty g^{(n)}(a) \left(\frac{i}{v}\right)^{n+\alpha} \frac{\Gamma(n+\alpha)}{n!}.$$

If $g(t)$ has a Taylor's series expansion about $t = a$ with finite radius of convergence T then since $n! = \Gamma(n+1)$ and $\Gamma(n+a)/\Gamma(n+b) \sim n^{a-b}$, $n \gg 1$, the terms of the series for $F_a(v)$ behave as

$$\frac{\sqrt{2\pi n}}{n^{1-\alpha}} \left(\frac{n}{e|v|T}\right)^n.$$

Combining the series for $F_a(v)$ and $F_b(v)$ gives, as $|v| \to \infty$,

$$\int_a^b dt \, (t-a)^{\alpha-1} g(t) e^{ivt} \quad \sim \quad -\frac{i e^{ivb}}{v} \sum_{r=0}^\infty \left(\frac{i}{v}\right)^r \frac{d^r}{db^r} \left[g(b)(b-a)^{\alpha-1}\right]$$
$$+ e^{iva} \left(\frac{i}{v}\right)^\alpha \sum_{r=0}^\infty \frac{\Gamma(r+\alpha)}{r!} \left(\frac{i}{v}\right)^r g^{(r)}(a). \qquad (14.40)$$

Compare this with equation 14.32 and observe that if $\alpha = 1$ they are the same, but for $\alpha < 1$ the effect of the $(t-a)^{-1+\alpha}$ term is to decrease the rate at which the function tends to zero as $|v| \to \infty$.

Exercise 14.17
Show that if $v \gg 1$,

$$\int_0^1 dt \, t^{1/2} e^{ivt} \sim \frac{\sqrt{\pi}}{2v^{3/2}} e^{3\pi i/4} - \frac{i}{v} e^{iv} \left(1 - 2 \sum_{k=1}^\infty \frac{(2k-2)!}{(k-1)!} \left(-\frac{i}{4v}\right)^k\right).$$

Show that this asymptotic expansion should be terminated at $k \simeq v$ for optimum accuracy.

Now we derive these results rigorously using integration by parts. Set

$$h(t) = (t-a)^{\alpha-1} e^{ivt} \quad \text{and} \quad f(t) = q_1(t) g(t),$$

so the formula 14.31 gives

$$F_a(v) \sim -\sum_{r=0}^{\infty} (-1)^r g^{(r)}(a) h^{(-r-1)}(a), \tag{14.41}$$

there being no contribution from $t = b$. The only problem is to evaluate $h^{(-r-1)}(a)$.

The evaluation of $h^{(-n)}(a)$ for $v > 0$

The functions $h^{(-n)}(a)$ are evaluated by careful integration and by noting that if t is complex and $0 < \arg t < \pi/2$ then e^{ivt} decreases exponentially as $\Im(t)$ increases. For $h^{(-1)}(t)$ we have, as may be verified by direct differentiation,

$$
\begin{aligned}
h^{(-1)}(t) &= -\int_t^{t+i\infty} ds\,(s-a)^{\alpha-1} e^{ivs} \\
&= -ie^{ivt} \int_0^{\infty} d\sigma\,(t-a+i\sigma)^{\alpha-1} e^{-v\sigma}, \quad s = t + i\sigma.
\end{aligned}
$$

At $t = a$ this integral may be evaluated in terms of the gamma function,

$$h^{(-1)}(a) = -\frac{\Gamma(\alpha)}{v^\alpha} \exp i\left(av + \frac{\pi}{2}\alpha\right).$$

The $(n+1)$th integral is obtained in the same manner; on integrating n times we obtain

$$h^{(-n-1)}(t) = (-1)^{n+1} \int_t^{t+i\infty} ds_n \int_{s_n}^{s_n+i\infty} ds_{n-1} \cdots \int_{s_1}^{s_1+i\infty} ds_0\,(s_0-a)^{\alpha-1} e^{ivs_0}.$$

Now set $s_m = s_{m+1} + i\sigma_m$, $m = 0, 1, \ldots, n$, with $s_{n+1} = t$ to cast this in the form

$$h^{(-n-1)}(t) = (-i)^{n+1} e^{ivt} \int_0^{\infty} d\sigma_n \int_0^{\infty} d\sigma_{n-1} \cdots \int_0^{\infty} d\sigma_0\,(t-a+i\Sigma_n)^{\alpha-1} \exp(-v\Sigma_n),$$

where $\Sigma_n = \sigma_0 + \sigma_1 + \cdots + \sigma_n$.

Finally, observe that the integrand depends only upon the sum Σ_n, which suggests introducing the integration variables $x_k = \sigma_0 + \sigma_1 + \cdots + \sigma_k$, $k = 0, 1, \ldots, n$, having the Jacobian determinant $\partial(\mathbf{x})/\partial(\boldsymbol{\sigma}) = 1$. Hence

$$
\begin{aligned}
h^{(-n-1)}(t) &= (-i)^{n+1} e^{ivt} \int_0^{\infty} dx_n \int_0^{x_n} dx_{n-1} \cdots \int_0^{x_1} dx_0\,(t-a+ix_n)^{\alpha-1} e^{-vx_n} \\
&= e^{ivt} \frac{(-i)^{n+1}}{n!} \int_0^{\infty} dx\,(t-a+ix)^{\alpha-1} x^n e^{-vx}.
\end{aligned}
$$

Thus at $t = a$ we have

$$h^{(-n-1)}(a) = \frac{(-1)^{n+1}\Gamma(n+\alpha)}{n! v^{n+\alpha}} \exp i\left(va + \frac{\pi}{2}(n+\alpha)\right), \tag{14.42}$$

where we have used the integral definition of the gamma function. With this, equations 14.39 and 14.41, we regain equation 14.40.

These results can be used to determine the asymptotic expansion of a wide variety of integrals, the most common of which will be considered in the next section.

We end this section by illustrating how these techniques may be used to derive an asymptotic expansion for the integral

$$F(v) = \int_a^b dt\, f(t) e^{ivh(t)}, \quad |v| \gg 1, \tag{14.43}$$

where both $f(t)$ and $h(t)$ have Taylor's series expansions about $t = a$ and b and $h'(t) \neq 0$ for $a \le t \le b$, so there are no stationary phase points: for convenience we initially assume that $h(t)$ is increasing, $h'(t) > 0$. First introduce two suitable neutralisers, to isolate the contributions from each end point, and write the function in the form $F(v) = F_a(v) + F_b(v)$. The contribution from the neighbourhood $t = a$ is found by introducing a new integration variable

$$u = h(t) - h(a) \ge 0,$$

which has a single valued inverse $t(u)$. This puts the integral in the familiar form, equation 14.32,

$$F_a(v) = e^{ivh(a)} \int_0^\infty du\, \frac{f(t)}{h'(t)} q(t) e^{ivu},$$

and integration by parts, equation 14.32, yields

$$F_a(v) \sim e^{ivh(a)} \frac{i}{v} \sum_{r=0}^\infty \left(\frac{i}{v} \right)^r \frac{d^r}{du^r} \left[\frac{f(t)}{h'(t)} \right]_{u=0}.$$

Similarly, $F_b(v)$ is re-written using the new variable $v = h(b) - h(t) \ge 0$ to give the final result

$$F(v) \sim \frac{i}{v} e^{ivh(a)} \sum_{r=0}^\infty \left(\frac{i}{v} \right)^r \frac{d^r}{du^r} \left[\frac{f(t)}{h'(t)} \right]_{u=0}$$
$$- \frac{i}{v} e^{ivh(b)} \sum_{r=0}^\infty \left(-\frac{i}{v} \right)^r \frac{d^r}{dv^r} \left[\frac{f(t)}{h'(t)} \right]_{v=0}. \tag{14.44}$$

This result remains true for $h'(t) < 0$, and if $h(t) = t$ it reduces to 14.32. The first two terms of this sum are relatively easy to determine and give

$$F(v) \sim -\frac{i}{v} \left[\frac{f(t)}{h'(t)} e^{ivh(t)} \right]_{t=a}^b + \frac{1}{v^2} \left[\left(\frac{f'(t)h'(t) - f(t)h''(t)}{h'(t)^3} \right) e^{ivh(t)} \right]_{t=a}^b + \cdots.$$

Often this is the most useful approximation because in many cases the higher-order differentials can be numerically large. If more terms are required, perhaps the easiest way of proceeding is to use Maple to express $u(t)$ and $v(t)$ as power series, invert these and then express $f(t)/h'(t)$ as power series in u and v.

For instance, consider the example

$$F(v) = \int_0^{\pi/4} dt\, \frac{e^{iv \tan t}}{1 + t^2}, \quad v \gg 1.$$

By introducing neutralisers q_1 and q_2 and making suitable changes of variable, this may be written in the form

$$F(v) = \int_0^\infty du\, \frac{\cos^2 t(u)}{1 + t(u)^2} q_1(u) e^{ivu} + e^{iv} \int_0^\infty dv\, \frac{\cos^2 t(v)}{1 + t(v)^2} q_2(v) e^{-ivv},$$

where $u = \tan t$ and $v = 1 - \tan t$. Expanding the pre-exponential parts of each integral about the origin,

$$\frac{\cos^2 t(u)}{1 + t(u)^2} = 1 - 2u^2 + \frac{11}{3}u^4 - \frac{293}{45}u^6 + \cdots$$

$$\frac{\cos^2 t(v)}{1 + t(v)^2} = \frac{8}{16 + \pi^2} + \frac{8(\pi^2 + 4\pi + 16)}{(16 + \pi^2)^2}v$$

$$+ \frac{4(\pi^4 + 12\pi^3 + 56\pi^2 + 192\pi + 128)}{(16 + \pi^2)^3}v^2 + \cdots,$$

and using the integral

$$\int_0^\infty du\, u^k e^{ivu} = k!\left(\frac{i}{v}\right)^{k+1}, \quad v \neq 0,$$

gives the required asymptotic expansion, the real part of which is

$$\int_0^{\pi/4} dt\, \frac{\cos(v\tan t)}{1 + t^2} \sim -\frac{0.4595}{v^2}\left(1 - \frac{2.802}{v^2}\right)\cos v + \frac{0.3092}{v}\left(1 - \frac{2.620}{v^2}\right)\sin v.$$

$$(14.45)$$

A comparison of this with the exact numerical integration of the integral, the thick line, is shown in the following figure.

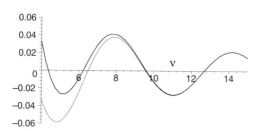

Figure 14.14   Comparison of the real part of $F(v)$, the thicker line, and its asymptotic expansion 14.45.

Exercise 14.18
Show that

$$\int_0^{\pi/4} dt\, \frac{\sin(v\tan t)}{1 + t^2} \sim \left(\frac{1}{v} + \frac{4}{v^3} + \frac{88}{v^5} + \cdots\right) - \frac{0.4598}{v^2}\left(1 - \frac{2.8012}{v^2} + \cdots\right)\sin v$$

$$- \frac{0.3098}{v}\left(1 - \frac{2.6169}{v^2} + \frac{0.0749}{v^4} + \cdots\right)\cos v.$$

Exercise 14.19

Show that if $a > 0$ and $v \gg 1$,

$$\int_0^1 dt \, \frac{t^{1/2}}{1+t^2} \cos(v(t + at^2)) \quad \sim \quad \frac{\sqrt{2\pi}}{4v^{3/2}} \left(-1 + \frac{15a}{4v} + \cdots\right)$$
$$+ \frac{\sin(v(1+a))}{2v(1+2a)} - \frac{1+6a}{4v^2(1+2a)^2} \cos(v(1+a)) + \cdots.$$

Why is this approximation poor for large a?

Exercise 14.20

Use equation 14.32 (page 546) to show that if $|\lambda/v| < 1$,

$$\int_0^\infty dt \, e^{-\lambda t} e^{ivt} \quad = \quad \frac{1}{\lambda - iv},$$

$$\int_0^\infty dt \, \cos \lambda t \, e^{ivt} \quad = \quad \frac{i}{v} \sum_{k=0}^\infty \left(\frac{\lambda}{v}\right)^{2r} = \frac{iv}{v^2 - \lambda^2}.$$

Compare these expression with those obtained using direct integration.

Exercise 14.21

Show that if $\alpha, \beta > -1$ and $h'(t) > 0$ for $a \le t \le b$,

$$\int_a^b dt \, (t-a)^\alpha (b-t)^\beta g(t) e^{ivh(t)} \sim B(b) - A(a),$$

where

$$A(a) \quad = \quad -\frac{\exp(ivh(a))}{|v|^{1+\alpha}} \exp\left(i\frac{\pi}{4}(1+\alpha)\operatorname{sgn}(v)\right) \sum_{k=0}^\infty g_a^{(k)}(0) \frac{\Gamma(1+k+\alpha)}{k!} \left(\frac{i}{v}\right)^k,$$

$$B(b) \quad = \quad \frac{\exp(ivh(b))}{|v|^{1+\beta}} \exp\left(i\frac{\pi}{4}(1+\beta)\operatorname{sgn}(v)\right) \sum_{k=0}^\infty g_b^{(k)}(0) \frac{\Gamma(1+k+\beta)}{k!} \left(\frac{i}{v}\right)^k,$$

and

$$g_a(s) \quad = \quad \left(\frac{t-a}{s}\right)^\alpha \frac{(b-t)^\beta}{h'(t)} g(t), \quad s = h(t) - h(a) > 0,$$

$$g_b(u) \quad = \quad \frac{(t-a)^\alpha}{-h'(t)} \left(\frac{b-t}{u}\right)^\beta g(t), \quad u = h(b) - h(t) > 0.$$

Exercise 14.22

If $h(t)$ has a Taylor's series expansion about $t = a$, is positive, monotonic increasing for $t \ge a$ and $h(a) \ne 0$, show that for $\alpha > 0$,

$$\int_a^\infty dt \, f(t) e^{iv(a-x)^\alpha h(t)} \sim \frac{\exp\left(i\frac{\pi}{2\alpha}\operatorname{sgn}(v)\right)}{\alpha |v|^{1/\alpha}} \sum_{k=0}^\infty g_a^{(k)}(0) \frac{\Gamma\left(\frac{1+k}{\alpha}\right)}{k!} \left(\frac{i}{v}\right)^{k/\alpha},$$

where

$$g_a(s) = \frac{f(t)}{ds/dt}, \quad s = (t-a)h(t)^{1/\alpha}.$$

14.7 Stationary phase approximation II

The methods described in the previous section can be trivially extended to find an asymptotic expansion for the integral

$$F(v) = \int_a^b dt\, f(t) e^{ivh(t)}, \quad |v| \gg 1, \tag{14.46}$$

where $f(t)$ and $h(t)$ are real and infinitely differentiable for $a \le t \le b$ and $h(t)$ has stationary points in the interval $a < t < b$. The cases in which $f(t)$ is singular at either, or both, end points, or when the stationary point is at one of the end points, can be dealt with using a combination of the methods described here and in the preceding section, so are left as exercises for the reader.

Here we consider the situation where $h(t)$ has a single stationary point at $t = t_s$ with $a < t_s < b$ and $h''(t_s) \ne 0$, in which case the asymptotic expansion of $F(v)$ comprises terms from the expansions about $t = t_s$ (the dominant term) and $t = a$ and $t = b$. We therefore proceed by isolating the contributions from these three points using suitable neutralisers and, to avoid confusion, assume that $h(t)$ has a local minimum at t_s, so $h''(t_s) > 0$.

The contribution from the end points is given by equation 14.44. The dominant contribution, however, comes from the neighbourhood of the minimum at $t = t_s$; near here we may define a new independent variable with the equation

$$w^2 = h(t) - h(t_s),$$

and the sign convention $w > 0$ for $t > t_s$ and $w < 0$ for $t < t_s$, allowing $w(t)$ to be expressed as a power series in $t - t_s$,

$$w = (t - t_s)\sqrt{\frac{h''(t_s)}{2}} + O\left((t - t_s)^2\right).$$

Using w as the integration variable, the contribution from the neighbourhood of the stationary point becomes

$$F_s(v) = e^{ivh(t_s)} \int_{-\infty}^{\infty} dw\, q(w) \left(\frac{2wf(t)}{h'(t)}\right) e^{ivw^2}, \tag{14.47}$$

where $q(w)$ is a suitable neutraliser. This integral may be evaluated using integration by parts, as in section 14.6; however, it is usually easier to use the formally equivalent method of expanding the pre-exponential part of the integral as a series in w. Thus, the expansion

$$\frac{2wf(t)}{h'(t)} = \sum_{k=0}^{\infty} \frac{a_k}{k!} w^k, \quad a_k = \lim_{w \to 0} \left(\frac{d^k}{dw^k}\left[\frac{2wf(t)}{h'(t)}\right]\right), \tag{14.48}$$

together with the integral

$$\int_{-\infty}^{\infty} dt\, t^{2n} e^{ivt^2} = \frac{\Gamma(n+1/2)}{|v|^{n+1/2}} \exp\left(i\frac{\pi}{2}\left(n+\frac{1}{2}\right)\mathrm{sgn}(v)\right),$$

yields

$$F_s(v) = \frac{1}{\sqrt{|v|}} \exp\left(ivh(t_s) + i\frac{\pi}{4}\operatorname{sgn}(v)\right) \sum_{k=0}^{\infty} a_{2k} \frac{\Gamma(k+1/2)}{(2k)!} \left(\frac{i}{v}\right)^k. \qquad (14.49)$$

If $h(t)$ has a maximum, rather than a minimum, then the natural transformation for the stationary phase contribution is

$$\zeta^2 = h(t_s) - h(t), \quad \Longrightarrow \quad \zeta = (t - t_s)\sqrt{-\frac{h''(t_s)}{2}} + O((t - t_s)^2)$$

and then

$$F_s(v) = \frac{1}{\sqrt{|v|}} \exp\left(ivh(t_s) - \frac{i\pi}{4}\operatorname{sgn}(v)\right) \sum_{k=0}^{\infty} b_{2k} \left(\frac{i}{v}\right)^k \frac{\Gamma(k+1/2)}{(2k)!}, \qquad (14.50)$$

where

$$b_k = \lim_{\zeta \to 0} \left(\frac{d^k}{d\zeta^k}\left[\frac{2\zeta f(t)}{-h'(t)}\right]\right).$$

14.7.1 Summary

When $h(t)$ has N isolated stationary points, t_k, $k = 1, 2, \ldots, N$, satisfying

$$a < t_1 < t_2 < \cdots < t_n < b, \quad h'(t_k) = 0, \ h''(t_k) \neq 0, \quad k = 1, 2, \ldots, N,$$

the asymptotic expansion of the integral 14.46 comprises a sum of contributions from each stationary phase point and from the end points,

$$\int_a^b dt\, f(t)e^{ivh(t)} \sim \sum_{k=1}^N F_s^{(k)}(v) + F_{\text{end}}(v).$$

The contribution from the end points is

$$F_{\text{end}}(v) \sim \frac{i}{v}e^{ivh(a)} \sum_{r=0}^{\infty} \left(\frac{i}{v}\right)^r \frac{d^r}{du^r}\left[\frac{f(t)}{h'(t)}\right]_{u=0}$$

$$-\frac{i}{v}e^{ivh(b)} \sum_{r=0}^{\infty} \left(-\frac{i}{v}\right)^r \frac{d^r}{dv^r}\left[\frac{f(t)}{h'(t)}\right]_{v=0}.$$

The contribution from the kth stationary phase point is given by equation 14.49 at a minimum, $h''(t_k) > 0$, or equation 14.50 at a maximum, $h''(t_k) < 0$.

The stationary phase contribution gives the terms $v^{-k-1/2}$, $k = 0, 1, \ldots$, and the end-point contributions the terms v^{-k-1}, $k = 0, 1, \ldots$.

The first two terms of this asymptotic expansion are

$$\int_a^b dt\, f(t)e^{ivh(t)} \sim \sum_k f(t_k)\sqrt{\frac{2\pi}{|vh''(t_k)|}} \exp i\left(vh(t_k) + \frac{\pi}{4}\operatorname{sgn}(vh''(t_k))\right)$$

$$-\frac{i}{v}\left[\frac{f(t)}{h'(t)}e^{ivh(t)}\right]_{t=a}^b.$$

The other terms are more complicated and are usually best derived as special cases as and when needed; Maple makes light work of this type of algebra.

Exercise 14.23
Show that

$$a_0 = b_0 = f(t_s)\sqrt{\frac{2}{|h''(t_s)|}},$$

where a_k and b_k are defined in equations 14.48 and 14.50.

Exercise 14.24
Show that if $|v| \gg 1$,

$$\int_{-1}^{1} dt \, e^{ivt^2} \sim \sqrt{\frac{\pi}{|v|}} \exp\left(i\frac{\pi}{4}\operatorname{sgn}v\right) - \frac{ie^{iv}}{v\sqrt{\pi}} \sum_{k=0}^{\infty} \Gamma\left(k + \frac{1}{2}\right)\left(-\frac{i}{v}\right)^k.$$

Show also that for optimum results this sum should be truncated at about $k = [v] + 1$, where $[v]$ denotes the integer part of v.

Compare this approximation with the exact evaluation of the integral for $v > 2$.

Exercise 14.25
Show that if $\beta > -1$ and $|v| \gg 1$,

$$\int_{0}^{1} dt \, t^\beta e^{ivt^2} \sim \frac{\Gamma\left(\frac{1+\beta}{2}\right)}{2|v|^{(1+\beta)/2}} \exp\left(i\frac{\pi}{4}(1+\beta)\operatorname{sgn}(v)\right) - \frac{ie^{iv}}{2v} \sum_{k=0}^{\infty} \left(-\frac{i}{v}\right)^k A_k,$$

where

$$A_k = \frac{\Gamma\left(k + \frac{1-\beta}{2}\right)}{\Gamma\left(\frac{1-\beta}{2}\right)} = (-1)^k \frac{\Gamma\left(\frac{1+\beta}{2}\right)}{\Gamma\left(\frac{1+\beta}{2} - k\right)}.$$

Show that this expression is exact if $\beta = 1, 3, 5, \ldots$.

Exercise 14.26
Show that if a and b are real numbers with $a \neq 0$, then

$$\int_{-\infty}^{\infty} dt \, \exp i\left(at^2 + bt\right) = \sqrt{\frac{\pi}{|a|}} \exp\left(-i\frac{b^2}{4a} + i\frac{\pi}{4}\operatorname{sgn}(a)\right).$$

Exercise 14.27
Show that if $x > 0$ and $|v| \gg 1$,

$$\int_{-\infty}^{\infty} dt \, \exp\left(iv\sinh^2 t\right) \sim \frac{\exp\left(i\frac{\pi}{4}\operatorname{sgn}(v)\right)}{\sqrt{\pi|v|}} \sum_{k=0}^{\infty} \frac{\Gamma(k + 1/2)^2}{k!}\left(-\frac{i}{v}\right)^k$$

and determine the number of terms that should be included in the series for optimum accuracy.

Exercise 14.28
Show that if $v \gg 1$,

$$\int_{0}^{\infty} dt \, \sin(at)\cos\left(vt^4(1 - t^2)\right) \sim \sqrt{\frac{\pi}{v}} \left\{\frac{a}{4\sqrt{2}} + \sqrt{\frac{3}{8}}\sin\frac{a\sqrt{6}}{3}\cos\left(\frac{4v}{27} - \frac{\pi}{4}\right)\right\} + O(v^{-3/2})$$

and find the next term in the expansion.

14.8 Coalescing stationary points: uniform approximations

It has been shown that the dominant contribution to oscillatory integrals comes from the neighbourhood of the real stationary phase points, and that if these points are sufficiently far apart the contributions are additive, equation 14.15 (page 540). But if the stationary points are too close this type of approximation is poor, and when two stationary points coalesce the stationary phase approximation is normally singular, see equation 14.10 (page 535) and the subsequent discussion.

It might be thought that the occurrence of close or coalescing stationary points is sufficiently rare that the problem may be ignored and the exceptions treated numerically. This, however, is not the case, for such situations occur in a wide variety of optical and wave-type phenomena, so its correct mathematical description is required. The approximations developed earlier in this chapter all involved reducing the integral to sums of simpler integrals of the form

$$\int_0^\infty dt\, f(t)e^{ivt} \quad \text{or} \quad \int_0^\infty dt\, f(t)e^{ivt^2}.$$

These methods fail when stationary points are too close,† for then we need the exponent to be a cubic in t and the required generic function is the Airy function, defined in chapter 11. The approximation described in this section is useful not only when two real stationary phase points are close but also when they are far apart — for then the new approximation reduces to the stationary phase approximation — and when they are complex and the integral is exponentially small. Because the approximation is accurate through the awkward region where the stationary phase approximation is singular, this type of approximation is named a *uniform approximation*. Before dealing with uniform approximations we provide a brief discussion of rainbows because this theory was the origin of such approximations; the next subsection contains no essential material.

14.8.1 Rainbows

Rainbows are of intrinsic interest, but also provided the first example of a particular type of integral useful in approximating other, more complicated integrals. The proper mathematical description of rainbows is due to Airy (1838, 1849) and in this work he introduced what is now called the Airy function. Rainbows are one of many naturally occurring optical phenomena that can be understood with relatively elementary ideas, and there are several excellent books describing such diverse and fascinating phenomena in fairly simple terms; see for example the books by Greenler (1991), Meinel and Meinel (1991) and Minnaert (1954). More detailed accounts of the rainbow may be found in the article by Walker (1976).

G. B. Airy
(1801–1892)

---

† They also fail when too close to the end points, but the methods described here cannot deal with this situation.

R. Descartes
(1596–1650)

The first 'correct' description of rainbows was due to Descartes in his 1637 essay *Les Météores*, which was part of *Discours de la Méthode*, but speculations as to the origins of rainbows are very much older; the interested reader will find it worthwhile reading Boyer's (1987) history of rainbows.

In order to understand why rainbows occur an elementary knowledge of reflection and refraction at boundaries is necessary. Indeed, Descarte's discovery of the law of refraction was essential for his description of rainbows. A ray of light reflects from a flat surface so that the angle of incidence, i, equals the angle of reflection, r, as shown in figure 14.15.

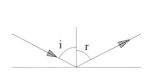

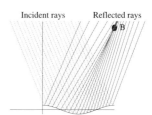

Figure 14.15   Reflection from a flat
surface.

Figure 14.16   Reflection from a
curved surface.

Reflections from an irregular surface, figure 14.16, will produce more complicated light patterns; in particular there are often regions where the density of rays is high and bright spots or lines are produced at B.

The bright regions, such as B in figure 14.16, are seen frequently; for instance on the underside of a bridge over a canal as the sun light is refected from the ruffled water surface, see for instance Plate IV of Minnaert (1954, page 32), and the type of glass commonly used in bathroom windows frequently creates such patterns.

Similar patterns are seen on the bottom of swimming pools, as seen on the front cover, or any clear shallow water, but these are caused by refraction whereby the direction of light rays is changed on passing through a boundary of two transparent media, such as air and water, in each of which the speed of light is different. In this case the angle of incidence i and the angle of refraction r are related by Snell's law, exercise 14.14 (page 543), which states that the ratio $\sin i / \sin r$ is constant for any pair of media. If the light passes from medium 1 to medium 2, as shown in figure 14.17, then

$$\frac{\sin i}{\sin r} = n, \quad \text{where} \quad n = \frac{v_1}{v_2}, \qquad \text{(Snell's law)}, \qquad (14.51)$$

where v_k is the velocity of light in medium k. For an air–water interface, with the light passing into the water the value of n varies between $n = 1.331$ for red light and $n = 1.343$ for blue light, so $i > r$.

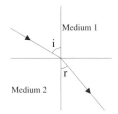

Figure 14.17   Diagram illustrating
equation 14.51.

In practice not all the light is refracted through a surface; a fraction, which depends upon i, is reflected; this is the reason why at most two rainbows are ever seen.

Now consider a ray of light entering a spherical drop at B, figure 14.18, suffering one internal reflection at C before being refracted out again at D.

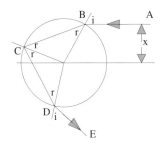

Figure 14.18   Passage of a ray through a
spherical drop with one internal reflection.

If θ is the acute angle between the incoming ray, AB, and the exiting ray, DE, and the angles of incidence and refraction at B are i and r respectively, simple geometry shows that $\theta = 4r - 2i$. Using Snell's law, equation 14.51, this gives θ in terms of i,

$$\theta(i) = 4\sin^{-1}\left(\frac{1}{n}\sin i\right) - 2i.$$

The function $\theta(i)$ has a local maximum and, for reasons that will soon become clear, the value of $\theta(i)$ at this maximum is named the *rainbow angle*, and is denoted by θ_R.

Exercise 14.29
Show that $\theta(i)$ has a local maximum at

$$\cos i = \sqrt{\frac{n^2 - 1}{3}}.$$

The value of n varies between 1.331 (red light) and 1.343 (blue light). Show that the maximum value of $\theta(i)$, the rainbow angle, varies between 42.37° and 40.65°.

The significance of the maximum value in $\theta(i)$ is that here rays accumulate to produce brighter light than at other values of θ. This can be seen by considering adjacent parallel rays AB and $A'B'$, incident on a water droplet of radius R. If the ray AB is a distance x from the parallel axis through the centre, so that $x/R = \sin i$, and is scattered through the angle $\theta(x)$; an adjacent ray at a distance $x + \delta x$ will be scattered through the angle $\theta(x + \delta x) \simeq \theta(x) + \theta'(x)\delta x$. Thus the light in the beam between the two rays will be spread into the angular distance $\theta'(x)\delta x$.

For a uniform beam of light the intensity of light scattered into an angle ϕ will be inversely proportional to the rate at which $\theta(x)$ is changing with respect to x when $\theta(x) = \phi$,

$$I(\phi) = A \left| \frac{d\theta}{dx} \right|^{-1}, \quad \theta(x) = \phi, \tag{14.52}$$

for some constant A. But $\frac{d\theta}{dx} = \frac{d\theta}{di} \frac{di}{dx}$ and $\frac{di}{dx} = 1/(R \cos i)$. Hence at the rainbow angle, where $\theta'(i) = 0$, the intensity, $I(\theta_R)$, is infinite. Thus light scattered in this direction is more intense than in other directions: this produces the rainbow. The important point to remember is that the rainbow is at the angle where $\theta(i)$ is stationary.

A schematic diagram of an Earth-bound observed O facing away from the Sun is shown in figure 14.19; light scattered by the rain drops will be more intense when coming from the rainbow angle, $\theta_R \simeq 42°$, as shown. Since the rays are symmetric with respect to rotations about the axis OS through the observer and the Sun (assumed infinitely far away) the rainbow comprises the light from all the water droplets lying on the surface of a cone, the apex of which is at the observer, the axis, along OS and the semi-vertical angle is $42°$. This explain why the rainbow has the form of an arc and why the angle between it and the Sun is always the same; this is essentially Descartes' 1637 explanation. The centre of the rainbow lies on the projection of SO so is below the horizion. If the Sun is higher than $42°$ no primary rainbow can be seen: near sunset the primary rainbow may comprise almost a complete semicircle.

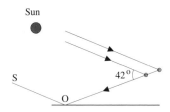

Figure 14.19 Diagram showing the rays
from the Sun being refracted to the
observer at O.

The spread of colours is due to the variation of the refractive index n with wave length. The calculations performed in exercise 14.29 show that red light is scattered through a larger angle than blue so the top of the rainbow at $42°$ is red and blue is at the bottom; the angular spread of the rainbow is about $1.72°$.

If the Sun is bright and there is sufficient rain one can observe two rainbows, the second outside the first at an angle of about $51°$ and with colour ordering reversed. This rainbow is produced by rays undergoing two internal reflections, so the angle of deflection is $\theta = 6r - 2i$ and the value of i at which θ is stationary is given by $\cos i = \sqrt{(n^2 - 1)/6}$. In principle a ray can undergo any number of internal reflections; for k reflections $\theta = k\pi + 2i - 2r(k + 1)$.

Exercise 14.30
Show that in the case of k internal reflections the value of i at which $\theta(i)$ is stationary is given by

$$\cos i = \sqrt{\frac{n^2 - 1}{k(k + 2)}}.$$

Find the rainbow angles for the first ten rainbows for red and for blue light.

Thus in principle one should see many more than two rainbows. More are not seen because the light intensity decreases at each internal reflection (and each refraction) with the result that in reality only the first two are sufficiently intense to observe in nature, though more can be observed in the laboratory; Walker (1976) discusses these and many other aspects of rainbows.

The ray explanation is sufficient to describe most commonly observed properties. It is derived assuming the Sun to be a point source, and the fact that it is not – the Sun's disk has an angular spread of $1/108$ radians or about $\frac{1}{2}°$ – means that the rainbow colours are mixed.

One distinct feature of this theory is that the results are independent of the rain drop size; this must be approximately true, otherwise rainbows would either not be seen or would vary much more, but there are size effects which cannot be explained with ray optics. For example, drops of diameter 1–2 mm emphasise violet and green; for diameters less than 0.2–0.3 mm there is no red; other differences are described by Minnaert (1954).

Another distinct feature of the ray description is the singularity in $I(\phi)$ at $\phi = \theta_R$. This is a consequence of ignoring the phase associated with each ray; the correct treatment is due to Airy, and this theory leads directly to the necessity of dealing with coalescing stationary points.

The complete theory of rainbows is based on the wave description of light and is complicated. However, the wavelength of light is normally far shorter than the radius of a water droplet and this allows use of the ray picture provided the phase along each ray is taken into account. But in this approximation the light rays are the lines along which the phase is stationary, section 14.2, and since the rainbow is formed when the direction of rays is stationary, equation 14.52, it may be shown that at a rainbow two stationary points of the phase coalesce. The correct description of a rainbow takes this into account, and for $\phi \simeq \theta_R$ gives the intensity as

$$I(\phi) = B \, |\mathrm{Ai}(-z)|^2, \tag{14.53}$$

where, for our purposes, B may be considered as a constant, $\mathrm{Ai}(x)$ is Airy's function and, if R is the radius of the water drop,

$$z = \left(\frac{4\pi}{3}\right)^{2/3} \left(\frac{3\lambda R^2}{\sin i}\right)^{1/3} \frac{\phi - \theta_R}{\lambda} \cos i. \tag{14.54}$$

An important difference between the predictions of this theory and the ray theory is that the maximum brightness of a particular colour is not at $\phi = \theta_R$ but at a slightly smaller angle, where $\mathrm{Ai}(-z)^2$ has its first maximum, that is, $z \simeq 1.02$.

14.8.2 Uniform approximations
The basic idea described in this section is a simple generalisation of the previous method. The details, however, can sometimes be quite complicated, so we proceed by giving a quick overview of the method, then apply it to a particular problem before describing the general theory.

  Consider the integral

$$F(v, x) = \int_{-\infty}^{\infty} dt\, f(t, x) e^{ivh(t,x)}, \quad |v| \gg 1, \tag{14.55}$$

where the functions $f(t, x)$ and $h(t, x)$ are analytic functions of both x and t. We assume also that there are only two stationary phase points $t_1(x)$ and $t_2(x)$, and that these depend upon the parameter x. The situation of interest is when there is a critical value of $x = x_c$, such that $t_1(x_c) = t_2(x_c)$, and the stationary points are real for $x > x_c$ and complex, with $t_1 = t_2^*$, for $x < x_c$, or vice versa. For $x \simeq x_c$ the stationary phase method fails, so our objective is to find a uniform approximation valid for $x \simeq x_c$. It transpires that the approximation is also valid when the stationary points are real and far apart and when they are complex.

  As x varies and passes through x_c the graph of $h(t, x)$ changes its shape, as shown in the following three figures.

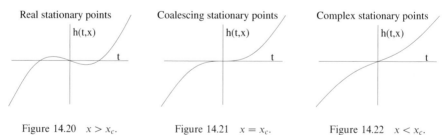

Real stationary points	Coalescing stationary points	Complex stationary points
Figure 14.20 $x > x_c$.	Figure 14.21 $x = x_c$.	Figure 14.22 $x < x_c$.

  The cubic $w^3/3 - \zeta w$ has a similar behaviour, with $\zeta > 0$ for the left hand figure, $\zeta = 0$ for the centre figure and $\zeta < 0$ for the right hand figure, which suggests defining a new variable $w(t)$ by the equation

$$\frac{1}{3}w^3 - \zeta(x)w + \xi(x) = h(t, x), \tag{14.56}$$

where ζ and ξ are functions of x chosen to ensure that $w(t)$ is analytic over a reasonable range of t, that includes the stationary points, and has an inverse; we shall see how this is achieved later. With this transformation the contribution to the original integral 14.55 from the region of stationary points is

$$F(v, x) \sim e^{iv\xi(x)} \int_{-\infty}^{\infty} dw \, g(t(w)) \frac{dt}{dw} \exp iv \left(\frac{1}{3} w^3 - \zeta(x) w \right),$$

and by expanding $g(t) dt/dw$ as a suitable power series in w we may evaluate the integral in terms of the Airy function and its first derivative. This technique provides an approximation to the original function where conventional stationary phase methods fail.

In order to carry through the analysis we require various integrals, all variants of the integral definition of the bounded Airy function, $\text{Ai}(x)$, defined in section 11.4, which may be expressed in the form

$$a^{-1/3} \text{Ai} \left(-x a^{-1/3} \right) = \frac{1}{2\pi} \int_{-\infty}^{\infty} dt \, \exp i \left(\frac{1}{3} a t^3 - xt \right). \tag{14.57}$$

By differentiating under the integral sign and using the differential equation, if necessary, the following useful relations are obtained:

$$\frac{1}{2\pi} \int_{-\infty}^{\infty} dt \, t \exp i \left(\frac{1}{3} a t^3 - xt \right) = -i a^{-2/3} \text{Ai}' \left(-x a^{-1/3} \right). \tag{14.58}$$

$$\frac{1}{2\pi} \int_{-\infty}^{\infty} dt \, t^2 \exp i \left(\frac{1}{3} a t^3 - xt \right) = a^{-4/3} x \text{Ai} \left(-x a^{-1/3} \right). \tag{14.59}$$

The Bessel function
The general method is most easily understood by applying it to the Bessel function because this has symmetries that simplify the analysis. It is also simpler to deal with the function $J_n(nx)$ having the integral representation

$$J_n(nx) = \frac{1}{2\pi} \int_{-\pi}^{\pi} dt \, e^{in(t - x \sin t)}, \tag{14.60}$$

so $h(t, x) = t - x \sin t$. In this case the critical value of x is at $x = 1$, and for $x > 1$ there are two real stationary phase points; these coalesce at $x = 1$ and are complex for $x < 1$. Because the phase is an odd function of t the new variable $w(t)$, equation 14.56, is defined by the equation

$$\frac{1}{3} w^3 - \zeta(x) w = t - x \sin t, \quad -\pi \leq t \leq \pi, \tag{14.61}$$

where $\zeta(x)$ is a real function of x; the other function, $\xi(x)$, equation 14.56, is identically zero because h is odd. This equation is used to determine both $\zeta(x)$ and $w(t)$. First concentrate on $\zeta(x)$, which is obtained by ensuring that $w(t)$ is well behaved. Differentiation gives

$$\frac{dt}{dw} = \frac{w^2 - \zeta(x)}{1 - x \cos t}, \tag{14.62}$$

and we ensure that dt/dw remains finite by chosing $\zeta(x)$ to make both numerator and denominator simultaneously zero. If $x > 1$ there are two real stationary points of $t - x \sin t$ at $\pm t_s(x)$, where $x \cos t_s = 1$, $0 \le t_s < \pi/2$, and $\zeta(x) > 0$. Choose ζ so that $w(t_s)^2 - \zeta(x) = 0$ when $1 = x \cos t_s$, that is, $\zeta(x) = w(t_s(x))^2$. Substituting this into equation 14.61 gives

$$\frac{2}{3}\zeta(x)^{3/2} = x \sin t_s - t_s = \sqrt{x^2 - 1} - \cos^{-1}\left(\frac{1}{x}\right), \qquad x \ge 1. \tag{14.63}$$

If $x < 1$ the stationary points of $t - x \sin t$ are complex; the relevant ones are

$$t_s = \pm i v(x), \quad v = \cosh^{-1}\left(\frac{1}{x}\right) = \ln\left(\frac{1 + \sqrt{1 - x^2}}{x}\right), \qquad x \ge 1.$$

When $x < 1$, $\zeta(x) < 0$, hence we have

$$\zeta(x) = -(3/2)^{2/3}\left(v(x) - x \sinh v(x)\right)^{2/3}, \qquad x \le 1. \tag{14.64}$$

Although $\zeta(x)$ is represented by different expressions for $x \le 1$ and $x \ge 1$ it is an analytic function of x. Its series representation is easily obtained from either 14.63 or 14.64, and is

$$\zeta(x) = 2^{1/3}(x - 1) - \frac{3 \cdot 2^{1/3}}{10}(x - 1)^2 + \frac{32 \cdot 2^{1/3}}{175}(x - 1)^3 + \cdots, \tag{14.65}$$

and for $x \gg 1$,

$$\frac{2}{3}\zeta^{3/2} = x - \frac{\pi}{2} + \frac{1}{2x} + O(x^{-3}).$$

For $x > 1$, $\zeta(x)$ is almost linear, as seen in the following graph.

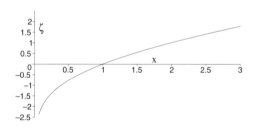

Figure 14.23   The function $\zeta(x)$, defined by
equations 14.63 and 14.64.

Now consider the function $t(w)$. If $x = 1$, $\zeta = 0$ and equation 14.61 becomes $\frac{1}{3}w^3 = t - \sin t = \frac{1}{6}t^3 + \cdots$, so $t = 2^{1/3}w + w^3/30 + O(w^5)$. For $x \simeq 1$ we may use the approximation 14.65 and invert 14.61 in terms of the series

$$t(w) = 2^{1/3}\left(1 - \frac{3(x - 1)}{10}\right)w + \left(\frac{1}{30} - \frac{19(x - 1)}{700}\right)w^3 + \cdots.$$

Higher-order terms of this series are readily found and suggest that $t(w)$ is close to linear. A clearer picture of $t(w)$ is obtained using `implicitplot` to draw graphs of the real solutions of equation 14.61 for fixed x. Three cases, $x = 0.9$, 1.1 and 2.0, are shown next.

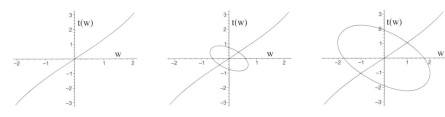

Figure 14.24 $x = 0.9$.   Figure 14.25 $x = 1.1$.   Figure 14.26 $x = 2.0$.

For $x = 0.9$, $\zeta = -0.13$ and the equation $\frac{1}{3}w^3 - \zeta w = t - x\sin t$ has one real root for each t. If $x > 1$ there are three real roots provided t is not too large: the approximate straight line is the solution we require and the the ovals in the centre and right figures are the other solutions.

Returning to the main analysis, we see that with this definition of $w(t)$ the asymptotic expansion of the integral 14.60 becomes†

$$J_n(nx) \sim \frac{1}{2\pi} \int_{-\infty}^{\infty} dw\, \frac{dt}{dw} \exp in\left(\frac{1}{3}w^3 - \zeta(x)w\right). \tag{14.66}$$

The complete asymptotic expansion is obtained from this integral by noting that $w(t)$ is an odd function of t, so dt/dw is an even function which is most conveniently expressed as the power series

$$\frac{dt}{dw} = \sum_{k=0}^{\infty} c_k(x)(w^2 - \zeta(x))^k, \tag{14.67}$$

which is an expansion about the stationary points, not the origin. The integrals

$$I_k = \int_{-\infty}^{\infty} dw \left(\frac{dg}{dw}\right)^k e^{ivg(w)}, \quad g(w) = \frac{1}{3}w^3 - \zeta w,$$

may be easily be expressed in terms of the Airy function and its derivative; in particular,

$$I_0 = 2\pi v^{-1/3} \text{Ai}\left(-v^{2/3}\zeta\right), \quad I_1 = 0, \quad I_2 = 4\pi v^{-5/3} \text{Ai}'\left(-v^{2/3}\zeta\right).$$

The coefficients $c_k(x)$, which are analytic functions of x, may be expressed in terms of the differentials of t with respect to w, evaluated at $w = \sqrt{\zeta}$, thus

$$c_0 = \lim_{w \to \sqrt{\zeta}} \frac{dt}{dw}, \qquad c_1 = \frac{1}{2\sqrt{\zeta}} \lim_{w \to \sqrt{\zeta}} \frac{d^2 t}{dw^2},$$

$$c_2 = \frac{1}{8\zeta}\left(\lim_{w \to \sqrt{\zeta}} \frac{d^3 t}{dw^3} - 2c_1\right), \quad c_3 = \frac{1}{48\zeta^{3/2}}\left(\lim_{w \to \sqrt{\zeta}} \frac{d^4 t}{dw^4} - 24c_2\sqrt{\zeta}\right).$$

These derivatives are most easily obtained by successive differentiation of equation 14.61, then setting $x\cos t = 1$ and $w = \sqrt{\zeta}$ to evaluate the required limits.

---

† At this point we should include a neutraliser, particularly if there are more stationary phase points, but the exposition is simpler without.

For example, the second differential gives

$$2w = (1 - x\cos t)\frac{d^2 t}{dw^2} + x\sin t\left(\frac{dt}{dw}\right)^2,$$

and hence

$$c_0(x) = \left(\frac{4\zeta(x)}{x^2 - 1}\right)^{1/4} = 2^{1/3} - \frac{2^{1/3}}{5}(x - 1) + \frac{3.2^{1/3}}{35}(x - 1)^2 + \cdots.$$

The nth differential is given by differentiating $n + 1$ times. The algebra is messy, but is ideally suited to Maple:

$$c_1(x) = \frac{\sqrt{2}(x^2 - 1)^{3/4} - \sqrt{2}\zeta^{3/4}}{6} \frac{1}{(x^2 - 1)\zeta^{3/4}} = \frac{1}{10} - \frac{37}{700}(x - 1) + \frac{361}{12\,600}(x - 1)^2 + \cdots,$$

$$c_2(x) = \frac{\sqrt{2}(4 + 6x^2)\zeta^{3/2} - 5(x^2 - 1)^{3/2}}{96} \frac{}{[\zeta(x^2 - 1)]^{7/4}} = \frac{2^{2/3}}{140} - \frac{2^{2/3}37}{6300}(x - 1) + \cdots.$$

The first term in the expansion 14.67 gives the approximation

$$J_n(nx) \simeq \frac{c_0 I_0}{2\pi} = \left(\frac{4\zeta(x)}{x^2 - 1}\right)^{1/4} n^{-1/3} \mathrm{Ai}\left(-n^{2/3}\zeta(x)\right). \tag{14.68}$$

The following figure shows a graph of the difference between the exact values of $J_n(nx)$ and this approximation for $n = 2$ and $0 < x < 10$.

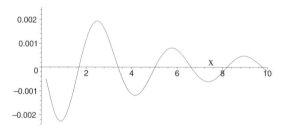

Figure 14.27   Absolute error of the approximation 14.68 for $n = 2$.

These comparisons show the uniform approximation to be good even for $n = 2$, that it remains accurate when $x \gg 1$, when the stationary phase points are well separated, and for $x < 1$ when they are complex, though in this case the relative error increases as $x \to 0$.

Exercise 14.31
Write a Maple procedure that evaluates the approximation 14.68 and use this to examine its error for various n; for $x > 1$ consider the absolute error and for $x < 1$ the relative error.

Exercise 14.32
Show that for $x > 1$ the stationary phase approximation to the integral 14.66 with dt/dw represented by the series 14.67 is the same as the stationary phase approximation to I_0 and to the original integral 14.60, given in equation 14.10 (page 535).

Exercise 14.33
Show that including the first three terms of the series 14.67 leads to the approximation

$$J_n(nx) \simeq c_0(x) n^{-1/3} \mathrm{Ai}\left(-n^{2/3}\zeta(x)\right) + 2c_2(x) n^{-5/3} \mathrm{Ai}'\left(-n^{2/3}\zeta(x)\right).$$

Show, graphically, that for $n \geq 2$ the absolute error in this approximation is significantly smaller than the zero-order approximation, equation 14.68.

General theory
The general theory for the approximation of

$$F(v, x) = \int_{-\infty}^{\infty} dt\, f(t, x) e^{ivh(t,x)}, \qquad |v| \gg 1 \tag{14.69}$$

is similar to the above special case, though normally the factor $f(t, x) dt/dw$ needs to be approximated in a different manner. Suppose that $t_1(x)$ and $t_2(x)$ are the two stationary points of $h(t, x)$ that are real and satisfy $t_1(x) < t_2(x)$ for $x > x_c$, coalesce at $x = x_c$ and are complex for $x < x_c$, with $t_1 = t_2^*$ and $\Im(t_2) > 0$. We shall also assume that for $x > x_c$, $h(t)$ has the shape shown in the diagram, so $h(t_1(x), x) > h(t_2(x), x)$.

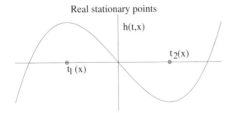

Real stationary points

$h(t,x)$

$t_2(x)$

$t_1(x)$

The new variable $w(t)$ is defined implicitly by the equation

$$\frac{1}{3}w^3 - \zeta(x)w + \xi(x) = h(t, x), \tag{14.70}$$

with $\zeta(x) > 0$ for $x > x_c$ and $\zeta(x) < 0$ for $x < x_c$. As before, differentiation with respect to t gives

$$\frac{dt}{dw} = \frac{w^2 - \zeta(x)}{h'(t)}, \qquad h'(t) = \frac{dh}{dt}, \tag{14.71}$$

and we ensure that $t'(w)$ exists in the neighbourhood of the stationary points by making numerator and denominator simultaneously zero. Thus at $t = t_2$, $\zeta(x) = w(t_2(x), x)^2$

and at $t = t_1$ we have $w(t_1) = -w(t_2)$. At the stationary points equation 14.70 becomes

$$-\frac{2}{3}w(t_2)^3 + \xi(x) = h(t_2(x), x),$$

$$+\frac{2}{3}w(t_1)^3 + \xi(x) = h(t_1(x), x), \quad \text{with} \quad w(t_2) = -w(t_1) = \zeta(x)^{1/2}.$$

Hence

$$\xi(x) = \frac{1}{2}(h(t_1(x), x) + h(t_2(x), x)),$$

$$\frac{2}{3}\zeta(x)^{3/2} = \frac{1}{2}(h(t_1(x), x) - h(t_2(x), x)).$$

When $x > x_c$, and the stationary points are real, since $h(t_1) > h(t_2)$ both $\xi(x)$ and $\zeta(x)$ are real. When $x < x_c$ we have $t_1 = t_2^*$ and $h(t_1) = h(t_2^*) = h(t_2)^*$ and hence

$$\xi(x) = \Re(h(t_2)), \quad \zeta(x) = -(3/2)^{2/3}|\Im(h(t_2))|^{2/3}, \quad x \le x_c.$$

With these definitions the uniform approximation to the integral becomes

$$F(v, x) \sim e^{iv\xi(x)} \int_{-\infty}^{\infty} dw\, f(t(w)) \frac{dt}{dw} \exp iv\left(\frac{1}{3}w^3 - \zeta(x)w\right). \tag{14.72}$$

Up to this point the analysis is similar to that used for the Bessel function, but now it differs because the expansion 14.67 is useful only if $f\, dt/dw$ is even in t. In general it is easier to approximate $f\, dt/dw$ using the linear function

$$f(t(w), x)\frac{dt}{dw} \simeq d_0(x) + d_1(x)w, \tag{14.73}$$

with the coefficients $d_0(x)$ and $d_1(x)$ obtained by fitting this at the stationary points, thus

$$d_0(x) = \begin{cases} \zeta^{1/4}\left(\dfrac{f(t_2)}{\sqrt{2h''(t_2)}} + \dfrac{f(t_1)}{\sqrt{-2h''(t_1)}}\right), & x \ge x_c, \\[3ex] 2(-\zeta)^{1/4}\,\Re\left(\dfrac{f(t_2)}{\sqrt{2h''(t_2)}}\right), & x < x_c, \end{cases} \tag{14.74}$$

$$d_1(x) = \begin{cases} \zeta^{-1/4}\left(\dfrac{f(t_2)}{\sqrt{2h''(t_2)}} - \dfrac{f(t_1)}{\sqrt{-2h''(t_1)}}\right), & x \ge x_c, \\[3ex] 2(-\zeta)^{-1/4}\,\Im\left(\dfrac{f(t_2)}{\sqrt{2h''(t_2)}}\right), & x < x_c. \end{cases} \tag{14.75}$$

With this approximation to $f\, dt/dw$ the integral 14.72 may be evaluated in terms of the Airy function and its derivative:

$$F(v) \sim 2\pi e^{iv\xi(x)}\left\{\frac{d_0(x)}{v^{1/3}}\mathrm{Ai}\left(-v^{2/3}\zeta(x)\right) - \frac{id_1(x)}{v^{2/3}}\mathrm{Ai}'\left(-v^{2/3}\zeta(x)\right)\right\}. \tag{14.76}$$

Exercise 14.34

Show that at $x = x_c$, that is, when $t_1 = t_2 = t_c$,

$$d_0(x_c) = f(t_c, x_c)\left(\frac{2}{h^{(3)}}\right)^{1/3}, \quad \text{where} \quad h^{(n)} = \left.\frac{d^n h}{dt^n}\right|_{t=t(x_c)},$$

$$d_1(x_c) = \frac{df}{dt}\left(\frac{2}{h^{(3)}}\right)^{2/3} - \frac{f(t_c, x_c)}{3.\,2^{1/3}}\frac{h^{(4)}}{\left(h^{(3)}\right)^{5/3}}.$$

Hint: use the original definition, 14.70, with $x = x_c$ and $\zeta(x_c) = 0$.

Exercise 14.35

Using the asymptotic expansions, valid as $u \to \infty$,

$$\text{Ai}(-u) \sim \frac{1}{u^{1/4}\pi^{1/2}}\sin\left(\frac{2}{3}u^{3/2} + \frac{\pi}{4}\right), \quad \text{Ai}'(-u) \sim -\frac{u^{1/4}}{\pi^{1/2}}\cos\left(\frac{2}{3}u^{3/2} + \frac{\pi}{4}\right),$$

obtained using a stationary phase approximation, show that the approximation 14.76 reduces to the stationary phase approximation of the original integral 14.69 if $t_2 - t_1$ is sufficiently large. Note that *both* terms are needed to obtain the lowest-order stationary phase approximation.

Exercise 14.36

Show that if $v \gg 1$,

$$\int_{-\infty}^{\infty} dt \exp iv\,(t - x\sinh t) \sim \frac{2\pi}{v^{1/3}}\left(\frac{4\zeta(x)}{1 - x^2}\right)^{1/4}\text{Ai}\left(-v^{2/3}\zeta(x)\right),$$

where

$$\zeta(x) = \begin{cases} \left(\frac{3}{2}\right)^{2/3}\left[\ln\left(\frac{1 + \sqrt{1-x^2}}{x}\right) - \sqrt{1-x^2}\right]^{2/3}, & x \le 1, \\ -\left(\frac{3}{2}\right)^{2/3}\left[\sqrt{x^2-1} - \cos^{-1}(1/x)\right]^{2/3}, & x \ge 1. \end{cases}$$

14.9 Exercises

Exercise 14.37

Use equation 14.32 to show that

$$\int_0^{\infty} dt\,J_0(t)e^{ivt} = \frac{i}{\sqrt{v^2-1}}, \quad v > 1.$$

Hint: use the Taylor's series expansion of $J_0(t)$, equation 11.62 (page 405), and the relations

$$(2k)! = \Gamma(2z), \ z = k + \frac{1}{2}, \quad \text{and} \quad \Gamma(2z) = \frac{4^z}{\sqrt{4\pi}}\Gamma(z)\Gamma\left(z + \frac{1}{2}\right).$$

Note that for $v < 1$ the value of the integral can be shown to be

$$\int_0^{\infty} dt\,J_0(t)e^{ivt} = \frac{1}{\sqrt{1-v^2}}, \quad v < 1.$$

There are similar, but more complicated results, for the Bessel function $J_n(t)$.

Exercise 14.38
Use the integral definition of the Bessel function to show that

$$J_n(n) \sim \frac{1}{3\Gamma(2/3)}\left(\frac{6}{n}\right)^{1/3} - \frac{\Gamma(2/3)\sqrt{3}}{2520\pi}\left(\frac{6}{n}\right)^{5/3} - \frac{1}{675}\frac{6^{1/3}}{\Gamma(2/3)n^{7/3}}$$
$$+ \frac{1213}{6\,142\,500}\frac{\Gamma(2/3)6^{2/3}\sqrt{3}}{\pi\,n^{11/3}} + \frac{151\,439}{654\,885\,000}\frac{6^{1/3}}{\Gamma(2/3)n^{13/3}} + \cdots$$

and find the next three terms.

This is Meissel's third expansion of $J_n(n)$ derived by Cauchy in 1854 and in a more complete form by Meissel in 1891, see Watson (1966, page 232).

Exercise 14.39
Show that

$$\mathscr{J}_v(v) \sim \frac{\Gamma\left(\frac{1}{3}\right)}{2\pi\sqrt{3}}\left(\frac{6}{v}\right)^{1/3} - \frac{\sqrt{3}\Gamma(2/3)}{2520\pi}\left(\frac{6}{v}\right)^{5/3} + \frac{\sin v\pi}{2\pi v} + \cdots.$$

Exercise 14.40
Use equation 14.32 to show that if $v \gg 1$,

$$\frac{1}{\pi}\int_0^1 dt\,\frac{\sin\pi t}{1+t^2}\cos vt \sim -\frac{\pi}{v^2}\left(1+\frac{1}{2}\cos v\right) - \frac{\pi\sin v}{v^3} - \frac{\pi}{v^4}\left(\pi^2+6+\frac{1}{2}\left(\pi^2-3\right)\cos v\right).$$

Exercise 14.41
Show that if $v \gg 1$,

(i) $$\int_0^1 dt\,\sin\pi t\cos(vt^2) \sim \frac{\pi(\pi^2-3\cos v)}{12v^2} - \frac{3\pi\sin v}{8v^3}$$
$$-\frac{\pi(\pi^6-(1575-105\pi^2)\cos v)}{1680v^4}+\cdots,$$

(ii) $$\int_{-1}^1 dt\,\frac{e^{iv\cosh t}}{\sqrt{1-t^2}} \sim \sqrt{\frac{2\pi}{v}}\left\{\exp(i(v+\pi/4)) + \frac{1}{\sqrt{\sinh(1)}}\exp(i(v\cosh(1)-\pi/4))\right\},$$

(iii) $$\int_0^\pi dt\,t^{1/2}e^{iv(t+\sin t)}\cos\sqrt{1+t} \sim -\frac{1.288}{v^{1/3}}\cos\left(v\pi-\frac{\pi}{6}\right) + \frac{0.7686}{v^{2/3}}\cos\left(v\pi-\frac{\pi}{3}\right).$$

Exercise 14.42
Show that

$$\int_0^1 dt\,\frac{e^{ivt^3}}{\sqrt{1-t^2}} \sim \frac{1}{3\sqrt{\pi}}\sum_{k=0}^\infty \frac{\Gamma(k+1/2)\Gamma(2k/3+1/3)}{k!}\left(\frac{i}{v}\right)^{(2k+1)/3}$$
$$+ \frac{e^{iv}}{3}\sum_{k=0}^\infty a_k\Gamma(k+1/2)\left(-\frac{i}{v}\right)^{k+1/2},$$

where a_k is the coefficient of u^k in the Taylor's series of

$$g(u) = u^{1/2}(1-u)^{-2/3}\left(1-(1-u)^{2/3}\right)^{-1/2}$$

about $u = 0$, $a_0 = \sqrt{6}/2$, $a_1 = -3\sqrt{6}/8$ and $a_2 = -71\sqrt{6}/1728$.

Exercise 14.43

Show that as $v \to \infty$,

$$\int_0^1 dt \, \frac{\sin\left(\pi t/2\right) \cos(v\sqrt{1-t^2})}{t^{3/4}(1-t)^{1/4}} \quad \sim \quad \frac{\pi}{2^{11/8}} \frac{\Gamma(5/8)}{v^{5/8}} \cos\left(v - \frac{5\pi}{16}\right) + \frac{\pi}{2^{23/8}} \frac{\Gamma(9/8)}{v^{9/8}} \cos\left(v - \frac{9\pi}{16}\right)$$
$$-\frac{\pi}{2^{11/8}} \left(\frac{1}{2} + \frac{\pi^2}{12}\right) \frac{\Gamma(13/8)}{v^{13/8}} \cos\left(v - \frac{13\pi}{16}\right) - \frac{\sqrt{\pi}}{2^{5/4} v^{3/2}},$$

and that the next term is $O(v^{-17/8})$.

Exercise 14.44

Show that as $v \to \infty$

$$\int_0^1 dt \, \sin \pi t \cos v t^4 \quad \sim \quad \frac{\pi}{4} \sum_{k=0}^\infty \frac{(-\pi^2)^k}{v^{k/2+1/2}} \frac{\Gamma\left(k/2 + 1/2\right)}{(2k+1)!} \cos\left(\frac{\pi}{4}(k+1)\right)$$
$$+ \sum_{k=0}^\infty \frac{a_k}{v^{k+1}} \sin\left(v - \frac{\pi}{2}k\right),$$

where $a_0 = \pi/16$, $a_1 = 9\pi/64$, $a_2 = \pi(111 - \pi^2)/256$ and $a_3 = 15\pi(119 - 2\pi^2)/1024$.

Exercise 14.45

Use the relation

$$\ln x = \lim_{\epsilon \to 0} \frac{x^\epsilon - 1}{\epsilon}$$

together with the integral definition of the gamma function (equation 20.4, page 834), to show that

$$\int_0^\infty dt \, e^{-vt} \ln t \;\;=\;\; -\frac{\gamma + \ln v}{v}, \quad v > 0,$$
$$\int_0^\infty dt \, e^{i\omega t} \ln t \;\;=\;\; -\frac{\pi}{2\omega} - i\left(\frac{\gamma + \ln \omega}{\omega}\right), \quad \omega > 0,$$

where γ is Euler's constant. Show that these results are consistent by putting $\omega = iv$.

By differentiating under the integral sign, or otherwise, show that

$$\int_0^\infty dt \, t^n e^{-vt} \ln t \;\;=\;\; \frac{n!}{v^{n+1}} \left(\psi(n+1) - \ln v\right), \quad v > 0,$$
$$\int_0^\infty dt \, t^n e^{i\omega t} \ln t \;\;=\;\; n! \left(\frac{i}{\omega}\right)^{n+1} \left(\psi(n+1) - \ln \omega + i\frac{\pi}{2}\right), \quad \omega > 0.$$

Hint: recall that the psi (or digamma) function is defined by $\psi(z) = \Gamma'(z)/\Gamma(z)$, that Euler's constant is $\gamma = \psi(1)$ and that

$$\psi(n) = -\gamma + \sum_{k=1}^{n-1} k^{-1}, \quad n \geq 2.$$

Exercise 14.46

Use the heuristic arguments introduced in deriving 14.40 to show that if $g(t)$ is an infinitely differentiable function and $v \gg 1$,

$$\int_a^b dt\, g(t) \ln(t-a) e^{ivt} \quad\sim\quad -i\frac{e^{ivb}}{v} \sum_{r=0}^{\infty} \left(\frac{i}{v}\right)^r \frac{d^r}{db^r} \left[g(b)\ln(b-a)\right]$$

$$+ i\frac{e^{iva}}{v} \sum_{r=0}^{\infty} \left(\frac{i}{v}\right)^r \left(\psi(r+1) - \ln v + i\frac{\pi}{2}\right) g^{(r)}(a).$$

Exercise 14.47

Show that if $x \geq 1$ and $v \gg 1$,

$$F_v(x) = \int_{-\infty}^{\infty} dt\, e^{iv(t - x\tanh t)} \quad\sim\quad \frac{2\pi}{v^{1/3}} \left(\frac{x\zeta(x)}{x-1}\right)^{1/4} \mathrm{Ai}\left(-v^{2/3}\zeta(x)\right),$$

where

$$\frac{2}{3}\zeta(x)^{3/2} = \sqrt{x(x-1)} - \ln(\sqrt{x} + \sqrt{x-1}).$$

Show also that the integral may be written in the form

$$F_v(x) \;=\; 2\int_0^T dt\, \cos v\,(t - x\tanh t) + R,$$

$$R \;=\; 2\int_T^{\infty} dt\, \cos v\,(t - x\tanh t) \sim -\frac{2\sin v\,T_1}{v\,T_2} + \frac{4x\sinh T \cos v\,T_1}{v^2(T_2\cosh T)^3},$$

where $T_1 = T - x\tanh T$ and $T_2 = 1 - x/\cosh(T)^2$, provided T is suitably chosen. Use this expression to evaluate the integral numerically and make a comparison of the two approximations for various values of v and x.

15

Uniform approximations for differential equations

15.1 Introduction

The idea of a uniform approximation, introduced in the previous chapter, can also be applied to some differential equations. Here we apply this technique to the approximation of the eigenvalues and eigenfunctions of the Sturm–Liouville problem

$$\frac{d^2y}{dx^2} + v^2(\lambda - f(x))y = 0, \tag{15.1}$$

where v is a large, positive parameter and λ the eigenvalue. It is shown in exercise 15.1 how most Sturm–Liouville problems can, in principle, be cast into this form. Such equations occur frequently in the description of all types of wave phenomenon, including quantum mechanical systems, and the theory of this chapter may be applied when the wave length, usually proportional to v^{-1}, is relatively small.

The study of these equations dates from the early part of the nineteenth century and seems to have been started by Liouville, who considered the equation J. Liouville
(1809–1882)

$$\frac{d^2y}{dx^2} + v^2g(x)y = 0, \tag{15.2}$$

that is now named a *Liouville equation*. Green (1838) also considered such equations in G. Green
(1793–1841) his studies of waves in canals of small but variable depth and width. In the 1920s the discovery of Schrödinger's equation led to further, independent, investigations of this type of equation by Wentzel (1926), Kramers (1926) and Brillouin (1926). These studies were important because they formed a connecting link between the old quantum theory of Bohr and the solutions of Schrödinger's equation and explained why the heuristic ideas of the old quantum theory worked; an excellent history of these developments is given by Jammer (1989). Moreover, this theory provides one of the essential links between quantum theory and its asymptotic limit, Newtonian dynamics. Since then the type of approximation developed in section 15.2 is frequently named a WKB approximation. The generalisation of the ideas presented here to many dimensional problems is difficult, important and far from properly understood.

The same type of equations arise in the description of boundary layers in fluid motion. A boundary layer is a narrow region where the solution of a differential equation changes rapidly, and this is normally caused by the coefficient of the highest-order derivative being relatively small. Setting $v^{-1} = \epsilon$, equation 15.2 becomes

$$\epsilon^2 \frac{d^2y}{dx^2} + g(x)y = 0.$$

In this form it is clear that a straightforward power series expansion in the small parameter ϵ cannot work, for when $\epsilon = 0$ the equation becomes $g(x)y = 0$ and is fundamentally different. Boundary layer theory is based on the assumption that there are different regions where the solution behaves in different ways: in a narrow region, near a boundary, the solution changes rapidly and elsewhere the variation is slower. The general idea is to find appropriate approximations in each region and match these in the interval of common applicability. An introduction to this type of method is given in sections 15.3.3 and 15.5. Boundary layer theory is not restricted to linear systems so is more general than the methods developed here; for more information on this type of problem the reader should consult Bender and Orszag (1978) or Holmes (1995). Another manifestation of this type of behaviour is seen in the relaxation oscillations studied in chapter 17.

The purpose of this chapter is to provide an introduction to some of the powerful methods that can be used to approximate the solutions of equations 15.1 and 15.2 for various types of functions $f(x)$ and $g(x)$. The analysis is not rigorous and, apart from a few comments in section 15.7, does not extend beyond first order, but in many cases this is adequate, moreover higher-order approximations are normally quite unwieldy. The reader requiring more mathematical rigor should consult Olver (1974).

Exercise 15.1
By making the transformation

$$y(x) = A(x)v(\xi(x)),$$

where $A(x)$, $\xi(x)$ and $v(\xi)$ are arbitrary functions, show that the Sturm–Liouville equation

$$\frac{d}{dx}\left(p(x)\frac{dy}{dx}\right) + \left(q(x) + \lambda w(x)\right)y = 0$$

becomes

$$pA\left(\frac{d\xi}{dx}\right)^2\frac{d^2v}{d\xi^2} + \frac{1}{A}\frac{d}{dx}\left(pA^2\frac{d\xi}{dx}\right)\frac{dv}{d\xi} + \left\{\frac{d}{dx}\left(p\frac{dA}{dx}\right) + (q + \lambda w)A\right\}v = 0.$$

By setting $p(x)\xi'(x)^2 = w(x)$ and choosing $A(x)$ to make the coefficient of $v'(\xi)$ zero, show that this equation may be cast in the form

$$\frac{d^2v}{d\xi^2} + (F(\xi) + \lambda)v = 0,$$

where $A(\xi) = [p(x(\xi))\,w(x(\xi))]^{-1/4}$ and

$$F(\xi) = \frac{q(\xi)}{w(\xi)} - A(\xi)\frac{d^2}{d\xi^2}\left(\frac{1}{A(\xi)}\right), \quad\text{and}\quad \xi(x) = \int dx\,\sqrt{\frac{w(x)}{p(x)}}.$$

Exercise 15.2
Use the transformation found in the previous exercise to show that Bessel's equation,

$$x^2\frac{d^2y}{dx^2} + x\frac{dy}{dx} + \left(x^2 - v^2\right)y = 0,$$

may be cast in the form

$$\frac{d^2v}{d\xi^2} + \left(e^{2\xi} - v^2\right)v = 0, \quad \text{where} \quad \xi = \ln x \quad \text{and} \quad y = v(\xi).$$

Exercise 15.3

The associated Laguerre polynomials satisfy the equation

$$x\frac{d^2y}{dx^2} + (1 + \alpha - x)\frac{dy}{dx} + ny = 0, \quad x \geq 0.$$

Show that this may be put in the standard self-adjoint form with

$$p(x) = x^{1+\alpha}e^{-x}, \quad w(x) = x^{\alpha}e^{-x} \quad \text{and} \quad q(x) = 0.$$

Show further that this may be transformed into

$$\frac{d^2v}{d\xi^2} + \left(\frac{1}{2}(1 + 2n + \alpha) - \frac{\xi^2}{16} + \frac{1 - 4\alpha^2}{4\xi^2}\right)v = 0,$$

where $\xi = 2\sqrt{x}$ and $y = x^{-1/4-\alpha/2}\exp(x/2)v(\xi)$.

Exercise 15.4

This exercise provides a simple example of a boundary layer, near $x = 0$. Find a solution of the linear equation

$$\epsilon\frac{d^2y}{dx^2} + 2\frac{dy}{dx} + y = 0, \quad y(0) = 0, \quad y(1) = 1,$$

where $0 < \epsilon \ll 1$. Plot the graph of the solution for $\epsilon = 0.02$ and show that for $x > 2\epsilon$ this is approximately the same as the solution of the equation

$$2\frac{dy}{dx} + y = 0, \quad y(1) = 1.$$

15.2 The primitive WKB approximation

In this section we consider the simplest approximation to the Sturm–Liouville problem

$$\frac{d^2y}{dx^2} + v^2g(x)y = 0, \quad v \gg 1, \tag{15.3}$$

where v is a large, positive number and $g(x)$ a sufficiently well behaved function. Normally there are boundary or initial conditions but, for the present, these will be ignored while we concentrate on obtaining approximations to the general solution.

If $g(x)$ is constant the nature of the solution depends upon its sign; for positive values, $g(x) = 1$, the two linearly independent solutions are the oscillatory functions $y(x) = \exp(\pm ivx)$; for negative values, $g(x) = -1$, the solutions are $y(x) = \exp(\pm vx)$ which change exponentially. These limiting cases suggest expressing the solution of 15.3 in the form

$$y(x) = \begin{cases} \exp(iv\phi(x)), & g(x) > 0, \\ \exp(v\phi(x)), & g(x) < 0, \end{cases} \tag{15.4}$$

for some function $\phi(x)$. Substituting into equation 15.3 gives the following nonlinear equations for $\phi(x)$

$$\frac{i}{v}\frac{d^2\phi}{dx^2} + \left(g(x) - \left(\frac{d\phi}{dx}\right)^2\right) = 0, \quad g(x) > 0,$$

$$\frac{1}{v}\frac{d^2\phi}{dx^2} + \left(g(x) + \left(\frac{d\phi}{dx}\right)^2\right) = 0, \quad g(x) < 0. \tag{15.5}$$

It may seem bizarre to replace a linear by a nonlinear equation, but frequently it is easier to find a solution, or an approximate solution, of the latter. Since $v \gg 1$ an approximate solution is obtained by ignoring the first term, so the simplest approximate solutions of 15.3 are

$$y(x) \sim \begin{cases} \exp\left(\pm iv \int dx \sqrt{g(x)}\right), & g(x) > 0, \\[2mm] \exp\left(\pm v \int dx \sqrt{-g(x)}\right), & g(x) < 0. \end{cases} \tag{15.6}$$

Equation 15.5 suggests that a more accurate solution to equation 15.3 may be obtained using the series expansion,

$$\phi(x) = \phi_0(x) + \frac{1}{v}\phi_1(x) + \frac{1}{v^2}\phi_2(x) + \cdots.$$

Substituting this into 15.5 and equating the coefficient of v^{-n} to zero gives an equation for $\phi_n(x)$ in terms of $\phi_k(x)$, $k = 1, 2, \ldots, n-1$. If $g(x) > 0$ we obtain

$$\left(\frac{d\phi_0}{dx}\right)^2 = g(x),$$

$$i\frac{d^2\phi_0}{dx^2} = 2\frac{d\phi_0}{dx}\frac{d\phi_1}{dx},$$

$$i\frac{d^2\phi_1}{dx^2} = 2\frac{d\phi_0}{dx}\frac{d\phi_2}{dx} + \left(\frac{d\phi_1}{dx}\right)^2, \tag{15.7}$$

$$\vdots \qquad\qquad \vdots$$

$$i\frac{d^2\phi_{n-1}}{dx^2} = \sum_{p=0}^{n}\frac{d\phi_p}{dx}\frac{d\phi_{n-p}}{dx}.$$

The first and second of these equations can be solved to give

$$\phi_0(x) = \pm\int dx\sqrt{g(x)}, \quad \phi_1(x) = \frac{i}{4}\ln g(x), \tag{15.8}$$

and hence

$$y \sim \frac{1}{g(x)^{1/4}}\exp\left(\pm iv\int dx\sqrt{g(x)}\right), \quad g(x) > 0.$$

Expressions for some of the remaining $\phi_k(x)$ may also be found, exercise 15.32 (page 624), but these do not concern the present analysis. However, we note that when $g(x) > 0$, $\phi_{2k+1}(x)$ is purely imaginary and $\phi_{2k}(x)$ is real. Moreover, $\phi_{2k}(x)$ may be expressed explicitly in terms of $g(x)$ and its derivatives.

Exercise 15.5
Show that if $g(x) < 0$ then

$$\phi_0 = \pm v \int dx \sqrt{-g(x)}, \quad \phi_1(x) = -\frac{1}{4} \ln(-g(x)),$$

and hence

$$y \sim \frac{1}{(-g(x))^{1/4}} \exp\left(\pm v \int dx \sqrt{-g(x)}\right), \quad g(x) < 0.$$

Using these expressions the approximate general solution of 15.3 is, for $g(x) > 0$,

$$y(x) \sim \frac{1}{g(x)^{1/4}} \left\{ A \exp\left(iv \int dx \sqrt{g(x)}\right) + B \exp\left(-iv \int dx \sqrt{g(x)}\right)\right\}, \quad (15.9)$$

and for $g(x) < 0$,

$$y(x) \sim \frac{1}{(-g(x))^{1/4}} \left\{ C \exp\left(v \int dx \sqrt{-g(x)}\right) + D \exp\left(-v \int dx \sqrt{-g(x)}\right)\right\}, \quad (15.10)$$

for some constants A, B, C and D.

The solutions 15.9 and 15.10 are useful provided $v^2 g(x)$ is not too small. If $g(x)$ has a simple zero, where $g(x) > 0$ the solution has the form 15.9 and elsewhere the form 15.10, and it is necessary to connect these solutions, that is to find relations between the constants (A, B) and (C, D). This is not straightforward because neither solution is valid in the neighbourhood of the zero of $g(x)$, indeed both approximations are singular at the zero.

The points where $g(x) = 0$ are therefore special and are named *transition* or *turning points*. This second name comes from quantum mechanics, because Schrödinger's equation for the wave function $\psi(x)$ of a particle of mass m moving in one dimension under the action of a potential $V(x)$ is

$$\frac{d^2\psi}{dx^2} + \frac{2m}{\hbar^2}\left[E - V(x)\right]\psi = 0,$$

where E is the energy. The coefficient of $\psi(x)$ is zero when $V(x) = E$; since the equivalent Newtonian motion satisfies the energy equation $\frac{1}{2}m\dot{x}^2 = E - V(x)$, at these points the particle stops and reverses its direction of motion. Because the Newtonian motion with energy E is confined to the region $E > V(x)$, the quantal wave function is often exponentially decreasing when $V(x) - E$ is positive and increasing. Turning points are important because on either side the solution behaves in distinctly different ways.

Exercise 15.6
Use the approximation 15.9 to show that if $v \gg 1$ the equation

$$\frac{d^2 y}{dx^2} + (v + x)y = 0, \quad y(0) = a, \quad y'(0) = b, \quad x \geq 0,$$

has the approximate solution

$$y(x) = \frac{1}{(1+x/v)^{1/4}} \left\{ a\cos\left(\frac{2}{3}v^{3/2}F(x)\right) + \frac{a+4bv}{4v^{3/2}}\sin\left(\frac{2}{3}v^{3/2}F(x)\right) \right\},$$

where $F(x) = (1+x/v)^{3/2} - 1$.

Find the exact solution in terms of Airy functions and make a graphical comparison for various values of v, a, b and ranges of x.

Exercise 15.7

The expansion defined by equations 15.7 is not the only type of expansion that can be used and sometimes others are more practical.

The Airy function $\mathrm{Ai}(x)$ is defined to be the solution of the equation

$$\frac{d^2 y}{dx^2} - v^2 xy = 0$$

that tends to zero as $x \to \infty$ and with $v = 1$. By applying the result found in exercise 15.5 to this problem, show that the leading term in the asymptotic expansion is $\exp(-2vx^{3/2}/3)$. Define $f(x)$ by the equation

$$y(x) = f(x)\exp(-2vx^{3/2}/3)$$

and show that it satisfies the differential equation

$$\frac{1}{v}\frac{d^2 f}{dx^2} = 2x^{1/4}\frac{d}{dx}\left(x^{1/4}f(x)\right).$$

By writing

$$f(x) = \frac{1}{x^{1/4}} + \sum_{n=1}^{\infty}\frac{f_n(x)}{v^n}, \quad \text{show that} \quad f_{n+1}(x) = -\frac{1}{2x^{1/4}}\int_x^{\infty}dx\,\frac{f_n''(x)}{x^{1/4}},$$

and deduce that

$$\mathrm{Ai}(x) \sim ax^{-1/4}\exp\left(-\frac{2}{3}vx^{3/2}\right)\sum_{k=1}^{\infty}\frac{c_k}{\zeta^k},$$

where a is an arbitrary constant (that cannot be determined without further information) and

$$c_n = -\frac{(6n-5)(6n-1)}{72n}c_{n-1}, \quad c_0 = 1, \quad \zeta = \frac{2}{3}x^{3/2}.$$

Exercise 15.8

Another expansion for equation 15.3 is obtained by writing the solution in the form

$$y(x) = u(x)^{-1/2}\exp\left(iv\int dx\,u(x)\right).$$

Show that $u(x)$ satisfies the equation

$$u^2 - g(x) = \frac{1}{v^2}u^{1/2}\frac{d^2}{dx^2}\left(u^{-1/2}\right).$$

By assuming an expansion of the form

$$u = g(x)^{1/2} + \sum_{k=1}^{\infty}\frac{1}{v^{2k}}u_{2k}(x), \quad \text{show that} \quad u_2(x) = \frac{1}{2g^{1/4}}\frac{d^2}{dx^2}\left(g^{-1/4}\right).$$

This expansion can be used to obtain the complete asymptotic expansion of the eigenvalues of some equations, see for instance equation 15.95 (page 622) and exercises 15.36 and 15.39.

15.3 Uniform approximations

15.3.1 General theory

There are two important techniques for connecting solutions across turning points. One involves moving round the turning point in the complex x-plane to find connections between the constants (A, B) and (C, D) of equations 15.9 and 15.10; this method is described by Heading (1962) and Dingle (1973), for instance. The other method, described here, does not involve the complex plane but uses a transformation in both the dependent and independent variable to re-express the original equation in a form that is approximately the same as an equation with a known solution. This method produces a *uniform approximation* accurate in the neighbourhood of the zeros of $g(x)$ and which reduces to 15.9 when $g(x) > 0$ and to 15.10 when $g(x) < 0$. The idea behind this method is very similar to that used in the previous chapter where integrands were transformed into generic types for which the integrals could be evaluated.

The objective is to transform the original equation

$$\frac{d^2y}{dx^2} + v^2 g(x)y = 0, \quad v \gg 1, \tag{15.11}$$

into a *comparison equation*

$$\frac{d^2\psi}{d\xi^2} + v^2 G(\xi)\psi = 0 \tag{15.12}$$

which has a known solution, or at least is close to an equation with a known solution.†
This means that $G(\xi)$ must be one of the small number of functions listed in the following table.

Table 15.1. *List of possible generic functions $G(\xi)$ and solutions. Here a and b are constants.*

Function $G(\xi)$	Solution to equation 15.12
± 1	Trigonometric or exponential functions
$\pm \xi$	Airy functions
$a \pm \frac{1}{4}\xi^2$	Parabolic cylinder functions or Hermite polynomials
$a - 2b \sin 2\xi$	Mathieu functions

The zeros of $g(x)$ determine the nature of the solution $y(x)$ and the zeros of $G(\xi)$ determine the nature of the solution $\psi(\xi)$ so it is clear that $y(x)$ and $\psi(\xi)$ can be similar, on given intervals of x and ξ, only if $g(x)$ and $G(\xi)$ have the same number and type of zeros. For instance, if $g(x)$ has one simple zero then $G(\xi)$ must also have only

---

† It is not necessary to include the factor v^2 in the comparison equation. Taking the equation to be $\psi'' + G(\xi)\psi = 0$ simply delays the introduction of this term until later.

one simple zero. We shall show how to use this information after first considering a general transformation.

Write the required solution in terms of two unknown functions $A(x)$ and $\xi(x)$,

$$y(x) = A(x)\psi(\xi(x)), \tag{15.13}$$

where $\psi(x)$ is a solution of equation 15.12. Differentiate this expression, substitute into 15.11 and substitute for $d^2\psi/d\xi^2$ from 15.12 to obtain the equation

$$A(x)\psi(x)v^2 \left[g(x) - \left(\frac{d\xi}{dx} \right)^2 G(\xi) \right] + \frac{\psi'(x)}{A(x)} \frac{d}{dx} \left(A(x)^2 \frac{d\xi}{dx} \right) + \frac{d^2A}{dx^2} \psi(\xi) = 0.$$

The first and second terms are made zero by defining

$$\left(\frac{d\xi}{dx} \right)^2 = \frac{g(x)}{G(\xi)}, \quad \text{and} \quad \frac{d}{dx} \left(A(x)^2 \frac{d\xi}{dx} \right) = 0. \tag{15.14}$$

The first of these equations is similar to equation 14.71 (page 567), and yields a single valued solution if the integration constant is chosen to make the zeros of $g(x)$ and $G(\xi)$ coincide, so the ratio is defined everywhere: we shall see later how this is done using particular examples. Having determined $\xi(x)$, the second equation may be integrated to give

$$A(x) = \left(\frac{d\xi}{dx} \right)^{-1/2} = \left(\frac{G(\xi(x))}{g(x)} \right)^{1/4}. \tag{15.15}$$

With this choice of $\xi(x)$ and $A(x)$ we see that the function $y(x) = A(x)\psi(\xi(x))$ satisfies the equation

$$\frac{d^2y}{dx^2} + v^2 \left(g(x) + \frac{1}{v^2 A(x)} \frac{d^2A}{dx^2} \right) = 0,$$

so the accuracy of the approximation depends upon how well the inequality

$$\frac{d^2A}{dx^2} \ll v^2 g(x)A(x)$$

is satisfied. In practice this inequality needs to be considered for each case, though in most practical applications $A(x)$ varies sufficiently slowly for the inequality to be satisfied even when v is not very large. Examples given later, for instance exercise 15.11 and figure 15.15 (page 609), illustrate this behaviour.

Exercise 15.9
Show that

$$\int_{x_1}^{x_2} dx \, g(x)|y(x)|^2 = \int_{\xi_1}^{\xi_2} d\xi \, G(\xi)|\psi(\xi)|^2,$$

where $\xi_k = \xi(x_k)$, $k = 1, 2$.

15.3.2 No turning points

The simplest example is when $g(x)$ is never zero, and then we should expect the above uniform approximation to reduce to equations 15.9 and 15.10 according to the sign of $g(x)$. Suppose that $g(x) > 0$, then a natural choice for $G(\xi)$ is $G = 1$ and the comparison equation 15.12 becomes

$$\frac{d^2\psi}{d\xi^2} + v^2\psi = 0, \quad \text{with solution} \quad \psi(\xi) = \exp iv\xi.$$

Equation 15.14 for ξ is simply

$$\frac{d\xi}{dx} = \pm\sqrt{g(x)} \quad \Longrightarrow \quad \xi(x) = \pm\int dx \sqrt{g(x)},$$

and equation 15.15 gives $A(x) = g(x)^{-1/4}$, yielding the approximate solution

$$y(x) = \frac{1}{g(x)^{1/4}} \left\{ A \exp\left(iv \int dx \sqrt{g(x)}\right) + B \exp\left(-iv \int dx \sqrt{g(x)}\right) \right\}, \quad g(x) > 0.$$
$$(15.16)$$

This is equation 15.9.

Exercise 15.10
Show that if $g(x) < 0$ the uniform approximation, with $G(\xi) = -1$, gives 15.10.

Exercise 15.11
The equation

$$\frac{d^2 y}{dx^2} + v^2 e^{2x} y = 0, \quad v > 0,$$

has linearly independent solutions $J_0(ve^x)$ and $Y_0(ve^x)$, where $J_0(z)$ and $Y_0(z)$ are the two zero-order Bessel functions.

Show that if $y(x)$ satisfies the boundary conditions $y(0) = y(1) = 0$, then v takes a set of discrete values defined by the positive roots of the equation

$$J_0(ev)Y_0(v) - J_0(v)Y_0(ev) = 0,$$

and find numerical approximations for the first ten of these.

Using the approximation 15.16 show that an approximation to these eigenvalues is

$$v_n = \frac{n\pi}{e - 1}, \quad n = 1, 2, \ldots,$$

and compare these with the exact values found previously. Also show that the approximate solution is

$$y_n(x) \simeq a e^{-x/2} \sin\left(\frac{e^x - 1}{e - 1} n\pi\right),$$

where a is a constant.

By fixing the arbitrary multiplicative constants to make the exact and approximate solutions agree at one point, make a graphical comparison between the two solutions for some values of n.

15.3.3 Isolated turning points

If $g(x)$ has only one simple zero, or if there are several but all are well separated — in a sense that will be made clear later — we may treat each on its own, in the same manner that isolated stationary phase points were treated in the previous chapter. With only one turning point, we chose $G(\xi) = \pm\xi$ so the comparison equation becomes Airy's equation.

Suppose that $g(b) = 0$ and $g(x) < 0$ for $x > b$, so $g'(b) < 0$, as is shown in the diagram.

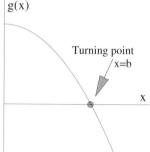

We choose $G(\xi) = -\xi$, and equation 15.14 for $\xi(x)$ becomes

$$\left(\frac{d\xi}{dx}\right)^2 = \frac{g(x)}{-\xi}. \tag{15.17}$$

Now choose the integration constant so that $\xi = 0$ when $x = b$, ensuring that $\xi(x)$ is defined for all x, and choose the signs to make the right hand side of 15.17 positive, that is, $\xi < 0$ when $x < b$ and vice versa. Thus

$$
\begin{aligned}
\frac{2}{3}(-\xi(x))^{3/2} &= \int_x^b dt\,\sqrt{g(t)}, & x \le b, \\
\frac{2}{3}\xi(x)^{3/2} &= \int_b^x dt\,\sqrt{-g(t)}, & x \ge b.
\end{aligned}
\tag{15.18}
$$

Near the turning point, $x = b$, a Taylor expansion of $g(x)$ shows that

$$\xi(x) = |g'(b)|^{1/3}(x - b) + O((x - b)^2).$$

Some higher-order terms of this expansion are given in exercise 15.33, (page 624). With $G(\xi) = -\xi$ the comparison equation 15.12 (page 579), is Airy's equation,

$$\frac{d^2\psi}{d\xi^2} - v^2\xi\psi = 0,$$

having the general solution

$$\psi(\xi) = \alpha\mathrm{Ai}(v^{2/3}\xi) + \beta\mathrm{Bi}(v^{2/3}\xi)$$

for some constants α and β. Thus the approximate solution of equation 15.11 is

$$y(x) \simeq \left(-\frac{\xi(x)}{g(x)}\right)^{1/4}\left\{\alpha\mathrm{Ai}\left(v^{2/3}\xi(x)\right) + \beta\mathrm{Bi}\left(v^{2/3}\xi(x)\right)\right\}, \tag{15.19}$$

with $\xi(x)$ defined in equation 15.18. In this solution both components are oscillatory for $x < b$, that is, $\xi < 0$. For $x > b$, ($\xi > 0$), the first Airy function, $\mathrm{Ai}(\zeta)$, is exponentially decreasing and the second Airy function, $\mathrm{Bi}(\zeta)$, is exponentially increasing.

Equation 15.19 is the fundamental approximate solution; we now show that sufficiently far from the turning point it reduces to the solutions 15.9 and 15.10 found previously. If $|x - b|$ is sufficiently large, the argument of the Airy functions will be large enough to use the asymptotic expansions

$$\mathrm{Ai}(z) \sim \frac{1}{2}\pi^{-1/2}z^{-1/4}\exp\left(-\frac{2}{3}z^{3/2}\right), \quad z \to \infty,$$

$$\mathrm{Ai}(-z) \sim \pi^{-1/2}z^{-1/4}\sin\left(\frac{2}{3}z^{3/2} + \frac{\pi}{4}\right), \quad z \to \infty,$$

$$\mathrm{Bi}(z) \sim \pi^{-1/2}z^{-1/4}\exp\left(\frac{2}{3}z^{3/2}\right), \quad z \to \infty,$$

$$\mathrm{Bi}(-z) \sim \pi^{-1/2}z^{-1/4}\cos\left(\frac{2}{3}z^{3/2} + \frac{\pi}{4}\right), \quad z \to \infty.$$

Thus for $x \gg b$, ($v^{2/3}\xi \gg 1$), we have

$$y(x) \sim \frac{c}{(-g(x))^{1/4}}\left(\frac{\alpha}{2}e^{-\zeta} + \beta e^{\zeta}\right), \quad \zeta(x) = v\int_b^x dt\,\sqrt{-g(t)},$$

and for $x \ll b$, ($v^{2/3}\xi \ll -1$),

$$y(x) \sim \frac{c}{g(x)^{1/4}}\left(\alpha\sin\left(\zeta + \frac{\pi}{4}\right) + \beta\cos\left(\zeta + \frac{\pi}{4}\right)\right), \quad \zeta(x) = v\int_x^b dt\,\sqrt{g(t)}, \qquad (15.20)$$

where $c = v^{-1/6}\pi^{-1/2}$. These solutions have the same form as those given in equations 15.9 (page 577) and 15.10, but now there are only two arbitrary constants, α and β, because the solutions either side of the turning point are connected.

An initial value problem
We illustrate this method by finding approximate solutions to the initial value problem

$$\frac{d^2y}{dx^2} + \left(B - \frac{1}{4}x^2\right)y = 0, \quad y(0) = 1, \quad y'(0) = 0, \quad B \gg 1, \quad x > 0. \qquad (15.21)$$

Comparing this with equation 15.11 we see that $g(x) = B - x^2/4$ and $v = 1$. Also for $x > 0$ there is one turning point at $x = b = 2\sqrt{B}$ and $B \gg 1$ plays the role of the large parameter. The general analysis shows that for $0 \le x \le 2\sqrt{B}$ the solution is oscillatory, and for larger x it is either exponentially increasing or decreasing. The initial conditions are most easily applied to the asymptotic expansion 15.20 by writing

$$\zeta(x) = \frac{1}{2}\int_x^{2\sqrt{B}} dt\,\sqrt{4B - t^2}, \quad x \le 2\sqrt{B},$$

$$= \frac{\pi B}{2} - x\sqrt{B} + O(x^2),$$

and expanding about $x = 0$ to eventually give

$$c\alpha = B^{1/4}\sin\left(\frac{\pi}{2}\left(B + \frac{1}{2}\right)\right), \quad c\beta = B^{1/4}\cos\left(\frac{\pi}{2}\left(B + \frac{1}{2}\right)\right).$$

Hence the approximation 15.19 becomes

$$y(x) \sim \sqrt{\pi}\left(\frac{4B\xi(x)}{x^2 - 4B}\right)^{1/4}\left\{\sin\phi\,\mathrm{Ai}(\xi(x)) + \cos\phi\,\mathrm{Bi}(\xi(x))\right\}, \qquad (15.22)$$

where $\phi = \pi B/2 + \pi/4$ and where equation 15.18 for $\xi(x)$ becomes

$$\xi(x) = \begin{cases} \left[\dfrac{3}{2}\left(\dfrac{1}{4}x\sqrt{x^2 - 4B} - B\ln\left(\dfrac{x + \sqrt{x^2 - 2B}}{2\sqrt{B}}\right)\right)\right]^{2/3}, & x \geq 2\sqrt{B}, \\[4mm] -\left[\dfrac{3}{2}\left(\dfrac{\pi B}{2} - \dfrac{x}{4}\sqrt{4B - x^2} - B\sin^{-1}\left(\dfrac{x}{2\sqrt{B}}\right)\right)\right]^{2/3}, & 0 \leq x \leq 2\sqrt{B}. \end{cases}$$

Observe that in this approximation the coefficient, $\cos\phi$, of the exponentially increasing Airy function, $\mathrm{Bi}(\xi)$, is zero if $B = 2n + \frac{1}{2}$, $n = 0, 1, \ldots$, and then the solution decreases exponentially to zero as $x \to \infty$; for this value of B equation 11.30 shows that the solution is proportional to $\exp(-x^2/4)H_{2n}(x/\sqrt{2})$. For other values of B, $|y(x)|$ increases exponentially. In figure 15.1 we show a graph, the solid line, of the approximate solution for $B = 4.499$, close to a value of B at which the solution tends to zero as $x \to \infty$; the circles superimposed on this line are the values obtained by numerical integration and provide some idea of the accuracy of this approximation.

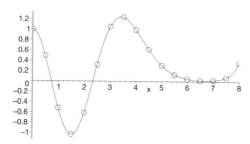

Figure 15.1   The approximate solution 15.22 with
$B = 4.499$.

Exercise 15.12
Explain why the solution 15.22 is not related to a Hermite polynomial when $B = n + \frac{1}{2}$, and n is an odd integer.

Exercise 15.13
Derive equation 15.22 and show that the approximate solution of equation 15.21 with the

initial conditions $y(0) = 0$, $y'(0) = 1$, is

$$y(x) \sim \sqrt{\pi} \left(\frac{4\xi(x)}{B(x^2 - 4B)} \right)^{1/4} \left\{ \sin \phi \, \mathrm{Bi}(\xi(x)) - \cos \phi \, \mathrm{Ai}(\xi(x)) \right\}$$

where $\phi = \pi B/2 + \pi/4$ and $\xi(x)$ is defined after equation 15.22.

Exercise 15.14
Write Maple procedures that (a) plot graphs of the solution to equation 15.21 with the initial conditions of exercise 15.13, and (b) evaluate the uniform approximation for this problem. Use these procedures to investigate the accuracy of the uniform approximation for various B and ranges of x.

For a turning point $x = a$ with $g'(a) > 0$ we choose $G(\xi) = \xi$, then equation 15.17 becomes

$$\left(\frac{d\xi}{dx} \right)^2 = \frac{g(x)}{\xi}, \quad \Longrightarrow \quad \begin{array}{ll} \dfrac{2}{3}\xi(x)^{3/2} & = \displaystyle\int_a^x dt \, \sqrt{g(t)}, \quad x \geq a, \\[3mm] \dfrac{2}{3}(-\xi(x))^{3/2} & = \displaystyle\int_x^a dt \, \sqrt{-g(t)}, \quad x \leq a, \end{array} \tag{15.23}$$

and the approximate solution is

$$y(x) \sim \left(\frac{\xi(x)}{g(x)} \right)^{1/4} \left(\gamma \mathrm{Ai}\left(-v^{2/3}\xi(x) \right) + \delta \mathrm{Bi}\left(-v^{2/3}\xi(x) \right) \right) \tag{15.24}$$

for some constants γ and δ. In this solution both components are oscillatory for $x > a$, that is, $\xi > 0$. For $x < a$, ($\xi < 0$), the first Airy function is exponentially decreasing and the second is exponentially increasing.

Two turning points
If $g(x)$ is positive for $a < x < b$, but is otherwise negative, as shown in the following diagram, the approximation 15.19 may be used in the neighbourhood of $x = b$ and the approximation 15.24 in the neighbourhood of $x = a$ provided $b - a$ is sufficiently large. Then these two solutions may be matched in the region $a \ll x \ll b$ using the asymptotic expansion of the Airy function to give an approximate solution covering the whole range of x.

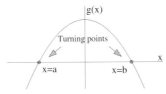

In order to illustrate this method we consider a problem with the boundary conditions $y(x) \to 0$ as $|x| \to \infty$. The solution 15.19 is valid in the neighbourhood of $x = b$, and because $y(x) \to 0$ as $x \to \infty$ the coefficient of $\mathrm{Bi}(\zeta)$ must be zero. Similarly the solution 15.24 is valid in the neighbourhood of $x = a$ and the boundary conditions give $\delta = 0$. Thus we have

$$y_b(x) \sim \alpha \left(-\frac{\xi_b(x)}{g(x)}\right)^{1/4} \text{Ai}\left(v^{2/3}\xi_b(x)\right), \quad x \gg a,$$

$$y_a(x) \sim \gamma \left(\frac{\xi_a(x)}{g(x)}\right)^{1/4} \text{Ai}\left(-v^{2/3}\xi_a(x)\right), \quad x \ll b,$$

where ξ_b and ξ_a are given respectively by equations 15.18 and 15.23 and α and γ are constants, to be determined.

In order to proceed it is necessary to use the asymptotic expansion of the Airy function, page 583, to obtain two simpler expressions for the solutions in the region $a \ll x \ll b$,

$$y_b(x) \sim \frac{\alpha}{\pi^{1/2}v^{1/6}} g(x)^{-1/4} \sin\left(\zeta_b(x) + \frac{\pi}{4}\right), \quad \zeta_b(x) = v \int_x^b dt \sqrt{g(t)}, \quad x \gg a,$$

$$y_a(x) \sim \frac{\gamma}{\pi^{1/2}v^{1/6}} g(x)^{-1/4} \sin\left(\zeta_a(x) + \frac{\pi}{4}\right), \quad \zeta_a(x) = v \int_a^x dt \sqrt{g(t)}, \quad x \ll b.$$

These two functions must be the same, and by writing

$$\zeta_a(x) = v \int_a^b dt \sqrt{g(t)} - \zeta_b(x),$$

the equation $y_a(x) = y_b(x)$ becomes

$$\alpha \sin\left(\zeta_b(x) + \frac{\pi}{4}\right) = -\gamma \sin\left(\zeta_b(x) + \frac{\pi}{4} - A - \frac{\pi}{2}\right), \quad A = v \int_a^b dt \sqrt{g(t)}.$$

Since this must be true for a range of x we have $\sin(A + \pi/2) = 0$, that is,

$$v \int_a^b dt \sqrt{g(t)} = \left(n + \frac{1}{2}\right)\pi, \quad n = 0, 1, \ldots \quad \text{and} \quad \alpha = (-1)^n \gamma. \tag{15.25}$$

This equation gives the approximate values of v for which solutions exist.

The procedure in which two solutions, here $y_a(x)$ and $y_b(x)$, are compared in a common interval of validity to find a unique solution valid over a larger interval is named a *matching* procedure. Normally, as in this application, the asymptotic expansions of the two solutions are matched, and then the method works only if the asymptotic expansions of both approximations are valid in the matching interval, which must therefore be sufficiently far from the turning points,

$$v^{2/3}\xi_b(x) \gg 1 \quad \text{and} \quad v^{2/3}\xi_a(x) \gg 1 \quad \text{for} \quad a \ll x \ll b.$$

The accuracy of the approximation is therefore dependent upon the magnitude of $\xi_a(x)$, $\xi_b(x)$ as well as v. If the above inequalities are satisfied, the turning points $x = a$ and b may be treated as isolated and the solution around each may be approximated by an Airy function, as in equation 15.19. But if either of the inequalities is not satisfied, use of the first term in the asymptotic expansion is not justified: now the turning points may not be treated as isolated and a comparison equation in which $G(\xi)$ is a quadratic in ξ is needed. The solution of such equations can be expressed in terms of the parabolic cylinder functions, so the properties of these are needed before coalescing turning points can be dealt with.

Exercise 15.15
Consider the Sturm–Liouville equation

$$\frac{d^2y}{dx^2} + v^2(\lambda - f(x))y = 0, \quad f(x) = \alpha^2\left(1 - e^{-\beta x}\right)^2,$$

with the boundary condition $y(x) \to 0$ as $|x| \to \infty$. Show that for sufficiently large v this system has a finite set of discrete eigenvalues and that the first term in their asymptotic expansion is

$$\lambda_n = \alpha^2 - \left(\alpha - \frac{\beta}{v}\left(n + \frac{1}{2}\right)\right)^2, \quad n = 0, 1, \ldots, N, \quad N = \left[\frac{v\alpha}{\beta} - \frac{1}{2}\right].$$

15.4 Parabolic cylinder functions

If two turning points are close then the function $G(\xi)$ in the comparison equation must also have two zeros. The simplest suitable function is a quadratic, and the resulting differential equation has parabolic cylinder functions for solutions. We consider these functions partly to illustrate use of the methods discussed earlier, partly to show what happens when two turning points become close, and finally because we shall need the results given here in the general study of nearby turning points in section 15.6.

Parabolic cylinder functions are solutions of the differential equation

$$\frac{d^2y}{dx^2} + (ax^2 + bx + c)y = 0,$$

which has the two distinct, real, standard forms,

$$\frac{d^2y}{dx^2} + \left(\frac{1}{4}x^2 - a\right)y = 0 \tag{15.26}$$

and

$$\frac{d^2y}{dx^2} - \left(\frac{1}{4}x^2 + a\right)y = 0. \tag{15.27}$$

The second of these equations may be obtained from the first by simultaneously replacing x by $e^{i\pi/4}x$ and a by $-ia$.

These are the next most complicated differential equations after Airy's equation. Changing the sign of x leaves each standard equation invariant, so the linearly independent solutions may be chosen to be the odd and even solutions, although linear combinations of these are often useful. Note that if $a = -n - 1/2$, n being a non-negative integer, the standard equation 15.27 becomes 11.30 and its bounded solution is $y = \exp(-x^2/4)H_n(x/\sqrt{2})$, where H_n is a Hermite polynomial.

It is more convenient, and less confusing, to deal with each of the standard equations separately.

15.4.1 The equation $y'' + (x^2/4 - a)y = 0$
The case $a > 0$:
This is the most interesting case because the solution displays different types of behaviour, depending on the value of x. For $|x| \ll 2\sqrt{a}$ the equation is approximately $y'' - ay = 0$, having solutions proportional to $\cosh(x\sqrt{a})$ and $\sinh(x\sqrt{a})$. For $|x| > 2\sqrt{a}$ the coefficient of y is positive and the solution is oscillatory.

The simplest solutions are power series and, although rarely of practical value, they provide a useful starting point. The even solutions behave like $\cosh(x\sqrt{a})$ for small x, suggesting a power series of the form

$$y_1(x) = 1 + b_1 \frac{x^2}{2!} + b_2 \frac{x^4}{4!} + \cdots + b_n \frac{x^{2n}}{(2n)!} + \cdots .$$

This fixes the arbitrary multiplicative constant. Substituting the series into the differential equation and equating the coefficients of x^{2n} to zero gives

$$y_1(x) = 1 + a\frac{x^2}{2!} + \left(a^2 - \frac{1}{2}\right)\frac{x^4}{4!} + a\left(a^2 - \frac{7}{2}\right)\frac{x^6}{6!} + \left(a^4 - 11a^2 + \frac{15}{4}\right)\frac{x^8}{8!} + \cdots, \quad (15.28)$$

where coefficients b_n of $x^{2n}/(2n)!$ are related by

$$b_n = ab_{n-1} - (n-1)\left(n - \frac{3}{2}\right)b_{n-2}, \quad b_0 = 1, \quad b_1 = a.$$

Similarly, the odd solution, being proportional to $\sinh(x\sqrt{a})$ for small x, is defined by the expansion

$$y_2(x) = x + a\frac{x^3}{3!} + \left(a^2 - \frac{3}{2}\right)\frac{x^5}{5!} + a\left(a^2 - \frac{13}{2}\right)\frac{x^7}{7!} + \cdots, \quad (15.29)$$

where the coefficients b_n of $x^{2n+1}/(2n+1)!$ are related by

$$b_n = ab_{n-1} - (n-1)\left(n - \frac{1}{2}\right)b_{n-2}, \quad b_0 = 1, \quad b_1 = a.$$

These solutions are also valid for $a < 0$ and then, for small $|x|$, we see that $y_1(x) \simeq \cos(x\sqrt{-a})$ and $y_1(x) \simeq \sin(x\sqrt{-a})/\sqrt{-a}$.

For a large and positive the two turning points at $x = \pm 2\sqrt{a}$ are well separated and the Airy function expansions, described in section 15.3.3, are valid. For $x \gg -2\sqrt{a}$ the expansion about $2\sqrt{a}$ gives, on using equation 15.24,

$$y(x) \sim \left(\frac{4\xi(x)}{x^2 - 4a}\right)^{1/4}\left\{\alpha \operatorname{Ai}\left(-\xi(x)\right) + \beta \operatorname{Bi}\left(-\xi(x)\right)\right\}, \quad (15.30)$$

where α and β are constants and, for $x \geq 2\sqrt{a}$,

$$\frac{2}{3}\xi(x)^{3/2} = F(x) = \frac{1}{2}\int_{2\sqrt{a}}^{x} dt\,\sqrt{t^2 - 4a} = \frac{1}{4}x\sqrt{x^2 - 4a} - a\ln\left(\frac{x + \sqrt{x^2 - 4a}}{2\sqrt{a}}\right), \quad (15.31)$$

and for $-2\sqrt{a} \ll x \leq 2\sqrt{a}$,

$$\frac{2}{3}(-\xi(x))^{3/2} = \frac{1}{2}\int_{x}^{2\sqrt{a}} dt\,\sqrt{4a - t^2} = \frac{\pi a}{2} - a\sin^{-1}\left(\frac{x}{2\sqrt{a}}\right) - \frac{x}{4}\sqrt{4a^2 - x^2}.$$

This solution may be associated with the series solutions found above by imposing relevant conditions at $x = 0$ and noting that $F(x) = \pi a/2 - x\sqrt{a} + O(x^2)$ to find an approximation near $x = 0$.

For the even solution, with conditions $y_1(0) = 1$, $y_1'(0) = 0$, using the asymptotic expansion of the Airy functions gives, after some analysis,

$$\alpha = \sqrt{\pi} a^{1/4} e^{\pi a/2} \quad \text{and} \quad \beta = \frac{\sqrt{\pi}}{2} a^{1/4} e^{-\pi a/2}.$$

Hence, an approximation to the even solution, $y_1(x)$, is

$$y_1(x) \sim \sqrt{\pi} \left(\frac{4a\xi(x)}{x^2 - 4a} \right)^{1/4} \left\{ e^{\pi a/2} \mathrm{Ai}(-\xi(x)) + \frac{1}{2} e^{-\pi a/2} \mathrm{Bi}(-\xi(x)) \right\} \quad (15.32)$$

$$\sim \left(\frac{4a}{x^2 - 4a} \right)^{1/4} \left\{ e^{\pi a/2} \sin G(x) + \frac{1}{2} e^{-\pi a/2} \cos G(x) \right\}, \quad x \gg 2\sqrt{a},$$

$$\sim \left(\frac{4a}{4a - x^2} \right)^{1/4} \cosh \left(\frac{1}{2} \int_0^x dt \sqrt{4a - t^2} \right), \quad |x| \ll 2\sqrt{a},$$

where $G(x) = F(x) + \pi/4$ and $F(x)$ is defined in equation 15.31. These approximations are needed later. Similarly, an approximation to the odd solution is

$$y_2(x) \sim \sqrt{\pi} \left(\frac{4\xi(x)}{a(x^2 - 4a)} \right)^{1/4} \left\{ e^{\pi a/2} \mathrm{Ai}(-\xi(x)) - \frac{1}{2} e^{-\pi a/2} \mathrm{Bi}(-\xi(x)) \right\} \quad (15.33)$$

$$\sim \left(\frac{4}{a(x^2 - 4a)} \right)^{1/4} \left\{ e^{\pi a/2} \sin G(x) - \frac{1}{2} e^{-\pi a/2} \cos G(x) \right\}, \quad x \gg 2\sqrt{a},$$

$$\sim \left(\frac{4}{a(4a - x^2)} \right)^{1/4} \sinh \left(\frac{1}{2} \int_0^x dt \sqrt{4a - t^2} \right), \quad |x| \ll 2\sqrt{a}.$$

Note that for $a \gg 1$, $y_1(x) \simeq \sqrt{a} y_2(x)$.

Figure 15.2 shows graphs of the odd and even functions, obtained by numerically solving the equation, for $a = 3$ and $x > 0$: for this value of a the differences between the exact and approximate solution, 15.32 and 15.33, are too small to be noticeable in such a graph, and are comparable to the errors shown in figure 15.1 (page 584). When $a = 3$, $e^{-\pi a/2} \simeq 0.01$, and both solutions are dominated by the $\mathrm{Ai}(-\xi)$ term, so $y_1(x) \simeq \sqrt{a} y_2(x)$ for most values of x.

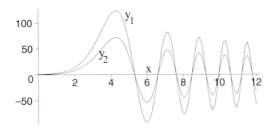

Figure 15.2   Graphs of $y_1(x)$ and $y_2(x)$ for $a = 3$.

Figure 15.3 shows these graphs with $a = 0.2$, for which $e^{-\pi a/2} \simeq 0.7$; with this smaller value of a the differences between $y_1(x)$ and $y_2(x)$ are seen to be more significant. Also the differences between the exact and approximate solutions are larger.

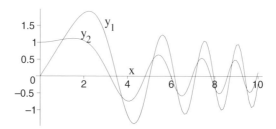

Figure 15.3   Graphs of $y_1(x)$ and $y_2(x)$ for $a = 0.2$.

These approximations show clearly how these parabolic cylinder functions behave. The solutions are oscillatory for $|x| > 2\sqrt{a}$, but behave like hyperbolic functions for smaller $|x|$. With increasing a the turning points move apart and consequently the amplitude of the oscillations in the regions $|x| > 2\sqrt{a}$ increase exponentially with a.

In the exact analysis a variety of notations is used to label various linear combinations of $y_1(x)$ and $y_2(x)$, see for instance Abramowitz and Stegun (1965, chapter 19). Some choices are made to force particular types of asymptotic behaviour, and here we introduce the two linearly independent solutions, $W(a, \pm x)$, that are needed later. These two functions are *defined* by the relation

$$W(a, \pm x) = 2^{-3/4} \left\{ \sqrt{|G|}\, y_1(x) \mp \sqrt{\frac{2}{|G|}}\, y_2(x) \right\}, \quad G(a) = \frac{\Gamma\left(\frac{1}{4} + \frac{1}{2}ia\right)}{\Gamma\left(\frac{3}{4} + \frac{1}{2}ia\right)}. \tag{15.34}$$

With this definition it can be shown that for $x > 2\sqrt{|a|}$,

$$W(a, x) \;\sim\; \sqrt{\frac{2k(a)}{x}} \cos\left(\frac{1}{4}x^2 - a\ln x + \frac{1}{2}\Phi(a) + \frac{\pi}{4}\right), \tag{15.35}$$

$$W(a, -x) \;\sim\; \sqrt{\frac{2}{k(a)x}} \sin\left(\frac{1}{4}x^2 - a\ln x + \frac{1}{2}\Phi(a) + \frac{\pi}{4}\right) \tag{15.36}$$

where

$$\Phi(a) = \arg\Gamma\left(\frac{1}{2} + ia\right), \quad k(a) = \sqrt{1 + e^{2\pi a}} - e^{\pi a} = \frac{1}{\sqrt{1 + e^{2\pi a}} + e^{\pi a}}, \tag{15.37}$$

and the function $\Phi(a)$ is made unique by choosing $\Phi(0) = 0$: higher-order terms in this asymptotic expansion may be found in Abramowitz and Stegun (1965, chapter 19). The phase $\Phi(a)$ is an odd function, because $\Gamma(z^*) = \Gamma(z)^*$, is zero at $a = \pm 2.703$, besides

$a = 0$, for small $|a|$ it behaves as $\Phi(a) \simeq -2a$ and for large $|a|$ as $a \ln |a| - a$. The next figure shows the graph of $\Phi(a)$.

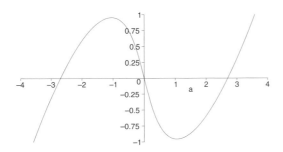

Figure 15.4   Graph of the phase $\Phi(a) = \arg \Gamma(1/2 + ia)$.

The difference between this asymptotic expansion and that of equations 15.32 and 15.33 — valid when the turning points are isolated — is seen by using the asymptotic expansion of the Airy function and the approximation $|G| \sim \sqrt{2/a}$ to give, for $a > 0$,

$$W(a, x) \sim \sqrt{\frac{2k}{x}} \cos\left(\frac{1}{4}x^2 - a \ln x + \frac{1}{2}a(\ln a - 1) + \frac{\pi}{4}\right),$$

$$W(a, -x) \sim \sqrt{\frac{2}{kx}} \sin\left(\frac{1}{4}x^2 - a \ln x + \frac{1}{2}a(\ln a - 1) + \frac{\pi}{4}\right), \quad x \gg 2\sqrt{a}.$$

Comparing these expressions with equations 15.35 and 15.36, we see that the phase $\Phi(a)$ is approximated by $a(\ln a - 1)$. This is the consequence of assuming the turning points to be isolated, clearly invalid if $|a|$ is small. The difference between $\Phi(a)$ and $a(\ln |a| - 1)$ is largest at $|a| = 0.18$ and here the relative error is about 30%.

Exercise 15.16
Use the property $\Gamma(z^*) = \Gamma(z)^*$ to show that $|G(a)|$ is an even function of a and the first term of the asymptotic expansion of the gamma function (equation 20.5, page 835), to show that

$$|G(a)| \sim \frac{\sqrt{2e}}{\left(\left(a^2 + \frac{1}{4}\right)\left(a^2 + \frac{9}{4}\right)\right)^{1/8}} \exp\left(-\frac{a}{2} \tan^{-1}\left(\frac{4a}{3 + 4a^2}\right)\right).$$

Plot the graph of $|G(a)|$ to compare this approximation with the exact values of $|G(a)|$.

Exercise 15.17
(i) Show that the small a expansion of $\Phi(a)$ is

$$\begin{aligned}
\Phi(a) &= -(\gamma + \ln 4)a + \frac{7}{3}\zeta(3)a^3 - \frac{31}{5}\zeta(5)a^5 + \frac{127}{7}\zeta(7)a^7 + \cdots \\
&\simeq -1.9635a + 2.8049a^3 - 6.4288a^5 + \cdots.
\end{aligned}$$

(ii) Show that the asymptotic expansion of $\Phi(a)$ is

$$\Phi(a) \quad \sim \quad \Phi_1(a) = a \ln \sqrt{a^2 + \frac{1}{4}} - a - \tan^{-1}\left(\frac{2a}{7 + 24a^2}\right) + O(a^{-3}) \qquad (15.38)$$

$$= \quad a \ln|a| - a + \frac{1}{24a} + O(a^{-3}),$$

and plot the graph of the difference $\Phi(a) - \Phi_1(a)$ for $0 < a < 4$.
Show also that

$$\Phi_1(a) = -1.979a + 2.987a^3 - 7.439a^5 + \cdots, \quad |a| \ll 1.$$

These results show that the relative error in the asymptotic expansion $\Phi_1(a)$ is less than 1% for all a.

The case $a < 0$:

In this case there are no turning points, and since the coefficient of y is positive for all x the solutions are oscillatory. Using equation 15.9 (page 577) and the appropriate conditions at $x = 0$ gives approximations to the even and odd solutions,

$$y_1(x) \quad = \quad \left(\frac{4b}{x^2 + 4b}\right)^{1/4} \cos\left(\frac{1}{2}\int_0^x dt \sqrt{t^2 + 4b}\right), \quad b = -a > 0, \qquad (15.39)$$

$$y_2(x) \quad = \quad \left(\frac{4}{b(x^2 + 4b)}\right)^{1/4} \sin\left(\frac{1}{2}\int_0^x dt \sqrt{t^2 + 4b}\right). \qquad (15.40)$$

For large x the phase in this approximation has the expansion

$$\frac{1}{2}\int_0^x dt \sqrt{t^2 + 4b} = \frac{1}{4}x^2 + b \ln x + \frac{1}{2}(b - b \ln b) + O(b^2/x^2),$$

giving the following asymptotic expansion for $W(a, \pm x)$, when $x > 0$,

$$W(a, x) \quad \sim \quad \sqrt{\frac{2}{x}} \cos\left(\frac{1}{4}x^2 - a \ln x + \frac{a}{2}\ln|a| - \frac{a}{2} + \frac{\pi}{4}\right),$$

$$W(a, -x) \quad \sim \quad \sqrt{\frac{2}{x}} \sin\left(\frac{1}{4}x^2 - a \ln x + \frac{a}{2}\ln|a| - \frac{a}{2} + \frac{\pi}{4}\right).$$

15.4.2 The equation $y'' - (x^2/4 + a)y = 0$

The linearly independent solutions of this equation may be chosen to be the odd and even solutions. Because this equation may be derived from that dealt with previously by the replacements $x \to e^{i\pi/4}x$, $a \to -ia$, the series representations of these solutions are easily derived from 15.28 and 15.29,

$$y_1(x) \quad = \quad 1 + a\frac{x^2}{2!} + \left(a^2 + \frac{1}{2}\right)\frac{x^4}{4!} + a\left(a^2 + \frac{7}{2}\right)a\frac{x^6}{6!} + \cdots, \qquad (15.41)$$

$$y_2(x) \quad = \quad x + a\frac{x^3}{3!} + \left(a^2 + \frac{3}{2}\right)\frac{x^5}{5!} + a\left(a^2 + \frac{13}{2}\right)\frac{x^7}{7!} + \cdots. \qquad (15.42)$$

The case $a > 0$:

If $a \geq 0$ the coefficient of y in the differential equation is always negative, so we may use equation 15.10 (page 577) with appropriate initial conditions to obtain

$$y_1(x) \quad \sim \quad \left(\frac{4a}{x^2 + 4a} \right)^{1/4} \cosh H(x), \tag{15.43}$$

$$y_2(x) \quad \sim \quad \left(\frac{4}{a(x^2 + 4a)} \right)^{1/4} \sinh H(x), \tag{15.44}$$

where $H(x)$ is the odd function

$$H(x) = \frac{1}{2} \int_0^x dt \, \sqrt{4a + t^2} = \frac{1}{4} x \sqrt{x^2 + 4a} + a \ln \left(\frac{x + \sqrt{x^2 + 4a}}{2\sqrt{a}} \right).$$

If a is large, since $|G(a)| \sim \sqrt{2/a}$,

$$W(a, \pm x) \sim \left(x^2 + 4a \right)^{-1/4} \exp(\mp H(x)).$$

The case $a < 0$:

If $a < 0$ there are two turning points, and the analysis of section 15.2 shows that if $-a$ is large enough the solutions are oscillatory for $|x| < 2\sqrt{|a|}$ and exponentially increasing or decreasing for $|x| \gg 2\sqrt{|a|}$. Expanding about the turning point at $x = 2\sqrt{|a|}$ and using equation 15.19 (page 582) with appropriate initial conditions gives, on setting $b = -a > 0$,

$$y_1(x) \quad \sim \quad \sqrt{\pi} \left(\frac{4b\xi(x)}{x^2 - 4b} \right)^{1/4} \{ \sin \phi \, \mathrm{Ai}(\xi) + \cos \phi \, \mathrm{Bi}(\xi) \}, \tag{15.45}$$

$$y_2(x) \quad \sim \quad \sqrt{\pi} \left(\frac{4\xi(x)}{b(x^2 - 4b)} \right)^{1/4} \{ \sin \phi \, \mathrm{Bi}(\xi) - \cos \phi \, \mathrm{Ai}(\xi) \}, \tag{15.46}$$

where $\phi = \pi(b/2 + 1/4)$. The function $\xi(x)$ is negative for $-2\sqrt{b} \leq x \leq 2\sqrt{b}$, and here it is given by

$$\frac{2}{3} (-\xi(x))^{3/2} = \frac{1}{2} \int_x^{2\sqrt{b}} dt \, \sqrt{4b - t^2} = \frac{\pi b}{2} - \frac{1}{4} x \sqrt{4b - x^2} - b \sin^{-1} \left(\frac{x}{2\sqrt{b}} \right),$$

and for $x \geq 2\sqrt{b}$, $\xi(x)$ is positive and given by

$$\frac{2}{3} \xi(x)^{3/2} = \frac{1}{2} \int_{2\sqrt{b}}^x dt \, \sqrt{t^2 - 4b} = \frac{1}{4} x \sqrt{x^2 - 4b} - b \ln \left(\frac{x + \sqrt{x^2 - 4b}}{2\sqrt{b}} \right).$$

If the coefficient of $\mathrm{Bi}(\xi)$ is zero the solutions tend to zero as $x \to \infty$. Thus if $-a = b = 2n + \frac{1}{2}$, $n = 0, 1, \ldots$, we obtain an approximation to the even Hermite polynomials $y_1 = C \exp(-x^2/4) H_{2n}(x/\sqrt{2})$, for some constant C, and if $-a = b = 2n - \frac{1}{2}$, $n = 1, 2, \ldots$, an approximation proportional to the odd Hermite polynomial $y_2(x) = C \exp(-x^2/4) H_{2n-1}(x/\sqrt{2})$.

15.5 Some eigenvalue problems

The Sturm–Liouville system

$$\frac{d^2y}{dx^2} + v^2 (\lambda - f(x)) y = 0, \tag{15.47}$$

where v is a large, fixed parameter and λ the eigenvalue, occurs in a variety of problems. For instance, such equations are common in the quantum mechanical description of a single particle moving in a separable potential. Indeed, all Sturm–Liouville systems may be cast in this form, as shown in exercise 15.1. Here we shall consider only the cases where $y(x)$ is defined on the whole real axis and is bounded as $|x| \to \infty$.

For such systems the eigenvalues of the operator $Ly = y'' - v^2 f(x)y$ may comprise discrete values $\{\lambda_1, \lambda_2, \ldots\}$ (a discrete spectrum), or the equation $Ly + \lambda y = 0$ may have solutions for all values of λ in an interval, for instance $(0, \infty)$ (a continuous spectrum) or there may be some discrete and some continuous eigenvalues (a mixed spectrum). The nature of the spectrum depends mainly upon the shape of the function $f(x)$.

In the following four diagrams are shown some typical functions $f(x)$: the statements following these diagrams will be justified later by constructing approximate solutions. A more rigorous treatment of such problems can be found in Kato (1976).

Case (a): discrete spectrum

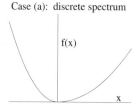

Case (b): continuous spectrum

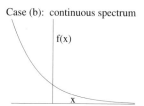

In case (a) the function $f(x)$ has a single minimum and tends to infinity as $|x| \to \infty$; the spectrum is discrete and given approximately by equation 15.48 below. Higher-order terms in the asymptotic expansion in λ may be computed using the method outlined in exercises 15.36 and 15.39. In case (b) the function $f(x)$ has no stationary points so there is only one turning point; the spectrum is continuous.

Case (c): mixed spectrum

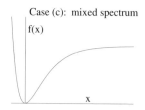

Case (d): continuous spectrum

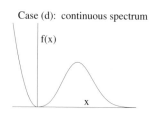

In case (c) the function has a single minimum and tends to infinity as $x \to -\infty$ but to a constant, $f(x) \to f_\infty$ as $x \to \infty$. For sufficiently large v, if $\min(f(x)) < \lambda < f_\infty$ the spectrum is discrete and also given approximately by equation 15.48, but for $\lambda > f_\infty$ it is continuous.

Case (d) is, perhaps, the most interesting. In principle the spectrum is continuous, but in practice for sufficiently large v, there are eigenvalues for which the eigenfunctions behave like the discrete spectrum of cases (a) and (b).

In addition to these cases $f(x)$ could be periodic, and then the eigenvalues and eigenfunctions will be like those of the Mathieu functions described in chapter 11, though there can be subtle complications if $f(x)$ has many stationary points in one period.

These five examples describe important types of possible behaviour, though others are possible, for instance the case where $f(x)$ has two or more minima is important in understanding the structure of some molecules. The reason why the eigenvalues behave in this manner is a simple consequence of the boundary conditions, and because the only two types of solution are oscillatory, 15.9 (page 577) and exponentially increasing or decreasing, 15.10, although the algebra needed to find an approximate solution can obscure this simplicity.

Case a: discrete spectrum

If $f(x)$ has a single minimum and $|f'(x)| > 0$, except at this minimum as in case (a), then $g(x) = \lambda - f(x)$ has two turning points, $a < b$, which depend upon λ. For $x > b$ and $x > a$ the general solution has one component that is exponentially decreasing, and another that is exponentially increasing which is zero only for the discrete set of λ, given approximately by equation 15.25 (page 586) which, for this problem, gives

$$v \int_{a(\lambda)}^{b(\lambda)} dx \sqrt{\lambda - f(x)} = \left(n + \frac{1}{2} \right) \pi, \quad n = 0, 1, \ldots, \quad \text{where} \quad f(a) = f(b) = \lambda. \quad (15.48)$$

This provides an approximation to the eigenvalues; note that the smallest eigenvalue must be larger than $\min(f)$. The eigenfunctions can be expressed in terms of Airy functions using equations 15.19 and 15.24, but a neater method, using Hermite polynomials, is given in section 15.6.1.

Exercise 15.18
Show that the eigenvalues of the equation

$$\frac{d^2 y}{dx^2} + \left(\lambda - x^2 \right) y = 0, \quad |y| \to 0 \quad \text{as} \quad |x| \to \infty,$$

as given by equation 15.48 are $\lambda_n = 2n + 1$, $n = 0, 1, \ldots$, and that these are exact.
Hint: this is equation 11.30.

The result found in this exercise is important. It shows that, in the case $f(x) = x^2$, the first term in the asymptotic expansion of the eigenvalue is exact. The superficial reason for this is that all the other terms in the asymptotic expansion of the eigenvalue (not the eigenfunction) are zero, but why this should be so is rather subtle: one explanation is given by Norcliffe (1973) and also Norcliffe *et al.* (1969).

The significance of this fact is that at minima most functions behave quadratically, so when $\lambda \simeq \min(f)$ and the integral 15.48 is dominated by the quadratic part of $f(x)$ the errors in λ_n are due mainly to the small, higher-order terms of $f(x)$, not the higher-order terms in the asymptotic expansion of λ_n (discussed on page 622). As λ increases (that is, as n increases) the non-quadratic terms of $f(x)$ become more significant, but now the higher-order terms in the asymptotic expansion are less significant. Thus for *all* λ the errors in λ due to using the first order asymptotic expansion are often relatively small.

Exercise 15.19
Show that the system

$$\frac{d^2 y}{dx^2} + \left(\lambda - |x|\right) y = 0, \quad |y| \to 0 \quad \text{as} \quad |x| \to \infty,$$

has the solution

$$y(x) = \begin{cases} c_1 \, \mathrm{Ai}(x - \lambda), & x \geq 0, \\ c_2 \, \mathrm{Ai}(-\lambda - x), & x \leq 0. \end{cases}$$

Using the fact that both $y(x)$ and $y'(x)$ are continuous at $x = 0$, deduce that:

(i) the eigenvalues of the odd eigenfunctions are given by the positive real roots of $\mathrm{Ai}(-\lambda) = 0$, and that $c_2 = -c_1$;

(ii) the eigenvalues of the even eigenfunctions are given by the positive real roots of $\mathrm{Ai}'(-\lambda) = 0$ and that $c_2 = c_1$.

Show that the approximation 15.48 gives

$$\lambda_n = \left[\frac{3}{4}\left(n + \frac{1}{2}\right)\pi\right]^{2/3}, \quad n = 0, 1, \ldots,$$

and make a numerical comparison between the exact and approximate eigenvalues.

Exercise 15.20
Show that the difference between two adjacent eigenvalues is approximately

$$\lambda_{n+1} - \lambda_n \simeq \frac{2\pi}{\nu}\left[\int_a^b dx \, \frac{1}{\sqrt{\bar{\lambda} - f(x)}}\right]^{-1},$$

where $f(a) = f(b) = \bar{\lambda} = (\lambda_{n+1} + \lambda_n)/2$.

Case b: continuous spectrum
This is the next simplest example because there can be only one turning point. Consider the situation shown in the diagram where $f(x)$ decreases monotonically, $f'(x) < 0$. If $a(\lambda)$ is the turning point, $f(a) = \lambda$, then for $x > a$, $g(x) = \lambda - f(x) > 0$ and the solution is oscillatory. The approximate solution, for all x, is given by equation 15.24 (page 585), with $\delta = 0$ to ensure that $y(x)$ is bounded as $x \to -\infty$. Thus

$$y(x) \sim \left(\frac{\xi(x)}{\lambda - f(x)}\right)^{1/4} \mathrm{Ai}\left(-\nu^{2/3}\xi(x)\right) \tag{15.49}$$

and equation 15.23 for $\xi(x)$ becomes

$$\frac{2}{3}\xi(x)^{3/2} = \int_{a(\lambda)}^{x} dt\,\sqrt{\lambda - f(t)}, \quad x \geq a,$$

$$\frac{2}{3}(-\xi(x))^{3/2} = \int_{x}^{a(\lambda)} dt\,\sqrt{f(t) - \lambda}, \quad x \leq a.$$

This solution is bounded and exists for all real values of λ for which $f(a) = \lambda$ has a real solution. Typically, the solution oscillates when $x > a(\lambda)$, with the first maximum to the right of $x = a$, and is exponentially decaying for $x < a$. In figure 15.5 is shown an eigenfunction in the case $f(x) = e^{-x}$ with $\lambda = e^{-1}$ and $v = 1$; this function and the line $y = \lambda$ are shown by the faint lines and the turning point is $x = a$, where these faint lines intersect. Notice that the first maximum of the eigenfunction is to the right of the turning point, that is, in the region $f(x) > \lambda$.

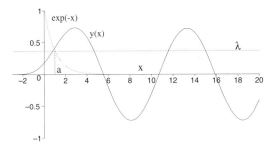

Figure 15.5 The eigenfunction when $f = e^{-x}$ for $\lambda = e^{-1}$ and $v = 1$.

Case c: mixed spectrum
Functions of type (c) combine the attributes of types (a) and (b). If $f(x) \to f_{\infty}$ as $x \to \infty$ and $\min(f) < \lambda < f_{\infty}$ there are two turning points and the discrete part of the spectrum is approximated by equation 15.48, which has solutions provided

$$A = v \int_{a}^{\infty} dx\,\sqrt{f_{\infty} - f(x)} > \frac{\pi}{2},$$

where a is defined by $f(a) = f_{\infty}$. If A is finite there are about A/π components to the discrete spectrum. For $\lambda > f_{\infty}$ the spectrum is continuous and the approximate eigenfunctions are given by equation 15.49.

Case d: quasi-discrete spectrum
In this case $f(x)$ has a local maximum $\max(f) = f_{\max}$. If $\lambda > f_{\max}$ then there is only one real turning point and the spectrum is continuous, case (b), and the eigenfunctions are given by equation 15.49. If $\lambda < f_{\max}$ then there are three turning points, the spectrum is also continuous but is more interesting. If $\lambda \simeq f_{\max}$ then the turning points near the

maximum of f are too close for the methods of this section to be used; we deal with this important problem in the next section.

If $\lambda < f_{\max}$ there are three well separated turning points $a(\lambda) < b(\lambda) < c(\lambda)$ which define the three regions I, II and III shown in the following figure.

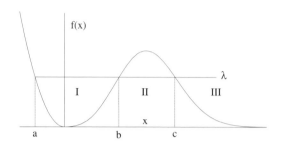

Figure 15.6 Diagram showing the turning points for case (d).

In regions I and III, $\lambda > f(x)$, so the solution will be oscillatory, and in region II it will comprise exponentially increasing and decreasing components. In addition, for $x < a$ the eigenfunction must be exponentially decreasing, because we are assuming that $y(x)$ is bounded as $|x| \to \infty$. Provided the turning points are sufficiently far apart, the solution is easily constructed from the components found previously; we use the approximation 15.24 (page 585) for the turning points at $x = a$ and c and the approximation 15.19 (page 582) for the turning point at $x = b$. These solutions are then matched in the interior of regions I and II, using the asymptotic expansion quoted on page 583. The analysis is fairly lengthy but straightforward provided figure 15.6 is kept in mind.

The solution around the turning point at $x = a$ is, from equation 15.24 (page 585) with $\delta = 0$,

$$y_a(x) = \alpha_a \left(\frac{\xi_a(x)}{\lambda - f(x)} \right)^{1/4} \mathrm{Ai}\left(-v^{2/3} \xi_a(x) \right), \tag{15.50}$$

where

$$\frac{2}{3} \xi_a(x)^{3/2} = \int_a^x dt \, \sqrt{\lambda - f(t)}, \quad x \geq a,$$

$$\frac{2}{3} (-\xi_a(x))^{3/2} = \int_x^a dt \, \sqrt{f(t) - \lambda}, \quad x \leq a.$$

This solution is exponentially decreasing for $x < a$, as required by the boundary conditions.

The solution around the turning point at $x = b$ is, from equation 15.19 (page 582),

$$y_b(x) = \left(-\frac{\xi_b(x)}{\lambda - f(x)} \right)^{1/4} \left\{ \alpha_b \, \mathrm{Ai}\left(v^{2/3} \xi_b(x) \right) + \beta_b \, \mathrm{Bi}\left(v^{2/3} \xi_b(x) \right) \right\}, \tag{15.51}$$

where

$$\frac{2}{3}(-\xi_b(x))^{3/2} = \int_x^b dt \sqrt{\lambda - f(t)}, \quad x \le b,$$

$$\frac{2}{3}\xi_b(x)^{3/2} = \int_b^x dt \sqrt{f(t) - \lambda}, \quad x \ge b.$$

The constants α_b and β_b may be expressed in terms of α_a by matching the asymptotic expansions of these two expressions in the region $a \ll x \ll b$. Using the asymptotic expansion given on page 583 we have, after some algebra,

$$y_a(x) \sim \frac{\alpha_a(\lambda - f(x))^{-1/4}}{\sqrt{\pi}\, v^{1/6}} \sin\left(F_a(x) + \frac{\pi}{4}\right),$$

$$y_b(x) \sim \frac{(\lambda - f(x))^{-1/4}}{\sqrt{\pi} v^{1/6}} \left\{\alpha_b \sin\left(A - F_a(x) + \frac{\pi}{4}\right) + \beta_b \cos\left(A - F_a(x) + \frac{\pi}{4}\right)\right\},$$

where

$$F_a(x) = v \int_a^x dt \sqrt{\lambda - f(t)}, \quad a \le x \le b, \quad \text{and} \quad A = F_a(b).$$

These two approximations are the same if

$$\alpha_a \sin\left(F_a(x) + \frac{\pi}{4}\right) = \alpha_b \sin\left(A + \frac{\pi}{4} - F_a(x)\right) + \beta_b \cos\left(A + \frac{\pi}{4} - F_a(x)\right).$$

This equation must be true for x in the range $a \ll x \ll b$, so the coefficients of both $\sin F_a(x)$ and $\cos F_a(x)$ on each side must be equal, hence:

$$\alpha_b = -\alpha_a \cos\left(A + \frac{\pi}{2}\right), \quad \beta_b = -\alpha_a \sin\left(A + \frac{\pi}{2}\right).$$

The solution around the turning point at $x = c$ is

$$y_c(x) = \left(\frac{\xi_c(x)}{\lambda - f(x)}\right)^{1/4} \left\{\alpha_c \operatorname{Ai}\left(-v^{2/3}\xi_c(x)\right) + \beta_c \operatorname{Bi}\left(-v^{2/3}\xi_c(x)\right)\right\}, \qquad (15.52)$$

where

$$\frac{2}{3}\xi_c(x)^{3/2} = \int_c^x dt \sqrt{\lambda - f(t)}, \quad x \ge c,$$

$$\frac{2}{3}(-\xi_c(x))^{3/2} = \int_x^c dt \sqrt{f(t) - \lambda}, \quad x \le c,$$

and the constants (α_c, β_c) are connected to the constants (α_b, β_b) by using the asymptotic expansions in region II,

$$y_b(x) \sim \frac{(f(x) - \lambda)^{-1/4}}{\sqrt{\pi} v^{1/6}} \left\{\frac{1}{2}\alpha_b \exp\left(-F_b(x)\right) + \beta_b \exp\left(F_b(x)\right)\right\},$$

$$y_c(x) \sim \frac{(f(x) - \lambda)^{-1/4}}{\sqrt{\pi} v^{1/6}} \left\{\frac{1}{2}\alpha_c \exp\left(F_b(x) - \pi B\right) + \beta_c \exp\left(\pi B - F_b(x)\right)\right\},$$

where

$$F_b(x) = v \int_b^x dt \sqrt{f(t) - \lambda}, \quad b \le x \le c, \quad \text{and} \quad \pi B = F_b(c).$$

The factor π multiplying B is introduced here for consistency with the later, more general theory. Equating these two equations gives

$$\frac{1}{2}\alpha_b e^{-F_b(x)} + \beta_b e^{F_b(x)} = \frac{1}{2}\alpha_c e^{-\pi B} e^{F_b(x)} + \beta_c e^{\pi B} e^{-F_b(x)}.$$

Hence

$$\alpha_c = 2\beta_b e^{\pi B} = 2\alpha_a e^{\pi B} \sin\left(A + \frac{\pi}{2}\right),$$

$$\beta_c = \frac{1}{2}\alpha_b e^{-\pi B} = -\frac{1}{2}\alpha_a e^{-\pi B} \cos\left(A + \frac{\pi}{2}\right),$$

where

$$A = v \int_a^b dt \, \sqrt{\lambda - f(t)} \quad \text{and} \quad \pi B = v \int_b^c dt \, \sqrt{f(t) - \lambda}.$$

The original equation is linear, so the constant α_a is the arbitrary multiplicative constant. Thus for every value of λ for which turning points exist, and are sufficiently far apart for the matching procedure to be valid, we have a solution; that is, the spectrum is continuous.

If v is large the constant πB will also be large, and for most values of λ, $|\alpha_c| \gg |\beta_c|$ and the solution in region III will be oscillatory with large amplitude. But if $\sin(A + \pi/2) = 0$, we regain equation 15.48,

$$A = v \int_{a(\lambda)}^{b(\lambda)} du \, \sqrt{\lambda - f(u)} = \left(n + \frac{1}{2}\right)\pi, \quad f(a) = f(b) = \lambda, \qquad (15.53)$$

then $\alpha_c = 0$ and $|\beta_c| \ll |\alpha_a|$. In this case the solution in region III will be oscillatory but with small amplitude. Further, if $\alpha_c = 0$ the solution in region II decreases exponentially as x increases from b to c. Thus for these particular values of λ the eigenfunction behaves like the solutions for type (a) systems: for $a < x < b$ the solution is oscillatory and for $x < a$ and $x > b$ it decreases exponentially. However, the exponential decrease stops at $x = c$ and there remains a residual exponentially small oscillatory component for $x > c$.

For other values of λ, for which $\sin(A + \pi/2) \neq 0$, since $e^{\pi B} \gg 1$, $|\alpha_c| \gg |\alpha_a|$. As before, the solution is oscillatory in $a < x < b$ and $x > c$, but now the amplitude in region III is exponentially larger than in I, contrary to the previous case. The values of λ at which $\alpha_c = 0$ are often named *resonances* and play an important role in quantum mechanics and hence in nature; another similar type of resonance is considered in exercise 15.38 (page 625).

For a specific example, suppose

$$f(x) = \frac{1}{x} + \frac{\beta}{1 + (x - \alpha)^2}, \quad x > 0.$$

For sufficiently large $\beta\alpha^3$ this has a single maximum at $x \simeq \alpha$ and, because $f(x) \to \infty$ as $x \to 0$, a minimum for $x < \alpha$. We choose $\beta = 5$ and $\alpha = 3$ so the graph of $f(x)$ is as shown below.

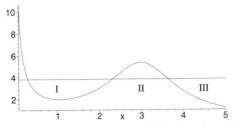

Figure 15.7　Graph of $f(x)$ with $\beta = 5$ and $\alpha = 3$.

The local minimum and maximum values of $f(x)$ are 1.99 and 5.33, so for $1.99 < \lambda < 5.33$, and large enough v, there will be values of λ for which $e^{\pi B}\sin(A + \pi/2) \ll 1$ and these eigenfunctions will be small in region III.

For this function the value of A varies between 0, when $\lambda = \min(f)$, to $3.55v$, when $\lambda = \max(f)$, so from equation 15.53 we see that the maximum value of n is about $\max(A)/\pi \simeq 1.1v$. For any given v the values of λ that satisfy equation 15.53 may be computed (though if v is large high accuracy is needed); when $v = 5$ and $n = 3$ we obtain $\lambda = 3.822\,620\,32$, $\pi B = 6.1$ and $\alpha_c = 1.1 \times 10^{-5}$, and this eigenfunction is shown in figure 15.8. The eigenfunction for the slightly different value $\lambda = 3.823$, for which $\alpha_c = 1.85$, is shown in figure 15.9. Note how a relatively small change in λ has dramatically altered the nature of the eigenfunction.

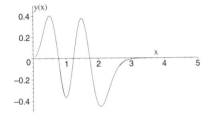

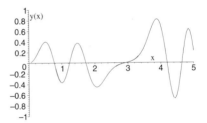

Figure 15.8　$\lambda = 3.822\,620\,32$.   Figure 15.9　$\lambda = 3.823$.

In the cases shown the turning points are, approximately, 0.31, 2.3 and 3.6, and we see that the eigenfunction between the first two turning points is practically unchanged by this small change in λ. For $x > 3$ however, the eigenfunctions are quite different because increasing λ by a factor of 1.0001 changes α_c by a factor of 1.7×10^5, from 1.1×10^{-5} to 1.85.

Near a resonance the shapes of the eigenfunctions are very sensitive to small changes in λ. For most values of λ the relative magnitude of the oscillations in region I is small, but for a narrow band around the resonance values the oscillations in region III have a relatively small amplitude. We may estimate the width of this band by expanding about the value of λ at which $A(\lambda) = (n + 1/2)\pi$; let this value be λ_n and set $\lambda = \lambda_n + \delta$, and expand the expression for α_c to give

$$\alpha_c \simeq -2\alpha_a(-1)^n e^{\pi B_n} A'(\lambda_n)\delta, \quad A'(\lambda_n) = \frac{v}{2}\int_{a(\lambda_n)}^{b(\lambda_n)} dt \, \frac{1}{\sqrt{\lambda_n - f(t)}},$$

and $\pi B_n = v \int_{b(\lambda_n)}^{c(\lambda_n)} dt \sqrt{f(t) - \lambda_n}$. For large v, $e^{\pi B_n}$ is very large so, approximately, the magnitude of α_c is negligible only if $\delta = O(e^{-\pi B_n})$, that is, the width of the resonance decreases exponentially with v. In the example of figure 15.8, for which $v = 5$, $\pi B_3 = 6.1$ and $e^{-\pi B_3} = 0.002$; if $v = 10$, $e^{-\pi B_3} \simeq 5 \times 10^{-6}$, so for large v an accurate determination of the resonance eigenfunction requires very accurate evaluation of the integrals. The approximate estimate of the resonance position and width is, however, relatively simple.

These resonances are isolated but important. In quantum mechanics they represent the energies at which particles may be trapped behind barriers; the inverse of the width of the resonance is tnen related to the mean time the particle is trapped and determines, for instance, the rate of nuclear decay. A similar physical phenomenon is considered in exercise 15.38.

Resonances exist when the value of λ is less that the local maximum of $f(x)$. For larger λ there is only one turning point and the eigenfunctions behave like those of type b. There can be no abrupt change in the behaviour of the eigenfunctions, and the resonances gradually become less pronounced as λ increases to $\max(f)$, because their width increases. In the current approximation we see that as λ increases the value of $B(\lambda)$ decreases because the turning points at $x = b$ and c approach each other. However, if $c - b$ is too small the asymptotic expansions used in region II to match $y_b(x)$ and $y_c(x)$ are no longer valid and the approximation used here breaks down. In particular the resonance position is given incorrectly.

When turning points are too close a different comparison equation is needed; this is the subject of section 15.6.

15.5.1 A double minimum problem

An illustration of the power of these methods is provided by the eigenvalue problem

$$\frac{d^2 y}{dx^2} + v^2(\lambda - f(x))y = 0, \quad y(x) \to 0 \quad \text{as} \quad |x| \to \infty,$$

where $f(x)$ has the shape shown in figure 15.10 below. This is an important problem as in quantum mechanics $f(x)$ will represent a potential energy with two equilibrium points (the minima) separated by a barrier. Such *double well* potentials occur in many molecular systems and for such systems the eigenvalues are closely related to the molecular spectrum; further, the characteristic behaviour of the eigenvalues can be used to infer the shape of $f(x)$, which is directly related to the forces between the atoms in the molecule.

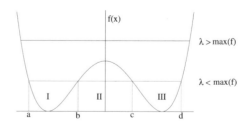

Figure 15.10

The function $f(x)$ has a single maximum and two minima which we assume to have the same value, $\min(f) = 0$, and $f(x) \to \infty$ as $|x| \to \infty$. For this problem the spectrum is discrete and, provided λ is not too close to $\max(f)$, the eigenvalues are given by matching the various uniform approximations in the three regions I, II and III.

If $\lambda > \max(f)$ there are only two turning points and the eigenvalues are approximated by equation 15.25, page 586. If $\lambda < \max(f)$ there are four turning points, $a < b < c < d$, and hence four uniform approximations. The solution around the turning points at $x = a$, b and c are given by equations 15.50 (page 598), 15.51 and 15.52 respectively; the solution round the turning point $x = d$ is similar to that around $x = b$, but because $y(x) \to 0$ as $x \to \infty$ the coefficient of the second Airy function must be zero and hence

$$y_d(x) = \alpha_d \left(\frac{\xi_d(x)}{f(x) - \lambda} \right)^{1/4} \text{Ai}\left(v^{2/3} \xi_d(x) \right), \tag{15.54}$$

where

$$\frac{2}{3}(-\xi_d(x))^{3/2} = \int_x^d dt \sqrt{\lambda - f(t)}, \quad x \leq d,$$

$$\frac{2}{3}\xi_d(x)^{3/2} = \int_d^x dt \sqrt{f(t) - \lambda}, \quad x \geq d.$$

Matching the solutions in region I gives, as before,

$$\alpha_b = -\alpha_a \cos\left(A_1 + \frac{\pi}{2} \right), \quad \beta_b = -\alpha_a \sin\left(A_1 + \frac{\pi}{2} \right), \quad A_1 = v \int_a^b dt \sqrt{\lambda - f(t)}, \tag{15.55}$$

and matching in region II gives

$$\alpha_c = 2\alpha_a e^{\pi B} \sin\left(A_1 + \frac{\pi}{2} \right), \quad \beta_c = -\frac{1}{2}\alpha_a e^{-\pi B} \cos\left(A_1 + \frac{\pi}{2} \right), \tag{15.56}$$

$$\pi B = v \int_b^c dt \sqrt{f(t) - \lambda}.$$

The new match in region III gives the third set of equations:

$$\alpha_c = -\alpha_d \cos\left(A_2 + \frac{\pi}{2} \right), \quad \beta_c = \alpha_d \sin\left(A_2 + \frac{\pi}{2} \right), \quad A_2 = v \int_c^d dt \sqrt{\lambda - f(t)}. \tag{15.57}$$

Equations 15.56 and 15.57 may be rearranged to give

$$\alpha_d \cos\left(A_2 + \frac{\pi}{2} \right) = -2\alpha_a e^{\pi B} \sin\left(A_1 + \frac{\pi}{2} \right),$$

$$\alpha_d \sin\left(A_2 + \frac{\pi}{2} \right) = -\frac{1}{2}\alpha_a e^{-\pi B} \cos\left(A_1 + \frac{\pi}{2} \right), \tag{15.58}$$

then division gives an equation for the eigenvalue λ,

$$\tan\left(A_1(\lambda) + \frac{\pi}{2} \right) \tan\left(A_2(\lambda) + \frac{\pi}{2} \right) = \frac{1}{4}e^{-2\pi B(\lambda)}. \tag{15.59}$$

In order to understand this equation we assume that $f(x)$ is even so $a = -d$, $b = -c$

and $A_1 = A_2 = A(\lambda)$, then this equation simplifies to

$$\tan\left(A(\lambda) + \frac{\pi}{2}\right) = \pm\frac{1}{2}e^{-\pi B(\lambda)}. \tag{15.60}$$

Remember that $\pi B \gg 1$ (if $\lambda \ll \max(f)$ and $v \gg 1$) so the right hand side is small and the zero-order approximation to this equation is simply

$$A(\lambda) = v\int_c^d dt\ \sqrt{\lambda - f(t)} = \left(n + \frac{1}{2}\right)\pi,$$

which is just equation 15.25 (page 586). If this approximation to the nth eigenvalue is denoted by $\lambda_n^{(0)}$, then first-order perturbation theory applied to equation 15.60 gives

$$\lambda_n = \lambda_n^{(0)} \pm \frac{1}{2}\left(\frac{\partial A}{\partial\lambda_n^{(0)}}\right)^{-1}\exp(-\pi B(\lambda_n^{(0)})).$$

Thus for each n there are two eigenvalues differing by an exponentially small quantity. With equation 15.60 satisfied the remaining constants are:

$$\alpha_b = -\alpha_a\cos\left(A + \frac{\pi}{2}\right) \simeq \alpha_a(-1)^n,$$

$$\beta_b = -\alpha_a\sin\left(A + \frac{\pi}{2}\right) \simeq \pm\frac{1}{2}(-1)^n\alpha_a e^{-\pi B} \ll 1,$$

$$\alpha_c = \mp\alpha_a\cos\left(A + \frac{\pi}{2}\right) \simeq \pm(-1)^n\alpha_a,$$

$$\beta_c = -\frac{1}{2}\alpha_a e^{-\pi B}\cos\left(A + \frac{\pi}{2}\right) \simeq (-1)^n\frac{1}{2}\alpha_a e^{-\pi B} \ll 1,$$

$$\alpha_d = \mp\alpha_a.$$

Thus $|\beta_b|$ and $|\beta_c|$ are exponentially small. This means that the eigenfunctions for each exponentially close pair of eigenvalues can be written approximately in the form $y_\pm(x) = \phi_1(x) \pm \phi_3(x)$, where $\phi_1(x)$ is the 'eigenfunction' for region I obtained by assuming it decreases to zero outside the turning points at $x = a$ and b. Similarly for $\phi_3(x)$. Thus the inner product of these two functions must be zero, $(\phi_1, \phi_3) = 0$, and if each is normalised to give $(\phi_1, \phi_1) = (\phi_3, \phi_3)$ it follows that $(y_+, y_-) = 0$, that is, the eigenfunctions of the exponentially close eigenvalues are orthogonal, as expected from Sturm–Liouville theory. This analysis also shows that the eigenfunction of the smallest eigenvalue is even.

The behaviour just described is shown in the following figures, where the eigenfunctions for $f(x) = (1 - x^2)^2$, having minima at $x = \pm1$, and $v = 10$ are shown: these are computed numerically using the method outlined in exercise 15.21. In figures 15.11 and 15.12 are shown the eigenfunctions for the two lowest eigenvalues, $\lambda = 0.194\,68 \pm 0.301 \times 10^{-5}$, for which $|y_0(x)|$ and $|y_1(x)|$ are indistinguishable on the scale of these graphs, however $y_0(x)$ is an even function and $y_1(x)$ an odd function.

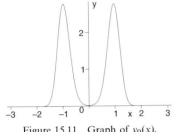

Figure 15.11   Graph of $y_0(x)$,
$\lambda_0 = 0.194\,672\,996$.

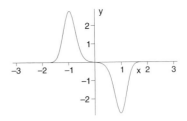

Figure 15.12   Graph of $y_1(x)$,
$\lambda_1 = 0.194\,679\,024$.

In this case the approximation of equation 15.60 gives $\lambda = 0.196\,107 \pm 0.284 \times 10^{-5}$.

The eigenfunctions for the next two eigenvalues are shown in the following two figures. In this case $\lambda = 0.559\,51 \pm 0.332 \times 10^{-3}$ and the approximation equation 15.60 gives $\lambda = 0.5616 \pm 0.331 \times 10^{-3}$.

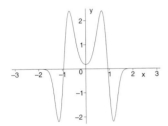

Figure 15.13   Graph of $y_2(x)$,
$\lambda_2 = 0.5591\,77$.

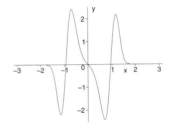

Figure 15.14   Graph of $y_3(x)$,
$\lambda_3 = 0.5598\,42$.

Exercise 15.21
Solutions to the Sturm–Liouville problem $y'' + v^2(\lambda - f(x))y = 0$ may be approximated numerically by assuming the solution can be represented as a truncated Fourier series,

$$y(x) = \sum_{n=-N}^{N} c_n e^{-inx},$$

provided the outer turning points are not too close to $\pm\pi$.

By substituting this series unto the differential equation derive the matrix equation $M\mathbf{c} = \lambda\mathbf{c}$, $\mathbf{c} = (c_{-N}, \dots, c_0, \dots, c_N)$, where M is the $(2N+1) \times (2N+1)$ matrix with elements

$$M_{nm} = \frac{n^2}{v^2}\delta_{nm} + \frac{1}{2\pi}\int_{-\pi}^{\pi} dx\, f(x)e^{i(n-m)x}.$$

Use this method to reproduce the results quoted above, for which $f(x) = (1 - x^2)^2$, and to draw the graphs of the eigenfunctions with eigenvalues $\lambda \simeq 1$ and $\lambda > 1$.
Note: the results quoted above were obtained with $N = 30$ and Digits$= 16$. Care is needed when numerically solving this type of problem because components of the solution that increase exponentially as $|x| \to \infty$ can cause problems. The method used in exercise 15.24 avoids this problem.

15.6 Coalescing turning points

If two turning points are so close that they cannot be treated as isolated the comparison equation needs to explicitly include both turning points, so $G(\xi)$ becomes a quadratic function having either a minimum, $G \sim \xi^2$, or a maximum, $G \sim -\xi^2$. In either case the solution of the comparison equation can be expressed in terms of parabolic cylinder functions.

15.6.1 Expansion about a minimum

If $f(x)$ has a single local minimum at x_m and if the solutions are bounded as $|x| \to \infty$, as in cases (a) and (c), then near this minimum,

$$\lambda - f(x) \simeq \lambda - f(x_m) - \frac{1}{2}f''(x_m)(x - x_m)^2, \quad \lambda > f(x_m),$$

which suggests using the comparison equation

$$\frac{d^2\psi}{d\xi^2} + \left(\mu^2 - \xi^2\right)\psi = 0, \quad \mu = \mu_n = \sqrt{2n+1}, \quad n = 0, 1, \ldots. \tag{15.61}$$

This is the equation for a parabolic cylinder function, and you will recall from section 15.4 that solutions are bounded as $|\xi| \to \infty$ only if $\mu^2 = 2n + 1$, and then the solutions can be written in terms of the Hermite polynomials,

$$\psi_n(\xi) = \exp(-\xi^2/2)H_n(\xi). \tag{15.62}$$

Now equation 15.14 (page 580) for $\xi(x)$ becomes, for the nth eigenvalue,

$$\left(\frac{d\xi}{dx}\right)^2 = \frac{\nu^2(\lambda - f(x))}{2n + 1 - \xi^2}. \tag{15.63}$$

If $a(\lambda) < b(\lambda)$ are the turning points, $f(a) = f(b) = \lambda$, then $\xi(x)$ is defined everywhere if the zeros of the numerator and denominator are made to coincide. For the nth eigenvalue this gives

$$\int_{-\sqrt{2n+1}}^{\sqrt{2n+1}} d\xi \sqrt{2n + 1 - \xi^2} = \nu \int_a^b dx \sqrt{\lambda - f(x)}, \tag{15.64}$$

which gives the required approximation to λ_n. Evaluating the integral on the left hand side gives

$$\nu \int_a^b dx \sqrt{\lambda - f(x)} = \left(n + \frac{1}{2}\right)\pi, \quad n = 0, 1, \ldots. \tag{15.65}$$

This is exactly the same as equation 15.25, which is surprising as that equation was derived assuming the turning points to be isolated, an approximation *not* made here. The reason why the approximations 15.25 and 15.65 are the same is partly to do with the result found in exercise 15.18 (page 595) and the subsequent discussion. The advantage of the present approximation is that it gives a reasonably simple approximation for the eigenfunctions, valid over the entire range of x.

The single valued function $\xi(x)$ is obtained by integrating the differential equation

$$\frac{d\xi}{dx} = v\sqrt{\frac{\lambda_n - f(x)}{2n+1-\xi^2}}, \qquad \xi(a) = -\sqrt{2n+1}. \tag{15.66}$$

This may be done either by numerically solving this equation,† though some care is necessary at $\xi = \pm\sqrt{2n+1}$ where both numerator and denominator are zero and numerical errors may produce singularities on the right hand side. With Maple the procedure dsolve, with the numeric option, will perform this task, though with this procedure there seems to be no mechanism available for a careful treatment of the turning points.

The alternative method is to perform the ξ-integral analytically. Thus for $x \geq b$ ($\xi \geq \sqrt{2n+1}$), we have

$$v\int_b^x dx \sqrt{f(x) - \lambda} = \frac{2n+1}{4}(\sinh 2u - 2u), \quad \cosh u = \frac{\xi}{\sqrt{2n+1}},$$

with other expressions for the regions $a \leq x \leq b$ and $x \leq a$. Now the integral needs to be evaluated numerically, for each value of x, and the equation inverted numerically to find $\xi(x)$. In practice this method is less convenient.

Once a representation of $\xi(x)$ has been found, an approximate solution for the nth eigenfunction of the equation

$$\frac{d^2y}{dx^2} + v^2(\lambda - f(x))y = 0,$$

with λ_n given by equation 15.65, is

$$y_n(x) \sim \left(\frac{2n+1-\xi(x)^2}{\lambda_n - f(x)}\right)^{1/4} \exp\left(-\frac{\xi(x)^2}{2}\right) H_n(\xi(x)).$$

As an example of this approximation we consider the eigenvalue problem

$$\frac{d}{dr}\left(r^2\frac{dR}{dr}\right) + \left[\lambda r - \frac{r^2}{4} - l(l+1)\right]R = 0, \quad r \geq 0, \tag{15.67}$$

the solution of which can be expressed in terms of associated Laguerre polynomials, equation 15.68 overpage. This equation is the radial part of Schrödinger's equation when applied to the electron of a hydrogen atom; the independent variable r is a scaled radius, see Landau and Lifshitz (1965b, section 36), l is the non-negative integer that determines the angular momentum of the electron and λ is the eigenvalue that needs to be found. The electron energy, $E = -1/2\lambda^2$, depends only upon this eigenvalue.

This equation has two linearly independent solutions which behave as r^l and r^{-l-1} as $r \to 0$; only the first of these is of physical interest, and the eigenvalue is determined by the condition that $R(r) \to 0$ as $r \to \infty$. With these boundary conditions it may be

† The sceptic would point out that we have merely replaced a differential equation that needs to be solved numerically by another that also needs to be solved numerically. The original equation, however, is a boundary value problem, while the current equation is an initial value problem, which is far easier to deal with.

shown that the solution can be written in terms of the associated Laguerre polynomials,

$$R(r) = r^l e^{-r/2} L_{n+l}^{2l+1}(r).$$ (15.68)

Now we show how to find the uniform approximation for this problem.

First it is necessary to express the equation in the standard form using the transformation found in exercise 15.1. In this case $p(r) = r^2$, $w(r) = r$ and $q(r) = -r^2/4 - l(l+1)$, so the new independent variable $x(r)$ is

$$x(r) = \int dr \sqrt{\frac{w(r)}{p(r)}} = 2r^{1/2} \quad \text{or} \quad r = \left(\frac{x}{2}\right)^2.$$ (15.69)

The overall scaling factor is

$$A(r) = \left(\frac{1}{pw}\right)^{1/4} \quad \text{or} \quad A = \left(\frac{2}{x}\right)^{3/2}.$$

Hence the equation becomes

$$\frac{d^2v}{dx^2} + (\lambda - f(x))v = 0, \quad f(x) = \frac{x^2}{16} + \frac{\alpha^4}{x^2}, \quad \alpha^4 = 4\left(l + \frac{1}{2}\right)^2 - \frac{1}{4}.$$ (15.70)

The function $f(x)$ has a single minimum at

$$x = 2\alpha \quad \text{at which} \quad f(2\alpha) = \frac{\alpha^2}{2},$$

and $f(x) \to \infty$ as $x \to 0$ and ∞.

An approximation to the eigenvalues is given by equation 15.65:

$$\left(n + \frac{1}{2}\right)\pi = \int_a^b dx \sqrt{\lambda - \frac{x^2}{16} - \frac{\alpha^4}{x^2}}, \quad a^2, b^2 = 8\lambda \pm 8\sqrt{\lambda^2 - \alpha^4/4},$$

$$= \frac{1}{2}\int_{\sqrt{a}}^{\sqrt{b}} \frac{dz}{z} \sqrt{-\alpha^4 + \lambda z - \frac{z^2}{16}} = \frac{\pi}{2}\left(2\lambda - \alpha^2\right).$$

Hence

$$\lambda_{nl} = n + \frac{1}{2} + \sqrt{\left(l + \frac{1}{2}\right)^2 - \frac{1}{16}} = n + l + 1 + O(l^{-1}).$$ (15.71)

The function $\xi(x)$ is defined by equation 15.63 which, in this context, can be written as

$$\frac{d\xi}{dx} = \sqrt{\left|\frac{\lambda_{nl} - f(x)}{2n + 1 - \xi^2}\right|}, \quad \xi(a) = -\sqrt{2n + 1}.$$

The modulus is introduced because the numerical value of $\xi(b)$ will not be exactly $\sqrt{2n + 1}$. In practice the Maple commands

```
>  eqn:=diff(y(x),x)=sqrt(abs((L(n,l)-f(x,l))/(G(y,n))));
>  s:=dsolve({eqn,y(a)=evalf(-sqrt(2*n1+1))},{y(x)},type=numeric);
```

are sufficient to solve this differential equation, and in most cases the integration procedure will integrate through the points $\xi(a)$ and $\xi(b)$, which are potential sources of numerical problems because here both numerator and denominator are zero. For the case $l = 10$, $n = 2$, $\lambda \simeq 13$ the function $\xi(x)$ is shown in the following figure, in which the two horizontal lines are $\xi = \pm\sqrt{2n+1}$ so intersect the graph of $\xi(x)$ at the turning points $a = 6.532$ and $b = 12.86$.

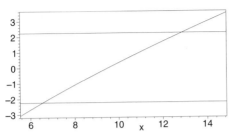

Figure 15.15   The function $\xi(x)$ for $l = 10$, $n = 2$.

In the next two figures we compare the uniform approximation

$$R_{nl}^{(u)}(x) = x^{-3/2} \left(\frac{2n + 1 - \xi(x)^2}{\lambda_{nl} - f(x)} \right)^{1/4} \exp\left(-\frac{\xi(x)^2}{2} \right) H_n(\xi(x))$$

as a function of $x = 2\sqrt{r}$ with the exact function defined in equation 15.68. The approximation is depicted by the solid lines and the exact function by the circles. Normalisation is chosen to make the exact function agree with the approximation at one point. On the left, figure 15.16, $l = 10$ and $n = 2$, as above, and on the right, figure 15.17 $l = 2$, $n = 1$. In both cases the agreement is remarkably good.

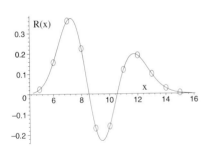

Figure 15.16 $l = 10$, $n = 2$.

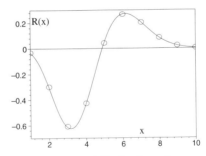

Figure 15.17 $l = 2$, $n = 1$.

15.6.2 *Expansion about a maximum*

If $f(x)$ has a local maximum at $x = x_m$ and $\lambda \simeq f(x_m)$, then near here

$$\lambda - f(x) \simeq \lambda - f(x_m) + \frac{1}{2}|f''(x_m)|(x - x_m)^2,$$

which suggests a comparison equation of the form

$$\frac{d^2\psi}{d\xi^2} + \left(\frac{1}{4}\xi^2 - B \right)\psi = 0, \tag{15.72}$$

where the constant B may be positive or negative. This is the equation for the parabolic cylinder functions dealt with in section 15.4.1.

Equation 15.14 (page 580) for $\xi(x)$ is then

$$\frac{d\xi}{dx} = v\sqrt{\frac{\lambda - f(x)}{\frac{1}{4}\xi^2 - B}}, \tag{15.73}$$

and the relation between B and λ is obtained by matching the turning points, although now we have to extend previous ideas to deal with the case $\lambda > \max(f)$, when the relevant roots of $f(x) = \lambda$ are complex. If $\lambda < \max(f)$ and the turning points are $x = b$ and c, integrating between these gives

$$v\int_b^c dx\,\sqrt{f(x) - \lambda} = \frac{1}{2}\int_{-2\sqrt{B}}^{2\sqrt{B}} d\xi\,\sqrt{4B - \xi^2} = \pi B(\lambda) > 0. \tag{15.74}$$

Normally $B(\lambda)$ is an analytic function of λ and the integral can be written as a contour integral in the complex x-plane round the contour $\mathscr{C}_x$ enclosing the two relevant roots of $f(x) = \lambda$:

$$2\pi B = v\oint_{\mathscr{C}_x} dz\,\sqrt{f(z) - \lambda}. \tag{15.75}$$

If $\lambda < \max(f)$, this expression reduces to the real form 15.74 and is zero if $\lambda = \max(f)$ and the turning points coincide; but if $\lambda > \max(f)$ the roots of $f(x) = \lambda$ become complex and $B < 0$. In practice, if v is large enough, the value of B is adequately approximated by expanding about $\lambda = \max(f)$, as in the following exercise.

Exercise 15.22
Show that for $\lambda \simeq \max(f)$,

$$B(\lambda) = v\,\frac{f(x_m) - \lambda}{\sqrt{2|f''(x_m)|}}.$$

Further terms of this expansion are given in exercise 15.37 (page 625).

Exercise 15.23
In the case $f(x) = (x^2 - 1)^2$ show that the expressions 15.74 and 15.75 for B become

$$\pi B(\lambda) = 2v\left(1 - \sqrt{\lambda}\right)\sqrt{1 + \sqrt{\lambda}}\int_0^{\pi/2} d\phi\,\cos^2\phi\sqrt{1 - \mu\sin^2\phi},$$

where $\mu = (1 - \sqrt{\lambda})/(1 + \sqrt{\lambda})$.

Use Maple to compute the above integrals — which may be evaluated in terms of complete elliptic integrals, with different expressions for $\lambda < 1$ and $\lambda > 1$ — and compare these results with the approximation derived in the previous exercise.

15.6.3 The double minimum problem

For the first application of this theory we consider the situation where $f(x)$ has the shape shown in figure 15.10 (page 602): so $f(x)$ is even, has a single maximum at $x = 0$, two minima, and $f(x) \to \infty$ as $|x| \to \infty$.

If $\lambda < \max(f)$ there are four turning points $a < b < c < d$ and three solutions to match, but in this case the solutions are either odd or even solutions, so we need only match in region III, $c \ll x \ll d$.

The solution of the comparison equation is either $\psi_1(\xi)$ (the even solution) or $\psi_2(\xi)$ (the odd solution) and these functions are defined in equations 15.28 and 15.29 (see also 15.32 and 15.33). It is more convenient to express these in terms of the functions $W(B, \pm\xi)$ defined in equation 15.34, which can be rearranged to give

$$\psi_{1,2}(\xi) = \gamma \left(W(B, -\xi) \pm W(B, \xi) \right) \tag{15.76}$$

for some, irrelevant, constant γ. The asymptotic expansions of these functions, equation 15.35 (page 590) and 15.36, show that in region III

$$\psi_{1,2}(\xi) \sim \gamma' \sqrt{\frac{2}{\xi}} \sin\left(\frac{1}{4}\xi^2 - B\ln\xi + \frac{\pi}{4} + \frac{1}{2}\Phi(B) \pm \delta(B)\right), \tag{15.77}$$

where

$$\tan\delta = k(B), \quad 0 \le \delta \le \frac{\pi}{2},$$

and where $\Phi(B)$ and $k(B)$ are defined in equation 15.37 (page 590), and also 15.79 below.

The function $\xi(x)$ is given by integrating equation 15.73. When the turning points are real, $\lambda < \max(f)$, and when $x > c$ we have

$$\nu \int_c^x dx \sqrt{\lambda - f(x)} = \frac{1}{2} \int_{2\sqrt{B}}^\xi d\xi \sqrt{\xi^2 - 4B}, \quad B \ge 0,$$

$$= \frac{1}{4}\xi^2 - B\ln\xi + \frac{1}{2}(B\ln B - B) + O(\xi^{-2}). \tag{15.78}$$

Alternatively, the solution in region III can be obtained by expanding about the turning point $x = d$, to obtain $y_d(x)$, equation 15.54 (page 603). The asymptotic expansion of this solution in region III is

$$y_d(x) \sim \frac{1}{\sqrt{\pi}\nu^{1/6}} \frac{\alpha_d}{(\lambda - f(x))^{1/4}} \sin\left(\nu \int_x^d dx \sqrt{\lambda - f(x)} + \frac{\pi}{4}\right).$$

The solutions derived from $\psi_{1,2}(\xi)$ in this region are

$$y_{1,2}(x) \sim \left(\frac{\xi^2 - 4B}{\lambda - f(x)}\right)^{1/4} \left[W(B, -\xi(x)) \pm W(B, \xi(x))\right],$$

and combining the asymptotic expansion 15.77 with 15.78 gives

$$y_{1,2}(x) \sim \frac{\gamma'}{(\lambda - f(x))^{1/4}} \sin\left(\nu \int_c^x dx \sqrt{\lambda - f(x)} + \frac{1}{2}\Phi(B) - \frac{B}{2}(\ln B - 1) + \frac{\pi}{4} \pm \delta(B)\right).$$

Equating the phases of these two approximations gives an equation for the eigenvalues,

$$\nu \int_c^d dx \sqrt{\lambda - f(x)} = \left(n + \frac{1}{2}\right)\pi - \frac{1}{2}\Phi(B) + \frac{B}{2}(\ln B - 1) \mp \tan^{-1} k(B), \tag{15.79}$$

where n is an integer,

$$\Phi(B) = \arg \Gamma\left(\frac{1}{2} + iB\right), \quad k(B) = \sqrt{1 + e^{2\pi B}} - e^{\pi B} = \frac{1}{\sqrt{1 + e^{2\pi B}} + e^{\pi B}},$$

and where

$$\pi B(\lambda) = v \int_b^c dx \sqrt{f(x) - \lambda}.$$

In this expression the upper sign gives even solutions, the lower sign odd solutions. Remember also that the turning points a, b, c and d depend upon λ.

For a given n these equations may be solved to find the eigenvalue λ_n. If $\lambda \ll \max(f)$, B is large and the equation reduces to 15.48 (page 595). If we ignore $\Phi(B) - B \ln B + B$ the equation becomes the same as equation 15.60, as it ought because these terms are the corrections due to the proximity of the turning points b and c, which were assumed infinitely far apart when deriving the earlier approximation.

If $\lambda > \max(f)$, $B < 0$ and the only turning points are at $x = \pm d$. Now $\xi(x)$ is given by

$$v \int_0^x dx \sqrt{\lambda - f(x)} = \frac{1}{2} \int_0^\xi d\xi \sqrt{\xi^2 - 4B}, \quad B < 0$$

$$= \frac{1}{4}\xi^2 - B \ln \xi + \frac{B}{2}(\ln|B| - 1) + O(\xi^{-2}),$$

and the same matching procedure leads to

$$v \int_0^d dx \sqrt{\lambda - f(x)} = \left(n + \frac{1}{2}\right)\pi - \frac{1}{2}\Phi(B) + \frac{B}{2}(\ln|B| - 1) \mp \tan^{-1} k(B), \quad B < 0. \quad (15.80)$$

This is essentially the same as equation 15.79 except the lower limit of the integral is at $x = 0$.

If $\lambda \gg \max(f)$ the value of $|B|$ will be large, then $\tan^{-1}(k(B)) \simeq \pi/4$, and in this limit equation 15.80 for λ_n becomes

$$v \int_{-d}^d dx \sqrt{\lambda - f(x)} = \left(2n + 1 \mp \frac{1}{2}\right)\pi,$$

which is identical to equation 15.48, because the right hand side has the values $(m + 1/2)\pi$ for integer m.

An elegant method of visualising how the eigenvalues change as λ increases through $\max(f)$ is obtained by writing equations 15.79 and 15.80 in the form

$$\cos\left(2vF(\lambda) + \chi(B)\right) = -\cos 2\delta(B) = -\frac{e^{\pi B}}{\sqrt{1 + e^{2\pi B}}}, \quad (15.81)$$

where $\chi(B) = \Phi(B) - B(\ln|B| - 1)$ and

$$F(\lambda) = \int_{\bar{c}}^d dx \sqrt{\lambda - f(x)}, \quad \bar{c} = \begin{cases} c & \text{if } \lambda \le \max(f), \\ 0 & \text{if } \lambda \ge \max(f). \end{cases}$$

The graphs of the left and right hand sides of this equation show how the eigenvalues vary.

A typical example is provided by the function†

$$f(x) = x^2 + \beta e^{-x^2}.$$

If $\beta > 1$, $f(x)$ has a single maximum at $x = 0$, with $f(0) = \beta$, and two minima at $x^2 = \ln \beta$ with $\min(f) = 1 + \ln \beta$. Thus for large enough β the eigenfunctions with eigenvalues in the interval $\min(f) < \lambda \ll \beta$ will behave like those shown in figures 15.11–15.14 (page 605).

In the following figure is shown graphs of each side of equation 15.81, the left hand side being the oscillatory curve. The eigenvalues are at the intersection of these curves.

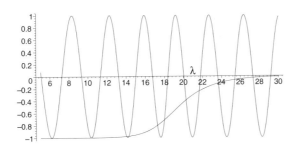

Figure 15.18   Graphs of the left (the oscillatory curve) and right hand side of equation 15.81, for $\beta = 18$ and $v = 1$.

For the small eigenvalues $\lambda \ll \max(f) = \beta$, $B \gg 1$ and the right hand side of the equation is approximately $-1 + \exp(-2\pi B)/2$, so the intersections are exponentially close to the roots of $vF(\lambda) = (n + 1/2)\pi$: these pairs of eigenvalues belong to similar odd and even eigenfunctions.

As λ increases through β the value of B decreases, being zero when $\lambda = \beta(= 18)$, and for $\lambda \gg \beta$, $B \ll -1$ and the right hand side of the equation decreases rapidly to zero, so now the eigenvalues are given, approximately, by the roots of $2vF(\lambda) \simeq (n + 1/2)\pi$.

In the table below is a comparison between the exact eigenvalues of this system, with $v = 1$ and $\beta = 18$, computed using the method outlined in exercise 15.24 and using the uniform approximation, equation 15.79. This comparison shows how accurate the uniform approximation can be, even for this relatively small value of v, and also how the character of the eigenvalues change as λ passes through $\max(f)$.

Exercise 15.24

The eigenvalues and eigenfunctions of the system

$$\frac{d^2 y}{dx^2} + (\lambda - f(x)) y = 0, \quad y(x) \to 0 \text{ as } |x| \to \infty,$$

with $f(x) = x^2 + \beta e^{-x^2}$ may be found by using the orthonormal set $\phi_n(x) = e^{-x^2/2} H_n(x)/\sqrt{h_n}$, where $h_n = 2^n n! \sqrt{\pi}$ and $H_n(x)$ is a Hermite polynomial that satisfies equation 11.30. By

---

† The reason for making this choice is that the eigenvalues of this function have been computed by various methods, see for instance Korsch and Laurent (1981), Paulsson *et al.* 1983, and also Pajunen and Child (1980) who have devised a scheme for deriving the function $f(x)$ from its spectrum.

Table 15.2. *Some eigenvalues for the function $f = x^2 + 18e^{-x^2}$ with $v = 1$.*

n	even parity Exact	even parity Uniform	odd parity Exact	odd parity Uniform
0	6.1508	6.1692	6.1570	6.1745
1	10.2766	10.2978	10.3287	10.3552
2	13.9428	13.9811	14.1968	14.2184
3	17.1524	17.2074	17.9358	17.9586
4	20.2447	20.2851	21.6485	21.6752
5	23.6597	23.6765	25.3913	25.4145
6	27.3241	27.3344	29.1799	29.1970

writing

$$y(x) = \sum_{k=0}^{N} c_k \phi_k(x)$$

show that the $N + 1$ Fourier coefficients $\mathbf{c} = (c_0, c_1, \ldots, c_N)$ and the eigenvalues of the differential equation are given by the matrix equation

$$M\mathbf{c} = \lambda \mathbf{c},$$

where the elements of the square, symmetric $(N + 1) \times (N + 1)$ matrix are

$$M_{ij} = (2i + 1)\delta_{ij} + \beta \int_{-\infty}^{\infty} dx\, e^{-2x^2} H_i(x)H_j(x).$$

Write a Maple procedure that constructs the matrix M and finds a list of eigenvalues λ_k of M, arranged in ascending order, for particular values of β. The integral may be evaluated using the relation

$$\int_{-\infty}^{\infty} dx\, e^{-2a^2x^2} H_n(x)H_m(x) = \frac{2^{(n+m-1)/2}}{a^{n+m+1}} \left(1 - 2a^2\right)^{(n+m)/2} \Gamma\left((n + m + 1)/2\right)$$

$$\times F\left(-m, -n; \frac{1 - n - m}{2}; \frac{a^2}{2a^2 - 1}\right),$$

which is valid if n and m have the same parity, where $F(a, b; c; z)$ is the hypergeometric function defined in chapter 6.

In addition find the eigenvectors of M and use these to construct and plot graphs of particular eigenfunctions of the original differential equation.

Note that you will need to experiment with the value of N; typically for $\beta = 18$ the first 20 eigenvalues are given to a reasonable accuracy with $N = 30$ and `Digits= 16`.

Exercise 15.25

Write a Maple program to compute the uniform approximation to the eigenvalues computed in exercise 15.24.

This is an involved computation with many subsidiary parts and you will need to write (and test) several procedures to evaluate various intermediate quantities. First you need the outer turning point $d(\lambda)$, which may initially be estimated using the perturbation expansion

$$d^2 = \lambda - \delta - \delta^2 - \frac{3}{2}\delta^3 - \frac{8}{3}\delta^4 + \cdots, \qquad \delta = \beta e^{-\lambda}.$$

Use this estimate as a first guess in `fsolve`. The inner turning point, $c(\lambda)$, is zero if $\lambda \geq \beta$ but needs to be found numerically if $\lambda < \beta$ using the knowledge that $0 < c < d$. The function $B(\lambda)$ may be approximated using the result given in exercise 15.22 (or exercise 15.37), or by numerical integration as for $A(\lambda)$. The function $A(\lambda)$ can, in principle, be evaluated directly using the `evalf(Int(*))` construction, but this is normally too slow for this type of use. Moreover, because the turning points are not known precisely the value of the integrand at a numerically computed turning point may be complex. It is therefore more efficient to note that the integrand has square root singularities at both ends and use either the appropriate Gauss quadrature method, described in chapter 11, or to write the integral in the form

$$A = \int_c^d dx\, \sqrt{(x-c)(d-x)}\sqrt{\frac{\lambda - f(x)}{(x-c)(d-x)}},$$

where c and d are the numerically generated numbers. The function inside the second square root is well behaved for $c \leq x \leq d$, and its value at $x = c$ and d is obtained using l'Hospital's rule. Now substitute $x = c\cos^2\phi + d\sin^2\phi$ and use Simpson's rule to evaluate the integral. The integral for $B(\lambda)$ may be computed in the same manner. For $|B| > 4$ the function $\chi(B)$ may be approximated by the asymptotic expansion $(24B)^{-1}$. Finally, the numerical solution of equation 15.79 may be found using the false position method described in chapter 3.

15.6.4 Mathieu's equation

We end this chapter by finding uniform approximations to the eigenvalues of Mathieu's equation,

$$\frac{d^2y}{dx^2} + (\lambda - 2q\cos 2x)y = 0,$$

valid through the difficult region $\lambda \simeq 2q$, where simple perturbation expansions fail. The analysis in this section is fairly lengthy, but it is described only briefly with many stages omitted.

First recall that the required eigenvalues are the values of λ that produce π- and 2π-periodic solutions to Mathieu's equation. In this example comparison with 15.1 shows that $f(x) = 2q\cos 2x$, so if $-2q < \lambda < 2q$ there are four turning points and if $\lambda > 2q$ there are none. The following diagram shows a graph of $f(x)$ and the turning point $x = a(\lambda)$ used in the subsequent analysis.

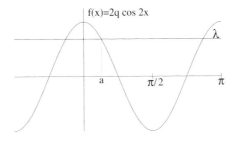

Figure 15.19   Graph of $f(x) = 2q\cos 2x$ showing the turning points for $\lambda < 2q$.

The ensuing analysis is made easier if the symmetries of Mathieu functions are used: these are summarised in the following table.

Table 15.3. *Symmetries of Mathieu functions.*

function	eigenvalue	period	parity about $x = 0$	parity about $x = \frac{\pi}{2}$
$ce_{2n}(x, q)$	$a_{2n}(q)$	π	even	even
$se_{2n+1}(x, q)$	$b_{2n+1}(q)$	2π	odd	even
$ce_{2n+1}(x, q)$	$a_{2n+1}(q)$	2π	even	odd
$se_{2n+2}(x, q)$	$b_{2n+2}(q)$	π	odd	odd

Recall also that

$$a_0(q) < b_1(q) < a_1(q) < b_2(q) < a_2(q) < \cdots, \quad q \neq 0.$$

Real turning points: $-2q < \lambda \leq 2q$

Because Mathieu functions have various symmetries we need consider only the interval $0 \leq x \leq \pi$ and an expansion about the turning point at $x = a$, shown in figure 15.19. It is convenient to write Mathieu's equation in the form $y'' + (4q \sin^2 x - 2q + \lambda)y = 0$; by expanding about $x = 0$ and comparing with equation 15.72 we see that $B > 0$ and that equation 15.74 (page 610) gives

$$\pi B = 2 \int_0^a dx \sqrt{2q - \lambda - 4q \sin^2 x}$$

$$= 4\sqrt{q} \left[E(s_a) - (1 - s_a^2)K(s_a) \right], \quad s_a = \sin a = \sqrt{\frac{2q - \lambda}{4q}}. \quad (15.82)$$

We need also the relation between x and ξ in the region $a < x < \pi - a$, where $\lambda - 2q \cos 2x > 0$; this is given by equation 15.73, which becomes

$$\int_a^x dx \sqrt{2q + \lambda - 4q \cos^2 x} = \frac{1}{4}\xi\sqrt{\xi^2 - 4B} - B \ln\left(\frac{\xi + \sqrt{\xi^2 - 4B}}{2\sqrt{B}} \right)$$

$$= \frac{1}{4}\xi^2 - B \ln \xi - \frac{B}{2}(1 - \ln B) + O(\xi^{-2}). \quad (15.83)$$

The even and odd solution of the comparison equation 15.72, here denoted respectively by $\psi_1(\xi, B)$ and $\psi_2(\xi, B)$, are defined in section 15.4.1. From equation 15.34 (page 590), we have

$$\psi_1(\xi, B) = c_1 [W(B, \xi) + W(B, -\xi)], \quad \text{even about } \xi = 0,$$
$$\psi_2(\xi, B) = c_2 [W(B, -\xi) - W(B, \xi)], \quad \text{odd about } \xi = 0,$$

where $c_{1,2}$ are constants that play no subsequent role. The asymptotic expansion of $W(B, \pm\xi)$, valid when $\xi \gg 2\sqrt{B}$, equations 15.35 and 15.36, then gives

$$\psi_{1,2}(\xi) \sim c\sqrt{\frac{2}{\xi}}\sqrt{k(B) + \frac{1}{k(B)}} \sin\left(\frac{1}{4}\xi^2 - B \ln \xi + \frac{1}{2}\Phi(B) + \frac{\pi}{4} \pm \delta(B) \right), \quad (15.84)$$

where

$$\Phi(B) = \arg \Gamma \left(\frac{1}{2} + iB \right), \quad k = \sqrt{1 + e^{2\pi B}} - e^{\pi B} \quad \text{and} \quad \tan 2\delta = e^{-\pi B}.$$

Using equation 15.13 (page 580), this provides the required approximation to $y(x)$ in the region where $\lambda - 2q \cos 2x > 0$,

$$y_{1,2}(x) = \frac{c}{(\lambda - 2q \cos 2x)^{1/4}} \sin \left(\int_a^x dt \sqrt{2q + \lambda - 2 \cos^2 t} + \chi(B) \pm \delta(B) \right), \quad (15.85)$$

where $\chi = \frac{1}{2}\Phi(B) + \frac{1}{2}B(1 - \ln B) + \frac{\pi}{4}$. Here $y_1(x)$ — the $+$ sign — is the even solution, associated with $ce_k(x, q)$, and $y_2(x)$ is the odd solution, associated with $se_k(x, q)$. In order to obtain equations for λ from this approximation we need only impose the symmetry conditions given in table 15.3.

First consider the π-periodic solutions: one of these, $ce_{2r}(x, q)$, is even and the other, $se_{2r}(x, q)$ is odd about both $x = 0$ and $\pi/2$. The periodicity condition is imposed by writing $F(\pi - x) = A - F(x)$, where

$$F(x) = \int_a^x dt \sqrt{2q + \lambda - 4q \cos^2 t} \quad \text{and}$$

$$A = \int_a^{\pi-a} dt \sqrt{2q + \lambda - 4q \cos^2 t}$$

$$= 4\sqrt{q} \left(E(c_a) - s_a^2 K(c_a) \right), \quad c_a = \cos a = \sqrt{\frac{2q + \lambda}{4q}}. \quad (15.86)$$

For future reference we note that A is just the integral 15.48 (page 595). This gives

$$y_{1,2}(\pi - x) = \frac{c}{(\lambda - 2q \cos 2x)^{1/4}} \sin \left(A + \chi(B) \pm \delta(B) - F(x) \right).$$

For the even, π-periodic solution, $y_1(-x) = y_1(\pi - x) = y_1(x)$, and for the odd, π-periodic solution, $y_1(-x) = y_1(\pi - x) = -y_1(x)$. These conditions imposed upon equations 15.85 and 15.86 give the following equations for λ:

$$A(\lambda) = \left(2k \pm \frac{1}{2} \right) \pi - \Phi(B) - B(1 - \ln B) \mp 2\delta(B), \quad k = 0, 1, 2, \ldots, \quad \begin{cases} a_{2k}(q) \\ b_{2k}(q) \end{cases}.$$

If $q \gg 1$ and $\lambda \simeq -2q$, B is large and, to a first approximation, this equation becomes $A = (2k \pm \frac{1}{2})\pi$, which is the same as given by equation 15.48. More precisely, in this limit $A \simeq \pi(2q + \lambda)/4\sqrt{q}$ so

$$a_{2k} \simeq -2q + 2(4k + 1)\sqrt{q}, \quad k = 0, 1, \ldots,$$

$$b_{2k} \simeq -2q + 2(4k - 1)\sqrt{q}, \quad k = 1, 2, \ldots.$$

For the 2π-periodic solutions we note that the even solution, ce_{2r+1}, is odd about $x = \frac{\pi}{2}$ and that the odd solution, se_{2r+1}, is even about $x = \frac{\pi}{2}$. The details of the calculation are left for the reader: the final result is

$$A(\lambda) = \left(2k \mp \frac{1}{2} \right) \pi - \Phi(B) - B(1 - \ln B) \mp 2\delta(B), \quad k = 0, 1, 2, \ldots, \quad \begin{cases} a_{2k-1}(q) \\ b_{2k+1}(q) \end{cases}.$$

In the limit $q \gg 1$ and $\lambda \simeq -2q$, $B \simeq 0$ and using the first-order expansion for $A(\lambda)$ given in exercise 15.27, we obtain

$$
\begin{aligned}
b_{2k+1} &\simeq -2q + 2(4k+1)\sqrt{q} \simeq a_{2k}, \quad k = 0, 1, \ldots, \\
a_{2k-1} &\simeq -2q + 2(4k-1)\sqrt{q} \simeq b_{2k}, \quad k = 1, 2, \ldots.
\end{aligned}
$$

This, and the similar result quoted above, is obtained by treating the regions $a < x < \pi - a$ in isolation, by setting $B = 0$, and neglecting the effect of the similar region $-(\pi - a) < x < -a$. In this limit the eigenvalues $a_r(q)$ and $b_{r+1}(q)$ are identical. In fact there is interference between these two region, because $B \neq 0$ — tunnelling in the language of quantum mechanics — and the eigenvalues of the π- and 2π-periodic solutions differ by a quantity that is exponentially small for large q and $\lambda \ll 2q$, equation 15.93 below.

No turning points: $\lambda > 2q$

When $\lambda > 2q$ the equation $\lambda - 2q \cos 2x = 0$ has no real roots and the relevant comparison equation is $\psi'' + (B + \xi^2/4)\psi = 0$, $B > 0$, where ξ and x are related through the equation

$$
\int_0^\xi ds \sqrt{B + \frac{1}{4}s^2} = \int_0^x dt \sqrt{\lambda - 2q + 4q \sin^2 t}. \tag{15.87}
$$

The integrands have square-root singularities at $s = \pm 2i\sqrt{B}$ and $t = \pm iv$, respectively, where $\sinh v = \sqrt{(\lambda - 2q)/4q}$, so that equation 15.75 becomes

$$
\begin{aligned}
\pi B &= 2 \int_0^v dw \sqrt{\lambda - 2q - 4q \sinh^2 w} \\
&= 2\sqrt{\lambda + 2q} \left(K(k) - E(k) \right), \quad k^2 = \frac{\lambda - 2q}{\lambda + 2q}. \tag{15.88}
\end{aligned}
$$

The required asymptotic expansions of the comparison equation solutions are

$$
\psi_{1,2}(B, \xi) \sim \frac{c}{\sqrt{\xi}} \sin\left(\frac{1}{4}\xi^2 + B \ln \xi + \frac{1}{2}\Phi(-B) + \frac{\pi}{4} \pm \delta(-B) \right).
$$

Relation 15.87, between ξ and x, can, for large ξ, be written as

$$
\frac{1}{4}\xi^2 + B \ln \xi + \frac{1}{2}B(1 - \ln B) + O(B^2/\xi^2) = \int_0^x dt \sqrt{\lambda - 2q + 4q \sin^2 t}, \quad \xi \gg 2\sqrt{B},
$$

which gives the required uniform approximation,

$$
\begin{aligned}
y_{1,2}(x) \sim \frac{c}{(\lambda - 2q \cos 2x)^{1/4}} \sin\Bigg(&\int_0^x dt \sqrt{\lambda - 2q + 4q \sin^2 t} \\
&+ \frac{1}{2}\Phi(-B) - \frac{1}{2}B(1 - \ln B) + \frac{\pi}{4} \pm \delta(-B) \Bigg). \tag{15.89}
\end{aligned}
$$

Equations for the eigenvalue λ are now obtained, as before, by imposing the relevant symmetry conditions summarised in table 15.3.

Summary and discussion

The final expressions for each type of eigenvalue may be written as a single expression, valid for $\lambda < 2q$ and $\lambda \geq 2q$. In ascending order of magnitude these are:

$$a_n(q): \qquad A(\lambda) = \left(n + \frac{1}{2}\right)\pi - \chi(B) - 2\delta(B),$$

$$b_{n+1}(q): \qquad A(\lambda) = \left(n + \frac{1}{2}\right)\pi - \chi(B) + 2\delta(B),$$

(15.90)

where $\chi(B) = \Phi(B) + B(1 - \ln|B|)$, $\Phi(B) = \arg\Gamma\left(1/2 + iB\right)$, $\tan 2\delta = e^{-\pi B}$,

$$A(\lambda) = \begin{cases} 4\sqrt{q}\left(E(c_a) - (1 - c_a^2)K(c_a)\right), & c_a = \sqrt{\dfrac{2q + \lambda}{4q}}, & -2q \leq \lambda < 2q, \\[4mm] 2\sqrt{\lambda + 2q}\, E\left(\sqrt{4q/(\lambda + 2q)}\right), & & \lambda \geq 2q \end{cases}$$

(15.91)

and

$$\pi B(\lambda) = \begin{cases} 4\sqrt{q}\left(E(s_a) - (1 - s_a^2)K(s_a)\right), & s_a = \sqrt{\dfrac{2q - \lambda}{4q}}, & -2q \leq \lambda < 2q, \\[4mm] -2\sqrt{\lambda + 2q}\left(K(k) - E(k)\right), & k = \sqrt{\dfrac{\lambda - 2q}{\lambda + 2q}}, & \lambda \geq 2q. \end{cases}$$

(15.92)

The function $B(\lambda)$ decreases from $4\sqrt{q}/\pi$ to zero as λ increases from $-2q$ to $2q$ and tends to $-\infty$ as λ increases beyond $2q$. Hence $\delta(B)$ is very small for $\lambda \ll 2q$, $(B \gg 1)$, and is close to $\pi/4$ for $\lambda \gg 2q$. Equations 15.90 therefore show that for $\lambda \ll 2q$, $b_{n+1} \sim a_n$ whilst for $\lambda \gg 2q$, $b_n \simeq a_n$, in accordance with the behaviour seen in figure 11.18 (page 414). The exponentially small differences between the eigenvalues in these regions is due to the exponentially small values of δ, for $\lambda \ll 2q$, and of $\pi/4 - \delta$ for $\lambda \gg 2q$.

The function $A(\lambda)$ increases monotonically with increasing λ: for $\lambda \simeq -2q$,

$$A(\lambda) = \frac{\pi(\lambda + 2q)}{4\sqrt{q}} + O((\lambda + 2q)^2), \quad \text{and for } \lambda \gg 2q, \quad A(\lambda) \sim \pi\sqrt{\lambda}.$$

Also $A(2q) = 4\sqrt{q}$.

For $\lambda \ll 2q$ the difference between close eigenvalues is given by

$$b_{n+1}(q) - a_n(q) \sim \frac{8}{\pi}q^{n/2 + 3/4}\left(\frac{32e}{2n + 1}\right)^{n + 1/2} e^{-4\sqrt{q}}, \quad \text{as} \quad q \to \infty.$$

(15.93)

For $\lambda \gg 2q$ the differences are

$$a_n(q) - b_n(q) \sim \frac{4n}{\pi}\left(\frac{e^2 q}{4n^2}\right)^n, \quad q \gg 1, \quad n \gg 2q \quad \text{and} \quad a_n \sim b_n \sim n^2.$$

(15.94)

Outline derivations of these results are given in exercises 15.27 to 15.29.

Numerical solutions of equations 15.90 are easily obtained because $A + \chi(B)/\pi \pm 2\delta(B)/\pi$ is a monotonic increasing function of λ and is numerically similar to $A(\lambda)$. In the following table we compare the values of $a_{2n}(q)$, for $q = 60$, computed using the numerical methods described in chapter 11 and by numerically solving equation 15.90.

Table 15.4. *Comparison of the uniform approximation to $a_{2n}(60)$ with exact values.*

n	0	1	2	3	4
a_{2n} exact	−104.8	−45.95	8.216	56.88	97.66
a_{2n} approximate	−104.7	−45.87	8.299	56.96	97.76

n	5	6	7	8
a_{2n} exact	124.5	157.6	205.5	263.2
a_{2n} approximate	124.6	157.7	205.6	263.2

In this case $q \gg 1$ and the relative accuracy of the uniform approximation is better than 1%. The difference between nearby eigenvalues is also given accurately, even if these differences are far less than the errors in the absolute value of the eigenvalues. This is illustrated in the following table, which gives values of $b_1(q) - a_0(q)$ for various q.

Table 15.5. *Values of the difference $\Delta(q) = b_1(q) - a_0(q)$. The top row is obtained using the numerical method described in chapter 11; the second row by numerically solving equation 15.90 for b_1 and a_0. For $q = 20$ the accuracy of these numerical methods was too poor to compute the difference: for the third row equation 15.93 was used.*

$\Delta \backslash q$	1	5	10	15	20
exact	3.49(−1)	9.97(−3)	4.27(−4)	3.43(−5)	3.90(−6)
equation 15.90	3.22(−1)	9.25(−3)	3.97(−4)	3.21(−5)	
equation 15.93	4.35(−1)	1.04(−2)	4.29(−4)	3.39(−5)	3.82(−6)

Exercise 15.26
Use the approximation $A \simeq \pi(2q + \lambda)/(4\sqrt{q})$, valid if $\lambda \simeq -2q$, and equations 15.90 to show that

$$a_r \sim b_{r+1} \sim -2q + 2(2r + 1)\sqrt{q}.$$

Exercise 15.27
Show that if $\lambda = -2q + \Delta$ with Δ small,

$$\pi A(\lambda) = \frac{\Delta}{4q^{1/2}} + \frac{\Delta^2}{128q^{3/2}} + O(\Delta^3)$$

$$B(\lambda) = \frac{4\sqrt{q}}{\pi} - \frac{\Delta}{4\pi\sqrt{q}}\left(1 + \ln\left(\frac{64q}{\Delta}\right)\right) + \cdots.$$

Using these expansions, equations 15.90, the approximation $\Phi(u) + u(1 - \ln|u|) \sim (24u)^{-1}$, $|u| \gg 1$, and ignoring $\delta(B)$, show that

$$\lambda = a_r \simeq b_{r+1} \simeq -2q + 2R\sqrt{q} - \frac{1}{8}\left(\frac{1}{3} + R^2\right) + \cdots, \quad R = 2r + 1.$$

Exercise 15.28

Using equations 15.90 show that when $q \gg 1$ and $B \gg 1$ the difference between $b_{n+1}(q)$ and $a_n(q)$ for $\lambda \ll 2q$ is given by

$$b_{n+1}(q) - a_n(q) \sim \frac{4\delta(B)}{A'(\lambda) + \chi'(B)B'(\lambda)},$$

where all quantities are evaluated at the value of λ derived in the previous question. Use this approximation to derive 15.93.

Exercise 15.29

Show that if $\lambda \gg 2q$,

$$A(\lambda) \sim \pi\sqrt{\lambda}\left(1 - \frac{q^2}{4\lambda^2} - \frac{15q^4}{64\lambda^4} - \frac{105q^6}{256\lambda^6} + \cdots\right),$$

$$\pi B(\lambda) \sim -\sqrt{\lambda}\left\{\ln\left(\frac{4\lambda}{e^2 q}\right) - \left(\frac{q}{2\lambda}\right)^2 \ln\left(\frac{4\lambda e^3}{q}\right) + \cdots\right\}.$$

Use these expansions to deduce that

$$a_n \sim b_n \sim n^2 + \frac{q^2}{2n^2} + \frac{5q^4}{32n^6} + \cdots,$$

and to derive equation 15.94.

Exercise 15.30

In exercise 12.18 the equation

$$\frac{d^2x}{dt^2} + \frac{dx}{dt}\sin t + (\omega^2 + a\cos t)x = 0$$

was considered. Using the results obtained in exercise 15.1 show that this may be cast in the form

$$\frac{d^2v}{d\tau^2} + (\lambda - p(\tau))v = 0, \quad p(\tau) = 4\left(a - \frac{1}{2}\right)\cos 2\tau - \frac{1}{2}\cos 4\tau,$$

where $t = 2\tau + \pi$, $\lambda = 4\omega^2 - \frac{1}{2}$ and $x = v(\tau)\exp\left(\frac{1}{2}\cos 2\tau\right)$.

Explain why the theory described in this section cannot be used to approximate the eigenvalues of this problem if $a = 1$. In this case use the method described in section 11.5.2 to find the eigenvalues for all types of solutions and show that the first ten eigenvalues are

$$\omega = 0, 0.284\,48, 0.777\,02, 1.0796, 1.0796, 1.5426, 1.5500, 2.0332, 2.0332, 2.5259.$$

15.7 Conclusions

The results presented in this chapter have been chosen to illustrate how the first term in the asymptotic expansion of the eigenvalues and eigenfunctions of the Sturm–Liouville system

$$\frac{d^2y}{dx^2} + v^2(\lambda - f(x))y = 0$$

may be found. Other applications and further developments of this theory may be found in the books by Dingle (1973), Heading (1962), Fröman and Fröman (1967) and

Olver (1974). We have concentrated on the first-order term in the asymptotic expansion of both the eigenvalues and eigenfunctions because this is the easiest term to find; it is also often the only term that can be found without significant labour. In the simple case of two isolated turning points, however, there is an elegant result, due to Dunham (1932) who showed that the complete asymptotic expansion for the eigenvalues can be obtained from the series

$$\oint_{\mathscr{C}} dz \sqrt{\lambda - f(z)} + \sum_{k=1}^{\infty} \frac{1}{v^{2k}} \oint_{\mathscr{C}} dz \, u_{2k}(z) = \frac{2\pi}{v} \left(n + \frac{1}{2} \right), \tag{15.95}$$

which is the appropriate generalisation of equation 15.48 (page 595). Here $u_{2k}(z)$ are the functions defined in exercise 15.8 (page 578); note that these functions also depend upon λ. A derivation of this expression is outlined in exercise 15.39. This series generally provides an asymptotic expansion to the nth eigenvalue. There are, however, some functions $f(x)$ for which the series terminates to give exact eigenvalues. A list of these functions and associated eigenvalues is given in the following table.

Table 15.6. *Functions for which the asymptotic expansion 15.95 yields exact eigenvalues, adapted from Rosenzweig and Krieger (1968).*

Function $f(x)$	Range of x	Eigenvalues
$\omega^2 x^2$	$(-\infty, \infty)$	$(2n+1)\omega/v$
$\dfrac{\beta^2}{x^2} - \dfrac{2\alpha}{x}$	$(0, \infty)$	$-\left(\dfrac{2\alpha v}{\beta v + n + 1/2} \right)^2$
$\omega^2 x^2 + \dfrac{\beta^2}{x^2}$	$(0, \infty)$	$2\omega \left(\beta + (2n+1)/v \right)$
$-\alpha^2 / \cosh^2(x/\beta)$	$(-\infty, \infty)$	$-\left(\alpha - \dfrac{n+1/2}{v\beta} \right)^2$
$\alpha^2 \tan^2(\pi x/2\beta)$	$(-\beta, \beta)$	$\dfrac{(2n+1)\pi}{4v\beta} \left(\dfrac{(2n+1)\pi}{4v\beta} + 2\alpha \right)$
$\alpha^2 \left(1 - e^{-\beta x} \right)^2$	$(-\infty, \infty)$	$\alpha^2 - \left(\alpha - (n+1/2)\beta/v \right)^2$

Even when the series does not terminate it sometimes provides a very accurate estimate of the eigenvalues, even for small values of n. The case $f(x) = x^4$ is treated in Bender and Orszag (1978, section 10.7) and there it is shown, for instance, that for $n = 1$ the first two terms of the asymptotic expansion give 10% accuracy and for $n = 2$ these terms are accurate to 1 part in 10^5; for $n = 6$ the first ten terms are accurate to 7 parts in 10^{11}. The first three terms of this expansion are derived in exercise 15.36. A more significant result of Voros (1983) shows that the asymptotic expansion 15.95 can, under certain circumstances, be re-summed by allowing v to be complex to give *exact* results.

Finding more terms in the asymptotic expansion of eigenfunctions is usually difficult. When there is only one isolated turning point, Olver (1974, chapter 11, sections 7 and 11) shows how the method used in section 15.5 may be extended to find the asymptotic expansion. In general, however, only the first term in the asymptotic expansion can be obtained, though this often provides a good simple approximation, adequate for many applications.

The generalisation of this one-dimensional theory to many-dimensional systems, for instance the partial differential equation

$$\frac{\partial^2 F}{\partial x^2} + \frac{\partial^2 F}{\partial y^2} + v^2(\lambda - f(x, y))F = 0,$$

with $F(x, y)$ satisfying given boundary conditions, is far from straightforward, except when the equation is separable.† Most systems are not separable, then the substitution $F = \exp(iv\phi(x, y))$ leads to a nonlinear, first-order partial differential equation. If this is integrable, then for some systems the method of Maslov (1972, see also Percival 1977) may be used to obtain the first two terms in the asymptotic expansion of the eigenvalue, but no general method exists for higher-order terms. If the system is not integrable, some solutions for $\phi(x, y)$ may be chaotic and these solutions affect the nature of the eigenvalues; discussions of this problem in relation to quantum mechanical systems may be found in Gutzwiller (1990), Ozorio de Almeida (1988), Reichl (1992) and Tabor (1989).

15.8 Exercises

Exercise 15.31
Throughout this chapter we have encountered integrals of the form

$$F(\epsilon) = \oint_{\mathscr{C}} dz \, \sqrt{g(z)}, \quad g(z) = 1 - z^2 + \epsilon f(z),$$

where $f(z)$ is real on the real axis and $g(z)$ has real zeros, x_+ close to ± 1, and possibly others that are far from the origin. The integrand has branch points at $z = x_\pm$ and the contour $\mathscr{C}$ encloses these and no other singularities; in other words the integrand has the same analytic structure as $\sqrt{1 - z^2}$, inside $\mathscr{C}$.

For real z, $\sqrt{g(z)}$ is real if $x_- \leq z \leq x_+$ and is negative on the upper side of the cut; it is imaginary outside this range with $\arg(g) = \pi/2$ for $z > x_+$ and $\arg(g) = -\pi/2$ for $z < x_-$.

For small $|\epsilon|$ we expect $F(\epsilon)$ to possess a Taylor's series. By considering the related function

$$G(\epsilon, \lambda) = \oint_{\mathscr{C}} dz \, \sqrt{\lambda - z^2 + \epsilon f(z)},$$

where λ is a real variable, and by noting that

$$\frac{1}{(\lambda - z^2)^{n-1/2}} = (-1)^{n-1} \frac{2^{2n-2}(n-1)!}{(2n-2)!} \frac{d^{n-1}}{d\lambda^{n-1}} (\lambda - z^2)^{-1/2}, \quad n \geq 2,$$

---

† For a separable system the solution may be expressed as the product $F = X(x)Y(y)$, with each component a function of one independent variable satisfying an ordinary differential equation. Whether or not a system is separable depends upon the equation *and* the boundary conditions.

show that

$$F(\epsilon) = \pi + \epsilon \int_{-\pi/2}^{\pi/2} d\phi \, f(\sin\phi) + \frac{1}{\sqrt{\pi}} \sum_{n=2}^{\infty} (4\epsilon)^n \frac{\Gamma(n+1/2)}{(2n)!} \frac{d^{n-1}}{d\lambda^{n-1}} \int_{-\pi/2}^{\pi/2} d\phi \, f(\sqrt{\lambda}\sin\phi)^n,$$

where λ is set to unity after all computations have been completed.

Exercise 15.32
Using the expressions given in equations 15.7 (page 576), show that, if $g(x) > 0$,

$$\phi_2(x) = -\frac{1}{8} \int dx \, \frac{1}{g(x)^{1/4}} \frac{d}{dx} \left(\frac{g'(x)}{g(x)^{5/4}} \right),$$

$$\phi_3(x) = -\frac{i}{16g(x)^{3/4}} \frac{d}{dx} \left(\frac{g'(x)}{g(x)^{5/4}} \right),$$

$$\phi_4(x) = \frac{i\phi_3'(x)}{\sqrt{g(x)}} - \frac{1}{128} \int dx \, \frac{1}{g(x)} \left[\frac{d}{dx} \left(\frac{g'(x)}{g(x)^{5/4}} \right) \right]^2,$$

$$\phi_5(x) = \frac{i}{2} \left(\frac{\phi_4'(x)}{\phi_0'(x)} - \frac{1}{2} \left(\frac{\phi_2'(x)}{\phi_0'(x)} \right)^2 \right).$$

Exercise 15.33
Substitute the Taylor expansion of $g(x)$ about $x = b$ into the integral 15.18 to show that near the turning point, $x = b$,

$$\xi(x) = |g'(b)|^{1/3}(x-b) \left\{ 1 - \frac{g''(b)}{10|g'(b)|}(x-b) - \left[\frac{g'''(b)}{42|g'(b)|} + \frac{2}{175} \left(\frac{g''(b)}{g'(b)} \right)^2 \right] (x-b)^2 + \cdots \right\}.$$

Exercise 15.34
Show that the system

$$\frac{d^2y}{dx^2} + \left(\lambda - |x^\alpha| \right) y = 0, \quad |y| \to 0 \quad \text{as} \quad |x| \to \infty,$$

where α is a positive real number, has eigenvalues given approximately by

$$\lambda_n = \left(\alpha\sqrt{\pi} \frac{\Gamma(3/2 + 1/\alpha)}{\Gamma(1/\alpha)} \right)^{2\alpha/(2+\alpha)} \left(n + \frac{1}{2} \right)^{2\alpha/(2+\alpha)}.$$

Exercise 15.35
Show that if $f(x) = x^4$ the approximate eigenvalues given by equation 15.48 (page 595) are

$$\lambda_n^{3/4} = \frac{3}{\sqrt{2\pi}} \Gamma\left(3/4\right)^2 \left(n + \frac{1}{2} \right), \quad n = 0, 1, \ldots.$$

Write a Maple program to find numerical approximations to the eigenvalues and show that the relative accuracy of the lowest three eigenvalues given by this formula are respectively 22%, 1% and $\frac{1}{2}\%$.

Hint: the numerical methods outlined in either exercise 15.21 or 15.24 may be used.

Exercise 15.36

For the function $f(x) = x^4$ show that the first two expansion coefficients defined in exercise 15.8 (page 578) are

$$u_2(x) = \frac{x^2(2x^4 + 3\lambda)}{2(\lambda - x^4)^{5/2}}, \quad u_4(x) = -\frac{112x^{12} + 666\lambda x^8 + 321\lambda^2 x^4 + 6\lambda^3}{8(\lambda - x^4)^{11/2}}$$

and find $u_6(x)$. Hence show that the asymptotic expansion 15.95 for this function is

$$n + \frac{1}{2} = \lambda^{3/4}\frac{\sqrt{2\pi}}{3\Gamma(3/4)^2} - \frac{\sqrt{2}\Gamma(3/4)^2}{8\pi^{3/2}\lambda^{3/4}} + \frac{11}{1536}\frac{\sqrt{2\pi}}{\Gamma(3/4)^2\lambda^{9/4}} + \cdots$$

and hence that

$$\lambda_n = \left[\frac{3\Gamma(3/4)^2}{\sqrt{2\pi}}N\right]^{4/3}\left(1 + \frac{1}{9\pi N^2} - \frac{11\pi^2 + 60\Gamma(3/4)^8}{7776\pi^2\Gamma(3/4)^8 N^4} + \frac{11(93\pi^4 - 40\Gamma(3/4)^8)}{34\,992\pi^2\Gamma(3/4)^8 N^6} + \cdots\right),$$

where $N = n + \frac{1}{2}$. Note that the expansion 15.95 is particularly easy to apply in this case because the eigenvalue λ may be scaled out of the integrals.

Exercise 15.37

Use the method described in exercise 15.31 to show that the first few terms of the Taylor's series of the function B, defined in equation 15.75 (page 610), are

$$\frac{B}{v} = \frac{\delta}{2a_2} + \left\{\frac{5f_3^2}{768a_2^7} + \frac{f_4}{128a_2^5}\right\}\delta^2$$

$$+ \left\{\frac{35f_4^2}{73\,728a_2^9} + \frac{385f_3^4}{884\,736a_2^{13}} + \frac{35f_3^2f_4}{24\,576a_2^{11}} + \frac{7f_3f_5}{9216a_2^9}\right\}\delta^3 + \cdots,$$

where $\delta = f(x_m) - \lambda$, f_k is the kth differential of $f(x)$ evaluated at the maximum, $x = x_m$ and $a_2 = \sqrt{|f_2|/2}$.

For the example dealt with in exercise 15.23 make a numerical comparison of this approximation with the exact results derived in that exercise.

Exercise 15.38

This exercise shows how the resonance phenomenon seen in figures 15.8 and 15.9 occurs in other circumstances.

Consider an elastic string of length L fixed at each end and free to to vibrate in a plane. A weight of mass M is firmly attached to the string a distance $a = L\delta$ from one end. The equation of motion for the string is, on neglecting gravity,

$$\frac{\partial^2 y}{\partial x^2} - \frac{1}{c^2}\frac{\partial^2 y}{\partial t^2} = 0, \quad T = \rho c^2,$$

where ρ is the string density, T the tension in the stationary string and $y(x, t)$ the amplitude of small oscillations. If $y_1(x, t)$ and $y_2(x, t)$ are respectively the solutions for $0 \le x \le a$ and $a \le x \le L$, the boundary conditions are $y_1(0, t) = y_2(L, t) = 0$, $y_1(a, t) = y_2(a, t)$, for all t, and the equation of motion for the mass gives

$$M \frac{\partial^2 y}{\partial t^2} = T \left(\frac{\partial y_2}{\partial x} - \frac{\partial y_1}{\partial x} \right), \quad \text{at} \quad x = a.$$

By assuming solutions of the form

$$y_1(x, t) = A \sin \frac{\omega}{c} x \cos(\omega t + \alpha), \quad 0 \le x \le a = L\delta,$$

$$y_2(x, t) = B \sin \frac{\omega}{c}(L - x) \cos(\omega t + \alpha), \quad a \le x \le L,$$

show that the frequencies Ω are given by the equation $\epsilon \Omega \sin \Omega \delta \sin \Omega (1 - \delta) = \sin \Omega$, where ϵ and Ω are the dimensionless parameters $\epsilon = M/(L\rho)$, $\Omega = \omega L/c$ and $B \sin \Omega (1 - \delta) = A \sin \Omega \delta$. Further, show that

(i) if $M = 0$ this reduces to the normal equation for such a string,
(ii) if $M \to \infty$ it reduces to the equation for a string of length $L - a$, fixed at $x = L\delta = a$ and $x = L$.

For the case $\epsilon = 0.2$, $\delta = 0.05$ use Maple to compute the first 50 eigen-frequencies and plot the graphs of some eigenvectors. In particular, show that the 19th and 21st eigenfunctions look like those shown below.

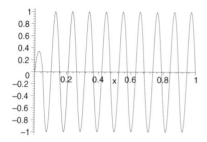

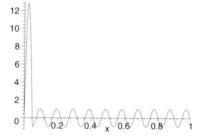

Figure 15.20   19th eigenfunction.   Figure 15.21   21st eigenfunction.

Show also that the low frequency solutions are practically unaffected by the mass and that for most high frequency eigenfunctions the displacement of the string for $0 \le x \le L\delta$ is relatively small.

Exercise 15.39
Consider the eigenvalue problem

$$\frac{d^2 y}{dx^2} + v^2(\lambda - f(x))y = 0$$

where $f(x)$ has a single minimum and $f(x) \to \infty$ as $|x| \to \infty$, so there are only two turning points, as in case (a) of section 15.5. The nth eigenfunction, $y(x)$, has n simple zeros, so Cauchy's residue theorem gives

$$\oint_\mathscr{C} dz \, \frac{y'(z)}{y(z)} = 2\pi i n,$$

where $\mathscr{C}$ is a contour containing both turning points. Using this and the expansion defined in exercise 15.8 (page 578), show that

$$2\pi n = v \oint_{\mathscr{C}} dz\, u(z) - \frac{1}{2i} \oint_{\mathscr{C}} dz\, \frac{u'(z)}{u(z)}.$$

Use the fact that $u(z)^2$ has two simple zeros inside $\mathscr{C}$, because it is similar to $\lambda - f(z)$, to deduce that the value of the second term in this expression is π and hence that

$$\oint_{\mathscr{C}} dz\, \sqrt{\lambda - f(z)} + \sum_{k=1}^{\infty} \frac{1}{v^{2k}} \oint_{\mathscr{C}} dz\, u_{2k}(z) = \frac{2\pi}{v}\left(n + \frac{1}{2}\right).$$

16

Dynamical systems I

16.1 Introduction

Differential equations are one of the principal means of describing changing phenomena and have consequently played a dominant role in the natural sciences since the time of Newton. In the century following Newton the understanding of differential equations and their solutions developed rapidly, and it was then that many of the methods and techniques now taught in conventional calculus courses were originally conceived. The brief history presented in this introduction serves partly to remind you of the various types of differential equations that exist and have solutions, and partly to put the theory described in this and the remaining chapters in context: some exercises on these standard equations are set at the end of this introduction. More detailed historical accounts may be found in the books by Barrow-Green (1996), Boyer (1968), Cajori (1991) and Kline (1972).

The seventeenth and eighteenth centuries saw a remarkable increase in our knowledge, and it is difficult for the modern student to appreciate the work of the natural philosophers of that era because so much of what they achieved is now taken for granted. Much of the analysis taught in schools and elementary undergraduate courses and almost all text books on dynamics written before 1970, for instance, contain work developed mainly during this period. The logarithm and exponential functions were defined and the trigonometric addition formulae, $\sin(x \pm y)$ for example, were discovered. Problems from the physical world provide a rich source of mathematical problems and the methods of Newton, Leibniz and subsequent mathematicians provided the tools which allowed some of them to be solved. Typical problems from this period are the nature of the vibrations of a violin string, the shape of an elastic beam or membrane under load, the shape of a hanging rope or chain, the amplitude dependence of the pendulum period and the accurate predications of lunar motion.

These problems require the understanding of phenomena changing in either space or time, or both, and naturally give rise to differential equations. But at the end of the seventeenth century this type of equation was new and its solutions poorly understood, indeed there was a vigorous debate as to what actually constituted a solution. Hence not only did mathematicians have to formulate the physical ideas in terms of new concepts but also to find acceptable solutions to these unfamiliar equations.

I. Newton
(1642–1727)
Newton in 1671 discussed fluxional equations† and classified them into three classes. In modern notation the first class is comprised of equations containing two fluxions

---

† In modern notation a fluxion is a differential of a fluent, the name Newton gave to dependent variables.

but only one fluent:

$$\frac{\dot{y}}{\dot{x}} = \frac{dy}{dx} = f(x) \quad \text{or} \quad \frac{dy}{dx} = f(y).$$

The second class is comprised of equations which contain two fluxions and two fluents:

$$\frac{\dot{y}}{\dot{x}} = \frac{dy}{dx} = f(x, y).$$

Note the use of the dot notation, originally due to Newton. The third class are now known as partial differential equations. Newton's general method of solving these equations was to expand the right hand side as a power series in x and y and then to develop the solution $y(x)$ as a power series in x. In finding solutions of this type Newton observed that there was always an arbitrary constant that could not be determined; and that for equations of the form $d^n y/dx^n = f(x)$ the solution was undetermined to within a polynomial of degree n.

James Bernoulli showed, in 1690, that the isochrone problem — the problem of finding a curve such that the time taken for a particle to slide from a point to the minimum is independent of the starting point† — is equivalent to solving the differential equation

James Bernoulli (1654–1705)

$$\frac{dy}{dx} = \frac{a^{3/2}}{\sqrt{b^2 y - a^3}},$$

where a and b are constants. He solved this problem and in the same paper proposed the problem of finding the shape of the curve assumed by a flexible, inextensible cord freely hung between two fixed points, that is, the catenary; in the same work the term *integral* appeared for the first time. In 1691 Leibniz, Huygens and John Bernoulli (brother of James) all published independent solutions. John Bernoulli used separation of variables, a technique independently discovered by Leibniz in 1691 when he showed that the equation

G. W. L. Leibniz (1647–1716)
C. Huygens (1629–1695)
John Bernoulli (1667–1748)

$$y\frac{dx}{dy} = X(x)Y(y)$$

can be solved by quadrature, that is reduced to an integral. He also discovered a method of solving the first-order homogeneous differential equations

$$\frac{dy}{dx} = g(y/x). \tag{16.1}$$

The equation

$$\frac{dy}{dx} + yP(x) = y^n Q(x), \quad \text{(Bernoulli's equation)}, \tag{16.2}$$

was proposed for solution by James Bernoulli in 1695. Leibniz (in 1696) pointed out that the change of variable $z = y^{1-n}$ reduced this to a linear equation; John Bernoulli found another method, described in Ince (1956, page 532).

The mathematical description of physical phenomena is probably the most important

---

† The isochrone problem was important in designing a pendulum clock with a pendulum that oscillated with a period independent of its amplitude. Huygens developed such a pendulum in 1673, a diagram of which is in Boyer (1968, page 413).

source of new mathematical problems. In particular, various types of differential equation have been introduced via physical problems: vibrating media both linear and nonlinear, fluid motion and planetary motion are a just a few examples. But geometry is also a rich source of interesting equations. In the 1720s the Italian Count Jacopo Riccati, in his investigations of two-dimensional curves with specified properties, was led to the consideration of differential equations of the general form $f(y, y', y'') = 0$. By regarding y as the independent variable, defining a new variable $p = dy/dx$ and using the relation $y'' = p\,dp/dy$, he cast these equations into the form $f(y, p, p\,dp/dy) = 0$, which is first-order in p rather than second-order in y. However, the particular equation for which Riccati is known is

J. F. Riccati
(1676–1754)

$$\frac{dy}{dx} = A(x) + B(x)y + C(x)y^2, \quad \text{(Riccati's equation)}. \tag{16.3}$$

Daniel Bernoulli
(1700–1782)

L. Euler
(1707–1783)

Riccati's equation was studied by many including the Bernoullis. Daniel Bernoulli (son of John) originally concealed his solution in an anagram yet to be deciphered, though he published the solution in 1724; Watson (1966, chapter 1) gives a short account of the history of this equation. Euler was the first to point out that if a particular solution, $v(x)$, is known, then the substitution $y = v(x) + 1/z$ converts this nonlinear equation into a linear equation for $z(x)$, so the general solution may be found, exercise 16.4.

Euler was responsible for many of the methods in common use today: the integration factor method, systematic methods of solving linear high-order equations with constant coefficients and the distinction between general and particular solutions are all, at least in part, due to Euler.

Much of the early work was concerned with techniques of solving equations, that is, of finding solutions in terms of known functions, but by the beginning of the nineteenth century the emphasis was changing with the realisation that it is impossible to integrate most differential equations in this manner. Cauchy in particular was asking more general questions, about existence and uniqueness of solutions for instance, which led to the creation of the theory of linear differential equations in the complex plane.

A.-L. Cauchy
(1789–1857)

This theory naturally arose from the linear partial differential equations used to describe a variety of physical problems, for example heat flow, vibrating strings and membranes and sound waves. These phenomena are described by linear equations because the disturbance considered is relatively small. They are second-order in time because of Newton's law of motion, and in space because of the nature of the media considered, although higher-order spatial derivatives are often important, for instance in the vibrations of elastic media, a problem first considered by Daniel Bernoulli. Linear partial differential equations could be tackled only by the method of separating variables and, if separable, gave rise to equations of the form

$$a(x)\frac{d^2y}{dx^2} + b(x)\frac{dy}{dx} + c(x)y = d(x), \tag{16.4}$$

with appropriate boundary conditions and where the coefficients are known functions. The study of these Sturm–Liouville problems, considered in chapter 10, gave rise not only to the theory of special functions, some of which are considered in chapter 11,

but also to the very powerful complex variable techniques of Cauchy and Riemann G. F. B. Riemann
and the theory of Fourier series. (1826–1866)

In parallel with this development was the study of dynamical systems, then manifest
in the problems of Celestial Mechanics, which also started with Newton's work. Prior
to the publication of the *Principia* in 1687 Celestial Mechanics had been an empirical
subject with predictions based on extrapolations from careful observations, the culmi-
nation being Kepler's three laws. Newton changed the emphasis of science: his three
laws of motion and the law of gravitational attraction solved a central problem of
astronomy and created the ethos of modern physics. A lively account of this work, both
the mathematics of Newton put in a modern context and the role of other scientists
involved, particularly Hooke, is provided by Arnold (1990). These basic laws enabled R. Hooke
Newton to derive the shape of the planetary orbits from fundamental principles. (1635–1702)
Newton's published derivation is based on geometric arguments, rather than the analysis he knew, and a modern attempt at reproducing this derivation is given in *Feynman's
Lost Lecture* (Goodstein and Goodstein 1997).

Halley,† who encouraged and financed the publication of the *Principia*, was appar- E. Halley
ently the first to use Newton's laws by noting that a comet had been seen in 1531, 1607 (1656–1742)
and 1682 and concluded that rather than separate events these were periodic appear-
ances of the same body and predicted the return in 1758. The French mathematician
Clairaut made approximate computations of the effect of Saturn and Jupiter on the A. C. Clairaut
orbit and predicted the date of return to its perihelion to be April 1759; its actual (1713–1765)
return in March, just within the errors quoted by Clairaut, provided early confirmation
that Newton's laws were more widely applicable than originally thought.

It is a minor understatement to say that Clairaut made approximate calculations,
and gives no idea of the importance attached to this prediction. It must be remembered
that even Newton was not clear about the general applicability of his laws, and that
the accurate prediction of the return date of this comet was their first major test. In
fact in 1748 Clairaut, Lalande and Lapaute sequestered themselves to calculate the
return date of the comet, last seen in 1682. Lalande wrote 'During the six months we
calculated from morning to night, sometimes even at meals; the consequence of which
was, that I contracted an illness which changed my constitution for the rest of my life.'
The difficulty was created by the need to account for the effects of Jupiter and Saturn
and the fact that probably the only calculating aids they had were logarithms, see Gear
and Skeel (1990) and Grossman (1997).

It was, however, the three-body problem, either two planets and the Sun or the
Sun–Earth–Moon system, that provided the most significant challenge. It led to major
advances in the techniques of solving nonlinear differential equations and ultimately
to the more general understanding of the behaviour of their solutions, a subject that
remains an active area of research; an excellent account of the history of this problem

---

† Edmond Halley is today best known by the comet named after him. He achieved far more and did
innovative research in geophysics, navigation, astronomy and mathematics; Halley's method of iteratively
solving equations is described in exercise 3.62 (page 130). He made many more discoveries of lasting
importance, see for example Grossman (1997, page 52), and for a biography setting Halley's work in the
context of his time, Cook (1998).

is given by Barrow-Green (1996, chapter 2). The interest in this problem was not just theoretical; the accurate prediction of lunar motion was considered the most likely method of accurate determination of longitude at sea, though it was the development of an accurate clock by Harrison that eventually solved this problem, see for instance Brown (1949 and 1960) and Sobel (1995).

By the mid-nineteenth century it had become clear that the solutions for most initial conditions could not be expressed in terms of finite series of known functions. For short times the series approximations that had been developed were adequate, though in 1877 Hill, the American mathematical astronomer, noted the discrepancy between the computed value of the lunar perigee† and the values derived from observation, and queried whether this was due to inaccuracies of the calculation or due to the presence of other forces that had been ignored. As a consequence Hill set about developing an entirely new method of calculation, which had a significant effect on Poincaré's work and today provides an important method of determining the stability of periodic orbits: examples of this use were discussed in sections 12.5 and 12.6.

G. W. Hill
(1838–1914)

H. Poincaré
(1854–1912)

The central question was whether or not the infinite series solutions could converge; a related and important question was that of determining whether the Solar System was stable, a problem discussed in the appendix of chapter 17. It transpired that these questions could not be answered with the available mathematics, and the major achievement of Poincaré was to develop important new ideas that helped resolve some of these problems. However, only in 1952 did Kolmogorov, Arnold and Moser prove that for some initial conditions and in some circumstances convergent perturbation expansions existed; the proof of this result, the KAM theorem, is very long and difficult but a brief description of it can be found in Arnold (1978, appendix 8, see also 1963 for a simple technical discussion and examples), Lichtenberg and Lieberman (1983) and Guckenheimer and Holmes (1983).

The three-body problem is described by nonlinear equations and these cannot be solved explicitly in terms of known functions, so the problem of stability cannot be solved by examining the solutions. Poincaré therefore attempted to understand the long-time behaviour of the solution by examining the equations of motion, and the theory he developed, called the qualitative theory of differential equations, was presented in a set of four papers in 1881, 1882, 1885 and 1886. The theory in this and the next chapter comes mainly from the 1881 paper.

Poincaré started his studies with the nonlinear equation

$$\frac{dy}{dx} = \frac{Y(x, y)}{X(x, y)} \tag{16.5}$$

and considered only real variables — a radical departure from contemporary practice — the motivation apparently being that equations of the form $\dot{x} = X(x, y)$, $\dot{y} = Y(x, y)$ are amongst the simplest nonlinear systems not having known solutions and that equation 16.5 is obtained from these by eliminating time. Such equations have subsequently been shown to describe a wide variety of phenomena and are worthy of

---

† The lunar apsides, the apogee and perigee, are the two points on the orbit that are furthest and closest, respectively, from the Earth.

study in their own right. Examples of such equations arise in population models (Lotka 1956, Maynard Smith 1977 and Murray 1989), in the description of some chemical reactions, astronomy and electrical circuits.

In this and the next chapter we present some of the techniques needed to understand the solutions of the equations $\dot{x} = X(x, y)$, $\dot{y} = Y(x, y)$, where x and y are real variables and (X, Y) are real sufficiently well behaved functions. These nonlinear equations present no problem to the modern numerical analyst and a small personal computer can be used to find accurate solutions numerically, at least for finite times. Why then is it necessary to understand the theory and techniques developed in an age before such computations were possible? For some problems, if the equations of motion and initial conditions are precisely defined, a numerical solution is often adequate and sometimes preferable, and there are many complicated problems for which this is the only method of attack. But it is worth remembering that there are also many common problems, where instabilities are present, for which most numerical solutions are necessarily wrong after quite short times. More generally, in most physical problems the equations of motion contain parameters and it is necessary to know how the solutions depend upon these as well as the initial conditions, and we may need to know how solutions behave as $|t| \to \infty$. Numerical solutions, whilst helpful, can rarely shed light on these problems, so some qualitative understanding of the system is necessary.

In practice one needs both numerical solutions and mathematical analysis and one aim of this chapter is to show, by example, why this is the case. The present chapter provides elementary background theory and starts with the simplest systems of all, first-order systems, in order to introduce some ideas which are then used in more complicated problems. The fixed point analysis introduced here is important as fixed points essentially organise the global behaviour of the system and are readily understood. In chapter 17 we consider in some detail the periodic orbits as these are the next level of complexity which can be understood and, when they exist, are important; in chaotic systems fixed points and periodic orbits are two of the features that allow us to understand the motion.

Exercise 16.1
Use the method of separation of variables to find the general solution of the equations

$$\text{(i) } \frac{dy}{dx} = x^3 y^2, \quad \text{(ii) } \frac{dy}{dx} = \frac{x(1-y)}{y(a-x)}, \quad \text{(iii) } \frac{dy}{dx} = \frac{x+ay}{ax-y},$$

where a is a positive constant.

Exercise 16.2
The homogeneous equation defined in equation 16.1 can be put in separable form by defining a new dependent variable $v(x)$ by $y = xv$. Use this method to find the general solution of the equations

$$\text{(i) } \frac{dy}{dx} = \frac{x-y}{x+y}, \quad \text{(ii) } \frac{dy}{dx} = \frac{x^2+y^2}{2xy}, \quad \text{(iii) } \frac{dy}{dx} = \frac{y-x+1}{y+x+5}, \quad \text{(iv) } \frac{dy}{dx} = \frac{xy}{x^2-y^2}.$$

Exercise 16.3

Bernoulli's equation

In equation 16.2 if $n = 0$ the equation can be solved using the integrating factor method. Use this to solve the equations

$$\text{(i)} \quad x\frac{dy}{dx} + y = x^2, \qquad \text{(ii)} \quad \frac{dy}{dx} + 2xy = 3e^{-x^2}.$$

If $n \geq 2$ show that the change of variable $z = y^{1-n}$ reduces equation 16.2 to the form

$$\frac{1}{1-n}\frac{dz}{dx} + zP(x) = Q(x),$$

and hence solve the equations

$$\text{(iii)} \quad 2\frac{dy}{dx} + xy = xy^2, \qquad \text{(iv)} \quad 2\frac{dy}{dx} = \frac{y}{x} + \frac{y^3}{x^3}.$$

Exercise 16.4

Riccati's equation

If, in equation 16.3, either $A(x) = 0$ or $B(x) = 0$, the equation reduces to Bernoulli's equation.

(i) Show that if $y(x)$ is expressed in terms of the new variable $u(x)$ by

$$y = -\frac{1}{uC(x)}\frac{du}{dx},$$

then u satisfies the linear second-order equation

$$-C\frac{d^2u}{dx^2} + \left(BC + \frac{dC}{dx}\right)\frac{du}{dx} - AC^2u = 0.$$

Hence find the general solution of the equations

$$\text{(a)} \quad \frac{dy}{dx} = -2 - 5y - 2y^2, \qquad \text{(b)} \quad x^2\frac{dy}{dx} + 2 - 2xy + x^2y^2 = 0.$$

(ii) If a particular solution $y = v(x)$ is known show that the equation satisfied by $z(x)$, where $y = v + 1/z$, is Bernoulli's equation,

$$\frac{dz}{dx} + (B + 2vC)z = -C.$$

16.2 First-order autonomous systems

16.2.1 *Fixed points and stability*

The simplest dynamical systems are those whose state can be represented by a single real variable x, which may be considered as a coordinate of a point in an abstract one-dimensional space, which we name the *phase space*. Examples of such systems are radioactive decay, simple chemical reactions, a light body falling through a very viscous fluid and the discharge of an electrical condenser through a resistance.

The motion of the system is represented by a function $x(t)$ of time satisfying a first-order differential equation

$$\frac{dx}{dt} = \dot{x} = v(x,t),\tag{16.6}$$

where $v(x,t)$ is a known sufficiently well-behaved *velocity function* of x and t; the value of $v(x,t)$ for a particular x and t is the *phase velocity*. Here we shall only consider *autonomous systems* for which the velocity function is independent of the time t.

If x_0 represents the state of the system at time t_0, then for an autonomous system equation 16.6 can be integrated directly to give t as a function of x:

$$t - t_0 = \int_{x_0}^{x} \frac{dx}{v(x)},\tag{16.7}$$

provided the integral exists. The inverse of $t(x)$ will give x as a function of the time and we note that this depends only upon the time difference $t - t_0$, not t and t_0 separately: this property is restricted to autonomous systems and consequently it is often convenient to set $t_0 = 0$.

Important information about the behaviour of $x(t - t_0)$ can be obtained without evaluating equation 16.7, and this is often useful because the integral may not exist in terms of known functions and, even if it does, the inverse of $t(x)$ may not be easy to find. The behaviour of $x(t)$ is controlled by the zeros, x_f, of the velocity function, $v(x_f) = 0$. If a system is initially at x_f then it remains there for all time; the points x_f represent the state of equilibrium and are named the *fixed points* of the system; at all other points the state of the system changes. A system starting in an open interval between two fixed points cannot pass either of them, so such open intervals, together with those that extend from a fixed point to infinity, are invariant. They represent *invariant sets of states* having the property that if the system is initially in such a state then it remains there for all times, past and future. Fixed points are also invariant sets. Usually only the elementary invariant sets that cannot be decomposed into smaller invariant sets are considered. The whole phase space of a system comprises invariant sets and they provide useful information about the behaviour over arbitrarily long periods of time.

The most important fixed points are at the simple zeros of the velocity function, that is, $v(x_f) = 0$ and $v'(x_f) \neq 0$, and there are only two types of *simple fixed points*, stable and unstable. A fixed point is *stable* if $v(x)$ is decreasing at x_f, $v'(x_f) < 0$, for if $x < x_f$ then $\dot{x} > 0$ and $x(t)$ increases towards x_f and if $x > x_f$, $\dot{x} < 0$ and $x(t)$ decreases towards x_f. Thus neighbouring points approach x_f, as shown schematically in figure 16.1. Similarly, a fixed point is *unstable* if $v(x)$ is increasing at x_f, $v'(x_f) > 0$, figure 16.2.

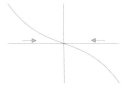   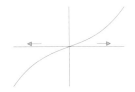

Figure 16.1   A stable fixed point.   Figure 16.2   An unstable fixed point.

Near a stable simple fixed point we may approximate the velocity function by a linear velocity function, $v(x) = -a(x - x_f)$, $a > 0$, so the motion nearby is given by the equation

$$\dot{x} = -a(x - x_f) \quad \Longrightarrow \quad x(t) = x_f + (x_0 - x_f)e^{-at},$$

where $x_0 = x(0)$. This solution shows that $x(t)$ does not reach the fixed point x_f in any finite time.

Other types of fixed points for which $v'(x_f) = 0$ can occur, and these produce different types of behaviour. Here we consider a particular example; another type is considered in section 16.2.2. If the velocity function is

$$v(x) = ax^2, \quad a > 0, \tag{16.8}$$

then $\dot{x} > 0$ for all $x \neq 0$ and there is a fixed point at $x = 0$. If $x > 0$ then $x(t)$ is moving away from the fixed point and if $x < 0$ then $x(t)$ is moving towards the fixed point; thus the fixed point is neither stable nor unstable.

This type of fixed point is unusual because it is *structurally unstable*, which means that if the velocity function is perturbed by an arbitrarily small perturbation $\epsilon w(x)$, where $w(x)$ is differentiable in the neighbourhood of the fixed point, then for most functions $w(x)$ the nature of the fixed point changes.† In the current example we may take $w(x) = 1$ to give the new velocity function $\tilde{v}(x) = ax^2 + \epsilon$, and if $\epsilon > 0$ there are no fixed points and if $\epsilon < 0$ there are two simple fixed points, as illustrated in the following graphs.

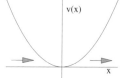

| Figure 16.3   Velocity function $v = x^2$. | Figure 16.4   Velocity function $v = x^2 + \epsilon$. | 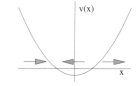 Figure 16.5   Velocity function $v = x^2 - \epsilon$. |

On the contrary, simple fixed points, for which $v'(x_f) \neq 0$, retain their nature when similarly perturbed: they are *structurally stable*.

Another property of the velocity function $v(x) = ax^2$ is that the motion *terminates* because $x(t)$ becomes infinite at a finite time. If $x(0) = x_0$ then

$$x(t) = \frac{x_0}{1 - ax_0 t}, \quad t < \frac{1}{ax_0}, \tag{16.9}$$

and the motion terminates at $t = 1/ax_0$; it is not defined beyond this time. This

---

† Equations describing physical systems normally have structurally stable fixed points because there are always small effects which are usually ignored in the mathematical description. Fixed points like that of equation 16.8 can exist if there is a fundamental symmetry forcing the velocity function to have this form.

example shows that not all velocity functions define motion for all times: in practice the equations of motion become invalid before the critical time is reached.

Exercise 16.5
Show that all the fixed points of the system with velocity function $v(x) = x(1 - x^2)$ are simple and classify them as stable or unstable.

Exercise 16.6
An autonomous first-order system has exactly two fixed points. Can both points be stable? Can both points be unstable? How many essentially different phase diagrams can there be for such a system?

Exercise 16.7
Show that for all positive integers $n \geq 2$ the motion with the velocity function $v(x) = x^n$ and initial condition $x(0) \neq 0$ is terminating.

16.2.2 *Natural boundaries*
If the variable x can take values on only a segment of the real line, then at the boundaries of this segment the phase space has a *natural boundary*. For instance, if x represents a population or a distance then $x \geq 0$ and there is a natural boundary at $x = 0$. Normally, for first-order systems the velocity function is zero at a natural boundary.

An example of a dynamical system with a natural boundary at $x = 0$ is

$$\frac{dx}{dt} = v(x) = -\sqrt{x}, \quad x \geq 0. \tag{16.10}$$

Here $\dot{x} < 0$, consequentially $x(t)$ decreases towards zero; in fact the solution is $x(t) = (\sqrt{x_0} - t/2)^2$, with $x_0 = x(0)$, and the origin is reached at the finite time $t = 2\sqrt{x_0}$. For $t > 2\sqrt{x_0}$, $x(t)$ exists, but is not the solution of the differential equation.

Exercise 16.8
Show that the system with velocity function

$$v(x) = \sqrt{1 - x^2}, \quad -1 \leq x \leq 1,$$

has natural boundaries at $x = \pm 1$, and find an approximation to the motion for large t.

16.2.3 *Rotations*
Phase space is not always the whole real line. For some types of motion, for instance a rotation about an axis, the phase space can be considered as a circle, in which case it is convenient to choose a coordinate θ in the range $[-\pi, \pi]$ with the phase space points $\theta = \pm\pi$ representing the same physical point. In this case the velocity function must be 2π-periodic in θ, that is, $v(\theta + 2\pi) = v(\theta)$ for all θ. If there are no fixed points then $\theta(t)$ either increases or decreases monotonically and the motion is periodic.

Exercise 16.9

Prove that if the motion is periodic then the period is given by the integral

$$T = \int_{-\pi}^{\pi} \frac{d\theta}{v(\theta)}.$$

Exercise 16.10

Consider the system on the circle with the equation of motion

$$\dot{\theta} = a + b \sin \theta, \quad a > 0, \quad b > 0.$$

Find the fixed points and the invariant sets of the motion when $a < b$, $a = b$ and $a > b$. For which values of a and b is the motion a rotation? Find the period of the rotational motion.

16.2.4 *The logistic equation*

When a population of a single species in some given region is sufficiently large it may be represented by a real variable $x \geq 0$. If we suppose that the population can be described by a single equation of the form $\dot{x} = v(x)$, then because $x = 0$ is a natural boundary we know that $v(0) = 0$.

In practice the population of a given region cannot increase without bound, because of competition for resources such as food and space, and the simplest way of allowing for this is to assume that there is a maximum sustainable population, $x = X$, at which the birth and death rates are the same. Then we must have $v(X) = 0$ with $v(x) > 0$ for $0 < x < X$. The simplest velocity function satisfying these conditions is $v(x) = ax(X - x)$, for some positive constant a, and this gives the *logistic equation*,

$$\frac{dx}{dt} = ax(X - x). \tag{16.11}$$

An alternative, discrete, form of the logistic equation, often called the logistic map, is

$$w_{k+1} = bw_k(1 - w_k), \quad b > 0, \quad k = 0, 1, 2, \ldots, \tag{16.12}$$

which may describe a population of a species having a short breeding season with w_k the population immediately after this season. Another connection between the differential and the discrete form of the logistic equation is derived in exercise 16.56 (page 670).

The logistic equation is solved by separating variables,

$$\int_{x_0}^{x} dx \, \frac{1}{x(X - x)} = a(t - t_0) \quad \Longrightarrow \quad \frac{x}{X - x} = \frac{x_0}{X - x_0} \exp\left(aX(t - t_0)\right), \tag{16.13}$$

where $x(t_0) = x_0$; rearranging gives the more convenient form

$$x(t) = \frac{Xx_0}{x_0 + (X - x_0)\exp(-aX(t - t_0))}. \tag{16.14}$$

The shape of this function is shown in the following graph:

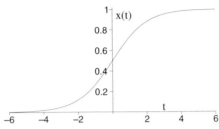

Figure 16.6   The function $x = 1/(1 + \exp(-t))$.

The solution 16.14 depends upon the explicit form chosen for the velocity function, but any velocity function with simple fixed points at $x = 0$ and X and which is positive in between must have a solution with the same shape as shown in figure 16.6, see exercise 16.11.

A nice illustration of this approximation to a population is obtained using the USA population data from 1790 to 1970, given in the *Historical Statistics of USA: Colonial times to 1970*, Table A6-8. The raw data are given in table 16.1. This type of analysis was apparently first done by Pearl and Reed (1920) using data from 1790 to 1910, and their calculation is reproduced by Lotka (1956, chapter 7), who also gives other examples showing that the logistic equation describes some types of populations quite accurately.

Table 16.1. *Population of USA, in millions, from 1790 to 1970.*

Year	1790	1800	1810	1820	1830	1840	1850	1860
Population	3.93	5.30	7.22	9.62	12.90	17.12	23.26	31.51
Year	1870	1880	1890	1900	1910	1920	1930	1940
Population	39.91	50.26	63.06	76.09	92.41	106.5	123.1	132.1
Year	1950	1960	1970					
Population	151.2	180.7	204.9					

Taking $t_0 = 1790$ and $x_0 = 3.93$ we can use the data at 1850 and 1910 to estimate X and a to give the approximation

$$x(t) = \frac{198}{1 + 49.4 \exp(56.19 - 0.0314t)}. \tag{16.15}$$

A graph of this function and the original data is shown in figure 16.7, where we see that the approximation is remarkably good up to about 1930, but subsequently underestimates the population.

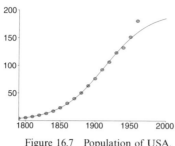

Figure 16.7   Population of USA,
in millions.

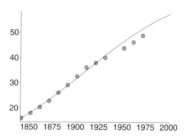

Figure 16.8   English and Scottish
population, in millions.

In figure 16.8 we show a similar comparison for the English and Scottish population data between 1841 and 1961, taken from Mitchell (1992), which is also fitted well by the solution of the logistic equation between 1841 and 1940, but subsequently not so well. Comparisons with many other types of populations are given by Lotka (1956, chapter 7).

Exercise 16.11

Consider the equation

$$\frac{dx}{dt} = A \sin \alpha x, \quad A > 0, \, \alpha > 0.$$

Show that the solution of this equation with $x(0) = x_0$, $0 < x_0 < \pi/\alpha$, is

$$x = \frac{2}{\alpha} \tan^{-1} \left(\tan \left(\frac{\alpha x_0}{2} \right) e^{\alpha A t} \right).$$

Make some graphical comparisons between this solution and the equivalent solution of the logistic equation, 16.14, where the parameters (a, X) are chosen to give the same behaviour near $x = 0$ and to make the other fixed point the same.

Exercise 16.12

Consider a time-dependent logistic equation

$$\frac{dx}{dt} = x(X(\epsilon t) - x), \quad 0 < x(0) < 1, \quad 0 \le \epsilon \ll 1,$$

where $X(\epsilon t)$ is a positive slowly varying function of time.
  Use Maple to numerically solve the equation in the particular case

$$X(w) = \frac{1}{1 + w},$$

for some representative values of ϵ and $x(0)$ and describe what you see. Do you think that the actual form of $X(w)$ is important? In this particular case can you find an approximate expression for the difference $x(t) - X(\epsilon t)$ as $t \to \infty$? Can you find an approximation to this difference for arbitrary $X(\epsilon t)$?
Hint: this time-dependent logistic equation is a Bernoulli equation.

16.2.5 Summary

Autonomous first-order systems have relatively simple solutions that are determined qualitatively by the fixed points. In the long-time limit a solution tends either to $\pm\infty$ or to a fixed point.

A time-dependent first-order system, with velocity function $v(x, t)$, can be considered as an autonomous second-order system by defining a new variable $y = t$ to give the new equations of motion $\dot{x} = v(x, y)$, $\dot{y} = 1$, so we next turn to a discussion of this type of system.

Exercise 16.13

Three first-order systems have the following velocity functions. Which of them is autonomous?

$$\text{(i)} \quad v = e^x, \qquad \text{(ii)} \quad v = xt, \qquad \text{(iii)} \quad v = \begin{cases} x, & t \le 0, \\ x^2, & t > 0. \end{cases}$$

Exercise 16.14

Draw the phase diagrams and find the fixed points and invariant sets of the systems with the following velocity functions:

$$\text{(i)} \quad v(x) = (a - x)(x - b), \qquad \text{(ii)} \quad v(x) = (a - x)(b - x),$$

where $0 < a < b$ and x can vary along the whole real line.

Without solving the equations of motion, describe the qualitative features of the motion for $t > 0$ when $a < x(0) < b$, in each case.

Exercise 16.15

Nitric oxide (NO) and oxygen (O_2) react to form NO_2 as follows:

$$2NO + O_2 \rightarrow 2NO_2.$$

If $C(t)$ represents the concentration of NO_2, it is found to satisfy the differential equation

$$\frac{dC}{dt} = k(\alpha - C)^2(2\beta - C), \quad C \ge 0, \quad C(0) = 0,$$

where k is a positive constant for the reaction and α and β are the initial concentrations of NO and O_2 respectively, both greater than zero.

Discuss the qualitative behaviour of the NO_2 concentration when $\alpha < 2\beta$ and $\alpha > 2\beta$.

16.3 Second-order autonomous systems

16.3.1 Introduction

Most dynamical systems are more complicated than the first-order systems of the previous section, but second-order autonomous systems have the virtue that the phase space, defined below, is two-dimensional and is often easily visualised using diagrams of the phase plane. The motion of such systems is relatively simple by comparison to some, nevertheless it is usually difficult to form a global picture of the behaviour; the examples in some of the following exercises should convince you of this. The methods developed in this and the next chapter are needed to understand the behaviour of solutions and are also necessary to understand the behaviour of more complicated systems which are impenetrable without these methods.

The state of a second-order dynamical system is represented by two real variables, (x, y), or as a single two dimensional vector $\mathbf{r} = (x, y)$, which may be considered as coordinates of an abstract two-dimensional *phase space*. The motion of the system is represented by a vector $\mathbf{r}(t) = (x(t), y(t))$ satisfying the first-order differential equation

$$\frac{d\mathbf{r}}{dt} = \mathbf{v}(\mathbf{r}, t), \tag{16.16}$$

or, in component form,

$$\frac{dx}{dt} = X(x, y, t), \quad \frac{dy}{dt} = Y(x, y, t), \tag{16.17}$$

where the vector velocity function $\mathbf{v}(\mathbf{r}, t) = (X(x, y, t), Y(x, y, t))$ is a sufficiently well behaved function of $\mathbf{r}$ and t. In this chapter we deal mainly with autonomous systems for which $\mathbf{v}$ does not depend upon t; for these systems solutions depend upon the time t and the initial time t_0 only through the difference $t - t_0$.

A particular motion of the system is obtained by starting at a given point (x_0, y_0) in phase space at a given time t_0. Most initial conditions determine a unique solution, $\mathbf{r}(t) = (x(t), y(t))$; the conditions on the velocity function for a solution to exist and to be unique are given in the appendix to this chapter. The solution $\mathbf{r}(t)$ traces out a continuous curve in phase space, named the *phase curve*, and the set of all possible motions is named the *phase flow*. For brevity we often use the terms *orbit* and *trajectory* for a phase curve.

Phase curves are given parametrically by the solution $\mathbf{r}(t)$ of the equations of motion when expressed as a function of t, or by eliminating t at the outset by dividing the two equations of motion to give a differential equation involving x and y only,

$$\frac{dy}{dx} = \frac{Y(x, y)}{X(x, y)}. \tag{16.18}$$

The solutions of this equation will be of the form $F(x, y, C) = 0$, where C is the integration constant.

Any second-order differential equation can be cast in the form of equation 16.17 by defining a second independent variable. For instance, if $x(t)$ satisfies the equation

$$\frac{d^2x}{dt^2} = F\left(x, \frac{dx}{dt}, t\right), \tag{16.19}$$

by defining a new dependent variable $y = \dot{x}$ we obtain the two coupled equations

$$\frac{dx}{dt} = y, \quad \frac{dy}{dt} = F(x, y, t),$$

which is in the form of equation 16.17 with velocity function $\mathbf{v} = (y, F(x, y, t))$. There are infinitely many other ways of converting second-order differential equations into coupled first-order systems, see for instance exercise 16.16 below. Second-order differential equations are common because Newton's second law relates accelerations to forces which often depend only upon the position of the particle, so a single particle confined to move along a line is usually described by this type of equation.

Finally, we note that if the velocity function $\mathbf{v}(\mathbf{r}, \alpha)$ of an autonomous system depends upon a parameter α, the nth differential of $\mathbf{v}$ with respect to α exists and if $\mathbf{v}(\mathbf{r}_0, \alpha_0) \neq 0$, the solution $\mathbf{r}_0$ is n times differentiable in α for sufficiently small $|t|$, $|\mathbf{r} - \mathbf{r}_0|$ and $|\alpha - \alpha_0|$, Arnold (1973, section 7.6) and Sànchez (1968, chapter 6).

Exercise 16.16
Convert the second-order differential equation $\ddot{x} + 6\dot{x} - x^2 + 4x = 0$ to the canonical form using the new variables

$$\text{(i)} \quad y = \dot{x}, \qquad \text{(ii)} \quad z = \dot{x} + 6x, \qquad \text{(iii)} \quad w = \dot{x} + 4x.$$

Exercise 16.17
Use the new variable $y = \dot{x} + F(x)$, with suitable choices of $F(x)$, to show that the second-order equation

$$\frac{d^2x}{dt^2} + f(x)\frac{dx}{dt} + x = 0$$

can be written in either of the equivalent forms

$$\text{(i)} \quad \begin{aligned} \dot{x} &= y, \\ \dot{y} &= -x - yf(x), \end{aligned} \qquad \text{(ii)} \quad \begin{aligned} \dot{x} &= y - F(x), \\ \dot{y} &= -x, \end{aligned}$$

where, in (ii), $F(x) = \int_0^x du\, f(u)$. The phase plane, (x, y), defined in equations (ii), is known as the Liénard plane and is sometimes more useful than the obvious phase plane defined in (i), particularly when $f(x)$ takes large values. We shall see why in the analysis of the damped linear oscillator on page 646 and, more important, in the treatment of the van der Pol oscillator and relaxation oscillations in the next chapter.

Exercise 16.18
Consider the second-order autonomous system

$$\frac{dx}{dt} = -y, \quad \frac{dy}{dt} = x.$$

Find the solution with initial conditions $\mathbf{r}(0) = (a, b)$ and from these expressions eliminate the time to find an equation for the phase curves.

Show also that the same expression is obtained by solving the equations

$$\frac{dy}{dx} = \frac{Y(x, y)}{X(x, y)} = -\frac{x}{y}.$$

What extra information is provided by the first solution?

Exercise 16.19
Show that the function $Ax^2 + 2Bxy + Cy^2 = D$, where A, B, C and D are constants, is a solution of the equation $dy/dx = -(Ax + By)/(Bx + Cy)$.

Exercise 16.20
Show that the function $F(\mathbf{r}) = a\tan^{-1}(y/x) + \ln\sqrt{x^2 + y^2}$ is a solution of the equation $dy/dx = (ay - x)/(y + ax)$, where a is a constant.

16.3.2 The phase portrait

For autonomous systems a phase curve cannot cross itself, consequently the phase flow
can be represented by drawing a representative sample of phase curves. Such a diagram
is named a *phase portrait* or a *phase diagram.*

An approximate picture of the phase portrait can be obtained, without solving the
equations of motion, by representing the velocity function, $\mathbf{v}(\mathbf{r})$, as a set of arrows
on a grid $\mathbf{r}_{ij}$ with the magnitude of the arrow centred at $\mathbf{r}_{ij}$ proportional to $|\mathbf{v}(\mathbf{r}_{ij})|$
and direction along $\mathbf{v}(\mathbf{r}_{ij})$. This construction can be carried out by hand but it is
tedious: Maple provides commands to remove this tedium which we illustrate with a
few examples.

The command `phaseportrait` will plot the arrows alone, but in practice I have
found it better to use `DEplot` as this allows one to include a few numerically generated
solutions, which is usually necessary in order to obtain a clearer understanding of the
phase portrait.

For the first example take the case of a particle falling freely in the vertical direction
in the Earth's gravitational field. If x is the height of the particle above the ground,
Newton's equation of motion is

$$\frac{d^2 x}{dt^2} = -g,$$

where g is the gravitational acceleration, 32.2 feet/sec^2 or 981 cm/sec^2; choose units to
give $g = 1$. Now define $y = \dot{x}$, the vertical velocity, to put this equation in the form

$$\frac{dx}{dt} = y, \quad \frac{dy}{dt} = -1, \tag{16.20}$$

or

$$\frac{dx}{dy} = -y. \tag{16.21}$$

The solution of equations 16.20 is

$$x(t) = x_0 + y_0 t - t^2/2, \quad y(t) = y_0 - t.$$

The solution of equation 16.21 is $x = x_1 - y^2/2$, which can also be obtained from the
first solution by eliminating the time t. The phase curves are therefore parabolas.

We may graph these using `DEplot`, which will both numerically solve equations 16.20
to draw phase curves and simultaneously plot representative phase arrows using the
phase velocity $\mathbf{v} = (y, -1)$. A typical use of this command is shown below. First, load
the necessary package:

```
>  restart;   with(DEtools):
```

Define the equations of motion and put these in a variable eqn:

```
>  eqn:=diff(x(t),t)=y(t), diff(y(t),t)=-1;
```

$$eqn := \frac{\partial}{\partial t} \, x(t) = y(t), \, \frac{\partial}{\partial t} \, y(t) = -1$$

It is sometimes useful to define the range of variation of the two coordinates (x, y); for this problem we restrict x to the range $(0, 0.6)$ and y to $(-1, 1)$.

```
>  rang:=x=0..0.6,y=-1..1;
```
$$rang := x = 0...6, \ y = -1..1$$

Now set up an expression sequence containing the various DEplot options.

```
>  conds:=arrows=medium,            # Defines the arrow type
>  colour=green,                    # Colour of the arrows
>  dirgrid=[15,15],                 # Grid size for the arrow positions
>  linecolour=black,        # Colour of numerically generated curves
>  stepsize=0.1:                # Step size for numerical integration
```

If no arrows are required set arrows=none, then the next two options in the above list are unnecessary.

Next it is necessary to define the initial conditions of the phase curves to be computed. These have to be in the form of a list of lists, with each member of the inner list of the form [x(0)=a,y(0)=b], where a and b are numbers. Less typing is required if we first define a list of lists of the type [a,b] and then convert this to the required form automatically. In this example we start all phase curves at ground level, $x = 0$, with the positive velocities 0.2, 0.4, ..., 1.0. The first list is therefore

```
>  ic:=[seq([0,0.2*k],k=1..5)];
```
$$ic := [[0, .2], [0, .4], [0, .6], [0, .8], [0, 1.0]]$$

Now convert this to the required form:

```
>  ic1:=[seq([x(0)=ic[k][1],y(0)=ic[k][2]],k=1..nops(ic))]:
```

Finally the command that creates both the arrows and the phase curves is as shown below: in this command you should note the purpose of the other arguments.

- The first two arguments define the equations and the dependent variables.

- The argument t=0..2 defines the maximum time interval of integration.

- The argument scene=[x,y], or equivalently scene=[x(t),y(t)], defines the picture to be shown and also the axis labels; for example if we require a graph of $x(t)$ the option scene=[t,x] should be used and the arrows are automatically not drawn, as shown in figure 16.10.

Figure 16.9, on the left, is created with the command:

```
>  DEplot([eqn],[x(t),y(t)],t=0..2,scene=[x,y],rang, ic1,conds);
```

and figure 16.10, on the right, is created with the command:

```
>  DEplot([eqn],[x(t),y(t)],t=0..2,scene=[t,x],rang, ic1,conds,thickness=1);
```

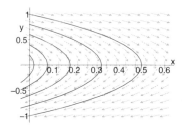
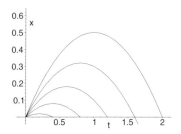

Figure 16.9   Phase curves for a particle   Figure 16.10   Variation of the height x
falling in gravity, equation 16.21; x is the   with time.
            height above ground.

The use of DEplot has been described in some detail; similar commands will allow you to create the phase portrait of any other system. In practice, however, if the system is at all complicated such a phase portrait on its own does not provide a clear enough picture to understand the global behaviour of the system, and the mathematical analysis of this and the next chapter is usually essential.

First we consider the behaviour of a few simple systems to provide some background examples with which to understand the mathematical analysis. Some simple systems are treated in the next few exercises, and then we consider in some detail the motion of a strongly damped linear oscillator and the vertical pendulum.

Exercise 16.21
Use DEplot to draw the phase portrait and the phase curves with initial conditions $x(0) = 0$, $\dot{x}(0) = 0.2k$, $k = 1, 2, \ldots, 5$, for the linear oscillator

$$\frac{d^2x}{dt^2} + x = 0.$$

Exercise 16.22
Use DEplot to draw the phase portrait for the system

$$\frac{d^2x}{dt^2} - x + x^3 = 0.$$

In particular plot the phase curves passing through the points $(0, 1/2)$, $(0, 1)$, $(\sqrt{2}, 0)$, $(-\sqrt{2}, 0)$, $(1/2, 0)$, $(-1/2, 0)$.

The strongly damped oscillator
Consider the linear differential equation

$$\frac{d^2x}{dt^2} + 2v\frac{dx}{dt} + x = 0, \quad v > 1, \tag{16.22}$$

v being a constant, which describes a strongly damped linear oscillator, for instance the small oscillations of a pendulum swinging in a viscous fluid. In the next chapter we shall allow v to vary with x, and it is helpful to introduce a few ideas with this simple example that can be solved exactly. The solution with the initial conditions $x(0) = 1$ and $\dot{x}(0) = 0$ is

$$x(t) = \left(\frac{\sqrt{v^2 - 1} + v}{2\sqrt{v^2 - 1}} \right) e^{-\lambda_1 t} - \left(\frac{v - \sqrt{v^2 - 1}}{2\sqrt{v^2 - 1}} \right) e^{-\lambda_2 t},$$

where $\lambda_1 = v - \sqrt{v^2 - 1}$ and $\lambda_2 = v + \sqrt{v^2 - 1}$. If $v \gg 1$, λ_1 is small and λ_2 is large, and the behaviour of the solution is more transparent when is expanded in inverse powers of v with only the lowest-order terms retained,

$$x(t) \simeq \exp(-t/2v) - \frac{1}{4v^2} \exp(-2vt), \quad v \gg 1. \tag{16.23}$$

In this limit the two terms are of different magnitude and have different time-scales. The first term is $O(1)$, decays very slowly and arises from the solution of the original equation when the acceleration term, $\ddot{x}$, is ignored, that is,

$$2v\dot{x} + x = 0 \quad \Longrightarrow \quad x = a\exp(-t/2v).$$

The second term, e^{-2vt}, is small, decays very rapidly and arises from the original equation when the restoring force, x, is ignored,

$$\ddot{x} + 2v\dot{x} = 0 \quad \Longrightarrow \quad x = b\exp(-2vt).$$

For almost all initial conditions the motion comprises the sum of these terms, but the large damping term very quickly brings the viscous and inertial forces into equilibrium, so after a short initial period the motion is described accurately by the equation $2v\dot{x} + x = 0$. The larger v the quicker the system reaches this equilibrium motion and the slower the system approaches the fixed point at the origin.

This type of time-scale separation occurs in other more complicated circumstances where the damping coefficient v can vary with the position, x. Then the subsequent motion is more complicated, but can be understood using the idea developed here. The trick is to find a phase space representation in which one coordinate represents the slow motion, the first term of equation 16.23, and the other is the fast term: it transpires that this is the Liénard plane, introduced in exercise 16.17 (page 643). Write the original equation in the form

$$\frac{d}{dt}\left(\frac{dx}{dt} + 2vx \right) + x = 0 \tag{16.24}$$

and assume that both x and $\dot{x}$ are $O(1)$ and that $v \gg 1$ so the term in the brackets is $O(v)$. Hence, we define a new variable $y = O(1)$ by $vy = \dot{x} + 2vx$ to give the new equations of motion,

$$\dot{x} = v(y - 2x), \quad \dot{y} = -x/v. \tag{16.25}$$

In this phase plane the motion is very simple when $v \gg 1$. Consider a point not on the line $y = 2x$, then $|\dot{y}| \ll 1$ and $|\dot{x}| \gg 1$, so the flow is practically horizontal, rapid and towards the line $y = 2x$. On the line $y = 2x$ the flow is towards the origin and slow. This behaviour is seen in the figure where a single solution is shown, the solid line, together with the phase arrows in the case $v = 2$.

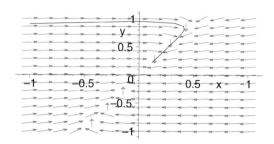

Figure 16.11   Phase flow of the damped oscillator, with $v = 2$, in the Liénard plane.

Now the two time-scales are clearly seen. The almost horizontal part of the solution represents the rapid motion, the e^{-2vt} part of equation 16.23, during which the viscous and inertial forces are rapidly brought into equilibrium. When the phase curve reaches the line $y = 2x$ the rapid x motion ceases and the phase curve moves slowly down the curve $y = 2x$. Indeed, if we set $y = 2x$ the second of equations 16.25 becomes $2v\dot{x} + x = 0$, the original equation less the acceleration term.

This type of analysis may seem rather involved for such a simple equation, but we shall see that the same ideas can be used to understand the solutions of equations that do not have elementary solutions.

Exercise 16.23
Consider the system defined by the equation

$$\frac{d^2x}{dt^2} + 2v\frac{dx}{dt} + x^3 = 0, \quad v > 0.$$

(i) Show that the equations of motion in the Liénard plane are

$$\frac{dx}{dt} = v(y - 2x), \quad \frac{dy}{dt} = -\frac{x^3}{v}.$$

(ii) Use Maple to plot some representative phase curves in the Liénard plane for $v = 2$.

(iii) Show that the phase curve approaches the origin slowly along the line $y = 2x$ with $x(t)$ given approximately by $\dot{x} = -x^3/2v$, and hence that $x^2 = vb^2/(v + b^2t)$, where b is some constant.

The vertical pendulum

G. Galileo
(1564–1642)

R. Hooke
(1635–1702)

The vertical pendulum is a simple problem of both historical interest and current importance. Galileo observed that the period of the pendulum was constant and independent of the amplitude† and used this to design a clock, but it was Hooke and

† Galileo considered only small amplitude motion.

Huygens who perfected the workings of the pendulum clock, and their clocks were far C. Huygens
more accurate than the other types of mechanical clocks of the period.† (1629–1695)

Another use of the pendulum was in the measurements of g, the acceleration due
to gravity, through the formula $T = 2\pi\sqrt{l/g}$ relating the period T to the pendulum
length l. Measurements of T at various points of the Earth thus provided a means of
estimating the shape of the Earth. In this manner Newton estimated that the ratio of
the equatorial to the polar radius was $1 + \frac{1}{230}$; the current estimate is $1 + \frac{1}{294}$, with the
equatorial and the polar radius being 3963.37 and 3949.92 miles respectively.

The physical pendulum is of practical use in time keeping and is generally confined to
small amplitude motion. The mathematical pendulum — that is, equation 16.26 below
— is important because it occurs in many nonlinear problems in which resonances
occur as a local approximation whenever the periods of two component motions
have a rational ratio. This feature of nonlinear systems often gives rise to chaotic
motion for reasons that will be briefly discussed in the following chapters, in particular
exercise 17.28 (page 697) and section 18.6. The motion of a perturbed pendulum can
also be surprising, as shown in exercise 16.29 below and in more detail in chapter 19.

The physical pendulum, depicted in figure 16.12, comprises a mass m at P joined
to a frictionless hinge at O by a light rod of length l, making an angle x with the
downward vertical, and is hinged so that it can freely swing over the top and execute
rotations about O.

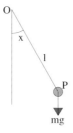

Figure 16.12   Diagram of the vertical
pendulum.

Newton's equation of motion for this system is independent of the mass, m, and is
given by

$$\frac{d^2x}{dt^2} + \alpha^2 \sin x = 0, \quad \alpha = \sqrt{\frac{g}{l}}. \tag{16.26}$$

The positions $x + 2n\pi$ and x are physically equivalent, so it is necessary only to consider
the interval $-\pi < x \le \pi$ and identify the points $x = \pi$ and $-\pi$. Casting this equation

† One reason why time-keeping was important was for the determination of the longitude of a ship at sea,
and because pendulum clocks were affected too much by the ship's motion they could not be used. The
absence of accurate clocks meant that an accurate determination of the Moon's position was necessary,
hence the need for accurate computations of the lunar orbit mentioned in the introduction. However, it
turned out easier to make an accurate clock than to understand the motion of the Moon in sufficient
detail, see the discussion on page 632.

in the canonical form of equation 16.17 gives

$$\frac{dx}{dt} = y, \quad \frac{dy}{dt} = -\alpha^2 \sin x. \tag{16.27}$$

The pendulum can execute two types of motion. In the normal pendulum motion, it oscillates about the downward vertical, $x = 0$: in the second type the pendulum rotates around O.

The oscillatory motion about the downward vertical, in which the velocity $\dot{x}$ changes sign, is often named *librational motion*. For this type of motion $|x(t)| < \pi$: if $|x(t)| \ll 1$ it is the usual motion of a clock pendulum and the period of this small amplitude motion is $T = 2\pi\sqrt{l/g}$, which is independent of the amplitude of the swing. In the *rotational motion* the angle $x(t)$ is always either increasing, that is $\dot{x} > 0$ for all t, or always decreasing, $\dot{x} < 0$ for all t.

A typical librational phase curve is depicted by curve B in figure 16.13: starting at the downward vertical, point a where $y = a > 0$, $x = 0$, the pendulum swings away from the vertical with $y = \dot{x} > 0$, but y decreasing until the maximum amplitude at $x = b$ is reached; this is the arc ab. At b it instantaneously stops before returning to the vertical at c, on the arc ac. The pendulum now retraces its path but in the opposite direction until it reaches the starting point $y = a$, $x = 0$, moving on the arc cda. Then it repeats this motion, so the phase curve is a closed loop and the actual motion is periodic.

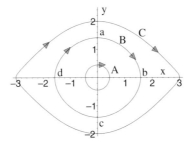

Figure 16.13 Librational motion for
the vertical pendulum.

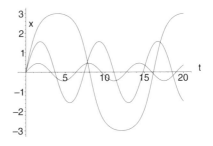

Figure 16.14 Motion $x(t)$ for the phase
curves shown.

The phase curves A and C represent other motions of this type; curve A depicts small amplitude motion and curve C large amplitude motion in which the maximum value of $x(t)$ is slightly less than π. Figure 16.14 shows graphs of $x(t)$ for the three phase curves shown. For the small amplitude motion, curve A, $x(t)$ is close to a sinusoidal function, as is the larger amplitude motion, curve B, but the large amplitude motion, curve C, is quite different because now the motion near $x = \pi$ is very slow and the pendulum takes a long time to pass here, which is why the maxima and minima of $x(t)$ are so broad.

In the rotational motion the angle $x(t)$ is always either increasing, that is $\dot{x} = y(t) > 0$ for all t, or always decreasing, $y(t) < 0$ for all t. Further, the faster the pendulum rotates the smaller the effect of gravity, so we should expect the variations in $y(t)$ to be relatively smaller as $|y|$ increases. Some typical phase curves of this rotational motion are shown in figure 16.15. Remember that $y = \dot{x}$ is the angular velocity and on the lowest curve shown y is very small near $x = \pm\pi$, so the pendulum is only just moving. On the upper curve y is large and almost constant.

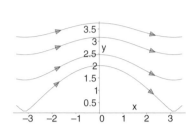

Figure 16.15   Rotational motion for   Figure 16.16   Motion $x(t)$ for the phase
the vertical pendulum.   curves shown.

In figure 16.16 is shown the solutions $x(t)$ as a function of the time t for each of the four phase curves of figure 16.15. The lowest, wiggly, line corresponds to the rotational motion with only just enough energy to go all the way round: near the top, $x = (2n + 1)\pi$, it is moving slowly so $\dot{x}$ is small. The upper line is a rapid rotation, the upper phase curve, and here $x(t)$ increases almost linearly with time; the pendulum is rotating almost as fast at the top as at the bottom of its swing. The other two solutions are intermediate between these two extremes.

We end this section with some exercises which use Maple and DEplot to graph solutions of various differential equations. In each of the following six exercises you are asked to solve some equations of motion with given initial conditions. These are important exercises and are designed to provide you with some idea of the range of types of behaviour that these relatively simple systems can display and, hopefully, to convince you that it is very difficult to form a global understanding of the system behaviour from numerical solutions alone. In each case particular initial conditions are suggested in order to demonstrate specific behaviour, but other conditions should be experimented with. We return to each of these systems later in order to understand the observed behaviour.

Exercise 16.24

Consider the equations

$$\frac{dx}{dt} = x(1 - y), \quad \frac{dy}{dt} = ay(x - 1),$$

where a is a constant. Write a Maple procedure which numerically solves these and will (i) plot the graphs of $x(t)$ and $y(t)$ as functions of t on the same graph, and (ii) plot the phase curve, depending upon the value of a parameter in the argument list. Use your procedure to investigate the behaviour for the five initial conditions, values of a, integration time T and step size δt given in the table.

	$x(0)$	$y(0)$	a	δt	T
i	0.5	0.5	1	0.1	10
ii	1.5	1.5	1	0.1	10
iii	3.0	3.0	1	0.1	10
iv	1.5	1.5	5	0.05	5
v	1.5	1.5	10	0.02	3

Exercise 16.25

For the system defined by the equations

$$\frac{dx}{dt} = y - x - \frac{1}{2}x^3, \quad \frac{dy}{dt} = -x + 3x^2 y - \frac{1}{2}y^3,$$

plot the graphs of $x(t)$ against t for the solutions with initial conditions $(x(0), y(0)) = (-1.893, 0)$, $(-1.894, 0)$ and $(-1.895, 0)$ for $0 < t < 18$.

This system is investigated further in exercise 16.48 (page 668).

Exercise 16.26

Use Maple to plot the phase curve of the system

$$\frac{dx}{dt} = y + \frac{1}{2}x^3, \quad \frac{dy}{dt} = -x - \alpha x^2 y + \frac{1}{2}y^3, \quad \alpha = 3,$$

for the initial condition $x(0) = 0.3$, $y(0) = 0$ and the time $T = 1000$ phase curve. The step size $\delta t = 0.1$ is recommended if DEplot is used.

Compare this phase curve with that of the same system and initial condition but with $\alpha = \frac{5}{2}$.

Exercise 16.27

The equation

$$\frac{d^2 x}{dt^2} + v(x^2 - 1)\frac{dx}{dt} + x = 0,$$

where v is a constant, describes the van der Pol oscillator, which occurs in a variety of circumstances and will be dealt with in chapter 17. Use Maple to plot the phase curves of the system for the following five initial conditions, values of v, integration time T and step size δt.

	$x(0)$	$y(0)$	v	T	δt
i	0.1	0.1	0.1	100	0.1
ii	0.1	0.1	1	30	0.1
iii	0.1	0.1	4	15	0.02
iv	0	12	1	30	0.02
v	0	12	4	20	0.01

Exercise 16.28

Consider the equations

$$\frac{dx}{dt} = y(1 - y^2), \quad \frac{dy}{dt} = -x + x^2 + y^2(x - \mu),$$

where μ is a constant.

For $\mu = 0.9$ plot the solutions $x(t)$ having the initial conditions $(0.458, 0)$ and $(0.459, 0)$ for $0 < t < 20$ and with a step size $\delta t = 0.05$.

For the same value of μ plot $x(t)$ for the initial conditions $(\sqrt{\mu}, 0.99)$ and $(0.99, -0.01)$ for $0 < t < 100$ and $\delta t = 0.1$. Also plot the phase curves of these solutions for $0 < t < 200$.

Plot the phase curves for $\mu = 0.9$, $0 < t < 20$, step size $\delta t = 0.1$ and the initial conditions $(0.1, 0.1)$, $(0.4, 0.4)$, $(0.5, 0)$, $(1, 0.05)$, $(2, 0)$, $(\sqrt{\mu}, 0.8)$ and $(\sqrt{\mu}, -0.8)$.

The reasons for the behaviour seen is examined in exercise 16.49 (page 668).

Exercise 16.29
A variation of the pendulum problem is obtained by making the point of support, O in figure 16.12, move periodically in the vertical direction. If the distance of O from a fixed point is $a \sin \Omega t$ then the equations of motion may be shown to be

$$\frac{dx}{dt} = y, \quad \frac{dy}{dt} = -\alpha^2 \left(1 + \frac{a\Omega^2}{g} \sin \Omega t \right) \sin x, \quad \alpha = \sqrt{\frac{g}{l}}.$$

If the period $2\pi/\Omega$ is comparable to the natural frequency of the motion this is a particularly efficient manner of creating chaotic motion. However, here we are interested in high frequency oscillations of the point of support, that is, when $a^2\Omega^2 > 2gl$, for then an approximate theory which also assumes that $a \ll l$ — too lengthy for inclusion here, see however Landau and Lifshitz (1965a, section 30) or Percival and Richards (1982, chapter 9), where a derivation of the equations of motion is given — shows that the upward vertical position becomes stable.

Write a Maple procedure that solves these equations and plot the solutions with the initial condition $y(0) = 0$ and $x(0) = \pi - \delta$, $\delta = 0.1$ and 0.01 for $\alpha = g = 1$, $a = 0.1$ and for various values of Ω in order to check this approximation. Note that you will need to use a stepsize that is small by comparison with $2\pi/\Omega$, and for these parameters the upward vertical becomes stable if $\Omega > 15$, approximately.

16.3.3 Integrals of the motion
An *integral of the motion* is a single valued, differentiable function $F(\mathbf{r})$ which is constant along all solutions of the equations of motion; that is, given a solution $\mathbf{r}(t)$, $F(\mathbf{r}(t))$ is constant. Using the chain rule we obtain

$$0 = \frac{d}{dt} F(\mathbf{r}(t)) = \dot{x} \frac{\partial F}{\partial x} + \dot{y} \frac{\partial F}{\partial y} = \mathbf{v}(\mathbf{r}) \cdot \operatorname{grad} F.$$

The above equation shows that $\operatorname{grad} F$ is perpendicular to the velocity function. The equation $F(\mathbf{r}) = \text{constant}$ defines a contour, or level curve, of the function $F(\mathbf{r})$, and these contours must be solutions of the equations of motion. Integrals of the motion are often called *constants of the motion*, particularly in mechanics where the energy and angular momentum are constants of the motion for conservative and spherically symmetric systems, respectively. For the vertical pendulum the function $H(x, y) = y^2/2 - \alpha^2 \cos x$ is an integral of the motion.

Integrals of the motion are often obtained by integrating the equations

$$\frac{dy}{dx} = \frac{Y(x, y)}{X(x, y)},$$

and hence are sometimes called *first integrals*. Such integration is always possible when coordinates (u, v) can be found such that component of the velocity function can be expressed as products of functions of u only and of v only. Thus if $X = f_1(u)g_1(v)$ and $Y = f_2(u)g_2(v)$ the differential equation becomes

$$\frac{dv}{du} = \frac{f_2(u)g_2(v)}{f_1(u)g_1(v)} \implies F(u,v) = \int dv \frac{g_1(v)}{g_2(v)} - \int du \frac{f_2(u)}{f_1(u)}.$$

The function $F(u, v)$ constructed in this manner need not be single valued, and then it is not an integral of the motion; examples of this type of system are given in exercises 16.33 and 16.42.

An important case for which integrals exist is when a function $H(x, y)$ exists such that the equations of motion can be written in the form

$$\frac{dx}{dt} = \frac{\partial H}{\partial y}, \quad \frac{dy}{dt} = -\frac{\partial H}{\partial x}. \tag{16.28}$$

Then $H(x, y)$ is an integral of the motion, exercise 16.32. This type of system occurs frequently in physics and is named a *Hamiltonian system*, and the function $H(x, y)$ is the *Hamiltonian function* or just the Hamiltonian, named after Hamilton who developed the theory of Hamiltonian dynamics; the function $H(x, y)$ defined in exercise 16.31 is the Hamiltonian function for the vertical pendulum. The Hamiltonian may be a function of time, in which case it is not constant along a solution; Hamiltonian systems of higher dimensions exist and are also very important.

W. R. Hamilton (1805–1865)

This passing reference to Hamiltonian systems does scant justice to one of the more important developments in mathematical physics and to one of the most important types of dynamical systems, which dominate physics. Hamilton was an Irish mathematician, born in Dublin at midnight of 13/14th August 1805, who, by the age of 13 was fluent in 13 languages, and whose interest in mathematics was awakened upon meeting the American calculating genius Zerah Colburn (1804–1839) when 12 years old. By 17 he was purported to have found an error in Lagrange's proof of the parallelogram law of forces and by 18 seems to have discovered the characteristic function, discussed in his influential paper on optics, published in 1828; a discussion of this can be found in Whittaker (1964, chapter 11). Hamilton extended this work on ray optics to dynamics in 1834, thereby laying the mathematical foundations of almost the whole of modern physics by reformulating Newtonian dynamics in a fundamental manner far more suited to the development of relativity in 1915 and quantum mechanics in the 1920s. The Hamiltonian formulation of dynamics was also used in the perturbation analysis developed by Poincaré in his studies of the three-body problem and which led, amongst other things, to the discovery of chaotic motion. Hamiltonian dynamics is considered in more detail in the books by Arnold (1978), Corben and Stehle (1994), Goldstein (1980) and Percival and Richards (1982).

Exercise 16.30
Show that if $F(\mathbf{r})$ is a constant of the motion then so are $CF(\mathbf{r})$ and $F(\mathbf{r}) + C$, where C is a constant. Show also that if $G(z)$ is any differentiable function then $G(F(\mathbf{r}))$ is a constant of the motion.

Exercise 16.31

Show that the value of the function $H(x, y) = y^2/2 - \alpha^2 \cos x$ is a constant along all the phase curves of the vertical pendulum.

Plot the contours of the function $H(x, y) = y^2/2 - \cos x$ for the values of $H = -0.9, 0$ and 0.99 and compare your results with the phase curves depicted in figure 16.13.

Exercise 16.32

If $H(x, y)$ is the Hamiltonian of a system, by differentiating $H(x, y)$ with respect to t and using the equations of motion 16.28 show that $H(x, y)$ is an integral of the motion.

Exercise 16.33

In exercises 16.19 and 16.20 (page 643), it was shown that for the equations

$$\frac{dy}{dx} = -\frac{Ax + By}{Bx + Cy}, \qquad \frac{dy}{dx} = \frac{Ay - x}{y + Ax},$$

where A, B, and C are constants, the functions

$$F(\mathbf{r}) = Ax^2 + 2Bxy + Cy^2, \qquad F(\mathbf{r}) = A\tan^{-1}(y/x) + \ln\sqrt{x^2 + y^2}$$

are respectively constants along the phase curves. Which, if any, of these functions is an integral of the motion?

Exercise 16.34

The three-dimensional vector $\mathbf{x} = (x_1, x_2, x_3)$ satisfies the vector equation $\dot{\mathbf{x}} = \boldsymbol{\omega}(t) \times \mathbf{x}$, where $\times$ is the vector cross product and $\boldsymbol{\omega}$ a given time-dependent vector. Show that the distance $x_1^2 + x_2^2 + x_3^2$ is a constant of the motion and hence that the end of the vector $\mathbf{x}$ moves on the surface of a sphere.

Further, show that if the direction of $\boldsymbol{\omega}(t)$ is constant, so that $\boldsymbol{\omega}(t) = f(t)\boldsymbol{\Omega}$ where $\boldsymbol{\Omega}$ is a constant vector, then $\boldsymbol{\Omega} \cdot \mathbf{x}$ is also a constant of the motion.

Exercise 16.35

Show that the equation

$$\frac{dy}{dx} = -\frac{x + axy}{y + bx^2 + cy^2}$$

has the solution $x^2 = Ay^2 + By + C$ if

$$A = -\frac{c}{a+b}, \quad B = \frac{2(c-a-b)}{(a+b)(a+2b)}, \quad C = \frac{a+b-c}{b(a+b)(a+2b)}.$$

Is the function $F(\mathbf{r}) = Ay^2 - x^2 + By$ an integral of the motion for the system $\dot{x} = y + bx^2 + cy^2$, $\dot{y} = -x - axy$? This system is considered in more detail in exercise 17.19 (page 690).
Hint: a function $F(\mathbf{r})$ is an integral of the motion if $F(\mathbf{r}) = $ constant on *all* solutions in a region of phase space.

16.3.4 Fixed points and stability

A point $\mathbf{r}_f$ at which the velocity function is zero, $\mathbf{v}(\mathbf{r}_f) = 0$, is named a fixed point and represents the system in equilibrium. Thus, for the vertical pendulum, equation 16.27,

the fixed points are where

$$\mathbf{v}(x, y) = (y, -\alpha^2 \sin x) = 0,$$

that is, at $y = 0$ and $x = 0$ and π, the downward and upward verticals respectively. Phase points at which $\mathbf{v}$ is defined and $\mathbf{v} \neq 0$ are called *ordinary points*: in most systems almost all points are ordinary points, though it is the exceptional fixed points that control the dynamics.

Synonyms for a fixed point are *equilibrium point* and *singular point*; the origin of the former is clear. The latter term comes from investigations of equation 16.18 (page 642), because $\lim_{\mathbf{r} \to \mathbf{r}_f} Y(\mathbf{r})/X(\mathbf{r})$ does not exist and the existence and uniqueness theorems for this equation which apply at ordinary points fail at singular points.

A system in equilibrium must stay there, but there is negligible chance of being exactly at equilibrium, so its behaviour in the neighbourhood of a fixed point is important: for the vertical pendulum we saw that the behaviour near the two fixed points was quite different. For second-order systems there is a greater variety of types of behaviour than for first-order systems, in fact there are ten common types of fixed points having distinctively different local behaviour. Before discussing these we must refine our notions of stability.

A fixed point $\mathbf{r}_f$ is an *attractor* for some motion $\mathbf{r}(t)$ if

$$\lim_{t \to \infty} \mathbf{r}(t) = \mathbf{r}_f, \tag{16.29}$$

and such an orbit may look like:

Figure 16.17   An attractor.

In this definition the time is increasing: a fixed point such that

$$\lim_{t \to -\infty} \mathbf{r}(t) = \mathbf{r}_f$$

is named a *repellor* of the motion.

A fixed point may be an attractor for some motion passing close to it but not others; for instance the point $x = \pi$, $y = 0$ in the vertical pendulum phase space is an attractor for the special motion which starts at $x = 0$, $y > 0$ with y chosen so that $x(t) \to \pi$ as $t \to \infty$, but for no other motion.

The fixed point $\mathbf{r}_f$ is *strongly* or *asymptotically stable* if it is an attractor for *all* phase curves passing through a neighbourhood $\mathbf{r}_f$. A typical strongly stable fixed point is shown in figure 16.18, because all curves passing through the circle tend to the fixed point at its centre.

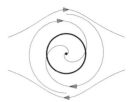

Figure 16.18 A strongly stable
fixed point.

Exercise 16.36
Is it possible for a single fixed point to be simultaneously an attractor and a repellor of the
same motion?

16.3.5 Classification of simple fixed points
In order to understand the behaviour of a system it is necessary to understand how it
behaves in the neighbourhood of the fixed points, because the nature and distribution
of these points play a significant role in controlling the overall behaviour of the system.
This type of analysis was first performed by Poincaré (1881) in his early work which
was a prelude to his profound analysis of the three-body problem in which chaotic
motion was first discovered, Poincaré (1890, 1892).

In order to understand and classify the motion in the vicinity of fixed points we need
to make some assumptions about the nature of the velocity function, otherwise there
are too many possibilities. We shall assume that its Taylor's series expansion about any
fixed point of interest exists and that we can ignore all second-order terms. The first
assumption is satisfied by many physical system that occur in practice and we shall
discuss the implications of the second assumption in section 16.3.6.

Expanding the velocity function, $\mathbf{v} = (X, Y)$, about the fixed point $\mathbf{r}_f$ gives

$$X(\mathbf{r} - \mathbf{r}_f) = (x - x_f)\frac{\partial X}{\partial x} + (y - y_f)\frac{\partial X}{\partial y} + O(|\mathbf{r} - \mathbf{r}_f|^2),$$

$$Y(\mathbf{r} - \mathbf{r}_f) = (x - x_f)\frac{\partial Y}{\partial x} + (y - y_f)\frac{\partial Y}{\partial y} + O(|\mathbf{r} - \mathbf{r}_f|^2),$$

where all derivatives are evaluated at the fixed point. We obtain the linearised equations
of motion of the nonlinear equations $\dot{\mathbf{r}} = \mathbf{v}(\mathbf{r})$ by ignoring the second-order terms:

$$\frac{d\mathbf{z}}{dt} = A\mathbf{z}, \quad \mathbf{z} = \mathbf{r} - \mathbf{r}_f, \quad A = \begin{pmatrix} \dfrac{\partial X}{\partial x} & \dfrac{\partial X}{\partial y} \\ \dfrac{\partial Y}{\partial x} & \dfrac{\partial Y}{\partial y} \end{pmatrix}, \tag{16.30}$$

the elements of A being evaluated at the fixed point. It is convenient to write

$$A = \begin{pmatrix} a & b \\ c & d \end{pmatrix}, \tag{16.31}$$

where all the elements are real.

A *simple* fixed point is one for which the determinant of the matrix A is nonzero. In the neighbourhood of a simple fixed point we shall show that the shape of the phase curves (of the linear system) depends only upon the nature of the eigenvalues of A which, in turn, depend only upon the trace and determinant of A. This is seen by transforming to a new coordinate system $\mathbf{u} = M\mathbf{z}$ where M is a constant, non-singular 2×2 matrix; if $\det(M) > 0$ the transformation is orientation preserving. In this coordinate system the linearised equations of motion, equation 16.30, become

$$\frac{d\mathbf{u}}{dt} = B\mathbf{u}, \quad B = MAM^{-1}. \tag{16.32}$$

The matrix M can always be chosen to cast B into one of three distinct types, the actual form being determined only by the eigenvalues of A,

$$\lambda = \frac{1}{2}\left[\mathrm{Tr}(A) \pm \sqrt{\mathrm{Tr}(A)^2 - 4\det(A)}\right], \tag{16.33}$$

where the trace and determinant are

$$\mathrm{Tr}(A) = a + d, \quad \det(A) = ad - bc.$$

Each of these three types produces distinctly different shaped phase curves.

Type 1: Eigenvalues of A, λ_1 and λ_2, real and distinct

In this case it is convenient to label the eigenvalues so that $\lambda_2 > \lambda_1$ and to choose M to be

$$M = \begin{pmatrix} c & \lambda_1 - a \\ |c| & (\lambda_2 - a)\mathrm{sgn}(c) \end{pmatrix}, \quad \text{giving} \quad B = \begin{pmatrix} \lambda_1 & 0 \\ 0 & \lambda_2 \end{pmatrix}. \tag{16.34}$$

With this choice $\det(M) > 0$. The linearised equations of motion in the $\mathbf{u}$-representation are

$$\dot{u}_1 = \lambda_1 u_1, \quad \dot{u}_2 = \lambda_2 u_2,$$

with solutions $u_1 = \alpha e^{\lambda_1 t}$, $u_2 = \beta e^{\lambda_2 t}$, from which t may be eliminated to give the equation $(u_1/\alpha)^{\lambda_2} = (u_2/\beta)^{\lambda_1}$ for the phase curves. There are now three situations to consider depending upon the signs of the eigenvalues.

If both eigenvalues are positive, $\lambda_2 > \lambda_1 > 0$, both $u_1(t)$ and $u_2(t)$ increase with increasing t and the fixed point is unstable; this type of fixed point is named an *unstable node* and some typical phase curves are shown in figure 16.19. Adjacent to these, figure 16.20, is shown some typical phase curves when transformed back into the **r**-representation.

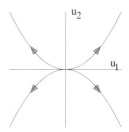

Figure 16.19   An unstable node in the
u-representation, $\lambda_2 > \lambda_1 > 0$.

Figure 16.20   An unstable node in the
r-representation, $\lambda_2 > \lambda_1 > 0$.

If both eigenvalues are negative the shape of the phase curves are similar but the flow is in the opposite direction, so the fixed point is strongly stable and is named a *stable node*. An example of this type of flow, for the case $\lambda_1 < \lambda_2 < 0$, is shown in figure 16.21, in the **u**-representation, and figure 16.22, in the original representation.

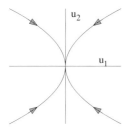

Figure 16.21   A stable node in the
u-representation, $\lambda_1 < \lambda_2 < 0$.

Figure 16.22   A stable node in the
r-representation, $\lambda_1 < \lambda_2 < 0$.

If the eigenvalues have opposite sign, $\lambda_1 < 0 < \lambda_2$, the u_1-subsystem is stable and the u_2-subsystem is unstable, hence the fixed point is unstable but not strongly unstable. Now the phase curves are given by an equation of the form $u_1^{\lambda_2} u_2^{|\lambda_1|} = $ constant and look like generalised hyperbolas, as shown in figures 16.23 and 16.24. This type of fixed point is named a *saddle point* or, in some texts, a *hyperbolic fixed point*.

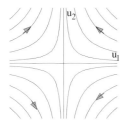

Figure 16.23   A saddle in the
u-representation, $\lambda_1 < 0 < \lambda_2$.

Figure 16.24   A saddle in the
r-representation, $\lambda_1 < 0 < \lambda_2$.

The eigenvalues determine the nature of the fixed point, but for saddles and nodes it is often useful to compute the eigenvectors as these show the direction of the motion flowing directly away or towards the fixed point; further, the orbits that leave the fixed point along these eigenvectors can determine important boundaries in phase space, as in exercises 16.25 and 16.48.

Type 2: Complex eigenvalues, $\lambda = \alpha + i\omega$, $\lambda^* = \alpha - i\omega$, $\mathrm{sgn}(\omega) = \mathrm{sgn}(c)$
In this case we take the transformation matrix M, equation 16.32, to be

$$M = \begin{pmatrix} c & \alpha - a \\ 0 & \omega \end{pmatrix}, \quad \text{giving} \quad B = \begin{pmatrix} \alpha & -\omega \\ \omega & \alpha \end{pmatrix} = |\lambda| \begin{pmatrix} \cos\phi & -\sin\phi \\ \sin\phi & \cos\phi \end{pmatrix}, \quad (16.35)$$

where ϕ is the phase of the eigenvalue, $\alpha + i\omega = \sqrt{\alpha^2 + \omega^2}e^{i\phi}$. The matrix B represents a rotation through the angle ϕ and a uniform scaling which is a contraction if $|\lambda| < 1$ and an expansion if $|\lambda| > 1$.

This geometric interpretation of B suggests that the equations of motion $\dot{\mathbf{u}} = B\mathbf{u}$ would be simpler if expressed in terms of the polar coordinates, $\mathbf{u} = (r\cos\theta, r\sin\theta)$. In this representation the equations of motion are

$$\frac{dr}{dt} = \alpha r, \quad \frac{d\theta}{dt} = \omega, \tag{16.36}$$

and have the solution

$$r(t) = r_0 e^{\alpha t}, \quad \theta(t) = \omega t + \theta_0.$$

Thus if the real part of the eigenvalue, α, is negative the fixed point is strongly stable. If, in addition, $\omega > 0$ the phase point rotates anticlockwise, as shown in figure 16.25; in the adjacent figure these curves are shown after transforming back into the **r**-representation. This type of fixed point is named a *stable spiral*, or in some texts a *focus*.

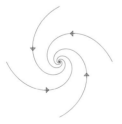

Figure 16.25   Stable spiral in the **u**-representation, $\alpha < 0$, $\omega > 0$.

Figure 16.26   Stable spiral in the **r**-representation, $\alpha < 0$, $\omega > 0$.

Changing the sign of ω, that is, c, changes the direction of motion, but not the stability of the fixed point.

On the other hand, changing the sign of α changes stability of the fixed point and if $\alpha > 0$ it is an *unstable spiral*; an example with $\alpha > 0$ and $\omega < 0$, unstable clockwise spirals, is shown in the next figure.

Figure 16.27   Unstable spiral in the **u**-representation, $\alpha > 0$, $\omega < 0$.

Figure 16.28   Unstable spiral in the **r**-representation, $\alpha > 0$, $\omega < 0$.

If $\alpha = 0$, so that $r(t) = $ constant, then the phase curves in the **u**-representation are circles, with the motion clockwise if $\omega < 0$: these circles become ellipses in the original **r**-representation, as seen in the adjacent figure. This type of fixed point is named a *centre* or an *elliptic fixed point*.

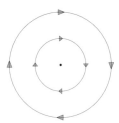

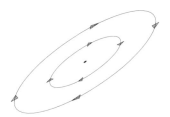

Figure 16.29   Centre in the **u** representation, $\alpha = 0$, $\omega < 0$.

Figure 16.30   Centre in the **r** representation, $\alpha = 0$, $\omega < 0$.

Type 3: Eigenvalues real and equal
The eigenvalues are real and equal if

$$\mathrm{Tr}(A)^2 = 4\det(A), \quad \text{that is,} \quad (a - d)^2 + 4bc = 0, \tag{16.37}$$

and this equation can be satisfied in two ways.

First, if $b = c = 0$ and $a = d$, A is already diagonal and the linearised equations of motion, 16.30, are

$$\dot{z}_1 = az_1, \quad \dot{z}_2 = az_2, \quad \text{with solutions} \quad \mathbf{z} = \mathbf{z}_0 e^{at}.$$

Thus both z_1 and z_2 increase $(a > 0)$, or decrease $(a < 0)$, at the same rate. All phase

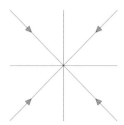

Figure 16.31   A stable star, $b = c = 0$,
$a = d < 0$.

curves are therefore the straight lines $z_1 = \alpha z_2$, as shown in the figure above.

This type of fixed point is named a *stable star* if $a = d < 0$ and an *unstable star* if $a = d > 0$: in the latter case the phase curves are the same as shown in the figure but the direction of the arrows is reversed. A stable star is strongly stable.

The second way in which equation 16.37 can be satisfied is if at least one of b and c is nonzero, then A has only one linearly independent eigenvector. The matrix M may be chosen to be

$$M = \begin{pmatrix} a - d & 2b \\ 2c & 0 \end{pmatrix}, \quad \text{giving} \quad B = \begin{pmatrix} \bar{a} & 0 \\ c & \bar{a} \end{pmatrix}, \quad \bar{a} = \frac{1}{2}(a + d),$$

when $c \neq 0$. The equations of motion, 16.30, are now

$$\dot{z}_1 = \bar{a} z_1, \quad \dot{z}_2 = c z_1 + \bar{a} z_2,$$

and have the solution

$$z_1 = \alpha e^{\bar{a}t}, \quad z_2 = (\beta + c\alpha t)e^{\bar{a}t}$$

for some constants $\alpha = z_1(0)$ and $\beta = z_2(0)$.

If $\mathrm{Tr}(A) = 2\bar{a} < 0$ the fixed point is stable and for sufficiently large times the β term may be neglected to give

$$z_1 = \alpha e^{\bar{a}t}, \quad z_2 = c\alpha t e^{\bar{a}t}, \quad \text{or} \quad z_2 = \frac{c}{|\bar{a}|} z_1 \ln(\alpha/z_1), \quad \bar{a} < 0, \quad \frac{z_1}{\alpha} > 0.$$

If $c > 0$ the phase curve crosses the z_1-axis at $z_1 = \alpha$ and approaches the origin along the positive z_2-axis as $t \to \infty$. As $t \to -\infty$, z_2 becomes negative so the phase curves go to infinity in the fourth quadrant, as shown in figure 16.32.

If $c > 0$ and $\alpha = z_1(0) < 0$ the phase curve approaches the origin along the positive z_2-axis as $t \to \infty$. As $t \to -\infty$, the phase curves go to infinity in the second quadrant. If $\bar{a} < 0$ and $c < 0$ the fixed point remains stable and the phase curves look like those shown in figure 16.33.

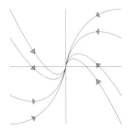

Figure 16.32   Stable improper node $\bar{a} < 0$
and $c > 0$.

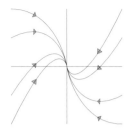

Figure 16.33   Stable improper node $\bar{a} < 0$
and $c < 0$.

This type of fixed point is named a *stable improper node* if $\mathrm{Tr}(A) < 0$ or an *unstable improper node* if $\mathrm{Tr}(A) > 0$.

If $\bar{a} > 0$ the fixed point is unstable but the phase curves have the same shape as those shown in the figure, though the direction of the arrows is reversed.

Summary of classifications

The nature of the phase curves in the neighbourhood of a simple fixed point depends only upon the eigenvalues of the linearisation matrix A, defined in equation 16.30, page 657. We have seen that there are ten distinctly different types of flow.

Eigenvalues real and different: $\mathrm{Tr}(A)^2 > 4\det(A)$

>   **Stable node** Eigenvalues real and negative.
>
>   **Unstable node** Eigenvalues real and positive.
>
>   **Saddle** Eigenvalues real and of different sign.

Eigenvalues complex: $\mathrm{Tr}(A)^2 < 4\det(A)$

>   **Stable spiral** Real part of eigenvalues negative.
>
>   **Unstable spiral** Real part of eigenvalues positive.
>
>   **Centre** Eigenvalues imaginary.

Eigenvalues the same: $\mathrm{Tr}(A)^2 = 4\det(A)$

>   **Stable star** A is diagonal, $b = c = 0$ and $\mathrm{Tr}(A) < 0$.
>
>   **Unstable star** A is diagonal, $b = c = 0$ and $\mathrm{Tr}(A) > 0$.
>
>   **Stable improper node** A is not diagonal and $\mathrm{Tr}(A) < 0$.
>
>   **Unstable improper node** A is not diagonal and $\mathrm{Tr}(A) > 0$.

This completes the classification of simple fixed points. The most important point of this analysis is that the nature of the linearised flow depends only upon the values of the eigenvalues of the matrix A, defined in equation 16.30, which in turn depend only upon the trace and determinant of A.

Exercise 16.37
Derive equation 16.36.

Exercise 16.38
Classify the fixed points of the following linear systems and state whether they are unstable, stable or strongly stable:

(i) $\dot{x} = 3x + 4y,$ $\dot{y} = 2x + y,$ (ii) $\dot{x} = 3x,$ $\dot{y} = 2x + y,$

(iii) $\dot{x} = x + 2y,$ $\dot{y} = -2x + 5y,$ (iv) $\dot{x} = x + 4y,$ $\dot{y} = -x + y,$

(v) $\dot{x} = 2x - y,$ $\dot{y} = -x - 2y,$ (vi) $\dot{x} = x - 2y,$ $\dot{y} = 2x - y,$

(vii) $\dot{x} = -2x,$ $\dot{y} = -2y,$ (viii) $\dot{x} = x + 3y,$ $\dot{y} = -2x - 4y.$

Exercise 16.39
Convert the following second-order differential equations to canonical form and find and classify their fixed points:

(i) $\ddot{x} + k(x^2 - 1)\dot{x} + x = 0, \quad k > 0,$
(ii) $\ddot{x} - \dot{x}^2 + x^2 - x = 0.$

If any fixed points are centres you should use Maple to examine some numerical solutions to guess its true character.

Exercise 16.40
If the matrix A is real and symmetric, that is, $b = c$, what type of isolated fixed points can the system have?

16.3.6 The linearisation theorem
The classification scheme described above used the linear system, $\dot{\mathbf{z}} = A\mathbf{z}$, equation 16.30 (page 657), obtained by ignoring all terms $O(|\mathbf{r} - \mathbf{r}_f|^2)$. We need to know whether this classification is valid for the original nonlinear equations $\dot{\mathbf{r}} = \mathbf{v}(\mathbf{r})$. The linearisation theorem provides the crucial link between the flows of these two related systems.

Linearisation theorem If the nonlinear system

$$\dot{\mathbf{r}} = \mathbf{v}(\mathbf{r})$$

has a simple fixed point at $\mathbf{r}_f$, then there is a neighbourhood of $\mathbf{r}_f$ in which the phase portrait of the system and its linearisation are qualitatively equivalent *provided* that the fixed point of the linearised system is *not a centre*.

The proof of this statement can be found in Hartman (1964, chapter 8).

This theorem is the justification for ignoring the second-order terms in equation 16.30, and it means that a fixed point of the nonlinear system is stable whenever the linearised system is strongly stable and unstable whenever the linearised system is unstable.

If the fixed point is a centre for the linearised system then $\mathrm{Tr}(A) = 0$ and any small perturbation, such as the ignored nonlinear terms, could make $\mathrm{Tr}(A)$ nonzero and hence change the centre into a spiral. For centres more investigation is needed to determine the true nature of the fixed point: we return to this problem in section 17.4. Thus when stating that a fixed point is a centre it is essential to also state whether the classification is for the original system or its linearisation.

Exercise 16.41

Show that the origin is a centre of the linearisation of the system

$$\dot{x} = y - x(x^2 + y^2), \quad \dot{y} = -x - y(x^2 + y^2).$$

Show that in polar coordinates, $x = r\cos\theta$, $y = r\sin\theta$, the equations of motion are

$$\dot{r} = -r^3, \quad \dot{\theta} = -1,$$

and hence deduce that the fixed point at the origin is not a centre of the nonlinear system but is strongly stable.

Exercise 16.42

Volterra's problem

Volterra (1926) formulated a simple population model for two species, A and B, in which A has an external food source, say vegetation (assumed available in unlimited quantities), whereas B feeds exclusively on A. The equations that approximate the populations of these species are

$$\frac{dN_A}{dt} = aN_A(1 - \alpha N_B), \quad N_A(t) \geq 0,$$

$$\frac{dN_B}{dt} = -bN_B(1 - \beta N_A), \quad N_B(t) \geq 0,$$

where the constants a, α, b and β are all positive. The parameter a represents the natural undisturbed growth of species A and αN_B the decrease of this growth due to the predator population B. The parameter b is the death rate of B in the absence of A, its sole food source. The rate of growth of the population B is assumed to depend upon the food supply and is taken to be $b(1 - \beta N_A)$.

Introduce the scaled populations $x = \beta N_A$, $y = \alpha N_B$ and the scaled time $\tau = bt$ to write the equations in the simple form, depending upon only one parameter,

$$\frac{dx}{d\tau} = \frac{a}{b}x(1 - y), \quad \frac{dy}{d\tau} = -y(1 - x).$$

Show that the fixed points are at $(0,0)$ and $(1,1)$ and that the latter is a centre.

Find an integral of the motion for this system and use this to show that all solutions of these equations are periodic.

The same equations were derived by Lotka (1920) to describe certain chemical reactions and Volterra derived the equations to approximate the population of fish in the Adriatic sea; hence these equations are often called the Volterra–Lotka equations.

16.3.7 The Poincaré index

The notion of an index was introduced by Poincaré in his 1881 paper in order to establish a *necessary* criterion for the existence of periodic orbits. The Poincaré index is an invariant of closed curves in phase space, which *need not* be orbits. They are important because they impose restrictions upon the way fixed points can be distributed and hence the manner in which the system can behave. For instance, if Γ is a periodic orbit inside of which there are only simple fixed points, N nodes, S_p spirals, S_a saddles and C centres, then we must have $N + S_p + C - S_a = 1$. In addition if the velocity function $\mathbf{v}(\mathbf{r}, a)$ depends upon a parameter a and has a fixed point at $\mathbf{r}_f(a)$ it can be shown that as *a* varies this fixed point can change its nature in only a few ways; for instance, it cannot just disappear and a centre cannot become a saddle.

Consider the general autonomous system $\dot{x} = X(x, y)$, $\dot{y} = Y(x, y)$. At every ordinary point† in the phase plane $\phi(\mathbf{r})$ is defined to be the angle between the vector $\mathbf{v} = (X(x, y), Y(x, y))$ and the *x*-axis, that is,

$$\tan \phi = \frac{Y(x, y)}{X(x, y)}. \tag{16.38}$$

If Γ is a closed curve on which there are no fixed points $\phi(\mathbf{r})$ varies continuously as $\mathbf{r}$ moves along the curve, and if traversed once the vector $\mathbf{v}$ returns to its initial direction, consequently the angle $\phi(\mathbf{r})$ must change by an integer multiple of 2π; for an anticlockwise traversal this multiple is named the *Poincaré index* and it may be zero, positive or negative. If the curve Γ coincides with a periodic orbit then its index must be $+1$, exercise 16.43.

The change in $\phi(\mathbf{r})$ along any curve $\mathscr{C}$ is

$$\Delta\phi = \int_{\mathscr{C}} d\phi = \int_{\mathscr{C}} \frac{X\,dY - Y\,dX}{X^2 + Y^2},$$

and hence the Poincaré index I_Γ along the closed curve Γ is

$$I_\Gamma = \frac{1}{2\pi} \int_\Gamma \frac{X\,dY - Y\,dX}{X^2 + Y^2}, \tag{16.39}$$

where the curve Γ is traversed anticlockwise. This is not a pleasant integral and it is somewhat surprising that its value is an integer: fortunately we do not need to evaluate it. The most important point to notice is that the denominator of the integrand is $X^2 + Y^2$ so that the integrand is not defined at a fixed point where $X = Y = 0$, hence we require that Γ does not pass through any fixed points, although there may be fixed points inside Γ. Observe that the value of I_Γ is unchanged if the signs of both X and Y are changed, that is if the direction of the motion is changed.

Consider two closed curves Γ_1 and Γ_2 such that Γ_1 can be continuously deformed into Γ_2 in such a manner that the number of fixed points inside the curve is constant and none appear on the boundary, so the integrand is always finite. The index must vary continuously with Γ and since its value is an integer it must be constant. This means that we can define the index of a fixed point, which need not be simple, to be the index along any curve which contains only that fixed point.

Further, if Γ contains no fixed points it may be deformed to an arbitrarily small curve, so I_Γ is arbitrarily small and hence must be zero.

It also follows that if the curve Γ contains a set of N fixed points, which need not be simple, then

$$I_\Gamma = \sum_{k=1}^{N} I_k, \tag{16.40}$$

where I_k is the index of the *k*th fixed point. The diagram in figure 16.34 shows why this result is true, in the case of two fixed points. On the left are two closed curves, *A*

† At an ordinary point both X and Y are defined and at least one is not zero.

and B each enclosing a single fixed point, and an outer curve C surrounding both of these.

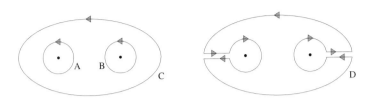

Figure 16.34   Figure 16.35

In figure 16.35 is a single closed curve D obtained from the three curves on the left by making the two cuts shown. The closed curve D contains no fixed points and so $I_D = 0$. On D the curves A and B are traversed clockwise, and since the contribution to the integral along the parallel lines, traversed in opposite directions, cancel, we have $I_D = I_C - I_A - I_B$ and consequently $I_C = I_A + I_B$. Clearly this result can be extended to include any number of fixed points.

Finally, note that if the velocity function depends continuously upon some parameter $\mathbf{a}$ then the index I_Γ is independent of $\mathbf{a}$ provided changes in $\mathbf{a}$ do not cause any fixed points to cross Γ. This is because I_Γ is continuous in $\mathbf{a}$, and being an integer must be constant.

To summarise, the properties of the Poincaré index are as follows.

 (i) If Γ_1 is a closed curve obtained from another closed curve Γ by a continuous transformation such that no fixed points cross the boundary then $I_\Gamma = I_{\Gamma_1}$.
 (ii) The index of a closed curve containing no fixed points is zero.
 (iii) The index of a periodic orbit is $+1$.
 (iv) The index of a node, spiral, star and a centre is $+1$.
 (v) The index of a saddle is -1.
 (vi) The index is additive, equation 16.40.
 (vii) The index of a curve is unchanged if the velocity function $\mathbf{v}$ is replaced by $-\mathbf{v}$; this is equivalent to changing the direction of time.

The index of a periodic orbit is $+1$, consequently a periodic orbit must contain fixed points with indices summing to $+1$: alternatively a system with no fixed points or with indices that do not sum to $+1$ has no periodic orbit.

Exercise 16.43
Show that the index of a centre is $+1$. More generally, show that the index of any periodic orbit is $+1$.

Exercise 16.44
Show that the index of a centre, spiral and node, including an improper node, is $+1$ and that the index of a saddle is -1.

Exercise 16.45

Show that the system

$$\frac{dx}{dt} = 1 - xy, \quad \frac{dy}{dt} = x + y^3$$

has no periodic orbits.

Exercise 16.46

Consider the system

$$\dot{x} = y, \quad \dot{y} = -vx - 2ky - x^3, \quad k > 0.$$

If $v > 0$ show that there is a saddle at $x = y = 0$ and show how this fixed point changes as v decreases and becomes negative. As v passes through zero confirm that the number of fixed points in the vicinity of the origin changes but that the index on a curve containing all these fixed points does not.

Exercise 16.47

Determine the index of the fixed point at the origin for the system $\dot{x} = y^2$, $\dot{y} = -x^3$.

Exercise 16.48

Use Maple to determine the position and nature of the fixed points of the system defined in exercise 16.25 (page 652), and explain the behaviour of the solution computed in that exercise.

Exercise 16.49

Determine the fixed points of the system defined in exercise 16.28 (page 652). Classify these for $0 < \mu < 1$ and explain the behaviour of the numerical solutions.

16.4 Appendix: existence and uniqueness theorems

Consider the autonomous system $\dot{\mathbf{r}} = \mathbf{v}(\mathbf{r})$: in the neighbourhood of an ordinary point $\mathbf{r}_0$ there exists a differentiable change of coordinates $\mathbf{u} = \mathbf{r}(\mathbf{u})$ such that the equations of motion become $\dot{\mathbf{u}} = (1, 0)$, provided $\mathbf{v}(\mathbf{r})$ is smooth and has an inverse which is also smooth in the neighbourhood; in particular this means that the neighbourhood must not contain any fixed points.

The existence of solutions in the neighbourhood of ordinary points follows from this result: the solution of the equation $\dot{\mathbf{u}} = (1, 0)$ is $\boldsymbol{\phi}(t) = (u_0 + t, v_0)$ for $|t|$ sufficiently small. Uniqueness also follows, for if $\boldsymbol{\phi}_1$ and $\boldsymbol{\phi}_2$ are two solutions satisfying the same initial conditions then $\boldsymbol{\phi}_1 - \boldsymbol{\phi}_2 = 0$. At a fixed point $\mathbf{v}(\mathbf{r}_0) = 0$ the solution through $\mathbf{r}_0$ is $\boldsymbol{\phi}(t) = \mathbf{r}_0$. Existence and uniqueness follow.

Note that these results are also true for systems of order n where $\mathbf{r} = (x_1, x_2, \ldots, x_n)$.

Counter examples are always helpful in understanding the value of such theorems. The first-order system $\dot{x} = |x|^{1/2}$, $x(0) = 0$, has two solutions $x(t) = 0$ and $x(t) = t^2/4$ because the velocity function is not differentiable at the origin.

The reader will have noted that the above results are valid only for short, or at most finite, times and in many applications we are interested in the long-time behaviour of solutions. If the velocity function $\mathbf{v}(\mathbf{r})$ is defined on a compact manifold then the system has a unique solution for all real times. Examples of compact manifolds are a circle, a sphere, and a torus and in these cases there can be no terminating motion.

16.5 Exercises

Exercise 16.50
Discuss the nature of the motion and give the invariant sets of the system with velocity functions

$$\text{(i)} \quad v(x) = x \sin x \quad \text{and} \quad \text{(ii)} \quad v(x) = x \cos x.$$

Exercise 16.51
For each of the following velocity functions, give a qualitative description of the motion of the system and obtain expressions for the motion in the vicinity of any fixed points.

$$\text{(i)} \ v(x) = a\sqrt{x - b}, \quad a > 0, \ x \geq b, \qquad \text{(ii)} \ v(x) = a\sqrt{b - x}, \quad a > 0, \ x \leq b.$$

Exercise 16.52
Here you will prove the rather unsurprising result that the larger the velocity function the shorter the time taken to travel between two points.

Consider the two velocity functions $v_1(x)$ and $v_2(x)$ such that for some constants A and B with $A < B$,

$$0 < v_1(x) \leq v_2(x) \quad \text{when} \quad A \leq x \leq B,$$

and that $v_1(x)$ and $v_2(x)$ are finite. If T_k is the time system k takes to go from $x = A$ to $x = B$ show that $T_2 \leq T_1$.

Exercise 16.53
A point moves round a circle in such a way that

$$\dot{\theta} = \sqrt{a + \cos \theta},$$

where θ is the angle between the radius to the point and a fixed radius and a is a constant with $a > -1$. For what values of a is the motion a rotation? Show that the period of this rotational motion is, for large a,

$$T = \frac{2\pi}{\sqrt{a}} \left(1 + \frac{3}{16a^2} + \frac{105}{1024a^4} + O(a^{-6}) \right).$$

Show also that $T(a)$ diverges logarithmically as $a \to 1$ from above.

Exercise 16.54
Air resistance can often be assumed to be proportional to the square of the speed. If V is the velocity of a body falling vertically in the Earth's gravitational field, then for $V \geq 0$,

$$\frac{dV}{dt} = g - kV^2, \quad k > 0,$$

where g is the acceleration due to gravity and k is a constant. If initially $V(0) = 0$, describe the behaviour of $V(t)$ and determine its limiting value as $t \to \infty$.

Exercise 16.55

Consider a first-order system with velocity function $v(x,t)$ which is homogeneous of degree zero, that is, $v(ax,at) = v(x,t)$ for all nonzero a.

Show that a change of variables exists in which the system $\dot{x} = v(x,t)$ reduces to

$$\frac{du}{d\tau} = v(u,1) - u.$$

Show that a fixed point of this autonomous system corresponds to uniform motion $x = kt$, for some constant k of the original system. Discuss the nature of the motion of the original system as $t \to \infty$ when the fixed point of the autonomous system is simple and stable.

Exercise 16.56

The simplest and crudest method of solving a first-order differential equation numerically is by taking a set of equally spaced points $t_k = k\delta t$ for the independent variable t and approximating the differential by a difference $\dot{x}(t_{k+1}) \simeq (x(t_{k+1}) - x(t_k))/\delta t$. Writing $x_k = x(t_k)$ use this approximation to show that the logistic equation 16.11 transforms to the logistic map and find w_k and b in terms of x_k, a, X and δt.

Exercise 16.57

(i) Show that the equation of motion 16.26 of the vertical pendulum can be integrated to give the energy equation

$$\frac{1}{2}\left(\frac{dx}{dt}\right)^2 = W + \alpha^2 \cos x \quad (\geq 0),$$

where W is a constant. Deduce that for $-\alpha^2 \leq W < \alpha^2$ the motion is librational with amplitude $X = \cos^{-1}(-W/\alpha^2)$, and for $W > \alpha^2$ the motion is rotational.

(ii) The smallest allowed value of W is $-\alpha^2$, and for $W \simeq -\alpha^2$ the amplitude of the motion is small. Put $W = -\alpha^2 + \epsilon^2/2$ and show that to lowest order the energy equation becomes $\dot{x}^2 + \alpha^2 x^2 = \epsilon^2$ with solution $x(t) = \epsilon \sin(\alpha t + \delta)$ and period $T = 2\pi\sqrt{l/g}$, where δ is a constant, depending upon the initial conditions. This represents simple harmonic motion and has the important property that the period is independent of the amplitude of the motion.

(iii) Show that the period of librational motion is given by

$$T = 2\sqrt{2}\int_0^X dx\, \frac{1}{\sqrt{W + \alpha^2 \cos x}} = \frac{4}{\alpha}K(1/k), \quad k^2 = \frac{2\alpha^2}{W + \alpha^2},$$

where K is the complete elliptic integral of the second kind, see Appendix II, and that if $W = \alpha^2(1 - 2\epsilon^2)$, $0 < \epsilon \ll 1$, then $T(W) \sim (4/\alpha)\ln(4/\epsilon)$.

(iv) Show that the period of the rotational motion is

$$T = \sqrt{2}\int_0^\pi dx\, \frac{1}{\sqrt{W + \alpha^2 \cos x}} = \frac{2k}{\alpha}K(k), \quad k^2 = \frac{2\alpha^2}{W + \alpha^2} > 0,$$

where $K(k)$ is the complete elliptic integral of the first kind.

Show also that if $W = \alpha^2(1 + 2\epsilon^2)$, where $0 < \epsilon \ll 1$, then $T \sim (2/\alpha)\ln(4/\epsilon)$.

(v) The pendulum motion with $W = \alpha^2$ is special because its phase curve separates the rotational and the librational regions of phase space. Show that on this phase curve the energy equation becomes $\dot{x} = \pm 2\alpha \cos(x/2)$, and if $x(0) = 0$ and $\dot{x}(0) = 2\alpha$ the subsequent motion is $x(t) = 4\tan^{-1}\tanh(\alpha t/2) \simeq \pi - 4e^{-\alpha t}$ as $t \to \infty$. This motion never reaches the upward vertical position; as $W \to \alpha^2$ the periodic motion approaches this asymptotic solution and the period of motion tends to infinity. For this system the solution with $W = \alpha^2$ is the only non-periodic motion.

Exercise 16.58
Find and classify the fixed points of the system

$$\frac{dx}{dt} = 2xy - 4y - 8, \quad \frac{dy}{dt} = 4y^2 - x^2.$$

Use Maple to plot the phase curves in the region $-5 \le x \le 3$, $-3 \le y \le 3$ and show, numerically, that solutions with initial conditions $x(0) = 0$, $\dot{x}(0) > 0$ either tend to $x = -1$ or to $-\infty$ as $t \to \infty$.

Exercise 16.59
Show that the fixed points of the second-order system defined by the equation

$$\frac{d^2x}{dt^2} = F\left(x, \frac{dx}{dt}\right)$$

all lie on the x-axis of the phase plane and that their x-coordinates are found by solving the equation $F(x, 0) = 0$.

If $F(x, \dot{x})$ is an even function of $\dot{x}$ show that the fixed points are all either centres or saddles.

Exercise 16.60
Show that the second-order autonomous system obtained from the first-order system $\dot{x} = v(x, t)$ has no fixed points.

Exercise 16.61
Find and classify the fixed points of the following system:

(i) $\dot{x} = x + 3y + 4$, $\dot{y} = -6x + 5y - 1$,
(ii) $\dot{x} = x + 3y + 1$, $\dot{y} = -6x + 5y + 1$,
(iii) $\dot{x} = 3x + y + 1$, $\dot{y} = -x + y - 6$.

Exercise 16.62
Given any differentiable function $F(x, y)$ we can form the *gradient system* by defining the velocity function to be $\mathbf{v} = \text{grad } F$, so the equations of motion are

$$\frac{dx}{dt} = \frac{\partial F}{\partial x}, \quad \frac{dy}{dt} = \frac{\partial F}{\partial y}.$$

Show that the fixed points of the system occur at the stationary points of $F(x, y)$ and that if the fixed point is simple the saddles of F are saddle points, and determine the nature of the fixed points at maxima and minima of F.

Exercise 16.63
Find and classify the fixed points of the following nonlinear systems:

(i) $\ddot{x} - \dot{x} + x^2 - 2x = 0$,   (ii) $\ddot{x} - \dot{x}^3 + x + 5 = 0$.

In each case use Maple to plot some representative phase curves.

Exercise 16.64

Determine all the possible types of fixed points of the system $\dot{\mathbf{x}} = A\mathbf{x}$ when the real, 2×2, matrix A is anti-symmetric, that is, $A_{12} = -A_{21}$.

Exercise 16.65

For autonomous systems the phase curves can be obtained directly from equation 16.18 (page 642). Show that the same method can be used for the non-autonomous system

$$\frac{dx}{dt} = f(t)X(x, y), \quad \frac{dy}{dt} = f(t)Y(x, y),$$

where $f(t)$ is a function of t only and both X and Y are independent of t. Show also that this system can be converted to an autonomous system.

Exercise 16.66

Show that the origin is a strongly stable fixed point of the system $\dot{x} = -x^2$, $\dot{y} = -y^2$, but not of the system $\dot{x} = -y^2$, $\dot{y} = -x^2$.

Exercise 16.67

A cell population consisting of a mixture of 2-chromosome and 4-chromosome cells is described approximately by

$$\frac{dx}{dt} = (\lambda - \mu)x, \quad \frac{dy}{dt} = \mu x + \nu y,$$

where x is the number of 2-chromosome cells, y is the number of 4-chromosome cells and λ, μ and ν are constants with $\lambda \neq \mu$ and $\nu > 0$.

Show that whatever the values of λ, μ and ν as $t \to \infty$ the proportion of 2-chromosome cells in the population tends to a value independent of the initial conditions. For what parameter values is this limiting ratio nonzero?

Exercise 16.68

Consider the system $\dot{x} = X(x, y)$, $\dot{y} = Y(x, y)$. Show that a necessary condition for the function $F(x, y)$ to be constant along all phase curves is that it should satisfy the partial differential equation

$$X(x, y)\frac{\partial F}{\partial x} + Y(x, y)\frac{\partial F}{\partial y} = 0.$$

For the case $(X, Y) = (xy, \ln x)$ show that the phase curves are given by $y^2 - (\ln x)^2 = $ constant and deduce that the solution of the equation

$$xy\frac{\partial f}{\partial x} + \frac{\partial f}{\partial y}\ln x = 0$$

is $f(x, y) = y^2 - (\ln x)^2$.

17

Dynamical systems II: periodic orbits

17.1 Introduction

Periodic solutions of differential equations are important partly because they organise phase space and partly because they are readily observable — periodic phenomena being easier to distinguish than the less ordered kind; recall, for instance, the discussion of Halley's comet on page 631. This chapter deals with some elementary aspects of periodic solutions, using traditional methods primarily based on perturbation theory. Perturbations, however, may produce a qualitative change to a periodic solution that can only be understood using non-perturbative methods, some of which are developed in the next chapter.

For autonomous two-dimensional systems there are two main types of such motion: the isolated periodic orbits, or limit cycles, and continuous families of periodic orbits, typical of systems with integrals of the motion. Given any two-dimensional autonomous system it is usually far from obvious that periodic orbits exist. Hence this chapter starts with statements of some theorems that provide conditions necessary for the existence, or otherwise, of periodic solutions. This is followed by a discussion of limit cycles, which normally enclose spiral fixed points and are often associated with relaxation oscillations. Families of periodic orbits, on the other hand, normally surround centres. The linearisation theorem quoted in chapter 16 shows that centres are delicate objects and can be destroyed by the higher-order terms of the system. A nonlinear system has a centre only if an integral of the motion exists in its neighbourhood, and this is defined in section 17.4 where we also provide a formal method of analysing the effects of the nonlinear terms on a centre of the linearised system

The chapter ends with a description of two standard perturbative methods of finding periodic solutions, both of which can be written as computer programs, given in appendices to this chapter.

Prior to the use of computer assisted algebra only low-order perturbation theory could be used for most problems, but now the limit is often set only by the available computer memory. However, it is not always sensible to carry such calculations as far as is now possible because the starting equations are normally only approximations: it is therefore important to have some idea of the magnitude of these errors and to match these with the approximation used for the solutions. A clear, qualitative exposition of this type of process is provided by an article of Poincaré on the stability of the Solar System, reproduced here as an appendix to this chapter because this is an important aspect of any approximation.

17.2 Existence theorems for periodic solutions

Here we consider some general theorems that help us understand how solutions can behave. The most general theorem shows that solutions are not too complicated: orbits of autonomous systems cannot cross themselves or other orbits and near an ordinary point all move in the same direction. Intuitively we expect that each orbit either:

  (a) goes to infinity, as $t \to \infty$;
  (b) approaches a fixed point, or is a periodic orbit;
  (c) approaches an isolated closed orbit.

The *Poincaré–Bendixson theorem* proves that this expectation is realised and effectively shows that the motion of a second-order autonomous system is not too complicated: in particular, it cannot be chaotic.

The Poincaré–Bendixson theorem If $\mathscr{D}$ is a closed bounded region of phase space and a solution of the equations of motion is such that $\mathbf{r}(t)$ remains in $\mathscr{D}$ for all $t \geq 0$, then the orbit is either a closed path, approaches a closed path as $t \to \infty$, or approaches a fixed point.

This theorem was proved by Bendixson (1901) by extending the ideas in Poincaré's 1881 paper; hence the name. A proof may be found in Hirsch and Smale (1974, chapter 11). The theorem gives a sufficient condition for the existence of a closed orbit. The main difficulty in using this theorem for finding periodic orbits is determining the region $\mathscr{D}$. One technique is to find an annular region surrounded by two closed curves, $\mathscr{A}$ and $\mathscr{B}$ as shown in the diagram, such that:

  (a) there are no fixed points inside $\mathscr{D}$, and
  (b) orbits enter $\mathscr{D}$ through *every* point of $\mathscr{A}$ and $\mathscr{B}$, or leave through every point.

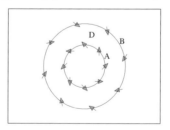

If the phase curves enter $\mathscr{D}$ through $\mathscr{A}$ and $\mathscr{B}$, as in the figure, then all orbits asymptotically approach a stable periodic orbit which lies entirely within $\mathscr{D}$; if they exit they approach a periodic orbit in $\mathscr{D}$ as $t \to -\infty$. Such a region may contain more than one periodic orbit, as is shown by the following exercise, which is followed by an exercise showing how this method may be used to prove the existence of a periodic orbit.

Exercise 17.1
Consider the system

$$\dot{x} = -y + xr^2 \sin \pi r, \quad \dot{y} = x + yr^2 \sin \pi r, \quad r^2 = x^2 + y^2.$$

Express the equation of motion in polar coordinates and hence show that:

(i) the origin is an unstable spiral;

(ii) the circles $r = 1, 3, 5 \cdots$ are stable periodic orbits and the circles $r = 2, 4, 6, \ldots$, are unstable periodic orbits.

Exercise 17.2

Consider the system

$$\frac{dx}{dt} = x - y - x(x^2 + 2y^2), \quad \frac{dy}{dt} = x + y - y(x^2 + y^2).$$

Show that in polar coordinates, $x = r \cos \theta$, $y = r \sin \theta$, the radial equation of motion is

$$\frac{dr}{dt} = r - \frac{1}{8} r^3 \left(9 - \cos 4\theta\right),$$

and deduce that there is at least one closed orbit inside the annulus $2/\sqrt{5} < r < 1$.

A simpler test for the existence of periodic orbits, due to Bendixson (1901), and sometimes called *Bendixson's negative criterion*, states that if $\mathscr{D}$ is a simply connected region of phase space then the autonomous system with velocity function $\mathbf{v} = (X(x, y), Y(x, y))$ has no periodic orbit wholly contained in $\mathscr{D}$ if

$$\operatorname{div} \mathbf{v} = \frac{\partial X}{\partial x} + \frac{\partial Y}{\partial y}$$

is of constant sign in $\mathscr{D}$. A simply connected region is a region with no 'holes' in it, as illustrated in the figure. In a simply connected region every closed curve can be continuously deformed into a point without leaving the region.

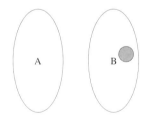

Figure 17.1   Region *A* is simply connected but *B* has a hole in it and is not simply connected.

An extension of this criterion is provided by Dulac's test, which states that if there exists a function $\rho(\mathbf{r})$ having continuous first differentials such that

$$\operatorname{div}(\rho \mathbf{v}) = \frac{\partial(\rho X)}{\partial x} + \frac{\partial(\rho Y)}{\partial y}$$

has constant sign in a simply connected region $\mathscr{D}$, then the system with velocity function (X, Y) has no periodic orbit in $\mathscr{D}$; proofs of both theorems are outlined in exercise 17.46 (page 724).

For particular categories of differential equations there are many theorems giving sufficient conditions for the existence of periodic orbits. Here we quote a few results that apply in quite general circumstances in order to provide a flavour of the subject; many more theorems are given in the compendious volume of Ye Yan-Qian *et al.* (1986).

One of the more general results concerns the system $\dot{x} = X(x, y)$, $\dot{y} = Y(x, y)$, where X is even in x and Y is odd in x,

$$X(x, y) = X(-x, y), \quad Y(x, y) = -Y(-x, y).$$

For this system, if the linearised system has a centre at the origin so does the nonlinear system.

In order to see this, first observe that all phase curves cross the y-axis at right angles because $Y(0, y) = 0$. Consider an orbit starting on the positive y-axis, at $(0, b)$, $b > 0$; if this orbit is $(s_1(t), s_2(t))$, with $(s_1(0), s_2(0)) = (0, b)$, then at some later time $T > 0$ the orbit must cross the negative y-axis, because the fixed point at the origin is either a centre or a spiral. Now consider the backward evolution of this orbit by setting $\tau = -t$ and also $u = -x$ and $v = y$ to give the equations

$$\frac{du}{d\tau} = X(u, v), \quad \frac{dv}{d\tau} = Y(u, v), \quad u(0) = 0, \quad v(0) = b.$$

These are exactly the same as the original equations, hence $u(\tau) = s_1(\tau)$ and at $-t = \tau = T$ the orbit is at $(0, s_2(T))$, the same point as the forward evolute. It follows that the orbit is closed and hence periodic.

If the fixed point at the origin is not simple, the application of these ideas is more complicated, but provides useful results, see Minorsky (1962, page 114).

Another general theorem concerns Liénard's equation,

$$\frac{d^2x}{dt^2} + f(x)\frac{dx}{dt} + g(x) = 0. \tag{17.1}$$

The following conditions are *sufficient* for a stable periodic orbit to exist.

(1) The functions $f(x)$ and $g(x)$ are continuous;
(2) $f(x)$ is an even function;
(3) $g(x)$ is odd and $xg(x) > 0$ for all $x \neq 0$;
(4) the function

$$F(x) = \int_0^x du\, f(u)$$

satisfies $F(x) < 0$ for $0 < x < a$ and $F(x) > 0$ for $x > a$, so the only zeros of $F(x)$ are at $x = 0$ and $\pm a$.

Exercise 17.3
Show that the equation 17.1 can have no periodic solution which lies entirely in a region where $f(x)$ is of one sign.

Exercise 17.4

Show that if $F(x) > 0$ for $0 < x < a$ and $F(x) < 0$ for $x > a$ then Liénard's equation has an unstable periodic orbit.

Exercise 17.5

Show that the equation

$$\frac{d^2x}{dt^2} + v(x^{2m} - 1)\frac{dx}{dt} + x^{2n-1} = 0, \quad v \geq 0,$$

where n and m are positive integers, has a stable periodic orbit.

Exercise 17.6

Consider the system

$$\frac{dx}{dt} = y, \quad \frac{dy}{dt} = -4x - ay + \frac{6y}{1 + x^2}, \quad a > 0,$$

which is a simplification of a particular vacuum-tube circuit, Andronov and Chaikin (1949).

Show that if $2 < a < 10$ the origin is a spiral point which is stable if $6 < a < 10$. Determine the nature of the fixed point if $a < 2$ or $a > 10$.

Show also that if $a < 6$ the system has a stable periodic orbit, and use Maple to draw some representative phase curves for the cases $a = 5$ and $a = 1$.

17.3 Limit cycles

17.3.1 Simple examples

A limit cycle is an isolated periodic orbit, which means that all adjacent orbits either spiral towards or away from it. A limit cycle is an invariant set and in its neighbourhood there can be no integral of the motion.

Limit cycles are important for reasons that we discuss after this elementary introduction which concentrates on simple, soluble examples. Consider the autonomous second-order system

$$\dot{x} = x + y - y(x^2 + y^2), \quad \dot{y} = x - y - x(x^2 + y^2), \tag{17.2}$$

having a single fixed point at the origin. In polar coordinates this system is separable (which is why it was chosen) and becomes

$$\dot{r} = r(1 - r^2), \quad \dot{\theta} = 1, \quad r \geq 0.$$

The radial motion is similar to the logistic equation, equation 16.11, with fixed points at $r = 0$ and 1, and a sketch of the radial velocity function, $r(1 - r^2)$, shows that the fixed point at the origin is an unstable spiral. At $r = 1$ the radial equation has a stable fixed point but θ is not constant, so in the original cartesian coordinates this orbit is the circle $r = 1$ and represents an invariant periodic orbit; moreover all adjacent orbits spiral towards this circle.

The invariant circle $r = 1$ is an example of an isolated periodic orbit and such an orbit is named a *limit cycle*, which is either an attracting or a repelling set and is usually structurally stable.

Stable limit cycles are important in the theory of nonlinear oscillators, particularly if they attract motion from a large area of phase space and are consequently clearly observable. Such behaviour is seen in exercise 16.27 (page 652), where you saw that solutions of the van der Pol oscillator, starting from widely different regions of phase space, tended towards the same periodic orbit; in this example it can be shown that *all* initial conditions, the origin excepted, lead to the same limit cycle.

For the more general autonomous second-order system,

$$\dot{x} = X(x, y), \quad \dot{y} = Y(x, y),$$

it is normally very difficult to determine whether limit cycles exist and, if they do, how many exist. Indeed, one of the classical problems posed by Hilbert, as part of his sixteenth problem, is to find an upper bound on the number of possible limit cycles of these equations when X and Y are polynomials of degree n; a modern discussion of this problem is given by Smale (1998). For quadratic polynomials it may be shown that there cannot be infinitely many limit cycles, and examples have been constructed with four limit cycles. A wealth of information about polynomial systems is contained in the article by Lloyd (1987) and the book by Ye Yan-Qian *et al.* (1986).

Exercise 17.7
By transforming to polar coordinates find the limit cycle of the system

$$\frac{dx}{dt} = -y + x(4 - x^2 - y^2), \quad \frac{dy}{dt} = x + y(4 - x^2 - y^2),$$

and investigate its stability.

Exercise 17.8
Show by numerical investigation that the system

$$\frac{d^2x}{dt^2} + v(x^2 - 1)\frac{dx}{dt} + x - \frac{1}{3}\epsilon x^3 = 0, \quad |\epsilon| \ll 1,$$

may have a stable limit cycle provided that ϵ is sufficiently small.
Hint: fix $v = 1$ and consider orbits starting near the origin, for example $(0.2, 0)$ and $0 < \epsilon \le 0.70$.

17.3.2 *Van der Pol's equation and relaxation oscillations*

B. van der Pol
(1889–1959)

In 1926 the Dutch physicist B. van der Pol published a paper in which he described the oscillations of the grid voltage in a simple electrical circuit containing a triode valve. The equation that he derived to describe this voltage is, after suitable scaling,

$$\frac{d^2x}{dt^2} + v(x^2 - 1)\frac{dx}{dt} + x = 0, \quad v \ge 0. \tag{17.3}$$

Subsequently, equations like this were shown to describe a wide variety of phenomena and to be one of the simplest types of equation to display *relaxation oscillations*; consequently this equation is now named after its founder. Many applications can be found in the collected works of van der Pol, Bremmer and Bouwkamp (1960), and also Grasman (1987).

Van der Pol's equation is nonlinear and it is not easy to guess how its solution will behave, so we start by showing that it has a stable limit cycle and then we examine some numerical solutions. Comparing 17.3 with 17.1 gives $g(x) = x$ and $f(x) = v(x^2 - 1)$, which satisfy conditions (1), (2) and (3); also $F(x) = vx(x^2/3 - 1)$ which has the correct behaviour if $v > 0$. Thus the van der Pol oscillator has a stable limit cycle. If $v < 0$ this is unstable, as may be seen by reversing the direction of time.

In figures 17.2 to 17.7 are shown typical solutions for $v = 0.2$, 1 and 5; on the left of each pair is the phase curve of the orbit and on the right the graph of $x(t)$. In all cases the initial condition is $x(0) = 0$, $\dot{x}(0) = 0.1$; different initial conditions produce qualitatively similar solutions.

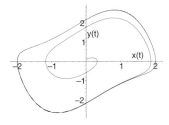

Figure 17.2   A phase curve for $v = 0.2$.

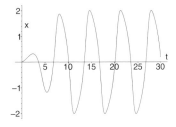

Figure 17.3   Graph of $x(t)$, $v = 0.2$.

The first two figures, $v = 0.2$, show that the solution takes several periods to approximate the limit cycle and that eventually $x(t)$ looks similar to the graph of a sine-curve: we show in section 17.6 that for this small value of v low-order perturbation theory provides a good approximation to the limit cycle.

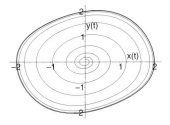

Figure 17.4   A phase curve for $v = 1$.

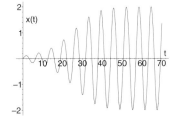

Figure 17.5   Graph of $x(t)$, $v = 1$.

In the second example $v = 1$ and the solution has lost the characteristic shape of a sine-curve and approaches the periodic orbit much faster. Nevertheless, a high-order perturbation expansion provides a good approximation to the limit cycle, as shown in figure 17.21 (page 707).

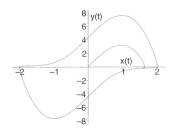

Figure 17.6   A phase curve for $v = 5$.

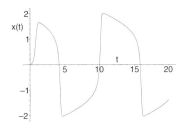

Figure 17.7   Graph of $x(t)$, $v = 5$.

Finally, for $v = 5$ the solution has adopted a quite different and distinct shape and the periodic orbit is approached very quickly, in less than a period: for this value of v the perturbation expansion does not converge. Thus we see a gradual change from a sinusoidal behaviour, at small v, to a distinctively different behaviour for the larger values of v. Observe also that the period of the limit cycle increases with v: if v is large we show below that $T \simeq 1.61v$.

Exercise 17.9

Use Maple to reproduce a few of these graphs and investigate the effect of changing the initial conditions and draw the equivalent graphs for $v = 10, 20$ and 30.

Exercise 17.10

Write a Maple procedure to compute, numerically the period, $T(v)$, of a limit cycle for any given value of v. Use this procedure to draw the graph of $T(v)$ for $0 < v < 10$.

Hint: this is not easy because the numerical integration packages available in Maple are limited. Observe that an orbit starting at $(0, y_1)$, $y_1 > 0$, next crosses the y-axis at $(0, -y_2)$ with $y_2 > 0$ and that for a periodic orbit $y_1 = y_2$. The first stage of the calculation is thus to write a procedure that finds y_2 in terms of y_1 and finds the time taken along this orbit. The next stage is to solve the equation $y_2(y_1) = y_1$, to find the periodic orbit.

A characteristic feature of the large v solution, figure 17.7, is that $x(t)$ decreases relatively slowly from 2 to 1, then changes abruptly to -2, increases slowly to -1 and then changes abruptly to 2, and so on. The larger v the longer the period between rapid changes and the more rapid they are.

The abrupt changes seen in figure 17.7 are described as *relaxation oscillations* because it is as if the system is slowly storing energy, until it becomes unstable and flips to another state. The term *self-excited oscillations* is also used to describe this phenomenon. A very simple model that behaves like this is the see-saw with a water reservoir on the right hand side filled by the drops, as shown in the diagram:

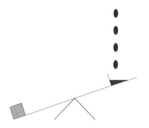

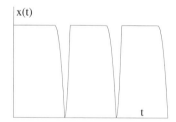

Figure 17.8   Model see-saw which executes relaxation oscillations.

Figure 17.9   Graph of the see-saw angle with time.

In this device, when the weight of water on the right exceeds the weight on the left the see-saw flips, empties and returns to its initial position. If the angle between the see-saw and the horizontal is $x(t)$ its graph will be as shown in figure 17.9.

Other simple examples are given by Grasman (1987), who also provides a full account of the subject. Van der Pol and van der Mark (1929) give a long list of phenomena which are typical relaxation oscillations; some instances are: the aeolian harp (consisting of a string against which a wind is blowing), a pneumatic hammer, the scratching noise of a knife on a plate, the waving of a flag in a wind, the humming noise sometimes made by a water-tap, the squeaking of a door and periodic re-occurrence of epidemics. The period of relaxation oscillations is determined by a diffusion or relaxation time, and has nothing to do with the natural period of the system when executing small sinusoidal oscillations. For instance, in the see-saw model if the bucket fills sufficiently slowly the period is determined almost entirely by the water flow and is unrelated to the period of natural oscillations. More detailed descriptions of relaxation oscillations in some instances can be found in Minorsky (1962, chapters 27 and 28) and Hagedorn (1988, chapter 3).

Other named equations with solutions behaving similarly are Rayleigh's equation and the more general Liénard equation:

$$\frac{d^2x}{dt^2} + \epsilon\left(\frac{1}{3}\left(\frac{dx}{dt}\right)^2 - 1\right)\frac{dx}{dt} + x = 0, \quad \text{(Rayleigh's equation);} \quad (17.4)$$

$$\frac{d^2x}{dt^2} + f(x)\frac{dx}{dt} + g(x) = 0, \quad \text{(Liénard's equation).} \quad (17.5)$$

Rayleigh's equation was derived phenomenologically in his investigations of vibrating systems (Rayleigh, 1883 and 1894, section 68a) and of the effect of external forces which depend upon the velocity of the vibrations. The simplest system considered has the equation $\ddot{u} + \kappa\dot{u} + \kappa'\dot{u}^3 + \omega^2 u = 0$, and Rayleigh deduced that for small oscillations no steady vibratory motion was possible unless κ is negative and κ' positive. This analysis is reproduced in exercise 17.56 (page 725). Rayleigh's equation is closely connected to van der Pol's equation, exercise 17.51.

Baron Rayleigh
J. W. Strutt
(1842–1919)

Consider the behaviour of the solution for large v: for this discussion it is more convenient to consider the slightly more general case by setting $x^2 - 1 = f(x)$, so equation 17.3 becomes

$$\frac{d^2x}{dt^2} + vf(x)\frac{dx}{dt} + x = 0. \quad (17.6)$$

In all that follows you should remember that $|x| = O(1)$. Recall that if $v \gg 1$ and $f =$ constant the solution has two time-scales, equation 16.23 (page 647), a fast one determined by the acceleration and damping term and a slow one determined by the equilibrium motion between the damping term and the restoring force. Here the damping term, $vf(x)\dot{x}$, can change sign and this produces the limit cycle.

Make the same change as in passing from equation 16.22 to 16.24 (pages 646, 647), to write 17.6 in the form

$$\frac{d}{dt}\left(\frac{dx}{dt} + vF(x)\right) + x = 0, \quad F(x) = \int_0^x du\, f(u). \quad (17.7)$$

For most values of x, $|vF(x)|$ will be large, so it is sensible to define a new variable y by $vy = \dot{x} + vF(x)$, so that $y = O(1)$, and the equation of motion becomes

$$\frac{dx}{dt} = v(y - F(x)), \quad \frac{dy}{dt} = -\frac{x}{v}. \quad (17.8)$$

This phase space is the Liénard plane, introduced in exercise 16.17 (page 643); these equations are the nonlinear equivalent of equations 16.25 (page 647). The advantage of using this representation is that $|\dot{y}|$ is always small and $|\dot{x}|$ is large, except close to the line defined by $y = F(x)$. The phase curves are therefore quite easy to understand: away from the curve $y = F(x)$ the flow is almost horizontal, moving rapidly to the left if $y > F(x)$, or to the right if $y < F(x)$, but on the curve $y = F(x)$ the flow is nearly vertical as $\dot{x} \simeq 0$. For the van der Pol oscillator, $f(x) = x^2 - 1$ and $F(x) = -x + x^3/3$, the phase flow is shown in the figures: the flow is shown by the arrows, the light solid line is an actual solution and the solid heavy line is the curve $y = F(x) = -x + x^3/3$.

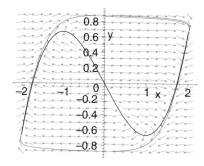

Figure 17.10   Phase curves for $v = 5$.

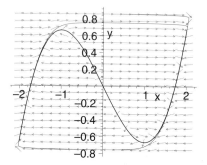

Figure 17.11   Phase curves for $v = 10$.

Using this representation we have, as for the damped oscillator, separated the motion into fast and slow components. Starting, for instance, on the positive y-axis, $\dot{x} \gg 1$ and $|\dot{y}| \sim 0$, the solution moves rapidly to the right, until it reaches the curve $y = F(x)$ where $\dot{x} \sim 0$ and $\dot{y} < 0$, and the solution moves relatively slowly down this curve until near the minimum at $F'(x) = 0$ at $x = 1$.

Here $\dot{y} < 0$ so the solution can no longer follow the curve $y = F(x)$, but once it leaves this curve it is swept away to the left to join the other branch of $y = F(x)$ at $x \simeq 2$, and the process continues.

Increasing v has two effects. First, the curve $y = F(x)$ is adhered to more closely, as seen by comparing figures 17.10 and 17.11, so the maximum amplitude of the motion in the infinite v limit is at the simple root of $F(x) = \max(F(x)) = \frac{2}{3}$, that is, $x = 2$. Second, from any initial condition the limit cycle is approached quicker with increasing v, as seen in figures 17.3 to 17.7 (page 679).

A consequence of this behaviour is that the period of the limit cycle in the large v limit is easily estimated. The crucial observation is that as $v \to \infty$ the proportion of time spent on the slow motion along $y = F(x)$ increases, so the period, T, is approximately twice the time taken to go from $x = 2$ to $x = 1$ along the curve $y = F(x)$, that is,

$$T \simeq 2 \int_2^1 \frac{dx}{\dot{x}}.$$

Since y and x are, in this approximation, constrained by the relation $y = F(x)$, we have $\dot{y} = F'(x)\dot{x} = f(x)\dot{x}$, so using the second equation of motion $v\dot{y} = -x$, we obtain

$vf(x)\dot{x} = -x$ and hence the period is

$$T(v) \simeq 2v \int_1^2 dx \, \frac{f(x)}{x} = (3 - 2\ln 2)v \simeq 1.61v, \quad v \gg 1. \tag{17.9}$$

This approximation is in fact the first term in an asymptotic expansion of $T(v)$. For any v the limit cycle is given by the solution of the equation

$$y\frac{dy}{dx} + v(x^2 - 1)y + x = 0$$

that is a closed curve. There are no analytic expressions for the whole curve, but there are analytic approximations for sections of it and the asymptotic expansion of the period and amplitude may be obtained by matching these. This calculation is complicated and given in Grasman (1987, section 2.2.2): the first few terms are

$$T = (3 - 2\ln 2)v + \frac{3\alpha}{v^{1/3}} - \frac{2}{3v}\ln v + O(v^{-1}), \tag{17.10}$$

$$A = 2 + \frac{\alpha}{3v^{4/3}} + O\left(\frac{\ln v}{v^2}\right),$$

where α is the first zero of the Airy function, $\alpha \simeq 2.338$.

For small v the period and amplitude may be estimated using the perturbation method developed in section 17.6; this gives

$$T = 2\pi\left(1 + \frac{v^2}{16} - \frac{5v^4}{3072} + \cdots\right), \quad A = 1 + \frac{v^2}{96} - \frac{1033v^4}{552\,960} + \cdots.$$

Exercise 17.11
Consider the equation

$$\frac{d^2x}{dt^2} + v(x^4 - 1)\frac{dx}{dt} + x = 0, \quad v > 0.$$

Show that this equation has a stable limit cycle and that in the large v limit the period, T, and amplitude, A, of this are given by

$$T \simeq \left(\frac{1}{2}(\beta^4 - 1) - 2\ln\beta\right)v, \quad A \simeq \beta,$$

where $\beta \neq -1$ is the real root of $x - x^5/5 = 4/5$. Use Maple to find β and to draw some phase curves and solutions.

Exercise 17.12
The system of differential equations

$$\frac{dx}{dt} = v(y - F(x, t)), \quad \frac{dy}{dt} = -\frac{x}{v},$$

where $F(x, t) = x(x + 1)(x + 2)(x + 3)(x - a(t))$ and $a(t) = 0.8 + 0.1\sin(0.1t)$, has the solution shown in figure 17.12.

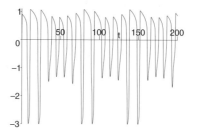

Figure 17.12   Solution for $v = 1$.

Noting that $a(t)$ is relatively slowly varying, explain why the amplitude of the motion varies in the manner shown.

17.4 Centres and integrals of the motion

When the fixed point is a centre of the linearised system it may or may not be a centre of the nonlinear system, and usually further analysis is required to establish the true nature of the fixed point. At a centre $\mathrm{Tr}(A) = 0$ and $\det(A) > 0$, A being the matrix defined in equation 16.30 (page 657), so an arbitrarily small perturbation is likely to change $\mathrm{Tr}(A)$, making the fixed point a spiral; it may, therefore, be thought that centres are structurally unstable and consequently play no significant role in real systems. Indeed, Minorsky (1962, page 38) goes so far as to state that in terrestrial problems a 'centre has never a real existence and appears merely as a convenient mathematical idealisation separating convergent and divergent spiral trajectories'. This view is probably correct but, as we shall see, not necessarily of practical significance.

Suppose that the nonlinear system has a centre, so all phase curves in the neighbourhood of the fixed point are closed. Then in this neighbourhood there is an integral of the motion; as may be proved by construction. At each ordinary point (x, y) of the neighbourhood we can define a function $\mathscr{A}(x, y)$ to be the area enclosed by the phase curve through (x, y); this function is, by definition, a constant of the motion. In addition it follows that a coordinate system can be found in which the system can be expressed in Hamiltonian form (equation 16.28, page 654), though the proof of this would take us too far from our theme.

Thus it would be helpful either to find an integral of the motion or to show that one does not exist. The original analysis of Poincaré can often be used to perform the latter task and can also be used to determine other useful information about the system. This technique is described in Minorsky (1962, chapter 1), but seems to have escaped inclusion in more recent texts although it is particularly easy to implement using Maple. Poincaré's method is interesting for two other reasons. First, it is a form of perturbation theory that approximates integrals of the motion rather than solutions and is a precursor of the more general perturbation methods stemming from Lindstedt's method, section 17.6, the methods based on the Hamilton–Jacobi equation, see Born (1960) and Nayfeh (1973), and the more elegant methods based on Lie series, Giacaglia

(1972) or Nayfeh (1973). Second, the method is related to the elementary method of averaging which isolates fast and slow time-scales and can be used to investigate the stability of limit cycles, see for instance Jordan and Smith (1999, chapter 4), Verhulst (1990, chapter 11), Nayfeh (1973, chapter 5) or Nayfeh and Mook (1979, chapter 3).

For simplicity, move the fixed point, which is a centre of the linearised system, to the origin, make a coordinate transformation and scale the independent variable to make the linearised velocity function $(y, -x)$, so the linearised phase curves are circular. Now assume that

$$
\begin{aligned}
X(x, y) &= y + X_2(x, y) + X_3(x, y) + \cdots \\
Y(x, y) &= -x + Y_2(x, y) + Y_3(x, y) + \cdots,
\end{aligned}
\tag{17.11}
$$

where (X_n, Y_n) are homogeneous polynomials of degree n in x and y; thus $X_2(x, y)$ has the form $\alpha x^2 + \beta xy + \gamma y^2$ for some constants α, β and γ.

Assume that an integral $F(x, y)$ exists and that this also can be written in the form

$$
F(x, y) = F_1(x, y) + F_2(x, y) + \cdots,
\tag{17.12}
$$

where $F_n(x, y)$ is, homogeneous polynomial of degree n in x and y. Since F is an integral of the motion we must have

$$
\frac{dF}{dt} = \dot{x}\frac{\partial F}{\partial x} + \dot{y}\frac{\partial F}{\partial y} = X\frac{\partial F}{\partial x} + Y\frac{\partial F}{\partial y} = 0,
$$

and our aim is either to solve this equation or to show that a single-valued solution does not exist. First substitute the series expansion of X, Y and F, and collect the powers of (x, y) to obtain

$$
\begin{aligned}
0 =\ & y\frac{\partial F_1}{\partial x} - x\frac{\partial F_1}{\partial y} \\
+\ & y\frac{\partial F_2}{\partial x} - x\frac{\partial F_2}{\partial y} + \left(X_2\frac{\partial F_1}{\partial x} + Y_2\frac{\partial F_1}{\partial y}\right) \\
+\ & y\frac{\partial F_3}{\partial x} - x\frac{\partial F_3}{\partial y} + \left(X_3\frac{\partial F_1}{\partial x} + Y_3\frac{\partial F_1}{\partial y}\right) + \left(X_2\frac{\partial F_2}{\partial x} + Y_2\frac{\partial F_2}{\partial y}\right) \\
& \vdots \\
+\ & y\frac{\partial F_n}{\partial x} - x\frac{\partial F_n}{\partial y} + \sum_{k=2}^{n}\left(X_k\frac{\partial F_{n+1-k}}{\partial x} + Y_k\frac{\partial F_{n+1-k}}{\partial y}\right) + \cdots.
\end{aligned}
\tag{17.13}
$$

The first line of this equation is a polynomial of degree one, the second line a polynomial of degree two and the nth a polynomial of degree n. Since the expression is zero for all (x, y) in the neighbourhood of the origin, each of these polynomials can be equated to zero to give a set of equations the nth of which expresses F_n in terms of $(F_1, F_2, \ldots, F_{n-1})$.

The first equation shows that $F_1 = 0$ because the most general degree-one polynomial is $F_1(x, y) = \alpha x + \beta y$, for some constants α and β, and substituting this into the equation gives

$$y\frac{\partial F_1}{\partial x} - x\frac{\partial F_1}{\partial y} = \alpha y - \beta x = 0.$$

The only solution valid for *all* x and y in a neighbourhood of the origin is $\alpha = \beta = 0$, thus $F_1 = 0$.

The second equation then gives

$$y\frac{\partial F_2}{\partial x} - x\frac{\partial F_2}{\partial y} = 0.$$

The most general quadratic is $F_2(x, y) = \alpha x^2 + 2\beta xy + \gamma y^2$, and on substituting this into the equation we find that $\beta = 0$ and $\alpha = \gamma$; now choose the arbitrary multiplicative constant to give $F_2 = x^2 + y^2$.

These two equations depend only on the fact that the fixed point is a centre and are independent of the nonlinear terms in the velocity function. The third equation is the first to depend upon these terms, for it becomes

$$x\frac{\partial F_3}{\partial y} - y\frac{\partial F_3}{\partial x} = 2(xX_2 + yY_2).$$

We need a solution of this which is single valued. Since both sides of the equation are of degree three the use of the polar coordinates $x = r\cos\theta$, $y = r\sin\theta$ will give an overall factor of r^3; further, in these coordinates $x\partial/\partial y - y\partial/\partial x = \partial/\partial\theta$, and the equation becomes

$$\begin{aligned}\frac{\partial F_3}{\partial\theta} &= 2r\left[X_2(r\cos\theta, r\sin\theta)\cos\theta + Y_2(r\cos\theta, r\sin\theta)\sin\theta\right] \\ &= r^3 H_3(\theta),\end{aligned} \qquad (17.14)$$

where $H_3(\theta)$ is a 2π-periodic function of θ only. This equation can be integrated to give

$$F_3(r, \theta) = r^3\int_0^\theta d\phi\, H_3(\phi).$$

If $F_3(x, y)$ is single valued, when expressed in terms of (r, θ) it must be 2π-periodic in θ, consequently $F_3(r, 0) = F_3(r, 2\pi)$, and since $F_3(r, 0) = 0$

$$F_3(r, 2\pi) = r^3\int_0^{2\pi} d\phi\, H_3(\phi) = 0.$$

That is, the mean value of $H_3(\theta)$ must be zero if $F_3(x, y)$ is to be single valued.

If the mean of H_3 is not zero then $F_3(r, \theta)$ will contain a term proportional to θ and hence will not be single valued and no integral of the motion exists. But if the mean is zero, $F_3(r, \theta)$ will be periodic in θ, and we may proceed to the next step.

At the nth stage, supposing that $(F_3, F_4, \ldots, F_{n-1})$ have been found, we have

$$\begin{aligned}\frac{\partial F_n}{\partial\theta} &= \sum_{k=2}^{n-1}\left(X_k\frac{\partial F_{n+1-k}}{\partial x} + Y_k\frac{\partial F_{n+1-k}}{\partial y}\right) \qquad (17.15) \\ &= r^n H_n(\theta),\end{aligned} \qquad (17.16)$$

where $H_n(\theta)$ is 2π-periodic in θ. As before, if the mean of $H_n(\theta)$ is zero, integration gives an expression for $F_n(r, \theta)$ which is 2π-periodic in θ and we may proceed to the next term.

Now the terms occurring in the right hand side of equation 17.15 are of the form $x^{n-k} y^k$, $k = 0, 1, \ldots, n$, so that if n is odd this expression can be written in the form $\cos\theta f_n(\cos^2\theta) + \sin\theta g_n(\cos^2\theta)$, where $f_n(z)$ and $g_n(z)$ are polynomials in z, hence the mean of $H_{2n+1}(\theta)$ is always zero; therefore the procedure can fail only at even values of n.

Suppose that the mean of $H_{2m}(\theta)$, $m \geq 2$, is the first to be nonzero; define the periodic function $\tilde{F}_{2m}(r, \theta)$ by the equation

$$\tilde{F}_{2m}(r, \theta) = r^{2m} \int_0^\theta d\phi \, (H_{2m}(\phi) - \mathscr{H}_{2m}), \quad \mathscr{H}_{2m} = \frac{1}{2\pi} \int_0^{2\pi} d\phi \, H_{2m}(\phi), \quad (17.17)$$

which is constructed to ensure that $\tilde{F}_{2m}(r, 0) = \tilde{F}_{2m}(r, 2\pi)$. Finally, define the function

$$G(x, y) = F_2(x, y) + F_3(x, y) + \cdots + \tilde{F}_{2m}(x, y), \quad (17.18)$$

which satisfies the equation

$$\frac{dG}{dt} = r^{2m} \mathscr{H}_{2m}.$$

If m is large dG/dt will be very small when $|r^{2m} \mathscr{H}_{2m}| \ll 1$, and there will be a region near the fixed point where the motion behaves like a centre for very long times, though eventually the spiral nature of the fixed point must become apparent. If the procedure fails at $\mathscr{H}_4$, the first term at which failure is possible, then the fixed point is not a centre, but a spiral. Its stability is determined by the sign of $\mathscr{H}_4$, being stable if $\mathscr{H}_4 < 0$ and unstable otherwise.

If the procedure fails at $\mathscr{H}_{2m}$, $m > 2$, the same conclusion holds and the spiral is stable if $\mathscr{H}_{2m} < 0$. But now, close to the origin, $|\dot{G}|$ is very small and $G(x, y)$ is an approximate constant of the motion for many spirals. In other words, although the fixed point is a spiral it behaves like a centre for relatively long times. In these circumstances the short-time dynamics can be approximated by an integrable system: an example of this type of system is treated after the next few exercises, in particular see figure 17.13. A real example of this type of approximation is the motion of the planets which, to a very good approximation, is described by a Hamiltonian system and both energy and angular momentum are conserved quantities; nevertheless, we know that there are dissipative forces due to tides and the elasticity of the planets, for instance, which cannot be incorporated into the Hamiltonian but which decrease the orbital energy significantly over times comparable to the age of the Solar System, about 4×10^9 years, but have negligible effects on the planetary motion on a human time-scale. A more detailed discussion of this effect and its relation to the long-time behaviour of the Solar System is given in the appendix to this chapter. The motion of atoms and molecules provides other examples, described by Hamilton's equations, where centres are important.

Exercise 17.13

Consider the system

$$\frac{dx}{dt} = y, \quad \frac{dy}{dt} = -x + ay^3,$$

where a is a constant.

Using the theory just described, show that $F_{2k+1} = 0$, $k = 0, 1, \ldots$, and that F_4 is given by

$$\frac{\partial F_4}{\partial \theta} = 2ay^4 = 2ar^4 \sin^4 \theta.$$

Deduce that the system does not possess an integral of the motion and that the fixed point at the origin is not a centre. Further, show that

$$G = x^2 + y^2 - \frac{1}{4}axy(3x^2 + 5y^2) \quad \text{and that} \quad \frac{dG}{dt} = \frac{3}{4}ar^4.$$

Exercise 17.14

Prove that the general single valued solution of the second-order component of equation 17.13 is $F_2 = A(x^2 + y^2)$, where A is a constant. In the text we set $A = 1$: what is the effect of choosing a different value of A?

Exercise 17.15

Prove that

$$\frac{dG}{dt} = r^{2m} \mathcal{H}_{2m},$$

where the integration is along an orbit and G is defined by equation 17.18.

In order to see how this idea works in practice we consider the example dealt with by Poincaré (1881) when he introduced the method; as with many of the methods of this type the algebra is lengthy and Maple is a significant help; it is of comfort to lesser mortals that the original analysis contained an algebraic error, albeit of no fundamental significance. The worksheet used to derive the results quoted here is given in section 17.9.1.

The system to be considered is

$$\frac{dx}{dt} = y - \frac{1}{2}\beta x^3, \quad \frac{dy}{dt} = -x + \alpha x^2 y - \frac{1}{2}\beta y^3, \tag{17.19}$$

which has the cubic perturbation with $X_3 = -\beta x^3/2$ and $Y_3 = \alpha x^2 y - \beta y^3/2$, where α and β are constants to be chosen to ensure that as many as possible of the functions F_k are single valued. For this system equation 17.15 becomes

$$\frac{\partial F_n}{\partial \theta} = X_3 \frac{\partial F_{n-2}}{\partial x} + Y_3 \frac{\partial F_{n-2}}{\partial y},$$

and since $F_1 = 0$ all the odd numbered members of the series are zero, that is, $F_{2k+1} = 0$. The equation for F_4 is

$$\begin{aligned}
\frac{\partial F_4}{\partial \theta} &= X_3 \frac{\partial F_2}{\partial x} + Y_3 \frac{\partial F_2}{\partial y} = 2xX_3 + 2yY_3 \\
&= -\beta x^4 + 2\alpha x^2 y^2 - \beta y^4 \\
&= -\frac{1}{4}r^4 \left((\beta + \alpha)\cos 4\theta + 3\beta - \alpha \right).
\end{aligned} \tag{17.20}$$

Thus the mean of $H_4(\theta)$ is zero only if $\alpha = 3\beta$, and then

$$F_4(x, y) = \beta xy(y^2 - x^2).\tag{17.21}$$

The mean of $H_6(\theta)$ is zero and

$$F_6 = \frac{\beta^2}{8}(y^2 - x^2)(y^4 + 10x^2y^2 + x^4).\tag{17.22}$$

At the next stage, the calculation for F_8, we find that the mean of H_8 is not zero but that $\mathcal{H}_8 = 9\beta^3 r^8/64$, thus we define $\tilde{F}_8$ using equation 17.17:

$$\tilde{F}_8 = \frac{\beta^3}{64}xy(15x^6 - 73x^4y^2 + 89x^2y^4 + 33y^6).\tag{17.23}$$

In figure 17.13 we show a phase curve for this system with $\beta = -1$, $\alpha = -3\beta$, which makes the origin a stable spiral. One orbit starting at $(0.3, 0)$ is shown for the time $0 < t < 1000$, or about 150 periods; it is seen that the orbit spirals into the fixed point very slowly and that for short times it is as if this fixed point were a centre. In the adjacent graph we show the variation of $G(t) = F_2 + F_4 + F_6 + \tilde{F}_8$, equation 17.18, along this orbit and note that it decreases very slowly: indeed $\dot{G} = -9r^8/64 \simeq -9 \times 10^{-6}$.

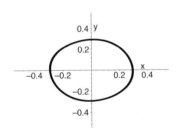

Figure 17.13   Orbit of equation
17.19 with $\beta = -1$, $\alpha = -3$.

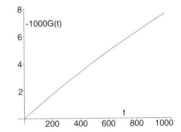

Figure 17.14   Graph of $-1000G(t)$ for the
same orbit.

In the next figure, for comparison, is shown the orbit having the same initial conditions, with $\beta = -1$ but $\alpha = -2.5$, parameters for which the series 17.12 terminates sooner, at F_4 and $\dot{G} \simeq 0.0014$.

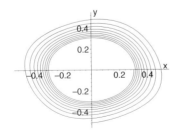

Figure 17.15   An orbit of equation 17.19
with $\beta = -1$, $\alpha = -2.5$, integrated up to
$t = 50$.

Exercise 17.16
Consider the system

$$\frac{dx}{dt} = y + x^2 - (1 + 3c + 6c^2)xy + cy^2, \quad \frac{dy}{dt} = -x + x^2 + (3 + 5c)xy - (2 + c + 2c^2)y^2.$$

Show that the linearised system has a centre at the origin. For various values of c, use Maple to plot some phase curves of the nonlinear system, in order to examine the true nature of the fixed point at the origin. Use the method described above to explain your results.

Exercise 17.17
Consider the system

$$\frac{dx}{dt} = y + x^2 + 2b_1xy + c_1y^2, \quad \frac{dy}{dt} = -x + x^2 + 2b_2xy + c_2y^2,$$

where b_1, b_2, c_1 and c_2 are constants.

Use Maple to compute the series 17.12 for the integral and show that if b_1 and b_2 are defined by

$$b_1 = \frac{1}{2}(5 + 3c_2), \quad b_2 = \frac{1}{2}(3 + 5c_1),$$

then $\mathcal{H}_6 = 0$, where $\mathcal{H}_{2n}$ is defined in equation 17.16. Find other relations between (b_1, b_2) and (c_1, c_2) that make $\mathcal{H}_6 = 0$.

Show further that for the values of b_1 and b_2 given above if, in addition, $c_2 = -2 - c_1(1 + 2c_1)$, then $\mathcal{H}_m = 0$ for $m \le 14$.

Exercise 17.18
In exercise 16.28 (page 652), the system

$$\frac{dx}{dt} = y(1 - y^2), \quad \frac{dy}{dt} = -x + x^2 + y^2(x - \mu)$$

was examined and the numerical evidence suggested that the fixed point at the origin is a centre. Find an approximation to the integral for the motion in the neighbourhood of the origin and show that $\mathcal{H}_m = 0$ for $m = 1, 2, \ldots, 14$. Plot some of the contours of this integral, including polynomials up to order ten, to reproduce the phase curves found in exercise 16.28.

Exercise 17.19
In exercise 16.35 (page 655), it was shown that the equations

$$\dot{x} = y + bx^2 + cy^2, \quad \dot{y} = -x - axy$$

had a solution which can be expressed in the form $g(\mathbf{r}) = x^2 - Ay^2 + By + C = 0$, where the constants A, B and C depend upon a, b and c.

In the particular case $a = b = c = 1$ determine the first ten terms of the series 17.12 for $F(\mathbf{r})$, showing that $\mathcal{H}_{2m} = 0$ for $m = 1, 2, \ldots, 5$. By plotting some contours of the integral $F(\mathbf{r})$, some phase curves obtained using DEplot and also the phase curve defined by the particular solution $g(\mathbf{r}) = 0$, or otherwise, determine the connection between the integral $F(\mathbf{r})$ and the function $g(\mathbf{r})$.

17.5 Conservative systems

In this section we consider solutions of the equation

$$m\frac{d^2x}{dt^2} + \frac{dV}{dx} = 0, \quad m > 0, \tag{17.24}$$

m being a positive constant. This is a variation of equation 17.1 (page 676), with $f(x) = 0$ and $g(x) = dV/dx$. Equations of this type frequently possess families of periodic orbits, each labelled by a continuous index. We assume that $V(x)$ is a sufficiently well behaved function; in practice V need only be piecewise differentiable. This type of equation normally arises when applying Newton's law of motion to the motion of a particle of mass m moving in one dimension under the action of a force $F(x) = -dV/dx$, where $V(x)$ is the *potential energy*, which is why we have introduced the constant m. We often refer to the function $V(x)$ simply as 'the potential'. Similar equations arise when describing the motion of a many-dimensional separable system.

The solutions of this type of equation are relatively easy to understand because the first integral

$$H(x, y) = \frac{1}{2m}y^2 + V(x), \quad y = m\frac{dx}{dt} \tag{17.25}$$

exists. In mechanical applications the function H is often the energy of the system, so we normally use the symbol E for values of $H(x, y)$ and name E the energy.

Exercise 17.20
Show, by differentiation, that the function $H(x, y)$ is an integral of the motion for the equation 17.24. Further, show that $H(x, y)$ is a Hamiltonian function for this system.

The form of the equation of motion 17.24 imposes severe constraints on the nature of the motion; in particular the only types of fixed point that this nonlinear system can possess are centres and saddles. This is seen by writing the equations of motion in the form

$$\dot{x} = \frac{y}{m}, \quad \dot{y} = -\frac{dV}{dx},$$

which shows that the fixed points are at $y = 0$ and the solutions of $V'(x) = 0$. If one of these points is x_f, put $x = x_f + w$ and expand to give the linearised equations

$$\dot{w} = \frac{y}{m}, \quad \dot{w} = -wV''(x_f).$$

Thus the fixed point is a centre if $V''(x_f) > 0$, a minimum of the potential, and a saddle if $V''(x_f) < 0$, a maximum of $V(x)$, and these are the only types of fixed points. Since there is an integral of the motion the centre of the linearised system is also a centre of the nonlinear system.

The phase portrait is most easily drawn, for a specific potential $V(x)$, by drawing the contours of $H(x, y) = E$ using the Maple command `contourplot`, for suitable values of the energy E. However, it is normally helpful to perform some preliminary analysis in order to understand the results produced by this 'black-box' routine. We describe the method by dealing with two particular cases which illustrate all the salient points.

First we observe that the integral defined in equation 17.25 can be cast in the form

$$y = \pm\sqrt{2m(E - V(x))}, \quad y = m\dot{x}, \tag{17.26}$$

and, since y is real, for any given energy x is restricted to the regions where $V(x) < E$, suggesting that we first draw the graph of the potential $V(x)$.

Consider first the example in which $V(x)$ has a single minimum at $x = 0$ and monotonically increases as $x \to \pm\infty$, as shown in the upper graph of figure 17.16.

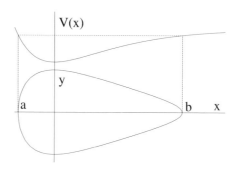

Figure 17.16   Diagram showing a potential, top curve, and the phase curve corresponding to the energy of the horizontal dashed line.

Consider the motion with energy E, the value of which can be any value taken by the potential. The horizontal line in the upper graph intersects the potential at only two points $x = a(E)$ and $b(E)$, where $V(x) = E$, so the motion is confined to the range $a \le x \le b$; the phase curve, shown in the lower figure, comprises the two branches $y = \pm\sqrt{2m(E - V(x))}$, with $y > 0$ when x is increasing and vice versa. The *turning points* at $x = a$, b are at the minimum and maximum values, respectively, of $x(t)$, and here $y = \dot{x} = 0$. At these points $|\frac{dy}{dx}|$ is unbounded, exercise 17.21, so the phase curve is perpendicular to the x-axis.

The phase curve is closed so the motion is periodic. The period, $T(E)$, is twice the time taken to travel from $x = a$ to $x = b$, because the time taken to travel from a to b on the upper branch is the same as the time taken to travel from b to a on the lower branch. Thus

$$T(E) = 2\int_0^{T/2} dt = 2\int_a^b \frac{dx}{\dot{x}} = \sqrt{2m}\int_{a(E)}^{b(E)} \frac{dx}{\sqrt{E - V(x)}}, \quad V(a) = V(b) = E. \tag{17.27}$$

This simple construction shows that in this case *all* solutions in the vicinity of the potential minimum are periodic.

Exercise 17.21
Show that at a turning point the phase curve of the conservative system defined by equation 17.24 crosses the x-axis at right angles.

Under what conditions do the phase curves cross the y-axis at an angle that is *not* a right angle?

Exercise 17.22
For most potentials the period T depends upon the energy E. In the particular case $V(x) = m\omega^2 x^2/2$ find the period and show that it is independent of the energy.

Exercise 17.23
For the potential $V(x) = m\omega^4 x^4/4$ show that the period $T(E)$ is proportional to $E^{-1/4}$.

Now consider the potential

$$V(x) = \frac{1}{2}x^2 - \frac{1}{3}x^3, \qquad (17.28)$$

which is sufficiently general to illustrate all the other important features, because it has both a maximum and a minimum. The fixed points are at $y = 0$ and the stationary points of the potential, that is, where $V'(x) = 0$. This cubic potential has two stationary points: at $x = 0$ it has a local minimum and at $x = 1$ a local maximum at which $V(1) = \frac{1}{6}$, as shown in figure 17.17. Thus at the origin there is a centre and at $x = 1$ a saddle. In the graph of the potential the horizontal solid lines are at the energies $E = \frac{1}{12}$ and $E = \frac{1}{4}$ and the dotted line in between is at $E = \frac{1}{6}$, the value of $V(x)$ at its local maximum.

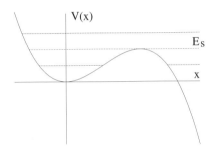

Figure 17.17   The potential 17.28.

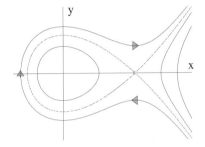

Figure 17.18   Phase curves at the energies shown on the left.

The phase curves, shown in figure 17.18, are constructed using the following considerations.

(1) For an energy $0 < E < \frac{1}{6} = V(1)$ the equation $V(x) = E$ has three real roots $x_1(E) < x_2(E) < x_3(E)$, the turning points, and the motion is confined either to the range $x_1 \leq x \leq x_2$ or $x > x_3$ depending upon the initial conditions.

The phase curves of the bound motion, $x_1 \leq x \leq x_2$, are constructed as in the previous example; this motion is periodic about the centre at $y = x = 0$ and has the period

$$T(E) = \sqrt{2m} \int_{x_1(E)}^{x_2(E)} dx \, \frac{1}{\sqrt{E - \frac{1}{2}x^2 + \frac{1}{3}x^3}}. \qquad (17.29)$$

On the other branch $x \geq x_3$ the motion is unbounded. Thus for $0 < E < \frac{1}{6}$ there are two possible types of motion.

(2) For $E < 0$ the equation $V(x) = E$ has only one real root which is positive and the motion is confined to the right of this point. This motion is unbounded and is similar to the unbounded motion discussed in (1). No example of negative energy motion is shown in figure 17.18.

(3) If $E > V(1) = \frac{1}{6}$ then again the equation $V(x) = E$ has only one real root which is now negative, and the unbounded motion is confined to the right of this point.

(4) The phase curve with energy $E = \frac{1}{6}$ is special. It is the phase curve which passes through the saddle and has two branches. To the left is the closed loop, on which the motion is clockwise, which is the dividing line between the unbounded motion with $E > \frac{1}{6}$ and the periodic motion with $0 < E < \frac{1}{6}$. To the right of the saddle the motion goes to infinity on the upper curve and comes from infinity on the lower curve. Because this special phase curve divides phase space into invariant regions containing different types of motion, it is given the name *separatrix*. The equation of the separatrix, in this case, is

$$
\begin{aligned}
y^2 &= 2m(E_s - V(x)), \quad E_s \text{ being the saddle energy,} \\
&= \frac{m}{3}(x-1)^2(2x+1),
\end{aligned}
$$

so near the saddle at $x = 1$ the phase curves are $y \sim \pm\sqrt{m}(1-x)$. This line does not intersect the x-axis at right angles like most other phase curves, because at the saddle $\mathbf{v} = 0$ and $\frac{dy}{dx} = \frac{\dot{y}}{\dot{x}}$ has a finite limit.

The motion on the separatrix near the saddle and moving towards it is given by

$$
\dot{x} = y = (1-x)\sqrt{m} \quad \Longrightarrow \quad x(t) = 1 - (1-x_0)e^{-t\sqrt{m}}.
$$

The solution is very slowly varying and never reaches the saddle: this has very important consequences which we discuss below, see also exercise 17.28.

On the separatrix the motion is not periodic and is exponentially slow as the saddle is approached. Periodic motion near the separatrix, that is, with $E \simeq \frac{1}{6}$, must reflect this behaviour and we should expect the period, $T(E)$, to diverge as $E \to \frac{1}{6}$ from below. In fact for potentials having a quadratic maxima at $E = E_s$, $T(E) \sim -\ln(|E - E_s|)$ as $E \to E_s$. In the next exercise equation 17.29 is cast into a form that helps make this clear for the potential $V = \frac{1}{2}x^2 - \frac{1}{3}x^3$, and following this we show that the result is true in general.

Exercise 17.24
Show that for the potential $V = \frac{1}{2}x^2 - \frac{1}{3}x^3$ equation 17.29 for the period can be written in the form

$$
T(E) = 2\sqrt{\frac{6m}{a+c}}K(k), \quad k^2 = \frac{a+b}{a+c}, \tag{17.30}
$$

where $K(k)$ is the complete elliptic integral of the first kind and where a, b and c are positive numbers, $-a(E)$ being the negative turning point and $b(E) \le c(E)$ the two positive turning points near the saddle. If $E = \frac{1}{6}$, $a = -\frac{1}{2}$ and $b = c = 1$.

Use Maple to find expressions for $a(E)$, $b(E)$ and $c(E)$ and to plot the graph of $T(E)/2\pi$ and the frequency $\omega(E) = 2\pi/T(E)$ for $0 \le E < \frac{1}{6}$.

Also solve the equations of motion numerically and plot the graphs of $x(t)$ for $0 < t < T(E)$ for $E = 0.02$, 0.05, 0.1, 0.16 and 0.16666. Can you explain the shape of the graph when $E = 0.16666$?

Using the first term in its series expansion of $K(k)$ about $k = 1$, given on page 843, show that

$$T(E) = 2\sqrt{\frac{6m}{a+c}} K(k) \sim -2\sqrt{3} \ln\left(\frac{1}{\sqrt{8}} \left(\frac{1}{6} - E\right)\right) \quad \text{as } E \to \frac{1}{6}.$$

Thus for the potential $V(x) = \frac{1}{2}x^2 - \frac{1}{3}x^3$ as $E \to E_s = \frac{1}{6}$, from below, the period diverges logarithmically. The same result holds for any potential with this shape. Consider a potential that supports bound motion and has a maximum at $x = x_m$ with $V(x_m) = E_s$. The motion is slowest near the maximum, where we may write $V(x) \simeq E_s - a^2(x_m - x)^2$, with $a^2 = \frac{1}{2}|V''(x_m)|$. If $E = E_s - \epsilon$, $0 < \epsilon \ll 1$, the integral for the period, equation 17.27, is dominated by the time taken to travel from some arbitrary point x_0 to the turning point $x_2 \simeq x_m$,

$$
\begin{aligned}
T(E) &\simeq \sqrt{2m} \int_{x_0}^{x_2} dx\, \frac{1}{\sqrt{a^2(x - x_m)^2 - \epsilon}} \\
&= \sqrt{2m}\, \cosh^{-1}\left(\frac{x_m - x_0}{x_m - x_2}\right) \sim -\ln(x_m - x_2),
\end{aligned}
$$

which proves the result since $a(x_m - x_2) = \sqrt{\epsilon} = \sqrt{|E - E_s|}$.

Because the phase point is moving slowly near the separatrix it can be significantly perturbed by any suitable small perturbation; this often produces chaotic motion, because the solution becomes very sensitive to the initial conditions. For instance, if the system is perturbed by a small periodic potential, $\delta x \sin \Omega t$, where $|\delta| \ll 1$, then provided the energy is close enough to the saddle energy it can be proved that the orbit becomes unstable; a numerical example of this effect is examined in exercise 17.28.

Exercise 17.25
One of the potentials used to approximate forces between atoms in molecules is the Morse potential, which can be written in the form

$$V(x) = U\left(1 - e^{-x/a}\right)^2,$$

where U and a are positive constants.

Plot the graph of this potential and show that it supports periodic motion for energies $0 < E < U$. Also draw some phase curves for energies in this range.

For the motion of a particle of mass m and energy E show that the turning points are at $x = -a \ln\left(1 \pm \sqrt{E/U}\right)$ and that the period is

$$T(E) = 2\pi a\sqrt{\frac{2m}{U - E}}.$$

Exercise 17.26

Consider the motion of a particle of unit mass in the potential

$$V(x) = \frac{1}{2}x^2 - \frac{1}{4}\epsilon x^4.$$

By sketching graphs of the potential show that:

 (i) if $\epsilon \leq 0$ motion is possible only for energies $E \geq 0$ and that all motion is periodic;
 (ii) if $\epsilon > 0$ motion is possible for all energies, but periodic motion occurs for $0 < E < 1/4\epsilon$.

For $\epsilon = \pm 1$ sketch some phase curves and for $\epsilon = 1$ show the separatrix. Explain why the phase curves are symmetric with respect to reflections in both the x- and y-axes.

Show that the period of the motion with amplitude a can be expressed in terms of the integral

$$
T(a) = \frac{4}{\sqrt{1 - \epsilon a^2/2}} \int_0^{\pi/2} d\phi \, \frac{1}{\sqrt{1 - k^2 \sin^2 \phi}}, \quad k^2 = \frac{\epsilon a^2}{2 - \epsilon a^2} < 1
$$

$$
= \frac{2}{\sqrt{1 - \epsilon a^2/2}} \sum_{p=0}^{\infty} \left(\frac{\Gamma(p + \frac{1}{2})}{p!} \right)^2 k^{2p}.
$$

Use the asymptotic expansion

$$\Gamma(az + b) \sim \sqrt{2\pi}\, e^{-az} (az)^{az+b-1/2}, \quad a > 0, \quad |\arg z| < \pi,$$

to rearrange the latter expression in the form

$$
T = \frac{2}{\sqrt{1 - \epsilon a^2/2}} \left[\pi - \ln\left(\frac{1 - \epsilon a^2}{1 - \epsilon a^2/2} \right) \right] + \frac{2\sqrt{2}}{\sqrt{2 - \epsilon a^2}} \sum_{p=1}^{\infty} \left[\left(\frac{\Gamma(p + \frac{1}{2})}{p!} \right)^2 - \frac{1}{p} \right] k^{2p},
$$

to demonstrates explicitly the logarithmic singularity at $\epsilon a^2 = 1$.

Write Maple procedures to evaluate the period (*a*) by expressing the integral in terms of the elliptic integral $K(k)$, and (*b*) by truncating the above sum at the Nth term. Compare these two expressions graphically for $0 \leq \epsilon a^2 < 1$ with $N = 1, 2$ and 4.

Exercise 17.27

Consider the motion of a particle of unit mass in the potential

$$V(x) = -\frac{1}{2}x^2 + \frac{1}{4}x^4.$$

Show that $V(x)$ has two minima at $x = \pm 1$, where $V(\pm 1) = -\frac{1}{4}$, and a maximum at the origin. By sketching the potential show that all motion is periodic but that there are three invariant regions in phase space corresponding to $(E < 0, x > 0)$, $(E < 0, x < 0)$ and $(E > 0)$. Write down the equation of the separatrix dividing the $E > 0$ from the $E < 0$ motion. Draw some representative phase curves showing all types of motion.

Show that the period of the motion of energy $E > 0$ and amplitude $a > \sqrt{2}$ can be written in the form

$$T(a) = \frac{4}{\sqrt{a^2 - 1}} K(k), \quad k^2 = \frac{a^2}{2(a^2 - 1)}.$$

Consider the limit of large a and show that $T(a) \simeq 4^{3/4}E^{-1/4}K\left(1/\sqrt{2}\right)$, and by comparison with the solution of exercise 17.23 deduce that

$$K(1/\sqrt{2}) = \frac{\sqrt{\pi}}{2^{3/2}}\Gamma(1/4)/\Gamma(3/4) = \frac{\pi^{3/2}}{2\Gamma(3/4)^2}.$$

Plot the graph of $T(a)$ for $2 < a^2 < 10$ and determine the behaviour of the period as $a \to \sqrt{2}$.

For energies in the range $-\frac{1}{4} \le E < 0$ if the turning points are $a < b$ show that the period can be written in the form

$$T(E) = 8\sqrt{2}\int_0^{\pi/2} d\phi\, \frac{1}{\sqrt{[3a + b - (b-a)\cos 2\phi][a + 3b - (b-a)\cos 2\phi]}}$$

and plot the graph of this for $-\frac{1}{4} \le E < 0$.

Exercise 17.28
The previous exercise dealt with motion in the potential $V = -x^2/2 + x^4/4$ and it was shown that there are three invariant regions with the separatrix

$$y = \pm x\sqrt{1 - x^2/2}$$

between the invariant regions with $E > 0$ and $E < 0$. It was also shown that on the separatrix and near $x = 0$ the motion is exponentially slow. In these circumstances we expect a suitable small perturbation to change the nature of the motion and the results obtained in this exercise demonstrate this.

Consider the modified potential

$$V(x, t) = -\frac{1}{2}x^2 + \frac{1}{4}x^4 + x\delta\sin(t + \phi),$$

where $|\delta|$ is a small constant and ϕ a constant phase, for which the equations of motion are

$$\dot{x} = y, \quad \dot{y} = x - x^3 - \delta\sin(t + \phi).$$

Use Maple to solve these equations for the initial conditions $x(0) = 1$, $y(0) = 1/\sqrt{2}$, that is, a point on the separatrix, and plot graphs of the solution $x(t)$ for $0 < t < 300$ with $\delta = 0.001$ and for $\phi = 0, 0.001, 0.01, 0.1$.

17.6 Lindstedt's method for approximating periodic solutions

The theories discussed so far have described conditions for the existence of periodic orbits, but provide no means of determining these orbits. There are a number of methods available for approximating these orbits, most of which are based upon some form of perturbation theory. However, these can become very cumbersome and error prone, and before the advent of computer-assisted algebra were rarely used beyond the second-order term; with Maple, however, one is limited only by the machine memory provided the method can be described as an algorithm. For example, the method described here has been used to compute the period of the van der Pol limit cycle using 164 orders of the perturbation expansion, Andersen and Geer (1982).

Here we describe a method developed by Lindstedt and Poincaré which is a double expansion in both the frequency and the solution and is therefore quite complicated.

A. Lindstedt
(1854–1939)

H. Poincaré
(1854–1912)

This complication is necessary in order to obtain periodic solutions, because for nonlinear systems the frequency depends upon the amplitude of the motion and neither is known in advance. The method approximates periodic orbits, but it is slightly easier to apply to conservative systems for which *all* orbits in a region of phase space are periodic, so we deal with these before considering limit cycles.

17.6.1 Conservative systems

The perturbation schemes described in this section provide expansions of the orbit and frequency about a potential minimum. In the neighbourhood of a minimum, which we move to the origin, it is assumed that the potential can be expanded as a Taylor's series, so that the equation of motion 17.24 (page 691) will be

$$m\frac{d^2x}{dt^2} + xV^{(2)} + \frac{1}{2}x^2V^{(3)} + \cdots + \frac{x^n}{n!}V^{(n+1)} + \cdots = 0, \tag{17.31}$$

where $V^{(k)}$ is the kth differential of the potential at the origin. The equation $\ddot{x} + \omega^2 x = 0$ is just the linear oscillator approximation with natural frequency $\omega = \sqrt{V^{(2)}/m}$; remember that $V^{(2)} > 0$. The case $V^{(2)} = V^{(3)} = 0$ is special and not amenable to the treatment of this section, but see exercise 17.42 (page 711).

If the amplitude of the motion being considered is a, a quantity having the same dimensions as x, then it is convenient to introduce the scaled time $\tau = \omega t$ and displacement $z = x/a$, so that equation 17.31 becomes

$$\frac{d^2z}{d\tau^2} + z + \sum_{k=2}^{\infty} \epsilon_k z^k = 0, \quad \epsilon_n = \frac{a^{n-1}V^{(n+1)}}{n!V^{(2)}}, \tag{17.32}$$

where the dimensionless numbers ϵ_n need to be small and to decrease with increasing n. With this rescaling the dependence upon the amplitude has been moved from the initial conditions to the perturbation parameter ϵ. If X is the radius of convergence of the Taylor's series for $V(x)$, then $X^nV^{(n)} \sim n!$ so $\epsilon_n \sim n(a/X)^{n+1}$, and if $a < X$, ϵ_n decreases rapidly with n.

Exercise 17.29
Derive equation 17.32.

Exercise 17.30
For the vertical pendulum, equation 16.26 (page 649), show that for motion with amplitude θ_m the equation of motion can be expanded in the form

$$\frac{d^2z}{d\tau^2} + z + \sum_{k=1}^{\infty}(-1)^k\delta_{2k+1}z^{2k+1} = 0, \quad \delta_{2n+1} = \frac{\theta_m^{2n}}{(2n+1)!},$$

where $z = \theta/\theta_m$, $\tau = \alpha t$ and $\alpha = \sqrt{g/l}$.
  Show also that if $\epsilon = \delta_3 = \theta_m^2/6$ then

$$\frac{\delta_{2k+1}}{\epsilon^k} = \frac{6^k}{(2k+1)!} \simeq \frac{1}{2k+1}\frac{1}{\sqrt{4\pi k}}\left(\frac{e\sqrt{6}}{2k}\right)^{2k}$$

and evaluate the first few of these ratios.

When applying perturbation theory to conservative systems two independent approximations are normally made. First, it is necessary to truncate the series in equation 17.32. Second, the solution is expanded as a perturbation series of a particular order. Both these approximations introduce errors which need to be comparable: for instance it would be silly to use a tenth-order perturbation expansion on the equation $z'' + z + \epsilon z^2 = 0$ unless the ignored terms had less effect than the last term of the perturbation expansion. It is important to remember this, as with computerised algebra it is easy to carry out higher-order perturbation expansions than is warranted.

Thus we start with a simple example and proceed in three stages. First we show why a naïve application of perturbation theory fails; second we introduce Lindstedt's method using conventional algebra in order to understand how the method works and finally we implement the method using Maple.

Consider the system with $m = 1$ and the even potential $V(x) = x^2/2 - \epsilon x^4/4$, dealt with in exercise 17.26; the equation of motion is

$$\frac{d^2x}{dt^2} + x - \epsilon x^3 = 0, \quad |\epsilon| \ll 1. \tag{17.33}$$

Recall that all solutions are periodic if $\epsilon \leq 0$, and if $\epsilon > 0$ some solutions with energy in the range $0 < E < 1/4\epsilon$ are periodic. Because the potential is even the phase curves are invariant with respect to reflections in both the x- and y-axes, and this makes our analysis slightly easier.

We search for solutions, $x(t)$, with $x(0) = A > 0$ and $\dot{x}(0) = 0$, so A is the amplitude of the motion. The solution has the following properties.

P1 The phase curves are invariant with respect to reflections in the x-axis, because the energy integral is an even function of $\dot{x}$. Thus if $\dot{x}(0) = 0$ the solution $x(t)$ is an even function of time and its Fourier series contains only the terms $\cos k\omega t$, where ω is the frequency.

P2 The phase curves are also invariant with respect to reflections in the y-axis, because the potential is an even function of x. If $\dot{x}(0) = 0$ then $x(T/4) = 0$ and $x(t)$ is odd about $t = T/4$, where T is the period. Hence the Fourier development of $x(t)$ contains only the components $\cos(2k + 1)\omega t$, $k = 0, 1, \ldots$.

P3 The function $z(t) = x(t)/A$, with $z(0) = 1$, $\dot{z}(0) = 0$, satisfies

$$\frac{d^2z}{dt^2} + z - \delta z^3 = 0, \quad \delta = \epsilon A^2, \tag{17.34}$$

which depends upon only one parameter δ; there are periodic solutions only if $\delta < 1$. As in the general case, we have moved the amplitude dependence to the perturbation parameter, δ.

Naïve perturbation theory

Now we explore the result of using a naïve perturbation theory such as described in chapter 9: this analysis is important because although the method fails it is helpful to know why and how. Assume a solution of the form

$$z(t) = z_0(t) + \delta z_1(t) + \delta^2 z_2(t) + \cdots, \qquad (17.35)$$

with $z_0(0) = 1$, $\dot{z}_0(0) = 0$ and $z_k(0) = \dot{z}_k(0) = 0$ for $k = 1, 2, \ldots$, substitute this into equation 17.34 and set the coefficients of the powers of δ to zero, to obtain equations of the form

$$\ddot{z}_n + z_n = f_n(z_0, z_1, \ldots, z_{n-1}), \quad n = 1, 2, \ldots.$$

The first few of these equations are

$$
\begin{aligned}
\ddot{z}_0 + z_0 &= 0, & z_0(0) &= 1, \ \dot{z}_0(0) = 0, \\
\ddot{z}_1 + z_1 &= z_0(t)^3, & z_1(0) &= \dot{z}_1(0) = 0, \\
\ddot{z}_2 + z_2 &= 3z_0(t)^2 z_1(t), & z_2(0) &= \dot{z}_2(0) = 0.
\end{aligned}
\qquad (17.36)
$$

This scheme looks promising because $z_n(t)$ is given recursively in terms of the previous terms. But the solutions for z_0 and z_1 are

$$z_0(t) = \cos t \quad \text{and} \quad z_1(t) = \frac{1}{32}(\cos t - \cos 3t + 12t \sin t),$$

and all subsequent terms contains powers of t; thus $z_n(t)$, $n \geq 1$, is not periodic and this scheme does not produce a satisfactory solution.

Exercise 17.31
Derive equations 17.36 and find solutions for the first three terms of the series. You will probably find it easier and quicker to use Maple to do all this exercise.

The unphysical terms in this solution which grow as powers of t are called *secular terms*.† The reason why these terms occur can be described either physically or mathematically; since it is important to understand their origin we give both. First consider the mathematics: because $z_0(t) = \cos t$ the right hand side of the equations for $(z_1, \ldots, z_n)$ will contain linear combinations of the functions $\{\cos t, \cos 3t\}, \ldots$, $\{\cos t, \cos 3t, \ldots, \cos(2n+1)t\}$. In order to understand the effects of these terms we need only consider the general equation

$$\frac{d^2z}{dt^2} + z = \cos \omega t, \quad z(0) = 0, \quad \dot{z}(0) = 0, \qquad (17.37)$$

having the solution

$$z(t) = \frac{\cos \omega t - \cos t}{1 - \omega^2}.$$

This solution is valid for all ω, but when $\omega = 1$ both numerator and denominator are

---

† The name secular is used in astronomy and celestial mechanics to describe the time variation of variables whose mean over a sufficiently long time changes monotonically; for instance if $u = a_0 t + a_1 \sin \omega_1 t + a_2 \sin \omega_2 t$, it is said to have a secular component $a_0 t$. Many approximation schemes, such as that described here, produce secular variations and the problem in practice is to know whether these are real or an artifact of the method. The distinction between secular and oscillatory variations in observables is clearly important, but not always apparent from observations made over relatively short times: for instance the variation of $u = a \sin \omega t$ will appear linear over times short by comparison to π/ω.

zero, so care is needed. The limit is obtained using l'Hospital's rule

$$z(t) = \lim_{\omega \to 1} \frac{\cos \omega t - \cos t}{1 - \omega^2} = \lim_{\omega \to 1} \frac{-t \sin \omega t}{-2\omega} = \frac{1}{2} t \sin t,$$

which reproduces the term linear in t.

The physical reason for this secular term is that equation 17.37 represents a linear oscillator with natural frequency of unity — the $\ddot{z} + z$ term — disturbed by a periodic force, $\cos \omega t$. Because the frequency of oscillations of a linear oscillator is independent of the amplitude, when the forcing term is in resonance, $\omega = 1$, it remains in resonance no matter how large the amplitude becomes. Thus whatever the amplitude of the motion the resonant force increases the energy of the oscillator and the amplitude steadily increases. For a nonlinear oscillator the frequency of the motion always depends upon the amplitude, so a resonant driving force cannot stay in resonance if it drives the amplitude to large values; this effect is considered in exercise 17.36. This crucial ingredient is missing from the naïve perturbation theory.

Lindstedt's method

In Lindstedt's method the frequency of the solution ω is treated as an unknown to be determined together with the solution. This is achieved by defining a new time

$$\tau = \omega t,$$

so equation 17.34 becomes

$$\omega^2 \frac{d^2 z}{d\tau^2} + z - \delta z^3 = 0, \tag{17.38}$$

and we attempt to find a solution which is 2π-periodic in τ, by expanding *both* ω and $z(\tau)$ as power series,

$$\begin{aligned} \omega &= \omega_0 + \delta \omega_1 + \delta^2 \omega_2 + \cdots, \\ z(\tau) &= z_0(\tau) + \delta z_1(\tau) + \delta^2 z_2(\tau) + \cdots, \end{aligned} \tag{17.39}$$

where $z_0(0) = 1$, $\dot{z}_0(0) = 0$ and $z_k(0) = \dot{z}_k(0) = 0$ for $k \geq 1$. When $\delta = 0$ we know that $\omega = 1$ so $\omega_0 = 1$, but all the remaining coefficients are unknown.

Substitute these series into equation 17.38 and equate the coefficients of the powers of δ to zero. The first two equations are

$$\begin{aligned} \frac{d^2 z_0}{d\tau^2} + z_0 &= 0, & z_0(0) = 1, \quad \dot{z}_0(0) = 0, \\ \frac{d^2 z_1}{d\tau^2} + z_1 &= -2\omega_1 \frac{d^2 z_0}{d\tau^2} + z_0^3, & z_1(0) = \dot{z}_1(0) = 0. \end{aligned} \tag{17.40}$$

The first equation gives $z_0(\tau) = \cos \tau$ and then the second becomes, after some rearrangement,

$$\frac{d^2 z_1}{d\tau^2} + z_1 = \left(2\omega_1 + \frac{3}{4} \right) \cos \tau + \frac{1}{4} \cos 3\tau.$$

This equation is similar to equation 17.36 but now we have the freedom to set the coefficient of $\cos \tau$ to zero, and hence remove the secular term, by choosing $\omega_1 = -\frac{3}{8}$. Then the solution is

$$z_1(\tau) = \frac{1}{32}(\cos \tau - \cos 3\tau),$$

Hence our approximate solution is, to $O(\delta)$,

$$z(t) \;=\; \cos \tau + \frac{\delta}{32}(\cos \tau - \cos 3\tau), \quad \tau = \omega t, \qquad (17.41)$$

$$\omega \;=\; 1 - \frac{3}{8}\delta.$$

On substituting $\delta = \epsilon A^2$ we see that the frequency expansion agrees with the first term in the series obtained in exercise 17.26.

This procedure in entirely automatic and readily carried out by Maple to an order that depends upon available computer memory. The following Maple commands determine the solution to $O(\delta^2)$.

We start with equation 17.38, and in the Maple code use t in place of τ and d in place of δ to save typing and make the Maple input neater. First we set $N = 2$ to find the second-order expansion; higher-order expansions are found simply by increasing N, but then some of the output ought to be suppressed by changing the terminators.

```
>  N:=2:
```

Define the perturbation expansion of the frequency and the solution, to $O(N)$ in d:

```
>  W:=1+add(d^k * w.k,k=1..N):
```

and the solution, here denoted by z,

```
>  z:=x0(t) + add(d^k * x.k(t),k=1..N);
```

$$z := x0(t) + d\,x1(t) + d^2\,x2(t)$$

Note that, in principle, the components w.k and x.k(t) could both be replaced by elements of a list, but using the list x[k](t) produces an error in dsolve when using Maple 5, but not with Maple 6.†

```
>  eqn:=series(W^2 * diff(z,t$2) + z - d*z^3,d=0,N+1):
```

```
>  for k from 0 to N do; eq[k]:=coeff(eqn,d,k)=0; od:
```

These are the equations that have to be solved; the first and second are equations 17.40. The solution of the unperturbed equation with the initial conditions $x_0(0) = 1$, $\dot{x}_0(0) = 0$, is

```
>  sol:=dsolve({eq[0],x0(0)=1,D(x0)(0)=0},x0(t));
```

$$sol := x0(t) = \cos(t)$$

```
>  x0:=unapply(rhs(sol),t);
```

$$x0 := \cos$$

With this expression for x0(t) the equation, eq[1], for x1(t) will contain a term that produces a secular variation in x1(t). This needs to be isolated and removed by choosing w1 appropriately. With Maple we perform this operation exactly as in

---

† Recall that in Maple 6 the concatenation operator **a.b** is different and a more comprehensible output is obtained if the list construction is used.

the previous analysis by converting the expression to multiple angle form, using the combine command. Remember that it is often necessary to use expand before using the combine command when converting to a multiple angle form, though in this particular problem it is not necessary.

Removal of the secular term means that the solution of eq[1], x1(t), is periodic. Substituting this into eq[2] then gives another equation from which secular terms have to be removed by an appropriate choice of w2: this ensures that x2(t) is periodic. In principle this process may be carried out to any order: the first N terms are given by the following loop construction.

```
>    for k from 1 to N do;
>    eq[k]:=combine(expand(eq[k]),trig):
>    secular:=select(has,lhs(eq[k]),cos(t)):
>    w.k:=solve(secular=0,w.k);
>    sol:=combine( dsolve( {eq[k],x.k(0)=0,D(x.k)(0)=0},x.k(t)), trig):
>    x.k:=unapply(rhs(sol),t):
>    od:
```

With $N = 6$ these commands give the following sixth-order approximations to the frequency:

$$\omega = 1 - \frac{3}{8}\delta - \frac{21}{256}\delta^2 - \frac{81}{2048}\delta^3 - \frac{6549}{262144}\delta^4 - \frac{37737}{2097152}\delta^5 - \frac{936183}{67108864}\delta^6 + O(\delta^7).$$

The second-order approximation to the solution is

$$z = \cos \omega t + \frac{\delta}{32}(\cos \omega t - \cos 3\omega t) + \frac{\delta^2}{1024}(23\cos \omega t - 24 \cos 3\omega t + \cos 5\omega t).$$

In terms of the original variables we have $x(t) = Az(t)$ and $\delta = \epsilon A^2$, where A is the amplitude of the motion.

Higher-order terms are readily computed, and in the next two graphs we compare the solutions given by tenth-order perturbation theory, the faint lines, and by numerical integration, the solid lines, for the motion with $\delta = 0.95$, which is close to the separatrix. On the left is the direct comparison of $z(t)$ over approximately one period; on the right we have used the power series for ω to obtain the [5/5] Padé approximant for ω and used this in the series for z. As can be seen, this results in a substantial improvement in accuracy.

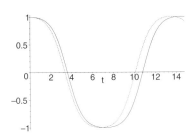

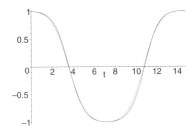

Figure 17.19 Comparison of the tenth-order expansion with the numerical solution, the solid line.

Figure 17.20 As in 17.19 but using the Padé approximant for the frequency.

Exercise 17.32

Use the Maple commands listed above to reproduce figures 17.19 and 17.20 and make similar comparisons for $\delta = 0.5$ and $\delta = -1$.

Exercise 17.33

In this problem you will find a perturbation approximation to the motion of the vertical pendulum with amplitude $\theta_m \simeq \pi/2$. Use the result of exercise 30 to show that $\delta_5 = 3\epsilon^2/10$, $\delta_7 = 3\epsilon^3/70$ and $\delta_9 = \epsilon^4/280$, where $\epsilon = \delta_3 = \theta_m^2/6$, and that if $\theta_m = \pi/2$, $\epsilon = 0.411$.

Use the approximate equation of motion

$$\frac{d^2z}{dt^2} + z - \epsilon z^3 + \frac{3}{10}\epsilon^2 z^5 - \frac{3}{70}\epsilon^3 z^7 = 0, \quad z(0) = 1, \quad \dot{z}(0) = 0,$$

and Lindstedt's method to find a third-order perturbation approximation to the motion and to show that

$$\frac{T(\epsilon)}{2\pi} = 1 + \frac{3\epsilon}{8} + \frac{33\epsilon^2}{256} + \frac{519\epsilon^3}{10\,240} + O(\epsilon^4),$$

$$z(t) = \cos(\tau) + \frac{\epsilon}{32}(\cos\tau - \cos 3\tau) + \frac{3\epsilon^2}{5120}(17\cos\tau - 20\cos 3\tau + 3\cos 5\tau) + O(\epsilon^3),$$

where $\tau = \omega t$, and find the third- and fourth-order terms of $z(t)$.

Compare this approximation with the solution obtained by numerical integration of the original equations of motion, $\ddot{\theta} + \sin\theta = 0$ for $\theta_m = \pi/2$ and $\theta_m = 3\pi/4$. For this comparison you should plot the graphs of both solutions and also the difference between the two over one period.

Exercise 17.34

Use equation 17.27 to show that the period of the librational motion, that is, the oscillatory motion with amplitude $\theta_m < \pi$, of the vertical pendulum described by the equation

$$\ddot{\theta} + \alpha^2 \sin\theta = 0$$

can be expressed in terms of the elliptic integral as follows:

$$T(\theta_m) = \frac{4}{\alpha}K\left(\sin(\theta_m/2)\right).$$

Show that the series expansion of this expression when expressed in terms of $\epsilon = \theta_m^2/6$ agrees with the series obtained in the previous exercise to $O(\epsilon^3)$.

Exercise 17.35

Use Lindstedt's method to find a fourth-order perturbation approximation for the frequency and the solutions of Volterra's equations, defined in exercise 16.42 (page 665),

$$\dot{x} = ax(1-y), \quad \dot{y} = -y(1-x), \quad a > 0,$$

about the fixed point at $(1,1)$.

Exercise 17.36

Consider the forced nonlinear oscillator

$$\frac{d^2x}{dt^2} + x + ax^3 = F\sin\Omega t,$$

which is a nonlinear version of equation 17.37 if $a = 0$ and $F = 1$.

Write a Maple procedure that will solve this equation numerically and plot the graph of the solution, for particular values of a, F, Ω and initial conditions. Specifically, for the initial conditions $(x(0), y(0)) = (0, 0.5)$ and with $F = 0.1$, consider the solutions with $a = 0$, $\Omega = 1.1$ and 1 and $a = 0.1$, $\Omega = 1.1$ and 1. In every case integrate over a sufficiently long time for the significant trends to become apparent.

17.6.2 Limit cycles

The use of Lindstedt's method to approximate limit cycles is very similar, though there is one important difference, namely the amplitude of the motion, besides the frequency, is unknown. This means that the initial conditions of each term in the perturbation expansion are initially unknown and are determined by removing components leading to secular solutions.

Consider the van der Pol equation: make the same substitution $\tau = \omega t$ with ω the unknown frequency and seek a solution 2π-periodic in τ. The equation becomes

$$\omega^2 \frac{d^2x}{d\tau^2} + v\omega(x^2 - 1)\frac{dx}{d\tau} + x = 0, \tag{17.42}$$

where v is the small expansion parameter. As before, we assume that

$$\begin{aligned}
\omega &= 1 + v\omega_1 + v^2\omega_2 + \cdots, \\
x(\tau) &= x_0(\tau) + vx_1(\tau) + v^2x_2(\tau) + \cdots,
\end{aligned} \tag{17.43}$$

where each $x_k(\tau)$ is 2π-periodic in τ. Since the points at which the orbit intersects either axis of phase space are not known we have to use the initial conditions

$$x_k(0) = a_k, \quad \dot{x}_k(0) = 0, \tag{17.44}$$

where the a_k are unknown numbers; compare these initial conditions with those in equation 17.40.

On substituting equations 17.43 into the differential equation and collecting powers of v we obtain

$$\begin{aligned}
\ddot{x}_0 + x_0 &= 0, \tag{17.45} \\
\ddot{x}_1 + x_1 &= \dot{x}_0(1 - x_0^2) + 2\omega_1\ddot{x}_0, \tag{17.46} \\
\ddot{x}_2 + x_2 &= \dot{x}_1(1 - x_0^2) - 2x_0x_1\dot{x}_0 - 2\omega_1\ddot{x}_1 - \left(\omega_1^2 - 2\omega_2\right)\ddot{x}_0. \tag{17.47}
\end{aligned}$$

These expressions can be obtained using Maple: the relevant worksheet is described in section 17.9.2, and this may be used to generate higher-order terms in the series.

The solution of equation 17.45, with the initial condition 17.44, is

$$x_0(\tau) = a_0 \cos \tau,$$

where a_0 is, at present, unknown. Substituting $x_0(\tau)$ into equation 17.46, we obtain, after some algebraic manipulation,

$$\ddot{x}_1 + x_1 = -2\omega_1 a_0 \cos\tau + a_0 \left(1 - \frac{1}{4}a_0^2\right)\sin\tau - \frac{1}{4}a_0^3 \sin 3\tau.$$

For x_1 to be periodic we must make the coefficients of both $\cos\tau$ and $\sin\tau$ zero. Thus we choose $a_0 = 2$ and $\omega_1 = 0$ and solve the equation with the initial condition 17.44 to give

$$x_1(\tau) = a_1 \cos\tau + \frac{3}{4}\sin\tau - \frac{1}{4}\sin 3\tau.$$

At the next stage, for $x_2(\tau)$ both a_1 and ω_2 can be determined in a similar manner, and so on; at the nth stage, for $x_n(\tau)$ we will determine a_{n-1} and ω_n. Thus the nth-order perturbation series will give the frequency to $O(v^n)$ and the solution $x(\tau)$ to $O(v^{n-1})$. The pattern of these calculations is clear and easily implemented using Maple: the relevant code is given in section 17.9.2. Tenth-order perturbation theory gives, for the frequency,

$$\omega = 1 - \frac{1}{16}v^2 + \frac{17}{3072}v^4 + \frac{35}{884\,736}v^6 - \frac{678\,899}{5\,096\,079\,360}v^8 + \frac{28\,160\,413}{2\,293\,235\,712\,000}v^{10} + O(v^{12}),$$

and the first few terms of the solution are

$$x(\tau) = 2\cos\tau + \frac{v}{4}(3\sin\tau - \sin 3\tau) - \frac{v^2}{8}\left(\cos\tau - \frac{3}{2}\cos 3\tau + \frac{5}{12}\cos 5\tau\right) + O(v^3).$$

The amplitude of the motion is just the sum of the components $a_k v^k$,

$$A = 2 + \frac{1}{96}v^2 - \frac{1\,033}{552\,960}v^4 + \frac{1\,019\,689}{55\,738\,368\,000}v^6 + \frac{9\,835\,512\,276\,689}{157\,315\,969\,843\,200\,000}v^8 + O(v^{10}).$$

One of the differences between this and conservative systems is that it seems to be impossible to find either the frequency or amplitude of the motion without also finding the motion, a consequence of there being no integral of the motion.

It is not possible to estimate the radius of convergence directly from this series for ω, because there are insufficient terms, but the [4/4] and [6/6] Padé approximants, the latter being obtained using 12th-order perturbation theory, have respectively four and six singularities at a minimum distance of 2.6 and 2.2 from the origin, suggesting that the radius of convergence is about 2. In fact Andersen and Geer (1982) have computed this series to $O(v^{164})$ and estimate that the radius of convergence is between 1.84 and 1.85.

In figures 17.21 and 17.22 we compare the solution $x(\tau)$ given by tenth-order perturbation theory and numerical integration with initial condition given by the value of the perturbation solution at $t = 0$. In the first figure $v = 1.4$ and there is no difference detectable in this graph; in the second figure, on the right, where $v = 1.6$ spurious high frequency oscillations occur in the perturbation series, which disappear if a higher-order expansion is used.

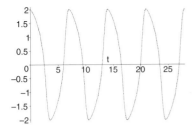

Figure 17.21   Comparison of the
tenth-order expansion and the numerical
solution for $v = 1.4$.

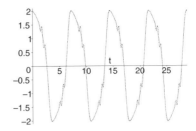

Figure 17.22   Comparison of the
tenth-order expansion and the numerical
solution for $v = 1.6$.

Exercise 17.37
Compare the values of period of the van der Pol oscillator given by perturbation theory and
by the numerical procedure obtained in exercise 17.10 (page 680). Also make a comparison
with the values given by the asymptotic expansion in equation 17.10 (page 683).

Exercise 17.38
Consider the equation

$$\frac{d^2x}{dt^2} + v(x^4 - 1)\frac{dx}{dt} + x = 0, \quad v \ge 0.$$

Show that for small v the period of the limit cycle is given by the perturbation expansion

$$\frac{T}{2\pi} = 1 + \frac{7}{48}v^2 - \frac{2981}{138\,240}v^4 - \frac{18\,059\,927}{3\,483\,648\,000}v^6 + \frac{307\,491\,710\,018\,051}{29\,496\,744\,345\,600\,000}v^8 + O\left(v^{10}\right),$$

and find the expansion for the amplitude of the motion. For $v = 0.8$ and 0.9 compare the
solution as given by eighth-order perturbation theory with the numerical solution.

17.7 The harmonic balance method

The harmonic balance method is a conceptually simple technique of approximating
periodic orbits, but in practice is difficult to apply except in the lowest order, unless a
computer is used.

Suppose that we seek a periodic solution of a nonlinear differential equation with
period ω. The solution can be approximated with a truncated Fourier series:

$$x(t) = \sum_{k=-N}^{N} c_k e^{-ik\omega t}, \quad c_{-k} = c_k^*, \tag{17.48}$$

in which there are $N + 1$ coefficients $(c_0, c_1, \ldots, c_N)$ comprising $2N + 1$ real numbers.
By substituting this series into the differential equation and setting the coefficients
of $e^{-ik\omega t}$, $k = -N, \ldots, N$, to zero we obtain $2N + 1$ equations for these $2N + 1$
constants. In principle these equations can be solved to provide an approximation to
the solution, though in practice solving the $2N + 1$ nonlinear algebraic equations is not
straightforward.

In order to see how this technique works consider the simple example treated in section 17.6, namely

$$\frac{d^2z}{dt^2} + z - \delta z^3 = 0, \quad \delta < 1. \tag{17.49}$$

In this example we know that, with a suitable choice of time-origin, the Fourier series will contain only the components $\cos(2k+1)\omega t$, so we assume a solution of the form

$$z(t) = a\cos\omega t + b\cos 3\omega t. \tag{17.50}$$

Substituting this expression into the equation of motion and rearranging gives

$$\left[a(1-\omega^2) - \frac{3}{4}a\delta\left(a^2 + ab + 2b^2\right) \right]\cos\omega t \tag{17.51}$$

$$+ \left[b(1-9\omega^2) - \frac{1}{4}b\delta\left(3b^2 + 6a^2\right) - \frac{1}{4}\delta a^3 \right]\cos 3\omega t \;=\; \text{higher harmonics.}$$

Equation 17.51 is valid for all t and therefore the coefficients of $\cos\omega t$ and $\cos 3\omega t$ are both zero, hence we have the two equations

$$a(1-\omega^2) \;=\; \frac{3}{4}a\delta\left(a^2 + ab + 2b^2\right), \tag{17.52}$$

$$b(1-9\omega^2) \;=\; \frac{1}{4}b\delta\left(3b^2 + 6a^2\right) + \frac{1}{4}a^3\delta. \tag{17.53}$$

In principle these can be solved to give (a,b) in terms of (ω,δ). But these are nonlinear equations and it is necessary to develop a method that will work when more terms are included in the trial solution. For this reason we consider a perturbation expansion.

In the unperturbed limit $\delta = 0$ we know that $b = 0$ and $\omega = 1$ and that $a > 0$, but is otherwise arbitrary. Because of this it is easier to solve the equations for (b,ω) assuming that (a,δ) are known, for we can write

$$\omega = 1 + \delta\omega_1,$$

then it is clear from equation 17.53 that $b = O(\delta)$, provided $a = O(1)$. Thus on ignoring terms $O(\delta^2)$ the first equation gives

$$-2a\omega_1 - \frac{3}{4}a^3 = O(\delta) \quad\Longrightarrow\quad \omega_1 = -\frac{3}{8}a^2,$$

and the second equation gives

$$8b = -\frac{1}{4}\delta b\left(6a^2 + 3b^2\right) - \frac{1}{4}\delta a^3,$$

which shows that $b = O(\delta)$ and that, to this order,

$$b = -\frac{a^3\delta}{32}.$$

Hence the approximate solution is

$$z(t) = a\cos\omega t - \frac{a^3\delta}{32}\cos 3\omega t, \quad \omega = 1 - \frac{3a^2}{8}\delta. \tag{17.54}$$

Exercise 17.39
Derive equations 17.51 and 17.52, without using Maple, and provide the details of the calculations leading to the solution 17.54.

Exercise 17.40
The solution 17.54 is not the same as that of equation 17.41 (page 702), obtained previously, because $z(0) = a - \delta a^3/32$. Find the value of a, to $O(\delta)$, which makes $z(0) = 1$ and hence show that the two solutions agree.

The inclusion of higher-order terms in the trial Fourier series leads to more equations of the type 17.51 and 17.53 and it is not immediately evident how these can be solved. For instance, adding the next term of the Fourier series 17.50 gives

$$z(t) = a \cos \omega t + b_3 \cos 3\omega t + b_5 \cos 5\omega t,$$

and using the same technique we obtain the following three equations:

$$a(1 - \omega^2) = \frac{3}{4}\delta \left(a^3 + a^2 b_3 + 2a(b_3 b_5 + b_3^2 + b_5^2) + b_3^2 b_5 \right), \qquad (17.55)$$

$$b_3(1 - 9\omega^2) = \frac{\delta}{4} \left[a^3 + 3 \left(a^2 b_5 + b_3 \right) + 6b_3 \left(a^2 + ab_5 + b_5^2 \right) \right], \qquad (17.56)$$

$$b_5(1 - 25\omega^2) = \frac{3}{4}\delta \left[ab_3 (a + b_3) + 2b_5 \left(a^2 + b_3 \right) + b_5^2 \right]. \qquad (17.57)$$

The left hand side of each of these equations comes from the linear part of the equations of motion, $\ddot{z} + z$, and the right hand side from the δz^3 term.

These equations can be solved using a perturbation expansion in δ and employing the fact that the harmonic $\cos 3\omega t$ appears only in first order, so $b_3 = O(\delta)$, and the $\cos 5\omega t$ in second order, giving $b_5 = O(\delta^2)$. Thus we write

$$\omega = 1 + \omega_1 \delta + \omega_2 \delta^2, \quad b_3 = b_{31}\delta + b_{32}\delta^2, \quad b_5 = b_{52}\delta^2.$$

Substituting these into equations 17.55 to 17.57 and collecting the coefficients of δ^k, $k = 0$, 1 and 2, gives the equations

$$\delta a \left(2\omega_1 + \frac{3}{4}a^2 \right) + a\delta^2 \left(2\omega_2 + \omega_1^2 + \frac{3}{4}ab_{31} \right) = 0, \qquad (17.58)$$

$$\delta \left(\frac{1}{4}a^3 + 8b_{31} \right) + \delta^2 \left(\frac{3}{2}a^2 b_{31} + 8b_{32} + 18\omega_1 b_{31} \right) = 0, \qquad (17.59)$$

$$\delta^2 \left(\frac{3}{4}a^2 b_{31} + 24b_{52} \right) = 0. \qquad (17.60)$$

Equating coefficients of δ^k to zero yields five equations for the five unknown parameters. The lowest-order coefficients of each of the above equations can be solved first to give

$$\omega_1 = -\frac{3}{8}a^2, \quad b_{31} = -\frac{a^3}{32},$$

and then the second-order terms may be solved to give

$$\omega_2 = -\frac{15}{256}a^4, \quad b_{32} = -\frac{21}{1024}a^5, \quad b_{52} = \frac{1}{1024}a^5,$$

so the approximate solution is

$$z(t) = a \cos \omega t - \frac{a^3 \delta}{32} \left(1 + \frac{21}{32} a^2 \delta \right) \cos 3\omega t + \frac{a^5 \delta^2}{1024} \cos 5\omega t,$$

with

$$\omega = 1 - \frac{3}{8} a^2 \delta - \frac{15}{256} a^4 \delta^2 + O(\delta^3).$$

It should now be clear how to proceed to higher-order approximations. With the addition of another term to the trial Fourier series we introduce another constant $b_7 = b_{73} \delta^3$, the coefficient of the $\cos 7\omega t$ term, and three new constants b_{33}, b_{53} and ω_3 to the power series for b_3, b_5 and ω respectively, to give a total of nine unknowns. Equations 17.58 to 17.60 change only by the addition of a term $O(\delta^3)$, but there is also an additional equation starting at this order in δ; so we gain an extra four equations for these new unknowns. All of these nine equations could be solved as a large set of linear equations, using the `solve` command, but this is particularly inefficient. It is better to proceed by first solving the two equations $O(\delta)$ for ω_1 and b_{31}, then the three equations $O(\delta^2)$, individually, and then the three equations $O(\delta^3)$, again individually.

In this manner it can be seen how to generalise this scheme to cope with N terms in the trial Fourier series. We set

$$z(t) = a \cos \omega t + \sum_{k=2}^{N} b_{2k-1} \cos(2k-1)\omega t, \tag{17.61}$$

and

$$
\begin{aligned}
\omega &= 1 + \omega_1 \delta + \omega_2 \delta^2 + \cdots + \omega_{N-1} \delta^{N-1}, \\
b_3 &= \delta b_{31} + \delta^2 b_{32} + \cdots + \delta^{N-1} b_{3N-1}, \\
\vdots\; &=\quad\quad \vdots \\
b_{2n+1} &= \delta^n b_{2n+1\ n} + \delta^{n+1} b_{2n+1\ n+1} + \cdots + \delta^{N-1} b_{2n+1\ N-1} \\
\vdots\; &=\quad\quad \vdots \\
b_{2N-1} &= \delta^{N-1} b_{2N-1\ N-1},
\end{aligned}
\tag{17.62}
$$

so there are now $(N-1)(N+2)/2$ unknown quantities. Substituting these into the equation of motion, collecting the coefficients of all the harmonics used and then the coefficients of δ^k for $k = 1, 2, \ldots, N-1$ gives a set of $(N-1)(N+2)/2$ equations for these coefficients. As in the particular cases these equations can be dealt with individually.

It is very difficult to implement this scheme without computer assistance, and even using Maple care is necessary otherwise the computational time becomes excessive. A working scheme is listed in section 17.9.3.

Exercise 17.41
Find the solution of the equation

$$\ddot{z} + z - \delta z^3 + A\delta^2 z^5 = 0, \quad z(0) = 1, \ \dot{z}(0) = 0,$$

where A is a constant, which includes the first eleven harmonics, that is, up to the term $\cos 11\omega t$.

Exercise 17.42

Consider the conservative system

$$\ddot{z} + z^3 = 0.$$

Show that all solutions are periodic and that, with the correct choice of the time origin, the Fourier series of the solutions contains only the harmonics $\cos(2n + 1)\omega t$, where ω is the frequency of the motion.

Show that a simple harmonic balance approximation yields

$$z(t) \simeq z_1(t) = \frac{2\omega}{\sqrt{3}} \cos \omega t.$$

Use Maple to numerically solve the differential equation with the initial conditions $z(0) = A$, $\dot{z}(0) = 0$ and compare this solution with the harmonic balance approximation for a range of amplitudes, A.

Extend the harmonic balance approximation by assuming an approximate solution of the form $z(t) = a \cos \omega t + b \cos 3\omega t$ and show that $a \simeq 1.1275\omega$ and $b = 0.0550\omega$. Compare this with numerical solutions of the original equation.

17.8 Appendix: On the Stability of the Solar System, by H. Poincaré

This article was published in *Nature*, Volume **58** (Number 1495), June 23 1898, and printed here with permission. It is a translation of an article published in *Annuaire du Bureau des Longitudes*, 1898.

All persons who interest themselves in the progress of celestial mechanics, but can only follow it in a general way, must feel surprised at the number of times demonstrations of the stability of the solar system have been made.

Lagrange was the first to establish it, Poisson then gave a new proof; afterwards other demonstrations came, and others will still come. Were the old demonstrations insufficient, or are the new ones unnecessary?

The astonishment of these persons would doubtless be increased if they were told that perhaps some day a mathematician would show by rigorous reasoning that the planetary system is unstable. This may happen, however; there would be nothing contradictory in it, and the old demonstrations would still retain their value.

The demonstrations are really but successive approximations; they do not pretend to strictly confine the elements of the orbits within narrow limits that they may never exceed, but they at least teach us that certain causes, which seemed at first to compel some of these elements to vary fairly rapidly, only produce in reality much slower variations.

The attraction of Jupiter, at an equal distance, is a thousand times smaller than that of the sun; the disturbing force is therefore small: nevertheless, if it always acted in the same direction it would not fail to produce appreciable effects. But the direction is not constant, and this is the point that Lagrange established. After a small number of years two planets, which act on each other, have occupied all possible positions in their orbits; in these diverse positions their mutual action is directed sometimes one way, sometimes in the opposite way, and that in such a fashion that after a short time there is almost exact compensation. The major axes of the orbits are not absolutely invariable, but their variations are reduced to oscillations of small amplitude about a mean value.

This mean value, it is true, is not rigorously fixed, but the changes which it undergoes are extremely slow, as if the force which produces them was not a thousand times, but a million times smaller than the solar attraction. One may, therefore, neglect these changes, which are of the order of the square of the masses. As to the other elements of the orbits, such as the eccentricities and the inclinations, these may acquire round their mean value wider and slower oscillations, to which, however, limits may easily be assigned.

This is what Lagrange and Laplace pointed out, but Poisson went further. He wished to study the slow changes experienced by the mean values — changes to which I have already referred, and which his predecessors had at first neglected. He showed that these changes reduced themselves again to periodic oscillations round a mean value which is only liable to variations a thousand times slower.

This was a step further, but it was still only an approximation. Since then further advance has been made, but without arriving at a complete definitive and rigorous demonstration. There is a case which seemed to escape the analysis of Lagrange and Poisson. If the two mean movements are commensurable among themselves, at the end of a certain number of revolutions, the two planets and the sun will be found in the same relative situation, and the disturbing force will act in the same direction as at first. The compensation, to which I have referred, will not any more be produced, and it might be feared that the effects of the disturbing forces will end by accumulating and becoming very considerable. More recent works, amongst others those of Delaunay, Tisserland, and Gyldén, have shown that this accumulation does not actually occur. The amplitude of the oscillations is slightly increased, but remains, nevertheless, very small. This particular case, therefore, does not escape the general rule.

The apparent exceptions have not only been dispensed with, but the real reasons of these compensations, which the founders of celestial mechanics had observed, have been better explained. The approximation has been pushed further than was done by Poisson, but it is still only an approximation.

It can be shown, in particular cases, that the elements of the orbit of one planet will return an infinite number of times to very nearly the initial elements, and that is also probably true in the general case; but it does not suffice. It should be shown that these elements will not only regain their original values, but that they will never deviate much from them.

This last demonstration has never been given in a definite manner, and it is even probable that the proposition is not strictly true. The statement that is true, is that the elements can only deviate extremely slowly from their original values, and this after a long interval of time. To go further, and affirm that these elements will remain not for *a very long time*, but always confined within narrow limits, is what we cannot do.

But the problem does not take this form.

The mathematician only considers fictitious bodies, reduced to simple material points, and subject to the exclusive action of their mutual attractions, which rigorously follows Newton's law. How would such a system behave, would it be stable? This is a problem which is as difficult as it is interesting for an analyst. But it is not one which actually occurs in nature. Real bodies are not material points, and they are subject to other forces than the Newtonian attraction. These complementary forces ought to have the effect of gradually modifying the orbits, even when the fictitious bodies, considered by the mathematician, possess absolute stability.

What we must ask ourselves then is, whether this stability will be more easily destroyed by the simple action of Newtonian attraction or by these complementary forces.

When the approximation shall be pushed so far that we are certain that the very slow variations, which the Newtonian attraction imposes on the orbits of the fictitious bodies, can only be very small during the time that suffices for the complementary forces to destroy the system: when, I say, the approximation shall be pushed as far as that, it will be useless to go further, at least from the point of view of application, and we must consider ourselves satisfied.

But it seems that this point is attained: without quoting figures, I think that the effects of these complementary forces are much greater than those of the terms neglected by the analysts in the most recent demonstrations on stability.

Let us see which are the most important of these complementary forces. The first idea which comes to mind is that Newton's law is, doubtless, not absolutely correct: that the attraction is not rigorously proportional to the inverse square of the distances, but to some other function of them. In this way Prof. Newcomb has recently tried to explain the movements of the perihelion of Mercury. But it is soon seen that this would not influence the stability. It is true, according to a theory of Jacobi, that there would be instability if the attraction were inversely proportionate to the cube of the distance. It is easy by rough reasoning to account for this; with such a law, the attraction would be great for the small distances and extremely feeble for great distances. If therefore, for any reason, the distance of one of the planets from the central body were to increase the attraction would diminish rapidly until it would not be capable of retaining the planet in its orbit. But that only takes place with laws very different from that of the square of the distances. All laws, near enough to that of Newton's to be acceptable, are equivalent from the stability point of view.

But there is another reason which opposes the theory that bodies move without ever deviating much from their original orbits. According to the second law of thermodynamics, known by the name of Carnot's Principle, there is a continual dissipation of energy, which tends to lose the form of mechanical work and to take the form of heat. There exists a certain function called entropy, which it is unnecessary to define here; entropy, according to this second law, either remains constant or diminishes, but can never increase. When once it has deviated from its original value, which it can only do by diminishing, it can never return again as it would have to increase. The world consequently could never return to this original state, or to a slightly different state, so soon as its entropy has changed. It is the contrary of stability.

But the entropy diminishes every time that an irreversible phenomenon takes place, such as the friction of two solids, the movement of a viscous liquid, the exchange of heat between two bodies of different temperatures, the heating of a conductor by the passage of a current. If we observe, then, that there is not in reality a reversible phenomenon, that the reversibility is only a limiting case — an ideal case which nature can more or less approach but can never attain — we shall be led to conclude that instability is the law of all natural phenomena.

Are the movements of the heavenly bodies the only ones to escape? One might believe it by seeing that they move in a vacuum, and are thus free from friction. But it is the interplanetary vacuum absolute, or do the bodies move in an extremely attenuated medium of which the resistance is extremely feeble, but nevertheless is capable of offering resistance?

Astronomers have only been able to explain the movement of Encke's comet by supposing the existence of such a medium. But the resisting medium which would account for the anomalies of this comet, if it exists, is confined to the immediate neighbourhood of the sun. This comet would penetrate it: but at the distances at which the planets are, the action of this medium would cease to make itself felt, or would become much more feeble. As an indirect effect, it would accelerate the movements of the planets; losing energy, they would tend to *fall* on the sun, and by reason of Kepler's third law the duration of the revolution would diminish at the same time as the distance to the central body. But it is impossible to form an idea of the rapidity with which this effect would be produced, as we have no notion of the density of this hypothetical medium.

Another cause to which I am going to refer must have, it seems, a more rapid action. It had for some time been imagined, but was first more especially brought to light by Delaunay, and afterwards by G. Darwin.

The tides, which are direct consequences of celestial movements, could only stop if these movements are ceased. But the oscillations of the seas are accompanied by friction, and consequently produce heat. This heat can only be borrowed from the energy which produces the tides - that is to say, to the *vis viva* of the celestial bodies. We can therefore foresee that, for this reason, this *vis viva* is gradually dissipated, and a little reflection will enable us to understand by what mechanism. The surface of the seas, raised by the tides, presents a kind of wave. If high tide took place at the time of the meridian passage of the moon, this surface would be that of an ellipsoid, the axis of which would pass through the moon. Everything would be symmetrical in relation to this axis, and the attraction of the moon on this wave could neither slow down nor accelerate the celestial rotation. This is what would happen if there were no friction: but in consequence of this friction, high tide is late on the Moon's meridian passage: symmetry ceases: the attraction of the moon on the wave no longer passes through the centre of the earth, and tends to slow down the rotation of our globe.

Delaunay estimated that, for this cause, the length of the sidereal day increases by one second in a hundred thousand years. It is thus he wishes to account for the secular acceleration of the moon's motion. The lunation would seem to us to become shorter and shorter, because the unit of time to which we ascribed it, the day, would become longer and longer.

Whatever we may think of the figures given by Delaunay, and the explanation which he proposes for the anomalies of the moon's movement, it is difficult to dispute the effect produced by the tides.

It is just this that may help us to understand a well-known but very surprising fact. It is known that the period of rotation of the moon is exactly equal to that of its revolution; in such a way that, if there were seas on this body, they would have no tides — at least, tides due to the attraction of the earth: because for an observer situated at a point on the surface of the moon, the earth would be always at the same height above the horizon. It is also known that Laplace tried to explain this curious coincidence. How can the two velocities be *exactly* the same? It is exceedingly improbable that this strict equality is due to mere chance. Laplace supposes that the moon has the form of an elongated ellipsoid; this ellipsoid behaves like a pendulum, which would be in equilibrium when the major axis is directed along the line joining the centres of the two bodies.

If the *initial* velocity of rotation differs slightly from that of revolution, the ellipsoid will oscillate about its position of equilibrium without ever deviating much from it. A pendulum which has received a slight impetus behaves in this way. The *mean* velocity of rotation is then exactly the same as that of the position of equilibrium round which the major axis oscillates; it is, therefore, the same as that of the straight line which joins the centres of the two bodies. It is, therefore, strictly equal to the velocity of revolution.

If, on the contrary, the initial velocity differs considerably from the velocity of revolution, the major axis will not oscillate any more round its position of equilibrium, like a pendulum which under a strong impulse describes a complete circle.

It suffices, therefore, that the velocity of revolution should be *almost* equal to the *initial* velocity of rotation, in order that it may be exactly equal to the *mean* velocity of rotation. A strict equality being no longer necessary, the paradox does not exist any more. The explanation is nevertheless incomplete. What is the reason of this approximate equality, of which the probability is no longer zero, it is true, but still very small? And, especially, why does not the moon undergo slight oscillations about its position of equilibrium (if we eliminate, of course, its numerous librations, due to other well known causes)? These oscillations must originally have existed: they must have become extinct by a kind of friction, and everything tends to make us believe that the mechanism of this friction is that which I have just analysed with respect to the ocean tides.

When the moon was not yet solid, and formed a fluid in the form of a spheroid, this spheroid must have experienced enormous tides, by reason of the proximity of the earth and of its mass. These tides could only have ceased when the oscillations became almost entirely extinct.

It seems that Jupiter's satellites, and the two planets nearest the sun, Mercury and

Venus, have also a rotation, the duration of which is the same as that of their revolution: it is doubtless for the same reason.

It might be thought that this tidal action has no connection with our subject. I have as yet only spoken of rotations, and in the studies relative to the stability of the solar system the movements of translation are only dealt with; but a little attention shows that the same action makes itself equally felt on the latter.

We have just seen that the attraction of the moon on the earth does not act exactly through the centre of the earth. The attraction of the earth on the moon, which is equal and exactly opposite, would not pass either through this centre; that is to say, through the focus of the lunar orbit. A disturbing force is the result, very small in reality, but sufficient to make the moon increase in energy. The active force of translation thus gained by the moon is evidently smaller than that of rotation, lost by the earth: because a part of the energy must be transformed into heat in consequence of the friction engendered by the tides. The period of revolution of the moon lasting about twenty-eight sidereal days, a very simple calculation shows that this body gains twenty-eight times less *vis viva* than the earth loses.

I have already explained the action of a resisting medium: I have shown how, by making the planets lose energy, their movements are accelerated: on the contrary the action of the tides, by increasing the energy of the moon, retards its movements: the month lengthens therefore as well as the day. Now if this cause acts alone, what is the final state towards which the system will tend? Obviously this action would only stop when the tides have ceased — that is to say, when the rotation of the earth would have the same duration.

This is not all: in the final state the orbit of the moon must have become circular, if it were otherwise, the variations of the distance of the moon to the earth would suffice to produce tides. As the movement of rotation would not have changed, it would be easy to calculate what angular velocity would be common to the earth and to the moon. One finds that, at the limit, the month, like the day, would last about sixty-five of our actual days.

Such would be the final state if there were no resisting medium, and if the earth and the moon existed alone.

But the sun also produces tides, the attraction of the planets likewise produces them on the sun. The solar system therefore would tend to a condition in which the sun, all the planets and their satellites, would move with the same velocity round the same axis, as if they were parts of one solid invariable body. The final angular velocity would, on the other hand, differ little from the velocity of revolution of Jupiter. This would be the final state of the solar system if there were not a resisting medium; but the action of this medium, if it exists, would not allow such a condition to be assumed, and would end by precipitating all the planets into the sun.

It must not be thought that a solid globe which was not covered by seas would, by the absence of tides, find itself free from actions analogous to those just mentioned, even by admitting that the solidification had reached the centre of the globe. This body, which we suppose solid, would not on that count be an invariable one; such bodies only exist in text books on rational "mechanics". It would be elastic and be subject, by the attraction of neighbouring celestial bodies, to deformations analogous to tides and of the same order of magnitude.

If the elasticity were perfect, these deformations would occur without loss of work, and without the production of heat. But perfectly elastic bodies do not exist. There would be in consequence development of heat, which would take place at the expense of the energy of rotation and translation of the bodies, and which will produce absolutely the same effects as the heat engendered by the friction of the tides.

This is not all: the earth is magnetic and very probably the sun and other planets are the same. The following well-known experiment is one which we owe to Foucault: a copper disc rotating in the presence of an electromagnetic field suffers a great resistance, and becomes heated when the electromagnetic field is brought into action. A moving conductor in a magnetic field is traversed by induction currents which heat it: the produced heat can only be derived from the *vis viva* of the conductor. We can therefore foresee that the electrodynamic actions of the electromagnet on the currents of induction must oppose the movement of the conductor. In this way Foucault's experiment is explained. The celestial bodies must undergo an analogous resistance because they are magnetic and conductors.

The same phenomenon, though much weakened by the distance, will therefore be produced: but the effects, being produced always in the same direction, will end by accumulating; they add themselves, besides, to those of the tides, and tend to bring the system to the same final state.

Thus the celestial bodies do not escape Carnot's law, according to which the world tends to a state of final repose. They would not escape it, even if they were separated by an absolute vacuum. Their energy is dissipated: and although this dissipation only takes place extremely slowly, it is sufficiently rapid that one need not consider terms neglected in the actual demonstrations of the stability of the solar system.

17.9 Appendix: Maple programs for some perturbation schemes

17.9.1 Poincaré's method for finding an integral of the motion

The following commands use the method described in section 17.4 to compute the approximate integral of the motion for the system defined in equation 17.19.

The function F_n is obtained from F_{n-2} in two stages from equation 17.15, which, in this case, is

$$\frac{\partial F_n}{\partial \theta} = X_3 \frac{\partial F_{n-2}}{\partial x} + Y_3 \frac{\partial F_{n-2}}{\partial y},$$

so $F_{2n-1} = 0$, $n = 1, 2, \ldots$. The following method works in more general cases.

At each stage we first evaluate the right hand side, using the procedure HN(f), and then check that the mean of this is zero. The procedure FN(h) then integrates this result. The first procedure computes $H_{2n}(\theta)$ (θ is denoted by p in the Maple script) given F_{2n-2}, denoted by f in the argument list, and the output is $r^{2n}H_{2n}(\theta)$ expressed in terms of multiple angles. Thus the constant of the motion is obtained by successive applications of these procedures after defining $F_2 = x^2 + y^2$: H4=HN(F2) $\rightarrow$

F4=FN(H4) → H6=HN(F4), etc., at each stage checking that the mean of $H_{2m}(x, y)$ is zero and, if necessary, making necessary modifications. This set of commands is not easily automated.

```
> restart;
> HN:=proc(f) local X3,Y3,h; global a,b;
> X3:=-b*x^3/2; Y3:=a*x^2*y-b*y^3/2;
> h:=X3*diff(f,x)+Y3*diff(f,y);
> h:=expand(subs(x=r*cos(p),y=r*sin(p),h)); h:=combine(h,trig);
> factor(h); end:
```

Next is a procedure for integrating $H_n(\theta)$ assuming it to be periodic in θ. The output $F_n(\theta)$ is given as a function of x and y:

```
> FN:=proc(h) local f;
> f:=int(h,p);
> f:=expand(f); f:=subs(cos(p)=x/r,sin(p)=y/r,r=sqrt(x^2+y^2),f);
> factor(f); end;
```

First define $F_2(x, y)$:

```
> F2:=x^2+y^2;
```

$$F2 := x^2 + y^2$$

Now use this to compute H_4:

```
> H4:=HN(F2);
```

$$H4 := -\frac{1}{4} r^4 (b \cos(4p) + 3b - a + a \cos(4p))$$

Now remove the constant term; in this case it is obvious that we need to put $a = 3b$, but it is useful to do this simple calculation using Maple: first find the constant term

```
> c4:=eval( subs(cos=0,H4) );
```

$$c4 := -\frac{1}{4} r^4 (3b - a)$$

and then solve this for a. (Note the following command does not work if a and b are local variables in the procedure HN(f).)

```
> a:=solve(c4=0,a);
```

$$a := 3b$$

So that H_4 becomes

```
> H4;
```

$$-b r^4 \cos(4p)$$

and we can express this in terms of x and y as in the next command. Here you should note the following.

(i) First it is necessary to expand H_4 in order to convert $\cos 4p$ into powers of $\cos p$;

(ii) The order of the substitutions is important.

```
> H4xy:=factor(subs(cos(p)=x/r,sin(p)=y/r,r=sqrt(x^2+y^2),expand(H4)));
```

$$H4xy := -b(-y^2 + 2xy + x^2)(-y^2 - 2xy + x^2)$$

The value of F_4 is

> F4:=FN(H4);

$$F4 := -b\,y\,x\,(x-y)(x+y)$$

Now $F_5 = 0$, and the value of H_6 is given by:

> H6:=HN(F4);

$$H6 := -\frac{3}{8}\,b^2\,r^6\,(\sin(6\,p) - \sin(2\,p))$$

The mean of this is zero and the consequent value of F_6 is:

> F6:=FN(H6);

$$F6 := -\frac{1}{8}\,b^2\,(x-y)(x+y)(y^4 + 10\,x^2\,y^2 + x^4)$$

Then H_8 is:

> H8:=HN(F6);

$$H8 := -\frac{3}{64}\,b^3\,r^8\,(6\cos(4\,p) - 3\cos(8\,p) - 3 - 4\cos(6\,p) - 4\cos(2\,p))$$

The mean of this is not zero and is given by:

> c8:=eval(subs(cos=0,H8));

$$c8 := \frac{9}{64}\,b^3\,r^8$$

Now re-define H_8 according to equation 17.17:

> H8:=simplify(H8-c8);

$$H8 := -\frac{9}{32}\,b^3\,r^8\cos(4\,p) + \frac{9}{64}\,b^3\,r^8\cos(8\,p) + \frac{3}{16}\,b^3\,r^8\cos(6\,p) + \frac{3}{16}\,b^3\,r^8\cos(2\,p)$$

giving the following expression for F_8:

> F8:=FN(H8);

$$F8 := \frac{1}{64}\,b^3\,y\,x\,(15\,x^6 - 73\,x^4\,y^2 + 89\,x^2\,y^4 + 33\,y^6)$$

17.9.2 Lindstedt's method applied to the van der Pol oscillator

The following Maple commands find a perturbation expansion for the limit cycle of the van der Pol oscillator,

$$\frac{d^2x}{dt^2} + \epsilon(x^2 - 1)\frac{dx}{dt} + x = 0,$$

using Lindstedt's method. The perturbation parameter is ϵ and the commands also determine expansions for the frequency and amplitude of the limit cycle.

In this application of Lindstedt's method the analysis of the first-order term is different from all subsequent terms because the the equation for a_0 is quadratic with two possible solutions: the choice between these is most easily done by hand. All subsequent equations have unique solutions and cause no problems.

> restart; N:=3:

The value of N is the order of the perturbation expansion, and increasing its value is all that is necessary to increase the order of the expansion produced. I have run this with $N = 20$, but for large N the memory required is large as is the computation time; there has been no attempt to improve the efficiency of this method, in particular the use of dsolve is lazy and slows the computation.

Now define the unkown frequency W and solution X as series in ϵ, note that here the variable t denotes τ:

```
>  W:=1+add(w[k]*e^k,k=1..N);
>  X:= x.0(t)+ add(x.k(t)*e^k,k=1..N) ;
```

$$W := 1 + w_1 \, e + w_2 \, e^2 + w_3 \, e^3$$

$$X := x0(t) + x1(t) \, e + x2(t) \, e^2 + x3(t) \, e^3$$

Substitute these into the equation and collect the coefficients of ϵ^k for $k = 0, 1, \ldots, N$. If N is large these terms should not be displayed.

```
>  eqn:=series( W^2*diff(X,t$2) + e*W*(X^2-1)*diff(X,t) + X,e=0,N+1):
>  for k from 0 to N do; eq[k]:=coeff(eqn,e,k)=0: od:
```

We now have to solve these: first define the appropriate initial conditions, solve the equation and then define the function $x_0(t)$:

```
>  ic:=x0(0)=a[0],D(x0)(0)=0:        # The initial conditions
>  dsolve({eq[0],ic},x0(t)):        # Solve the equations
>  x0:=unapply(rhs(%),t);           # Define the function x0(t)
```

$$x0 := t \to a_0 \cos(t)$$

The first-order equation eq[1] needs to be expressed in multiple angle form:

```
>  eq[1]:=combine(eq[1],trig);
```

$$eq_1 := x1(t) + (\tfrac{\partial^2}{\partial t^2} x1(t)) - 2 \, w_1 \, a_0 \cos(t) - \frac{1}{4} \, a_0^3 \sin(3 \, t) - \frac{1}{4} \, a_0^3 \sin(t) + a_0 \sin(t) = 0$$

and we set the coefficients of $\cos t$ and $\sin t$ to zero in order to remove the secular terms. The equations for this are:

```
>  c[1]:=0=select(has,lhs(eq[1]),cos(t));
```

$$c_1 := 0 = -2 \, w_1 \, a_0 \cos(t)$$

```
>  s[1]:=0=select(has,lhs(eq[1]),sin(t));
```

$$s_1 := 0 = -\frac{1}{4} \, a_0^3 \sin(t) + a_0 \sin(t)$$

and their solution is

```
>  sol:=[solve({c[1],s[1]},{a[0],w[1]})];
```

$$sol := [\{a_0 = 0, \, w_1 = w_1\}, \, \{w_1 = 0, \, a_0 = 2\}, \, \{w_1 = 0, \, a_0 = -2\}]$$

In this case there are three solutions and it is easiest to chose the appropriate one, $a_0 = 2$, $\omega_1 = 0$, by hand:

```
>  a[0]:=2: w[1]:=0:
```

so that the equation for the first-order corrections becomes:

```
>  eq[1];
```

$$x1(t) + (\tfrac{\partial^2}{\partial t^2} x1(t)) - 2 \sin(3 \, t) = 0$$

The remaining part of the calculation is automatic so we solve this and the higher-order equations in a loop:

```
>  for k from 1 to N do;
>  ic:=x.k(0)=a[k],D(x.k)(0)=0;        # Define the initial conditions
>  dsolve({eq[k],ic},x.k(t));          # Solve the equations
>  u1:=combine(rhs(%),trig);
>  x.k:=unapply(u1,t);
```

Now the secular term has to be isolated and removed; this is necessary only for the $k = 1, 2, \ldots, N - 1$ terms of the series.

```
>  if k=N then break fi;
>  eq[k+1]:=combine(eq[k+1],trig);
>  s:=select(has,lhs(eq[k+1]),sin(t));    # coefficients of sin(t)
>  c:=select(has,lhs(eq[k+1]),cos(t));    # coefficients of cos(t)
>  sol:=solve({s=0,c=0},{a[k],w[k+1]});   # Remove the secular terms
>  assign(sol);
>  od:
```

The final solution is therefore

```
>  X:=collect( combine(simplify(X,{e^N=0}), trig),e);
```

$$X := (-\frac{5}{96}\cos(5t) + \frac{3}{16}\cos(3t) - \frac{1}{8}\cos(t))e^2 + (-\frac{1}{4}\sin(3t) + \frac{3}{4}\sin(t))e + 2\cos(t)$$

Remember that the t here denotes the τ used in the text. The series for the frequency is

```
>  W;
```

$$1 - \frac{1}{16}e^2$$

The solution $x(t)$ in terms of the original variable t are obtained by the substitution $\tau = \omega t$, and $y(t)$ is obtained by differentiating this:

```
>  Xt:=subs(t=W*t,X):  Yt:=diff(Xt,t):
```

The series for the amplitude of the motion is

```
>  A:=add(a[k]*e^k,k=0..N-1);
```

$$A := 2 + \frac{1}{96}e^2$$

17.9.3 Harmonic balance method

Here we apply an Nth-order harmonic balance approximation to the equation

$$\ddot{z} + z - dz^3 = 0,$$

using the perturbation scheme defined in equations 17.61 and 17.62. This equation is defined in line #1 of the Maple script below and is easily changed. In this example we set $N = 2$, but this can be increased simply by changing the value of N in the first command below; most output has been suppressed so to understand how all the commands work you may need to re-instate the output.

```
>  restart;  N:=2:
```

With the solution denoted by z the approximate solution, equation 17.61, is

```
>  z:=a*cos(t) + add( add(d^j*b[k,j],j=k..N)*cos((2*k+1)*t),k=1..N):
```

where we have used equation 17.62 in the definition of the coefficients. Note that $\tau = \omega t$ is here denoted by t. The perturbation expansion of the frequency, denoted by W, is

```
>  W:=1+add(w[k]*d^k,k=1..N):
```

Substitute these two expansions into the equation of motion and expand as a series in d; the quickest method of doing this is to use the series command.

```
>  eqn:=series( W^2*diff(z,t$2) + z - d*z^3,d=0,N+1):                    # 1
```

This expression will contain powers of trigonometric functions, which are converted to the multiple angle form with the following command. For large N this is the slowest part of the calculation.

```
>  eqn:=map(combine,eqn,trig):
```

Now collect together all the required equations. There are two loops in the next set of commands. The outer loop collects the coefficients of d^k for $k = 1, 2, \ldots, N$. For each of these we collect the coefficients of $\cos(2j - 1)t$, where j runs from 1 to $k + 1$, and the array eq[k,j] contains the coefficient of $d^k \cos(2j - 1)t$.

```
>  for k from 1 to N do;
>  cd[k]:=coeff(eqn,d,k);
>  for j from 1 to k+1 do; n:=2*j-1;
>  eq3:=collect(cd[k],cos(n*t));
>  eq[k,j]:=coeff(eq3,cos(n*t));
>  od: od:
```

The following two loops print these equations and should be omitted if N is increased; the commands in the next four lines are not necessary for the final result.

```
>  for k from 1 to N do;
>  for j from 1 to k+1 do;
>  print('k,j=',k,j,'    eq[k,j]=',eq[k,j]);
>  od:od:
```

$$k,j =, 1, 1, \quad eq[k,j] =, -2 w_1 a - \frac{3}{4} a^3$$

$$k,j =, 1, 2, \quad eq[k,j] =, -8 b_{1,1} - \frac{1}{4} a^3$$

$$k,j =, 2, 1, \quad eq[k,j] =, -a w_1{}^2 - \frac{3}{4} a^2 b_{1,1} - 2 a w_2$$

$$k,j =, 2, 2, \quad eq[k,j] =, -8 b_{1,2} - 18 w_1 b_{1,1} - \frac{3}{2} a^2 b_{1,1}$$

$$k,j =, 2, 3, \quad eq[k,j] =, -24 b_{2,2} - \frac{3}{4} a^2 b_{1,1}$$

These have the pattern described in the text so may be solved recursively rather than as a large set of linear equations, which is much slower when N is large.

```
>  for k from 1 to N do; for j from 1 to k+1 do;
>  if j=1 then
>  w[k]:=solve(eq[k,j]=0,w[k]) else
>  b[j-1,k]:=solve(eq[k,j]=0,b[j-1,k]) fi;
>  od;od;
```

The solution is therefore

> `z:=map(factor,z);`

$$z := a\cos(t) - \frac{1}{1024}\, d\, a^3\, (32 + 21\, d\, a^2)\cos(3\, t) + \frac{1}{1024}\, d^2\, a^5\cos(5\, t)$$

and the frequency is

> `W;`

$$1 - \frac{3}{8}\, d\, a^2 - \frac{15}{256}\, a^4\, d^2$$

But the value of $z(0)$ is not unity, as it is for the perturbation method, so to compare the results we need to find the value of a that makes $z(0) = 1$. Let $z(0) = x_0$,

> `x0:=collect(eval(subs(t=0,z)),d):`

and define a series $A = 1 + \sum_{k=1}^{N} A_k d^k$,

> `A:=1+add(A.k * d^k,k=1..N): z0:=series(subs(a=A,x0),d=0,N+1):`

Now solve this to ensure that $x_0 = 1$:

> `for k from 1 to N do;  A.k:=solve( coeff(z0,d,k)=0, A.k); od:`

Thus the required solution is:

> `x:=convert(series(subs(a=A,z),d=0,N+1),polynom);`

$$x := \cos(t) + \left(\frac{1}{32}\cos(t) - \frac{1}{32}\cos(3\, t)\right) d + \left(\frac{23}{1024}\cos(t) - \frac{3}{128}\cos(3\, t) + \frac{1}{1024}\cos(5\, t)\right) d^2$$

and the frequency of this motion is:

> `Wa:=convert(series(subs(a=A,W),d=0,N+1),polynom);`

$$Wa := 1 - \frac{3}{8}\, d - \frac{21}{256}\, d^2$$

17.10   Exercises

Exercise 17.43
Show that the system

$$\frac{dx}{dt} = 2x - xy^2 + \cos y, \qquad \frac{dy}{dt} = -y - x^2 y + \sin x$$

has no closed orbit inside the unit circle.

Exercise 17.44
Show that the following systems have no periodic solutions:

(i)	$\dot{x} = y,$	$\dot{y} = a + x^2 - (b-x)y,$	$a > 0,\ b > 0,$
(ii)	$\dot{x} = (x-a)^3 + xy^2,$	$\dot{y} = y + y^3,$	$a > 0,$
(iii)	$\dot{x} = 4xy + ax^3,$	$\dot{y} = -bx^2 + y - 2y^2 + y^3,$	$a > 0,\ b > 0,$
(iv)	$\dot{x} = x,$	$\dot{y} = 1 + x + y^2.$	

Exercise 17.45

Find a function $g(x)$ of x only that can be used in Dulac's test to show that the system

$$\dot{x} = y, \quad \dot{y} = x^2 + y^2 - x - ay, \quad a > 0$$

has no periodic solutions.

Exercise 17.46

Green's theorem relates an integral over a simply connected region $\mathscr{D}$ in the plane to an integral over the boundary curve, $\mathscr{C}$,

$$\oint_{\mathscr{C}} P(x, y)dx + Q(x, y)dy = \iint_{\mathscr{D}} dxdy \left(\frac{\partial Q}{\partial x} - \frac{\partial P}{\partial y} \right),$$

where $P(x, y)$ and $Q(x, y)$ have continuous first derivatives and the line integral along the boundary is taken in the anticlockwise direction.

By assuming that a periodic orbit exists and chosing $\mathscr{C}$ to be this closed curve, use this theorem to prove Bendixson's negative criterion and Dulac's extension.

Exercise 17.47

Show that the equation

$$\ddot{x} + v(x^2 - 1)\dot{x} + \tanh \alpha x = 0, \quad v > 0, \quad \alpha > 0,$$

has a stable limit cycle.

Exercise 17.48

Are there any periodic orbits for the system

$$\dot{x} = y(1 + x - y^2), \quad \dot{y} = x(1 + y - x^2), \quad x \geq 0, \quad y \geq 0 ?$$

Exercise 17.49

Show that the system

$$\dot{x} = -y - x(x^2 + y^2 - 2x - 3), \quad \dot{y} = x - y(x^2 + y^2 - 2x - 3)$$

has at least one periodic orbit in the annulus $1 < r < 3$. Use Maple to provide evidence that there is only one stable orbit in this annulus.

Exercise 17.50

Show that the equation

$$\frac{d^2 z}{d\tau^2} + \gamma(z^2 - a^2)\frac{dz}{d\tau} + \omega^2 z = 0, \quad a > 0,$$

where γ, a and ω are constants, can be written in the canonical form, equation 17.48, by introducing new variables $x = z/\alpha$ and $t = \tau/\beta$ and choosing the constants α and β appropriately.

Exercise 17.51
Determine the relation between Rayleigh's and van der Pol's equations.

Exercise 17.52
Find the period of bound motion for a particle of unit mass in the potentials

(i) $V(x) = -U/(\alpha|x|)$,
(ii) $V(x) = U \tan^2 \alpha x$,
(iii) $V(x) = -U/\cosh^2(\alpha x)$,
(iv) $V(x) = U(\alpha x)^{2n}$,

where U and α are positive constants and n a positive integer.

Exercise 17.53
Consider the system

$$\frac{d^2 x}{dt^2} + v(x^2 - 1)\left(\frac{dx}{dt}\right)^3 + x = 0, \quad v > 0.$$

Show that the only fixed point is at the origin, which is a centre of the linearised system, but an unstable spiral of the nonlinear system.

Use Maple to find numerical solutions of this system and use these solutions to suggest that the system has a stable limit cycle.

Use Lindstedt's method to show that the small v expansion of the period and amplitude of this limit cycle are

$$\frac{T}{2\pi} = 1 - \frac{81}{64}v^2 + \frac{130\,977}{20\,480}v^4 - \frac{106\,237\,175\,103}{1\,835\,008\,000}v^6 + O\left(v^8\right),$$

$$A = \sqrt{6}\left(1 - \frac{783}{1280}v^2 + \frac{7\,289\,917\,353}{1\,605\,632\,000}v^4 + O(v^6)\right).$$

Exercise 17.54
Using one of the theorems quoted in the introduction show that the nonlinear system

$$\dot{x} = xy, \quad \dot{y} = a_0 + a_1 x - a_2 x^2 + a_3 y^2, \quad a_0 > 0, \quad a_2 > 0,$$

has two centres on the x-axis.

Use Lindstedt's method to find a sixth-order perturbation series approximation to the frequency of the motion at each fixed point in the case $a_0 = 2$, $a_1 = a_2 = 1$ and $a_3 = b$, an arbitrary parameter.

Exercise 17.55
Consider the equation

$$\frac{d^2 x}{dt^2} + v(x^4 - 1)\frac{dx}{dt} + x = 0, \quad v > 0,$$

dealt with in exercise 17.11 (page 683). Use Lindstedt's method to find a series expansion of the period of the limit cycle and compare the sixth-order perturbation approximation of the solution with the numerical solution of the equation for various values of $v < 1$.

Exercise 17.56
Rayleigh (1883) considered the conditions necessary for a system subject to dissipative

forces to exhibit permanent oscillations. He noted that if the displacement is described by the equation

$$\frac{d^2x}{dt^2} + \epsilon\frac{dx}{dt} + \omega^2 x = 0, \quad \epsilon > 0,$$

periodic motion is impossible as all solutions decay to zero. In order that vibrations are maintained the system must be in connection with an energy source which, for small oscillations, will depend upon the displacement x and the velocity $\dot{x}$. Consequently he considered the system described by the equation

$$\frac{d^2x}{dt^2} + \epsilon\frac{dx}{dt} + \epsilon k\left(\frac{dx}{dt}\right)^3 + x = 0, \quad \epsilon > 0,$$

in order to determine conditions necessary for maintained vibrations.

Use the harmonic balance method with the trial solution

$$x = A\sin\omega t + \epsilon(B\sin 3\omega t + C\cos 3\omega t), \quad \omega = 1 + \epsilon\omega_1,$$

to approximate the solutions of this equation. Show that $B = \omega_1 = 0$ and that there are real solutions only if $k < 0$, and then

$$A = \frac{2}{\sqrt{-3k}}, \quad C = \frac{kA^3}{32}.$$

Exercise 17.57

The periodic function $\sin(z\sin x)$ has the Fourier series expansion

$$\sin(z\sin x) = 2\sum_{k=0}^{\infty} J_{2k+1}(z)\sin(2k+1)x,$$

where $J_n(x)$ is an ordinary Bessel function.

Use this expansion together with the trial function $\theta = A\sin\omega t$ to show that an approximation to the motion of the vertical pendulum, described by the equation $\ddot{\theta} + \alpha^2\sin\theta = 0$, with amplitude $A = a$, is $\theta = a\sin\omega t$, where $\omega^2 = 2\alpha^2 J_1(a)/a$.

Obtain a power series expansion for $T(a)/2\pi$, T being the period, in terms of the amplitude a and show that this differs from the series given by the exact result obtained in exercise 17.34 by $a^4/3072 + O(a^6)$.

18

Discrete Dynamical Systems

18.1 Introduction

In this chapter we study orbits produced by maps rather than differential equations. These are obtained by repeated applications of a discrete transformation in phase or state space: a point $\mathbf{x} = (x_1, x_2, \ldots, x_n)$ is mapped to a point $\mathbf{T}(\mathbf{x}) = (T_1(\mathbf{x}), T_2(\mathbf{x}), \ldots, T_n(\mathbf{x}))$ by some suitably defined functions $\mathbf{T}(\mathbf{x})$. An orbit is a sequence of points $\mathbf{x}_0, \mathbf{x}_1, \ldots, \mathbf{x}_N, \ldots$ with $\mathbf{x}_{k+1} = \mathbf{T}(\mathbf{x}_k)$.

Our main reason for studying maps is that orbits of differential equations and maps behave in similar ways, but it is far easier to understand the behaviour of maps, partly because it is much easier to compute map orbits and partly because the analysis of maps is easier. Thus the study of maps provides an easier route to the understanding of the solutions of differential equations. Maps, however, have a significance of their own: they arise as a natural description of various processes and as important approximations to solutions of differential equations.

The description of populations with short breeding seasons leads naturally to maps. The combination of population biology and population genetics leads to a wide variety of maps, see for instance the review by May (1987) and the books by Maynard Smith (1977) and Murray (1989). Linear maps also arise in the description of some Markov processes, Cox and Miller (1965, chapter 3). In mathematics iterative processes are frequently used to find approximate solutions, usually numerical, of a variety of types of equations, for instance the iterative schemes of Newton and Halley, described in chapter 9 and exercise 18.52 (page 779). The study of iterations, particularly in the complex plane, is an important area of study and has been for over a century: much of the recent work on fractals, see for instance Barnsley (1988) and Peitgen *et al.* (1992), arises from iterations in the complex plane.

Maps also arise naturally in the study of certain types of differential equations. The monodromy matrix introduced in chapter 12 is one example of a linear map; an example of a nonlinear map is discussed at the end of this chapter, section 18.8. Maps can also be used to approximate the dynamics described by differential equations. This is important because, even with modern computers, the numerical solution of many ordinary differential equations is difficult and time consuming, particularly if the system has two or more widely differing time-scales and integration over long times is necessary: an example is the Solar System where the period of the largest planet, Jupiter, is about fifty times longer than that of Mercury, 11.9 and 0.24 years respectively. These problems are compounded if it is necessary to study the dependence of the solutions on many parameters or initial conditions. There are two notable successes of this type of approximation. First is the approximation of a periodically driven one-dimensional system. If the driving frequency is larger than the natural frequency of the system the dynamics is well approximated by a map: this method was introduced by Casati *et al.* (1987) in studies of the hydrogen atom affected by a strong periodic electric field, see also Jensen *et al.* (1991) and Richards (1997); the method was subsequently generalised to arbitrary systems by Dando and Richards (1990). It is interesting to note that for weak perturbations the map produced by this type of approximation can always be transformed into the standard map, defined in exercise 18.63, regardless of the original system. Thus the standard map plays the same role to periodically driven systems as does the linear oscillator to motion near equilibrium. The second success is the work of Wisdom (1982) in his work on asteroids near a 3:1 resonance — that is, asteroids with a period $\frac{1}{3}$ of Jupiter's period – in which the change in the orbital elements of an asteroid is computed at each period of Jupiter. This method provides a powerful algorithm that is over 1000 times faster than conventional numerical integration and allows the investigation of the motion over much longer times than is otherwise possible, typically 2×10^6 years. Such investigations provide an explanation of the origin of the Kirkwood gaps in the asteroid belt. Another interesting consequence of these studies (Wisdom, 1983), is the observation that the eccentricity of chaotic trajectories of some asteroids becomes so large, $\epsilon \simeq 0.6$, that their orbits cross the Earth's orbit: such resonant excitation of asteroids by Jupiter is now considered to be one of the mechanisms for producing meteorites.

This chapter provides an introduction to maps and deals only with systems of one and two dimensions: far more detail may be found in the books by Devaney (1992), Ott (1993), Peitgen *et al.* (1992) and Rasband (1990). The simplest maps are one-dimensional, and these are used to introduce a variety of ideas and phenomena, particularly period doubling. We then move to two-dimensional maps but limit attention to area-preserving maps, that is, those for which the Jacobian determinant of $\mathbf{T}(\mathbf{x})$ is unity: this type of map is closely connected with certain Hamiltonian systems, and here we concentrate on features needed in the next chapter.

18.2 First-order maps

18.2.1 *Fixed points and periodic orbits*

A point x_f is a fixed point of the first-order map $x_{n+1} = F(x_n)$ if

$$x_f = F(x_f).$$

The stability, or otherwise, of a fixed point is determined by considering the behaviour of an orbit starting near x_f, so we set $x = x_f + \xi$, with $|\xi|$ small, and make a Taylor expansion to obtain

$$\xi_{n+1} = F'(x_f)\xi_n + O(\xi_n^2). \tag{18.1}$$

The *linearised* map is obtained by ignoring the second-order term,

$$\xi_{n+1} = F'(x_f)\xi_n \implies \xi_n = F'(x_f)^n \xi_0. \tag{18.2}$$

The value of $F'(x_f)$ thus determines the behaviour of ξ_n, and in most cases the behaviour of x_n, as $n \to \infty$.

- If $|F'(x_f)| < 1$ the difference $|x_n - x_f|$ tends to zero exponentially as $x_n \to x_f$: the fixed point is *stable*.
- If $|F'(x_f)| = 1$ linearisation provides no information about the nature of the fixed point.
- If $|F'(x_f)| > 1$ for most initial conditions the orbit moves away from the fixed point, which is *unstable*.
- If $|F'(x_f)| = 0$ the fixed point is *superstable*, that is $|\xi_n| \to 0$ faster than the exponential convergence of equation 18.2, exercise 18.46 (page 779). The iterative schemes of Newton and Halley are important examples of superstable systems, exercise 18.52.

For one-dimensional systems a stable fixed point *attracts* all sufficiently nearby points and is therefore called an *attractor*; whereas an unstable fixed point repels almost all points† in its neighbourhood and is therefore called a *repellor*.

Of equal interest to fixed points are orbits that return repeatedly to their original position after a set number of iterations. Such an orbit is said to be a *periodic orbit*. If it takes N, but no fewer than N, steps to return to its original position the orbit is said to have *period N* and the orbit is called a period-N or N-period orbit. A fixed point is a period-1 orbit.

Thus any of the N points, $(x_0, x_1, \ldots, x_{N-1})$, of an N-period orbit satisfies the equation

$$x = F^N(x) \quad \text{but} \quad x \neq F^k(x) \quad k = 1, 2, \ldots, N-1,$$

where $F^k(x)$ is the function $F(x)$ composed with itself k times,

$$F^2(x) = F(F(x)), \ F^3(x) = F(F(F(x))) \text{ and, in general, } F^k(x) = F(F^{k-1}(x)).$$

It is convenient to name $F^k(x)$ the period-k map or the kth iterate of the point x. With this notation it is helpful and consistent to define $F^0(x)$ to be the identity, $F^0(x) = x$ for all x.

This notation must be distinguished from that used to denote the nth differential of a function, $F^{(n)}(x)$, and the nth power of a function, $F(x)^n$. If F is denoted by the Maple function F then F^n is (F@@n).

A period-N orbit is thus a fixed point of $F^N(x)$: but not all the fixed points of $F^N(x)$ need be N-period orbits. For instance, if $N = 6$ the 1-, 2- and 3-period orbits that are the fixed points of $F(x)$, $F^2(x)$ and $F^3(x)$ are also fixed points of $F^6(x)$. In general, if n divides N the fixed points of $F^n(x)$ are also fixed points of $F^N(x)$ but are not period-N orbits.

Exercise 18.1

If $F(x) = x^2 + c$, for some constant c, show that $F^2(x) = x^4 + 2cx^2 + c(1 + c)$. What degree polynomial is $F^k(x)$?

---

† The fixed point itself is not repelled, and if $F'(x_f) < -1$ there may be iterates of the nonlinear map that land on the fixed point after only a finite number of iterations.

Exercise 18.2

Show that if $f(x)$ is an even function then so is $f^n(x)$, $n \geq 1$.

Exercise 18.3

Show that if a is on a period-N orbit of $F(x)$ then so is $b = F^n(a)$ for any integer n.

The same reasoning that gave equation 18.2 also shows that the N-period orbit, $(a_0, a_1, \ldots, a_{N-1})$, is stable if

$$\left| \frac{d}{dx} F^N(a_n) \right| < 1, \quad 0 \leq n \leq N - 1. \tag{18.3}$$

Using the chain rule the differential can be expressed as the product

$$\frac{d}{dx} F^N(x) = F'(x) F'(F(x)) F'(F^2(x)) \cdots F'(F^{N-1}(x))$$

$$= \prod_{k=0}^{N-1} F'(F^k(x)).$$

Hence another way of writing the differential 18.3 is

$$\frac{d}{dx} F^N(a_n) = \prod_{k=0}^{N-1} F'(a_k), \quad a_{j+1} = F(a_j),$$

which shows that the value of the differential in 18.3 is independent of n.

Exercise 18.4

The *logistic map*, studied in the next section, can be written in the form

$$x_{n+1} = F(x_n) = bx_n(1 - x_n), \quad 0 < b \leq 4.$$

Show that it has two fixed points, $x_f = 0$, $1 - 1/b$, and that

(i) the fixed point $x_f = 0$ is stable for $b \leq 1$ and unstable for $b > 1$,

(ii) the fixed point $x_f = 1 - 1/b$ is stable for $1 \leq b \leq 3$ and unstable for $3 < b \leq 4$. Note the case $b = 3$ is special and dealt with in exercise 18.50.

Exercise 18.5

(i) Show that the fixed points of $F^2(x)$, where $F(x) = bx(1 - x)$, are at

$$x_0 = 0, \quad x_1 = 1 - \frac{1}{b}, \quad x_{2,3} = \frac{\sqrt{b+1}}{2b} \left(\sqrt{b+1} \mp \sqrt{b-3} \right).$$

(ii) Show that the fixed point x_1 is a stable fixed point of $F(x)$ and $F^2(x)$ if $b \leq 3$ and unstable if $b > 3$.

(iii) Show that the points x_2 and x_3 are stable fixed points of $F^2(x)$ for $3 \leq b \leq 1 + \sqrt{6}$ and unstable for $b > 1 + \sqrt{6}$.

Exercise 18.6

Show that if $b = 2$ the expansion of the logistic map about the fixed point at $x_f = \frac{1}{2}$ leads to the map $\xi_{n+1} = -2\xi_n^2$, which is superstable.

For the general map $x_{n+1} = F(b, x_n)$ show that the fixed point x_f is superstable if $F(b, x)$ has a maximum at $x = x_f$.

Exercise 18.7

If x_f is a fixed point of a map $x_{n+1} = F(x_n)$, show that it is also a fixed point of $F^m(x)$, for any positive integer m, and that if x_f is a stable (unstable) fixed point of $F(x)$ it is a stable (unstable) fixed point of $F^m(x)$,

18.2.2 The logistic map

The *logistic map* is a quadratic first-order iterative system: it is important because it is the simplest map of its type, so is the easiest to analyse, yet its orbits behave in a similar manner to those of most similar maps. It can be written in the form

$$x_{n+1} = F(x_n) = bx_n(1 - x_n), \quad 0 < x_0 < 1, \quad 0 < b \le 4. \tag{18.4}$$

The value of the constant b is restricted otherwise x is not confined to the interval $(0, 1)$. With this restriction all iterations are confined to a finite interval inside $[0, 1]$ and it is this, together with the fact that the inverse of $F(x)$ is multivalued, that produces interesting behaviour. The initial conditions $x = 0$ and 1 are excluded because they lead to uninteresting orbits.

The behaviour of the iterates x_n depends upon the value of b and some typical results are shown in the next few figures; in all cases $x_0 = 0.1$ though the value of x_0 does not affect the long-time behaviour of the orbits. In the first figure $b = 2$ and the orbit settles, very quickly, onto the stable fixed point at $x_f = \frac{1}{2}$, exercise 18.6; in the second figure, $b = 2.99$, the orbit seems to oscillate between two points, but after a sufficiently long time it settles on to the fixed point $x_f = 1 - 1/b = 0.66 \ldots$; in the third case, in which $b = 3.01$ is only slightly larger, the iterates approach a period-2 orbit. Notice that for short times there is no apparent difference between the $b = 2.99$ and $b = 3.01$ orbits, though they are distinctly different for large n.

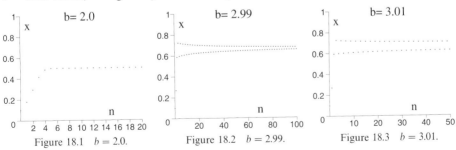

Figure 18.1 $b = 2.0$. Figure 18.2 $b = 2.99$. Figure 18.3 $b = 3.01$.

In the first of the next three figures $b = 3.5$ and the iterates approach a period-4 orbit. In the second figure, where $b = 3.84$, a period-3 orbit is produced, though this is not apparent for the first 20 iterations. In the third figure, where $b = 3.9$, there is no apparent order to the motion.

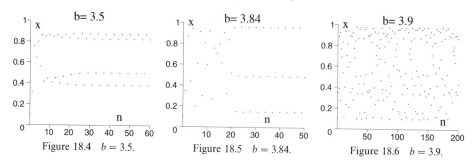

Figure 18.4 $b = 3.5$.   Figure 18.5 $b = 3.84$.   Figure 18.6 $b = 3.9$.

These few examples provide some idea of the variety of orbits that this relatively simple dynamical system can produce. In the rest of this section we shall try and understand why such variety exists, and shall determine some of the critical values of b at which the orbits change their nature. However, before this you should do the next two exercises in order to see for yourself how the system behaves.

Exercise 18.8

Write a Maple program that can reproduce the figures shown above and use it to investigate the behaviour of the logistic map for various values of b. For instance, consider the intervals $3.828 \leq b \leq 3.841$, $3.7017 \leq b \leq 3.7025$, $3.7385 \leq b \leq 3.742$, and $b = 1, 2, 3$ and 4.

A graphical interpretation of this type of iteration was described in section 9.3, and on page 316 we list a procedure that creates an animated sequence of graphs showing how the iterations proceed. The command `anim(3.84,0.1,100);` will animate a graphical representation of 100 iterations, for $b = 3.84$, starting at $x_0 = 0.1$.

Exercise 18.9

Use the procedure listed on page 316 to look at some of the iterations performed in exercise 18.8.

The behaviour depicted in figures 18.1–18.6 is mostly explained by the nature of the fixed points and periodic orbits of the map, some of which were considered in exercises 18.4 and 18.5, where it was found that $F(x)$ has two fixed points, at $x = 0$ and $x = 1 - 1/b$. It is convenient to divide the interval $0 < b \leq 4$ into smaller non-intersecting intervals; the behaviour in the first three intervals is relatively simple.

- $0 < b < 1$: The fixed point at $x = 0$ is stable and that at $x = 1 - 1/b$ is unstable. For any initial condition, $x_n \to 0$ as $n \to \infty$: for large n the fixed point is approached exponentially rapidly.

- $b = 1$: The fixed point at $x = 0$ is stable, but the convergence is slower and $x_n = O(1/n)$ as $n \to \infty$, exercise 18.49 (page 779).

- $1 < b \leq 3$: The fixed point at $x = 0$ is unstable and that at $x = 1 - 1/b$ is stable. For any initial condition, $x_n \to 1 - 1/b$ as $n \to \infty$. For $b \neq 2$ or 3 convergence to the fixed point is exponential.

  * For $b = 2$, figure 18.1, the fixed point at $x_f = \frac{1}{2}$ is superstable, exercise 18.6, though the behaviour for $b \simeq 2$ is not noticeably different in a graphical comparison.

* For $b = 3$ the convergence is slower and $|x - x_f| = O(n^{-1/2})$ as $n \to \infty$, exercise 18.50 (page 779).

* $3 < b \leq 1 + \sqrt{6} = 3.4495\ldots$: For $b > 3$ both fixed points of $F(x)$ are unstable, but there exist stable period-2 orbits if $b \leq 1 + \sqrt{6}$, exercise 18.5. All initial conditions lead to a period-2 orbit oscillating between the points

$$\frac{\sqrt{1+b}}{2b}\left(\sqrt{b+1} \pm \sqrt{b-3}\right),$$

as seen in figure 18.3, where $b = 3.01$.

Bifurcations

It is not a coincidence that the period-1 orbit becomes unstable at $b = 3$ precisely where a stable period-2 orbit is born. In the following three figures are shown the graphs of $y = x$, $y = F(x)$ (curve [1]) and $y = F^2(x)$ (curve [2]) in the neighbourhood of the fixed point for $b = 2.8$, 3.0 and 3.2.

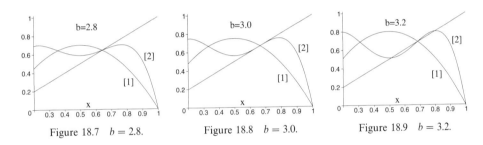

Figure 18.7 $b = 2.8$.   Figure 18.8 $b = 3.0$.   Figure 18.9 $b = 3.2$.

As b increases from 2.8 to 3.2 the gradient of $F^2(x)$ at $x_f = 1 - 1/b$ is seen to increase through unity at $b = 3$. The reason for this is simply that

$$\frac{d}{dx}F^2(x) \;=\; F'(x)\,F'(F(x))$$
$$\;=\; F'(x_f)^2 \quad \text{at} \quad x = x_f = F(x_f).$$

Thus as b passes through a critical value where the fixed point of $F(x)$ becomes unstable, $|F'(x_f)| = 1$, the gradient of $F^2(x)$ must also be unity. Moreover, $F^2(x)$ develops two new fixed points as b increases through 3: because these are not fixed points of $F(x)$ a period-2 orbit is born at $b = 3$, precisely where the period-1 orbit becomes unstable.

In general the period-2^n orbit becomes unstable at $b = b_n$, and for $b_n < b < b_{n+1}$ the period-2^{n+1} orbit is stable. This process is named *bifurcation* and is conveniently depicted in a graph showing all the fixed points of $F^n(x)$, $n = 1, 2, 4, \ldots$, $x_f(b)$, as a function of b. In the following diagram, figure 18.10, the fixed points of $F^n(x)$, $n = 1, 2$ and 4 are shown: this graph is easily drawn using `implicitplot` with the equations $F(x) = x$ and $(F^{2n}(x) - x)/(F^{2n-2}(x) - x) = 0$, $n = 1$ and 2. Another, more general method of producing bifurcation diagrams is described in exercise 18.12.

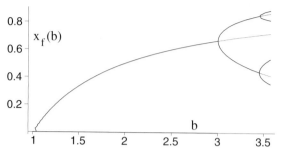

Figure 18.10   Bifurcation diagram for $F^n(x)$, $n = 1$, 2 and 4. Here
$b_1 = 3$ and $b_2 = 3.4495$.

In this diagram the stable fixed points are shown by the black lines, the unstable fixed points by the lighter lines. The period-1 orbit bifurcates at $b = 3$ and the period-2 orbit at $b = 1 + \sqrt{6} = 3.4495$.... At each bifurcation the local shape of fixed points has the same characteristic form and is conventionally named the *pitchfork bifurcation*. If the bifurcation forms at a critical value b_c of b the equation for $b(x_f)$ for x_f after the bifurcation is the parabola

$$b = b_c + \alpha(x - x_f(b_c))^2 + \cdots, \quad \alpha > 0.$$

The bifurcation diagram 18.10 shows the first two bifurcations at $b = b_1 = 3$ and $b_2 = 1 + \sqrt{6}$; there is also the point $b = 1$ at which the fixed point at $x_f = 0$ becomes unstable and the stable fixed point at $x_f = 1 - 1/b$ is born.

The remainder of this section is devoted to analysing the conditions required for such bifurcations, to showing that there are infinitely many bifurcations at b_n, $n = 1, 2, 3, \ldots$, and that the difference $b_{n+1} - b_n$ decreases so rapidly that $b_\infty \simeq 3.57$....

The values b at which the 2^n-period orbit becomes unstable are denoted by b_n and some of these are shown in table 18.1: these numbers provide a framework for the following analysis.

Table 18.1. *Values of b_n at which the 2^n-period orbit becomes unstable. The top row is the result of a numerical calculation, see for instance Rasband (1990, chapter 2) or Peitgen et al. (1992, chapter 11). The second row shows the approximate values given by the sequence defined by equation 18.15.*

n	1	2	3	4
b_n	3.0	3.449 490	3.544 090	3.564 407
	3.0	3.449 490	3.539 585	3.557 261

n	5	6	7	8
b_n	3.568 759	3.569 692	3.569 891	3.569 934
	3.560 715	3.561 389	3.561 521	3.561 547

The values of b_n in this table show that convergence to a limit is very rapid. If the limit is denoted by b_∞ it is shown below, for a particular map, that

$$b_n = b_\infty - \frac{c}{\delta^n}, \quad n \to \infty, \tag{18.5}$$

for some constants c and $\delta > 1$.

It transpires that the constant δ has the same value for all maps with a single quadratic maximum, that is, $f(x) - f(x_m) \sim (x - x_m)^2$, near the maximum at x_m. For such quadratic maps $\delta = 4.669\,202\ldots$. This constant is now named *Feigenbaum's constant* after Feigenbaum's original 1978 paper. Further discussion of this aspect of one-dimensional maps may be found in Devaney (1992), Peitgen *et al.* (1992) and Rasband (1990).

In order to understand the behaviour of $F^2(x)$ for b near $b_1 = 3$ and x near the fixed point $x_f = 1 - 1/b$ of $F(x)$ we make a Taylor expansion by putting $x = x_f(b) + y$, $b = b_1 + c = 3 + c$ and define the new function $F^2(x_f(b) + y) - x_f(b)$; expanding to third order then gives an approximation to the iterates of F^2:

$$y_{n+1} = G(y_n) = (1 + c)^2 y_n - 3cy_n^2 - 18y_n^3, \tag{18.6}$$

where $x_{2n} = x_f(b) + y_n$. This expansion is most easily achieved using mtaylor.† Notice that the coefficient of y_n^2 is small, and zero at the bifurcation point $c = 0$. Note also that the coefficient of y_n^3 is negative with large magnitude and consequently dominates the behaviour of the iterates when $c > 0$ — to see the relative significance of the cubic and quadratic terms ignore the former, find the fixed points and compare the result you get with those quoted in the next equation.

The map $G(y)$ has three fixed points at

$$y = 0, \quad -\frac{c}{12} \pm \frac{1}{12}\sqrt{16c + 9c^2}.$$

The fixed point at $y = 0$, $(x = x_f)$ is stable if $c < 0$ and unstable if $c > 0$, in accordance with the general result obtained above. At the other fixed points the gradient of $G(y)$ is

$$G'(y_f) = 1 - 4c \pm c^{3/2} + O(c^2),$$

so these are stable when $c > 0$, $(b > 3)$ and unstable when $c < 0$, $(b < 3)$.

The astute reader will have noted that the map 18.6 has three fixed points — and hence that a period-2 orbit is born — *only* because the coefficient of y_n^3 is negative. It transpires that this is a general property of many functions and is related to the fact that the Schwarzian derivative of the map function $F(x)$,

$$SF(x) = \frac{F'''(x)}{F'(x)} - \frac{3}{2}\left(\frac{F''(x)}{F'(x)}\right)^2, \tag{18.7}$$

is negative. The reason for this is discussed after exercises 18.10 and 18.11.

Thus, precisely at the point where a stable period-2 orbit is born the period-1 orbit becomes unstable. For the same reason a period-4 orbit is born when the period-2 orbit becomes unstable, at $b = 1 + \sqrt{6}$. In the following three figures are shown graphs of $y = x$, $y = F^2(x)$ (curve [2]), and $y = F^4(x)$ (curve [4]), in the neighbourhood of $b = 1 + \sqrt{6}$ and one of the fixed points of $F^2(x)$. Apart from the compressed scale, the similarity with figures 18.7 to 18.9 is clear.

† In Maple 5 **mtaylor** needs to be loaded using the **readlib(mtaylor)** command. This is unnecessary in Maple 6.

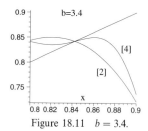

Figure 18.11 $b = 3.4$.

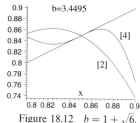

Figure 18.12 $b = 1 + \sqrt{6}$.

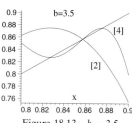

Figure 18.13 $b = 3.5$.

For $b = 3.4$ the fixed point x_f, given in exercise 18.5, of $F^2(x)$ is stable and $F^4(x)$ has no other fixed points in the vicinity of x_f. When $b > 1 + \sqrt{6}$ the gradient of $F^2(x)$ at the fixed point is greater than -1 and $F^4(x)$ has developed two more fixed points nearby.

As before, the Taylor expansion of $F^4(x_f(b) + y) - x_f(b)$ about $y = 0$ and $b = 1 + \sqrt{6}$ can be found and this gives a local approximation to the map:

$$y_{n+1} \;=\; H(y_n) = \left(1 + 4\sqrt{6}c + 26c^2\right) y_n - 4c\left(6 + 3\sqrt{3} + 9\sqrt{2} + \sqrt{6}\right) y_n^2$$
$$-20\left(14 + 7\sqrt{2} + 4\sqrt{3} + 4\sqrt{6}\right) y_n^3. \tag{18.8}$$

This map should be compared with equation 18.6: the coefficient of y_n increases through unity as c increases through zero, the coefficient of y_n^2 is $O(c)$ and the coefficient of y_n^3 is independent of c, negative with large magnitude that is even larger than before.

The continuation of this bifurcation process depends upon two ingredients. First, that a stable (unstable) fixed point of $F^n(x)$ is also a stable (unstable) fixed point of $F^{2n}(x)$, as proved in exercise 18.7. Second, that the coefficient of the cubic term in the expansion of $F^{2n}(x)$ about the fixed point if $F^n(x)$ at the critical point is negative. That this is true follows from the results derived in the following two exercises.

Exercise 18.10

If x_f is a fixed point of a map $F(x)$ show that the Taylor expansion of $F^2(x)$ about x_f leads to the following cubic approximation to the period-2 map near x_f:

$$y_{n+2} \;=\; F'(x_f)^2 y_n + \frac{1}{2}F''(x_f)F'(x_f)\left(1 + F'(x_f)\right) y_n^2$$
$$+\left\{\frac{1}{6}F'(x_f)F'''(x_f)\left(1 + F'(x_f)^2\right) + \frac{1}{2}F'(x_f)F''(x_f)^2\right\} y_n^3, \tag{18.9}$$

where $x_n = x_f + y_n$ and $x_f = F(x_f)$.

Hence show that if $F'(x_f) = -1$,

$$y_{n+1} = y_n - \frac{1}{3}\left\{F'''(x_f) + \frac{3}{2}F''(x_f)^2\right\} y_n^3, \tag{18.10}$$

and check that this reduces to equations 18.6 and 18.8 when $F(x) = bx(1-x)$ and appropriate expressions for x_f are used.

Exercise 18.11
The Schwarzian derivative of a function $f(x)$ is defined to be

$$S f(x) = \frac{f'''(x)}{f'(x)} - \frac{3}{2}\left(\frac{f''(x)}{f'(x)}\right)^2 = -2\sqrt{f'(x)}\frac{d^2}{dx^2}\left(\frac{1}{\sqrt{f'(x)}}\right). \tag{18.11}$$

(i) Show that for the logistic map, $F(x) = bx(1-x)$,

$$S F(x) = -\frac{6}{(2x-1)^2} < 0.$$

(ii) Show that if $f(x)$ and $g(x)$ both have negative Schwarzian derivatives, $S f(x) < 0$ and $S g(x) < 0$, then the Schwarzian derivative of their composition is also negative, that is, $S f(g(x)) < 0$.

These two exercises show that if a map $F(x)$ has a negative Schwarzian derivative the expansion of $F^2(x)$ about a fixed point x_f of $F(x)$ when $F'(x_f) = -1$ has the general form $y_{n+1} = y_n - \alpha y_n^3$, for some positive constant α. This shows that a period-2 orbit is born as $F'(x_f)$ decreases through -1.

Period doubling
In order to estimate the critical values, b_n, of b where the fixed points of $F^{2^n}(x)$ become unstable, we first compare expansions of $F(x) = bx(1-x)$ and $F^2(x)$ about their fixed points. Expanding $F(x)$ about $x_f = 1 - 1/b$,

$$y_{k+1} = (2-b)y_k - A_1(b)y_k^2, \quad y = x - x_f, \tag{18.12}$$

for some (irrelevant) constant $A_1(b)$. A similar expansion of $F^2(x)$ about $x_f = ((1+b) + \sqrt{(b-3)(b+1)})/2b$ leads to

$$z_{k+1} = (4 + 2b - b^2)z_k + A_2(b)z_k^2 + O(z_k^3), \tag{18.13}$$

for some constant $A_2(b)$.

It is not possible to carry this type of analysis further because the fixed points of $F^4(x)$ cannot be obtained in terms of simple formulae, so an approximation is needed. First observe that a rescaling $z_n = \alpha w_n$ in equation 18.13 does not affect the linear terms, so we may re-write this as a logistic map,

$$w_{k+1} = (4 + 2b - b^2)w_k(1 - w_k), \tag{18.14}$$

with a different constant. In order to approximate the expansion of $F^4(x)$ about its fixed point we take 18.14 as the basis map and define a new constant β_1 by the equation $2 - \beta_1 = 4 - 2b - b^2$, so it becomes

$$w_{k+1} = (2 - \beta_1)w_k(1 - w_k), \quad b = 1 + \sqrt{3 + \beta_1}.$$

This is an approximation to the period-2 map, but has the same form as the original map 18.12, so may be used in the same manner to approximate the period-4 map:

$$u_{k+1} = (2 - \beta_2)u_k(1 - u_k), \quad \beta_1 = 1 + \sqrt{3 + \beta_2}.$$

This process may be carried on indefinitely to give

$$v_{k+1} = (2 - \beta_n)v_k(1 - v_k), \quad \beta_{n-1} = 1 + \sqrt{3 + \beta_n}, \tag{18.15}$$

for the approximation to the period-2^n map.

Now consider the critical values of b. The period-2 orbit becomes unstable when $\beta_1 = 3$, that is, $b = b_1 = 1 + \sqrt{6}$; the period-4 orbit becomes unstable when $\beta_2 = 3$, so $\beta_1 = 1 + \sqrt{6}$ and $b_2 \simeq 3.539\ldots$: the period-8 orbit becomes unstable when $\beta_3 = 3$, which gives $b_3 \simeq 3.557$, and so on. The limit point, b_∞, is a solution of the equation

$$B = 1 + \sqrt{3 + B}, \quad \text{that is,} \quad b_\infty \simeq B = 3.5616, \tag{18.16}$$

which compares with the exact value of $3.5699\ldots$, obtained numerically. Moreover, since starting with $\beta_N = 3$ gives an approximation to $b_N \simeq b_0$, Feigenbaum's constant, defined in equation 18.5, is given approximately by

$$\delta \simeq \frac{1}{g'(B)} = 5.123, \quad g(z) = 1 + \sqrt{3 + z},$$

which compares with the exact value of $\delta = 4.669\ldots$.

This approximate analysis shows how period doubling occurs and the magnitude of δ shows why the approach to the limit b_∞ is so rapid.

The complications of the logistic map, and all similar maps, do not stop here. For (approximately) $3.828 \le b \le 3.841$ the logistic map has a stable period-3 orbit, exercise 18.53. This is important because of a theorem published by Yorke in 1975 which proves that if a map has a period-3 orbit then periodic orbits of *all* periods exist, although these need not be stable: indeed, one of the bifurcation diagrams drawn as part of exercise 18.12 suggests that only the period-3 orbit is stable. This theorem and the more general result of Sarkovskii, published in Russian in 1964, are discussed by Devaney (1992, chapter 11).

Finally, we briefly mention the special case $b = 4$, for then the transformation $x = \sin^2 \pi\xi$, $0 \le \xi \le 1$, converts the logistic map into the *tent map*,

$$\xi_{n+1} = T(\xi_n) = \begin{cases} 2\xi_n, & 0 \le \xi_n \le \dfrac{1}{2}, \\ 2(1 - \xi_n), & \dfrac{1}{2} \le \xi_n \le 1. \end{cases} \tag{18.17}$$

The name arises from the shape of the function $T(\xi)$. For this map it can be shown that:

- there are periodic orbits of every period;
- there are infinitely many dense orbits, that is, orbits that pass arbitrarily close to any point in $[0, 1]$;
- all information about an orbit with an initial condition defined with finite accuracy is lost after a finite number of iterations.

An orbit with these properties is said to be chaotic. The third property shows that despite the system being deterministic its long-time behaviour is not predictable. This property is often described as *extreme sensitivity to initial conditions*.

Exercise 18.12

Bifurcation diagram

The method used to draw the bifurcation diagram 18.10 is useful for small values of n only: in most cases numerical methods are necessary. The conventional method of drawing a bifurcation diagram for $A \leq b \leq B$ is illustrated in the following Maple commands. Here a set of $N+1$ values of b, $b_k = A + k(B-A)/N$, $k = 0, 1, \ldots, N$, is chosen. For each b_k an orbit starting at x_0 (the precise value of which is immaterial) is iterated a given number of times (here 100) in order to remove the transients and the next 200 iterations are plotted. Finally, these $N+1$ graphs are combined.

```
>  restart; with(plots):
>  gr:=proc(A,B,N) local pg,b,db,x,pts,j,cond;
>  cond:=style=point,symbol=point,view=[A..B,0..1],colour=black:
>  db:=evalf( (B-A)/N ); pg:=NULL:
>     for b from A to B by db do;
>     x:=0.6;
>        for j from 1 to 100 do;
>        x:=evalhf( x*b*(1-x) ); od;              # remove transients
>        pts:=NULL: for j from 1 to 200 do;
>        x:=evalhf( x*b*(1-x) );
>        pts:=pts,[b,x]; od;
>     pg:=pg,plot([pts],cond);
>     od:
>  display(pg);
>  end:
```

Use this procedure to investigate the bifurcation diagram for various ranges of b, in particular consider the ranges $3.825 \leq b \leq 3.865$, $3.73 \leq b \leq 3.75$, $3.70 \leq b \leq 3.705$ and $3.95 \leq b \leq 3.97$, using suitably small increments in b.

18.2.3 Lyapunov exponents for one-dimensional maps

The stability of an orbit, as opposed to a fixed point, may be examined by computing the rates at which orbits with nearby initial conditions separate. Consider two orbits of a one-dimensional map with neighbouring initial conditions x_0 and $x_0 + \delta$; after n iterations the distance between the orbits is proportional to

$$d_n(x_0) = |F^n(x_0 + \delta) - F^n(x_0)| \, / \delta. \tag{18.18}$$

This quantity depends upon both n and the initial condition x_0, and is weakly dependent upon the initial distance δ, if small enough.

For the logistic map, $F(x) = bx(1 - x)$, if $b < 3$ we know that $F^n(x) \to 1 - 1/b$ as $n \to \infty$ for all x_0, so $d_n \to 0$ as $n \to \infty$. Similarly, for any value of b for which all orbits are stable we should expect $d_n \to 0$ as $n \to \infty$, provided δ is sufficiently small. But for values of b for which there are no stable orbits we should expect d_n to increases with n; clearly the distance cannot increase for all n because $|F^n(x) - F^n(y)| \leq 1$. The following two graphs show some examples of the function $d_n(x_0)$ for some values of b. Here we plot $\log(\langle d_n(x_0) \rangle)$, where $\langle \ \rangle$ denotes the mean over x_0 in the interval $(0, 1)$; for these graphs we set $\delta = 0.001$, but the results are insensitive to the value of δ. In figure 18.14 is shown the logarithm of the mean distance for $b = 2.5$, 2.9 and 3.4; in all cases $\langle d_n(x_0) \rangle$ decreases exponentially, until numerical errors dominate. In figure 18.15

we show the same variable for $b = 3.6$, 3.8 and 3.99, and in this case the mean distance increases exponentially for small n.

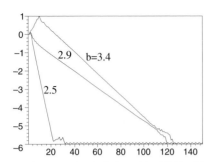

Figure 18.14   Variation of $\log(\langle d_n\rangle)$ with n for some stable orbits.

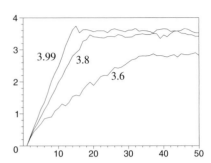

Figure 18.15   Variation of $\log(\langle d_n\rangle)$ with n for some unstable orbits

Exercise 18.13

Write a Maple procedure to draw the graph of $\log(d_n(x_0))$, for various x_0, and use this to show, graphically, that the gradient of the lines shown in the above figures is relatively insensitive to changes in x_0 and to δ.

These graphs suggest that for $|\delta| \ll 1$ and n not too large we may write

$$d_n(x_0) \simeq e^{n\lambda}, \tag{18.19}$$

for some constant exponent λ which is positive if the orbit is unstable and negative if stable. Now take the limit $\delta \to 0$ in equation 18.18 to obtain

$$d_n(x_0) = \left| \frac{d}{dx_0} F^n(x_0) \right| = \prod_{j=0}^{n-1} |F'(x_j)|,$$

where x_j are the n points on the orbit, $x_j = F(x_{j-1})$. Equation 18.19 now becomes:

$$\lambda(x_0) = \frac{1}{n} \ln\left(d_n(x_0)\right) = \frac{1}{n} \sum_{j=0}^{n-1} \ln |F'(x_j)|.$$

The *Lyapunov exponent* is defined by taking the limit $n \to \infty$:

$$\lambda_L = \lim_{n\to\infty} \frac{1}{n} \sum_{j=0}^{n-1} \ln |F'(x_j)|. \tag{18.20}$$

Note that the two limits $n \to \infty$ and $\delta \to 0$ are not interchangeable and that we have assumed λ_L to be independent of x_0. This important parameter is named after the Russian mathematician Lyapunov† who made important contributions to the theory of nonlinear systems.

A. M. Lyapunov
(1857–1918)

† The name Lyapunov is also anglicised as Liapunov.

If the iterates of an initial point tend towards a period-N orbit $(a_0, a_1, \ldots, a_{N-1})$, the Lyapunov exponent is just the finite sum

$$\lambda_L = \frac{1}{N} \sum_{j=0}^{N-1} \ln |f'(a_j)|, \quad \text{(period-}N\text{ orbit)},$$

because the first few transients have a negligible contribution to the infinite sum. In this case λ_L is clearly independent of x_0, the initial value of x.

Thus for the logistic map and $1 < b < 3$, since $f^n(x_0) \to 1 - 1/b$, for all x_0,

$$\lambda_L = \ln(|2 - b|), \quad \text{(logistic map, } 1 < b < 3).$$

Notice that $\lambda_L \to -\infty$ as $b \to 2$ because when $b = 2$ the fixed point at $x_f = \frac{1}{2}$ is superstable, exercise 18.6 (page 730). If $3 < b < 1 + \sqrt{6}$ all initial conditions lead to the 2-period orbit oscillating between $x_2(b)$ and $x_3(b)$, exercise 18.5, so

$$\lambda_L = \frac{1}{2} \ln |4 + 2b - b^2|.$$

Again λ_L is clearly independent of the initial value of x.

For values of b where the orbit is not periodic most initial conditions lead to orbits that are arbitrary close to all points in the interval $(0, b)$, and the value of λ_L is the same for almost all values of x_0. This is not true for higher-dimensional systems where arbitrarily close orbits can have Lyapunov exponents with different signs.

For non-periodic orbits it is normally necessary to evaluate λ_L numerically using the definition 18.20. In this case it is expedient to ignore the first few iterations, say the first 100, so that any initial transients are not included in the sum: this is particularly important for periodic orbits.

The Lyapunov exponent is a function of the system parameters. For the logistic map a graph of $\lambda_L(b)$ shows clearly the regions where the iterates are stable and where they are unstable: with sufficient resolution such a graph also shows the position of any superstable orbits for which $\lambda_L \to -\infty$, exercise 18.16. Such graphs provide a method of obtaining an overall view of the system behaviour relatively painlessly. The following figure was obtained using Maple by iterating a single orbit with $x_0 = 0.2$, ignoring the first 100 iterations and performing the sum over the next 200 iterations: the increment in b was chosen to be small, 0.001, otherwise some fine detail is missed.

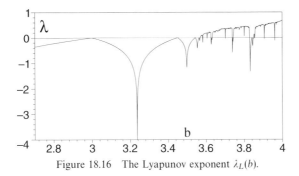

Figure 18.16   The Lyapunov exponent $\lambda_L(b)$.

Exercise 18.14

Show that for the linear map defined by the function $F(x) = bx$, the Lyapunov exponent is $\lambda_L = \ln b$.

Exercise 18.15

Show that for the logistic map, $F(x) = bx(1 - x)$, with $3 < b < 1 + \sqrt{6}$, the Lyapunov exponent is

$$\lambda_L(b) = \frac{1}{2} \ln |4 + 2b - b^2|.$$

Explain why $\lambda_L \to -\infty$ as $b \to 1 + \sqrt{5} = 3.2361\ldots$ and why $\lambda_L(3) = 0$.

Exercise 18.16

If x_f is a fixed point of $F^N(x)$, where $F(x) = bx(1 - x)$, show that the resulting N-period orbit is superstable if $x_f = 1/2$. By solving the equation

$$g_N(b) = F^N(1/2) - 1/2 = 0$$

for b show that superstable orbits exist for the following combinations of N and b:

N	3	4	5
		3.2361...	3.7389...
b	3.831 87...	3.4996...	3.9057...
		3.9603...	3.9902...

Exercise 18.17

Consider the function

$$\lambda(x_0, n) = \frac{1}{n+1} \sum_{j=J}^{n+J} \ln |F'(x_j)|, \quad x_{j+1} = F(x_j),$$

where $F(x) = bx(1 - x)$ and J is some large number, typically $J = 100$.

(i) Plot graphs of $\lambda(x_0, n)$ for $0 < x_0 < 1$ and various n, between 100 and 5000, and b between 3 and 4, to examine the dependence of this function on x_0.

(ii) For fixed x_0 plot graphs for $100 \leq n \leq 5000$ to examine the convergence of $\lambda(x_0, n)$ towards λ_L.

Exercise 18.18

Write a Maple program to plot the graph of the Lyapunov exponent, $\lambda_L(b)$, for the logistic map over a given range of b. Use your procedure to look in more detail at the intervals $3.8 \leq b \leq 3.9$, with $\delta b = 0.0001$, and $3.92 \leq b \leq 3.98$, with $\delta b = 0.00002$.

18.2.4 Probability densities

For most values of $b > 3.56\ldots$ the iterates x_k, $k = 0, 1, \ldots$, do not adhere to any well defined pattern, and because the orbits are unstable exact numerical calculations are not feasible. In these circumstances knowledge of the distribution of the iterates is often more useful than actual orbits.

Consider a long segment of an orbit $(x_0, x_1, x_2, \ldots, x_N)$, $N \gg 1$, and divide the interval $(0, 1)$ into $M (\ll N)$ equal length subintervals $[\xi_{j-1}, \xi_j)$, where $\xi_j = j/M$ and $j = 1, 2, \ldots, M$. The relative frequency that the iterates fall into the jth interval is

$$p_j = \frac{1}{N} \sum_{k=1}^{N} \left[H(x_k - \xi_{j-1}) - H(x_k - \xi_j) \right], \quad j = 1, 2 \ldots, M \ll N, \quad (18.21)$$

where $H(x)$ is the Heaviside unit function. For a fixed number of iterations, N, and large enough M, the value of p_j is approximately proportional to the length of the subintervals, $\Delta \xi = 1/M$; the more subintervals the fewer iterates fall in each. It is therefore better to define a density function

$$\rho(\bar{x}_j) = p_j / \Delta \xi, \quad \bar{x}_j = \frac{1}{2}(\xi_{j-1} + \xi_j),$$

which is not sensitive to small variations in M. Formally, we may take the limit as $N \to \infty$, replace the difference in equation 18.21 by a differential and let $\bar{x}_j \to x$ to obtain the function $\rho(x)$.

$$\rho(x) = \lim_{N \to \infty} \frac{1}{N} \sum_{k=0}^{N} \delta(x_k - x), \quad x_j = F^j(x_0). \quad (18.22)$$

This is named the probability density function. For the numerical approximation of $\rho(x)$ this equation is less useful than the original approximation 18.21, which is easily computed using the iterates and a suitably chosen grid.

The quantity $\rho(x)\delta x$ is proportional to the number of iterates falling in the interval $(x - \delta x/2, x + \delta x/2)$, so the graph of $\rho(x)$ provides a succinct global picture of the motion. Some examples for the logistic map are shown in the next four figures.

In figure 18.17 $b = 3.5$ and the iterates settle down to a period-4 orbit, so the density comprises a sum of four delta functions, as seen in the figure. In the next case the value of b has been increased slightly to 3.6 and we see that the iterates now lie in two bands; the density is not uniform in either band and is strongly peaked at the edges.

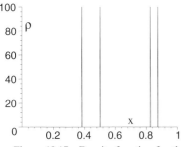

Figure 18.17 Density function for the logistic map with $b = 3.5$.

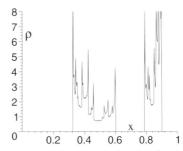

Figure 18.18 Density function for the logistic map with $b = 3.6$.

For still larger $b = 3.8$ the iterates lie in one band $0.18 < x < 0.95$ but the distribution is far from uniform and is weighted towards the edges of the band. For the largest value of $b = 4$ all structure has disappeared and the density is a smooth function given, in fact, by equation 18.28, with square root singularities at $x = 0$ and 1.

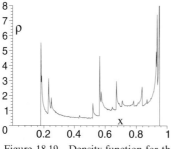

Figure 18.19   Density function for the logistic map with $b = 3.8$.

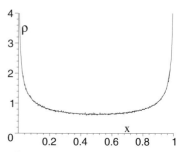

Figure 18.20   Density function for the logistic map with $b = 4.0$.

These graphs were all obtained using $M = 500$, iterating one orbit $N = 100 + 10^6$ times and ignoring the first 100 iterations.

For the logistic map the values of p_j, equation 18.21, are independent of the initial position for almost all values of x_0, because most orbits sample almost all of the accessible values of x; this is not true of dynamical systems in more than one dimension, in which some initial conditions lead to unstable orbits and others to stable orbits.

The mean value of any function $g(x)$ along an orbit is defined to be

$$\langle g \rangle = \lim_{N \to \infty} \frac{1}{N} \sum_{j=0}^{N-1} g(x_j), \quad x_j = F^j(x_0), \tag{18.23}$$

and this may be written as the integral over the probability density $\rho(x)$, provided ρ is independent of x_0,

$$\langle g \rangle = \int_0^1 dx\, \rho(x) g(x). \tag{18.24}$$

The equality between the spatial average 18.24 and the time-average 18.23 is an example of ergodic behaviour. This means that the motion is sufficiently complicated for almost all orbits to explore the entire allowed region of phase space. For complicated systems, a gas for example, it is often very helpful to assume ergodic behaviour as this allows drastic simplification by replacing a complicated time-average with the far simpler spatial average; when made this assumption is known as the ergodic hypothesis. An ergodic dynamical system is one for which the equality between time and spatial averages for almost all orbits can be proved. Ergodicity is necessary but not sufficient for chaos: integrable systems are not ergodic.

With the definition 18.24 the Lyapunov exponent, equation 18.20, is just the mean of $\ln |F'(x)|$, and consequently another expression for λ_L is

$$\lambda_L = \int_0^1 dx\, \rho(x) \ln |F'(x)|. \tag{18.25}$$

The probability density $\rho(x)$ defined in equation 18.22 is the asymptotic density, approached only after an infinity of iterations. The density $\rho_{n+1}(x)$ after $n+1$ iterations will depend upon the density $\rho_n(x)$ at n iterations and $F(x)$, the iteration function. The relation between successive densities is

$$\rho_{n+1}(x) \;=\; \int_0^1 dy\,\rho_n(y)\delta(x - F(y)), \tag{18.26}$$

$$=\; \sum_i \frac{\rho_n(y_i)}{|F'(y_i)|}.$$

The second of these equations follows from the general relation, equation 7.13 (page 230). A geometric interpretation of this relation is as follows: the value of the density $\rho_{n+1}(x)$ at x is determined by those values of y, at the nth iteration, which can lead to x, that is, $x = F(y)$, as shown in the diagram.

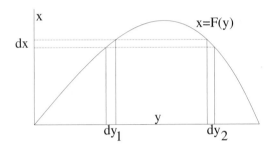

Figure 18.21   A geometric interpretation of equation 18.26.

In this example two values of y contribute to a given value of x and the density $\rho_{n+1}(x)$ is just the weighted sum of $\rho_n(x)$ at these two values of y:

$$\rho_{n+1}(x)dx \;=\; \rho_n(y_1)dy_1 + \rho_n(y_2)dy_2, \quad y_{1,2} = F(x),$$

$$=\; \left(\frac{\rho_n(y_1)}{|F'(y_1)|} + \frac{\rho_n(y_2)}{|F'(y_2)|} \right) dx.$$

We expect $\rho_n(x)$ to converge to an invariant function, independent of n and of the initial density, as $n \to \infty$. This limit is just $\rho(x)$, and hence another equation for the asymptotic density $\rho(x)$ is obtained by putting $\rho_n = \rho_{n+1} = \rho$,

$$\rho(x) = \int_0^1 dy\,\rho(y)\delta(x - F(y)). \tag{18.27}$$

The graphs shown in figures 18.17 to 18.20 show that the solution of this integral equation is, except when $b = 4$, complicated: if $b = 4$ we have, exercise 18.20,

$$\rho(x) = \frac{1}{\pi\sqrt{x(1 - x)}}, \quad (b = 4). \tag{18.28}$$

The graph of this function is also shown in figure 18.20, although it is almost indistinguishable from the numerically generated result.

Exercise 18.19
Use equations 18.25 and 18.28 to show that the Lyapunov exponent of the logistic map for $b = 4$ is $\lambda_L(4) = \ln 2$.

Exercise 18.20
Show that for the logistic map with $b = 4$ equation 18.27 becomes

$$\rho(y_1) + \rho(y_2) = 4\sqrt{1-x}\,\rho(x), \quad y_{1,2}(x) = \frac{1}{2} \pm \frac{1}{2}\sqrt{1-x}.$$

Further, show that this functional equation is satisfied by the function

$$g(x) = \frac{A}{\sqrt{x(1-x)}}$$

for some constant A. Using the fact that $\int_0^1 dx\,\rho(x) = 1$, obtain 18.28.

18.3 Area-preserving second-order systems

18.3.1 Fixed points and stability

Two-dimensional, area-preserving maps are special and simpler to understand than more general two-dimensional maps because the behaviour of their solutions is constrained. But, because many Hamiltonian systems can be expressed as area-preserving maps they are important, and the theory presented here underpins some of the phenomena described in the next chapter.

The two-dimensional map

$$\mathbf{x}_{n+1} = \mathbf{T}(\mathbf{x}_n) \quad \text{or} \quad \begin{aligned} x_{n+1} &= X(x_n, y_n), \\ y_{n+1} &= Y(x_n, y_n), \end{aligned} \tag{18.29}$$

taking a point $\mathbf{x}_n = (x_n, y_n)$ in the phase-plane to another, unique point $\mathbf{x}_{n+1}$ is area-preserving when the determinant of the Jacobian matrix is unity:

$$\frac{\partial(X, Y)}{\partial(x, y)} = \begin{vmatrix} \dfrac{\partial X}{\partial x} & \dfrac{\partial X}{\partial y} \\ \dfrac{\partial Y}{\partial x} & \dfrac{\partial Y}{\partial y} \end{vmatrix} = 1, \quad \text{for all } x \text{ and } y.$$

Area-preserving maps are therefore invertible, so $\mathbf{x}_n$ is uniquely determined by $\mathbf{x}_{n+1}$, allowing backwards as well as forwards iteration, in contrast to the first-order systems discussed in the previous section.

Fixed points of the map are given by the roots of the equation $\mathbf{x} = \mathbf{T}(\mathbf{x})$, or

$$x = X(x, y), \quad y = Y(x, y). \tag{18.30}$$

Normally the fixed points are isolated: the condition for this is that the Jacobian matrix of the function $\mathbf{T}(\mathbf{x}) - \mathbf{x}$ at the fixed point is non-singular. Important exceptions to this rule occur, but these cases have to be treated individually, as in section 18.3.3.

The nature of a fixed point $\mathbf{x}_f$ is usually determined by making a first-order Taylor expansion about $\mathbf{x}_f$: if $\mathbf{x} = \mathbf{x}_f + \boldsymbol{\xi}$ this gives

$$\boldsymbol{\xi}_{n+1} = L(\mathbf{x}_f)\boldsymbol{\xi}_n, \quad L = \begin{pmatrix} \dfrac{\partial X}{\partial x} & \dfrac{\partial X}{\partial y} \\[2ex] \dfrac{\partial Y}{\partial x} & \dfrac{\partial Y}{\partial y} \end{pmatrix}, \tag{18.31}$$

where $L(\mathbf{x}_f)$ is the Jacobian, or linearisation, matrix of the transformation evaluated at the fixed point; because the map is area-preserving $\det(L(\mathbf{x})) = 1$ for all $\mathbf{x}$ and hence L is non-singular.

With this *linearisation* of the map the nth iterate of the point (ξ_0, η_0) is simply

$$\begin{pmatrix} \xi_n \\ \eta_n \end{pmatrix} = L(\mathbf{x}_f)^n \begin{pmatrix} \xi_0 \\ \eta_0 \end{pmatrix}, \tag{18.32}$$

and the stability of the fixed point is largely determined by the eigenvalues, λ_1 and λ_2, of $L(\mathbf{x}_f)$. Because $\mathbf{T}$ is area-preserving $\det(L) = 1$ so $\lambda_1\lambda_2 = 1$, and since $\mathbf{T}$ is real there are just two cases to consider:

Elliptic fixed point: λ_1 and λ_2 are complex conjugates and lie on the unit circle: it is convenient to include the two degenerate cases $\lambda = \pm 1$ in this category.

Hyperbolic fixed point: the eigenvalues are real and either both positive or both negative; λ_1 and λ_2 are reciprocal points on the real axis, $\lambda_1 = 1/\lambda_2$.

A fixed point x_f of $\mathbf{T}(\mathbf{x})$ is said to be *stable* if for *every* neighbourhood U of x_f there exists a neighbourhood V whose images $\mathbf{T}^k(V)$ lie in U for all integers. Otherwise the fixed point is *unstable*.

Thus, a hyperbolic fixed point is unstable. Almost all elliptic fixed points are stable, but there are exceptional cases $\lambda = \pm 1$, $\pm i$ and $\exp(\pm 2\pi i/3)$ for which stability cannot be determined from the linearised map. Because the eigenvalues of L are given by the equation

$$\lambda^2 - \mathrm{Tr}(L)\lambda + 1 = 0,$$

a necessary condition for stability is $|\mathrm{Tr}(L)| < 2$.

For complex eigenvalues $\lambda = e^{\pm i\theta}$, $0 < \theta < \pi$, there is a coordinate transformation $\mathbf{u} = M\boldsymbol{\xi}$ (equation 16.35, page 660), allowing each iteration to be represented by a rotation through an angle θ:

$$\mathbf{u}_{n+1} = \begin{pmatrix} \cos\theta & \sin\theta \\ -\sin\theta & \cos\theta \end{pmatrix} \mathbf{u}_n, \quad \cos\theta = \frac{1}{2}\mathrm{Tr}(L). \tag{18.33}$$

In the original coordinate system the iterates $(\boldsymbol{\xi}_0, \boldsymbol{\xi}_1, \ldots)$ move round an ellipse: the mean rate of rotation round the ellipse is θ, and the shape of the ellipse is determined by the matrix M.

Elliptic fixed points are normally stable, though in the special cases mentioned above linearisation is not sufficient to categorise the fixed point. An example of what can happen when $\lambda = e^{\pm 2\pi i/3}$ is considered in exercises 18.27 (page 750) and 18.62 (page 782).

Hyperbolic fixed points are unstable. A transformation $\mathbf{u} = M\boldsymbol{\xi}$ may be found (equation 16.34, page 658), so that for small n the iterates in the $\mathbf{u}$-representation are

$$\mathbf{u}_n = \begin{pmatrix} \lambda^n & 0 \\ 0 & \lambda^{-n} \end{pmatrix} \mathbf{u}_0, \tag{18.34}$$

that is, the iterates move on the branches of a hyperbola.

For a fixed point with $\lambda > 0$ successive iterates remain on the same branch and it is named a *hyperbolic* fixed point. If $\lambda < 0$ successive iterates are in opposite quadrants and the fixed point is named a *hyperbolic with reflection* or *inversion hyperbolic*.

A more useful parameter than $\mathrm{Tr}(L)$, that distinguishes between all three types of fixed point, is the *residue*, defined by

$$R = \frac{1}{4}\left(2 - \mathrm{Tr}(L)\right). \tag{18.35}$$

- The fixed point is unstable and hyperbolic if $R < 0$, $(\lambda > 0)$.
- The fixed point is unstable and hyperbolic with reflection if $R > 1$, $(\lambda < 0)$.
- The fixed point is stable and elliptic if $0 < R < 1$ provided $R \neq \frac{1}{2}$ or $\frac{3}{4}$. If $\lambda = e^{\pm i\theta}$ the residue for an elliptic fixed point is $R = \sin^2(\theta/2)$. The exceptional cases $R = 0, \frac{1}{2}, \frac{3}{4}$ and 1, corresponding to $\theta = 0, \pi/2, 2\pi/3$ and π, respectively, have to be considered individually.

Exercise 18.21
Show that if $\mathbf{x}_f$ is an inversion hyperbolic fixed point of the map $\mathbf{T}$ then it is an ordinary hyperbolic fixed point of $\mathbf{T}^2(\mathbf{x}) = \mathbf{T}(\mathbf{T}(\mathbf{x}))$.

Exercise 18.22
If R_1 is the residue of a fixed point of $\mathbf{T}$ show that the residue of the same fixed points of $\mathbf{T}^n$, $n = 2$, 3 and 5 are, respectively,

$$R_2 = 4R_1(1 - R_1), \quad R_3 = R_1(3 - 4R_1)^2, \quad R_5 = R_1(16R_1^2 - 20R_1 + 5)^2.$$

Show also that if $R_1 = \frac{3}{4}$, $R_n = 0$ if n is a multiple of n, otherwise $R_n = \frac{3}{4}$.

Hint: $\mathrm{Tr}(L^n) = \lambda_1^n + \lambda_2^n$ and $\lambda_1\lambda_2 = 1$, where (λ_1, λ_2) are the eigenvalues of L.

In the following five exercises you will explore aspects of the behaviour of the map

$$\mathbf{T}: \begin{array}{rcl} x_{n+1} & = & 2cx_n - y_n + 2x_n^2 \\ y_{n+1} & = & x_n \end{array} \tag{18.36}$$

for some values of the constant c and some initial conditions. The aim of these exercises is to provide first hand experience of how a typical area-preserving map behaves and to build a framework for further analysis. Maple is needed to do some of these exercises and you should experiment with values of c and initial conditions other than those suggested.

Exercise 18.23

Show that the map **T**, defined in equation 18.36, is area-preserving and that the Jacobian matrix at a point (a, b) is

$$L(\mathbf{x}_f) = \begin{pmatrix} 2c + 4a & -1 \\ 1 & 0 \end{pmatrix}. \tag{18.37}$$

Show also that the trace and residue of L are $\mathrm{Tr}(L) = 2c + 4a$, $R = (1 - c - 2a)/2$.

Exercise 18.24

(i) Show that the fixed points of **T**, equation 18.36, are at $x = y = 0$ and $x = y = 1 - c$ with residues $R(0, 0) = \frac{1}{2}(1 - c)$, $R(1 - c, 1 - c) = \frac{1}{2}(c - 1)$.

(ii) Show that the fixed point at the origin is elliptic if $|c| < 1$, except possibly if $c = 0$ or $-\frac{1}{2}$. Show also that this fixed point is hyperbolic if $c > 1$ and hyperbolic with reflection if $c < -1$.

(iii) Show that the fixed point $(1 - c, 1 - c)$ is elliptic if $1 < c < 3$, except possibly if $c = \frac{5}{2}$, and hyperbolic if $c < 1$ and hyperbolic with reflection if $c > 3$.

Exercise 18.25

Write a Maple procedure that plots the first N iterates of **T**, equation 18.36, starting with a given initial condition and a given value of c. You may limit the x- and y-ranges to $(-\frac{1}{2}, \frac{1}{2})$, or smaller in some cases, and you should ensure that the iterations stop when the iterates are sufficiently far outside these limits, say $|x| + |y| > 1$.

Check your procedure by reproducing figures 18.22 and 18.23, obtained using about 200 iterations for each of the following initial conditions:

figure 18.22: $c = -0.2$, $x_0 = y_0 = 0.03, 0.1, 0.15, 0.3, 0.38$;
figure 18.23: $c = -0.9$, $x_0 = -y_0 = 0.02 + 0.04j$, $j = 1, 2, \dots, 7$.

In the following figures are shown orbits for $c = -0.2$ and -0.9 with the initial conditions defined in the previous exercise.

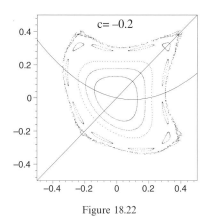

Figure 18.22

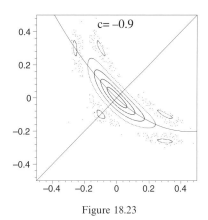

Figure 18.23

In figure 18.22, $c = -0.2$ and five orbits surrounding the stable fixed point at the origin are shown: for future reference we also show the lines $y = x$ and $y = x(c + x)$, which we note pass through some fixed points. Three types of orbit are seen in this figure. Moving out from the origin the first three orbits lie on closed curves surrounding the fixed point; these are reminiscent of the closed contours surrounding a centre of the integrable system in figure 17.18 (page 693). The fourth orbit, with initial condition $\mathbf{x}_0 = (0.3, 0.3)$, is different, comprising a set of eleven closed curves: this type of formation is colloquially called an *island chain* and, as will be shown later, is produced by a period-11 orbit. The final orbit, outside the island chain, starting at $(0.38, 0.38)$, does not lie on a closed curve and is unstable: in figures 18.40 and 18.42 (page 776) it is shown, numerically, that the largest Lyapunov exponent of the island orbit is zero and that of the outer orbit is positive. There are other unstable orbits closer to the origin but, for the sake of clarity, none are shown.

In figure 18.23, $c = -0.9$, the stable fixed point at the origin still exists but the islands surrounding it are far narrower and a quite different shape from that in figure 18.22; we should expect the orbits to change shape as $c \to -1$ because here the origin changes from a stable to an unstable fixed point, so when $c = -1$ the central island must disappear. On the outer edge of this island is another island chain surrounding a 5-period orbit, and beyond that most orbits are unbounded. There is also a 12-period orbit through $(-0.170\,886\,230\,3, 0.183)$ which is surrounded by an exceedingly small island chain, exercise 18.38.

Exercise 18.26
Use the procedure written in exercise 18.25 to examine other cases. For instance, consider the following values of c and initial conditions:

   (i) $c = -0.9 + 0.05k$, $k = 1, 2, \ldots, 4$: $(x_0, y_0) = (0.02j, -0.02j)$, $j = 1, 2, \ldots, 12$,
   (ii) $c = -1.05$, $(x_0, y_0) = (0.001, -0.001)$ and $(0.30, -0.2)$,
   (iii) $c = 0.5$, $(x_0, y_0) = (0.31, 0.31)$ and $(0.39, 0.36)$, with about 500 iterations,
   (iv) $c = -0.3$, $(x_0, y_0) = (0.2, 0.2)$, $(0.35, 0.35)$ and $(0.3123, 0)$,
   (v) $c = -0.5$, $(x_0, y_0) = (0.002, 0.002)$ and $(0.003, 0.003)$.

It was pointed out above that if the value of the residue is $\frac{3}{4}$ an elliptic fixed point is not necessarily stable. In the next exercise the numerical behaviour of the system as R passes through $\frac{3}{4}$, that is, c near $-\frac{1}{2}$, is investigated.

Exercise 18.27

The case $R = \frac{3}{4}$:
Plot the iterates of **T** for the following values of c and initial conditions; in most cases you will need to display about 1000 iterations.

   (i) $c = -0.49$, $(x_0, y_0) = (0.001, -0.001)$, $(0.010\,425\,7, 0.010\,425\,7)$.
   (ii) $c = -0.5$, $(x_0, y_0) = (a, -a)$ with $a = 0.002j$, $j = 1, \ldots, 4$.
   (iii) $c = -0.51$, $(x_0, y_0) = (0.001, -0.001)$.

The reason for the observed behaviour is explored in exercise 18.62 (page 782).

18.3.2 Periodic orbits

A period-N orbit is a set of N distinct points

$$\mathbf{a}, \quad \mathbf{T}(\mathbf{a}), \quad \mathbf{T}^2(\mathbf{a})), \quad \ldots, \quad \mathbf{T}^{N-2}(\mathbf{a}), \quad \mathbf{T}^{N-1}(\mathbf{a}), \tag{18.38}$$

where

$$\mathbf{a} = \mathbf{T}^N(\mathbf{a}) \quad \text{and} \quad \mathbf{a} \neq \mathbf{T}^k(\mathbf{a}), \quad k = 1, 2, \ldots, N-1, \tag{18.39}$$

and where $\mathbf{T}^k(\mathbf{x})$ represents k applications of the map $\mathbf{T}$ to the point $\mathbf{x}$, that is,

$$\mathbf{T}^k(\mathbf{x}) = \mathbf{T}(\mathbf{T}^{k-1}(\mathbf{x})).$$

A period-N orbit is thus a fixed point of $\mathbf{T}^N(\mathbf{x})$ that is not a fixed point of $\mathbf{T}^k(\mathbf{x})$ for $0 \leq k \leq N-1$. Fixed points of $\mathbf{T}(\mathbf{x})$ may be considered as orbits with period-1.

The notation $\mathbf{T}^k$ may, as in the one-dimensional case, be extended: $\mathbf{T}^0(\mathbf{x})$ is defined to be the identity, $\mathbf{T}^0(\mathbf{x}) = \mathbf{x}$ and, because the map is invertible, we define $\mathbf{T}^{-k}(\mathbf{x})$ to be $\mathbf{T}$ iterated backwards k times. Thus if $\mathbf{y} = \mathbf{T}^{-1}(\mathbf{x})$ then $\mathbf{x} = \mathbf{T}(\mathbf{y})$ and, more generally, if $\mathbf{y} = \mathbf{T}^{-k}(\mathbf{x})$, $\mathbf{x} = \mathbf{T}^k(\mathbf{y})$.

The stability of a periodic orbit is examined in exactly the same manner as the stability of a fixed point, though in practice the details are more complicated. Consider the iterates of orbits adjacent to the N-periodic orbit defined in equation 18.38: for this we need the first-order Taylor expansion of the map $\mathbf{T}$ about an arbitrary point, $\mathbf{z}$:

$$\mathbf{T}(\mathbf{z} + \mathbf{u}) = \mathbf{T}(\mathbf{z}) + L(\mathbf{z})\mathbf{u} + \cdots, \tag{18.40}$$

where $L(\mathbf{z})$ is the Jacobian matrix of the map at the point $\mathbf{z}$.

$$L(\mathbf{z}) = \begin{pmatrix} \dfrac{\partial X}{\partial x} & \dfrac{\partial X}{\partial y} \\ \dfrac{\partial Y}{\partial x} & \dfrac{\partial Y}{\partial y} \end{pmatrix}, \quad \text{where} \quad \mathbf{T} = (X(x, y), Y(x, y)).$$

We also need the first-order Taylor expansion of any power of $\mathbf{T}$. The Taylor's series of $\mathbf{T}^2$ is given by two applications of the above formula:

$$\begin{aligned} \mathbf{T}^2(\mathbf{z} + \mathbf{u}) &= \mathbf{T}(\mathbf{T}(\mathbf{z} + \mathbf{u})) = \mathbf{T}(\mathbf{T}(\mathbf{z}) + L(\mathbf{z})\mathbf{u}) \\ &= \mathbf{T}^2(\mathbf{z}) + L(\mathbf{T}(\mathbf{z})) L(\mathbf{z})\mathbf{u}, \end{aligned}$$

and similarly,

$$\begin{aligned} \mathbf{T}^k(\mathbf{z} + \mathbf{u}) &= \mathbf{T}^k(\mathbf{z}) + L(\mathbf{T}^{k-1}(\mathbf{z})) L(\mathbf{T}^{k-2}(\mathbf{z})) \cdots L(\mathbf{T}(\mathbf{z})) L(\mathbf{z}) \mathbf{u} + \cdots, \\ &= \mathbf{T}^k(\mathbf{z}) + \left(\prod_{r=0}^{k-1} L\left(\mathbf{T}^r(\mathbf{z})\right) \right) \mathbf{u} + \cdots. \end{aligned} \tag{18.41}$$

The order of this product of matrices is important.

If $\mathbf{a}$ is on a period-N orbit this equation shows that the Nth iterate of a neighbouring point $\mathbf{x} = \mathbf{a} + \mathbf{u}_0$ is $\mathbf{a} + \mathbf{u}_1$, where

$$\mathbf{u}_1 = M(\mathbf{a})\mathbf{u}_0, \quad M(\mathbf{a}) = \prod_{r=0}^{N-1} L(\mathbf{T}^r(\mathbf{a})). \tag{18.42}$$

Similarly, after m periods, that is, mN iterations of $\mathbf{T}$, the point $\mathbf{a} + \mathbf{u}_0$ is mapped to $\mathbf{a} + \mathbf{u}_m$, where

$$\mathbf{u}_m = M(\mathbf{a})^m \mathbf{u}_0.$$

Thus the stability of the periodic orbit through $\mathbf{a}$ depends upon the eigenvalues of the matrix $M(\mathbf{a})$. Since the determinant of a product of matrices is the product of the determinants of the individual matrices it follows that $\det(M) = 1$ and the classification scheme discussed above also applies to periodic orbits.

An orbit which is a fixed point of $\mathbf{T}^N$, and not $\mathbf{T}^k$ for $1 \leq k \leq N - 1$, comprises N distinct points. If the fixed point is stable there are adjacent orbits that lie on a chain of N closed curves surrounding these points; an example of this behaviour is shown in figure 18.22. The appearance is of a ring of islands, named an *island chain*: usually there is a region between these islands containing unstable motion. These islands are also named *resonance islands* because they are produced by resonances between frequencies of two component motions.

More complicated arrangements than that seen in figure 18.22 exist and are common. An example is shown in the next three figures, which depict *one* orbit of the map 18.36 with increasing degrees of magnification. Here $c = -1.05$ and a period-2 orbit oscillates between $(0.25, -0.2)$ and $(-0.2, 0.25)$, as shown in exercise 18.40 (page 764). The orbit shown starts at $(0.346, -0.2)$ and the first figure, on the left, shows 1000 iterates of the whole orbit and the lines $y = x$ and $y = x(c + x)$, which will be shown to pass through one or more of the stable fixed points at the island centres. Superficially this diagram seems to comprise a set of 33 islands. On the right is shown an enlargement of the region contained in the rectangle which seems to confirm that each of the 11 groups seen on the left comprises three separate islands.

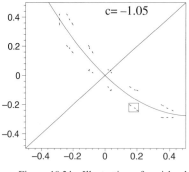

Figure 18.24   Illustration of an island chain.

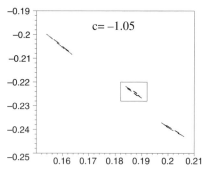

Figure 18.25   Enlargement region around $(0.2, -0.2)$.

However, enlarging the rectangle in figure 18.25 we see, figure 18.26, that each comprises five separate islands. On the right, figure 18.27, is shown two orbits. One orbit is that used in the previous three figures and produces the islands from the extreme left group in figure 18.24, but magnified we see that it comprises a chain of five islands, surrounding a 165-period orbit. The other orbit, with initial condition $(-0.2876, 0.386)$,

produces a separate island at the centre of this group. Also shown in this figure is the
line $y = x(c + x)$.

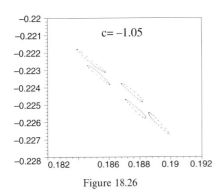

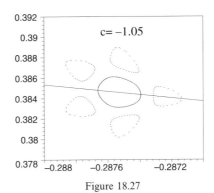

Figure 18.26   Figure 18.27

These figures show that island chains can be complicated and that they can surround other island chains. This behaviour is typical of nonlinear systems. With each
magnification of scale similar structures are seen: we used this feature to examine the
period doubling sequence of the logistic map and we shall show that the same type of
analysis may be applied to two-dimensional maps, but now each stable periodic orbit
is surrounded by an island of stability.

The twist map theorem described in the next section provides a clear description of
how the motion in the neighbourhood of an elliptic fixed point (and hence also a stable
periodic island) is changed when perturbed, and partly explains the behaviour seen in
the previous four figures.

Exercise 18.28
If the set of points $(\mathbf{a}, \mathbf{T}(\mathbf{a}), \mathbf{T}(\mathbf{T}(\mathbf{a})), \ldots, \mathbf{T}^{N-1}(\mathbf{a}))$ with $\mathbf{T}^N(\mathbf{a}) = \mathbf{a}$ is an N-period orbit, show
that the set of points $(\mathbf{b}, \mathbf{T}(\mathbf{b}), \mathbf{T}(\mathbf{T}(\mathbf{b})), \ldots, \mathbf{T}^{N-1}(\mathbf{b}))$, where $\mathbf{b} = \mathbf{T}^p(\mathbf{a})$ for any integer p,
represents the same orbit.

Exercise 18.29
(i) If $\mathbf{x}$ and λ are respectively the eigenvector and eigenvalue of the product AB of two
    matrices A and B, show that $B\mathbf{x}$ is an eigenvalue of BA with the same eigenvalue.
(ii) Use this result, and the solution to exercise 18.28, to show that when $\mathbf{a}$ and $\mathbf{b}$ are
    different points on the same periodic orbit then the matrices $M(\mathbf{a})$ and $M(\mathbf{b})$, defined in
    equation 18.42, have the same eigenvalues. Conclude that, as would be expected, it does
    not matter which point of a periodic orbit is used to determine its stability.

18.3.3 Twist maps
In the vicinity of an elliptic fixed point the linearised motion is particularly simple,
but the effect of the nonlinear terms on this motion is far from simple. The twist map
theorem describes some of these complications, and it is important because it applies
to local regions of phase space of many Hamiltonian systems and explains why the
motion of many systems is so complicated.

A *twist map* is conventionally written in the form[†]

$$\mathbf{T}_\epsilon : \begin{array}{rcl} I_{n+1} & = & I_n + \epsilon f(\theta_n), \quad |\epsilon| \ll 1 \\ \theta_{n+1} & = & \theta_n + \omega(I_{n+1}). \end{array} \qquad (18.43)$$

It has its origin in Hamiltonian dynamics and it is from this theory that the notation[‡] arises: here we need only note that θ is an angle, so the points $\theta \pm 2n\pi$, $n = 0, 1, \ldots$, are physically equivalent and the perturbation, $\epsilon f(\theta)$, must be a periodic function of θ; normally this is chosen to have zero mean value. Geometrically the coordinates (θ, I) may be interpreted as defining points on the surface of a cylinder, with I being the distance along the axis and θ the angle on a circle perpendicular to the axis. The unperturbed motion, $\epsilon = 0$, is on one of these circles.

An alternative representation is obtained by interpreting $\sqrt{I}$ as the distance from the origin in a two-dimensional plane and θ as the polar angle, then the unperturbed iterations are on concentric circles with their centre at the origin. This interpretation corresponds more closely to the iterates surrounding the origin in figures 18.22 and 18.23 and we use it in the following discussion.

The two crucially important properties of the twist map are:

P 1: that it is area-preserving:

$$\frac{\partial(I_{n+1}, \theta_{n+1})}{\partial(I_n, \theta_n)} = \begin{vmatrix} 1 & \epsilon f'(\theta_n) \\ \omega'(I_{n+1}) & 1 + \epsilon \omega'(I_{n+1}) f'(\theta_n) \end{vmatrix} = 1,$$

P 2: that $\omega(I)$ varies with I.

Orbits of the unperturbed map $\mathbf{T}_0$ are simple:

$$I_n = I_0 = \text{constant}, \quad \theta_n = \theta_0 + n\omega(I_0),$$

that is, all iterations lie on circles. If $\omega(I_0)/2\pi$ is irrational the circle is eventually filled, in the sense that given any point (α, I_0) on a circle there are iterates in every neighbourhood of α. But if $\omega(I_0)/2\pi$ is rational, $\omega(I_0)/2\pi = r/s$ with r and s coprime integers, the iterates of $\mathbf{T}_0$ are the s distinct points $I = I_0$, $\theta_k = \theta_0 + 2\pi r k/s$, $k = 0, 1, \ldots, s - 1$, so after s iterations of $\mathbf{T}_0$ the angle θ has increased by $2\pi r$: that is *all* points on the circle $I = I_0$ are fixed points of $\mathbf{T}_0^s$. When $\omega(I_0)/2\pi$ is rational the fixed points are *not* isolated. Because all previous analysis assumed that fixed points are isolated it may be thought that lines of fixed points are exceptional. In fact they are common and examples are given in the next chapter.

H. Poincaré
(1854–1912)

G. D. Birkhoff
(1884–1944)

The Poincaré–Birkhoff twist map theorem shows that these lines of fixed points are destroyed by an arbitrarily small perturbation and become a *finite* set of isolated fixed points. Specifically, it shows that for some integer k, $\mathbf{T}_\epsilon^s$ has $2ks$ fixed points close to the unperturbed phase curve which are alternately elliptic and hyperbolic fixed points, so there are k of each: this behaviour is shown schematically in the following diagram.

[†] In some texts this is known as the radial twist map: a more general form of the twist map is given in exercise 18.30.

[‡] (θ, I) are the angle-action variable of an integrable Hamiltonian system.

The proof is elementary and uses the fact that the map is area-preserving: it may be found in Arnold and Avez (1968, chapter 4).

Perturbation of a rational phase curve

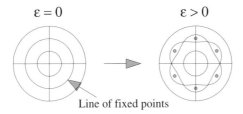

Line of fixed points

This change in the nature of the map is fundamental because the topology of the phase curves when $\epsilon = 0$ and $\epsilon \neq 0$ is different. It means, for instance, that perturbation theory has to be used with great care and is why it normally provides an asymptotic expansion rather than a convergent power series when applied to this type of problem, which includes most Hamiltonian systems.

This picture of the perturbed motion is very complicated; at every rational frequency of the unperturbed system the rotational phase curves are broken up into sets of resonance islands. Moreover, at every irrational frequency the rotational phase curves are unchanged in form; the proof of this last statement is very long and difficult and is one of the triumphs of modern mathematics. Because every irrational number is approximated to arbitrary accuracy by a rational number, the perturbed motion is extremely complicated and difficult to imagine.

The first impression is that an arbitrarily small perturbation will destroy all the order of the unperturbed system; but this is not true. The reason is that most rational numbers have large denominators, so the vast majority of resonances belong to rational frequencies $\omega/2\pi = r/s$ with $s \gg 1$, and these are associated with periodic orbits with very long periods. Thus for any *finite* time only a finite number of resonances can be distinguished from non-periodic orbits.

Further, the area of phase space associated with most elliptical islands is relatively very small. For many systems it transpires that the area of the r/s resonance is typically $O(e^{-s})$ and the total area of all these islands is significantly less than the whole phase space, provided the perturbation is small enough.†

Now consider the generation of islands within islands as seen in figures 18.26 and 18.27. A rigorous analysis of this important feature is not possible here so, instead, we present a heuristic argument that shows how they arise. Consider the twist map 18.43 with $f(\theta) = \sin \theta$ and an expansion in I about the point at which $\omega = \pi$, so that two applications of the unperturbed map gives the identity. If $\omega(A) = \pi$, we put $J = I - A$ to obtain

$$\mathbf{T}_1: \qquad J_{n+1} = J_n + \epsilon \sin \theta_n, \quad \theta_{n+1} = \theta_n + \pi + \alpha J_{n+1}, \quad \alpha = \omega'(A),$$

---

† The exponential behaviour is due to the behaviour of the high harmonics of the Fourier series. For infinitely differentiable functions it was shown in chapter 8 that the nth harmonic decreases faster than any power of n.

and two applications give

$$\mathbf{T}_2 : \begin{array}{rcl} J_{n+2} &=& J_n + \epsilon \left\{ \sin \theta_n - \sin \left(\theta_n + \alpha(J_n + \epsilon \sin \theta_n) \right) \right\}, \\ \theta_{n+2} &=& \theta_n + \alpha \left(J_n + J_{n+2} + \epsilon \sin \theta_n \right). \end{array} \qquad (18.44)$$

Linearising this about the origin gives

$$\begin{pmatrix} \theta_{n+2} \\ J_{n+2} \end{pmatrix} = L_2 \begin{pmatrix} \theta_n \\ J_n \end{pmatrix}, \quad \text{where} \quad L_2 = \begin{pmatrix} 1 + \alpha\epsilon - (\alpha\epsilon)^2 & \alpha(2 - \alpha\epsilon) \\ -\alpha\epsilon^2 & 1 - \alpha\epsilon \end{pmatrix},$$

with eigenvalues $\lambda = e^{\pm i\psi}$, $\sin(\psi/2) = \alpha\epsilon/2$.

Successive iterates of L_2 move points on an ellipse around the origin, so it is natural to make a coordinate transformation $\mathbf{y} = M \begin{pmatrix} \theta \\ J \end{pmatrix}$, where the matrix M converts L_2 to the standard form, equation 16.35 (page 660),

$$ML_2M^{-1} = B(\psi) = \begin{pmatrix} \cos \psi & \sin \psi \\ -\sin \psi & \cos \psi \end{pmatrix}.$$

In this coordinate system the $\mathbf{T}_2$ map is

$$\mathbf{y}_{n+2} = B(\psi)\mathbf{y}_n + \begin{pmatrix} g_1(M^{-1}\mathbf{y}_n) \\ g_2(M^{-1}\mathbf{y}_n) \end{pmatrix}, \qquad (18.45)$$

where $g_k(\theta, J)$ are the nonlinear parts of $\mathbf{T}_2$. For the present purposes the exact form of these is irrelevant, but we note that because the map is area-preserving these two functions cannot be completely independent.

Since the linear part of this map represents a clockwise rotation through an angle ψ it is natural to re-express the map in terms of the polar coordinates $y_1 = \sqrt{R}\cos\phi$, $y_2 = \sqrt{R}\sin\phi$: this form of the transformation is chosen so that its Jacobian determinant is constant. The linear part of the map 18.45 is just $R_{n+2} = R_n$, $\phi_{n+2} = \phi_n - \psi \simeq \phi_n - \epsilon\alpha$. The addition of the nonlinear terms changes this to $R_{n+2} = R_n + F_1(\phi_n, R_n)$, $\phi_{n+2} = \phi_n - \psi + F_2(\phi_n, R_n)$, where F_1 and F_2 are two (related) periodic function of ϕ. If the mean, $\langle F_2 \rangle$, of F_2 is nonzero then the map has the form of the more general twist map given in exercise 18.30. Thus if $(\langle F_2 \rangle - \psi)/2\pi$ is rational the perturbation due to the remaining part of the map will produce islands within the original islands. This process can be carried on indefinitely. An example of these island structures is considered in exercise 18.31.

Exercise 18.30

The most general form for the twist map is

$$\begin{array}{rcl} I_{n+1} &=& I_n + \epsilon F(I_{n+1}, \theta_n), \\ \theta_{n+1} &=& \theta_n + \omega(I_{n+1}) + \epsilon G(I_{n+1}, \theta_n). \end{array}$$

Show that this is area-preserving if

$$\frac{\partial F}{\partial I_{n+1}} + \frac{\partial G}{\partial \theta_n} = 0.$$

Exercise 18.31
Show that the map $\mathbf{T}_2$, equation 18.44, has an elliptic fixed point at the origin. Plot some iterates of $\mathbf{T}_2$ for $\epsilon = 0.2$ and $\alpha = 4$ for initial conditions that produce an island surrounding the origin. In particular, consider the initial condition $(0, 0.048)$; show that this produces islands within the original islands of $\mathbf{T}_1$ and investigate, numerically, the size of this island.

18.3.4 Symmetry methods

Periodic orbits are important so, consequently, are methods of finding them. Conventional numerical techniques for finding solutions of the nonlinear equation $\mathbf{T}^n(\mathbf{x}) = \mathbf{x}$ involve a two-dimensional search which can be time consuming, particularly if n is large, and do not guarantee that all solutions are found. The theory presented here simplifies this numerical task by replacing a two-dimensional by a one-dimensional search; it becomes particularly important when the algebraic map $\mathbf{T}$ is replaced by the solution of a differential equation. The method relies upon finding the symmetries of the map and uses the fact that periodic orbits lie on lines defined by these symmetries, hence limiting the search to these lines.

A map $\mathbf{R}$ is named a *reflection* if $\mathbf{R}^2$ is the identity, $\mathbf{R}^2 = \mathbf{I}$, *and* the Jacobian determinant of $\mathbf{R}$ is everywhere negative. It can be shown that the fixed points of $\mathbf{R}$ form a curve. An example is the map $\mathbf{R}(x, y) = (x, -y)$ representing a reflection in the x-axis which is the line of fixed points.

We say that $\mathbf{S}$ is a *symmetry* of a map $\mathbf{T}$ if both $\mathbf{S}$ and $\mathbf{TS}$ are reflections. The curve of fixed points of $\mathbf{S}$ is called a *symmetry curve* of $\mathbf{T}$. The following properties are easily derived.

P 1: If both $\mathbf{S}$ and $\mathbf{TS}$ are reflections then $\mathbf{T}^{-1} = \mathbf{STS}$ and $\mathbf{S} = \mathbf{TST}$.
P 2: If $\mathbf{S}$ is a reflection *and* $\mathbf{S} = \mathbf{TST}$ then $\mathbf{S}$ is a symmetry of $\mathbf{T}$.
P 3: If $\mathbf{S}$ is a symmetry of $\mathbf{T}$, then it is also a symmetry of $\mathbf{T}^n$, for all integers n.
P 4: If $\mathbf{S}$ is a symmetry of $\mathbf{T}$ then $\mathbf{S}' = \mathbf{TS}$ is also a symmetry of $\mathbf{T}$.

The map $\mathbf{S}' = \mathbf{TS}$ is named the symmetry of $\mathbf{T}$ complementary to $\mathbf{S}$, and the fixed curve of $\mathbf{S}'$ is the complementary symmetry curve. Since $\mathbf{S}'\mathbf{S} = \mathbf{T}$, the two symmetries are said to factor $\mathbf{T}$ and its fixed points may be found using one or more of the following properties.

P 5: If $\mathbf{S}$ and $\mathbf{S}'$ are complementary symmetries of $\mathbf{T}$ and $\mathbf{x}$ lies at the intersection of the two symmetry curves, then $\mathbf{x}$ is a fixed point of $\mathbf{T}$.
P 6: If $\mathbf{x}$ is on the $\mathbf{S}$-symmetry line then the forward and backward iterates of $\mathbf{T}$ are reflections of each other:

$$\mathbf{ST}^{-k}(\mathbf{x}) = \mathbf{T}^k(\mathbf{x}), \quad k = \pm 1, \pm 2, \ldots. \tag{18.46}$$

P 7: If $\mathbf{x}$ and $\mathbf{T}^n(\mathbf{x}) \neq \mathbf{x}$ lie on the same symmetry curve, then $\mathbf{x}$ is a fixed point of $\mathbf{T}^{2n}$.
P 8: If $\mathbf{x}$ lies at the intersection of the symmetry curves of $\mathbf{S}$ and $\mathbf{T}^n\mathbf{S}$ then $\mathbf{x}$ is a fixed point of $\mathbf{T}^n$.
P 9: If $\mathbf{x}$ lies on the symmetry line of $\mathbf{S}$ and $\mathbf{T}^n(\mathbf{x})$ on the symmetry line of $\mathbf{S}'$, then $\mathbf{x}$ is a fixed point of $\mathbf{T}^{2n-1}$.

Most of these properties are easily verified, and **P1**–**P4** we leave as exercises for the reader and provide outline proofs of **P5**–**P9**. In the following we denote the symmetry line of **S** by $\mathscr{S}$.

For **P5** we have

$$\mathbf{S}(\mathbf{x}) = \mathbf{x} \quad \Longrightarrow \quad \mathbf{TS}(\mathbf{x}) = \mathbf{T}(\mathbf{x}) \quad \Longrightarrow \quad \mathbf{S}'(\mathbf{x}) = \mathbf{T}(\mathbf{x}),$$

but $\mathbf{S}'(\mathbf{x}) = \mathbf{x}$ and hence $\mathbf{T}(\mathbf{x}) = \mathbf{x}$, so $\mathbf{x}$ is a fixed point of **T**.

For **P6** we note that since $\mathbf{S}(\mathbf{x}) = \mathbf{x}$, **P1** gives

$$\mathbf{x} = \mathbf{TST}(\mathbf{x}) \quad \Longrightarrow \quad \mathbf{T}^{-1}(\mathbf{x}) = \mathbf{ST}(\mathbf{x}) \quad \Longrightarrow \quad \mathbf{ST}^{-1}(\mathbf{x}) = \mathbf{T}(\mathbf{x}),$$

which is equation 18.46 for $k = \pm 1$. Similarly

$$\mathbf{T}^{-k}(\mathbf{x}) = \underbrace{\mathbf{STS\,STS}\cdots\mathbf{STS}}_{k \text{ times}}(\mathbf{x}) = \mathbf{ST}^{k}(\mathbf{x}),$$

and hence $\mathbf{ST}^{-k}(\mathbf{x}) = \mathbf{T}^{k}(\mathbf{x})$, because $\mathbf{S}(\mathbf{x}) = \mathbf{x}$ and $\mathbf{S}^2 = \mathbf{I}$. Operating with **S** gives $\mathbf{T}^{-k}(\mathbf{x}) = \mathbf{ST}^{k}(\mathbf{x})$: this proves **P6**.

For **P7** consider the point $\mathbf{x}$ on $\mathscr{S}$, $\mathbf{S}(\mathbf{x}) = \mathbf{x}$, and its iterates, $\mathbf{T}^{k}(\mathbf{x})$, $k = 1, 2, \ldots$. Assume that n is the smallest integer such that $\mathbf{T}^{n}(\mathbf{x})$ is on $\mathscr{S}$,

$$\mathbf{ST}^{n}(\mathbf{x}) = \mathbf{T}^{n}(\mathbf{x}) \quad \text{and} \quad \mathbf{S}(\mathbf{x}) = \mathbf{x}.$$

Typical sets of iterates are shown in the figures below. Here $\mathbf{T}^{3}(\mathbf{x})\,[= \mathbf{T}^{-3}(\mathbf{x})]$ is on $\mathscr{S}$ and we show the two possibilities: on the left $\mathbf{x} \neq \mathbf{T}^{3}(\mathbf{x})$ and on the right $\mathbf{x} = \mathbf{T}^{3}(\mathbf{x})$.

$2n$-periodic orbit $(n = 3)$ $(2n + 1)$-periodic orbit $(n = 1)$

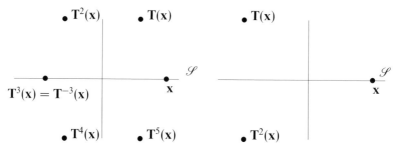

If $\mathbf{T}^{n}(\mathbf{x}) \neq \mathbf{x}$ then equation 18.46 gives $\mathbf{T}^{-n}(\mathbf{x}) = \mathbf{T}^{n}(\mathbf{x})$ and hence $\mathbf{x}$ is on a $2n$-periodic orbit.

If $\mathbf{T}^{n}(\mathbf{x}) = \mathbf{x}$ then n must be odd, for if $n = 2m$, $\mathbf{T}^{m}(\mathbf{x})$ must lie on $\mathscr{S}$ since

$$\mathbf{ST}^{-m}(\mathbf{x}) = \mathbf{T}^{m}(\mathbf{x}) \quad \text{and} \quad \mathbf{x} = \mathbf{T}^{2m}(\mathbf{x}) \quad \Longrightarrow \mathbf{ST}^{m}(\mathbf{x}) = \mathbf{T}^{m}(\mathbf{x}).$$

This contradicts the assumption that $n\,(> m)$ is the smallest integer for which $\mathbf{T}^{n}(\mathbf{x})$ is on $\mathscr{S}$.

For **P8**, since, from **P3**, both **S** and $\mathbf{T}^{n}\mathbf{S}$ are reflections and $\mathbf{x}$ is on the symmetry lines of both, that is,

$$\mathbf{S}(\mathbf{x}) = \mathbf{x} \quad \text{and} \quad \mathbf{T}^{n}\mathbf{S}(\mathbf{x}) = \mathbf{x} \quad \text{and hence} \quad \mathbf{T}^{n}(\mathbf{x}) = \mathbf{x}.$$

Finally, for **P 9**, if **x** is on $\mathscr{S}$ and $\mathbf{T}^n(\mathbf{x})$ on the $\mathbf{S}' = \mathbf{TS}$ symmetry line, we have

$$\mathbf{S}(\mathbf{x}) = \mathbf{x} \quad \text{and} \quad \mathbf{TST}^n(\mathbf{x}) = \mathbf{T}^n(\mathbf{x}),$$

then using equation 18.46 we obtain $\mathbf{TT}^{-n}(\mathbf{x}) = \mathbf{T}^n(\mathbf{x})$ and hence **x** is a fixed point of $\mathbf{T}^{2n-1}$.

These results may be used to find some fixed points by searching along a symmetry line. Further discussion of these methods is given by Greene *et al.* (1981), here we illustrate this method by applying it to the area-preserving map

$$\mathbf{W} : \quad \begin{cases} u_{n+1} & = & -v_n + f(u_n) \\ v_{n+1} & = & u_n - f(u_{n+1}). \end{cases} \tag{18.47}$$

where $f(u)$ is an arbitrary function: this is sometimes called the general DeVogelaere map. It is easily seen that if

$$\mathbf{S}_1 : \quad \begin{cases} u' & = & u \\ v' & = & -v \end{cases} \quad \text{and} \quad \mathbf{S}_2 : \quad \begin{cases} u' & = & v + f(u) \\ v' & = & u - f(u'), \end{cases} \tag{18.48}$$

then both $\mathbf{S}_k$, $k = 1, 2$, are reflections, $\mathbf{W} = \mathbf{S}_2\mathbf{S}_1$ and $\mathbf{W}^{-1} = \mathbf{S}_1\mathbf{S}_2$. The $\mathbf{S}_1$-symmetry line is the u-axis and the $\mathbf{S}_2$-symmetry line is $v = u - f(u)$.

The area-preserving transformation

$$x = u, \quad y = v + f(u)$$

gives an alternative form for this map:

$$\mathbf{T} : \quad \begin{cases} x_{n+1} & = & -y_n + 2f(x_n) \\ y_{n+1} & = & x_n. \end{cases} \tag{18.49}$$

With $f(x) = x(c + x)$ this is the map considered in exercises 18.23 to 18.27 and also figures 18.24–18.27, some of which include these symmetry lines, $\mathbf{S}_1$ being $y = f(x)$ and $\mathbf{S}_2$ being $y = x$.

Exercise 18.32
Show that $\mathbf{S}_1$ and $\mathbf{S}_2$, defined in equation 18.48, are reflections and satisfy the relations $\mathbf{W} = \mathbf{S}_2\mathbf{S}_1$ and $\mathbf{W}^{-1} = \mathbf{S}_1\mathbf{S}_2$.

Exercise 18.33
Show that the maps 18.47 and 18.49 are area-preserving and also that the inverse of 18.47 is

$$\mathbf{W}^{-1} : \quad \begin{cases} u_n & = & v_{n+1} + f(u_{n+1}) \\ v_n & = & f(u_n) - u_{n+1}. \end{cases}$$

For the map 18.47 the $\mathbf{S}_1$-symmetry line is the u-axis, $v = 0$, and we consider the points $\mathbf{x} = (a, 0)$ on this line. Iterate such a point n times to obtain $\mathbf{W}^n(a, 0) = (F_1(a), G_1(a))$, which is on the symmetry line if $G_1(a) = 0$. Generally this equation will have many roots each of which will give either a 2n- or an n-periodic orbit, **P 7**.

The S_2-symmetry curve is the line $v = u - f(u)$ and this may be used in the same manner. Now the equations are $\mathbf{W}^n(a, a - f(a)) = (F_2(a), G_2(a))$ and we must search for the zeros of $G_2(a) = F_2(a) + f(F_2(a))$.

In the (x, y)-coordinate system the S_1 and S_2 symmetry curves are $y = f(x)$ and $y = x$ respectively. These are drawn in figures 18.24 to 18.27, where it is seen that each passes through fixed points.

It is a relatively trivial matter to numericaly compute the vector $(F_k(a), G_k(a))$ over a given range of a for any n, provided n is not too large. If n is large there will be many roots and between these both $|F_k|$ and $|G_k|$ are large, so great care is needed if all roots are required. If n is small algebraic methods may suffice.

First, as a simple illustration, consider $n = 1$. If $f(x) = x(c + x)$ we have, for a point initially on $\mathscr{S}_1$, $\mathbf{W}(a, 0) = (f(a), -a - f(f(a)))$ and this iterate returns to $\mathscr{S}_1$ if $a = f(f(a))$, that is,

$$a - f(f(a)) = a(a + c - 1)\left(a^2 + a(c + 1) + c + 1\right) = 0. \tag{18.50}$$

One set of solutions of $a = f(f(a))$ is clearly $a = f(a)$; these account for the first two factors and give periodic orbits. The third factor, $a^2 + a(c + 1) + c + 1$ gives a 2-period orbit.

When $c = -1$ this equation becomes $a^3(a - 2) = 0$ and has a triple root at $a = 0$. If $c = -1 - \delta$ and $0 < \delta \ll 1$ there is one real root near the origin and for $-1 \ll \delta < 0$ there are three real roots. This is consistent with the observation made in exercise 18.24 (page 749) that when c decreases through -1 the fixed point of $\mathbf{T}$ (and hence $\mathbf{W}$) at the origin changed from elliptic to hyperbolic with reflection.

Exercise 18.34

Show that for the map $\mathbf{T}$, equation 18.49, a point on the S_2-symmetry line gives $\mathbf{T}(a, a) = (2f(a) - a, a)$, and a point on the S_1-symmetry line gives $\mathbf{T}(a, f(a)) = (f(a), a)$. These equations show that a periodic orbit is given by the roots of $a = f(a)$. Is this consistent with equation 18.50, which has more roots?

Exercise 18.35

Show that the factor $a^2 + a(c + 1) + c + 1$ of equation 18.50 gives a fixed point $\mathbf{x} = (a, 0)$, of $\mathbf{W}^2$, where $a = \left(\delta \pm \sqrt{4\delta + \delta^2}\right)/2$, $c = -1 - \delta$, and that its residue is $R = \delta(4 + \delta)$.

For large n it is normally easier to perform the analysis numerically and we illustrate how one proceeds for $n = 2$, the case treated in the previous exercise. If $\mathbf{W}^n(a, 0) = (F_1(a), G_1(a))$, $n = 2$, the fixed points of $\mathbf{W}^2$ or $\mathbf{W}^4$ are given by the roots of $G_1(a) = 0$. In figure 18.28 we show the graph of this function for $c = -1.23$ and see that there are roots at $a = -0.37818$ and $a = 0.60818$: these are also roots when $n = 1$ so are period-2 orbits. In the right-hand figure are shown 200 iterations of $\mathbf{W}$ for an orbit starting near $a = -0.37818$, indicating that this root is indeed a 2-period orbit. The root near $a = -0.8$ gives an unstable orbit.

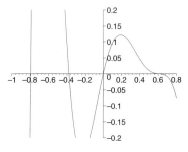

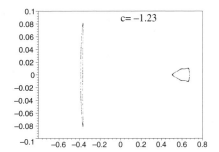

Figure 18.28 $G_1(a)$ for $c = -1.23$, $n = 2$.

Figure 18.29   An orbit starting near the zero at $x \simeq -0.378\,18$.

Figure 18.28 shows that $G_1(a)$ is almost tangential to the a-axis at $a \simeq 0.6$. The reason for this is that when $c = 1 - \sqrt{5} = -1.236\ldots$, $G_1(a)$ has a double root at $a = (\sqrt{5} - 1)/2 \simeq 0.68$, exercise 18.36. As c passes through $1 - \sqrt{5}$ the fixed point bifurcates.

As $c \to 1 - \sqrt{5}$ from below the island centred at $(-0.378\ldots, 0)$ becomes long and thin and eventually collapses in the same manner as the island around the origin, when $c = 1$. In the following two figures are shown the graph of $G_1(a)$ for $c = -1.25$ and $n = 4$.

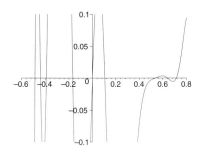

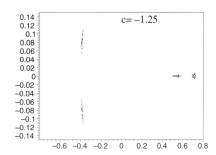

Figure 18.30 $G_1(a)$ for $c = -1.25$, $n = 4$.

Figure 18.31   An orbit starting near $(0.54, 0)$.

Near $a = 0.6$ there are now three real roots, which are also roots for $n = 2$. The central root, $a = 0.640\,39$, represents an unstable period-2 orbit (residue $R = 1.06$) while the roots $a = \{0.5364, 0.7136\}$ are two points on a stable period-4 orbit. In figure 18.31 is shown a single orbit starting at $\mathbf{x} = (0.5364 + 0.0125, 0)$ (note, this particular orbit in unbounded). The islands associated with this stable period-4 orbit are seen to be exceedingly small and two of these islands lie on the S_2 symmetry line.

Exercise 18.36

Show that the fixed points, $\mathbf{x}_f = (a, 0)$, of $\mathbf{W}^2$ that lie on the S_1-symmetry line are given by the roots of

$$f(a) = f(2f(f(a)) - a).$$

For the case $f(x) = x(c + x)$, by factoring out the fixed points of $\mathbf{W}$, show that the period-2 orbits are given by the real roots of $g(a) = 0$, where

$$g(a) = \left(a^2 + a(c+1) + c + 1\right)\left(2a^2(a+c)^2 + 2ca(a+c) + c\right).$$

Hence show that the equation $g(a) = 0$ has double roots when $c = -1, 0, 2, 3, 1 \pm \sqrt{5}$.

Exercise 18.37

Plot the graph of $G_1(a)$, where $\mathbf{T}^4(a, 0) = (F_1(a), G_1(a))$ and $f(x) = x(x + c)$, in the range $0.4 < a < 0.8$ for various values of c in the interval $-1.265 \le c \le -1.255$, and show that the stable period-4 orbit becomes unstable at $c \simeq -1.262$, at which point a period-8 orbit is born.

In the case $c = -1.265$ find a point on the abscissa through which this orbit passes and plot the iterates of a neighbouring stable orbit. Note that the island surrounding this periodic orbit is very small and can be seen only by limiting the view to a small area around one of the islands.

Exercise 18.38

Consider the map $\mathbf{W}$ defined in equation 18.47, with $f(x) = x(x - 0.9)$ and the symmetry line S_1, $\mathbf{u} = (a, 0)$. Plot the graph of $G(a)$, where $\mathbf{W}^6(a, 0) = (F(a), G(a))$ and show that $G(a) = 0$ at $a = a_1 = -0.170\,886\ldots$ and that $\mathbf{u} = (a_1, 0)$ lies on a stable 12-period orbit.

Using a similar method with $\mathbf{W}^5(a, 0)$ show that $\mathbf{u} = (a_2, 0)$, with $a_2 = -0.260\,75$, lies on a 5-period orbit.

18.4 The Poincaré index for maps

The results shown in figures 18.28–18.31 show that the fixed points of a map can bifurcate when a parameter changes through critical values. If there is a symmetry and the position of a fixed point is determined by the solution of an equation in one variable, figures 18.28 and 18.30 suggest that the fixed points are normally born in pairs. The theory presented in this section shows that this feature is quite general and also put constraints on the types of bifurcation that are possible.

The Poincaré index was introduced in chapter 16 and applied to solutions of differential equations. The same notion can be applied to a map $\mathbf{T}(\mathbf{x})$ by considering the difference $\mathbf{v}(\mathbf{x}) = \mathbf{T}(\mathbf{x}) - \mathbf{x}$, which defines a vector everywhere except at the fixed points where $\mathbf{v}(\mathbf{x}) = 0$. The index of a closed curve Γ, that passes through no fixed point of $\mathbf{T}$, is defined as the number of times the vector $\mathbf{v}(\mathbf{x})$ encircles the origin as $\mathbf{x}$ traverses Γ. This number is defined to be positive if the encirclement and transversal are in the same direction: otherwise it is negative.

The properties of the Poincaré index for a map are the same as for the vector field of a differential equation or, indeed, any vector field, and we list them below; more detail is given in section 16.3.7.

(i) If Γ_1 is a closed curve obtained from another closed curve Γ by a continuous transformation such that no fixed points cross the boundary then $I_\Gamma = I_{\Gamma_1}$.

(ii) The index of a closed curve containing no fixed points is zero.

(iii) The index of an elliptic fixed point is $+1$.

(iv) The index of a hyperbolic fixed point is -1.

(v) The index of hyperbolic fixed point with reflection is $+1$.

(vi) The index is additive, equation 16.40 (page 666).

Exercise 18.39

Find the index of the following linear maps:

(i) $\mathbf{T} = (-y, x)$, (elliptic fixed point);
(ii) $\mathbf{T} = (x, -y)$, (hyperbolic fixed point);
(iii) $\mathbf{T} = (-x, -y)$, (hyperbolic fixed point with reflection).

An example of the creation of a stable period-2 orbit is given in exercise 18.24 (page 749), where we consider the map $\mathbf{T}(x, y) = (2cx - y + 2x^2, x)$ having a fixed point at the origin. Consider a closed curve, $\mathscr{C}$, enclosing the origin and all the relevant fixed points of $\mathbf{T}$ and $\mathbf{T}^2$. For $|c| < 1$ the fixed point is elliptic with residue $R = \frac{1}{2}(1 - c)$. As c decreases through -1, R increases through $+1$, the fixed point of $\mathbf{T}$ becomes hyperbolic with reflection and the index is unchanged, consistent with there being no new fixed point of $\mathbf{T}$.

The index of $\mathbf{T}^2$ must also be constant, but the index of its fixed point at the origin changes from $+1$ to -1. One way the value of the index of $\mathscr{C}$ can be preserved is the creation of two elliptic fixed points of $\mathbf{T}^2$, each with index $+1$. Note also that the residue of $\mathbf{T}^2$ is $R_2 = 4R(1 - R) = 1 - c^2$, exercise 18.22, and this becomes negative as $|c|$ increases through 1.

In general if the residue of a fixed point of $\mathbf{T}^n$ increases through unity the index of the fixed point does not change, but that of $\mathbf{T}^{2n}$ changes from $+1$ to -1. One way the balance can be redressed is with the birth of two $2n$-periodic orbits.

Other changes are possible. For instance, the map defined in equation 18.36 (page 748) has, for $c < 1$, an elliptic fixed point (index $+1$) at the origin and a hyperbolic fixed point (index -1) at $(1 - c, 1 - c)$. As c increases through unity these change roles and the index remains constant.

If the residue decreases though zero the index of the fixed point changes from $+1$ to -1, and two possible ways of maintaining the value of the index are:

(i) the birth of two elliptic fixed points; or
(ii) the absorption of two hyperbolic fixed points.

These processes are summarised in the following bifurcation diagram.

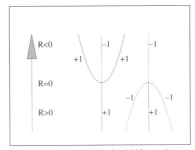

Figure 18.32 A typical bifurcation diagram with index changing between $+1$ and -1.

Other configurations are possible, Greene *et al.* (1981), but less common.

18.5 Period doubling

In the previous section we found examples of periodic orbits becoming unstable with
a stable periodic orbit of twice the period being born as a parameter passed through a
critical value. The same behaviour was seen in one-dimensional maps and it was noted
that the value of the parameter b_k where a period-2^k orbit is born is approximately
$b_k = b_\infty + A/\delta^k$, section 18.2.2, and that the number δ is the same for all quadratic maps.
Area preserving maps behave in exactly the same way: we show this by considering an
approximation to a particular example, though the phenomenon is universal.

We shall use the map defined in equation 18.49 (page 759), but before embarking
upon the analysis it would be helpful to do the following exercise, which derives some
results used later.

Exercise 18.40
Consider the area-preserving map

$$\mathbf{T}: \quad x_{n+1} = -y_n + 2f(x_n), \quad y_{n+1} = x_n. \tag{18.51}$$

(i) Show that

$$\mathbf{T}^2: \quad x_{n+2} = -x_n + 2f(f(x_n) - y_n), \quad y_{n+2} = 2f(x_n) - y_n \tag{18.52}$$

and deduce that the fixed points of $\mathbf{T}^2$ are given by the solutions of the equation

$$x = f(f(x)) \quad \text{with} \quad y = f(x).$$

(ii) In the case $f(x) = x(c + x)$, show that with the fixed points of $\mathbf{T}$ factored out these
equations become

$$x^2 + (c + 1)x + (c + 1) = 0 \quad \text{and} \quad y = -x - c - 1. \tag{18.53}$$

(iii) Deduce that fixed points of $\mathbf{T}^2$ exist for $c > 3$ or $c < -1$ and are given by

$$\mathbf{x}_1 = \begin{pmatrix} f_1 \\ f_2 \end{pmatrix}, \ \mathbf{x}_2 = \begin{pmatrix} f_2 \\ f_1 \end{pmatrix}, \ \text{where } f_k = -\frac{1}{2}\left\{c + 1 + (-1)^k \Delta\right\} \tag{18.54}$$

and where $\Delta = \sqrt{(c + 1)(c - 3)}$.

(iv) Show that the linearisation of $\mathbf{T}^2$ about (f_1, f_2) is

$$\begin{pmatrix} x_{n+2} \\ y_{y+2} \end{pmatrix} = L_2 \begin{pmatrix} x_n \\ y_y \end{pmatrix}, \quad \text{where} \quad L_2 = \begin{pmatrix} 15 + 8c - 4c^2 & 2(1 + \Delta) \\ 2(\Delta - 1) & -1 \end{pmatrix},$$

and deduce that the period-2 orbit becomes unstable as c decreases through $1 - \sqrt{5}$.

The map 18.51 can be written in the form

$$x_{n+1} + x_{n-1} = 2cx_n + 2x_n^2, \tag{18.55}$$

which is more convenient for this analysis. Now expand about the period-2 orbit by
setting $x_n = g_n + z_n$, where $g_{2k} = f_2$ and $g_{2k+1} = f_1$, to obtain

$$z_{n+1} + z_{n-1} = 2(c + 2g_n)z_n + 2z_n^2. \tag{18.56}$$

Replacing n with $2m+1$ and $2m-1$ gives two equations that can be added to give

$$z_{2m+2} + z_{2m-2} = 2(c + 2f_1)(z_{2m+1} + z_{2m-1}) - 2z_{2m} + 2\left(z_{2m+1}^2 + z_{2m-1}^2\right). \qquad (18.57)$$

But from equation 18.56,

$$z_{2m+1} + z_{2m-1} = 2(c + 2f_2)z_{2m} + 2z_{2m}^2,$$

and hence

$$
\begin{aligned}
z_{2m+2} + z_{2m-2} &= 2\left\{2(c + 2f_1)(c + 2f_2) - 1\right\}z_{2m} + 4(c + 2f_1)z_{2m}^2 \\
&\quad + 2\left(z_{2m-1}^2 + z_{2m+1}^2\right).
\end{aligned}
\qquad (18.58)
$$

Apart from the last term this is similar to the original map, equation 18.56, hence it can be used to approximate the fixed points of $\mathbf{T}^2$ and the value of c at which they become unstable. Moreover, it is clear that if we ignore the last term, it would be possible, by rescaling z, to convert this map to the same form as the original map and then we could repeat the process to form a quadratic map connecting z_{4m-4}, z_{4m} and z_{4m+4} and so on.

Because z_{2m} is small we may approximate the sum $z_{2m-1}^2 + z_{2m+1}^2$ by a term proportional to the square of the central term, z_{2m}^2, by writing

$$z_{2m-1}^2 + z_{2m+1}^2 = \beta z_{2m}^2 + O(z_{2m}^3)$$

for some positive constant β, the value of which turns out to be unimportant. With this approximation equation 18.58 becomes

$$
\begin{aligned}
z_{2m+2} + z_{2m-2} &= 2c_1 z_{2m} + 2z_{2m}^2(2c + 4f_1 + \beta) + O(z_{2m}^3), \\
c_1 &= 2(c + 2f_1)(c + 2f_2) - 1.
\end{aligned}
$$

Because z is small the third-order term may be ignored. Now rescale this equation by defining

$$w_{2m} = \alpha z_{2m} \quad \text{with} \quad \alpha = 2c + 4f_1 + \beta$$

to give

$$w_{2m+2} + w_{2m-2} = 2c_1 w_{2m} + 2w_{2m}^2, \qquad (18.59)$$

which is identical to the original map, equation 18.56, except that c has been replaced by c_1 and n by $2m$.

Exercise 18.41

Using relation 18.54 for f_n show that

$$c_1 = 7 + 4c - 2c^2. \qquad (18.60)$$

The rest is now clear: treat the new map, equation 18.59, in precisely the same way as before, to obtain a succession of similar maps the fixed points of which are the periodic orbits with period 2^n.

For instance, the map 18.59 bifurcates when $c_1 = -1$; the stable period-2 orbit

Discrete Dynamical Systems

becomes unstable and a period-4 orbit is born. When $c_1 = -1$ the value of c is given by the negative root of 18.60, that is

$$2c^2 - 4c - 8 = 0 \quad \text{or} \quad c = 1 - \sqrt{5}.$$

This is the value obtained in exercise 18.40 for the value of c at which the stable fixed point of $\mathbf{T}^2$ becomes unstable. Here we have shown that this change coincides with the birth of a stable fixed point of $\mathbf{T}^4$.

But on repeating the above analysis with the map 18.59 we see that its stable fixed points become unstable and a period-8 orbit is born when $c_1 = 1 - \sqrt{5}$, that is, when c satisfies the equation

$$-2c^2 + 4c + 7 = 1 - \sqrt{5} \quad \text{or} \quad c = 1 - \sqrt{4 + \frac{1}{2}\sqrt{5}} = -1.2623\ldots.$$

In general, at the kth stage of this process we will obtain a map

$$z_{2m+2}^{(k)} + z_{2m-2}^{(k)} = 2c_k z_{2m}^{(k)} + 2(z_{2m}^{(k)})^2, \tag{18.61}$$

where $z_{2m}^{(0)} = z_{2m}$, $z_{2m}^{(1)} = w_{2m}$ and the constants c_k and c_{k-1} are connected by

$$c_k = 7 + 4c_{k-1} - 2c_{k-1}^2, \quad c_0 = c. \tag{18.62}$$

All the maps of equation 18.61, for $k = 0, 1, 2, \ldots$, are the same, so all bifurcate at $c_k = -1$. By setting $c_N = -1$ and iterating equation 18.62 backwards we can obtain the value of c_0, which we denote by C_N, at which the period-2^N orbit becomes unstable and a period-2^{N+1} orbit is born. Since we are interested in the value of C_N it is more convenient to rewrite equation 18.62 as:

$$c_{k-1} = 1 - \sqrt{\frac{1}{2}(9 - c_k)}. \tag{18.63}$$

Some values are shown in table 18.2.

Table 18.2. *Approximate value of C_N at which the period-2^N orbit becomes unstable, obtained by setting $c_N = -1$ and using 18.63 to find $c_0 = C_N$.*

N	Doubling sequence	Value of C_N	Numerical value
0	$1 \to 2$	-1	-1
1	$2 \to 2^2$	$1 - \sqrt{5}$	$-1.236\,068$
2	$2^2 \to 2^3$	$1 - \sqrt{4 + \frac{1}{2}\sqrt{5}}$	$-1.262\,307$
3	$2^3 \to 2^4$	$1 - \sqrt{4 + \frac{1}{2}\sqrt{4 + \frac{1}{2}\sqrt{5}}}$	$-1.265\,205$
4	$2^4 \to 2^5$	$-$	$-1.265\,525$
5	$2^5 \to 2^6$	$-$	$-1.265\,560$

This iterative process converges to C_∞, that is, the limiting value of c, which is given by the root of

$$C_\infty = 7 + 4C_\infty - 2C_\infty^2 \quad \text{or} \quad C_\infty = \frac{3 - \sqrt{65}}{4} = -1.265\,564. \tag{18.64}$$

Numerical computations have been performed on the original map 18.51, Greene *et al.* (1981), see also Rasband (1990, chapter 7), and these give the more exact value $C_\infty = -1.266\,311$, the differences being due to the approximation used to derive equation 18.61.

Finally, you will notice from the table that the convergence to C_∞ is very rapid; the numerical values suggest that

$$C_k \simeq C_\infty + A\delta^{-k} \tag{18.65}$$

for some constants A and δ. The exact value of δ, obtained numerically, is $8.721\,097\ldots$, and its magnitude explains why the convergence to the limit is so rapid. The value of this parameter is the same for all such area-preserving maps.

Exercise 18.42
By putting $c_k = C_\infty + d_k$ in equation 18.63 show that

$$d_{k-1} = \frac{d_k}{2\sqrt{2(9 - C_\infty)}},$$

and deduce that for large N,

$$C_N \simeq C_\infty + \frac{A}{\delta^N}, \quad \delta = 2\sqrt{2(9 - C_\infty)} \simeq 9.06,$$

for some constant A.

18.6 Stable and unstable manifolds

In this section we analyse the motion in the vicinity of hyperbolic fixed points, which is important because, for the first time, it shows how the motion can become extraordinarily complicated and how chaos develops.

In the vicinity of a hyperbolic fixed point at $\mathbf{x} = \mathbf{x}_f$ the map can be expressed in the form

$$\mathbf{z}_{n+1} = L(\mathbf{x}_f)\mathbf{z}_n + \mathbf{g}(\mathbf{z}_n), \quad \mathbf{z}_n = \mathbf{x}_n - \mathbf{x}_f, \tag{18.66}$$

where L is the linearised matrix, equation 18.31 (page 747), and $\mathbf{g}(\mathbf{z}) = O(\mathbf{z}^2)$ represents the nonlinear terms. Assume that the eigenvalues of L are positive and $0 < \lambda_1 < 1 < \lambda_2$ with eigenvectors $\mathbf{e}_1$ and $\mathbf{e}_2$, so these vectors define two straight lines along which orbits of the linearised system move towards $\mathbf{x}_f$ along $\mathbf{e}_1$ or away from $\mathbf{x}_f$ along $\mathbf{e}_2$.

Consider the iterates of the short, straight line segment $\mathbf{x} = \mathbf{x}_f + \delta\mathbf{e}_2$, $0 < \delta \ll 1$, originating at $\mathbf{x}_f$ and pointing in the direction $\mathbf{e}_2$. Applying $\mathbf{T}^n$ to this line will produce another line, also starting at $\mathbf{x}_f$. For small n we may replace $\mathbf{T}$ by the linear map L to obtain the straight line $\mathbf{x} = \mathbf{x}_f + \delta\lambda_2^n\mathbf{e}_2$, with a length increasing exponentially with n. Similarly, the backward iterate of the line $\mathbf{x} = \mathbf{x}_f + \delta\mathbf{e}_1$ gives another line, which for small n is the straight line $\mathbf{x} = \mathbf{x}_f + \delta\lambda_1^{-n}\mathbf{e}_1$.

This analysis ignores all but first-order terms, but is a useful starting point: we proceed further by defining the local *stable* and *unstable manifolds*, which are curves tangent to $\mathbf{e}_1$ and $\mathbf{e}_2$, respectively, at $\mathbf{x} = \mathbf{x}_f$.

Let U be a neighbourhood of $\mathbf{x}_f$: the *local stable manifold* $W^s_{loc}(\mathbf{x}_f)$ is the set of all points in U which when iterated forwards in time to the infinite future reach $\mathbf{x}_f$, without leaving U. Formally,

$$W_{\text{loc}}^{\text{s}}(\mathbf{x}_f) \quad = \quad \{\mathbf{x} \in U \mid \mathbf{T}^n(\mathbf{x}) \to \mathbf{x}_f \quad \text{as} \quad n \to \infty,$$
$$\text{and} \quad \mathbf{T}^n(\mathbf{x}) \in U \quad \text{for all} \quad n \geq 0 \}. \qquad (18.67)$$

Similarly, the *local unstable manifold* $W_{\text{loc}}^{\text{u}}(\mathbf{x}_f)$ is defined as the set of all points in U which, when iterated backwards in time to the infinite past reach $\mathbf{x}_f$, without leaving U. Formally,

$$W_{\text{loc}}^{\text{u}}(\mathbf{x}_f) \quad = \quad \{\mathbf{x} \in U \mid \mathbf{T}^n(\mathbf{x}) \to \mathbf{x}_f \quad \text{as} \quad n \to -\infty,$$
$$\text{and} \quad \mathbf{T}^n(\mathbf{x}) \in U \quad \text{for all} \quad n \leq 0 \}. \qquad (18.68)$$

For sufficiently small neighbourhoods U these manifolds† are close to the straight lines through the fixed point in the direction of the eigenvectors of the linearisation matrix $L(\mathbf{x}_f)$, as illustrated in figure 18.33.

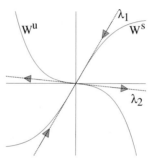

Figure 18.33   Diagram showing the local stable
and unstable manifolds; the straight lines denote
the directions of the eigenvectors $\mathbf{e}_1$ and $\mathbf{e}_2$.

These local definitions are extended into the whole phase space by defining the *stable manifold* and the *unstable manifold*; the extension is straightforward and mimics exactly the way one constructs these manifolds numerically on a computer. First, consider the stable manifold: to define it, take a point on the local stable manifold $W_{\text{loc}}^{\text{s}}(\mathbf{x}_f)$ and iterate it *backwards* for ever, that is, away from the fixed point; do the same with *all* points on $W_{\text{loc}}^{\text{s}}(\mathbf{x}_f)$ to form a curve, named the stable manifold $W^{\text{s}}(\mathbf{x}_f)$. In other words, the stable manifold is formed by iterating the whole local stable manifold backwards for ever. Formally,

$$W^{\text{s}}(\mathbf{x}_f) = \bigcup_{n \geq 1} \mathbf{T}^{-n} \{W_{\text{loc}}^{\text{s}}(\mathbf{x}_f)\}. \qquad (18.69)$$

Similarly, for the unstable manifold take a point on the local unstable manifold $W_{\text{loc}}^{\text{u}}(\mathbf{x}_f)$

---

† The term 'manifold' simply means a curve and is taken from the higher-dimensional case where the corresponding objects are not curves.

and iterate it *forwards* for ever, again away from the fixed point; do the same with all points on $W_{\text{loc}}^{\text{u}}(\mathbf{x}_{\text{f}})$ to form a curve, named the unstable manifold $W^{\text{u}}(\mathbf{x}_{\text{f}})$. Formally,

$$W^{\text{u}}(\mathbf{x}_{\text{f}}) = \bigcup_{n \geq 1} \mathbf{T}^n \{W_{\text{loc}}^{\text{u}}(\mathbf{x}_{\text{f}})\}. \tag{18.70}$$

Every point on the stable manifold reaches the fixed point asymptotically under forward iteration, and every point of the unstable manifold reaches the fixed point asymptotically under backward iteration. Moreover, since $W^{\text{s}}(\mathbf{x}_{\text{f}})$ and $W^{\text{u}}(\mathbf{x}_{\text{f}})$ are defined by repeated iteration of the map $\mathbf{T}$, if a point $\mathbf{x}$ lies on either of them so must $\mathbf{T}^n(\mathbf{x})$, $n = \pm 1, \pm 2, \dots$.

In order to make these definitions clear we shall use them to construct a picture of an unstable manifold of the area-preserving map

$$\mathbf{T} : \begin{cases} y_{n+1} &= y_n + 2cx_n(x_n - 1) \\ x_{n+1} &= x_n + y_{n+1}, \end{cases} \tag{18.71}$$

which has an elliptic fixed point at the origin if $0 < c < 4$ and a hyperbolic fixed point at $(1,0)$ if $c > 0$ or $c < -4$. The linearisation matrix at $(1,0)$ is

$$L = \begin{pmatrix} 1 + 2c & 1 \\ 2c & 1 \end{pmatrix}. \tag{18.72}$$

Assuming $c \geq 0$, the eigenvalues and eigenvectors are

$$\begin{aligned}
\lambda_1 &= 1 + c - \sqrt{2c + c^2} < 1, & \mathbf{e}_1^T &= (c - \sqrt{2c + c^2}, 2c), \\
\lambda_2 &= 1 + c + \sqrt{2c + c^2} > 1, & \mathbf{e}_2^T &= (c + \sqrt{2c + c^2}, 2c).
\end{aligned}$$

The unstable manifold will be tangential to $\mathbf{e}_2$ at $(1,0)$, so to draw W^{u} we take a large number of initial points,

$$\mathbf{a}_k = \mathbf{x}_{\text{f}} + \delta_k \mathbf{e}_2, \quad k = 1, 2, \dots, M, \quad \mathbf{x}_{\text{f}} = (1,0),$$

where δ_k are small numbers, negative in this case, and iterate each of these N times. The set of $N \times M$ points then lies close to W^{u} and if $|\delta_k|$ are sufficiently small and N not too large these points will provide an approximation to the line W^{u}. In practice it is necessary to chose $|\delta_k|$ so small that there is no need for the first few iterations to be plotted. Figure 18.34 shows part of the unstable manifold for $c = \frac{3}{4}$.

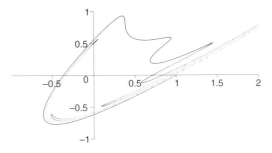

Figure 18.34   An unstable manifold of 18.71 for $c = \frac{3}{4}$.

This unstable manifold emanates from the hyperbolic fixed point at $(1, 0)$ in a south-westerly direction, curving round to approach the point $(0.5, 1.0)$; after this the curve starts oscillating wildly but, on average, heads back towards the fixed point. You will observe that the computer-generated curve becomes ragged towards its end; this is an example of the exponential dependence on initial conditions and is due to the use of a finite number of points.

Figure 18.34 shows the eighth to fifteenth iterates of 800 initial conditions starting on the $\mathbf{e}_2$ vector with $\delta_k = -10^{-7}k$, and was drawn using the following Maple commands. The first few lines set up the initial parameters: here M is the number of initial conditions and Ns the number of iterates performed before plotting graphs.

```
>  ticks:=xtickmarks=[seq(k/2,k=-1..4)], ytickmarks=[seq(k/2,k=-2..2)]:
>  conds:=style=point,symbol=point,view=[-1..2,-1..1],colour=black,ticks:
>  c:=0.75:
>  c1:=(c+sqrt(2*c+c^2)):
>  M:=800: Ns:=8:
>  sg:=array(1..M):
```

Now iterate all M initial points Ns times: note the use of evalhf which speeds up the computation.

```
>  for k from 1 to M do;                    # Loop round initial conditions
>  dk:=-1.0e-07*k:
>  x:= 1+ dk*c1; y:=2*c*dk;                 # Define initial conditions
>      for j from 1 to Ns do;                   # Perform Ns iterations
>      y:=evalhf(y+2*c*x*(x-1)); x:=evalhf(x + y); od:          #3
>  sg[k]:=[x,y]:
>  od:
```

Finally iterate each of these points in the array sg eight times, plotting a graph after each iteration. It is interesting to note that over 95% of the computational time† is taken in plotting the graph.

```
>  for n from 1  to 8 do;                   # Loop round next 8 iterations
>      for k from 1 to M do;                # Loop round initial conditions
>      x:=sg[k][1]: y:=sg[k][2]:
>      y:=evalhf(y+2*c*x*(x-1)); x:=evalhf(x + y);                 #4
>      sg[k]:=[x,y]; od:                          # update array
>  pg.n:=plot(convert(sg,list),conds):           # Create graph
>  od:
```

After running this set of commands all the graphs can be displayed together, as in figure 18.34, with the command display(seq(pg.j,j=1..8)); but it is also instructive to view the graphs as an animated sequence with the command

```
display([seq(pg.j,j=1..8)],insequence=true);
```

which shows how the initial segment moves and expands as the number of iterations increases.

In figure 18.35 below we show the stable manifold, of the same fixed point, drawn in the same way; the backward iteration is performed by rearranging equation 18.71 to express $\mathbf{x}_n$ in terms of $\mathbf{x}_{n+1}$:

$$x_n = x_{n+1} - y_{n+1}, \quad y_n = y_{n+1} - 2cx_n(x_n - 1),$$

---

† There is a bug in both Maple 5 and 6 that makes the time to plot N points increase more rapidly than N, which is particularly annoying for $N > 500$.

that is, by replacing lines #3 and #4 by

```
>  x:=evalhf(x-y): y:=evalhf(y - 2*c*x*(x-1));
```

In addition it is necessary to redefine c1 as c1:=c-sqrt(2*c+c^2): and to change the sign of dk.

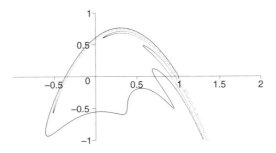

Figure 18.35   A stable manifold of 18.71 for $c = \frac{3}{4}$.

The fact that W^s can be transformed into W^u by an anticlockwise 90° rotation and a reflection in the line $x = y$ is a property of the particular map chosen. The backwards iterates along W^s leave the hyperbolic fixed point in a northwesterly direction. On superimposing the stable and unstable manifolds, by eye, we see that they intersect many times, even for the relatively few iterations shown here. We emphasise that in order to draw these curves it is necessary to iterate only a few times; with more iterations the points become too spread for the continuous curve to be distinguished.

This definition of the stable and unstable manifolds assumed only that the eigenvalues of the linearisation matrix satisfy $0 < \lambda_1 < 1 < \lambda_2$, and did not assume that the map was area-preserving.

Exercise 18.43
The method used to construct the pictures 18.34 and 18.35 of the unstable and stable manifolds is the conventional method, but with Maple there is an alternative that provides clearer figures if the number of iterations is not too large.

Write a procedure to produce a symbolic expression for $\mathbf{T}^n(\mathbf{x})$ for particular values of n and where $\mathbf{T}$ is defined in equation 18.71. For $c = \frac{3}{4}$ and for $n = 4, 5, \ldots, 10$ substitute $\mathbf{x} = \mathbf{x}_f + \delta\mathbf{e}_2$ in this expression to form the two functions $(u_n(\delta), v_n(\delta))$ and plot the curve defined parametrically by $x = u_n(\delta)$, $y = v_n(\delta)$ for δ small and negative.

Use this technique to investigate the shape of the unstable manifold for different values of c, in the range $0 < c < 2$, using $\mathbf{T}^8(\mathbf{x})$.

Exercise 18.44
Show that the map

$$\mathbf{T} : x_{n+1} = -y_n + 2x_n^2, \quad y_{n+1} = x_n$$

is area-preserving and has a stable fixed point at the origin and an unstable fixed point at

$x = y = 1$, and that the normalised eigenvectors of the linearised map at $x = y = 1$ are

$$\mathbf{u}_{\pm} = \frac{1}{2}\begin{pmatrix} \sqrt{\lambda_{\pm}} \\ \sqrt{\lambda_{\mp}} \end{pmatrix}, \quad \lambda_{\pm} = 2 \pm \sqrt{3}.$$

Use the method described in the previous exercise to plot the unstable manifolds passing through $(1, 1)$. Use the inverse map,

$$\mathbf{T}^{-1} : x_n = y_{n+1}, \quad y_n = -x_{n+1} + 2y_{n+1}^2,$$

to plot the stable manifold. In addition plot iterates of some orbits starting near the origin to see how stable and unstable manifolds coexist with better behaved orbits..

18.6.1 Homoclinic points and the homoclinic tangle

The examples of unstable and stable manifolds shown in figures 18.34 and 18.35 show these lines intersecting many times and that at each crossing W^u and W^s are not tangential: such crossings are said to be transversal. This behaviour is common and we now consider its implications and in particular shall see how these crossings cause chaotic motion.

If a point $\mathbf{r}$ is on both W^s *and* W^u it follows that all the points $\mathbf{r}_n = \mathbf{T}^n(\mathbf{r})$, $n = \pm 1, \pm 2, \ldots$, are on *both* curves; these points are distinct, so if the curves cross just once they must cross infinitely many times.

Consider the points $\mathbf{r}_n = \mathbf{T}^n(\mathbf{r})$ on the stable manifold $W^s(\mathbf{x}_f)$; for sufficiently large n these will be close to $\mathbf{x}_f$ and the distance between successive points will decrease geometrically. For example, sufficiently near $\mathbf{x}_f$ the iterates will satisfy

$$\xi_{n+1} = \lambda_1 \xi_n, \quad \lambda_1 < 1,$$

where ξ is the coordinate along the stable eigenvector of $L(\mathbf{x}_f)$; so $\xi_{n+N} = \lambda_1^N \xi_n$.

At each point $\mathbf{r}_n$ the curves $W^s(\mathbf{x}_f)$ and $W^u(\mathbf{x}_f)$ intersect transversally, so near $\mathbf{x}_f$ the curve $W^u(\mathbf{x}_f)$ will have the form shown below.

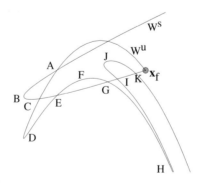

Figure 18.36

Since the map is area-preserving the action of **T** on a region preserves its orientation, so the region ABC in figure 18.36 must be mapped onto region EFG or another similar region, but not regions similar to CDE; then the region EFG will be mapped onto another region IJK, and so on. Since **T** is area-preserving the areas of the loops ABC, EFG and IJK, etc. are the same, but since the distances along $W^s(\mathbf{x}_f)$, EG, IK, etc. are decreasing, the lengths of the loops EFG and IJK must be increasing in the same proportion.

If we consider the repeated images of two neighbouring points, such as those at A and B, then it is clear that neighbouring orbits must separate exponentially with the number of iterations.

Similarly for the iterates $\mathbf{T}^{-n}(\mathbf{r})$, which approach the fixed point in the distant past, the same construction shows that the stable manifold, $W^s(\mathbf{x}_f)$, must oscillate wildly about the unstable manifold in a similar manner. It can be shown that the loops of $W^s(\mathbf{x}_f)$ and $W^u(\mathbf{x}_f)$ must cross each other near $\mathbf{x}_f$, to produce a complicated tangle, named the *homoclinic tangle*. Drawing such an infinitely tangled web is clearly quite impossible.

Some idea of the complexity of the motion can be gleaned by doing the following exercise in which Maple is used to construct segments of the homoclinic tangle

Exercise 18.45

Homoclinic tangle:
Use the Maple commands listed on page 770 to draw the unstable manifold of the map defined in equation 18.71, for $c = \frac{3}{4}$, using up to $N = 2000$ points and between $n = 10$ and 25 iterations. View these all together and as an animated sequence, to see how the tangle develops.
Note: with this many points Maple takes a long time to draw the graphs.

In figures 18.37 and 18.38 are shown two of the graphs obtained in the previous exercise, for $n = 18$ and 22 iterations.

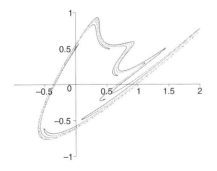

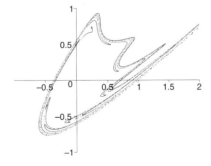

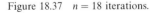

Figure 18.37 $n = 18$ iterations.   Figure 18.38 $n = 22$ iterations.

The case of 15 iterations is shown in figure 18.34; comparing this with figure 18.37 and this with 18.38 we see that additional iterations simply add extra homoclinic points and longer loops. Since W^u cannot intersect itself this loop is forced round the inside of the outer boundary. In some cases these extra loops are so long they encircle the stable fixed point at the origin.

More iterations add little at this resolution, but we can imagine some of the complications: the extra intersections approach the hyperbolic point rapidly, thus as the number of tendrils increases so does their length, but the total area enclosed remains finite. Since there is a stable fixed point at the origin there is a region of stable invariant curves surrounding the origin, so all the tendrils are confined to the annulus between these and the outer boundary.

Starting with two neighbouring orbits near either $W^s(\mathbf{x}_f)$ or $W^u(\mathbf{x}_f)$, then no matter how close they are initially the above discussion shows that for the first few iterations — just how many depends upon the size of λ_2 — the orbits separate as λ_2^n, that is, exponentially with n, and after many iterations they will no longer be close enough to be considered neighbouring.

18.7 Lyapunov exponents

The Lyapunov exponent, λ, of a one-dimensional map provides a measure of the mean rate at which adjacent orbits separate: if $\lambda < 0$ neighbouring orbits approach each other in the future and if $\lambda < 0$ they separate exponentially, at least for short times. For maps in two dimensions there are two Lyapunov exponents, so the situation is a little more complicated.

Consider orbits with nearly identical initial conditions $\mathbf{x}_0$ and $\mathbf{x}_0 + \boldsymbol{\delta}$. After n iterations of the map $\mathbf{T}$ the difference between the two orbits is

$$\boldsymbol{\delta}_n = \mathbf{T}^n(\mathbf{x}_0 + \boldsymbol{\delta}) - \mathbf{T}^n(\mathbf{x}_0).$$

For small $|\boldsymbol{\delta}|$ and n not too large this may be approximated by

$$\boldsymbol{\delta}_n = L_n(\mathbf{x}_0)\boldsymbol{\delta} = L(\mathbf{x}_{n-1})L(\mathbf{x}_{n-2})\cdots L(\mathbf{x}_0)\boldsymbol{\delta}.$$

If the map is area-preserving, $\det(L) = 1$, the eigenvalues of L_n are either real and typically have the form $\{n\mu, n/\mu\}$, for some real number μ, or are complex with the form $\exp(\pm in\theta)$, for some real θ.

Consider the effect of L_n on a small circle surrounding $\mathbf{x}_0$. If the eigenvalues of each $L(\mathbf{x}_k)$ are real (and independent of $\mathbf{x}_k$), this circle is transformed into an ellipse with the length of the major axis increasing exponentially with n and that of the minor axis decreasing exponentially with n. For large n the ellipse is essentially a straight line pointing in the direction of the eigenvector belonging to the numerically largest eigenvalue.

Alternatively, if the eigenvalues of each $L(\mathbf{x}_k)$ are complex and independent of $\mathbf{x}_k$ the action of L_n on $\boldsymbol{\delta}$ will be a rotation, through an angle proportional to n, plus a shear the magnitude of which is independent of n.

The length, δ_n, of $\boldsymbol{\delta}_n$ is given by

$$\delta_n^2 = \boldsymbol{\delta}_n^\top \boldsymbol{\delta}_n = \boldsymbol{\delta}^\top H_n(\mathbf{x}_0)\boldsymbol{\delta}, \quad H_n(\mathbf{x}_0) = L_n(\mathbf{x}_0)^\top L_n(\mathbf{x}_0),$$

where $A^\top$ is the transpose of A. The matrix H_n is real, symmetric and $\mathrm{Tr}(H_n) > 0$, so has real, positive eigenvalues, exercise 18.60. In addition, for area-preserving maps $\det(H_n) = 1$. Note that $L_n(\mathbf{x}_0)$ is a product of n matrices, but H_n cannot be decomposed in the same manner. The following two simple cases illustrate the most important types of behaviour.

First, if L represents a pure rotation, independent of $\mathbf{x}_0$, as near an elliptic fixed point, then

$$L = \begin{pmatrix} \cos\theta & \sin\theta \\ -\sin\theta & \cos\theta \end{pmatrix}, \quad \text{so} \quad L_n = \begin{pmatrix} \cos n\theta & \sin n\theta \\ -\sin n\theta & \cos n\theta \end{pmatrix} \quad (18.73)$$

and $H_n = I$, for all θ, and $\delta_n = |\boldsymbol{\delta}|$ for all n.

Second, linearisation near a hyperbolic fixed point leads to matrices of the form

$$L = \begin{pmatrix} \mu & 0 \\ 0 & \mu^{-1} \end{pmatrix}, \quad \text{so} \quad L^n = \begin{pmatrix} \mu^n & 0 \\ 0 & \mu^{-n} \end{pmatrix}, \quad \mu > 1, \quad (18.74)$$

and $H_n = L^{2n}$: if $\boldsymbol{\delta} = (\cos\phi, \sin\phi)$ we have $\delta_n^2 = \mu^{2n}\cos^2\phi + \mu^{-2n}\sin^2\phi$. For almost all values of ϕ and n large, δ^n increases exponentially with n.

In general, for an area-preserving map $H_n(\mathbf{x}_0)$ has real, positive eigenvalues $h_{n1} \geq h_{n2} = 1/h_{n1}$ and orthonormal eigenvectors $\mathbf{e}_{nj}$ that depend upon n. Setting $\boldsymbol{\delta} = \alpha\mathbf{e}_{n1} + \beta\mathbf{e}_{n2}$ then gives $\delta_n^2 = \alpha^2 h_{n1} + \beta^2 h_{n2}$. For unstable orbits we expect h_{n1} to behave as μ^{2n}, where $\mu > 1$, and so we define the two Lyapunov exponents by the limit

$$\lambda_k = \lim_{n \to \infty} \frac{1}{2n} \ln(h_{nk}), \quad \lambda_1 \geq \lambda_2. \quad (18.75)$$

It is assumed that the limits exist. Near a hyperbolic fixed point $\lambda_1 > 0$ and $\lambda_2 < 0$ and near an elliptic fixed point $\lambda_1 = \lambda_2 = 0$.

In the next two sets of figures are shown examples of the behaviour of $\ln(h_{n1})/2n$ for the map

$$\mathbf{W}: \quad x_{n+1} = -y_n + f(x_n), \quad y_{n+1} = x_n - f(x_{n+1}), \quad f(x) = x(x+c),$$

a transformed version of which was used to draw figures 18.22 to 18.27 (pages 749 to 753). Here we set $c = -0.2$ and investigate two of the orbits shown in figure 18.22. In this representation the linearised map is

$$L(\mathbf{x}) = \begin{pmatrix} c + 2x & -1 \\ 1 - c^2 + 2cy - (2c^2 + 2c - 4y)x - 6cx^2 - 4x^3 & c - 2y + 2cx + 2x^2 \end{pmatrix}.$$

At each point $\mathbf{x}_k$ of the orbit the matrix $L(\mathbf{x}_k)$ is found together with the product $L_n(\mathbf{x}_0)$ and the largest eigenvalue of $H_n(\mathbf{x}_0)$, an expression for which is given in exercise 18.60.

A representative stable orbit, with $\mathbf{x}_0 = (0.3, 0.27)$, is shown in figure 18.39 and the variation of $\ln(h_{n1})/2n$ in figure 18.40. In the latter figure the faint line is $\langle \ln(h_{n1}) \rangle / 2n$, where the mean is taken over the last few hundred iterations. These graphs suggest that $\lambda_1 = 0$ and that the orbit is stable, as expected. The approach to the limit is slow and care is needed when interpreting such data.

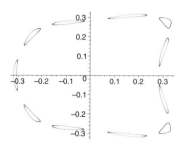

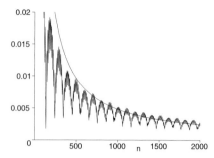

Figure 18.39   The orbit for $c = -0.2$,
$\mathbf{x}_0 = (0.3, 0.27)$.

Figure 18.40 $\ln(h_{1n})/2n$.

In the next example the initial point is $\mathbf{x}_0 = (0.375, 0.3116)$. The graph on the right and the lack of agreement with the faint line, constructed as before, suggests that for this orbit $\lambda_1 > 0$.

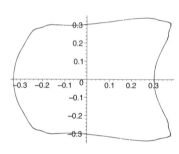

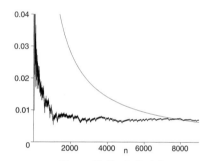

Figure 18.41   The orbit for $c = -0.2$,
$\mathbf{x}_0 = (0.375, 0.3116)$.

Figure 18.42 $\ln(h_{1n})/2n$.

18.8 An application to a periodically driven, nonlinear system

The theory and results presented above have important consequences for the solutions of many types of differential equations. We illustrate how using a particular example, the forced, nonlinear oscillator described by the Hamiltonian

$$H(y, x, t) = \frac{1}{2}y^2 + \frac{1}{2}x^2 - \frac{1}{3}x^3 + \epsilon x \cos \Omega t, \tag{18.76}$$

giving the equations of motion

$$\frac{dx}{dt} = y, \quad \frac{dy}{dt} = -x + x^2 - \epsilon \cos \Omega t. \tag{18.77}$$

The unperturbed limit, $\epsilon = 0$, was dealt with in chapter 17, and the perturbed system is considered in the next chapter. Here our aim is to show why the theory of this chapter applies to this type of system. The phase curves of the unperturbed motion are reproduced in figure 18.43.

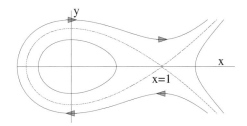

Figure 18.43   Representative phase curves of the unperturbed
system, $\epsilon = 0$.

If necessary the reader should refer to section 17.5 (page 691) for a discussion of this
unperturbed motion. For our purposes the most important feature in this figure is the
separatrix, the dashed curve through the saddle at $(1,0)$, which is the boundary between
the periodic and the unbounded motion. On this curve and near the saddle

$$x(t) \simeq 1 - (1 - x_0)e^{-t},$$

that is, the unperturbed motion is very slow in the neighbourhood of the saddle.

We need to understand the effect of the perturbation on the motion near the
separatrix, and there are two ostensibly conflicting observations, as follows.

(1) Because motion in the vicinity of the saddle is very slow the perturbation has a
long time to act and therefore we should expect it to have a significant effect.

(2) A naive first-order perturbation theory suggests that the periodic force will
induce small periodic oscillations on this slow motion, so we should expect the
mean effect of these oscillations to be negligible.

These two observations are not in conflict: the first observation is correct for phase
curves that are very close to the separatrix; the second is true for most other phase
curves. The theory of this chapter helps us understand the motion near the separatrix,
but first it is necessary to see the relation between the differential equation 18.77 and
an area-preserving map.

This Hamiltonian system is converted to an area-preserving map by the simple device
of viewing the actual motion $(x(t), y(t))$ using a stroboscope with period $T = 2\pi/\Omega$, so
we observe only the set of points

$$(x_n, y_n) = (x(t_0 + nT), y(t_0 + nT)), \quad n = 0, 1, \ldots, \quad T = \frac{2\pi}{\Omega},$$

where t_0 is some arbitrary initial time; because the value of t_0 is of no fundamental
significance it is normally set to zero. The point $\mathbf{x}_{n+1} = (x_{n+1}, y_{n+1})$ is determined
uniquely by the previous point $\mathbf{x}_n$ by solving the equations of motion 18.77 with the
initial conditions $x(nT) = x_n$, $y(nT) = y_n$. Since the equations of motion are invariant
under the translation $t \to t + T$ in time, we can write $\mathbf{x}_{n+1} = \mathbf{T}(\mathbf{x}_n)$, where the function
$\mathbf{T}(\mathbf{x})$ is independent of n. Similarly, we may integrate backwards in time to obtain $\mathbf{x}_n$
from $\mathbf{x}_{n+1}$; thus the inverse $\mathbf{T}^{-1}$ exists. Finally, we note that because the equations of

motion are Hamiltonian the map **T** is area-preserving; a proof is given in Percival and Richards (1982, chapter 4). Moreover, it is shown in section 19.4 that reflections in the *x*-axis are a symmetry of **T**.

The unstable fixed point of the unperturbed Hamiltonian system, at $(1,0)$, is also an unstable fixed point of the unperturbed map. The closed loop of the separatrix is both an unstable and a stable manifold of this hyperbolic fixed point, and in this conservative limit is special because these branches of W^s and W^u coincide everywhere, unlike those in figures 18.34 and 18.35.

Now consider the question of what happens to the separatrix under the action of the period force. There are three possibilities.

- First, it may be merely perturbed slightly, but otherwise remain similar to that shown in figure 18.43: this cannot happen unless the perturbed Hamiltonian is also conservative.
- Second, the stable and unstable manifolds may separate to become distinct, *non-intersecting* curves. This can occur, but only if a sufficiently large frictional term is added, in which case the system is no longer Hamiltonian. An example of this type of system is

$$\ddot{x} + x - x^2 = \epsilon \cos \Omega t - \delta \dot{x}, \quad \delta > 0. \tag{18.78}$$

- Third, the stable and unstable manifolds separate into two distinct *intersecting* curves: if this happens then the consequences are as described in the previous section, in particular figure 18.36. It may be proved, using Melnikov's method described in, for instance, Arrowsmith and Place (1990, chapter 3), that for any $|\epsilon| > 0$ the stable and unstable manifolds cross tranversally, so in the vicinity of the separatrix the perturbed ($\epsilon = 0$) and the unperturbed ($\epsilon \neq 0$) motion are qualitatively different. It can also be shown, though not with the same degree of rigor, that the area of phase space associated with the homoclinic tangle is small and typically $O(\exp(-1/\epsilon))$, see for instance Sagdeev *et al.* (1988, chapter 5) and Zaslavsky (1998, chapter 2).

  It is also interesting to note that the presence of damping, equation 18.78, inhibits the creation of the homoclinic tangle, which is delayed until the perturbation is sufficiently strong,

$$|\epsilon| > \delta \frac{\sinh \pi\Omega}{10\pi\Omega^2},$$

a result that may also be proved by an application of Melnikov's method.

This brief discussion shows that an arbitrarily small perturbation, $\epsilon x \cos \Omega t$ in this case, changes the motion near the separatrix qualitatively: globally it changes an integrable system to a non-integrable system for any $\epsilon > 0$. This type of effect is quite general and occurs in most dynamical systems, not just those that are periodically forced: this is why most systems are not integrable. It is also worth noting that in our example, provided ϵ is not too large, motion far from the sepatrix can be accurately described using perturbation theory, at least for short times; that is, there are regions of phase space in which the system behaves as if it were integrable. Again this type of behaviour is typical.

18.9 Exercises

Exercise 18.46
Show that the first-order iterative scheme,

$$x_{n+1} = cx_n^p, \quad p > 1, \quad x_0 > 0, \quad c > 0,$$

where c is a constant, has the solution

$$x_n = dx_0^{p^n}, \quad \ln d = \frac{p^n - 1}{p - 1} \ln c,$$

and hence that x_n converges to zero more rapidly than the exponential convergence in equation 18.2.

Show also that in the limit $p \to 1$ this solution becomes $x_n = c^n x_0$.

Exercise 18.47
Explain why the linear map

$$\mathbf{x}_n = \begin{pmatrix} \alpha & \beta \\ \gamma & \delta \end{pmatrix} \mathbf{x}_{n-1}, \quad \mathbf{x}_n = \begin{pmatrix} x_n \\ x_{n-1} \end{pmatrix}$$

can represent the two-term recurrence relation $x_{n+2} + 2bx_{n+1} + cx_n = 0$ only if $\alpha = -2b$, $\beta = -c$, $\gamma = 1$ and $\delta = 0$.

Exercise 18.48
Consider the map $x_{n+1} = a \ln x_n$, $a > 0$. Show that if $a < e$ there are no fixed points and if $a > e$ there are two fixed points. In the latter case, if the fixed points are $\xi_1 < \xi_2$, show that ξ_1 is unstable and ξ_2 is stable.

Exercise 18.49
Logistic map with $b = 1$:
Consider the map with $x_{n+1} = x_n - x_n^2$. By approximating the difference $x_{n+1} - x_n$ by the differential dx/dn show that an approximate solution is $x_n \simeq u_n = x_0/(1 + nx_0)$. Make a graphical comparison of x_n and u_n, where x_n are the exact iterates for various values of x_0.

Exercise 18.50
Logistic map with $b = 3$:
Consider the logistic map with $b = 3$ and make the expansion $x = x_f + y$ about the fixed point $x_f = 1 - 1/b$ to show that $y_{n+1} = -y_n - 3y_n^2$ and hence that $z_{n+1} = z_n - 18z_n^3 - 27z_n^4$, where $z_n = y_{2n}$. If $|z_n| \ll 1$ show that an approximate solution to this equation is $z_n = z_0(1 + 36z_0^2 n)^{-1/2}$. Deduce that $x_n \to x_f$ as $n^{-1/2}$ and hence show that $\lambda_L(3) = 0$.

Exercise 18.51
Show that the Schwarzian derivatives of the functions $b \sin \pi x$, $bx(1 - x^n)$ and $xe^{b(1-x)}$, where b and n are positive constants, are all negative.

Exercise 18.52
Newton's method
Show that the fixed points of the map

$$x_{n+1} = x_n - \frac{f(x_n)}{f'(x_n)}$$

are the roots of $f(x) = 0$, provided these are simple. Show also that these fixed points are superstable.

Examine the behaviour of this iterative scheme in the neighbourhood of a point $x = a$, where $f(a) = 0$ and $f^{(k)}(a) = 0$, $k = 1, 2, \ldots, N$ for some finite N.

Exercise 18.53

(i) Plot the curve in the (x, b) plane defined by the equation $g(x, b) = 0$, where $g(x, b) = (f^3(x) - x)/(f(x) - x)$, where $f(x) = bx(1 - x)$, for $0 < x < 1$ and $3 < b < 4$.
(ii) Solve the equations $g(x, b) = \partial g/\partial b = 0$ to show that a period-3 orbit appears when $b \simeq 3.828\,427\ldots$.
(iii) Show also that this periodic orbit remains stable for approximately $3.828\,427 \le b \le 3.484\,15$.

Exercise 18.54

Show that the map $x_{n+1} = -b \exp(x_n)$, $b > 0$, has a single stable fixed point $\xi_1(b)$ if $0 < b < e$ which becomes unstable as b increases through e. For small and large b show that

$$\xi_1(b) \quad = \quad -b + b^2 - \frac{3}{2}b^3 + O(b^4),$$

$$= \quad -\left(1 - \frac{\ln(\ln b)}{1 + \ln b}\right) \ln b, \quad b \gg 1.$$

Use the Schwarzian derivative, exercises 18.10 and 18.11 (pages 736, 737), to show that as b increases through e a stable 2-period orbit is born.

Investigate whether or not other bifurcations occur as b increases.

Exercise 18.55

Consider the map $x_{n+1} = F(x_n) = b \sin \pi x_n$, $0 < x_0 < 1$, $0 < b \le \pi$, and show that:

(i) for $\pi b \le 1$ the only fixed point is $x = 0$, which is stable;
(ii) if $\pi b > 1$ there is another fixed point x_1 that becomes unstable at $b = b_1$, where $\pi x_1 + \tan \pi x_1 = 0$, $\pi b_1 = -1/\cos \pi x_1$, that is, $b_1 = 0.719\,962$, $x_1 = 0.645\,774$;
(iii) that the stable 2-period orbit born at $b = b_1$ bifurcates at $b = b_2 \simeq 0.833\,266$;
(iv) there is a superstable fixed point of F at $x = b = \frac{1}{2}$ and a superstable 2-period orbit at $b = b_s = 0.777\,733\,8$ that cycles between $x = b_s$ and $x = \frac{1}{2}$.

Plot the graph of the Lyapunov exponent for $0 \le b \le 1$ in order to find the ranges of b where stable orbits exist.

Exercise 18.56
Bernoulli shift
The one-dimensional map

$$x_{n+1} = B(x_n) = \begin{cases} 2x_n, & 0 \le x_n < \dfrac{1}{2}, \\[2mm] 2x_n - 1, & \dfrac{1}{2} \le x_n < 1, \end{cases}$$

is named the Bernoulli shift after J. Bernoulli's study of random events with only two possible outcomes. The function $B(x)$ maps $[0, 1]$ onto $[0, 1]$ and has the following simple interpretation: express the number x as a binary number

$$x = \sum_{k=1}^{\infty} \frac{d_k}{2^k} = [d_1, d_2, \ldots],$$

where the d_k are either 0 or 1. Show that $B(x) = [d_2, d_3, \ldots]$. Use this result to prove the following results.

(i) If x_0 has binary representation of finite length N so $x_0 = M2^{-N}$ for some integer M, for example $x_0 = [1, 0, 1, 0, 1] = 21/2^5$, then $x_p = 0$ for $p \geq N$.
(ii) A binary number comprising a recurrent sequence of length N gives rise to an N-period orbit. Show that $x_0 = \frac{1}{3}$ produces a 2-period orbit and $x_0 = \frac{1}{7}$ a 3-period orbit.
(iii) Consider the effect of the map on two initially close points x_0 and y_0 such that the first N elements in their binary representations are identical. For how many iterations do these orbits remain close?

Exercise 18.57
Möbius sequences
The Möbius transformation

$$f(x) = \frac{ax + b}{cx + d}, \quad ad - bc \neq 0,$$

where a, b, c and d are real numbers, may be used to define a Möbius sequence

$$x_{n+1} = f(x_n) = \frac{ax_n + b}{cx_n + d}.$$

(i) Show that provided $(a - d)^2 + 4bc > 0$ there are two fixed points α and β which satisfy

$$f'(\alpha) = \frac{c\beta + d}{c\alpha + d}, \quad f'(\beta) = \frac{c\alpha + d}{c\beta + d},$$

and that

$$\frac{x_n - \alpha}{x_n - \beta} = \frac{x_0 - \alpha}{x_0 - \beta} f'(\alpha)^n, \quad n = 0, 1, 2, \ldots.$$

Deduce that one of the fixed points is stable.
(ii) Also show that if $(a - d)^2 + 4bc = 0$ then $f'(\alpha) = 1$ and

$$u_n = u_0 + \frac{2nc}{a + d}, \quad \text{where} \quad u = \frac{1}{x - \alpha}.$$

Exercise 18.58
Consider the map $x_{n+1} = F(x_n) = bx_n(1 - x_n^2)$, $0 < b \leq 3\sqrt{3}/2 = 2.598\ldots$, and show that:

(i) there is a fixed point at $x = 0$ which is stable if $b \leq 1$ and unstable for $b > 1$;
(ii) there is a fixed point at $x = \sqrt{1 - 1/b}$ which is stable if $1 < b \leq 2$ and unstable for $b > 2$;
(iii) there is a stable 2-period orbit for $2 < b \leq \sqrt{5}$.

Plot the graph of the Lyapunov exponent for $2 \leq b \leq 3\sqrt{3}/2$ to find the ranges of b where stable orbits exist, and compute the density function $\rho(x)$, equation 18.22, for various values of b.

Exercise 18.59

Consider the mean values M_p of the logistic map,

$$M_p = \lim_{N \to \infty} \frac{1}{N} \sum_{k=1}^{N} x_k^p, \quad x_{n+1} = bx_n(1 - x_n),$$

where p is a positive integer.

Show that if $b \leq 1$ then $M_p = 0$ and that for $b > 1$,

$$M_1 = \frac{b}{b-1} M_2, \quad M_2 = \frac{b^2}{b^2-1}(2M_3 - M_4), \quad M_3 = \frac{b^3}{b^3-1}(3M_4 - 3M_5 + M_6),$$

and in general

$$M_p = \frac{p! \, b^p}{b^p - 1} \sum_{r=1}^{p} \frac{(-1)^r M_{p+r}}{r! \, (p-r)!}.$$

Show also that $\sum_{k=1}^{N} x_k^2 = x_1 + x_N(1 - x_N)$ if $b = 1$ and deduce that $x_n \to 0$ faster than $n^{-1/2}$ as $n \to \infty$.

Exercise 18.60

If L is a real, 2×2 matrix with $\det(L) = 1$ and

$$L = \begin{pmatrix} a & b \\ c & d \end{pmatrix}, \quad \text{show that} \quad H = L^{\mathsf{T}} L = \begin{pmatrix} a^2 + c^2 & ab + cd \\ ab + cd & b^2 + d^2 \end{pmatrix},$$

and deduce that the eigenvalues of H are real and positive and that the eigenvectors are orthogonal.

Show that the eigenvalues of H are λ and $1/\lambda$, where

$$\lambda = \frac{1}{2}\left(T + \sqrt{T^2 - 4}\right), \quad T = \operatorname{Tr}(H) = a^2 + b^2 + c^2 + d^2.$$

Exercise 18.61

Consider the map

$$\mathbf{T}: \begin{array}{rcl} x_{n+1} &=& 2cx_n - y_n + 2x_n^2, \\ y_{n+1} &=& x_n, \end{array}$$

used to draw figures 18.22 and 18.23 (page 749). Show that $\mathbf{T}^2$ has fixed points at the roots of $2x(x + c - 1)(x^2 + x(c + 1) + c + 1) = 0$, $y = x(c + x)$, and that the linear approximation to $\mathbf{T}^2$ is

$$L_2(\mathbf{x}) = \begin{pmatrix} 4(2x + c)(4x^2 + 4cx + c - 2y) - 1 & -2(c + 4xc + 4x^2 - 2y) \\ 2(c + 2x) & -1 \end{pmatrix}.$$

For the 2-period orbit show that $\operatorname{Tr}(L_2) = 14 + 8c - 4c^2$ and deduce that this orbit is stable if $1 - \sqrt{5} < c < -1$ or $3 < c < 1 + \sqrt{5}$.

Exercise 18.62

The case $R = \frac{3}{4}$:

If $\mathbf{T}(\mathbf{x})$ is the map defined in the previous question, use Maple to show that an approximation to $\mathbf{T}^3$ is

$$\mathbf{T}^3(\mathbf{x}) \simeq \mathbf{U}(\mathbf{x}) = \begin{pmatrix} (1 + 2d)x + 2dy - 2x(x + 2y) \\ -4dx + y(1 - 2d) + 2y(y + 2x) \end{pmatrix},$$

where $c = -\frac{1}{2} + d$ and where terms of third order in x, y and d have been ignored.

Show that $\mathbf{U}(\mathbf{x})$ has the four fixed points $(0,0)$, (d,d), $(d,-2d)$ and $(-2d,d)$ and at each the residue of the linearised map is, to the order quoted, zero.

Using these approximations as a basis for a perturbation expansion, show that the fixed points of $\mathbf{U}(\mathbf{x})$ are

$$(d + 4d^2 + 24d^3, d + 4d^2 + 24d^3),$$
$$(d + 4d^2 + 24d^3, -2d - 4d^2 - 24d^3),$$
$$(-2d - 4d^2 - 24d^3, d + 4d^2 + 24d^3),$$

and that their residues are $R = -9d^2 + 6d^3$. Note that for $d = 0.1$ the fixed point near (d,d) is at (a,a), where $a = 0.010\,425\,7$, the value used in exercise 18.27.

Show also that the residue of $\mathbf{T}^3$ at the origin is

$$R(0,0) = \frac{1}{2}(1-c)(1+2c)^2 = 3d^2 - 2d^3 + \cdots,$$

which is zero at $c = -\frac{1}{2}$ and $R = 1$ when $c = -1$ and $\frac{1}{2}$.

Plot some iterates of $\mathbf{T}$ for small $|d|$ and initial conditions inside the triangle having vertices (d,d), $(d,-2d)$ and $(-2d,d)$.

Exercise 18.63
The standard map
The standard map is defined by

$$\mathbf{T}: \begin{array}{rcl} I_{n+1} & = & I_n - k\sin\theta_n, \quad k \geq 0, \\ \theta_{n+1} & = & \theta_n + I_{n+1}. \end{array}$$

This system is clearly 2π-periodic in θ, so the coordinates (I,θ) may be interpreted as defining points on the surface of a cylinder. In some applications the system is also considered to be 2π-periodic in I, so the iterates lie on the surface of a torus: for the consequences of this interpretation see Lichtenberg and Lieberman (1983, chapter 3).

Show that the standard map is area-preserving and that it has an unstable fixed point at $I = 0$, $\theta = \pi$, for all $k > 0$, and another fixed point $I = \theta = 0$ with residue $R = k/4$. Determine the position and nature of the fixed points of $\mathbf{T}^2$ and plot some representative orbits to check these calculations.

Exercise 18.64
Show that the map

$$\mathbf{T}: x_{n+1} = -y_n + 2x_n^3, \quad y_{n+1} = x_n$$

is area-preserving, has a stable fixed point at the origin, unstable fixed points at $x = y = \pm 1$ and that the normalised eigenvectors of the linearised map at $x = y = 1$ are

$$\mathbf{u}_\pm = \frac{1}{\sqrt{6}}\left(\begin{array}{c}\sqrt{\lambda_\pm} \\ \sqrt{\lambda_\mp}\end{array}\right), \quad \lambda_\pm = 3 \pm 2\sqrt{2}.$$

Use the method described in section 18.6 to plot the stable and unstable manifolds passing through $(1,1)$. For the unstable manifold, parallel to $\mathbf{u}_+$ near the fixed point, you will need to use M initial points

$$\mathbf{x}_k = \left(1 - k\delta, 1 - k\delta/\lambda_+\right), \quad k = 1, 2, \ldots, M, \quad 0 < \delta \ll 1.$$

For $\delta \sim 10^{-8}$ about 15 iterations suffice.

For the stable manifold use the inverse map

$$\mathbf{T}^{-1} : x_n = y_{n+1}, \quad y_n = -x_{n+1} + 2y_{n+1}^3,$$

with the initial conditions

$$\mathbf{x}_k = \left(1 - k\delta, 1 - k\delta/\lambda_-\right), \quad k = 1, 2, \dots, M, \quad 0 < \delta \ll 1.$$

In addition to these stable and unstable manifolds you should also plot iterates of some orbits starting near the origin to see how the chaotic tangle can coexist with well behaved orbits.

The unstable manifold produced by this map differs from that shown in figure 18.34 (page 769), because it intersects the stable manifold of the other unstable fixed point at $(-1, -1)$. Intersections of manifolds of different fixed points are name heteroclinic points.

19.1 Introduction: Duffing's equation

Resonances in dynamical systems are important: they occur when frequency ratios are rational and produce a wide variety of phenomena, for instance meteorites, the rings around Saturn, sounds from many musical instruments and those annoying rattles in mechanical systems. Resonances also produce periodic solutions and, as shown in the previous chapter, these help organise phase space.

In this chapter we concentrate on one dynamical system, although the methods and ideas discussed are applicable to many types of problem. Our aim is partly to illustrate how the ideas introduced in the previous chapter apply to differential equations and partly to understand the effect of a periodic perturbation on a nonlinear system by using a combination of numerical methods and analytic approximations. The story is complicated because a periodically driven nonlinear system can behave in a wide variety of ways, and a good understanding of this type of system is obtained only by solving illustrative problems, some of which are set as exercises. After some preliminary analysis we present sample surface-of-sections for a few carefully chosen parameters. The most significant features of these graphs are then explained using the harmonic balance approximation. It is also shown how these simple approximations may be used in conjunction with Floquet theory to analyse the stability of periodic orbits. Finally, we describe an averaging method that provides an approximation to a wider class of solutions than the harmonic balance method.

We investigate the behaviour of Duffing's equation, which is one of the simplest periodically driven nonlinear systems. Duffing's original 1918 study was an investigation into the behaviour of electrical machinery and various experiments on simple electro-mechanical systems which displayed a nonlinear response to periodic forcing, that is, the frequency of the forced oscillations is dependent upon the amplitude of the motion. The 1918 paper considered various forms of the periodically driven nonlinear system

$$\frac{d^2x}{dt^2} + 2\mu\frac{dx}{dt} + \omega^2 x + \alpha x^2 + \beta x^3 = F\sin\Omega t,$$

as well as the driven vertical pendulum, $\ddot{x} + \omega^2\sin x = F\sin\Omega t$, and systems driven by two frequencies, $F_1\sin\Omega_1 t + F_2\sin\Omega_2 t$. However, the equation that is now known as

Duffing's equation† is, after suitable rescaling of lengths and times,

$$\frac{d^2x}{dt^2} + 2\mu\frac{dx}{dt} + x - \epsilon x^3 = F\cos\Omega t. \tag{19.1}$$

This system is damped (the $2\mu\dot{x}$ term), nonlinear (the x^3 term) and driven by the periodic force $F\cos\Omega t$ with constant magnitude $F \geq 0$. It has four independent parameters:

- the damping strength, $2\mu \geq 0$;
- the strength of the nonlinear term, ϵ, which may be positive or negative;
- the magnitude of the driving term, $F \geq 0$;
- the frequency of the driving term, $\Omega > 0$.

In most real systems the driving force is switched on and off, over a finite time, and one is not always free to chose the time origin so that the maximum of the applied force is at time $t = 0$. In a practical situation it is therefore often necessary to replace the right hand side of equation 19.1 by $FA(t)\cos(\Omega t + \delta)$ where $0 \leq A(t) \leq 1$ is an envelope function representing the switching on and off of the disturbance and δ is a phase; the envelope normally changes slowly by comparison with $\cos\Omega t$, that is, $|\dot{A}| \ll A\Omega$. The presence of the envelope $A(t)$ can significantly change the response of the system for given initial conditions although it does not qualitatively alter the possible types of behaviour. However, it is necessary to ignore its effect in any initial investigation, otherwise little progress can be made.

Our objective is to develop techniques that help us understand the behaviour of the solutions of Duffing's equation in as much of parameter space as possible. Clearly this is a difficult task and we shall very quickly find that different regions of this space require quite different techniques and that some regions are far more difficult to understand than others. In this introduction selection is necessary in order to keep the discussion to a reasonable length, so we mainly study the undamped case, $\mu = 0$.

As is usual with any problem with many parameters, it is advisable to first consider limiting cases that are easily understood and then to consider those values of the parameters for which the solutions should be fairly simple. Thus in this presentation we follow Duffing's example and first consider well understood limiting cases, the forced linear oscillator $\epsilon = 0$, and the undamped, unperturbed nonlinear oscillator, $\mu = F = 0, \epsilon \neq 0$.

19.2 The forced linear oscillator

When $\epsilon = 0$ equation 19.1 is linear:

$$\frac{d^2x}{dt^2} + 2\mu\frac{dx}{dt} + x = F\cos\Omega t. \tag{19.2}$$

If $0 < \mu < 1$ the unforced system, $F = 0$, executes decaying oscillations with frequency $v = \sqrt{1 - \mu^2}$,

$$x(t) = ae^{-\mu t}\sin(vt + \delta), \quad 0 \leq \mu < 1, \quad (F = 0),$$

where a and δ are constants depending upon the initial conditions.

---

† There is no generally accepted definition of what Duffing's equation is. Some texts use the name to signify equation 19.1 with $F = 0$ and others with $\mu = 0$.

Because the system is linear the forcing term merely adds another oscillatory term to this, giving the general solution

$$x(t) = ae^{-\mu t}\sin(vt + \delta) + F\frac{(1 - \Omega^2)\cos\Omega t + 2\mu\Omega\sin\Omega t}{(1 - \Omega^2)^2 + 4\mu^2\Omega^2}, \quad 0 \le \mu < 1. \tag{19.3}$$

The first, decaying, term is the *transient* and, after a time long compared with μ^{-1}, it becomes negligible, leaving only the oscillations caused by the periodic force. The amplitude, $A(\Omega)$, of these forced oscillations is

$$A(\Omega) = \begin{cases} \dfrac{F}{\sqrt{(\Omega^2 - 1 + 2\mu^2)^2 + 4\mu^2(1 - \mu^2)}}, & \mu \ne 0, \\ \dfrac{F}{|1 - \Omega^2|}, & \mu = 0. \end{cases} \tag{19.4}$$

- If $0 < \mu^2 < \frac{1}{2}$ the amplitude $A(\Omega)$ has a maximum at $\Omega^2 = 1 - 2\mu^2$ with

$$\max(A) = \frac{F}{2\mu\sqrt{1 - \mu^2}}, \quad 0 < \mu^2 < \frac{1}{2},$$

  and a local minimum at $\Omega = 0$, figure 19.1.
- If $\mu^2 > \frac{1}{2}$ the amplitude has a maximum at $\Omega = 0$ and decreases monotonically with increasing Ω, figure 19.2.
- If $\mu = 0$ then $A \to \infty$ as $\Omega \to 1$.

Some typical graphs of $A(\Omega)$ for $\mu^2 < \frac{1}{2}$ and $\mu^2 > \frac{1}{2}$ with $F = 1, 2, 3$ and 4 are shown below: such curves are named *resonance* or *response* curves and contain important information about the system behaviour in simple graphical form.

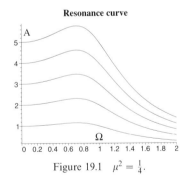

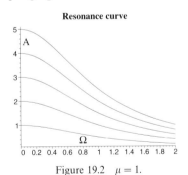

Figure 19.1 $\mu^2 = \frac{1}{4}$.   Figure 19.2 $\mu = 1$.

The amplitude of these forced oscillations is finite for all Ω provided $\mu \ne 0$, but when $\mu = 0$ and at resonance, $\Omega = v = 1$, the general solution of equation 19.2 is

$$x(t) = a'\sin(t + \delta') + \frac{1}{2}Ft\sin t, \tag{19.5}$$

and the amplitude of the oscillations increases linearly with time. This is simply because the frequency of the unperturbed motion is independent of the amplitude, so no matter how much energy is absorbed by the oscillator its motion remains in resonance with the applied force. The solution 19.5 can be obtained from 19.3, but not by simply putting $\mu = 0$ and $\Omega = 1$, see exercise 19.41. For a nonlinear system the frequency of the unperturbed motion is amplitude dependent, so a weak periodic driving force does not normally produce a large amplitude response.

Exercise 19.1
The case $\mu > 1$

(i) Show that if $\mu > 1$ the general solution of equation 19.2 is

$$x(t) = ae^{\lambda_+ t} + be^{\lambda_- t} + F\frac{(1 - \Omega^2)\cos\Omega t + 2\mu\Omega\sin\Omega t}{(1 - \Omega^2)^2 + 4\mu^2\Omega^2},$$

where $\lambda_\pm = -\mu \pm \sqrt{\mu^2 - 1}$ and a and b are constants.

(ii) If $\mu = 1$ show that the general solution is

$$x(t) = (a + bt)e^{-t} + F\frac{(1 - \Omega^2)\cos\Omega t + 2\Omega\sin\Omega t}{(1 + \Omega^2)^2},$$

where a and b are constants.

Exercise 19.2
The linear oscillator driven by the sum of two or more frequencies is described by the equation

$$\frac{d^2x}{dt^2} + 2\mu\frac{dx}{dt} + x = \sum_k F_k \cos\Omega_k t.$$

Show that if $\mu < 1$ the general solution is

$$x(t) = ae^{-\mu t}\sin(vt + \delta) + \sum_k F_k\frac{(1 - \Omega_k^2)\cos\Omega_k t + 2\mu\Omega_k\sin\Omega_k t}{(1 - \Omega_k^2)^2 + 4\mu^2\Omega_k^2}, \quad 0 < \mu < 1.$$

Exercise 19.3
Put the equation

$$\frac{d^2x}{dt^2} + 2\mu\frac{dx}{dt} + x = f(t), \quad x(0) = \dot{x}(0) = 0,$$

in self-adjoint form, find the Green's function and hence show that the solution may be written in the form

$$x(t) = \begin{cases} \dfrac{1}{v}\displaystyle\int_0^t d\tau\, f(\tau)e^{-\mu(t-\tau)}\sin v(t - \tau), & 0 < \mu < 1, \quad v = \sqrt{1 - \mu^2}, \\[2ex] \displaystyle\int_0^t d\tau\, f(\tau)e^{-(t-\tau)}(t - \tau), & \mu = 1, \\[2ex] \dfrac{1}{v}\displaystyle\int_0^t d\tau\, f(\tau)e^{-\mu(t-\tau)}\sinh v(t - \tau), & \mu > 1, \quad v = \sqrt{\mu^2 - 1}. \end{cases}$$

19.3 Unperturbed nonlinear oscillations

The simplest limit of Duffing's equation is the undamped, $\mu = 0$, unforced, $F = 0$, case when the system is conservative and the motion can be described in terms of standard functions. If the system is just unforced, $F = 0$, $\mu > 0$, simple solutions do not exist, but the qualitative behaviour of the small amplitude motion, normally considered, is easily understood. This section is devoted to a discussion of these limiting cases because the more complicated periodically forced system cannot be understood without knowledge of these solutions. The most important aspect of these solutions is the dependence of the natural frequency on the amplitude of the oscillations, shown graphically in figures 19.4 and 19.9. The Fourier series representation of the solutions, equations 19.10 and 19.14, are also important because the magnitude of the harmonics determines the effectiveness of the coupling between the unperturbed motion and a weak periodic perturbation. The conservative case $F = \mu = 0$ is dealt with first.

Conservative systems were introduced in section 17.5, and comparing Duffing's equation with equation 17.24 (page 691), we see that $m = 1$ and that the potential is even, $V(x) = \frac{1}{2}x^2 - \frac{1}{4}\epsilon x^4$. The fixed points are at the roots of $V'(x) = 0$, exercise 19.26, that is, $x = 0$ and $x^2 = 1/\epsilon$. If $\epsilon < 0$ all motion is bound and if $\epsilon > 0$ only some motion with energy $0 < E < 1/4\epsilon$ is bound, therefore it is best to consider these two cases separately.

19.3.1 Conservative motion, $F = \mu = 0$: $\epsilon < 0$

When $\epsilon < 0$ the force towards the origin, $dV/dx = x + (-\epsilon)x^3$, is larger than for the linear force, $\epsilon = 0$, so in the context of mechanical systems this is known as a hard spring.

The algebra is clearer if we define $\delta = -\epsilon > 0$. The theory of section 17.5 shows that the phase curves are the contours of the Hamiltonian function which, in this case, is also the energy of the system,

$$H(x, y) = \frac{1}{2}y^2 + \frac{1}{2}x^2 + \frac{1}{4}\delta x^4 = E \geq 0, \quad y = \frac{dx}{dt}. \tag{19.6}$$

There is only one fixed point: it is at the origin and is a centre. If $\delta = 0$ the contours are circles of radius $\sqrt{2E}$; if $\delta > 0$ the small energy contours remain close to these circles, but with increasing energy the quartic term becomes more significant and the contours approach those defined by $\frac{1}{2}y^2 + \frac{1}{4}\delta x^4 = E$. This gradual change is shown in figure 19.3 where equally spaced contours are shown.

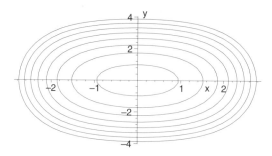

Figure 19.3   Contours of the Hamiltonian 19.6 with $\delta = \frac{1}{4}$ and
spacing $\Delta E = 1$; the first contour has energy $E = \frac{1}{2}$.

The period of the motion can be expressed in terms of the complete elliptic integral
of the first kind, (equation 21.7, page 841). If a is the amplitude of the motion, related
to the energy by $E = \frac{1}{2}a^2 + \frac{1}{4}\delta a^4$, the period depends only on the parameter $A = a\sqrt{\delta}$
and is

$$T(A) = \frac{4}{\sqrt{1 + A^2}} K(k), \quad k^2 = \frac{A^2}{2(1 + A^2)} < 1, \quad A = a\sqrt{\delta}. \tag{19.7}$$

The derivation of this result is sketched out in exercise 19.44 (page 828). The angular
frequency of the unperturbed motion, $\omega = 2\pi/T$, is particularly simple in the case of
small and large amplitude motion:

$$\omega(A) = \frac{2\pi}{T(A)} \sim \begin{cases} 1, & A \to 0, \quad \text{(small amplitude)} \\ A\, \dfrac{2\pi^{3/2}}{\Gamma(1/4)^2}, & A \to \infty, \quad \text{(large amplitude)}. \end{cases} \tag{19.8}$$

The first of these limits is just the frequency of the unperturbed linear oscillator. The
frequency in the large amplitude limit is proportional to the amplitude and is derived
using the identity $K\left(1/\sqrt{2}\right) = \Gamma(1/4)^2/(4\sqrt{\pi})$. This is the frequency of the motion in
the potential $V(x) = \frac{1}{4}\delta x^4$, exercise 19.8.

The frequency is a monotonic increasing function of the $A = a\sqrt{\delta}$, with $\omega = 1$ for
$A = 0$; the graph of $\omega(A)$ is shown in figure 19.4.

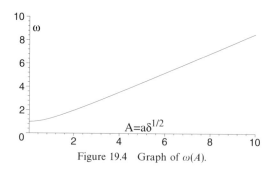

Figure 19.4   Graph of $\omega(A)$.

The next three figures show how the position $x(t)$ varies over one period when $\delta = \frac{1}{2}$ and for the amplitude $a = 0.1$, 1 and 10. Here we have plotted $x(t)/a$ against $t/T(a)$ in order that the scales remain the same, exercise 19.8 (page 793); observe that all three cases are similar and that the two cases $a = 1$ and $a = 10$ are practically indistinguishable.

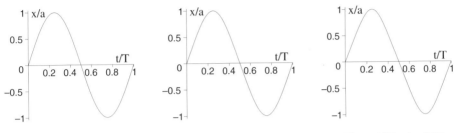

Figure 19.5 $A = 0.0707$.   Figure 19.6 $A = 0.707$.   Figure 19.7 $A = 7.07$.

The motion, being periodic, may be represented by a Fourier series, most easily obtained using the general results given in Whittaker and Watson (1965, chapter 22) pertaining to the Jacobi elliptic functions: a short desciption of these functions is provided in Appendix II. For motion with amplitude a and $x(0) = 0$, the time is given in terms of the position by the integral

$$t = \frac{1}{\sqrt{1 + A^2}} \int_0^{x/a} dz \, \frac{1}{\sqrt{(1 - z^2)(k'^2 + k^2 z^2)}}, \quad k^2 = \frac{A^2}{2(1 + A^2)}, \tag{19.9}$$

where $k'^2 + k^2 = 1$; note that $0 \le k^2 < \frac{1}{2}$. For small amplitude motion this reduces to

$$t = \int_0^{x/a} dz \, \frac{1}{\sqrt{1 - z^2}} \quad \text{with inverse} \quad \frac{x}{a} = \sin t.$$

More generally, the inverse is given in terms of the Jacobi elliptic function† $c_n(z, k)$, defined in Appendix II (equation 21.15, page 843). This gives

$$
\begin{aligned}
x(t) &= a \operatorname{cn}\left(\sqrt{1 + A^2}\, (T/4 - t), k \right), \quad x(0) = 0, \\
&= \frac{2\pi a}{kK(k)} \sum_{n=0}^{\infty} (-1)^n \frac{q^{n+1/2}}{1 + q^{2n+1}} \sin(2n + 1)\omega t, \tag{19.10}
\end{aligned}
$$

where T is the period, given in 19.7, and ω the frequency. The quantity q, named the *gnome*, is defined by the equation

$$q = \exp\left(-\frac{\pi K(k')}{K(k)} \right), \quad k' = \sqrt{1 - k^2},$$

and since k depends only upon $A = a\sqrt{\delta}$, q also depends only upon A. By expressing $K(k)$ in terms of the frequency, equation 19.7, the Fourier series for x may be written

† This function is denoted in Maple by **JacobiCN(z,k)**.

in the form

$$x(t)\sqrt{\delta} = 2\sqrt{2}\,\omega \sum_{n=0}^{\infty} (-1)^n \frac{\sin(2n+1)\omega t}{\cosh Q_n}, \qquad Q_n = -\left(n+\frac{1}{2}\right)\ln q.$$

For small A, $q(A) = A^2/32 + O(A^4)$ and the Fourier series reduces to the linear oscillator limit $x = a\sin t$. As A increases, $k^2 \to \frac{1}{2}$ and $q(A)$ increases monotonically to its large A limit $q \to e^{-\pi} = 0.043\ldots$, and is within 5% of this limit when $A > 5$. Thus, for all A the magnitude of the successive Fourier components decreases rapidly and for large A the relative values of the Fourier coefficients change little; this explains the similarity of figures 19.6 and 19.7. Another explanation of this behaviour is given in exercise 19.8 (page 793).

Summary
Two useful observations can be made.

(1) The natural frequency of this system, a hard spring, is always larger than unity. Hence a periodic driving force can resonate, in first order, only if its frequency is larger than unity. Lower frequency forces can resonant but only in higher order, so their effect is significant only for stronger forces, see for instance figure 19.16 (page 800).

(2) The harmonics of the Fourier series decay rapidly with n, and since a high frequency force couples to the system, in first order, via the nearest harmonic, the effect of a high frequency force on this system is relatively weak.

Exercise 19.4
Derive equations 19.9 and 19.10.

Exercise 19.5
For small $A = a\sqrt{\delta}$ show that

$$q(A) = \frac{A^2}{32} - \frac{3A^4}{128} + \frac{149A^6}{8192} + O(A^8),$$

and hence that the Fourier series 19.10 reduces to the linear oscillator as $A \to 0$.

Show also that in the opposite limit, $A \gg 1$, $k \simeq k' \simeq 1/\sqrt{2}$ and hence that $q \to e^{-\pi}$ as $A \to \infty$ and that the first two terms of the asymptotic expansion of $q(A)$ are

$$q(A) \sim \exp\left(-\pi - \frac{\pi K'(1/\sqrt{2})}{A^2\sqrt{2K(1/\sqrt{2})}}\right) = e^{-\pi}\left(1 - \frac{1.435}{A^2}\right).$$

Plot the graph of $q(A)$ and examine the values of the coefficients $q^{n+1/2}/(1+q^{2n+1})$ for $n = 0, 1, \ldots, 5$ and for various values of A.

Exercise 19.6
Show that in the limit of large frequencies equation 19.10 becomes

$$x(t)\sqrt{\delta} \sim 2\sqrt{2}\,\omega \sum_{n=0}^{\infty} (-1)^n \frac{\sin(2n+1)\omega\tau}{\cosh\left(n+\frac{1}{2}\right)\pi}$$

$$\simeq \omega\left(1.1272\sin\omega\tau - 0.0508\sin 3\omega\tau + 0.0022\sin 5\omega\tau + \cdots\right).$$

Exercise 19.7
Use the method of harmonic balance with the trial function $x = a \sin \omega t$ to show that the frequency of the motion with amplitude a is approximately

$$\omega = \sqrt{1 + \frac{3}{4}A^2}, \quad A = a\sqrt{\delta},$$

and compare the small and large A expansion of this with the exact result, equation 19.7.

Further, show that for high frequency this approximation gives $x\sqrt{\delta} = 1.155\omega \sin \omega t$ and compare this with the exact solution found in exercise 19.6.

Exercise 19.8
Show that the period of the motion of a particle, of unit mass, moving in the potential $V(x) = \frac{1}{4}\delta x^4$ with amplitude a, is

$$T(a) = \frac{\sqrt{2}}{A} \int_0^1 dy \, \frac{1}{y^{3/4}\sqrt{1-y}} = \frac{\Gamma(1/4)^2}{A\sqrt{\pi}}, \quad A = a\sqrt{\delta}.$$

Show also that if $\xi = x/a$ and $\tau = t/T(a)$ the equation of motion becomes

$$\left(\frac{d\xi}{d\tau}\right)^2 = c(1 - \xi^4), \quad c = \frac{\Gamma\left(\frac{1}{4}\right)^4}{2\pi}.$$

Deduce that in these scaled variables the motion is independent of the amplitude.

19.3.2 Conservative motion, $F = \mu = 0$: $\epsilon > 0$
When $\epsilon > 0$ the force towards the origin, $dV/dx = x - \epsilon x^3$, is smaller than for the linear force, $\epsilon = 0$, so this is known as a soft spring. The phase curves are the contours of the Hamiltonian

$$H(x, y) = \frac{1}{2}y^2 + \frac{1}{2}x^2 - \frac{1}{4}\epsilon x^4 = E, \quad y = \frac{dx}{dt}. \tag{19.11}$$

Now there are three fixed points: there is a centre at $x = y = 0$ and saddles at $y = 0$, $x = \pm 1/\sqrt{\epsilon}$. There is motion for all energies but bound, periodic motion only for $0 \le E < (4\epsilon)^{-1}$, as may be seen by plotting the graph of the potential. This bound periodic motion is separated from the unbound motion by the separatrix, section 17.5, which is the phase curve with the energy of the saddle, $E = E_s = 1/4\epsilon$: it has the equation

$$y(x) = \pm\frac{1}{\sqrt{2\epsilon}}(1 - \epsilon x^2)$$

and is depicted by the solid lines in figure 19.8. The motion on the separatrix is neither unbounded nor periodic, but on it and near $x = \pm 1/\sqrt{\epsilon}$ the motion is exponentially slow, equation 19.16. Because of this very slow motion most disturbances, no matter how small, destroy the separatrix to produce a homoclinic tangle.

The phase curves near the origin are close to circles of radius $\sqrt{2E}$ but become more and more distorted as $E \to E_s$. Some typical phase curves, equally spaced in energy, are shown in the figure for the case $\epsilon = \frac{1}{2}$.

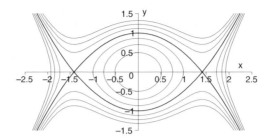

Figure 19.8 Contours of the Hamiltonian 19.6 with $\epsilon = \frac{1}{2}$ and $\Delta E = \frac{1}{8}$: the first contour has energy $E = \frac{1}{8}$.

The period of the motion with amplitude $a < 1/\sqrt{\epsilon}$ is

$$T(a) = \frac{4\sqrt{2}}{\sqrt{2 - \epsilon a^2}} K(k), \quad k^2 = \frac{a^2 \epsilon}{2 - a^2 \epsilon}. \tag{19.12}$$

The derivation of this expression is outlined in exercise 19.44, (page 828).

As $a^2 \epsilon \to 1$, $E \to E_s$, $k \to 1$ and hence $T(a) \to \infty$, which is to be expected because the motion on the separatrix is exponentially slow as $x^2 \epsilon \to 1$, equation 19.16. Thus in the two limiting cases the frequency behaves as follows:

$$\omega(a) = \frac{2\pi}{T(a)} \sim \begin{cases} 1, & a^2 \epsilon \to 0, & \text{(small amplitude)} \\ 2\pi \left[2\sqrt{2}\ln\left(\frac{8}{1 - a^2 \epsilon}\right)\right]^{-1}, & a^2 \epsilon \to 1, & \text{(large amplitude)}. \end{cases} \tag{19.13}$$

The second of these limits shows that $\omega(a) \to 0$ as $a\sqrt{\epsilon} \to 1$, but that the approach to this limit is exceedingly slow; for instance $\omega = 0.1$ when $a\sqrt{\epsilon} \simeq 1 - 1.8 \times 10^{-9}$ and has decreased by a factor of 10, to $\omega = 0.01$, when $a\sqrt{\epsilon} = 1 - 2.7 \times 10^{-96}$. Hence the graph of $\omega(a)$, shown below, has an almost vertical drop near $a\sqrt{\epsilon} = 1$, and for most values of $a\sqrt{\epsilon}$ the frequency is between $\frac{1}{2}$ and 1. This logarithmic behaviour is caused by the quadratic maximum in the potential at $x = \pm 1/\sqrt{\epsilon}$ and is typical of such systems. One important consequence of this behaviour is that when perturbed the separatrix normally splits into a stable and an unstable manifold, and an area of phase space surrounding the unperturbed separatrix is filled by the heteroclinic tangle, but because ω changes rapidly as $a\sqrt{\epsilon}$ decreases from unity the area of phase space filled by chaotic orbits is $O(\exp(-1/\epsilon))$ for weak perturbations.

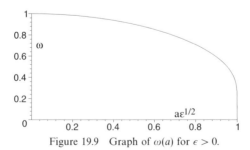

Figure 19.9 Graph of $\omega(a)$ for $\epsilon > 0$.

The next three figures show how $x(t)$ varies over one period for $\epsilon = \frac{1}{2}$ and $a = 0.1$, 1.0 and 1.414. In the first case the amplitude of the motion is small and the motion indistinguishable from that of the linear oscillator; the intermediate case, $a = 1$, has a longer period and the extrema are slightly broader. The third case shows motion very close to the separatrix, $E = 0.9994E_s$, which lingers for a long time in the vicinity of $x = \pm a$, near the unstable fixed points at $x = \pm\sqrt{2}$.

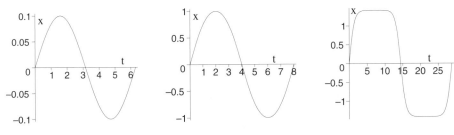

Figure 19.10 $a\sqrt{\epsilon} = 0.071$.   Figure 19.11 $a\sqrt{\epsilon} = 0.71$.   Figure 19.12 $a\sqrt{\epsilon} = 0.99985$.

The Fourier series of this motion is obtained by first expressing the time as an integral over the position in the form

$$t = \frac{1}{\sqrt{1 - \frac{1}{2}a^2\epsilon}} \int_0^{x/a} dw \, \frac{1}{\sqrt{(1 - w^2)(1 - k^2w^2)}}, \quad k^2 = \frac{\epsilon a^2}{2 - \epsilon a^2},$$

and then using the definition of the Jacobi elliptic function† $\mathrm{sn}(u, k)$, given in Appendix II (equation 21.14, page 843), to give

$$
\begin{aligned}
x(t) &= a\,\mathrm{sn}\left(t\sqrt{1 - \frac{1}{2}\epsilon a^2}, k\right), \quad k^2 = \frac{\epsilon a^2}{2 - a^2\epsilon}, \quad x(0) = 0, \\
&= \frac{2\pi a}{kK(k)} \sum_{n=0}^{\infty} \frac{q^{n+1/2}}{1 - q^{2n+1}} \sin(2n+1)\omega t, \quad q = \exp\left(-\frac{\pi K(k')}{K(k)}\right), \quad (19.14)
\end{aligned}
$$

where $k'^2 + k^2 = 1$: note that here k is different from that in 19.9. This looks similar to the previous Fourier series, equation 19.10, but the changed sign in the denominator of the Fourier coefficient and the fact that $k \to 1$ as $a\sqrt{\epsilon} \to 1$ makes the behaviour of this series much more interesting.

On the separatrix, $E = E_s$, the motion is not periodic but is bounded; on the upper branch $x\sqrt{\epsilon} \sim 1 - 2e^{-t\sqrt{2}}$, as $t \to \infty$, exercise 19.13. As $E \to E_s$ the frequency tends to zero and the Fourier series must mimic this limiting behaviour. In order to do this *all* the Fourier components must become significant as $E \to E_s$. We examine this limit by setting $\epsilon a^2 = 1 - \delta^2$, for some small positive number δ, then

$$q \simeq \exp\left(-\frac{\pi^2}{\ln(8/\delta^2)}\right), \quad \text{hence} \quad \frac{q^{n+1/2}}{1 - q^{2n+1}} \simeq \frac{1}{2\sinh\left(\frac{\pi^2(n+1/2)}{\ln(8/\delta^2)}\right)}.$$

† This function is denoted in Maple by **JacobiSN(z,k)**.

Note that as $\delta \to 0$, $q \to 1$ very slowly, and $q > 0.9$ only when $\delta < 1.3 \times 10^{-20}$. The nth Fourier coefficients become small (less than about 0.1) when $n > N = \ln(8/\delta^2)/\pi^2$, which increases as δ decreases, but the π^2 term ensures that it is small except when δ is exceedingly small. For instance, if $\delta = 10^{-6}$, $N \sim 3$. Nevertheless, as $E \to E_s$ the number of significant terms in the Fourier series increases, in contrast to equation 19.10 in which the number of significant Fourier components is always small.

Summary
The character of the motion for $\epsilon > 0$ is quite different from that when $\epsilon < 0$, in particular unbound solutions exist; this has the following two effects.

(1) The natural frequency of this system, a soft spring, is always smaller than unity. Hence a periodic driving force can resonate, in first order, only if its frequency is smaller than unity. Higher frequency forces can resonate but only in higher order, so their effect is significant only for stronger forces.

(2) The effect of almost all perturbations, no matter how small, is to split the separatrix between the bound and unbound motion into a stable and an unstable manifold, with chaotic neighbouring orbits. In this example most of these chaotic orbits are eventually unbound, but if the motion either side of the separatrix is bound the chaotic motion results in transport between regions. An example of such a system is the potential $V(x) = x^4/4 - x^2/2$ driven by a periodic force.

Exercise 19.9
Examine the values of $q^{n+1/2}/(1 - q^{2n+1})$ for $n = 1, 2, \ldots, 15$ and $\delta = 10^{-6}$, 10^{-9} and 10^{-12}.

Exercise 19.10
Show that the harmonic balance method with the trial function $x(t) = a \cos \omega t$ gives the frequency of the motion with amplitude a to be

$$\omega^2 = 1 - \frac{3}{4} \epsilon a^2.$$

Explain the physical significance of the fact that $\omega = 0$ when $a = 2/\sqrt{3\epsilon}$.

19.3.3 *Damped motion:* $F = 0$, $\mu > 0$
When damping is present the centre at $x = \dot{x} = 0$ changes to a stable spiral if $0 < \mu < 1$ or a stable node if $\mu > 1$; if the initial conditions are close to the origin the system tends towards the stable fixed point $(0,0)$ as $t \to \infty$. In general, when $\mu > 0$ the rate of change of the energy, E, defined in equation 19.6, is never positive, for we have

$$\frac{dE}{dt} = -2\mu\dot{x}^2 \le 0. \tag{19.15}$$

It follows that any motion starting with initial conditions such that the conservative motion, $\mu = 0$, would be periodic must decay towards the stable fixed point at the origin.

Exercise 19.11

Use Maple to numerically generate some solutions of the equation

$$\frac{d^2x}{dt^2} + 2\mu\frac{dx}{dt} + x - \epsilon x^3 = 0,$$

with initial conditions $x(0) = 0$, $\dot{x}(0) > y_0$ and with $\mu = \epsilon = 0.1$. Demonstrate that there is a critical value of y_0 above which the solution is unbound and below which the solutions spiral into the origin.

Exercise 19.12

Prove equation 19.15. If $0 < \mu \ll 1$ the damping terms has little effect during one period of the undamped motion. By replacing the right hand side of 19.15 by its mean over one unperturbed period, show that for small amplitude motion

$$\frac{dE}{dt} \simeq -2\mu E.$$

A method of extending this result to include motion with larger amplitude is given in exercise 19.37.

Exercise 19.13

Motion on the separatrix

Show that the motion on the separatrix of the Hamiltonian 19.11 (page 793), starting at $x = 0$ when $t = 0$ with $\dot{x} > 0$, is given by

$$x\sqrt{\epsilon} = \frac{1 - e^{-t\sqrt{2}}}{1 + e^{-t\sqrt{2}}}, \qquad (19.16)$$

and that $x\sqrt{\epsilon} \simeq 1 - 2e^{-t\sqrt{2}}$ as $t \to \infty$ and $x\sqrt{\epsilon} \simeq -1 + 2e^{t\sqrt{2}}$ as $t \to -\infty$.

Exercise 19.14

Show that if

$$x(t) = \sum_{n=0}^{\infty} a_n \sin(2n+1)\omega t \quad \text{then} \quad a_n = \frac{4\omega}{\pi} \int_0^{T/4} dt\, x(t) \sin(2n+1)\omega t.$$

Use the expression obtained in the previous exercise for $x(t)$ on the separatrix to derive the following approximation, accurate as $\omega \to 0$,

$$a_n\sqrt{\epsilon} \simeq \frac{4}{\pi} \int_0^{\infty} du \tanh\left(\frac{u}{\omega\sqrt{2}}\right) \sin(2n+1)u \simeq \frac{4}{(2n+1)\pi}.$$

Compare this expression with the exact result, equation 19.14, in the limit $a\sqrt{\epsilon} \to 1$. Note that this method of approximating the Fourier coefficients does not depend upon the details of the potential.

19.4 Maps and general considerations

Most analytic techniques available for approximating the solutions of Duffing's, or similar, equations work for specific types of orbits or particular parameter regimes, for instance the neighbourhood of periodic orbits. In general there is such great variety in the behaviour of solutions that no single analytic approximation can suffice in all circumstances. It is therefore difficult to obtain a good overall understanding of the system behaviour, and normally both analytic and numerical methods are needed.

One of the major problems with numerical solutions is to understand the results and to distinguish patterns. Modern graphical methods are of some help but do not overcome the problems inherent in trying to understand the behaviour of functions of many variables. In chapter 17 we saw that the global behaviour of a two-dimensional, conservative Hamiltonian system is easily understood by plotting contours of the Hamiltonian. There are two types of more complicated systems that can also be understood using two-dimensional plots.

(1) The motion of an autonomous Hamiltonian system with two degrees of freedom, that is, a four-dimensional phase space. This motion can be represented by a two-dimensional area-preserving map constructed by computing the points where an orbit, which is confined to the three-dimensional surface defined by the Hamiltonian $H(\mathbf{q}, \mathbf{p}) = E$, intersects a fixed plane, for example the (q_1, p_1)-plane. This map is named the *Poincaré section* or the *surface-of-section*. Details of its construction are given in Lichtenberg and Lieberman (1983, chapter 3) or Hand and Finch (1998, chapter 11), for example.

(2) The motion of a periodically driven system with one degree of freedom. The construction of this type of map is discussed in section 18.8; recall that if the system is Hamiltonian then the resulting map is area-preserving. This type of map is also named a *period map* as well as a Poincaré section or a surface-of-section.

These constructions are important because the resulting two-dimensional plots provide a global picture of the system behaviour that is otherwise extremely difficult to obtain. Here we consider only the second type of map applied to Duffing's equation. The Maple procedure `poincare` will find the surface-of-section for a given Hamiltonian system.

The construction of the period map is relatively simple, though the choice of orbits requires some time consuming experimentation. Numerical integration over one period may be performed using the Maple procedure `dsolve`. Using the default value of `Digits` this is about seven times slower than a Fortran program using double precision floating point arithmetic (on the same computer). However, increasing the size of `Digits` dramatically increases the computational time of `dsolve`: for instance setting `Digits` to 11 and 12 increases the ratio to 17 and 27 respectively. The default value of `Digits`, however, seems sufficient for most of the examples presented here; but, because the Maple procedures are too slow for the figures presented here I used a Fortran routine taken from Press *et al.* (1987, chapter 15) and Maple to graph the output.

The choice of t_0 is arbitrary and not important so we set $t_0 = 0$. The uniqueness of solutions of differential equations ensures that if $\mathbf{x}_0$ and $\mathbf{x}_1$ are distinct points so are $\mathbf{T}(\mathbf{x}_0)$ and $\mathbf{T}(\mathbf{x}_1)$, and also that the inverse $\mathbf{T}^{-1}(\mathbf{x})$ exists.

The fixed points of the map $\mathbf{T}$ are the solutions of the equation $\mathbf{x} = \mathbf{T}(\mathbf{x})$ and these represent T-period orbits. More generally, if $\mathbf{x}_0 \neq \mathbf{T}^k(\mathbf{x}_0)$, $k = 1, 2, \ldots, N - 1$, and $\mathbf{x}_0 = \mathbf{T}^N(\mathbf{x}_0)$ then the orbit starting at $\mathbf{x}_0$ is NT-periodic. That is, in a Poincaré section a T-periodic solution is represented by a single point and an NT-periodic orbit by N distinct points. The stability of a periodic solution is the same as the stability of the fixed point it generates.

An example of a periodic orbit and its period-map is shown in the three figures below. This orbit has period $6T$: the system parameters are $\epsilon = 1/6$, $F = 0.2$, $\Omega = 0.975$ and the initial condition is $\mathbf{x}_0 = (0.367\,742\ldots, 0)$, which is computed using the method discussed later in this section. Since $\mathbf{x}_6 = \mathbf{T}^6(\mathbf{x}_0)$, although $\mathbf{x}_k \neq \mathbf{x}_0$, $k = 1, 2, \ldots, 5$; the Poincaré section of this orbit comprises only the six points $\mathbf{x}_k$, $k = 0, 1, \ldots, 5$. For completeness we show the actual orbit in the following figures: on the left is the function $x(t)$, in the centre is the function $y(t) = \dot{x}$, and on the right is shown the orbit in phase space with the dots marking the six distinct iterates $\mathbf{T}^k(\mathbf{x}_0)$, $k = 0, 1, \ldots, 5$.

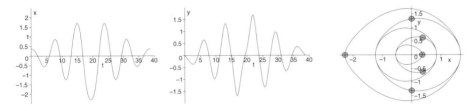

Figure 19.13 $x(t)$ vs t.   Figure 19.14 $y(t)$ vs t.   Figure 19.15   Phase space.

In the unforced limit $F = 0$ the iterates of the unperturbed map, $\mathbf{T}_0$, lie on a phase curve of the autonomous system; in the conservative, undamped case ($\mu = 0$), the bound motion is represented by concentric closed phase curves, figures 19.3 (page 790) and 19.8 (page 794). The frequency of the unperturbed motion, $\omega(E)$, depends upon the energy, and if the ratio $\omega(E)/\Omega$ is irrational the iterates of $\mathbf{T}_0$ fill the phase curve, in the sense that there is an iterate arbitrarily close to any given point on the phase curve. But if the frequency ratio is rational, $\omega(E)/\Omega = r/s$ for coprime integers r and s, $\mathbf{x} = \mathbf{T}_0^{(s)}(\mathbf{x})$ for *all* points on the phase curve: these phase curves consist entirely of fixed points. When $\mu = 0$ and $F > 0$ coordinates may be found such that $\mathbf{T}$ becomes a twist map, as defined in section 18.3.3. Thus for $F > 0$ the rational phase curves, $\omega(E)/\Omega = r/s$, change their nature and are perturbed into a set of ks elliptic points interleaved with ks hyperbolic points for some integer k.

Since there are rational numbers arbitrarily close to every irrational number the smallest perturbation will completely change the nature of the Poincaré section. In practice, however, the changes are rarely so dramatic because the area of the elliptic islands created by a small perturbation is normally relatively small. The reason for this is that the Fourier components of the unperturbed motion normally decrease exponentially, see for example equations 19.14 and 19.10, and it is the size of these components that determines the size of the islands. Not all systems behave like this, however. For example, a particle moving in the square well potential

$$V(x) = \begin{cases} 0, & |x| < a, \\ \infty, & |x| > a, \end{cases}$$

instantaneously reverses direction at $x = |a|$, so $\dot{x}$ is discontinuous here, and from the general theory of chapter 8 the nth Fourier component of $x(t)$ decreases as n^{-2}; this behaviour is demonstrated explicitly in exercise 19.45. Another important example is the one-dimensional Coulomb potential, $V = 1/x$, $x > 0$, which is singular at $x = 0$; in this case the results quoted in the appendix to chapter 7 show that the nth Fourier component of $x(t)$ decreases as $n^{-1/3}$.

In general, if $\omega/\Omega = r/s$ and s is large the iterates $\mathbf{T}^n$, $n = 0, 1, \ldots, N$, can resolve the island structure only when $N \gg s$: for smaller times there is no observable difference between the motion on a rational or an irrational curve.

Now consider a few Poincaré sections in order to obtain some idea of what type of patterns emerge and how the unperturbed motion and periodic driving force interact in different circumstances to produce different types of behaviour. We shall see that in many cases the dominant features of the sections are produced by the perturbation, $F \cos \Omega t$, resonating with particular unperturbed orbits. Recall that if $\epsilon < 0$ the unperturbed frequency ω is always larger than unity, figure 19.4 (page 790), and all solutions are bound. If $\epsilon > 0$ the unperturbed frequency ω is always smaller than unity, figure 19.9 (page 794), and unbounded solutions exist. Thus, in each case we should expect the $\Omega < 1$ and the $\Omega > 1$ Poincaré sections to be quite different so it is convenient to treat the $\epsilon < 0$ and $\epsilon > 0$ cases separately.

The hard spring: $\epsilon < 0$

If $\epsilon < 0$ all motion of the conservative system, $F = 0$, is bound and the unperturbed frequency is larger than unity. In general, for $F > 0$ the force towards the origin increases so rapidly with increasing $|x|$ that all motion is bound. In figure 19.16 is shown a typical Poincaré section for Duffing's equation with $\Omega = \frac{3}{4}$, when there are no unperturbed orbits directly in resonance with the driving force. Here $F = \frac{1}{2}$ and $\epsilon = -\frac{1}{2}$, both fairly large values; most orbits were integrated for 100 periods, a few for 200 or 300 periods.

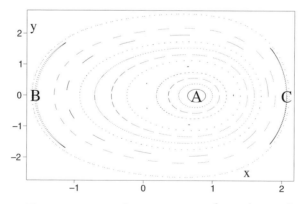

Figure 19.16   Poincaré sections with $\Omega = \frac{3}{4}$, $F = \frac{1}{2}$, $\epsilon = -\frac{1}{2}$.

There are a number of points to notice:

- There is one clear fixed point, $A = (0.774\,899\,4\ldots,0)$, surrounded by points lying on concentric closed curves. This fixed point represents a stable T-periodic orbit which has evolved from the periodic linear oscillator orbit given in equation 19.3 (page 787). A simple harmonic balance approximation to this periodic solution is derived in exercise 19.28 (page 812), and for these parameters is

$$x(t) = 0.760 \cos \Omega t + 0.015 \cos 3\Omega t,$$

which suggests that the fundamental of the driving force dominates the motion. In figure 19.17 is shown $\log(|\mathscr{F}(\omega)|)$, where

$$\mathscr{F}(\omega) = \int_0^{NT} dt\, x(t) e^{i\omega t}$$

is the Fourier transform of the periodic orbit through A over N periods. This confirms that the motion is dominated by the fundamental of the driving force, but also contains a small 3Ω component.

The Fourier transforms shown below were computed using the Maple fast Fourier transform procedure described in the appendix to this chapter. Typically $N = 52$ and a sample of either 10 or 20 points per period was used, and the orbit obtained by numerical integration. If an orbit is known to be periodic with a known frequency it is quicker and easier to compute the Fourier coefficients using the conventional formula, equation 8.13 or 8.15, using N equally spaced points as described in section 8.7.

- There are two other fixed points, at $B = (-1.569\,59\ldots,0)$ and $C = (2.066\,25\ldots,0)$, each of which is surrounded by closed curves topologically distinct from those surrounding A. These fixed points are the elliptic fixed points due to the break-up of the unperturbed phase curve with frequency $\omega = 2\Omega = \frac{3}{2}$ and amplitude $a = 1.85$. The approximation of these periodic orbits is difficult because their Fourier series contain many significant harmonics, as shown by their Fourier transform, figure 19.18 and also exercise 19.20.

- Figure 19.19 shows the Fourier transform of a 'typical' orbit surrounding the fixed point A; its initial condition is $(-0.5, 0)$. The three largest peaks are at the frequencies $\frac{5}{3}\Omega$, Ω and $\frac{7}{3}\Omega$ respectively; the fourth largest is at the frequency 0.35Ω, and is not clearly related to Ω. This solution is multiply periodic.

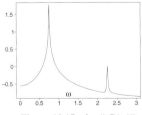

Figure 19.17 $\log(|\mathscr{F}(\omega)|)$ for orbit A.

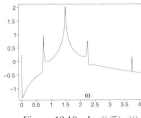

Figure 19.18 $\log(|\mathscr{F}(\omega)|)$ for orbit B.

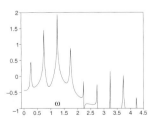

Figure 19.19 $\log(|\mathscr{F}(\omega)|)$ for a non-periodic orbit.

The Poincaré section for the higher frequency, $\Omega = \frac{3}{2}$, figure 19.20, looks quite different because there are unperturbed orbits that can resonate directly with the driving force.

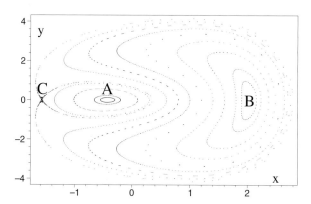

Figure 19.20   Poincaré sections with $\Omega = \frac{3}{2}$, $F = \frac{1}{2}$, $\epsilon = -\frac{1}{2}$.

- The stable fixed point at $A = (-0.423\,18, 0)$ is the perturbed, periodic linear oscillator orbit given in equation 19.3 with $a = 0$ (page 787); this is equivalent to orbit A in figure 19.16. The harmonic balance approximation, exercise 19.28 (page 812), is

$$x(t) = -0.423 \cos \Omega t - 0.0005 \cos 3\Omega t, \qquad \left(\Omega = \frac{3}{2} \right),$$

  showing that the fundamental frequency dominates the motion. The Fourier transform of this orbit is therefore dominated by a single peak at $\Omega = \frac{3}{2}$.

- There is a centre at $B = (2.031\,99, 0)$ and a saddle at $C = (-1.588\,13, 0)$: both are caused by the resonance between the applied force and the unperturbed orbit with frequency $\Omega = \frac{3}{2}$ which, from equation 19.7 (page 790), is seen to have amplitude $a \simeq 1.85$. In the absence of the driving force all orbits on circle of radius 1.85 have frequency $\frac{3}{2}$, so this is an example of the twist map theorem in operation, whereby particular points on a resonant invariant curve of the unperturbed map become elliptic and hyperbolic points when perturbed.

  The harmonic balance method gives the following approximation to these periodic orbits:

$$\begin{aligned} x(t) &= 1.97 \cos \Omega t + 0.059 \cos 3\Omega t, & \text{orbit } B, \\ x(t) &= -1.56 \cos \Omega t - 0.027 \cos 3\Omega t, & \text{orbit } C. \end{aligned}$$

  The derivation of these equations is part of exercise 19.28, (page 812). In these cases the 3Ω harmonic, though small, is larger than for the solution A. Note that these fixed points are produced by a quite different mechanism than that which produces the fixed points B and C in figure 19.16.

- In the neighbourhood of the unstable fixed point at C a homoclinic tangle develops, though it cannot be seen on the scale of this graph.

- The Fourier transforms of the stable periodic orbits through B and A both have a dominant peak at $\Omega = \frac{3}{2}$, as in figure 19.17. In contrast the Fourier transform of the unstable periodic orbit through C, whilst also having a dominant peak at $\Omega = \frac{3}{2}$, has many other components which are very sensitive to the exact value of the initial conditions, because this orbit is unstable. The Fourier transform of the orbits through B and C are shown in the following figures. Note that the unstable orbit through C can be followed numerically only for relatively short times and the small-scale structure in the Fourier transform of this orbit is sensitive to the integration time and the initial condition; here the orbit was integrated for 52 periods, with $x_0 = -1.588\,132\,9$; if $x_0 = -1.588$ a far noisier Fourier transform is produced.

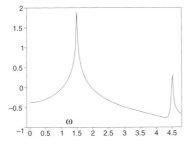

Figure 19.21 $\log(|\mathcal{F}(\omega)|)$ for the stable periodic orbit through B.

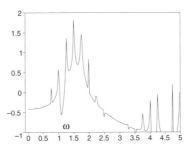

Figure 19.22 $\log(|\mathcal{F}(\omega)|)$ for an unstable periodic orbit near C.

Soft spring: $\epsilon > 0$

When $\epsilon > 0$ unbounded motion is possible, consequently it is normally necessary to limit the magnitude of ϵ and F. In the next two examples we set $\epsilon = \frac{1}{6}$ and choose the driving frequency close to resonance, $\Omega = 0.975$. At this frequency the system is more sensitive to the driving force so F needs to be much smaller than previously.

The Poincaré section for $F = 0.005$ is shown in figure 19.23. This section has a similar structure to that in figure 19.20, though the $y = \dot{x}$ variation is smaller and now solutions beyond the outer curve are unbounded.

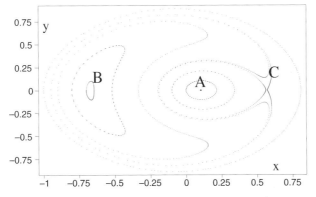

Figure 19.23 Poincaré sections with $\Omega = 0.975$, $\epsilon = \frac{1}{6}$, $F = 0.005$.

The following points should be noted.

- The stable fixed point at $A = (0.104\,118, 0)$ is the perturbed, periodic linear oscillator orbit given in equation 19.3 with $a = 0$ (page 787), and is equivalent to orbit A in figure 19.20. The simple harmonic balance approximation derived in section 19.5.1 gives the approximation

$$x(t) = 0.104\,123 \cos \Omega t \quad (\Omega = 0.975)$$

for this solution, which is in good agreement with the exact solution. Approximations for orbits near this periodic solution are derived in section 19.7.1, in particular compare this with figure 19.34 (page 819).

- The centre at $B = (-0.673\,16, 0)$ and the saddle at $C = (0.569\,48, 0)$ are both caused by the resonance between the applied force and the unperturbed orbit with angular frequency $\omega = 0.975$ which, from equation 19.12, can be seen to have amplitude $a = 0.629$. This is another example of the twist map theorem.

  The Fourier transforms of the orbits through A, B and C are dominated by a single peak at $\Omega = 0.975$.

- As in figure 19.20, the stable and unstable manifolds at C develop into a homoclinic tangle.

The positions of the fixed points A, B and C in figure 19.23 depend upon the magnitude of the force, F. With increasing F the fixed points at A and C move towards each other, eventually coalescing when $F \simeq 0.012$, section 19.5.1, and B moves to the left. In the next example the force is much stronger, $F = 0.2$; the bound motion is confined to the interior of the jagged outer curve, the fixed points at A and C have disappeared and that at B has moved to $(-1.274\,82, 0)$, but remains stable. For this stronger force resonance islands surrounding the period-6 orbit, passing through $(-0.6148, 0)$, are clearly visible, but these can be distinguished only after about 400 periods. This period-6 orbit is fairly complicated and its graph is shown in figure 19.26. With increasing F the area of phase space containing bound motion decreases: the fate of the T-periodic orbit is interesting and is explored in some detail in exercise 19.19. Briefly, it becomes unstable at $F \simeq 1.11$ but not with the birth of a $2T$-periodic orbit.

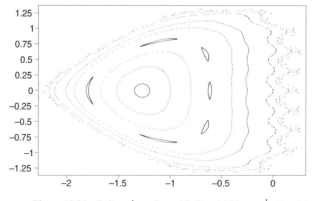

Figure 19.24   Poincaré section with $\Omega = 0.975$, $\epsilon = \frac{1}{6}$, $F = 0.2$.

Periodic orbits

It is clear from the Poincaré sections of figures 19.16, 19.16 and 19.24 that periodic orbits are important because they provide a focus around which other phase curves are organised; methods of finding them are therefore important. There are no foolproof techniques that can be applied to all problems and which, for any problem, yield all the periodic orbits; but the symmetry method discussed in chapter 18 can often help and here we show how it may be applied to the Duffing oscillator, for which the potential is even.

Observe that if $\tau = -t$ and $v = -y(= \dot{x})$ the equations of motion for $(x(t), y(t))$ are identical to those for $(x(\tau), v(\tau))$. Thus if $\mathbf{S}$ represents reflections in the x-axis, $\mathbf{S}(x, y) = (x, -y)$, the effect of the transformation $\mathbf{S}$ followed by $\mathbf{T}^{-1}$ is the same as $\mathbf{T}$ followed by $\mathbf{S}$, that is,

$$\mathbf{T}^{-1}\mathbf{S}(\mathbf{x}) = \mathbf{S}\mathbf{T}(\mathbf{x}),$$

so $\mathbf{S}$ is a symmetry curve of $\mathbf{T}$, since $(\mathbf{TS})^2 = \mathbf{I}$.

Exercise 19.15
Write Maple procedures that numerically compute the two transformations $\mathbf{T}^{-1}\mathbf{S}(\mathbf{x})$ and $\mathbf{S}\mathbf{T}(\mathbf{x})$ and use these to demonstrate numerically the identity $\mathbf{T}^{-1}\mathbf{S}(\mathbf{x}) = \mathbf{S}\mathbf{T}(\mathbf{x})$ for various values of $\mathbf{x}$.

It follows that we may use the theory of section 18.3.4 to restrict the search for some periodic orbits to the x-axis. Thus for nT- or $2nT$-period orbits we start with the initial condition $(x, 0)$ and integrate forward through one period to form

$$\mathbf{T}^n(x, 0) = (g(x), f(x)). \tag{19.17}$$

The periodic orbit is given by the values of x for which $f(x) = 0$. In practice we evaluate $f(x)$ at a suitable number of points and plot its graph to provide a rough guide of the position of the periodic orbits; then we can use a standard technique, for instance the false position method, described in chapter 3, to obtain a more accurate estimate of a root.

As an example consider a period-6 orbit. In figure 19.25 we show the graph of $f(x)$, where $\mathbf{T}^6(x, 0) = (g(x), f(x))$ in the case $\Omega = 0.975, \epsilon = \frac{1}{6}$ and $F = 0.2$ treated in figure 19.24. It is seen that $f(x) = 0$ has four roots in the range shown; the zero at $x = -1.2748\ldots$ is the T-periodic orbit, the two at $x = -1.801$ and -0.6148 belong to the $6T$-periodic orbit and the root at $x = 0.3677$ is the orbit shown in figure 19.15 (page 799). On the right the phase curve of the period-6 orbit passing through $(-0.6148, 0)$ is shown.

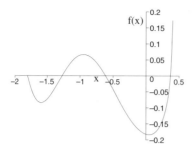

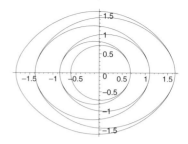

Figure 19.25   Graph of $f(x)$ where
$\mathbf{T}^6(x,0) = (g,f)$.

Figure 19.26   Phase curve of the period-6
orbit through $(-0.6148, 0)$.

Exercise 19.16

The Poincaré section in figure 19.24 suggests that long-period orbits exist in the neighbour-hood of $x \simeq 0.2$, and the shape of the plot suggest that some of these high-period orbits must intersect the x-axis near $x = -2.1$.

Write a procedure to compute $f(x)$, where $\mathbf{T}^n(x,0) = (g(x), f(x))$ and use this to plot the graph of $f(x)$ for $-2.2 < x < -1.2$ and $n = 25$.

Write another procedure, based on the false position method described in chapter 3, to show that this period-25 orbit intersects the x-axis at $x = -2.0789\ldots$.

Exercise 19.17

Use the procedures developed in the previous question to show that the initial condition $\mathbf{x}_0 = (0.377\,373\ldots, 0)$ gives a 20-period orbit and plot the phase curves of this orbit.

Note: care is needed in this case; compute $\mathbf{T}^{10}(x,0)$ for $0.37 < x < 0.38$.

Exercise 19.18

Use a modification of the procedure given in the appendix to this chapter to compute $\log(|\mathscr{F}(\omega)|)$ in conjunction with dsolve to plot graphs of the Fourier transform of some orbits. In particular consider the initial conditions $\mathbf{x}_0 = (x_0, 0)$ with the parameters

(i) $\Omega = 3/4$, $F = 1/2$, $\epsilon = -1/2$, $x_0 = -1.569\,59$, 0.7749, $2.066\,25$, (figure 19.16);
(ii) $\Omega = 3/2$, $F = 1/2$, $\epsilon = -1/2$, $x_0 = -1.588\,13$, $-0.423\,18$, $2.031\,99$, (figure 19.20);
(iii) $\Omega = 0.975$, $F = 0.005$, $F = 0.005$, $\epsilon = 1/6$, $x_0 = -0.673\,16$, $0.104\,118$, $0.569\,48$,
      (figure 19.23);
(iv) $\Omega = 0.975$, $F = 0.2$, $\epsilon = 1/6$, $x_0 = -1.274\,82$, 0.6148, (figure 19.24).

Exercise 19.19

Figure 19.24 shows that for $\Omega = 0.975$, $\epsilon = 1/6$ and $F = 0.2$ there is a stable T-periodic orbit through $\mathbf{x} = (-1.274\,81, 0)$. This orbit exist for higher values of F, but at $F \simeq 1.11$ it bifurcates. This can be shown by the method used above, figure 19.25.

If $\mathbf{T}(x,0) = (g(x), f(x))$ some of the roots of $f(x) = 0$ are shown in the following table.

F	Some roots of $f(x) = 0$		
1.00		−2.0377137	
1.10		−2.0977908	
1.11		−2.1035901	
1.111	−2.1328704,	−2.1041740	−2.0740086
1.12	−2.2070246	−2.1093652	−1.9922467
1.13	−2.2485935	−2.1151010	−1.9422144

Assuming that the periodic orbit has the Fourier series

$$x(t) = a_0 + \sum_{k=1}^{\infty} a_k \cos k\Omega t,$$

compute the Fourier coefficients for these solutions and observe the changes as F increases through 1.11.

This bifurcation is examined further in exercises 19.55 and 19.56.

Exercise 19.20
Show that the periodic orbit B shown in the surface-of-section 19.16 has the Fourier components:

a_0	a_1	a_2	a_3	a_4	a_5	a_6	a_7
0.01994	0.1278	−1.780	0.09871	−0.002944	0.01549	−0.04329	0.005454

where the coefficients a_k are defined in the previous exercise.

19.5 Duffing's equation: forced, no damping

19.5.1 Approximations to T-periodic orbits
Although there are infinitely many periodic orbits, generally the most important are those with short periods. These are sometimes relatively easy to approximate because their Fourier series representation contains few significant harmonics, as seen in figures 19.17–19.19. In this case the harmonic balance method may be used to good effect. An advantage of this technique is that the resulting algebra sometimes has a clear physical interpretation: another is that it often works when perturbation theory is invalid.

A T-periodic solution of the equation

$$\frac{d^2x}{dt^2} + x - \epsilon x^3 = F \cos \Omega t, \quad T = \frac{2\pi}{\Omega}, \tag{19.18}$$

can be represented as a truncated Fourier series. From the previous discussion the initial point will lie on the x-axis, so the Fourier series has the form

$$x(t) = a_0 + \sum_{k=1}^{N} a_k \cos k\Omega t.$$

The unperturbed solution, $F = 0$, has a Fourier series with only odd harmonics —

because the force $-x + \epsilon x^3$ is an odd function of x — and has zero mean value, so $a_{2k} = 0$, $k = 0, 1, \ldots$. Therefore for small F we assume the truncated Fourier series,

$$x(t) = \sum_{k=1}^{N} a_{2k+1} \cos(2k + 1)\Omega t. \tag{19.19}$$

Exercises 19.19, 19.55 and 19.56 show that this assumed form can be invalid for large F.

A natural first approximation is obtained by ignoring all but the first term, the fundamental, substituting $x = a \cos \Omega t$ into the equation and setting the coefficient of $\cos \Omega t$ to zero to obtain the following cubic equation for a:

$$a(1 - \Omega^2) - \frac{3\epsilon}{4} a^3 = F. \tag{19.20}$$

The amplitude of this motion is $|a|$. As before, it is easier to consider the cases $\epsilon < 0$ and $\epsilon > 0$ separately, though, apart from various sign changes, the algebra is identical. In the next two subsections we show how this approximation can be used to understand the dominant features of figures 19.16 and 19.20.

Hard spring: $\epsilon < 0$

For $\epsilon < 0$ set $\delta = -\epsilon > 0$ and define a new variable b as follows:

$$a = b \frac{4\sqrt{|1 - \Omega^2|}}{3\sqrt{\delta}}.$$

With these definitions equation 19.20 becomes

$$4b^3 + 3b \;=\; g = \frac{9F\sqrt{\delta}}{4(1 - \Omega^2)^{3/2}} > 0, \quad \Omega < 1, \tag{19.21}$$

$$4b^3 - 3b \;=\; g = \frac{9F\sqrt{\delta}}{4(\Omega^2 - 1)^{3/2}} > 0. \quad \Omega > 1, \tag{19.22}$$

so, as expected, a different response is obtained according as $\Omega < 1$ or $\Omega > 1$. The case $\Omega = 1$ is special. The qualitative behaviour of these equations is most easily seen by drawing graphs of $y = 4b^3 \pm 3b$.

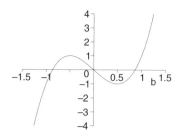

Figure 19.27   Graph of $y = 4b^3 - 3b$.

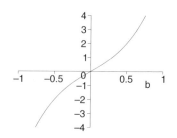

Figure 19.28   Graph of $y = 4b^3 + 3b$.

The function $y = 4b^3 - 3b$ has two stationary points with a local maximum of unity at $b = -1/2$, so the equation $4b^3 - 3b = g$ has three real roots if $0 < g < 1$ and one if $g > 1$. The function $y = 4b^3 + 3b$ is monotoic increasing so the equation $4b^3 + 3b = g$ has one real root for all g. These roots are readily found using trigonometric identities, exercises 19.27 and 19.42. The real solution of equation 19.21 is

$$b = \sinh\left(\frac{1}{3}\ln\left(g + \sqrt{1+g^2}\right)\right), \quad \Omega < 1, \tag{19.23}$$

and the real solutions of 19.22 are

$$b = \begin{cases} \cos\left(\frac{1}{3}\cos^{-1}g\right), & \cos\left(\frac{2\pi}{3} \pm \frac{1}{3}\cos^{-1}g\right), & 0 < g < 1, \quad \Omega > 1, \\ \cosh\left(\frac{1}{3}\ln(g + \sqrt{g^2 - 1})\right), & g > 1. \end{cases} \tag{19.24}$$

When $\Omega < 1$ there is only one real root of the cubic 19.21, because there are no unperturbed solutions having the same frequency as the driving force, figure 19.4. If F (and g) is small the easiest method of approximating this root is to use first-order perturbation theory, so equation 19.21 becomes $3b = g$, or

$$a = \frac{F}{1 - \Omega^2} + O(F^2).$$

The same result may be obtained by expanding 19.23. In this limit, because the cubic term is negligible, the approximation reproduces the second term of equation 19.3 (page 787), with $\mu = 0$. Equation 19.23 shows that the perturbation series converges only for $|g| < 1$, that is, $9F\sqrt{|\epsilon|} < 4|1 - \Omega^2|^{3/2}$, so is of limited value, particularly if $\Omega \simeq 1$. For instance, with the parameters used in figure 19.16 (page 800), perturbation theory is invalid, though the harmonic balance method provides a good approximation to the orbit through A.

Exercise 19.21
Show that the parameters $F = \frac{1}{2}$, $\epsilon = -\frac{1}{2}$ and $\Omega = \frac{3}{4}$, used to construct figure 19.16 (page 800), are such that the perturbation series does not converge. In this case use equation 19.23 to show that an approximation to the periodic orbit through A is

$$x = 0.763 \cos\left(\frac{3}{4}t\right).$$

Use the method described at the end of the last section to show that a more accurate estimate for a is 0.7749 and compare the exact, numerically computed, orbit with that given above.
Note: the Fourier transform of the exact orbit, figure 19.17 (page 801), shows that the magnitude of the third harmonic is about 60 times smaller than that of the first.

If $\Omega > 1$ resonances with unperturbd orbits are possible, and equation 19.22 has three real roots if

$$g < 1, \quad \text{that is,} \quad F\sqrt{\delta} < \frac{4}{9}(\Omega^2 - 1)^{3/2}, \quad (\Omega > 1),$$

otherwise there is only one. For F small perturbation theory shows that the three real roots are approximately $b \simeq -g/3$ and $b \simeq \pm\sqrt{3}/2 + g/6$. The small root gives

$$a = -\frac{F}{\Omega^2 - 1} - \frac{3F^3\delta}{4(\Omega^2 - 1)^4} + O(F^5),$$

which is the equivalent of the periodic solution found above for $\Omega < 1$ and it approximates orbit A in figure 19.20 (page 802). The other two roots are just the perturbed T-periodic solutions of the unforced nonlinear oscillator and give

$$a = \pm\frac{2\sqrt{\Omega^2 - 1}}{\sqrt{3\delta}} + \frac{F}{2(\Omega^2 - 1)} \mp \frac{3F^2\sqrt{3\delta}}{16(\Omega^2 - 1)^{5/2}} + \cdots.$$

This very simple analysis explains very easily and quickly some of the important features of the Poincaré section shown in figures 19.16 (page 800) and 19.20 (page 802), though, without further work, it does not provide information about the stability of the orbits and hence the nature of the fixed points of $\mathbf{T}$; this is done in the sections 19.6 and 19.7.

Exercise 19.22
Show that for the parameters $F = \frac{1}{2}$, $\epsilon = -\frac{1}{2}$ and $\Omega = \frac{3}{2}$, used to construct figure 19.20 (page 802), the harmonic balance approximation gives the three periodic solutions

$$x_C(t) = -1.577\cos\Omega t, \quad x_A(t) = -0.4226\cos\Omega t \quad \text{and} \quad x_B(t) = 2\cos\Omega t,$$

corresponding to the three periodic solutions shown in that figure.

Soft spring, low frequency: $\epsilon > 0$, $\Omega < 1$
For a soft spring, $\epsilon > 0$, the natural frequency of the unforced system is less than unity, so first-order resonances occur only for $\Omega < 1$. Also, unbounded motion is possible so near a resonance the magnitude of the force, F, must be small for bound solutions to exist.

If $\epsilon > 0$, equation 19.20 can be written as

$$\begin{aligned} 4c^3 - 3c = g, && \Omega < 1, \\ 4c^3 + 3c = g, && \Omega > 1, \end{aligned} \tag{19.25}$$

where $a = -4c\sqrt{|1 - \Omega^2|}/3\sqrt{\epsilon}$ and $g = 9F\sqrt{\epsilon}/\left(4\left|1 - \Omega^2\right|^{3/2}\right)$, which are the same as equation 19.21 and 19.22. For the example in figure 19.23, $\Omega = 0.975$, $\epsilon = \frac{1}{6}$ and $F = 0.005$, we have $g = 0.419$ and equation 19.25 has three real roots corresponding to the three periodic orbits seen in that figure.

Exercise 19.23
For the parameters used in figure 19.23, $\Omega = 0.975$, $\epsilon = \frac{1}{6}$ and $F = 0.005$, show that the harmonic balance equation has the three roots $a = \{-0.674, 0.104\,12, 0.5699\}$. Use the procedures developed in exercise 19.16 to show that the exact equivalents of these are $x_0 = \{-0.673\,16, 0.104\,12, 0.569\,48\}$ and make a graphical comparison of the exact and approximate solutions.

Exercise 19.24
Repeat the above exercise for the parameters used in figure 19.24, $\Omega = 0.975$, $\epsilon = \frac{1}{6}$ and $F = 0.2$, but note that now there is only one real root of the harmonic balance equation, $a = -1.2819$, and that the exact equivalent of this is $x_0 = -1.2748$. Determine how a and x_0 change as F increases to 1.2, taking care when $F \simeq 1.11$.

The cubic equations 19.22 ($\epsilon < 0$) and 19.25 ($\epsilon > 0$) have three real roots if $0 < g < 1$, but only one for $g > 1$. As $g \to 1$ from below the two roots approach each other and coalesce when $g = 1$. The Poincaré sections is figures 19.20 ($\epsilon < 0$) and 19.23 ($\epsilon > 0$) show that one of these periodic orbits is stable and that the other is unstable (hyperbolic) so when they coalesce the Poincaré index of a curve enclosing both is unchanged at zero.

It is possible to determine whether or not periodic orbits of the exact dynamics coalesce and annihilate each other simply by plotting the graphs of $f(x)$ where $\mathbf{T}(x, 0) = (g(x), f(x))$. Consider the case $\Omega = \frac{3}{4}$, $\epsilon = 0.2$, so the critical force according to the harmonic balance approximation, obtained by setting $g = 1$, is $F_c = 0.287$. In the following three figures are shown graphs of $f(x)$ for $F = 0.28$, 0.29 and 0.30 and we observe that the two fixed points of $\mathbf{T}(\mathbf{x})$ do indeed coalesce approximately where predicted.

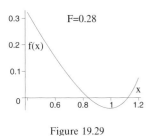

Figure 19.29

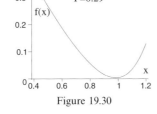

Figure 19.30

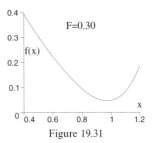

Figure 19.31

Exercise 19.25
Repeat the calculations that produce figures 19.29–19.31 but for $\Omega = \frac{3}{2}$ and $\epsilon = -\frac{1}{2}$, and show that the exact periodic orbits coalesce at $F_c \simeq 0.875$ compared with the value $F_c = 0.878$ predicted by the harmonic balance approximation.

Exercise 19.26
Equation 19.20 changes its form when $\Omega = 1$ and for $\Omega \simeq 1$ the perturbation expansion is of little value, the radius of convergence being very small. If $\Omega = 1$ the numerical method discussed in the previous section can be used to find the amplitude of the periodic orbit; some typical values of x_0, where $\mathbf{T}(x_0, 0) = (g(x_0), 0)$, for $\epsilon = -\frac{1}{2}$ and $\frac{1}{6}$ are given below.

	F	0.001	0.003	0.01	0.1	0.5	1.0
$\epsilon = -1/2$	x_0	0.1387	0.2001	0.2990	0.6465	1.1162	1.4200
$\epsilon = 1/6$	x_0	−0.2000	−0.2884	−0.4306	−0.9256	−1.5745	−1.9752

Compare the values given in this table with the solution of equation 19.20, that is, $a = -(4F/3|\epsilon|)^{1/3}\mathrm{sgn}(\epsilon)$.

This periodic orbit is stable for $F < 1$ — as may be seen by plotting the surface-of-sections of neighbouring orbits, or by the methods described in section 19.6. However, we should expect it to become unstable for some critical value of F: the nature of this bifurcation is examined in exercise 19.55.

Exercise 19.27
Use the identity $4\sinh^3 z + 3\sinh z = \sinh 3z$ to show that the solution of equation 19.21 is the expression given in 19.23. Use this solution to obtain the perturbation expansion

$$a = \frac{F}{1 - \Omega^2} - \frac{3F^3\delta}{4(1 - \Omega^2)^4} + \frac{27F^5\delta^2}{16(1 - \Omega^2)^7} + O(\delta^3),$$

for the amplitude of a T-periodic orbit.

Exercise 19.28
Using the trial function

$$x(t) = a\cos\Omega t + b\cos 3\Omega t$$

in Duffing's equation, $\ddot{x} + x - \epsilon x^3 = F\cos\Omega t$, show that the equations for a and b are

$$(1 - \Omega^2)a - \frac{3}{4}\epsilon a\left(a^2 + ab + 2b^2\right) = F,$$

$$b(1 - 9\Omega^2) - \frac{1}{4}\epsilon\left(3b^3 + a^3 + 6a^2b\right) = 0.$$

By solving these equations numerically show that the periodic orbits through A, B and C in figure 19.20 (page 802), for which $F = \frac{1}{2}$, $\Omega = \frac{1}{2}$ and $\epsilon = -\frac{1}{2}$, are given approximately by

$$x_A(t) = -0.42268\cos\Omega t - 0.0005\cos 3\Omega t,$$

$$x_B(t) = 1.9716\cos\Omega t + 0.0587\cos 3\Omega t,$$

$$x_C(t) = -1.5604\cos\Omega t - 0.0273\cos 3\Omega t.$$

For these values of (F, Ω, ϵ) there is a stable T-periodic orbit with initial condition $\mathbf{x}_0 = (-0.422\,463\ldots, 0)$. Use Maple to compute this solution numerically and compare it with $x_A(t)$ defined above.

The Poincaré section in figure 19.16 was drawn using the parameters $F = \frac{1}{2}$, $\Omega = \frac{3}{4}$ and $\epsilon = -\frac{1}{2}$; show that in this case the harmonic balance approximation to the periodic orbit through A is

$$x(t) = 0.7595\cos\Omega t + 0.0151\cos 3\Omega t.$$

In this case the stable T-periodic solution has the initial condition $\mathbf{x}_0 = (0.774\,90\ldots, 0)$. Compare the numerically generated solution with this approximation.

Exercise 19.29
The equations for (a, b) derived in exercise 19.28 are nonlinear and often cannot be solved using a direct perturbation expansion. However, usually $|a| \gg |b|$, which suggests putting $a = a_0 + c$, where a_0 is a solution of the first equation with $b = 0$, expanding both equations to first order in (c, b) and then solving the resulting linear equations. Show that this procedure gives the approximations

$$x_A(t) = -0.422\,68\cos\Omega t - 0.0005\cos 3\Omega t,$$

$$x_B(t) = 1.9728\cos\Omega t + 0.0590\cos 3\Omega t,$$

$$x_C(t) = -1.5609\cos\Omega t - 0.0273\cos 3\Omega t.$$

19.5.2 Subharmonic response

Linear systems respond to a periodic driving force at the driving frequency, Ω. Nonlinear systems are different and can also respond at the subharmonic frequencies Ω/n, $n = 2, 3,\ldots$. In any particular system only certain frequencies are allowed, and these are determined by the form of the nonlinearity. Such behaviour is important because it provides a mechanism for changing a high input frequency to a lower frequency response.

For the Duffing equation the most important subharmonic is $\Omega/3$ because the nonlinear term is cubic. This connection is made transparent by setting $x = a\cos(\Omega t/3)$ to give

$$x^3 = \frac{1}{4}a^3 \left(3\cos\Omega t + 4\cos(\Omega t/3)\right).$$

The first term has the same frequency as the driving force. Also, the unperturbed motion ($F = 0$) has natural frequency $\omega < 1$ if $\epsilon > 0$, figure 19.9 (page 794), hence if $\Omega/3 < 1$ there can also be a resonance between the driving force and the unperturbed motion through the x^3 term.

This analysis suggests using a harmonic balance method with the trial function

$$x(t) = a\cos\left(\frac{1}{3}\Omega t\right) + b\cos\Omega t, \quad \text{with} \quad \Omega \simeq 3, \tag{19.26}$$

which has period $6\pi/\Omega \simeq 2\pi$. Substituting this into the equation of motion $\ddot{x}+x-\epsilon x^3 = F\cos\Omega t$, equating the coefficients of $\cos(\Omega t/3)$ and $\cos\Omega t$ to zero and assuming $a \neq 0$ gives the following equations:

$$\begin{aligned} 1 - \frac{\Omega^2}{9} - \frac{3\epsilon}{4}\left(a^2 + ab + 2b^2\right) &= 0, \\ b(1 - \Omega^2) - \frac{\epsilon}{4}\left(a^3 + 6a^2b + 3b^3\right) &= F. \end{aligned} \tag{19.27}$$

These equations are easier to understand if we write $a^2 + ab + 2b^2 = 2(b+a/4)^2 + 7a^2/8$ and define a new variable $c = b + a/4$, which simplifies the first equation:

$$\frac{3\epsilon}{2}\left(c^2 + \frac{7}{16}a^2\right) = 1 - \frac{\Omega^2}{9}, \tag{19.28}$$

which defines an ellipse if $\epsilon > 0$ and $\Omega < 3$ — (or $\epsilon < 0$ and $\Omega > 3$) — as expected from the previous discussion. This type of analysis is valid only if a and b are both $O(1)$, hence we must have $|\Omega - 3| = O(\epsilon)$, so the second of 19.27 can be approximated by the straight line

$$b = c - \frac{a}{4} = -\frac{F}{\Omega^2 - 1} + O(\epsilon). \tag{19.29}$$

This intersects the ellipse 19.28 if

$$F < F_{\text{crit}} = \frac{4}{3}(\Omega^2 - 1)\sqrt{\frac{9 - \Omega^2}{21\epsilon}}, \quad |\Omega - 3| = O(\epsilon). \tag{19.30}$$

It follows that if $F > F_{\text{crit}}$ there are no subharmonic solutions, and if $F < F_{\text{crit}}$ there are two. Since these two solutions coalesce and disappear as F increases through F_{crit}, consideration of the Poincaré index suggests that one of these period-3 orbits is stable and the other unstable. In section 19.6 this will be shown directly, exercise 19.33.

The predictions made by this simple approximation are readily checked by numerical integration. The solutions of equations 19.28 and 19.29 will give the value of $x(0) = c + 3a/4$, and this quantity may be computed independently using the method described in section 19.4 for finding the fixed points of $\mathbf{T}^3$. Some typical values of the relevant quantities are given in the table.

Table 19.1. *Values of a and c, the roots of equations 19.28 and 19.29, and the values of $x(0)$ computed from these solutions and by numerically finding the fixed points of $\mathbf{T}^3$. Here $\Omega = 2.85$ and $\epsilon = \frac{1}{6}$.*

F	a	c	$x(0)$ approximate	$x(0)$ exact
1.0	0.934 −0.798	0.093 −0.334	0.7929 −0.9323	0.7935 −0.9318
2.0	0.942 −0.668	−0.412 −0.441	0.6654 −0.9423	0.6665 −0.9416
3.0	0.901 −0.483	−0.187 −0.537	0.4882 −0.8988	0.4894 −0.8985
4.0	0.776 −0.207	−0.356 −0.609	0.2261 −0.7644	0.2268 −0.7653

The following figure shows the values of $x(0)$ obtained by numerically determining the fixed points of $\mathbf{T}^3$ for $\Omega = 2.85$ and $\epsilon = \frac{1}{6}$, for which 19.30 gives $F_{\text{crit}} = 4.76$.

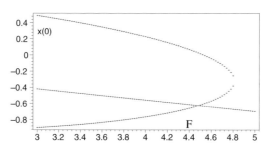

Figure 19.32   Locus of the initial conditions for the 1- and 3-period orbits for $\Omega = 2.85$, $\epsilon = \frac{1}{6}$.

In this figure the straight line is a fixed point of $\mathbf{T}(\mathbf{x})$; the curved line shows the fixed points of $\mathbf{T}^3(\mathbf{x})$; the two branches of this curve coalesce at $F \simeq 4.812$, close to the estimate 4.76 of F_{crit} given by equation 19.30.

In addition the fixed points of $\mathbf{T}$ and $\mathbf{T}^3$ coincide at $F \simeq 4.48$; in the harmonic balance approximation this occurs when $a = 0$, that is when the two equations

$$F = b(1 - \Omega^2) - \frac{3\epsilon}{4}b^3, \quad \frac{3}{2}\epsilon b^2 = 1 - \frac{\Omega^2}{9},$$

are satisfied. For $\epsilon = \frac{1}{6}$ and $\Omega = 2.85$ these give $b = x(0) = -0.62$ and $F = 4.48$, which are close to the exact values.

From these comparisons we conclude that the harmonic balance method provides a good description of the $\frac{1}{3}$ subharmonic. The comparisons made in the following exercise will confirm this.

Exercise 19.30
Use the values of a, c and the initial conditions given in table 19.1 to plot graphs of some of these subharmonic solutions and to compare the exact solutions with those given by approximation 19.26.

19.6 Stability of periodic orbits: Floquet theory

Periodic orbits may be stable or unstable and can change character as parameters change, so methods of characterising orbits are important. Stability is determined by the behaviour of neighbouring solutions, so we make an expansion about the periodic orbit, to give Hill's equation, and use Floquet theory, chapter 12, to determine whether or not the original solution is stable.

Let $p(t)$ be the periodic orbit; write the nearby solution in the form $x(t) = p(t) + \xi(t)$, where $\xi(t)$ is an unknown function sufficiently small that all terms $O(\xi^2)$ may be ignored. On substituting $x(t)$ into Duffing's equation 19.19, expanding and using the fact that $p(t)$ also satisfies the equation, the following equation for $\xi(t)$ is obtained:

$$\frac{d^2\xi}{dt^2} + \left(1 - 3\epsilon p(t)^2\right)\xi = 0. \tag{19.31}$$

This is Hill's equation, which has both stable and unstable solutions, depending upon the values of Ω, ϵ and F. The stability problem is thus reduced to finding the boundaries between these regions.

In the simplest application we use a harmonic balance approximation with one term to approximate $p(t)$; that is, we set $p(t) = A \cos \Omega t$, where $A(\epsilon, \omega, F)$ is a known function. Then equation 19.31 reduces to Mathieu's equation:

$$\frac{d^2\xi}{d\tau^2} + (a - 2q \cos 2\tau)\xi = 0, \quad q = \frac{3A^2|\epsilon|}{4\Omega^2}, \quad a = \frac{1}{\Omega^2} - 2q \operatorname{sgn}(\epsilon), \tag{19.32}$$

which has stable solutions if $a_r(q) \le a \le b_{r+1}(q)$, where $a_r(q)$ and $b_{r+1}(q)$ are the eigenvalues of the periodic Mathieu functions, defined in section 11.5.1, see also figure 11.18 (page 414) and figure 12.4 (page 455). Note, the variable a used here should not be confused with that used in the previous section to denote the amplitude of the motion.

As an example consider the case shown in the Poincaré section 19.20 (page 802), where $\Omega = \frac{3}{2}$, $F = \frac{1}{2}$ and $\epsilon = -\frac{1}{2}$; for these parameters the solution of equation 19.20 (page 808) gives the three values for the pairs (q, a):

Label of orbit in figure 19.20	A	a	q	stability
C	-1.577	0.415	1.27	unstable
A	-0.423	0.0298	0.504	stable
B	2.00	0.667	1.778	stable

In the following figure graphs of a_0, b_1, and a_1 are drawn and the position of the pairs (q, a) for each of the orbits A, B and C is labelled appropriately: for these small values of q the series defined in exercise 11.45 (page 424), suffice. This diagram shows the orbits A and B are stable and C is unstable, in agreement with the Poincaré section 19.20.

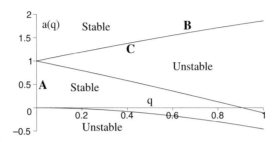

Figure 19.33   Stability regions for equation 19.32 and the 3-period orbits in the case $\Omega = \frac{3}{2}$, $F = \frac{1}{2}$ and $\epsilon = -\frac{1}{2}$.

Exercise 19.31
Derive equations 19.31 and 19.32.

Exercise 19.32
When $\epsilon > 0$ and $\Omega < 1$ the equation $a(1 - \Omega^2) - 3\epsilon a^3/4 = F$ has three roots given in equation 19.24. For the parameters $\Omega = 0.975$, $\epsilon = 1/6$ and $F = 0.005$, used to construct figure 19.23, show that two of these periodic orbits are stable and one unstable and that this classification is consistent with that figure.

When the expression for the periodic solution $p(t)$ is more complicated, or is generated numerically, the variational equation 19.31 cannot be solved in terms of Mathieu functions and then it is usually necessary to solve equations 19.31 numerically in order to construct the monodromy matrix.

Exercise 19.33
Show that if $\Omega = 2.85$, $\epsilon = 1/6$ and $F = 2$, equations 19.27 have two real solutions leading to the approximate $3T$-periodic solutions

$$x_1(t) = 0.9416 \cos\left(\frac{1}{3}\Omega t\right) - 0.2808 \cos \Omega t,$$

$$x_2(t) = -0.6609 \cos\left(\frac{1}{3}\Omega t\right) - 0.2808 \cos \Omega t, \quad \Omega = 2.85.$$

(i) Compare these approximations with the exact solution obtained by numerical integration using initial conditions defined by the approximations.
(ii) Show that the monodromy matrices generated by these two approximate solutions are, respectively,

$$E_1 = \begin{pmatrix} 0.8817 & -0.3972 \\ 0.5605 & 0.8817 \end{pmatrix}, \quad E_2 = \begin{pmatrix} 1.056 & 0.2622 \\ 0.4408 & 1.056 \end{pmatrix},$$

and deduce that $x_1(t)$ is a stable solution and $x_2(t)$ an unstable solution.

19.7 Averaging methods

Averaging methods are important and useful when fast and slow motions coexist. Physical examples of dynamical system having fast and slow motions are abundant. For instance, the length of a pendulum in a clock varies slowly by comparison with the pendulum period as the temperature changes. The motion of the planets round the Sun is rapid by comparison with the slowly changing mass of the Sun, a relative change of about 10^{-13} per year. The simplicity of the gas laws is due to the motion of the gas molecules being far faster than the relative motion of the containing walls.

Averaging methods are important because they decrease the number of dimensions of the dynamics by removing the fastest component, and under very general circumstances provide an approximation which remains close to the exact motion for long, but not necessarily all, times. In general they give the first term in an asymptotic expansion, though other terms are usually very difficult to find; we consider only the first term.

The slow motion considered near a resonance is produced by viewing the system in a reference frame moving with a particular (approximate) periodic solution; the neighbouring solutions will then appear to be moving relatively slowly. The transformation to a moving reference frame usually produces rapidly oscillating terms in the equations of motion that can be averaged to produce simpler, solvable equations.

In this section, we first illustrate the general idea by expanding about a periodic solution of Duffing's equation, we then state a general theorem before considering other examples.

19.7.1 Averaging over periodic solutions

First consider the case where the driving frequency Ω is similar to the natural frequency of the unperturbed linear system, $\Omega \simeq 1$; the aim is to approximate solutions in the neighbourhood of the T-periodic solution, where $T = 2\pi/\Omega$. There are several ways of applying averaging methods to this problem, and here we proceed in a manner convenient for the application of the general theory that follows; alternative, but equivalent, formulations can be found in Jordan and Smith (1999, chapter 4) or Nayfeh (1973, chapter 5).

The phase space coordinates are x and $y = \dot{x}$, so Duffing's equation, with $F = \epsilon f$, can be written in the form

$$\frac{d\mathbf{x}}{dt} = A\mathbf{x} + \epsilon \mathbf{g}(\mathbf{x}, t), \qquad (19.33)$$

where

$$\mathbf{x} = \begin{pmatrix} x \\ y \end{pmatrix}, \quad A = \begin{pmatrix} 0 & 1 \\ -1 & 0 \end{pmatrix}, \quad \mathbf{g} = \begin{pmatrix} 0 \\ x^3 + f \cos \Omega t \end{pmatrix}.$$

The solution, $\mathbf{x}_0(t)$, of the unperturbed problem, $\epsilon = 0$, is simply a clockwise rotation in phase space, with unit angular velocity,

$$\mathbf{x}_0(t) = \begin{pmatrix} \cos t & \sin t \\ -\sin t & \cos t \end{pmatrix}.$$

The T-periodic orbit has the approximate solution $a(\cos \Omega t, -\Omega \sin \Omega t)^\top$, $\Omega \simeq 1$, which traces out an ellipse in phase space. We thus move to a coordinate system which is rotating with angular speed Ω, $\mathbf{y} = R(t)^{-1}\mathbf{x}$, where R is the rotation,

$$R(t) = \begin{pmatrix} \cos \Omega t & \sin \Omega t \\ -\sin \Omega t & \cos \Omega t \end{pmatrix}.$$

The equation of motion in the $\mathbf{y}$ representation is obtained by substituting $\mathbf{x} = R\mathbf{y}$ into equation 19.33 to give

$$\dot{R}\mathbf{y} + R\dot{\mathbf{y}} = AR\mathbf{y} + \epsilon \mathbf{g}\left(R(t)\mathbf{y}, t\right). \tag{19.34}$$

Since R satisfies the equation $\dot{R} = \Omega AR$ and $R^{-1}AR = A$, this becomes

$$\frac{d\mathbf{y}}{dt} = (1-\Omega)A\mathbf{y} + \epsilon R(t)^{-1}\mathbf{g}\left(R(t)\mathbf{y}, t\right). \tag{19.35}$$

In coordinate form, with $\mathbf{y} = (u, v)^\top$, the equations of motion are

$$\begin{aligned}
\dot{u} &= (1-\Omega)v - \epsilon \sin \Omega t \left[(u\cos \Omega t + v\sin \Omega t)^3 + f\cos \Omega t\right], \\
\dot{v} &= -(1-\Omega)u + \epsilon \cos \Omega t \left[(u\cos \Omega t + v\sin \Omega t)^3 + f\cos \Omega t\right].
\end{aligned} \tag{19.36}$$

Note that no approximations have been made in deriving these equations.

If $\Omega = 1 + O(\epsilon)$ the rates of change of u and v are $O(\epsilon)$, which is much slower than the rates of change of $\cos \Omega t$. Hence, during one oscillation of $\cos \Omega t$ neither u nor v change significantly; it is therefore plausible to derive approximate equations of motion by averaging the equations 19.36 over a period $2\pi/\Omega$ while keeping u and v constant. This gives the *averaged* equations of motion or the *mean-motion* equations,

$$\begin{aligned}
\dot{U} &= (1-\Omega)V - \tfrac{3}{8}\epsilon V\left(U^2 + V^2\right), \\
\dot{V} &= -(1-\Omega)U + \tfrac{3}{8}\epsilon U\left(U^2 + V^2\right) + \tfrac{1}{2}\epsilon f,
\end{aligned} \tag{19.37}$$

where we have used the integrals

$$\frac{1}{2\pi}\int_{-\pi}^{\pi} d\theta \, \sin^4 \theta = \frac{1}{2\pi}\int_{-\pi}^{\pi} d\theta \, \cos^4 \theta = \frac{3}{8}, \quad \frac{1}{2\pi}\int_{-\pi}^{\pi} d\theta \, \sin^2 \theta \cos^2 \theta = \frac{1}{8}.$$

The variables (U, V) are closely related to (u, v) by equations of the form $u(t) = U(t) + \epsilon\alpha(t)$ where $\alpha(t) = O(1)$ and is T-periodic with zero mean value, or at worst a mean value that increases only very slowly with t, section 19.7.2.

The fixed points of this system are at $V = 0$ and the roots of

$$2(1-\Omega)U - \frac{3}{4}\epsilon U^3 = F \quad (\text{recall} \quad F = \epsilon f),$$

which, provided $\Omega = 1 + O(\epsilon)$, is the same, to $O(\epsilon^2)$, as equation 19.20 (page 808).

The mean-motion equations 19.37 are Hamiltonian because the original system is Hamiltonian. If $K(U, V)$ is the mean-motion Hamiltonian then $\dot{U} = \partial K/\partial V$ and $\dot{V} = -\partial K/\partial U$, with

$$K(U, V) = \frac{1}{2}(1-\Omega)(U^2 + V^2) - \frac{3}{32}\epsilon\left(U^2 + V^2\right)^2 - \frac{1}{2}\epsilon f U. \tag{19.38}$$

Further, since

$$x(t) \simeq U(t)\cos\Omega t + V(t)\sin\Omega t, \quad y(t) \simeq -U(t)\sin\Omega t + V(t)\cos\Omega t,$$

the values of $\mathbf{x}(t)$ at the times $t_n = nT$, used to construct a surface-of-section plot, are $\mathbf{x}(t_n) = (U(t_n), V(t_n))$. But these points lie on the contours of $K(U, V)$, hence the contours of the mean-motion Hamiltonian approximate the Poincaré section. For instance, using the same parameters as in figures 19.23 and 19.24 (pages 803 and 804), $\Omega = 0.975$, $\epsilon = \frac{1}{6}$, gives the contours of $K(U, V)$ shown in the following two figures. On the left $f = 0.03$, $(F = 0.005)$, and we see a close similarity between the contours of $K(U, V)$ and the Poincaré section. On the right $f = 1.2$, $(F = 0.2)$; even for this very strong field the simple averaging approximation provides a reasonable approximation near the stable fixed point. Naturally the period-6 orbit is missing, as is the chaotic motion on the outer edge of 19.24.

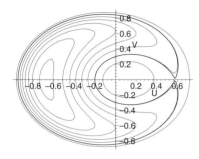

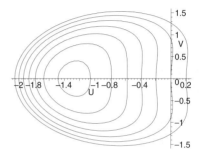

Figure 19.34   Contours of $K(U, V)$,
equation 19.38, for $\Omega = 0.975$, $\epsilon = \frac{1}{6}$, $F = 0.005$.

Figure 19.35   Contours of $K(U, V)$,
equation 19.38, for $\Omega = 0.975$, $\epsilon = \frac{1}{6}$,
$F = 0.2$.

19.7.2 General theory

There are many versions of averaging theorems; here we quote a result proved in Verhulst (1990, chapter 11) where other theorems valid under more general conditions are also proved. Other treatments may be found in Guckenheimer and Holmes (1983, chapter 4) and Arnold (1978, chapter 10).

Consider the N-dimensional initial value problem

$$\frac{d\mathbf{x}}{dt} = \epsilon \mathbf{f}(\mathbf{x}, t) + \epsilon^2 \mathbf{g}(\mathbf{x}, t, \epsilon), \quad \mathbf{x}(0) = \mathbf{x}_0, \tag{19.39}$$

where $\mathbf{x}$ is a real N-dimensional vector in $\mathbf{R}^N$, ϵ is a small quantity and where $\mathbf{f}$ and $\mathbf{g}$ satisfy the following conditions.

(i) The function $\mathbf{f}(\mathbf{x}, t)$ is T-periodic in t with mean value

$$\bar{\mathbf{f}}(\mathbf{x}) = \frac{1}{T} \int_0^T dt\, \mathbf{f}(\mathbf{x}, t). \tag{19.40}$$

(ii) The vectors $\mathbf{f}$, $\mathbf{g}$ and $\partial \mathbf{f}/\partial x_k$, $k = 1, 2, \ldots, N$ exist and are continuous for all $t > 0$ and in some region $\mathscr{D}$ of $\mathbf{R}^N$.

(iii) The vector $\mathbf{g}$ satisfies the Lipschitz condition, $|\mathbf{g}(\mathbf{x}_1, t, \epsilon) - \mathbf{g}(\mathbf{x}_2, t, \epsilon)| \leq L|\mathbf{x}_1 - \mathbf{x}_2|$ for all $\mathbf{x}_1$ and $\mathbf{x}_2$ in $\mathscr{D}$ and where L is a constant. Note that functions satisfying the Lipschitz condition are continuous but the converse is not true.

The *averaged* or *mean-motion* equations for the system 19.39 are defined to be

$$\frac{d\mathbf{y}}{dt} = \epsilon \bar{\mathbf{f}}(\mathbf{y}), \quad \mathbf{y}(0) = \mathbf{x}_0. \tag{19.41}$$

If $\mathbf{y}(t)$ remains inside $\mathscr{D}$ and $\mathbf{x}(t)$ is the solution of 19.39 then it may be proved that $|\mathbf{y}(t) - \mathbf{x}(t)| = O(\epsilon)$ for times $0 < t < O(1/\epsilon)$.

This theorem guarantees that the solution of the averaged system is a good approximation to the exact solution for time $O(1/\epsilon)$. In practice the approximations are often far more useful that this suggests. For instance, the phase curves of the mean-motion Hamiltonian 19.38 remain close to iterates of the map $\mathbf{T}$, that is, the exact solution, in some regions of phase space for *all* time. Note, however, that this does *not* mean that $|\mathbf{y}(t) - \mathbf{x}(t)| = O(\epsilon)$ for all times because the frequencies of the mean and exact motion differ by $O(\epsilon^2)$.

Exercise 19.34
Consider the one-dimensional system $\dot{x} = \epsilon x \sin t$. Show that the mean-motion equations are $\dot{y} = 0$ and that $|\mathbf{y}(t) - \mathbf{x}(t)| = O(\epsilon)$ for all time.

Exercise 19.35
Consider the one-dimensional system

$$\frac{dx}{dt} = 2\epsilon x \sin^2 t.$$

Show that the mean-motion equations are $\dot{y} = \epsilon y$ and that $|y(t) - x(t)| = O(\epsilon)$ for $0 < t < 1/\epsilon$ if $\epsilon > 0$ but for all t if $\epsilon < 0$.

Exercise 19.36
Show that for the one-dimensional system

$$\frac{dx}{dt} = \epsilon x(1 - x) \sin^2 t, \quad 0 < x(0) < 1,$$

the solutions of the mean-motion and exact equations differ by $O(\epsilon)$ for all t.

Averaging of a damped autonomous system
As an example we apply this general theory to a system having no time-dependent forcing term but with damping and a quadratic nonlinearity. The equation of motion is

$$\frac{d^2 x}{dt^2} + x + 2\epsilon \mu \frac{dx}{dt} - \epsilon x^2 = 0, \quad 0 \leq \mu < 1, \tag{19.42}$$

and we derive an approximation to the small amplitude motion. The undamped, $\mu = 0$, limit of this system was considered in section 17.5, and there it is shown that there is a centre at the origin, a saddle at $(1/\epsilon, 0)$ and that bound periodic motion exists for energies $0 \le E < 1/6\epsilon^2$.

As before we write the equation in the form

$$\frac{d\mathbf{x}}{dt} = A\mathbf{x} - \epsilon\mathbf{g}(\mathbf{x}), \quad \mathbf{g} = \begin{pmatrix} 0 \\ 2\mu y - x^2 \end{pmatrix}, \tag{19.43}$$

where A is defined after equation 19.33. Now change to a phase space coordinate system rotating with the unperturbed motion; in this case the rotation matrix is

$$R(t) = \begin{pmatrix} \cos t & \sin t \\ -\sin t & \cos t \end{pmatrix},$$

and if $\mathbf{x} = R(t)\mathbf{u}$ equation 19.43 becomes $\dot{\mathbf{u}} = -\epsilon R(t)^{-1}g(R(t)\mathbf{u})$. In component form this is

$$\begin{aligned} \frac{du}{dt} &= \epsilon \sin t \left\{ 2\mu(v\cos t - u\sin t) + (u\cos t + v\sin t)^2 \right\}, \\ \frac{dv}{dt} &= -\epsilon \cos t \left\{ 2\mu(v\cos t - u\sin t) + (u\cos t + v\sin t)^2 \right\}. \end{aligned} \tag{19.44}$$

The averaged equations of motion are formed by taking the mean of the right hand side, keeping u and v fixed:

$$\begin{aligned} \frac{dU}{dt} &= -\mu\epsilon U, & \Longrightarrow & \quad U(t) = U(0)\exp(-\mu\epsilon t), \\ \frac{dV}{dt} &= -\mu\epsilon V. & \Longrightarrow & \quad V(t) = V(0)\exp(-\mu\epsilon t). \end{aligned} \tag{19.45}$$

In this case the mean-motion equations are linear because the mean over the nonlinear terms is zero; this is not always the case, as seen in exercise 19.38. Thus the approximate general solution of the original system is

$$x(t) = \alpha e^{-\mu\epsilon t}\cos(t + \beta), \quad y(t) = -\alpha e^{-\mu\epsilon t}\sin(t + \beta), \tag{19.46}$$

where α and β are constants depending upon the initial conditions.

This type of approximation assumes that the unperturbed motion is $\alpha(\cos t, -\sin t)$, that is, that the initial phase point is close enough to the origin. As the amplitude of the initial motion increases the frequency of the unperturbed motion, ($\mu = 0$), decreases, so it is clear that approximation 19.46 cannot be correct. It is normally difficult to find good approximations to large amplitude motion.

Exercise 19.37

(i) For the system defined by equation 19.42 the energy E is defined by the equation

$$E = \frac{1}{2}\dot{x}^2 + \frac{1}{2}x^2 - \frac{1}{3}\epsilon x^3.$$

Show that for small ϵ

$$\frac{dE}{dt} \simeq -2\epsilon\mu \left\langle \dot{x}^2 \right\rangle,$$

where $\langle \ \rangle$ denotes the mean over one period of the undamped motion, $\mu = 0$, $\epsilon \ne 0$.

(ii) Show that the period, T, of the undamped motion may be written in terms of the integrals

$$T = 2 \int_a^b dx \, \frac{1}{\sqrt{\lambda - x^2 + \frac{2}{3}\epsilon x^3}} = \oint_{\mathscr{C}} dz \, \frac{1}{\sqrt{\lambda - z^2 + \frac{2}{3}\epsilon z^3}}, \qquad \lambda = 2E,$$

where $\mathscr{C}$ is the contour in the complex plane that encloses the turning points a and b, but no other singularity of the integrand.

Further, show that

$$\left\langle \dot{x}^2 \right\rangle = \frac{1}{T} \int_0^T dt \, \dot{x}^2 = \frac{2}{T} \int_a^b dx \, \dot{x} = \frac{1}{T} \oint_{\mathscr{C}} dz \, \sqrt{\lambda - z^2 + \frac{2}{3}\epsilon z^3}.$$

By first differentiating this result with respect to λ, show that

$$\left\langle \dot{x}^2 \right\rangle = \frac{2}{T} \int_0^\lambda d\lambda \, T(\lambda).$$

(iii) Use this expression and the series for $T(\lambda)$, given in exercise 19.47 (page 829), to show that

$$\frac{dE}{dt} = -2\epsilon\mu E \left(1 - \frac{5}{12}\epsilon^2 E - \frac{155}{108}\epsilon^4 E^2 + \cdots \right).$$

Exercise 19.38

Show that the mean-motion equations for the weakly damped, unforced Duffing equation,

$$\frac{d^2x}{dt^2} + 2\mu\epsilon \frac{dx}{dt} + x - \epsilon x^3 = 0, \qquad 0 < \epsilon \ll 1,$$

are

$$\frac{dU}{dt} = -\epsilon\mu U - \frac{3}{8}\epsilon V \left(U^2 + V^2 \right), \qquad \frac{dV}{dt} = -\epsilon\mu V + \frac{3}{8}\epsilon U \left(U^2 + V^2 \right).$$

Exercise 19.39

(i) Show that the mean-motion equations for the weakly damped and forced Duffing equation,

$$\frac{d^2x}{dt^2} + 2\mu\epsilon \frac{dx}{dt} + x - \epsilon x^3 = \epsilon f \cos \Omega t, \qquad 0 < \epsilon \ll 1,$$

are

$$\frac{dU}{dt} = (1 - \Omega)V - \epsilon\mu U - \frac{3}{8}\epsilon V \left(U^2 + V^2 \right),$$

$$\frac{dV}{dt} = -(1 - \Omega)U - \epsilon\mu V + \frac{3}{8}\epsilon U \left(U^2 + V^2 \right) + \frac{1}{2}\epsilon f.$$

(ii) If $\Omega^2 = 1 - \epsilon v$ show that the fixed points of this system are at the roots of

$$R^2 \left(4\mu^2 + \left(v - \frac{3}{4}R^2 \right)^2 \right) = f^2, \qquad \text{where} \quad R^2 = U^2 + V^2.$$

(iii) Show that the eigenvalues of the linearisation matrix are

$$\lambda = -\epsilon\mu \pm \frac{\epsilon}{8}\sqrt{(3R^2 - 4v)(4v - 9R^2)}.$$

19.7.3 Averaging over a subharmonic

A driving force with frequency $\Omega \simeq 3$ will excite a subharmonic with frequency close to unity, the natural period of the system. The application of the averaging method to this subharmonic resonance, equation 19.26 (page 813), is more difficult than to the main resonance because the periodic solution has two components and it is therefore more difficult to cast the equations in the required form.

The fastest term, $\cos \Omega t$, comes directly from the applied force and, for small $|\epsilon|$, this term is given approximately by the solution of the linear equation $\ddot{x} + x = F \cos \Omega t$, that is, $\frac{F}{1-\Omega^2} \cos \Omega t$. We remove this component by defining a new variable, ξ, by the relation

$$x = \xi - \frac{F}{\Omega^2 - 1} \cos \Omega t,$$

and then the equations of motion become

$$\frac{d^2 \xi}{dt^2} + \xi - \epsilon \left(\xi - \frac{F}{\Omega^2 - 1} \cos \Omega t \right)^3 = 0,$$

or, in vector form,

$$\frac{d\boldsymbol{\xi}}{dt} = A\boldsymbol{\xi} + \epsilon \mathbf{g}(\boldsymbol{\xi}), \quad \text{where} \quad \mathbf{g} = \begin{pmatrix} 0 \\ \left(\xi - \frac{F \cos \Omega t}{\Omega^2 - 1} \right)^3 \end{pmatrix}, \tag{19.47}$$

and where A is defined after 19.33 (page 817). Notice that the $\cos \Omega t$ terms are now $O(\epsilon)$.

Now proceed as before and move to a reference frame rotating clockwise with angular frequency $\Omega/3 \simeq 1$, so the new coordinates are $\mathbf{y} = R(t)^{-1} \boldsymbol{\xi}$, where

$$R(t) = \begin{pmatrix} \cos(\Omega t/3) & \sin(\Omega t/3) \\ -\sin(\Omega t/3) & \cos(\Omega t/3) \end{pmatrix}.$$

In this moving reference frame the equations of motion are

$$\frac{d\mathbf{y}}{dt} = \left(1 - \frac{\Omega}{3} \right) A\mathbf{y} + \epsilon R(t)^{-1} \mathbf{g}(R(t)\boldsymbol{\xi}). \tag{19.48}$$

The right hand side of these equations are $3T$-periodic and provided $\Omega = 3 + O(\epsilon)$ satisfy the condition of the averaging theorem, section 19.7.2. Averaging these equations, most easily accomplished using Maple, gives the mean-motion equations

$$\frac{du}{dt} = \left(1 - \frac{\Omega}{3} \right) v - \frac{3}{8} \epsilon v(u^2 + v^2) - \frac{3}{32} \epsilon F uv - \frac{3}{256} \epsilon F^2 v, \tag{19.49}$$

$$\frac{dv}{dt} = -\left(1 - \frac{\Omega}{3} \right) u + \frac{3}{8} \epsilon u(u^2 + v^2) + \frac{3\epsilon}{64} F(v^2 - u^2) + \frac{3\epsilon}{256} F^2 u. \tag{19.50}$$

In deriving these we have set $\Omega = 3 + O(\epsilon)$ in the nonlinear terms and ignored all terms $O(\epsilon^2)$. These are Hamiltonian equations, as they must be, with the Hamiltonian

$$H(u,v) = \frac{1}{2} \left(1 - \frac{\Omega}{3} \right) (u^2 + v^2) - \frac{3\epsilon}{32} (u^2 + v^2)^2 - \frac{3\epsilon F}{64} u \left(v^2 - \frac{u^2}{3} \right) - \frac{3\epsilon F^2}{512} (u^2 + v^2). \tag{19.51}$$

In the following figure some representative contours of this Hamiltonian are shown.

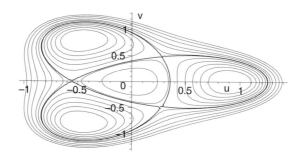

Figure 19.36   Some representative contours of $H(u, v)$ for $F = 3$, $\epsilon = \frac{1}{6}$,
$\Omega = 2.85$.

Note that there is a centre at the origin, corresponding to a stable T-periodic orbit, and three other centres associated with a stable $3T$-periodic orbit. There are also three saddles, corresponding to an unstable $3T$-periodic orbit, and the separatrix through these is depicted by the solid line.

The phase curves of these three orbits are shown in the following graphs.

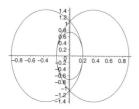   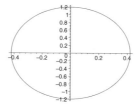   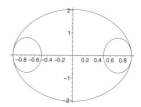

Figure 19.37   Unstable period-3   Figure 19.38   Stable period-1   Figure 19.39   Stable period-3
        orbit.   orbit.   orbit.

Exercise 19.40
With the trial function

$$x = a\cos(\Omega t/3) + b\cos\Omega t$$

use the harmonic balance method to show that a and b satisfy equations 19.27. For the parameters used in figure 19.36 use `implicitplot` to plot the graphs of these two equations and to demonstrate numerically that there are two real solutions, which give the approximate solutions

$$x_1(t) = -0.412\cos\Omega t + 0.901\cos\left(\frac{1}{3}\Omega t\right), \quad \Omega = 2.85,$$

$$x_2(t) = -0.416\cos\Omega t - 0.483\cos\left(\frac{1}{3}\Omega t\right).$$

Plot the graphs of these approximate solutions and compare these with the graphs shown in figures 19.39 and 19.37. Use the Floquet method, outlined in section 19.6, to show that $x_1(t)$ is stable and $x_2(t)$ is unstable.

19.8 Appendix: Fourier transforms

The Fourier transform procedure provided by Maple is useful but both the input and output data require some manipulation before it can be used. However, the Maple help files do not state clearly what the fast Fourier transform procedure actually returns; this appendix provides the information necessary to use this procedure.

Suppose that we need to evaluate the finite Fourier transform

$$F(\omega) = \frac{1}{\sqrt{2\pi}} \int_a^b dx \, e^{-i\omega x} f(x), \tag{19.52}$$

and that $f(x)$ is real so that $F(-\omega) = F^*(\omega)$, and only non-negative frequencies are required. If $-a = b = \infty$ the inverse is

$$f(x) = \frac{1}{\sqrt{2\pi}} \int_{-\infty}^{\infty} d\omega \, e^{i\omega x} F(\omega),$$

which is the reason for this choice of external factor. It is necessary to know the values of $f(x)$ at a set of N equally spaced points,

$$x_j = a + \frac{j}{N}(b - a), \quad j = 0, 1, \ldots, N - 1, \quad x_0 = a, \quad (x_N = b),$$

where $N = 2^n$ for some n. The discrete approximation to 19.52 can be written in the form

$$F(\omega) = \frac{e^{-i\omega a}}{\sqrt{N}} \sum_{j=0}^{N-1} z_j \exp\left(-i\omega(b - a)j/N\right), \quad z_j = \frac{b - a}{\sqrt{2\pi N}} f(x_j). \tag{19.53}$$

The definition of the discrete Fourier transform $\mathbf{Z} = [Z_0, Z_1, \ldots, Z_{N-1}]$ of a list of N complex numbers $\mathbf{z} = [z_0, z_1, \ldots, z_{N-1}]$ is

$$Z_k = \alpha \sum_{j=0}^{N-1} z_j \exp\left(-\frac{2\pi i jk}{N}\right), \tag{19.54}$$

with the inverse

$$z_j = \beta \sum_{k=0}^{N-1} Z_k \exp\left(\frac{2\pi i jk}{N}\right),$$

where α and β are real normalisation constants that satisfy $\alpha\beta = N$. The fast Fourier transform procedure in Maple uses the convention that $\alpha = 1$ and $\beta = N$. The Maple procedure FFT(n,x,y), where x and y are arrays of floating point numbers of length 2^n with $z_j = x[j] + iy[j]$, returns the complex numbers Z_k with $Z_k = x[k] + iy[k]$, that is, the original data is over-written. The computational time for this calculation varies as $N \ln N$ rather than N^2, as would be the case if conventional integration methods were used. Details of how this method works may be found in Powell (1981, chapter 13).

The transform returns the values of $F(\omega)$ at 2^n values of the frequency. By comparing 19.53 and 19.54 we see that the values of the positive frequencies are

$$\omega_k = \frac{2\pi k}{b - a}, \quad k = 0, 1, \ldots, 2^{n-1} - 1, \tag{19.55}$$

and that the required values of $F(\omega_k)$ are

$$F_k = F(\omega_k) = \frac{e^{-i\omega_k a}}{\sqrt{N}} Z_k, \quad k = 0, 1, \ldots, 2^{n-1} - 1. \tag{19.56}$$

There are four points to notice.

(1) The number of data points N *must* be of the form $N = 2^n$; often this necessitates truncating the data set.
(2) The frequency spacing is, from 19.55,

$$\Delta\omega = \frac{2\pi}{b-a}, \quad \text{hence} \quad \Delta x \Delta\omega = \frac{2\pi}{N},$$

where Δx is the spacing of the original data points, equation 19.53. Note that for fixed a and b the frequency spacing is independent of the number of points N.
(3) The first 2^{n-1} elements Z_k, $k = 0, 1, 2, \ldots, 2^{n-1} - 1$, give $F(\omega_k)$, and the remaining 2^{n-1} elements Z_k, $k = 2^n - 1 - p$, $p = 0, 1, 2, \ldots, 2^{n-1} - 1$, give the Fourier transform at the negative frequencies $F(\omega_{-p})$.
(4) The maximum frequency is given by

$$\max(\omega) = \frac{\pi N}{b-a} = \frac{\pi}{\Delta x}.$$

When computing the finite Fourier transform the values of a, b and N must be chosen to provide appropriate values of $\Delta\omega$ and $\max(\omega)$. The following Maple procedure uses FFT to compute $F(\omega_k)$ given a, b, n and the array x.

The following procedure uses the above analysis and the Maple FFT procedure, which is accessed with the `readlib(FFT)` command, to find the Fourier transform of a real function $f(x)$, equation 19.52, and plot the graphs of the real part of $F(\omega)$ over the range $(0, \omega_m)$. It assumes that the values of $f(x)$ are in the array f, that the kth element of f is $[x_k, f(x_k)]$ and that there are $N = 2^n$ elements in the array, for some positive integer n. The fourth argument of the procedure, wmax$= \omega_m$, defines the upper limit of the final graphs and has nothing to do with the calculation of the Fourier transform. The first few lines of the procedure define necessary constants: the value of $N = 2^n$ is calculated; the values of a and b in the integral 19.52 are computed from the input data assuming the increments in x are all equally spaced.

```
>   readlib(FFT):
>   F:=proc(f,n,wmax)
>   local a,b,c1,Dx,Fi,Fr,h,k,N,pi2,w,xf,yf,z;
>   N:=2^n;
>   a:=f[1][1]; Dx:=f[2][1]-f[1][1];
>   b:=evalf(N*Dx+a);
>   pi2:=evalf(2.0*Pi): c1:=evalf( (b-a)/sqrt(pi2*N)):
>   h:=(b-a)/N;
>   printf("Max frequency component is %8.4f and increment
>                is %10.5f \n", evalf(Pi/h),evalf(2*Pi/(b-a)));
>   xf:=array(1..N):
>   yf:=array(1..N,[0.0$N]):        # Assume function f(x) is real
```

Compute the array defined in equation 19.53,

```
>   for k from 1 to N do;
>         xf[k]:=evalf(c1*f[k][2]);
>   od:
```

Find the finite Fourier transform,

```
>   FFT(n,xf,yf);
```

Now scale the data from the FFT, equation 19.56, and plot the graph for $\omega > 0$.

```
>   Fi:=NULL: Fr:=NULL:
>   for k from 0 to 2^(n-1)-1 do;
>         w:=evalhf( pi2*k/(b-a) );
>         z:=exp(-I*a*w)*(xf[k+1]+I*yf[k+1])/sqrt(N);
>         Fr:=Fr,[w,evalf(Re(z))];
>         Fi:=Fi,[w,evalf(Im(z))];
>   od:
>   plot([[Fr]],0..wmax,colour=black):
>   end:
```

As an example consider the Fourier transform of $f(x) = \exp(-A(x-B)^2)$, $A, B > 0$, which is, provided $a \ll B$ and $b \gg B$,

$$F(\omega) = \frac{e^{-i\omega B}}{\sqrt{2A}} \exp\left(-\frac{\omega^2}{4A}\right).$$

For $A = \frac{1}{4}$ and $B = 20$ we take $a = 0$ and $b = 60\pi$ and form the array s comprising 2^8 pairs $[x_k, f(x_k)]$, $k = 1, 2, \ldots, 2^8$

```
>   f:=x-> exp(-A*(x-B)^2):              # Define the function
>   A:=0.25: B:=20:                          # Define A   and B
>   a:=0: b:=evalf(60*Pi):                   # Define a and b
>   n:=8: N:=2^n:             # Define the number of points to be used
>   s:=NULL:
>   for k from 0 to N-1 do; # Begin a loop that defines the array of points
>         x:=evalf( a + k*(b-a)/N );
>         s:=s,[x,evalhf(f(x))];
>   od: s:=[s]:
```

For these parameters $\Delta\omega = 1/30$, $\Delta x \simeq 0.736$ and $\max(\omega) \simeq 4.3$, and the graph for $0 < \omega < 2$ is produced with the command F(s,n,2). This graph is shown in the following figure together with the graph of the real part of the exact Fourier transform; the differences are barely noticeable.

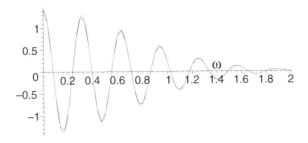

Figure 19.40   Fourier transform of $f(x) = \exp(-(x-20)^2/4)$.

19.9 Exercises

Exercise 19.41

By considering the integral $\int_0^1 dt\,(1-t^4)^{-1/2}$, show that

$$K\left(\frac{1}{\sqrt{2}}\right) = \frac{1}{2}\frac{\Gamma(1/4)}{\Gamma(3/4)}\sqrt{\frac{\pi}{2}} = \frac{\Gamma(1/4)^2}{4\sqrt{\pi}}.$$

Exercise 19.42

Show that the solution of equation 19.2 with $\mu = 0$, $\Omega \neq 1$ and initial conditions $x(0) = a$, $\dot{x}(0) = b$ is

$$x(t) = a\cos t + b\sin t + \frac{F}{1-\Omega^2}(\cos\Omega t - \cos t).$$

By taking the limit as $\Omega \to 1$ show that for $\Omega = 1$ the solution is

$$x = a\cos t + b\sin t + \frac{1}{2}Ft\sin t.$$

Exercise 19.43

Use the trigonometric identity $\cos 3\theta = 4\cos^3\theta - 3\cos\theta$ to show that the cubic equation $4w^3 - 3w = g$ has the solutions given in equation 19.24.

Exercise 19.44

Period of conservative Duffing system

Show that the period of the unperturbed Duffing oscillator, with amplitude a, may be written in terms of the integral

$$T(a) = 4\int_0^a dx\,\frac{1}{\sqrt{(a^2-x^2)\left(1-\frac{\epsilon}{2}(a^2+x^2)\right)}} = 4\int_0^{\pi/2}d\phi\,\frac{1}{\sqrt{1-\frac{1}{2}\epsilon a^2 - \frac{1}{2}\epsilon a^2\sin^2\phi}},$$

where $x = a\sin\phi$. Consider the cases $\epsilon < 0$ and $\epsilon > 0$ separately and derive equations 19.7 and 19.12 respectively.

Exercise 19.45

One of the simplest oscillating systems is a particle moving on the x-axis bouncing elastically from two walls situated at $x = \pm a$, so for $|x| < a$ the particle moves freely, $x(t) = A + vt$, with velocity v, but at $x = \pm a$ its velocity instantaneously changes sign, but not magnitude. It may be shown that that the discontinuous potential

$$V(x) = \begin{cases} 0, & |x| < a \\ \infty, & |x| \geq a \end{cases}$$

produces such a motion.

Show that if the speed of the particle is v then the frequency of the motion is $\omega = \pi v/2a$, and that for a particle with the initial conditions $x(0) = 0$, $\dot{x}(0) = v > 0$ the Fourier series of the motion is

$$x(t) = \frac{8a}{\pi^2}\sum_{k=0}^{\infty}\frac{(-1)^k}{(2k+1)^2}\sin(2k+1)\omega t.$$

Compare this Fourier series with that of equation 19.10 (page 791), and explain why it converges more slowly.

Exercise 19.46
Consider the system defined by the equation

$$\frac{d^2x}{dt^2} + U\operatorname{sgn}(x) = F\cos\Omega t, \quad U > 0.$$

(i) Show that when $F = 0$ harmonic balance gives the approximate solution

$$x = a\cos\omega t, \quad a = \frac{4U}{\pi\omega^2},$$

and that the exact relation between the amplitude a and frequency ω is $a = \pi^2 U/8\omega^2$.

(ii) Show that if $F > 0$ the harmonic balance approximation for the solution with period $2\pi/\Omega$ is

$$x = a\cos\Omega t, \quad a = \frac{4U - \pi F}{\pi\Omega^2}.$$

(iii) Examine the exact solution with the initial conditions $x(0) = a > 0$, $\dot{x}(0) = 0$ for $F < U$ and $F > U$ and determine why this harmonic balance approximation breaks down for $F \sim U$.

Exercise 19.47
Consider the system

$$\frac{d^2x}{dt^2} + x - \epsilon x^2 = 0.$$

(i) Show that the period of the motion with energy E and turning points $a < b$ may be expressed in the following forms:

$$
\begin{aligned}
T(E) &= 2\int_{a(E)}^{b(E)} dx \, \frac{1}{\sqrt{2E - x^2 + 2\epsilon x^3/3}} \\
&= 2\sum_{p=0}^{\infty} \left(\frac{8\epsilon}{3}\right)^{2p} (2E)^p \frac{\Gamma(2p+1/2)\Gamma(3p+1/2)}{(4p)!\,p!} = 2\pi\left(1 + \frac{5}{6}\epsilon^2 E + \cdots\right).
\end{aligned}
$$

Hint: convert the integral for T to a contour integral and use the method described in exercise 15.31 (page 623).

(ii) Show also that for the unforced Duffing oscillator with $\epsilon > 0$ the period, equation 19.12, can be expanded in terms of ϵE, rather than $\epsilon^2 E$, and is

$$T(E) = 2\pi\left(1 + \frac{3}{4}\epsilon E + \frac{105}{64}(\epsilon E)^2 + \cdots\right).$$

Explain why the expansion parameter is different.

(iii) Use the trial function $x(t) = A + B\cos\omega t$ with the harmonic balance method to show that an approximate solution with frequency ω is

$$x(t) = \frac{1-\omega^2}{2\epsilon} + \sqrt{\frac{1-\omega^4}{2\epsilon^2}}\cos\omega t.$$

Hence show that an approximate solution with energy E is

$$x(t) = \epsilon E + \sqrt{2E}\cos\omega t, \quad \omega = 1 - \epsilon^2 E,$$

and explain why the relation between ω and $\epsilon^2 E$ differs from that found in part (i).

(iv) Use Lindstedt's method to show that the solution with initial conditions $x(0) = a$, $\dot{x}(0) = 0$ is

$$x = a \cos \tau + \frac{1}{6} \epsilon a^2 \left(3 - 2 \cos \tau - \cos 2\tau\right) - \frac{\epsilon^2 a^3}{144} \left(48 - 29 \cos \tau - 16 \cos 2\tau - 3 \cos 3\tau\right),$$

where $\tau = \omega t$, $\omega = 1 - 5\epsilon^2 a^2/16$. Show that this solution has energy

$$E = \frac{1}{2} a^2 \left(1 - \frac{2}{3} \epsilon a\right) + O(\epsilon^3),$$

and by expressing a in terms of E show that this approximation gives

$$x = \epsilon E + \sqrt{2E} \cos \tau - \frac{1}{3} \epsilon E \cos 2\tau + \cdots .$$

Exercise 19.48
Use the harmonic balance method to show that the approximate periodic solution of the equation

$$\frac{d^2 x}{dt^2} + x - \epsilon x^2 = \epsilon f \cos \Omega t, \quad \Omega^2 = 1 - \epsilon v,$$

where f and v are $O(1)$, is $x(t) = \epsilon f^2/(2v) + (f/v) \cos \Omega t$.
 Show that if $v \neq 0$ Lindstedt's method applied to this problem gives

$$x(t) = \frac{\epsilon f^2}{2v} + \frac{f}{v} \cos \Omega t + \frac{\epsilon f^2}{6v^4} \left(5f \cos \Omega t - v^2 \cos 2\Omega t\right) + O(\epsilon^2).$$

Exercise 19.49
Show that the mean-motion equations of the problem defined in the previous exercise are

$$\frac{dU}{dt} = (1 - \Omega)V, \quad \frac{dV}{dt} = -(1 - \Omega)U + \frac{1}{2}\epsilon f,$$

and that these are Hamiltonian equations with the Hamiltonian

$$H(U, V) = \frac{1}{2}(1 - \Omega) \left(U^2 + V^2\right) - \frac{1}{2}\epsilon f U.$$

Explain why the periodic solution given by this approximation differs from that given in the previous exercise by a constant term.

Exercise 19.50
Show that the system

$$\frac{dx}{dt} = y, \quad \frac{dy}{dt} = -x + \epsilon(1 - ax^2 - by^2)y,$$

where a and b are positive constants, has a periodic solution and that the radial component of the mean-motion equations is

$$\frac{dR}{dt} = \frac{1}{8}\epsilon R \left(1 - (a + 3b)R^2\right).$$

Deduce that there is a stable limit cycle with radius $2/\sqrt{a + 3b}$.

Exercise 19.51
Apply the averaging method to the van der Pol equation

$$\frac{d^2x}{dt^2} + \epsilon \left(x^2 - 1\right) \frac{dx}{dt} + x = 0$$

to obtain the mean-motion equations

$$\frac{dU}{dt} = \frac{\epsilon U}{8} \left(4 - U^2 - V^2\right), \quad \frac{dV}{dt} = \frac{\epsilon V}{8} \left(4 - U^2 - V^2\right),$$

and show that in polar coordinates the radial mean-motion equation is $\dot{R} = \epsilon R(4 - R^2)/8$. Deduce that there is a stable limit cycle.

Exercise 19.52
Show that Mathieu's equation, $\ddot{x} + (a - 2\epsilon \cos 2t)x = 0$, may be written in the form

$$\dot{\mathbf{x}} = A\mathbf{x} - \epsilon \begin{pmatrix} 0 & 0 \\ a_1 - 2\cos 2t & 0 \end{pmatrix} \mathbf{x}, \quad \mathbf{x} = \begin{pmatrix} x \\ \dot{x} \end{pmatrix},$$

where A is the matrix defined in equation 19.33 (page 817), and $a = 1 + \epsilon a_1$.

Show that averaging method applied to this equation gives the mean-motion equations

$$\frac{d}{dt} \begin{pmatrix} U \\ V \end{pmatrix} = \epsilon \begin{pmatrix} 0 & (1 + a_1)/2 \\ (1 - a_1)/2 & 0 \end{pmatrix} \begin{pmatrix} U \\ V \end{pmatrix},$$

and deduce that if $|\epsilon|$ is small the solutions of Mathieu's equation are stable if $|a_1| < 1$.

Exercise 19.53
Consider the periodically forced van der Pol equation

$$\frac{d^2x}{dt^2} + \epsilon \left(x^2 - 1\right) \frac{dx}{dt} + x = \epsilon f \cos \Omega t, \quad f > 0, \quad 0 < \epsilon \ll 1.$$

(i) Show that if $\Omega^2 = 1 - \epsilon v$ the harmonic balance method gives the approximate solution $x = a \cos \Omega t + b \sin \Omega t$, where a and b satify the equations

$$g_1(a, b) = a \left(a^2 + b^2 - 4\right) - 4bv, \quad g_2(a, b) = b \left(a^2 + b^2 - 4\right) + 4av - 4f.$$

Show that if $(a, b) = (\alpha, \beta)$ is a solution of these equations then $(-\alpha, \beta)$ is a solution when v changes sign.

(ii) Show that these equations may be rearranged to give

$$r\sqrt{16v^2 + (r^2 - 4)^2} = 4f, \quad r^2 = a^2 + b^2.$$

Deduce that for $v^2 > 1/3$ there is only one real root for all values of f, but for smaller v^2 there may be either one or three roots, depending on the value of f.

(iii) Show also that the mean-motion equations in the neighbourhood of the periodic orbit are

$$\frac{dU}{dt} = -\frac{\epsilon}{8} g_1(U, V), \quad \frac{dV}{dt} = -\frac{\epsilon}{8} g_2(U, V).$$

Show that the linearisation matrix about the fixed point defined in part (i) is proportional to the matrix

$$A = \begin{pmatrix} 4 - 3a^2 - b^2 & 4v - 2ab \\ -4v - 2ab & 4 - 3b^2 - a^2 \end{pmatrix},$$

having $\mathrm{Tr}(A) = 4(2 - r^2)$, $\det(A) = 3r^4 - 16r^2 + 16(1 + v^2)$ and with eigenvalues

$$\lambda = 2(2 - r^2) \pm \sqrt{r^4 - 16v^2}.$$

(iv) Show that in the (r^2, v)-plane, with $v > 0$:
   (a) if $r^2 < 2v$ the fixed points are spirals, stable if $r^2 > 2$ and unstable if $r^2 < 2$.
   (b) if $\det(A) < 0$ the fixed points are saddles.
   (c) if $\det(A) > 0$ the fixed points are nodes, stable if $r^2 > 2$ and unstable if $r^2 < 2$.
   Plot these regions in the (r^2, v)-plane.

Exercise 19.54

Show that the mean-motion equations for the system

$$\frac{dx}{dt} = y + \epsilon(xy - 1)\sin t, \quad \frac{dy}{dt} = x, \quad 0 < \epsilon \ll 1,$$

about the period-2π solution are

$$\frac{dU}{dt} = \frac{1}{8}\left(V^2 - U^2\right), \quad \frac{dV}{dt} = -\frac{1}{4}(2 + UV).$$

Find and classify the fixed points of the mean-motion equations and hence show that the original system has a stable limit cycle.

Exercise 19.55

Consider the Duffing equation

$$\frac{d^2x}{dt^2} + x - \epsilon x^3 = F\cos t, \quad 0 < \epsilon \ll 1, \quad (\Omega = 1).$$

(i) Show that the harmonic balance method with the trial function $x(t) = m - a\cos t$, $m \neq 0$, gives

$$3\epsilon a\left(a^2 + 4m^2\right) = 4F, \quad m^2 + \frac{3}{2}a^2 = \frac{1}{\epsilon},$$

   and sketch the curves in the (a, m)-plane defined by these equations.
(ii) Show that if F is small these equations have two real solutions, giving

$$x = \pm\sqrt{\frac{1}{\epsilon} - \frac{F}{3}}\cos t.$$

   By expanding the original equations about $x = \sqrt{1/\epsilon}$ and linearising the resulting equations, explain the origin of these solutions.
(iii) Show that there are four real solutions of these equations if $F > (6\epsilon)^{-1/2}$.
(iv) If $\mathbf{T}(\mathbf{x})$ is the period map, plot the graph of $f(x)$, where $\mathbf{T}(x, 0) = (g(x), f(x))$ for $\epsilon = 1/6$, $-2.3 < x < -2.0$ and F between 1.20 and 1.22. Hence show that for x in this interval $f(x) = 0$ has the following roots.

F	Roots of $f(x) = 0$		
1.200		-2.0959831	
1.202		-2.0971182	
1.203		-2.0976854	
1.204	-2.1064319	-2.0982583	-2.0899630
1.210	-2.1814011	-2.1016461	-2.0100093
1.220	-2.2313304	-2.1072770	-1.9517631

(v) For $F = 1.20$ and 1.21 plot surface-of-sections with $\epsilon = 1/6$ and for the initial conditions $\mathbf{x} = (x_0, 0)$ with $x_0 = \{-2.2, -2.1, -2.05, -2.0, -1.9\}$ and $x_0 = \{-2.2, -2.1, -2.102, -2.05, -2.0, -1.9\}$ respectively, and confirm numerically that the bifurcation involves the creation of two stable T-period orbits.

(vi) Assuming that the periodic orbits have the Fourier series

$$x(t) = a_0 + \sum_{k=1}^{\infty} a_k \cos kt,$$

compute the Fourier components of the orbits listed in the table in part (iv) above. In particular show that for $F \leq 1.203$, $a_0 = 0$ and all the Fourier components of the even harmonics are zero, and that for $F \geq 1.204$ all the Fourier components of the stable orbits are nonzero.

Exercise 19.56
Extend the analysis in the previous question to deal with the case $\Omega^2 = 1 - \epsilon v$, $v > 0$.

Exercise 19.57
If $f(x) = O(x^3)$ and is such that the equation

$$\frac{d^2 x}{dt^2} + x + \epsilon f'(x) = 0, \quad |\epsilon| \ll 1,$$

has periodic solutions for small x, show that the period of the oscillation with energy E can be expressed in terms of the infinite series

$$T = \frac{2}{\sqrt{\pi}} \sum_{k=0}^{\infty} (8\epsilon)^k \frac{\Gamma(k + 1/2)}{(2k)!} \frac{d^k}{d\lambda^k} \int_{-\pi/2}^{\pi/2} d\theta\, f(\sqrt{\lambda}) \sin \theta)^k, \quad \lambda = 2E.$$

20

Appendix I: The gamma and related functions

The gamma function

The gamma function occurs frequently in physical and statistical problems and is unusual in that it is not a solution of a differential equation with rational coefficients. It was first defined by Euler as the infinite product

$$\Gamma(z) = \lim_{n \to \infty} \frac{n! \, n^z}{z(z+1)(z+2) \cdots (z+n)}, \quad z \neq 0, -1, -2, \ldots, \tag{20.1}$$

and Weierstrass defined it by

$$\frac{1}{\Gamma(z)} = z e^{\gamma z} \prod_{n=1}^{\infty} \left[\left(1 + \frac{z}{n} \right) e^{-z/n} \right], \tag{20.2}$$

where γ is Euler's constant, denoted by gamma in Maple,

$$\gamma = \lim_{n \to \infty} \left(1 + \frac{1}{2} + \frac{1}{3} + \cdots + \frac{1}{n} - \ln n \right) = 0.577\,215\,7\ldots .$$

It follows from this definition that the gamma function is analytic everywhere except at $z = 0, -1, -2, \ldots$, where it has simple poles. It also follows that

$$\Gamma(z+1) = z\Gamma(z) \tag{20.3}$$

and since, from the first definition $\Gamma(1) = 1$, it follows that if $z = n$ is an integer $\Gamma(n+1) = n!$.

The gamma function may also be defined by the integral

$$\Gamma(z+1) = \int_0^\infty dt \, t^z e^{-t}, \quad \Re(z) > -1, \tag{20.4}$$

which can be shown to be equivalent to the Euler product 20.1 (Whittaker and Watson, 1965, section 12.2). In this integral, and elsewhere, t^z is defined to be $t^z = \exp(z \ln t)$ for real $t > 0$. Using this representation and 20.3 we see that

$$\Gamma\left(\frac{1}{2}\right) = \sqrt{\pi}, \quad \Gamma\left(\frac{3}{2}\right) = \frac{1}{2}\sqrt{\pi}, \quad \ldots, \quad \Gamma\left(n + \frac{1}{2}\right) = \frac{1.3.5.7 \cdots (2n-1)}{2^n} \Gamma\left(\frac{1}{2}\right).$$

Other similar expressions are

$$\Gamma\left(n + \frac{1}{3}\right) = \frac{1.4.7.10 \cdots (3n-2)}{3^n} \Gamma\left(\frac{1}{3}\right),$$

$$\Gamma\left(n + \frac{1}{4}\right) = \frac{1.5.9.13 \cdots (4n-3)}{4^n} \Gamma\left(\frac{1}{4}\right).$$

Other useful properties of the gamma function are:

Reflection formula

$$\Gamma(z)\Gamma(1-z) = -z\Gamma(-z)\Gamma(z) = \frac{\pi}{\sin \pi z}$$

$$= \int_0^\infty dt \, \frac{t^{z-1}}{1+t}, \quad 0 < \Re(z) < 1.$$

Duplication formula

$$\Gamma(2z) = \frac{2^{2z-1}}{\sqrt{\pi}} \Gamma(z)\Gamma\left(z + \frac{1}{2}\right).$$

Gauss' multiplication formula

$$\Gamma(nz) = \sqrt{\frac{2\pi}{n}} \frac{n^{nz}}{(2\pi)^{n/2}} \prod_{k=0}^{n-1} \Gamma\left(z + \frac{k}{n}\right).$$

The asymptotic expansion of the gamma function may be derived from the integral 20.4, see chapter 13 and in particular exercise 13.29 for more terms in this series,

$$\Gamma(z) = \sqrt{\frac{2\pi}{z}} \left(\frac{z}{e}\right)^z \left[1 + \frac{1}{12z} + \frac{1}{288z^2} - \frac{139}{51\,840z^3} + \cdots\right], \quad |\arg z| < \pi. \qquad (20.5)$$

From this the following useful formulae may be derived:

$$\Gamma(z+a) \sim \sqrt{\frac{2\pi}{z}} \left(\frac{z}{e}\right)^z z^a, \quad |z| \to \infty, \qquad (20.6)$$

and

$$\frac{\Gamma(z+a)}{\Gamma(z+b)} \sim z^{a-b} \left(1 + \frac{(a-b)(a+b-1)}{2z} + O(z^{-2})\right). \qquad (20.7)$$

The psi or digamma function

Derivatives of the gamma function are defined in terms of the psi, or digamma, function, $\Psi(z)$,

$$\Psi(z) = \frac{\Gamma'(z)}{\Gamma(z)} = \frac{d}{dz} \ln \Gamma(z).$$

A recurrence formula for the psi function is

$$\Psi(z+1) = \Psi(z) + \frac{1}{z},$$

and this leads to some special values of this function and shows its relation with the harmonic series,

$$\Psi(1) = -\gamma, \qquad \Psi(n) = -\gamma + \sum_{k=1}^{n-1} \frac{1}{k}, \quad n \geq 2,$$

$$\Psi\left(\tfrac{1}{2}\right) = -\gamma - 2\ln 2, \qquad \Psi\left(n + \frac{1}{2}\right) = -\gamma - 2\ln 2 + 2\sum_{k=1}^{n} \frac{1}{2k-1}, \quad n \geq 1.$$

The reflection and duplication formulae are, respectively,

$$\Psi(1-z) = \Psi(z) + \frac{\pi}{\tan \pi z}, \quad \Psi(2z) = \frac{1}{2}\Psi(z) + \frac{1}{2}\Psi\left(z + \frac{1}{2}\right) + \ln 2.$$

The asymptotic expansion of the psi function as $|z| \to \infty$ is

$$\Psi(z) \quad \sim \quad \ln z - \frac{1}{2z} - \sum_{k=1}^{\infty} \frac{B_{2k}}{2kz^{2k}}, \quad |\arg(z)| < \pi, \tag{20.8}$$

$$= \quad \ln z - \frac{1}{2z} - \frac{1}{12z^2} + \frac{1}{120z^4} - \frac{1}{252z^6} + \cdots, \tag{20.9}$$

where B_{2k} are the Bernoulli numbers.

The beta function

The beta function is a function of two complex variables, defined by the integrals

$$B(z, w) \quad = \quad \int_0^1 dt \, t^{z-1}(1-t)^{w-1} = \int_0^\infty dt \, \frac{t^{z-1}}{(1+t)^{z+w}}, \quad \Re(z), \Re(w) > 0,$$

$$= \quad 2\int_0^{\pi/2} d\theta \, (\sin\theta)^{2z-1}(\cos\theta)^{2w-1},$$

$$= \quad \frac{\Gamma(z)\Gamma(w)}{\Gamma(z+w)}.$$

The latter identity is a generalisation of Wallis' formula

$$\frac{2}{\pi}\int_0^{\pi/2} d\theta \, \sin^{2n}\theta \quad = \quad \frac{2}{\pi}\int_0^{\pi/2} d\theta \, \cos^{2n}\theta$$

$$= \quad \frac{1.3.5\cdots(2n-1)}{2.4.6\cdots(2n)} = \frac{(2n)!}{(2^n n!)^2} = \frac{\Gamma(n+1/2)}{n!\sqrt{\pi}}$$

$$\sim \quad \frac{1}{\sqrt{\pi n}}\left(1 - \frac{1}{8n} + \frac{1}{128n^2} + \frac{5}{1024n^3} + \cdots\right).$$

The relative error of the asymptotic expansion truncated beyond the n^{-3} term is better than 0.2% for $n \geq 1$.

Some useful integrals

The asymptotic expansions derived in chapters 13 and 14 require definite integrals involving powers and exponentials, which can usually be evaluated in terms of the gamma function. In the following, the symbols a, b, c and x represent real variables and w and z complex variables, all unrestricted unless otherwise stated.

$$\int_0^\infty dt \, t^z e^{-wt} = \frac{\Gamma(1+z)}{|w|^{1+z}} \exp\left(-i(1+z)\arg w\right), \quad \Re(z) > -1, \quad \Re(w) > 0, \tag{20.10}$$

and useful special cases are

$$\int_0^\infty dt\, t^z e^{-xt} = \frac{\Gamma(z+1)}{x^{z+1}} = \frac{z\Gamma(z)}{x^{z+1}}, \quad \Re(z) > -1, \quad x > 0, \tag{20.11}$$

$$\int_0^\infty dt\, t^c e^{-at^b} = \frac{1}{b}\left(\frac{1}{a}\right)^{\frac{1+c}{b}} \Gamma\left(\frac{1+c}{b}\right), \quad c > -1, \ a, b > 0, \tag{20.12}$$

$$\int_{-\infty}^\infty dt\, t^{2n} e^{-at^2} = \left(\frac{1}{a}\right)^{n+\frac{1}{2}} \Gamma\left(n+\frac{1}{2}\right), \quad a > 0, \quad (n \text{ an integer}),$$

$$= \frac{1.3.5\cdots(2n-1)}{(2a)^n}\sqrt{\frac{\pi}{a}}, \tag{20.13}$$

$$\int_{-\infty}^\infty dt\, e^{-at^2+zt} = \sqrt{\frac{\pi}{a}}\exp\left(\frac{z^2}{4a}\right), \quad a > 0. \tag{20.14}$$

If $\Re(w) = 0$, $w = -ix$ for x real and $x \neq 0$, then the integral 20.10 is oscillatory and care is needed. It may be shown that

$$\int_0^\infty dt\, t^z e^{ixt} = \frac{\Gamma(1+z)}{|x|^{1+z}}\exp\left(\frac{i\pi}{2}(1+z)\mathrm{sgn}\,x\right) = \Gamma(1+z)\left(\frac{i}{x}\right)^{1+z}, \tag{20.15}$$

which transforms to

$$\int_0^\infty dt\, t^z e^{ixt^\beta} = \frac{\Gamma(w)}{\beta|x|^w}\exp\left(\frac{i\pi}{2}w\,\mathrm{sgn}\,x\right), \quad w = \frac{1+z}{\beta}, \quad \beta > 0. \tag{20.16}$$

When $z = n$ is an integer equation 20.15 is a special case of the identity derived in chapter 7, that is,

$$\int_0^\infty dt\, t^n e^{ixt} = n!\left(\frac{i}{x}\right)^{1+n} + \pi\delta^{(n)}(x), \quad n = 0, 1, \ldots.$$

Useful particular cases are

$$\int_0^\infty dt\, e^{ixt^2} = \frac{1}{2}\sqrt{\frac{\pi}{|x|}}\exp\left(i\frac{\pi}{4}\mathrm{sgn}x\right), \tag{20.17}$$

$$\int_0^\infty dt\, t^{2n} e^{ixt^2} = \frac{\Gamma(n+1/2)}{2|x|^{n+1/2}}\exp\left(i\frac{\pi}{2}\left(n+\frac{1}{2}\right)\mathrm{sgn}x\right). \tag{20.18}$$

The binomial expansion
The binomial expansion is

$$(1+z)^\mu = \sum_{k=0}^\infty b_k z^k, \quad |z| < 1,$$

where

$$b_k = \frac{\Gamma(1+\mu)}{k!\,\Gamma(1+\mu-k)} \quad \mu \neq -1, -2, -3, \ldots \tag{20.19}$$

$$= \frac{(-1)^k \Gamma(k-\mu)}{k!\,\Gamma(-\mu)} \quad \mu \neq 0, 1, 2, \ldots. \tag{20.20}$$

If $\mu < 0$ the following version is sometimes useful:

$$(1+z)^{-a} = \frac{1}{\Gamma(a)} \sum_{k=0}^{\infty} \frac{\Gamma(k+a)}{k!} (-z)^k, \quad |z| < 1.$$

21

Appendix II: Elliptic functions

An important property of the circular functions, $\sin z$, $\cos z, \ldots$, is their periodicity: if $f(z)$ denotes any of them,

$$f(z + 2\pi) = f(z) \quad \text{and} \quad f(z + 2n\pi) = f(z)$$

for all integers n. These are *singly-periodic* functions. An *elliptic function*, $f(z)$, is a single valued, *doubly-periodic* function,

$$f(z + \omega_1) = f(z), \quad f(z + \omega_2) = f(z),$$

where the ratio of the two periods, ω_1 and ω_2, is not real.† In addition it is analytic, except at its poles, and has no other singularities other than poles in the finite part of the complex plane.

Elliptic functions occur in many applications of both pure and applied mathematics and have had a significant influence on the developement of the general theory of functions of a complex variable.

In the theory of integration it is a standard result that if $R(x)$ is a linear or quadratic function of the real variable x then the integral $\int dx\, F(x, \sqrt{R(x)})$, $F(x, y)$ being a rational function of x and y, is an elementary function, see 13.2.1. But if $R(x)$ is a cubic or a quartic, with no repeated factors, such integrals are new functions. These were named *elliptic integrals*, because one important application is the computation of the arc length along an ellipse.

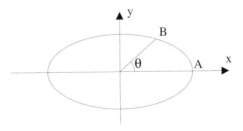

The ellipse in the diagram has the parametric equations $x = a\cos\theta$, $y = b\sin\theta$, with

† If ω_1/ω_2 is real the function reduces to a singly-periodic function if the ratio is rational and a constant if it is irrational.

$a > b$, and the length of the arc AB is

$$s(\theta) = a \int_0^\theta d\phi \sqrt{1 - \epsilon^2 \sin^2 \phi}, \quad \epsilon = \sqrt{1 - (b/a)^2}$$

$$= a \int_0^{\sin^{-1} \theta} dt \sqrt{\frac{1 - \epsilon^2 t^2}{1 - t^2}}.$$

The first systematic study of these integrals was made by Legendre, who proved, among other results, that any elliptic integral can be reduced to the sum of elementary functions and of constant multiples of integrals in three standard forms with the two parameters k (the *modulus*) and n (the *parameter*),

$$\int \frac{dx}{\sqrt{(1 - x^2)(1 - k^2 x^2)}}, \quad \int dx \sqrt{\frac{1 - k^2 x^2}{1 - x^2}}, \quad \int \frac{dx}{(1 + nx^2)\sqrt{(1 - x^2)(1 - k^2 x^2)}}.$$

These are named Legendre's standard elliptic integrals of the first, second and third kind, respectively. This work was published in 1825 and 1826.

As Legendre finished his work, Abel and Jacobi revolutionised the subject by writing, for example,

$$u = \int_0^x dt \, \frac{1}{\sqrt{(1 - t^2)(1 - k^2 t^2)}},$$

and by considering x as a function of u, rather than following Legendre by considering u as a function of x. When $k = 0$, this gives $x = \sin u$, but for other values of k it gives one of Jacobi's elliptic functions, $\mathrm{sn}(u, k)$.

These three functions are the simplest examples of the general class of elliptic functions. Because these functions are doubly-periodic in the complex plane it suffices to consider a parallelogram of periods, that is, any parallelogram with vertices a, $a + \omega_1$, $a + \omega_1 + \omega_2$, $a + \omega_2$ for arbitrary a. Alternatively, an elliptic function may be considered as an analytic function on the torus formed by identifying opposite sides of a parallelogram of periods. Liouville proved the following general properties.

- An elliptic function $f(z)$ that has no pole in a parallelogram of periods is constant.
- The sum of residues of $f(z)$ at the poles in a parallelogram is zero.
- The function $f(z)$ has at least two poles, or a multiple pole, in a parallelogram.
- The number of poles in a parallelogram is equal to the number of zeros and is equal to the number of points at which the function has any assigned value.
- The sum of the values of z at the poles in a parallelogram is either equal to the sum of the values of z at the zeros or differs from it by some period.

It is apparent from this brief summary that much is known about elliptic functions in general and the Jacobi elliptic functions that arise in physical problems. In the following we provide a review of the most useful results involving these functions; for further results and proofs the reader should refer to Whittaker and Watson (1965, chapters 20–22) or Lawden (1989). Other useful lists of formulae can be found in Abramowitz and Stegun (1965) and Gradshteyn and Ryzhik (1965).

Theta functions

Related to Jacobi's elliptic functions are four theta functions $\theta_i(z, q)$, $i = 1, 2, 3, 4$. Two of these arise in the solution of the heat, or diffusion, equation,

$$\frac{\partial \theta}{\partial t} = \kappa \frac{\partial^2 \theta}{\partial z^2}, \quad 0 \leq z \leq \pi.$$

For the boundary conditions $\theta(0, t) = \theta(\pi, t) = 0$ and the initial condition $\theta(z, 0) = \pi \delta(z - \pi/2)$ the solution is

$$\theta = \theta_1(z, q) = 2 \sum_{n=0}^{\infty} (-1)^n q^{(n+1/2)^2} \sin(2n+1)z, \quad q = e^{-4\kappa t}, \tag{21.1}$$

which is the first theta function.

For the boundary conditions $\partial \theta / \partial z = 0$ at $z = 0$, π, that is, no conduction across the boundary but the same initial conditions, we obtain the fourth theta function,

$$\theta = \theta_4(z, q) = 1 + 2 \sum_{n=0}^{\infty} (-1)^n q^{n^2} \cos 2nz, \quad q = e^{-4\kappa t}. \tag{21.2}$$

The theta function $\theta_1(z, q)$ is defined by the series 21.1 by allowing q and z to be complex with $|q| < 1$. Similarly for $\theta_4(z, q)$.

The function $\theta_1(z, q)$ has period 2π in z, and incrementing z by a quarter period gives the second theta function, $\theta_2(z, q) = \theta_1(z + \pi/2, q)$. Similarly $\theta_3(z, q) = \theta_4(z + \pi/2, q)$. Alternative representations of the theta functions are obtained by replacing the trigonometric functions by exponentials. This gives

$$\theta_1(z, q) = -i \sum_{n=-\infty}^{\infty} (-1)^n q^{(n+1/2)^2} e^{i(2n+1)z}$$

$$= 2 \sum_{n=0}^{\infty} (-1)^n q^{(n+1/2)^2} \sin(2n+1)z, \tag{21.3}$$

$$\theta_2(z, q) = \sum_{n=-\infty}^{\infty} q^{(n+1/2)^2} e^{i(2n+1)z} = 2 \sum_{n=0}^{\infty} q^{(n+1/2)^2} \cos(2n+1)z, \tag{21.4}$$

$$\theta_3(z, q) = \sum_{n=-\infty}^{\infty} q^{n^2} e^{2inz} = 1 + 2 \sum_{n=0}^{\infty} q^{n^2} \cos 2nz, \tag{21.5}$$

$$\theta_4(z, q) = \sum_{n=-\infty}^{\infty} (-1)^n q^{n^2} e^{2inz} = 1 + 2 \sum_{n=0}^{\infty} (-1)^n q^{n^2} \cos 2nz. \tag{21.6}$$

Complete elliptic integrals

The two complete elliptic integrals are defined in terms of Legendre standard integrals. The integral of the first kind is defined to be

$$K(k) = \int_0^1 dt \, \frac{1}{\sqrt{(1-t^2)(1-k^2 t^2)}} = \int_0^{\pi/2} d\theta \, \frac{1}{\sqrt{1 - k^2 \sin^2 \theta}}, \quad |k| < 1. \tag{21.7}$$

The complete elliptic integral of the second kind is

$$E(k) = \int_0^1 dt \, \sqrt{\frac{1 - k^2 t^2}{1 - t^2}} = \int_0^{\pi/2} d\theta \, \sqrt{1 - k^2 \sin^2 \theta}, \quad |k| < 1. \tag{21.8}$$

In some texts k^2 is replaced by m, $K(k) \to K(m)$ and all subsequent equations change accordingly.

These definitions show that

$$E(0) = K(0) = \frac{\pi}{2}, \quad E(1) = 1,$$

and that $K(1)$ is undefined.

The series representation of these functions is obtained by expanding the integrand and term-by-term integration,

$$K(k) = \frac{\pi}{2} \left(1 + \left(\frac{1}{2} \right)^2 k^2 + \left(\frac{1 \cdot 3}{2 \cdot 4} \right)^2 k^4 + \left(\frac{1 \cdot 3 \cdot 5}{2 \cdot 4 \cdot 6} \right)^2 k^6 + \cdots \right), \tag{21.9}$$

$$E(k) = \frac{\pi}{2} \left(1 - \left(\frac{1}{2} \right)^2 k^2 - \frac{1}{3} \left(\frac{1 \cdot 3}{2 \cdot 4} \right)^2 k^4 - \frac{1}{5} \left(\frac{1 \cdot 3 \cdot 5}{2 \cdot 4 \cdot 6} \right)^2 k^6 + \cdots \right). \tag{21.10}$$

In the theory of these functions k is named *the parameter* and $k' = \sqrt{1 - k^2}$ the *complementary parameter*, and the notation

$$K(k') = K'(k), \quad E(k') = E'(k),$$

is used. In this context $K'(k)$ *does not* mean dK/dk.

The gnome (or nome), q, is defined by the relation

$$q(k) = \exp \left(-\frac{\pi K(k')}{K(k)} \right) = \left(\frac{k}{4} \right)^2 + 8 \left(\frac{k}{4} \right)^4 + 84 \left(\frac{k}{4} \right)^6 + \cdots.$$

In terms of the gnome we have

$$K(k) = \frac{\pi}{2} + 2\pi \sum_{s=1}^{\infty} \frac{q^s}{1 + q^{2s}}.$$

Both $K(k)$ and $K(k')$ are linearly independent solutions of the differential equation, Lawden (1989, chapter 3),

$$k(1 - k^2)\frac{d^2 y}{dk^2} + (1 - 3k^2)\frac{dy}{dk} - ky = 0,$$

whilst $E(k)$ is one of the solutions of

$$k(1 - k^2)\frac{d^2 y}{dk^2} + (1 - k^2)\frac{dy}{dk} + ky = 0.$$

These equations allow the domain of k to be extended from $|k| < 1$ to the whole complex plane, see Lawden (1989, chapter 8).

For $k \sim 1$ the series 21.9 and 21.10 are of little practical value and the following series are more useful:

$$K(k') = \ln\left(\frac{4}{k}\right) + \frac{k^2}{4}\left(\ln\left(\frac{4}{k}\right) - 1\right) + \frac{k^4}{64}\left(9\ln\left(\frac{4}{k}\right) - \frac{21}{2}\right)$$
$$+ \frac{k^6}{256}\left(25\ln\left(\frac{4}{k}\right) - \frac{185}{6}\right) + \cdots, \tag{21.11}$$

$$E(k') = 1 + \frac{k^2}{2}\left(\ln\left(\frac{4}{k}\right) - \frac{1}{2}\right) + \frac{k^4}{16}\left(3\ln\left(\frac{4}{k}\right) - \frac{11}{4}\right)$$
$$+ \frac{k^6}{128}\left(15\ln\left(\frac{4}{k}\right) - \frac{61}{4}\right) + \cdots, \tag{21.12}$$

where $k' = \sqrt{1 - k^2}$.

Jacobi elliptic functions

The incomplete elliptic integral of the first kind,

$$u = \int_0^y dt \, \frac{1}{\sqrt{(1 - t^2)(1 - k^2 t^2)}} = \int_0^\phi d\theta \, \frac{1}{\sqrt{1 - k^2 \sin^2 \theta}}, \quad y = \sin\phi, \tag{21.13}$$

can be used to define y as a function of u (and k); this function is denoted by

$$y = \text{sn}(u, k) = \sin\phi, \tag{21.14}$$

and is one of the Jacobi elliptic functions; the angle ϕ is named the *amplitude*. Two of the other elliptic functions are defined by the relations

$$u = \int_y^1 dt \, \frac{1}{\sqrt{(1 - t^2)(k'^2 + k^2 t^2)}} \quad \Longrightarrow \quad y = \cos\phi = \text{cn}(u, k), \tag{21.15}$$

and

$$u = \int_y^1 dt \, \frac{1}{\sqrt{(1 - t^2)(t^2 - k'^2)}} \quad \Longrightarrow \quad y = \text{dn}(u, k). \tag{21.16}$$

From these definitions and those of the complete elliptic integrals we have

$$\text{sn}(K, k) = 1, \quad \text{cn}(K, k) = 0, \quad \text{dn}(K, k) = k',$$

and for small k,

$$\text{sn}(u, k) = \sin u - \frac{k^2}{4}(u - \sin u \cos u)\cos u + O(k^4),$$

$$\text{cn}(u, k) = \cos u + \frac{k^2}{4}(u - \sin u \cos u)\sin u + O(k^4),$$

$$\text{dn}(u, k) = 1 - \frac{k^2}{2}\sin^2 u + O(k^4).$$

Elliptic funtions are doubly-periodic and the Jacobi functions can be shown to satisfy the relations

$$\text{sn}\left(u + 2mK(k) + 2inK(k'), k\right) = (-1)^m \text{sn}(u, k),$$
$$\text{cn}\left(u + 2mK(k) + 2inK(k'), k\right) = (-1)^{m+n} \text{cn}(u, k),$$
$$\text{dn}\left(u + 2mK(k) + 2inK(k'), k\right) = (-1)^n \text{dn}(u, k).$$

Applying Pythagoras' theorem to formulae 21.14 and 21.15 gives $\text{cn}(u, k)^2 + \text{sn}(u, k)^2 = 1$; in addition to this similarity with trigonometric functions there are several addition formulae, some of which are:

$$\text{sn}(u + v, k) = \frac{\text{sn}(u, k)\,\text{cn}(v, k)\,\text{dn}(v, k) + \text{sn}(v, k)\,\text{cn}(u, k)\,\text{dn}(u, k)}{1 - k^2 \text{sn}^2(u, k)\,\text{sn}^2(v, k)},$$

$$\text{cn}(u + v, k) = \frac{\text{cn}(u, k)\,\text{cn}(v, k) - \text{sn}(u, k)\,\text{sn}(v, k)\,\text{dn}(u, k)\,\text{dn}(v, k)}{1 - k^2 \text{sn}^2(u, k)\,\text{sn}^2(v, k)},$$

$$\text{dn}(u + v, k) = \frac{\text{dn}(u, k)\,\text{dn}(v, k) - k^2 \text{sn}(u, k)\,\text{sn}(v, k)\,\text{cn}(u, k)\,\text{cn}(v, k)}{1 - k^2 \text{sn}^2(u, k)\,\text{sn}^2(v, k)}.$$

On the real line $\text{dn}(u, k)$ is $2K(k)$-periodic, whilst $\text{sn}(u, k)$ and $\text{cn}(u, k)$ are $4K(k)$-periodic. The Fourier series representations were found by Jacobi and are:

$$\text{sn}(u, k) = \frac{2\pi}{kK(k)} \sum_{n=0}^{\infty} \frac{q^{n+1/2}}{1 - q^{2n+1}} \sin \frac{(2n+1)\pi}{2K(k)}, u$$

$$\text{cn}(u, k) = \frac{2\pi}{kK(k)} \sum_{n=0}^{\infty} \frac{q^{n+1/2}}{1 + q^{2n+1}} \cos \frac{(2n+1)\pi}{2K(k)} u,$$

$$\text{dn}(u, k) = \frac{\pi}{2K(k)} + \frac{2\pi}{K(k)} \sum_{n=1}^{\infty} \frac{q^n}{1 + q^{2n}} \cos \frac{n\pi}{K(k)} u.$$

For small values of u these functions have the following Taylor's series:

$$\text{sn}(u, k) = u - (1 + k^2)\frac{u^3}{3!} + \left(1 + 14k^2 + k^4\right)\frac{u^5}{5!} + \cdots,$$

$$\text{cn}(u, k) = 1 - \frac{u^2}{2!} + (1 + 4k^2)\frac{u^4}{4!} + \cdots,$$

$$\text{dn}(u, k) = 1 - k^2\frac{u^2}{2!} + k^2(4 + k^2)\frac{u^4}{4!} + \cdots.$$

22

References

22.1 References

Abramowitz M. and Stegun I. A. 1965 *Handbook of Mathematical Functions* (Dover, NY)

Acton F. S. 1970 *Numerical Methods that Work* (Harper Row, NY)

Airy G. B. 1838 On the Intensity of Light in the Neighbourhood of a Caustic *Trans. Camb. Phil. Soc.* **6** 379–403

Airy G. B. 1849 Supplement to a Paper 'On the Intensity of Light in the Neighbourhood of a Caustic' *Trans. Camb. Phil. Soc.* **8** 595–599

Andersen C. M. and Geer J. F. 1982 *SIAM J. App. Math.* **42** 678–693

Andronov A. A. and Chaikin C. E. 1949 *Theory of Oscillations* (Princeton University Press)

Apostol T. M. 1963 *Mathematical Analysis* (Addison-Wesley)

Apostol T. M. 1976 *Analytic Number Theory* (Springer-Verlag)

Arfken G. B. and Weber H. J. 1995 *Mathematical Methods for Physicists*, Fourth edition (Academic Press)

Arnold V. I. 1963 Small Denominators and Problems of Stability of Motion in Classical and Celestial Mechanics *Russian Mathematical Surveys* **18** 85–191

Arnold V. I. 1973 *Ordinary Differential Equations*, Translated and edited from the Russian by R. A. Silverman (MIT Press)

Arnold V. I. 1978 *Mathematical Methods of Classical Mechanics* (Springer-Verlag)

Arnold V. I. 1990 *Huygens and Barrow, Newton and Hooke* (Birkhäuser)

Arnold V. I. and Avez A. 1968 *Ergodic Problems of Classical Mechanics* (Benjamin)

Arrowsmith D. K. and Place C. M. 1990 *Dynamical Systems: Differential Equations Maps and Chaotic Behaviour* (Chapman and Hall)

Baker G. A. 1975 *Essentials of Padé Approximants* (Academic Press)

Barnsley M. 1988 *Fractals Everywhere* (Academic Press)

Barrow-Green J. 1996 *Poincaré and the Three Body Problem* (American Mathematical Society)

Bender C. M. and Orszag S. A. 1978 *Advanced Mathematical Methods for Scientists and Engineers* (McGraw-Hill)

Bendixson I. 1901 *Acta Math.* **24** 1–88

Berry M. V. 1987 *The Bakerian Lecture: Quantum Chaology* in *Dynamical Chaos*, pages 183–198, Eds M. V. Berry, I. C. Percival and N. O. Weiss (The Royal Society of London)

845

Blackham D. E. 1978 The Physics of the Piano, in *The Physics of Music*, Readings from Scientific American (W. H. Freeman and Co)

Bleistein N. and Handelsman R. A. 1975 *Asymptotic Expansions of Integrals* (Dover)

Born M. 1960 *The Mechanics of the Atom* (Frederick Ungar)

Boyer C. B. 1968 *A History of Mathematics* (John Wiley)

Boyer C. B. 1987 *The Rainbow: from Myth to Mathematics* (Princeton University Press)

Bremmer H. and Bouwkamp C. J. 1960 *Balthasat van der Pol: Selected Scientific Papers* Volumes I and II (North-Holland)

Brezinski, C. 1981 *The Long History of Continued Fractions and Padé Approximants, Padé Approximation and its Applications*, Lecture Notes in Math. 888 Eds M. G. de Bruin and H. van Rossum (Springer-Verlag, Berlin, New York), 1–27

Brezinski C. 1985 *The Birth and Early Developments of Padé Approximants*, in Lecture Notes in Pure and Appl. Math. 100 (New York), 105–121

Brezinski C. 1996 Extrapolation Algorithms and Padé Approximations: a Historical Survey, *Appl. Numer. Math.* **20** (3), 299–318

Brillouin L. 1926 *Comptes Rendus* **183** 24–26

Brink D. M. and Satchler G. R. 1971 *Angular Momentum* (Clarendon Press, Oxford Library of the Physical Sciences)

Brown L. A. 1949 *The Story of Maps* (Little, Brown and Company)

Brown L. A. 1960 The Longitude, published in *The World of Mathematics* Volume **2**, pages 780–819 (George Allen and Unwin). This article comprises extracts from Brown 1949

Bryant H. C. 1995 *Am. J. Phys.* **63** 14

Cajori F. 1991 *A History of Mathematics* (Chelsea Publishing Company, originally published in 1919 by Macmillan)

Cannell D. M. 1993 *George Green* (Athlone Press)

Casati G., Chirikov B. V., Shepelyansky D. L. and Guarneri I. 1987 *Phys. Rep.* **154** 77–123

Cesari L. 1963 *Asymptotic Behavior and Stability Problems in Ordinary Differential Equations* (Academic Press)

Cook A. 1998 *Edmond Halley: Charting the Heavens and the Seas* (Clarendon Press, Oxford)

Copson E. T. 1967 *Asymptotic Expansions* (Cambridge University Press)

Corben H. C. and Stehle. P 1994 *Classical Mechanics* (Dover)

Courant R. and Hilbert D. 1965 *Methods of Mathematical Physics* Volume I (John Wiley and Sons)

Cox D. R. and Miller H. D. 1965 *The Theory of Stochastic Processes* (Methuen)

Dando P. A. and Richards D. 1990 *J. Phys.* **B23** 3179–3204

Davenport J. H. 1981 *On the Integration of Algebraic Functions*, Lecture Notes in Computer Science, (Springer-Verlag, Berlin, New York)

Davenport J. H., Siret Y. and Tournier E. 1993 *Computer Algebra* (Academic Press)

Dean C. 1999 *Am. J. Phys.* **67**(10) 928–929

de Bruijn N. G. 1961 *Asymptotic Methods in Analysis* (North-Holland)

Devaney R. L. 1992 *A First Course in Chaotic Dynamical Systems* (Addison-Wesley)

Dingle R. B. 1973 *Asymptotic Expansions: Their Derivation and Interpretation* (Academic Press, London)

Duffing G. 1918 *Erzwungene Schwingungen bei veränderlicher Eigenfrequenz und ihre technische Bedeutung* (Braunschweig, Druck und Verlag von Friedr. Vieweg und Sohn)

Dunham J. L. 1932 *Phys. Rev.* **41** 721–731

Edmonds A. R. 1974 *Angular Momentum in Quantum Mechanics* (Princeton University Press)

English L. Q. and Winters R. R. 1997 *Am. J. Phys.* **65**(5) 390–393

Erdélyi A. 1956 *Asymptotic Expansions* (Dover)

Feigenbaum M. J. 1978 *J. Stat. Phys.* **19** 25–52

Feynman R. P. and Hibbs A. R. 1965 *Quantum Mechanics and Path Integrals* (McGraw-Hill)

Forsyth A. R. 1935 Old Tripos Days at Cambridge *The Math. Gazette* **19** 162–179

Frobenius G. 1881 *J. für Math.* (Crelle) **90** 1–17

Fröman N. and Fröman P. O. 1967 *JWKB-Approximation, Contributions to the Theory* (North-Holland, Amsterdam)

Gandy R. O. 1973 Bertrand Russell, as Mathematician *Bull. London Math. Soc.* **5** 342–348

Gathen J. von zur and Gerhard J. 1999 *Modern Computer Algebra* (Cambridge University Press)

Gear C. W. and Skeel R. D. 1990 in *A History of Numerical Computing* (Addison-Wesley)

Geddes K. O., Czapor S. R. and Labahn G. 1992 *Algorithms for Computer Algebra* (Kluwer Academic Publishers)

Giacaglia G. E. O. 1972 *Perturbation Methods in Non-Linear Systems* (Springer-Verlag)

Gibbs J. W. 1898 *Nature* **59** 200, reprinted in *The Collected Works of J. Willard Gibbs* (Yale University Press)

Gibbs J. W. 1899 *Nature* **59** 606, reprinted in *The Collected Works of J Willard Gibbs* (Yale University Press)

Goldstein H. 1980 *Classical Mechanics* (Addison-Wesley)

Goodstein D. L. and Goodstein J. R. 1997 *Feynman's Lost Lecture: the Motion of Planets around the Sun* (Vintage)

Gradshteyn I. S. and Ryzhik I. M. 1965 *Tables of Integrals, Series and Products* (Academic Press)

Grasman J. 1987 *Asymptotic Methods for Relaxation Oscillations and Applications* (Springer-Verlag, Applied Mathematical Sciences 63)

Gratton-Guinness I. 1970 *The Development of the Foundations of Mathematical Analysis from Euler to Riemann* (MIT press)

Green G. 1828 *An Essay on the Application of Mathematical Analysis to the Theories of Electricity and Magnetism*, originally published at Nottingham in 1828, reprinted in *The Scientific Papers of George Green* Volume 1, by The George Green Memorial Committee (Nottingham) (1993)

Green G. 1838 On the Motion of Waves in a Variable Canal of Small Depth and

Width *Camb. Phil. Trans.* **6** 457–462, reprinted in *The Scientific Papers of George Green* Volume 3, by The George Green Memorial Committee (Nottingham) (1993)

Greene J. M., MacKay R. S., Vivaldi F. and Feigenbaum M. J. 1981 *Physica* **3D** 468–86 (Reprinted in *Hamiltonian Dynamical Systems* Eds R. S. MacKay and J. D. Meiss (Adam Hilger 1987)

Greenler R. 1991 *Rainbows, Halos and Glories* (Cambridge University Press)

Grossman N. 1997 *The Sheer Joy of Celestial Mechanics* (Birkhäuser)

Guckenheimer J. and Holmes P. 1983 *Nonlinear Oscillations, Dynamical Systems, and Bifurcations of Vector Fields* (Springer-Verlag)

Gutzwiller M. C. 1990 *Chaos in Classical and Quantum Mechanics* (Springer-Verlag)

Hagedorn P. 1988 *Non-Linear Oscillations* (Oxford Science Publications)

Hand L. N. and Finch J. D. 1998 *Analytic Mechanics* (Cambridge University Press)

Harding R. D. and Quinney D. A. 1989 *A Simple Introduction to Numerical Analysis: Interpolation and Approximation* Volume 2 (Adam Hilger, Bristol)

Hardy G. H. 1948 The Case Against the Mathematical Tripos, *The Math. Gazette* **32** 134–145

Hardy G. H. 1966 *The Integration of Functions of a Single Variable*, Cambridge Tracts in Mathematics and Mathematical Physics (Cambridge University Press)

Hardy G. H. 1967 *A Course of Pure Mathematics* (Cambridge University Press)

Hartman P. 1964 *Ordinary Differential Equations* (John Wiley)

Heading J. 1962 *Phase Integral Methods* (Methuen, London)

Heck A. 1996 *Introduction to Maple*, Second edition (Springer)

Hinch E. J. 1991 *Perturbation Methods* (Cambridge University Press)

Hirsch M. W. and Smale S. 1974 *Differential Equations, Dynamical Systems and Linear Algebra* (Academic Press)

Holmes M. H. 1995 *Introduction to Perturbation Methods* (Springer-Verlag)

Holstein B. R. 1997 *Am. J. Phys.* **65**(12) 1133–1135

Hoskins R. F. 1979 *Generalised Functions* (Ellis Horwood)

Ince E. L. 1956 *Ordinary Differential Equations* (Dover)

Jammer M. 1989 *The Conceptual Development of Quantum Mechanics* (Tomash publishers) American Institute of Physics (The History of Modern Physics **12**)

Jensen R. V., Susskind S. M. and Saunders M. M. 1991 Chaotic Ionization of Highly Excited Hydrogen Atoms *Phys. Rep.* **201** 1–56

Jones D. S. 1966 *The Theory of Generalised Functions* (Cambridge University Press)

Jordan D. W. and Smith P. 1999 *Nonlinear Ordinary Differential Equations*, Third edition (Clarendon Press, Oxford Applied and Engineering Mathematics)

Kato T. 1976 *Perturbation Theory for Linear Operators* (Springer-Verlag)

Keating J. 1993 in *The Nature of Chaos* Ed. T. Mullin (Oxford University Press)

Khinchin A. Y. 1964 *Continued Fractions* (Chicago University Press)

Kline M. 1972 *Mathematical Thought from Ancient to Modern Times* (Oxford University Press)

Körner T. W. 1988 *Fourier Analysis* (Oxford University Press)

Korsch J. H. and Laurent H. 1981 *J. Phys.* **B 14** 4213–4230

Kramers H. A. 1926 *Zeitschrift für Physik* **39** 828–840

Krantz R. J. 1998 *Am. J. Phys.* **66**(4) 276–277

Kreider D. L., Kuller R. G., Ostberg D. R. and Perkins F. W. 1966 *An Introduction to Linear Analysis* (Addison-Wesley Publishing Company)

Lanczos C. 1966 *Discourse on Fourier Series* (Oliver and Boyd)

Landau L. D. and Lifshitz E. M. 1965a *Mechanics Course of Theoretical Physics*, Volume 1 (Pergamon Press)

Landau L. D. and Lifshitz E. M. 1965b *Quantum Mechanics Course of Theoretical Physics*, Volume 3 (Pergamon Press)

Lawden D. F. 1989 *Elliptic Functions and Applications* (Springer-Verlag)

Lichtenberg A. J. and Lieberman M. A. 1983 *Regular and Stochastic Motion* (Springer-Verlag)

Li T.-Y. and Yorke J. 1975 Period Three Implies Chaos *Am. Math. Monthly* **82** 985–992

Lighthill M. J. 1962 *Introduction to Fourier Analysis and Generalised Functions* (Cambridge University Press)

Lloyd N. G. 1987 Limit Cycles of Polynomial Systems, in *New Directions in Dynamical Systems* Eds T. Bedford and J. Swift, London Mathematical Lecture Note Series **127** (Cambridge University Press)

Lopez-Lopez F. J. 1995 *Am. J. Phys.* **63**(7) 583–584

Longhurst R. S. 1967 *Geometrical and Physical Optics* (Longmans)

Lotka A. J. 1920 Undamped Oscillations Derived from the Law of Mass Action *J. Amer. Chem. Soc.* **42** 1595–1599

Lotka A. J. 1956 *Elements of Mathematical Biology* (Dover, New York)

Lützen J. 1990 *Joseph Liouville 1809–1882: Master of Pure and Applied Mathematics* (Springer-Verlag).

MacCallum M. and Wright F. 1991 *Algebraic Computing with Reduce* (Oxford University Press)

MacDonald N. 1994 *Reduce for Physicists* (Institute of Physics)

Mandel L. and Wolf E. 1995 *Optical Coherence and Quantum Optics* (Cambridge University Press)

Maslov V. 1972 *Théories des Perturbations* (Dunod, Paris)

May R. M. 1987 Chaos and the Dynamics of Biological Populations in *Dynamical Chaos*, pages 27–44, Eds M. V. Berry, I. C. Percival and N. O. Weiss (The Royal Society of London)

Maynard Smith J. 1977 *Mathematical Ideas in Biology* (Cambridge University Press)

Melzak Z. A. 1973 *Companion to Concrete Mathematics* (John Wiley)

Meinel A. and Meinel M. 1991 *Sunsets, Twilights and Evening Skies* (Cambridge University Press)

Minnaert M. 1954 *The Nature of Light and Colour in the Open Air* (Dover)

Minorsky N. 1962 *Nonlinear Oscillations* (D. van Nostrand)

Mitchell B. R. 1992 *International Historical Statistics: Europe 1750–1988* (MacMillan)

Monk R. 1992 *Bertrand Russell: the Spirit of Solitude* (Jonathan Cape)

Moore C. G. 1964 *An Introduction to Continued Fractions* (National Council of Teachers of Mathematics, Washington D.C., USA)

Moses J. 1971 *Comm. ACM* **14** 548–560

Murray J. D. 1984 *Asymptotic Analysis* (Springer-Verlag)

Murray J. D. 1989 *Mathematical Biology* (Springer-Verlag)

Nayfeh A. 1973 *Perturbation Methods* (Wiley)

Nayfeh A. and Mook D. T. 1979 *Nonlinear Oscillations* (Wiley)

Norcliffe A. 1973 *Case Studies in Atom. Phys.* **4**(Nov) 1–55

Norcliffe A., Percival I. C. and Roberts M. J. 1969, *J. Phys.* **B2** 578–589

Observer's Handbook 1995 Ed. R L Bishop (The Royal Astronomical Society of Canada)

Olver F. W. J. 1974 *Asymptotics and Special Functions* (Academic Press)

Ott E. 1993 *Chaos in Dynamical Systems* (Cambridge University Press)

Ozorio de Almeida A. M. 1988 *Hamiltonian Systems: Chaos and Quantisation* (Cambridge University Press)

Padé H. 1892 (Thesis) Sur la Représentation Approchée d'une Fonction pour des Fractions Rationnelles *Ann. Sci. École Norm. Sup. Suppl.* [3] **9** 1–93

Pajunen P. and Child M. S. 1980 *Mol. Phys.* **40** 597–604

Paulsson R., Karlsson F. and Le Roy R. J. 1983 *J. Chem. Phys.* **79**(9) 4346–4354

Pearl R. and Reed L. J. 1920 *Proc. Natl. Acad. Sci.* **6** 275

Pearson K. 1936 Old Tripos Days at Cambridge, as Seen from Another Viewpoint, *The Math. Gazette* **20** 27–36

Peitgen H.-O., Jürgens H. and Saupe D. 1992 *Chaos and Fractals* (Springer)

Percival I. C. 1977 Semiclassical Theory of Bound States *Adv. Chem. Phys.* **36** 1–61

Percival I. C. and Richards D. 1982 *Introduction to Dynamics* (Cambridge University Press)

Phelps F. M. 1995 *Am. J. Phys.* **63**(7) 584–585

Pickett T. J. 1997 *Am. J. Phys.* **65**(6) 461–462

Poincaré H. 1880 *C. R. Acad. des Sciences* **90** 673–675

Poincaré H. 1881 *J. de Math.* (3) **7** 375–422

Poincaré H. 1882 *J. de Math.* (3) **8** 251–296

Poincaré H. 1885 *J. de Math.* (4) **1** 90–161

Poincaré H. 1886 *J. de Math.* (4) **2** 151–217

Poincaré H. 1886a *Acta Math.* **6**, 295–344

Poincaré H. 1890 *Acta Math.* **13** 1–270

Poincaré H. 1892 *Les Méthodes Nouvelles de la Mécaniques Céleste* **I** (Gauthiers-Villars, republished by Blanchard, Paris, 1987)

Powell M. J. D. 1981 *Approximation Theory and Methods* (Cambridge University Press)

Press W. H., Flannery B. P., Teukolsky S. A. and Vetterling W. T. 1987 *Numerical Recipes* (Cambridge University Press)

Rasband S. N. 1990 *Chaotic Dynamics of Nonlinear Systems* (John Wiley)

Rayleigh J. W. S. 1883 *Phil. Mag.* **15** 229–235

Rayleigh J. W. S. 1894 *The Theory of Sound*, Volume 1 (Reprinted by Dover, 1945)

Reichl L. E. 1992 *The Transition to Chaos in Conservative Classical Systems: Quantum Manifestations* (Springer-Verlag)

Richards D. 1997 The Periodically Driven Excited Hydrogen Atom, in *Classical, Semiclassical and Quantum Dynamics in Atoms* Eds H. Friedrich and B. Eckhardt (Springer)

Risch R. H. 1969 *Trans. AMS* **139** 167–189

Rosenzweig C. and Krieger J. B. 1968 *J. Math. Phys.* **9** 849–60

Rudin W. 1976 *Principles of Mathematical Analysis*, Third edition (McGraw-Hill)

Sagdeev, R. Z. and Zaslavsky, G. M. 1988 *Nonlinear Physics: from the Pendulum to Turbulence and Chaos*, Translated from Russian by I. R. Sagdeev (Harwood Academic Publishers).

Sanmartín J. R. 1984 *Am. J. Phys.* **52** 937–945

Sànchez D. A. 1968 *Ordinary Differential Equations and Stability Theory* (W H Freeman and Company)

Sawyer W. W. 1967 *A First Look at Numerical Functional Analysis* (Oxford University Press)

Shank D. 1955 *J. Math. Phys.* **34**, 1–42

Shiu P. 1996 *Math. Gazette*, **80**, 54–70

Smale S. 1998 *The Mathematical Intelligencer*, **20**(2), 7–15

Sobel D. 1995 *Longitude* (Fourth Estate)

Srinivasan T. P. 1992 *Am. J. Phys.* **60**(5) 461–462

Sullivan J. 1978 *Am. J. Phys.* **46**(5) 489–494

Tabor M. 1989 *Chaos and Integrability in Nonlinear Dynamics* (John Wiley and Sons)

Townes C. H. and Schawlow A. L. 1975 *Microwave Spectroscopy* (Dover)

van der Pol B. 1920 *Radio Rev. London* **1** 701–710, 754–762

van der Pol B. 1926 *Phil. Mag.* **2** 978–992

van der Pol B. and van der Mark M. J. 1929 *Arch. Néerl. Physiol.* **14** 418–443

Verhulst F. 1990 *Nonlinear Differential Equations and Dynamical Systems* (Springer-Verlag)

Volterra V. 1926 Variazioni e Fluttuazioni del Nemero d'Individui in Specie Animali Conviventi *Mem. Accad. Naz. Lincei* **2**, 31–113. English translation in Chapman R. N. 1931 *Animal Ecology* pages 409–448 (McGraw-Hill, New York)

Voros A. 1983 *Annales de l'Institut Henri Poincaré* section A **39**, 211–338

Walker J. 1977 *The Flying Circus of Physics* (John Wiley)

Walker J. D. 1976 *Am. J. Phys.* **44**(5) 421–455

Watson G. N. 1966 *A Treatise on the Theory of Bessel Functions* (Cambridge University Press)

Wentzel G. 1926 *Zeitschrift für Physik* **38**, 518–529

Wester M. J. 1999 *Computer Algebra Systems: A Practical Guide* (John Wiley)

Whittaker E. T. 1964 *A Treaties on the Analytical Dynamics of Particles and Rigid Bodies* (Cambridge University Press)

Whittaker E. T. and Watson G. N. 1965 *A Course of Modern Analysis* (Cambridge University Press)

Wisdom J. 1982 *Astron. J.* **87**(2) 577–593

Wisdom J. 1983 *J. Meteonics* **18** 422–423

Wynn P. 1956 *Proc. Camb. Phil. Soc.* **52** 663–671

Yakubovich V. A. and Starzhinskii V. M. 1975 *Linear Differential Equations with Periodic Coefficients* (John Wiley and Sons; Israel program for scientific translations)

Ye Yan-Qian, Cai Sui-Lin and Ma Zhi-En 1986 *Theory of Limit Cycles* (American Mathematical Society) Translated by Chi Y. Lo.

References

Young N. 1988 *An Introduction to Hilbert Space* (Cambridge University Press)
Zaslavsky G. M. 1998 *Physics of Chaos in Hamiltonian Systems* (Imperial College Press)
Zygmund A. 1990 *Trigonometric Series*, Second edition, Volumes I and II combined
(Cambridge University Press)

22.2 Books about computer-assisted algebra

There are many books that describe Maple and/or use Maple in a specific context. The following is a selected list of those books and also texts on related background material that I have found useful. A more comprehensive list may be found on the Maple website, www.maplesoft.com

Bauldry W. C., Evans B., Johnson J. 1995 *Linear Algebra with Maple* (John Wiley)

Burkhardt W. 1994 *First Steps in Maple* (Springer-Verlag)

Coombes K. R., Hunt B. R., Lipsman R. L., Osborn J. E. and Stuck G. J. 1996 *Differential Equations with Maple* (John Wiley)

Corless R. M. 1995 *Essential Maple* (Springer-Verlag)

Davenport J. H., Siret Y. and Tournier E. 1993 *Computer Algebra Systems and Algorithms for Algebraic Computation* (Academic Press)

Fowkes N. D. and Mahony J. J. 1996 *An Introduction to Mathematical Modelling* (John Wiley)

Gander W. and Hrebicek J. 1991 *Solving Problems in Scientific Computing Using Maple and Matlab* (Springer)

Gathen J. von zur and Gerhard J. 1999 *Modern Computer Algebra* (Cambridge University Press)

Geddes K. O., Czapor S. R. and Labahn G. 1992 *Algorithms for Computer Algebra* (Kluwer Academic Publishers)

Greene R. L. 1995 *Classical Mechanics with Maple* (Springer)

Heck A. 1996 Introduction to Maple (Springer)

Holmes M. H., Ecker J. G., Boyce W. E. and Siegmann W. L. 1993 *Exploring Calculus with Maple* (Addison-Wesley)

Horbatsch M. 1995 *Quantum Mechanics Using Maple* (Springer)

Kofler M. 1997 *Maple: an Introduction and Reference* (Addison-Wesley)

MacCallum M. and Wright F. 1991 *Algebraic Computing with Reduce* (Oxford University Press)

MacDonald N. 1994 *Reduce for Physicists* (Institute of Physics)

Nicholaides R. and Walkington N. 1996 *Maple: a Comprehensive Introduction* (Cambridge University Press)

Redfern D. 1995 *The Practical Approach: Utilities for Maple* (Springer)

Redfern D. and Chandler E. 1995 *Maple ODE Lab. Book* (Springer)

Wester M. J. 1999 *Computer Algebra Systems: a Practical Guide* (John Wiley)

Index

Maple commands are **bold** in index and footnotes

$, 48
%, 19

Abel N. H., 172, 485, 840
Abel's test for uniform convergence, 172
about, 55
abs, 19
absolute convergence, 139, 173
absolutely convergent series, 139
acceleration due to gravity, 644
add, 13, 14
adjoint system, 475
Airy function, 107, 406, 510, 578
    asymptotic expansions, 583
Airy J. B., 407, 557
AiryAi, 107, 406
AiryBi, 406
Aitken A. C., 161
Aitken's Δ^2 process, 161
algebraic functions, 486
alias, 107, 180, 254, 327
allvalues, 9, 109
alternating harmonic series, 148
alternating series, 148
anames, 15
Anger C. T., 532
Anger's function, 496, 532
AngerJ, 532
animation, 81
animation of graphs, 42
antisymmetric matrix, 45
aphelion, 258
apogee, 632
approximation in the mean, 270
argument, 19
Aryabhata, 196
assign, 8
assume, 55
    constant, 106
    showassumed, 56
asteroid, 728
asympt, 186
asymptotic expansion
    Airy functions, 407
    Bernoulli numbers, 179, 506
    Bessel function, 255, 408
    Bessel function zeros, 409
    definition, 184
    differentiation of, 188
    digamma function, 836

error function, 192
exponential integral, 494
gamma function, 188, 506, 835
Hermite polynomials, 392
integration of, 188
Laguerre polynomials, 398
Mathieu functions, 427
modified Bessel function, 499
psi function, 836
Sine integral, 193
Sturm–Liouville system, 373
uniqueness of, 186
asymptotic sequence, 189
asymptotically stable, 656
attractor, 656, 729
autonomous systems, 635, 642
average of periodic functions, 278
averaged equations of motion, 818, 820
averaging, 817
    over a subharmonic, 823

back-quotes, 11
    special characters, 36
Bendixson's negative criterion, 675, 724
Bernoulli
    Daniel, 132, 630
    James, 629
    John, 629
bernoulli, 148, 180
Bernoulli equation, 629, 634, 640
Bernoulli numbers, 143, 148, 179
    asymptotic expansion, 179, 506
Bernoulli polynomials, 179, 192
    generating function, 179
Bernoulli shift, 780
Bessel function, 66, 107, 404
    addition formula, 295
    and planetary motion, 258
    asymptotic expansion, 255, 408, 563
    differential equation, 254
    generating function, 405
    integral representation, 254
    modified, 499
    series representation, 254
    spherical, 406
    zeros, 351
Bessel's equation, 350, 404
Bessel's inequality, 271
BesselJ, 254, 327, 404
BesselJZeros, 351, 409

855

BesselY, 404
BesselYZeros, 409
beta function, 836
bifurcation, 733
    diagram, 739, 763
binomial expansion, 166, 837
Birkhoff G. D., 754
bound variable, 7
boundary conditions
    mixed, 355, 360
    periodic, 345, 355
boundary layer, 573
break, 30, 31

canal equation, 346
catenary, 629
Cauchy A.-L., 154, 630
Cauchy's
    integral, 382
    test, 144
    theorem, 507
centre, 661
cfrac, 198
chaos, 767
characteristic
    exponents, 450
    multipliers, 449
    numbers, 449
Chebychev, 387
Christoffel–Darboux formula, 380, 385
circle, 153
circle of convergence, 151
Cissoid of Diocles, 488
Clairaut A. C., 631
coeff and Fourier coefficients, 83
coeff and **coeffs**, 82
collect, 82, 83
combine, 72
    symbolic option, 73
    trig, 73
commensurable numbers, 351
comparison equation, 579
comparison test, 144
complete
    orthogonal system, 271
    orthonormal system, 357
completeness relation, 271
complex conjugate, 269
compose, 157, 158
composition of functions, 100
concatenation, 33
Conchoid of Nicomedes, 489
conditionally convergent series, 139
conjugate, 19
conjugate variables, 468
conservative systems, 691
constants of the motion, 653
context-sensitive plotting, 22
continue, 31
continued fraction
    canonical representation, 203
    convergence of, 201
    definition, 195

elements of, 195
evaluation of, 199
finding from a given number, 200
hypergeometric function, 220
of order n, 195
periodic, 202
remainder, 201
using convert command, 74
continuous spectrum, 594
contourplot, 23, 109, 691
contours option in plots, 25
convergence
    in the mean, 272
    of a sequence, 135
convergent series, 138
convert, 74, 157
    Padé approximants, 214
    confrac, 198
    continued fraction, 74
    ratpoly, 212, 214
    to a string, 78
coords option in plot, 23
cosine series, 284
Cotes R., 223
critical point, 508, 527
cycloid, 22

D, 103
    and total derivatives, 105
d'Alembert J. R., 131
d'Alembert's ratio test, 145
damped linear oscillator, 646
day, definition of, 197
de Broglie wavelength, 527
De Moivre A., 179
degenerate eigenvalue, 350
delta function, 228
    derivatives of, 229
    periodic array of, 236
    sequences for, 235
delta function decomposition, 359
denom, 68
density function, 743
dependence upon a parameter, 643
DEplot, 644
Descartes R., 558
det, 46
DEtools, 324
DeVolgelaere map, 759
diagonal matrix, 45
diff, 48, 103
differentiation under the integral sign, 479
digamma function, 835
Digits, default value, 16
diophantine equation, 196, 197
Diophantus, 196
Dirac, 228
directory specification, 123
Dirichlet's test, 146
discrete spectrum, 594
display, 42, 79
divergent series, 138
dot notation, 629

double integrals, 520
double well, 602, 610
dsolve, 119, 324, 331, 608, 798
   classical option, 456
   maxfun option, 456
   numpoints option, 456
Duffing's equation, 786
Dulac's test, 675, 724

Earth, direction of rotation, 197, 259
Earth's radius, 649
eccentric anomaly, 259
eccentricity, 259
   of Earth's orbit, 260, 338
eigenfunction, 342
eigenfunction expansion, 365
eigenvalue, 342
   degenerate, 350
eigenvalues, 420
eigenvectors, 416
Ei(n,x), 182, 184
elementary functions, 486
elements of a continued fraction, 195
elif, 29
elliptic fixed point, 661, 747
elliptic function, 839
elliptic integrals, 426, 839
EllipticK and **EllipticE**, 426
equilibrium point, 656
erf, 180
ergodic hypothesis, 744
error function, 180, 192
   asymptotic expansion, 192
Euclid's algorithm, 196
Euler L., 54, 132, 198, 630, 834
Euler transform, 210, 214
   and Padé approximant, 214
Euler's constant, 15, 143, 834
Euler–Maclaurin expansion, 148
eval, 17, 43, 310
   to remove order symbol, 76, 187
evalc, 19
evalf, 15, 16
   and **map**, 99
   and **int**, 53, 490
evalhf, 32
evalm, 43
evaln, 11, 114
eval
   with diff, 50
existence theorems, 668
exp and e^x in Maple, 73
exp of a matrix, 87
expand, 63
   Bessel functions, 66
   operation on functions, 66
expansion coefficients, 270
explicit algebraic functions, 486
exponential integral, 182, 494
   asymptotic expansion, 494
exponential of a matrix, 473
expression sequence, 33

factor, 13, 67
   difference between **normal**, 69
factorial function
   asymptotic expansion, 503, 506
False Position Method, 115
fast Fourier transform, 801, 825
Feigenbaum's constant, 735
Fermat's principle, 544
FFT, 825
Fibonacci numbers, 117
figure justification, 22
first integral, 653
fixed point, 635
   classification, 663
   for $1d$-maps, 728
   for $2d$-maps, 746
   for second-order systems, 655
   index of, 666
   simple, 635, 658
   superstable, 729
floor, 201
Floquet G., 443
Floquet theory, 815
Floquet–Lyapunov theorem, 451
fnormal, 20
focus, 660
folder specification, 123
for loop, 29
forget, 78
forward-quotes, 11
Fourier coefficients, 273, 358
   behaviour of, 277
Fourier integrals, 496
   convergence of, 496
Fourier J. B. J., 132, 346
Fourier series, 273, 358
   and the wave equation, 132
   as Laurent series, 74
   differentiation of, 282
   integration of, 281
Fourier transform, 801, 825
Fourier's theorem, 275
free variables, 7
Fresnel integral, 527, 531
Frobenius F. G., 212
fsolve, 18, 27, 95, 152, 315, 401
function
   composition, 100
   definition, 91
   of many variables, 94
   piecewise, 102
function composition, 100
function of a matrix, definition, 100
functions known to Maple, 91
fundamental matrix, 438
Fundamental Theorem of Calculus, 478

g, acceleration due to gravity, 644
Galileo G., 648
GAMMA, 15, 67
gamma, 15, 143
gamma function, 15, 67, 834
   asymptotic expansion, 188, 506, 835

Gaussian quadrature, 400
gcd, 68
generalised functions, 232
generating function, 380, 433
    Bernoulli polynomials, 179
    Bessel functions, 405
    Hermite polynomials, 391
    Laguerre polynomials, 397
    Legendre polynomials, 383
geometric series, 141
Gibbs J. W., 287
Gibbs phenomenon, 286
global variables, 113, 117
gnome, 791, 842
gradient system, 671
Gram–Schmidt orthogonalisation, 297
Green G., 227, 261, 573
Green's function, 227, 361, 362
    and linearity, 240
    for the linear oscillator, 248
    generalised, 367
    series representation, 365
Green's theorem, 724
Gregory's series, 155
grid, option in plot, 24

Hadamard product, 485
half-range Fourier series, 285
Halley E., 631
Halley's method, 130, 260, 327
    with matrices, 335
    with power series, 335
Hamilton W. R., 654
Hamilton's equations, 468
Hamiltonian
    function, 468
    system, 654
hard spring, 789, 800, 808
harmonic balance, 707, 808
harmonic series, 138, 835
has in **select**, 39
Heaviside, 230
Heaviside function, 230, 287
    series representation, 238
Hermite polynomials, 264, 389
    asymptotic expansion, 392, 593
heteroclinic points, 784
heteroclinic tangle, 794
Hilbert's sixteenth problem, 678
Hill G. W., 411, 632
Hill's equation, 411, 452, 815
homoclinic tangle, 773, 793, 802, 804
Hooke R., 631
horn equation, 346
Horowitz method, 483
Huygens C., 197, 529, 629
hyperbolic
    fixed point, 659, 747, 748
    with reflection, 748
hypergeom, 219
hypergeometric function
    continued fractions, 220
    differential equation for, 219

integral representation, 507
series definition, 219

idempotency operator, 485
identity matrix, 45
if conditional, 26
ifactor, 5
Im, 19
implicit algebraic functions, 486
implicitplot, 23, 24, 127
improper node, 663
impulsive forces and Green's functions, 241
incoherent sum, 529
index, 666
inifcn, 91, 377
initial value problems, 248
inner product, 269, 354, 356
input prompt, 2
insequence=true option, 42, 81
Int, 51
int, 9, 51
    and **evalf**, 53
    definite, 51
integral of the motion, 653
integral test, 146
integration
    by parts, 493
    of periodic functions, 278
integration problem, 484
interface and **showassumed**, 56
interference, 529
invariant sets of states, 635
inverse, 157
inverse, different meanings of, 157
inverse square law, 258, 383
inversion hyperbolic, 748
irreducible polynomial, 486
island chain, 750, 752
isochrone problem, 629
isprime, 5
iterate, 729
iterative methods, 315

Jacobi C. G. J., 840
Jacobi elliptic function, 791
Jacobian
    determinant, 746
    matrix, 746, 747
JacobiCN(z,k), 791
JacobiSN(z,k), 795
justification of figures, 22

Kapteyn series, 255
Kapteyn W., 255
Kepler ellipse, 197
    precession, 197
Kepler's
    equation, 117, 259, 338
    laws, 631
    third law, 259
Kirchhoff G. R., 530
Kirkwood gap, 728
Kronecker delta, 269

labels, 97, 491
Lagrange J.-L., 258
Lagrange's identity, 354
Laguerre polynomials, 395, 608
    asymptotic expansion, 398, 609
Laplace integrals, 382, 495
Laplace P. S., 180, 485
Legendre A. M., 840
Legendre functions of second kind, 382
Legendre polynomials, 119, 220, 238, 381
    asymptotic expansion, 386
Leibniz G. W. L., 629
Leibniz's criterion, 148
lemniscate of Bernoulli, 23
Liénard
    equation, 466, 676, 681
    plane, 643, 647
Liapunov, *see* Lyapunov
library packages, 23
librational motion, 650
limit, 87, 95
limit cycle, 677
limit of a sequence, 135
Lindstedt A., 697
Lindstedt's method, 422, 701, 705
Linear Algebra in Maple 6, 46
linear dependence, 438
linear oscillator, 248, 265, 391, 646, 786
    Green's function for, 788
linearisation matrix, 747
linearisation theorem, 664
linearised equations of motion, 657
linearised map, 728
Liouville equation, 573
Liouville J., 342, 439, 484, 573, 840
Liouville's principle, 487
Lipschitz–condition, 820
Lissajou's figures, 23
list, 27, 34
local variables, 113
local variables, 113
logarithm
    of a matrix, 475
    principal branch of, 450
logistic
    equation, 638
    map, 38, 315, 638, 730, 731
loops, 26
    controlled by lists, 36
Lotka A. J., 665
lunar apsides, 632
Lyapunov A. M., 740
Lyapunov exponent, 740, 774

Maclaurin series, 153
manifold, stable and unstable, 768
map
    linearised, 728, 747
    one-dimensional, 728
    two-dimensional, 746
map, 39, 85
    and **evalf**, 99
    with function notation, 99

Maple 6, 29, 32, 34, 46
Maple settings, 56
Maplev5.ini file, 56
Maplev5.ini file, 56
matching procedure, 586
Mathieu E., 411
Mathieu functions, 410, 615
    asymptotic expansion, 427, 619
Mathieu's equation, 411, 452, 454, 815
    damped, 457
matrix
    exponential of, 473
    logarithm of, 475
    partitioned, 469
matrix, 43
    exponential of, 87
    multiplication, 44
matrizant, 439
Maxwell J. C., 161
mean of periodic functions, 278
mean square difference, 268
mean value theorem, 233
mean-motion equations, 818, 820
Meissner's equation, 474
Melnikov's method, 778
Mercury, 260
Mersenne primes, 57
meteorites, 728
Méziriac C. G. B. de, 197
Michelson A. A., 286
minimax polynomial, 387
mixed spectrum, 594
Möbius transformation, 781
modp, 30
monodromy matrix, 439, 449
Morse potential, 695
mtaylor, 735
multiply, 157

name in Maple, 18
natural boundary, 637
Neumann series, 257
neutraliser, 547
Newton I., 179, 325, 628, 631
Newton's equations, 239
Newton's law, 691
Newton's method, 117, 133, 260, 325, 328, 779
    for division, 332
    with differential equations, 329
    with matrices, 335
    with power series, 332
next, 31
nextprime, 5
nodal lines, 351
node, 658
nome, 842
non-degenerate eigenvalue, 350
nops, 35, 39
norm of a function, 269
normal, 68
    difference between **factor**, 69
    with generalised polynomials, 70
normalised functions, 269

NULL sequence, 31
NULL sequence, 31
numer, 68
numerical integration, 53
numpoints, option in plot, 23

O Botafumeiro, 446
odeplot, 119, 324
op, 35, 39
option in procedures, 117
orbit, 642
Order
  an environment variable, 76
  and **solve**, 77
order notation, 136
order symbol, removal in Maple, 76, 187
ordinary point, 656
orrery, 197
orthogonal
  functions, 269
  system, 269
orthonormal system, 269
oscillation theorem, 357

packages, 23
Padé approximant, 212
  and poles, 216
Padé H. E., 209
pade, 215
parabolic cylinder functions, 587
parametric pumping, 444
parse, 78
Parseval's theorem, 276, 358
partial derivaties, 108
partial sum, 138, 169
Pauli matrices, 47
Pell's equation, 197
perigee, 632
perihelion, 258
period doubling
  in 1d maps, 737
  in 2d maps, 764
period map, 798
periodic continued fraction, 202
periodic orbit, 805
  of 1d map, 729
  of 2d map, 751
  stability of, 730
periodic solutions
  conditions for, 676
phase, 526
  curve, 642
  diagram, 644
  flow, 642
  portrait, 644
  velocity, 635
phase space, 634
phaseportrait, 644
piecewise functions, 102
pitchfork bifurcation, 734
Planck's constant, 527
plot, 21
  parametric, 22

  scaling option, 22
plot3d, 25, 109
plotsetup, 125
Pluto, 260
Poincaré
  –Bendixson theorem, 674
  index, 666, 762, 811
  section, 798
Poincaré H., 180, 185, 632, 665, 697, 754
point (font size), 79
pointplot, 152
pointwise convergence, 169
Poisson S. D., 346
Poisson summation formula, 291
population of England and Scotland, 640
population of USA, 639
PostScript output, 124
potential energy, 691
powadd, 157
powcreate, 157
power series, 150
powseries, 156
powsolve, 159
prime number theorem, 189
prime numbers and zeta function, 142
print, 13
  for arrays, 41
  for matrices, 43
printf, 17
probability density, 743
procedure, 112
  general form, 117
product of series, 139
prompt, for input, 2
propagator, 439
psi function, 835

quadrature, 629
quartic potential, 622, 625
quotes, 18

Raabe's test, 145
radial twist map, 754
radius of convergence, 151
rainbow angle, 559
rand, 38
Raphson J., 325
rational function, 209, 483, 485
ratpoly, 212
Rayleigh's equation, 681, 725
Re, 19
read, 122
readdata, 123
recurrence relation
  Bessel functions, 408
  orthogonal polynomials, 379
recursive procedure, 118
reduction formulae, 493
reflection map, 757
reflexive polynomials, 470
regular perturbations, 301
relaxation oscillations, 678
remainder of series, 138

remember option, 119
remove, 39
repellor, 656, 729
residue, 748
resonance, 600, 602, 625, 785
   curve, 787
   island, 752, 799, 804
response curve, 787
restart, 5
RETURN, 115
reversion, 158, 165
Riccati J., 630
Riccati's equation, 630, 634
Richardson's extrapolation, 164
Riemann
   hypothesis, 143
   –Lebesgue lemma, 496
   zeta function, 142
Riemann G. F. B., 54, 142, 507, 631
Riesz–Fischer theorem, 276
Rodrigues O., 379
Rodrigues' formula
   general case, 379
   Hermite polynomials, 390
   Laguerre polynomials, 396
   Legendre polynomials, 381
rotational motion, 650

saddle point, 509, 659
save, 122
saw tooth, mechanical, 299
scaling in plot, 22
Schläfli's integral, 382
Schrödinger's equation, 437, 577, 607
Schwarz inequality, 269
Schwarzian derivative, 735, 737
scientific notation, 78
second-order dynamical system, 642
secular terms, 700
Seidel P. L. V., 171
select, 39, 120
self-adjoint operators, 354
self-excited oscillations, 680
separating variables, 347
separatrix, 694
   and stable manifold, 777
   motion on, 797
seq, 13, 26, 36
sequence, 131, 134
sequence operator, 48
series, 138
   alternating, 148
   asymptotic, 184
   convergent, 138
   divergent, 138
   geometric, 141
   Gregory's, 155
   harmonic, 138
   infinite, 131
   power, 150
   remainder of, 138
series, 75, 137, 156, 186, 214, 310, 311
   with **solve**, 77

setoptions, 26
setoptions3d, 26
sets in Maple, 8, 38
sgn function, 235
Shanks transformation, 161
showassumed in **interface**, 56
side-relations, in simplify, 70
siderial day and year, 197
signature of a matrix, 543
signum, 235
simple fixed point, 635, 658
simple harmonic motion, 670
simplify, 70
   hypergeom, 219
   side-relations, 70
   symbolic option, 72
simplifying, problem of, 63
Simpson's rule, 128
Sine integral, 193, 289
   asymptotic expansion, 193
sine series, 284
singular perturbations, 301
singular point, 656
singular Sturm–Liouville system, 350
singularities and Padé approximants, 216
singularity, 154, 166
Snell W., 544
Snell's law, 543, 558
soft spring, 793, 803, 810
Solar System
   age of, 687
   stability of, 711
solve, 20
   with **series**, 77, 311
sort, 38, 46, 71
sparse matrix, 45
special functions, 481
   available in Maple, 91
speed of light, 286
spherical Bessel functions, 406
spiral, 660
stability of periodic orbits, 730
stability of the Solar System, 711
stable fixed point, 635, 729
   of a map, 747
stable manifold, 767, 768
standard map, 728, 783
star, 662
stationary phase approximation, 526
steepest descent, 506, 507
Stirling J., 179
Stirling's approximation, 127, 179, 188, 503, 835
Stokes G. G., 171
string, 18
string, vibrating, 131
strongly stable, 656
structurally unstable, 636
Sturm J. C. F., 342
Sturm–Liouville eigenvalues, 594
Sturm–Liouville system, 352, 575
   regular, 352
   singular, 350, 353
Sturm–Liouville theorem, 356

subharmonics, 813
subs, 8, 9
    comparison with **eval**, 17
    for matrices, 44
    to remove order symbol, 76, 187
sum, 12, 14, 22
superstable, 729
surface-of-section, 798
symbol in Maple, 18
symbolic
    option of **combine**, 73
    option of **simplify**, 72
symmetric matrix, 45
symmetry
    method, 757, 805
    of a map, 757
symmetry curve, 757

taylor, 154, 156, 186
Taylor polynomial, 153
Taylor's series, 153
    with Maple, 154
Tchebychev P. L., 387
Tchebychev polynomials, 387
    of second kind, 389
tent map, 738
terminating motion, 636
terminator, 3
text added to figures, 79
textplot, 79
three-body problem, 631
tickmarks, 80
tickmarks, 80
timelimit, 61, 492
titlefont, 79
total derivatives, 105
tpsform, 157
trace, 46
trace facility, 118
trajectory, 642
transient, 787
transition points, 577
trapezoidal rule, 129
trigonometric series, 273, 358
trun, 201
turning points, 577, 692
    coalescing, 606
    isolated, 582
twist map, 754, 799, 802, 804

unapply, 96, 104
unassign, 12

unbounded variables, 7
uniform approximation, 557, 579
uniform convergence, 171, 173
    of power series, 172
uniqueness theorems, 668
unstable fixed point, 635, 729, 747
unstable manifold, 767, 768
untrace, 118

value function, 49
value, as used by Maple, 8
van der Pol B., 678
van der Pol oscillator, 652
vector, 41
velocity function, 635
vertical pendulum, 649, 670
Volterra–Lotka equations, 665

Wallis J., 196
Wallis' formula, 494, 836
Watson's lemma, 499
wave equation, 131, 346
    for a membrane, 347
    for a string, 370, 625
Weber equation, 391
Weber function, 496, 539
Weierstrass M-test, 171
Weierstrass K., 171, 834
Weierstrass' approximation theorem, 272, 296, 377
weight function, 356, 377
whattype, 75
while loop, 29, 30
Wilkinson's polynomial, 313
with, 23
    DEtools, 644
    linalg, 46, 418
    numapprox, 215
    numtheory, 198
    orthopoly, 238, 264, 377
    plots, 23, 79, 109, 259, 316
    plottools, 153, 259, 316
    powseries, 156
WKB approximation, 573, 622
Wronskian, 364, 375, 439, 472

xtickmarks, 80

ytickmarks, 80

zero equivalence problem, 63
zero-eigenvalue problem, 366
zeta function, 54, 142, 189, 295